CHEMICAL ENGINEERING COMPUTATION WITH MATLAB®

CHEMICAL ENGINEERING COMPUTATION WITH MATLAB®

Yeong Koo Yeo

CRC Press
Taylor & Francis Group
Boca Raton London New York

CRC Press is an imprint of the
Taylor & Francis Group, an **Informa** business

Second edition published 2021
by CRC Press
6000 Broken Sound Parkway NW, Suite 300, Boca Raton, FL 33487-2742

and by CRC Press
2 Park Square, Milton Park, Abingdon, Oxon, OX14 4RN

© 2021 Taylor & Francis Group, LLC

First edition published by CRC Press 2017

CRC Press is an imprint of Taylor & Francis Group, LLC

Library of Congress Cataloging-in-Publication Data
Names: Yeo, Yeong-Koo, author.
Title: Chemical engineering computation with MATLAB / by Yeong Koo Yeo.
Description: Second edition. | Boca Raton, FL : CRC Press/Taylor & Francis   Group, LLC, [2021] | Includes bibliographical references and index. |
Summary: "This new edition continues to present basic to advanced levels of problem-solving techniques using MATLAB. It provides even more examples and problems extracted from core chemical engineering subject areas and all code is updated to MATLAB version 2020. It also includes a new chapter on computational intelligence. This essential textbook readies engineering students, researchers, and professionals to be proficient in the use of MATLAB to solve sophisticated real-world problems within the interdisciplinary field of chemical engineering"-- Provided by publisher.
Identifiers: LCCN 2020043323 (print) | LCCN 2020043324 (ebook) | ISBN 9780367547820 (hardback) | ISBN 9780367547844 (pbk) | ISBN 9781003090601 (ebook)
Subjects: LCSH: Chemical engineering--Data processing. | MATLAB.
Classification: LCC TP184 .Y46 2021 (print) | LCC TP184 (ebook) | DDC 660.0285--dc23
LC record available at https://lccn.loc.gov/2020043323
LC ebook record available at https://lccn.loc.gov/2020043324

ISBN: 978-0-367-54782-0 (hbk)
ISBN: 978-1-003-09060-1 (ebk)

Typeset in Times
by MPS Limited, Dehradun

Access the Support Materials: https://www.routledge.com/9780367547820

Contents

Preface

This book focuses on the presentation of problem-solving techniques in the main subject areas of chemical engineering. Basics to advanced levels of problem solving are covered utilizing MATLAB®, which is a high-performance language for technical computing. This book provides numerous examples and exercises, as well as extensive problem-solving instruction and solutions, for various chemical engineering problems. Solutions are developed using fundamental principles to construct mathematical models, and an equation-oriented approach is used to generate numerical results. For efficient problem solving, various numerical methods—including linear and nonlinear algebraic equations, finite difference methods, ordinary differential equations, boundary-value problems, partial differential equations, curve fitting, and linear and nonlinear regressions—are employed in the examples.

Most problems encountered in chemical engineering are very sophisticated and interdisciplinary. Thus, calculations for chemical engineering problem solving by portable calculators or hand calculations are limited to only a few simple applications and problems. Since the introduction of personal computers and mathematical software packages in the early 1980s, emphasis has gradually shifted to computer-based problem solving. It is increasingly becoming important for today's engineering students, researchers, and professionals to be proficient in the use of software tools for problem solving.

MATLAB is one such tool that has been found suitable for implementing algorithms required in engineering education, graduate research, advanced mathematics, and industrial operation-aid systems. It is distinguished by its ability to perform calculations in vector-matrix form, a large library of built-in functions, strong structural language, and a rich set of graphical visualization tools. Furthermore, MATLAB integrates computations, visualization, and programming in an intuitive, user-friendly environment. These useful features of MATLAB can be effectively applied in chemical engineering problem-solving activities. Examples and problems introduced in this book demonstrate how some of these special capabilities of MATLAB can be best used for effective and efficient problem solving.

This book is designed for chemical engineering students and industrial professionals. MATLAB is adopted as the computation environment throughout this book. This book provides examples and problems extracted from core chemical engineering subject areas and presents a basic instruction in the use of MATLAB for problem solving. Emphasis is placed on setting up problems systematically and obtaining required solutions efficiently. Additionally, since this book presents examples and problems commonly encountered in most chemical engineering subject areas, it can serve as a reference in various scenarios and environments. Since all examples and problems presented in this book are solved with MATLAB, this book can also be a resource in developing proficiency in the use and application of MATLAB. Practicing engineers and students alike can learn to write MATLAB programs for various chemical engineering applications.

Examples presented in this book demonstrate the implementation of various problem-solving approaches and methodologies for problem formulation, problem solving, analysis, and presentation—as well as visualization and documentation of results. This book also provides aid with advanced problems that are often encountered in graduate research and industrial operations, such as nonlinear regression, parameter estimation in differential systems, two-point boundary-value problems, and partial differential equations and optimization.

This book is intended for those with interest in learning how to use MATLAB to solve chemical engineering problems using computers. It can be used as a textbook in a one-semester course for students in chemical engineering and related disciplines. For undergraduate students, this book can be a resource for learning how to classify and analyze problems according to the numerical methods that facilitate efficient and effective computations. It can also be utilized as a reference for chemical engineering researchers and engineers, particularly in computer-aided problem solving.

Features in the Second Edition
There have been some major changes in some chapters. Major new features are as follows:

- All programs provided are rewritten in MATLAB 2020a. Most of these programs are also compatible with MATLAB 2019a/b or lower.
- Many new examples and problems have been added throughout.
- A new section on Simulink has been added in Chapter 1.
- Typical numerical analysis algorithms with MATLAB programs are presented in Chapter 2.
- In Chapter 3, the section about water and steam has been revised and all the relevant programs have been rewritten.
- In Chapter 6, examples and problems about nonisothermal catalytic reactions have been added. Calculation methods for unsteady-state nonisothermal reactions have been introduced. Cell growth models are introduced with ample examples and problems.
- Chapters 7 and 8, concerning mass and heat transfer, are equipped with new examples and problems.
- A new chapter concerning computational intelligence has been added. In this new chapter, topics on fuzzy logic, artificial neural network, and machine learning are presented with relevant MATLAB programs.

Yeong Koo Yeo

MATLAB® is a registered trademark of The MathWorks, Inc. For product information, please contact:
The MathWorks, Inc.
3 Apple Hill Drive
Natick, MA, 01760-2098 USA
Tel: 508-647-7000
Fax: 508-647-7001
E-mail: info@mathworks.com
Web: www.mathworks.com

Acknowledgments

This book is a result of my research and teaching career in chemical engineering, and I gratefully acknowledge the contributions of my students to the development of many of the ideas contained herein.

I especially thank the late Professor Chang Kyun Choi at Seoul National University, whose passion, knowledge, and discipline in this field have been a tremendous inspiration for my work.

I am indebted to Professor Kyu Yong Choi at the University of Maryland, who gave valuable comments and suggestions and provided continuous support and encouragement in the preparation of this book.

My acknowledgments are due to Sand Duck Lee at McGraw-Hill and Allison Shatkin at CRC Press/Taylor & Francis Group for their strong backing, support, and encouragement with this book effort.

I want to thank Adrián Parodi at the Institute of Research and Development in Process Engineering and Applied Chemistry, Argentina, who tested the MATLAB programs on water and steam properties and gave valuable comments and suggestions.

I thank Gabrielle Vernachio, editorial assistant at Taylor & Francis Group, for finding mistakes, and inconsistencies, and for her attention to detail in working with the copyedited proof.

Finally, my special gratitude goes to my wife, Myoung, my son, Jonathan, and my daughter-in-law, Grace, who have been my personal inspiration for so many years and have shared the burden of this effort.

Author

Yeong Koo Yeo teaches chemical engineering at the College of Engineering of Hanyang University, Seoul, South Korea. He earned a BA (1979) and an MS (1982) in chemical engineering at Seoul National University and a PhD (1986) in chemical engineering at Auburn University. Before joining the Hanyang faculty in 1993, Professor Yeo was a senior researcher at KIST. He has taught various courses, such as process control, plant design, process analysis, optimization, computational process design, and model predictive control. A leading authority on process control and application of MATLAB, he is the author of four books. *Introduction and Application of MATLAB* (5th ed., 2019) has played a key role in establishing technical computing as an important field. *Modern Process Control Engineering* (3rd ed., 2020) has played a key role in establishing practical control education. Professor Yeo has also authored more than 149 articles in leading academic journals, such as *Industrial and Engineering Chemistry Research*, *IEEE Transactions on Automatic Control*, *Chemical Engineering Communication*, *Japanese Journal of Chemical Engineering*, *Korean Journal of Chemical Engineering*, and *Hwahak Gonhak*. His research interests include process control, process modeling and simulation, process artificial intelligence, and process optimization.

1 Introduction to MATLAB®

1.1 STARTING MATLAB

You can start MATLAB as you would any other software application. In Windows, you access it via the Start menu. Alternatively, you may have a desktop icon that enables you to start MATLAB with a simple double click. When MATLAB is started, a window opens in which the main part is the Command Window. Figure 1.1 contains an example of a newly launched MATLAB desktop. In the Command Window, you will see the prompt (»). If the Command Window is active, the prompt will be followed by a cursor. That is the place where you will enter MATLAB commands. If the Command Window is not active, just click in it anywhere.

In addition to the Command Window, there are couple of other windows that may be opened by default. The layout can always be customized. To the left of the Command Window is the Current Folder Window. The folder that is set as the Current Folder is where files will be saved. This window shows the files that are stored in the Current Folder. To the right of the Command Window are the Workspace Window on top and the Command History Window on the bottom. The Command History Window shows commands that have been entered. The configuration of each window can be altered by clicking the down arrow at the top right corner of the window. This will show a menu of options including closing and undocking that window. Alternatively, hitting the down arrow under Layout in the HOME tab allows for customization of the window within the desktop environment, as shown in Figure 1.2.

1.1.1 ENTERING COMMANDS IN THE COMMAND WINDOW

Click in the Command Window to make it active. When a window becomes active, its title bar darkens and a blinking cursor appears after the prompt. Now you can begin entering commands. Try typing a=[2 4 7] and pressing the Enter or Return key:

```
>> a = [2 4 7]
a =
  2   4   7
```

MATLAB assigns the vector [2 4 7] to a variable of your choice (for example, a). The spacing between the MATLAB commands or expressions and the results can be controlled by entering "format compact". As another example, type 3 + 5 and press the Enter key. Then try typing factor(46738921) and sin(80) Figure 1.3 shows the results displayed in the Command Window.

Entering a command and manipulating it require us to master the following rules:

- MATLAB is case sensitive, which means that there is a difference between upper- and lowercase letters.
- A semicolon (;) at the end of a command withholds displaying the answer.
- A command in a preceding line cannot be changed; to correct or repeat an executed command, press the up arrow key (↑). Also, an old command can be recalled by typing the first few characters followed by the up arrow.
- You can type "help topic" to access online help on the command, function, or symbol topic.

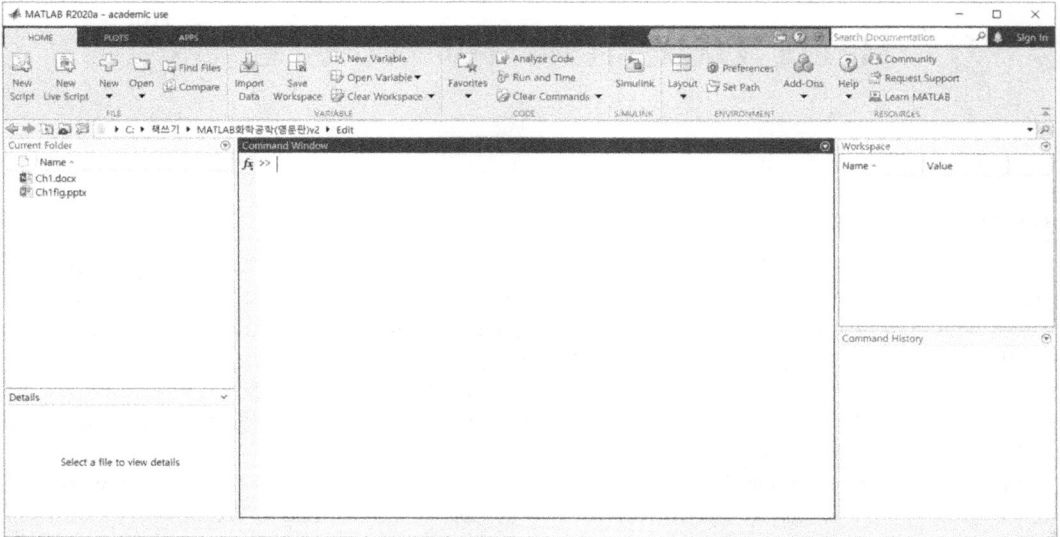

FIGURE 1.1 A MATLAB desktop.

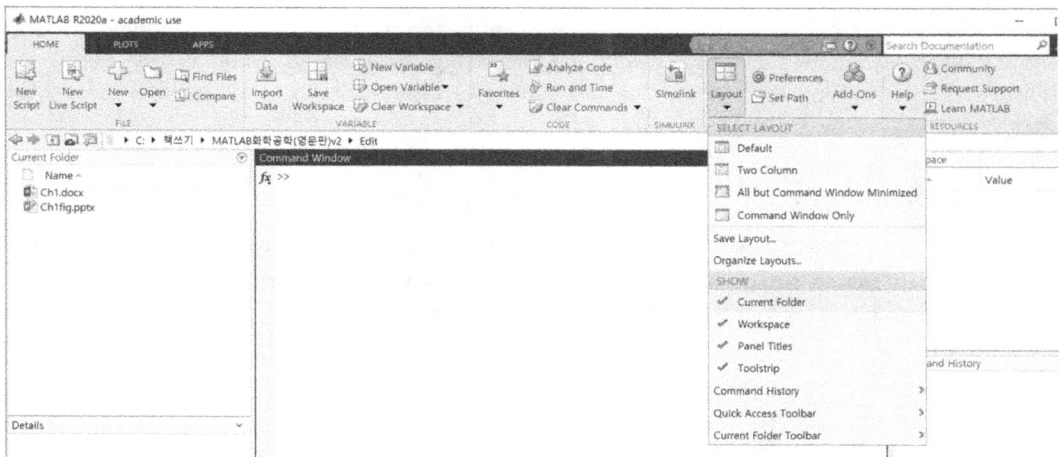

FIGURE 1.2 Customization of the windows using the Layout menu.

- If you press the Tab key after partially typing a function or variable name, MATLAB will attempt to complete it, offering you a selection of choices if there is more than one possible completion.
- MATLAB uses parentheses (), square brackets [], and curly braces { }, and these are not interchangeable.
- You can quit MATLAB by typing "exit" or "quit" or clicking the *x* icon (×) located at the upper right corner of the MATLAB window.

1.1.2 HELP COMMAND

The "help" command can be used to identify MATLAB functions and how to use them. To find out what a particular function or command does and how to call it, type "help" and then

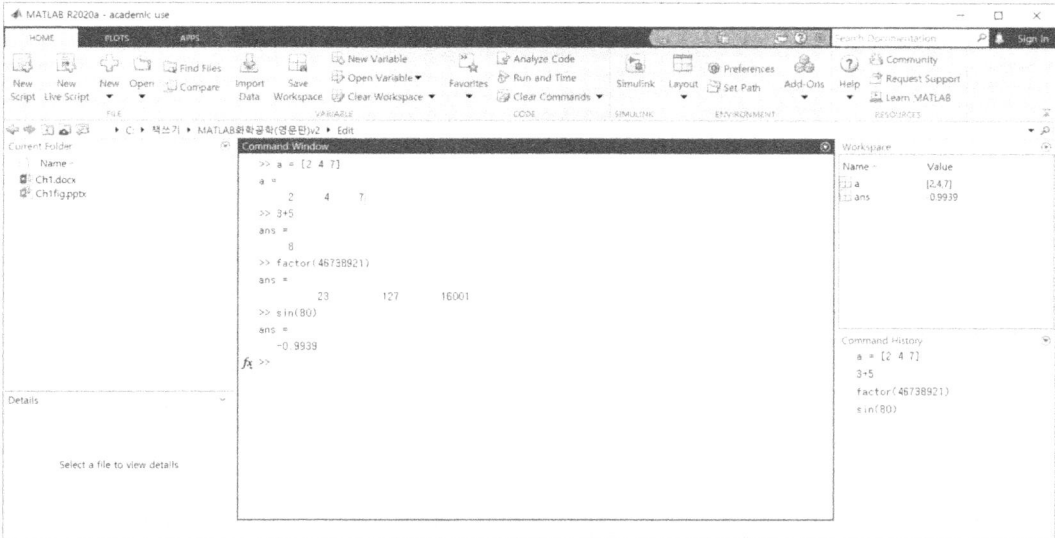

FIGURE 1.3   MATLAB desktop with several commands evaluated.

the name of the function or command. The explanations appear immediately in the Command Window. Alternatively, hitting the down arrow under "Help" on the toolstrip of the RESOURCES group of the desktop HOME tab allows a choice of various options, as shown in Figure 1.4.

For example, the following will give a description of the linspace command, as shown in Figure 1.5.

```
>> help linspace
```

1.1.3 EXITING MATLAB

To exit MATLAB, enter either "quit" or "exit" at the prompt or click the exit button (×) located at the upper right corner of the MATLAB window. It is important to save your files or graphics before exiting MATLAB.

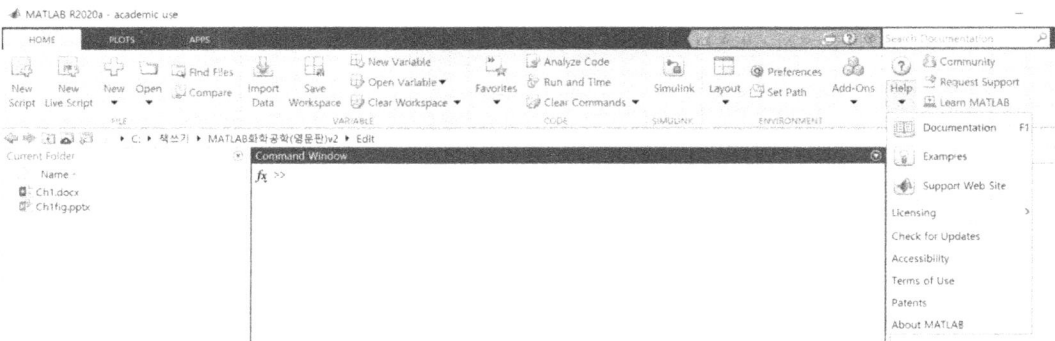

FIGURE 1.4   Submenus of the "Help" menu.

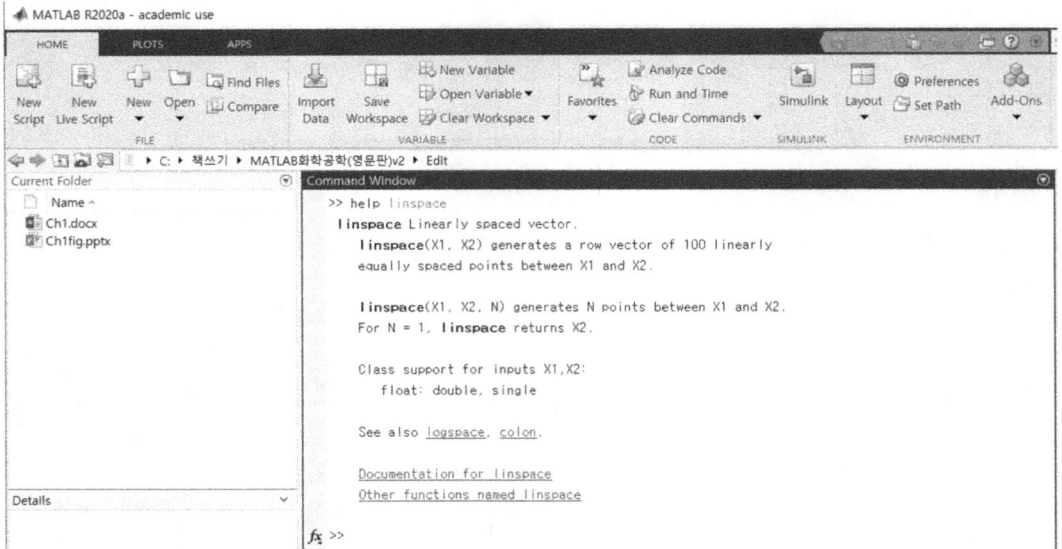

FIGURE 1.5   Description of the linspace command.

1.2 OPERATIONS AND ASSIGNMENT OF VARIABLES

MATLAB can be used as a hand calculator to perform arithmetic operations. MATLAB uses the symbols +, -, *, /, and ^ for addition, subtraction, multiplication, division, and exponentiation (powers) of scalars, respectively. For example:

```
>> 6^3 - (7 + 5)/2 + 9*4
ans =
 246
>> (-2 + sqrt(6))/2
ans =
  0.2247
>> 2^(-48)
ans =
 3.5527e-15
```

MATLAB assigns the most recent answer to a variable called `ans`, which is an abbreviation for "answer." You can use the variable `ans` for further calculations. For example:

```
>> ans^3 + sqrt(ans)
ans =
 5.9605e-08
```

You can use variables to write mathematical expressions. Instead of using the default variable `ans`, you can assign the result to a variable of your own choice. The Workspace Window shows variables that have been created and their values. MATLAB uses the assignment operator (=) to create a variable. For example:

```
>> u = cos(18)
u =
  0.6603
```

```
>> v = sin(18)
v =
  -0.7510
>> u^2 + v^2
ans =
   1
```

The MATLAB trigonometric functions `sin(x)`, `cos(x)`, `tan(x)`, etc., use radian measure. Trigonometric functions ending in d (`sind(x)`, `cosd(x)`, `tand(x)`, etc.) take the argument x in degrees. Variable names must begin with a letter; the rest of the name may contain letters, digits, and underscore characters. Variable names can be no longer than 63 characters. MATLAB is case sensitive, which means that there is a difference between upper- and lowercase letters. The variable pi is a permanent variable with value π.

```
>> y = tan(pi/6)
y =
   0.5774
```

Table 1.1 shows the basic arithmetic operators used in MATLAB. Array operations are defined to act elementwise and are generally obtained by preceding the symbol with a dot.

1.2.1 ERRORS IN INPUT

If you make an error in an input line, MATLAB will normally print an error message. For example, here is what happens when you try to evaluate $4x^4$:

```
>> 4x^4
 4x^4
 ↑
Error: Invalid expression. Check for missing multiplication operator, missing
or unbalanced
delimiters, or other syntax error. To construct matrices, use brackets instead
of parentheses.
```

The error is a missing multiplication operator *. The correct input would be `4*x^4`. Note that MATLAB places a marker (an up arrow) at the place where it thinks the error might be. You can edit an input line by using the up arrow key (↑) to redisplay the previous commands, editing the specific command using the left arrow key (←) and right arrow key (→), and then pressing Return or Enter.

TABLE 1.1
Basic Arithmetic Operators Used in MATLAB

Symbol	Operation	Symbol	Operation
+	Addition	.^	Array exponentiation
-	Subtraction	\	Backslash, left division
*	Multiplication	/	Slash, right division
.*	Array multiplication	.\	Array left division
^	Exponentiation	./	Array right division

1.2.2 ABORTING CALCULATIONS

If MATLAB gets hung up in a calculation, or seems to be taking too long to perform an operation, you can usually abort it by typing Ctrl+C, that is, holding down the Ctrl key and pressing C. While not foolproof, this is the method of choice when MATLAB is not responding.

1.3 VECTORS AND MATRICES

1.3.1 VECTORS

In MATLAB, a vector is simply a list of scalars, whose order of appearance in the list might be significant. You can create a vector of any length in MATLAB by typing a list of numbers, separated by commas and/or spaces, inside square brackets. For example:

```
>> u = [-2 3 4 8 15]
u =
  -2   3   4   8   15
>> u = [-2,3,4,8,15]
u =
  -2   3   4   8   15
>> v = [-6 3 -11 9 12 -2 0 5]
v =
  -6   3   -11   9   12   -2   0   5
```

If the values in the vector are regularly spaced, the colon operator (:) can be used to iterate through these values. For example:

```
>> t = 0:7
t =
  0   1   2   3   4   5   6   7
>> x = -4:3
x =
  -4   -3   -2   -1   0   1   2   3
```

Suppose that you want to create a row vector of values running from 1 to 10. Here is how to do it without typing each number:

```
>> w = 1:10
w =
  1   2   3   4   5   6   7   8   9   10
```

The increment can be specified as the middle of three arguments. For example, to create a vector with all integers from 8 to 24 in steps of 2:

```
>> y = 8:2:24
y =
  8   10   12   14   16   18   20   22   24
```

Increments can be fractional or negative. For example:

```
>> g = 3:3:10, h = 4:-0.75:0
g =
  3   6   9
h =
  4.0000   3.2500   2.5000   1.7500   1.0000   0.2500
```

The elements of the vector w can be extracted as w(1), w(2), etc. For example:

```
>> w = 5:12
w =
     5    6    7    8    9   10   11   12
>> w(4)
ans =
     8
```

A subset of a vector, which would be a vector itself, can also be obtained using the colon operator. For example, the following statement would get the third through sixth elements of the vector w and store the result in a vector variable *v*:

```
>> v = w(3:6)
v =
   7    8    9   10
```

Any row vector created using any method can be transposed to result in a column vector. In MATLAB, the apostrophe (') is built in as the transpose operator. For example:

```
>> v'
ans =
   7
   8
   9
  10
```

You can perform mathematical operations on vectors. For example, to square the elements of the vector x, type

```
>> x = 1:2:13
x =
     1    3    5    7    9   11   13
>> x.^2
ans =
     1    9   25   49   81  121  169
```

The dot (period) in this expression says that the numbers in *x* should be squared individually, or element byelement. Typing x^2 would tell MATLAB to use matrix multiplication to multiply *x* by itself and would produce an error message in this case. Similarly, you must type .* or ./ if you want to multiply or divide vectors element by element. For example, to multiply the elements of the vector a by the corresponding elements of the vector b, type

```
>> a = 2:3:14
a =
     2    5    8   11   14
>> b = [-3 2 -4 1 8]
b =
    -3    2   -4    1    8
>> a.*b
ans =
    -6   10  -32   11  112
```

The linspace function creates a linearly spaced vector: linspace(a,b,n) creates a vector with n variables in the inclusive range from a to b. If n is omitted, the default is n = 100. For example:

```
>> linspace(-2,2,8)
ans =
  -2.0000  -1.4286  -0.8571  -0.2857  0.2857  0.8571  1.4286  2.0000
```

MATLAB has many mathematical functions that operate in the array sense when given a vector or matrix argument. These include `exp`, `log`, `sqrt`, `sin`, `cos`, `tan`, etc. For example:

```
>> a = [0.2 1.5 3]
a =
     0.2000   1.5000   3.0000
>> exp(a)
ans =
     1.2214   4.4817  20.0855
>> log(ans)
ans =
     0.2000   1.5000   3.0000
>> sqrt(a)
ans =
     0.4472   1.2247   1.7321
```

If x and y are column vectors, the dot product or inner product of these two vectors is accomplished using the * operator and transposing the first vector, or using the `dot` function in MATLAB. The cross product or outer product of two vectors x and y is defined only when both x and y have three elements. It can be defined as a matrix multiplication of a matrix composed from the elements of x in a particular manner and the column vector y. MATLAB has a built-in function `cross` to accomplish this operation. For example:

```
>> x = [-2 0 2]', y = [3 5 7]'
x =
        -2
         0
         2
y =
         3
         5
         7
>> x'*y
ans =
         8
>> dot(x,y)
ans =
         8
>> cross(x,y)
ans =
       -10
        20
       -10
```

The result of the outer product operation x*y' becomes a matrix:

```
>> Z = x*y'
Z =
      -6  -10  -14
       0    0    0
       6   10   14
```

1.3.2 MATRICES

A matrix is a rectangular array of numbers. Row and column vectors are examples of matrices. Consider the 3×4 matrix given by

$$A = \begin{bmatrix} -2 & 1 & 3 & 5 \\ 4 & 0 & 6 & -1 \\ 3 & 5 & 2 & 7 \end{bmatrix}$$

It can be entered in MATLAB with the command

```
>> A = [-2 1 3 5; 4 0 6 -1; 3 5 2 7]
A =
   -2    1    3    5
    4    0    6   -1
    3    5    2    7
```

The matrix elements in any row are separated by commas or spaces, and the rows are separated by semicolons. If two matrices A and B are the same size, their element-by-element sum is obtained by entering A + B. You can also add a scalar c to a matrix A by typing A + c. Likewise, A-B represents the difference of A and B, and A-c subtracts the number c from each element of A.

If A and B are multiplicatively compatible, that is, if A is $n \times m$ and B is $m \times l$, then their product A*B is $n \times l$. Recall that the element of A*B in the ith row and jth column is the sum of the products of the elements from the ith row of A times the elements from the jth column of B. The product of a number c and the matrix A is given by c*A. A simple illustration is given by the matrix product of the 3×4 matrix A above by the 4×1 column vector x:

```
>> x = [2 -3 4 5]'
x =
    2
   -3
    4
    5
>> A*x
ans =
   30
   27
   34
```

The result is a 3×1 matrix, in other words, a column vector. A' represents the conjugate transpose of A. Consider two 3×3 matrices A and B given by

$$A = \begin{bmatrix} 3 & 1 & 2 \\ -1 & 0 & 1 \\ 6 & 2 & 4 \end{bmatrix}, \quad B = \begin{bmatrix} 2 & 0 & 1 \\ 1 & -3 & 2 \\ -2 & 1 & 1 \end{bmatrix}$$

Then the transpose of A (A'), addition C = A + B, subtraction D = A − B and product P = A*B are given by

```
>> A = [3 1 2; -1 0 1; 6 2 4]
A =
    3    1    2
   -1    0    1
    6    2    4
>> A'
```

```
ans =
   3   -1    6
   1    0    2
   2    1    4
>> B = [2 0 1;1 -3 2;-2 1 1]
B =
   2    0    1
   1   -3    2
  -2    1    1
>> C = A+B
C =
   5    1    3
   0   -3    3
   4    3    5
>> D = A-B
D =
   1    1    1
  -2    3   -1
   8    1    3
>> P = A*B
P =
   3   -1    7
  -4    1    0
   6   -2   14
```

The function sum returns the sum of a vector or sums of the columns of a matrix. To find the sum of all elements of a matrix A, we can use sum(sum(A)):

```
>> sum(A)
ans =
   8    3    7
>> sum(sum(A))
ans =
  18
```

The function diag returns the diagonal in a square matrix:

```
>> diag(A)
ans =
   3
   0
   4
>> sum(diag(A))
ans =
   7
```

For the exponentiation operator, typing A^2 would tell MATLAB to use matrix multiplication to multiply A by itself, A*A. But typing A.^2 returns the matrix formed by raising each of the elements of A to the power 2. For example:

```
>> W = [2 1; -3 4]
W =
   2    1
  -3    4
>> W^2
```

```
ans =
    1    6
  -18   13
>> W.^2
ans =
    4    1
    9   16
```

In exponentiation, a vector or a matrix can be used as a power. For example:

```
>> x = [3 1 2], y = [4 2 3], z = [2 1; 4 3]
x =
    3    1    2
y =
    4    2    3
z =
    2    1
    4    3
>> x.^y
ans =
   81    1    8
>> 2.^x
ans =
    8    2    4
>> 2.^z
ans =
    4    2
   16    8
```

As stated before, the conjugate transpose of the matrix A is obtained from A'. If A is real, this is simply the transpose. The transpose without conjugation is obtained with A.'. The functional alternatives `ctranspose(A)` and `transpose(A)` are sometimes more convenient. Let Z be a complex matrix given by

$$Z = \begin{bmatrix} -1 & i \\ 2+i & 1+2i \end{bmatrix}$$

```
>> Z = [-1 i;2+i 1+2i]
Z =
  -1.0000 + 0.0000i   0.0000 + 1.0000i
   2.0000 + 1.0000i   1.0000 + 2.0000i
>> Z'
ans =
  -1.0000 + 0.0000i   2.0000 - 1.0000i
   0.0000 - 1.0000i   1.0000 - 2.0000i
>> Z.'
ans =
  -1.0000 + 0.0000i   2.0000 + 1.0000i
   0.0000 + 1.0000i   1.0000 + 2.0000i
>> ctranspose(Z)
ans =
  -1.0000 + 0.0000i   2.0000 - 1.0000i
   0.0000 - 1.0000i   1.0000 - 2.0000i
>> transpose(Z)
```

```
ans =
 -1.0000 + 0.0000i   2.0000 + 1.0000i
  0.0000 + 1.0000i   1.0000 + 2.0000i
```

The functions rank, det, and inv calculate, respectively, the rank, the determinant, and the inverse of a matrix. The inverse of the matrix B can also be obtained by B^-1. For example:

```
>> B = [2 0 1; 1 -2 3; -2 1 1]
B =
   2    0    1
   1   -2    3
  -2    1    1
>> rank(B)
ans =
   3
>> det(B)
ans =
  -13
>> inv(B)
ans =
  0.3846   -0.0769   -0.1538
  0.5385   -0.3077    0.3846
  0.2308    0.1538    0.3077
>> B^-1
ans =
  0.3846   -0.0769   -0.1538
  0.5385   -0.3077    0.3846
  0.2308    0.1538    0.3077
```

The element in row 2 and column 3 of matrix B can be accessed as B(2,3). The second column can be extracted by B(:,2), and the third row can be extracted by B(3,:). For example:

```
>> B(2,3)
ans =
   3
>> B(:,2)
ans =
   0
  -2
   1
>> B(3,:)
ans =
  -2   1   1
```

The submatrix C formed by taking the second and third columns of the matrix B is given by

```
>> C = B(:,2:3)
C =
   0   1
  -2   3
   1   1
```

MATLAB has several commands that generate special matrices. The command eye(n) represents the $n \times n$ identity matrix, and the commands zeros(n,m) and ones(n,m) generate $n \times m$ matrices of zeros and ones, respectively. The first argument (n) denotes the row and the second argument (m) denotes the column. The commands `eye(3)` and `eye(3,3)` produce the same identity matrix.

```
>> I3 = eye(3,3), Y = zeros(3,5), Z = ones(2)
I3 =
   1   0   0
   0   1   0
   0   0   1
Y =
   0   0   0   0   0
   0   0   0   0   0
   0   0   0   0   0
Z =
   1   1
   1   1
```

The function `rand` creates an array of random numbers between 0 and 1, and the function `randn` creates an array of normal random numbers with mean 0 and variance 1.

```
>> F = rand(3), G = randn(1,6)
F =
   0.8147   0.9134   0.2785
   0.9058   0.6324   0.5469
   0.1270   0.0975   0.9575
G =
   2.7694  -1.3499   3.0349   0.7254  -0.0631   0.7147
```

1.3.3 COMPLEX NUMBER

In MATLAB, i and j are used as the imaginary unit $\sqrt{-1}$. The imaginary unit can also be represented by `sqrt(-1)` or by the function `complex(0,1)`. For example, the complex number $3 - 2i$ can be entered as `3-2i`, `3-2*i`, `3-2*sqrt(-1)`, or `complex(3,-2)`. MATLAB supports complex arithmetic, with `conj`, `real`, and `imag` taking the conjugate and the real and imaginary parts from a complex number, respectively. The functions `abs` and `angle` compute the magnitude and the phase angle of a complex number, respectively. For example:

```
>> w = (-1)^0.36
w =
 0.4258 + 0.9048i
>> z = conj(w)
z =
 0.4258 - 0.9048i
>> [real(z) imag(z)]
ans =
 0.4258  -0.9048
>> exp(i*pi)
ans =
 -1.0000 + 0.0000i
>> angle(w)
ans =
 1.1310
>> abs(w)
ans =
 1
```

The Hermitian transpose of a complex number is obtained by the transpose operator ('). For example, the Hermitian transpose of $x = [3 - 2i \ -5 + 4i]$ is given by x'. But x.' produces a simple transpose of x. For example:

```
>> x = [3-2i -5+4i]
x =
 3.0000 - 2.0000i  -5.0000 + 4.0000i
>> x'
ans =
 3.0000 + 2.0000i
-5.0000 - 4.0000i
>> x.'
ans =
 3.0000 - 2.0000i
-5.0000 + 4.0000i
```

1.3.4 SUPPRESSION OF SCREEN OUTPUT

Putting a semicolon (;) at the end of an input line suppresses printing of the output of the MATLAB command. It can be used in any situation where the MATLAB output need not be displayed. For example, putting a semicolon at the end of the definition of the matrix A suppresses the display of matrix A on the screen. The result of the operation A*x is not shown in the screen:

```
>> A = [1 2 3 4; 5 6 7 8; 9 10 11 12];
>> A*x;
```

The semicolon should generally be used when defining large vectors or matrices (for example, x = -1:0.1:2;).

1.4 NUMERICAL EXPRESSIONS

By default, MATLAB carries out all its arithmetic operations in double-precision floating-point arithmetic, which is accurate to approximately 14 decimal digits. In other words, MATLAB stores floating-point numbers and carries out elementary operations to an accuracy of about 16 significant decimal digits. However, MATLAB displays only four significant decimal digits by default. The "format" command can be used to specify the output format of expressions. To display more digits, type "format long". Then all subsequent numerical outputs will have 14 decimal digits displayed. Type "format short" to return to a four decimal-digit display. For example:

```
>> format long
>> v = [1 2 3];
>> sqrt(v)
ans =
1.000000000000000  1.414213562373095  1.732050807568877
>> format short
>> sqrt(v)
ans =
 1.0000  1.4142  1.7321
```

The "format" command can also be used to control the spacing between the MATLAB command or expression and the result. For example:

```
>> 6.2*4
ans =
  24.8000
```

```
>> format compact
>> 6.2*4
ans =
   24.8000
```

Table 1.2 shows numerical display formats used in MATLAB.

The command vpa can be used to do variable-precision arithmetic. For example, to print 50 digits of $\sqrt{5}$, enter:

```
>> vpa(sqrt(5), 50)
ans =
2.2360679774997896964091736687312762354406183596115
```

If you don't specify the number of digits, the default setting is 32. You can change the default with the command digits.

MATLAB has some rounding and remainder functions that are very useful. Table 1.3 shows some of these functions.

For example:

```
>> 53/7
ans =
  7.5714
>> round(53/7)
ans =
  8
>> ceil(53/7)
ans =
  8
>> floor(53/7)
```

TABLE 1.2
Numerical Display Formats

MATLAB Command	Display Format	Example
format	Default: same as format short	
format bank	2 real decimal digits	3.47
format compact	Suppresses redundant line	theta = pi/6 theta = 0.5236
format long	14 decimal digits	3.14159265358979
format short	4 decimal digits	3.1416
format rat	Fractional form	377/211

TABLE 1.3
Rounding and Remainder Functions

Function	Description
round	Rounds to the nearest integer
ceil	Rounds to the nearest integer toward positive infinity
floor	Rounds to the nearest integer toward negative infinity
fix	Rounds to the nearest integer toward zero

```
ans =
   7
>> fix (53/7)
ans =
   7
```

1.5 MANAGING VARIABLES

1.5.1 CLEAR COMMAND

The "clc" command clears the command window, leaving a blank page for you to work on. However, this command does not delete from memory the actual variables you have created. The "clear" command deletes all of the saved variables. The action of the "clear" command is reflected in the Workspace Window. If you want to delete a specific variable, type the variable name right after the "clear" command. For example, if you want to delete the variable x, enter

```
>> clear x
```

Table 1.4 shows some options of "clear" command.

1.5.2 COMPUTATIONAL LIMITATIONS AND CONSTANTS

MATLAB includes functions to identify the largest real numbers and the largest integers it can process. MATLAB also keeps some constants, which are values that are known ahead of time and cannot possibly change. An example of a constant value would be pi (π), which is 3.14159265... Table 1.5 shows some computational limitations and constants.

TABLE 1.4
Some Options of the "clear" Command

Option	Description
clear, clear variables	Delete all variables from the Workspace Window
clear global	Delete all global variables
clear functions	Delete all code files compiled and link to mex files
clear all	Delete all variables, global variables, functions, mex links, and Java package import list
clear import	Delete Java package import list (cannot be used within functions)
clear classes	Delete classes

TABLE 1.5
Typical Computational Limitations and Constants

Limits and Constants	Description	Value
eps	Returns the distance from 1.0 to the next largest floating-point number	2.2204e-16
realmax	Returns the largest possible floating-point number	1.7977e+308
realmin	Returns the smallest possible floating-point number	2.2251e-308
pi	π	3.1415926535897
i, j	Imaginary unit	$\sqrt{-1}$
inf	Infinite number	∞
NaN	Not a Number	

1.5.3 "WHOS" COMMAND

The "whos" command shows variables that have been defined in the Command Window. This command shows more information on the variables compared to the "who" command.

```
>> whos
Name    Size         Bytes Class    Attributes
B       3x3          72 double
a       1x3          24 double
b       1x3          24 double
x       3x3          72 double
```

1.6 SYMBOLIC OPERATIONS

1.6.1 CREATION OF SYMBOLIC VARIABLES

Symbolic operations mean doing mathematical operations on symbols. The symbolic math functions are included in the Symbolic Math Toolbox in MATLAB. Enter "help symbolic" to check whether the Symbolic Math Toolbox is installed in your system or not. Simple symbolic variables can be created with the "sym" and "syms" commands. For example, to create a symbolic variable y, enter

```
>> syms y
>> a = 8;
>> a^3 - 4*a*y + y
ans =
512 - 31*y
```

The following creates symbolic variables x and y and defines z as a function of these symbolic variables:

```
>> syms x y
>> z = x^3 - 4*x*y + y^2
z =
x^3 - 4*x*y + y^2
>> 3*y*z
ans =
3*y*(x^3 - 4*x*y + y^2)
```

The expand function will multiply out terms, and the factor function will do the reverse. For example:

```
>> syms x y
>> (x - y)*(x - y)^2
ans =
(x-y)^3
>> expand(ans)
ans =
x^3 - 3*x^2*y + 3*x*y^2 - y^3
>> factor(ans)
ans =
(x-y)^3
```

The simplify function simplifies each part of an expression or equation. For example:

```
>> simplify((x^3 - y^3)/(x-y))
```

```
ans =
x^2 + x*y + y^2
>> syms p x y
>> y = ((x^p)^(p+1))/x^(p-1);
>> simplify(y)
ans =
x*(x^p)^p
```

Both `syms x` and `x = sym('x')` set the character "x" equal to the symbolic variable x. The "`syms`" command is particularly convenient, because it can be used to create multiple symbolic variables at the same time. To set the imaginary unit as a symbolic variable, use `sym(sqrt(-1))` or `sym(i)`. For example:

```
>> syms x y sym(i)
>> z = (x-3*i)*(y+4*i)
z =
- (4*i + y)*(3*i - x)
>> expand(z)
ans =
4*i*x - 3*i*y + x*y - 12*i^2
```

1.6.2 SUBSTITUTION IN SYMBOLIC EQUATIONS

Sometimes it is necessary to substitute numerical values or other symbolic expressions for one or more symbolic variables. The `subs` function will substitute a value for a symbolic variable in an expression. For example, in the following commands, `subs(w, u, 2)` substitutes 2 for the symbolic variable u in the symbolic expression w:

```
>> d = 1, syms u v
d =
   1
>> w = u^2 - v^2
w =
u^2 - v^2
>> subs(w,u,2)
ans =
4 - v^2
>> subs(w,v,d)
ans =
u^2 - 1
>> subs(w,v,u+v)
ans =
u^2 - (u+v)^2
>> simplify(ans)
ans =
-v*(2*u + v)
```

1.7 CODE FILES

Code files are ordinary text files containing MATLAB commands. You can create and modify them using any text editor or word processor that is capable of saving files as plain ASCII text. You can use the built-in editor, which you can start by typing `edit`, either by itself (to edit a new file) or followed by the name of an existing code file in the current directory. The Editor Window can also be started by hitting the New menu on the toolstrip of the FILE group of the desktop HOME tab, as shown in Figure 1.6. In order to run the code file created, the Current Folder should be set to the directory where the code file is stored. There are two different kinds of code files: script code files and function code files.

1.7.1 Script Code Files

A script code file is simply a list of MATLAB statements that are saved in a file with a ".m" file extension. A script code file contains a sequence of MATLAB commands to be run in order. Open the MATLAB editor and create a file containing the following lines:

```
format long
x = [1.5 0.4 0.06];
y = sin(x)./x
```

Save this file with the name *scex1.m* in your current directory, or in some directory in your path. You can name the file any way you like, but the ".m" suffix is mandatory. You can run this script by entering *scex1* in the Command Window. The output will be displayed in the Command Window. The code file can easily be modified. For example, if you wish to calculate $\sin(0.02)/0.02$, you can modify the code file as follows:

```
format long
x = [1.5 0.4 0.06 0.02];
y = sin(x)./x
```

Suppose that you want to treat results of examinations with MATLAB functions such as sort, mean, median, and std. In the MATLAB editor, type the following commands into a code file and save this file with the name *mdat.m*:

```
graddat = [15 0 4 26 75 6 48];
exsort = sort(graddat)
exmean = mean(graddat)
```

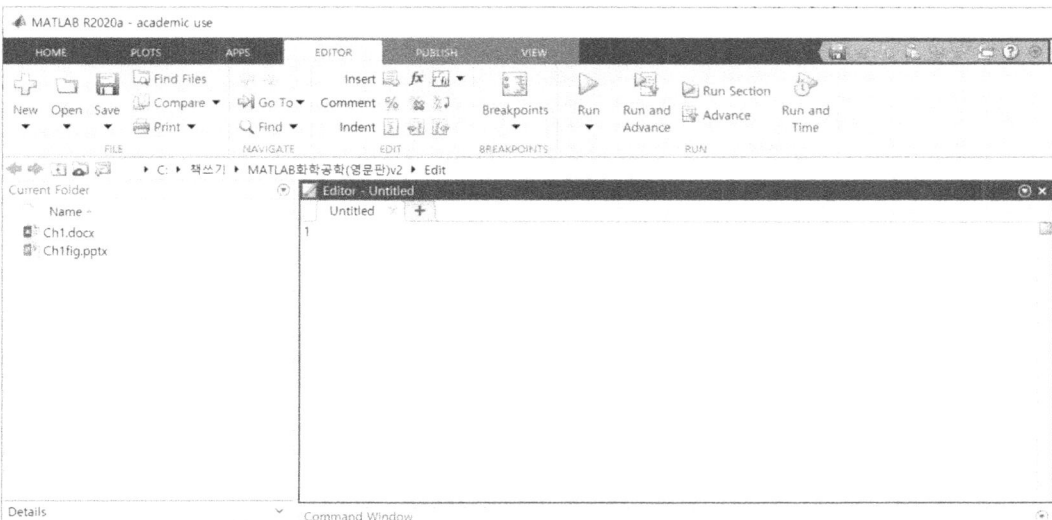

FIGURE 1.6 MATLAB Editor Window.

```
exmed = median(graddat)
exstd = std(graddat)
```

You can run this script by entering `mdat` in the Command Window:

```
>> mdat
exsort =
   0    4    6   15   26   48   75
exmean =
 24.8571
exmed =
  15
exstd =
 27.5586
```

1.7.2 Adding Comments

A comment is an explanatory line for the reader that is ignored by MATLAB when the script is executed. To put in a comment, simply type the % character at the beginning of a line or select the comment lines and then click on the % symbol on the desktop EDITOR tab, as shown in Figure 1.7. In the MATLAB Editor, the comments will be displayed in green.

The following script is a modified *scex1.m* file with new comments:

```
format long    % 14 decimal digits
x = [1.5 0.4 0.06 0.02];
y = sin(x)./x
% the value of y=sin(x)/x approaches to 1
% as x approaches to zero.
```

Longer comments, called comment blocks, consist of everything in between %{ and %}, which must be alone on separate lines. For example:

```
format long    % 14 decimal digits
x = [1.5 0.4 0.06 0.02];
y = sin(x)./x
%{
  Example of comment block: this is long comments
  the value of y=sin(x)/x approaches to 1
  as x approaches to zero.
%}
```

FIGURE 1.7   Adding comments.

1.7.3 Function Code Files

Function code files, unlike script code files, allow you to specify input values when you run them from the MATLAB command line or from another code files. Like a script code file, a function code file is a plain text file that should reside in your current directory or elsewhere in your MATLAB path.

As an example, let's create a function code file that calculates some values of $\sin(x)/x$ with $x = 10^{-a}$ for several values of a. Here is a function code file called $sinx$ designed to execute these calculations; this file is stored in a file called $sinx.m$.

```
function y = sinx(z)
% sinx calculates sin(x)/x for x = 10^(-a),
% where a = 1, ..., z.
format long
a = 1:z;
x = 10.^(-a);
y = (sin(x)./x)';
end
```

The first line of the file starts with function, which identifies the file as a function code file. The Editor colors this special word blue. The first line of the code file specifies the name of the function and describes both input arguments (or parameters) and output values. In this example, the function is called sinx. The file name without the ".m" extension and the function name should match. It is good practice to follow the first line of a function code file with one or more comment lines explaining what the code file does. If you do, "help" will automatically retrieve this information. For example:

```
>> help sinx
sinx calculates sin(x)/x for x = 10^(-a),
where a = 1, ..., z.
```

The following is an example of a call to this function in which z is set to 3 and the value returned is stored in the default variable ans:

```
>> sinx(3)
ans =
  0.998334166468282
  0.999983333416666
  0.999999833333342
```

1.8 FUNCTIONS

MATLAB contains a wide variety of built-in functions. However, you will often find it useful to create your own MATLAB functions.

1.8.1 Built-In Functions

MATLAB has many built-in functions, such as sqrt, cos, sin, tan, log, exp, atan, gamma, erf, and besselj. The function log is the natural logarithm, which is written as "ln." The base 10 logarithm is represented by the function log10 in MATLAB. For example:

```
>> log(exp(5))
ans =
  5
>> log10(exp(5))
```

```
ans =
  2.1715
>> sin(2*pi/5)
ans =
  0.9511
```

Symbolic representation can also be used:

```
>> sin(sym(2*pi/5))
ans =
(2^(1/2)*(5^(1/2) + 5)^(1/2))/4
```

MATLAB also contains many linear algebra functions. Consider a system of linear equations given by

$$- x_1 - 3x_3 = -2$$

$$5x_1 + 2x_2 - 6x_3 = 1$$

$$- 4x_1 + x_2 + 8x_3 = 3$$

Now let

$$A = \begin{bmatrix} -1 & 0 & -3 \\ 5 & 2 & -6 \\ -4 & 1 & 8 \end{bmatrix}, \quad x = \begin{bmatrix} x_1 \\ x_2 \\ x_3 \end{bmatrix}, \quad b = \begin{bmatrix} -2 \\ 1 \\ 3 \end{bmatrix}$$

Then the linear system can be expressed as

$$Ax = b$$

The solution to this system can be obtained by using the backslash function (\):

```
>> A = [-1 0 -3; 5 2 -6; -4 1 8]
A =
  -1    0   -3
   5    2   -6
  -4    1    8
>> b = [-2; 1; 3]
b =
  -2
   1
   3
>> x = A\b
x =
  0.4754
  0.8361
  0.5082
```

1.8.2 USER-DEFINED FUNCTIONS

There are three ways to create functions:

1. Use the inline command.
2. Create an anonymous function using @ operator.
3. Create a separate code file.

Anonymous or inline functions are most useful for defining simple functions that can be expressed in one line and for turning the output of a symbolic command into a function. Function code files are useful for defining functions that require several intermediate commands to compute the output. A separate code file can be created by using the MATLAB Editor, as mentioned before. As an example, let's define the function $f(x) = x^3$. We can use the `inline` command:

```
>> fl = inline('x^3', 'x')
fl =
   Inline function:
   fl(x) = x^3
```

We can also use the @ operator to define $f(x)$ as an anonymous function:

```
>> f = @(x) x^3
f =
 function_handle with value:
   @(x)x^3
```

Once the function is created, we can use it by providing the input value:

```
>> fl(5)
ans =
 125
>> f(5)
ans =
 125
```

Vectors and matrices can also be used as input arguments. Thus, it is good practice to put a dot symbol (.) right before mathematical operators such as *, /, and ^. The function $f(x) = x^3$ can be redefined using the dot symbol as:

```
>> f = @(x) x.^3
f =
 function_handle with value:
   @(x)x.^3
```

   or
```
>> fl = inline('x.^3', 'x')
fl =
   Inline function:
   fl(x) = x.^3
```

The function defined in this manner can accept a vector as an input argument. For example:

```
>> f(2:8)
ans =
  8   27   64   125   216   343   512
```

We can also define a function of more than two independent variables. For example:

```
>> g = @(x, y) x^3 + y^2;
>> g(1,2)
ans =
  5
>> gl = inline('sqrt(x) + y^3', 'x', 'y');
>> gl(1,2)
```

```
ans =
   9
```

If we want to allow vector-matrix operations in the function, we can define the function $g(x)$ as

```
>> g = @(x, y) x.^3 + y.^2;
```

For example, we can get function values at two points:

```
>> g([1 2], [3 4])
ans =
   10   24
```

1.9 LOOPS

MATLAB has four flow control structures: the `if` statement, the `for` loop, the `while` loop, and the `switch` statement.

1.9.1 IF STATEMENT

The simplest form of the `if` statement is

```
if (expression)
   (statements)
end
```

where the `statements` are executed if the elements of `expression` are all nonzero. For example, the following code swaps x and y if x is greater than y:

```
if x > y
   temp = y;
   y = x;
   x = temp;
end
```

One or more further tests can be added with `else if`. There must be no space between `else` and `if`.

1.9.2 FOR LOOP

The basic form of the `for` loop is

```
for (variable) = (expression)
   (statements)
end
```

For example, the sum of the first 25 terms of the harmonic series $1/k$ is calculated by

```
>> s = 0;
>> for k = 1:25, s = s + 1/k; end, s
s =
   3.8160
```

Multiple `for` loops can be nested. The `expression` in the `for` loop can be a matrix, in which case `variable` is assigned the columns of `expression` from first to last.

The `while` loop has the form

```
while (expression)
  (statements)
end
```

The `statements` are executed as long as `expression` is true.

1.10 GRAPHICS

1.10.1 PLOTTING WITH EZPLOT

The simplest way to graph a function of one variable is with `ezplot`. The `ezplot` function expects a string, a symbolic expression, or an anonymous function, representing the function to be plotted. For example, to plot $x^3 - 4x + 3$ on the interval -3 to 3 using the string form of `ezplot`, enter

```
>> ezplot('x^3 - 4*x + 3', [-3 3])
```

The plot will appear on the screen in a new window labeled "Figure 1." Using a symbolic expression, you can produce the plot in Figure 1.8 with the following input:

```
>> syms x, ezplot('x^3 - 4*x + 3', [-3 3])
```

An anonymous function can also be used as the argument to `ezplot` as in

```
>> ezplot(@(x) x.^3 - 4*x + 3, [-3 3])
```

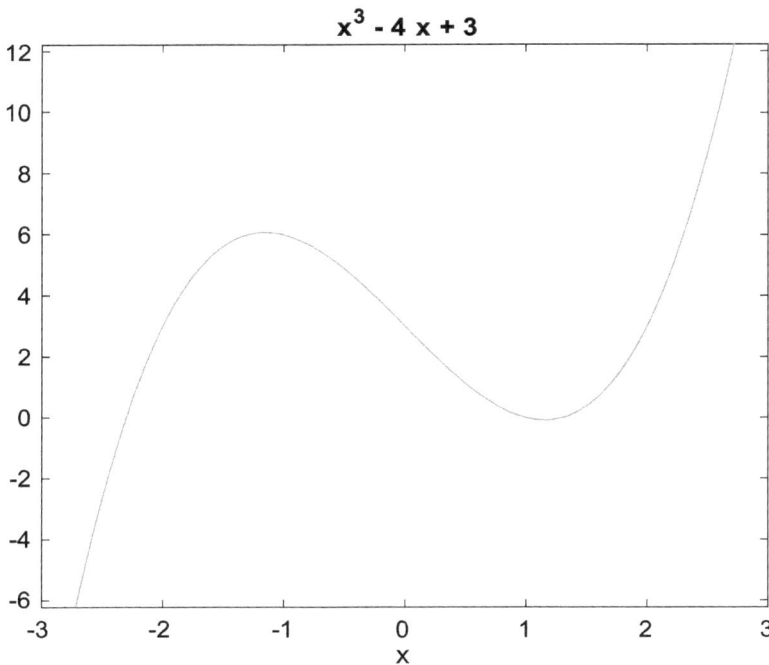

FIGURE 1.8 Plot of $x^3 - 4x + 3$ on the interval [-3 3].

1.10.2 Modifying Graphs

A graph can be modified in a number of ways. You can change the title above the graph in Figure 1.8 by typing in the Command Window:

```
>> title 'Plot of a 3rd-order polynomial'
```

The same change can be made directly in the figure window by selecting Axes Properties... from the Edit menu at the top of the figure window, as shown in Figure 1.9. You can just type the new title in the box marked "Title."

You can add a label on the vertical axis with y label or change the label on the horizontal axis with x label. Also, you can change the horizontal and vertical ranges of the graph with the axis command. For example, to confine the horizontal range to the interval from -2 to 2 and the vertical range to the interval from -2 to 8, type

```
>> axis ([-2 2 -2 8])
```

The first two numbers are the range of the horizontal axis, and the last two numbers are the range of the vertical axis. Both ranges must be included even if only one is changed. To make the shape of the graph square, type axis square. This command also makes the scale the same on both axes if the x and y ranges have equal length. For ranges of any length, you can force the same scale on both axes without changing the shape by typing axis equal.

1.10.3 Graphing with *plot*

The *plot* function works on vectors of numerical data. The basic syntax is plot (X, Y), where X and Y are vectors of the same length. For example:

```
>> X = [1 2 5]; Y = [4 2 8]; plot(X,Y)
```

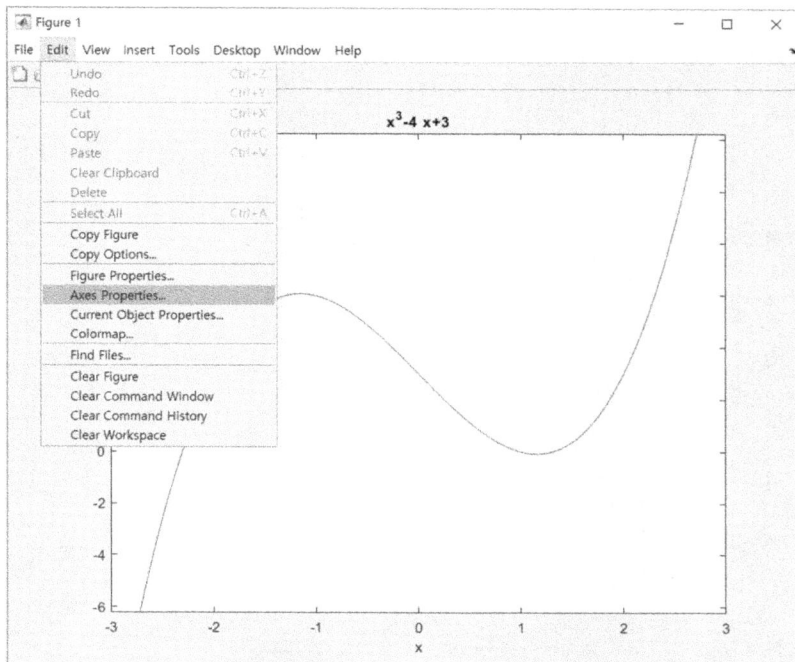

FIGURE 1.9   Edit menu of the Figure window.

The plot function considers the vectors X and Y to be lists of the x and y coordinates of successive points on a graph, and connects the points with line segment. So in Figure 1.10, MATLAB connects (1,4) to (2,2) to (5,8).

In general, the `plot` function generates a two-dimensional (2D) graph:

```
>> t = 0:0.002:1; z = exp(9.8*t.*(t-1)).*sin(10*pi*t);
>> plot(t,z)
```

Entering `plot(t,z)` uses a basic solid line to join the points t(i) with z(i) to create the curve shown in Figure 1.11. The figure window can be closed by typing `close` at the prompt.

To plot $x^3 - 4x + 3$ on the interval [-2, 2], we specify the range of the independent variable x, compute $y = x^3 - 4x + 3$, and then plot the results. The function `grid` introduces a light horizontal and vertical hashing that extends from the axis ticks.

The following code produces the plot shown in Figure 1.12.

```
>> x = -2:0.1:2;
>> y = x.^3 -4*x +3;
>> plot(x,y)
>> grid
>> title('x^3 -4*x +3')
>> xlabel('x'), ylabel('y')
```

1.10.4 PLOTTING MULTIPLE CURVES

Each time you execute a plotting command, MATLAB erases the old plot and draws a new one. If you want to overlay two or more plots, use `hold on`. This command instructs MATLAB to retain the old graphics and draw any new graphics on top of the old ones. It remains in effect until you type `hold off`. Do not forget to enter `hold off` after graphing whenever you use `hold on`. The following is an example using the `ezplot` function to create the plot shown in Figure 1.13:

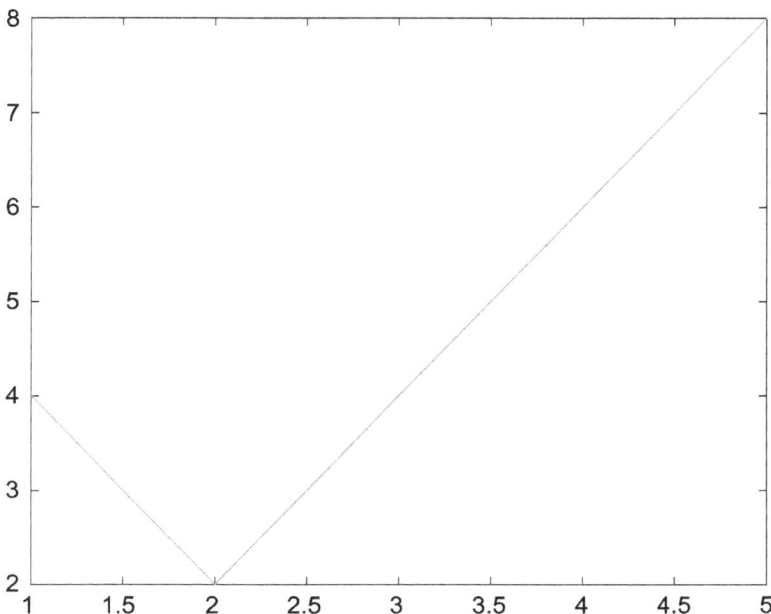

FIGURE 1.10 Plot of line segments.

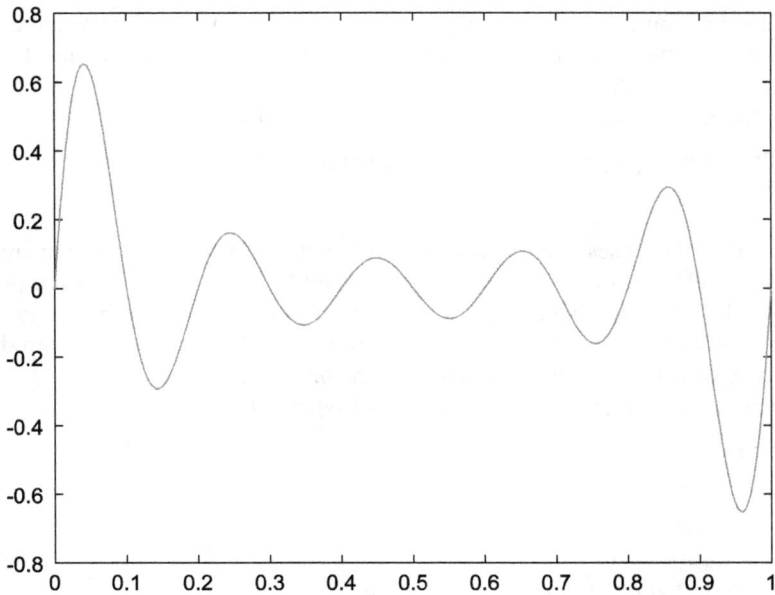

FIGURE 1.11   2D graph by plot function.

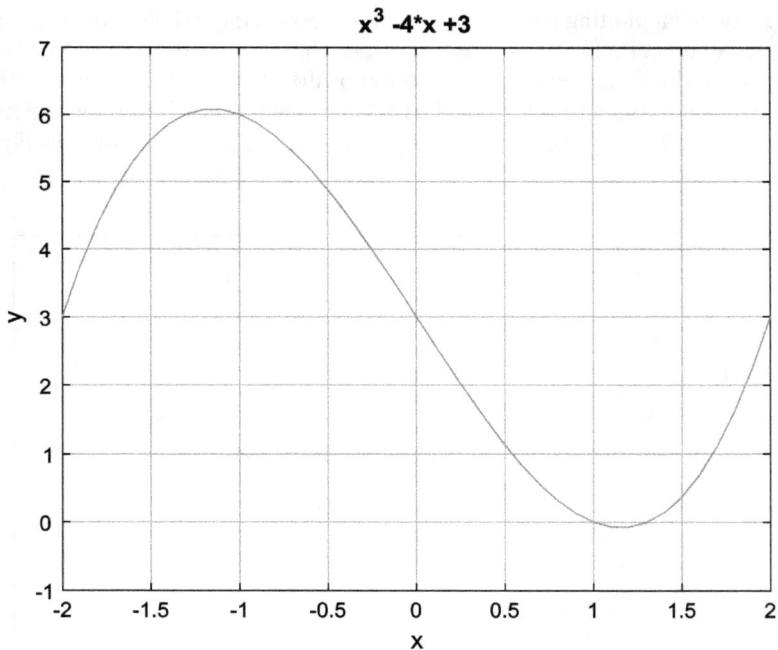

FIGURE 1.12   Graph of $x^3 - 4x + 3$ on the interval [-2, 2].

```
>> ezplot('exp(-x^1.2)', [0 6])
>> hold on
>> ezplot('cos(2*x)', [0 6])
>> hold off
>> title 'exp(-x^{1.2}) and cos(2*x)'
```

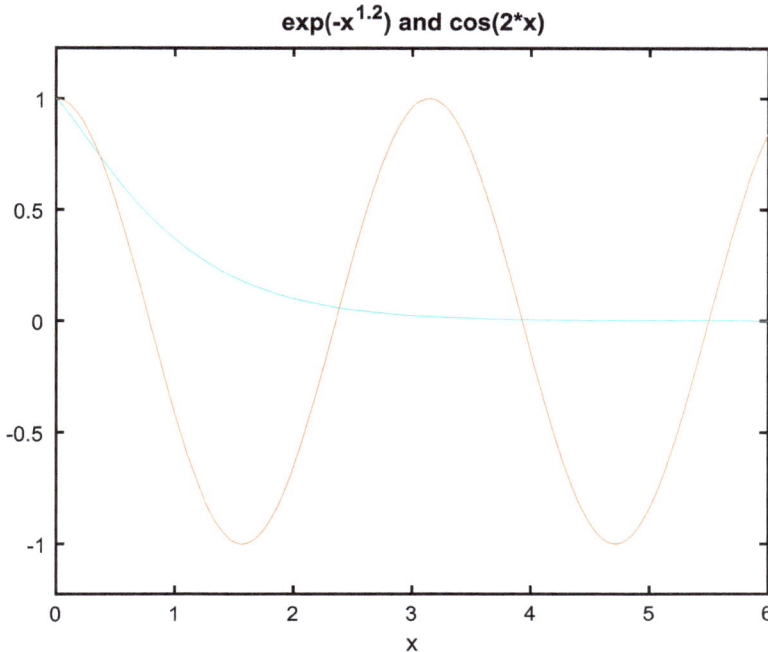

FIGURE 1.13   Plots of $e^{-x^{1.2}}$ and $\cos(2x)$.

The plot function can also be used to generate the curves shown in Figure 1.13:

```
>> x = 0:0.01:6; plot(x, exp(-x.^1.2), x, cos(2*x))
```

The command `hist` produces a histogram that shows the distribution of data by intervals:

```
>> hist(randn(1000,1))
```

Here, `hist` is given 1000 points from the normal (0,1) random number generator `randn` to create the histogram shown in Figure 1.14.

1.10.5 THREE-DIMENSIONAL PLOTS

For plotting curves in three-dimensional space, the basic command is `plot 3`. As an example, plot the following functions on the interval $0 \le t \le 8\pi$ using the function `plot 3`:

$$x = 0.8 \sin(t), \quad y = 1.2 \cos(t)$$

The following commands generate the three-dimensional (3D) curve shown in Figure 1.15:

```
>> t = [0:0.1:8*pi];
>> x = 0.8*sin(t);
>> y = 1.2*cos(t);
>> plot3(x, y, t)
>> xlabel('x '), ylabel('y '), zlabel('t')
>> grid
```

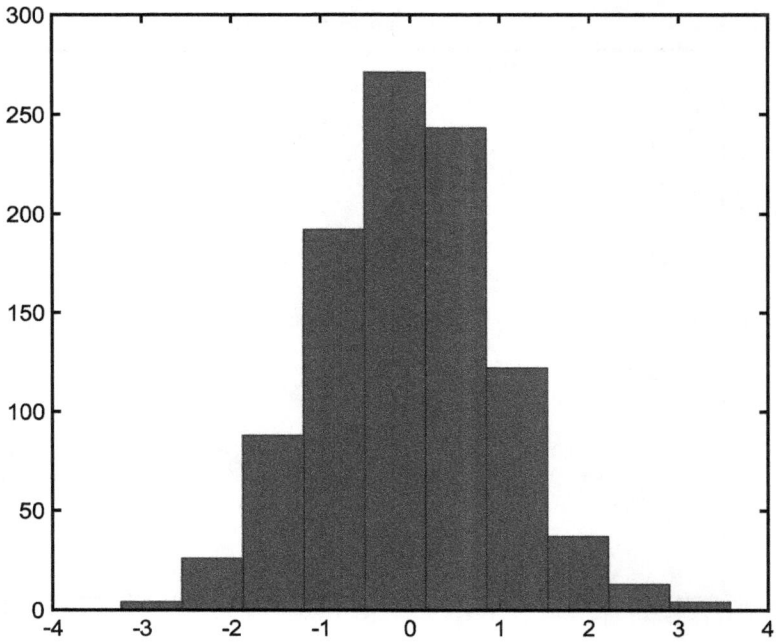

FIGURE 1.14   Histogram by the function hist.

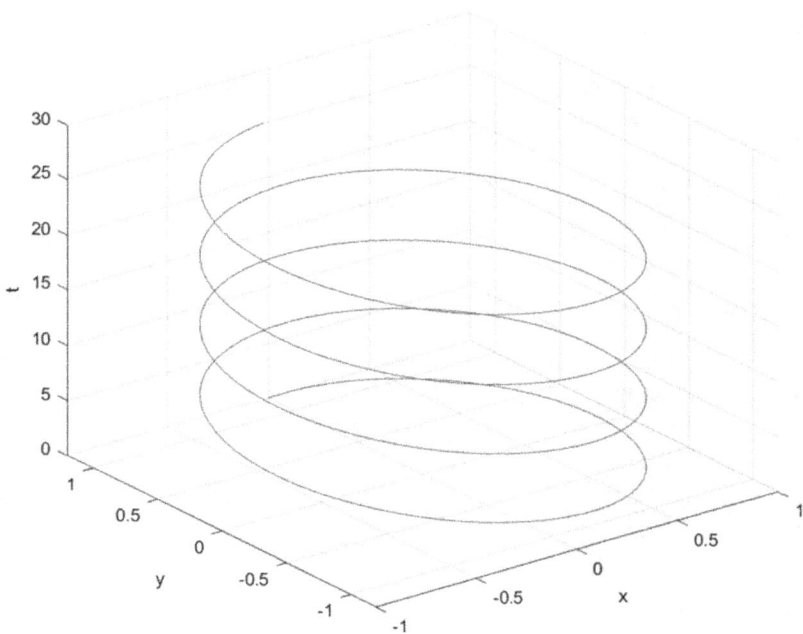

FIGURE 1.15   Example of a 3D curve.

1.11 ORDINARY DIFFERENTIAL EQUATIONS

MATLAB has a range of functions for solving ordinary differential equations (ODEs). Each of MATLAB's ODE solvers is designed to be efficient in specific circumstances. The function ode45 implements an adaptive Runge-Kutta algorithm and is typically the most efficient solver. As an example to use the function ode45, let's solve the following initial-value ODE on the interval $0 \leq t \leq 3$:

$$\frac{dy(t)}{dt} = -y(t) - 5e^{-t}\sin(5t)$$

The function file *ode1.m* defines the ODE equation to be solved:

```
function dy = ode1(t,y)
% dy = myode1(t,y) solves ODE.
dy = - y - 5*exp(-t)*sin(5*t);
end
```

Then, in the Command Window, type:

```
>> tinv = [0 3]; y0 = 1;
>> [t,y] = ode45(@ode1,tinv,y0);
>> plot(t,y)
```

where tinv defines the time interval $0 \leq t \leq 3$ and y 0 = 1 represents the initial condition. The solution is shown in Figure 1.16.

The ODE can also be defined as an anonymous function. The following code will produce the same results:

```
>> tinv = [0 3]; y0 = 1;
>> dy = @(t,y) - y - 5*exp(-t)*sin(5*t); % define ODE as an anonymous function
>> [t,y] = ode45(dy,tinv,y0);
>> plot(t,y)
```

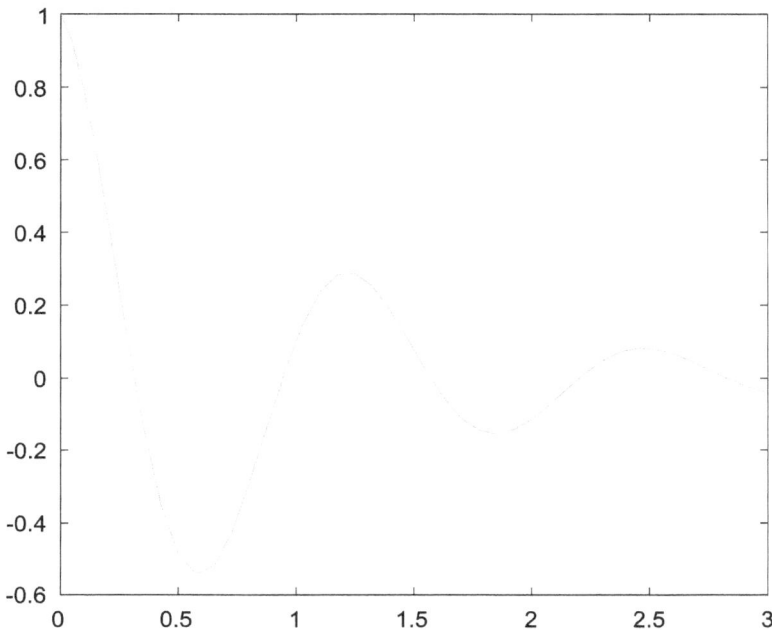

FIGURE 1.16 Solution of a simple ODE.

1.12 CODE FILE EXAMPLES

1.12.1 POPULATION GROWTH MODEL

A population growth model can be expressed as $x_{k+1} = ax_k(1 - x_k)$. The following code file defines this model:

```
function [t, x] = pgmodel(a, xinit, n)
x(1) = xinit; t(1) = 0;
for k = 2:n+1;
  t(k) = k-1;
  x(k) = a*x(k-1)*(1-x(k-1));
end
end
```

This code file should be stored as *pmod.m*. In the Command Window, specify the values of a (=2.9), the initial value (=0.2), and the computing time (=25), and call the function pmod. Figure 1.17 shows the resultant plot.

```
>> [tv, xv] = pgmodel (2.9, 0.2, 25);
>> plot(tv, xv), xlabel('Time'), ylabel('Population')
>> title('Population growth model')
```

1.12.2 RANDOM FIBONACCI SEQUENCE

A random Fibonacci sequence $\{x_n\}$ is generated by choosing x_1 and x_2 and setting

$$x_{n+1} = x_n \pm x_{n-1} \ (n \geq 2)$$

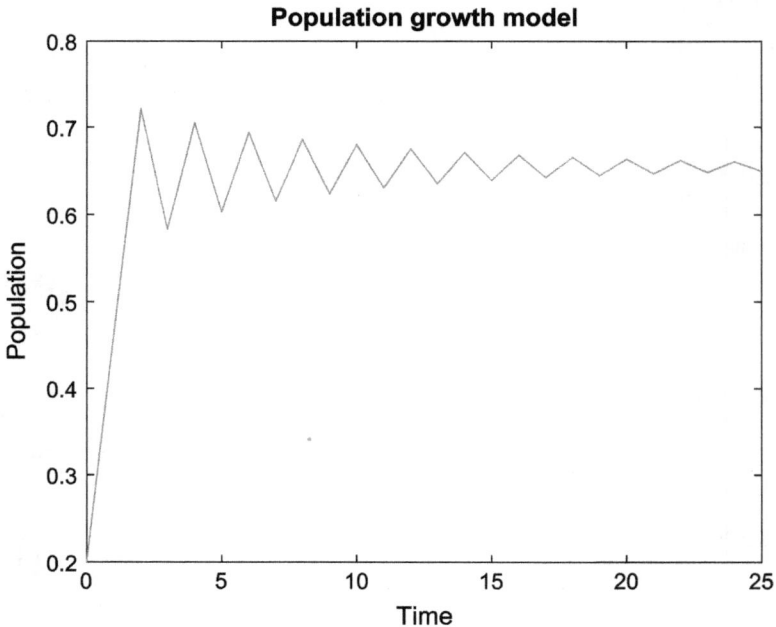

FIGURE 1.17   Population growth curve.

where ± indicates that + and − must have equal probability of being chosen. For large n, the quantity $|x_n|$ increases like a multiple of d^n, where $d = 1.13198824....$ In the MATLAB Editor Window, create the script code file as following and save it (the file name is *ranfib.m*):

```
clear; rand('state',100)    % Set random number state.
m = 1000;        % number of iterations
x = [1 2];       % initial condition
for n = 2:m-1    % for loop
  x(n+1) = x(n) + sign(rand-0.5)*x(n-1);
end
semilogy(1:1000,abs(x))
d = 1.13198824;
hold on
semilogy(1:1000, d.^[1:1000])
hold off
```

Here, the `for` loop stores a random Fibonacci sequence in the array x. MATLAB automatically extends x each time a new element x(n + 1) is assigned. The `semilogy` function then plots n on the x-axis against abs(x) on the y-axis, with logarithmic scaling for the y-axis. Typing `hold on` tells MATLAB to superimpose the next picture on top of the current one. The second `semilogy` function produces a line of slope d. Now type

```
>> ranfib
```

at the command line. This will create at the command line. This will create the graph shown in Figure 1.18.

1.12.3 GENERATION OF A 3D OBJECT

Let's generate a volume-swept 3D object as shown in Figure 1.19. The script *twinobj.m* uses the command `surf(X,Y,Z)` to create a 3D surface where the heights Z are specified at points X and Y in the x-y plane.

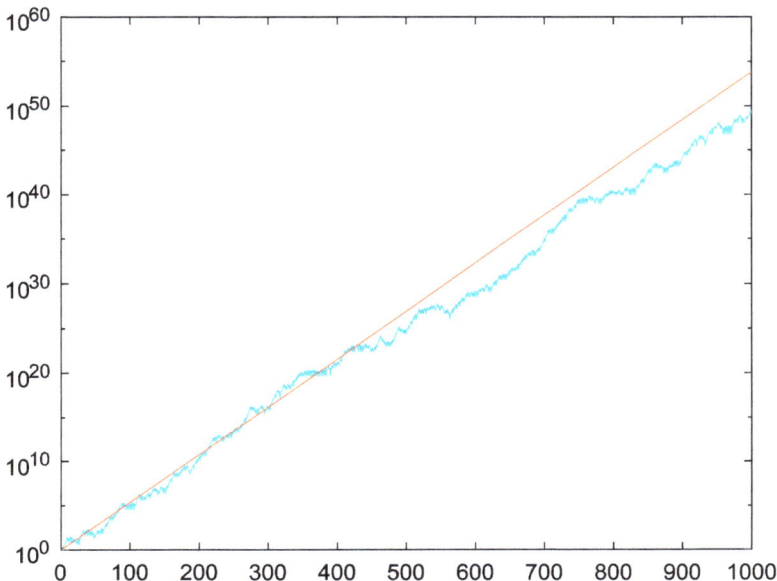

FIGURE 1.18 Random Fibonacci sequence.

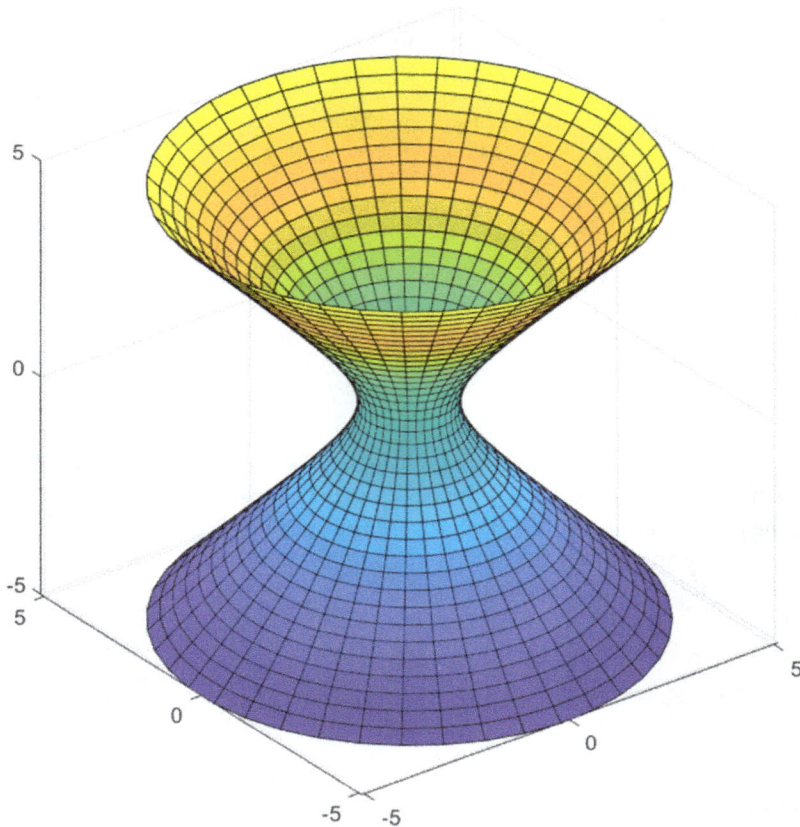

FIGURE 1.19   A volume-swept 3D object.

```
% twinobj.m : generate a volume-swept three-dimensional (3D) object
clear all;
N = 40;  % number of increments
z = linspace(-5,5,N)';
radius = sqrt(1+z.^2);
theta = 2*pi*linspace(0,1,N);
X = radius*cos(theta);
Y = radius*sin(theta);
Z = z(:,ones(1,N));
surf(X,Y,Z)
axis equal
```

1.13 SIMULINK®

1.13.1 SIMULINK BLOCKS

Simulink is a block diagram environment for model-based design and multi-domain simulation. It supports system-level design, automatic code generation, simulation, and continuous test and verification of embedded systems. Simulink provides a graphical editor, customizable block libraries, and solvers for modeling and simulating dynamic systems. Simulink is built on top of MATLAB and integrated with it. Thus, you can incorporate MATLAB algorithms into models and export simulation results to MATLAB for further analysis. You can use Simulink to model a system and then simulate the dynamic behavior of that system. The basic techniques you use to create a simple model in this tutorial are the same as those you use for more complex models.

A Simulink block is a model element that defines a mathematical relationship between its input and output. A block can represent a physical component, a small system, or a function. The Simulink graphical interface enables you to position the blocks, resize them, label them, specify block parameters, and interconnect the blocks to describe complex systems for simulation. Simulink provides block libraries that are collections of blocks grouped by functionality.

1.13.2 CREATION OF A SIMPLE SIMULINK MODEL

To start Simulink, type Simulink in the Command Window or click on the Simulink icon under the HOME tab. The Simulink Start Page opens, as shown in Figure 1.20. If you click on the Blank Model option, an untitled model window will open, as shown in Figure 1.21(a). In the untitled model window, select the Library Browser under the View menu or click on the Library Browser icon in the menu-bar strip ⊞. The Simulink Library Browser window opens, as shown in Figure 1.21(b).

The Simulink blocks are located in these libraries displayed under the Simulink heading in Figure 1.21(b). To select a block from the Simulink Library Browser, click on the name of the appropriate library and a list of blocks within that library appears. To bring a block to the model window, click on the block name or icon, hold the mouse button down, drag the block to the model window, and release the button. As an example, let's take the Sine Wave block from the Source library. You can access "help" for the block by right-clicking on its name or icon and selecting "Help" from the drop-down menu that appears, as shown in Figure 1.22.

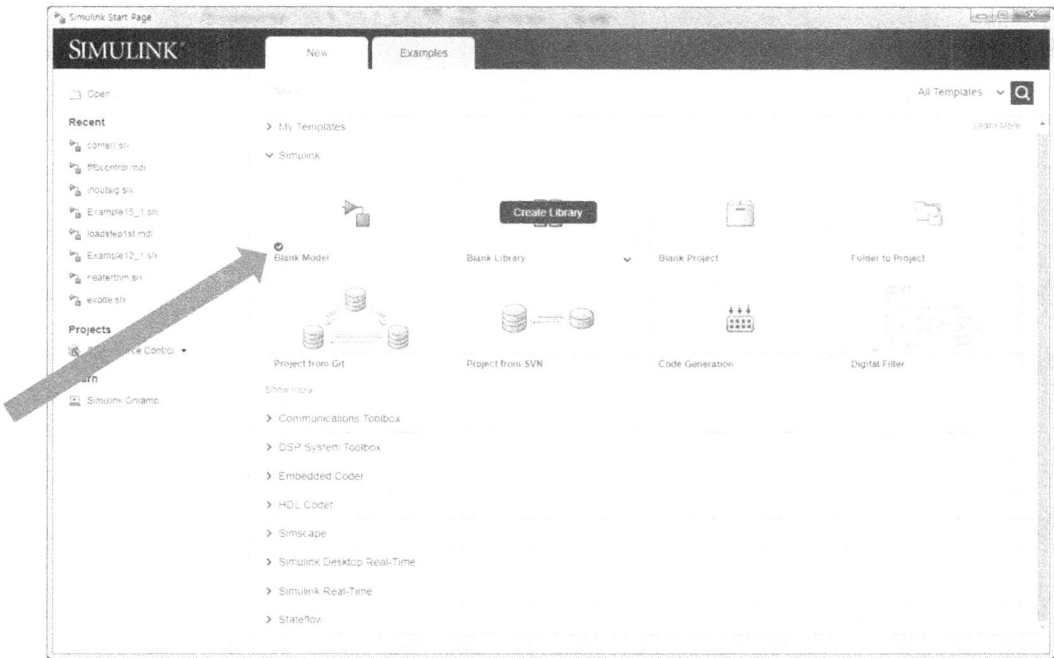

FIGURE 1.20 Simulink Start Page.

(a)

(b)

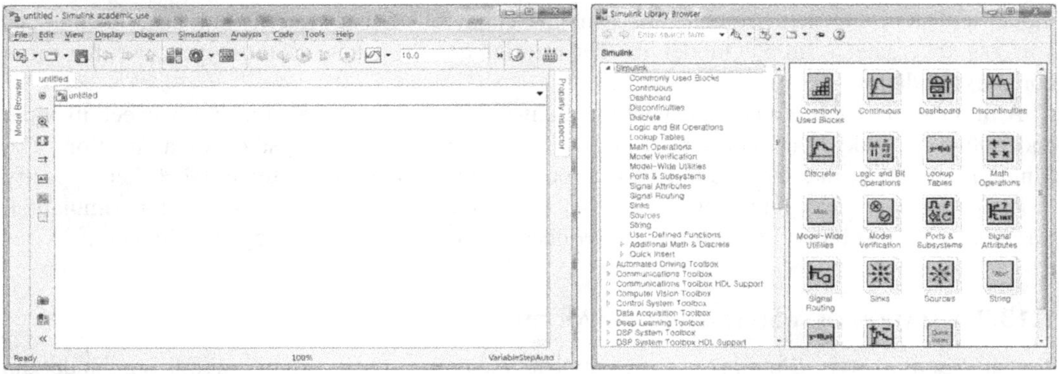

FIGURE 1.21   (a) Untitled model window and (b) Simulink Library Browser.

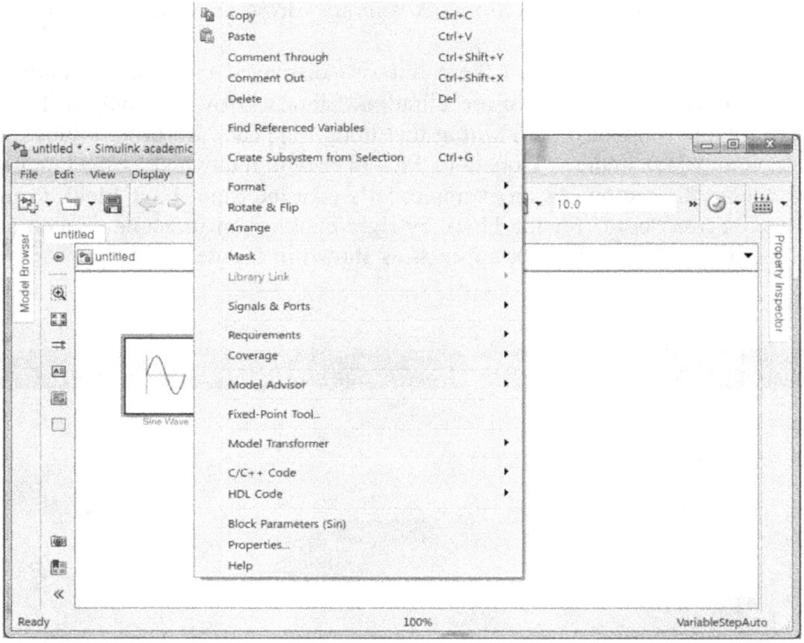

FIGURE 1.22   Drop-down menu for a block.

Let's now use Simulink to build a model to solve the following initial-value problem for $0 \leq t \leq 10$:

$$\frac{dy}{dt} = 5 \sin t, \quad y(0) = 0$$

Step 1: Start Simulink and open a new model window as shown previously.

Step 2: Select the Sine Wave block from the Source library and place it in the model window. Double-click on the block to open the Block Parameters Window shown in Figure 1.23(a). Set the amplitude to 1, the bias to 0, the frequency (rad/sec) to 1, the phase (rad) to 0, and the sample time to 0, and then click the OK button.

(a)   (b)   (c)

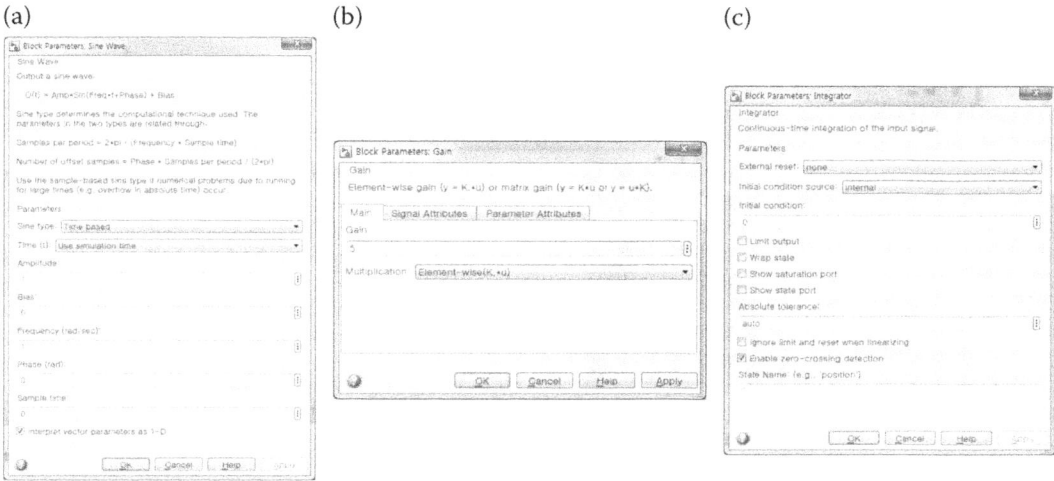

FIGURE 1.23 Block Parameters Windows for (a) Sine Wave block, (b) Gain block, and (c) Integrator block.

Step 3: Select the Gain block from the Math Operations library and place it in the model window. Double-click on the block to open the Block Parameters window shown in Figure 1.23(b). Set the gain value to 5 and then click the OK button. You can see that the value 5 then appears in the Gain block.

Step 4: Select the Integrator block from the Continuous library and place it in the model window. Double-click on the block to open the Block Parameters window shown in Figure 1.23(c). Set the initial condition to 0 and then click the OK button.

Step 5: Select and place the Scope block from the Sinks library in the model window.

Step 6: Connect the input port on each block to the output port on the preceding block. To do this, move the cursor to an input port or an output port, hold the mouse button down, and drag the cursor to a port on another block to be connected. If you release the mouse button, Simulink will connect those blocks with an arrow pointing at the input port. The model should look like that shown in Figure 1.24.

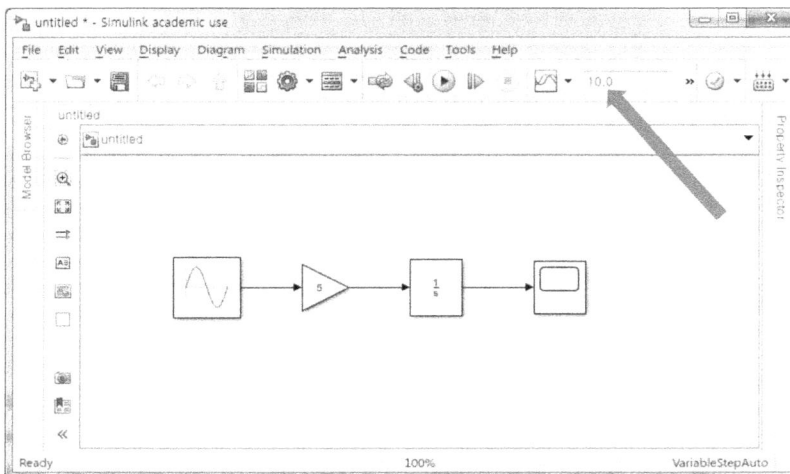

FIGURE 1.24 The Simulink model window showing the blocks connected.

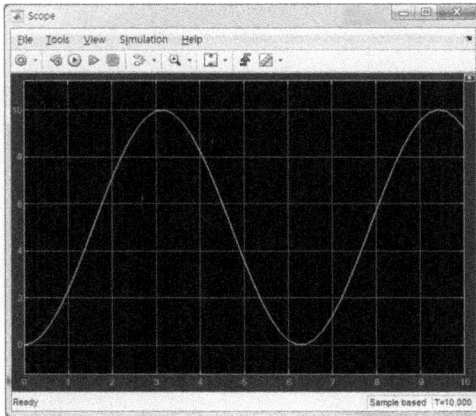

FIGURE 1.25 The Scope block window showing the simulation results.

Step 7: Enter 10.0 for the simulation stop time, as indicated by the long arrow in Figure 1.24. This time value has no unit; time unit in Simulink depends on how the equations are constructed.

Step 8: Run the simulation by clicking on the Run icon (▶) on the toolbar. Double-click on the Scope block to view the response. You can see an oscillating curve with an amplitude of 5 and a period of 2π, as shown in Figure 1.25.

To save the model created, click on the File menu and select the Save As submenu. In the File name text box, enter a name for your model and click Save. The model is saved with the file extension ".slx."

BIBLIOGRAPHY

Attaway, S., *MATLAB*, 3rd ed., Butterworth-Heinemann, Kidlington, Oxford, UK, 2013.
Burstein, L., *MATLAB in Quality Assurance Sciences*, Woodhead Publishing, Sawston, Cambridge, UK, 2015.
Gdeisat, M. and F. Lilley, *MATLAB by Example: Programming Basics*, Elsevier, Waltham, MA, 2013.
Gilat, A., *MATLAB: An Introduction with Applications*, 4th ed., John Wiley & Sons, Inc., Hoboken, NJ, 2011.
Hanselman, D. and B. Littlefield, *Mastering MATLAB 7*, Pearson Education International, Inc., Upper Saddle River, NJ, 2005.
Higham, D. J. and N. J. Higham, *MATLAB Guide*, 2nd ed., Siam, New York, NY, 2005.
Hunt, B. R., R. L. Lipsman and J. M. Rosenberg, *A Guide to MATLAB*, 2nd ed., Cambridge Press, Cambridge, UK, 2006.
Knight, A., *Basics of MATLAB and Beyond*, Chapman & Hall/CRC, Boca Raton, FL, 2000.
McMahon, D., *MATLAB Demystified*, McGraw-Hill, New York, NY, 2007.
Moore, H., *MATLAB for Engineers*, 2nd ed., Pearson Education, Inc., Upper Saddle River, NJ, 2009.
Palm III, W. J., *Introduction to MATLAB for Engineers*, 4th ed., McGraw-Hill, New York, NY, 2019.
Yeo, Y. K., *Introduction to MATLAB Programming* (in Korean), 5th ed., Ajin, Seoul, Korea, 2019.

2 Numerical Methods with MATLAB®

2.1 LINEAR SYSTEMS

Linear algebraic equations arise from the analysis of linear systems whose responses are proportional to the inputs. When chemical processes are modeled mathematically, they are occasionally described by systems of linear algebraic equations. In particular, linear equation sets occurring in many chemical engineering problems are often very large and consume a lot of computational resources. In this section, we will examine how such equation systems are solved. In most cases, it is possible to reduce the computational run time and the storage requirements by employing some special properties of linear algebra. Let's consider the simultaneous linear equations with n unknowns ($x_1, x_2, \ldots, x_n$) and m equations given by

$$a_{11}x_1 + a_{12}x_2 + \cdots + a_{1n}x_n = b_1$$
$$a_{21}x_1 + a_{22}x_2 + \cdots + a_{2n}x_n = b_2$$
$$\vdots$$
$$a_{m1}x_1 + a_{m2}x_2 + \cdots + a_{mn}x_n = b_m$$

These equations can be expressed in terms of vectors and matrices as

$$Ax = b$$

where
   A is an $m \times n$ matrix of known coefficients
   x is the column vector of n unknowns
   b is a column vector of m known coefficients as shown below:

$$A = \begin{bmatrix} a_{11} & a_{12} & \cdots & a_{1n} \\ a_{21} & a_{22} & \cdots & a_{2n} \\ \vdots & \vdots & \ddots & \vdots \\ a_{m1} & a_{m2} & \cdots & a_{mn} \end{bmatrix}, x = \begin{bmatrix} x_1 \\ x_2 \\ \vdots \\ x_n \end{bmatrix}, b = \begin{bmatrix} b_1 \\ b_2 \\ \vdots \\ b_m \end{bmatrix}$$

There are many numerical methods and algorithms dedicated to the solution of large sets of linear equations, each one tailored to a specific form of the coefficient matrix.

2.1.1 GAUSS ELIMINATION METHOD

The Gauss elimination method consists of the elimination phase and the solution phase. In the elimination phase, the system $Ax = b$ is transformed into the form $Ux = c$, where U is an upper triangular matrix. If the ith row is a typical row below the pivot equation to be transformed, we multiply the pivot row by $\lambda = a_{ik}/a_{kk}$ and subtract it from the ith row:

$$a_{ij} \leftarrow a_{ij} - \lambda a_{kj}, \, b_i \leftarrow b_i - \lambda b_k \, (j = k, \, k+1, \, \ldots, n)$$

In the solution phase, the solution is obtained by back substitution as follows:

$$x_k = \frac{1}{a_{kk}}\left(b_k - \sum_{j=k+1}^{n} a_{kj}x_j\right)(k = n - 1, n - 2, ..., 1)$$

The function *gausselm* performs the elimination and the back substitution phases.

```
function x = gausselm(A,b)
% Solves Ax=b by Gauss elimination method (A: square(m=n))
b = (b(:).').'; % b must be a column vector
n = length(b);
% Elimination phase
for k = 1:n-1
  for i = k+1:n
    if A(i,k) ~= 0
      c = A(i,k)/A(k,k); A(i,k+1:n) = A(i,k+1:n) - c*A(k,k+1:n);
      b(i) = b(i) - c*b(k);
    end
  end
end
% Solution phase: back substitution
for k = n:-1:1, b(k) = (b(k) - A(k,k+1:n)*b(k+1:n))/A(k,k); end
x = b; % vector b contains the solution
end
```

2.1.2 GAUSS-SEIDEL METHOD

Extracting the term containing x_i, the linear system $Ax = b$ can be rewritten as

$$a_{ii}x_i + \sum_{j=1, j\neq i}^{n} a_{ij}x_j = b_i (i = 1, 2, ..., n)$$

Solving for x_i, we obtain

$$x_i = \frac{1}{a_{ii}}\left(b_i - \sum_{j=1, j\neq i}^{n} a_{ij}x_j\right)(i = 1, 2, ..., n)$$

From this equation we have the following iterative formula:

$$x_i \leftarrow \frac{\rho}{a_{ii}}\left(b_i - \sum_{j=1, j\neq i}^{n} a_{ij}x_j\right) + (1 - \rho)x_i (i = 1, 2, ..., n)$$

where ρ is called the relaxation factor. In the Gauss-Seidel method, the convergence is guaranteed only if the coefficient matrix is diagonally dominant (each diagonal element is larger than the sum of the other elements in the same row). The function *GaussSeidel* implements the Gauss-Seidel method with relaxation.

```
function x = GaussSeidel(A,b,rho)
% Solves Ax=b by Gauss-Seidel iterative method (A should be square and diag-
onally dominant)
% input:
%   A: coefficient matrix
%   b: constant vector
%   rho: relaxation factor
```

```
% Initialization
itmax = 500; tol = 1e-8;
b = (b(:).')'; % b should be column vector
n = length(b); [nr nc] = size(A);
if nr ~= nc, error('Matrix A is not square.'); end
if nr ~= n, error('Matrix A and vector b are not consistent.'); end
if det(A) == 0, fprintf('\n Rank = %7.3g\n',rank(A)); error('A is sin-
gular.'); end
% Iteration
M = A;
for k = 1:n, M(k,k) = 0; x(k) = 0; end % initial values
for k = 1:n % check diagonal dominancy of A
  if sum(M(k,:)) > A(k,k), disp('A is not diagonally dominant.'); return; end
end;
for k = 1:n, M(k,:) = M(k,:)/A(k,k); s(k) = b(k)/A(k,k); end
iter = 0; x = (x(:).')';
while (1)
  x0 = x;
  for k = 1:n
    x(k) = rho*(s(k) - M(k,:)*x) + (1-rho)*x(k);
    if x(k) ~= 0, err(k) = abs((x(k) - x0(k))/x(k)); end
  end
  iter = iter + 1;
  if max(err) <= tol | iter >= itmax, break; end
end
end
```

2.1.3 Conjugate Gradient Method

Let's consider the problem of finding the vector x that minimizes the quadratic function given by

$$f(x) = \frac{1}{2}x^T A x - b^T x$$

where A is a positive definite symmetric matrix. The value of $f(x)$ is minimized when its first derivative $\nabla f = Ax - b$ is zero. Thus we can see that the minimization problem is equivalent to solving $Ax = b$. The conjugate gradient method starts with an initial vector x_0. At each iterative cycle k, the solution vector x_{k+1} is updated by

$$x_{k+1} = x_k + \alpha_k d_k$$

The step length α_k, selected so that x_{k+1} minimizes $f(x)$ in the search direction d_k, is given by

$$\alpha_k = \frac{d_k{}^T s_k}{d_k{}^T A d_k}$$

where $s_k = b - Ax_k$. The search direction is updated by $d_{k+1} = s_{k+1} + \beta_k d_k$, where the constant β_k is given by

$$\beta_k = -\frac{s_{k+1}{}^T A d_k}{d_k{}^T A d_k}$$

The conjugate gradient method can handle large and sparse linear systems effectively, but convergence is guaranteed only if the coefficient matrix is diagonally dominant. The function *congrad* implements the conjugate gradient method.

```
function x = congrad(A,x0,b)
% Solves Ax=b by conjugate gradient method (A is square)
function x = congrad(A,x0,b)
% Solves Ax=b by conjugate gradientmethod (A is square)
% input:
%   A: coefficient matrix
%   b: constant vector (A*x=b)
%   x0: initial guess
% Initialization
itmax = 500; tol = 1e-6;
b = (b(:).')'; x0 = (x0(:).')'; % b and x0 should be column vector
if det(A) == 0
  fprintf('\n Rank = %7.3g\n',rank(A)); error('Matrix A is singular.');
end
n = length(b); x = x0; M = A;
for k = 1:n % check diagonal dominancy of A
  if sum(M(k,:)) > A(k,k), disp('A is not diagonally dominant.'); return; end
end;
% Iteration
s = b - A*x; d = s;
for k = 1:n
  v = A*d; alpa = dot(d,s)/dot(d,v); x = x + alpa*d; s = b - A*x;
  if sqrt(dot(s,s)) < tol, return;
  else, beta = -dot(s,v)/dot(d,v); d = s + beta*d; end
end
end
```

Example 2.1 Solution of a Linear System

Use Gauss elimination, Gauss-Seidel, and conjugate gradient methods to solve $Ax = b$ where

$$A = \begin{bmatrix} 2 & -1 & 0 & 0 & 0 & 0 \\ -1 & 2 & -1 & 0 & 0 & 0 \\ 0 & -1 & 2 & -1 & 0 & 0 \\ 0 & 0 & -1 & 2 & -1 & 0 \\ 0 & 0 & 0 & -1 & 2 & -1 \\ 0 & 0 & 0 & 0 & -1 & 2 \end{bmatrix}, b = \begin{bmatrix} 0 \\ 0 \\ 0 \\ 0 \\ 0 \\ 4 \end{bmatrix}$$

Use $\rho = 0.8$ and $x_0 = [0\ 0\ 0\ 0\ 0\ 0]^T$.

Solution

The script *trilin* defines the given linear system and finds the solution for the system using three methods.

```
% trilin.m: solve a linear system
% Gauss elimination, Gauss-Seidel and conjugate gradient methods are used.
A = zeros(6,6); for i = 1:6, A(i,i) = 2; end, for i = 1:5, A(i,i+1) = -1; A(i+1,i) =
-1; end
b = zeros(6,1); b(6,1) = 4; rho = 1; x0 = zeros(1,6);
xg = gausselm(A,b); xs = GaussSeidel(A,b,rho); xc = congrad(A,x0,b);
display('x by Gauss elimination method: '), display(xg');
display('x by Gauss-Seidel method: '), display(xs');
display('x by conjugate gradient method: '), display(xc');
```

```
>> trilin
x by Gauss elimination method:
  0.5714   1.1429   1.7143   2.2857   2.8571   3.4286
x by Gauss-Seidel method:
  0.5714   1.1429   1.7143   2.2857   2.8571   3.4286
x by conjugate gradient method:
  0.5714   1.1429   1.7143   2.2857   2.8571   3.4286
```

2.1.4 USE OF MATLAB BUILT-IN FUNCTIONS

If the coefficient matrix A is square and nonsingular, the solution of the linear system is given by

$$x = A^{-1}b$$

If matrix A is not square (i.e., $A \in R^{m \times n}$, $m > n$), we cannot compute A^{-1} and the solution cannot be obtained from the above relation. If we multiply A^T on both sides of the linear system, we have

$$A^T A x = A^T b$$

The product $A^T A$ is square, so we can find the inverse of the product. Thus we obtain

$$x = (A^T A)^{-1} A^T b$$

Letting $A^+ = (A^T A)^{-1} A^T$, we have

$$x = A^+ b$$

The matrix A^+ is called Moore-Penrose pseudo-inverse of A. We can use MATLAB's built-in function *pinv* to compute A^+. In most cases, the solution of linear systems can be found by using the MATLAB backslash operator (\). Table 2.1 summarizes the MATLAB built-in functions that can be used to solve the linear system $Ax = b$.

Example 2.2 Pseudo-Inverse of a Matrix

Use MATLAB's built-in function *pinv* to compute the pseudo-inverse of

$$A = \begin{bmatrix} -2 & -1 & 3 \\ 4 & -5 & 7 \\ 6 & 2 & -5 \\ -3 & 2 & 1 \end{bmatrix}$$

TABLE 2.1
MATLAB Built-In Solvers for the Linear System $Ax = b$

Name	Syntax	Description
\	x = A\b	Backslash or left matrix divide operator
pcg	x = pcg(A,b)	Uses preconditioned conjugate gradients method
minres	x = minres(A,b)	Finds a minimum norm residual solution
bicg	x = bicg(A,b)	Uses biconjugate gradients method
gmres	x = gmres(A,b)	Uses generalized minimum residual method
qmr	x = qmr(A,b)	Uses quasi-minimal residual method

Solution

```
>> A = [-2 -1 3; 4 -5 7; 6 2 -5; -3 2 1];
>> Ap = pinv(A)
Ap =
  -0.0109   0.0948   0.1437   0.0878
   0.0506   0.0615   0.2103   0.4689
   0.0620   0.1294   0.0744   0.2802
```

Example 2.3 Solution of a Linear System Using the Backslash Operator (\)

Use the backslash operator to solve the following equation system:

$$4x_1 + 2x_2 - x_3 = 8$$
$$-3x_1 + x_2 + 2x_3 = -6$$
$$2x_1 - 4x_2 + x_3 = 12$$

Solution

The solution is given by x = A\c. In this problem, the coefficient matrix A is nonsingular and the inverse of A can be used to find the solution.

```
>> A = [4 2 -1; -3 1 2; 2 -4 1]; b = [8; -6; 12];
>> x = A\b
x =
   3
  -1
   2
>> x = inv(A)*b
x =
   3
  -1
   2
```

Example 2.4 Heat Conduction by Conjugate Gradients Method

Figure 2.1 shows a flat square plate with its sides held at constant temperatures. Find the temperature at each node x_1, x_2, x_3, and x_4 using the conjugate gradients method. Each dot represents a node, and the temperature at each node is assumed to be given by the average temperatures of adjacent nodes.

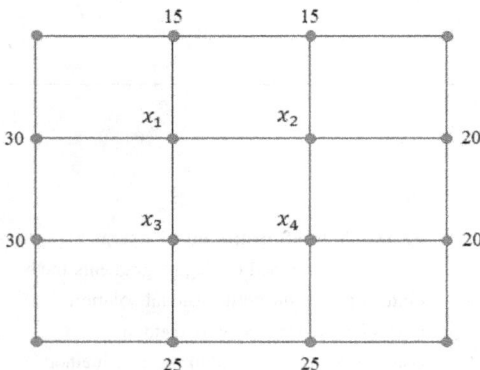

FIGURE 2.1 Flat square plate with heat transfer.

Solution

$$x_1 = \frac{1}{4}(30 + 15 + x_2 + x_3), \; x_2 = \frac{1}{4}(x_1 + 15 + 20 + x_4), \; x_3 = \frac{1}{4}(30 + x_1 + x_4 + 25),$$

$$x_4 = \frac{1}{4}(x_3 + x_2 + 20 + 25)$$

This relationship can be rearranged in the form of a system of linear equations as

$$4x_1 - x_2 - x_3 = 45, \; -x_1 + 4x_2 - x_4 = 35, \; -x_1 + 4x_3 - x_4 = 55, \; -x_2 - x_3 + 4x_4 = 45$$

or

$$\begin{bmatrix} 4 & -1 & -1 & 0 \\ -1 & 4 & 0 & -1 \\ -1 & 0 & 4 & -1 \\ 0 & -1 & -1 & 4 \end{bmatrix} \begin{bmatrix} x_1 \\ x_2 \\ x_3 \\ x_4 \end{bmatrix} = \begin{bmatrix} 45 \\ 35 \\ 55 \\ 45 \end{bmatrix}$$

We can use the built-in *pcg* function.

```
>> A = [4 -1 -1 0; -1 4 0 -1; -1 0 4 -1; 0 -1 -1 4]; b = [45 35 55 45]';
>> x = pcg(A,b)
x =
  22.5000
  20.0000
  25.0000
  22.5000
```

Example 2.5 Two-Dimensional Heat Transfer[1]

Consider the cross section of a rectangular flue with steady heat conduction along the x- and y-axes as shown in Figure 2.2. Since both ends of the section are symmetrical, we can consider only 1/8 of the section. From this section, we can construct a node network consisting of small squares.

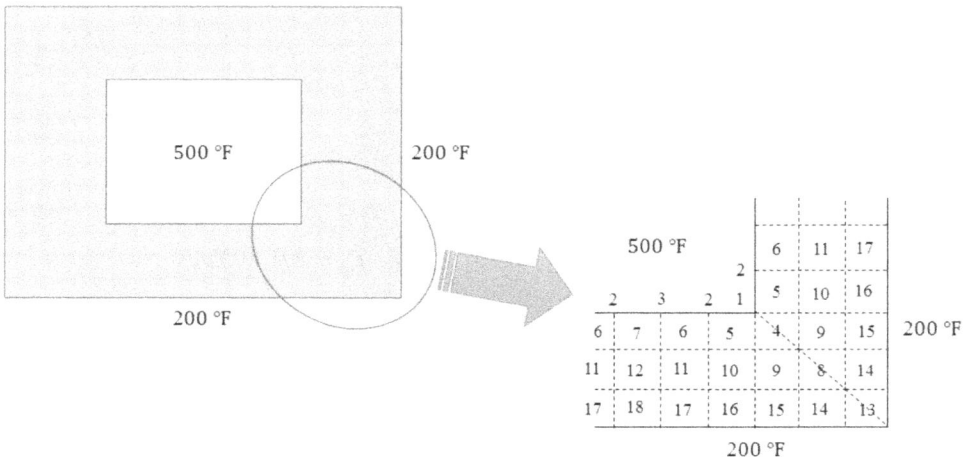

FIGURE 2.2 Cross section of rectangular flue. (From Kapuno, R. R. A., *Programming for Chemical Engineers*, Infinity Science Press, Hingham, MA, 2008, p. 122.)

A heat balance on node 4 gives

$$q_4 = \frac{kA}{\Delta x}(T_5 - T_4) + \frac{kA}{\Delta x}(T_9 - T_4) + \frac{kA}{\Delta y}(T_5 - T_4) + \frac{kA}{\Delta y}(T_9 - T_4)$$

Since each node is square, $\Delta x = \Delta y$, and by rearranging we have

$$\frac{q_4 \Delta x}{kA} = 2T_5 + 2T_9 - 4T_4$$

At steady state, q_4/kA is assumed to be zero. Find temperatures T_4, T_5 ..., T_{12} at nodes 4 through 12.

Solution

From the heat balance at each node, and considering the boundary conditions, we have

$$T_1 = T_2 = T_3 = 500°F, \; T_{13} = T_{14} = T_{15} = T_{16} = T_{17} = T_{18} = 200°F$$

At node 4: $-4T_4 + 2T_5 + 2T_9 = 0$, At node 5 : $T_4 - 4T_5 + T_6 + T_{10} = -500$
At node 6: $T_5 - 4T_6 + T_7 + T_{11} = -500$, At node 7 : $2T_6 - 4T_7 + T_{12} = -500$
At node 8: $-4T_8 + 2T_9 = -400$, At node 9 : $T_4 + T_8 - 4T_9 + T_{10} = -200$
At node 10: $T_5 + T_9 - 4T_{10} + T_{11} = -200$, At node 11 : $T_6 + T_{10} - 4T_{11} + T_{12} = -200$
At node 12: $T_7 + 2T_{11} - 4T_{12} = -200$

These equations can be solved by using the backslash operator:

```
>> A = [-4 2 0 0 0 2 0 0 0;1 -4 1 0 0 0 1 0 0;0 1 -4 1 0 0 0 1 0; ...
   0 0 2 -4 0 0 0 0 1;0 0 0 0 -4 2 0 0 0;1 0 0 0 1 -4 1 0 0;...
   0 1 0 0 0 1 -4 1 0;0 0 1 0 0 0 1 -4 1;0 0 1 0 0 0 0 2 -4];
>> b = [0 -500 -500 -500 -400 -200 -200 -200 -200]';
>> x = A\b
x =
 311.9011
 369.4600
 387.6418
 391.7217
 227.1712
 254.3423
 278.2969
 289.3855
 291.6032
```

The results are summarized as follows:

$T_4 = 311.9011°F$, $T_5 = 369.4600°F$, $T_6 = 387.6418°F$, $T_7 = 391.7217°F$, $T_8 = 227.1712°F$,

$T_9 = 254.3423°F$, $T_{10} = 278.2969°F$, $T_{11} = 289.3855°F$, $T_{12} = 291.6032°F$

2.2 NONLINEAR EQUATIONS

2.2.1 POLYNOMIAL EQUATIONS

In MATLAB, a polynomial is represented by a row vector whose elements denote each coefficient sorted by descending power. The solution of the polynomial equation

$$f(x) = a_n x^n + a_{n-1} x^{n-1} + \cdots + a_1 x + a_0 = 0$$

is given by MATLAB's built-in function *roots*. If we define the coefficient vector as $p = [a_n \; a_{n-1} \; \cdots \; a_1 \; a_0]$, the solution is obtained from

```
x = roots(p)
```

Example 2.6 Solution of a Polynomial Equation

Find the roots of the following equation:

$$f(x) = x^5 - 3x^4 + 3x^3 - 2x^2 - 4x + 1 = 0$$

Solution

We can use the built-in function *roots*. But first we have to construct a coefficient vector with the coefficients sorted in a descending order:

```
>> c = [1 -3 3 -2 -4 1]; % coefficient vector
>> x = roots(c)
x =
 2.3600 + 0.0000i
 0.5980 + 1.4065i
 0.5980 - 1.4065i
-0.7865 + 0.0000i
 0.2306 + 0.0000i
```

Example 2.7 Specific Volume of CO_2

The van der Waals equation of state is given by

$$\left(P + \frac{a}{v^2}\right)(v - b) = RT$$

In this equation, $v = (V/n)$ (n: number of moles), $R = 0.082054 \; liter \cdot atm/(mol \cdot K)$, and $a = 3.592$ and $b = 0.04267$ for CO_2. Find the specific volume (*liter/mol*) of CO_2 when $P = 12 \; atm$ and $T = 315.6 \; K$.

Solution

The van der Waals equation can be rearranged as

$$Pv^3 - (bP + RT)v^2 + av - ab = 0$$

which is a 3rd-order polynomial with respect to v. Therefore, the specific volume can be obtained by using the *roots* function:

```
>> P = 12; T = 315.6; R = 0.08205; a = 3.592; b = 0.04267;
>> roots([P -(b*P+R*T) a -a*b])
ans =
 2.0582 + 0.0000i
 0.0712 + 0.0337i
 0.0712 - 0.0337i
```

Since v should be a real number, we take $v = 2.0582 \; liter/mol$.

2.2.2 ZEROS OF NONLINEAR EQUATIONS

Several numerical methods can be used to find zeros of nonlinear functions of a single or multiple variables. The following is an overview of several numerical techniques for finding zeros of a nonlinear equation of one variable given by, $f(x) = 0$ where $x \in R^1$.

2.2.2.1 Bisection Method

In this method, an initial interval $[x_1, x_2]$ is selected first so that $f(x_1)f(x_2) < 0$. The midpoint m is calculated by $m = (x_1 + x_2)/2$. If $f(x_1)f(m) < 0$, the root is in $[x_1, m]$ and set $x_2 = m$. Otherwise, the root is in $[m, x_2]$ and set $x_1 = m$. This process is repeated until the length of the search interval is less than the prescribed small value. The bisection method can be summarized as follows:

1. Determine two initial values x_1 and x_2 so that $f(x_1)f(x_2) < 0$.
2. Find the midpoint $x_3 = (x_1 + x_2)/2$ and evaluate $f(x_3)$.
3. If $f(x_1)f(x_3) > 0$, replace x_1 by x_3 (i.e., $x_1 = x_3$). Otherwise, replace x_2 by x_3 (i.e., $x_2 = x_3$).
4. Check for convergence. If it has not occurred, go back to the second step.

This iterative procedure continues until the size of the interval $|x_2 - x_1|$ shrinks below a certain tolerance level. The bisection method is also known as the interval halving method, because the size of the search interval is halved in each iteration, as shown in Figure 2.3. This method never fails when the initial interval contains a root. But it converges slowly and is not easily extended to multivariable systems.

The function *bisectn* implements the bisection method to find the root of $f(x) = 0$.

```
% input
%   fun: function handle that returns f(x)
%   x1,x2: interval limits containing the root
% Output
%   x: zero of f(x)
tol = 1e-8; kmax = ceil(log(abs(x2-x1)/tol)/log(2));
f1 = feval(fun,x1); f2 = feval(fun,x2);
if f1 == 0, x = x1; return; end
if f2 == 0, x = x2; return; end
if f1*f2 > 0, error('The root is not located in [x1, x2].'); end
for k = 1:kmax
   x3 = (x1+x2)/2; f3 = feval(fun,x3);
```

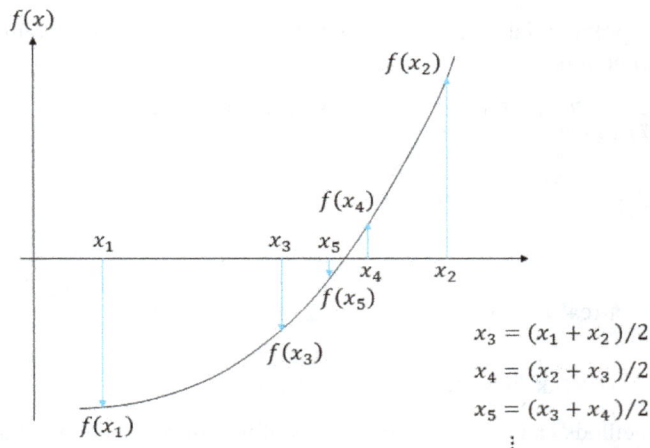

FIGURE 2.3   Illustration of the bisection method.

```
  if abs(f3) <= tol, x = x3; break; end
  if f2*f3 < 0, x1 = x3; f1 = f3; else x2 = x3; f2 = f3; end
end
x = (x1+x2)/2;
fprintf('Number of iterations: %g\n', k);
end
```

2.2.2.2 False Position Method

The false position method, or the regula falsi method, is similar to the bisection method. Two points $(x_0, f(x_0))$ and $(x_1, f(x_1))$ are selected first so that $f(x_0)f(x_1) < 0$. The intermediate point x_{k+1} is calculated by

$$x_{k+1} = x_k - \frac{x_k - x_{k-1}}{f(x_k) - f(x_{k-1})} f(x_k)(k > 0)$$

If $f(x_{k-1})f(x_{k+1}) < 0$, set $x_k = x_{k+1}$; otherwise, set $x_{k-1} = x_{k+1}$. This iteration process is terminated when the root is estimated adequately.

2.2.2.3 Newton-Raphson Method

Let the Taylor series expansion of $f(x)$ be zero:

$$f(x + \Delta x) = f(x) + \frac{\partial f(x)}{\partial x}\Delta x + \frac{\partial^2 f(x)}{\partial x^2}\frac{(\Delta x)^2}{2!} + \frac{\partial^3 f(x)}{\partial x^3}\frac{(\Delta x)^3}{3!} + \cdots = 0$$

By neglecting all terms of order 2 and higher, we have

$$f(x) + \frac{\partial f(x)}{\partial x}\Delta x = f(x) + f'(x)\Delta x = 0$$

from which

$$\Delta x = -\frac{f(x)}{f'(x)}$$

Let $\Delta x_{k+1} = x_{k+1} - x_k$. Then

$$\Delta x_{k+1} = x_{k+1} - x_k = -\frac{f(x_k)}{f'(x_k)} = -\frac{f_k}{f_k'}$$

The Newton-Raphson method is expressed as

$$x_{k+1} = x_k - \frac{f_k}{f_k'}$$

Figure 2.4 shows the graphical representation of the Newton-Raphson method.
    The function *newtrap* implements the Newton-Raphson method to find the root of $f(x) = 0$.

```
function x = newtrap(f,df,x0)
% Implements Newton-Raphson method to find root(s) of an equation
% input
%   f: function handle that returns f(x)
%   df: function handle that returns df/dx
%   x0: initial guess
```

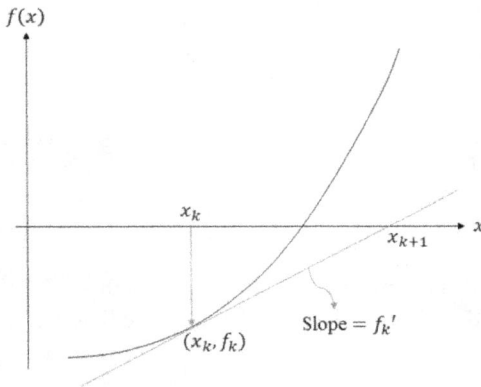

FIGURE 2.4 Graphical illustration of the Newton-Raphson method.

```
% Output
%   x: zero of f(x)
tol = 1e-8; kmax = 1e4; f0 = feval(f,x0); df0 = feval(df,x0);
if f0 == 0, x = x0; return; end
for k = 1:kmax
   x1 = x0 - f0/df0; f1 = feval(f,x1); df1 = feval(df,x1);
   if abs(f1) <= tol || abs(x1-x0) <= tol, break; end
   if k >= kmax, disp('The Newton-Raphson method not converged.'); break; end
   x0 = x1; f0 = f1; df0 = df1;
end
x = x1; fprintf('Number of iterations: %g\n', k);
end
```

2.2.2.4 Secant Method

Let x_{k-1} and x_k be two approximations to the root of $f(x) = 0$, and let $f_{k-1} = f(x_{k-1})$ and $f_k = f(x_k)$. We construct a straight line (called a secant or chord) through the points (x_{k-1}, f_{k-1}) and (x_k, f_k). This line can be expressed as

$$y = \left(\frac{f_k - f_{k-1}}{x_k - x_{k-1}} \right)x + \frac{f_{k-1}x_k - f_k x_{k-1}}{x_k - x_{k-1}}$$

As the next approximation x_{k+1}, we take the point of intersection of the straight line with the x-axis. If $y = 0$, we have

$$x = x_{k+1} = \frac{f_k x_{k-1} - f_{k-1}x_k}{f_k - f_{k-1}}$$

The secant method is illustrated graphically in Figure 2.5.

The function *secant* implements the secant method to find the root of $f(x) = 0$. The iteration process is terminated when the root is estimated adequately.

```
function x = secant(fun,x1,x2)
% Implements secant method to find root(s) of an equation
% input
%   fun: function handle that returns f(x)
%   x1,x2: interval limits containing the root
% Output
```

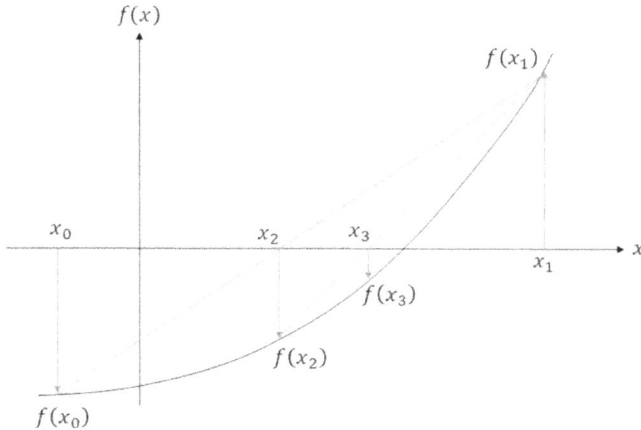

FIGURE 2.5   Illustration of the secant method.

```
%   x: zero of f(x)
tol = 1e-8; kmax = 1e4; f1 = feval(fun,x1); f2 = feval(fun,x2);
if f1 == 0, x = x1; return; end
if f2 == 0, x = x2; return; end
for k = 1:kmax
  x3 = x2 - f2*(x2 - x1)/(f2 - f1); f3 = feval(fun,x3);
  if abs(f3) <= tol, break; end
  if k >= kmax, disp('The secant method not converged.'); break; end
  x1 = x2; x2 = x3; f1 = f2; f2 = f3;
end
x = x3; fprintf('Number of iterations: %g\n', k);
end
```

2.2.2.5 Muller Method

For three given points $(x_0, f(x_0))$, $(x_1, f(x_1))$, and $(x_2, f(x_2))$, we compute

$$m_1 = \frac{f(x_1) - f(x_0)}{x_1 - x_0}, \ m_2 = \frac{f(x_2) - f(x_1)}{x_2 - x_1}, \ n_1 = \frac{m_2 - m_1}{x_2 - x_0}, \ s = m_2 + n_1(x_2 - x_1)$$

The next approximation is given by

$$x = x_2 - \frac{2f(x_2)}{s + \mathrm{sign}(s)\sqrt{s^2 - 4n_1 f(x_2)}}$$

Example 2.8 Reduction of an Iron Ore

In a reduction experiment of iron ore by hydrogen, the time $t\,(sec)$ spent in the reduction zone is given by the function of unreacted particle core $r\,(cm)$ as

$$t = 9.496 \times 10^3(1 - 12r^2 + 16r^3)$$

Find r when $t = 30\,min$ using the bisection, secant, and Newton-Raphson methods. As the initial search interval, use [0, 0.5], and as an initial guess, let $x_0 = 0.3$.

Solution

```
>> f = @(r) 9.496e3*(1-12*r^2 + 16*r^3) - 1800;
>> df = @(r) 9.496e3*(-24*r + 48*r^2); % 1st derivative of f(r) (=df/dr)
```

```
>> x = bisectn(f,0,0.5) % bisection method
Number of iterations: 26
x =
   0.3607
>> x = secant(f,0,0.5) % secant method
Number of iterations: 7
x =
   0.3607
>> x = newtrap(f,df,0.3) % Newton-Raphson method
Number of iterations: 4
x =
   0.3607
```

2.2.3 Solution of Nonlinear Equations of Several Variables

Some methods used to find the zeros of a single-variable nonlinear equation can be extended to a system of nonlinear equations of multiple variables of a form $F(x) = 0$, where $x \in R^{n \times 1}$ and $F \in R^{n \times 1}$. The following is an overview of some numerical techniques for finding zeros of a nonlinear equation of several variables.

2.2.3.1 Newton-Raphson Method

Let a multivariable system be represented by

$$f_1(x_1, x_2, \ldots, x_n) = 0$$
$$f_2(x_1, x_2, \ldots, x_n) = 0$$
$$\vdots$$
$$f_n(x_1, x_2, \ldots, x_n) = 0$$

Let $x = [x_1 \quad x_2 \quad \cdots \quad x_n]^T$. The Taylor expansion for each f_i after neglecting the 2nd- and higher-order terms gives

$$f_i(x + \Delta x) = f_i(x) + \sum_{j=1}^{n} \frac{\partial f_i(x)}{\partial x_j} \Delta x_j = 0$$

This equation can be expressed in vector-matrix form as

$$f(x) + J\Delta x = 0$$

where $f = \begin{bmatrix} f_1 & f_2 & \cdots & f_n \end{bmatrix}^T$ and J is the Jacobian matrix given by

$$J = \begin{bmatrix} \frac{\partial f_1}{\partial x_1} & \frac{\partial f_1}{\partial x_2} & \cdots & \frac{\partial f_1}{\partial x_n} \\ \vdots & \vdots & \ddots & \vdots \\ \frac{\partial f_n}{\partial x_1} & \frac{\partial f_n}{\partial x_2} & \cdots & \frac{\partial f_n}{\partial x_n} \end{bmatrix}$$

Now we have

$$\Delta x = -J^{-1} f(x)$$

Thus the guess for x at iteration k + 1, x^{k+1} is given by

$$x^{k+1} = x^k - J^{-1} f(x^k)$$

The function *newtrapmv* implements the Newton-Raphson method to find the root of a system of nonlinear equations.

```
function z = newtrapmv(fun,x0)
% Implements Newton-Raphson method to find roots of a system of equations
% input
% fun: function handle that returns f(x) = [f1,f2,...,fn]
% x0: vector of initial guesses
% Output
% z: zeroes of f(x)
tol = 1e-8; kmax = 1e3;
x = x0; if size(x,1) = = 1, x = x'; end % x should be column vector
for k = 1:kmax
[J,f] = jacob(fun,x);
if sqrt(dot(f,f)/length(x)) < tol, z = x; return; end
dx = J\(-f); x = x + dx;
if sqrt(dot(dx,dx)/length(x)) < tol*max(abs(x),1), z = x; return; end
if k > = kmax, disp('The Newton-Raphson method does not converge.'); break; end
end
fprintf('Number of iterations: %g\n', k);
end
function [J,f] = jacob(fun,x)
% Computes the Jacobian matrix J and f(x)
hx = 1e-4; n = length(x); J = zeros(n); f0 = feval(fun,x); f = f0;
for k = 1:n
tempx = x(k); x(k) = tempx + hx; f1 = feval(fun,x);
x(k) = tempx; J(:,k) = (f1-f0)/hx;
end
end
```

2.2.3.2 Secant Method

The secant method is an alternative to the Newton-Raphson method that does not require computation of the Jacobian. The Jacobian is replaced by a matrix A that satisfies $A(x_1 - x_0) = f(x_1) - f(x_0)$. The matrix A is updated by

$$A_n = A + \frac{(z - As)s^T}{\| s \|^2}$$

where $z = f(x_n) - f(x)$ and $s = x_n - x$.

2.2.3.3 Fixed-Point Iteration Method for Nonlinear Systems

In this method, the nonlinear system $F(x) = 0$ is converted to the fixed-point form $x = g(x)$. If there is a region Q such that $g(Q) \subset Q$ and the Jacobian G of g satisfies $\| G(z) \|_\infty < 1$, then the iteration $x(k + 1) = g(x(k))$ will converge.

Example 2.9 Nonlinear Equation System

Determine a solution of the following system of equations using the Newton-Raphson method. Let the initial guess be $x_0 = [1 \ 1 \ 1]^T$.

$$\cos(x) + y^2 + \ln z = 8$$
$$4x + 3^y - z^3 = -2$$
$$x + y + z = 6$$

Solution

Let $x(1) = x$, $x(2) = y$, and $x(3) = z$.

```
>> f = @(x) [cos(x(1)) + x(2)^2 + log(x(3)) - 8; 4*x(1) + 3^x(2) - x(3)^3 + 2; x(1)
+ x(2) + x(3) - 6];
>> x0 = [1 1 1]'; z = newtrapmv(f,x0)
z =
0.7542
2.5024
2.7434
```

Example 2.10 Catalytic Dehydrogenation Reactions of Ethane C_2H_6

Ethylene (C_2H_4) and acetylene (C_2H_2) are obtained from the following catalytic dehydrogenation reactions of ethane (C_2H_6):

Reaction 1: $C_2H_6 \leftrightarrow C_2H_4 + H_2$ (extent of reaction: η_1)

Reaction 2: $C_2H_6 \leftrightarrow C_2H_2 + 2H_2$ (extent of reaction: η_2)

The equilibrium constants K_1 and K_2 for each reaction may be expressed in terms of the extents of reaction η_1 and η_2 as follows:

$$K_1 = \frac{\eta_1(\eta_1 + 2\eta_2)}{(1 - \eta_1 - \eta_2)(1 + \eta_1 + 2\eta_2)}, \quad K_2 = \frac{\eta_2(\eta_1 + 2\eta_2)^2}{(1 - \eta_1 - \eta_2)(1 + \eta_1 + 2\eta_2)^2}$$

At the reaction condition of 980°C and 1atm, the equilibrium constants were measured to be $K_1 = 3.78$ and $K_2 = 0.137$. Determine the extents of reaction η_1 and η_2.

Solution

Let $x(1) = \eta_1$ and $x(2) = \eta_2$.

```
>> K1 = 3.78; K2 = 0.137; x0 = [0.80 0.10];
>> f = @(x) [x(1)*(x(1) + 2*x(2))/(1 - x(1) - x(2))/(1 + x(1) + 2*x(2)) - K1; x
(2)*(x(1) + 2*x(2))^2/(1 - x(1) -x(2))/(1 + x(1) + 2*x(2))^2 - K2];
>> z = newtrapmv(f,x0)
z =
0.8310
0.0617
```

We can see that $\eta_1 = 0.8310$ and $\eta_2 = 0.0617$.

2.2.4 USE OF MATLAB BUILT-IN FUNCTIONS

MATLAB has a built-in function *fzero* which can be used to find the zeros of a single-variable nonlinear equation $f(x) = 0$. One form of its syntax is

```
x = fzero(@fun, x0)
```

where *fun* is the name of the function $f(x)$ and *x0* is a user-defined guess for the zero. The *fzero* function returns a value of x that is near *x0*. This function only identifies the points where the function touches the axis.

The zeros of a system of nonlinear equations can be obtained by using MATLAB's built-in function *fsolve*. Usage of this function is similar to that of *fzero*:

```
x = fsolve(@fun, x0) or
[x, fval] = fsolve(@fun, x0)
```

where *fun* denotes the name of the function defining a system of nonlinear equations and *fval* is the values of the function evaluated at the solution vector *x*.

Example 2.11 Zeros of a Nonlinear Equation
Find the zeros of

$$f(x) = x^3 - x\sin x + 1 = 0$$

Solution
We can use the *fzero* function to find the zero near −1:

```
>> fzero(@(x) x^3-x*sin(x)+1, -1)
ans =
-0.7725
```

Example 2.12 Friction Factor Using the Colebrook Equation
The Colebrook equation is given by

$$\frac{1}{\sqrt{f}} = -0.86\ln\left(\frac{\epsilon/D}{3.7} + \frac{2.51}{N_{Re}\sqrt{f}}\right)$$

Find the friction factor f for $N_{Re} = 6.5 \times 10^4$ and $\epsilon/D = 0.00013$. As the first guess for f, use $f_0 = 0.1$.

Solution
The equation should be rearranged in the form $f(x) = 0$ as

$$\frac{1}{\sqrt{f}} + 0.86\ln\left(\frac{\epsilon/D}{3.7} + \frac{2.51}{N_{Re}\sqrt{f}}\right) = 0$$

We can find f by using the function *fzero* as follows:

```
>> eD = 1.3e-4; Nre = 6.5e4; f0 = 0.1; % data (f0: initial guess)
>> Cf = @(f) 1/sqrt(f) + 0.86*log(eD/3.7 + 2.51/Nre/sqrt(f)); % define equation
>> f = fzero(Cf,f0)
f =
0.0206
```

Example 2.13 SRK Equation of State
The Soave-Redlich-Kwong (SRK) equation of state is given by

$$P = \frac{RT}{(V-b)} - \frac{a}{V(V+b)\sqrt{T}}$$

where $a = 0.42748\, R^2 T_c^{5/2}/P_c$, $b = 0.08664\, RT_c/P_c$, T_c and P_c denote the critical temperature (K) and critical pressure (*atm*), respectively, and V is the specific volume (*liter*/*mol*). Find the specific volume of 1-butene at 415 K and 21 MPa. The gas constant is $R = 0.082054$ *liter·atm*/(*mol·K*) and the critical properties of 1-butene are $T_c = 419.6\ K$ and $P_c = 40.2\ MPa$.

Solution

First, we have to rearrange the SRK equation in the form of $f(x) = 0$ as follows:

$$f(V) = \frac{RT}{(V - b)} - \frac{a}{V(V + b)\sqrt{T}} - P = 0$$

Since 1 $MPa = 9.86923$ *atm*, $P = 207.2538$ *atm*, and $P_c = 396.743$ *atm*. The script *srkeqn* uses the built-in solver *fzero* to solve the equation with the initial guess of 0.1.

```
% srkeqn.m: solves the SRK equation
R = 0.082054; Tc = 419.6; Pc = 396.743; T = 415; P = 207.2538; % data
a = 0.42748*R^2*Tc^2.5/Pc; b = 0.08664*R*Tc/Pc; % coefficients a and b
f = @(V) R*T./(V-b) - a./(V.*(V+b)*sqrt(T)) - P; % define the equation
V0 = 0.1; V = fzero(f,V0) % use the built-in solver fzero (V0: initial guess)

>> srkeqn
V =
0.1291
```

We can see that $V = 0.1291$ *liter*/*mol*.

Example 2.14 Bubble Point of a Mixture

The vapor pressure P_{sat} (*mmHg*) of benzene and toluene can be represented by the Antoine equation:

$$\ln P_{sat} = A - \frac{B}{t(^\circ C) + C}$$

Determine the bubble point temperature $t(^\circ C)$ of a mixture containing benzene and toluene at a pressure $P = 1\ atm$. P is given by

$$P\,(mmHg) = x_B P_{sat,B} + x_T P_{sat,T}$$

where x_B and x_T are mole fractions of benzene and toluene, respectively.
Data: $x_B = 0.4$; $x_T = 0.6$; benzene: $A = 15.90085$, $B = 2788.507$, $C = 220.790$; toluene: $A = 16.01066$, $B = 3094.543$, $C = 219.377$.

Solution

The bubble point temperature $t(^\circ C)$ can be found by solving the following nonlinear equation:

$$f(t) = x_B P_{sat,B}(t) + x_T P_{sat,T}(t) - P = 0$$

The script *btbubbleT* defines the equation and uses the built-in solver *fzero* to get the required bubble point temperature.

```
% btbubbleT.m: bubble point temperature of beznene-toluene mixture
% 1: benzene, 2: toluene
A = [15.90085, 16.01066]; B = [2788.507, 3094.543]; C = [220.790, 219.377];
P = 760; % total pressure
x = [0.4, 0.6]; % mole fraction
f = @(t) x(1)*exp(A(1)-B(1)/(t+C(1))) + x(2)*exp(A(2)-B(2)/(t+C(2))) - P;
t = fzero(f,50); % initial guess = 50 deg.C
```

```
fprintf('The bubble point temperature = %g deg.C\n', t);

>> btbubbleT
The bubble point temperature = 95.1417 deg.C
```

Example 2.15 A System of Nonlinear Equations

Find the zeros of the following three equations:

$$\sin(x) + y^2 + \ln z = 7, \; 3x + 2y - z^3 = -1, \; x + y + z = 5$$

Solution

We start with the definition of the function containing equations in vector form. Let $x(1) = x$, $x(2) = y$, and $x(3) = z$. The initial estimate of a common solution should be set in advance. We can use the built-in solver *fsolve* to solve the set of equations.

```
>> fun = @(x) [sin(x(1))+x(2)^2+log(x(3))-7; 3*x(1)+2*x(2)-x(3)^3+1; x(1)+x
(2)+x(3)-5];
>> x0 = [0 2 2]'; x = fsolve(fun,x0) % use the built-in solver fsolve
Equation solved.
fsolve completed because the vector of function values is near zero
as measured by the default value of the function tolerance, and
the problem appears regular as measured by the gradient.
<stopping criteria details>
x =
0.6331
2.3934
1.9735
```

Example 2.16 Composition of the Equilibrium Mixture[2]

Ethane reacts with steam to form hydrogen over a cracking catalyst at a temperature of $T = 1000\,K$ and pressure of $P = 1\,atm$. The feed contains 4 moles of steam per mole of ethane. The components present in the equilibrium mixture are shown in Table 2.2. The Gibbs energies of formation *(kcal/gmol)* of the various components at the reaction temperature $(1000\,K)$ are also given in Table 2.2, as is the initial guess for each component. The equilibrium composition of the effluent mixture is to be calculated using the data given in Table 2.2.

This problem can be regarded as an optimization problem, which minimizes the total Gibbs energy. The objective function to be minimized is given by

---

TABLE 2.2
Components Present in Effluent of Ethane-Steam Cracking Reactor

i	Component	Gibbs Energy *(kcal/gmol)*	Feed *(gmol)*	Effluent Initial Guess
1	CH_4	4.61		0.001
2	C_2H_4	28.249		0.001
3	C_2H_2	40.604		0.001
4	CO_2	-94.61		0.993
5	CO	-47.942		1
6	O_2	0		0.0001
7	H_2	0		5.992
8	H_2O	-46.03	4	1
9	C_2H_6	26.13	1	0.001

$$\min_{n_i} \frac{G}{RT} = \sum_{i=1}^{c} n_i \left(\frac{G_i^0}{RT} + \ln \frac{n_i}{\Sigma\, n_i} \right)$$

where
  n_i is the number of moles of component i
  c is the total number of components
  R is the gas constant
  G_i^0 is the Gibbs energy of pure component i at temperature T
  Oxygen, hydrogen, and carbon balances should be set to find n_i.
  Oxygen: $f_1 = 2n_4 + n_5 + 2n_6 + n_8 - 4 = 0$
  Hydrogen: $f_2 = 4n_1 + 4n_2 + 2n_3 + 2n_7 + 2n_8 + 6n_9 - 14 = 0$
        Carbon: $f_3 = n_1 + 2n_2 + 2n_3 + n_4 + n_5 + 2n_9 - 2 = 0$
  These balance equations are constraints that can be introduced into the objective function using Lagrange multipliers λ_1, λ_2, and λ_3. The extended objective function is given by

$$\min_{n_i, \lambda_j} F = \sum_{i=1}^{c} n_i \left(\frac{G_i^0}{RT} + \ln \frac{n_i}{\Sigma\, n_i} \right) + \sum_{j=1}^{3} \lambda_j f_j$$

All the partial derivatives of the function F with respect to n_i and λ_j vanish at the minimum point. For example, the partial derivative of F with respect to n_i is given by

$$\frac{\partial F}{\partial n_i} = \frac{G_i^0}{RT} + \ln \frac{n_i}{\Sigma_{k=1}^{c} n_k} + 1 - \frac{n_i}{\Sigma_{k=1}^{c} n_k} + \sum_{j=1}^{3} \lambda_j \left(\frac{\partial f_j}{\partial n_i} \right) = 0 (i = 1, 2, ..., c), f_j = 0 (j = 1, 2, 3)$$

Solution
  From $f_j = 0 (j = 1, 2, 3)$, we have

$$f_1 = 2n_4 + n_5 + 2n_6 + n_8 - 4 = 0$$

$$f_2 = 4n_1 + 4n_2 + 2n_3 + 2n_7 + 2n_8 + 6n_9 - 14 = 0$$

$$f_3 = n_1 + 2n_2 + 2n_3 + n_4 + n_5 + 2n_9 - 2 = 0$$

Let $n_s = \Sigma_{k=1}^{c} n_k$ in the relation $\partial F / \partial n_i = 0 (i = 1, 2, ..., c)$, and transform logarithms into exponential functions to avoid taking logarithms of very small numbers. Applying the Gibbs energy data given in Table 2.2, we have the following nonlinear equations:

$$n_1 - n_s \exp \left[-\left(\frac{4.61 \times 10^3}{RT} + 1 - \frac{n_1}{n_s} + 4\lambda_2 + \lambda_3 \right) \right] = 0$$

$$n_2 - n_s \exp \left[-\left(\frac{28.249 \times 10^3}{RT} + 1 - \frac{n_2}{n_s} + 4\lambda_2 + 2\lambda_3 \right) \right] = 0$$

$$n_3 - n_s \exp \left[-\left(\frac{40.604 \times 10^3}{RT} + 1 - \frac{n_3}{n_s} + 2\lambda_2 + 2\lambda_3 \right) \right] = 0$$

$$n_4 - n_s \exp \left[-\left(\frac{-94.61 \times 10^3}{RT} + 1 - \frac{n_4}{n_s} + 2\lambda_1 + \lambda_3 \right) \right] = 0$$

$$n_5 - n_s \exp\left[-\left(\frac{-47.942 \times 10^3}{RT} + 1 - \frac{n_5}{n_s} + \lambda_1 + \lambda_3\right)\right] = 0, \quad n_6 - n_s \exp\left[-\left(1 - \frac{n_6}{n_s} + 2\lambda_1\right)\right] = 0$$

$$n_7 - n_s \exp\left[-\left(1 - \frac{n_7}{n_s} + 2\lambda_2\right)\right] = 0, \quad n_8 - n_s exp\left[-\left(\frac{-46.03 \times 10^3}{RT} + 1 - \frac{n_8}{n_s} + \lambda_1 + 2\lambda_2\right)\right] = 0$$

$$n_9 - n_s \exp\left[-\left(\frac{26.13 \times 10^3}{RT} + 1 - \frac{n_9}{n_s} + 6\lambda_2 + 2\lambda_3\right)\right] = 0, \quad R = 1.9872 (cal/(gmol \cdot K))$$

There are 12 unknown variables—n_i ($i = 1, 2, ..., 9$) and λ_j ($j = 1, 2, 3$)—and 12 equations. First we construct the MATLAB function *rxnfun*, which defines the system of nonlinear equations in vector form. In this function, all the unknown variables are represented in terms of x_i ($i = 1, 2, ..., 12$)—i.e., $x_i = n_i$ ($i = 1, 2, ..., 9$) and $x_i = \lambda_i$ ($i = 10, 11, 12$).

```
function f = rxnfun(x,T)
R = 1.9872; ns = sum(x(1:9));
f(1,1)  = 2*x(4)+x(5)+2*x(6)+x(8)-4;
f(2,1)  = 4*x(1)+4*x(2)+2*x(3)+2*x(7)+2*x(8)+6*x(9)-14;
f(3,1)  = x(1)+2*x(2)+2*x(3)+x(4)+x(5)+2*x(9)-2;
f(4,1)  = x(1)-ns*exp(-(4.61e3/(R*T)+1-x(1)/ns+4*x(11)+x(12)));
f(5,1)  = x(2)-ns*exp(-(28.249e3/(R*T)+1-x(2)/ns+4*x(11)+2*x(12)));
f(6,1)  = x(3)-ns*exp(-(40.604e3/(R*T)+1-x(3)/ns+2*x(11)+2*x(12)));
f(7,1)  = x(4)-ns*exp(-(-94.61e3/(R*T)+1-x(4)/ns+2*x(10)+x(12)));
f(8,1)  = x(5)-ns*exp(-(-47.942e3/(R*T)+1-x(5)/ns+x(10)+x(12)));
f(9,1)  = x(6)-ns*exp(-(1-x(6)/ns+2*x(10)));
f(10,1) = x(7)-ns*exp(-(1-x(7)/ns+2*x(11)));
f(11,1) = x(8)-ns*exp(-(-46.03e3/(R*T)+1-x(8)/ns+x(10)+2*x(11)));
f(12,1) = x(9)-ns*exp(-(26.13e3/(R*T)+1-x(9)/ns+6*x(11)+2*x(12)));
end
```

As the initial estimate, we choose x_0 = [0.001 0.001 0.001 0.993 1 0.0001 5.992 1 0.001 10 10 10]. Zeros of the nonlinear equation system can be found by using the built-in function *fsolve*. The script file *ethanrxn* shows the solution procedure.

```
% ethanrxn.m: ethane-steam cracking reaction
x0 = [0.001 0.001 0.001 0.993 1 0.0001 5.992 1 0.001 10 10 10]'; T = 1000;
[x, fval] = fsolve(@rxnfun, x0, [],T); % use fsolve to solve the equation system
comp = {'CH4','C2H4','C2H2','CO2','CO','O2','H2','H2O','C2H6'};
lamda = {'lambda1','lambda2','lambda3'};
fprintf('\n i\tComp. \t Initial Val.\t\tFinal val.\n');
for k = 1:length(x)-3, fprintf('%d\t%s \t%12.9f\t%12.9f\n',k,comp
{k},x0(k),x(k)); end
fprintf('\n i\tLambda\tInitial Val.\tFinal val.\n');
for i = k+1:length(x), fprintf('%g\t%s\t\t%4.1f\t%15.9f\n',i,lamda{i-length
(comp)},x0(i),x(i)); end
```

Running this script gives

```
>> ethanrxn
Equation solved.
```

```
fsolve completed because the vector of function values is near zero
as measured by the default value of the function tolerance, and
the problem appears regular as measured by the gradient.
<stopping criteria details>

i    Comp. Initial Val.    Final val.
1    CH4         0.001000000         0.081717614
2    C2H4        0.001000000         0.000000172
3    C2H2        0.001000000         0.000000000
4    CO2         0.993000000         0.664820888
5    CO          1.000000000         1.253460254
6    O2          0.000100000         0.000000000
7    H2          5.992000000         5.419665106
8    H2O         1.000000000         1.416897971
9    C2H6        0.001000000         0.000000450
i    Lambda      Initial Val.        Final val.
10   lambda1              10.0       24.051830977
11   lambda2              10.0        0.051093350
12   lambda3              10.0        1.168410847
```

2.3 REGRESSION ANALYSIS

2.3.1 INTRODUCTION TO STATISTICS

2.3.1.1 Elementary Statistics

Let N designate the total number of items in the population under consideration, n the number of items contained in the sample taken from the population ($0 \leq n \leq N$), and m_j ($j = 1,2, ...,M$) the number of times the value x_j occurs. The probability of occurrence is defined by the number of occurrences of x_j divided by the total number of observations:

$$Pr(X = x_j) = \lim_{n \to N} \frac{m_j}{m} = p(x_j)$$

For a discrete random variable, $p(x_j)$ is called the probability function, and it has the following properties:

$$0 \leq p(x_j) \leq 1, \ \sum_{j=1}^{M} p(x_j) = 1$$

The expected value of a discrete random variable is defined as

$$\mu = E[X] = \sum_{j=1}^{M} x_j p(x_j)$$

The expected value corresponds to the center of gravity of the probability density distribution. For the entire population, the expected value is given by the arithmetic average of the random variable:

$$\mu = E[X] = \frac{1}{N} \sum_{i=1}^{N} x_i$$

The sample mode of a sample observation is the value that occurs most frequently, and the sample mean (or arithmetic average) is the value obtained by dividing the sum of observations by the total tally:

$$\bar{x} = \frac{1}{n} \sum_{i=1}^{n} x_i$$

The expected value of the sample mean is given by

$$E[\bar{x}] = E\left[\frac{1}{n} \sum_{i=1}^{n} x_i\right] = \frac{1}{n} \sum_{i=1}^{n} E[\bar{x}] = \frac{1}{n} \sum_{i=1}^{n} \mu = \mu$$

The population variance is defined as the expected value of the square of the deviation of random variable X from its expectation:

$$\sigma^2 = E[(X - E[X])^2] = E[(X - \mu)^2]$$

For a discrete random variable, this definition is equivalent to

$$\sigma^2 = \sum_{j=1}^{M} (x_j - \mu)^2 p(x_j) = \frac{1}{N} \sum_{i=1}^{N} (x_i - \mu)^2$$

So the population standard deviation is defined as the positive square root of the population variance:

$$\sigma = \sqrt{\sigma^2}$$

The sample variance s^2 is defined as the arithmetic average of the square of the deviations of x_i from the population mean μ:

$$s^2 = \frac{1}{n} \sum_{i=1}^{n} (x_i - \mu)^2$$

If μ is not known, $\bar{x}$ is used as an estimate of μ, and the sample variance is given by

$$s^2 = \frac{1}{n-1} \sum_{i=1}^{n} (x_i - \bar{x})^2$$

In fact, the sample variance is an unbiased estimate of the population variance:

$$E[s^2] = \sigma^2$$

The positive square root of the sample variance is called the sample standard deviation:

$$s = \sqrt{s^2} = \sqrt{\frac{1}{n-1} \sum_{i=1}^{n} (x_i - \bar{x})^2}$$

The median is the value in the middle of the data if the number of data points is odd.

In MATLAB, the sample mean is calculated by the built-in function *mean*, and the sample mode is calculated by the built-in function *mode*. The variance and the standard deviation are calculated by the built-in functions *var* and *std*, respectively. The built-in function *median* returns the median value of the elements of the data.

Example 2.17 Elementary Statistics

Table 2.3 shows the measurements of thermal expansion coefficients of a certain metal. Find the mean, median, mode, variance, and standard deviation of these data.

Solution

```
>> tc = [5.394 5.564 5.654 5.465 5.495 5.404 5.524 5.414...
5.514 5.335 5.614 5.455 5.534 5.524 5.475 5.295...
5.384 5.614 5.554 5.675 5.455 5.554 5.504 5.584];
>> fprintf('Mean = %9.6f', mean(tc))
>> fprintf('\nMedian = %9.6f', median(tc))
>> fprintf('\nMode = %9.6f', mode(tc))
>> fprintf('\nVariance = %9.6f', var(tc))
>> fprintf('\nStandard deviation = %9.6f\n', std(tc))
Mean = 5.499333
Median = 5.509000
Mode = 5.455000
Variance = 0.009405
Standard deviation = 0.096977
```

2.3.1.2 Probability Distribution

A Bernoulli trial is an experiment that generates only two mutually exclusive and exhaustive outcomes. If a Bernoulli trial is repeated independently n times and the probability of success in a single trial is p, the probability of k successes out of n independent trials is given by the binomial probability distribution:

$$P|_k = C_k^n p^k (1 - p)^{n-k}$$

A density curve, or probability density curve, is a smooth curve under which the area is 1. The mathematical representation that describes the shape of the density curve is called the density function $f(x)$, which satisfies

$$\int_{-\infty}^{\infty} f(x)dx = 1$$

The normal probability density, a continuous unimodal function symmetric about the mean of the distribution μ, is defined by

---

TABLE 2.3
Thermal Expansion Coefficients

Order	Coefficient	Order	Coefficient	Order	Coefficient
1	5.394	9	5.514	17	5.384
2	5.564	10	5.335	18	5.614
3	5.654	11	5.614	19	5.554
4	5.465	12	5.455	20	5.675
5	5.495	13	5.534	21	5.455
6	5.404	14	5.524	22	5.554
7	5.524	15	5.475	23	5.504
8	5.414	16	5.295	24	5.584

$$f(x) = \frac{1}{\sqrt{2\pi}\sigma}e^{-(x-\mu)^2/2\sigma^2}$$

The standard normal distribution, called a normal curve $N(0, 1)$, has a probability density given by

$$f(x) = \frac{1}{\sqrt{2\pi}}e^{-z^2/2}$$

where z is a normally distributed variable with $\sigma = 1$ and $\mu = 0$. A variable x that exhibits a normal distribution can be rescaled as $z = (x - \mu)/\sigma$ to produce a standard normal distribution. This is known as the z-statistic and is a measure of the standard deviations between x and μ. For a continuous distribution, the cumulative distribution function Φ is defined as

$$\Phi(y) = \int_{-\infty}^{y} f(x)dx = P(x \le y)$$

For large sample sizes, the variance s can be used as a reliable estimate of σ. For small sample sizes, t-curves (commonly known as Student's t distribution) can be used as a modification of the normal curve. Suppose that the population distribution is estimated using sample statistics $N(\bar{x}, s)$. Then the confidence level for the population μ can be obtained using the t-statistic, which follows the t distribution and is given by

$$t = \frac{\bar{x} - \mu}{s/\sqrt{n}}$$

The shape of the t distribution depends on the sample size n and approaches the normal distribution as $n \to \infty$. Significance tests may be classified as one-tailed or two-tailed hypothesis tests. In the one-tailed test, only one of the two ends of the probability distribution function is covered. Both tails of the probability distribution function are covered in the two-tailed test.

The MATLAB built-in functions *binopdf* and *normpdf* compute the binomial probability distribution and the normal probability density. For example, $z = binopdf(x,n,p)$ computes the binomial probability distribution at each of the values in x using the corresponding number of trials in n and probability of success for each trial in p. The function *normpdf* can be used to generate the normal probability density function. For example, $f = @(x)normpdf(x,0.2,1)$ generates the function handle f with $\mu = 0.2$ and $\sigma = 1$. The built-in function *normcdf* computes the cumulative probability distribution for a normally distributed random variable. The inverse cumulative distribution function for a probability distribution can be obtained by the built-in function *norminv*.

The MATLAB built-in function *tcdf* computes the cumulative probability (i.e., the area under the t-curve) for values of t, and *tinv* computes the critical value of the t-statistic. The function *ttest* performs a paired t test and returns whether the null hypothesis should be rejected at the significance level of 5%.

Example 2.18 Normal Probability
We can define the normal probability density $f(x)$ with $\mu = 0$ and $\sigma = 1$ using the built $-$ in function *normpdf*. Let $f z = (x - \mu)/\sigma$.
(1) Determine the probability of z assuming any value between -0.8 and +0.8.
(2) What is the probability of z lying between -0.8 and +0.8?
(3) Determine the probability of observing $z \ge 2$.

Solution
The script *pbfun* use the built-in function *normcdf* to get the results.

```
% pbfun.m: normal probability density function
clear all;
```

```
mu = 0; sigma = 1; lx = -0.8; ux = 0.8; zl = 2;
fx = @(x) normpdf(x,mu,sigma); % normal probability density function f(x)
P = quad(fx,lx,ux); % probability assuming any value between lx and ux
Pr = normcdf(ux,mu,sigma) - normcdf(lx,mu,sigma); % probability of z lying
between lx and ux
Pz = 1 - normcdf(zl, mu,sigma); % probability of observing z> = zl
fprintf('Probability of z assuming any value between %g and %g  =  %g
\n',lx,ux,P);
fprintf('Probability of z lying between %g and %g = %g\n',lx,ux,Pr);
fprintf('Probability of observing z > = %g = %g\n',zl,Pz);

>> pbfun
Probability of z assuming any value between -0.8 and 0.8  =  0.576289
Probability of z lying between -0.8 and 0.8  =  0.576289
Probability of observing z > = 2  =  0.0227501
```

Example 2.19 *t* Test

In a sequence of light absorbance experiments to determine the concentration of a hemoglobin solution, the individual absorbance values measured are [0.72, 0.54, 0.62, 0.80, 0.76, 0.64, 0.75, 0.94, 0.44]. The absorbance data are approximately normally distributed. Determine if the mean absorbance value is 0.86 (i.e., the null hypothesis is $H_0: \mu = 0.86$).

Solution

The script *ttestex* uses the built-in function *ttest* to perform a *t* test.

```
% ttestex.m: perform t-test
abdat = [0.72, 0.54, 0.62, 0.80, 0.76, 0.64, 0.75, 0.94, 0.85, 0.44];
avg = mean(abdat); stdv = std(abdat); % mean and standard deviation
mu = 0.86; % null hypothesis
cf = 0.05; % significance level
[h,p] = ttest(abdat,mu,cf,'both'); % both: both tails
fprintf('h (result of the hypothesis test) = %g\n',h);
fprintf('p (probability) = %g\n',p);

>> ttestex
h (result of the hypothesis test) = 1
p (probability) = 0.00966083
```

Since $h = 1$, the null hypothesis H_0 is rejected. If $h = 0$, H_0 is not rejected at the significance level *cf*. Also, since $p < 0.05$, the null hypothesis is rejected.

2.3.2 GENERATION OF RANDOM NUMBERS

MATLAB provides various built-in functions for generating random numbers. The function *rand* generates uniformly distributed random numbers between 0 and 1. The function *randn* generates normally distributed random numbers having a mean of 0 and a standard deviation of 1. The following commands show how to generate an $m \times n$ matrix consisting of random numbers:

```
x = rand(m,n)
x = randn(m,n)
```

To generate a random number x within $[x_a x_b]$ (i.e., $x_a < x < x_b$), we use

$$x = x_a + (x_b - x_a)_* \mathrm{rand}(m, n)$$

A normally distributed $m \times n$ random matrix having a mean of $\bar{x}$ and a standard deviation of s can be generated by

$$x = \bar{x} + s_*\text{randn}(m, n)$$

Random integers can be generated by the built-in function *randi*. Let's construct a 2×3 matrix consisting of random integers from 1 to 30:

```
>> M = randi([1 30], 2, 3)
M =
    25    4    19
    28   28    3
```

Example 2.20 Generation of Random Numbers

(1) Generate a 2×3 matrix x of random numbers distributed uniformly between 0 and 1.
(2) Generate a 2×3 matrix y of normally distributed random numbers.
(3) Generate a 3×3 matrix z of random numbers distributed uniformly between 3 and 5.

Solution

(1)
```
>> x = rand(2,3)
x =
    0.2785    0.9575    0.1576
    0.5469    0.9649    0.9706
```

(2)
```
>> y = randn(2,3)
y =
    0.7254    0.7147    -0.1241
   -0.0631   -0.2050     1.4897
```

(3)
```
>> z = 3+(5-3)*rand(3,3)
z =
    4.5844    3.0714    4.3575
    4.9190    4.6983    4.5155
    4.3115    4.8680    4.4863
```

2.3.3 LINEAR REGRESSION ANALYSIS

A linear relationship can be represented as

$$y^* = a_0 + a_1 x$$

where
   y^* is the dependent variable
   x is the independent variable
   If we let X be the vector of observations of the independent variable and y be the vector of observations of the dependent variable, then the linear model can be rewritten as

$$y = \alpha + X\beta + u$$

where
   u is the vector of disturbance terms
   α is the y-intercept of the line
   β is the slope of the line

This relation can be extended to include more than one independent variable as follows:

$$y = X_1\beta_1 + X_2\beta_2 + \cdots + X_m\beta_m + u$$

where X_1, X_2, ..., X_m are the vectors of observations of m independent variables. The vector X_1 can be taken as a vector whose elements are all equal to 1. In this case, β_1 becomes the y-intercept. We can rewrite this relation in vector-matrix form as

$$y = X\beta + u$$

where $y \in R^{n\times1}$, $X \in R^{n\times m}$, $\beta \in R^{m\times1}$, $u \in R^{n\times1}$, and n denotes the number of observations.

If b denotes an m-element vector which is an estimate of the parameter vector β, then we can define a vector of residuals as

$$\epsilon = y - Xb = y - \hat{y}$$

Each element of ϵ is a difference between the experimental observation y and the calculated value of $\hat{y}$ using the estimated vector b. Let Φ be a sum of the squared residuals, given by

$$\Phi = \epsilon^T\epsilon = (y - Xb)^T(y - Xb)$$

Taking the partial derivative of Φ with respect to b and setting the result to 0, we have

$$\frac{\partial\Phi}{\partial b} = (-X)^T(y - Xb) + (y - Xb)^T(-X) = -2X^T(y - Xb) = 0 \Rightarrow (X^TX)b = X^Ty$$

Solving the resulting equation for vector b, we have the least-squares solution given by

$$b = (X^TX)^{-1}X^Ty$$

Vector b obtained from this equation minimizes the objective function Φ. The expected value of b is given by

$$E[b] = E[\beta] + (X^TX)^{-1}X^T E[u] = E[\beta] = \beta$$

We can see that b is the unbiased estimate of β. The weighted objective function Φ_w can be defined as

$$\Phi_w = (y - Xb)^T W (y - Xb)$$

where W is the weighting matrix. From the minimization of Φ_w, we can get the weighted least-squares solution b_W:

$$b_W = (X^TWX)^{-1}X^TWy$$

2.3.4 POLYNOMIAL REGRESSION ANALYSIS

2.3.4.1 Data Fitting by Least-Squares Method

Data obtained from experiments or plant operations typically contain a significant amount of noise due to measurement errors. If this noise is confined to dependent variables, the least-squares fitting method is most commonly used to approximate data. Suppose that the measured data $(x_i, y_i)(i = 1, 2, ...,n)$ are approximated by the fitting function $f(x)$. The residual is defined by $r_i = y_i - f(x_i)$, and least-squares fitting minimizes the residual function defined by

$$\Phi(a_1, a_1, ..., a_m) = \sum_{i=1}^{n} r_i^2 = \sum_{i=1}^{n} \left\{ y_i - f(x_i) \right\}^2$$

If the fitting function $f(x)$ is chosen as a linear combination of m specified functions $f_j(x)(j = 1, 2, ..., m)$, and $f_j(x)$ is given by $f_j(x) = x^{j-1}$, $f(x)$ can be represented as

$$f(x) = a_1 f_1(x) + a_2 f_2(x) + ... + a_m f_m(x) = \sum_{j=1}^{m} a_j x^{j-1}$$

The optimal parameters are given by the solution of the equations

$$\frac{\partial \Phi}{\partial a_j} = 0 (j = 1, 2, ..., m)$$

Using vector-matrix notation, these equations can be rewritten as $Ap = b$ where

$$A = \begin{bmatrix} n & \sum x_i & \cdots & \sum x_i^{m-1} \\ \sum x_i & \sum x_i^2 & \cdots & \sum x_i^m \\ \vdots & \vdots & \ddots & \vdots \\ \sum x_i^{m-1} & \sum x_i^m & \cdots & \sum x_i^{2(m-1)} \end{bmatrix}, p = \begin{bmatrix} a_1 \\ a_2 \\ \vdots \\ a_m \end{bmatrix}, b = \begin{bmatrix} \sum y_i \\ \sum x_i y_i \\ \vdots \\ \sum x_i^{m-1} y_i \end{bmatrix}$$

where $\sum()$ stands for $\sum_{i=1}^{n}()$. The function *lspolfit* implements the least-squares polynomial fitting method.

```
function p = lspolfit(x,y,m)
% Implements least-squares polynomial fitting
% input
% x: vector of independent variables
% y: vector of measured dependent variables
% m: order of polynomial
% Output
% p: coefficient vector of mth order polynomial
n = length(x); m = m+1;
if n ~= length(y), error('x and y should be the same length.'); end
A = zeros(m,m);
% Define matrix A
for k = 1:2*(m-1), s(k) = sum(x.^k); end;
A(1,1) = n; A(1,2:m) = s(1:m-1);
for k = 2:m, A(k,:) = s(k-1:k-2+m); end;
% Define vector b
for k = 1:m, b(k) = sum((x.^(k-1)).*y); end; b = b';
% Determine parameter using backslash operator
c = A\b; p = c(end:-1:1); % coefficients in descending order
end
```

Example 2.21 Least-Squares Polynomial Fitting

The data shown in Table 2.4 were taken in an isothermal batch reactor where reactant A is decomposed to product B. Fit 2nd-, 3rd-, and 4th-order polynomials to the data.

Solution

The script *fitrxndata* fits 2nd-, 3rd-, and 4th-order polynomials to the reaction data and generates plots to compare fitting results, as shown in Figure 2.6.

TABLE 2.4

Concentration of Reactant A (C_A) as a Function of Time (t) in a Batch Reactor

t (min)	0	2	5	6	13	20	24	30	35	41	50
C_A (mol/liter)	0.86	0.61	0.47	0.39	0.25	0.18	0.15	0.12	0.10	0.09	0.08

```
% fitrxndata.m: fit polynomials to the reaction data
t = [0,2,5,6,13,20,24,30,35,41,50];
c = [0.86,0.61,0.47,0.39,0.25,0.18,0.15,0.12,0.10,0.09,0.08];
tmin = min(t); tmax = max(t); tint = linspace(tmin,tmax,100);
g = {'--',':','.-'}; m = [3 4 5]; % 2nd-, 3rd-, and 4th-order polynomials
for k = 1:3, p = lspolfit(t,c,m(k)); cint{k} = polyval(p,tint); end
% Plot data and fit polynomials
plot(t,c,'o',tint,cint{1},g{1},tint,cint{2},g{2},tint,cint{3},g{3})
xlabel('t(min)'), ylabel('C_A(mol/liter)')
legend('Data','2nd-order','3rd-order','4th-order','location','best')
```

2.3.4.2 Use of MATLAB Built-In Functions

MATLAB's built-in function *polyfit* implements least-squares polynomial regression analysis. The calling syntax is

```
p = polyfit(x, y, n)
```

where x is the vector of independent variables, y is the vector of dependent variables, and n is the order of the polynomial to be fitted. Vector p contains the polynomial's resulting coefficients in descending order.

The built-in function *lscov* performs weighted least-squares calculations for Xb = y. This function can be called as:

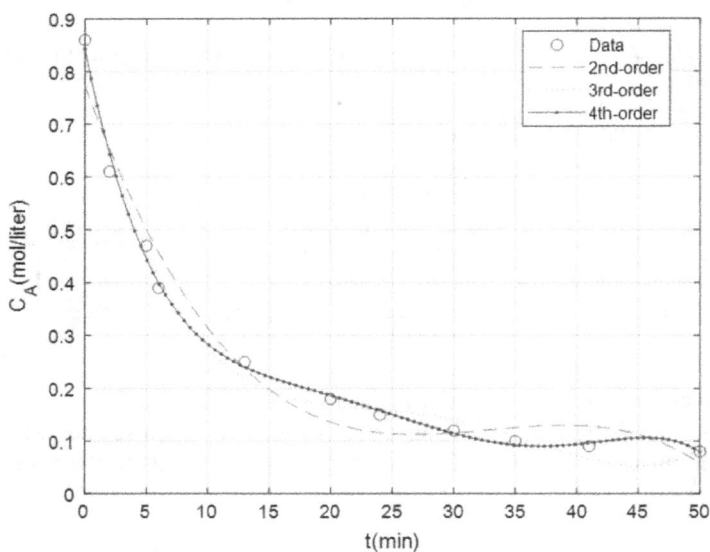

FIGURE 2.6 Polynomial fitting to data.

```
b = lscov(X, y, v)
```

where X is the matrix defined by independent variables, y is the vector of dependent variables, and v denotes the inverse of the weighting matrix W.

Example 2.22 Reaction Temperature Profile

Table 2.5 shows measurements of reaction temperature versus time. Determine the 1st-, 2nd-, and 4th-order polynomials to represent this data.

Solution

The script *rxntemfit* uses the built-in function *polyfit* to perform polynomial regressions.

```
% rxntemfit.m: polynomial fitting by polyfit function
t = [1 2 3 4 5 6 7 8]; tp = 1:0.1:8;
T = [50.8 56.4 55.1 60.6 61.5 59.5 54.1 53.8]; % temperature data
p1 = polyfit(t,T,1), p2 = polyfit(t,T,2), p4 = polyfit(t,T,4) % polynomial
fitting
T1 = polyval(p1,tp); % Calculate temperature by 1st-order polynomial
T2 = polyval(p2,tp); % Calculate temperature by 2nd-order polynomial
T4 = polyval(p4,tp); % Calculate temperature by 4th-order polynomial
plot(t,T,'o',tp,T1,':',tp,T2,'.-',tp,T4,'-'),  xlabel('t(hr)'),  ylabel
('T(C)'), grid
legend('Data','1st-order','2nd-order','4th-order')

>> rxntemfit
p1 =
0.2810 55.2107
p2 =
-0.6643 6.2595 45.2464
p4 =
0.0481 -0.8684 4.5996 -5.8696 53.5357
```

The graphs of these lines and the data points are shown in Figure 2.7. From the results, we can see that the 4th-order polynomial

$$T(t) = 0.0481t^4 - 0.8684t^3 + 4.5996t^2 - 5.8696t + 53.5357$$

is the best-fitting polynomial.

Example 2.23 Virial Coefficients[3]

Table 2.6 shows the density (ρ) of N_2 as a function of pressure (P) at $T = 200\ K$. Using these data, estimate the virial coefficients of the virial equation of state given by

$$Z = \frac{P}{\rho RT} = 1 + B\rho + C\rho^2 + D\rho^3$$

where R is the gas constant $(= 0.08206\ liter \cdot atm/(mol \cdot K))$.

TABLE 2.5
Reaction Temperature versus Time

$t\,(hr)$	1	2	3	4	5	6	7	8
$T\,(°C)$	50.8	56.4	55.1	60.6	61.5	59.5	54.1	53.8

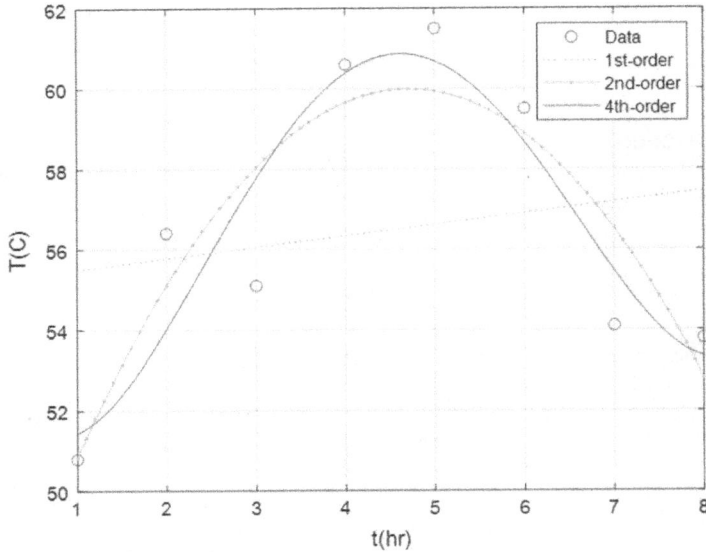

FIGURE 2.7 Polynomial representation of temperature data.

TABLE 2.6
Density of N_2 at $T = 200\ K$

$P\,(atm)$	3.2	6.0	9.0	12.0	14.0	17.0	19.0	21.0
$\rho\,(mol/liter)$	0.1995	0.3825	0.5895	0.8108	0.9685	1.2257	1.4159	1.6281

Solution

The problem is equivalent to 3rd-order polynomial fitting. In the script *N2denfit*, the built-in function *polyfit* is used with n = 3.

```
% N2denfit.m: density of N2
clear all; T = 200; R = 0.08206; % gas constant
P = [3.2 6.0 9.0 12.0 14.0 17.0 19.0 21.0]; % P(atm)
rho = [0.1995 0.3825 0.5895 0.8108 0.9685 1.2257 1.4159 1.6281]; % rho(mol/l)
Z = P./(R*rho*T); Vc = polyfit(rho, Z, 3)
x = [0.15:0.01:1.8]; Zcal = polyval(Vc, x); Pcal = R*Zcal.*x*T;
plot(x,Pcal,rho,P, 'o'), grid, xlabel ('\rho(mol/liter)'), ylabel ('P(atm)')
legend('Fitting curve', 'Data','location','best')

>> N2denfit
Vc =
   0.0001   -0.0130   -0.1105   0.9999
```

We can see that $D = 0.0001$, $C = -0.013$ and $B = -0.1105$. Figure 2.8 is a plot of the data and the fitting curve.

2.3.5 TRANSFORMATION OF NONLINEAR FUNCTIONS TO LINEAR FUNCTIONS

Nonlinear functions can be used in lieu of polynomial functions for fitting functions. Some nonlinear functions can be transformed into linear functions by taking a logarithm or inversion.

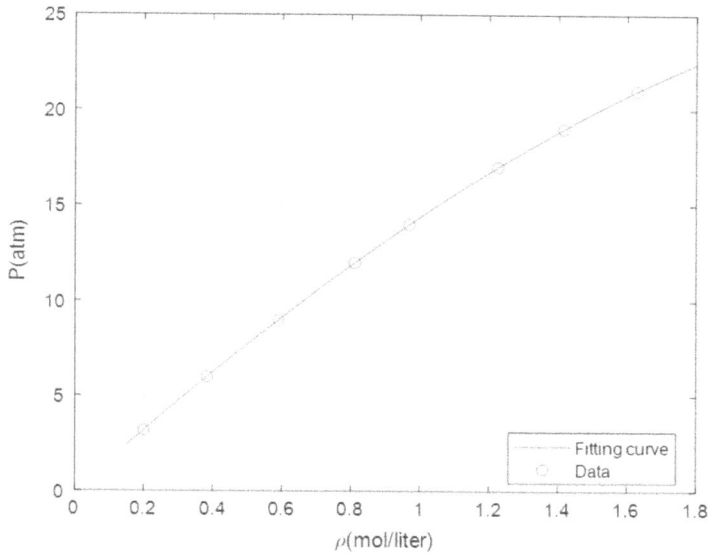

FIGURE 2.8 Density of N_2 versus pressure.

The resulting linear relation can be used in the linear regression analysis. MATLAB provides a built-in function *nlinfit* that can be used to fit the nonlinear model. Table 2.7 summarizes examples of linear transformation of some nonlinear models.

TABLE 2.7
Linearization of Nonlinear Models

Nonlinear Model	Linearization Method	Linear Model
$y = ae^{bx}$	Take natural logarithm on both sides	$\ln y = \ln a + bx$
$y = ax^b$	Take logarithm on both sides	$\log y = \log a + b\log x$
$y = \frac{ax}{b+x}$	Invert both sides	$\frac{1}{y} = \frac{1}{a} + \left(\frac{1}{b}\right)\frac{1}{x}$

Example 2.24 Michaelis-Menten Enzyme Kinetics

An enzyme behaves as a catalyst in a living cell. To represent enzyme-catalyzed reactions, the Michaelis-Menten equation is widely used:

$$r = \frac{a[S]}{b + [S]}$$

where
r is the reaction rate
$[S]$ denotes the concentration of the substrate S
a is the maximum initial reaction rate
b is a constant given by combination of rate constants

Table 2.8 shows experimental data of reaction rates versus substrate concentrations. Assuming that reaction rates can be represented by the Michaelis-Menten equation, determine parameters a and b of the equation.

TABLE 2.8
Enzyme Reaction Rates versus Substrate Concentrations

[S]	1.2	1.6	3.2	4.3	5.8	7.6	8.8
r	0.06	0.12	0.24	0.27	0.33	0.34	0.34

Solution

Taking the inverse of the Michaelis-Menten equation $r = a[S]/(b + [S])$, we have

$$\frac{1}{r} = \frac{1}{a} + \frac{b}{a}\frac{1}{S} \implies y = p_1 x + p_0$$

Coefficients p_0 and p_1 of the linear model can be obtained by using the built-in function *polyfit* with n = 1. From the relations $p_0 = 1/a$ and $p_1 = b/a$, we have

$$a = \frac{1}{p_0}, \; b = p_1 a = \frac{p_1}{p_0}$$

The script *mmmdl* uses *polyfit* to find the model parameters.

```
% mmmdl.m: Michaelis-Menten equation
S = [1.2 1.6 3.2 4.3 5.8 7.6 8.8]; r = [0.06 0.12 0.24 0.27 0.33 0.34 0.34];
p = polyfit(1./S,1./r,1); % p(1) = p1, p(2) = p0
a = 1/p(2), b = p(1)*a
Sv = S(1):0.1:S(end); rv = a*Sv./(b+Sv); % Reaction rates by Michaelis-Menten
model
plot(Sv,rv,S,r,'o'),xlabel('[S]'),ylabel('r'),legend('Michaelis-Menten
model','Experimental data')

>> mmmdl
a =
   -10.6918
b =
  -187.4149
```

We can see that the experimental data can be approximated by

$$r = \frac{-10.6918[S]}{[S] - 187.4149}$$

The graph of this line and the data points is shown in Figure 2.9.

Example 2.25 Vapor Pressure by Antoine Equation

The Antoine equation is widely used to represent the relationship between vapor pressure and temperature:

$$\log P_v = A + \frac{B}{T + C}$$

where

P_v is vapor pressure (mmHg)
T is temperature (°C)
A, B and C are parameters
Use the vapor pressure data given in Table 2.9 to find parameters A, B, and C.

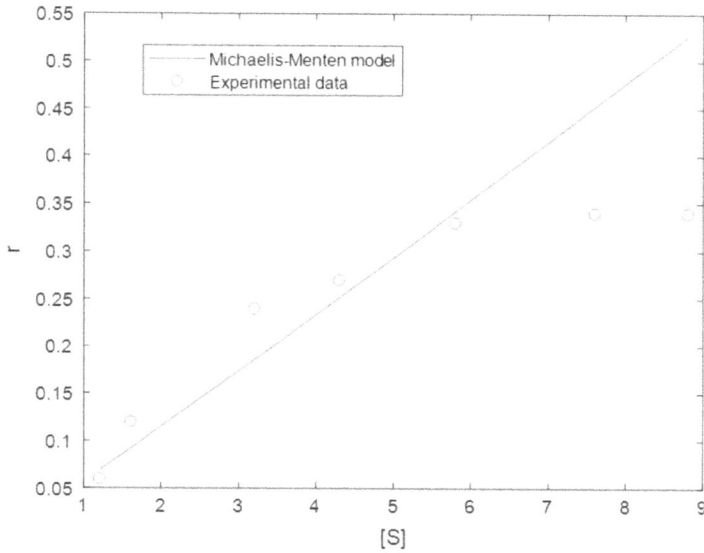

FIGURE 2.9 Michaelis-Menten model.

TABLE 2.9
Vapor Pressure versus Temperature

$T\,(°C)$	**25**	**27**	**30**	**31**	**35**	**36**	**37**
$P_v\,(mmHg)$	15.8	26.43	39.56	94.91	428.35	861.71	1851.24

Solution

Multiplying $(T + C)$ on both sides of the Antoine equation, we have

$$(T + C)\log P_v = A(T + C) + B$$

By rearranging, we have

$$T\log P_v = AT + (AC + B) - C\log P_v$$

And dividing both sides by T, we get

$$\log P_v = A + (AC + B)\left(\frac{1}{T}\right) - C\left(\frac{\log P_v}{T}\right) = \left[1\ \frac{1}{T}\ \frac{\log P_v}{T}\right]\left[\begin{array}{c} A \\ AC + B \\ -C \end{array}\right]$$

Let

$$y = \begin{bmatrix} \log P_{v1} \\ \log P_{v2} \\ \vdots \\ \log P_{vn} \end{bmatrix},\ X = \begin{bmatrix} 1 & 1/T_1 & \log P_{v1}/T_1 \\ 1 & 1/T_2 & \log P_{v2}/T_2 \\ \vdots & \vdots & \vdots \\ 1 & 1/T_n & \log P_{vn}/T_n \end{bmatrix},\ b = \begin{bmatrix} A \\ AC + B \\ -C \end{bmatrix}$$

where b is given by

$$b = (X^T X)^{-1} X^T y$$

The script *antvp* performs the required least-squares calculation.

```
% antvp.m: vapor pressure by Antoine eqn.
T = [25 27 30 31 35 36 37]; Pv = [15.8 26.43 39.56 94.91 428.35 861.71 1851.24];
X = [ones(1,length(T)); 1./T; log10(Pv)./T]';
b = inv(X'*X)*X'*log10(Pv)'; A = b(1), C = -b(3), B = b(2)-A*C
Ti = T(1):0.1:T(end); Pvi = 10.^(A + B./(Ti+C));
plot(Ti,Pvi,T,Pv,'o'),    xlabel('T(C)'),    ylabel('Pv(mmHg)'),    legend
('Fitting','Data')
>> antvp
A =
    -0.0644
C =
    -44.2767
B =
    -24.9991
```

We can see that the Antoine equation is given by

$$\log P_v = -0.0644 - \frac{24.9991}{T - 44.2767}$$

The graph of this line and the data points is shown in Figure 2.10.

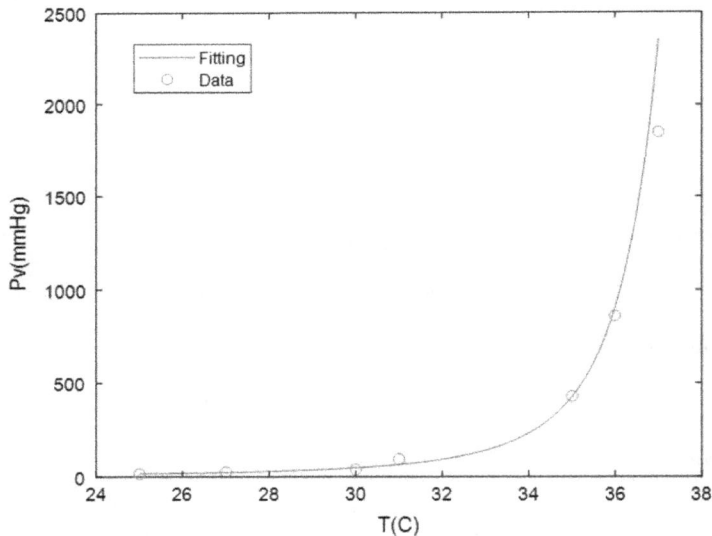

FIGURE 2.10   Antoine vapor pressure model.

Example 2.26 Nonlinear Regression by Least Squares and Built-In Function *nlinfit*

Assume that the experimental data shown in Table 2.10 can be represented by a nonlinear model

$$v = \frac{Ae^{-t}}{1 + Bt}$$

Use the least-squares method and the built-in function *nlinfit* to estimate A and B. Compare the nonlinear model and the results of nonlinear regressions by plotting the fitting curves and data on the same graph.

TABLE 2.10

Experimental Data

t	2.1	2.2	2.4	2.5	2.6	2.8	3.0	3.1
v	3.091	2.699	1.801	1.698	1.412	1.297	0.702	0.597

Solution

The nonlinear model can be rearranged as

$$\frac{1}{v} = \frac{1}{A}e^t + \frac{B}{A}te^t = C_1e^t + C_2te^t, \; A = \frac{1}{C_1}, \; B = AC_2$$

The parameters C_1 and C_2 can be determined by solving the following system of linear equations:

$$Y = \begin{bmatrix} 1/v_1 \\ 1/v_2 \\ \vdots \\ 1/v_n \end{bmatrix} = \begin{bmatrix} e^{t_1} & te^{t_1} \\ e^{t_2} & te^{t_2} \\ \vdots & \vdots \\ e^{t_n} & te^{t_n} \end{bmatrix} \begin{bmatrix} C_1 \\ C_2 \end{bmatrix} = XC \Rightarrow C = \begin{bmatrix} C_1 \\ C_2 \end{bmatrix} = (X^TX)^{-1}X^TY$$

The script *nonreg* compares the least-squares fitting by this equation with the results obtained from the built-in function *nlinfit*. The function *nlinfit* requires an initial guess for the parameters, which is set to $[1 \; 1]^T$ here.

```
% nonreg.m: nonlinear regression
clear all;
t = [2.1 2.2 2.4 2.5 2.6 2.8 3.0 3.1]; z = [min(t):0.01:max(t)];
v = [3.091 2.699 1.801 1.698 1.412 1.297 0.702 0.597];
X = [exp(t); t.*exp(t)]'; Y = [1./v]';
C = inv(X'*X)*X'*Y; A = 1/C(1), B = A*C(2) % regression by least squares
yz = A*exp(-z)./(1 + B*z);
f = @(D,t) D(1)*exp(-t)./(1 + D(2)*t); % model to be used in the built-in
function nlinfit
D0 = [1 1]'; % initial guess for parameter vector
D = nlinfit(t,v,f,D0); % use of built-in function nlinfit
plot(z,yz,z,f(D,z),'--',t,v,'o'), xlabel('t'), ylabel('v')
legend('Nonlinear  regression','Built-in  fun  nlinfit','Data','location',
'best')

>> nonreg
A =
-20.9206
B =
-0.8105
```

Results are shown in Figure 2.11. We can see that the experimental data can be reasonably approximated by the nonlinear model

$$v = \frac{-20.9206e^{-t}}{1 - 0.8105t} = \frac{20.9206e^{-t}}{0.8105t - 1}$$

Example 2.27 Nonlinear Regression by Built-In Function *nlinfit*[4]

A wooden slab with thickness $z = 0.03 \; m$ is dried from both sides by hot air. Table 2.11 shows the data on the free moisture in the wood, $x \, (kgH_2O/kg \; dry \; wood)$, obtained from the

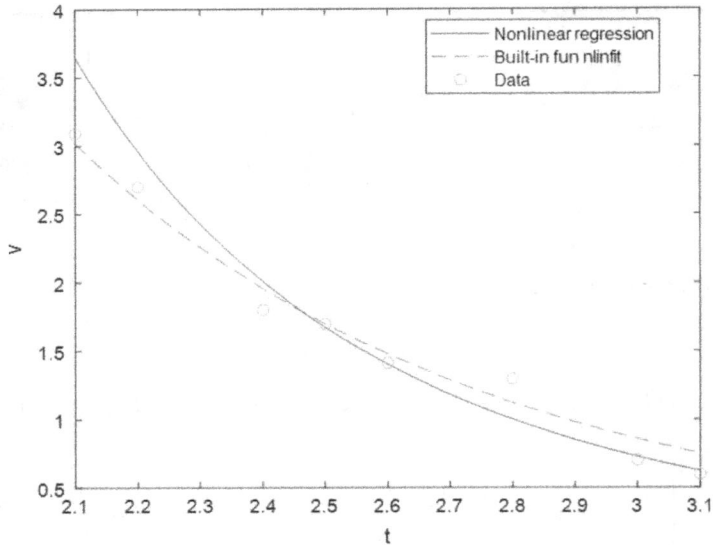

FIGURE 2.11 Comparison of nonlinear model with data.

drying experiment. The following model can be used to approximate the moisture content of the wood:

$$x(t) = \frac{8x_0}{\pi^2} \left\{ e^{-Dt\left(\frac{\pi}{2z}\right)^2} + \frac{1}{9} e^{-9Dt\left(\frac{\pi}{2z}\right)^2} \right\}$$

where x_0 is the initial free moisture of the wood and $D\,(m^2/hr)$ is the diffusivity of water in the wood. Estimate x_0 and D using the built-in function *nlinfit*.

TABLE 2.11

Free Moisture in the Wood as a Function of Time

$t\,(hr)$	5	10	15	20	25	30
x	0.245	0.230	0.211	0.197	0.187	0.176

Solution

   The script *diffwood* uses the built-in function *nlinfit* to estimate x_0 and D and compares results of fitting and data, as shown in Figure 2.12.

```
% diffwood.m: estimation of water diffusivity in the wood
clear all; t = [5:5:30]; % time(hr)
x = [0.245, 0.230, 0.211, 0.197, 0.187, 0.176]; % free moisture in the wood
z = 0.03; % wood thickness
f   = @(C,t) 8*C(1)*(exp(-C(2)*t*(pi/2/z)^2) + exp(-9*C(2)*t*(pi/2/z)^2)/
9)/pi^2;
C0 = [0.3; 0]; % initial guess (C(1) = x0, C(2) = D)
C = nlinfit(t,x,f,C0); % nonlinear regression by built-in function nlinfit
x0 = C(1); D = C(2); tv = 0:0.1:30;
fprintf('Initial free moisture content of the wood = %g\n', x0)
fprintf('Diffusivity of water in the wood = %g\n', D)
```

```
plot(tv,f(C,tv),t,x,'o'),   grid,   xlabel('t(hr)'),   ylabel('x(kg   H_2O/
kg wood)')
legend('Fitting by built-in fun nlinfit','Data','location','best')

>> diffwood
Initial free moisture content of the wood = 0.299382
Diffusivity of water in the wood = 4.01675e-06
```

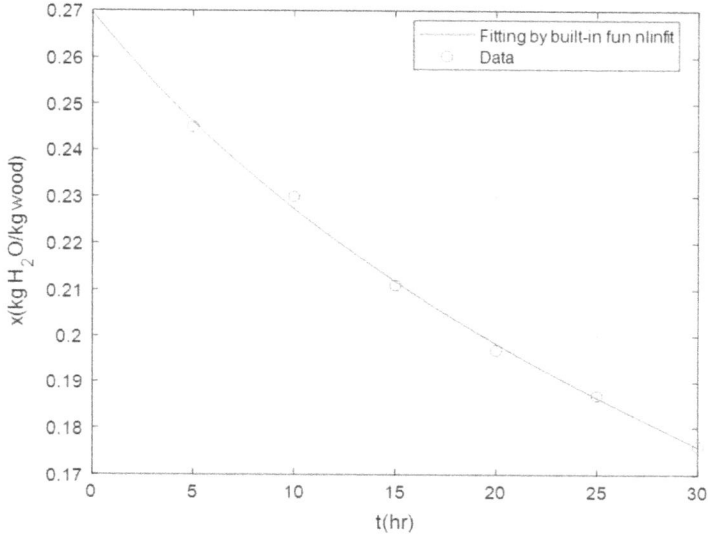

FIGURE 2.12 Nonlinear fitting by the built-in function *nlinfit*.

2.4 INTERPOLATION

2.4.1 POLYNOMIAL INTERPOLATION

Sometimes we need to represent a function based on knowledge about its behavior at a set of discrete points. Interpolation is a method that produces a function that best matches the given data while also providing a good approximation to the unknown values at intermediate points. Suppose that values of a function $f(x_i)$ are known at a set of independent variables x_i, as shown in Table 2.12.

The general objective in developing interpolating polynomials is to choose a polynomial $P_n(x)$ of the form

$$P_n(x) = a_0 + a_1 x + \cdots + a_n x^n$$

TABLE 2.12
Known Function Values

x_i	$f(x_i)$
x_1	$f(x_1)$
x_2	$f(x_2)$
$\vdots$	$\vdots$
x_n	$f(x_n)$

so that this equation best fits the base points of the function and connects these points with a smooth curve. In short, we find $P_n(x)$ such that

$$P_n(x) = f(x_i)(i = 0, 1, \cdots, n)$$

The resulting polynomial $P_n(x)$ can be used to estimate data values at intermediate points.

Example 2.28 Determination of Interpolating 2nd-Order Polynomial

Table 2.13 shows entropy data of saturated steam at three different temperatures. Identify the 2nd-order polynomial that best fits the data, and estimate the entropy at $T = 373.15K$ using this polynomial.

TABLE 2.13

Entropy of Saturated Steam

$T(K)$	Entropy $(kJ/(kg \cdot K))$
313.15	8.2583
363.15	7.4799
413.15	6.9284

Solution

We have to choose the coefficients of the quadratic equation

$$f(x) = a_2 x^2 + a_1 x + a_0$$

so that this equation best fits the data points. The procedure shown in Table 2.14 can be represented as

$$\begin{bmatrix} 313.15^2 & 313.15 & 1 \\ 363.15^2 & 363.15 & 1 \\ 413.15^2 & 413.15 & 1 \end{bmatrix} \begin{bmatrix} a_2 \\ a_1 \\ a_0 \end{bmatrix} = \begin{bmatrix} 8.2583 \\ 7.4799 \\ 6.9284 \end{bmatrix}$$

The coefficients can be calculated by the backslash operator.

```
>> format long
>> T = [313.15 363.15 413.15]'; A = [T.^2 T ones(3,1)]; b = [8.2583 7.4799
6.9284]';
>> x = A\b
x =
0.000045380000000
-0.046258494000000
18.294051973049942
```

TABLE 2.14

Determination of 2nd-Order Polynomial to Fit Entropy Data

i	x_i	$f(x_i) = f_i = a_2 x_i^2 + a_1 x_i + a_0$
1	313.15	$8.2583 = a_2(313.15)^2 + a_1(313.15) + a_0$
2	363.15	$7.4799 = a_2(363.15)^2 + a_1(363.15) + a_0$
3	413.15	$6.9284 = a_2(413.15)^2 + a_1(413.15) + a_0$

We can see that the entropy of saturated steam can be represented as

$$S(T) = 0.00004538T^2 - 0.0462585T + 18.294052$$

At $T = 373.15K$, the entropy is calculated as

```
>> s = x(1)*373.15^2 + x(2)*373.15+x(3)
s =
7.351448
```

From the steam table, we obtain s = 7.3554 $kJ/(kg \cdot K)$ at $T = 373.15\ K$. Thus, we can see that the entropy of the saturated steam can be adequately estimated by the 2nd-order polynomial.

2.4.2 LAGRANGE METHOD

It is always possible to determine a polynomial $P_{n-1}(x)$ of degree $n - 1$ that passes through n distinct data points. The Lagrange formula is one means of obtaining $P_{n-1}(x)$:

$$P_{n-1}(x) = \sum_{i=1}^{n} L_i(x)f_i$$

where n is the number of data points and $L_i(x)$ is called the cardinal function, defined by

$$L_i(x) = \frac{x - x_1}{x_i - x_1} \cdot \frac{x - x_2}{x_i - x_2} \cdots \frac{x - x_{i-1}}{x_i - x_{i-1}} \cdot \frac{x - x_{i+1}}{x_i - x_{i+1}} \cdots \frac{x - x_n}{x_i - x_n} = \prod_{j=1, j \neq i}^{n} \frac{x - x_j}{x_i - x_j} (i = 1, 2, \ldots, n)$$

The function *lagrintp* implements the Lagrange interpolation method.

```
function yi = lagrintp(x,y,xi)
% Implements Lagrange interpolation method
% input
% x: vector of independent variables
% y: vector of dependent variables
% xi: vector of independent variable at which the interpolation is performed
% Output
% yi: vector of interpolated values at xi
% Determine coefficients of Lagrange polynomial
n = length(x); m = length(xi);
if n ~ = length(y), error('x and y should be the same length.'); end
for k = 1:n
dx(k) = 1;
for j = 1:n
if j ~ = k, dx(k) = dx(k)*(x(k) - x(j)); end
a(k) = y(k)/dx(k);
end
end
% Interpolation using Lagrange polynomial
for i = 1:m, yi(i) = 0;
for j = 1:n, q(j) = 1;
for k = 1:n, if (j ~ = k), q(j) = q(j)*(xi(i) - x(k)); end; end
yi(i) = yi(i) + a(j)*q(j);
end
end
end
```

2.4.3 NEWTON METHOD

The polynomial $P_{n-1}(x)$ of degree $n-1$ that passes through n distinct data points can be written in the form

$$P_{n-1}(x) = b_1 + b_2(x - x_1) + b_3(x - x_1)(x - x_2) + \cdots + b_n(x - x_1)(x - x_2)\cdots(x - x_{n-1})$$

$$P_0(x) = b_n, \ P_k(x) = b_{n-k} + (x - x_{n-k})P_{k-1}(x)(k = 1, 2, \cdots, n-1)$$

The function *newtintp* implements the Newton interpolation method.

```
function yi = newtintp(x,y,xi)
% Implements Newton interpolation method
% input
% x: vector of independent variables
% y: vector of dependent variables
% xi: vector of independent variable at which the interpolation is performed
% Output
% yi: vector of interpolated values at xi
% Determine coefficients of Newton polynomial
n = length(x); m = length(xi);
if n ~ = length(y), error('x and y should be the same length.'); end
a(1) = y(1);
for k = 1:n-1, df(k,1) = (y(k+1) - y(k))/(x(k+1) - x(k)); end
for j = 2:n-1
for k = 1:n-j, df(k,j) = (df(k+1,j-1) - df(k,j-1))/(x(k+j) - x(k)); end
end
for k = 2:n, a(k) = df(1,k-1); end
% Interpolation using Newton polynomial
for k = 1:m
s(1) = 1; yi(k) = a(1);
for j = 2:n, s(j) = (xi(k) - x(j-1))*s(j-1); yi(k) = yi(k) + a(j)*s(j); end
end
end
```

2.4.4 CUBIC SPLINES

Cubic splines are most widely used because they provide the simplest representation with desired smoothness. Figure 2.13 shows cubic splines that span n data points, with $s_{i,i+1}(x)$ representing the 3rd-order cubic polynomial that spans the segment between points i and $i + 1$. Continuity of the second derivatives requires that

$$s''_{i-1,i}(x_i) = s''_{i,i+1}(x_i) = m_i(i = 2, 3, \cdots, n-1), \ m_1 = m_n = 0$$

Using Lagrange two-point interpolation, the second derivative of the cubic spline $s''_{i,i+1}(x)$, which is linear with respect to x, can be given by

$$s''_{i,i+1}(x) = \frac{m_i(x - x_{i+1}) - m_{i+1}(x - x_i)}{x_i - x_{i+1}}$$

Integrating twice with respect to x, we have

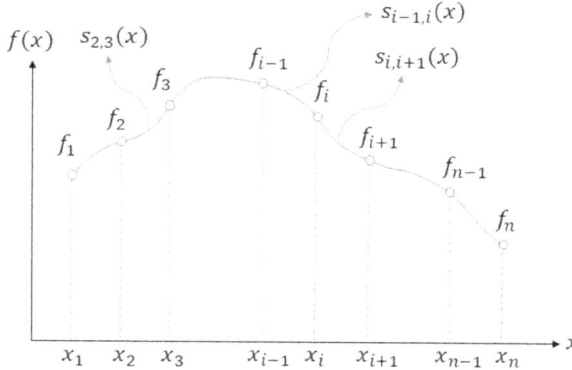

FIGURE 2.13 Cubic splines.

$$s_{i,i+1}(x) = \frac{m_i(x - x_{i+1})^3 - m_{i+1}(x - x_i)^3}{6(x_i - x_{i+1})} + a(x - x_{i+1}) - b(x - x_i)$$

where a and b are integration constants. Using the conditions $s_{i,i+1}(x_i) = f(x_i) = f_i$ and $s_{i,i+1}(x_{i+1}) = f(x_{i+1}) = f_{i+1}$, we obtain

$$a = \frac{f_i}{x_i - x_{i+1}} - \frac{m_i}{6}(x_i - x_{i+1}), \ b = \frac{f_{i+1}}{x_i - x_{i+1}} - \frac{m_{i+1}}{6}(x_i - x_{i+1})$$

Substitution of these relationships into the equation for $s_{i,i+1}(x)$ yields

$$s_{i,i+1}(x) = \frac{m_i}{6}\left\{\frac{(x - x_{i+1})^3}{x_i - x_{i+1}} - (x - x_{i+1})(x_i - x_{i+1})\right\} - \frac{m_{i+1}}{6}\left\{\frac{(x - x_i)^3}{x_i - x_{i+1}} - (x - x_i)(x_i - x_{i+1})\right\}$$

$$+ \frac{f_i(x - x_{i+1}) - f_{i+1}(x - x_i)}{x_i - x_{i+1}}$$

Using the slope continuity conditions $s'_{i-1,i}(x_i) = s'_{i,i+1}(x_i)(i = 2, 3, \cdots, n - 1)$ and some algebraic manipulations, we have

$$m_{i-1}(x_{i-1} - x_i) + 2m_i(x_{i-1} - x_{i+1}) + m_{i+1}(x_i - x_{i+1}) = 6\left(\frac{f_{i-1} - f_i}{x_{i-1} - x_i} - \frac{f_i - f_{i+1}}{x_i - x_{i+1}}\right)(i$$

$$= 2, \cdots, n - 1)$$

The function *cspintp* implements the cubic spline interpolation method.

```
function yi = cspintp(x,y,xi)
% Implements cubic spline interpolation method
% input
% x: vector of independent variables
% y: vector of dependent variables
% xi: vector of independent variable at which the interpolation is performed
% Output
% yi: vector of interpolated values at xi
% Formulation of tridiagonal system
```

```
n = length(x); m = length(xi);
if n ~ = length(y), error('x and y should be the same length.'); end
for k = 1:n-1 % parameters to form tridiagonal system
h(k) = x(k+1) - x(k); w(k) = (y(k+1) - y(k))/h(k);
end
for k = 1:n-2 % right-hand side and diagonal elements
v(k) = w(k+1) - w(k); d(k) = 2*(h(k) + h(k+1));
end
for k = 1:n-3, U(k) = h(k+1); L(k+1) = U(k); end
L(1) = 0; U(n-2) = 0;
% Solve tridiagonal system to find coefficients of spline functions
v(1) = v(1)/d(1); U(1) = U(1)/d(1);
for k = 2:n-3
dn = d(k)-L(k)*U(k-1); U(k) = U(k)/dn; v(k) = (v(k)-L(k)*v(k-1))/dn;
end
v(n-2) = (v(n-2) - L(n-2)*v(n-3))/(d(n-2) - L(n-2)*U(n-3)); a(n-2) = v(n-2);
for k = n-3:-1:1, a(k) = v(k) - U(k)*a(k+1); end; % coefficient vector a
b(1) = y(1)/h(1); c(1) = y(2)/h(1) - a(1)*h(1); % coefficient vectors b and c
for k = 2:n-2, b(k) = y(k)/h(k) - a(k-1)*h(k); c(k) = y(k+1)/h(k) - a(k)*h
(k); end
b(n-1) = y(n-1)/h(n-1) - a(n-2)*h(n-1); c(n-1) = y(n)/h(n-1);
a = [0 a]; % a(1) should be 0
% Interpolation at xi
for k = 1:m
for j = 1:n-1, if xi(k) > = x(j), id = j; end; end
if xi(k) > x(n), error('xi > max(x): cannot interpolate.'); end
if xi(k) < x(1), error('xi < min(x): cannot interpolate.'); end
h = x(id+1) - x(id);
if id = = 1
s(k) = a(2)*(xi(k)-x(1))^3/h + b(1)*(x(2)-xi(k)) + c(1)*(xi(k)-x(1));
elseif id = = n-1
s(k)  =  a(n-1)*(x(n)-xi(k))^3/h + b(n-1)*(x(n)-xi(k)) + c(n-1)*(xi(k)-x
(n-1));
else
s(k) = (a(id)*(x(id+1)-xi(k))^3 + a(id+1)*(xi(k)-x(id))^3)/h...
+ b(id)*(x(id+1)-xi(k)) + c(id)*(xi(k)-x(id));
end
end
yi = s; % vector of interpolated values at xi
end
```

Example 2.29 Isothermal Batch Reactor

The data shown in Table 2.15 were taken in an isothermal batch reactor where reactant A is decomposed to product B. Determine C_A at $t\,(min)$ = 8, 15, 25 and 32 using Lagrange, Newton, and cubic spline interpolation methods.

TABLE 2.15

Concentration of Reactant A (C_A) as a Function of Time (t) in a Batch Reactor

$t\,(min)$	0	2	5	6	13	20	24	30	35	41	50
$C_A\,(mol/liter)$	0.86	0.61	0.47	0.39	0.25	0.18	0.15	0.12	0.10	0.09	0.08

Solution

The script *compintp* calculates C_A at $t\,(min)$ = 8, 15, 25 and 32 using Lagrange, Newton, and cubic spline interpolation methods, and generates plots to compare interpolation results as shown in Figure 2.14.

```
% compintp.m: compare performance of various interpolation methods
t = [0,2,5,6,13,20,24,30,35,41,50];
c = [0.86,0.61,0.47,0.39,0.25,0.18,0.15,0.12,0.10,0.09,0.08];
ti = [8,15,25,32];
cL = lagrintp(t,c,ti); % Lagrange method
cN = newtintp(t,c,ti); % Newton method
cC = cspintp(t,c,ti); % cubic spline method
fprintf('Lagrange method: C = '); fprintf('%g ', cL);
fprintf('\nNewton method: C = '); fprintf('%g ', cN);
fprintf('\nCubic spline method: C = '); fprintf('%g ', cC);
plot(t,c,'o-',ti,cL,'*',ti,cN,'<',ti,cC,'kd')
xlabel('t(min)'), ylabel('C_A(mol/l)')
legend('Data','Lagrange','Newton','Cubic spline','location','best')

>> compintp
Lagrange method: C = 0.255609 0.271843 0.157431 0.0661202
Newton method: C = 0.255609 0.271843 0.157431 0.0661202
Cubic spline method: C = 0.281332 0.240582 0.144551 0.111026
```

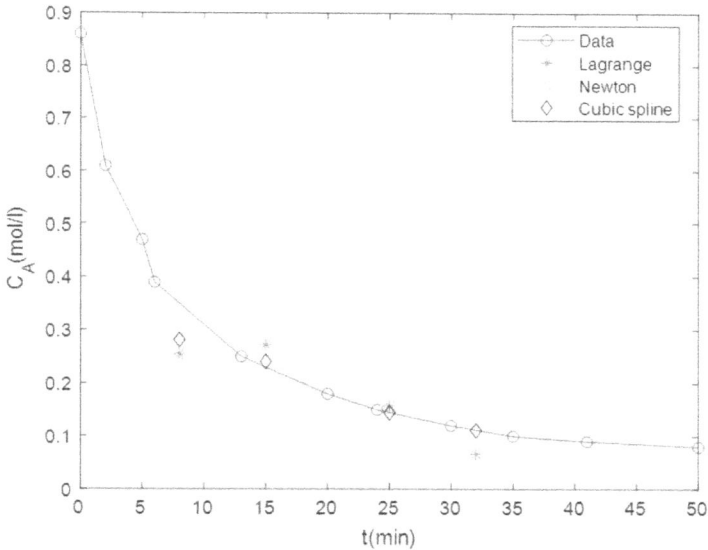

FIGURE 2.14 Interpolation of concentrations.

2.4.5 Use of MATLAB Built-In Functions

2.4.5.1 Polynomial and Cubic Spline Regression

The built-in function *polyfit* can be implemented as

```
p = polyfit(x, y, n)
```

where p is a vector of coefficients sorted by descending power, x is a vector of independent variables, y is a vector of dependent variables, and n is the order of the polynomial. When using *polyfit*, the number of data points should be greater than that of polynomial coefficients.

Function values at intermediate points can be calculated using the built-in function *polyval.* For example, the function value at an intermediate point z can be found by

```
f = polyval(p, z)
```

where p is the vector of coefficients sorted by descending power, as determined by the function *polyfit*; f and z may be vectors.

Cubic spline calculations can be performed by using the built-in function *spline.* The basic calling syntax is

```
yi = spline(x, y, xi)
```

where x and y are vectors of independent and dependent variables, respectively, and yi is the vector of the values of the interpolant at xi.

Example 2.30 Polynomial Regression by *polyfit*

Table 2.16 presents the enthalpy of saturated steam versus temperature. Determine a 2nd-order polynomial that fits these enthalpy data, and estimate the enthalpy at $T = 350.15\ K$ using the polynomial.

TABLE 2.16

Enthalpy of Saturated Steam

$T(K)$	283.15	303.15	323.15	363.15	393.15	413.15
Enthalpy *(kJ/kg)*	2519.9	2556.4	2592.2	2660.1	2706.0	2733.1

Solution

The script *enth2nd* finds coefficients of the 2nd-order polynomial using the built-in function *polyfit*.

```
% enth2nd.m: interpolation of enthalpy for saturated steam
format long;
T = [283.15 303.15 323.15 363.15 393.15 413.15]; % T(K)
H = [2519.9 2556.4 2592.2 2660.1 2706.0 2733.1]; % enthalpy (kJ/kg)
p = polyfit(T, H, 2)
Tv = 280:0.1:415; Hv = polyval(p,Tv); % enthalpy values by the 2nd polynomial
plot(Tv,Hv,T,H,'o'), xlabel('T(K)'), ylabel('H(kJ/kg)')
legend('2nd-order interpolation','Steam table','location','best')

>> enth2nd
p =
1.0e+03 *
-0.000002096326670 0.003108694505382 1.807163153926106
```

We can see that the enthalpy of the saturated steam shown in Table 2.16 can be represented by the quadratic polynomial

$$H(T) = -0.002096327T^2 + 3.108695T + 1807.163154$$

The graph of this curve and the enthalpy data points is shown in Figure 2.15. The enthalpy of the saturated steam at $T = 350.15\ K$ can be obtained as

```
>> f = polyval(p, 350.15)
f =
2.638652356432598e+03
```

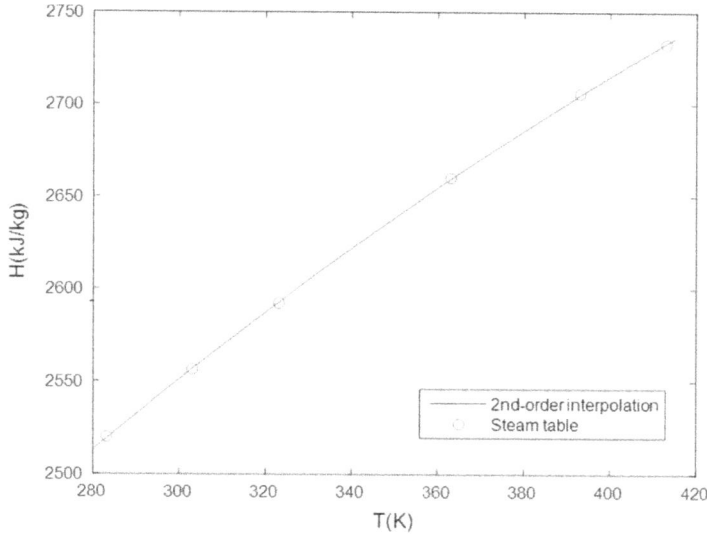

FIGURE 2.15 Interpolation of the enthalpy value.

We have $H = 2638.65 \, kJ/kg$. From the steam table, we get $H = 2638.7 \, kJ/kg$ at $T = 350.15 \, K$. Thus, we can see that 2nd-order polynomials can adequately represent the enthalpy of the saturated steam.

Example 2.31 Cubic Spline Interpolation

Table 2.17 shows experimental data on a pressure drop (kPa) according to flow rates $(liter/sec)$ in a filter. Perform cubic spline interpolation using the built-in function *spline*.

TABLE 2.17

Pressure Drop versus Flow Rates

Flow Rate (liter/sec)	Pressure Drop (kPa)	Flow Rate (liter/sec)	Pressure Drop (kPa)
0	0	31.7	1.695
9.7	0.28	35.2	2.306
15.6	0.524	38.4	2.781
21.3	0.998	42.9	3.205

Solution

The script *filtintp* produces the graph of the interpolation and data points shown in Figure 2.16.

```
% filtintp.m: cubic spline interpolation by the built-in spline function
x = [0 9.7 15.6 21.3 31.7 35.2 38.4 42.9];
y = [0 0.28 0.524 0.998 1.695 2.306 2.781 3.205];
xi = min(x):0.1:max(x); yi = spline(x,y,xi);
plot(xi,yi,x,y,'o'), xlabel('Flow rate(liter/sec)'), ylabel('Pressure
drop(kPa)')
legend('Cubic spline interpolation','Experimental data','Location','Best')
```

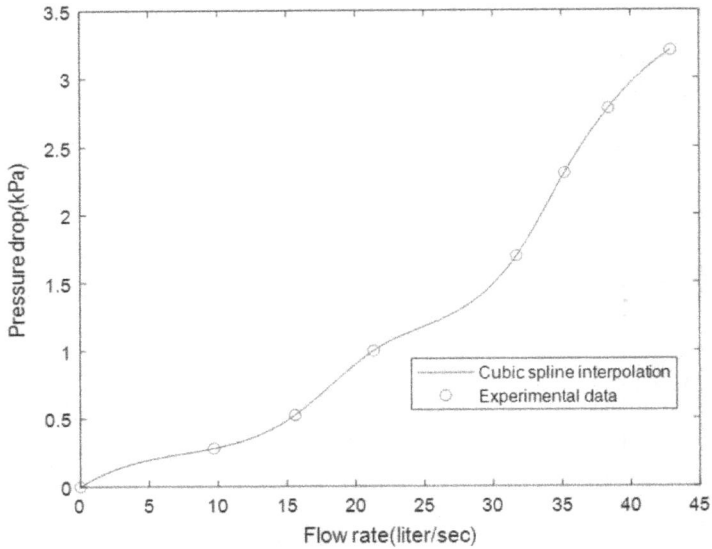

FIGURE 2.16 Cubic spline interpolation.

2.4.5.2 Interpolation in One Dimension

For one-dimensional data, the vector of an independent variable x must be monotonically increasing or decreasing. MATLAB's built-in function *interp1* finds the interpolated value yi for the specified value xi (xi and yi are vectors), based on data given in the vectors x and y. The calling syntax is

```
yi = interp1(x, y, xi, 'method')
```

where 'method' is the desired interpolation method; the default method is linear interpolation. MATLAB provides several methods, as shown in Table 2.18. These methods can also be used in two- and three-dimensional interpolation except *pchip*. As with cubic splines, the *pchip* method uses cubic polynomials to connect data points with continuous first derivatives. In this method, the second derivatives are not necessarily continuous.

Example 2.32 One-Dimensional Interpolation

Determine y at $x = 0.45$ using the benzene-toluene equilibrium data shown in Table 2.19. Try various interpolation methods.

TABLE 2.18

Interpolation Methods Used in the Function *interp1*

Method	Features
nearest	Nearest neighbor interpolation. This method sets the value of an interpolated point to the value of the nearest existing data point.
linear	Linear interpolation. This method uses straight lines to connect the points.
spline	Piecewise cubic spline. This is identical to the spline function.
pchip	Piecewise cubic Hermite interpolation.

TABLE 2.19
Benzene-Toluene Equilibrium Data at 1 atm

x	0.0	0.1	0.2	0.3	0.4	0.5	0.6	0.7	0.8	0.9	1.0
y	0.00	0.21	0.38	0.51	0.62	0.71	0.79	0.86	0.91	0.96	1.00

Solution

The script *bte1dintp* implements required one-dimensional interpolation using the built-in function *interp1*.

```
% bte1dintp.m: 1D interpolation for B-T equilibrium data
x = [0 .1 .2 .3 .4 .5 .6 .7 .8 .9 1.0];
y = [0.00 0.21 0.38 0.51 0.62 0.71 0.79 0.86 0.91 0.96 1.00];
lv = interp1(x,y,0.45,'linear'); % Linear interpolation
pv = interp1(x,y,0.45,'pchip'); % Piecewise cubic Hermite interpolation
sv = interp1(x,y,0.45,'spline'); % Piecewise cubic spline
nv = interp1(x,y,0.45,'nearest'); % Nearest neighbor interpolation
fprintf('linear:   %6.4f\npchip:   %6.4f\nspline:   %6.4f\nnearest:   %6.4f
\n',lv,pv,sv,nv);

>> bte1dintp
linear: 0.6650
pchip: 0.6668
spline: 0.6670
nearest: 0.7100
```

Example 2.33 One-Dimensional Fitting
Table 2.20 shows the time series of measurements of reaction temperature. Use MATLAB's built-in function *interp1* to fit these data with the *pchip* (piecewise cubic Hermite) option.

TABLE 2.20
Reaction Temperature versus Time

Time (min)	0	18	42	55	70	82	86	95	102	115
Temperature (°C)	15	23	24	36	78	80	98	96	127	126

Solution

The script *rxnt1dintp* implements one-dimensional interpolation and plots the *pchip* interpolation shown in Figure 2.17 using the built-in function *interp1*.

```
% rxnt1dintp.m: performs 1D interpolation and plots the pchip interpolation
tm = [0 18 42 55 70 82 86 95 102 115]; TC = [15 23 24 36 78 80 98 96 127 126]; % data
tinv = linspace(0,115); yp = interp1(tm,TC,tinv,'pchip'); % 1D interpolation
using the pchip method
plot(tm,TC,'o',tinv,yp), xlabel('Time(min)'), ylabel('Temperature(deg.C)')
axis([0 115 0 130]), legend('Data','Cubic Hermite interpolation','
location','best')
```

Example 2.34 One-Dimensional Fitting of Heat Capacity Data
Table 2.21 shows the heat capacity of nitrogen (C_p) at 1 atm. Determine C_p at $T = 580\,K$. Try various one-dimensional interpolation methods.

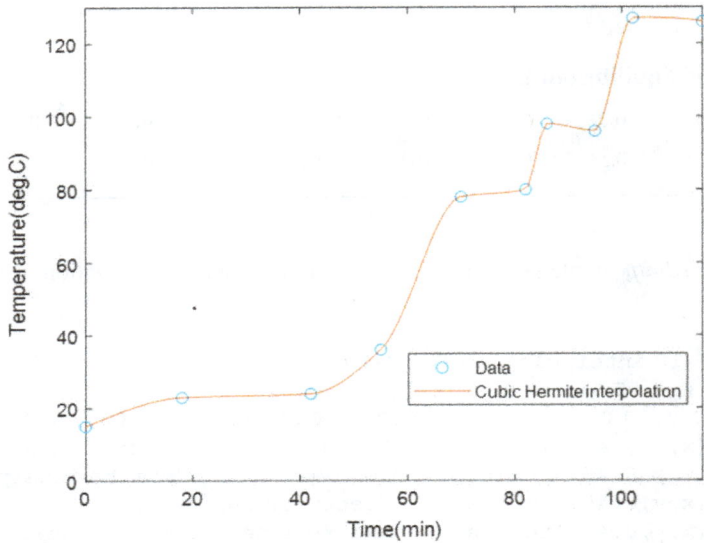

FIGURE 2.17   One-dimensional interpolation (use of *interp1*).

TABLE 2.21

Heat Capacity of Nitrogen at 1 atm

$T(K)$	373	473	573	673	773	873
$C_p(J/(mol \cdot K))$	29.189	29.291	29.462	29.678	29.971	30.269

Solution

The script *cp1dintp* implements various one-dimensional interpolations using the built-in function *interp1*.

```
% cp1dintp.m: 1-D interpolation for Cp data
T = 373:100:873; Cp = [29.189 29.291 29.462 29.678 29.971 30.269]; % data
lCp = interp1(T, Cp, 580, 'linear'); pCp = interp1(T, Cp, 580, 'pchip');
sCp = interp1(T, Cp, 580, 'spline'); nCp = interp1(T, Cp, 580, 'nearest');
fprintf('linear: %6.4f\npchip: %6.4f\nspline: %6.4f\nnearest: %6.4f\n',
lCp,pCp,sCp,nCp);

>> cp1dintp
linear: 29.4771
pchip: 29.4755
spline: 29.4754
nearest: 29.4620
```

2.4.5.3 Interpolation in Multiple Dimensions

Two-dimensional interpolation deals with determining intermediate values for functions of two independent variables, $z = f(x_i, y_i)$. Suppose that we have values at four different points: $z_{11} = f(x_1, y_1)$, $z_{21} = f(x_2, y_1)$, $z_{12} = f(x_1, y_2)$, $z_{22} = f(x_2, y_2)$. Let's interpolate between these points to estimate the function value $z = f(x_i, y_i)$ at an intermediate point $(x_i, y_i)(x_1 \le x_i \le x_2, y_1 \le y_i \le y_2)$. Using the Lagrange interpolation method, the function value at (x_i, y_1) is

$$z_{i1} = f(x_i, y_1) = \frac{x_i - x_2}{x_1 - x_2}z_{11} + \frac{x_i - x_1}{x_2 - x_1}z_{21}$$

and at (x_i, y_2) is

$$z_{i2} = f(x_i, y_2) = \frac{x_i - x_2}{x_1 - x_2}z_{12} + \frac{x_i - x_1}{x_2 - x_1}z_{22}$$

These points can then be used to linearly interpolate along the y-axis to yield

$$z = f(x_i, y_i) = \frac{y_i - y_2}{y_1 - y_2}z_{i1} + \frac{y_i - y_1}{y_2 - y_1}z_{i2}$$

$$= \frac{x_i - x_2}{x_1 - x_2}\frac{y_i - y_2}{y_1 - y_2}z_{11} + \frac{x_i - x_1}{x_2 - x_1}\frac{y_i - y_2}{y_1 - y_2}z_{21} + \frac{x_i - x_2}{x_1 - x_2}\frac{y_i - y_1}{y_2 - y_1}z_{12} + \frac{x_i - x_1}{x_2 - x_1}\frac{y_i - y_1}{y_2 - y_1}z_{22}$$

The MATLAB built-in function *interp2* implements two-dimensional piecewise interpolation. A simple syntax representation of *interp2* is

```
zi = interp2(x, y, z, xi, yi, 'method')
```

where x and y are matrices containing coordinates of the points at which the values in matrix z are given, zi is a matrix containing the results of the interpolation as evaluated at the points in the matrices xi and yi, and 'method' is the desired interpolation method. The methods are identical to those used by *interp1*, that is, nearest, linear, and spline. As with *interp1*, if the 'method' argument is omitted, the default is linear interpolation.

Three-dimensional interpolation deals with determining intermediate values for functions of three independent variables $v = f(x, y, z)$. The MATLAB built-in function *interp3* implements three-dimensional piecewise interpolation. The calling syntax of *interp3* is

```
Vi = interp3(X, Y, Z, V, Xi, Yi, Zi)
```

where X, Y, and Z are matrices containing the coordinates of the points at which the values in the matrix V are given; Vi is a matrix containing the results of the interpolation as evaluated at the points in the matrices Xi, Yi, and Zi; and 'method' is the desired interpolation method.

Example 2.35 Two-Dimensional Interpolation

Temperatures are measured at various points on a heated metal plate (Table 2.22). Estimate the temperature at $x_i = 6.4$ and $y_i = 5.2$ using two-dimensional piecewise cubic spline interpolation.

Solution

The commands to implement two-dimensional piecewise cubic spline interpolation are

```
>> x = [2 10]; y = [1 8]; z = [80 78; 75 90]; % data
>> zi = interp2(x, y, z, 6.4, 5.2, 'spline')
zi =
81.5100
```

TABLE 2.22

Temperature at Various Points on a Heated Metal Plate

x	y	$T = f(x, y)$
2	1	80
2	8	75
10	1	78
10	8	90

Example 2.36 Interpolation of Humidity and Dew Point

Table 2.23 presents the absolute humidity (H) and the dew point (DP) of the air as a function of relative humidity (RH). Estimate H and DP when RH = 58.4 and T(dry bulb temperature) = 46.8°C using two-dimensional piecewise cubic spline interpolation.

TABLE 2.23
Absolute Humidity and Dew Point

Relative Humidity (%RH) Dry Bulb (°C)		10%	30%	50%	70%	90%
51	H,g/m^3	8.27	24.75	41.30	58.10	73.51
	DP,°C	10.10	26.89	37.01	43.21	47.80
44	H,g/m^3	6.52	19.58	32.70	45.75	58.78
	DP,°C	6.40	23.34	32.18	38.32	42.81

```
% hum2dintp.m: performs 2D interpolation
RH = [10 30 50 70 90]; T = [51 44];
H = [8.27 24.75 41.30 58.10 73.51; 6.52 19.58 32.70 45.75 58.78];
DP = [10.10 26.89 37.01 43.21 47.80; 6.40 23.34 32.18 38.32 42.81];
Hv = interp2(RH,T,H,58.4,46.8,'spline') % absolute humidity
DPv = interp2(RH,T,DP,58.4,46.8,'spline') % dew point

>> hum2dintp
Hv =
42.2701
DPv =
36.9549
```

We can see that the absolute humidity is 42.27 g/m^3 and the dew point is 36.95°C when the relative humidity is 58.4% and the dry bulb temperature is 46.8°C.

Example 2.37 Two-Dimensional Interpolation of Steam Table Data

Table 2.24 shows the enthalpy $H(kJ/kg)$ of superheated steam excerpted from the steam table. Estimate H at $T = 380°C$ and $P = 260 \, kPa$.

Solution

The built-in function *interp2* may be used to estimate the enthalpy. The script *ss2dintp* implements two-dimensional interpolation using the cubic spline method.

TABLE 2.24
Enthalpy of Superheated Steam

$P(kPa)$	$T(°C)$ 300	350	400	450	500
150	3073.3	3174.7	3277.5	3381.7	3487.6
200	3072.1	3173.8	3276.7	3381.1	3487.0
225	3071.5	3173.3	3276.3	3380.8	3486.8
250	3070.9	3172.8	3275.9	3380.4	3486.5
275	3070.3	3172.4	3275.5	3380.1	3486.2
300	3069.7	3171.9	3275.2	3379.8	3486.0

```
% ss2dintp.m: 2D interpolation for steam table enthalpy data
T = [300 350 400 450 500]; P = [150 200 225 250 275 300];
H = [3073.3 3174.7 3277.5 3381.7 3487.6; 3072.1 3173.8 3276.7 3381.1 3487.0;
3071.5 3173.3 3276.3 3380.8 3486.8; 3070.9 3172.8 3275.9 3380.4 3486.5;
3070.3 3172.4 3275.5 3380.1 3486.2; 3069.7 3171.9 3275.2 3379.8 3486.0];
Hi = interp2(T,P,H,380,260,'spline') % use cubic spline method

>> ss2dintp
Hi =
3.2343e+03
```

2.5 DIFFERENTIATION AND INTEGRATION

2.5.1 DIFFERENTIATION

The derivative of a function can be estimated by using the function values at a set of discrete points. The difference formulas to estimate the first derivatives are based on the interpolation of a given data set using a straight line. Suppose that a function $f(x)$ and three data points $(x_{i-1}, f(x_{i-1}))$, $(x_i, f(x_i))$, and $(x_{i+1}, f(x_{i+1}))$ are given. The derivatives of $f(x)$ at x_i can be estimated by difference formulas. In general, the difference formulas are based on a straight line to interpolate the given data. The derivative can be approximated by using forward, central, or backward differences. An approximation to the 1st- and 2nd-order derivatives at $x = x_i$ are given as follows ($f_i = f(x_i)$ and $h = x_{i+1} - x_i$):

1. Forward difference formula:

$$\frac{df(x_i)}{dx} \approx \frac{f(x_{i+1}) - f(x_i)}{x_{i+1} - x_i} = \frac{f_{i+1} - f_i}{h}$$

$$\frac{d^2f(x_i)}{dx^2} \approx \frac{f(x_{i+2}) - 2f(x_{i+1}) + f(x_i)}{(x_{i+2} - x_{i+1})(x_{i+1} - x_i)} = \frac{f_{i+2} - 2f_{i+1} + f_i}{h^2}$$

2. Central difference formula:

$$\frac{df(x_i)}{dx} \approx \frac{f(x_{i+1}) - f(x_{i-1})}{x_{i+1} - x_{i-1}} = \frac{f_{i+1} - f_{i-1}}{2h}$$

$$\frac{d^2f(x_i)}{dx^2} \approx \frac{f(x_{i+1}) - 2f(x_i) + f(x_{i-1})}{(x_{i+1} - x_i)(x_i - x_{i-1})} = \frac{f_{i+1} - 2f_i + f_{i-1}}{h^2}$$

3. Backward difference formula:

$$\frac{df(x_i)}{dx} \approx \frac{f(x_i) - f(x_{i-1})}{x_i - x_{i-1}} = \frac{f_i - f_{i-1}}{h}$$

$$\frac{d^2f(x_i)}{dx^2} \approx \frac{f(x_i) - 2f(x_{i-1}) + f(x_{i-2})}{(x_i - x_{i-1})(x_{i-1} - x_{i-2})} = \frac{f_i - 2f_{i-1} + f_{i-2}}{h^2}$$

MATLAB has two built-in functions to determine the derivatives of data: *diff* and *gradient*. The function *diff* returns a vector containing the differences between adjacent elements. Let x and y be independent and dependent variables, respectively.

`diff(y)` computes the differences between the adjacent elements of y, and
`diff(x)` computes the differences between the adjacent elements of x.
The derivatives of data are given by
`d = diff(y)./diff(x)`

The *gradient* function evaluates derivatives at the data values themselves, rather than at the intervals between values. A simple expression of its syntax is

`df = gradient(f, h)`

where f is a one-dimensional vector and h is the interval (constant) between data points.

Example 2.38 Differentiation by *diff*

Differentiate the function

$$f(x) = 0.3 + 20x - 180x^2 + 650x^3 - 880x^4 + 360x^5$$

from $x = 0$ to 1 using the *diff* function. Compare your results with the exact solution given by

$$f'(x) = 20 - 360x + 1950x^2 - 3520x^3 + 1800x^4$$

Solution

The script *fxdiff* estimates the differentiation of $f(x)$ and compares the results with the exact solution graphically, as shown in Figure 2.18.

```
% fxdiff.m: differentiation using diff function
f = @(x) 0.3+20*x-180*x.^2+650*x.^3-880*x.^4+360*x.^5; % define function
x = 0:0.1:1.0; n = length(x); % range of x
y = f(x); dr = diff(y)./diff(x) % use diff to find f'(x) numerically
xm = (x(1:n-1) + x(2:n))/2; % the midpoint of each interval
xp = 0:0.01:1; yp = 20-360*xp+1950*xp.^2-3520*xp.^3+1800*xp.^4; % exact
differentiation
plot(xp,yp,xm,dr,'o'), xlabel('x'), ylabel('df(x)/dx')
legend('Exact differentiation','Numerical differentiation','location','best')
```

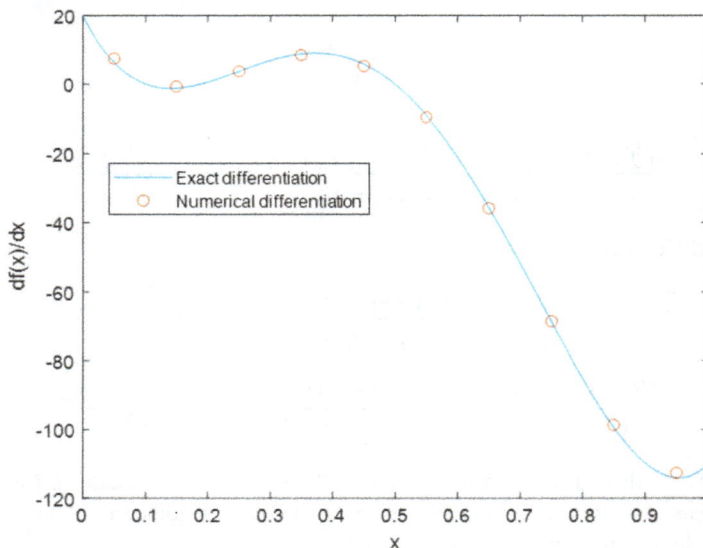

FIGURE 2.18 Differentiation using the built-in function *diff*.

```
>> fxdiff
dr =
7.6560  -0.5840  3.8960  8.6160  5.4160  -9.5440  -35.7840  -68.5040  -98.5840
-112.5840
```

Example 2.39 : Differentiation of CO_2 Concentration by diff[5]

In a transient mass transfer of CO2 through a membrane the concentration profile of CO2 at a certain time across the membrane wall is represented by the equation

$$C(x) = -1.3 \times 10^6 x^4 + 7.1 \times 10^3 x^3 - 14x^2 - 0.364x + 0.001 (-0.001 \le x \le 0.001)$$

where $C(kgmol/m^3)$ is the concentration of CO_2 and $x(m)$ is the distance from the center of the membrane. The mass transfer flux $N_{CO_2} kgmol/(sec \cdot m^2)$ at $x(m)$ from the center of the membrane is given by

$$N_{CO_2} = -D \frac{dC(x)}{dx}$$

where $D = 3.26 \times 10^{-8} m^2/sec$ is the diffusivity of CO_2 through the membrane. At time t, CO_2 accumulates within the membrane if the net flux (difference between input flux and output flux) is positive $(N_{CO_2}|_{x=-0.001} - N_{CO_2}|_{x=0.001} > 0)$ and is depleted from the membrane if the net flux is negative $(N_{CO_2}|_{x=-0.001} - N_{CO_2}|_{x=0.001} < 0)$. Is CO_2 accumulating or depleting in the membrane?

Solution

The numerical differentiation of $C(x)$ can be performed by using the built-in function diff. The script mflux performs numerical differentiation and calculates the net mass transfer flux.

```
% mflux.m
D = 3.26e-8; x = -1e-3:1e-4:1e-3;
C = @(x) -1.3e6*x.^4 + 7.1e3*x.^3 - 14*x.^2 - 0.364*x + 0.001;
f = C(x); numf = diff(f)./diff(x); % numerical differentiation
NCO2 = -D*numf; % flux of CO2
nfl = NCO2(1); nfr = NCO2(end); netf = nfl - nfr; % net flux
fprintf('Flux at x = -0.001: %g, Flux at x = 0.001: %g, Net flux: %g\n', nfl,
nfr, netf)

>> mflux
Flux at x = -0.001: 1.02262e-08, Flux at x = 0.001: 1.2252e-08, Net flux:
-2.02581e-09
```

Since the net flux is negative, we can see that depletion of CO_2 takes place.

Example 2.40 Differentiation of Slurry Data by diff[6]

The equation for constant pressure filtration in a plate-and-frame press is given by

$$\frac{dt}{dV} = \frac{\mu c_s}{A^2 (-\Delta P)} \alpha V + \frac{\mu}{A(-\Delta P)} R_m$$

where
A is the filter area
c_s is the slurry concentration
μ is the viscosity of water

α is the specific cake resistance

R_m is the resistance of the filter medium to filtrate flow

Table 2.25 shows data for filtration of $CaCO_3$ slurry in water at 298 K at a constant pressure ($-\Delta P$) of $3 \times 10^5 kg/(m \cdot sec^2)$. Determine $\alpha (m/kg)$ and R_m (m^{-1}) by using Table 2.25 and the data after it.

Data: $A = 0.04 \ m^2$, $c_s = 20 \ kg/m^3$, $\mu = 8.937 \times 10^{-4} \ kg/(m \cdot sec)$, $-\Delta P = 3 \times 10^5 \ kg/(m \cdot sec^2)$

TABLE 2.25

Data for Filtration of $CaCO_3$ Slurry in Water

t (sec)	4.4	9.5	16.3	24.6	34.7	46.1	59.0	73.6	89.4	107.3
V (liter)	0.498	1.000	1.501	2.000	2.498	3.002	3.506	4.004	4.502	5.009

Solution

Differentiation of the filtration data by using the built-in function *diff* yields dt/dV. We assume that dt/dV are derivatives at the midpoint of each time interval. Thus we use values of V at the midpoint of each interval, V_m, in the regression to find α and R_m. The filtration equation can be rearranged as follows:

$$\frac{dt}{dV} = k_1 \alpha V_m + k_2 R_m$$

$$Y = \begin{bmatrix} (dt/dV)_1 \\ (dt/dV)_2 \\ \vdots \\ (dt/dV)_n \end{bmatrix} = \begin{bmatrix} k_1 V_{m,1} & k_2 \\ k_1 V_{m,2} & k_2 \\ \vdots & \vdots \\ k_1 V_{m,n} & k_2 \end{bmatrix} \begin{bmatrix} \alpha \\ R_m \end{bmatrix} = XC \Rightarrow C = \begin{bmatrix} \alpha \\ R_m \end{bmatrix} = (X^T X)^{-1} X^T Y$$

where $k_1 = \mu c_s/(A^2(-\Delta P))$ and $k_2 = \mu/(A(-\Delta P))$. The script *filt* performs numerical differentiation and calculates α and R_m by regression.

```
% filt.m
t = [4.4 9.5 16.3 24.6 34.7 46.1 59.0 73.6 89.4 107.3]; % t (sec)
V = 1e-3*[0.498 1.0 1.501 2.0 2.498 3.002 3.506 4.004 4.502 5.009]; % V (m^3)
A = 0.04; cs = 20; vis = 8.937e-4; dp = 3e5; % data
dtV = diff(t)./diff(V); % numerical differentiation using diff
n = length(V); Vm = (V(1:n-1) + V(2:n))/2; % midpoint of each interval
k1 = vis*cs/(A^2*dp); k2 = vis/(A*dp); % define k1 and k2
Y = dtV'; X = [k1*Vm; k2*ones(1,n-1)]'; C = inv(X'*X)*X'*Y; % regression
fprintf('alpha = %g, Rm = %g\n', C(1), C(2))

>> filt
alpha = 1.65993e+11, Rm = 7.7766e+10
```

We can see that $\alpha = 1.65993 \times 10^{11} \ m/kg$ and $R_m = 7.7766 \times 10^{10} \ m^{-1}$.

Example 2.41 Differentiation by *gradient*

Differentiate the function

$$f(x) = 0.3 + 20x - 180x^2 + 650x^3 - 880x^4 + 360x^5$$

from $x = 0$ to 1 using the *gradient* function. Compare your results with the exact solution given by

$$f'(x) = 20 - 360x + 1950x^2 - 3520x^3 + 1800x^4$$

Solution

The script *dfgrad* estimates the differentiation of $f(x)$ and computes the exact solution.

```
% dfgrad.m
f = @(x) 0.3+20*x-180*x.^2+650*x.^3-880*x.^4+360*x.^5;
df = @(x) 20-360*x+1950*x.^2-3520*x.^3+1800*x.^4;
x = 0:0.1:1.0; y = f(x);
dr = gradient(y, 0.1), dy = df(x) % dr: estimation by gradient, dy: exact solution

>> dfgrad
dr =
7.6560 3.5360 1.6560 6.2560 7.0160 -2.0640 -22.6640 -52.1440 -83.5440
-105.5840 -112.5840
dy =
20.0000 0.1600 0.7200 7.0400 8.8000 0 -21.0400 -51.6800 -84.9600 -109.6000
-110.0000
```

2.5.2 INTEGRATION

2.5.2.1 Definite Integrals

A definite integral can be approximated by a weighted sum of function values at points within the specified integration interval $[a\ b]$. A numerical integration formula to approximate $\int_a^b f(x)dx$ has the form

$$\int_a^b f(x)dx \approx \sum_{i=0}^n c_i f(x_i) = I$$

where the parameters c_i depend on the particular numerical method.

Let the integral region $[a\ b]$ be divided into $n-1$ subintervals each of width h. The trapezoid rule, one of the simplest methods to estimate a definite integral, approximates $f(x)$ by a straight line in each subinterval. The area of the subinterval I_i is approximated by

$$I_i = \frac{h}{2}\{f_i + f_{i+1}\}$$

where $h = (b-a)/n$ and $f_i = f(x_i)$. The total area representing $\int_a^b f(x)dx$ is estimated by

$$I = \sum_{i=1}^{n-1} I_i = \frac{h}{2}\sum_{i=1}^{n-1}(f_i + f_{i+1}) = \frac{h}{2}\{f_1 + 2f_2 + 2f_3 + \cdots + 2f_{n-1} + f_n\}$$

The function *trapezoid* implements the trapezoidal rule to approximate $\int_a^b f(x)dx$; Figure 2.19 shows the trapezoidal rule. The function *trapezoidat* performs integration when data are given instead of the function $f(x)$.

```
function z = trapzoid(f,a,b,n)
% Implements trapezoidal rule to integrate f(x) from a to b
% input
% f: function handle to be integrated
% a,b: limits of integration
% n: number of subinterval points
% Output
% z: integral of f(x)
```

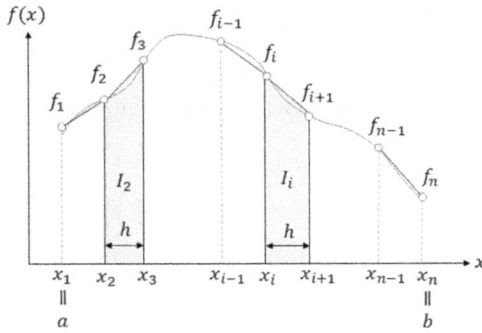

FIGURE 2.19   Trapezoidal rule.

```
h = (b-a)/n; s = f(a);
for k = 1:n-1, x = a + h*k; s = s + 2*f(x); end
s = s + f(b); z = h*s/2;
end

function s = trapzoidat(x,y)
% Implements trapezoidal rule to integrate data set (x,y)
% input
% x: vector of independent variables
% y: vector of dependent variables
% Output
% s: integral of y
n = length(x); s = 0;
for k = 1:n-1, s = s + (y(k) + y(k+1))*(x(k+1) - x(k))/2; end
end
```

The widely used Simpson rule approximates the function to be integrated by a quadratic polynomial. The Simpson 1/3 rule can be obtained by using a parabolic interpolant through three adjacent points, as shown in Figure 2.20. The integral in the interval $[x_i, x_{i+2}]$ is approximated by

$$\int_{x_i}^{x_{i+2}} f(x)dx \cong \frac{h}{3}\{f_i + 4f_{i+1} + f_{i+2}\}$$

The integral $\int_a^b f(x)dx$ is then estimated by

$$\int_a^b f(x)dx \cong \frac{h}{3}\{f_1 + 4f_2 + 2f_3 + 4f_4 + \cdots + 2f_{n-2} + 4f_{n-1} + f_n\} \, (n : \text{odd})$$

In this rule, n should be odd (i.e., the number of subintervals is even). If n is even, integration over the first three (or the last three) subintervals is approximated by the Simpson 3/8 rule as

$$I = \frac{3h}{8}\{f_1 + 3f_2 + 3f_3 + f_4\}$$

and the Simpson 1/3 rule is applied for the remaining subintervals.

The function *simps* implements the Simpson 1/3 rule to approximate $\int_a^b f(x)dx$.

```
function z = simps(f,a,b,n)
% Implements Simpson 1/3 rule to integrate f(x) from a to b
% input
% f: function handle to be integrated
% a,b: limits of integration
```

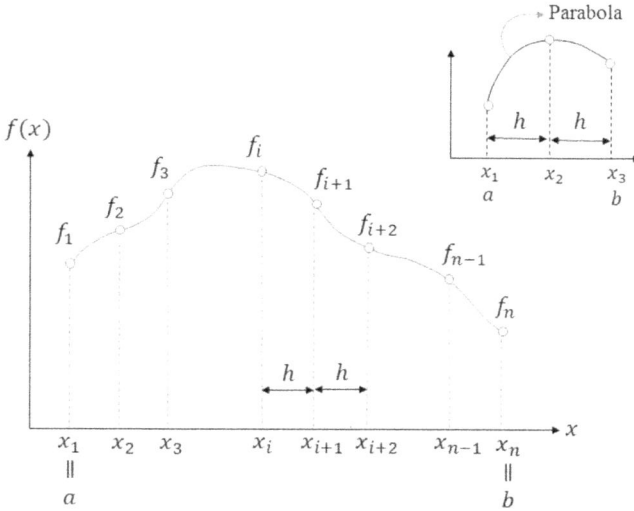

FIGURE 2.20 Simpson 1/3 rule.

```
% n: number of subinterval points (x1,...,xn)
% Output
% z: integral of f(x)
if ceil(n/2) - floor(n/2) < 1, n = n+1; end % n must be odd
h = (b-a)/(n-1); s = f(a);
if n <= 2, display('Too few subintervals.'); return; end
if n == 3, z = h*(s + 4*f((a+b)/2) + f(b))/3; return; end
for k = 2:2:n-1, x = a + h*(k-1); s = s + 4*f(x); end
for k = 3:2:n-2, x = a + h*(k-1); s = s + 2*f(x); end
s = s + f(b); z = h*s/3;
end
```

The midpoint rule uses only function evaluations at points within the integral interval. The midpoint rule is given by

$$\int_a^b f(x)dx \approx h \sum_{i=0}^{n} f(x_i), \; x_i = a + \left(i - \frac{1}{2}\right)h$$

Example 2.42 Numerical Integration
Evaluate $\int_0^{1.5} \frac{1}{1+x^2}dx$ using the trapezoidal rule and the Simpson 1/3 rule.

Solution
```
>> f = @(x) 1/(1+x^2); a = 0; b = 1.5; n = 10;
>> z = trapzoid(f,a,b,n) % trapezoidal rule
z =
    0.9823
>> z = simps(f,a,b,n) % Simpson 1/3 rule
z =
0.9828
```

MATLAB provides built-in functions that evaluate definite integrals for a given data set. The *trapz* function implements the trapezoidal rule. A simple representation of its syntax is

```
z = trapz(x, y)
```

where x and y are vectors of independent and dependent variables, respectively.

The *cumtrapz* function evaluates the cumulative integral. It has the general syntax

```
z = cumtrapz(x, y)
```

where x and y are vectors of independent and dependent variables, respectively.

The *integral* and *quad* functions implement adaptive Simpson quadrature, and the function *quadl* performs Lobatto quadrature. Simple expressions of their syntax are

```
z = integral(fun, a, b)
z = quad(fun, a, b, tol)
z = quadl(fun, a, b, tol)
```

where z is the integral of the function fun over the range from a to b and tol is the desired absolute error tolerance. If tol is not specified, a default value of 1×10^{-6} is used.

Example 2.43 Use of *trapz* and *cumtrapz*

Table 2.26 shows a series of time spot measurements of the velocity of a falling sphere. Determine the distance traveled when $t = 2.5 sec$. Find the cumulative distance traveled at each time spot.

TABLE 2.26
Velocity of a Falling Mass

$t (sec)$	0	0.5	1.2	1.6	2.5	3.1	4.8	6.9
$v (m/sec^2)$	0	5	12	15	23	28	38	47

Solution

The distance traveled to a certain time spot is given by the function *trapz*, and the distance traveled at each time spot can be calculated by the function *cumtrapz*. The following script estimates the distances:

```
>> t = [0 0.5 1.2 1.6 2.5 3.1 4.8 6.9]; t1 = [0 0.5 1.2 1.6 2.5];
>> v = [0 5 12 15 23 28 38 47]; v1 = [0 5 12 15 23];
>> z1 = trapz(t1,v1)
z1 =
    29.7000
>> z = cumtrapz(t,v)
z =
    0 1.2500 7.2000 12.6000 29.7000 45.0000 101.1000 190.3500
```

Table 2.27 summarizes the results.

TABLE 2.27
Distance Traveled

$t (sec)$	0	0.5	1.2	1.6	2.5	3.1	4.8	6.9
$v (m/sec^2)$	0	5	12	15	23	28	38	47
Distance (m)	0	1.25	7.2	12.6	29.7	45.0	101.1	190.35

Example 2.44 Use of *integral, quad,* and *quadl*

MATLAB provides a demonstrative function *humps,* defined as

$$f(x) = \frac{1}{(x - q)^2 + 0.01} + \frac{1}{(x - r)^2 + 0.04} + s$$

Estimate $\int_a^b f(x)dx$ when $a = 0$, $b = 1$, $q = 0.2$, $r = 0.7$ and $s = 4$. Use the built-in functions *integral, quad,* and *quadl* and compare the results.

Solution

The following script estimates the integrals:

```
>> format long
>> q = 0.2; r = 0.7; s = 4;
>> hfun = @(x) 1./((x-q).^2+0.01)+1./((x-r).^2+0.04)-s;
>> integral(hfun,0,1)
ans =
   32.912352455607831
>> quad(hfun,0,1)
ans =
   32.912352767792420
>> quadl(hfun,0,1)
ans =
   32.912352455611952
```

Example 2.45 Interpolation and Numerical Integration

Table 2.28 shows the heat capacity C_p ($J/(mol \cdot °C)$) of a gas as a function of temperature $t(°C)$. Compute the enthalpy change $\Delta H(J) = n \int_{t_1}^{t_2} C_p(t)dt$ for $n = 6.5$ mol of this gas heated from $t_1 = 55°C$ to $t_2 = 185°C$.

TABLE 2.28

Heat Capacity of a Gas as a Function of Temperature

$t(°C)$	20.1	49.8	81.5	109.8	140.2	171.1	201.4	229.5
$C_p(J/(mol \cdot °C))$	28.98	29.11	29.96	29.47	29.67	29.91	30.02	30.14

Solution

The built-in function *trapz*, which implements the trapezoidal rule, can be used to determine ΔH. The script numint first determines values of C_p at $t_1 = 55°C$ and $t_2 = 185$ by interpolation. Then numerical integration is performed on the subintervals $[t_1 \; t_2]$ and $[C_p(t_1) \; C_p(t_2)]$ by the function *trapz*.

```
% numint.m
t = [20.1 49.8 81.5 109.8 140.2 171.1 201.4 229.5]; % temp.(deg.C)
Cp = [28.98 29.11 29.96 29.47 29.67 29.91 30.02 30.14]; % Cp (J/mol/deg.C)
n = 6.5; t1 = 55; t2 = 185; % data
% step 1: find Cp(t1) and Cp(t2) by interpolation using piecewise cubic spline
method.
Cp1 = interp1(t, Cp, t1,'spline'); Cp2 = interp1(t, Cp, t2, 'spline'); % Cp at t
= t1 and t = t2
% step 2: numerical interpolation by trapz on the subintervals [t1 t2] and
[Cp1 Cp2]
```

```
ts  = [t1 81.5 109.8 140.2 171.1 t2]; % subinterval of t: t1 < = t < = t2
Cps = [Cp1 29.96 29.47 29.67 29.91 Cp2]; % subinterval of Cp: Cp1 < = Cp < = Cp2
delH = n*trapz(ts,Cps) % perform numerical integration
```

```
>> numint
delH =
    2.5100e+04
```

2.5.2.2 Multiple Integrals

The MATLAB built-in functions *integral2* and *dblquad* implement double integration

$$I = \int_a^b \int_c^d f(x, y)dydx = \int_a^b \left\{ \int_c^d f(x, y)dy \right\} dx = \int_c^d \left\{ \int_a^b f(x, y)dx \right\} dy$$

Simple expressions of the syntax for *integral2* and *dblquad* are

```
z = integral2(fun, xa, xb, yc, yd, tol)
z = dblquad(fun, xa, xb, yc, yd, tol)
```

where z is the double integral of the function fun over the ranges from xa to xb and yc to yd. If the desired absolute error tolerance tol is not specified, a default value of 1×10^{-6} is used. Here is an example of how *integral2* and *dblquad* can be used to evaluate the double integral

$$I = \int_0^{0.8} \int_0^{0.8} \frac{1}{1 - xy}dydx$$

```
>> f = @(x,y) 1./(1-x.*y);
>> I2 = integral2(f, 0, 0.8, 0, 0.8)
I2 =
    0.7900
>> I2 = dblquad(f, 0, 0.8, 0, 0.8)
I2 =
    0.7900
```

The built-in function *quad2d* can also be used to evaluate the double integral. This function can be used when the range of one of the two independent variables is a function of the other independent variable. As an example, the double integral

$$\int_1^2 \int_{x^2}^{x^4} x^2 y \, dy \, dx$$

can be estimated by using *integral2* and *quad2d*:

```
>> f = @(x,y) x.^2.*y;
>> I2 = integral2(f, 1, 2, @(x) x.^2, @(x) x.^4)
I2 =
83.9740
>> Id = quad2d(f, 1, 2, @(x) x.^2, @(x) x.^4)
I =
83.9740
```

The built-in functions *integral3* and *triplequad* implement triple integration

$$I = \int_a^b \int_c^d \int_g^h f(x, y, z)dzdydx = \int_a^b \left\{ \int_c^d \left\{ \int_g^h f(x, y, z)dz \right\} dy \right\} dx$$

Simple expressions of the syntax for *integral3* and *triplequad* are

```
w = integral3 (fun, xa, xb, yc, yd, ze, zf, tol)
w = triplequad(fun, xa, xb, yc, yd, ze, zf, tol)
```

where w is the triple integral of the function fun over the ranges from xa to xb, yc to yd, and ze to zf. If the desired absolute error tolerance tol is not specified, a default value of 1×10^{-6} is used. Here is an example of how *integral3* and *triplequad* can be used to compute the triple integral

$$I = \int_0^1 \int_0^1 \int_0^1 64xy(1-x)^2 z\,dz\,dy\,dx$$

```
>> f = @(x,y,z) 64*x.*y.*(1-x).^2.*z;
>> I2 = integral3(f, 0, 1, 0, 1, 0, 1)
I2 =
    1.3333
>> Id = triplequad(f, 0, 1, 0, 1, 0, 1)
I =
    1.3333
```

Example 2.46 Double integral
Evaluate the double integral

$$I = \int_0^5 \int_0^7 f(x, y)\,dx\,dy$$

for the function

$$f(x, y) = 3xy + x - 1.2x^2 - 3y^2 + 25$$

Solution
```
>> fxy = @(x,y) 3*x.^y+x-1.2*x.^2-3*y.^2+25;
>> I2 = integral2(fxy, 0, 7, 0, 5)
I2 =
    3.2878e+04
>> Id = dblquad(fxy, 0, 7, 0, 5)
Id =
    3.2878e+04
```

Table 2.29 summarizes MATLAB's built-in integration functions.

TABLE 2.29
MATLAB's Built-In Integration Functions

Built-In Function	Description
integral, quad, int	Lower-order numerical integration
quadl	Higher-order numerical integration
trapz	Integration by trapezoidal rule
integral2, dblquad	Numerical double integration
integral3, triplequad	Numerical triple integration

2.6 ORDINARY DIFFERENTIAL EQUATIONS

Differential equations are composed of an unknown function and its derivatives. The equation is called an ordinary differential equation (ODE) when this unknown function involves only one independent variable, and a partial differential equation (PDE) when more than one independent variable is involved.

2.6.1 INITIAL-VALUE PROBLEMS

2.6.1.1 Explicit Euler Method

The most basic method for finding the approximate solution of a first-order ODE is the Euler method. Consider an ordinary differential equation of the form

$$\frac{dy}{dt} = f(t, y)$$

A forward difference approximation of this equation yields

$$\frac{dy}{dt} \approx \frac{y_{k+1} - y_k}{t_{k+1} - t_k} = \frac{y_{k+1} - y_k}{\Delta t} = f(t_k, y_k)$$

where $\Delta t = h = t_{k+1} - t_k$ is the step size. Rearrangement of the approximation gives

$$y_{k+1} = y_k + hf(t_k, y_k) = y_k + h\left(\frac{dy}{dt}\right)_{t_k, y_k}$$

The explicit Euler method is graphically illustrated in Figure 2.21.

The function *eulerde* implements the explicit Euler method to solve $dy/dt = f(t, y)$.

```
function [t,y] = eulerde(f,tspan,y0,n)
% Implements explicit Euler method to solve dy/dt = f(t,y)
% input
%    f: dy/dt
%    tspan: vector of initial and final values of independent variable
%    y0: initial value of dependent variable
%    n: number of subintervals
% Output
%    t: vector of independent variable
%    y: vector of dependent variable (solution vector)
t0 = tspan(1); tf = tspan(2); h = (tf-t0)/n;
t = (t0:h:tf)'; nt = length(t); y = y0*ones(nt,1);
for k = 2:nt, y(k) = y(k-1) + h*f(t(k-1),y(k-1)); end
end
```

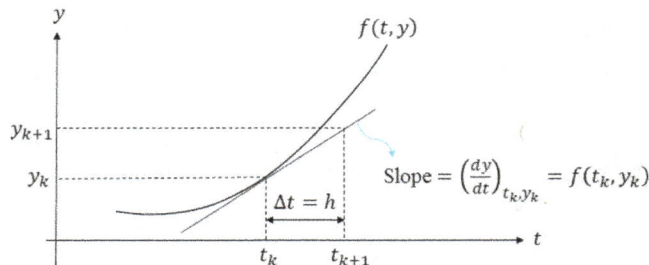

FIGURE 2.21   Illustration of the explicit Euler method.

2.6.1.2 Implicit Euler Method

For an ordinary differential equation of the form

$$\frac{dy}{dt} = f(t, y)$$

a backward difference approximation is given by

$$\frac{dy}{dt} \approx \frac{y_{k+1} - y_k}{t_{k+1} - t_k} = \frac{y_{k+1} - y_k}{h} = f(t_{k+1}, y_{k+1})$$

Rearranging, we obtain

$$y_{k+1} = y_k + hf(t_{k+1}, y_{k+1}) = y_k + h\left(\frac{dy}{dt}\right)_{t_{k+1}, y_{k+1}}$$

This equation indicates that we have to evaluate the derivative at the next time step t_{k+1}. In general, the implicit Euler method requires solution of nonlinear algebraic equations.

2.6.1.3 Heun Method

The ordinary differential equation to be solved can be written as

$$\frac{dy}{dt} = y'(t) = f(t, y)$$

By integrating from $t = t_k$ to $t = t_{k+1}$, we obtain

$$\int_{t_k}^{t_{k+1}} f(t, y)dt = \int_{t_k}^{t_{k+1}} y'(t)dt = y(t_{k+1}) - y(t_k) = y_{k+1} - y_k$$

Rearranging, we have

$$y_{k+1} = y_k + \int_{t_k}^{t_{k+1}} f(t, y)dt$$

The definite integral term in this equation can be approximated by using an appropriate approximating scheme. If the trapezoidal rule with step size $\Delta t = h = t_{k+1} - t_k$ is used, we obtain

$$y_{k+1} = y_k + \frac{h}{2}\{f(t_k, y_k) + f(t_{k+1}, y_{k+1})\}$$

If the explicit Euler method is used, we get

$$y_{k+1} = y_k + \frac{h}{2}\{f(t_k, y_k) + f(t_{k+1}, y_k + hf(t_k, y_k))\}$$

This equation can be described as a predictor-corrector algorithm as

$$P_{k+1} = y_k + hf(t_k, y_k), \quad t_{k+1} = t_k + h$$

$$y_{k+1} = y_k + \frac{h}{2}\{f(t_k, y_k) + f(t_{k+1}, P_{k+1})\}$$

2.6.1.4 Runge-Kutta Methods

The well-known Runge-Kutta methods are based on the determination of a sequence of approximations to the change $y_{i+1} - y_i$, followed by a combination of these to compute a new value of y. We would like to solve an ordinary differential equation of the form

$$\frac{dy}{dt} = f(t, y)$$

These methods give results that match the higher-order Taylor series expansions without requiring evaluation of derivatives of $f(t, y)$.

The optimal 2nd-order Runge-Kutta method is given by

$$k_1 = hf(t_i, y_i), \ k_2 = hf\left(t_i + \frac{2}{3}h, \ y_i + \frac{2}{3}k_1\right)$$

$$y_{i+1} = y_i + \frac{1}{4}(k_1 + 3k_2)$$

and the classic 4th-order Runge-Kutta method is given by

$$k_1 = hf(t_i, y_i), \ k_2 = hf\left(t_i + \frac{1}{2}h, \ y_i + \frac{1}{2}k_1\right),$$

$$k_3 = hf\left(t_i + \frac{1}{2}h, \ y_i + \frac{1}{2}k_2\right), \ k_4 = hf\left(t_i + h, \ y_i + k_3\right)$$

$$y_{i+1} = y_i + \frac{1}{6}(k_1 + 2k_2 + 2k_3 + k_4)$$

The function *rk4th* implements the 4th-order Runge-Kutta method to solve $dy/dt = f(t, y)$.

```
function [t,y] = rk4th(f,tspan,y0,n)
% Implements 4th-order Runge-Kutta method to solve dy/dt = f(t,y)
% input
%     f: dy/dt
%     tspan: vector of initial and final values of independent variable
%     y0: initial value of dependent variable
%     n: number of subintervals
% Output
%     t: vector of independent variable
%     y: vector of dependent variable (solution vector)
t0 = tspan(1); tf = tspan(2); h = (tf-t0)/n;
t = (t0:h:tf)'; nt = length(t); y = y0*ones(nt,1);
for k = 1:nt-1
    k1 = f(t(k),y(k)); k2 = f(t(k)+h/2, y(k)+h*k1/2);
    k3 = f(t(k)+h/2,y(k)+h*k2/2); k4 = f(t(k)+h, y(k)+h*k3);
    y(k+1) = y(k) + h*(k1 + 2*k2 + 2*k3 + k4)/6;
end
end
```

2.6.1.5 *n*th-Order ODE

An nth-order ODE can be transformed into a set of 1st-order ODEs. The general form of an nth-order differential equation can be expressed as

$$f\left(t, y, \frac{dy}{dt}, \frac{d^2y}{dt^2}, \ldots, \frac{d^{n-1}y}{dt^{n-1}}\right) = f(t, y, y', y'', \ldots, y^{(n-1)}) = \frac{d^ny}{dt^n} = y^{(n)}$$

The state function can be defined as

$$s(t) = \begin{bmatrix} y(t) \\ y'(t) \\ y''(t) \\ \vdots \\ y^{(n-1)}(t) \end{bmatrix} = \begin{bmatrix} s_1(t) \\ s_2(t) \\ s_3(t) \\ \vdots \\ s_n(t) \end{bmatrix}$$

If we substitute these relations into the nth-order differential equation, we have the equivalent system of simultaneous ordinary differential equations, which can be represented by

$$\frac{ds(t)}{dt} = \begin{bmatrix} y'(t) \\ y''(t) \\ y^{(3)}(t) \\ \vdots \\ y^{(n)}(t) \end{bmatrix} = \begin{bmatrix} y'(t) \\ y''(t) \\ y^{(3)}(t) \\ \vdots \\ f(t, y, y', \ldots, y^{(n-1)}) \end{bmatrix} = \begin{bmatrix} s_2(t) \\ s_3(t) \\ s_4(t) \\ \vdots \\ f(t, s_1(t), s_2(t), \ldots, s_n(t)) \end{bmatrix}$$

The solution of this system requires that n initial conditions be known at the starting value of t.

Example 2.47 Numerical Solution of ODE
Find the solution of the differential equation

$$\frac{dy}{dt} = 5e^{0.6t} - 2y, \ y(0) = 1.5 \ (0 \le t \le 3)$$

using the explicit Euler method and the 4th-order Runge-Kutta method. The number of sub-intervals is $n = 5$.

Solution
```
>> f = @(t,y) 5*exp(0.6*t) - 2*y; y0 = 1.5; tspan = [0 3]; n = 5;
>> [t,y] = eulerde(f,tspan,y0,n); disp([t y]) % explicit Euler method
0 1.5000
0.6000    2.7000
1.2000    3.7600
1.8000    5.4113
2.4000    7.7518
3.0000    11.1117
>> [t,y] = rk4th(f,tspan,y0,n); disp([t y]) % 4th-order Runge-Kutta method
0 1.5000
0.6000    2.6270
1.2000    3.9173
1.8000    5.6631
2.4000    8.1325
3.0000    11.6615
```

2.6.1.6 Use of MATLAB Built-In Functions
MATLAB provides many built-in functions for solving ordinary differential equations. In order to use them, the given differential equation should be transformed into the equivalent system of 1st-order differential equations. Table 2.30 presents some of the built-in functions.

TABLE 2.30
Some of MATLAB's Built-In Functions for Solving Ordinary Differential Equations

Function	Solution Method
ode23	Explicit Runge-Kutta lower-order (2nd and 3rd order)
ode45	Explicit Runge-Kutta higher-order (4th and 5th order)
ode113	Adams-Bashforth-Moulton (explicit linear multistep)

A stiff system involves both rapidly changing components and slowly changing ones. In some cases, the rapidly varying components die away quickly and the solution becomes dominated by the slowly varying components. MATLAB provides a number of built-in functions for solving stiff systems of ODEs. Table 2.31 shows some of these solvers.

MATLAB's built-in functions for solving ODEs can be called in a number of different ways. A simple representation of the syntax is

```
[t, y] = solver(odefun,tspan,y0,options)
```

where solver is the name of MATLAB's built-in function, odefun is the name of the function defining the system of 1st-order ODEs to be solved, tspan is the span of the independent variable, y0 is the vector of initial values, and options is a data structure created with the *odeset* function to control features of the solution.

TABLE 2.31
Some of MATLAB's Built-In Functions for Stiff Systems

Function	Solution Method
ode23s	Modified 2nd-order Rosenbrock
ode23t	Implicit trapezoidal rule (2nd and 3rd order)
ode23tb	Implicit Runge-Kutta (2nd and 3rd order)
ode15s	Variable-order implicit multistep method

Example 2.48 Solution of an ODE
Solve

$$\frac{dy}{dt} = e^{-t}, \; y(0) = -1$$

from $t = 0$ to 1.

Solution
The following script uses the built-in *ode45* function to solve the differential equation and generate a plot of the solution, as shown in Figure 2.22:

```
>> tspan = [0 1]; y0 = -1; dy = @(t,y) exp(-t); % define ODE
>> [t y] = ode45(dy, tspan, y0); plot(t,y), grid, xlabel('t'), ylabel('y(t)')
```

Example 2.49 van der Pol Equation
The van der Pol equation can be expressed as

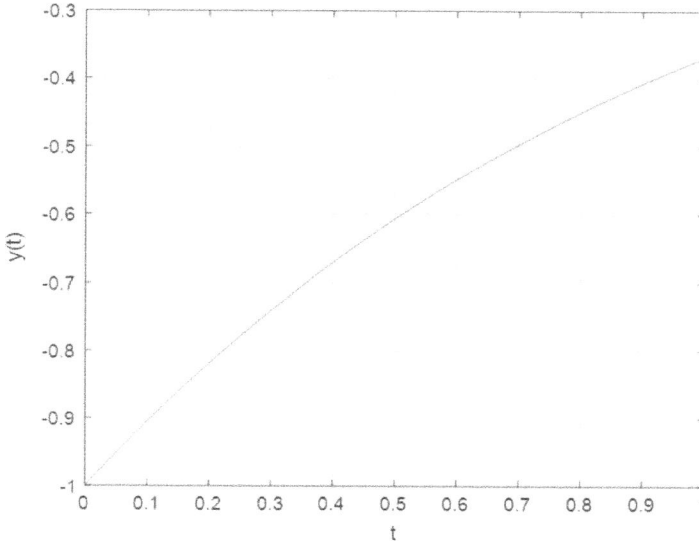

FIGURE 2.22 Plot of the solution of $dy/dt = e^{-t}$, $y(0) = -1$.

$$\frac{d^2 y_1}{dt^2} - \mu(1 - y_1^2)\frac{dy_1}{dt} + y_1 = 0$$

This equation can be transformed into a set of 1st-order differential equations as follows:

$$\frac{dy_1}{dt} = y_2, \quad \frac{dy_2}{dt} = -y_1 + \mu(1 - y_1^2)y_2$$

Plot the changes of y_1 and y_2 with respect to time t from $t = 0$ to 25. $\mu = 1$ and the initial conditions are $y_1(0) = y_2(0) = 1$.

Solution

The following script uses the built-in *ode45* function to integrate the van der Pol equation and generate a plot of the solution, as shown in Figure 2.23:

```
>> mu = 1; tspan = [0 25]; y0 = [1 1]; dy = @(t,y) [y(2); -y(1) + mu*(1-y(1)^2)*y
(2)]; % define ODE
>> [t y]   =   ode45(dy, tspan, y0); plot(t,y(:,1),t,y(:,2),':'), legend
('y_1','y_2'), xlabel('t'), ylabel('y')
```

Example 2.50 Well-Mixed Tanks

Figure 2.24 shows a series of three well-mixed tanks. From mass balance equations, we have

$$\frac{dV_1}{dt} = q_0 + m - q_1, \quad V_1\frac{dx_1}{dt} = q_0(x_0 - x_1) - mx_1, \quad V_2\frac{dx_2}{dt} = q_1(x_1 - x_2), \quad V_3\frac{dx_3}{dt} = q_2(x_2 - x_3)$$

where x_i ($i = 1,2,3$) is the concentration (*mol/liter*) of the solution contained in the tank i. Under normal steady-state operation, m is maintained at 0 and q_i ($i = 1,2,3$) is kept constant. At a certain time ($t = 0$), m is suddenly increased to 12 *liter/min*. Plot the concentrations in the three tanks from $t = 0$ to 2. The initial conditions are $x_0 = 0.15$ *mol/liter* and $q_0 = 15$ *liter/min*, and the initial volume of each tank is 20 *liter*.

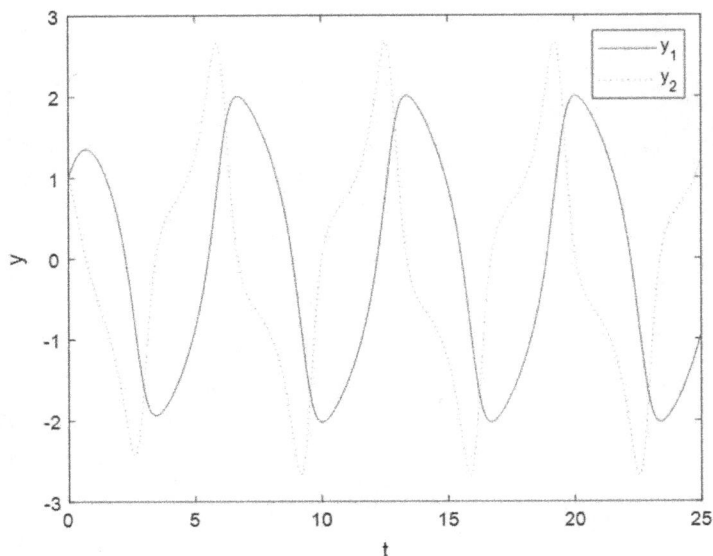

FIGURE 2.23 Plot of the solution of the van der Pol equation.

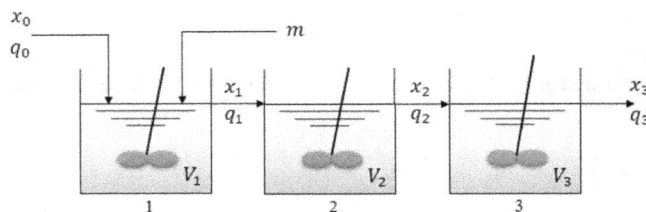

FIGURE 2.24 A series of three well-mixed tanks.

Solution

When $m = 0$, $V_1 = V_2 = V_3 = 20$ liter and q_i ($i = 1, 2, 3$) is constant ($q_i = 15$ liter/min). Thus V_2 and V_3 are constant and independent with changes in m. Substituting the given data, we have

$$\frac{dx_1}{dt} = \frac{2.25 - 27x_1}{V_1}, \quad \frac{dx_2}{dt} = \frac{15}{20}(x_1 - x_2), \quad \frac{dx_3}{dt} = \frac{15}{20}(x_2 - x_3), \quad \frac{dV_1}{dt} = 12$$

The system of differential equations can be defined as an anonymous function where $y_1 = x_1$, $y_2 = x_2$, $y_3 = x_3$ and $y_4 = V_1$. The following script uses the *ode45* solver to integrate the equation system and generate a plot of the solution, as shown in Figure 2.25:

```
>> tspan = [0 2]; y0 = [0.15 0.15 0.15 20];
>> dy = @(t,y) [(2.25-27*y(1))./y(4); 15*(y(1)-y(2))/20; 15*(y(2)-y(3))/20;
12]; % equation system
>> [t y] = ode45(dy, tspan, y0); plot(t,y(:,1),t,y(:,2),'.-',t,y(:,3),':'),
xlabel('t(min)'), ylabel('x(t)')
>> legend('x_1', 'x_2', 'x_3', 'location', 'best')
```

Example 2.51 Plug-Flow Reactor

A plug-flow reactor is used to carry out the reaction $A \rightarrow B$. The reaction rate is known to be represented by a Langmuir-Hinshelwood model given by

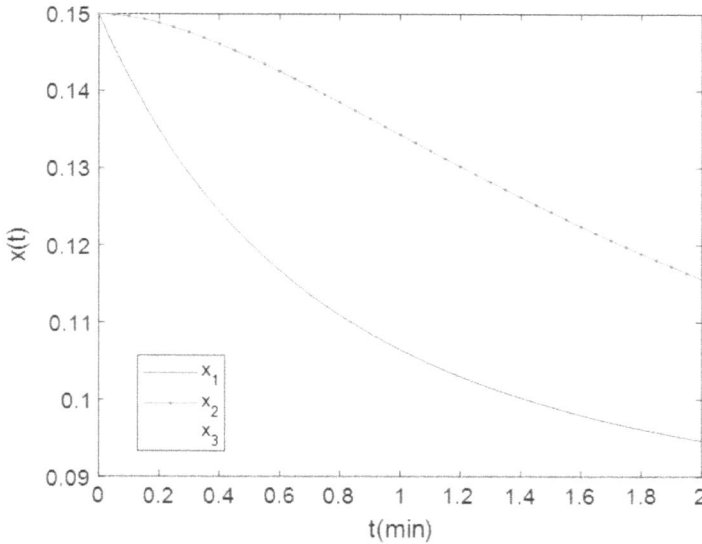

FIGURE 2.25 Concentration changes in mixed tanks.

$$\frac{dC_A}{dz} = \left(\frac{A_c}{q}\right)\frac{dF_A}{dV} = -\left(\frac{A_c}{q}\right)\frac{kC_A}{\sqrt{1 + k_r C_A^2}}$$

where z is the length of the reactor, V is the reactor volume, q is the inlet flow rate, A_c is the cross-sectional area of the reactor, and k and k_r are kinetic parameters. The initial concentration of A is $C_{A0} = 1\ mol/m^3$, the inlet flow rate is $q = 0.12\ m^3/sec$, $A_c = 0.26\ m^2$, and the kinetic parameters are $k = 2.1\ sec^{-1}$ and $k_r = 0.98\ mol^2/m^6$. F_A is given by $F_A = F_{A0}(1 - x_A)$ where $F_{A0} = qC_{A0}$.

(1) Generate the profiles of conversion and concentration of A for $0 \le z \le 0.5m$.
(2) Find the reactor volume required for 80% conversion of A.

Solution
(1) The concentration C_A can be obtained by using the built-in function *ode45*. The conversion is given by x = $(C_{A0}\text{-}C_A)/C_{A0}$. The script *pfrLHmodel* produces the required profiles, as shown in Figure 2.26.

(2) Since $F_A = F_{A0}(1 - x_A)$ and $C_A = C_{A0}(1 - x_A)$, we have

$$\frac{dF_A}{dV} = -F_{A0}\frac{dx_A}{dV} = -\frac{kC_{A0}(1 - x_A)}{\sqrt{1 + k_r C_{A0}^2(1 - x_A)^2}}$$

Rearrangement yields

$$\frac{dV}{dx_A} = \left(\frac{q}{k}\right)\frac{\sqrt{1 + k_r C_{A0}^2(1 - x_A)^2}}{(1 - x_A)}$$

The script *pfrLHmodel* integrates the differential equation to give the reactor volume required to achieve the conversion of $x_A = 0.8$.

```
% pfrLHmodel.m: PFR reaction calculation by Langmuir-Hinshelwood model
```

```
clear all;
q = 0.12; Ac = 0.26; k = 2.1; kr = 0.98; Ca0 = 1; % data
% (1) Profiles of concentration and conversion
zspan = [0 0.5];
dC = @(z,Ca) -(Ac/q)*k*Ca/sqrt(1 + kr*Ca^2);
[z Ca] = ode45(dC, zspan, Ca0); x = (Ca0 - Ca)/Ca0;
subplot(1,2,1), plot(z,Ca), grid, xlabel('z(m)'), ylabel('C_A(mol/m^3)')
subplot(1,2,2), plot(z,x), grid, xlabel('z(m)'), ylabel('x(conversion)')
% (2) Calculation of reactor volume
V0 = 0; xd = 0.8; xspan = [0 xd]; % conversion of A = 80%
dV = @(x,V) (q/k)*sqrt(1 + kr*Ca0^2*(1-x)^2)/(1-x);
[x V] = ode45(dV, xspan, V0);
fprintf('Reactor volume for 80 percent conversion = %g m^3\n', V(end));

>> pfrLHmodel
Reactor volume for 80 percent conversion = 0.104087 m^3
```

(a)   (b)

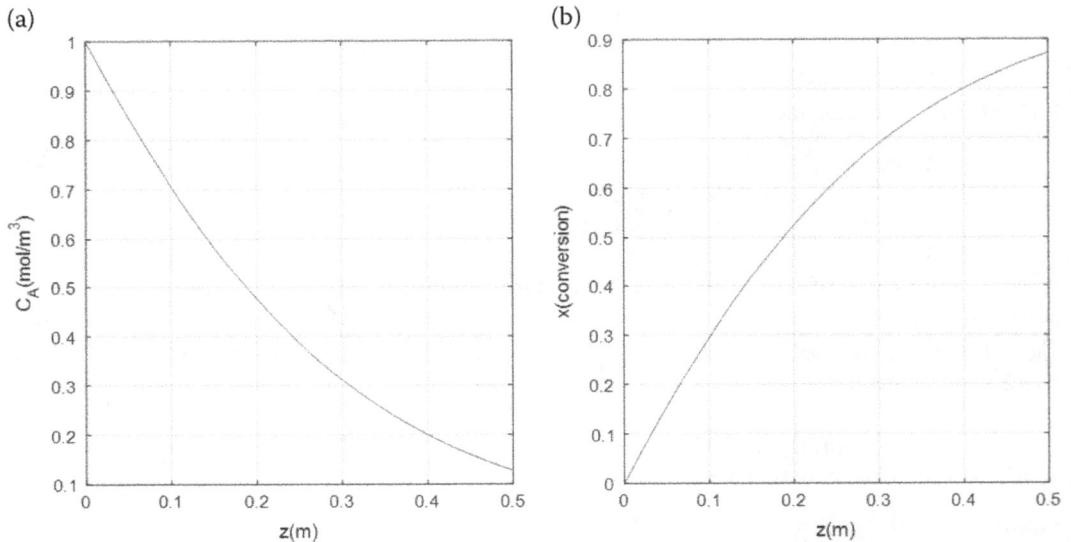

FIGURE 2.26   Profiles of (a) concentration and (b) conversion.

Example 2.52 Penicillin Production Reaction

Penicillin is produced in a batch reactor by fermentation. The reaction model is given by

$$\frac{dx_1}{dt} = a_1x_1 - \frac{a_1}{a_2}x_1^2, \frac{dx_2}{dt} = a_3x_1$$

where x_1 is the dimensionless cell concentration and x_2 is the dimensionless penicillin concentration. From experiments, it was found that $a_1 = 13.2$, $a_2 = 0.95$ and $a_3 = 1.76$. At $t = 0$, $x_1(0) = 0.028$ and $x_2(0) = 0.0$. Generate profiles of x_1 and x_2 as a function of dimensionless time t ($0 \le t \le 1$).

Solution

The script *penrxn* uses the built-in function *ode45* to solve the differential equations and produce profiles, as shown in Figure 2.27.

```
% penrxn.m: Penicillin production reaction
```

```
a = [13.2 0.95 1.76]; x0 = [0.028 0]; tspan = [0 1];
dx  =  @(t,x) [a(1)*x(1) - (a(1)/a(2))*x(1)^2; a(3)*x(1)]; [t x]  =  ode45
(dx,tspan,x0);
% Generate profiles
plot(t,x(:,1),t,x(:,2),'--'), grid, legend('x_1','x_2','location','best')
xlabel('t'), ylabel('x_1 and x_2')
```

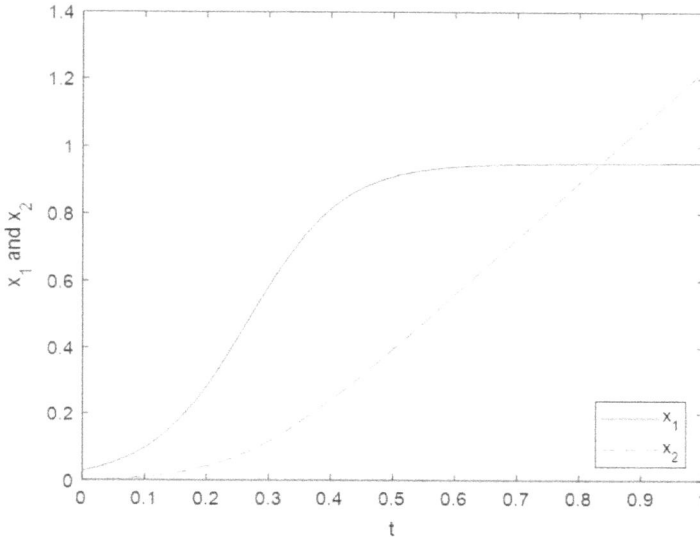

FIGURE 2.27 Profiles of x_1 and x_2.

Example 2.53 Growth of a Biomass from Substrate[7]

A biological process involving the growth of a biomass from substrate can be represented as

$$\frac{dB}{dt} = \frac{kBS}{K+S}, \frac{dS}{dt} = -\frac{0.75kBS}{K+S}$$

where B and S are the biomass and substrate concentrations, respectively. Solve these differential equations from $t = 0$ to 20. At $t = 0$, S and B are 5 and 0.05, respectively. The reaction kinetics are $k = 0.3$ and $K = 0.000001$.

Solution

This is a typical stiff system problem, and we can use the built-in functions for stiff equations. In the script *stiffode*, the stiff differential equations are defined as an anonymous function and the *ode15s* solver is called to integrate the differential equations and generate a plot of change of substrate and biomass concentrations, as shown in Figure 2.28. We can also use other solvers, such as *ode23s*, *ode23t*, or *ode23tb*, to solve the stiff system.

```
% stiffode.m: solution of stiff differential equations
tspan = [0 20]; y0 = [0.05 5]; k = 0.3; K = 1e-6; % data
dy  =  @(t,y) [k*y(1)*y(2)/(K+y(2)); -0.75*k*y(1)*y(2)/(K+y(2))]; % define
equations
[t y]  = ode15s(dy, tspan, y0); % use the built-in solver ode15s
plot(t,y(:,1),t,y(:,2),':'), grid, xlabel('t(min)'), ylabel('y(t)'), legend
('B(t)','S(t)','location','best')
```

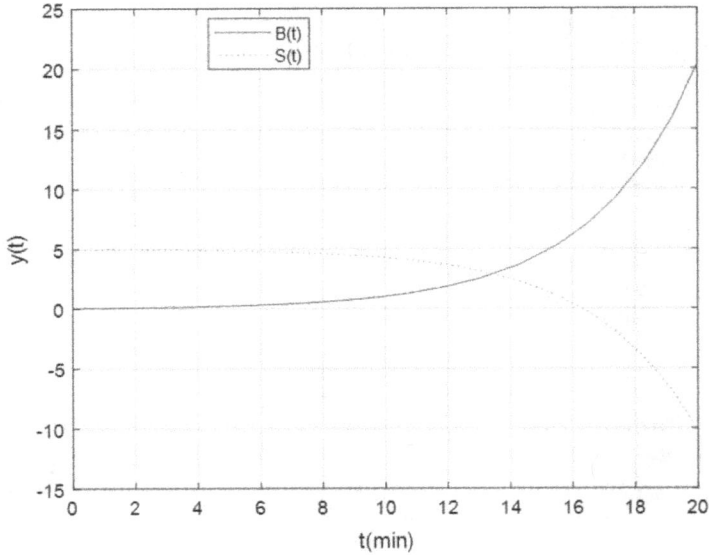

FIGURE 2.28   Change of substrate and biomass concentrations.

Example 2.54 Fluidized Packed Bed Catalytic Reactor[8]

The irreversible gas-phase catalytic reaction $A \rightarrow B$ is to be carried out in a fluidized packed bed reactor. Material and energy balances for this reactor yield

$$\frac{dP}{d\tau} = P_e - P + H_g(P_p - P), \quad \frac{dT}{d\tau} = T_e - T + H_T(T_p - T) + H_W(T_W - T)$$

$$\frac{dP_p}{d\tau} = \frac{H_g}{A}\{P - P_p(1 + K)\}, \quad \frac{dT_p}{d\tau} = \frac{H_T}{C}\{(T - T_p) + FKP_p\}, \quad K = 6 \times 10^{-4}\exp\left(20.7 - \frac{1000}{T_p}\right)$$

where
   $T(°R)$ is the temperature of the reactant
   $P(atm)$ is the partial pressure of the reactant
   $T_p(°R)$ is the temperature of the reactant at the surface of catalyst
   $P_p(atm)$ is the partial pressure of the reactant at the surface of catalyst
   K is the rate constant (dimensionless)
   τ is time (dimensionless)
and the subscript e is the inlet condition. The parameters and constants used in the model equations are

$$H_g = 320, \ T_e = 600, \ H_T = 266.67, \ H_W = 1.6, \ T_W = 720, \ F = 8000, \ A = 0.17142,$$

$$C = 205.74, \ P_e = 0.1$$

Solve the differential equations from $\tau = 0$ to 1500 and plot the changes of dependent variables. Initial conditions are $P(0) = 0.1$, $T(0) = 600$, $P_p = 0$, and $T_p = 761$.

Solution

The script *flupb* defines the differential equation system as an anonymous function and uses the built-in *ode15s* function for stiff systems to integrate the equation system and generate plots of the solutions, as shown in Figure 2.29.

(a)

(b)

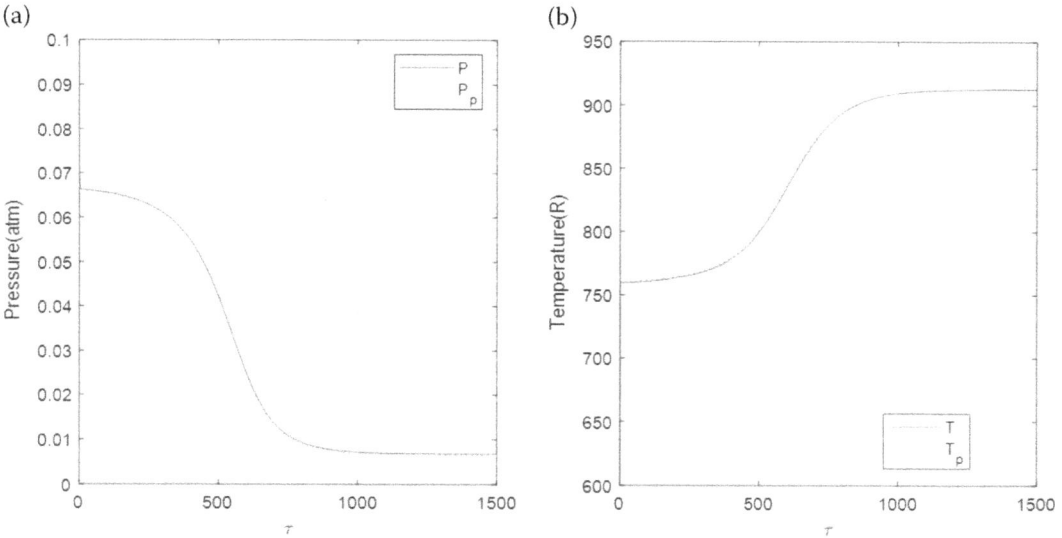

FIGURE 2.29 Change of (a) reaction pressure and (b) temperature.

```
% flupb.m: fluidized packed bed catalytic reactor
A = 0.17142; C = 205.74; F = 8000; Hg = 320; Ht = 266.67; Hw = 1.6; % data
Pe = 0.1; Te = 600; Tw = 720; % data and parameters
% define the differential equation system as an anonymous function
dy = @(t,y) [Pe - y(1) + Hg*(y(3) - y(1)); Te - y(2) + Ht*(y(4) - y(2)) + Hw*(Tw -
y(2));
Hg*(y(1) - y(3)*(1 + (6e-4 * exp(20.7-15000/y(4))))))/A;
Ht*((y(2) - y(4)) + F*(6e-4 * exp(20.7-15000/y(4)))*y(3))/C];
tspan = [0 1500]; y0 = [0.1 600 0 761]; % time range and initial conditions
[t y] = ode15s(dy,tspan,y0); % use the stiff system solver ode15s
subplot(1,2,1),plot(t,y(:,1),t,y(:,3),':'),grid,xlabel('\tau'),ylabel
('Pressure(atm)'),legend('P','P_p')
subplot(1,2,2),plot(t,y(:,2),t,y(:,4),':'),grid,xlabel('\tau'),ylabel
('Temperature(R)'),legend('T','T_p')
```

2.6.2 BOUNDARY-VALUE PROBLEMS

Problems where the value of an unknown function or its derivative is constrained at two different points are known as boundary-value problems. MATLAB provides the built-in function *bvp4c* to solve boundary-value problems. *bvp4c* produces a solution that is continuous on [a, b] and has a continuous first derivative there. A simple expression of the syntax for *bvp4c* is

```
sol = bvp4c(odefun,bcfun,solinit)
```

where *odefun* is a function handle that evaluates the differential equations, *bcfun* is a function handle that computes the residual in the boundary conditions, and *solinit* is a structure containing the initial guess for a solution. The procedure to use *bvp4c* consists of four steps:

Step 1: Transform the set of differential equations into a system of 1st-order differential equations, and create a function to hold the equation system (let's call it *bcprob.m*). It can have the form

```
dy = bcprob(x,y)
```

For a scalar x and a column vector y, *bcprob(x,y)* returns a column vector, dy, representing $f(x, y)$.

Step 2: Arrange the boundary conditions in the form y, $bcprob(x, y)$ and save it as a function (let's call it *bcval.m*). For two-point boundary-value conditions of the form bc(y(a),y(b)), *bcfun* can have the form

```
res = bcval(ya,yb)
```

where ya and yb are column vectors corresponding to y(a) and y(b). The output res is a column vector.

Step 3: Initialize the value of the dependent variable y. We assume values of y at several points in [a b] (let's call it *initsol*).

Step 4: Using the functions and initial values, call the built-in function *bvp4c*. The function *bvp4c* returns a structure (let's call it result) which has several fields as shown in Table 2.32.

TABLE 2.32

Fields of the Structure Generated by *bvp4c*

Field	Description
result.x	Values of the independent variables selected by *bvp4c*
result.y	Approximation to y(x) at the mesh points of x
result.yp	Approximation to y'(x) at the mesh points of x
result.stats	Computational cost statistics
result.solver	*bvp4c*

As an example, we now solve the boundary-value problem defined by

$$y'' + |y| = 0, \ y(0) = 0, \ y(4) = -2$$

using the built-in function *bvp4c*.[9]

Step 1: We need to rewrite the differential equation as a system of two 1st-order ordinary differential equations:

$$z_1' = y' = z_2, \ z_2' = y'' = -|y| = -|z_1|$$

where $z_1 = y$ and $z_2 = y'$. Create a function to hold these equations (let's call it *zode.m*):

```
function dz = zode(x,z)
dz = [z(2); -abs(z(1))];
end
```

Step 2: The boundary conditions can be rearranged as $y(0) = 0, \ y(4) + 2 = 0$ or $z_1(0) = 0, \ z_1(4) + 2 = 0$. Create a function to hold these conditions (let's call it *zobc.m*):

```
function res = zobc(za,zb)
res = [za(1); zb(1)+2];
end
```

Step 3: The range of the independent variable x is $x = [a \ b] = [0 4]$. Divide this range into 10 subintervals and initialize the values of dependent variable y at these points. Let $z_1(x) = y(x) = 1$ and $z_2(x) = y'(x) = 0$ (i.e., $[z_1 z_2] = [1 0]$). These conditions can be fed into the function *bvpinit* as

```
zinit = bvpinit(linspace(0,4,10),[1 0]);
```

Step 4: Solve the problem using *bvp4c*:

```
sol = bvp4c(@zode,@zobc,zinit);
```

The structure sol returned by *bvp4c* has the following fields:

```
sol =
    solver: 'bvp4c'
         x: [1x20 double]
         y: [2x20 double]
        yp: [2x20 double]
     stats: [1x1 struct]
```

We can evaluate the numerical solution at 100 equally spaced points and plot y(x) as

```
x = linspace(0,4,100); y = deval(sol,x); plot(x,y(1,:)), xlabel('x'), ylabel('y')
```

The whole solution procedure can be summarized as a single file. The script *bvp4cex* plots the solutions of this two-point boundary-value problem, as shown in Figure 2.30.

```
% bvp4cex.m
zinit = bvpinit(linspace(0,4,10),[1 0]); sol = bvp4c(@zode,@zobc,zinit);
x = linspace(0,4,100); y = deval(sol,x); plot(x,y(1,:)), grid, xlabel('x'), ylabel('y')
% define differential equations
function dz = zode(x,z)
dz = [z(2); -abs(z(1))];
end
% boundary conditions
function res = zobc(za,zb)
res = [za(1); zb(1)+2];
end
```

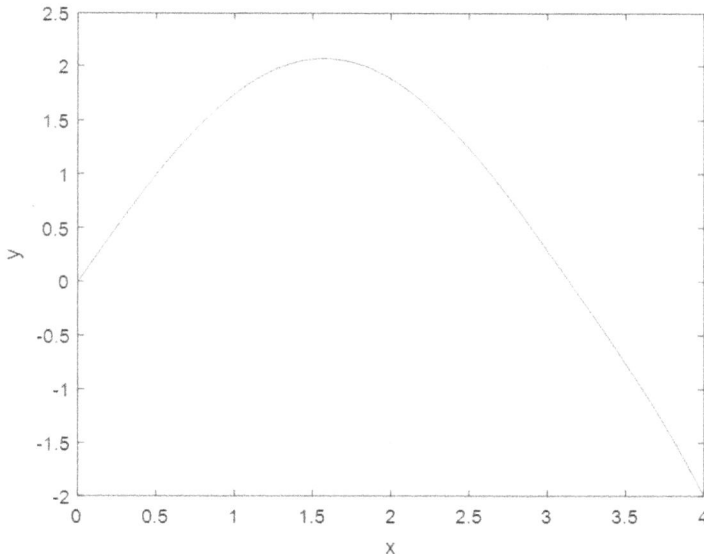

FIGURE 2.30   Plot of solutions of two-point boundary-value problem.

Example 2.55 Boundary-Value Problem

Solve the equation

$$x^2 \frac{d^2y}{dx^2} - 6y = 0$$

The range of x is $1 \le x \le 2$, and the boundary conditions are $y(1) = 1$, $y(2) = 1$.

Solution

The differential equation can be rearranged as

$$\frac{dz_1}{dx} = z_2, \quad \frac{dz_2}{dx} = \frac{6z_1}{x^2}$$

where $z_1 = y$ and $z_2 = \frac{dy}{dx}$. We can see that $z_1(1) = y(1) = 1$, $z_1(2) = y(2) = 1$.

Step 1: Create a function that holds the differential equations $z_1' = z_2$ and $z_2' = 6z_1/x$ (let's call it *bcprob*).

```
function deset = bcprob(x,z)
deset = [z(2); 6*z(1)./x.^2];
end
```

Step 2: The boundary conditions $y(1) = 1$ and $y(2) = 1$ can be rewritten as $y(1)-1 = 0$ and $y(2)-1 = 0$ or $z_1(1)-1 = 0$ and $z_2(1)-1 = 0$. Denote the boundary condition at $a = 1$ as $z_1(1) = z_1(a) = za$ and the condition at $b = 2$ as $z_1(2) = z_1(b) = zb$. We can create the function that holds these boundary conditions as

```
function bcset = bcval(za,zb)
bcset = [za(1)-1; zb(1)-1];
end
```

Step 3: Suppose that the interval $[a\ b] = [1\ 2]$ is divided into 10 subintervals and that the value of the dependent variable y is constant as 1 at all subinterval points. Since $z_1(x) = 1$ and $z_2(x) = z_1' = 0$ at the interval $[1\ 2]$, the initial guesses can be written as $[z_1(x)\ z_2(x)] = [1\ 0]$:

```
>> initsol = bvpinit(linspace(1,2,10), [1 0]);
```

Step 4: Solve the equation by calling the function *bvp4c* as

```
>> res = bvp4c(@bcprob, @bcval, initsol);
```

The script *qdbvp4c* shows the solution procedure and generates the plot shown in Figure 2.31:

```
% qdbvp4c.m
initsol = bvpinit(linspace(1,2,10), [1 0]); res = bvp4c(@bcprob, @bcval,
initsol);
x = linspace(1,2); y = deval(res,x); plot(x,y(1,:)), grid, xlabel('x'),
ylabel('y')
% define differential equations
function deset = bcprob(x,z)
deset = [z(2); 6*z(1)./x.^2];
end
% boundary conditions
function bcset = bcval(za,zb)
bcset = [za(1)-1; zb(1)-1];
end
```

`deval(res,x)` evaluates values of the solution structure *res* at x. The structure *res* returned by *bvp4c* has the following fields:

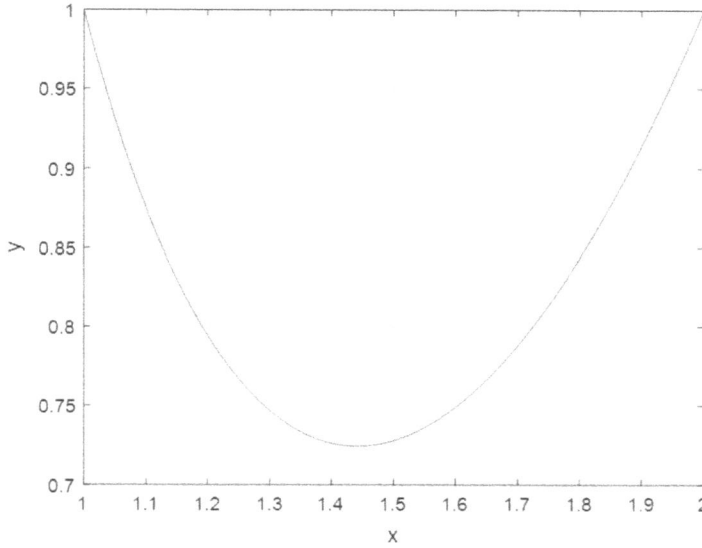

FIGURE 2.31 Solution of a boundary-value problem by *bvp4c*.

```
>> res
res =
struct with fields:
solver: 'bvp4c'
x: [1 1.1111 1.2222 1.3333 1.4444 1.5556 1.6667 1.7778 1.8889 2]
y: [2×10 double]
yp: [2×10 double]
stats: [1×1 struct]
```

Example 2.56 Temperature Distribution in a Rod[10]

A metal rod of length 1 *m* is placed between two tanks, one containing boiling water and the other containing ice. The rod is exposed to the air, and the temperature distribution in the rod can be expressed as

$$\frac{d^2T}{dx^2} = \frac{4h}{kD}(T - T_a),\ T(0) = 100°C,\ T(1) = 0°C$$

where x is the length of the rod, h is the heat transfer coefficient between the rod and air, k is the thermal conductivity of the rod, D is the diameter of the rod, and T_a is the ambient temperature. Data are given as $h = 50\ W/(m^2 \cdot K)$, $D = 0.04\ m$, $k = 390\ W/(m \cdot K)$, and $T_a = 25°C$. Produce the temperature profile as a function of x.

Solution

We first define differential equations. Let $y_1 = T$ and $y_2 = dT/dx$. Then the given equation becomes

$$\frac{dy_1}{dx} = y_2,\ \frac{dy_2}{dx} = \frac{4h}{kD}(y_1 - T_a),\ T(0) = 100°C,\ T(1) = 0°C$$

The subfunction *metfun* defines these differential equations. The boundary conditions can be re-written as $y_1(0) - 100 = 0$ and $y_1(1) = 0$. Let $y_1(0) = y_1(a) = ya$ and $y_1(1) = y_1(b) = yb$. The subfunction metbc defines these boundary conditions. We divide the interval [$a\ b$] = [0 1] into some subintervals and guess initial values for y_1 and y_2. For example, we can divide the interval [0 1] into 20 subintervals and set $y0 = [100\ 0]$. Next, solve the equation using the built-in function *bvp4c* and produce the profile of the temperature $T(= y_1)$ as a function of the length x. This procedure can be implemented by the function *metT*. Execution of the function *metT* yields the profile shown in Figure 2.32.

```
function metT
y0 = [100 0]; metin = bvpinit(linspace(0,1,20),y0); % divide the interval [0 1]
into 20 subintervals
y = bvp4c(@metfun, @metbc, metin);
xv = linspace(0,1); yv = deval(y,xv); plot(xv,yv(1,:)), xlabel('x'), ylabel
('T(degC)'), grid
end
% Define differential equations
function dy = metfun(x,y)
h = 50; D = 0.04; k = 390; Ta = 25;
dy = [y(2); 4*h*(y(1)-Ta)/(D*k)];
end
% Define boundary conditions
function mb = metbc(ya,yb)
mb = [ya(1)-100; yb(1)];
end
```

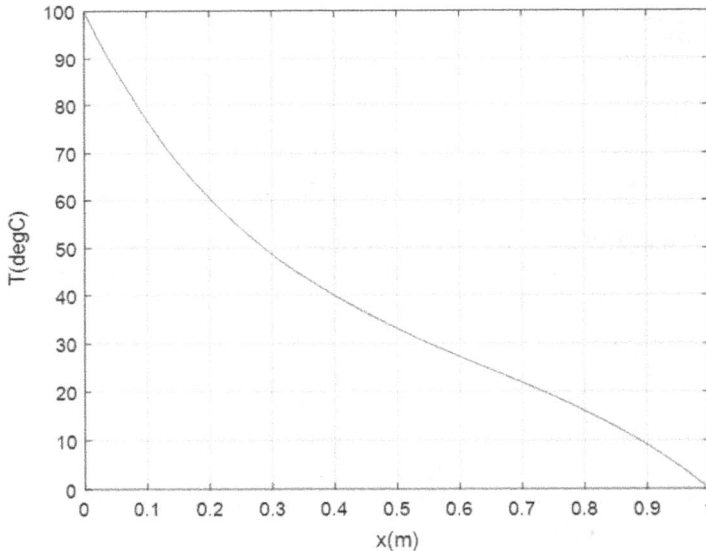

FIGURE 2.32 Temperature profile in a metal rod.

2.6.3 Differential Algebraic Equations (DAEs)

A system of differential algebraic equations (DAEs) can be represented as

$$\frac{dz}{dt} = f(t, x, z),\ 0 = g(t, x, z)$$

where $z \in R^m$ is a vector of differential variables and $x \in R^{n-m}$ are algebraic variables. In general, a system of DAEs can be expressed as

$$M\frac{dy}{dt} = f(t, y)$$

where M is defined as

$$M = \begin{bmatrix} I_m & 0_{m \times l} \\ 0_{l \times m} & 0_{l \times l} \end{bmatrix}$$

and $n = m + l$. The MATLAB built-in function *ode15s* can be used to solve the system of DAEs.

Example 2.57 Heterogeneous Reactor[11]

A 1st-order reaction $A \to B$ is carried out in a heterogeneous reactor. The reactor model can be represented as

$$u\frac{dC_A}{dz} = -k_g a (C_A - C_{As}), \quad 0 = k_g (C_A - C_{As}) - kC_{As}$$

where
u is the inlet velocity
a is the surface area to volume ratio
C_{As} is the surface concentration
k_g is the mass transfer coefficient
Produce the axial profiles of concentration C_A and C_{As} in the reactor when $a = 200$, $k = 0.02$, $k_g = 0.01$, $u = 1$, $C_{A0} = 1$, and L(reactor length) $= 1$.

Solution

The equations can be rearranged as

$$M\frac{dy}{dz} = f(z, y)$$

where

$$M = \begin{bmatrix} 1 & 0 \\ 0 & 0 \end{bmatrix}, \quad y = \begin{bmatrix} C_A \\ C_{As} \end{bmatrix}, \quad f(z, y) = \begin{bmatrix} -\frac{k_g a}{u}(C_A - C_{As}) \\ k_g (C_A - C_{As}) - kC_{As} \end{bmatrix}$$

For given C_{A0}, C_{As0} is given by

$$C_{As0} = \frac{k_g C_{A0}}{k + k_g}$$

The function *hrdae* defines the system of differential algebraic equations to be solved. The script *daerxn* uses the built-in function *ode15s* and produces the profiles of C_A and C_{As} shown in Figure 2.33.

```
% daerxn.m: solution of DAE (differential algebraic equations) system
clear all;
a = 200; k = 0.02; kg = 0.01; u = 1; Ca0 = 1; L = 1; % data
Cas0 = kg*Ca0/(k+kg); M = [1 0;0 0]; opt = odeset('Mass',M);
zspan = [0 L]; y0 = [Ca0; Cas0]; % initial conditions
[z,y] = ode15s(@hrdae,zspan,y0,opt,a,u,k,kg);
plot(z,y(:,1),z,y(:,2),'--'), xlabel('Location(z)'), ylabel('Concentration')
legend('C_A','C_{As}')
```

```
function dy = hrdae(z,y,a,u,k,kg)
% Differential algebraic equations - heterogeneous reactor (1st-order reaction)
Ca = y(1); Cas = y(2);
dy = [-(kg*a/u)*(Ca - Cas); % differential equation
kg*(Ca - Cas) - k*Cas]; % algebraic equation
end
```

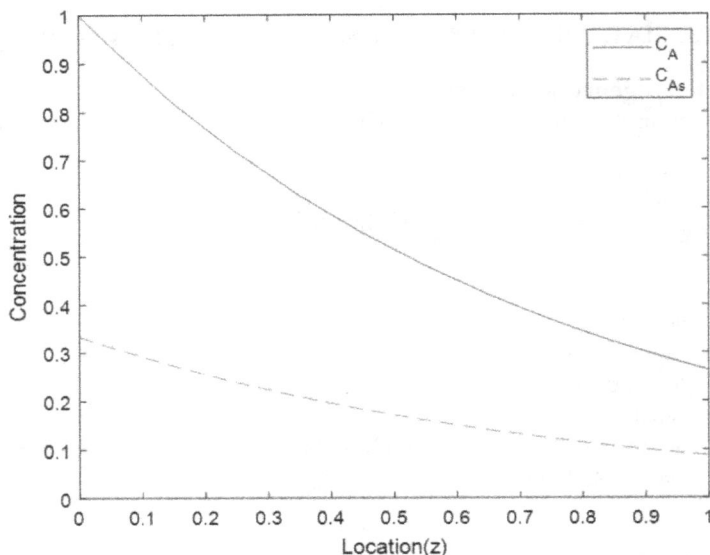

FIGURE 2.33   Concentration profiles.

2.7 PARTIAL DIFFERENTIAL EQUATIONS

2.7.1 CLASSIFICATION OF PARTIAL DIFFERENTIAL EQUATIONS

Partial differential equations (PDEs) are classified according to their order, linearity, and boundary conditions.

2.7.1.1 Classification by Order

The order of a partial differential equation is determined by the highest-order partial derivative in that equation. Examples of 1st-, 2nd-, and 3rd-order partial differential equations are

$$1st - order: \frac{\partial u}{\partial x} - \alpha\frac{\partial u}{\partial y} = 0$$

$$2nd - order: \frac{\partial^2 u}{\partial x^2} + u\frac{\partial u}{\partial y} = 0$$

$$3rd - order: \left(\frac{\partial^3 u}{\partial x^3}\right)^2 + \frac{\partial^2 u}{\partial x\partial y} + \frac{\partial u}{\partial y} = 0$$

2.7.1.2 Classification by Linearity

Partial differential equations can be classified into linear, quasilinear, and nonlinear equations. Let's consider the 2nd-order partial differential equation

$$a(\cdot)\frac{\partial^2 u}{\partial y^2} + 2b(\cdot)\frac{\partial^2 u}{\partial x \partial y} + c(\cdot)\frac{\partial^2 u}{\partial x^2} + d(\cdot) = 0$$

Linear: The coefficients are constants or functions of the independent variables only.
Quasilinear: The coefficients are functions of the dependent variable.
Nonlinear: The coefficients are functions of derivatives of the same order as the equation.

2.7.1.3 Classification of Linear 2nd-Order Partial Differential Equations
The general form of partial differential equations we consider is

$$a\frac{\partial^2 u}{\partial y^2} + 2b\frac{\partial^2 u}{\partial x \partial y} + c\frac{\partial^2 u}{\partial x^2} + d\frac{\partial u}{\partial x} + e\frac{\partial u}{\partial y} + fu + g = 0$$

Elliptic: $b^2 - ac < 0$ (example: Laplace equation $\partial^2 u/\partial x^2 + \partial^2 u/\partial y^2 = 0$)
Parabolic: $b^2 - ac = 0$ (example: heat conduction, or diffusion equation $\alpha \partial^2 u/\partial x^2 = \partial u/\partial t$)
Hyperbolic: $b^2 - ac > 0$ (example: wave equation $\alpha^2 \partial^2 u/\partial x^2 = \partial^2 u/\partial t^2$)

2.7.1.4 Classification by Initial and Boundary Conditions
The initial and boundary conditions associated with the partial differential equations must be specified in order to obtain unique solutions to these equations. In general, the boundary conditions for partial differential equations can be divided into three categories. Let's consider the one-dimensional unsteady-state heat conduction equation

$$\alpha \frac{\partial^2 T}{\partial x^2} = \frac{\partial T}{\partial t}$$

Dirichlet conditions (1st kind): The values of the dependent variable are given at fixed values of the independent variable. Examples of Dirichlet conditions for the heat conduction equation are

$$T = f(x)(t = 0, 0 \leq x \leq 1) \text{ or } T = T_0(t = 0, 0 \leq x \leq 1)$$

Boundary conditions of the 1st kind are represented as

$$T = f(x)(x = 0, t > 0), T = T_1(x = 1, t > 0)$$

Neumann conditions (2nd kind): The derivative of the dependent variable is given as a constant or as a function of the independent variable. For example,

$$\frac{\partial T}{\partial x} = 0 \ (x = 1, t \geq 0)$$

Cauchy conditions: These conditions are a combination of Dirichlet and Neumann conditions.
   Robbins conditions (3rd kind): The derivative of the dependent variable is given as a function of the dependent variable itself. For example,

$$k\frac{\partial T}{\partial x} = h\left(T - T_f\right) \ (x = 0, t \geq 0)$$

where h is the heat transfer coefficient of the fluid being considered.

2.7.2 SOLUTION OF PARTIAL DIFFERENTIAL EQUATIONS BY FINITE DIFFERENCE METHODS

2.7.2.1 Parabolic PDE

Consider the one-dimensional heat equation

$$\frac{\partial u}{\partial t} = \alpha \frac{\partial^2 u}{\partial x^2}$$

over a domain $0 \le t \le t_f$ and $0 \le x \le 1$. The initial and boundary conditions are given as

$$u(x, 0) = f(x)(0 \le x \le 1), \, u(0, t) = g_0(t), \, u(1, t) = g_1(t)(0 \le t \le t_f)$$

We divide the t domain into m sections and the x domain into n sections such that $\Delta t = t_f/m$ and $\Delta x = 1/n$. Then the interior points are given by $t_j = j\Delta t$ ($j = 1, 2, \cdots, m\text{-}1$) and $x_i = i\Delta x$ ($i = 1, 2, \cdots, n - 1$). Let the solution at the point (x_i, t_j) be represented as $u_{i,j}$. Figure 2.34 shows the points that are involved in the calculations at time steps $j - 1, j$ and $j + 1$. The partial differential terms can be approximated by finite differences as

$$\frac{\partial u}{\partial t} \cong \frac{1}{\Delta t}(u_{i,j+1} - u_{i,j}), \, \frac{\partial^2 u}{\partial x^2} \cong \frac{1}{(\Delta x)^2}(u_{i-1,j} - 2u_{i,j} + u_{i+1,j})$$

Replacing the partial differential terms by the finite differences gives a linear system of equations for u at the grid points:

$$u_{i,j+1} - u_{i,j} = \frac{\alpha \Delta t}{(\Delta x)^2}(u_{i-1,j} - 2u_{i,j} + u_{i+1,j}) = r(u_{i-1,j} - 2u_{i,j} + u_{i+1,j})$$

where $r = \alpha\Delta t/(\Delta x)^2$. Solving for $u_{i,j+1}$, we have

$$u_{i,j+1} = ru_{i-1,j} + (1 - 2r)u_{i,j} + ru_{i+1,j}(i = 2, 3, \cdots, n - 2)$$

$$u_{1,j+1} = rg_{0,j} + (1 - 2r)u_{1,j} + ru_{2,j}(i = 1)$$

$$u_{n-1,j+1} = ru_{n-2,j} + (1 - 2r)u_{n-1,j} + rg_{1,j}(i = n - 1)$$

In order for the solution to be stable, it is required that $r \le 0.5$.

The function *parapde* implements the finite difference method to solve the one-dimensional heat equation

$$\frac{\partial u}{\partial t} = \alpha \frac{\partial^2 u}{\partial x^2}$$

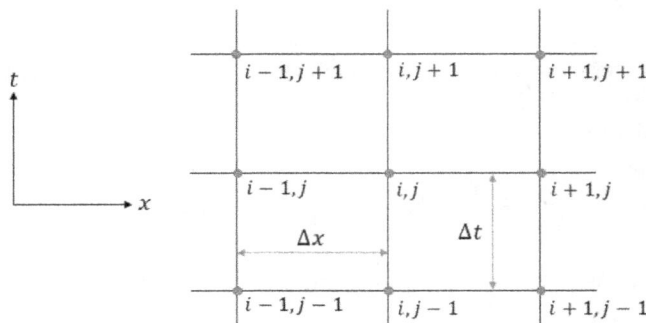

FIGURE 2.34   Grid points involved in the finite differences.

```
function u = parapde(f,g0,g1,tf,nx,nt,alpa)
% Solution of parabolib PDE (u_t = u_xx) using explicit method
% input
%     f: initial condition u(x,0) = f(x)
%     g0,g1: boundary conditions u(0,t) = g0(t), u(1,t) = g1(t)
%     tf: final time
%     nx,nt: number of subintervals
% Output
%     u: solution matrix
h = 1/nx; d = tf/nt; r = alpa*d/h^2; % r should be less than 1/2
x = 0:h:1; t = 0:d:tf;
u(1:nx+1,1) = f(x)'; % initial condition: u(x,0) = f(x)
u(1,1:nt+1) = g0(t); u(nx+1,1:nt+1) = g1(t); % boundary conditions
for k = 1:nt
    u(2:nx,k+1) = r*u(1:nx-1,k) + (1-2*r)*u(2:nx,k) + r*u(3:nx+1,k);
end
u = u'; surf(x,t,u) % display results (x from left to right)
colormap(gray); xlabel('x'), ylabel('t'), zlabel('T(t,x)')
end
```

Example 2.58 Temperature Distribution in a Rod

The temperature distribution in a rod of unit length can be given by

$$\frac{\partial u}{\partial t} = \alpha \frac{\partial^2 u}{\partial x^2} (0 \leq t \leq t_f, 0 \leq x \leq 1)$$

The initial and boundary conditions are given by

$$u(x, 0) = x^3 - 2x^2 + 1.5x \, (0 \leq x \leq 1)$$

$$u(0, t) = 0, \, u(1, t) = 2 (0 \leq t \leq t_f)$$

Plot the temperature profile in the rod using $t_f = 0.1$, $\alpha = 0.8$, $m = 50$, and $n = 10$.

Solution

The initial and boundary conditions are $f(x) = x^3 - 2x^2 + 1.5x$, $g_0(t) = 0$, $g_1(t) = 2$. The resultant temperature profile is shown in Figure 2.35.

```
>> f = @(x) x.^3 - 2*x.^2 + 1.5*x; g0 = @(t) 0; g1 = @(t) 2; nx = 10; nt = 50; alpa
= 0.8; tf = 0.1;
>> u = parapde(f,g0,g1,tf,nx,nt,alpa);
```

2.7.2.2 Hyperbolic PDE

Consider the one-dimensional wave equation

$$\frac{\partial^2 u}{\partial t^2} = \alpha \frac{\partial^2 u}{\partial x^2}$$

over a domain $0 \leq t \leq t_f$ and $0 \leq x \leq 1$. The initial and boundary conditions are given as

$$u(x, 0) = f_1(x), \, \frac{\partial u}{\partial t}(x, 0) = f_2(x) (0 \leq x \leq 1)$$

$$u(0, t) = g_0(t), \, u(1, t) = g_1(t) (0 \leq t \leq t_f)$$

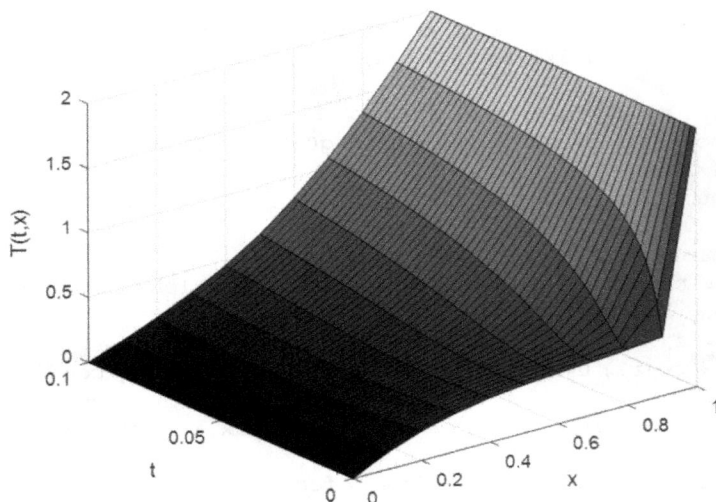

FIGURE 2.35 Temperature profile in a rod.

We divide the t domain into m sections and the x domain into n sections such that $\Delta t = t_f/m$ and $\Delta x = 1/n$. Then the interior points are given by $t_j = j\Delta t$ $(j = 1, 2, \cdots, m\text{-}1)$ and $x_i = i\Delta x$ $(i = 1, 2, \cdots, n - 1)$. Let the solution at a point (x_i, t_j) be represented as $u_{i,j}$. The partial differential terms can be approximated by finite differences as follows:

$$\frac{\partial^2 u}{\partial t^2} \cong \frac{1}{(\Delta t)^2}(u_{i,j-1} - 2u_{i,j} + u_{i,j+1}), \quad \frac{\partial^2 u}{\partial x^2} \cong \frac{1}{(\Delta x)^2}(u_{i-1,j} - 2u_{i,j} + u_{i+1,j})$$

Replacing the partial differential terms in the PDE by the finite differences gives a linear system of equations for u at the grid points:

$$u_{i,j-1} - 2u_{i,j} + u_{i,j+1} = \frac{\alpha(\Delta t)^2}{(\Delta x)^2}(u_{i-1,j} - 2u_{i,j} + u_{i+1,j}) = q(u_{i-1,j} - 2u_{i,j} + u_{i+1,j})$$

where $q = \alpha(\Delta t)^2/(\Delta x)^2$. In order for the solution to be stable, it is required that $q \leq 1$. Solving for $u_{i,j+1}$, we have

$$u_{i,j+1} = -u_{i,j-1} + qu_{i-1,j} + 2(1 - q)u_{i,j} + qu_{i+1,j} (i = 2, 3, \cdots, n - 2)$$

$$u_{1,j+1} = qg_{0,j} + (1 - 2q)u_{1,j} + qu_{2,j} (i = 1)$$

$$u_{n-1,j+1} = qu_{n-2,j} + (1 - 2r)u_{n-1,j} + qg_{1,j} (i = n - 1)$$

$$u_{i,1} - u_{i,-1} = 2(\Delta t)f_2(x_i) (j = 0)$$

We can see that $u_{i,-1} = u_{i,1} - 2(\Delta t)f_2(x_i)$ when $j = 0$. Then for $j = 0$, $u_{i,j+1} = u_{i,1}$ is given by

$$u_{i,1} = -u_{i,-1} + qu_{i-1,0} + 2(1 - q)u_{i,0} + qu_{i+1,0} = -u_{i,1} + 2(\Delta t)f_2(x_i) + qu_{i-1,0} + 2(1 - q)u_{i,0}$$
$$+ qu_{i+1,0}$$

Rearranging this equation gives

$$u_{i,1} = \frac{q}{2}u_{i-1,0} + (1-q)u_{i,0} + \frac{q}{2}u_{i+1,0} + (\Delta t)f_2(x_i)$$

The function *hypbpde* implements the finite difference method to solve the one-dimensional hyperbolic equation

$$\frac{\partial^2 u}{\partial t^2} = \alpha\frac{\partial^2 u}{\partial x^2}$$

```
function [u q] = hypbpde(f1,f2,g0,g1,xspan,tspan,nx,nt,alpa)
% 1-dimensional hyperbolic PDE (wave equation: u_tt = alpa*u_xx)
% Outputs:
%     u: solution matrix [u(x,t)]
% Inputs:
%     f1, f2: initial conditions f1(x) and f2(x) (at t = 0)
%     g0, g1: boundary conditions g0(t) and g1(t) (at x = 0,1)
%     nx, nt: number of subintervals along x and t directions
%     xspan, tspan: range of x and t
%     alpa: coefficient
% Output
%     u: solution matrix
x0 = xspan(1); xf = xspan(2); t0 = tspan(1); tf = tspan(2);
dx = (xf - x0)/nx; dt = (tf - t0)/nt; x = [0:nx]'*dx; t = [0:nt]*dt;
q = alpa*(dt/dx)^2; q1 = q/2; q2 = 2*(1-q);
u(:,1) = f1(x);
for k = 1:nt+1, u([1 nx+1],k) = [g0(t(k)); g1(t(k))]; end
u(2:nx,2)     =    q1*u(1:nx-1,1)   +   (1-q)*u(2:nx,1)   +   q1*u(3:nx+1,1)   +
dt*f2(x(2:nx));
for k = 3:nt+1
   u(2:nx,k)   =   q*u(1:nx-1,k-1) + q2*u(2:nx,k-1) + q*u(3:nx+1,k-1) - u
(2:nx,k-2);
end
surf(t,x,u) % display results
colormap(gray); xlabel('t'), ylabel('x'), zlabel('u(t,x)')
end
```

Example 2.59 Motion of a Vibrating String

The motion of a vibrating string with both ends held fixed can be described by

$$\frac{\partial^2 u}{\partial t^2} = \alpha\frac{\partial^2 u}{\partial x^2}\,(0 \le t \le t_f, 0 \le x \le 1)$$

The initial and boundary conditions are given by

$$u(x, 0) = x(1-x), \frac{\partial u(x, 0)}{\partial t} = 0(0 \le x \le 1)$$

$$u(0, t) = 0, u(1, t) = 0(0 \le t \le t_f)$$

Plot the position profile of the string using $t_f = 1$, $\alpha = 1$, $m = 40$ and $n = 20$.

Solution

The initial and boundary conditions are $f_1(x) = x(1-x), f_2(x) = 0, g_0(t) = 0, g_1(t) = 0$. The resultant position profile is shown in Figure 2.36.

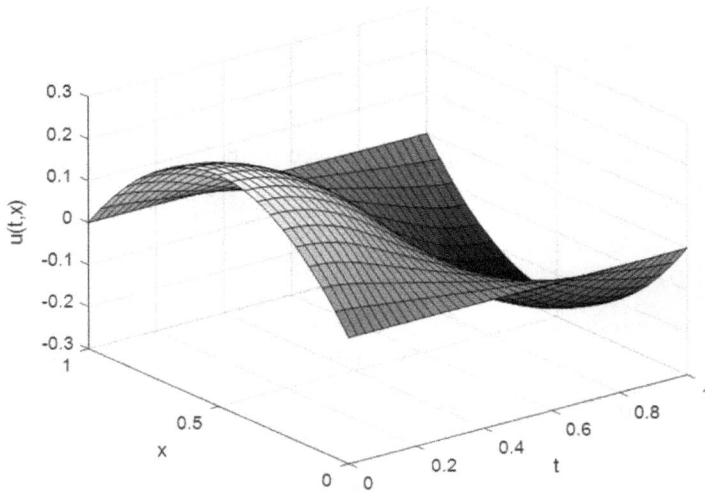

FIGURE 2.36 Position profile of a string.

```
>> f1 = @(x) x.*(1-x); f2 = @(x) 0; g0 = @(t) 0; g1 = @(t) 0; xspan = [0 1]; tspan
= [0 1];
>> nx = 20; nt = 40; alpa = 1; [u r] = hypbpde(f1,f2,g0,g1,xspan,tspan,nx,
nt,alpa);
```

2.7.2.3 Elliptic PDE

Consider the Helmholtz equation

$$\frac{\partial^2 u}{\partial x^2} + \frac{\partial^2 u}{\partial y^2} + g(x, y)u(x, y) = f(x, y)$$

over a domain $x_0 \le x \le x_f$ and $y_0 \le y \le y_f$. The boundary conditions are given as

$$u(x_0, y) = q_{x_0}(y), \ u(x_f, y) = q_{x_f}(y), \ u(x, y_0) = q_{y_0}(x), \ u(x, y_f) = q_{y_f}(x)$$

We divide the x domain into m sections and the y domain into n sections such that $\Delta x = (x_f - x_0)/m$ and $\Delta y = (y_f - y_0)/n$. The second derivatives can be replaced by the three-point central difference approximations as follows:

$$\left.\frac{\partial^2 u}{\partial x^2}\right|_{i,j} \cong \frac{1}{(\Delta x)^2}(u_{i+1,j} - 2u_{i,j} + u_{i-1,j}), \ \left.\frac{\partial^2 u}{\partial y^2}\right|_{i,j} \cong \frac{1}{(\Delta y)^2}(u_{i,j+1} - 2u_{i,j} + u_{i,j-1})$$

where $u_{i,j}$ is the function value at the point $x_i = x_0 + i(\Delta x)$ and $y_j = y_0 + j(\Delta y)$. For every interior point (x_i, y_j) with $1 \le i \le m - 1$ and $1 \le j \le n - 1$, we obtain the following finite difference equation:

$$\frac{1}{(\Delta x)^2}(u_{i+1,j} - 2u_{i,j} + u_{i-1,j}) + \frac{1}{(\Delta y)^2}(u_{i,j+1} - 2u_{i,j} + u_{i,j-1}) + g_{i,j}u_{i,j} = f_{i,j}$$

where $u_{i,j} = u(x_i, y_j)$, $f_{i,j} = f(x_i, y_j)$ and $g_{i,j} = g(x_i, y_j)$. Rearranging this equation, we have

$$u_{i,j} = b_y(u_{i+1,j} + u_{i-1,j}) + b_x(u_{i,j+1} + u_{i,j-1}) + b_{xy}(g_{i,j}u_{i,j} - f_{i,j})$$

$$u_{0,j} = q_{x_0}(y_j), \; u_{m,j} = q_{x_f}(y_j), \; u_{i,0} = q_{y_0}(x_i), \; u_{i,n} = q_{y_f}(x_i)$$

where

$$b_x = \frac{(\Delta x)^2}{2\{(\Delta x)^2 + (\Delta y)^2\}}, \; b_y = \frac{(\Delta y)^2}{2\{(\Delta x)^2 + (\Delta y)^2\}}, \; b_{xy} = \frac{(\Delta x)^2(\Delta y)^2}{2\{(\Delta x)^2 + (\Delta y)^2\}}$$

The function *helpde* solves this difference equation.

```
function [u,x,y] = helpde(f,g,qx0,qxf,qy0,qyf,R,m,n,crit,kmax)
% Solve the Helmholtz equation: u_xx + u_yy + g(x,y)u(x,y) = f(x,y)
% over the region R = [x0 xf y0 yf]
% Boundary conditions:
% u(x0,y) = qx0(y), u(xf,y) = qxf(y), u(x,y0) = qy0(x), u(x,yf) = qyf(x)
% m: number of subintervals along x-axis
% n: number of subintervals along y-axis
% crit: tolerance
% kmax: the maximum number of iterations
x0 = R(1); xf = R(2); y0 = R(3); yf = R(4);
dx = (xf - x0)/m; x = x0 + [0:m]*dx;
dy = (yf - y0)/n; y = y0 + [0:n]*dy;
dx2 = dx*dx; dy2 = dy*dy; dxy2 = 2*(dx2 + dy2);
bx = dx2/dxy2; by = dy2/dxy2; bxy = bx*dy2;
m1 = m + 1; n1 = n + 1;
% Boundary conditions
for k = 1:n1, u(k,[1 m1]) = [qx0(y(k)) qxf(y(k))]; end % at x = x0 and x = xf
for k = 1:m1, u([1 n1], k) = [qy0(x(k)) qyf(x(k))]; end % at y = y0 and y = yf
% Initialization
u(2:m,2:n) = 0;
for i = 1:m
    for j = 1:n, fv(i,j) = f(x(i),y(j)); gv(i,j) = g(x(i),y(j)); end
end
% Solve the difference equation by successive substitution method
for k = 1:kmax % perform iterative calculation
    for i = 2:m
        for j = 2:n
            u(i,j) = by*(u(i+1,j)+u(i-1,j)) + bx*(u(i,j+1)+u(i,j-1)) + bxy*(gv
(i,j)*u(i,j)-fv(i,j));
        end
    end
    if k>1 & max(max(abs(u-u0)))<crit, break; end
    u0 = u; % back substitution
end
end
```

Example 2.60 Steady-State Temperature Distribution over a Square Plate[12]

The steady-state temperature distribution over a square plate can be described by the Laplace equation

$$\nabla^2 u = \frac{\partial^2 u}{\partial x^2} + \frac{\partial^2 u}{\partial y^2} = 0 \ (0 \le x \le 4, \ 0 \le y \le 4)$$

Let the boundary conditions be given by

$$u(0, y) = e^y - \cos y, \ u(4, y) = e^y \cos 4 - e^4 \cos y, \ u(x, 0) = \cos x - e^x, \ u(x, 4) = e^4 \cos x - e^x \cos 4$$

Plot the temperature profile over the plate.

Solution

We divide the x domain into 40 sections and the y domain into 40 sections. The script *sstem* sets the boundary conditions and executes the function *helpde* to generate the temperature profile shown in Figure 2.37.

```
% sstem.m: steady-state temperature distribution
% u_xx + u_yy = 0
R = [0 4 0 4]; f = @(x,y) 0; g = @(x,y) 0;
qx0 = @(y) exp(y) - cos(y); qxf = @(y) exp(y)*cos(4) - exp(4)*cos(y);
qy0 = @(x) cos(x) - exp(x); qyf = @(x) exp(4)*cos(x) - exp(x)*cos(4);
m = 40; n = 40; crit = 1e-6; kmax = 1000;
[u,x,y] = helpde(f,g,qx0,qxf,qy0,qyf,R,m,n,crit,kmax);
mesh(x,y,u), xlabel('x'), ylabel('y'), zlabel('u(x,y)'), colormap(gray),
axis([0 4 0 4 -100 100])
```

2.7.3 METHOD OF LINES

A partial differential equation can be converted into a set of ordinary differential equations by discretizing in space, which is the basic idea of the method of lines. Consider a hyperbolic partial differential equation of the type

$$\frac{\partial u}{\partial t} + \alpha \frac{\partial u}{\partial x} = f(u)$$

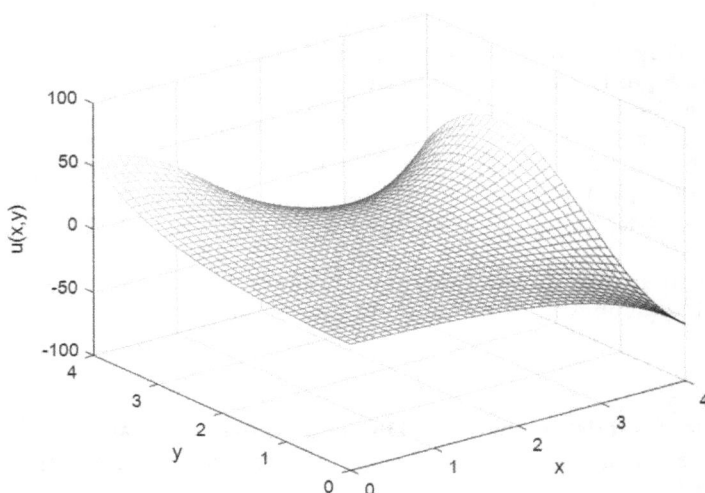

FIGURE 2.37 Steady-state temperature distribution over a plate.

Suppose that the spatial domain of length L is divided into n subintervals. Then the step size is $\Delta x = h = L/n$. Using the central difference formula, the equation is converted to a set of n coupled ordinary differential equations as follows:

$$\frac{du_i}{dt} = -\alpha\left(\frac{u_{i+1} - u_{i-1}}{2h}\right) + f(u_i)$$

Example 2.61 Material Balance on a Plug-Flow Reactor

The material balance on a plug-flow reactor can be expressed as a 1st-order hyperbolic PDE

$$\frac{\partial C}{\partial t} + v\frac{\partial C}{\partial x} = -kC, \, C(0, x) = C_0, \, C(t, 0) = C_{in}$$

where v is the inlet velocity.

(1) Use the method of lines to produce the steady-state concentration profile as a function of x for $0 \leq x \leq L\,(m)$ where L is the length of the reactor. Compare the result with the steady-state solution.

(2) Plot the exit concentration as a function of time $(0 \leq t \leq 10\,min)$. Use $L = 0.5\,m$, $v = 0.4\,m/min$ and $k = 0.2\,min^{-1}$.

Solution

Application of the method of lines yields

$$\frac{dC_i}{dt} = -v_i\left(\frac{C_{i+1} - C_{i-1}}{2h}\right) - kC_i\,(i = 2, 3, \cdots, n - 1)$$

$$\frac{dC_i}{dt} = -v_i\left(\frac{C_{i+1} - C_{in}}{2h}\right) - kC_i\,(i = 1)$$

$$\frac{dC_i}{dt} = -v_i\left(\frac{C_i - C_{i-1}}{h}\right) - kC_i\,(i = n)$$

At steady-state,

$$v\frac{dC}{dx} = -kC$$

from which we obtain

$$C = C_{in}e^{-kx/v}$$

The function *pfrconc* defines the system of difference equations to be integrated.

```
function dC = pfrconc(t,C,pf)
% Retrieve data
k = pf.k; v = pf.v; C0 = pf.C0; L = pf.L; n = pf.n;
% Initialization
h = L/n; dC = zeros(n,1);
% Difference equations
for m = 1:n
if m = = 1, s = (v/2/h)*(C(m+1) - C0);
elseif m = = n, s = (v/h)*(C(m) - C(m-1));
else, s = (v/2/h)*(C(m+1) - C(m-1)); end
dC(m) = -s - k*C(m);
end
end
```

The script *pfrmol* sets the data and parameters and employs the built-in function *ode45* to obtain the concentrations as a function of time and reactor length. Figure 2.38 shows the steady-state concentration profiles and the exit concentration as a function of time for $n = 20$.

```
% pfrmol.m: calculation of PFR using the method of lines (MoL)
clear all;
n = 20; pf.k = 0.18; pf.v = 0.5; pf.C0 = 1; pf.L = 0.5; % Data and parameters
pf.n = n; h = pf.L/n; C0 = pf.C0;
Z0 = ones(n,1)*C0; tspan = [0 10]; % initialize
[t,C] = ode45(@pfrconc,tspan,Z0,[],pf); % solve the set of ODEs
Cs = [C0, C(end,:)]; % steady-state by MoL
x = [0:h:pf.L]; Cm = C0*exp(-pf.k*x/pf.v); % steady-state by exact solution
% Plot results
subplot(1,2,1),   plot(x,Cs,x,Cm,'--'),   xlabel('x(m)'),   ylabel('C(mol/
liter)')
legend('St-st by MoL','St-st by exact solution')
subplot(1,2,2), plot(t,C(:,end)), xlabel('t(min)'), ylabel('C(mol/liter)')
```

2.7.4 USE OF THE MATLAB BUILT-IN FUNCTION *PDEPE*

The MATLAB built-in function *pdepe* solves initial-boundary value problems for parabolic/elliptic partial differential equations of the form[13]

$$g\left(x, t, u, \frac{\partial u}{\partial x}\right)\frac{\partial u}{\partial t} = x^{-m}\frac{\partial}{\partial x}\left(x^m f\left(x, t, u, \frac{\partial u}{\partial x}\right)\right) + r\left(x, t, u, \frac{\partial u}{\partial x}\right)$$

where
  $f(x, t, u, \partial u/\partial x)$ is a flux term
  $r(x, t, u, \partial u/\partial x)$ is a source term

(a)   (b)

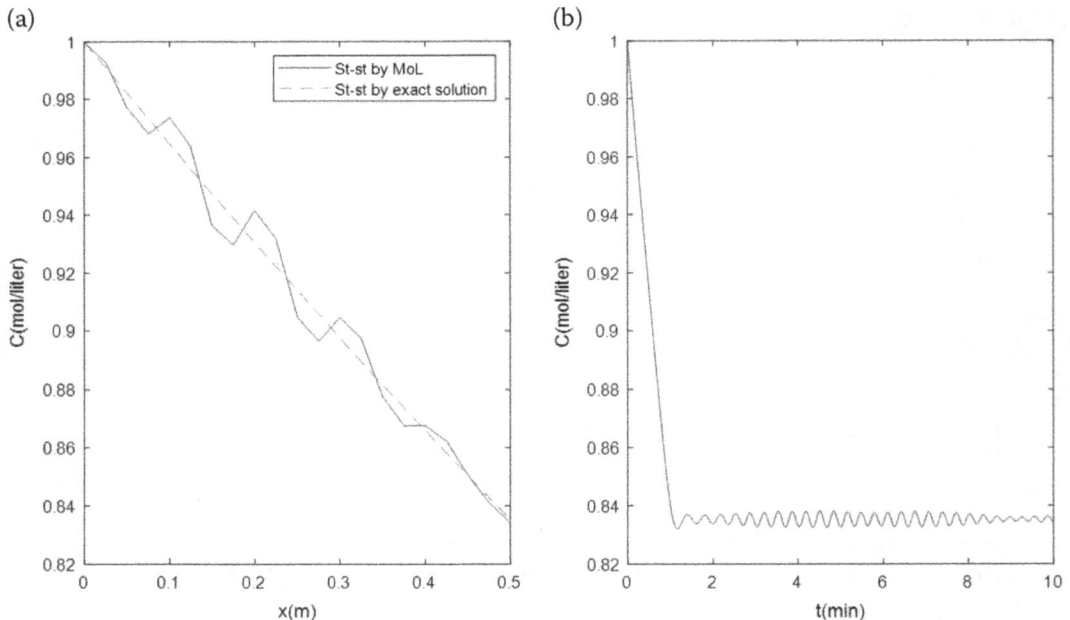

FIGURE 2.38   (a) Steady-state concentration profile and (b) exit concentration.

$g(x, t, u, \partial u/\partial x)$ is a diagonal matrix

The diagonal elements of this matrix are either identically zero or positive. An element that is identically zero corresponds to an elliptic equation, and otherwise to a parabolic equation. The value of m can be 0, 1, or 2, corresponding to slab, cylindrical, or spherical symmetry, respectively. If $m > 0$, then a must be greater than zero. The partial differential equations hold for $t_0 \leq t \leq t_f$ and $a \leq x \leq b$. The interval $[a\ b]$ must be finite.

The initial condition at $t = t_0$ is represented as

$$u(x, t_0) = u_0(x)$$

and the boundary condition at $x = a$ or $x = b$ is given by

$$p(x, t, u) + q(x, t)f\left(x, t, u, \frac{\partial u}{\partial x}\right) = 0$$

A simple expression of the syntax for *pdepe* is

```
sol = pdepe(m,pdefun,icfun,bcfun,xmesh,tspan)
```

where m is a parameter corresponding to the symmetry of the problem, *pdefun* is a handle to a function that defines the components of the partial differential equation, *icfun* is a handle to a function that defines the initial conditions, *bcfun* is a handle to a function that defines the boundary conditions, xmesh is a vector specifying the points at which a numerical solution is requested for every value in tspan, and tspan is a vector specifying the points at which a solution is requested for every value in xmesh. The function *pdepe* returns values of the solution (sol) on a mesh provided in xmesh.

The procedure to use *pdepe* consists of four steps.

Step 1: Specify the system of partial differential equations. Define m, g, f, and r using the given data and conditions, and create a function to hold these functions (let's call it *pdeeq.m*).

Step 2: Specify initial conditions. Define $u(x, t_0) = u_0(x)$ at $t = t_0$ and create a function to hold it (let's call it *pdeic.m*).

Step 3: Set boundary conditions. At $x = a$ or $x = b$, specify p, q, and f in the equation $p(x, t, u) + q(x, t)f(x, t, u, \partial u/\partial x) = 0$ and create a function to hold these functions (let's call it *pdebc.m*).

Step 4: Solve the partial differential equation using *pdepe*:

```
sol = pdepe(m, @pdeeq, @pdeic, @pdebc, x,t)
```

Example 2.62 One-Dimensional Parabolic PDE

The temperature $u(x, t)$ in a wall of unit length can be described by the one-dimensional heat equation

$$\frac{\partial u}{\partial t} = \alpha\frac{\partial^2 u}{\partial x^2}$$

The thickness of the wall is 1 m and the initial profile of the temperature in the wall at $t = 0$ sec is uniform at $T = 90°C$. At time $t = 0$, the ambient temperature is suddenly changed to 15°C and held there. If we assume that there is no convection resistance, the temperature of both sides of the wall is also held constant at 15°C. Determine the temperature distribution graphically within the wall from $t = 0$ to $t = 21,600$ *sec*. The wall property can be assumed as $\alpha = 4.8 \times 10^{-7}\ m/sec^2$.

Solution

Step 1: From the heat equation, we have $(1/\alpha)\partial u/\partial t = \partial^2 u/\partial x^2$. Thus, we can see that $m = 0$, $g = 1/\alpha$, $f = \partial u/\partial x$ and $r = 0$. Create a function that holds g, f and r:

```
function [g,f,r] = pdeTde(x,t,u,DuDx)
alpha = 4.8e-7; g = 1/alpha; f = DuDx; r = 0;
end
```

Step 2: Set the initial conditions. At $t = 0$, $u(x, t_0) = u(x, 0) = u_0(x) = 90$. We can write a function that defines initial conditions as

```
function u0 = pdeTic(x)
u0 = 90;
end
```

Step 3: Specify the boundary conditions $p(x, t, u) + q(x, t)f(x, t, u, \partial u/\partial x) = 0$. Since the temperature at both sides of the wall is 15°C, $p = pl = ul$-15 and $q = ql = 0$ at $x = 0$, and $p = pr = ur$-15 and $q = qr = 0$ at $x = 1$. Create a function that holds these boundary conditions.

```
function [pl,ql,pr,qr] = pdeTbc(xl,ul,xr,ur,t)
pl = ul-15; ql = 0; pr = ur-15; qr = 0;
end
```

Step 4: Set the intervals for x and t, and call the function pdepe to solve the equation. Suppose that the range of x is divided into 20 subintervals and the time span into 180 subintervals. The whole procedure can be summarized as the script *temprof*. The script *temprof* generates the temperature profile in the wall shown in Figure 2.39:

```
% temprof.m
m = 0; x = linspace(0,1,20); t = linspace(0,21600,54);
sol = pdepe(m,@pdeTde,@pdeTic,@pdeTbc,x,t);
T   =   sol(:,:,1); surf(x,t,T), xlabel('x'), ylabel('t(sec)'), zlabel
('T(deg C)')
function [g,f,r] = pdeTde(x,t,u,DuDx) % define g, f, and r
alpha = 4.8e-7; g = 1/alpha; f = DuDx; r = 0;
end
function u0 = pdeTic(x) % define initial conditions
```

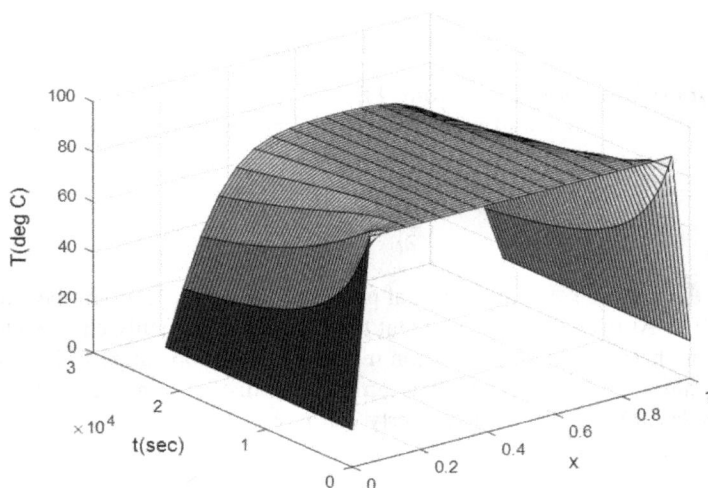

FIGURE 2.39   One-dimensional temperature profile.

```
u0 = 90;
end
function [pl,ql,pr,qr] = pdeTbc(xl,ul,xr,ur,t) % define boundary conditions
pl = ul-15; ql = 0; pr = ur-15; qr = 0;
end
```

2.8 HISTORICAL DEVELOPMENT OF PROCESS ENGINEERING SOFTWARE

The early development of chemical process engineering as a discipline was largely experimental and empirical. In typical process operations, the interaction of process variables with one another and their effect on the performance of the operation were not fully understood. Therefore, the approach adopted was to perform a number of experiments with the variables covering a wide operation range and to correlate the data using certain statistical methods and dimensional analyses. The resulting empirical model could be used for the design and analysis of chemical process systems.

Alongside the development of chemical process engineering, the use of computer power has been exploited by many process engineers. The history of simulations using computers started in 1946 with the first general-purpose computer (ENIAC). The first deployment of this computer was managed by John von Neumann to model the process of nuclear detonation during the Manhattan project.[14] Initial attempts toward computer simulation of chemical process engineering systems were made in the early 1950s, and one of the most significant advances in digital simulation took place in the late 1950s with the development of the FORTRAN programming language. The earliest process engineering programs were mostly automated versions of standard chemical engineering textbook modeling procedures for the common unit operations including reactor, distillation, heat exchanger, etc., using the empirical correlations. Derived chemical process models took considerable skill and time to develop into computer programs.

By the end of the 1950s, computer science was emerging as a promising field providing new simulation tools. However, up to the 1960s, most chemical engineers lacked the tools to design and simulate a complete process. In 1964, the digital computer program PACER, one of the ancestors of today's chemical process simulators, received its first practical test. It was written in FORTRAN for an IBM 7090 computer and consists of more than 1000 statements. Also in 1964, Imperial College in London released SPEEDUP (Simulation Program for the Economic Evaluation and Design of Unsteady-State Processes), the first process simulator based on an equation-oriented approach. For most chemical process engineers, these programs remained a "black box" model, and the continued use of manual calculations persisted even towards the end of the 1960s. In fact, early-generation computers were not adequate to solve some large and sophisticated chemical engineering problems, due to their limited memory and slow speed. In some cases, drastic simplifying assumptions had to be made in order to reduce the complexity of the problem, and this tended to give a distorted picture of the system behavior.[15] In 1966, Simulation Sciences commercialized a computer program for simulating distillation columns, which was the core of the flowsheeting package PROCESS. In 1969, ChemShare released DESIGN, which is a flowsheeting program for gas and oil applications.

During the 1970s, large high-speed digital computers became available for solving complex chemical engineering problems, with a considerable increase in the scope and usability of many computer programs. In the early 1970s, integrated process flowsheet package programs like PACER, Monsanto's FLOWTRAN, Kellogg's Flexible Flowsheet, and ICI's FLOWPACK appeared. With these programs, chemical engineers were able to model complex flowsheets on the computer and simulate the effect of interactions between various unit operations at different stages of the process. The traditional concept of designing a string of unit operations for a chemical process gave way to a new generation of computer-aided design techniques, where processes are designed as integrated units using the process flowsheeting approach. In 1976, the U.S. Department of Energy and MIT launched jointly the Aspen project, which eventually wound up with Aspen Plus, one of the most widely used simulators in the world.

In the 1980s, various specialized packages became available, such as DESIGN II, ASPEN, SIMSCI (PROII), HYSYS, and CHEMCAD, which started appearing on chemical engineering desktops. The improvement of computer technology allowed better data analysis and understanding of the processes and provided solutions to more complex problems. In 1985, AspenTech released Aspen Plus. During the late 1980s, PRO/II was upgraded from PROCESS by Simulation Sciences, and major packages migrated to PCs.[16]

In the early 1990s, all major software packages underwent periodic upgrades, and advanced applications such as pinch analysis were released. In the mid-1990s, major vendors converted the graphical user interface into a central part in the software development, and HYSIM became NYSYS. During this period of time, the process simulation market underwent severe transformations. The few systems that survived include CHEMCAD, Aspen Plus, Aspen HYSYS, PRO/II, ProSimPlus, SuperPro Designer, and gPROMS. Nowadays, most of the objected-oriented chemical process simulators are developed based on a combination of sequential modular and equation-oriented approaches using languages like C++, C#, MATLAB, and Java. Table 2.33 summarizes the evolution of commercial process simulators.[17]

For both the chemical engineering student at university and the chemical process engineer in the chemical industry, increasing emphasis has been placed both on understanding basic physical and

TABLE 2.33
Process System Engineering Software

Software	Year	Developer	Applications	Type
PACER	1963	Paul T. Shannon		Sequential modular
PROCESS	1966	Simulation Sciences	Steady-state simulation	Sequential modular
DESIGN	1969	ChemShare	Gas and oil applications	Sequential modular
CHESS	1969	Univ. of Houston	Flowsheeting, sizing, and costing	Sequential modular
FLOWTRAN	1970	Monsanto	Flowsheeting, sizing, and costing	Sequential modular
FLOWPACK II	1972	ICI	Flowsheeting	Sequential modular
DYSCO	1981	Institute of Paper Chemistry	Dynamic simulation	Dynamic modular simulation
ASCEND	1970–1980	Carnegie Mellon Univ.	Dynamic simulation	Equation oriented
Aspen Plus	1976	DOE, MIT	All-purpose flowsheeting	
TISFLO	1970–1980	DSM		Equation oriented
FLOWSIM	Early 1980s	Univ. of Connecticut		Equation oriented
HYSIM HYSYS	Mid 1990s	Hyprotech	Steady-state/dynamic simulation	
SpeedUp	1986	Imperial College	Dynamic simulation	Equation oriented
ProSimPlus	1989	ProSim	Steady-state simulation	Sequential modular
DESIGN II		WinSim	Flowsheeting	
Quasilin	Late 1980s	CADCentre	Simulation/optimization	Equation oriented
gPROMS		Imperial College, PSE Ltd.	Steady-state/dynamic modeling	Equation oriented
BatchPro Designer		Intelligen	Batch scheduling and design	
COCO simulator		AmsterCHEM	Flowsheeting	Sequential modular

Sources: https://en.wikipedia.org/wiki/List_of_chemical_process_simulators; Martin, M. M., *Introduction to Software for Chemical Engineers*, CRC Press, pp.291–293, 2015.

chemical principles and on process modeling, and the development of process engineering software has not entirely replaced the early approach of calculating individual unit operations. The practicing chemical engineer will notice that, to some degree, operation and optimization of chemical plants and the design of minor plant modifications are still largely empirical, backed by operating experience. By skillful combination of the analytical aspect and experimental empiricism, a framework can be developed for an economical, quantitative, and scientific approach to technical problems in the process industry. In this context, computer-aided design, process modeling, and simulation become essential tools for the chemical process engineer. One of the most important requirements of a process engineer is not just the ability to use the library of process engineering programs but the capacity to develop one's own program, either as a stand-alone program or appended to existing software to share some of the library programs. A process engineer with this ability may understand and appropriately respond to error or warning messages that appear in process engineering programs or computer-aided plant operation.

PROBLEMS

LINEAR SYSTEMS

2.1 In the photosynthesis reaction, water reacts with carbon dioxide to give glucose and oxygen. This reaction can be expressed as

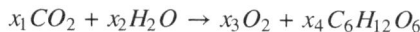

$$x_1 CO_2 + x_2 H_2O \rightarrow x_3 O_2 + x_4 C_6H_{12}O_6$$

Determine the values of coefficients x_1, x_2, x_3, and x_4 to balance the equation. Is it possible to determine these values? If not, under what conditions can the solutions be found?

2.2 Four reactors are connected by pipes, where directions of flow are depicted by means of arrows as shown in Figure P2.2.[18] The flow rate of the key component is given by the volumetric flow rate $Q (liter/sec)$ multiplied by the concentration $C (g/liter)$ of the component. The incoming flow rate is assumed to be equal to the outgoing rate. Using the flow rates given, calculate the concentration at each reactor:

$$Q_{13} = 75 liter/sec, \ Q_{24} = 20 liter/sec, \ Q_{33} = 60 liter/sec,$$

$$Q_{21} = 25 liter/sec, \ Q_{32} = 45 liter/sec, \ Q_{43} = 30 liter/sec$$

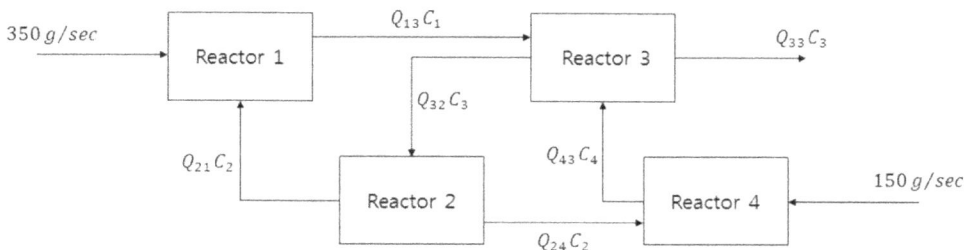

FIGURE P2.2 A reactor network.

2.3 Para-xylene, styrene, toluene, and benzene are to be separated with the array of distillation columns shown in Figure P2.3.[19] Determine the molar flow rates ($kgmol/min$) of D_1, D_2, B_1, and B_2.

2.4 Figure P2.4 shows a flat square plate the sides of which are held at constant temperatures (200°C and 500°C). Find the temperatures at inner nodes (i.e., $T_7 - T_9$, $T_{12} - T_{14}$, $T_{17} - T_{19}$ 2

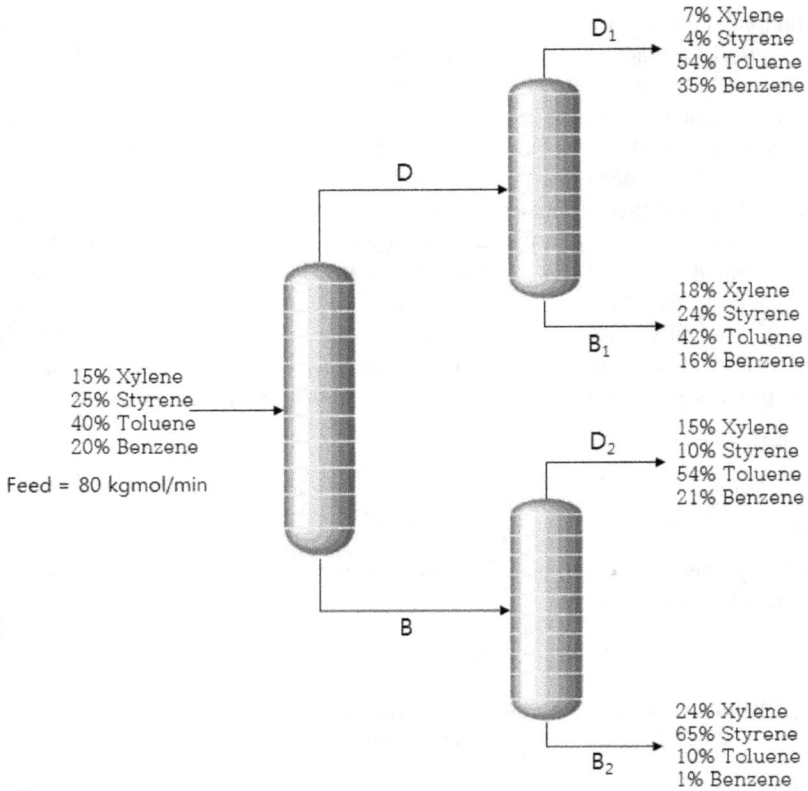

7% Xylene
4% Styrene
54% Toluene
35% Benzene

D_1

D

18% Xylene
24% Styrene
42% Toluene
16% Benzene

B_1

15% Xylene
25% Styrene
40% Toluene
20% Benzene

Feed = 80 kgmol/min

D_2

15% Xylene
10% Styrene
54% Toluene
21% Benzene

B

24% Xylene
65% Styrene
10% Toluene
1% Benzene

B_2

FIGURE P2.3   Array of distillation columns.

at constant temperatures (200). The temperature at each inner node is assumed to be given by the average of temperatures of adjacent nodes.

2.5 Figure P2.5 shows an ideal multi-component flash drum. The feed mixture of flow rate F consists of three isomers of xylene: *o*-xylene(1), *m*-xylene(2), and *p*-xylene(3). The feed contains mole fractions z_i of each component at temperature T_f and pressure P_f. In the flash drum,

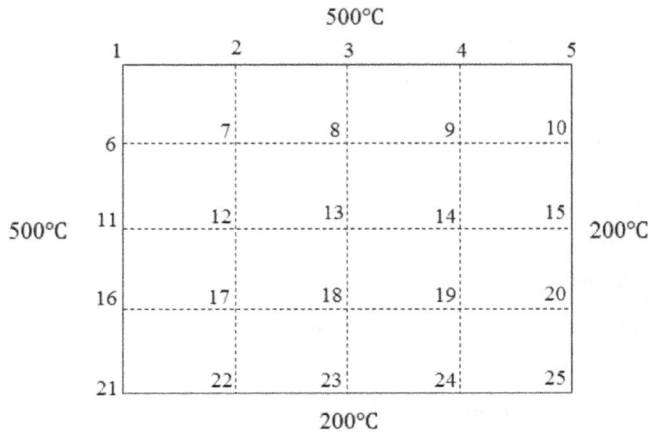

FIGURE P2.4   Temperature distribution in a flat square.

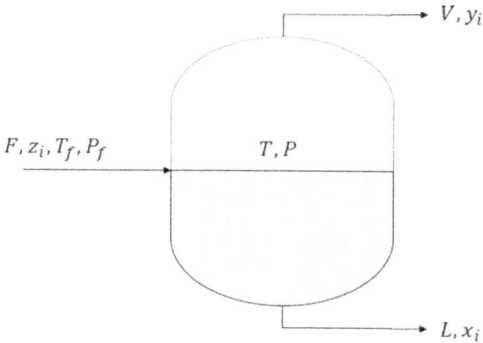

FIGURE P2.5 Illustration of ideal multi-component flash drum.

vapor-liquid equilibrium is achieved at T and P with a liquid flow rate L and vapor flow rate V. The vapor pressure of each component is assumed to be represented by the Antoine equation

$$\log_{10}P_i^{sat}(mmHg) = A_i - \frac{B_i}{T(°C) + C_i}$$

where A_i, B_i, and C_i are the Antoine coefficients for species i. Table P2.5 lists the Antoine coefficients for the three isomers of xylene. Assume that $P = 760\ mmHg$, $F = 1\ mol/sec$, and $L = 0.2mol/sec$. Generate a plot showing the range of operating temperature T as a function of the mole fraction of o-xylene z_1 ($0.1 \leq z_1 \leq 0.9$).[20]

TABLE P2.5

Antoine Coefficients for Three Isomers of Xylene

Component	A_i	B_i	C_i	Boiling point(°C)
o-xylene(1)	6.99891	1474.679	213.69	144.4
m-xylene(2)	7.00908	1462.266	215.11	139.1
p-xylene(3)	6.99052	1453.430	215.31	138.4

2.6 Consider the simplified process flow diagram shown in Figure P2.6.[21] In the diagram, $m_i(i = 1,2, \cdots, 12)$ represents the flow rate of stream i. Assume that no mass accumulations or chemical reactions take place in the process units. The feed flow m_1 is maintained at $100\ kg/min$, $m_3 = 0.7\ m_1 - m_2$, $m_6 = (m_7 + m_8)/3.2$, $m_7 = 0.84\ m_{12} - m_4$, $m_8 = 0.2\ m_5$, $m_{10} = 0.2\ m_9$, $m_9 = 0.85\ m_2 - m_{11}$, and $m_{12} = 0.55\ m_1 - m_9$. Determine the flow rates $m_i(i = 2,3, \cdots, 12)$.

2.7 The process shown in Figure P2.7 consists of a reactor and a separator. The reactants A and B are fed into the reactor with flow rates A_1 and B_1, respectively. The following two reactions are taking place in the reactor:

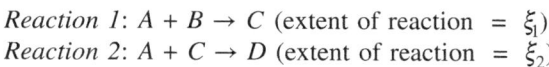

> *Reaction 1*: $A + B \rightarrow C$ (extent of reaction $= \xi_1$)
> *Reaction 2*: $A + C \rightarrow D$ (extent of reaction $= \xi_2$)

The intermediate product C produced by *Reaction 1* needs to be converted to the desired product D by *Reaction 2*. The single-pass conversion of the reactor is 90%, with 30% selectivity for *Reaction 2*. In the separator, the flow B_2 is evenly split between the product stream (stream 3) and the recycle stream (stream 4), 65% of D and 85% of C fed into the separator are recycled through stream 4, and 10% of flow A_2 is lost to the product stream (stream 3).[22]

The feed flow rates are $A_1 = 10mol/sec$ and $B_1 = 20mol/sec$. Determine flow rates A_i, B_i, C_i, $D_i(i = 2,3,4)$ and extents of reaction ξ_1 and ξ_2 for *Reaction 1* and 2.

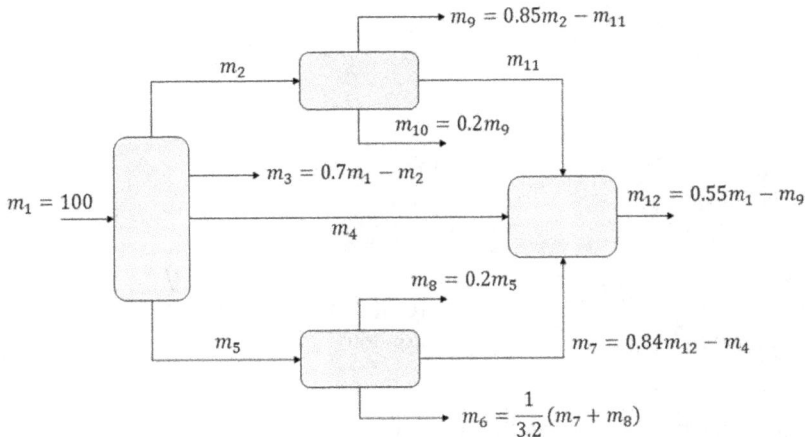

FIGURE P2.6 Simplified process flow diagram.

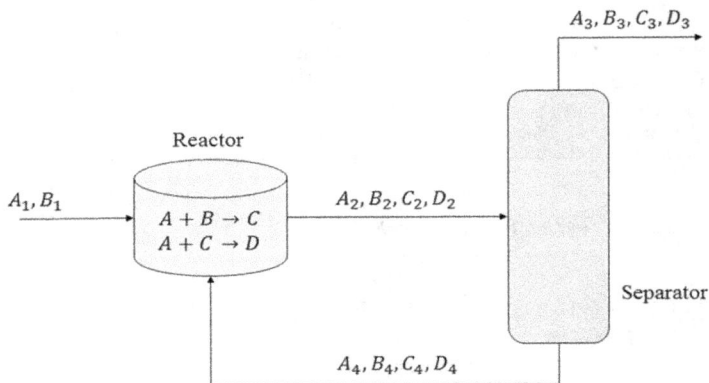

FIGURE P2.7 A process consisting of a reactor and a separator with recycle.

NONLINEAR EQUATIONS

2.8 The volume fraction of red blood cells in blood is called hematocrit. The core region hematocrit (H_c) is given by

$$\frac{H_c}{H_0} = 1 + \frac{(1 - \sigma^2)^2}{\sigma^2 \{2(1 - \sigma^2) + \sigma^2(1 - \alpha H_c)\}}, \quad \sigma = 1 - \frac{\delta}{R}, \quad \alpha = 0.07\exp\left(2.49H_c + \frac{1107}{T}e^{-1.69H_c}\right)$$

where
  H_0 is the hematocrit at the inlet of the blood vessel
  $\delta(\mu m)$ is the thickness of the plasma layer
  $R(\mu m)$ is the radius of the blood vessel
  $T(K)$ is the temperature
  Find H_c if $\delta = 2.94\ \mu m$, $R = 16\ \mu m$, $T = 315\ K$, and $H_0 = 0.45$.

2.9 The total number of unbound receptors present on a cell surface at equilibrium is given by

$$\frac{R_t}{R_{eq}} = 1 + \nu\left(\frac{L_0}{K_D}\right)(1 + K_x R_{eq})^{f-1}$$

where

R_t is the total number of receptors present on the cell surface
R_{eq} is the equilibrium concentration of unbound receptors present on the cell surface
ν is the number of binding sites
L_0 is the ligand concentration
K_D is the dissociation constant
K_x is the cross-linking equilibrium constant
f is the total number of binding sites available for binding to a single cell
Determine the equilibrium concentration R_{eq} using the following data:

Data: $R_t = 10692$, $\nu = 17$, $L_0 = 2.\,1 \times 10^{-9}M$, $K_D = 7.\,76 \times 10^{-5}M$, $K_x = 5.\,82 \times 10^{-5}$, $f = 9$.

2.10 The vapor pressure (mmHg) of *n*-pentane (A) and *n*-hexane (B) can be calculated from the Antoine equation (T:°C)[23]:

$$\log P_A = 6.85221 - \frac{1064.63}{T + 232.0}, \ \log P_B = 6.87776 - \frac{1171.53}{T + 224.366}$$

(1) Calculate the bubble point temperature and equilibrium composition associated with a liquid mixture of 10 mol% *n*-pentane and 90 mol% *n*-hexane at 1 atm.
(2) Repeat the calculations for liquid mixtures containing 0 mol% up to 100 mol% of *n*-pentane. Plot the bubble point temperature and mol % of *n*-pentane in the vapor phase as a function of the mol% in the liquid phase.
2.11 The force-extension behavior of DNA can be represented by[24]

$$\frac{Fb}{k_B T} = \frac{1}{2}\left(1 - \frac{x}{L}\right)^{-2} + \frac{2x}{L} - \frac{1}{2}$$

where

F is the force required to extend DNA to a distance x
k_B is the Boltzmann constant
T is the temperature
b is the Kuhn length that represents one step of the random walk
L is the total length of the DNA chain
Determine the extension x for $F = 30 \times 10^{-12}N$.

Data: $k_B = 1.38 \times 10^{-23} J/(mol \cdot K)$, $b = 10^{-7} m$, $T = 298\ K$, $L = 16.8\ \mu m$.

2.12 Consider an ethylbenzene(1)-toluene(2) mixture at equilibrium. The liquid and vapor phases are assumed to be ideal. The equilibrium condition requires that

$$x_i P_i^{sat} = p_i$$

where

x_i is the mole fraction of species i in the liquid phase
P_i^{sat} is the saturation pressure of species i
p_i is its partial pressure in the vapor phase
The saturation pressure can be approximated by using the Antoine equation

$$\log_{10} P_i^{sat} (mmHg) = A_i - \frac{B_i}{T\,(°C) + C_i}$$

The Antoine coefficients are shown in Table P2.12. The vapor pressures for ethylbenzene and toluene are $p_1 = 250\ mmHg$ and $p_2 = 343\ mmHg$, respectively. Find the liquid-phase mole fraction of ethylbenzene and the equilibrium temperature T.[25]

TABLE P2.12
Antoine Coefficients for Ethylbenzene and Toluene

Species	A	B	C
Ethylbenzene(1)	6.95719	1424.26	213.21
Toluene(2)	6.95464	1344.80	219.48

2.13 For a packed bed, the pressure drop per unit bed length $(\Delta P/L)$ can be given by the Ergun equation as

$$\frac{\Delta P}{L} = 150 \left(\frac{\eta v_0}{D_p^2} \right) \frac{(1-\varepsilon)^2}{\varepsilon^3} + 1.75 \left(\frac{\rho v_0^2}{D_p} \right) \frac{1-\varepsilon}{\varepsilon^3}$$

where
  η is the fluid viscosity
  v_0 is the superficial velocity
  D_p is the particle diameter
  ρ is the fluid density
  ε is the void fraction of the bed
  Determine ε when $\eta = 1.1\ cP$, $v_0 = 0.1\ m/sec$, $D_p = 4.9\ cm$, $\rho = 1.98\ g/cm^3$, $L = 1.6\ m$ and $\Delta P = 416\ Pa$.

2.14 The vapor pressure of component i in a binary mixture containing benzene and toluene can be represented by the Antoine equation

$$\ln P_{sat_i}(T) = A_i - \frac{B_i}{T + C_i} \quad (T : °C,\ P_{sat} : mmHg)$$

Antoine coefficients are
  Benzene: $A = 15.90085, B = 2788.507, C = 220.790$
  Toluene: $A = 16.01066, B = 3094.543, C = 219.377$
  The bubble point T_b can be determined by solving

$$P = x_B P_{sat_B}(T_b) + x_T P_{sat_T}(T_b)$$

where
  B denotes benzene
  T denotes toluene
  x_B and x_T are mole fractions of benzene and toluene, respectively
  (1) Calculate the bubble temperature T_b of a mixture containing $x_B = 0.65$ and $x_T = 0.35$ at a pressure of $P = 1\ atm$ (760 $mmHg$).
  (2) Plot the bubble point temperature T_b at a pressure $P = 1atm$ as a function of x_B $(0 \leq x_B \leq 1)$.
2.15 The yearly cost of a heat exchanger, C_d, due to depreciation can be determined from

$$C_d = (C_e - S_b) \frac{i}{(1 + i)^n - 1}$$

where

C_e is the initial cost of the heat exchanger

S_b is the scrap value of the heat exchanger

i is the annual compound interest rate

n is the lifetime of the heat exchanger

Heat exchanger A costs $6000 with a useful life of 9.8 years and a scrap value of $450, and heat exchanger B costs $10,500 with a useful life of 15 years and a scrap value of $1200.

(1) At what annual compound interest rate do the two heat exchangers (A and B) have the same yearly depreciation cost?

(2) Plot the yearly depreciation cost of each heat exchanger as a function of the annual compound interest rate ($0 \leq i \leq 0.1$) and confirm the result obtained from (1).

2.16 The virial equation of state can be represented as

$$\frac{P}{\rho RT} = 1 + B\rho + C\rho^2 + D\rho^3$$

where

$\rho\,(mol/liter)$ is the density

$P\,(atm)$ is the pressure

$T\,(K)$ is the temperature

$R = 0.08206\ liter \cdot atm/(mol \cdot K)$ is the gas constant

For nitrogen at $T = 200K$, $B = -0.0361 liter/mol$, $C = 2.7047 \times 10^{-3}(liter/mol)^2$, and $D = -4.4944 \times 10^{-4}(liter/mol)^3$. Find the density of nitrogen at $200K$ and $10\ atm$.

2.17 An exothermic reaction is carried out in an adiabatic continuous-stirred tank reactor with a fluid residence time of τ. From the material balance, the conversion of the reactant, x_m, is given by

$$x_m = \frac{k\tau}{1 + k\tau}, \quad k = Ae^{-\frac{E}{RT}}$$

where

A is the frequency factor

R is the gas constant

T is the temperature of the product stream

E is the activation energy

From the energy balance, the conversion of the reactant, x_e, is given by

$$x_e = \frac{C_{pf}\,(T - T_f)}{-\Delta H}, \quad \Delta H = \Delta H_R^0 + \Delta C_p\,(T - T_R)$$

where

C_{pf} is the heat capacity of the reactant

ΔH is the enthalpy change of reaction measured at T

T_f is the feed temperature

ΔH_R^0 is the standard heat of reaction at the reference temperature T_R

ΔC_p is the overall change in the heat capacity

(1) Plot x_m and x_e as a function of $T\,(260 \leq T \leq 360)$.

(2) Find values of T at which $x_m = x_e$.

Data: $\tau = 0.27\ hr$, $A = 4.711 \times 10^9\ sec^{-1}$, $R = 8.314\ J/(mol \cdot K)$, $E = 7.536 \times 10^4\ J/mol$, $C_{pf} = 1675\ J/(mol \cdot K)$, $T_f = 273.15\ K$, $\Delta H_R^0 = -1.216 \times 10^5\ J/mol$, $\Delta C_p = -29\,31\ J/(mol \cdot K)$, $T_R = 293.15K$

2.18 Figure P2.18 shows a tower used to produce CO_2. Limestone ($CaCO_3$) is calcinated continuously in the tower, where a flue gas flows in direct contact with it. The $CaCO_3$ enters the tower

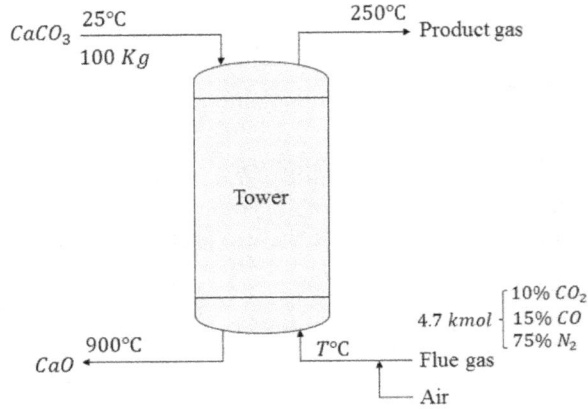

FIGURE P2.18 CO_2 production tower.

at 25°C and the calcinated material (CaO) leaves the tower at 900°C. The product gas leaves the tower at 250°C. Every 100 kg of $CaCO_3$ feed consumes 4.7 kmol of flue gas, which contains 10% CO_2, 15% CO, and 75% N_2 in volume. We assume that the gas mixture is burnt with the theoretical amount of air which enters the tower at 25°C, the combustion is complete, and energy losses and the water content in the air are negligible. The reaction enthalpy for each reaction is

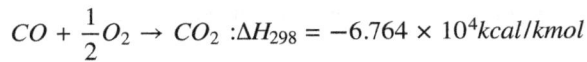

$$CaCO_3 \rightarrow CaO + CO_2 \;:\Delta H_{298} = 4.379 \times 10^4 kcal/kmol$$

$$CO + \frac{1}{2}O_2 \rightarrow CO_2 \;:\Delta H_{298} = -6.764 \times 10^4 kcal/kmol$$

The heat capacity ($kcal/kmol/K$) for each species at $T(K)$ is

$$C_{p,CO_2} = \int_{298}^{T} (6.339 + 0.01014T - 3.415 \times 10^{-6}T^2)dT$$

$$C_{p,CO} = \int_{298}^{T} (6.350 + 1.811 \times 10^{-3}T - 2.675 \times 10^{-7}T^2)dT$$

$$C_{p,N_2} = \int_{298}^{T} (6.457 + 1.389 \times 10^{-3}T - 6.9 \times 10^{-8}T^2)dT$$

$$C_{p,O_2} = \int_{298}^{T} (6.117 + 3.167 \times 10^{-3}T - 1.005 \times 10^{-6}T^2)dT$$

and $C_{p,CaCO_3}(T = 25°C) = 0.18$ and $C_{p,CaO}(T = 900°C) = 0.23$. Determine the temperature of the flue gas entering the tower.

2.19 For a binary liquid mixture containing acetone (1) and methanol (2), the following equations can be obtained from the basic thermodynamic relations:

$$y_1 P = x_1 \gamma_1 P_{sat,1}, \; (1 - y_1)P = (1 - x_1)\gamma_2 P_{sat,2}$$

where
  P is the total pressure (kPa)
  γ_i is the activity coefficient for component i
  x and y are the mole fractions of the liquid and vapor phases, respectively
  P_{sat} is the vapor pressure (kPa)
  γ_i is given by

$$\gamma_1 = \exp\{0.64(1 - x_1)^2\}, \ \gamma_2 = \exp\{0.64x_1^2\}$$

and $P_{sat,i}$ at $T\,(°C)$ can be determined from Antoine equation

$$\ln P_{sat,i}(T) = A_i - \frac{B_i}{T + C_i}\,(i = 1, 2; T : °C, P_{sat,i} : kPa)$$

Parameters for the Antoine equation are given in Table P2.19.

(1) Calculate the bubble point pressure $T\,(°C)$ of a liquid mixture containing 30 mol% acetone and 70 mol% methanol at $P = 101.325\,kPa\,(1\,atm)$. What is the vapor-phase mole fraction of acetone (y_1) at this temperature?

(2) Plot $T(°C)$ as a function of mole fractions x_1 and y_1.

TABLE P2.19
Parameters for Antoine Equation

	A	B	C
Acetone(1)	14.3916	2795.82	230.00
Methanol(2)	16.5938	3644.30	239.76

2.20 Ethylene and acetylene can be produced by the following catalytic dehydrogenation of ethane:

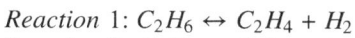

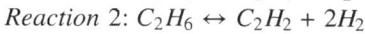

Reaction 1: $C_2H_6 \leftrightarrow C_2H_4 + H_2$
Reaction 2: $C_2H_6 \leftrightarrow C_2H_2 + 2H_2$

The equilibrium constants of *Reaction 1* (K_1) and *Reaction 2* (K_2) can be expressed in terms of the extents of the two reactions $(\xi_1$ and $\xi_2)$:

$$K_1 = \frac{\xi_1(\xi_1 + 2\xi_2)}{(1 - \xi_1 - \xi_2)(1 + \xi_1 + 2\xi_2)}, \ K_2 = \frac{\xi_2(\xi_1 + 2\xi_2)^2}{(1 - \xi_1 - \xi_2)(1 + \xi_1 + 2\xi_2)^2}$$

At 977°C and 1 *atm*, K_1 and K_2 were found to be 3.75 and 0.135, respectively. Determine the composition of the product stream.

REGRESSION ANALYSIS

2.21 Table P2.21 shows the weights of 60 students. Find the mean, median, mode, range, variance, and standard deviation for the given weight data set.

2.22 A manufacturer of memory chips produces batches of 500 chips for shipment to electronic companies. To determine if the chips meet specifications, the manufacturer tests all 500 chips in each batch. If 5% of the chips have defects, determine the probability that the manufacturer will find no defective chips in any batch. What is the most likely number of defective chips the manufacturer will find?

2.23 Table P2.23 shows experimental data of the reaction constant (k) for a 1st-order reaction as a function of reaction temperature (T). The reaction constant can be represented by the Arrhenius equation

$$k = Ae^{-E/RT}$$

where
A is the frequency factor

TABLE P2.21
Weights of Students

Student ID	Weight(*lb*)	Student ID	Weight(*lb*)	Student ID	Weight(*lb*)
1A	140	1U	170	2O	160
1B	145	1V	157	2P	155
1C	162	1W	130	2Q	124
1D	190	1X	185	2R	135
1E	155	1Y	190	2S	142
1F	165	1Z	154	2T	140
1G	150	2A	170	2U	164
1H	189	2B	155	2V	150
1I	194	2C	212	2W	145
1J	138	2D	150	2X	190
1K	159	2E	145	2Y	180
1L	155	2F	155	2Z	140
1M	153	2G	150	3A	150
1N	145	2H	155	3B	155
1O	170	2I	150	3C	146
1P	175	2J	155	3D	148
1Q	175	2K	150	3E	150
1R	170	2L	180	3F	155
1S	180	2M	162	3G	130
1T	135	2N	135	3H	160

R is the gas constant
E is the activation energy
Estimate values of A and E. Plot the fitting curve and data on the same graph.

TABLE P2.23
Reaction Constant for a 1st-Order Reaction

$T(K)$	302.9	311.2	313.8	319.1	325.2
$k(sec^{-1})$	0.00529	0.01128	0.01901	0.03649	0.07268

2.24 Table P2.24 presents the viscosity ($Pa \cdot sec$) of liquid propane as a function of temperature T(K). A nonlinear correlation is proposed to represent the viscosity data:

$$\log(\nu) = A + \frac{B}{T} + C\log T + DT^2$$

Determine parameters A, B, C, and D that best fit these viscosity data.[26]
2.25 Table P2.25 presents reaction rate data for a heterogeneous catalytic reaction given by $A \rightarrow B$. The following model has been suggested to correlate the data:

$$r = \frac{k_1 P_A}{(1 + K_A P_A + K_B P_B)^2}$$

where k_1, K_A, and K_B are coefficients to be determined by regression. Calculate the parameters using regression, and plot the reaction rate represented by this model and the rate data.[27]

TABLE P2.24
Viscosity of Liquid Propane

T	$10^4 \nu (Pa \cdot sec)$	T	$10^4 \nu (Pa \cdot sec)$
100	38.2	190	3.36
110	22.9	200	2.87
120	15.3	220	2.24
130	10.8	240	1.80
140	8.32	260	1.44
150	6.59	270	1.30
160	5.46	280	1.17
170	4.53	290	1.05
180	3.82	300	0.959

TABLE P2.25
Heterogeneous Catalytic Reaction Rate Data

n	$r \times 10^3$	Partial Pressure (atm)	
		P_A	P_B
1	5.11	1.00	0.00
2	5.42	0.91	0.09
3	5.55	0.82	0.18
4	5.85	0.69	0.31
5	6.02	0.61	0.39
6	6.15	0.52	0.48
7	6.31	0.41	0.59
8	6.45	0.29	0.71

TABLE P2.26
Mole Fraction of Benzene

x	0.0	0.1	0.2	0.3	0.4	0.5	0.6	0.7	0.8	0.9	1.0
y	0.000	0.211	0.378	0.512	0.623	0.714	0.791	0.856	0.911	0.959	1.000

2.26 Table P2.26 shows the mole fractions of benzene at a certain temperature and pressure in the benzene-toluene liquid mixture. x and y are the mole fractions of benzene in the liquid and vapor phases, respectively. Find the polynomial to fit these data.

2.27 Data of heat capacity $C_p(kJ/(kg \cdot mol \cdot K))$ versus temperature $T(K)$ for gaseous propane are presented in Table P2.27.[28] Find the polynomial that can be used to represent the variation of C_p as a function of T.

TABLE P2.27
Heat Capacity of Gaseous Propane

T	C_p	T	C_p
50	34.06	600	128.70
100	41.30	700	142.67
150	48.79	800	154.77
200	56.07	900	163.35
273.15	68.74	1000	174.60
298.15	73.60	1100	182.67
300	73.93	1200	189.74
400	94.01	1300	195.85
500	112.59	1400	201.21

2.28 The Margules equation, which describes the activity coefficients of a binary system, can be represented as

$$\ln\gamma_1 = x_2^2(2B - A) + 2x_2^3(A - B), \quad \ln\gamma_2 = x_1^2(2A - B) + 2x_1^3(B - A)$$

where x_1 and x_2 are the mole fractions of components 1 and 2, respectively. For this specific binary mixture, parameters A and B are constant. The activity coefficients for the binary mixture of benzene(1) and n-heptane(2) are given in Table P2.28. Estimate parameters A and B of the Margules equation using these data.

2.29 The van Laar equations for correlation of binary activity coefficients can be expressed as

$$\ln\gamma_1 = \frac{A}{\left(1 + \frac{x_1 A}{x_2 B}\right)^2}, \quad \ln\gamma_2 = \frac{B}{\left(1 + \frac{x_2 B}{x_1 A}\right)^2}$$

TABLE P2.28
Activity Coefficients of Benzene/n-Heptane Mixture

x_1	γ_1	γ_2
0.0464	1.2968	0.9985
0.0861	1.2798	0.9998
0.2004	1.2358	1.0068
0.2792	1.1988	1.0159
0.3842	1.1598	1.0359
0.4857	1.1196	1.0676
0.5824	1.0838	1.1096
0.6904	1.0538	1.1664
0.7842	1.0311	1.2401
0.8972	1.0078	1.4038

where x_1 and x_2 are the mole fractions of components 1 and 2, respectively. Parameters A and B are constant for this particular binary mixture. The activity coefficients for the binary mixture of benzene(1) and n-heptane(2) are given in Table P2.28. Estimate the parameter values of A and B of the van Laar equation.

2.30 Table P2.30 presents data of vapor pressure $P\,(Pa)$ versus temperature $T\,(K)$ for benzene.[29] Correlate the data using the Clapeyron equation

$$\log P = A + \frac{B}{T}$$

TABLE P2.30
Vapor Pressure of Benzene

$T\,(K)$	$P\,(Pa)$
302.39	15388.0
318.69	30464.0
330.54	47571.0
338.94	63815.0
346.24	81275.0
353.47	102040.0
358.87	120140.0

2.31 Table P2.31 shows the experimental data of relative vapor pressure (P_{rv}) of a glycerol-water mixture at 20°C. Find a polynomial that best fits the vapor pressure as a function of the weight fraction (x) of glycerol.[30]

2.32 The Wagner equation can be used to estimate vapor pressure between the triple point and the critical point:

$$\ln P_r = \frac{1}{T_r}(a_1\tau + a_2\tau^{1.5} + a_3\tau^3 + a_4\tau^6)$$

where $T_r = T/T_c$, $P_r = P/P_c$, and $\tau = 1 - T_r$. Table P2.32 presents ethane vapor pressure (Pa) versus temperature (K).[31] Correlate the data using the Wagner equation. For ethane, $T_c = 305.32\,K$ and $P_c = 4.872 \times 10^6\,Pa$.

2.33 Data of thermal conductivity $k\,(W/(m{\cdot}K))$ versus temperature $T\,(K)$ for gaseous propane are presented in Table P2.33.[32] The nonlinear equation $k = cT^n$ is known to be used to describe the relation between k and T. Estimate the parameters c and n that best fit the data.

TABLE P2.31
Relative Vapor Pressure of a Glycerol-Water Mixture at 20°C

x	0.00	0.20	0.35	0.48	0.60	0.65	0.75	0.83	0.92	1.0
P_{rv}	1.0	0.94	0.89	0.83	0.74	0.70	0.59	0.45	0.28	0.0

TABLE P2.32
Vapor Pressure of Ethane

$T\,(K)$	$10^{-4}P\,(Pa)$	$T\,(K)$	$10^{-4}P\,(Pa)$
92	0.00017	230	70.0000
100	0.0011	240	96.7000
110	0.0075	250	130.1000
120	0.0350	260	171.2000
130	0.1300	270	221.0000
140	0.3800	276	255.5000
150	0.9700	280	280.6000
158	1.8000	284	307.5000
170	4.3000	288	336.3000
180	7.9000	290	351.4000
192	14.9000	294	383.4000
200	21.7000	298	417.6000
210	33.4000	300	435.6000
220	49.2000	302	454.3000

TABLE P2.33
Thermal Conductivity of Gaseous Propane

$T\,(K)$	k	$T\,(K)$	k
231.07	0.0114	400	0.0306
240	0.0121	420	0.0334
260	0.0139	440	0.0363
280	0.0159	460	0.0393
300	0.0180	480	0.0424
320	0.0202	500	0.0455
340	0.0226	520	0.0487
360	0.0252	540	0.0520
380	0.0278	560	0.0553

2.34 Data of heat of vaporization of propane $\Delta H_v(J/kmol)$ versus temperature $T\,(K)$ are presented in Table P2.34.[33] ΔH_v is known to be represented as a nonlinear function of T as $\Delta H_v = A\,(T_c - T)^n$ (for propane, $T_c = 369.83\,K$). Estimate values of A and n.

INTERPOLATION

2.35 The vapor pressure (P) of water and the specific volume (v) of saturated water vapor are shown in Table P2.35. Find P and v at $T = 8$ and $14°C$. Use the cubic spline method.

2.36 Several models have been proposed to describe the rates of the reversible catalytic reforming reaction[34]

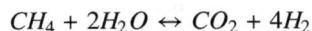

$$CH_4 + 2H_2O \leftrightarrow CO_2 + 4H_2$$

TABLE P2.34
Heat of Vaporization of Propane

$T(K)$	$10^{-7}\Delta H_v$	$T(K)$	$10^{-7}\Delta H_v$
90	2.46	220	1.93
100	2.42	231.07	1.88
110	2.37	240	1.83
120	2.33	250	1.78
130	2.29	260	1.73
140	2.25	270	1.67
150	2.21	280	1.61
160	2.17	290	1.54
170	2.13	300	1.46
180	2.09	310	1.38
190	2.05	320	1.28
200	2.01	330	1.18
210	1.97	340	1.05

TABLE P2.35
Vapor Pressure of Water and Specific Volume of Saturated Water Vapor

T (°C)	3.0	6.0	9.0	12.0	15.0	18.0
P (kPa)	0.758	0.935	1.148	1.402	1.705	2.064
v (m^3/kg)	168.13	137.73	113.39	93.78	77.93	65.04

One of several models for this catalytic reaction (model 1), in which methane is adsorbed on the catalyst surface, is given by

$$r_{CO_2} = k_s K_{CH_4} \left(P_{CH_4} P_{H_2O}^2 - \frac{P_{CO_2} P_{H_2}^4}{K_P} \right) \frac{1}{1 + K_{CH_4} P_{CH4}}$$

Another model (model 2) is a simple form of reversible reaction in which there is no component adsorption on the catalyst:

$$r_{CO_2} = k_1 \left(P_{CH_4} P_{H_2O}^2 - \frac{P_{CO_2} P_{H_2}^4}{K_P} \right)$$

where $K_P = 5.051 \times 10^{-5} atm^2$. Table P2.36 shows experimental results of reaction rate as a function of the partial pressure of the products at 350°C.

(1) Estimate the values of parameters k_s, K_{CH_4}, and k_1 of models 1 and 2.
(2) Determine graphically which model best represents the given experimental data.

2.37 Suppose you have measured heights at a number of coordinates on the surface of a mountainous area. The area is 4 km wide (x) and 5 km long (y).

```
>> x = 0:0.5:4; % x(km)
>> y = 0:0.5:5; % y(km)
>> z = [95 93 96 94 97 94 93 94 96; 96 94 93 94 97 93 96 94 94;
```

TABLE P2.36
Reaction Rate Data for Catalytic Reforming Reaction

n	$r_{CO_2} \times 10^3$ (gmol/hr/gr)	Partial Pressure (atm)			
		CH_4	H_2O	CO_2	H_2
1	0.13717	0.06298	0.23818	0.00420	0.01669
2	0.15584	0.03748	0.26315	0.00467	0.01686
3	0.20028	0.05178	0.29557	0.00542	0.02079
4	0.05700	0.04978	0.23239	0.00177	0.07865
5	0.20150	0.04809	0.29491	0.00655	0.02464
6	0.07887	0.03849	0.24171	0.00184	0.06873
7	0.14983	0.03886	0.26048	0.00381	0.01480
8	0.15988	0.05230	0.26286	0.05719	0.01635
9	0.26194	0.05185	0.33529	0.00718	0.02820
10	0.14426	0.06432	0.24787	0.00509	0.02055
11	0.20195	0.09609	0.28457	0.00652	0.02627

200	200	200	200	200	200	200	200	200	200	200	200
200	227.2	254.4	279.5	289.8	293	293	289.8	279.5	254.4	227.2	200
200	254.4	312	369.6	387.9	392.2	392.2	387.9	369.6	312	254.4	200
200	279.5	369.6	500	500	500	500	500	500	369.6	279.5	200
200	289.8	387.9	500	500	500	500	500	500	387.9	289.8	200
200	293	392.2	500	500	500	500	500	500	392.2	293	200
200	293	392.2	500	500	500	500	500	500	392.2	293	200
200	289.8	387.9	500	500	500	500	500	500	387.9	289.8	200
200	279.5	369.6	500	500	500	500	500	500	369.6	279.5	200
200	254.4	312	369.6	387.9	392.2	392.2	387.9	369.6	312	254.4	200
200	227.2	254.4	279.5	289.8	293	293	289.8	279.5	254.4	227.2	200
200	200	200	200	200	200	200	200	200	200	200	200

FIGURE P2.38　Temperature distribution for the cross section of a flue.

```
93 94 92 91 96 93 97 96 98; 95 93 92 91 93 95 96 96 93;
97 96 93 92 95 98 99 97 98; 97 98 97 96 98 99 98 96 95;
93 97 95 96 98 102 101 100 98; 91 94 95 96 98 101 102 101 95;
95 97 98 96 97 98 97 95 94; 98 99 101 98 96 95 96 94 93;
94 95 96 96 95 94 93 94 95];
```

Estimate the height at $x = 1.7$ and $y = 3.6$.

2.38 Consider a long, rectangular flue with steady heat conduction along the x and y axes. Figure P2.38 shows data of temperature distribution for the entire cross section of the rectangular flue, at 11 ft wide (x) and 11 ft long (y).[35] Divide the ranges of x and y into 55 subintervals (i.e., the distance between adjacent nodes is 0.2 ft) and estimate the temperature at each node. Display the results graphically.

DIFFERENTIATION AND INTEGRATION

2.39 Table P2.39 shows experimental data for a gas-phase reaction.[36] It was found that the reaction rate can be represented by the equation

$$\frac{dP}{dt} = k(3P_0 - P)^m$$

where k is the reaction rate constant, P_0 is the initial pressure, and m is the reaction order. Determine k and m.

TABLE P2.39
Experimental Data for a Gas-Phase Reaction

$t\,(min)$	0.0	2.5	5.0	10.0	15.0	20.0
$P\,(mmHg)$	7. 8	10.5	12.7	15.8	17.9	19.4

2.40 The equation of state for hard-chain fluids can be represented in term of the residual Helmholtz energy (a^{res}):[37]

$$a^{res} = m\frac{4\gamma - 3\gamma^2}{(1 - \gamma)^2} + (1 - m)\ln\frac{1 - \gamma/2}{(1 - \gamma)^3}$$

where m is the number of segments in a chain and γ is the reduced density. The compressibility factor $Z\,(=PV/RT)$ of any fluid can be obtained from a^{res} using the following relation:

$$Z = 1 + \gamma\frac{\partial a^{res}}{\partial \gamma}$$

Plot Z versus γ $(0 < \gamma < 0.5)$ when $m = 5$.

2.41 Estimate the amount of heat (cal) required to heat 1 gram mole of propane from 200°C to 700°C at 1 atm. The heat capacity of propane is given by

$$C_p = 2.41 + 0.057195T - 4.3 \times 10^{-6}T^2 (T : K)$$

ORDINARY DIFFERENTIAL EQUATIONS

2.42 A 2nd-order reaction is taking place in an isothermal continuous-stirred tank reactor. The reactor is well mixed and the volume V is maintained constant. The reaction rate is given by $r = kC^2$, where k is constant. The inlet concentration suddenly shows a periodic oscillation of the form

$$C_i = C_0\{1 + A\sin(2\pi ft)\}$$

where
  f is the frequency of oscillations
  C_0 is a constant

$A (<1)$ represents the magnitude of oscillations

The mass balance for the reactor can be written as

$$V\frac{dC}{dt} = q(C_i - C) - kVC^2, \; C(0) = 0 \, (0 \le t \le V/q)$$

Produce a plot of the dimensionless concentration $\theta = C/C_0$ as a function of the dimensionless time $\tau = tq/V$ for various values of f ($f = 0.1q/V, 4q/V$ and $8q/V$).[38] Assume that $A = 0.5$ and the Damköhler number $Da = kVC_0/q = 100$.

2.43 Solve the system of the Lorenz equations from $t = 0$ to 50:

$$\frac{d}{dt}y_1(t) = 10(y_2(t) - y_1(t))$$

$$\frac{d}{dt}y_2(t) = 28y_1(t) - y_2(t) - y_1(t)y_3(t)$$

$$\frac{d}{dt}y_3(t) = y_1(t)y_2(t) - \frac{8}{3}y_3(t)$$

The initial condition is $y(0) = \begin{bmatrix} 0 \\ 1 \\ 0 \end{bmatrix}$.

2.44 Figure P2.44 shows three continuous-stirred heating tanks connected in series. Steam is used to heat the solution in each tank. The mass flow rate w_i ($i = 0, 1, 2, 3$) at each tank is maintained at a constant value of $w \; kg/min$. If the heat transfer area A is constant for each tank, the amount of heat transferred from the steam pipe to the solution is given by

$$Q_i = UA(T_s - T_i)(i = 1, 2, 3)$$

The temperature at each tank before heating is T_0. Using the data given, plot the temperature profile at each tank as a function of time. What is the steady-state temperature of the 3rd tank?

Data: Heat capacity of the solution $C_p = 1.75 \; kJ/(kg \cdot °C)$; $UA = 12 \; kJ/(min \cdot °C)$; $T_0 = 25°C$; mass of solution at each tank $M = 1400 \; kg$; steam temperature $T_s = 200°C$; flow rate of the solution $w = 120 \; kg/min$; time span $t = [0 \, 200]$.

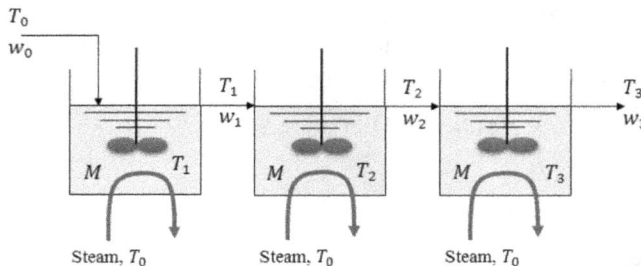

FIGURE P2.44   Continuous-stirred heating tanks.

2.45 The Monod model for biomass reactions can be represented as

$$\frac{dS}{dt} = -\frac{kSx}{k_s + S}, \frac{dx}{dt} = \frac{k\gamma Sx}{k_s + S} - aS$$

where

S is the growth-limiting substrate concentration

x is the biomass concentration
k is the maximum specific uptake rate of the substrate
k_s is the half saturation constant for growth
γ is the yield coefficient
a is the decay coefficient
Using the data given, generate profiles for S and x as a function of time t ($0 \leq t \leq 0.02$).
  Data: $S(0) = 980 \, kmol/m^3$, $x(0) = 112 \, kmol/m^3$, $k = 4.6 \, hr^{-1}$, $k_s = 18 \, kmol/m^3$, $\gamma = 0.05$,
$a = 0.01 \, hr^{-1}$

2.46 The system of Robertson ordinary differential equations models a reaction between three chemical components[39]:

$$\frac{dy_1(t)}{dt} = -0.04y_1(t) + 10^4 y_2(t)y_3(t)$$

$$\frac{dy_2(t)}{dt} = 0.04y_1(t) - 10^4 y_2(t)y_3(t) - 3 \times 10^7 y_2(t)^2$$

$$\frac{dy_3(t)}{dt} = 3 \times 10^7 y_2(t)^2$$

The Robertson reaction model is a typical example of the stiffness problem. Solve the Robertson ODE system from $t = 0$ to 3 and plot $y_2(t)$ versus time t. The initial conditions are $y_1(0) = 1$ and $y_2(0) = y_3(0) = 0$.

2.47 Consider a catalytic fluidized bed in which an irreversible gas-phase reaction $A \to B$ occurs. The mass and energy balances and kinetic rate constant for this system are given as[8,40]:

$$\frac{dP}{d\tau} = 0.1 + 320P_p - 321P, \quad \frac{dT}{d\tau} = 1752 - 269T + 267T_p,$$

$$\frac{dP_p}{d\tau} = 1.88 \times 10^3 \{P - P_p(1 + K)\}, \quad \frac{dT_p}{d\tau} = 1.3(T - T_p) + 1.04 \times 10^4 KP_p,$$

$$K = 6 \times 10^{-4} \exp\left(20.7 - \frac{1500}{T_p}\right)$$

where
  T is the absolute temperature (°R)
  P is the partial pressure of the reactant in the fluid (atm)
  T_p is the temperature of the reactant at the catalyst surface (°R)
  P_p is the partial pressure of the reactant at the catalyst surface (atm)
  K is the dimensionless reaction rate constant
  τ is the dimensionless time
Solve the equation system in the range of $0 \leq \tau \leq 1500$ and plot changes of dependent variables as functions of time. Use the initial values of $P(0) = 0.1$, $T(0) = 600$, $P_p = 0$, and $T_p = 761$.

2.48 A reaction $A \to B$ is taking place in a heterogeneous reactor. The reactor model can be represented as[41]

$$u\frac{dC_A}{dz} = -k_g a(C_A - C_{As})$$

$$0 = k_g(C_A - C_{As}) - r_A(C_{As}), \quad r_A(C_A) = \frac{kC_A}{\sqrt{1 + K_r C_A^2}}$$

where

    u is the inlet velocity

    a is the ratio of surface area to volume

    C_{As} is the surface concentration

    k_g is the mass transfer coefficient

Produce the axial profiles of concentration C_A and C_{As} in the reactor when $a = 380\, m_{cat}^2/m^3$, $k = 0.0045\, m^3/(m_{cat}^2 \cdot sec)$, $K_r = 1.1\, m^6/mol^2$, $k_g = 0.012\, m/sec$, $u = 1\, m/sec$, $C_{A0} = 1\, mol/m^3$ and L(reactor length) $= 1\, m$.

PARTIAL DIFFERENTIAL EQUATIONS

2.49 A thin metal plate of dimensions $1\, m \times 1\, m$ is subjected to four different heat sources which maintain the temperature $T\,(°C)$ on its four edges at $T\,(0, y) = 250°C$, $T\,(1, y) = 100°C$, $T\,(x, 0) = 500°C$, and $T\,(x, 1) = 25°C$ (see Figure P2.49). The flat sides of the plate are insulated and no heat is transferred through these sides. The steady-state temperature distribution over the plate can be described by the Laplace equation

$$\nabla^2 T = \frac{\partial^2 T}{\partial x^2} + \frac{\partial^2 T}{\partial y^2} = 0 (0 \le x \le 1, 0 \le y \le 1)$$

Plot the temperature profiles within the plate.[42]

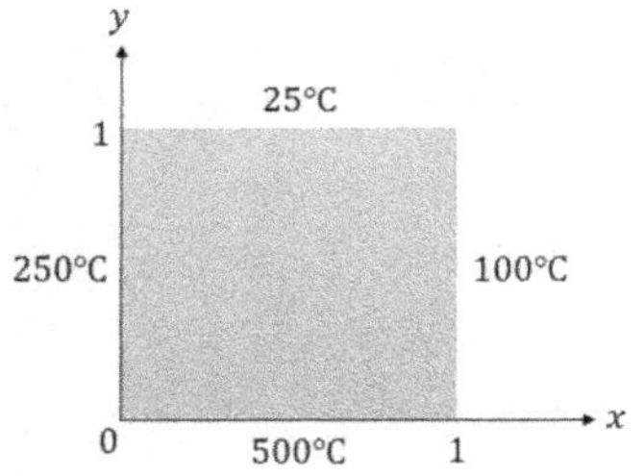

FIGURE P2.49   A metal plate subjected to four heat sources.

2.50 A solid plate of dimensions $1\, m \times 1\, m$ is in contact with atmosphere and the moisture goes into the solid (see Figure P2.50). The boundary conditions are $C\,(0, y) = 0$, $C\,(1, y) = 0$, $C\,(x, 0) = 0$, and $C\,(x, 1) = 32\, g/m^3$. The steady-state moisture distribution over the plate can be described by the Laplace equation

$$\frac{\partial^2 C}{\partial x^2} + \frac{\partial^2 C}{\partial y^2} = 0 (0 \le x \le 1, 0 \le y \le 1)$$

Plot the moisture profiles within the plate.

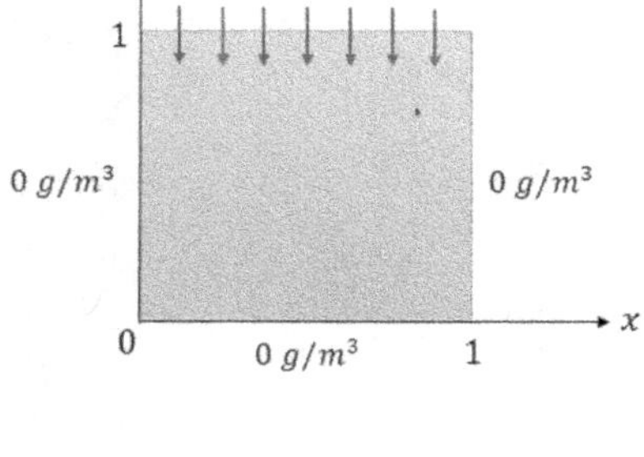

FIGURE P2.50   A solid plate contacting atmosphere.

REFERENCES

1. Kapuno, R. R. A., *Programming for Chemical Engineers*, Infinity Science Press, Hingham, MA, p. 122, 2008.
2. Cutlip, M. B. and M. Shacham, Problem Solving in Chemical and Biochemical Engineering with POLYMATH, Excel, and MATLAB, 2nd ed., Prentice-Hall, Boston, MA, p. 144, 2008.
3. Adidharma, H. and V. Temyanko, *Mathcad for Chemical Engineers*, Trafford publishing, Victoria, BC, Canada, p. 72, 2007.
4. Adidharma, H. and V. Temyanko, *Mathcad for Chemical Engineers*, Trafford publishing, Victoria, BC, Canada, p. 73, 2007.
5. Adidharma, H. and V. Temyanko, *Mathcad for Chemical Engineers*, Trafford publishing, Victoria, BC, Canada, p. 80, 2007.
6. Adidharma, H. and V. Temyanko, *Mathcad for Chemical Engineers*, Trafford publishing, Victoria, BC, Canada, p. 81, 2007.
7. Cutlip, M. B. and M. Shacham, Problem Solving in Chemical and Biochemical Engineering with POLYMATH, Excel, and MATLAB, 2nd ed., Prentice-Hall, Boston, MA, p. 203, 2008.
8. Aiken, R. C. and L. Lapidus, An effective integration method for typical stiff systems, *AIChE Journal*, 20(2), p. 368, 1974.
9. Shampine, L. F., I. Gladwell and S. Thompson, *Solving ODEs with MATLAB*, Cambridge Univ. Press, Cambridge, UK, pp. 168–169, 2003.
10. Victor J. Law, *Numerical Methods for Chemical Engineers*, CRC Press, Taylor & Francis Group, Boca Raton, FL, pp. 123–124, 2013.
11. Niket S. Kaisare, *Computational Techniques for Process Simulation and Analysis Using MATLAB*, CRC Press, Taylor & Francis Group, Boca Raton, FL, p. 343, 2018.
12. Won Y. Yang, Wenwu Cao, Tae-Sang Chung and John Morris, *Applied Numerical Methods Using MATLAB*, Wiley Interscience, John Wiley & Sons, Inc., Hoboken, NJ, p. 404, 2005.
13. Skeel, R. D. and M. Berzins, A method for the spatial discretization of parabolic equations in one space variable, *SIAM Journal on Scientific and Statistical Computing*, 11, pp. 1–32, 1990.
14. Martin, M. M., *Introduction to Software for Chemical Engineers*, CRC Press, Boca Raton, FL, p. 290, 2015.
15. Raman, R., *Chemical Process Computations*, Elsevier Applied Science Publishers, Barking, Essex, UK, p. 2, 1985.
16. Kano, M. and M. Ogawa, The state of the art in chemical process control in Japan: Good practice and questionnaire survey, *Journal of Process Control*, 11, pp. 969–982, 2010.
17. Martin, M. M., *Introduction to Software for Chemical Engineers*, CRC Press, Boca Raton, FL, pp. 291–293, 2015.
18. Kapuno, R. R. A., *Programming for Chemical Engineers*, Infinity Science Press, Hingham, MA, p. 112, 2008.
19. Cutlip, M. B. and M. Shacham, *Problem Solving in Chemical and Biochemical Engineering with POLYMATH, Excel, and MATLAB*, 2nd ed., Prentice-Hall, Boston, MA, pp. 23–24, 2008.
20. Kevin D. Dorfman and Prodromos Daoutidis, *Numerical Methods with Chemical Engineering Applications*, Cambridge, pp. 72–77, 2017.
21. Kevin D. Dorfman and Prodromos Daoutidis, *Numerical Methods with Chemical Engineering Applications*, Cambridge, pp. 132–133, 2017.
22. Kevin D. Dorfman and Prodromos Daoutidis, *Numerical Methods with Chemical Engineering Applications*, Cambridge, pp. 91–92, 2017.
23. Cutlip, M. B. and M. Shacham, Problem Solving in Chemical and Biochemical Engineering with POLYMATH, Excel, and MATLAB, 2nd ed., Prentice-Hall, Boston, MA, p. 41, 2008.
24. Kevin D. Dorfman and Prodromos Daoutidis, *Numerical Methods with Chemical Engineering Applications*, Cambridge, pp. 161–162, 2017
25. Kevin D. Dorfman and Prodromos Daoutidis, *Numerical Methods with Chemical Engineering Applications*, Cambridge, pp. 177–179, 2017.
26. Cutlip, M. B. and M. Shacham, Problem Solving in Chemical and Biochemical Engineering with POLYMATH, Excel, and MATLAB, 2nd ed., Prentice-Hall, Boston, MA, p. 686, 2008.
27. Cutlip, M. B. and M. Shacham, Problem Solving in Chemical and Biochemical Engineering with POLYMATH, Excel, and MATLAB, 2nd ed., Prentice-Hall, Boston, MA, p. 93, 2008.
28. Cutlip, M. B. and M. Shacham, Problem Solving in Chemical and Biochemical Engineering with POLYMATH, Excel, and MATLAB, 2nd ed., Prentice-Hall, Boston, MA, p. 685, 2008.

29. Cutlip, M. B. and M. Shacham, Problem Solving in Chemical and Biochemical Engineering with POLYMATH, Excel, and MATLAB, 2nd ed., Prentice-Hall, Boston, MA, p. 25, 2008.
30. Adidharma, H. and V. Temyanko, *Mathcad for Chemical Engineers*, Trafford publishing, Victoria, BC, Canada, p. 75, 2007.
31. Ingham, H., D. G. Friend, and J. F. Ely, Thermophysical Properties of Ethane, *Journal of Physical and Chemical Reference Data*, 20(2), pp. 275–347, 1991.
32. Younglove, B. A. and J. F. Ely, Thermophysical properties of fluids. II. Methane, ethane, propane, isobutene and n-butane, *Journal of Physical and Chemical Reference Data*, 16, p. 577, 1987.
33. Haynes, W. M. and R. D. Goodwin, Thermophysical Properties of Propane from 85 to 700 K at Pressures to 70 MPa, National Bureau of Standards Monograph 170, Boulder, CO, 1982.
34. Cutlip, M. B. and M. Shacham, *Problem Solving in Chemical and Biochemical Engineering with POLYMATH, Excel, and MATLAB*, 2nd ed., Prentice-Hall, Boston, MA, pp. 87–88, 2008.
35. Kapuno, R. R. A., *Programming for Chemical Engineers*, Infinity Science Press, Hingham, MA, pp. 240–241, 2008.
36. Adidharma, H. and V. Temyanko, *Mathcad for Chemical Engineers*, Trafford publishing, Victoria, BC, Canada, p. 86, 2007.
37. Adidharma, H. and V. Temyanko, *Mathcad for Chemical Engineers*, Trafford publishing, Victoria, BC, Canada, p. 84, 2007.
38. Kevin D. Dorfman and Prodromos Daoutidis, *Numerical Methods with Chemical Engineering Applications*, Cambridge, pp. 233–234, 2017.
39. Higham, D. J. and N. J. Higham, *MATLAB Guide*, 2nd ed., SIAM, Philadelphia, PA, pp. 184–187, 2005.
40. Cutlip, M. B. and M. Shacham, *Problem Solving in Chemical and Biochemical Engineering with POLYMATH, Excel, and MATLAB*, 2nd ed., Prentice-Hall, Boston, MA, pp. 207–208, 2008.
41. Niket S. Kaisare, *Computational Techniques for Process Simulation and Analysis Using MATLAB*, CRC Press, Taylor & Francis Group, Boca Raton, FL, p. 351, 2018.
42. Alkis Constantinides and Navid Mostoufi, *Numerical Methods for Chemical Engineers with MATLAB Applications*, Prentice Hall PTR, Upper Saddle River, NJ, p. 382, 1999.

3 Physical Properties

3.1 WATER AND STEAM

Many studies have been conducted on the thermodynamic properties of water and steam. Various formulations that approximate the properties of water and steam have been introduced through the years. In this section, we will focus on the computational procedures for properties of water and steam based on the formula depicted in the IAPWS-IF97 (International Association for Properties of Water and Steam–Industrial Formulation 1997)[1] revised in 2007. The range of validity covered in the IAPWS-IF97 is

$$273.15K \leq T \leq 1073.15K : P \leq 100MPa$$
$$1073.15K < T \leq 2273.15K : P \leq 50MPa$$

3.1.1 DIVISION OF PRESSURE-TEMPERATURE RANGE

The entire pressure-temperature range of validity of the IAPWS-IF97 is divided into five regions, as shown in Figure 3.1. Each region is individually covered by fundamental equations for the properties.

Region 1, which represents subcooled liquid under 623.15 K, and region 2, which represents superheated steam under 1073.15 K, are individually covered by a fundamental equation for the specific Gibbs free energy $g(P, T)$. Region 3 represents subcooled liquid and superheated steam located between regions 1 and 2 and is covered by a fundamental equation for the specific Helmholtz free energy $f(\rho, T)$. Region 4 is the saturated region, and is covered by a fundamental equation in the form of $T_s(P)$. Region 5 represents superheated steam between 1073.15 and 2273.15 K and is covered by a fundamental equation for the specific Gibbs free energy $g(P, T)$. The maximum pressure of regions 1, 2, and 3 is 100 MPa, and that of region 5 is 50 MPa.

3.1.2 PROPERTY EQUATIONS

3.1.2.1 Parameters and Auxiliary Equations
The values of parameters and reference constants used in the IAPWS-IF97 are

Specific gas constant: $R = 0.461526\,KJ/(kg{\cdot}K)$
Critical temperature: $T_c = 647.096\,K$
Critical pressure: $P_c = 22.064\,MPa$
Critical density: $\rho_c = 322\,kg/m^3$

The specific internal energy and specific entropy of the saturated liquid at the triple point are set equal to 0. This condition is met at the temperature and pressure of the triple point $T_t = 273.16\,K$, $P_t = 611.657\,Pa$ and at the enthalpy of the triple point $H_t = 0.611783\,J/kg$.

The boundary between regions 2 and 3 as shown in Figure 3.1 is defined by the following quadratic pressure-temperature relation:

$$\pi = n_1 + n_2\theta + n_3\theta^2$$

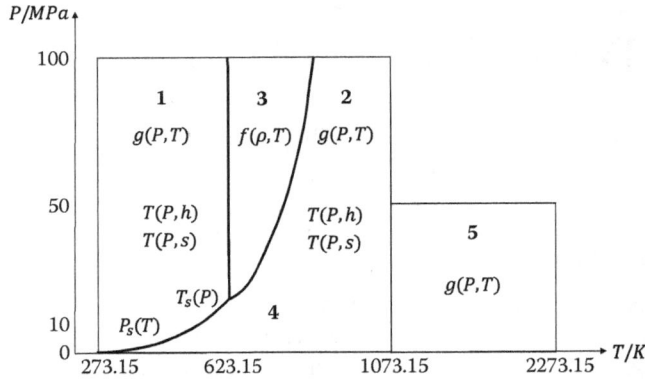

FIGURE 3.1   Pressure-temperature diagram divided into five regions, with property equations.

where $\pi = P/P^*$, $\theta = T/T^*$, $P^* = 1Mpa$, $T^* = 1K$, and n_1, n_2, and n_3 are constants. This equation roughly describes an isentropic line. The IAPWS-IF97 involves many constants, and MATLAB programs presented in this book contain these constants.

3.1.2.2 Basic Equation for Region 1

The basic equation for region 1 is a fundamental dimensionless equation for the specific Gibbs free energy g given:

$$\frac{g(P,T)}{RT} = \gamma(\pi, \tau) = \sum_{i=1}^{34} n_i (7.1 - \pi)^{I_i} (\tau - 1.222)^{J_i}$$

where $\pi = P/P^*$, $\tau = T^*/T$, $P^* = 16.53MPa$, and n_i and J_i are constants. Other thermodynamic properties can be obtained from this equation and its derivatives. For example, derivatives with respect to π and τ are used:

$$\left.\frac{\partial \gamma}{\partial \pi}\right|_\tau = -\sum_{i=1}^{34} n_i I_i (7.1 - \pi)^{I_i-1} (\tau - 1.222)^{J_i}, \quad \left.\frac{\partial \gamma}{\partial \tau}\right|_\pi = \sum_{i=1}^{34} n_i (7.1 - \pi)^{I_i} J_i (\tau - 1.222)^{J_i-1}$$

3.1.2.3 Basic Equation for Region 2

The basic equation for region 2 is a fundamental equation for the specific Gibbs free energy g. This equation can be expressed as a combination of an ideal gas part γ^o and a residual part γ^T:

$$\frac{g(P,T)}{RT} = \gamma(\pi, \tau) = \gamma^o(\pi, \tau) + \gamma^T(\pi, \tau)$$

$$\gamma^o = \ln\pi + \sum_{i=1}^{9} n_i^o \tau^{J_i^o}, \quad \gamma^T = \sum_{i=1}^{43} n_i \pi^{I_i} (\tau - 0.5)^{J_i}, \quad \gamma_\pi^o = \frac{1}{\pi}, \quad \gamma_\tau^o = \sum_{i=1}^{9} n_i^o J_i^o \tau^{J_i^o-1}$$

where $\pi = P/P^*$, $\tau = T^*/T$, $P^* = 1\,MPa$, $T^* = 540\,K$, and n_i, I_i, and J_i are constants. Derivatives of the ideal gas and residual parts with respect to π and τ are used in the calculation of physical properties:

$$\left.\frac{\partial \gamma^T}{\partial \pi}\right|_\tau = \sum_{i=1}^{43} n_i I_i \pi^{I_i-1} (\tau - 0.5)^{J_i}, \quad \left.\frac{\partial \gamma^T}{\partial \tau}\right|_\pi = \sum_{i=1}^{43} n_i \pi^{I_i} J_i (\tau - 0.5)^{J_i-1}$$

3.1.2.4 Basic Equation for Region 3

The basic equation for region 3 is a fundamental equation for the specific Helmholtz free energy f. This equation can be expressed in dimensionless form as

$$\frac{f(\rho, T)}{RT} = \Phi(\delta, \tau) = n_1 \ln\sigma + \sum_{i=2}^{40} n_i \delta^{l_i} \tau^{J_i}$$

where $\delta = \rho/\rho^*$, $\tau = T^*/T$, $\rho^* = \rho_c$, and $T^* = T_c$.

3.1.2.5 Basic Equation for Region 4

The saturation line can be expressed as a dimensionless quadratic equation, which can be solved in terms of saturation pressure P_s and saturation temperature T_s:

$$\beta^2\vartheta^2 + n_1\beta^2\vartheta + n_2\beta^2 + n_3\beta\vartheta^2 + n_4\beta\vartheta + n_5\beta + n_6\vartheta^2 + n_7\vartheta + n_8 = 0$$

where $\beta = (P_s/P)^{1/4}$, $\vartheta = T_s/T + n_9/((T_s/T^*) - n_{10})$, $P^* = 1\,MPa$, and $T^* = 1K$. The solution of this equation for the saturation pressure yields

$$\frac{P_s}{P^*} = \left[\frac{2C}{-B + \sqrt{B^2 - 4AC}}\right]^4$$

where $A = \vartheta^2 + n_1\vartheta + n_2$, $B = n_3\vartheta^2 + n_4\vartheta + n_5$, $C = n_6\vartheta^2 + n_7\vartheta + n_8$

Similarly, the solution of the quadratic equation for the saturation temperature yields

$$\frac{T_s}{T^*} = \frac{n_{10} + D - \sqrt{(n_{10} + D)^2 - 4(n_9 + n_{10}D)}}{2}$$

where

$$D = \frac{2G}{-F - \sqrt{F^2 - 4EG}}, E = \beta^2 + n_3\beta + n_6, F = n_1\beta^2 + n_4\beta + n_7, G = n_2\beta^2 + n_5\beta + n_8$$

3.1.2.6 Basic Equations for Region 5

The basic equation for high-temperature region 5 is a fundamental equation for the specific Gibbs free energy:

$$\frac{g(P, T)}{RT} = \gamma(\pi, \tau) = \gamma^o(\pi, \tau) + \gamma^T(\pi, \tau), \gamma^o = \ln\pi + \sum_{i=1}^{6} n_i^o \tau^{J_i^o}, \gamma^T = \sum_{i=1}^{6} n_i\pi^{l_i}\tau^{J_i}$$

where $\pi = P/P^*$, $\tau = T^*/T$, $P^* = 1MPa$, $T^* = 1000K$, and n_i, I_i and J_i are constants.

3.1.3 Properties of Saturated Steam

In this section we will focus on the thermodynamic properties of saturated steam valid from 273.15 to 647.15 K. The following description is based on the revised IAPWS Formulation 1995 for the Thermodynamic Properties of Ordinary Water Substance for General and Scientific Use.[2]

At a given temperature $T(K)$, the saturated vapor pressure can be approximated as

$$p = p_c \exp\left[\frac{T_c}{T}(a_1\tau + a_2\tau^{1.5} + a_3\tau^3 + a_4\tau^{3.5} + a_5\tau^4 + a_6\tau^{7.5})\right]$$

where $\tau = 1 - T/T_c$, p_c and T_c are critical pressure and critical temperature, respectively, and $a_i(i = 1, 2, \cdots, 6)$ are constants. Differentiation of the vapor pressure with respect to the saturation temperature gives

$$\frac{dp}{dT} = \left(-\frac{p}{T}\right)\left[7.5a_6\tau^{6.5} + 4a_5\tau^3 + 3.5a_4\tau^{2.5} + 3a_3\tau^2 + 1.5a_2\tau^{0.5} + a_1 + \ln\left(\frac{p}{p_c}\right)\right]$$

Alternatively, the saturation temperature can be calculated at a given pressure by converting the pressure equation into a nonlinear equation. By taking the logarithm of the pressure equation, we have

$$\ln\left(\frac{p}{p_c}\right) = \frac{T_c}{T}(a_1\tau + a_2\tau^{1.5} + a_3\tau^3 + a_4\tau^{3.5}a_5\tau^4 + a_6\tau^{7.5})$$

$$= \frac{1}{1-\tau}(a_1\tau + a_2\tau^{1.5} + a_3\tau^3 + a_4\tau^{3.5}a_5\tau^4 + a_6\tau^{7.5})$$

Rearrangement of this equation gives

$$f(\tau) = a_1\tau + a_2\tau^{1.5} + a_3\tau^3 + a_4\tau^{3.5}a_5\tau^4 + a_6\tau^{7.5} - (1-\tau)\ln\left(\frac{p}{p_c}\right) = 0$$

At a given temperature $T(K)$, the densities of the saturated liquid and steam can be represented, respectively, as

$$\rho_L = \rho_c[1 + b_1\tau^{1/3} + b_2\tau^{2/3} + b_3\tau^{5/3} + b_4\tau^{16/3} + b_5\tau^{43/3} + b_6\tau^{110/3}]$$

$$\rho_G = \rho_c\exp(c_1\tau^{1/3} + c_2\tau^{2/3} + c_3\tau^{4/3} + c_4\tau^3 + c_5\tau^{37/6} + c_6\tau^{71/6})$$

where $c_i(i = 1, 2, \cdots, 6)$ are constants.

The specific enthalpy of the saturated liquid and steam at a given temperature $T(K)$ can be obtained, respectively, from the following relations:

$$h_L = \alpha + \frac{T}{\rho_L}\frac{dp}{dT}, \, h_G = \alpha + \frac{T}{\rho_G}\frac{dp}{dT}, \, \alpha = \alpha_0[d_\alpha + d_1\theta^{-19} + d_2\theta + d_3\theta^{4.5} + d_4\theta^5 + d_5\theta^{54.5}]$$

where $d_i(i = 1, 2, \cdots, 5)$ are constants.

The specific entropy of the saturated liquid and steam at a given temperature $T(K)$ can be obtained, respectively, from the following equations:

$$s_L = \phi + \frac{1}{\rho_L}\frac{dp}{dT}, \, s_G = \phi + \frac{1}{\rho_G}\frac{dp}{dT}$$

$$\phi = \phi_0\left[d_\phi + \frac{19}{20}d_1\theta^{-20} + d_2\ln\theta + \frac{9}{7}d_3\theta^{3.5} + \frac{5}{4}d_4\theta^4 + \frac{109}{107}d_5\theta^{53.5}\right]$$

3.1.4 CALCULATION OF H₂O PROPERTIES BY MATLAB PROGRAMS

3.1.4.1 Properties of H₂O

The function *H2Oprop* estimates specific values of enthalpy, entropy, internal energy, and volume of H_2O at the given pressure $P(kPa)$ and temperature $T(°C)$. This function is written based on the revised release on the IAPWS-IF97.[1] The ranges of P and T are $0 \le P \le 100,000\,kPa$ and $0 \le T \le 2000°C$. The output is a structure containing estimated specific values of enthalpy $H(kJ/Kg)$, entropy $S(kJ/(Kg \cdot K))$, internal energy $U(kJ/Kg)$, and volume $V(cm^3/g)$. The function $H2Oprop$ and its subfunctions can be found in the Appendix at the end of this chapter.

```
>> help H2Oprop
  H2Oprop.m: properties (H,S,U,V) of H2O (water and steam)
  Based on the revised release on the IAPWS Industrial formulation 1997 for
  the Thermodynamic Properties of Water and Steam, August 2007 (Ref[1]).
  To find properties of saturated H2O, use satH2Oprop.m

Basic syntax: w = H2Oprop(dat1,dat2)
Inputs:
       dat1: pressure(kPa) (range: 0 ~ 100,000 kPa)
       dat2: temperature(deg.C) (range: 0 ~ 2000 deg.C)
Output:
       w- structure of property values (H,S,U,V) of H2O
       w.H: specific enthalpy (kJ/kg)
       w.S: specific entropy (kJ/(kg-K))
       w.U: specific internal energy (kJ/kg)
       w.V: specific volume (cm^3/g)

Examples:
  x = H2Oprop(150,350) returns the property structure x at 150 kPa and 350 deg.C
```

The script *wsprop* requests the user to input values of pressure $P\,(kPa)$ and temperature $T\,(°C)$, and calls the function $H2Oprop$ to estimate properties at the given conditions.

```
% wsprop.m: calculate properties (H,S,U,V) of H2O (water and steam) using
H2Oprop.m
P = input('Pressure (kPa) = ');
T = input('Temperature (deg.C) = ');
w = H2Oprop(P,T);
fprintf('Specific enthalpy = %g kJ/kg\n', w.H)
fprintf('Specific entropy = %g kJ/(kg-K)\n', w.S)
fprintf('Specific internal energy = %g kJ/kg\n', w.U)
fprintf('Specific volume = %g cm^3/g\n', w.V)
```

Example 3.1 Calculation of H_2O Properties

Calculate the specific volume, internal energy, enthalpy, and entropy of H_2O at the specified temperatures and pressures. Compare the results with those recorded in steam tables that can be found in standard thermodynamic texts.

1. 750 kPa, 167.76 °C
2. 2100 kPa, 214.85 °C

Solution

Results of calculations by the script *wsprop* are shown here. These results match the values in steam tables.

```
>> wsprop
Pressure (kPa) = 750
Temperature (deg.C) = 167.76

Specific enthalpy = 2765.65 kJ/kg
Specific entropy = 6.68356 kJ/(kg-K)
Specific internal energy = 2574.02 kJ/kg
Specific volume = 255.506 cm^3/g
```

We can see that $H = 2765.65\,kJ/Kg$, $S = 6.68356\,kJ/(Kg{\cdot}K)$, $U = 2574.02\,kJ/Kg$, and $v = 255.506\,cm^3/g$ when $P = 750\,kPa$ and $T = 167.76°C$.

```
>> wsprop
Pressure (kPa) = 2100
Temperature (deg.C) = 214.85
Specific enthalpy = 919.92 kJ/kg
Specific entropy = 2.46999 kJ/(kg-K)
Specific internal energy = 917.44 kJ/kg
Specific volume = 1.181 cm^3/g
```

We can see that $H = 919.92\,kJ/Kg$, $S = 2.46999\,kJ/(Kg{\cdot}K)$, $U = 917.44\,kJ/Kg$, and $v = 1.181cm^3/g$ when $P = 2100\,kPa$ and $T = 214.85°C$.

You can directly call the function *H2Oprop* to get the desired results, as follows:

```
>> w = H2Oprop(2100, 214.85)
w =
  struct with fields:
      H: 2.7657e+03
      S: 6.6836
      U: 2.5740e+03
      V: 255.5062
>> w = H2Oprop(2100, 214.85)
w =
  struct with fields:
      H: 919.9202
      S: 2.4700
      U: 917.4401
      V: 1.1810
```

3.1.4.2 Properties of Saturated H₂O

The function *satH2Oprop* estimates specific values of enthalpy $H(kJ/Kg)$, entropy $S(kJ/(Kg{\cdot}K))$, internal energy $U(kJ/Kg)$, and volume $V(cm^3/g)$ of H_2O, and saturated temperature (°C) or pressure (kPa) at the given conditions. The function *satH2Oprop* and its subfunctions can be found in the Appendix at the end of this chapter. This function is written based on the revised release of the IAPWS-IF97.[1] The basic syntax is

```
w = satH2Oprop(dtype, phH2O, dvalue)
```

where *dtype* is the data type (*'P'* if the pressure is specified, or *'T'* is the temperature is specified), *phH2O* is the state of H₂O (*'V'* for vapor state, or 'L' for liquid state), and *dvalue* is the value of pressure (kPa) or temperature (°C). The output *w* is a structure containing saturation pressure (kPa), saturation temperature (°C), and specific values of enthalpy $H(kJ/Kg)$, entropy $S(kJ/(Kg{\cdot}K))$, internal energy $U(kJ/Kg)$, and volume $V(cm^3/g)$.

```
>> help satH2Oprop
   satH2Oprop.m: properties of saturated H2O (water and steam)
   Based on the revised release on the IAPWS Industrial formulation 1997 for
   the Thermodynamic Properties of Water and Steam, August 2007 (Ref[1]).
   To find properties of superheated steam, use H2Oprop.m

Basic syntax: w = satH2Oprop(dtype, phH2O, dvalue)
Inputs:
        dtype: data type( 'P' = pressure, 'T' = temperature)
        phH2O: phase of H2O ('V' = vapor, 'L' = liquid)
        dvalue: value of pressure(kPa) or temperature(deg.C)
Output:
        w- structure of property values (T,P,H,S,U, or V) of saturated H2O
        w.P: saturation pressure (kPa)
        w.T: saturation temperature (deg.C)
        w.H: specific enthalpy (kJ/kg)
        w.S: specific entropy (kJ/(kg-K))
        w.U: specific internal energy (kJ/kg)
        w.V: specific volume (cm^3/g)

Examples:
   x = satH2Oprop('P','L',10420) returns the properties of saturated
       liquid at P = 10420 kPa
   x = satH2Oprop('T','V',220) returns the properties of saturated vapor
       at T = 220 degC
```

The script *satprop* requests the user to input the state of H_2O (vapor or liquid) and values of pressure $P(kPa)$ or temperature $T(°C)$, and calls the function *satH2Oprop* to estimate properties at the given conditions.

```
% satprop.m: properties of saturated H2O (vapor and liquid)

% Basic syntax: w = satH2Oprop(dtype, phH2O, dvalue)
% Inputs:
%          dtype: data type( 'P' = pressure, 'T' = temperature)
%          phH2O: phase of H2O ('V' = vapor, 'L' = liquid)
%          dvalue: value of pressure(kPa) or temperature(deg.C)
% Output:
%          w- structure of property values (T,P,H,S,U, or V) of saturated H2O
%          w.P: saturation pressure (kPa)
%          w.T: saturation temperature (deg.C)
%          w.H: specific enthalpy (kJ/kg)
%          w.S: specific entropy (kJ/(kg-K))
%          w.U: specific internal energy (kJ/kg)
%         w.V: specific volume (cm^3/g)

dtype = input('Data type (p(Pressure, kPa) or t(temperature, deg.C)) = ');
dvalue = input('Data value = ');
phH2O = input('Phase (V: vapor, L: liquid) = ');
w = satH2Oprop(dtype, phH2O, dvalue);
fprintf('\nSaturation pressure = %g kPa\n', w.P)
fprintf('Saturation temperature = %g deg.C\n', w.T)
fprintf('Specific enthalpy = %g kJ/kg\n', w.H)
fprintf('Specific entropy = %g kJ/(kg-K)\n', w.S)
fprintf('Specific internal energy = %g kJ/kg\n', w.U)
fprintf('Specific volume = %g cm^3/g\n', w.V)
```

Example 3.2 Physical Properties of the Saturated Liquid

Calculate the properties of the saturated liquid at $P = 10540$ kPa and those of saturated water vapor at $=198$ °C.

Solution
Direct call of the function *satH2Oprop* produces the following results:

```
>> x = satH2Oprop('P','L',10540)
x =
    struct with fields:
        P: 10540
        T: 314.8886
        H: 1.4310e+03
        S: 3.3983
        U: 1.4155e+03
        V: 1.4718
>> x = satH2Oprop('T','V',198)
x =
    struct with fields:
        T: 198
        P: 1.4907e+03
        H: 2.7908e+03
        S: 6.4453
        U: 2.5933e+03
        V: 132.4971
```

Or you can use the script *satprop* to get the desired results.

```
>> satprop
Data type (p(Pressure, kPa) or t(temperature, deg.C)) = 'p'
Phase (V: vapor, L: liquid) = 'l'
Data value = 10540

Saturation pressure = 10540 kPa
Saturation temperature = 314.889 deg.C
Specific enthalpy = 1430.96 kJ/kg
Specific entropy = 3.39835 kJ/(kg-K)
Specific internal energy = 1415.45 kJ/kg
Specific volume = 1.47182 cm^3/g

>> satprop
Data type (p(Pressure, kPa) or t(temperature, deg.C)) = 't'
Phase (V: vapor, L: liquid) = 'v'
Data value = 198

Saturation pressure = 1490.69 kPa
Saturation temperature = 198 deg.C
Specific enthalpy = 2790.82 kJ/kg
Specific entropy = 6.44527 kJ/(kg-K)
Specific internal energy = 2593.31 kJ/kg
Specific volume = 132.497 cm^3/g

Specific entropy = 6.44527 kJ/(kg-K)
```

```
Specific internal energy = 2593.31 kJ/kg
Specific volume = 132.497 cm^3/g
```

Example 3.3 Physical Properties of Saturated Steam

Calculate the specific volume and the enthalpy and entropy of saturated steam and liquid at 150°C. Compare the results with those recorded in steam tables that be found in standard thermodynamic texts.

Solution

Properties of saturated steam can be obtained from the script *satprop*, which calls the function *satH2Oprop*. Or you can directly call the function *satH2Oprop* to get the results. Outputs are summarized in Table 3.1.

```
>> x = satH2Oprop('T','V',150) % saturated vapor
x =
 struct with fields:
  T: 150
  P: 476.1014
  H: 2.7459e+03
  S: 6.8370
  U: 2.5590e+03
  V: 392.5024
>> x = satH2Oprop('T','L',150) % saturated liquid
x =
 struct with fields:
  T: 150
  P: 476.1014
  H: 632.2516
  S: 1.8420
  U: 631.7324
  V: 1.0905
```

3.2 HUMIDITY

3.2.1 RELATIVE HUMIDITY

Relative humidity (H_R) is defined as the ratio of the pressure exerted by water vapor to the vapor pressure of liquid water taken at that temperature. In general, relative humidity is represented in terms of percentage (%):

$$H_R = \frac{100 p_v}{P_w} (\%)$$

where
p_v is the partial pressure of water vapor

TABLE 3.1
Results of Physical Property Calculations of Saturated Steam

	Pressure	Liquid Volume	Steam Volume	Liquid Enthalpy	Steam Enthalpy	Liquid Entropy	Steam Entropy
	(kPa)	(cm^3/g)	(cm^3/g)	(kJ/Kg)	(kJ/Kg)	$(kJ/(Kg\cdot K))$	$(kJ/(Kg\cdot K))$
Equations	476.159	1.0905	392.50	632.25	2745.9	1.8420	6.8370
Steam table[3]	476	1.091	392.4	632.1	2745.4	1.8416	6.8358

P_w is the vapor pressure of water at the given temperature
This relation can also be expressed as

$$H_R = \frac{100 P_a}{P_s}$$

where
P_a is the actual vapor pressure
P_s is the saturation vapor pressure at the given temperature
At a given temperature $T\,(K)$, the saturated vapor pressure P_s can be approximated as

$$P_s = P_c \exp\left[\frac{T_c}{T}(a_1\tau + a_2\tau^{1.5} + a_3\tau^3 + a_4\tau^{3.5} + a_5\tau^4 + a_6\tau^{7.5})\right]$$

where $\tau = 1 - T/T_c$, P_c and T_c are critical pressure and critical temperature, respectively, and $a_i\,(i = 1,2,\cdots,6)$ are constants.

3.2.2 Absolute Humidity

Absolute humidity (H_A) is defined as the mass of water vapor per unit mass of vapor-free air. When the total pressure (sum of the partial pressure of vapor and air) is constant, absolute humidity depends solely on the pressure exerted by the water vapor in the vapor-air mixture:

$$H_A = \frac{M_w p_a}{M_a p_w} = \frac{M_w p_a}{M_a(P_T - p_a)}$$

where
M_a and M_w are the mol fractions of air and water vapor, respectively
p_a and p_w are the partial pressures of air and water vapor, respectively
P_T is the total pressure and can be taken as atmospheric pressure
H_A can be estimated by using the ideal gas equation of state. Since H_A can be represented as the density of water vapor in the air, we have from the ideal gas equation

$$H_A = \frac{p_a}{R_w T}$$

where $R_w = 461.5\,J/(kg{\cdot}K)$.

Calculation of absolute humidity is performed by the MATLAB function *humidest*. This function computes absolute humidity and the vapor pressure using relative humidity $H_R\,(\%)$ and the dry bulb temperature $T_d\,(°C)$. The basic function call is

```
hum = humidest(Hr,Td)
```

where Hr is relative humidity (%) and Td is the dry bulb temperature (°C). This function returns a structure that contains actual vapor pressure (hum.pres, *Pa*) and absolute humidity (hum.Ha, g/m^3).

```
function hum = humidest(Hr,Td)
% Hr: relative humidity
% Td: dry bulb temperature

% critical properties
Tc = 647.096; % critical temperature (K)
```

```
Pc = 22064000; % critical pressure (Pa)
% constants
Rw = 461.512244565;
a = [-7.85951783 1.84408259 -11.7866497 22.6807411 -15.9618719 1.80122502];
% temperature parameters
Td = Td+273.15; theta = Td/Tc; tau = 1 - theta;
% saturated pressure
tw = (Tc/Td)*(a(1)*tau + a(2)*tau^1.5 + a(3)*tau^3+a(4)*tau^3.5 + a(5)*tau^4 +
a(6)*tau^7.5);
Ps = Pc*exp(tw);
% results
hum.pres = (Hr/100)*Ps; % actual vapor pressure
hum.Ha = hum.pres*1000/((Td)*Rw); % absolute humidity
fprintf('Actual vapor pressure: %g Pa\n', hum.pres);
fprintf('Absolute humidity = %g g/m^3\n', hum.Ha);
```

Example 3.4 Humidity of Saturated Water Vapor

Calculate the actual vapor pressure (Pa) and absolute humidity (g/m^3 when the dry bulb temperature is 40 °C and relative humidity is 50%.

Solution

Call the function *humidest* with Hr = 50 and Td = 40:

```
>> hum = humidest(50,40);
Actual vapor pressure: 3692.56 Pa
Absolute humidity = 25.55 g/m^3
```

3.3 DENSITY OF SATURATED LIQUIDS

3.3.1 YAWS CORRELATION

Densities of saturated liquids at any given temperature can be given by the correlation[4]

$$\rho_L = AB^{-(1-T_r)^{2/7}}$$

where
  ρ_L is the saturated liquid density (g/cm^3)
  A and B are the correlation constants for the chemical compound being considered
  T_c is the critical temperature (°C)
  $T_r = T/T_c$ is the reduced temperature
  Table 3.2 lists typical components whose constants A and B and critical temperatures are known. Each component in Table 3.2 is assigned a specific identification number by the MATLAB function *compID*.

```
function ind = compID(cname)
% Assignment of identification number
cname = upper(cname);
switch cname
    case {'FLUORINE','F2'}, ind = 1;          case {'CHLORINE','CL2'}, ind = 2;
  case {'SULFUR DIOXIDE','SO2'}, ind = 3;     case {'CARBON MONOXIDE','CO'}, ind = 4;
```

TABLE 3.2

Typical Chemical Compounds with Known Constants A and B

1	Fluorine(F_2)	10	Hydrogen(H_2)	19	Aniline(C_6H_7N)
2	Chlorine(Cl_2)	11	Nitrogen(N_2)	20	Phenol(C_6H_6O)
3	Sulfur dioxide(SO_2)	12	Oxygen(O_2)	21	Cyclopropane(C_3H_6)
4	Carbon monoxide(CO)	13	Ethylene(C_2H_4)	22	Cyclohexane(C_6H_{12})
5	Carbon dioxide(CO_2)	14	Methane(CH_4)	23	1,3 Butadiene(C_4H_6)
6	Hydrogen chloride(HCl)	15	Ethane(C_2H_6)	24	Methanol(CH_4O)
7	Ammonia(NH_3)	16	Propane(C_3H_8)	25	Chloroform($CHCl_3$)
8	Water(H_2O)	17	Benzene(C_6H_6)	26	Carbon
9	Hydrogen peroxide(H_2O_2)	18	Toluene(C_7H_8)		tetrachloride(CCl_4)

Source: C.L. Yaws et al., *Physical Properties*, A Chemical Engineering publication, McGraw-Hill New York, 1977, p.235.

```
    case {'CARBON DIOXIDE','CO2'}, ind = 5;    case {'HYDROGEN CHLORIDE','HCL'}, ind = 6;
    case {'AMMONIA','NH3'}, ind = 7;           case {'WATER','H2O'}, ind = 8;
    case {'HYDROGEN PEROXIDE','H2O2'}, ind = 9; case {'HYDROGEN','H2'}, ind = 10;
    case {'NITROGEN','N2'}, ind = 11;          case {'OXYGEN','O2'}, ind = 12;
    case {'ETHYLENE','C2H4'}, ind = 13;        case {'METHANE','CH4'}, ind = 14;
    case {'ETHANE','C2H6'}, ind = 15;          case {'PROPANE','C3H8'}, ind = 16;
    case {'BENZENE','C6H6'}, ind = 17;         case {'TOLUENE','C7H8'}, ind = 18;
    case {'ANILINE','C6H7N'}, ind = 19;        case {'PHENOL','C6H6O'}, ind = 20;
    case {'CYCLOPROPANE','C3H6'}, ind = 21;    case {'CYCLOHEXANE','C6H12'}, ind = 22;
    case {'1,3 BUTADIENE','C4H6'}, ind = 23;   case {'METHANOL','CH4O'}, ind = 24;
    case {'CHLOROFORM','CHCL3'}, ind = 25;
    case {'CARBON TETRACHLORIDE','CCL4'}, ind = 26;
end
```

The MATLAB function *denL* calculates the density of saturated liquids at a given temperature. This function specifies constants A and B and critical temperatures for the compounds listed in Table 3.3. This function has the general syntax

```
rhoL = denL(T, cname)
```

where T is a given temperature (a scalar) or temperature range (a row vector) and cname is the name of the component, which can be specified either as a chemical formula (such as C6H6O) or a common name (such as Phenol). The parameters of this function are not case sensitive. This function calls the function *compID* to assign a unique identification number to the compound cname.

```
function rhoL = denL(T,cname)
% density of saturated liquid
% input
%    T: temperature(C) (scalar or row vector)
%    Tc: critical temperature(C)
%    cname: name or chemical formula of the compound
% output
%    rhoL: density of saturated liquid(g/cm^3)
% coefficients A and B of the density correlation equation
coefAB = [0.5649   0.2828   -129.0000;   0.5615   0.2720   144.0000;
```

TABLE 3.3
Vapor Pressure of Chloroform

$T\,(°C)$	$P_v\,(mmHg)$	$T\,(°C)$	$P_v\,(mmHg)$
0.001	65.4	80	1317.7
10	101.0	90	1733.9
20	159.3	100	2243.7
30	242.6	110	2859.2
40	357.9	120	3593.0
50	513.6	130	4457.9
60	718.7	140	5466.4
70	983.2	150	6631.0

```
        0.5164  0.2554   157.6000;    0.2931  0.2706  -140.1000;
        0.4576  0.2590    31.1000;    0.4183  0.2619   51.5000;
        0.2312  0.2471   132.4000;    0.3471  0.2740  374.2000;
        0.0     0.0        0.0 % missing component (H2O2)
        0.0315  0.3473  -240.2000;    0.3026  0.2763  -146.8000;
        0.4227  0.2797  -118.5000;    0.2118  0.2784    9.9000;
        0.1611  0.2877   -82.6000;    0.2202  0.3041   32.3000;
        0.2204  0.2753    96.7000;    0.3051  0.2714  288.9400;
        0.2883  0.2624   318.8000;    0.3392  0.2761  426.0000;
        0.4094  0.3246   420.0000;    0.2614  0.2826  124.9000;
        0.2729  0.2727   280.3000;    0.2444  0.2710  152.0000;
        0.2928  0.2760   239.4000;    0.5165  0.2666  263.4000;
        0.5591  0.2736   283.2000];
ind = compID(cname); % identification number of the component
% compute density
Tr = T./coefAB(ind,3); epn = -(1 - Tr).^(2/7); rhoL = coefAB(ind,1) * coefAB
(ind,2).^epn;
fprintf('Density = %g g/cm^3\n', rhoL)
end
```

Example 3.5 Density of Water

Calculate the density of saturated water at 150 °C.

Solution

```
>> rw = denL(150,'Water');
Density = 1.06207 g/cm^3
```

3.3.2 COSTALD Method

For the estimating liquid densities, the COSTALD method can be used.[5] This method can be applied for T_r within the range $0.25 < T_r < 1$.

$$v_s = \frac{1}{\rho} = v_c \cdot V_R^{(0)}\left(1 - \omega V_R^{(\delta)}\right)$$

$$V_R^{(0)} = 1 + a(1 - T_r)^{1/3} + b(1 - T_r)^{2/3} + c(1 - T_r) + d(1 - T_r)^{4/3}$$

$$V_R^{(\delta)} = \frac{e + fT_r + gT_r^2 + hT_r^3}{T_r - 1.00001}$$

where
    $a = -1.52816$, $b = 1.43907$, $c = -0.81446$, $d = 0.190454$
    $e = -0.296123$, $f = 0.386914$, $g = -0.0427458$, $h = -0.0480645$
    v_s is the specific volume (cm^3/mol)
    ω is the acentric factor
    v_c is the critical volume (cm^3/mol)
The function *costald* performs the COSTALD method. The basic calling syntax is

```
rhoL = costald(T,Tc,vc,w,Mw)
```

where T is the given temperature, Tc is the critical temperature, vc is the critical volume, w is the acentric factor, Mw is the molecular weight (g/mol), and rhoL is the estimated density (kg/m^3).

```
function rhoL = costald(T,Tc,vc,w,Mw)
% estimation of liquid density by COSTALD method
% input:
%   T,Tc: temperature and critical temperature (K)
%   vc: critical volume
%   w: acentric factor
%   Mw: molecular weight (g/mol)
% output
%   rhoL: estimated density (kg/m^3)
a = -1.52816; b = 1.43907; c = -0.81446; d = 0.190454;
e = -0.296123; f = 0.386914; g = -0.0427458; h = -0.0480645;
Tr = T./Tc; Vr0 = 1+a.*(1-Tr).^(1/3)+b.*(1-Tr).^(2/3)+c.*(1-Tr)+d.*(1-Tr).^
(4/3);
Vrd = (e + f*Tr + g*Tr.^2 + h*Tr.^3)./(Tr - 1.00001);
spV = vc.*Vr0.*(1 - w.*Vrd); % specific volume(cm^3/mol)
rhoL = 1000*Mw/spV; % density(kg/m^3)
fprintf('Density = %g g/cm^3\n', rhoL)
end
```

Example 3.6 Density of *n*-Hexane[6]

Estimate the liquid density of *n*-hexane at $T = 293.15\,K$ using the COSTALD method. For n $-$ hexane, $T_c = 507.8\,K$, $v_c = 386.8\ cm^3/mol$, $\omega = 0.3002$, and $M_w = 86.178\ g/mol$.

Solution

```
>> T=293.15; Tc=507.8; vc=386.8; w=0.3002; Mw = 86.178; rhoL = costald
(T,Tc,vc,w,Mw);
Density = 629.643 g/cm^3
```

The experimental value is $659.4\ kg/m^3$.

3.3.3 Gunn-Yamada Method[7]

For saturated liquid molar volume, the Gunn-Yamada method can be used:

$$\frac{V}{V_{sc}} = V_r^0(1 - \omega\Gamma)$$

where V is the liquid specific volume and V_{sc} is the scaling parameter defined in terms of the volume at $T_r = T/T_c = 0.6$:

$$V_{sc} = \frac{V_{0.6}}{0.3862 - 0.0866}$$

where $V_{0.6}$ denotes the saturated liquid molar volume at $T_r = 0.6$. If $V_{0.6}$ is not available, V_{sc} can be estimated by

$$V_{sc} = \frac{RT_c}{P_c}(0.2920 - 0.0967\omega)$$

where ω is the acentric factor. If the saturated liquid molar volume is available at any temperature, V_{sc} can be eliminated and V_r^0 and Γ can be expressed as a function of the reduced temperature T_r:

$$V_r^0 = \begin{cases} 0.33593 - 0.33953T_r + 1.51941T_r^2 - 2.02512T_r^3 \\ \quad + 1.11422T_r^4 \end{cases} : 0.2 \le T_r \le 0.8$$
$$\begin{cases} 1 + 1.3(1 - T_r)^{1/2}\log(1 - T_r) - 0.50879(1 - T_r) \\ \quad - 0.91534(1 - T_r)^2 \end{cases} : 0.8 < T_r < 1.0$$

$$\Gamma = 0.29607 - 0.09045T_r - 0.04842T_r^2 : 0.2 \le T_r < 1.0$$

The density is given by

$$\rho = \frac{M_w}{V}$$

where M_w is the molecular weight. The density of a mixture of n components, $\bar{\rho}$, can be obtained from the densities of pure components:

$$\frac{1}{\bar{\rho}} = \sum_{j=1}^{n} \frac{x_j}{\rho_j}$$

The function *gunyam* performs the Gunn and Yamada method. The basic calling syntax is
rhoL = gunyam(Tc,Pc,w,Mw,T)

where T(°C) is the given temperature, Tc(°C) is the critical temperature, w is the acentric factor, Mw (g/mol) is the molecular weight, and rhoL (kg/m^3) is the estimated density.

```
function rhoL = gunyam(Tc,Pc,w,Mw,T)
% saturated liquid molar volume by the Gunn and Yamada method
% Tc: deg,C, Pc: bar, Mw: g/mol
R = 0.08314; % liter*atm/(mol*K)
Tr = (T+273.15)/(Tc+273.15);
    if Tr <= 0.8
        Vr0 = 0.33593 - 0.33953*Tr + 1.51941*Tr^2 - 2.02512*Tr^3 + 1.11422*Tr^4;
    else
```

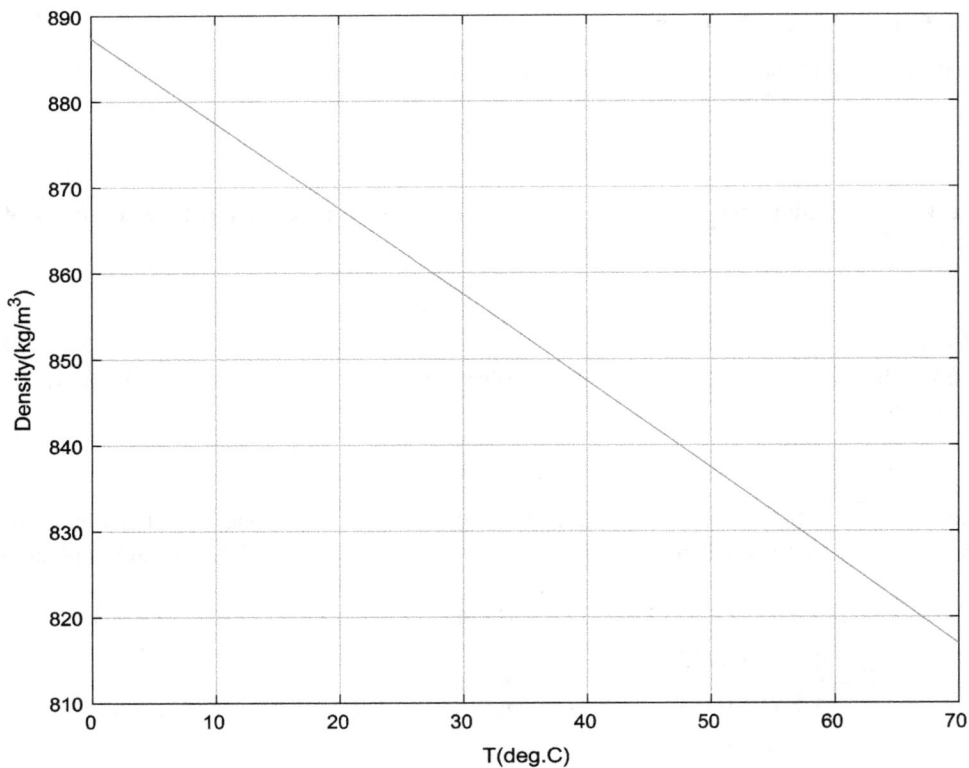

FIGURE 3.2   Plot of density of liquid benzene at 1 atm.

```
    Vr0 = 1 + 1.3*sqrt(1-Tr)*log10(1-Tr) - 0.50879*(1-Tr) - 0.91534*(1-Tr)^2;
  end
Gam = 0.29607 - 0.09045*Tr - 0.04842*Tr^2; Vsc = (R*(Tc+273.15)/Pc)*(0.2920 - 0.0967*w);
V = Vsc*Vr0.*(1 - w*Gam); rhoL = Mw./V; % density: kg/m^3
end
```

Example 3.7 Density of Benzene[8]

Plot the density of benzene as a function of temperature T at 1 atm $(0 \le T \le 70)$. For benzene, $T_c = 288.93°C$, $P_c = 49.24\ bar$, $\omega = 0.212$ and $M_w = 78$, and the gas constant $R = 0.08314\ l{\cdot}atm/(mol{\cdot}K)$.

Solution

The script *denlqbz* uses the function *gunyam* to calculate the density of the liquid benzene. The resultant plot of the density as a function of temperature is shown in Figure 3.2.

```
% denlqbz.m: density of liquid benzene
Tc = 288.93; Pc = 49.24; w = 0.212; % Tc: deg.C, Pc: bar
Mw = 78; % molecular weight (g/mol)
T = 0:70; n = length(T);
for k = 1:n
   denBz(k) = gunyam(Tc,Pc,w,Mw,T(k)); % density: kg/m^3
end
plot(T,denBz), grid, xlabel('T(deg.C)'), ylabel('Density(kg/m^3)')
```

3.4 VISCOSITY

3.4.1 VISCOSITY OF LIQUIDS

The viscosity of a saturated liquid can be expressed as a function of temperature:

$$\log \mu_L = A + \frac{B}{T} + CT + DT^2$$

where

μ_L is the viscosity of a saturated liquid (cP)

A, B, C, and D are correlation constants

T is temperature (K)

Viscosities of liquids decrease with temperature, and this variation is almost linear over a wide range of temperatures from the freezing point to the boiling point. This phenomenon can be represented by the Andrade correlation[9]:

$$\ln \mu_L = \ln A + \frac{B}{T}$$

The function *visL* calculates the viscosity of liquids. This function specifies constants A, B, C, and D for typical compounds listed in Table 3.2. This function can be used as follows:
mu = visL(T, cname)

where T is temperature (K); cname is the name of the component, which can be specified as chemical formula or common name; and mu is the resultant viscosity of the saturated liquid (cP). T can be a scalar or a row vector representing a temperature range. This function calls the function *compID* to assign an identification number to the compound cname.

```
function mu = visL(T,cname)
% Calculates viscosity of a saturated liquid (cP)
% input
%    T: temperature(K) (scalar or a row vector)
%    cname: common name or chemical formula of the compound
% output
%    mu: the viscosity of the saturated liquid (cP)
% correlation constants (A, B, C, D)
wv = 1.0e+03 * [-0.001576  0.085630  -0.0000004073  -0.000000002725
     -0.0007681  0.151400  -0.00000080650  0.000000000407500
     -0.0026700  0.406700  0.00000614100  -0.000000012540000
     -0.0023460  0.105200  0.00000461300  -0.000000019640000
     -0.0013450  0.021220  0.00001034000  -0.000000034050000
     -0.0015150  0.194600  0.00000306700  -0.000000013760000
     -0.0085910  0.876400  0.00002681000  -0.000000036120000
     -0.0107300  1.82800   0.00001966000  -0.000000014660000
     0.0         0.0       0.0            0.0   % missing component(H2O2)
     -0.0048570  0.025130  0.00014090000  -0.000002773000000
     -0.0121400  0.376100  0.00012000000  -0.000000470900000
     -0.0020720  0.093220  0.00000603100  -0.000000027210000
     -0.0077060  0.468100  0.00003725000  -0.000000076330000
     -0.0116700  0.499300  0.00008125000  -0.000000226300000
     -0.0044440  0.290100  0.00001905000  -0.000000041640000
```

```
       -0.0033720  0.313500  0.00001034000 -0.000000020260000
       0.0020030   0.064660 -0.00001105000 0.000000009648000
       0.0         0.0       0.0       0.0    % missing component (C6H6)
       0.0     0.0     0.0     0.0   % missing component (C6H6O)
       -0.0080390  1.889000  0.00001055000 -0.000000006718000
       -0.0013350  0.1162000 0.0000001108  -0.000000000038360
       -0.0019100  0.5992000 -0.0000006749  0.000000000502600
       -0.0026370  0.4345000 0.0000019370  -0.000000002907000
       -0.0026970  0.7009000 0.0000026820  -0.000000004917000
       -0.0018120  0.3975000 0.0000011740  -0.000000001784000
       -0.0056580  0.9945000 0.0000101600  -0.000000008733000];
ind = compID(cname); % identification number of the component
% compute viscosity
pn = wv(ind,1) + wv(ind,2)./T + wv(ind,3).*T + wv(ind,4).*T.^2; mu = 10.^pn;
fprintf('Viscosity = %g cP\n', mu)
end
```

Example 3.8 Viscosity of Phenol

Estimate the viscosity of liquid phenol at 150 °C.

Solution

```
>> T = 150+273.15; mu = visL(T,'C6H6O');
Viscosity = 0.485817 cP
```

3.4.2 VISCOSITY OF GASES

The viscosity of gases at low pressures can be expressed by[10]

$$\mu_G = 0.0001 \times (A + BT + CT^2)$$

where

μ_G is the viscosity of the gas at low pressure (cP)
T is temperature (K)
A, B, and C are correlation constants for a chemical compound

Viscosities of gas mixtures at low pressures can be estimated by the method of Wilke,[11] with Herning-Zipperer approximation[12]:

$$\mu_m = \frac{\sum y_i \mu_i (MW_i)^{0.5}}{\sum y_i (MW_i)^{0.5}}$$

where

μ_m is the gas mixture viscosity (cP)
y_i is the mole fraction of component i
μ_i is the gas viscosity of component i
MW_i is the molecular weight of component i

The function *visG* calculates the viscosity of gases listed in Table 3.2. This function specifies constants A, B, and C for typical compounds listed in Table 3.2. The basic calling syntax is
mu = visG(T, cname)

where T is temperature (K), cname is the name of the component (which can be specified as a chemical formula or common name), and mu is the resultant viscosity of the gas (*cP*). T can be a scalar or a row vector representing a temperature range. This function calls the function *compID* to assign an identification number to the compound cname.

```
function mu = visG(T,cname)
% Calculates viscosity of a gas (cP)
% input
% T: temperature(K) (scalar or a row vector)
% cname: common name or chemical formula of the compound
% output
% mu: the viscosity of the gas (cP)
% correlation constants (A, B, C)
cv = [22.0900     76.9000-211.6000; 5.1750  45.6900 -88.5400;
      -3.7930     46.4500 -72.7600; 32.2800  47.4700 -96.4800;
      25.4500     45.4900 -86.4900; -9.5540  54.4500 -96.5600;
      -9.3720     38.9900 -44.0500; -31.8900 41.4500  -8.2720;
       5.3810     28.9800  38.4000; 21.8700  22.2000 -37.5100;
      30.4300     49.8900-109.3000; 18.1100  66.3200-187.9000;
       3.5860     35.1300 -80.5500; 15.9600  34.3900 -81.4000;
       5.5760     30.6400 -53.0700; 4.9120   27.1200 -38.0600;
     -15.7600     32.4500 -72.3200; -8.4210  27.1100 -40.1800;
     -14.9800     29.0300  -1.1160; -16.4100 32.0000        0;
      -7.7870     34.7800 -81.3000; -4.7050  26.3200 -44.1000;
     -10.6700     34.3200 -80.8000; -5.6360  34.4500  -3.3400;
      -6.6880     37.2600 -50.8700; 5.6980   32.7300 -40.2800];
wv = [cv(:,1) cv(:,2)*1e-2 cv(:,3)*1e-6];
ind = compID(cname); % identification number of the component
% compute viscosity
pn = wv(ind,1) + wv(ind,2).*T + wv(ind,3).*T.^2; mu = pn*1e-4;
fprintf('Viscosity = %g cP\n', mu)
end
```

Example 3.9 Viscosity of Methane

Estimate the viscosity of methane gas at 100°C.

Solution

```
>> T = 100+273.15; mu = visG(T,'methane');
Viscosity = 0.0132952 cP
```

3.5 HEAT CAPACITY

3.5.1 HEAT CAPACITY OF LIQUIDS

3.5.1.1 Polynomial Correlation

A liquid's heat capacity can be expressed as polynomial of the form

$$C_p^L = A + BT + CT^2 + DT^3$$

where

C_p^L is the heat capacity of the saturated liquid ($cal/(g \cdot °C)$)

T is the temperature (K)

A, B, C, and D are correlation constants for a chemical compound

Liquid heat capacities are not strongly dependent upon temperatures except in the range of $0.7 \leq T_r \leq 0.8$, where T_r is the reduced temperature.[13] But at high T_r, liquid heat capacities become large and strongly dependent upon temperature. At the boiling point of most organic compounds, heat capacities are between 0.4 and 0.5 $cal/(g \cdot K)$.

The function *hcapL* calculates liquid heat capacities for the compounds listed in Table 3.2. This function specifies constants A, B, C, and D for typical compounds listed in Table 3.2. This function can be used as

```
hcp = hcapL(T,cname)
```

where T is the temperature (K); cname is the name of the component, which can be specified as a chemical formula or common name; and hcp is the resultant heat capacity of the liquid component ($cal/(g \cdot °C)$). T can be either a scalar or a row vector representing a temperature range. This function calls the function *compID* to assign an identification number to the compound cname.

```
function hcp = hcapL(T,cname)
% Calculates liquid heat capacity (cal/g/C)
% input
% T: temperature(K) (scalar or a row vector)
% cname: common name or chemical formula of the compound
% output
% hcp: liquid heat capacity(cal/g/C)

% correlation constants (A, B, C, D)
cv = 1.0e+04 * [-0.0000288   0.02528  -0.0331 0.1464
   -0.000013220   0.0004720  -0.0020370   0.002894000
   -0.000057370   0.0010340  -0.0040280   0.005285000
    0.000056450   0.0004798  -0.0143700   0.091195000
   -0.001930000   0.0254600  -0.1095500   0.157330000
   -0.000011210   0.0007048  -0.0035310   0.006621000
   -0.000192300   0.0031100  -0.0110900   0.013760000
    0.000067410   0.0002825  -0.0008371   0.000860100
    0.000044400   0.0001199  -0.0002738   0.000261500
    0.000379000  -0.0329800   1.2170900  -0.243480000
   -0.000106400   0.0059470  -0.0768700   0.335730000
   -0.000045870   0.0032340  -0.0395100   0.157570000
   -0.000034020   0.0006218  -0.0050120   0.012630000
    0.000123000  -0.0010330   0.0072000  -0.010730000
    0.000013880   0.0008481  -0.0056540   0.012610000
    0.000033260   0.0002332  -0.0013360   0.003016000
   -0.000148100   0.0015460  -0.0043700   0.004409000
   -0.000014610   0.0004584  -0.0013460   0.001425000
    0.000014070   0.0002467  -0.0006085   0.000592700
   -0.000068960   0.0008218  -0.0018420   0.001447000
   -0.000002618   0.0006913  -0.0034770   0.005990000
   -0.000128400   0.0013390  -0.0035100   0.003227000
    0.000037850   0.0001049  -0.0005761   0.001374000
    0.000083820  -0.0003231   0.0008296  -0.000016890
   -0.000009154   0.0003149  -0.0010640   0.001240000
   -0.000001228   0.0002058  -0.0007040   0.000861000];
wv = [cv(:,1) cv(:,2)*1e-3 cv(:,3)*1e-6 cv(:,4)*1e-9];
ind = compID(cname); % identification number of the component
```

```
% computes liquid heat capacity
hcp = wv(ind,1) + wv(ind,2).*T + wv(ind,3).*T.^2 + wv(ind,4).*T.^3;
fprintf(' Heat capacity = %g cal/g/C \n', hcp)
end
```

Example 3.10 Heat Capacity of Phenol

Estimate the heat capacity of liquid phenol at 150°C.

Solution
```
>> T = 150+273.15; hcp = hcapL(T, 'C6H6O');
Heat capacity = 0.585993 cal/g/C
```

3.5.1.2 ROWLINSON-BONDI METHOD

The Rowlinson-Bondi method can be used to estimate liquid heat capacities. This method is based on the ideal gas heat capacity and the corresponding-states principle[14]:

$$C_p^L = C_p^{id} + 1.45R + \frac{0.45R}{1 - T_r} + 0.25wR\left(17.11 + \frac{25.2(1 - T_r)^{1/3}}{T_r} + \frac{1.742}{1 - T_r}\right)$$

where C_p^{id} is the ideal gas heat capacity.

The function *hcRB* calculates liquid heat capacities using the Rowlinson-Bondi correlation method. This function can be invoked as

```
cpL = hcRB(T,Tc,w,Cpi)
```

where T is temperature (K), Tc is the critical temperature (K), w is the acentric factor, Cpi is the ideal gas heat capacity ($J/(mol \cdot K)$), and cpL is the estimated liquid heat capacity ($J/(mol \cdot K)$).

```
function cpL = hcRB(T,Tc,w,Cpi)
% Estimation of liquid heat capacity by Rowlinson/Bondi method
% input:
%   T,Tc: temperature and critical temperature (K)
%   w: acentric factor
%   Cpi: ideal gas heat capacity (J/mol/K)
% output
%   cpL: estimated liquid heat capacity (J/mol/K)
Tr = T./Tc; R = 8.3143;
cpL = Cpi + 1.45*R + 0.45*R./(1-Tr) + 0.25*w*R*(17.11 +...
   25.2*(1-Tr).^1/3 ./ Tr + 1.742./(1-Tr)); % (J/mol/K)
fprintf(' Heat capacity = %g J/mol/K \n', cpL)
end
```

Example 3.11 Heat Capacity of MEK

Use the Rowlinson-Bondi method to estimate the specific heat capacity of methyl ethyl ketone (MEK) at $T = 100°C$. Use $C_p^{id} = 1.671 \, J/(g \cdot K) = 120.496/(mol \cdot K)$, $T_c = 535.55 \, K$, $w = 0.323$, $M_w = 72.11 \, g/mol$, and $R = 8.3143 \, J/(mol \cdot K)$.

Solution

```
>> T = 373.15; Tc = 535.55; w = 0.323; Cpi = 120.496; cpL = hcRB(T,Tc,w,Cpi);
Heat capacity = 162.689 J/mol/K
```

3.5.2 Heat Capacity of Gases

The heat capacity of an ideal gas C_p^0 at low pressure can be expressed as a 3rd-degree polynomial function of temperature:

$$C_p^0 = A + BT + CT^2 + DT^3$$

where

C_p^0 is the heat capacity of the ideal gas at low pressure $(cal/(gmol \cdot K))$
T is the temperature (K)
A, B, C, and D are correlation constants

The function *hcapG* calculates heat capacities of gases listed in Table 3.2. This function defines constants A, B, C, and D in the range of $298\,K \le T \le 1500\,K$. This function can be used as
hcp = hcapG(T, cname)

where T is temperature (K); cname is the name of the component, which can be specified as a chemical formula or common name; and hcp is the resultant heat capacity of the gas $(cal/(gmol \cdot K))$. T can be either a scalar or a row vector representing a temperature range.

```
function hcp = hcapG(T,cname)
% Calculates gas heat capacity (cal/gmol/K)
% input
%   T: temperature(K) (scalar or a row vector)
%   cname: common name or chemical formula of the compound
% output
% hcp: gas heat capacity (cal/gmol/K)
% correlation constants (A, B, C, D)
cv = [5.8900    6.9400   -5.4800 1.5200; 7.0000    5.0500 -4.3900     1.3000;
    5.8500   15.4000 -11.1000  2.9100; 6.9200   -0.6500  2.8000    -1.1400;
    5.1400   15.4000  -9.9400  2.4200; 7.2400   -1.7600  3.0700    -1.0000;
    6.0700    8.2300   -0.1600 -0.6600; 8.1000   -0.7200  3.6300    -1.1600;
    5.5200   19.8000 -13.9000  3.7400; 6.8800   -0.0220  0.2100     0.1300;
    7.0700   -1.3200   3.3100  -1.2600; 6.2200    2.7100 -0.3700    -0.2200;
    0.9340   36.9000 -19.3000  4.0100; 5.0400    9.3200  8.8700    -5.3700;
    2.4600   36.1000  -7.0000  -0.4600; -0.5800  69.9000 -32.9000   6.5400;
   -8.7900  116.0000 -76.1000 18.9000; -9.3400 138.5000 -87.2000  20.6000;
   -8.1100  143.4000 -107.0000 30.7000; -5.6800 126.0000 -85.4000  20.5000;
   -6.2800   79.8000 -50.5000 12.2000;
    0.0       0.0       0.0      0.0; % missing component (C6H12)
   -0.5600   81.1000 -53.5000 13.6000; 3.6200   24.9000 -7.0500     0;
    7.7100   34.3000 -26.4000  7.2900; 12.6000  33.2000 -29.3000   8.5800];
wv = [cv(:,1) cv(:,2)*1e-3 cv(:,3)*1e-6 cv(:,4)*1e-9];
ind = compID(cname); % identification number of the component
% computes gas heat capacity
hcp = wv(ind,1) + wv(ind,2).*T + wv(ind,3).*T.^2 + wv(ind,4).*T.^3;
fprintf(' Heat capacity = %g cal/gmol/K\n', hcp)
end
```

Example 3.12 Heat Capacity of Carbon Dioxide

Estimate the heat capacity of carbon dioxide at 300 °C.

Solution

```
>> T = 300+273.15; v = hcapG(T,'CO2');
Heat capacity = 11.1568 cal/gmol/K
```

3.6 THERMAL CONDUCTIVITY

3.6.1 THERMAL CONDUCTIVITY OF LIQUIDS

The thermal conductivity of saturated liquids can be estimated by

$$k_L = A + BT + CT^2$$

where

k_L is the thermal conductivity of a saturated liquid ($\mu cal/(sec{\cdot}cm{\cdot}°C)$)

T is the temperature (K)

A, B, and C are correlation constants

Values of k_L for most common organic liquids range between 250 and 400 $\mu cal/(sec{\cdot}cm{\cdot}°C)$ at temperatures below the boiling point.

The function *condL* calculates thermal conductivities of liquid components listed in Table 3.2. This function defines constants A, B, and C. This function has the syntax

```
k = condL(T,cname)
```

where T is the temperature (K); cname is the name of the component, which can be specified as a chemical formula or common name; and k is the estimated thermal conductivity of the specified liquid ($\mu cal/(sec{\cdot}cm{\cdot}°C)$). T can be either a scalar or a row vector representing a temperature range.

```
function k = condG(T,cname)
% Calculates thermal conductivities of gases (microcal/cm/sec/K)
% input
% T: temperature(K) (scalar or a row vector)
% cname: common name or chemical formula of the compound
% output
% k: liquid thermal conductivity(microcal/cm/sec/C)

% correlation constants (A, B, C)
cv = 1000[0.6215   -0.1623  -0.1184 0.5590  -0.0483   -0.0152
  2.1408  -0.7837   0.0714; 0.4755   0.0033  -0.2143;
  0.9721  -0.2015  -0.0230; 1.0717  -0.0184  -0.0658;
  2.5513  -0.3766  -0.0294; -0.9166   1.2547  -0.1521;
 -0.4666   0.8059  -0.0876; -0.0201   2.4737  -5.3473;
  0.6280  -0.3689  -0.0226; 0.5838  -0.2105  -0.0483;
  0.8515  -0.2289  -0.0047; 0.7227  -0.1444  -0.0764;
  0.6993  -0.1659  -0.0049; 0.6235  -0.1268  -0.0021;
  0.4243   0.0011  -0.0090; 0.4851  -0.0538  -0.0006;
  0.5375  -0.0304  -0.0015; 0.4409   0.0867  -0.0070;
  0.3968  -0.0421  -0.0067; 0.3883  -0.0227  -0.0033;
  0.7183  -0.1872   0.0117; 0.7701  -0.1142   0.0028;
  0.3902  -0.0206  -0.0051];
wv = [cv(:,1) cv(:,2)*1e-2 cv(:,3)*1e-4];
ind = compID(cname); % identification number of the component
```

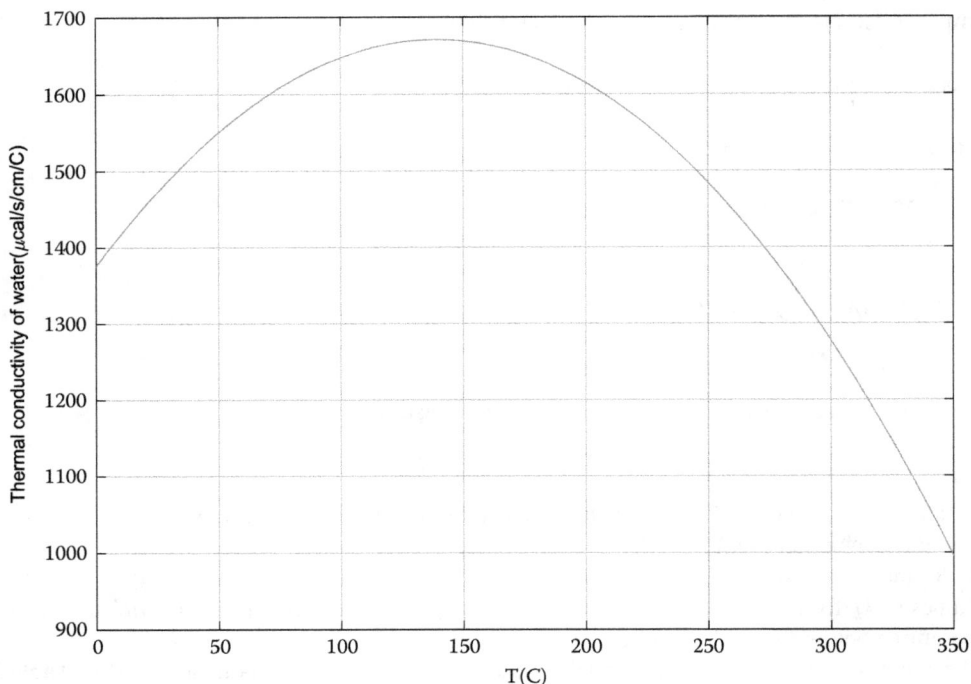

FIGURE 3.3   Plot of the thermal conductivity of water.

```
% Computes liquid thermal conductivity
k = wv(ind,1) + wv(ind,2).*T + wv(ind,3).*T.^2;
fprintf(' Thermal conductivity = %g microcal/cm/sec/C \n', k)
end
```

Example 3.13 Thermal Conductivity of Water

Calculate the thermal conductivity of water at the range of $0°C \leq T \leq 350°C$, and plot the results as a function of temperature.

Solution

The following commands compute the thermal conductivity and plot the results as shown in Figure 3.3:

```
>> T = [0:350]+273.15; k = condL(T,'water'); plot(T-273.15, k), xlabel('T(C)')
>> ylabel('Thermal conductivity of water(\mucal/s/cm/C)'), grid on
```

3.6.2 Thermal Conductivity of Gases

The thermal conductivities of low-pressure gases increase with temperature. The thermal conductivity k_G of a gas can be expressed as a function of temperature as

$$k_G = A + BT + CT^2 + DT^3$$

where

k_G is the thermal conductivity of the gas at low pressure $(\mu cal/(sec\cdot cm\cdot K))$
T is the temperature (K)
A, B, C, and D are correlation constants
    The function *condG* calculates thermal conductivities of gas components listed in Table 3.2. This function defines constants A, B, C, and D. A simple representation of its syntax is
k = condG(T,cname)

where T is temperature (K); cname is the name of the gas component, which can be specified as a chemical formula or common name; and k is the estimated thermal conductivity of the specified gas $(\mu cal/(sec\cdot cm\cdot K))$. T can be either a scalar or a row vector repr senting a temperature range.

```
function k = condG(T,cname)
% Calculates thermal conductivities of gases (microcal/cm/sec/K)
% input
%     T: temperature(K) (scalar or a row vector)
%     cname: common name or chemical formula of the compound
% output
%     k: bas thermal conductivity (microcal/cm/sec/K)

% correlation constants (A, B, C, D)
cv = [1.8654   19.7900    1.2400 -17.7700; 3.2500    5.8000    0.2100  -1.2500;
     -19.3100   15.1500   -0.3300   0.5500; 1.2100   21.7900   -0.8416   1.9580;
     -17.2300   19.1400    0.1308  -2.5140; -0.2600   12.6700   -0.2500   0.1600;
       0.9100   12.8700    2.9300  -8.6800; 17.5300   -2.4200    4.3000 -21.7300;
     -21.0700   16.9700    0.1700  -1.5600; 19.3400  159.7400   -9.9300  37.2900;
       0.9359   23.4400   -1.2100   3.5910; -0.7816   23.8000   -0.8939   2.3240;
     -42.0400   28.6500    0.7963  -3.2620; -4.4630   20.8400    2.8150  -8.6310;
     -75.8000   52.5700   -4.5930  39.7400; 4.4380   -1.1220    5.1980 -20.0800;
     -20.1900    8.6400    2.3400  -9.6900; 18.1400   -9.5700    5.6600 -22.2200;
     -26.3900   11.8900    1.5500  -4.3000; -31.8700   15.2600    1.7400  -4.4000;
     -20.4600    9.7400    3.7700 -16.2800; -20.5700    4.4500    4.0700 -17.3100;
     -67.9200   29.9600    1.7400 -12.2000; -18.6200    9.9500    2.9000 -12.3800;
      -5.7320    6.2900    0.5904  -3.3520; -0. 4161    4.0670    0.6115  -3.5660];
wv = [cv(:,1) cv(:,2)*1e-2 cv(:,3)*1e-4 cv(:,4)*1e-8];
ind = compID(cname); % % identification number of the component
% Computes gas thermal conductivity
k = wv(ind,1) + wv(ind,2)*T + wv(ind,3)*T.^2 + wv(ind,4)*T.^3;
fprintf(' Thermal conductivity = %g microcal/cm/sec/K \n', k)
end
```

Example 3.14 Thermal Conductivity of Propane

Estimate the thermal conductivity of propane gas at the range of $25°C \leq T \leq 900°C$, and plot the results as a function of temperature.

Solution
    The following commands generate a plot of the results as shown in Figure 3.4.

```
>> T = [25:900]+273.15; k = condG(T,'propane'); plot(T-273.15, k), xlabel
('T(C)'), axis tight, grid on
>> ylabel('Thermal conductivity of propane(\mucal/s/cm/K)')
```

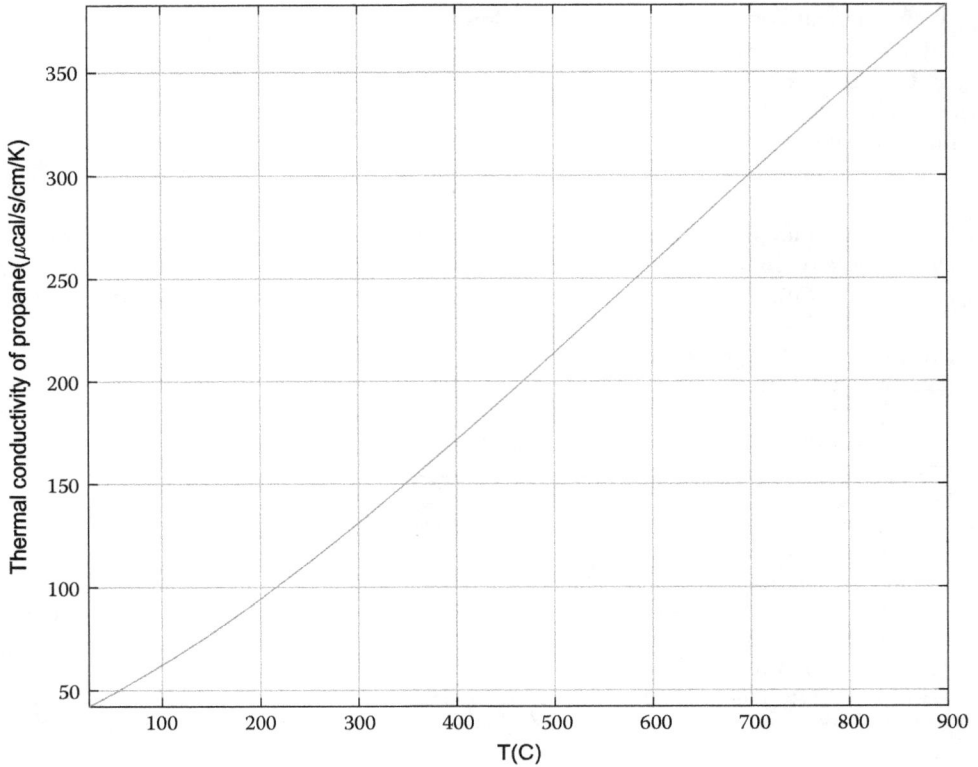

FIGURE 3.4   Plot of the thermal conductivity of propane.

3.7 SURFACE TENSION

3.7.1 SURFACE TENSION OF LIQUIDS

The surface tension of liquids can be expressed as[15]

$$\sigma = \sigma_1 \left(\frac{T_c - T}{T_c - T_1} \right)^r$$

$$\frac{\sigma}{\sigma_1} = \left(\frac{T_c - T}{T_c - T_1} \right)^r$$

where
     σ_1 is the surface tension ($dyne/cm$) at $T_1(K)$
     T_c is the critical temperature (K)
     r is a correlation parameter
     For water, the temperature range for which the surface tension is valid is

$$0°C\!\sim\!100°C : T_1 = 298.16\,K,\ r = 0.8105,\ \sigma_1 = 71.97$$

$$100°C\!\sim\!374.2°C : T_1 = 373.16\,K,\ r = 1.1690,\ \sigma_1 = 58.91$$

The function *surtL* calculates surface tensions of components listed in Table 3.2. This function defines the value of r. Its basic syntax is

```
sg = surtL(T,cname)
```

where T is temperature (°C); cname is the name of the component, which can be specified as a chemical formula or common name; and sg is the ratio of the estimated surface tension to the surface tension at T_1. If cname is water, sg is the value of the estimated surface tension (*dyne/cm*).

```
function sg = surtL(T,cname)
% Calculates liquid surface tension (dyne/cm)
% input
% T: temperature(C) (scalar or a row vector)
% cname: common name or chemical formula of the compound
% output
% sg: ratio of surface tension at T to that at T (sg = sigma/sigma1)

% data (T1, Tc) and correlation constant (r)
wv = [ -200.0000 -129.0000 0.8811; 20.0000 144.0000 1.0508; 30.0000 157.6000 1.1768;
 -193.0000 -140.1000 1.1441; 20.0000 31.1000 1.3015; -93.0000 51.5000 1.0972;
 -45.0000 132.4000 1.1548; 25.0000 374.2000 0.8105; 18.2000 455.0000 0.9141;
 -256.0000 -240.2000 1.1012; -203.0000 -146.8000 1.2123; -202.0000 -118.5000 1.1933;
 -120.0000 9.9000 1.2760; -168.1600 -82.6000 1.3941; -120.0000 32.3000 1.2060;
 -90.0000 96.7000 1.1982; 20.0000 288.9400 1.2243; 20.0000 318.8000 1.2364;
 20.0000 426.0000 1.1022; 60.0000 420.0000 1.0725; 10.0000 124.9000 1.3201;
 20.0000 280.3000 1.4246; 40.0000 152.0000 1.2055; 20.0000 239.4000 0.8115;
 25.0000 263.4000 1.1824; 30.0000 283.2000 1.2278];
ind = compID(cname); % identification number of the component
% compute ratio of surface tension
T0 = 273.15; T = T + T0; T1 = wv(ind,1) + T0; Tc = wv(ind,2) + T0;
if ind ~= 8
     sg = ((Tc - T)./(Tc-T1)).^(wv(ind,3));
else % ind=8: water
     sg = [];
     for j = 1:length(T)
          if T(j) >= T0 && T(j)-T0 <= 100+T0 % 0<=T<=100 (C)
               sg1 = 71.97;
               sgv = sg1*((wv(ind,2) + T0 - T(j))./(wv(ind,2) +T0-(wv(ind,1)
               + T0))).^(wv(ind,3));
          else % 100 <= T(C) <= 374.2
               wv(ind,:) = [100 374.2 1.169]; sg1 = 58.91;
               sgv = sg1*((wv(ind,2) + T0 - T(j))./(wv(ind,2) +T0-(wv(ind,1)
               + T0))).^(wv(ind,3));
               end
               sg = [sg sgv];
          end
     end
fprintf('Ratio of surface tension(sigma/sigma1) at T = %g C is %g\n', T-T0, sg);
end
```

Example 3.15 Surface Tension of Benzene

Estimate the ratio of the surface tension of benzene at $T = 60°C$.

Solution

```
surtL(60,'benzene');
Ratio of surface tension(sigma/sigma1) at T = 60 C is 0.82107
```

3.7.2 SURFACE TENSION BY CORRELATIONS

For cryogenic liquids, the surface tension can be estimated by the equation proposed by Sprow and Prausnitz[16]:

$$\sigma = \sigma_0(1 - T_r)^p$$

In this equation, σ_0 and p can be determined by the least-squares analysis of the measured data.

For nonpolar liquids, the corresponding-states correlation can be used to estimate surface tension[17]:

$$\sigma = P_c^{2/3} T_C^{1/3} Q(1 - T_r)^{11/9}$$

$$Q = 0.1196\left(1 + \frac{T_{br}\ln(P_c/1.03125)}{1 - T_{br}}\right) - 0.279$$

where

P_c is the critical pressure (*bar*)

T_c is the critical temperature (*K*)

T_b is the normal boiling point (*K*), and $T_{br} = T_b/T_c$

Pitzer proposed a series of relations for σ in terms of P_c, T_c and ω that together lead to the following corresponding-states relation for σ[18]:

$$\sigma = P_c^{2/3} T_C^{1/3} \frac{1.86 + 1.18\omega}{19.05}\left(\frac{3.75 + 0.91\omega}{0.291 - 0.08\omega}\right)^{2/3}(1 - T_r)^{11/9}$$

where ω is the acentric factor.

The function *corstsg* calculates the surface tension using the corresponding-states correlation. The basic syntax is

```
sg = corstsg(Pc,Tc,Tb,T)
```

where Pc is critical pressure (*bar*), Tc is critical temperature (*K*), Tb is the normal boiling point (K), T is temperature (*K*), and sg is the estimated surface tension (*dyne/cm*).

```
function sg = corstsg(Pc,Tc,Tb,T)
% Estimation of surface tension using the corresponding states correlation
% input
% Pc: critical pressure (bar)
% Tc: critical temperature (K)
% Tb: normal boiling point (K)
% T: temperature (K)
% output
% sg: surface tension (dyne/cm)
Tr = T./Tc; Tbr = Tb./Tc; Q = 0.1196*(1 + Tbr.*log(Pc/1.01325)./(1-Tbr))
- 0.279;
sg = Pc.^(2/3).*Tc.^(1/3).*Q.*(1-Tr).^(11/9);
fprintf('Surface tension = %g dyne/cm \n', sg);
end
```

The function *pitzersg* uses the Pitzer relation to estimate the surface tension. The basic syntax is
sg = pitzersg(w,Pc,Tc,T)

where w is the acentric factor, Pc is the critical pressure (*bar*), Tc is the critical temperature (*K*), T is the temperature (*K*), and sg is the estimated surface tension (*dyne/cm*).

```
function sg = pitzersg(w,Pc,Tc,T)
% Estimation of surface tension using Pitzer's relation
% input
% Pc: critical pressure (bar)
% Tc: critical temperature (K)
% w: acentric factor
% T: temperature (K)
% output
% sg: surface tension (dyne/cm)

Tr = T./Tc;
sg = (Pc.^(2/3)).*(Tc.^(1/3)).*((1.86+1.18*w)/19.05).*((3.75+...
  0.91*w)./(0.291-0.08*w)).^(2/3).*(1-Tr).^(11/9);
fprintf('Surface tension = %g dyne/cm \n', sg);
end
```

Example 3.16 Surface Tension of Ethanethiol

Use the corresponding-states correlation to estimate the surface tension of ethanethiol (ethyl mercaptan) at 30°C. For ethanethiol, $P_c = 54.9\,bar$, $T_c = 499\,K$, and $T_b = 308.15\,K$.

Solution
The function *corstsg* is used to evaluate the surface tension:

```
>> T = 303.15; Tc = 499; Tb = 308.15; Pc = 54.9; sg = corstsg(Pc,Tc,Tb,T);
Surface tension = 22.3398 dyne/cm
```

3.8 VAPOR PRESSURE

3.8.1 ANTOINE EQUATION

The Antoine equation is often used to estimate the vapor pressure P_v:

$$\ln P_v\,(mmHg) = A - \frac{B}{T\,(°C) + C}$$

where A, B, and C are Antoine parameters. This equation is applicable for pressures with ranges from 10 to 1500 *mmHg*. Multiplying both sides by $(T + C)$ to convert into a linear form, then rearranging, we have

$$y = a_1 + a_2 x_1 + a_3 x_2$$

where
  $y = \ln P_v$, $x_1 = \frac{1}{T}$, $x_2 = \frac{\ln P_v}{T}$, $a_1 = A$, $a_2 = AC - B$, $a_3 = -C$
  The function *prVp* calculates the vapor pressure of components listed in Table 3.2. This function defines correlation constants A, B, C, D, and E. The basic syntax is
pv = prVp(T,cname)

where T is the temperature (°C); cname is the name of the component, which can be specified as either a chemical formula or a common name; and pv is the estimated vapor pressure (*mmHg*) at T.

```
function pv = prVp(T,cname)
% Estimation of vapor pressure of the saturated liquid (mmHg)
% input
% T: temperature(C) (scalar or a row vector)
% cname: common name or chemical formula of the compound
% output
% pv: vapor pressure(mmHg) at T
% identification number of the component
ind = compID(cname);
% correlation constants (A, B, C, D, E)
cv = 1.0e+04 * [
    0.0021480 -0.051651 -0.00071218 0.00143550 0; 0.0042262 -0.20098 -0.001396300
    0.00093705 0;
    0.0046554 -0.24563 -0.001516900 0.00090026 0; 0.0032863 -0.06069 -0.001296900
    0.00275510 0;
    0.0047544 -0.17922 -0.001655900 0.00138330 0;
    0.0136050 -0.30473 -0.005841600 0.00954960 -0.0058507;
    0.0038440 -0.20662 -0.001210500 0.00077768 0;
    0.0016373 -0.28186 -0.000169080 -0.00057546 0.0004007;
    0.0044791 -0.40227 -0.001307600 0.00045627 0; 0.0005237 -0.00463 -0.000044809
    0.00252900 0;
    0.0021623 -0.04556 -0.000751070 0.00172140 0;
    0.0005648 -0.04113 0.000181180 -0.00250420 0.0062610;
    0.0030895 -0.11968 -0.001015300 0.00099351 0; 0.0022573 -0.06562 -0.000739420
    0.00118960 0;
    0.0016316 -0.10748 -0.000314340 0.00045534 0.0010373;
    0.0036007 -0.17372 -0.001166600 0.00085187 0; 0.0051204 -0.32457 -0.001640300
    0.00075400 0;
    0.0115210 -0.49181 -0.004346700 0.00385480 -0.0013496;
    0.0018893 -0.31040 -0.000347140 0.00000274 0;
    0.0672710 -2.21973 -0.026667000 0.02264300 -0.0073731;
    0.0038450 -0.18652 -0.001257800 0.00089375 0; 0.0064753 -0.36192 -0.002175300
    0.00107420 0;
    0.0041401 -0.21812 -0.001345100 0.00084524 0;
    -0.0042629 -0.11862 0.002327900 -0.00350820 0.00175780;
    0.0026828 -0.22926 -0.000718600 0.00031365 0; 0.0050612 -0.31357 -0.001631300
0.00078036 0];
wv = [cv(:,1) cv(:,2) cv(:,3) cv(:,4)*1e-3 cv(:,5)*1e-6];
% compute vapor pressure
T0 = 273.15; T = T + T0;
logp = wv(ind,1) + wv(ind,2)./T + wv(ind,3).*log10(T) + wv(ind,4).*T + wv
(ind,5).*T.^2; pv = 10.^logp;
fprintf('Vapor pressure = %g mmHg \n', pv);
end
```

Example 3.17 Vapor Pressure of Chloroform[19]

Table 3.3 shows the vapor pressure P_v (*mmHg*) of chloroform as a function of temperature T (°C). The following Antoine equation is known to be adequate to represent the vapor pressure of chloroform:

$$P_v(mmHg) = \exp\left(A - \frac{B}{T(°C) + C}\right)$$

Determine Antoine parameters A, B, and C. Plot values of P_v calculated by Antoine equation and P_v data on the same graph for $0 \leq T \leq 150(°C)$.

Solution

The Antoine equation can be rearranged as

$$\ln P_v = A - \frac{B}{T + C}$$

Multiplying $T + C$ on both sides and rearranging, we have

$$T\ln P_v = AT + (AC - B) - C\ln P_v \Rightarrow \ln P_v = A + (AC - B)\frac{1}{T} - C\frac{\ln P_v}{T}$$

This relation can be rewritten as

$$\ln P_v = y = k_1 + k_2\frac{1}{T} + k_3\frac{\ln P_v}{T} = \begin{bmatrix} 1 & \frac{1}{T} & \frac{\ln P_v}{T} \end{bmatrix}\begin{bmatrix} k_1 \\ k_2 \\ k_3 \end{bmatrix}$$

where $k_1 = A$, $k_2 = AC - B$ and $k_3 = -C$. The Antoine parameters are given by solving the following set of linear equations:

$$Y = \begin{bmatrix} \ln P_{v1} \\ \ln P_{v2} \\ \vdots \\ \ln P_{vn} \end{bmatrix} = \begin{bmatrix} 1 & 1/T_1 & (\ln P_{v1})/T_1 \\ 1 & 1/T_2 & (\ln P_{v2})/T_2 \\ \vdots & \vdots & \vdots \\ 1 & 1/T_n & (\ln P_{vn})/T_n \end{bmatrix}\begin{bmatrix} k_1 \\ k_2 \\ k_3 \end{bmatrix} = XK \Rightarrow K = \begin{bmatrix} k_1 \\ k_2 \\ k_3 \end{bmatrix} = (X^T X)^{-1} X^T Y$$

The Antoine parameters are given by $A = k_1$, $C = -k_3$ and $B = AC - k_2$. The script *vpchform* calculates the Antoine parameters and produces the required plot.

```
% vpchform.m: vapor pressure of chloroform
clear all;
t = [0.001 10:10:150]; % temp.(deg.C)
Pv = [65.4 101.0 159.3 242.6 357.9 513.6 718.7 983.2 1317.7 1733.9 ...
   2243.7 2859.2 3593.0 4457.9 5466.4 6631.0]; % vapor pressure (mmHg)
n = length(t); Y = (log(Pv))'; X = [ones(n,1) (1./t)' (log(Pv)./t)'];
K = inv(X'*X)*X'*Y; A = K(1), C = -K(3), B = A*C - K(2)
ti = 0:0.1:150; Pvi = exp(A - B./(ti + C)); % Generate points for plot
plot(ti,Pvi,t,Pv,'o'), xlabel('t(deg.C)'), ylabel('P_v(mmHg)')
legend('Antoine eqn.','Data','location','best')

>> vpchform
A =
   17.5056
C =
  278.2849
B =
   3.7082e+03
```

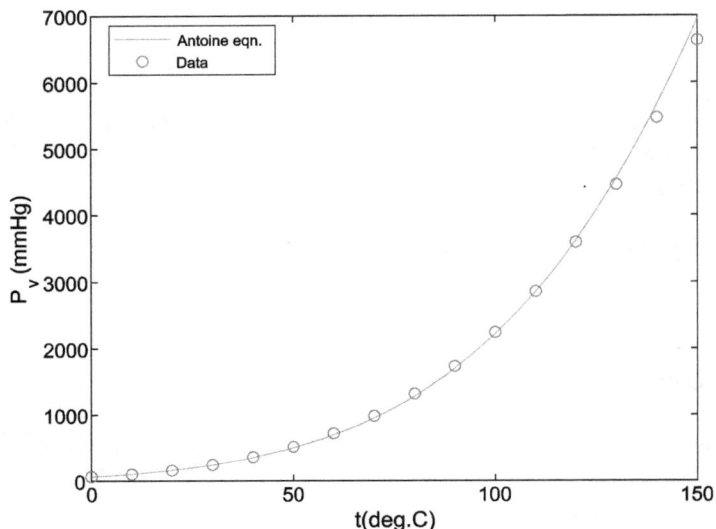

FIGURE 3.5 Plot of the vapor pressure of chloroform versus temperature.

Figure 3.5 shows the plot generated by the script *vpchform*. We can see that the vapor pressure of chloroform is represented adequately by the Antoine equation

$$P_v(mmHg) = \exp\left(17.5056 - \frac{3708.2}{T\,(°C) + 278.2849}\right)$$

Example 3.18 Vapor Pressure of Water

Estimate the vapor pressure of water at $T = 85°C$.

Solution

```
>> T = 85; pv = prVp(T,'H2O');
Vapor pressure = 434.184 mmHg
```

3.8.2 EXTENDED ANTOINE EQUATION

The vapor pressure of the saturated liquid can be expressed as the following extended Antoine equation[20]:

$$\log P_v = A + \frac{B}{T} + C\log T + DT + ET^2$$

where
   P_v is the vapor pressure of the saturated liquid ($mmHg$)
   T is the temperature (K)
   A, B, C, D, and E are correlation constants
   The following Antoine equation can be used to estimate the vapor pressure at high temperatures[21]:

$$\log_{10} P_{vb} = A - \frac{B}{T + C - 273.15} + 0.43429x^n + Ex^8 + Fx^{12}$$

where

P_{vb} is the vapor pressure (*bar*)

T is the temperature (*K*)

$x = (T - t_0 - 273.15)/T_c$

A, B, C, n, E, F, and t_0 are constants

The Riedel equation for vapor pressure is expressed as[22,23]

$$\ln P_v = A + \frac{B}{T} + C \ln T + DT^6$$

where A, B, C, and D are parameters. This equation can be written in a linear form as

$$y = A + Bx_1 + Cx_2 + Dx_3$$

where $y = \ln P_v$, $x_1 = 1/T$, $x_2 = \ln T$, and $x_3 = T^6$.

The Harlacher-Braun equation[24] is reasonably accurate from low vapor pressure up to the critical pressure:

$$\ln P_v = A + \frac{B}{T} + C \ln T + D \frac{P_v}{T^2}$$

where A, B, C, and D are parameters. This equation can be expressed in a linear form:

$$y = A + Bx_1 + Cx_2 + Dx_3$$

where $y = \ln P_v$, $x_1 = 1/T$, $x_2 = \ln T$, and $x_3 = P_v/T^2$.

Example 3.19 Vapor Pressure of 2,2,4-Trimethylpentane[25]

Table 3.4 shows the vapor pressure of liquid 2,2,4-trimethylpentane at various temperatures. Use the data given in Table 3.4 to determine the parameters for the Antoine equation, the Riedel equation, and the Harlacher-Braun equation.

Solution

For the Antoine equation, let

$$y = \begin{bmatrix} \ln P_{v1} \\ \ln P_{v2} \\ \vdots \\ \ln P_{vn} \end{bmatrix}, \quad X = \begin{bmatrix} 1 & 1/T_1 & \ln P_{v1}/T_1 \\ 1 & 1/T_2 & \ln P_{v2}/T_2 \\ \vdots & \vdots & \vdots \\ 1 & 1/T_n & \ln P_{vn}/T_n \end{bmatrix}, \quad b = \begin{bmatrix} A \\ AC - B \\ -C \end{bmatrix}$$

and use the relation

$$b = (X^T X)^{-1} X^T y$$

TABLE 3.4

Vapor Pressure (P_v) of Liquid 2,2,4-Trimethylpentane

T (°C)	-15	-4.3	7.5	20.7	29.1	40.7	58.1	78.0	99.2
P_v (kPa)	0.667	1.333	2.666	5.333	8.000	13.33	26.66	53.33	101.32

to calculate A, B, and C. For the Riedel and Harlacher-Braun equations, suitably define the matrix X and find the parameter vector $b = [A\ B\ C\ D]^T$. The script *compVPeqn* calculates the parameters of these equations. For the Harlacher-Braun equation, the nonlinear equation

$$A + \frac{B}{T} + C\ln T + \frac{Dx}{T^2} - \ln x = 0$$

is solved at a given T to give the vapor pressure x. The script *compVPeqn* generates the plot illustrated in Figure 3.6 showing vapor pressure values obtained from various equations whose parameters are estimated based on the data given in Table 3.4.

```
% compVPeqn.m: comparison of vapor equations
clear all; clc;
T = [-15 -4.3 7.5 20.7 29.1 40.7 58.1 78.0 99.2] + 273.15;
Pv = [0.667 1.333 2.666 5.333 8.000 13.33 26.66 53.33 101.32];
Xa = [ones(1,length(T)); 1./T; log(Pv)./T]'; ba = inv(Xa'*Xa)*Xa'*log(Pv)';
Aa = ba(1); Ca = -ba(3); Ba = Aa*Ca-ba(2); % Antoine eq.
Xr = [ones(1,length(T)); 1./T; log(T); T.^6]'; br = inv(Xr'*Xr)*Xr'*log(Pv)';
Ar = br(1); Br = br(2); Cr = br(3); Dr = br(4); % Riedel eq.
Xh  =  [ones(1,length(T)); 1./T;  log(T);  Pv./T.^2]';  bh  =  inv
(Xh'*Xh)*Xh'*log(Pv)';
Ah = bh(1); Bh = bh(2); Ch = bh(3); Dh = bh(4); % Harlecher-Braun eq
Ti = T(1):T(end); Pa = exp(Aa - Ba./(Ti+Ca));
Pr = exp(Ar + Br./Ti + Cr*log(Ti) + Dr*Ti.^6); Ph = [];
```

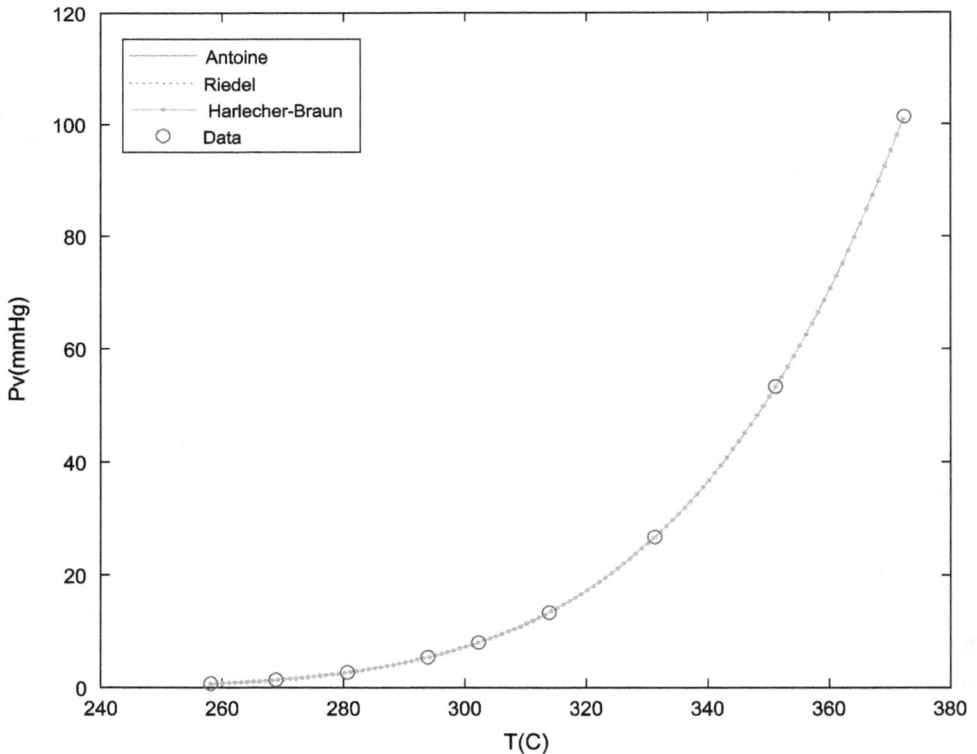

FIGURE 3.6   Vapor pressure of 2,2,4-trimethylpentane.

```
for k = 1:length(Ti)
  P0 = 100/length(Ti) * k;
  Phf = @(x) Ah + Bh/Ti(k) + Ch*log(Ti(k)) + Dh*x/Ti(k)^2 - log(x);
  Phv = fzero(Phf,P0); Ph = [Ph Phv];
end
plot(Ti,Pa,Ti,Pr,':',Ti,Ph,'.-',T,Pv,'o'), xlabel('T(C)'), ylabel('Pv(mmHg)')
legend('Antoine','Riedel','Harlecher-Braun','Data','location','best')
```

3.8.3 WAGNER EQUATION

The Wagner equation is a very useful correlation to estimate vapor pressures.[26] This equation correlates the whole vapor pressure curve from the triple point to the critical point:

$$\ln\frac{P_v}{P_c} = \frac{1}{T_r}[A(1 - T_r) + B(1 - T_r)^{1.5} + C(1 - T_r)^{2.5} + D(1 - T_r)^5]$$

where $T_r = T/T_c$. Reasonable ranges of parameters are $-9 \leq A \leq -5$, $-10 \leq B \leq 10$, $-10 \leq C \leq 10$, and $-20 \leq D \leq 20$.

The function *vpwagner* uses the Wagner equation to estimate vapor pressure. The basic syntax is
```
Pv = vpwagner(T,Tc,Pc,C)
```

where T is the temperature (K), Tc is the critical temperature (K), Pc is the critical pressure (MPa), C is the parameter vector, and Pv is the estimated vapor pressure (MPa).

```
function Pv = vpwagner(T,Tc,Pc,C)
% Estimation of vapor pressure using the Wagner equation
% input:
% T,Tc: temperature (K) and critical temperature (K)
% Pc: critical pressure (MPa)
% C: parameter vector of the Wagner equation
% output
% Pv: estimated vapor pressure (MPa)
Tr = T./Tc; a = C(1); b = C(2); c = C(3); d = C(4);
Pv = Pc.*exp((a*(1-Tr) + b*(1-Tr).^1.5 + c*(1-Tr).^2.5 + d*(1-Tr).^5)/Tr);
fprintf('Vapor pressure = %g MPa \n', Pv);
end
```

Example 3.20 Vapor Pressure of Acetone

Estimate the vapor pressure (MPa) of acetone at 273.15 K. For acetone, $T_c = 508.1\,K$, $P_c = 4.6924\,MPa$, and values of parameters of the Wagner equation are $A = -7.670734$, $B = 1.965917$, $C = -2.445437$, and $D = -2.899873$.

Solution

```
>> z = [-7.670734 1.965917 -2.445437 -2.899873]; T = 273.15; Tc = 508.1;
Pc=4.6924; Pv = vpwagner(T,Tc,Pc,z);
Vapor pressure = 0.00930124 MPa
```

3.8.4 HOFFMANN-FLORIN EQUATION

The vapor pressure can be calculated by a simple estimation procedure based on the equation of Hoffmann and Florin.[27] The Hoffmann-Florin equation has two adjustable parameters α and β:

$$\ln\frac{P_v}{P_0} = \alpha + \beta f(T)$$

where

$$f = \frac{1}{T} - 7.9151 \times 10^{-3} + 2.6726 \times 10^{-3}\log(T) - 0.8625 \times 10^{-6}T \ (T:K)$$

$$\alpha = \ln\frac{P_1^v}{P_0} - \ln\frac{P_1^v}{P_2^v}\cdot\frac{f(T_1)}{f(T_1) - f(T_2)}, \ \beta = \ln\frac{P_1^v/P_2^v}{f(T_1) - f(T_2)}$$

3.8.5 RAREY-MOLLER EQUATION

The Rarey-Moller equation is based on a new group contribution estimation method with the normal boiling point.[28] In this equation, the C parameter of the Antoine equation is correlated with the normal boiling point:

$$\ln P_v = B'\left(\frac{T - T_b}{T + 2.65 - (T_b^{1.485}/135)}\right) + D'\ln\frac{T}{T_b}$$

$$B' = 9.42208 + \sum_i v_i\Delta B_i + n_A \sum_j v_j\Delta B_j + \sum_k \Delta B_k + \frac{1}{2}\sum_i\sum_j GI_{ij}, \ D' = D + \frac{1}{n_A}\sum_i v_i\Delta E_i$$

where
  P_v is the vapor pressure (*bar*)
  T is the temperature (*K*)
  ΔB_i are the group contributions of the structural groups
  v_i is the frequency of group i in the molecule
  The GI_{ij} are the group interaction contributions, with $G_ij = G_ji$, D' denoting a correction term for aliphatic alcohols and carboxylic acids consisting of group contributions ΔD_i and a constant D, and n_A is the number of atoms except for hydrogen.
  The function *vpRM* performs the vapor pressure estimation by using the Rarey-Moller equation. This function can be called as
Pv = vpRM(T,Tb,nu,dB,GI,Dp)

where T and Tb are the temperature and normal boiling point (*K*), respectively; nu is the frequency of groups; dB and GI denote the ΔB_i and GI_{ij} of each group, respectively; Dp is D'; and Pv is the estimated vapor pressure (*bar*).

```
function Pv = vpRM(T,Tb,nu,dB,GI,Dp)
% Estimation of vapor pressure by using the Rarey/Moller equation
% input
%   T,Tb: temperature and normal boiling point (K)
%   nu: frequency of groups
%   dB, GI: delta B and GIij of each group
%   Dp: D prime
% output
```

TABLE 3.5

Groups and Frequencies

Group	v_i(Frequency)	ΔB_i
CH_2	2	0.07545
CH_3	1	-0.00227
C	1	0.11192
CH	5	0.01653

```
%  Pv: vapor pressure (bar)
sumdBi = sum(nu.*dB); sumGI = sum(sum(GI)); Bp = 9.42208 + sumdBi + sumGI;
Pv = exp(Bp*(T-Tb)./(T+2.65-(Tb.^1.485)/135) + Dp*log(T./Tb)); % bar
fprintf('Vapor pressure = %g bar \n', Pv);
end
```

Example 3.21 Vapor Pressure of *n*-Propylbenzene[29]

Estimate the vapor pressure of *n*-propylbenzene ($C(CH)_5(CH_2)_2CH_3$) at 100 °C and 200 °C using the Rarey-Moller equation. The normal boiling point of n-propyl benzene is $T_b = 159.22$°C. Each group and corresponding frequency of *n*-propylbenzene molecules are shown in Table 3.5.

Solution

The function *vpRM* returns the vapor pressure for each temperature.

```
>> Tb = 159.22+273.15; nu = [2 1 1 5]; Dp = 0; GI = []; dB = [0.07545 -0.00227
0.11192 0.01653];
>> T = 100 + 273.15; Pv = vpRM(T,Tb,nu,dB,GI,Dp);
Vapor pressure = 0.159473 bar
>> T = 200 + 273.15; Pv = vpRM(T,Tb,nu,dB,GI,Dp);
Vapor pressure = 2.61062 bar
```

3.8.6 VAPOR PRESSURE ESTIMATION BY CORRELATIONS

For some materials, vapor pressures can be effectively estimated using correlations. Table 3.6 shows correlations for vapor pressures (P_v, *mmHg*) of some substances as a function of temperature (T, °C).

Example 3.22 Vapor Pressure of Cyclohexanethiol

Estimate the vapor pressure of cyclohexanethiol at $T = 100$°C. Compare the result with the experimental value of $P_v = 132.9 \, mmHg$.

Solution

We can use the correlation shown in Table 3.6.

```
>> T = 100; Pv = 1.35175e-6*T^4 - 2.8e-5*T^3 - 0.0053375*T^2 + 1.1674*T - 38.62
Pv =
 131.9200
```

TABLE 3.6
Vapor Pressure Correlations for Typical Substances[30,31]

Substance	Vapor Pressure Correlation (P_v: mmHg)	Temperature Range (T:°C)	Data Source
Benzenethiol	$P_v = 8.615 \times 10^{-4}T^3 - 0.240482T^2 + 26.77258T - 1058.01046$	$115 \leq T \leq 210$	1
n-Butane	$P_v = 7.65 \times 10^{-6}T^4 + 0.00292T^3 + 0.43044T^2 + 29.2473T + 774.3643$	$-75 \leq T \leq 15$	2
1-Butanethiol	$P_v = 0.0013T^3 - 0.1277T^2 + 9.0858T - 162.398$	$55 \leq T \leq 130$	1
2-Butanethiol	$P_v = 0.0014T^3 - 0.0812T^2 + 6.6279T - 68.7349$	$40 \leq T \leq 120$	1
Cyclohexanethiol	$P_v = 1.35175 \times 10^{-6}T^4 - 2.8 \times 10^{-5}T^3 - 0.0053375T^2 + 1.1674T - 38.62$	$85 \leq T \leq 200$	1
Cyclopentanethiol	$P_v = 0.001T^3 - 0.181T^2 + 15.36T - 445.5201$	$85 \leq T \leq 170$	1
n-Decane	$P_v = 1.818 \times 10^{-6}T^4 - 3.3234 \times 10^{-4}T^3 + 0.03765T^2 - 1.95725T + 41.31365$	$60 \leq T \leq 200$	2
2,3-Dimethyl-2-butanethiol	$P_v = 1.6943 \times 10^{-6}T^4 + 1.37 \times 10^{-4}T^3 - 0.003516T^2 + 0.99033T - 12.7844$	$60 \leq T \leq 165$	1
Ethane	$P_v = 3.14 \times 10^{-5}T^5 + 0.01952T^3 + 4.5753T^2 + 479.3767T + 18959.3722$	$-140 \leq T \leq -75$	2
Ethanethiol	$P_v = 0.0023T^3 + 0.1327T^2 + 9.009T + 182.9919$	$5 \leq T \leq 60$	1
1-Heptanethiol	$P_v = 1.555 \times 10^{-6}T^4 - 2.385 \times 10^{-4}T^3 + 0.021794T^2 - 0.7481T + 7.2382$	$105 \leq T \leq 195$	1
n-Hexane	$P_v = 3.6871 \times 10^{-6}T^4 + 0.000657T^3 + 0.0539T^2 + 2.419385T + 45.5556$	$-25 \leq T \leq 90$	2
1-Hexanethiol	$P_v = 1.74 \times 10^{-6}T^4 - 9.24 \times 10^{-5}T^3 + 0.0033T^2 + 0.57904T - 21.625577$	$85 \leq T \leq 190$	1
2-Methyl-1-butanethiol	$P_v = 2.142 \times 10^{-6}T^4 + 1.314 \times 10^{-4}T^3 + 0.0017T^2 + 0.863076T - 7.7885$	$55 \leq T \leq 155$	1
2-Methyl-1-propanethiol	$P_v = 0.0014T^3 - 0.0961T^2 + 7.395T - 96.2743$	$45 \leq T \leq 120$	1
2-Methyl-2-butanethiol	$P_v = 0.0012T^3 - 1051T^2 + 7.8451T - 134.1219$	$55 \leq T \leq 135$	1
3-Methyl-2-butanethiol	$P_v = 2.15 \times 10^{-6}T^4 + 2.306 \times 10^{-4}T^3 + 0.002915T^2 + 1.0029236T - 2.3$	$45 \leq T \leq 145$	1
2-Methyl-2-pentanethiol	$P_v = 1.827 \times 10^{-6}T^4 + 1.352 \times 10^{-4}T^3 - 0.00431T^2 + 1.05128T - 14.817$	$60 \leq T \leq 160$	1
2-Methyl-2-propanethiol	$P_v = 0.0016T^3 - 0.0029T^2 + 5.1029T + 31.6341$	$25 \leq T \leq 95$	1
n-Pentane	$P_v = 5.094 \times 10^{-6}T^4 + 0.001393T^3 + 0.154777T^2 + 8.35808T + 183.30147$	$-50 \leq T \leq 55$	2
1-Pentanethiol	$P_v = 0.0011T^3 - 0.1831T^2 + 14.9687T - 412.9671$	$80 \leq T \leq 160$	1
1-Propanethiol	$P_v = 0.0017T^3 - 0.0321T^2 + 5.6123T + 6.9356$	$25 \leq T \leq 100$	1
2-Propanethiol	$P_v = 0.0018T^3 + 0.0396T^2 + 5.7869T + 79.6555$	$15 \leq T \leq 80$	1

Example 3.23 Vapor Pressure of 2-Methyl-1-Butanethiol

Estimate the vapor pressure of 2-methyl-1-butanethiol at $T = 112.6°C$. Compare the result with the experimental value of $P_v = 634 \, mmHg$.
Solution
   We can use the correlation shown in Table 3.6.

```
>> T = 112.6; Pv = 2.142e-6*T^4 + 1.314e-4*T^3 + 0.001*T^2 + 0.863076*T - 7.7885
Pv =
 633.9915
```

3.9 ENTHALPY OF VAPORIZATION

3.9.1 WATSON EQUATION

The enthalpy of vaporization, ΔH_v, is also termed the latent heat of vaporization. ΔH_v is the difference between the enthalpy of the saturated vapor and that of the saturated liquid at the same temperature. At a given temperature $T\,(K)$, ΔH_v can be estimated by the Watson correlation[32]

$$\Delta H_v = \Delta H_{v1}\left(\frac{T_c - T}{T_c - T_1}\right)^r$$

where
   ΔH_v is the heat of vaporization (cal/g) at a given temperature
   ΔH_{v1} is the heat of vaporization (cal/g) at $T_1(K)$
   T_c is the critical temperature (K)
   r is the characteristic constant for the component (usually $r = 0.38$)
   The function *hvapn* calculates the enthalpy of vaporization of components listed in Table 3.2. This function defines ΔH_{v1}, T_1, T_c, and r. The basic syntax is
hv = hvapn(T,cname)

where T is the temperature (°C); cname is the name of the component, which can be specified as a chemical formula or common name; and hv is the estimated heat of vaporization (cal/g) at T.

```
function hv = hvapn(T, cname)
% Estimation of heat of vaporization (cal/g)
% input
%   T: temperature(C) (scalar or a row vector)
%   cname: common name or chemical formula of the compound
% output
%   hv: estimated heat of vaporization at T (cal/g)
% parameters (dhv1, T1(C), Tc(C), r)
cv = [41.1000 -188.1000 -129.000 0.3800; 69.6000 -34.0600 144.0000 0.3800;
   93.0000 -10.0000 157.6000 0.3800; 51.6000 -191.5000 -140.1000 0.3800;
   56.1000 0 31.1000 0.3800; 105.8000 -85.0300 51.5000 0.3800;
   327.4000 -33.4300 132.4000 0.3800; 538.7000 100.0000 374.2000 0.3800;
   321.9000 150.2000 455.0000 0.3800; 107.0000 -252.8000 -240.2000 0.2370;
   47.5000 -195.8000 -146.8000 0.3800; 50.9000 -183.0000 -118.5000 0.3800;
   115.4000 -103.7000 9.9000 0.3800; 121.7000 -161.5000 -82.6000 0.3800;
   116.7000 -88.2000 32.3000 0.3800; 101.8000 -42.1000 96.7000 0.3800;
   94.1000 80.1000 288.9400 0.3800; 86.1000 110.6000 318.8000 0.3800;
```

```
    112.4000 184.4000 426.0000 0.3800;  116.4000 181.8000 420.0000 0.3800;
    113.8000 -32.8000 124.9000 0.3800;  0.0 0.0 0.0 0.0; % missing component (C6H12)
    100.2000 -4.4100 152.0000 0.3800;  260.1000 64.7000 239.4000 0.4000;
    58.9000 61.3000 263.4000 0.3800;  46.5500 76.7000 283.2000 0.3800];
ind = compID(cname); % identification number of the component
% computes heat of vaporization
T0 = 273.15; T = T + T0; Tc = cv(ind,3) + T0; T1 = cv(ind,2) + T0;
hv = cv(ind,1)*((Tc - T)./(Tc - T1)).^(cv(ind,4));
end
```

Example 3.24 Heat of Vaporization of Water

Estimate the heat of vaporization of water (cal/g) at the range of $0 \le T \le 350(°C)$, and plot the result as a function of temperature.

Solution

The following code produces the plot shown in Figure 3.7.

```
>> T = 0:350; hv = hvapn(T,'H2O');
>> plot(T,hv), xlabel('T(C)'), ylabel('Heat of vaporization of water(cal/
g)'), grid on
```

3.9.2 PITZER CORRELATION

The enthalpy of vaporization, ΔH_v, can be estimated by the correlation proposed by Pitzer et al.[33]:

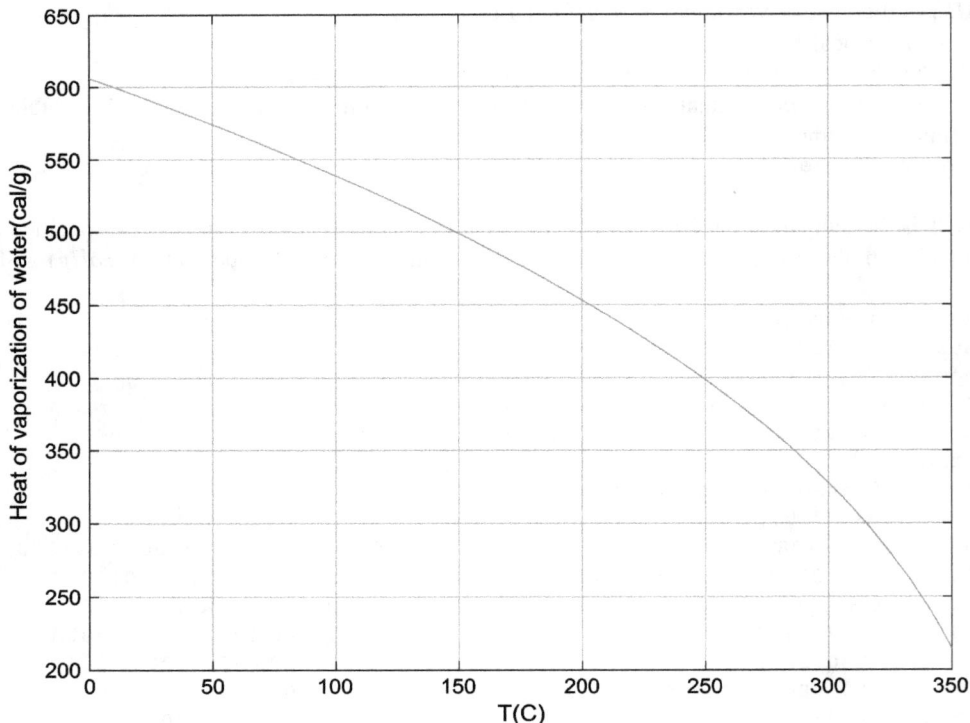

FIGURE 3.7 Heat of vaporization of water as a function of temperature.

$$\frac{\Delta H_v}{RT_c} = 7.08(1 - T_r)^{0.354} + 10.95\omega(1 - T_r)^{0.456}$$

where ω is the acentric factor.

3.9.3 Clausius-Clapeyron Equation

The derivative of the vapor pressure with respect to temperature in the Clausius-Clapeyron equation can be replaced with a suitable relation to be used in the estimation of enthalpy of vaporization:

$$\Delta H_V = -R\frac{d\ln P_v}{d(1/T)} = T(v^V - v^L)\frac{dP_v}{DT} \quad (T_r < 0.75)$$

Use of the Antoine equation: If the Antoine equation $\log P_v = A - B/(T + C)$ is used, the derivative of the vapor pressure with respect to temperature can be given by

$$\frac{d\ln P_v}{d(1/T)} = -\frac{2.3026B(T + 273.15)^2}{(T + C)^2}$$

Therefore, we have

$$\Delta H_V = \frac{2.3026RB(T + 273.15)^2}{(T + C)^2}$$

where
T is the temperature (°C)
B and C are parameters of the Antoine equation
Use of the Wagner equation: The Wagner correlation can be used to evaluate the derivative dP_v/dT. Differentiation of the Wagner equation

$$\ln\frac{P_v}{P_c} = \frac{1}{T_r}[A(1 - T_r) + B(1 - T_r)^{1.5} + C(1 - T_r)^{2.5} + D(1 - T_r)^5]$$

with respect to T gives[34]

$$\frac{dP_v}{dT} = -\frac{P_v}{T}\left[\ln\frac{P_v}{P_c} + A + 1.5B(1 - T_r)^{0.5} + 2.5C(1 - T_r)^{1.5} + 5D(1 - T_r)^4\right]$$

Substitution of this equation into the Clausius-Clapeyron equation yields

$$\Delta H_V = -P_v(v^V - v^L)\left[\ln\frac{P_v}{P_c} + A + 1.5B(1 - T_r)^{0.5} + 2.5C(1 - T_r)^{1.5} + 5D(1 - T_r)^4\right]$$

The function *vhwagner* estimates the enthalpy of vaporization using the Wagner equation. The syntax for this function is
dHv = vhwagner(T,Tc,Pc,vL,vV,C)

where T is the temperature (K), Tc is the critical temperature (K), Pc is the critical pressure (MPa), C is the parameter vector of the Wagner equation, and dHv is the estimated enthalpy of vaporization (J/mol).

```
function dHv = vhwagner(T,Tc,Pc,vL,vV,C)
% Estimation of enthalpy of vaporization using the Wagner equation
% input:
%   T,Tc: temperature and critical temperature (K)
%   Pc: critical pressure (MPa)
%   C: parameter vector of the Wagner equation
% output
%   dHv: estimated enthalpy of vaporization (J/mol)
Tr = T./Tc; a = C(1); b = C(2); c = C(3); d = C(4);
Pv = Pc.*exp((a*(1-Tr) + b*(1-Tr).^1.5 + c*(1-Tr).^2.5 + d*(1-Tr).^5)/Tr);
wd = log(Pv./Pc) + a + 1.5*b*(1-Tr).^0.5 + 2.5*c*(1-Tr).^1.5 + 5*d*(1-Tr).^4;
dHv = -Pv*(vV - vL).*wd*1e6; % J/mol
end
```

Example 3.25 Enthalpy of Vaporization of Acetone

Estimate the enthalpy of vaporization of acetone at $T = 273.15$ K($0\,°C$). For acetone, $v^L = 7.145 \times 10^{-5} m^3/mol$, $v^V = 0.2453 m^3/mol$, $T_3 = 508.1$K, $P_c = 4.6924$MPa, $M_w = 58.08$ g/mol, and the parameters of the Wagner equation are $A = -7.670734$, $B = 1.965917$, $C = -2.445437$, and $D = -2.899873$.

Solution

```
>> C = [-7.670734 1.965917 -2.445437 -2.899873];
>> T=273.15; Tc=508.1; Pc=4.6924; vL=7.145e-5; vV=0.2453; dHv = vhwagner
(T,Tc,Pc,vL,vV,C)
dHv =
   3.3015e+04
```

The result is $\Delta H_v = 33,015\,J/mol$. The experimental value is reported to be $32,460.9\,J/mol$.[35] We can see that the deviation is about 1.7%.

3.10 HEAT OF FORMATION FOR IDEAL GASES

The correlation of the heat of formation, ΔH_f^0, of the ideal gas at low temperature can be expressed as[36]

$$\Delta H_f^0 = A + BT + CT^2$$

where
ΔH_f^0 is the heat of formation of an ideal gas at low temperature ($kcal/gmol$)
T is the temperature (K)
A, B, and C are correlation constants

The function *hform* calculates the heat of formation of gases listed in Table 3.2. This function defines constants A, B, and C in the range of $298 \le T \le 1500(K)$. The basic syntax is

hf = hform(T,cname)

where T is the temperature (°C); cname is the name of the component, which can be specified as a chemical formula or common name; and hf is the estimated heat of formation ($kcal/gmol$) at T.

```
function hf = hform(T,cname)
```

```
% Estimation of heat of formation of gases (kcal/gmol) at low temperature
% input
%     T: temperature (C) (scalar or row vector)
%     cname: name or chemical formula of the compound
% output
%     hf: heat of formation (kcal/gmol)
%     correlation constants (A, B, C)
cv = [0.0 0.0 0.0; % missing component (F2)
     0.0 0.0 0.0; % missing component (Cl2)
     -69.6000 -5.2900 0; -26.5000 0.8400 -1.0500; -93.9000 -0.4300 0; -21.9000 -0.6100 0;
     -9.3400 -6.2000 2.3600; -57.4000 -1.7900 0; -31.8000 -3.0400 1.1900;
     0.0 0.0 0.0; % missing component (H2)
     0.0 0.0 0.0; % missing component (N2)
     0.0 0.0 0.0; % missing component (O2)
     0.0 0.0 0.0; % missing component (C2H4)
     -15.4000 -9.5900 3.5000; -16.4000 -14.8000 6.1300; -20.0000 -19.1000 8.1500;
     23.7000 -15.3000 6.2700; 16.8000 -19.0000 7.8400; 25.2000 -17.6000 8.9800;
     -19.3000 -14.6000 7.1800; 16.9000 -16.7000 7.9000; -21.6000 -32.0000 15.8000;
     29.0000 -10.1000 4.1300; -44.9600 -11.9000 4.9800; -24.7000 0.0334 0;
     -24.0000 2.4200 0];
wv = [cv(:,1) cv(:,2)*1e-3 cv(:,3)*1e-6];
ind = compID(cname); % identification number of the component
% computes heat of formation
T0 = 273.15; T = T + T0;
if ind ~= 3
  hf = wv(ind,1) + wv(ind,2)*T + wv(ind,3)*T.^2;
else % ind = 3: sulfur dioxide
  hf = [];
  for j = 1:length(T)
    if T(j) >= 298 && T(j) <= 717 % temperature range: 298K~717K
       ind = 1; hfv = wv(ind,1) + wv(ind,2)*T(j) + wv(ind,3)*T(j).^2;
    else % temperature range: 717K~1500K
       wv(ind,:) = [-86.9 0.32*1e-3 0]; hfv = wv(ind,1) + wv(ind,2)*T(j) + wv
(ind,3)*T(j).^2;
    end
    hf = [hf hfv];
  end
end
end
```

Example 3.26 Standard Heat of Formation of Methane

Estimate the heat of formation of methane (*kcal/gmol*) at 500°C.

Solution

```
>> T = 500; hf = hform(T,'methane')
hf =
-20.7223
```

3.11 GIBBS FREE ENERGY

The Gibbs free energy of formation, ΔG_f^0, is defined as

$$\Delta G_f^0 = \Delta H_f^0 - T\Delta S_f^0$$

where ΔS_f^0 is the entropy of formation. The Gibbs free energy of formation of the ideal gas at low pressure can be expressed as a linear relationship in temperature as[37]

$$\Delta G_f^0 = A + BT$$

where

ΔG_f^0 is the Gibbs free energy of formation of an ideal gas at low pressure (*kcal/gmol*)
T is the temperature (K)
A and B are correlation constants

The function *gfree* calculates the Gibbs free energy of formation of components listed in Table 3.2. This function defines constants A and B. This function is used as
gf = gfree(T,cname)

where T is the temperature (K); cname is the name of the component, which can be specified as a chemical formula or common name; and gf is the estimated Gibbs free energy of formation (*kcal/gmol*) at T. T can be either a scalar or a row vector representing a temperature range.

```
function gf = gfree(T,cname)
% Estimation of Gibbs free energy of gases (kcal/gmol) at low pressure
% input
%   T: temperature(K) (scalar or row vector)
%   cname: name or chemical formula of the compound
% output
%   gf: Gibbs free energy of formation (kcal/gmol)
% correlation constants (A, B)
cv = [ 0.0 0.0; % missing component (F2)
  0.0 0.0; % missing component (Cl2)
  -71.9000 0.2500;  -26.5000 -21.3000;  -94.2000 -0.4200;  -22.3000 -1.7200;
  -12.3000 27.2000;  -58.6000 12.7000;  -33.2000 26.4000;
  0.0 0.0; % missing component (H2)
  0.0 0.0; % missing component (N2)
  0.0 0.0; % missing component (O2)
  0.0 0.0; % missing component (C2H4)
  -20.1000 24.9000;  -23.3000 49.7000;  -28.8000 74.7000;  16.7000 45.9000;
  7.8000 68.7000;  18.5000 70.6000;  -25.0000 56.5000;  10.3000 47.7000;
  -35.1000 139.0000;  24.2000 38.6000;  -50.2000 36.5000;
  -24.8000 26.9000;  -22.6000 32.1000];
wv = [cv(:,1) cv(:,2)*1e-3];
ind = compID(cname); % identification number of the component
% computes Gibbs free energy
if ind ~= 3, gf = wv(ind,1) + wv(ind,2)*T;
else % ind = 3: sulfur dioxide
  gf = [];
  for j = 1:length(T)
    if T(j) >= 298 && T(j) <= 717 % temperature range: 298K~717K
      ind = 1; gfv = wv(ind,1) + wv(ind,2)*T(j);
    else % temperature range: 717K~1500K
      wv(ind,:) = [-86.8 17.7*1e-3]; gfv = wv(ind,1) + wv(ind,2)*T(j);
```

```
      end
      gf = [gf gfv];
   end
end
end
```

Example 3.27 Gibbs Free Energy of Formation of Benzene

Estimate the Gibbs free energy of formation of benzene at 60 °C.

Solution

```
>> T = 60+273.15; gfree(T, 'Benzene')
ans =
  31.9916
```

3.12 DIFFUSION COEFFICIENTS

3.12.1 LIQUID-PHASE DIFFUSION COEFFICIENTS

Diffusion coefficients in the liquid phase depend on the concentration and are valid for dilute solutions with solute concentrations less than 10 $mol\%$. For a binary mixture of solute A dissolved in solvent B, the diffusion coefficient can be represented as D_{AB}^0 for concentrations of A up to 10 $mol\%$. The Wilke-Chang method can be used to estimate D_{AB}^0[38]:

$$D_{AB}^0 = \frac{7.4 \times 10^{-8} T \sqrt{\phi M_{wB}}}{\mu_B \times V_A^{0.6}}$$

where
D_{AB}^0 is the diffusion coefficient of solute A at very low concentration in B (cm^2/sec)
M_{wB} is the molecular weight of solvent B
T is the temperature (K)
μ_B is the viscosity of solvent B (cP)
V_A is the molal volume of solute A at its normal boiling point ($cm^3/gmol$)
ϕ is the dimensionless association factor of solvent B: 2.6 for water, 1.9 for methanol, 1.5 for ethanol, and 1.0 for unassociated solvents
    The function *gdiffc* estimates diffusion coefficients in the liquid phase. The basic syntax of this function is
`Df = gdiffc(T,M,phi,mu,v)`

where T is the temperature (K; a scalar or a vector), M is the molecular weight, phi is the association factor of solvent B, mu is the viscosity of solvent B (cP), v is the molal volume of solute A at its normal boiling point ($cm^3/gmol$), and Df is the estimated diffusion coefficient in the liquid phase (cm^2/sec).

```
function Df = gdiffc(T,M,phi,mu,v)
% Estimation of diffusion coefficients in liquid phase
% input
%   T: temperature(C) (scalar or row vector)
%   M: molecular weight
```

```
% phi association factor of solvent B
% mu: viscosity of solvent B (cP)
% v: molal volume of solute A at its normal boiling point (cm^3/gmol)
% output
% Df: estimated diffusion coefficient in liquid phase (cm^2/s)
Df = 7.4e-8*T.* sqrt(phi.*M)./(mu.*v.^0.6);
end
```

3.12.2 Gas-Phase Diffusion Coefficients

Diffusion coefficients for nonpolar gases can be estimated from the Fuller-Schettler-Giddings method, which is expressed as[39]

$$D_{AB}^0 = \frac{10^{-3}T^{1.75}\sqrt{\frac{M_{wA}+M_{wB}}{M_{wA}M_{wB}}}}{P[(\Sigma v)_A^{1/3} + (\Sigma v)_B^{1/3}]^2}$$

where

D_{AB}^0 is the diffusion coefficient for a binary mixture of gases A and B (cm^2/sec)
M_{wA} and M_{wB} are the molecular weights of A and B, respectively
v_A and v_B are the atomic diffusion volumes of A and B, respectively
T is the temperature (K)
P is the pressure (atm)

The summation of the diffusion volume coefficients for components A and B, Σv, is shown in Table 3.7.

TABLE 3.7
Atomic Diffusion Volumes for Use in the Fuller-Schettler-Giddings Method[40]

Atomic and Structural Diffusion Volume Increments, v

C	16.5	Cl	19.5
H	1.98	S	17.0
O	5.48	Aromatic ring	-20.2
N	5.69	Heterocyclic ring	-20.2

Diffusion Volumes for Simple Molecules,

H_2	7.07	CO	18.9
D_2	6.70	CO_2	26.9
He	2.88	N_2O	35.9
N_2	17.9	NH_3	14.9
O_2	16.6	H_2O	12.7
Air	20.1	CCl_2F_2	114.8
Ar	16.1	SF_6	69.7
Kr	22.8	Cl_2	37.7
Xe	37.9	Br_2	67.2
SO_2	41.1		

3.13 COMPRESSIBILITY FACTOR OF NATURAL GASES

The compressibility factor Z of natural hydrocarbon gases can be estimated using the revised Awoseyin method[41]:

$$Z = F_1 \left[\frac{1}{1 + \frac{A_6 \times P \times 10^{1.785 S_g}}{T^{3.825}}} + F_2 \times F_3 \right] + F_4 + F_5$$

where
  T is the temperature (R)
  P is the pressure $(Kpsia)$

$F_1 = P(0.251 S_g - 0.15) - 0.202 S_g + 1.106$

$F_2 = 1.4 e^{-0.0054(T-460)}$

$F_3 = A_1 P^5 + A_2 P^4 + A_3 P^3 + A_4 P^2 + A_5 P$

$F_4 = (0.154 - 0.152 S_g) P^{(3.18 S_g - 1.0)} e^{-0.5P} - 0.02$

$F_5 = 0.35(0.6 - S_g) e^{-1.039(P-1.8)^2}$

and the values of the constants are

$$A_1 = 0.001946, \ A_2 = -0.027635, \ A_3 = 0.136315,$$

$$A_4 - 0.23849, \ A_5 = 0.105168, \ A_6 = 3.44 \times 10^8$$

The specific gravity, S_g, of a natural gas can be calculated from its density or molecular weight:

$$S_g = \frac{(density\ of\ gas)}{(density\ of\ air)} = \frac{\rho_{gas,60°F}}{\rho_{air,60°F}} \ or \ S_g = \frac{molecular\ weight\ of\ gas}{molecular\ weight\ of\ air} = \frac{M_{w,gas}}{M_{w,air}}$$

The function *ngasZ* estimates the compressibility factor Z of natural hydrocarbon gases. This function can be used as
nz = ngasZ(T,P,Sg)

where T is temperature (°F; a scalar or a vector), P is the pressure (psia; a scalar or a vector), Sg is the specific gravity of the natural gas, and nz is the estimated compressibility factor of the natural gas at T and P. The units of T and P are automatically converted to R and $Kpsia$, respectively, within the function.

```
function nz = ngasZ(T,P,Sg)
% Estimates the compressibility factor Z of natural gases
% input
%  T: temperature (F) (scalar or vector)
%  P: pressure (psia)
%  Sg: specific gravity of the natural gas
% output
%  nz: estimated compressibility factor Z of natural gases
P = P/1000; T = T + 460;
A1 = 0.001946; A2 = -0.027635; A3 = 0.136315; A4 = -0.23849; A5 = 0.105168; A6 =
3.44e8;
```

```
F1 = P.*(0.251*Sg-0.15) - 0.202*Sg + 1.106; den = 1 + A6*P.*10.^(1.785*Sg)./
(T.^3.825);
F2 = 1.4*exp(-0.0054*(T-460)); F3 = A1*P.^5 + A2*P.^4 + A3*P.^3 + A4*P.^2 + A5*P;
F4 = (0.154-0.152*Sg).*P.^(3.18*Sg-1).*exp(-0.5*P) - 0.02;
F5 = 0.35*(0.6-Sg).*exp(-1.039*(P-1.8).^2);
nz = F1.*(1./den + F2.*F3) + F4 + F5;
end
```

Example 3.28 Compressibility Factor of Natural Gases[42]

Estimate compressibility factors of natural gases at 60 °F and at the pressure range of $100 \le P \le 5000 (psia)$ when the specific gravity is 0.5, 0.6, 0.7, and 0.8. Plot the results as a function of pressure and specific gravity.

Solution

The script *comfacng* uses the function *ngasZ* to calculate the compressibility factor and produce the curves shown in Figure 3.8.

```
% comfacng.m
T = 60; P = 100:10:5000; nz = [];
for k = 1:4, Sg = 0.5+(k-1)*0.1; nzv = ngasZ(T,P,Sg); nz = [nz nzv']; end
plot(P,nz(:,1),P,nz(:,2),':',P,nz(:,3),'.-',P,nz(:,4),'--'), grid
axis([100  5000  0  1.1]),  legend('Sg=0.5','Sg=0.6','Sg=0.7','Sg=0.8','
location','best')
xlabel('P(psia)'), ylabel('Compressibility factor, Z')
```

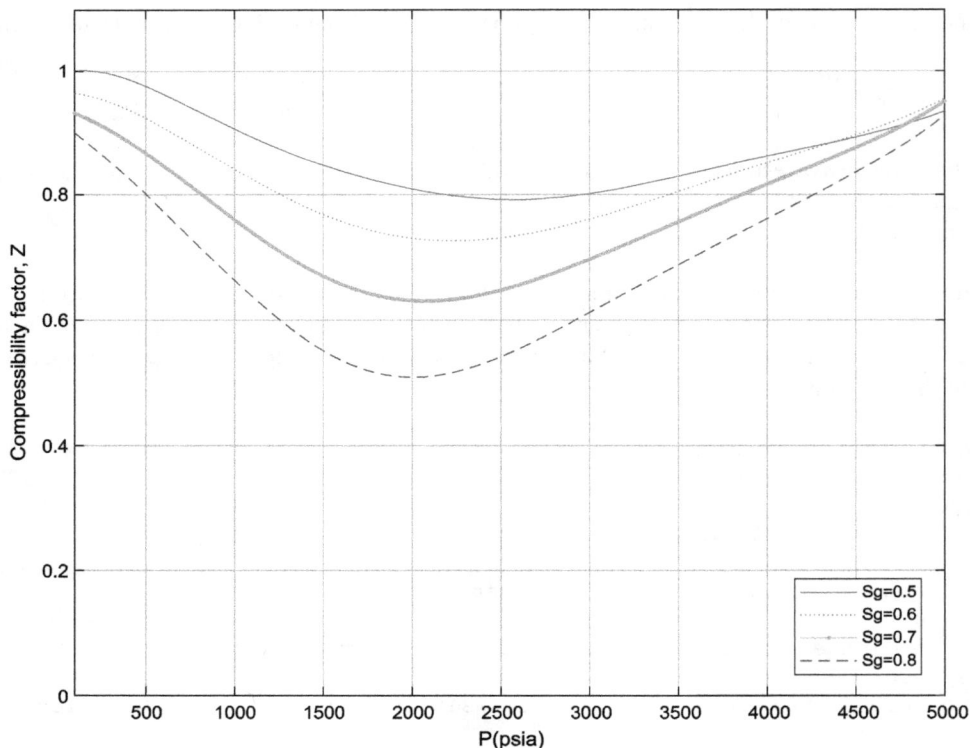

FIGURE 3.8 Plot of compressibility factors of natural gases as a function of pressure and specific gravity.

PROBLEMS

3.1 Estimate the physical properties of saturated steam for $200 \leq T \leq 300(°C)$ at 20°C intervals and tabulate the results.

3.2 Calculate values of absolute humidity (g/m^3) when the dry bulb temperature (T_d) increases from 10 °C to 40 °C at 10 °C intervals and the relative humidity varies from 20% to 80% at 20% intervals.

3.3 In a storage room, the initial air temperature is 28 °C and relative humidity is 73%. After a while, the air temperature and the relative humidity decrease to 16 °C and 55%, respectively. Assuming that the door of the room was kept closed, calculate the amount of water condensed. The volume of air in the room is 150 m^3.

3.4 Calculate the density of water for $0 \leq T \leq 350(°C)$ at 1°C intervals and plot the results as a function of temperature.

3.5 Calculate the viscosity of saturated water at 150 °C.

3.6 Calculate the viscosity of water for $0 \leq T \leq 350(°C)$ and plot the results as a function of temperature.

3.7 Estimate the viscosity of ethane for $25 \leq T \leq 900(°C)$ at 1 °C intervals and plot the results as a function of temperature.

3.8 Estimate the heat capacity of water at 150 °C.

3.9 Estimate the heat capacity of water for $0 \leq T \leq 350(°C)$ and plot the results as a function of temperature.

3.10 Estimate the heat capacity of carbon dioxide for $300 \leq T \leq 1500(K)$ at 1 K intervals and plot the results as a function of temperature.

3.11 Estimate the thermal conductivity of phenol for $0 \leq T \leq 350(°C)$ at 1°C intervals and plot the results as a function of temperature.

3.12 Estimate the thermal conductivities of carbon dioxide and methane for $20 \leq T \leq 500(°C)$ and plot the results as a function of temperature.

3.13 Calculate the surface tension of water (dyne/cm) for $0 \leq T \leq 350(°C)$ and plot the results as a function of temperature.

3.14 Estimate the vapor pressure of water at the range of $0 \leq T \leq 350(°C)$, and plot the results as a function of temperature.

3.15 Calculate the vapor pressures of benzene and phenol (mmHg) for $0 \leq T \leq 300(°C)$ and plot the results as a function of temperature.

3.16 Estimate the vapor pressure of RE-218 (CF_3-O-CF_2-CF_3) at 50.3 °C and at -45°C using the Rarey-Moller method. The normal boiling point is $T_b = -23.7$°C. Compare the results with experimental values (11.319 bar at 50.3 °C and 0.372 bar at –45 °C).[43] Each group and frequency of RE-218 molecules are shown in Table P3.16.

TABLE P3.16

Groups and Frequencies of RE-218 Molecules

Group	ν(Frequency)	ΔB_i
C	3	-0.0896
F_3	2	0.09402
F_2	1	0.1054
O	1	0.15049
no H	1	-0.19373

3.17 Estimate the vapor pressure of 1-propanethiol at $T = 79°C$. Compare the result with the experimental value of $P_v = 1074\,mmHg$.

3.18 Estimate the vapor pressure of 2-methyl-2-pentanethiol at $T = 71°C$. Compare the result with the experimental value of $P_v = 132.93\,mmHg$.

3.19 Estimate the enthalpy of vaporization of acetone at 273.15 K using the derivative relation based on an Antoine equation. The parameters of the Antoine equation are $A = 7.11714$, $B = 1210.595$, and $C = 229.664$.

3.20 Estimate the standard heat of formation of CO and CO_2 for $25 \le T \le 1200(°C)$ and plot the results as a function of temperature.

3.21 Calculate the Gibbs free energy of formation $(kcal/gmol)$ of ethane for $300 \le T \le 1500(K)$ and plot the results as a function of temperature.

3.22 Determine the value of the infinite-dilution diffusion coefficient of propane (A) in chlorobenzene (B) at 0 °C using the Wilke-Chang relation. The molecular weight of chlorobenzene is 112.56, $\mu = 1.05cP$, and the molar volume of propane is $74.5\,cm^3/gmol$, $T = 273.15\,K$, and $\varphi = 1.0$.

3.23 Estimate the diffusion coefficient of allyl chloride in air at 25 °C and 1 atm. Compare the result with the experimental value of $0.0975\,cm^2/sec$.[44]

REFERENCES

1. International Association for the Properties of Water and Steam, *Revised Release of the IAPWS Industrial Formulation 1997 for the Thermodynamic Properties of Water and Steam*, IAPWS, Lucerne, Switzerland, August 2007.
2. International Association for the Properties of Water and Steam, Revised Release on the IAPWS Formulation 1995 for the Thermodynamic Properties of Ordinary Water Substance for General and Scientific Use, IAPWS, Lucerne, Switzerland, September 2016.
3. Smith, J. M., H. C. Van Ness, and M. M. Abbott, *Introduction to Chemical Engineering Thermodynamics*, 7th ed., McGraw-Hill, New York, NY, p. 719, 2005.
4. Yaws, C. L. et al., *Physical Properties*, A Chemical Engineering Publication, McGraw-Hill, New York, NY, 1977.
5. Gmehling, J., B. Kolbe, M. Kleiber, and J. Rarey, *Chemical Thermodynamics for Process Simulation*, Wiley-VCH Verlag GmbH & Co., Weinheim, Germany, pp. 95–96, 2012.
6. Gmehling, J., B. Kolbe, M. Kleiber, and J. Rarey, *Chemical Thermodynamics for Process Simulation*, Wiley-VCH Verlag GmbH & Co., Weinheim, Germany, p. 96, 2012.
7. Nayef Ghasem, *Computer methods in chemical engineering*, CRC Press, Taylor & Francis Group, Boca Raton, FL, pp. 17–19, 2012.
8. Nayef Ghasem, *Computer methods in chemical engineering*, CRC Press, Taylor & Francis Group, Boca Raton, FL, p. 19, 2012.
9. Andrade, E. N. d. C., Properties of dense gases and liquids, *Philosophical Magazine*, 17, 497, p. 698, 1934.
10. Coker, A. K., *Chemical Process Design, Analysis and Simulation*, Gulf Publishing Company, Houston, Texas, p. 121, 1995.
11. Poling, B. E., J. M. Prausnitz, and J. P. O'Connell, *The Properties of Gases and Liquids*, 5th ed., McGraw-Hill, New York, NY, pp. 9–21, 2001.
12. Poling, B. E., J. M. Prausnitz, and J. P. O'Connell, *The Properties of Gases and Liquids*, 5th ed., McGraw-Hill, New York, NY, pp. 9–22, 2001.
13. Coker, A. K., *Chemical Process Design, Analysis and Simulation*, Gulf Publishing Company, Houston, Texas, p. 109, 1995.
14. Gmehling, J., B. Kolbe, M. Kleiber, and J. Rarey, *Chemical Thermodynamics for Process Simulation*, Wiley-VCH Verlag GmbH & Co., Weinheim, Germany, p. 111, 2012.
15. Coker, A. K., *Chemical Process Design, Analysis and Simulation*, Gulf Publishing Company, Houston, Texas, p. 110, 1995.
16. Coker, A. K., *Chemical Process Design, Analysis and Simulation*, Gulf Publishing Company, Houston, Texas, p. 111, 1995.

17. Poling, B. E., J. M. Prausnitz, and J. P. O'Connell, *The Properties of Gases and Liquids*, 5th ed., McGraw-Hill, New York, NY, pp. 12–18, 2001.
18. Poling, B. E., J. M. Prausnitz, and J. P. O'Connell, *The Properties of Gases and Liquids*, 5th ed., McGraw-Hill, New York, NY, pp. 13–16, 2001.
19. Adidharma, H. and V. Temyanko, *Mathcad for Chemical Engineers*, Trafford publishing, Victoria, BC, Canada, p.77, 2007.
20. Coker, A. K., *Chemical Process Design, Analysis and Simulation*, Gulf Publishing Company, Houston, Texas, p. 112, 1995.
21. Poling, B. E., J. M. Prausnitz, and J. P. O'Connell, *The Properties of Gases and Liquids*, 5th ed., McGraw-Hill, New York, NY, p. 7–7, 2001.
22. Poling, B.E., J. M. Prausnitz and J. P. O'Connell, *The Properties of Gases and Liquids*, 5th ed., McGraw-Hill, pp. 7.9–7.10, 2001.
23. Sandler, Stanley I., *Chemical, Biochemical and Engineering Thermodynamics*, 4th ed., Wiley, p. 320, 2006.
24. Sandler, Stanley I., *Chemical, Biochemical and Engineering Thermodynamics*, 4th ed., Wiley, p. 321, 2006.
25. Sandler, Stanley I., *Chemical, Biochemical and Engineering Thermodynamics*, 4th ed., Wiley, p. 319, 2006.
26. Gmehling, J., B. Kolbe, M. Kleiber, and J. Rarey, *Chemical Thermodynamics for Process Simulation*, Wiley-VCH Verlag GmbH & Co., Weinheim, Germany, pp. 84–85, 2012.
27. Gmehling, J., B. Kolbe, M. Kleiber, and J. Rarey, *Chemical Thermodynamics for Process Simulation*, Wiley-VCH Verlag GmbH & Co., Weinheim, Germany, p. 86, 2012.
28. Gmehling, J., B. Kolbe, M. Kleiber, and J. Rarey, *Chemical Thermodynamics for Process Simulation*, Wiley-VCH Verlag GmbH & Co., Weinheim, Germany, p. 89, 2012.
29. Gmehling, J., B. Kolbe, M. Kleiber, and J. Rarey, *Chemical Thermodynamics for Process Simulation*, Wiley-VCH Verlag GmbH & Co., Weinheim, Germany, pp. 89–90, 2012.
30. Osborn, A. G. and Douglin, D. R., *J.Chem.Eng.Data*, 11(4), pp. 502–509 (1966).
31. Thermodynamics Research Center API44 Hydrocarbon Project, *Selected values of properties of hydrocarbon and related compounds*, Texas A&M University, College Station, Texas (1978).
32. Poling, B. E., J. M. Prausnitz, and J. P. O'Connell, *The Properties of Gases and Liquids*, 5th ed., McGraw-Hill, New York, NY, pp. 7–24, 2001.
33. Pitzer, K. S., D. Lippman, R. F. Curl, C. M. Higgins, and D. E. Peterson, A new correlation method for enthalpies of vaporization, *Journal of the American Chemical Society*, 77, p. 3433, 1955.
34. Gmehling, J., B. Kolbe, M. Kleiber, and J. Rarey, *Chemical Thermodynamics for Process Simulation*, Wiley-VCH Verlag GmbH & Co., Weinheim, Germany, pp. 101–102, 2012.
35. Gmehling, J., B. Kolbe, M. Kleiber, and J. Rarey, *Chemical Thermodynamics for Process Simulation*, Wiley-VCH Verlag GmbH & Co., Weinheim, Germany, pp. 102–103, 2012.
36. Coker, A. K., *Chemical Process Design, Analysis and Simulation*, Gulf Publishing Company, Houston, Texas, p. 115, 1995.
37. Coker, A. K., *Chemical Process Design, Analysis and Simulation*, Gulf Publishing Company, Houston, Texas, pp. 118–119, 1995.
38. Poling, B. E., J. M. Prausnitz, and J. P. O'Connell, *The Properties of Gases and Liquids*, 5th ed., McGraw-Hill, New York, NY, pp. 11.21–11.22, 2001.
39. Coker, A. K., *Chemical Process Design, Analysis and Simulation*, Gulf Publishing Company, Houston, Texas, p. 123, 1995.
40. Coker, A. K., *Chemical Process Design, Analysis and Simulation*, Gulf Publishing Company, Houston, Texas, p. 124, 1995.
41. Coker, A. K., *Chemical Process Design, Analysis and Simulation*, Gulf Publishing Company, Houston, Texas, p. 125, 1995.
42. Coker, A. K., *Chemical Process Design, Analysis and Simulation*, Gulf Publishing Company, Houston, Texas, pp. 126–127, 1995.
43. Gmehling, J., B. Kolbe, M. Kleiber, and J. Rarey, *Chemical Thermodynamics for Process Simulation*, Wiley-VCH Verlag GmbH & Co., Weinheim, Germany, p. 91, 2012.
44. Coker, A. K., *Chemical Process Design, Analysis and Simulation*, Gulf Publishing Company, Houston, Texas, p. 127, 1995.

APPENDIX

A1. H2Oprop

```
function w = H2Oprop(dat1,dat2)
% H2Oprop.m: properties (H,S,U,V) of H2O (water and steam)
% Based on the revised release on the IAPWS Industrial formulation 1997 for
% the Thermodynamic Properties of Water and Steam, August 2007 (Ref[1]).
% To find properties of saturated H2O, use satH2Oprop.m
%
% Basic syntax: w = H2Oprop(dat1,dat2)
% Inputs:
%          dat1: pressure(kPa) (range: 0 ~ 100,000 kPa)
%          dat2: temperature(deg.C) (range: 0 ~ 2000 deg.C)
% Output:
%          w- structure of property values (H,S,U,V) of H2O
%          w.H: specific enthalpy (kJ/kg)
%          w.S: specific entropy (kJ/(kg-K))
%          w.U: specific internal energy (kJ/kg)
%          w.V: specific volume (cm^3/g)
%
% Examples:
%    x = H2Oprop(150,350) returns the property structure x at 150 kPa and 350 deg.C

P = dat1/1000; % kPa -> MPa
T = dat2 + 273.15; % deg.C -> K
rgn = regionPT(P,T);
switch rgn
    case 1
        w.H = H1pt(P,T); w.S = S1pt(P,T);
        w.U = U1pt(P,T); w.V = V1pt(P,T)*1e3;
    case 2
        w.H = H2pt(P,T); w.S = S2pt(P,T);
        w.U = U2pt(P,T); w.V = V2pt(P,T)*1e3;
    case 3
        w.H = H3pt(P,T);
        hs = H3pt(P,T); rhos = 1/V3ph(P,hs); w.S = S3rt(rhos,T);
        w.U = U3rt(rhos,T); w.V = V3ph(P,hs)*1e3;
    case 4
        w.H = NaN; w.S = NaN; w.U = NaN; w.V = NaN;
    case 5
        w.H = H5pt(P,T); w.S = S5pt(P,T);
        w.U = U5pt(P,T); w.V = V5pt(P,T)*1e3;
    otherwise
        w.H = NaN; w.S = NaN; w.U = NaN; w.V = NaN;
end
end

% Identify regions for given P and T ------------------------------------
function rgn = regionPT(P,T)
if T > 1073.15 && P < 10 && T < 2273.15 && P > 0.000611, rgn = 5;
elseif T <= 1073.15 && T > 273.15 && P <= 100 && P > 0.000611
  if T > 623.15
        if P > rgn23(T), rgn = 3;
```

```
                        if T < 647.096, val4 = rgn4T(T);
                                if abs(P - val4) < 1e-5, rgn = 4; end
                        end
                else, rgn = 2; end
        else
                val4 = rgn4T(T);
                if abs(P - val4) < 1e-5, rgn = 4;
                elseif P > val4, rgn = 1;
                else rgn = 2;
                end
        end
else
        rgn = 0; % Not in valid region
end
end


% Equations for region 1 ------------------------------------------------
function H1 = H1pt(P,T)
% Eqn.(7) (Ref[1])
% Parameters: Table 2 (Ref[1])
Ii = [0, 0, 0, 0, 0, 0, 0, 0, 1, 1, 1, 1, 1, 1, 2, 2, 2, 2, 2, 3, 3,...
  3, 4, 4, 4, 5, 8, 8, 21, 23, 29, 30, 31, 32];
Ji = [-2, -1, 0, 1, 2, 3, 4, 5, -9, -7, -1, 0, 1, 3, -3, 0, 1, 3, 17,...
  -4, 0, 6, -5, -2, 10, -8, -11, -6, -29, -31, -38, -39, -40, -41];
ni = [0.14632971213167, -0.84548187169114, -3.756360367204, 3.3855169168385,...
    -0.95791963387872, 0.15772038513228, -0.016616417199501, 8.1214629983568e-04,...
    2.8319080123804e-04, -6.0706301565874e-04, -0.018990068218419,...
    -0.032529748770505, -0.021841717175414, -5.283835796993e-05,...
    -4.7184321073267e-04, -3.0001780793026e-04, 4.7661393906987e-05,...
    -4.4141845330846e-06, -7.2694996297594e-16, -3.1679644845054e-05,...
    -2.8270797985312e-06, -8.5205128120103e-10, -2.2425281908e-06,...
    -6.5171222895601e-07, -1.4341729937924e-13, -4.0516996860117e-07,...
    -1.2734301741641e-09, -1.7424871230634e-10, -6.8762131295531e-19,...
    1.4478307828521e-20, 2.6335781662795e-23, -1.1947622640071e-23,...
    1.8228094581404e-24, -9.3537087292458e-26];
R = 0.461526; % kJ/(kg-K)
pv = P/16.53; tau = 1386/T; % pv = P/P*, tau = T*/T, P*=16.53 Mpa, T*=1386K
gt = 0;
for i = 1:34, gt = gt + (ni(i)*(7.1 - pv)^Ii(i) * Ji(i)*(tau-1.222)^(Ji(i)-
1)); end
H1 = R*T*tau*gt;
end


function S1 = S1pt(P,T)
% Eqn.(7) (Ref[1])
% Parameters: Table 2 (Ref[1])
Ii = [0, 0, 0, 0, 0, 0, 0, 0, 1, 1, 1, 1, 1, 1, 2, 2, 2, 2, 2, 3, 3, 3,...
      4, 4, 4, 5, 8, 8, 21, 23, 29, 30, 31, 32];
Ji = [-2, -1, 0, 1, 2, 3, 4, 5, -9, -7, -1, 0, 1, 3, -3, 0, 1, 3, 17,...
      -4, 0, 6, -5, -2, 10, -8, -11, -6, -29, -31, -38, -39, -40, -41];
ni = [0.14632971213167, -0.84548187169114, -3.756360367204,...
      3.3855169168385, -0.95791963387872, 0.15772038513228,...
      -0.016616417199501, 8.1214629983568e-04, 2.8319080123804e-04,...
      -6.0706301565874e-04, -0.018990068218419, -0.032529748770505,...
      -0.021841717175414, -5.283835796993e-05, -4.7184321073267e-04,...
```

```
       -3.0001780793026e-04, 4.7661393906987e-05, -4.4141845330846e-06,...
       -7.2694996297594e-16, -3.1679644845054e-05, -2.8270797985312e-06,...
       -8.5205128120103e-10, -2.2425281908e-06, -6.5171222895601e-07,...
       -1.4341729937924e-13, -4.0516996860117e-07, -1.2734301741641e-09,...
       -1.7424871230634e-10, -6.8762131295531e-19, 1.4478307828521e-20,...
       2.6335781662795e-23, -1.1947622640071e-23, 1.8228094581404e-24,...
      -9.3537087292458e-26];
R = 0.461526; %kJ/(kg-K)
pv = P/16.53; tau = 1386/T; % pv = P/P*, tau = T*/T, P*=16.53 Mpa, T*=1386K
gam = 0; gamt = 0;
for i = 1:34
     gamt = gamt + (ni(i)*(7.1-pv)^Ii(i) * Ji(i) * (tau-1.222)^(Ji(i) - 1));
     gam = gam + ni(i)*(7.1-pv)^Ii(i) * (tau-1.222)^Ji(i); % g(P,T)/(RT)
end
S1 = R*tau*gamt - R*gam;
end

function U1 = U1pt(P,T)
% Eqn.(7) (Ref[1])
% Parameters: Table 2 (Ref[1])
Ii = [0, 0, 0, 0, 0, 0, 0, 0, 1, 1, 1, 1, 1, 1, 2, 2, 2, 2, 2, 3, 3, 3,...
     4, 4, 4, 5, 8, 8, 21, 23, 29, 30, 31, 32];
Ji = [-2, -1, 0, 1, 2, 3, 4, 5, -9, -7, -1, 0, 1, 3, -3, 0, 1, 3, 17,...
     -4, 0, 6, -5, -2, 10, -8, -11, -6, -29, -31, -38, -39, -40, -41];
ni = [0.14632971213167, -0.84548187169114, -3.756360367204,...
     3.3855169168385, -0.95791963387872, 0.15772038513228,...
     -0.016616417199501, 8.1214629983568e-04, 2.8319080123804e-04,...
     -6.0706301565874e-04, -0.018990068218419, -0.032529748770505,...
     -0.021841717175414, -5.283835796993e-05, -4.7184321073267e-04,...
     -3.0001780793026e-04, 4.7661393906987e-05, -4.4141845330846e-06,...
     -7.2694996297594e-16, -3.1679644845054e-05, -2.8270797985312e-06,...
     -8.5205128120103e-10, -2.2425281908e-06, -6.5171222895601e-07,...
     -1.4341729937924e-13, -4.0516996860117e-07, -1.2734301741641e-09,...
     -1.7424871230634e-10, -6.8762131295531e-19, 1.4478307828521e-20,...
     2.6335781662795e-23, -1.1947622640071e-23, 1.8228094581404e-24,...
     -9.3537087292458e-26];
R = 0.461526; % kJ/(kg-K)
pv = P/16.53; tau = 1386/T; % pv = P/P*, tau = T*/T, P*=16.53 Mpa, T*=1386K
gt = 0; gp = 0;
for i = 1:34
  gp = gp - ni(i)*Ii(i)*(7.1-pv)^(Ii(i)-1) * (tau-1.222)^Ji(i);
  gt = gt + (ni(i)*(7.1-pv)^Ii(i) * Ji(i) * (tau-1.222)^(Ji(i)-1));
end
U1 = R*T*(tau*gt - pv*gp);
end

function V1 = V1pt(P,T)
% Eqn.(7) (Ref[1])
% Parameters: Table 2 (Ref[1])
Ii = [0, 0, 0, 0, 0, 0, 0, 0, 1, 1, 1, 1, 1, 1, 2, 2, 2, 2, 2, 3, 3, 3,...
     4, 4, 4, 5, 8, 8, 21, 23, 29, 30, 31, 32];
Ji = [-2, -1, 0, 1, 2, 3, 4, 5, -9, -7, -1, 0, 1, 3, -3, 0, 1, 3, 17,...
     -4, 0, 6, -5, -2, 10, -8, -11, -6, -29, -31, -38, -39, -40, -41];
ni = [0.14632971213167, -0.84548187169114, -3.756360367204,...
     3.3855169168385, -0.95791963387872, 0.15772038513228,...
```

```
        -0.016616417199501, 8.1214629983568e-04, 2.8319080123804e-04,...
        -6.0706301565874e-04, -0.018990068218419, -0.032529748770505,...
        -0.021841717175414, -5.283835796993e-05, -4.7184321073267e-04,...
        -3.0001780793026e-04, 4.7661393906987e-05, -4.4141845330846e-06,...
        -7.2694996297594e-16, -3.1679644845054e-05, -2.8270797985312e-06,...
        -8.5205128120103e-10, -2.2425281908e-06, -6.5171222895601e-07,...
        -1.4341729937924e-13, -4.0516996860117e-07, -1.2734301741641e-09,...
        -1.7424871230634e-10, -6.8762131295531e-19, 1.4478307828521e-20,...
        2.6335781662795e-23, -1.1947622640071e-23, 1.8228094581404e-24,...
        -9.3537087292458e-26];
R = 0.461526; %kJ/(kg K)
pv = P/16.53; tau = 1386/T; % pv = P/P*, tau = T*/T, P*=16.53 Mpa, T*=1386K
gp = 0;
for i = 1:34, gp = gp - ni(i)*Ii(i)*(7.1-pv)^(Ii(i)-1) * (tau-1.222)^Ji(i); end
V1 = R*T*pv*gp/P/1000;
end


% Equations for region 2 -------------------------------------------------
function H2 = H2pt(P,T)
% Parameters: Table 10 and 11 (Ref[1])
Ji0 = [0, 1, -5, -4, -3, -2, -1, 2, 3];
ni0 = [-9.6927686500217, 10.086655968018, -0.005608791128302,...
        0.071452738081455, -0.40710498223928, 1.4240819171444,...
        -4.383951131945, -0.28408632460772, 0.021268463753307];
Ii = [1, 1, 1, 1, 1, 2, 2, 2, 2, 2, 3, 3, 3, 3, 3, 4, 4, 4, 5, 6, 6, 6,...
        7, 7, 7, 8, 8, 9, 10, 10, 10, 16, 16, 18, 20, 20, 20, 21, 22, 23,...
        24, 24, 24];
Ji = [0, 1, 2, 3, 6, 1, 2, 4, 7, 36, 0, 1, 3, 6, 35, 1, 2, 3, 7, 3, 16,...
        35, 0, 11, 25, 8, 36, 13, 4, 10, 14, 29, 50, 57, 20, 35, 48, 21, 53,...
        39, 26, 40, 58];
ni = [-1.7731742473213e-03, -0.017834862292358, -0.045996013696365,...
        -0.057581259083432, -0.05032527872793, -3.3032641670203e-05,...
        -1.8948987516315e-04, -3.9392777243355e-03, -0.043797295650573,...
        -2.6674547914087e-05, 2.0481737692309e-08, 4.3870667284435e-07,...
        -3.227767723857e-05, -1.5033924542148e-03, -0.040668253562649,...
        -7.8847309559367e-10, 1.2790717852285e-08, 4.8225372718507e-07,...
    2.2922076337661e-06, -1.6714766451061e-11, -2.1171472321355e-03,...
    -23.895741934104, -5.905956432427e-18, -1.2621808899101e-06,...
    -0.038946842435739, 1.1256211360459e-11, -8.2311340897998,...
    1.9809712802088e-08, 1.0406965210174e-19, -1.0234747095929e-13,...
    -1.0018179379511e-09, -8.0882908646985e-11, 0.10693031879409,...
    -0.33662250574171, 8.9185845355421e-25, 3.0629316876232e-13,...
    -4.2002467698208e-06, -5.9056029685639e-26, 3.7826947613457e-06,...
    -1.2768608934681e-15, 7.3087610595061e-29, 5.5414715350778e-17,...
    -9.436970724121e-07];
R = 0.461526; % kJ/(kg-K)
pv = P; tau = 540/T; % pv = P/P*, tau = T*/T, P*=1 Mpa, T*=540 K
% Eqns in Table 14 (Ref[1])
gt0 = 0;
for i = 1:9, gt0 = gt0 + ni0(i)*Ji0(i)*tau^(Ji0(i)-1); end
gtr = 0;
for i = 1:43, gtr = gtr + ni(i) * pv^Ii(i) * Ji(i)*(tau-0.5)^(Ji(i)-1); end
H2 = R*T*tau*(gt0 + gtr); % Specific enthalpy (Table 12, Ref[1])
end
```

```
function S2 = S2pt(P,T)
% Parameters: Table 10 and 11 (Ref[1])
Ji0 = [0, 1, -5, -4, -3, -2, -1, 2, 3];
ni0 = [-9.6927686500217, 10.086655968018, -0.005608791128302,...
       0.071452738081455, -0.40710498223928, 1.4240819171444,...
       -4.383951131945, -0.28408632460772, 0.021268463753307];
Ii = [1, 1, 1, 1, 1, 2, 2, 2, 2, 2, 3, 3, 3, 3, 3, 4, 4, 4, 5, 6, 6, 6,...
      7, 7, 7, 8, 8, 9, 10, 10, 10, 16, 16, 18, 20, 20, 20, 21, 22, 23,...
      24, 24, 24];
Ji = [0, 1, 2, 3, 6, 1, 2, 4, 7, 36, 0, 1, 3, 6, 35, 1, 2, 3, 7, 3, 16,...
      35, 0, 11, 25, 8, 36, 13, 4, 10, 14, 29, 50, 57, 20, 35, 48, 21, 53,...
      39, 26, 40, 58];
ni = [-1.7731742473213e-03, -0.017834862292358, -0.045996013696365,...
      -0.057581259083432, -0.05032527872793, -3.3032641670203e-05,...
      -1.8948987516315e-04, -3.9392777243355e-03, -0.043797295650573,...
      -2.6674547914087e-05, 2.0481737692309e-08, 4.3870667284435e-07,...
      -3.227767723857e-05, -1.5033924542148e-03, -0.040668253562649,...
      -7.8847309559367e-10, 1.2790717852285e-08, 4.8225372718507e-07,...
      2.2922076337661e-06, -1.6714766451061e-11, -2.1171472321355e-03,...
      -23.895741934104, -5.905956432427e-18, -1.2621808899101e-06,...
      -0.038946842435739, 1.1256211360459e-11, -8.2311340897998,...
      1.9809712802088e-08, 1.0406965210174e-19, -1.0234747095929e-13,...
      -1.0018179379511e-09, -8.0882908646985e-11, 0.10693031879409,...
      -0.33662250574171, 8.9185845355421e-25, 3.0629316876232e-13,...
      -4.2002467698208e-06, -5.9056029685639e-26, 3.7826947613457e-06,...
      -1.276608934681e-15, 7.3087610595061e-29, 5.5414715350778e-17,...
      -9.436970724121e-07];
R = 0.461526; %kJ/(kg-K)
pv = P; tau = 540/T; % pv = P/P*, tau = T*/T, P*=1 Mpa, T*=540 K
% Eqns in Table 12 and 13 (Ref[1])
g0 = log(pv); gt0 = 0;
for i = 1:9
  g0 = g0 + ni0(i)*tau^Ji0(i); gt0 = gt0 + ni0(i)*Ji0(i)*tau^(Ji0(i)-1);
end
gr = 0; gtr = 0;
for i = 1:43
  gr = gr + ni(i)*pv^Ii(i) * (tau-0.5)^Ji(i);
  gtr = gtr + ni(i)*pv^Ii(i) * Ji(i)*(tau-0.5)^(Ji(i) - 1);
end
S2 = R*(tau*(gt0 + gtr) - (g0 + gr)); % Specific entropy (Table 12, Ref[1])
end


function U2 = U2pt(P,T)
% Parameters: Table 10 and 11 (Ref[1])
Ji0 = [0, 1, -5, -4, -3, -2, -1, 2, 3];
ni0 = [-9.6927686500217, 10.086655968018, -0.005608791128302,...
       0.071452738081455, -0.40710498223928, 1.4240819171444,...
       -4.383951131945, -0.28408632460772, 0.021268463753307];
Ii = [1, 1, 1, 1, 1, 2, 2, 2, 2, 2, 3, 3, 3, 3, 3, 4, 4, 4, 5, 6, 6, 6,...
      7, 7, 7, 8, 8, 9, 10, 10, 10, 16, 16, 18, 20, 20, 20, 21, 22, 23,...
      24, 24, 24];
Ji = [0, 1, 2, 3, 6, 1, 2, 4, 7, 36, 0, 1, 3, 6, 35, 1, 2, 3, 7, 3, 16,...
      35, 0, 11, 25, 8, 36, 13, 4, 10, 14, 29, 50, 57, 20, 35, 48, 21, 53,...
      39, 26, 40, 58];
ni = [-1.7731742473213e-03, -0.017834862292358, -0.045996013696365,...
```

```
        -0.057581259083432, -0.05032527872793, -3.3032641670203e-05,...
        -1.8948987516315e-04, -3.9392777243355e-03, -0.043797295650573,...
        -2.6674547914087e-05, 2.0481737692309e-08, 4.3870667284435e-07,...
        -3.227767723857e-05, -1.5033924542148e-03, -0.040668253562649,...
        -7.8847309559367e-10, 1.2790717852285e-08, 4.8225372718507e-07,...
        2.2922076337661e-06, -1.6714766451061e-11, -2.1171472321355e-03,...
        -23.895741934104, -5.905956432427e-18, -1.2621808899101e-06,...
        -0.038946842435739, 1.1256211360459e-11, -8.2311340897998,...
        1.9809712802088e-08, 1.0406965210174e-19, -1.0234747095929e-13,...
        -1.0018179379511e-09, -8.0882908646985e-11, 0.10693031879409,...
        -0.33662250574171, 8.9185845355421e-25, 3.0629316876232e-13,...
        -4.2002467698208e-06, -5.9056029685639e-26, 3.7826947613457e-06,...
        -1.2768608934681e-15, 7.3087610595061e-29, 5.5414715350778e-17,...
        -9.436970724121e-07];
R = 0.461526; %kJ/(kg-K)
pv = P; tau = 540/T; % pv = P/P*, tau = T*/T, P*=1 Mpa, T*=540 K
% Eqns in Table 12 and 13 (Ref[1])
gp0 = 1/pv; gt0 = 0;
for i = 1:9, gt0 = gt0 + ni0(i)*Ji0(i)*tau^(Ji0(i)-1); end
gpr = 0; gtr = 0;
for i = 1:43  % Table 14 (Ref[1])
  gpr = gpr + ni(i)*Ii(i)*pv^(Ii(i)-1) * (tau-0.5)^Ji(i);
  gtr = gtr + ni(i)*pv^Ii(i)*Ji(i)*(tau-0.5)^(Ji(i)-1);
end
U2 = R*T*(tau*(gt0 + gtr) - pv*(gp0 + gpr)); % Specific internal energy (Table 12,
Ref[1])
end


function V2 = V2pt(P,T)
% Parameters: Table 10 and 11 (Ref[1])
Ji0 = [0, 1, -5, -4, -3, -2, -1, 2, 3];
ni0 = [-9.6927686500217, 10.086655968018, -0.005608791128302,...
       0.071452738081455, -0.40710498223928, 1.4240819171444,...
       -4.383951131945, -0.28408632460772, 0.021268463753307];
Ii = [1, 1, 1, 1, 1, 2, 2, 2, 2, 2, 3, 3, 3, 3, 3, 4, 4, 4, 5, 6, 6, 6,...
      7, 7, 7, 8, 8, 9, 10, 10, 10, 16, 16, 18, 20, 20, 20, 21, 22, 23,...
      24, 24, 24];
Ji = [0, 1, 2, 3, 6, 1, 2, 4, 7, 36, 0, 1, 3, 6, 35, 1, 2, 3, 7, 3, 16,...
      35, 0, 11, 25, 8, 36, 13, 4, 10, 14, 29, 50, 57, 20, 35, 48, 21, 53,...
      39, 26, 40, 58];
ni = [-1.7731742473213e-03, -0.017834862292358, -0.045996013696365,...
      -0.057581259083432, -0.05032527872793, -3.3032641670203e-05,...
      -1.8948987516315e-04, -3.9392777243355e-03, -0.043797295650573,...
      -2.6674547914087e-05, 2.0481737692309e-08, 4.3870667284435e-07,...
      -3.227767723857e-05, -1.5033924542148e-03, -0.040668253562649,...
      -7.8847309559367e-10, 1.2790717852285e-08, 4.8225372718507e-07,...
       2.2922076337661e-06, -1.6714766451061e-11, -2.1171472321355e-03,...
      -23.895741934104, -5.905956432427e-18, -1.2621808899101e-06,...
      -0.038946842435739, 1.1256211360459e-11, -8.2311340897998,...
       1.9809712802088e-08, 1.0406965210174e-19, -1.0234747095929e-13,...
      -1.0018179379511e-09, -8.0882908646985e-11, 0.10693031879409,...
      -0.33662250574171, 8.9185845355421e-25, 3.0629316876232e-13,...
      -4.2002467698208e-06, -5.9056029685639e-26, 3.7826947613457e-06,...
      -1.2768608934681e-15, 7.3087610595061e-29, 5.5414715350778e-17,...
      -9.436970724121e-07];
```

```
R = 0.461526; %kJ/(kg-K)
pv = P; tau = 540/T; % pv = P/P*, tau = T*/T, P*=1 Mpa, T*=540 K
gp0 = 1/pv; gpr = 0;
for i = 1:43, gpr = gpr + ni(i)*Ii(i)*pv^(Ii(i)-1) * (tau-0.5)^Ji(i); end
V2 = R*T*pv*(gp0 + gpr)/P/1000; % Specific volume (Table 12, Ref[1])
end

% Auxiliary equation for the boundary between regions 2 and 3 -----------
function bval = rgn23(T) % Eqn (5) (Ref[1])
bval = 348.05185628969 - 1.1671859879975*T + 1.0192970039326e-3 * T^2;
end

function thet = bn23(P)  % Eqn (6) (Ref[1])
thet = 572.54459862746 + sqrt((P-13.91883977887)/1.0192970039326e-03);
end

% Equations for region 3 ------------------------------------------------
function H3 = H3pt(P,T)
if P < 22.06395      % Below the triple point
 Ts = Ts4(P);        % Saturation temperature (region 4)
 if T <= Ts          % Liquid side
     Ub = H4Lp(P); Lb = H1pt(P,623.15);
 else
    Lb = H4Vp(P); Ub = H2pt(P,bn23(P));
  end
else
 Lb = H1pt(P,623.15); Ub = H2pt(P,bn23(P));
end
Ts = T+1;
while abs(T-Ts) > 1e-5
  Hm = (Lb + Ub)/2; Ts = T3ph(P, Hm);
  if Ts > T, Ub = Hm; else, Lb = Hm; end
end
H3 = Hm;
end

function S3 = S3rt(r,T)
% Parameters: Table 30 (Ref[1])
Ii = [0, 0, 0, 0, 0, 0, 0, 0, 1, 1, 1, 1, 2, 2, 2, 2, 2, 2, 3, 3, 3, 3, 3,...
    4, 4, 4, 4, 5, 5, 5, 6, 6, 6, 7, 8, 9, 9, 10, 10, 11];
Ji = [0, 0, 1, 2, 7, 10, 12, 23, 2, 6, 15, 17, 0, 2, 6, 7, 22, 26, 0, 2,...
    4, 16, 26, 0, 2, 4, 26, 1, 3, 26, 0, 2, 26, 2, 26, 2, 26, 0, 1, 26];
ni = [1.0658070028513, -15.732845290239, 20.944396974307, -7.6867707878716,...
    2.6185947787954, -2.808078114862, 1.2053369696517, -8.4566812812502e-03,...
    -1.2654315477714, -1.1524407806681, 0.88521043984318, -0.64207765181607,...
    0.38493460186671, -0.85214708824206, 4.8972281541877, -3.0502617256965,...
    0.039420536879154, 0.12558408424308, -0.2799932969871, 1.389979956946,...
    -2.018991502357, -8.2147637173963e-03, -0.47596035734923,...
    0.0439840744735, -0.44476435428739, 0.90572070719733, 0.70522450087967,...
    0.10770512626332, -0.32913623258954, -0.50871062041158,...
    -0.022175400873096, 0.094260751665092, 0.16436278447961,...
    -0.013503372241348, -0.014834345352472, 5.7922953628084e-04,...
    3.2308904703711e-03, 8.0964802996215e-05, -1.6557679795037e-04,...
    -4.4923899061815e-05];
R = 0.461526; %kJ/(Kg-K)
```

```
Tc = 647.096; Pc = 22.064; rhoc = 322; % Critical parameters: Tc(K),Pc(MPa),rhoc(kg/m^3)
delt = r/rhoc; tau = Tc/T;
% Eqns in Table 32 (Ref[1])
phi = 0; phit = 0;
for i = 2:40
    phi = phi + ni(i)*delt^Ii(i) * tau^Ji(i);
    phit = phit + ni(i)*delt^Ii(i) * Ji(i)*tau^(Ji(i)-1);
end
phi = phi + ni(1)*log(delt);
S3 = R*(tau*phit - phi); % Specific entropy (Table 32, Ref[1])
end


function U3 = U3rt(r,T)
% Parameters: Table 30 (Ref[1])
Ii = [0, 0, 0, 0, 0, 0, 0, 0, 1, 1, 1, 1, 2, 2, 2, 2, 2, 2, 3, 3, 3, 3, 3,...
    4, 4, 4, 4, 5, 5, 5, 6, 6, 6, 7, 8, 9, 9, 10, 10, 11];
Ji = [0, 0, 1, 2, 7, 10, 12, 23, 2, 6, 15, 17, 0, 2, 6, 7, 22, 26, 0, 2,...
    4, 16, 26, 0, 2, 4, 26, 1, 3, 26, 0, 2, 26, 2, 26, 2, 26, 0, 1, 26];
ni = [1.0658070028513, -15.732845290239, 20.944396974307, -7.6867707878716,...
    2.6185947787954, -2.808078114862, 1.2053369696517, -8.4566812812502e-03,...
    -1.2654315477714, -1.1524407806681, 0.88521043984318, -0.64207765181607,...
    0.38493460186671, -0.85214708824206, 4.8972281541877, -3.0502617256965,...
    0.039420536879154, 0.12558408424308, -0.2799932969871, 1.389979956946,...
    -2.018991502357, -8.2147637173963e-03, -0.47596035734923,...
    0.0439840744735, -0.44476435428739, 0.90572070719733, 0.70522450087967,...
    0.10770512626332, -0.32913623258954, -0.50871062041158,...
    -0.022175400873096, 0.094260751665092, 0.16436278447961,...
    -0.013503372241348, -0.014834345352472, 5.7922953628084e-04,...
    3.2308904703711e-03, 8.0964802996215e-05, -1.6557679795037e-04,...
    -4.4923899061815e-05];
R = 0.461526; %kJ/(Kg-K)
Tc = 647.096; Pc = 22.064; rhoc = 322; % Critical parameters: Tc(K),Pc(MPa),rhoc
(kg/m^3)
delt = r/rhoc; tau = Tc/T; phit = 0;
for i = 2:40, phit = phit + ni(i)*delt^Ii(i) * Ji(i)*tau^(Ji(i)-1); end
U3 = R*T*tau*phit; % Specific internal energy (Table 32, Ref[1])
end


function V3 = V3ph(P,H)
% Backward equations T(p,h) and v(p,h) for subregions 3a and 3b (Ref[1])
h3ab = 2014.64004206875 + 3.74696550136983*P - 2.19921901054187e-02*P^2 +...
    8.7513168600995e-05*P^3;
if H < h3ab % Subregion 3a
  Ii = [-12, -12, -12, -12, -10, -10, -10, -8, -8, -6, -6, -6, -4, -4,...
    -3, -2, -2, -1, -1, -1, -1, 0, 0, 1, 1, 1, 2, 2, 3, 4, 5, 8];
  Ji = [6, 8, 12, 18, 4, 7, 10, 5, 12, 3, 4, 22, 2, 3, 7, 3, 16, 0, 1,...
    2, 3, 0, 1, 0, 1, 2, 0, 2, 0, 2, 2, 2];
  ni = [5.29944062966028e-03, -0.170099690234461, 11.1323814312927,...
    -2178.98123145125, -5.06061827980875e-04, 0.556495239685324,...
    -9.43672726094016, -0.297856807561527, 93.9353943717186,...
    1.92944939465981e-02, 0.421740664704763, -3689141.2628233,...
    -7.37566847600639e-03, -0.354753242424366, -1.99768169338727,...
    1.15456297059049, 5683.6687581596, 8.08169540124668e-03,...
    0.172416341519307, 1.04270175292927, -0.297691372792847,...
    0.560394465163593, 0.275234661176914, -0.148347894866012,...
```

```
        -6.51142513478515e-02, -2.92468715386302, 6.64876096952665e-02,...
        3.52335014263844, -1.46340792313332e-02, -2.24503486668184,...
        1.10533464706142, -4.08757344495612e-02];
    ps = P/100; hs = H/2100; vs = 0;
    for i = 1:32, vs = vs + ni(i)*(ps+0.128)^Ii(i) * (hs-0.727)^Ji(i); end
    V3 = vs * 0.0028;
else % Subregion 3b
  Ii = [-12, -12, -8, -8, -8, -8, -8, -8, -6, -6, -6, -6, -6, -6, -4,...
        -4, -4, -3, -3, -2, -2, -1, -1, -1, -1, 0, 1, 1, 2, 2];
  Ji = [0, 1, 0, 1, 3, 6, 7, 8, 0, 1, 2, 5, 6, 10, 3, 6, 10, 0, 2, 1,...
        2, 0, 1, 4, 5, 0, 0, 1, 2, 6];
  ni = [-2.25196934336318e-09, 1.40674363313486e-08, 2.3378408528056e-06,...
        -3.31833715229001e-05, 1.07956778514318e-03, -0.271382067378863,...
        1.07202262490333, -0.853821329075382, -2.15214194340526e-05,...
        7.6965608822273e-04, -4.31136580433864e-03, 0.453342167309331,...
        -0.507749535873652, -100.475154528389, -0.219201924648793,...
        -3.21087965668917, 607.567815637771, 5.57686450685932e-04,...
        0.18749904002955, 9.05368030448107e-03, 0.285417173048685,...
        3.29924030996098e-02, 0.239897419685483, 4.82754995951394,...
        -11.8035753702231, 0.169490044091791, -1.79967222507787e-02,...
        3.71810116332674e-02, -5.36288335065096e-02, 1.6069710109252];
  ps = P/100; hs = H/2800; vs = 0;
  for i = 1:30, vs = vs + ni(i)*(ps+0.0661)^Ii(i)*(hs-0.72)^Ji(i); end
  V3 = vs*0.0088;
end
end

function T3 = T3ph(P,H)
% Backward equations T(p,h) and v(p,h) for subregions 3a and 3b
h3ab = 2014.64004206875 + 3.74696550136983*P - 2.19921901054187e-02*P^2 +...
       8.7513168600995e-05*P^3;
if H < h3ab % Subregion 3a
  Ii = [-12, -12, -12, -12, -12, -12, -12, -12, -10, -10, -10, -8, -8,...
        -8, -8, -5, -3, -2, -2, -2, -1, -1, 0, 0, 1, 3, 3, 4, 4, 10, 12];
  Ji = [0, 1, 2, 6, 14, 16, 20, 22, 1, 5, 12, 0, 2, 4, 10, 2, 0, 1, 3,...
        4, 0, 2, 0, 1, 1, 0, 1, 0, 3, 4, 5];
  ni = [-1.33645667811215e-07, 4.55912656802978e-06, -1.46294640700979e-05,...
        6.3934131297008e-03, 372.783927268847, -7186.54377460447,...
        573494.7521034, -2675693.29111439, -3.34066283302614e-05,...
        -2.45479214069597e-02, 47.8087847764996, 7.64664131818904e-06,...
        1.28350627676972e-03, 1.71219081377331e-02, -8.51007304583213,...
        -1.36513461629781e-02, -3.84460997596657e-06, 3.37423807911655e-03,...
        -0.551624873066791, 0.72920227710747, -9.92522757376041e-03,...
        -0.119308831407288, 0.793929190615421, 0.454270731799386,...
        0.20999859125991, -6.42109823904738e-03, -0.023515586860454,...
        2.52233108341612e-03, -7.64885133368119e-03, 1.36176427574291e-02,...
        -1.33027883575669e-02];
    ps = P/100; hs = H/2300; Ts = 0;
    for i = 1:31, Ts = Ts + ni(i)*(ps+0.24)^Ii(i) * (hs-0.615)^Ji(i); end
    T3 = Ts*760;
else % Subregion 3b
    Ii = [-12, -12, -10, -10, -10, -10, -10, -8, -8, -8, -8, -8, -6, -6,...
          -6, -4, -4, -3, -2, -2, -1, -1, -1, -1, -1, -1, 0, 0, 1, 3, 5, 6, 8];
    Ji = [0, 1, 0, 1, 5, 10, 12, 0, 1, 2, 4, 10, 0, 1, 2, 0, 1, 5, 0, 4,...
          2, 4, 6, 10, 14, 16, 0, 2, 1, 1, 1, 1, 1];
```

```
        ni = [3.325457364492e-05, -1.27575556587181e-04, -4.75851877356068e-04,...
            1.56183014181602e-03, 0.105724860113781, -85.8514221132534,...
            724.140095480911, 2.96475810273257e-03, -5.92721983365988e-03,...
            -1.26305422818666e-02, -0.115716196364853, 84.9000969739595,...
            -1.08602260086615e-02, 1.54304475328851e-02, 7.50455441524466e-C2,...
            2.52520973612982e-02, -6.02507901232996e-02, -3.07622221350501,...
            -5.74011959864879e-02, 5.03471360939849, -0.925081888584834,...
            3.91733882917546, -77.314600713019, 9493.08762098587,...
            -1410437.19679409, 8491662.30819026, 0.861095729446704,...
            0.32334644281172, 0.873281936020439, -0.436653048526683,...
            0.286596714529479, -0.131778331276228, 6.76682064330275e-03];
        hs = H/2800; ps = P/100; Ts = 0;
        for i = 1:33, Ts = Ts + ni(i)*(ps+0.298)^Ii(i) * (hs-0.72)^Ji(i); end
        T3 = Ts*860;
    end
end


function P3s = P3sath(H)
% Boundary equations psat(h) & psat(s) for the saturation line of region 3
Ii = [0, 1, 1, 1, 1, 5, 7, 8, 14, 20, 22, 24, 28, 36];
Ji = [0, 1, 3, 4, 36, 3, 0, 24, 16, 16, 3, 18, 8, 24];
ni = [0.600073641753024, -9.36203654849857, 24.6590798594147,...
    -107.014222858224, -91582131580576.8, -8623.32011700662,...
    -23.5837344740032, 2.52304969384128e+17, -3.89718771997719e+18,...
    -3.33775713645296e+22, 35649946963.6328, -1.48547544720641e+26,...
    3.30611514838798e+18, 8.13641294467829e+37];
hs = H/2600; ps = 0;
for i = 1:14, ps = ps + ni(i)*(hs-1.02)^Ii(i) * (hs-0.608)^Ji(i); end
P3s = ps*22;
end


% Equations for region 4 -----------------------------------------------
function H4L = H4Lp(P)
if (P > 0.000611657 & P < 22.06395) == 1
    Ts = Ts4(P);
    if P < 16.529, H4L = H1pt(P,Ts);
    else
        Lb = 1670.858218; Ub = 2087.23500164864; ps = -1000;
        while abs(P-ps) > 1e-5
            hm = (Lb + Ub)/2; ps = P3sath(hm);
            if ps > P, Ub = hm; else, Lb = hm; end
        end
        H4L = hm;
    end
else
    H4L = -1e5;
end
end


function H4V = H4Vp(P)
if (P > 0.000611657 & P < 22.06395) == 1
  Ts = Ts4(P);
  if P < 16.529, H4V = H2pt(P,Ts);
  else
```

```
    Lb = 2087.23500164864; Ub = 2568.592004; ps = -1000;
    while abs(P-ps) > 1e-6
       hm = (Lb + Ub)/2; ps = P3sath(hm);
       if ps < P, Ub = hm; else, Lb = hm; end
    end
    H4V = hm;
  end
else
  H4V = -1e5;
end
end


function Ts = Ts4(P)
% Eqn.(31) (saturation-temperature equation) (Ref[1])
% Valid pressure range: 611.213 Pa <= P <= 22.064 MPa
beta = P^0.25; % Eqn.(29a) (Ref[1]), parameters: Table 34 (Ref[1])
E = beta^2 - 17.073846940092*beta + 14.91510861353;
F = 1167.0521452767*beta^2 + 12020.82470247*beta - 4823.2657361591;
G = -724213.16703206*beta^2 - 3232555.0322333*beta + 405113.40542057;
D = 2*G / (-F - (F^2 - 4*E*G)^0.5);
Ts = (650.17534844798 + D - ((650.17534844798 + D)^2 -...
  4*(-0.23855557567849 + 650.17534844798*D))^0.5) / 2;
end


function v4 = rgn4T(T)
x = T - 0.23855557567849 / (T - 650.17534844798);
a = x^2 + 1167.0521452767*x - 724213.16703206;
b = -17.073846940092*x^2 + 12020.82470247*x - 3232555.0322333;
c = 14.91510861353*x^2 - 4823.2657361591*x + 405113.40542057;
v4 = (2*c / (-b + sqrt(b^2 - 4*a*c)))^4;
end


% Equation for region 5 -------------------------------------------------
function H5 = H5pt(P,T)
% Parameters: Tables 37 and 38 (Ref[1])
Ji0 = [0, 1, -3, -2, -1, 2];
ni0 = [-13.179983674201, 6.8540841634434, -0.024805148933466,...
    0.36901534980333, -3.1161318213925, -0.32961626538917];
Iir = [1, 1, 1, 2, 3]; Jir = [0, 1, 3, 9, 3];
nir = [-1.2563183589592e-04, 2.1774678714571e-03, -0.004594282089991,...
    -3.9724828359569e-06, 1.2919228289784e-07];
R = 0.461526; % kJ/(kg-K)
tau = 1000/T; pv = P; gt0 = 0;
for i = 1:6, gt0 = gt0 + ni0(i)*Ji0(i)*tau^(Ji0(i)-1);
end
gt = 0;
for i = 1:5, gt = gt + nir(i)*pv^Iir(i) * Jir(i)*tau^(Jir(i)-1); end
H5 = R*T*tau*(gt0 + gt); % Specific enthalpy (Table 39, Ref[1])
end


function S5 = S5pt(P,T)
% Parameters: Tables 37 and 38 (Ref[1])
Ji0 = [0, 1, -3, -2, -1, 2];
ni0 = [-13.179983674201, 6.8540841634434, -0.024805148933466,...
    0.36901534980333, -3.1161318213925, -0.32961626538917];
```

```
Iir = [1, 1, 1, 2, 3]; Jir = [0, 1, 3, 9, 3];
nir = [-1.2563183589592e-04, 2.1774678714571e-03, -0.004594282089991,...
       -3.9724828359569e-06, 1.2919228289784e-07];
R = 0.461526; % kJ/(kg-K)
tau = 1000/T; pv = P; g0 = log(pv); gt0 = 0;
for i = 1:6
     gt0 = gt0 + ni0(i)*Ji0(i)*tau^(Ji0(i)-1); g0 = g0 + ni0(i)*tau^Ji0(i);
end
gr = 0; gtr = 0;
for i = 1:5
    gr = gr + nir(i)*pv^Iir(i) * tau^Jir(i);
    gtr = gtr + nir(i)*pv^Iir(i) * Jir(i)*tau^(Jir(i)-1);
end
S5 = R*(tau*(gt0 + gtr) - (g0 + gr)); % Specific entropy (Table 39, Ref[1])
end


function U5 = U5pt(P,T)
% Parameters: Tables 37 and 38 (Ref[1])
Ji0 = [0, 1, -3, -2, -1, 2];
ni0 = [-13.179983674201, 6.8540841634434, -0.024805148933466,...
       0.36901534980333, -3.1161318213925, -0.32961626538917];
Iir = [1, 1, 1, 2, 3]; Jir = [0, 1, 3, 9, 3];
nir = [-1.2563183589592e-04, 2.1774678714571e-03, -0.004594282089991,...
       -3.9724828359569e-06, 1.2919228289784e-07];
R = 0.461526; % kJ/(kg-K)
tau = 1000/T; pv = P; gp0 = 1/pv; gt0 = 0;
for i = 1:6, gt0 = gt0 + ni0(i)*Ji0(i)*tau^(Ji0(i)-1); end
gp = 0; gt = 0;
for i = 1:5
    gp = gp + nir(i)*Iir(i)*pv^(Iir(i)-1) * tau^Jir(i);
    gt = gt + nir(i)*pv^Iir(i) * Jir(i)*tau^(Jir(i)-1);
end
U5 = R*T*(tau*(gt0 + gt) - pv*(gp0 + gp)); % Specific internal energy (Table 39, Ref[1])
end


function V5 = V5pt(P,T)
% Parameters: Tables 37 and 38 (Ref[1])
Ji0 = [0, 1, -3, -2, -1, 2];
ni0 = [-13.179983674201, 6.8540841634434, -0.024805148933466,...
       0.36901534980333, -3.1161318213925, -0.32961626538917];
Iir = [1, 1, 1, 2, 3]; Jir = [0, 1, 3, 9, 3];
nir = [-1.2563183589592e-04, 2.1774678714571e-03, -0.004594282089991,...
       -3.9724828359569e-06, 1.2919228289784e-07];
R = 0.461526; % kJ/(kg-K)
tau = 1000/T; pv = P; gp0 = 1/pv; gp = 0;
for i = 1:5, gp = gp + nir(i)*Iir(i)*pv^(Iir(i)-1) * tau^Jir(i); end
V5 = R*T*pv*(gp0 + gp)/ P/1000; % Specific volume (Table 39, Ref[1])
end
```

A2. satH2Oprop

```
function propvalue = satH2Oprop(dtype, phH2O, dvalue)
% satH2Oprop.m: properties of saturated H2O (water and steam)
% Based on the revised release on the IAPWS Industrial formulation 1997 for
```

```
% the Thermodynamic Properties of Water and Steam, August 2007 (Ref[1]).
% To find properties of superheated steam, use H2Oprop.m
%
% Basic syntax: w = satH2Oprop(dtype, phH2O, dvalue)
% Inputs:
%        dtype: data type ( 'P' = pressure, 'T' = temperature)
%        phH2O: phase of H2O ('V' = vapor, 'L' = liquid)
%        dvalue: value of pressure(kPa) or temperature(deg.C)
% Output:
%        w- structure of property values (T,P,H,S,U, or V) of saturated H2O
%        w.P: saturation pressure (kPa)
%        w.T: saturation temperature (deg.C)
%        w.H: specific enthalpy (kJ/kg)
%        w.S: specific entropy (kJ/(kg-K))
%        w.U: specific internal energy (kJ/kg)
%        w.V: specific volume (cm^3/g)
%
% Examples:
%    x = satH2Oprop('P','L',10420) returns the properties of saturated
%        liquid at P = 10420 kPa
%    x = satH2Oprop('T','V',220) returns the properties of saturated vapor
%        at T = 220 degC

dtype = upper(dtype); % dtype = 'P' or 'T'
phH2O = upper(phH2O); % phH2O = 'V' or 'L'
switch dtype
     case 'P' % Properties at P(kPa)
          P = dvalue/1000; % kPa -> MPa
          propvalue.P = P*1e3;
          if P > 0.000611657 && P < 22.06395
           propvalue.T = Ts4(P) - 273.15; % Saturation temperature(deg.C) at P
     else
          propvalue.T = NaN; propvalue.H = NaN; propvalue.S = NaN;
          propvalue.U = NaN; propvalue.V = NaN; return;
     end
     switch phH2O
        case 'V' % Properties of saturated vapor at P (kPa)
            propvalue.H = H4Vp(P); % Saturated vapor enthalpy at P
            if P < 16.529
                propvalue.S = S2pt(P,Ts4(P)); % Saturated vapor entropy at P
                propvalue.U = U2pt(P,Ts4(P)); % Saturated vapor internal energy at P
     propvalue.V = V2pt(P,Ts4(P))*1e3; % Saturated vapor volume at P
   else
       propvalue.S = S3rt(1/(V3ph(P,H4Vp(P))),Ts4(P)); % Saturated vapor entropy at P
       propvalue.U = U3rt(1/(V3ph(P,H4Vp(P))),Ts4(P)); % Saturated vapor internal
       energy at P
       propvalue.V = V3ph(P,H4Vp(P))*1e3; % Saturated vapor volume at P
   end
 case 'L' % Properties of saturated liquid at P (kPa)
     propvalue.H = H4Lp(P); % Saturated liquid enthalpy at P
  if P < 16.529
     propvalue.S = S1pt(P,Ts4(P)); % Saturated liquid entropy at P
     propvalue.U = U1pt(P,Ts4(P)); % Saturated liquid internal energy at P
     propvalue.V = V1pt(P,Ts4(P))*1e3; % Saturated liquid volume at P
  else
```

```
            propvalue.S = S3rt(1/(V3ph(P,H4Lp(P))),Ts4(P)); % Saturated liquid entropy at P
            propvalue.U = U3rt(1/(V3ph(P,H4Lp(P))),Ts4(P)); % Saturated liquid internal
            energy at P
            propvalue.V = V3ph(P,H4Lp(P))*1e3; % Saturated liquid volume at P
       end
end
  case 'T' % Properties at T(deg.C)
        T = dvalue + 273.15; % deg.C -> K
        propvalue.T = dvalue;
        if T < 647.096 && T > 273.15
           propvalue.P = Ps4(T)*1e3; % Saturation pressure(kPa) at T(deg.C)
   else
           propvalue.P = NaN; propvalue.H = NaN; propvalue.S = NaN;
           propvalue.U = NaN; propvalue.V = NaN; return;
   end
     switch phH2O
        case 'V' % Properties of saturated vapor at T(deg.C)
           propvalue.H = H4Vp(Ps4(T)); % Saturated vapor enthalpy at T
           if T <= 623.15
           propvalue.S = S2pt(Ps4(T),T); % Saturated vapor entropy at T
           propvalue.U = U2pt(Ps4(T),T); % Saturated vapor internal energy at T
           propvalue.V = V2pt(Ps4(T),T)*1e3; % Saturated vapor volume at T

      else
       propvalue.S = S3rt(1/(V3ph(Ps4(T),H4Vp(Ps4(T)))),T); % Saturated vap. entropy at T
       propvalue.U = U3rt(1/(V3ph(Ps4(T),H4Vp(Ps4(T)))),T); % Saturated vap.
       internal energy at T
       propvalue.V = V3ph(Ps4(T),H4Vp(Ps4(T)))*1e3; % Saturated vapor
       volume at T
       end
  case 'L' % Properties of saturated liquid at P(kPa)
     propvalue.H = H4Lp(Ps4(T)); % Saturated liquid enthalpy at T
      if T <= 623.15
     propvalue.S = S1pt(Ps4(T),T); % Saturated liquid entropy at T
     propvalue.U = U1pt(Ps4(T),T); % Saturated liquid internal energy at T
     propvalue.V = V1pt(Ps4(T),T)*1e3; % Saturated liquid volume at T
 else
       propvalue.S = S3rt(1/(V3ph(Ps4(T),H4Lp(Ps4(T)))),T); % Saturated liq. entropy at T
       propvalue.U = U3rt(1/(V3ph(Ps4(T), H4Lp(Ps4(T)))), T); % Saturated liq. internal
       energy at T
       propvalue.V = V3ph(Ps4(T),H4Lp(Ps4(T)))*1e3; % Saturated liquid volume at T
           end
       end
end
end


% Equations for region 1 ------------------------------------------------
function H1 = H1pt(P,T)
% Eqn.(7) (Ref[1])
% Parameters: Table 2 (Ref[1])
Ii = [0, 0, 0, 0, 0, 0, 0, 0, 1, 1, 1, 1, 1, 1, 2, 2, 2, 2, 2, 3, 3,...
    3, 4, 4, 4, 5, 8, 8, 21, 23, 29, 30, 31, 32];
Ji = [-2, -1, 0, 1, 2, 3, 4, 5, -9, -7, -1, 0, 1, 3, -3, 0, 1, 3, 17,...
    -4, 0, 6, -5, -2, 10, -8, -11, -6, -29, -31, -38, -39, -40, -41];
ni = [0.14632971213167, -0.84548187169114, -3.756360367204, 3.3855169168385,...
```

```
      -0.95791963387872, 0.15772038513228, -0.016616417199501, 8.1214629983568e-04,...
       2.8319080123804e-04, -6.0706301565874e-04, -0.018990068218419,...
      -0.032529748770505, -0.021841717175414, -5.283835796993e-05,...
      -4.7184321073267e-04, -3.0001780793026e-04, 4.7661393906987e-05,...
      -4.4141845330846e-06, -7.2694996297594e-16, -3.1679644845054e-05,...
      -2.8270797985312e-06, -8.5205128120103e-10, -2.2425281908e-06,...
      -6.5171222895601e-07, -1.4341729937924e-13, -4.0516996860117e-07,...
      -1.2734301741641e-09, -1.7424871230634e-10, -6.8762131295531e-19,...
       1.4478307828521e-20, 2.6335781662795e-23, -1.1947622640071e-23,...
       1.8228094581404e-24, -9.3537087292458e-26];
R = 0.461526; % kJ/(kg-K)
pv = P/16.53; tau = 1386/T; % pv = P/P*, tau = T*/T, P*=16.53 Mpa, T*=1386K
gt = 0;
for i = 1:34, gt = gt + (ni(i)*(7.1 - pv)^Ii(i) * Ji(i)*(tau-1.222)^(Ji(i)-
1)); end
H1 = R*T*tau*gt;
end


function S1 = S1pt(P,T)
% Eqn.(7) (Ref[1])
% Parameters: Table 2 (Ref[1])
Ii = [0, 0, 0, 0, 0, 0, 0, 0, 1, 1, 1, 1, 1, 1, 2, 2, 2, 2, 2, 3, 3, 3,...
      4, 4, 4, 5, 8, 8, 21, 23, 29, 30, 31, 32];
Ji = [-2, -1, 0, 1, 2, 3, 4, 5, -9, -7, -1, 0, 1, 3, -3, 0, 1, 3, 17,...
      -4, 0, 6, -5, -2, 10, -8, -11, -6, -29, -31, -38, -39, -40, -41];
ni = [0.14632971213167, -0.84548187169114, -3.756360367204,...
      3.3855169168385, -0.95791963387872, 0.15772038513228,...
      -0.016616417199501, 8.1214629983568e-04, 2.8319080123804e-04,...
      -6.0706301565874e-04, -0.018990068218419, -0.032529748770505,...
      -0.021841717175414, -5.283835796993e-05, -4.7184321073267e-04,...
      -3.0001780793026e-04, 4.7661393906987e-05, -4.4141845330846e-06,...
      -7.2694996297594e-16, -3.1679644845054e-05, -2.8270797985312e-06,...
      -8.5205128120103e-10, -2.2425281908e-06, -6.5171222895601e-07,...
      -1.4341729937924e-13, -4.0516996860117e-07, -1.2734301741641e-09,...
      -1.7424871230634e-10, -6.8762131295531e-19, 1.4478307828521e-20,...
      2.6335781662795e-23, -1.1947622640071e-23, 1.8228094581404e-24,...
      -9.3537087292458e-26];
R = 0.461526; %kJ/(kg-K)
pv = P/16.53; tau = 1386/T; % pv = P/P*, tau = T*/T, P*=16.53 Mpa, T*=1386K
gam = 0; gamt = 0;
for i = 1:34
  gamt = gamt + (ni(i)*(7.1-pv)^Ii(i) * Ji(i) * (tau-1.222)^(Ji(i) - 1));
  gam = gam + ni(i)*(7.1-pv)^Ii(i) * (tau-1.222)^Ji(i); % g(P,T)/(RT)
end
S1 = R*tau*gamt - R*gam;
end


function U1 = U1pt(P,T)
% Eqn.(7) (Ref[1])
% Parameters: Table 2 (Ref[1])
Ii = [0, 0, 0, 0, 0, 0, 0, 0, 1, 1, 1, 1, 1, 1, 2, 2, 2, 2, 2, 3, 3, 3,...
      4, 4, 4, 5, 8, 8, 21, 23, 29, 30, 31, 32];
Ji = [-2, -1, 0, 1, 2, 3, 4, 5, -9, -7, -1, 0, 1, 3, -3, 0, 1, 3, 17,...
      -4, 0, 6, -5, -2, 10, -8, -11, -6, -29, -31, -38, -39, -40, -41];
ni = [0.14632971213167, -0.84548187169114, -3.756360367204,...
```

```
          3.3855169168385, -0.95791963387872, 0.15772038513228,...
         -0.016616417199501, 8.1214629983568e-04, 2.8319080123804e-04,...
         -6.0706301565874e-04, -0.018990068218419, -0.032529748770505,...
         -0.021841717175414, -5.283835796993e-05, -4.7184321073267e-04,...
         -3.0001780793026e-04, 4.7661393906987e-05, -4.4141845330846e-06,...
         -7.2694996297594e-16, -3.1679644845054e-05, -2.8270797985312e-06,...
         -8.5205128120103e-10, -2.2425281908e-06, -6.5171222895601e-07,...
         -1.4341729937924e-13, -4.0516996860117e-07, -1.2734301741641e-09,...
         -1.7424871230634e-10, -6.8762131295531e-19, 1.4478307828521e-20,...
          2.6335781662795e-23, -1.1947622640071e-23, 1.8228094581404e-24,...
         -9.3537087292458e-26];
R = 0.461526; % kJ/(kg-K)
pv = P/16.53; tau = 1386/T; % pv = P/P*, tau = T*/T, P*=16.53 Mpa, T*=1386K
gt = 0; gp = 0;
for i = 1:34
  gp = gp - ni(i)*Ii(i)*(7.1-pv)^(Ii(i)-1) * (tau-1.222)^Ji(i);
  gt = gt + (ni(i)*(7.1-pv)^Ii(i) * Ji(i)*(tau-1.222)^(Ji(i)-1));
end
U1 = R*T*(tau*gt - pv*gp);
end


function V1 = V1pt(P,T)
% Eqn.(7) (Ref[1])
% Parameters: Table 2 (Ref[1])
Ii = [0, 0, 0, 0, 0, 0, 0, 0, 1, 1, 1, 1, 1, 1, 2, 2, 2, 2, 2, 3, 3, 3,...
      4, 4, 4, 5, 8, 8, 21, 23, 29, 30, 31, 32];
Ji = [-2, -1, 0, 1, 2, 3, 4, 5, -9, -7, -1, 0, 1, 3, -3, 0, 1, 3, 17,...
       -4, 0, 6, -5, -2, 10, -8, -11, -6, -29, -31, -38, -39, -40, -41];
ni = [0.14632971213167, -0.84548187169114, -3.756360367204,...
      3.3855169168385, -0.95791963387872, 0.15772038513228,...
      -0.016616417199501, 8.1214629983568e-04, 2.8319080123804e-04,...
      -6.0706301565874e-04, -0.018990068218419, -0.032529748770505,...
      -0.021841717175414, -5.283835796993e-05, -4.7184321073267e-04,...
      -3.0001780793026e-04, 4.7661393906987e-05, -4.4141845330846e-06,...
      -7.2694996297594e-16, -3.1679644845054e-05, -2.8270797985312e-06,...
      -8.5205128120103e-10, -2.2425281908e-06, -6.5171222895601e-07,...
      -1.4341729937924e-13, -4.0516996860117e-07, -1.2734301741641e-09,...
      -1.7424871230634e-10, -6.8762131295531e-19, 1.4478307828521e-20,...
       2.6335781662795e-23, -1.1947622640071e-23, 1.8228094581404e-24,...
      -9.3537087292458e-26];
R = 0.461526; %kJ/(kg K)
pv = P/16.53; tau = 1386/T; % pv = P/P*, tau = T*/T, P*=16.53 Mpa, T*=1386K
gp = 0;
for i = 1:34, gp = gp - ni(i)*Ii(i)*(7.1-pv)^(Ii(i)-1) * (tau-1.222)^Ji(i); end
V1 = R*T*pv*gp/P/1000;
end


% Equations for region 2 ---------------------------------------------------
function H2 = H2pt(P,T)
% Parameters: Table 10 and 11 (Ref[1])
Ji0 = [0, 1, -5, -4, -3, -2, -1, 2, 3];
ni0 = [-9.6927686500217, 10.086655968018, -0.005608791128302,...
        0.071452738081455, -0.4710498223928, 1.4240819171444,...
      -4.383951131945, -0.28408632460772, 0.021268463753307];
Ii = [1, 1, 1, 1, 1, 2, 2, 2, 2, 2, 3, 3, 3, 3, 3, 4, 4, 4, 5, 6, 6, 6,...
```

```
    7, 7, 7, 8, 8, 9, 10, 10, 10, 16, 16, 18, 20, 20, 20, 21, 22, 23,...
    24, 24, 24];
Ji = [0, 1, 2, 3, 6, 1, 2, 4, 7, 36, 0, 1, 3, 6, 35, 1, 2, 3, 7, 3, 16,...
    35, 0, 11, 25, 8, 36, 13, 4, 10, 14, 29, 50, 57, 20, 35, 48, 21, 53,...
    39, 26, 40, 58];
ni = [-1.7731742473213e-03, -0.017834862292358, -0.045996013696365,...
    -0.057581259083432, -0.05032527872793, -3.3032641670203e-05,...
    -1.8948987516315e-04, -3.9392777243355e-03, -0.043797295650573,...
    -2.6674547914087e-05, 2.0481737692309e-08, 4.3870667284435e-07,...
    -3.227767723857e-05, -1.5033924542148e-03, -0.040668253562649,...
    -7.8847309559367e-10, 1.2790717852285e-08, 4.8225372718507e-07,...
    2.2922076337661e-06, -1.6714766451061e-11, -2.1171472321355e-03,...
    -23.895741934104, -5.905956432427e-18, -1.2621808899101e-06,...
    -0.038946842435739, 1.1256211360459e-11, -8.2311340897998,...
    1.9809712802088e-08, 1.0406965210174e-19, -1.0234747095929e-13,...
    -1.0018179379511e-09, -8.0882908646985e-11, 0.10693031879409,...
    -0.33662250574171, 8.9185845355421e-25, 3.0629316876232e-13,...
    -4.2002467698208e-06, -5.9056029685639e-26, 3.7826947613457e-06,...
    -1.276608934681e-15, 7.3087610595061e-29, 5.5414715350778e-17,...
    -9.436970724121e-07];
R = 0.461526; % kJ/(kg-K)
pv = P; tau = 540/T; % pv = P/P*, tau = T*/T, P*=1 Mpa, T*=540 K
% Eqns in Table 14 (Ref[1])
gt0 = 0;
for i = 1:9, gt0 = gt0 + ni0(i)*Ji0(i)*tau^(Ji0(i)-1); end
gtr = 0;
for i = 1:43, gtr = gtr + ni(i) * pv^Ii(i) * Ji(i)*(tau-0.5)^(Ji(i)-1); end
H2 = R*T*tau*(gt0 + gtr); % Specific enthalpy (Table 12, Ref[1])
end


function S2 = S2pt(P,T)
% Parameters: Table 10 and 11 (Ref[1])
Ji0 = [0, 1, -5, -4, -3, -2, -1, 2, 3];
ni0 = [-9.6927686500217, 10.086655968018, -0.005608791128302,...
    0.071452738081455, -0.40710498223928, 1.4240819171444,...
    -4.383951131945, -0.28408632460772, 0.021268463753307];
Ii = [1, 1, 1, 1, 1, 2, 2, 2, 2, 2, 3, 3, 3, 3, 3, 4, 4, 4, 5, 6, 6, 6,...
    7, 7, 7, 8, 8, 9, 10, 10, 10, 16, 16, 18, 20, 20, 20, 21, 22, 23,...
    24, 24, 24];
Ji = [0, 1, 2, 3, 6, 1, 2, 4, 7, 36, 0, 1, 3, 6, 35, 1, 2, 3, 7, 3, 16,...
    35, 0, 11, 25, 8, 36, 13, 4, 10, 14, 29, 50, 57, 20, 35, 48, 21, 53,...
    39, 26, 40, 58];
ni = [-1.7731742473213e-03, -0.017834862292358, -0.045996013696365,...
    -0.057581259083432, -0.05032527872793, -3.3032641670203e-05,...
    -1.8948987516315e-04, -3.9392777243355e-03, -0.043797295650573,...
    -2.6674547914087e-05, 2.0481737692309e-08, 4.3870667284435e-07,...
    -3.227767723857e-05, -1.5033924542148e-03, -0.040668253562649,...
    -7.8847309559367e-10, 1.2790717852285e-08, 4.8225372718507e-07,...
    2.2922076337661e-06, -1.6714766451061e-11, -2.1171472321355e-03,...
    -23.895741934104, -5.905956432427e-18, -1.2621808899101e-06,...
    -0.038946842435739, 1.1256211360459e-11, -8.2311340897998,...
    1.9809712802088e-08, 1.0406965210174e-19, -1.0234747095929e-13,...
    -1.0018179379511e-09, -8.0882908646985e-11, 0.10693031879409,...
    -0.33662250574171, 8.9185845355421e-25, 3.0629316876232e-13,...
    -4.2002467698208e-06, -5.9056029685639e-26, 3.7826947613457e-06,...
```

```
       -1.2768608934681e-15, 7.3087610595061e-29, 5.5414715350778e-17,...
       -9.436970724121e-07];
R = 0.461526; %kJ/(kg-K)
pv = P; tau = 540/T; % pv = P/P*, tau = T*/T, P*=1 Mpa, T*=540 K
% Eqns in Table 12 and 13 (Ref[1])
g0 = log(pv); gt0 = 0;
for i = 1:9
  g0 = g0 + ni0(i)*tau^Ji0(i); gt0 = gt0 + ni0(i)*Ji0(i)*tau^(Ji0(i)-1);
end
gr = 0; gtr = 0;
for i = 1:43
  gr = gr + ni(i)*pv^Ii(i) * (tau-0.5)^Ji(i);
  gtr = gtr + ni(i)*pv^Ii(i) * Ji(i)*(tau-0.5)^(Ji(i) - 1);
end
S2 = R*(tau*(gt0 + gtr) - (g0 + gr)); % Specific entropy (Table 12, Ref[1])
end


function U2 = U2pt(P,T)
% Parameters: Table 10 and 11 (Ref[1])
Ji0 = [0, 1, -5, -4, -3, -2, -1, 2, 3];
ni0 = [-9.6927686500217, 10.086655968018, -0.005608791128302,...
  0.071452738081455, -0.40710498223928, 1.4240819171444,...
  -4.383951131945, -0.28408632460772, 0.021268463753307];
Ii = [1, 1, 1, 1, 1, 2, 2, 2, 2, 2, 3, 3, 3, 3, 3, 4, 4, 4, 5, 6, 6, 6,...
    7, 7, 7, 8, 8, 9, 10, 10, 10, 16, 16, 18, 20, 20, 20, 21, 22, 23,...
    24, 24, 24];
Ji = [0, 1, 2, 3, 6, 1, 2, 4, 7, 36, 0, 1, 3, 6, 35, 1, 2, 3, 7, 3, 16,...
    35, 0, 11, 25, 8, 36, 13, 4, 10, 14, 29, 50, 57, 20, 35, 48, 21, 53,...
    39, 26, 40, 58];
ni = [-1.7731742473213e-03, -0.017834862292358, -0.045996013696365,...
    -0.057581259083432, -0.05032527872793, -3.3032641670203e-05,...
    -1.8948987516315e-04, -3.9392777243355e-03, -0.043797295650573,...
    -2.6674547914087e-05, 2.0481737692309e-08, 4.3870667284435e-07,...
    -3.227767723857e-05, -1.5033924542148e-03, -0.040668253562649,...
    -7.8847309559367e-10, 1.2790717852285e-08, 4.8225372718507e-07,...
     2.2922076337661e-06, -1.6714766451061e-11, -2.1171472321355e-03,...
    -23.895741934104, -5.905956432427e-18, -1.2621808899101e-06,...
    -0.038946842435739, 1.1256211360459e-11, -8.2311340897998,...
     1.9809712802088e-08, 1.0406965210174e-19, -1.0234747095929e-13,...
    -1.0018179379511e-09, -8.0882908646985e-11, 0.10693031879409,...
    -0.33662250574171, 8.9185845355421e-25, 3.0629316876232e-13,...
    -4.2002467698208e-06, -5.9056029685639e-26, 3.7826947613457e-06,...
    -1.2768608934681e-15, 7.3087610595061e-29, 5.5414715350778e-17,...
    -9.436970724121e-07];
R = 0.461526; %kJ/(kg-K)
pv = P; tau = 540/T; % pv = P/P*, tau = T*/T, P*=1 Mpa, T*=540 K
% Eqns in Table 12 and 13 (Ref[1])
gp0 = 1/pv; gt0 = 0;
for i = 1:9, gt0 = gt0 + ni0(i)*Ji0(i)*tau^(Ji0(i)-1); end
gpr = 0; gtr = 0;
for i = 1:43  % Table 14 (Ref[1])
  gpr = gpr + ni(i)*Ii(i)*pv^(Ii(i)-1) * (tau-0.5)^Ji(i);
  gtr = gtr + ni(i)*pv^Ii(i)*Ji(i)*(tau-0.5)^(Ji(i)-1);
end
```

```
U2 = R*T*(tau*(gt0 + gtr) - pv*(gp0 + gpr)); % Specific internal energy (Table 12,
Ref[1])
end


function V2 = V2pt(P,T)
% Parameters: Table 10 and 11 (Ref[1])
Ji0 = [0, 1, -5, -4, -3, -2, -1, 2, 3];
ni0 = [-9.6927686500217, 10.086655968018, -0.005608791128302,...
        0.071452738081455, -0.40710498223928, 1.4240819171444,...
        -4.383951131945, -0.28408632460772, 0.021268463753307];
Ii = [1, 1, 1, 1, 1, 2, 2, 2, 2, 2, 3, 3, 3, 3, 3, 4, 4, 4, 5, 6, 6, 6,...
      7, 7, 7, 8, 8, 9, 10, 10, 10, 16, 16, 18, 20, 20, 20, 21, 22, 23,...
      24, 24, 24];
Ji = [0, 1, 2, 3, 6, 1, 2, 4, 7, 36, 0, 1, 3, 6, 35, 1, 2, 3, 7, 3, 16,...
      35, 0, 11, 25, 8, 36, 13, 4, 10, 14, 29, 50, 57, 20, 35, 48, 21, 53,...
      39, 26, 40, 58];
ni = [-1.7731742473213e-03, -0.017834862292358, -0.045996013696365,...
      -0.057581259083432, -0.05032527872793, -3.3032641670203e-05,...
      -1.8948987516315e-04, -3.9392777243355e-03, -0.043797295650573,...
      -2.6674547914087e-05, 2.0481737692309e-08, 4.3870667284435e-07,...
      -3.227767723857e-05, -1.5033924542148e-03, -0.040668253562649,...
      -7.8847309559367e-10, 1.2790717852285e-08, 4.8225372718507e-07,...
      2.2922076337661e-06, -1.6714766451061e-11, -2.1171472321355e-03,...
      -23.895741934104, -5.905956432427e-18, -1.2621808899101e-06,...
      -0.038946842435739, 1.1256211360459e-11, -8.2311340897998,...
      1.9809712802088e-08, 1.0406965210174e-19, -1.0234747095929e-13,...
      -1.0018179379511e-09, -8.0882908646985e-11, 0.10693031879409,...
      -0.33662250574171, 8.9185845355421e-25, 3.0629316876232e-13,...
      -4.2002467698208e-06, -5.9056029685639e-26, 3.7826947613457e-06,...
      -1.2768608934681e-15, 7.3087610595061e-29, 5.5414715350778e-17,...
      -9.436970724121e-07];
R = 0.461526; %kJ/(kg-K)
pv = P; tau = 540/T; % pv = P/P*, tau = T*/T, P*=1 Mpa, T*=540 K
gp0 = 1/pv; gpr = 0;
for i = 1:43, gpr = gpr + ni(i)*Ii(i)*pv^(Ii(i)-1) * (tau-0.5)^Ji(i); end
V2 = R*T*pv*(gp0 + gpr)/P/1000;  % Specific volume (Table 12, Ref[1])
end


% Equations for region 3 -------------------------------------------------
function P3s = P3sath(H)
% Boundary equations psat(h) & psat(s) for the saturation line of region 3
Ii = [0, 1, 1, 1, 1, 5, 7, 8, 14, 20, 22, 24, 28, 36];
Ji = [0, 1, 3, 4, 36, 3, 0, 24, 16, 16, 3, 18, 8, 24];
ni = [0.600073641753024, -9.36203654849857, 24.6590798594147,...
      -107.014222858224, -91582131580576.8, -8623.32011700662,...
      -23.5837344740032, 2.52304969384128e+17, -3.89718771997719e+18,...
      -3.33775713645296e+22, 35649946963.6328, -1.48547544720641e+26,...
      3.30611514838798e+18, 8.13641294467829e+37];
hs = H/2600; ps = 0;
for i = 1:14, ps = ps + ni(i)*(hs-1.02)^Ii(i) * (hs-0.608)^Ji(i); end
P3s = ps*22;
end
```

```
function S3 = S3rt(r,T)
% Parameters: Table 30 (Ref[1])
Ii = [0, 0, 0, 0, 0, 0, 0, 0, 1, 1, 1, 1, 2, 2, 2, 2, 2, 2, 3, 3, 3, 3, 3,...
    4, 4, 4, 4, 5, 5, 5, 6, 6, 6, 7, 8, 9, 9, 10, 10, 11];
Ji = [0, 0, 1, 2, 7, 10, 12, 23, 2, 6, 15, 17, 0, 2, 6, 7, 22, 26, 0, 2,...
    4, 16, 26, 0, 2, 4, 26, 1, 3, 26, 0, 2, 26, 2, 26, 2, 26, 0, 1, 26];
ni = [1.0658070028513, -15.732845290239, 20.944396974307, -7.6867707878716,...
    2.6185947787954, -2.808078114862, 1.2053369696517, -8.4566812812502e-03,...
    -1.2654315477714, -1.1524407806681, 0.88521043984318, -0.64207765181607,...
    0.38493460186671, -0.85214708824206, 4.8972281541877, -3.0502617256965,...
    0.039420536879154, 0.12558408424308, -0.2799932969871, 1.389979956946,...
    -2.018991502357, -8.2147637173963e-03, -0.47596035734923,...
    0.0439840744735, -0.44476435428739, 0.90572070719733, 0.70522450087967,...
    0.10770512626332, -0.32913623258954, -0.50871062041158,...
    -0.022175400873096, 0.094260751665092, 0.16436278447961,...
    -0.013503372241348, -0.014834345352472, 5.7922953628084e-04,...
    3.2308904703711e-03, 8.0964802996215e-05, -1.6557679795037e-04,...
    -4.4923899061815e-05];
R = 0.461526; %kJ/(Kg-K)
Tc = 647.096; Pc = 22.064; rhoc = 322; % Critical parameters: Tc(K),Pc(MPa),rhoc
(kg/m^3)
delt = r/rhoc; tau = Tc/T;
% Eqns in Table 32 (Ref[1])
phi = 0; phit = 0;
for i = 2:40
  phi = phi + ni(i)*delt^Ii(i) * tau^Ji(i);
  phit = phit + ni(i)*delt^Ii(i) * Ji(i)*tau^(Ji(i)-1);
end
phi = phi + ni(1)*log(delt);
S3 = R*(tau*phit - phi); % Specific entropy (Table 32, Ref[1])
end

function U3 = U3rt(r,T)
% Parameters: Table 30 (Ref[1])
Ii = [0, 0, 0, 0, 0, 0, 0, 0, 1, 1, 1, 1, 2, 2, 2, 2, 2, 2, 3, 3, 3, 3, 3,...
    4, 4, 4, 4, 5, 5, 5, 6, 6, 6, 7, 8, 9, 9, 10, 10, 11];
Ji = [0, 0, 1, 2, 7, 10, 12, 23, 2, 6, 15, 17, 0, 2, 6, 7, 22, 26, 0, 2,...
    4, 16, 26, 0, 2, 4, 26, 1, 3, 26, 0, 2, 26, 2, 26, 2, 26, 0, 1, 26];
ni = [1.0658070028513, -15.732845290239, 20.944396974307, -7.6867707878716,...
    2.6185947787954, -2.808078114862, 1.2053369696517, -8.4566812812502e-03,...
    -1.2654315477714, -1.1524407806681, 0.88521043984318, -0.64207765181607,...
    0.38493460186671, -0.85214708824206, 4.8972281541877, -3.0502617256965,...
    0.039420536879154, 0.12558408424308, -0.2799932969871, 1.389979956946,...
    -2.018991502357, -8.2147637173963e-03, -0.47596035734923,...
    0.0439840744735, -0.44476435428739, 0.90572070719733, 0.70522450087967,...
    0.10770512626332, -0.32913623258954, -0.50871062041158,...
    -0.022175400873096, 0.094260751665092, 0.16436278447961,...
    -0.013503372241348, -0.014834345352472, 5.7922953628084e-04,...
    3.2308904703711e-03, 8.0964802996215e-05, -1.6557679795037e-04,...
    -4.4923899061815e-05];
R = 0.461526; %kJ/(Kg-K)
Tc = 647.096; Pc = 22.064; rhoc = 322; % Critical parameters: Tc(K),Pc(MPa),rhoc
(kg/m^3)
delt = r/rhoc; tau = Tc/T; phit = 0;
```

```matlab
for i = 2:40, phit = phit + ni(i)*delt^Ii(i) * Ji(i)*tau^(Ji(i)-1); end
U3 = R*T*tau*phit; % Specific internal energy (Table 32, Ref[1])
end


function V3 = V3ph(P,H)
% Backward equations T(p,h) and v(p,h) for subregions 3a and 3b (Ref[1])
h3ab = 2014.64004206875 + 3.74696550136983*P - 2.19921901054187e-02*P^2 +...
  8.7513168600995e-05*P^3;
if H < h3ab % Subregion 3a
  Ii = [-12, -12, -12, -12, -10, -10, -10, -8, -8, -6, -6, -6, -4, -4,...
      -3, -2, -2, -1, -1, -1, -1, 0, 0, 1, 1, 1, 2, 2, 3, 4, 5, 8];
  Ji = [6, 8, 12, 18, 4, 7, 10, 5, 12, 3, 4, 22, 2, 3, 7, 3, 16, 0, 1,...
      2, 3, 0, 1, 0, 1, 2, 0, 2, 0, 2, 2, 2];
  ni = [5.29944062966028e-03, -0.170099690234461, 11.1323814312927,...
      -2178.98123145125, -5.06061827980875e-04, 0.556495239685324,...
      -9.43672726094016, -0.297856807561527, 93.9353943717186,...
      1.92944939465981e-02, 0.421740664704763, -3689141.2628233,...
      -7.37566847600639e-03, -0.354753242424366, -1.99768169338727,...
      1.15456297059049, 5683.6687581596, 8.08169540124668e-03,...
      0.172416341519307, 1.04270175292927, -0.297691372792847,...
      0.560394465163593, 0.275234661176914, -0.148347894866012,...
      -6.51142513478515e-02, -2.92468715386302, 6.64876096952665e-02,...
      3.52335014263844, -1.46340792313332e-02, -2.24503486668184,...
      1.10533464706142, -4.08757344495612e-02];
  ps = P/100; hs = H/2100; vs = 0;
  for i = 1:32, vs = vs + ni(i)*(ps+0.128)^Ii(i) * (hs-0.727)^Ji(i); end
  V3 = vs * 0.0028;
else % Subregion 3b
  Ii = [-12, -12, -8, -8, -8, -8, -8, -8, -6, -6, -6, -6, -6, -6, -4,...
      -4, -4, -3, -3, -2, -2, -1, -1, -1, -1, 0, 1, 1, 2, 2];
  Ji = [0, 1, 0, 1, 3, 6, 7, 8, 0, 1, 2, 5, 6, 10, 3, 6, 10, 0, 2, 1,...
      2, 0, 1, 4, 5, 0, 0, 1, 2, 6];
  ni = [-2.25196934336318e-09, 1.40674363313486e-08, 2.3378408528056e-06,...
      -3.31833715229001e-05, 1.07956778514318e-03, -0.271382067378863,...
      1.07202262490333, -0.853821329075382, -2.15214194340526e-05,...
      7.6965608822273e-04, -4.31136580433864e-03, 0.453342167309331,...
      -0.507749535873652, -100.475154528389, -0.219201924648793,...
      -3.21087965668917, 607.567815637771, 5.57686450685932e-04,...
      0.18749904002955, 9.05368030448107e-03, 0.285417173048685,...
      3.29924030996098e-02, 0.239897419685483, 4.82754995951394,...
      -11.8035753702231, 0.169490044091791, -1.79967222507787e-02,...
      3.71810116332674e-02, -5.36288335065096e-02, 1.6069710109252];
  ps = P/100; hs = H/2800; vs = 0;
  for i = 1:30, vs = vs + ni(i)*(ps+0.0661)^Ii(i)*(hs-0.72)^Ji(i); end
  V3 = vs*0.0088;
end
end


% Equations for region 4 -------------------------------------------------
function Ps = Ps4(T)
% Eqn.(30) (saturation-pressure equation) (Ref[1])
% Valid temperature range: 273.15 K <= T <= 647.096 K
thet = T - 0.23855557567849/(T - 650.17534844798);
A = thet^2 + 1167.0521452767*thet - 724213.16703206;
B = -17.073846940092*thet^2 + 12020.82470247*thet - 3232555.0322333;
```

```
C = 14.91510861353*thet^2 - 4823.2657361591*thet + 405113.40542057;
Ps = (2*C/(-B + (B^2 - 4*A*C)^0.5))^4;
end


function Ts = Ts4(P)
% Eqn.(31) (saturation-temperature equation) (Ref[1])
% Valid pressure range: 611.213 Pa <= P <= 22.064 MPa
beta = P^0.25; % Eqn.(29a) (Ref[1]), parameters: Table 34 (Ref[1])
E = beta^2 - 17.073846940092*beta + 14.91510861353;
F = 1167.0521452767*beta^2 + 12020.82470247*beta - 4823.2657361591;
G = -724213.16703206*beta^2 - 3232555.0322333*beta + 405113.40542057;
D = 2*G / (-F - (F^2 - 4*E*G)^0.5);
Ts = (650.17534844798 + D - ((650.17534844798 + D)^2 -...
  4*(-0.23855557567849 + 650.17534844798*D))^0.5) / 2;
end


function H4V = H4Vp(P)
if (P > 0.000611657 & P < 22.06395) == 1
  Ts = Ts4(P);
  if P < 16.529, H4V = H2pt(P,Ts);
  else
     Lb = 2087.23500164864; Ub = 2568.592004; ps = -1000;
     while abs(P-ps) > 1e-6
       hm = (Lb + Ub)/2; ps = P3sath(hm);
       if ps < P, Ub = hm; else Lb = hm; end
     end
     H4V = hm;
  end
else
  H4V = -1e5;
end
end


function H4L = H4Lp(P)
if (P > 0.000611657 & P < 22.06395) == 1
  Ts = Ts4(P);
  if P < 16.529, H4L = H1pt(P,Ts);
  else
     Lb = 1670.858218; Ub = 2087.23500164864; ps = -1000;
     while abs(P-ps) > 1e-5
       hm = (Lb + Ub)/2; ps = P3sath(hm);
       if ps > P, Ub = hm; else Lb = hm; end
     end
     H4L = hm;
  end
else
  H4L = -1e5;
end
end
```

4 Thermodynamics

4.1 EQUATION OF STATE

4.1.1 VIRIAL STATE EQUATION

The virial equation of state is expressed as[1]

$$Z = \frac{PV}{RT} = 1 + \frac{BP}{RT}$$

where

P, is the pressure (*atm*),
V, is the molar volume (*liter/gmol*),
T, is the temperature (*K*),
R, is the gas constant (0.08206 *atm·liter/(gmol·K)*),
B, is the second virial coefficient (*liter/gmol*), given by[2]

$$B = \frac{RT_c}{P_c}(B^0 + wB^1), \, B^0 = 0.083 - \frac{0.422}{T_r^{1.6}}, \, B^1 = 0.139 - \frac{0.172}{T_r^{4.2}}$$

where

T_c, is the critical temperature (*K*),
P_c, is the critical pressure (*atm*),
w, is the acentric factor
T_r, and P_r, are reduced temperature and pressure, defined as

$$T_r = \frac{T}{T_c}, \, P_r = \frac{P}{P_c}$$

The function *virialEOS* evaluates the virial equation of state. This function computes the compressibility factor and molar volume (*liter/gmol*), at a given temperature (*K*), and pressure (*atm*). A simple expression of its syntax is

```
[Z V] = virialEOS(P,T,Pc,Tc,w)
```

Here, P and Pc are the pressure and critical pressure (*atm*), respectively; T and Tc are the temperature and critical temperature (*K*), respectively; w is the acentric factor; and Z and V are the compressibility factor and molar volume, respectively.

```
function [Z V] = virialEOS(P,T,Pc,Tc,w)
% Estimation of compressibility factor and molar volume
% at given T and P using the virial equation of state
% inputs:
%         P,Pc: pressure and critical pressure (atm)
%         T,Tc: temperature and critical temperature (K)
%         w: acentric factor
% outputs:
%         Z: compressibility factor
```

```
%           V: molar volume
R = 0.08206; % atm-liter/(gmol-K)
Tr = T./Tc; Pr = P./Pc; B0 = 0.083 - 0.422./(Tr.^1.6); B1 = 0.139 - 0.172./(Tr.^4.2);
B = R*Tc.*(B0 + w.*B1)./Pc; Z = 1 + B.*P./(R*T); V = R*Z.*T./P;
end
```

Example 4.1 Compressibility Factor and Molar Volume of Ethane[3]

Use the virial equation of state to determine the compressibility factor and the molar volume of ethane at 50 °C, and 15 bar. For ethane, $T_c = 305.3\,K$, $P_c = 48.08\,atm$, and $w = 0.1$.

Solution
Since 1 bar = 0. 986923atm, $P = (15)(0.986923) = 14.8atm$. The following commands produce the compressibility factor and the molar volume of ethane:
```
>> P = 14.8; T = 323.15; Pc = 48.08; Tc = 305.3; w = 0.1; [Z V] = virialEOS
(P,T,Pc,Tc,w)
Z =
   0.9122
V =
   1.6344
```

Example 4.2 Density of N_2 by Virial Equation[4]

The virial equation of state can be expressed as

$$\frac{P}{\rho RT} = 1 + B\rho + C\rho^2 + D\rho^3$$

where ρ is the molar density, R, is the gas constant (= 0.08206liter·atm/(mol· K)), and B, C, and D, are virial coefficients. For N_2, at $200\,K$, $B = -0.0361\,liter/mol$, $C = 2.7047 \times 10^{-3}(liter/mol)^2$, and $D = -4.4944 \times 10^{-4}(liter/mol)^3$. Plot the density of N_2, at $200\ K$, as a function of pressure $(1 \le P \le 30\,atm)$.

Solution
The density can be determined by solving the nonlinear equation

$$f(\rho) = 1 + B\rho + C\rho^2 + D\rho^3 - \frac{P}{\rho RT} = 0$$

The script *denvir* uses the built-in nonlinear solver *fsolve* to get the results and generate the required plot as shown in Figure 4.1.

```
% denvir.m
B = -0.0361; C = 2.7047e-3; D = -4.4944e-4; % virial constants
R = 0.08206; T = 200; n = 300; P = linspace(1,30,n); x0 = 0.5*ones(1,n);
f = @(x) 1 + B*x + C*x.^2 + D*x.^3 - P./(x*R*T); % define nonlinear equation
rho = fsolve(f,x0); plot(P,rho), xlabel('P(atm)'), ylabel('Density(mol/
liter)'), grid
```

4.1.2 LEE-KESLER EQUATION

The state equations in which the compressibility factor Z is represented as a function of T_r, and P_r, can be applied to most gases and are said to be generalized. Lee and Kesler[5] generalized the BWR

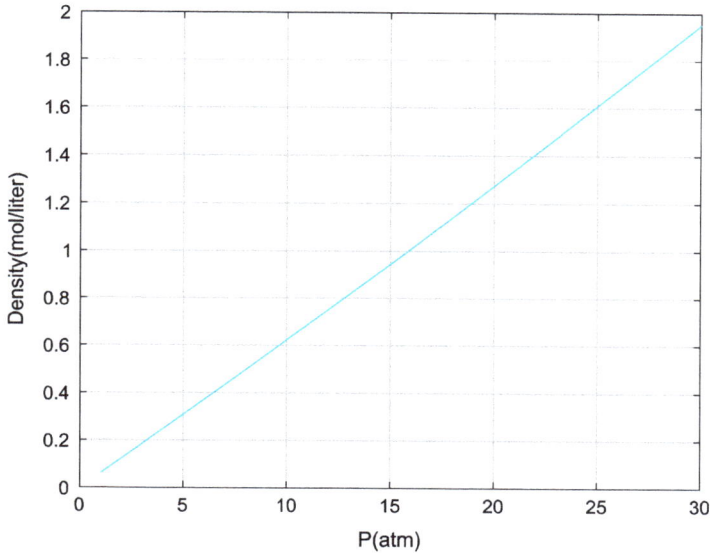

FIGURE 4.1 Plot of the density of N_2, versus pressure.

(Benedict-Webb-Rubin) equation using the corresponding-state principle to predict the properties of a variety of compounds including nonhydrocarbons. They extended the Curl-Pitzer method, which is based on the corresponding-state principle, to a lower temperature range and expressed the compressibility factor of a liquid in terms of the compressibility factor of the simple fluid ($w = 0$), $Z^{(0)}$ and that of a reference fluid (n-octane), $Z^{(r)}$. They used the revised BWR equation to express $Z^{(0)}$, and $Z^{(r)4}$:

$$Z = Z^{(0)} + \frac{w}{w^{(r)}}(Z^{(r)} - Z^{(0)}),\ Z = \left(\frac{P_r V_r}{T_r}\right) = 1 + \frac{B}{V_r} + \frac{C}{V_r^2} + \frac{D}{V_r^5} + \frac{c_4}{T_r^3 V_r^2}\left(\beta + \frac{\gamma}{V_r^2}\right)\exp\left(-\frac{\gamma}{V_r^2}\right)$$

where

$$B = b_1 - \frac{b_2}{T_r} - \frac{b_3}{T_r^2} - \frac{b_4}{T_r^3},\ C = c_1 - \frac{c_2}{T_r} + \frac{c_3}{T_r^3},\ D = d_1 + \frac{d_2}{T_r}$$

The constants b_i, c_i ($i = 1, 2, 3, 4$), d_1, d_2, β, and γ, are shown in Table 4.1.

The function zLK calculates the compressibility factor Z using the Lee-Kesler equation. This function can be called as

Z = zLK(T,Tc,P,Pc,w)

where T is the temperature (K), Tc is the critical temperature (K), P is the pressure (MPa), Pc is the critical pressure (MPa), and w is the acentric factor. The function zLK first computes the constants of the Lee-Kesler equation—B, C, and D—using the subfunction $compBCD$ followed by determination of the roots of the BWR equation. The BWR equation is defined by the subfunction $BWReq$.

```
function Z = zLK(T,Tc,P,Pc,w)
% zLK.m: calculates compressibility factor using the Lee-Kesler equation
% Input:
%    T,Tc: temperature(K)
```

TABLE 4.1
Constants of the Lee-Kesler Equation[6]

Constant	Simple Fluid	Reference Fluid
b_1	0.1181193	0.2026579
b_2	0.265728	0.331511
b_3	0.154790	0.027665
b_4	0.030323	0.203488
c_1	0.0236744	0.0313385
c_2	0.0186984	0.0503618
c_3	0.0	0.041577
c_4	0.042724	0.041577
$10^4 d_1$	0.155488	0.48736
$10^4 d_2$	0.623689	0.0740336
β	0.65392	1.226
γ	0.060167	0.03754

```
%    P,Pc: pressure (MPa)
%    w: acentric factor
%    constants and parameters
% Output:
%    Z: compressibility factor

b = [0.1181193 0.2657280 0.1547900 0.0303230; 0.2026579 0.3315110 0.0276550 0.2034880];
c = [0.0236744 0.0186984 0.0000000 0.0427240; 0.0313385 0.0503618 0.0169010 0.0415770];
d = 1e-4*[0.155488 0.623689; 0.487360 0.0740336];
beta = [0.653920 1.22600]; gam = [0.060167 0.03754]; wr = 0.3978;
Pr = P/Pc; Tr = T/Tc; % reduced pressure and temperature
% Calculation of Z0 of simple fluid (ind = 1)
ind = 1; [B C D] = compBCD(Tr,b,c,d,ind);
Vr0 = Tr./Pr; Vr = fzero(@BWReq,Vr0,[],B,C,D,c,Tr,Pr,ind,beta,gam); Z0 = Pr*Vr./Tr;
% Calculation of Zr of reference fluid (ind = 2)
ind = 2; [B C D] = compBCD(Tr,b,c,d,ind);
Vr = fzero(@BWReq,Vr,[],B,C,D,c,Tr,Pr,ind,beta,gam); Zr = Pr*Vr/Tr;
Z1 = 1/wr * (Zr - Z0);
Z = Z0 + w*Z1; % compressibility factor of real fluid
fprintf('Compressibility factor Z = %f\n', Z);
end

function [B C D] = compBCD(Tr,b,c,d,ind)
% ind 1: Simple fluids, ind 2: Reference fluids
B = b(ind,1) - b(ind,2)/Tr - b(ind,3)/Tr^2 - b(ind,4)/Tr^3;
C = c(ind,1) - c(ind,2)/Tr + c(ind,3)/Tr^3; D = d(ind,1) + d(ind,2)/Tr;
end

function f = BWReq(Vr,B,C,D,c,Tr,Pr,ind,beta,gam)
f = 1 + B/Vr + C/Vr^2 + D/Vr^5 + c(ind, 4)/(Tr^3*Vr^2)*(beta(ind)+...
gam(ind)/Vr^2)*exp(-gam(ind)/Vr^2) - Pr*Vr/Tr;
end
```

Example 4.3   Compressibility Factor of 1-Butene

Determine the compressibility factor of 1-butene at $400\ K$, and $20\ MPa$. For 1-butene, the critical temperature is $419.6\ K$, the critical pressure is 40.2 MPa, and the acentric factor is 0.191.

Solution

The function zLK produces the compressibility factor of 1-butene at given conditions.

```
>> T = 400; Tc = 419.6; P = 20; Pc = 40.2; w = 0.191; Z = zLK(T,Tc,P,Pc,w);
Compressibility factor Z = 0.753883
```

4.1.3   CUBIC EQUATIONS OF STATE

For vapor, the cubic equation of state is written as

$$V = \frac{RT}{P} + b - \frac{a(T)}{P}\frac{V - b}{(V + \epsilon b)(V + \sigma b)}$$

and for liquid, it is written as

$$V = b + (V + \epsilon b)(V + \sigma b)\left[\frac{RT + bP - VP}{a(T)}\right]$$

These equations can be rearranged as a 3rd-order polynomial equation with respect to V, thus the cubic equation. Here, a, and b, are given by

$$a(T) = \Psi\frac{\alpha(T_r)R^2T_c^2}{P_c},\ b = \Omega\frac{RT_c}{P_c}$$

Equations for Z, equivalent to these equations can be obtained by substituting the compressibility factor $Z = PV/(RT)$. For the vapor equation, we have

$$Z = 1 + \beta - q\beta\frac{Z - \beta}{(Z + \epsilon\beta)(Z + \sigma\beta)}$$

and for liquid,

$$Z = \beta + (Z + \epsilon\beta)(Z + \sigma\beta)\left(\frac{1 + \beta - Z}{q\beta}\right)$$

where $= bP/(RT) = \Omega P_r/T_r$, $q = a(T)/(bRT) = \Psi\alpha(T_r)/(\Omega T_r)$. These equations can be rearranged to give a 3rd-order polynomial equation:

$$Z^3 + \{(\sigma + \epsilon)\beta - (1 + \beta)\}Z^2 + \beta\{q + \epsilon\sigma\beta - (1 + \beta)(\sigma + \epsilon)\}Z - \beta^2\{q + (1 + \beta)\epsilon\sigma\} = 0$$

Parameters for each state equation type are given in Table 4.2.

The function $cubicEOSZ$ implements the cubic equation of state. This function uses the built-in function $fzero$ to estimate the compressibility factor and molar volume. A simple calling syntax is
`[Z V] = cubicEOSZ(state, eos, T, P, Tc, Pc, w)`

TABLE 4.2
Parameters for Equations of State[7]

Equation of State	$\alpha(T_r)$,	σ,	ϵ,	Ω,	Ψ,
van der Waals	1	0	0	0.12500	0.42188
Redlich-Kwong	$T_r^{-1/2}$,	1	0	0.08664	0.42748
Soave-Redlich-Kwong	α_{SRK},	1	0	0.08664	0.42748
Peng-Robinson	α_{PR},	$1 + \sqrt{2}$,	$1 - \sqrt{2}$,	0.07780	0.45724

$\alpha_{SRK}(T_r) = [1 + (0.480 + 1.574w - 0.176w^2)(1 - \sqrt{T_r})]^2$,

$\alpha_{PR}(T_r) = [1 + (0.37464 + 1.54226w - 0.26992w^2)(1 - \sqrt{T_r})]^2$,

Here, state denotes the state of the fluid: 'l' or 'L' when the fluid is liquid, 'v' or 'V' when the fluid is vapor. eos is the equation of state being used: 'VDW' when the van der Waals equation is used, 'RK' when the Redlich-Kwong equation is used, 'SRK' when the Soave-Redlich-Kwong equation is used, and 'PR' when the Peng-Robinson equation is used (both capital and lowercase letters are permitted). T and P are the temperature (K), and pressure (bar), respectively; Tc and Pc are the critical temperature (K), and pressure (bar), respectively; and w is the acentric factor. Z and V are estimated values of the compressibility factor and the molar volume (cm^3/mol).

```
function [Z V] = cubicEOSZ(state,eos,T,P,Tc,Pc,w)
% Estimation of Z and V using cubic equations of state
% input
%       state: fluid state (liquid: L, vapor: V)
%       eos: type of equation being used (VDW, RK, SRK, PR)
%       T,P: temperature (K) and pressure (bar)
%       Tc,Pc: critical temperature (K) and pressure (bar)
%       w- acentric factor
% Tr and Pr (reduced T and P)
Tr = T/Tc; Pr = P/Pc; nc = length(w); R = 83.14; % cm^3*bar/mol/K
eos = upper(eos);
switch eos
  case {'VDW'}
     sm = 0; ep = 0; om = 0.125; ps = 0.42188; mx = [0];
  case{'RK'}
     ep = 0; sm = 1; om = 0.08664; ps = 0.42748;
     al = 1./sqrt(Tr); mx = (Tr.^(-1/4) - 1)./(1 - sqrt(Tr));
  case{'SRK'}
     ep = 0; sm = 1; om = 0.08664; ps = 0.42748;
     al = (1+(0.48+1.574*w-0.176*w.^2).*(1-sqrt(Tr))).^2;
     mc = [0.48 1.574 0.176]; mx = [ones(nc,1) w -w.^2]*mc';
  case{'PR'}
     ep = 1 - sqrt(2); sm = 1 + sqrt(2); om = 0.07780; ps = 0.45724;
     mc = [0.37464 1.54226 0.26992]; mx = [ones(nc,1) w -w.^2]*mc';
end
% calculation of alpha, beta and q
al = (1 + mx.*(1-sqrt(Tr))).^2; beta = om*Pr./Tr; q = ps*al./(om*Tr);
% calculation of Z and V
state = upper(state);
```

```
c(1) = 1; c(2) = (sm +ep)*beta - (1+beta); c(3) = beta*(q + ep*sm*beta - (1+beta)*(sm+ep));
c(4) = -beta^2*(q +(1+beta)*ep*sm); Z = roots(c);
switch state
     case 'V', Z = max(Z);
     case 'L', Z = min(Z);
end
V = Z*R*T/P;
end
```

Example 4.4 Molar Volume of *n*-Butane[8]

Determine the molar volumes of saturated vapor and saturated liquid *n*-butane using the cubic equations of state (van der Waals, Redlich-Kwong, Soave-Redlich-Kwong equation, and Peng-Robinson). For *n*-butane, the vapor pressure at 350 K, is 9.4573 bar, $T_c = 425.1K$, $P_c = 37.96bar$, and $w = 0.2$.

Solution

The following commands show the calculation procedure for the saturated vapor using the van der Waals equation:

```
>> T = 350; P = 9.4573; Tc = 425.1; Pc = 37.96; w = 0.2; state = 'v'; eos = 'vdw';
>> [Z V] = cubicEOSZ(state,eos,T,P,Tc,Pc,w)
Z =
  0.8667
V =
  2.6669e+03
```

The calculation procedure for the saturated liquid using the Redlich-Kwong equation is as follows:

```
>> T = 350; P = 9.4573; Tc = 425.1; Pc = 37.96; w = 0.2; state = 'L'; eos = 'rk';
>> [Z V] = cubicEOSZ(state,eos,T,P,Tc,Pc,w)
Z =
  0.0433
V =
133.2663
```

Results of calculations for each equation of state are summarized in Table 4.3.

Example 4.5 Vapor Pressure by Peng-Robinson Equation[10]

The Peng-Robinson equation of state (EOS) is given by

$$P = \frac{RT}{V - b} - \frac{a}{V^2 + 2bV - b^2}$$

$$a = \frac{0.45724R^2T_c^2}{P_c}\{1 + (0.37464 + 1.54226\omega - 0.26992\omega^2)(1 - \sqrt{T_r})\}^2, b = 0.0778\frac{RT_c}{P_c}$$

Figure 4.2 shows the pressure-volume (PV), plot for CO_2, calculated by the Peng-Robinson equation of state when $T = 288.15K$ (15 °C). Graphically, the vapor pressure is a pressure at which the area of the region I shown in the PV, plot is exactly equal to the area of region II. This fact implies that the vapor pressure, P_v^{sat}, should satisfy the following relation:

TABLE 4.3

Results of Calculations (Molar Volume, cm^3/mol)

State	Experimental Data[9]	Equation of State			
		VDW	RK	SRK	PR
Saturated vapor	2482	2666.9	2555.3	2520.3	2486.4
Saturated liquid	115.0[1]	190.981	133.266	127.813	112.602

---

1 VDW = van der Waals; RK = Redlich-Kwong; SRK = Soave-Redlich-Kwong; PR = Peng-Robinson.

$$P_v^{sat}(V_G - V_L) = \int_{V_L}^{V_G} P dV$$

V_G, and V_L, are the roots of the nonlinear equation

$$f(V) = \frac{RT}{V - b} - \frac{a}{V^2 + 2bV - b^2} - P = 0$$

Determine P_v^{sat}, for CO_2, by applying appropriate numerical methods, and compare the result with that obtained from the extended Antoine equation

$$\log P_v = A + \frac{B}{T} + c\log T + DT \ (P_v : mmHg, \ T : K)$$

For CO_2, $T_c = 304.2K$, $P_c = 73.83 bar$, $\omega = 0.224$, $A = 47.544$, $B = -1792.2$, $C = -16.559$, and $D = 0.013833$.

Solution

Let

$$f(V) = \frac{RT}{V - b} - \frac{a}{V^2 + 2bV - b^2} - P$$

and

$$g_i = P_v^{sat}(V_{Gi} - V_{Li}) - \int_{V_{Li}}^{V_{Gi}} P dV$$

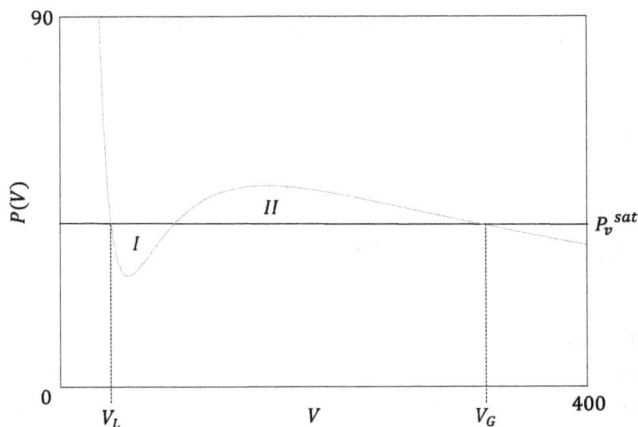

FIGURE 4.2   Pressure-volume plot for CO_2, by Peng-Robinson EOS.

The solution procedure is as follows:

1) $[P_a \quad P_b]$, and let $P_m = (P_a + P_b)/2$. Set initial values for V_L, and V_G. Here, we can set $[P_a \quad P_b] = [43 \quad 55]$, and $[V_{L0} \quad V_{G0}] = [60 \quad 300]$.

2) For each $P = P_i(i = a, b, m)$, solve $f(V) = 0$, to find the corresponding V_{Li}, and $V_{Gi},(i = a, b, m)$.

3) Calculate $g_i,(i = a, b, m)$. The built-in function quad can be used to evaluate the integral.

4) If $g_a * g_m < 0$, set $P_b = P_m$, and $g_b = g_m$. If $g_b * g_m < 0$, set $P_a = P_m$, and $g_a = g_m$. Calculate g_m.

5) If $|P_a - P_b| < \epsilon(=0.001)$, stop the calculation procedure. Otherwise go to step 4).

We use $R = 83.14 cm^3 \cdot bar/(mol \cdot K)$. The script *vppreos* performs this calculation using the bisection method.

```
% vppreos.m: Vp by PR EOS
R = 83.14; w = 0.224; Tc = 304.2; Pc = 73.83; % data for CO2
T = 288.15; Tr = T/Tc;
a = 0.45724*R^2*Tc^2*(1+(0.37464+1.54226*w-0.26992*w^2)*(1-sqrt(Tr)))^2/Pc;
b = 0.0778*R*Tc/Pc;
V = 40:0.5:400; P = R*T./(V - b) - a./(V.^2 + 2*b*V - b^2); % P by PR EOS
plot(V,P), axis([40 400 0 100]), grid, xlabel('V(cm^3/mol'), ylabel('P(bar)')
% Determine Pv using bisection method
x0l = 60; x0g = 300; % initial guess for Vl and Vg
xa = 43; fa = @(x) R*T./(x - b) - a./(x.^2 + 2*b*x - b^2) - xa;
xb = 55; fb = @(x) R*T./(x - b) - a./(x.^2 + 2*b*x - b^2) - xb;
xm = (xa + xb)/2; fm = @(x) R*T./(x - b) - a./(x.^2 + 2*b*x - b^2) - xm;
Ppr = @(x) R*T./(x - b) - a./(x.^2 + 2*b*x - b^2);
Vla = fsolve(fa,x0l); Vga = fsolve(fa,x0g); da = quad(Ppr,Vla,Vga) - xa*(Vga - Vla);
Vlb = fsolve(fb,x0l); Vgb = fsolve(fb,x0g); db = quad(Ppr,Vlb,Vgb) - xb*(Vgb - Vlb);
Vlm = fsolve(fm,x0l); Vgm = fsolve(fm,x0g); dm = quad(Ppr,Vlm,Vgm) - xm*(Vgm - Vlm);
crit = abs(xa - xb);
while crit > 1e-3 % Pv by bisection method
    if da*dm < 0, xb = xm; db = dm;
    elseif dm*db < 0, xa = xm; da = dm; end
    xm = (xa + xb)/2; Vlm = fsolve(fm,x0l); Vgm = fsolve(fm,x0g);
    dm = quad(Ppr,Vlm,Vgm) - xm*(Vgm - Vlm); crit = abs(xa - xb);
end
Pv = xm

>> vppreos
Pv =
 50.8219
```

The vapor pressure by the extended Antoine equation is given by
```
>> T = 288.15; Pv = 10^(47.544-1792.2/T -16.559*log10(T) + 0.013833*T)/
750.0615
Pv =
 50.8604
```

The result obtained from the extended Antoine equation is divided by 750.0615 because $1 bar = 750.0615 mmHg$. We can see that the difference between these two values is $0.0385 bar$.

TABLE 4.4
Thermodynamic State Equation Models[11]

Type	Name	Phase(s)
Ideal	Ideal gas law	V
Virial EOS	Nothnagel	V
	Hayden-O'Connell	V
	Benedict-Webb-Rubin	V,L[1]
	Lee-Kesler (LK)	V,L
Cubic EOS	Redlich-Kwong (RK)	V
	Soave-Redlich-Kwong (SRK)	V,L
	Peng-Robinson (PR)	V,L
	PR-Wong-Sandler (WS)	V,L
Mixing rules	Schwartzentruber-Renon	V,L
	Predictive SRK mixing rules	V,L
	WS mixing rules	V,L
	Modified Huron-Vidal mixing rules	V,L

1   V = vapor; L = liquid.

4.1.4 THERMODYNAMIC STATE MODELS

In choosing the correct property method, we have to consider the following factors: (1) the type of the thermodynamic property of interest, (2) the composition of the mixture, (3) the range of temperatures and pressures, and (4) the availability and reliability of the parameters. Table 4.4 shows the thermodynamic state equation models commonly available in commercial process software packages.

4.2 THERMODYNAMIC PROPERTIES OF FLUIDS

4.2.1 ENTHALPY CHANGE

The heat capacity, C_p, depends on temperature and can be expressed as a function of temperature:

$$\frac{C_p}{R} = A + BT + CT^2 + DT^{-2}$$

For some materials C, or D, may be 0. The enthalpy change, ΔH, and the mean heat capacity, $<C_p>_H$, are given by

$$\Delta H = Q = \int_{T_1}^{T_2} C_p dT, \ <C_p>_H = \frac{\Delta H}{T_2 - T_1}$$

The function *delH* estimates ΔH, and $<C_p>_H$, using the coefficients of the heat capacity relation at the given temperature range.

```
function [Q, mc] = delH(C,T1,T2)
```

```
% Calculates enthalpy change and avg. heat capacity of pure material
% input:
%      C: coefficients of Cp relation (C=[A,B,C,D])
%      T1,T2: temperature range (lower and upper limits of integral)
% output:
%      Q: delta H
R = 8.314; % J/(mol-K)
fH = @(T) C(1) + C(2)*T + C(3)*T.^2 + C(4)*T.^(-2); %T(K)
intCp = quadl(fH,T1,T2); Q = R*intCp; % J
mc = intCp/(T2-T1); % (Cp)h/R
end
```

Example 4.6 Enthalpy Change of Methane[12]

Determine the mean heat capacity, $<C_p>_H/R$, and the heat required to raise the temperature of 1 mol of methane from 260 °C, to 600 °C, in a steady flow process at a sufficiently low pressure that methane may be considered an ideal gas. For methane, the heat capacity is given by $C_p = 1.702 + 9.081 \times 10^{-3}T - 2.164 \times 10^{-6}T^2$.

Solution
The function *delH* produces the desired outputs:
```
>> [Q mc] = delH([1.702 9.081e-3 -2.164e-6 0], 533.15, 873.15)
Q =
  1.9778e+04
mc =
  6.9965
```

4.2.2 DEPARTURE FUNCTION

The departure function implies the departure of the actual fluid property from the same ideal gas property at the same temperature and pressure. The changes in thermodynamic properties of an actual fluid are equal to those for an ideal gas undergoing the same change of state plus the departure from ideal gas behavior of the initial state. Once the fluid equation of state is known, the departure function can be evaluated.

The change in enthalpy is expressed as

$$\Delta H = H_2 - H_1 = (H_2 - H_2^{ig}) + (H_2^{ig} - H_1^{ig}) - (H_1 - H_1^{ig})$$

Here, $H_2 - H_2^{ig}$, and $H_1 - H_1^{ig}$, are calculated by using the departure function, and $H_2^{ig} - H_1^{ig}$, is given by

$$H_2^{ig} - H_1^{ig} = \int_{T_1}^{T_2} C_p dT = \int_{T_1}^{T_2} (A + BT + CT^2 + DT^3)dT$$

Similarly, the change in entropy is written as

$$\Delta S = S_2 - S_1 = (S_2 - S_2^{ig}) + (S_2^{ig} - S_1^{ig}) - (S_1 - S_1^{ig})$$

where $S_2^{ig} - S_1^{ig}$, is obtained from the following relationship:

$$S_2^{ig} - S_1^{ig} = \int_{T_1}^{T_2} \frac{C_p}{T}dT + R\ln\frac{P_2}{P_1} = \int_{T_1}^{T_2} \left(\frac{A}{T} + B + CT + DT^2\right)dT + R\ln\frac{P_2}{P_1}$$

4.2.2.1 Departure Function from the Virial Equation of State

The virial equation of state may be used to represent the enthalpy and entropy departure functions:

$$\frac{H - H^{ig}}{RT} = -P_r \left[\frac{1.0972}{T_r^{2.6}} - \frac{0.083}{T_r} + w\left(\frac{0.8944}{T_r^{5.2}} - \frac{0.139}{T_r} \right) \right], \frac{S - S^{ig}}{R}$$

$$= -P_r \left(\frac{0.675}{T_r^{2.6}} + w\frac{0.722}{T_r^{5.2}} \right)$$

4.2.2.2 Departure Function from the VDW (van der Waals) Equation of State

Using the VDW (van der Waals) equation of state, we obtain the expressions for the enthalpy and entropy departure functions:

$$\frac{H - H^{ig}}{RT} = Z - 1 - \frac{3.375\beta}{ZT_r}, \frac{S - S^{ig}}{R} = \ln(Z - \beta)$$

where $\beta = bP/(RT)$.

4.2.2.3 Departure Function from the RK (Redlich-Kwong) Equation of State

Using the RK (Redlich-Kwong) equation of state, we obtain the expressions for the enthalpy and entropy departure functions:

$$\frac{H - H^{ig}}{RT} = (Z - 1) - \frac{1.5\Psi}{\sigma T_r^{1.5}}\ln\left(1 + \frac{b}{V} \right), \frac{S - S^{ig}}{R} = \ln\left(Z - \frac{bP}{RT} \right) - \frac{\Psi}{2\sigma T_r^{1.5}}\ln\left(1 + \frac{b}{V} \right)$$

4.2.2.4 Departure Function from the SRK (Soave-Redlich-Kwong) Equation of State

If the SRK (Soave-Redlich-Kwong) equation of state is used, the enthalpy and entropy departure functions are represented as

$$\frac{H - H^{ig}}{RT} = Z - 1 - \frac{A}{\beta}\left(\frac{\kappa\sqrt{T_r}}{1 + \kappa\left(1 - \sqrt{T_r} \right)} + 1 \right)\ln\left(1 + \frac{\beta}{Z} \right)$$

$$\frac{S - S^{ig}}{R} = \ln(Z - \beta) - \frac{A}{\beta}\left(\frac{\kappa\sqrt{T_r}}{1 + \kappa\left(1 - \sqrt{T_r} \right)} \right)\ln\left(1 + \frac{\beta}{Z} \right)$$

where $A = aP/(R^2T^2)$.

4.2.2.5 Departure Function from the PR (Peng-Robinson) Equation of State

The PR (Peng-Robinson) equation of state may be used to represent departure functions. The enthalpy departure function is given by

$$\frac{H - H^{ig}}{RT} = Z - 1 - \frac{A}{\beta\sqrt{8}}\left(1 + \frac{\kappa\sqrt{T_r}}{\sqrt{\alpha}}\right)\ln\left[\frac{Z + (1 + \sqrt{2})\beta}{Z + (1 - \sqrt{2})\beta}\right]$$

The entropy departure function is given by

$$\frac{S - S^{ig}}{R} = \ln(Z - \beta) - \frac{A}{\beta\sqrt{8}}\left(\frac{\kappa\sqrt{T_r}}{\sqrt{\alpha}}\right)\ln\left[\frac{Z + (1 + \sqrt{2})\beta}{Z + (1 - \sqrt{2})\beta}\right]$$

The internal energy departure function is given by

$$\frac{U - U^{ig}}{RT} = -\frac{A}{\beta\sqrt{8}}\frac{\kappa\sqrt{T_r}}{\sqrt{\alpha}}\ln\left[\frac{Z + (1 + \sqrt{2})\beta}{Z + (1 - \sqrt{2})\beta}\right]$$

The Gibbs free energy departure function is given by

$$\frac{G - G^{ig}}{RT} = Z - 1 - \ln(Z - \beta) - \frac{A}{\beta\sqrt{8}}\ln\left[\frac{Z + (1 + \sqrt{2})\beta}{Z + (1 - \sqrt{2})\beta}\right]$$

The function *deptfun* calculates departure functions. The basic syntax is

```
[Z V dH dS] = deptfun(state,eos,T,P,Tc,Pc,w)
```

where state denotes the state of the fluid (liquid: L, vapor: V); eos is the equation of state being used (VR, VDW, RK, SRK, PR); T and P are the temperature (K), and pressure (*bar*), respectively; Tc and Pc are the critical temperature (K), and pressure (*bar*), respectively; w is the acentric factor; Z is the compressibility factor; V is the molar volume (cm^3/mol); dH is the change in enthalpy (J/mol); and dS is the change in entropy ($J/(mol \cdot K)$).

```
function [Z V dH dS] = deptfun(state,eos,T,P,Tc,Pc,w)
% Calculation of departure functions using the virial and cubic EOS
% inputs
%       state: fluid state (liquid: L, vapor: V)
%       eos: type of the equation of state (VR, VDW, RK, SRK, PR)
%       T,P: temperature (K) and pressure (bar)
%       Tc,Pc: critical temperature (K) and critical pressure (bar)
%       w- acentric factor
% outputs:
%       Z: compressibility factor
%       V: molar volume
%       dH: enthalpy departure
%       dS: entropy departure
% Tr and Pr (reduced T and P)
Tr = T/Tc; Pr = P/Pc; R = 83.14; % cm^3*bar/mol/K
eos = upper(eos);
switch eos
    case 'VDW', al = 1; sm = 0; ep = 0; om = 0.125; ps = 0.42188; kappa = 0;
    case 'RK', al = 1./sqrt(Tr); sm = 1; ep = 0; om = 0.08664; ps = 0.42748;
    case 'SRK', kappa = 0.480 + 1.574*w - 0.176*w^2;
      al = (1 + kappa*(1-sqrt(Tr))).^2; sm = 1; ep = 0; om = 0.08664; ps = 0.42748;
    otherwise % PR, or VR
    kappa = 0.37464 + 1.54226*w - 0.26992*w^2; al = (1 + kappa*
    (1-sqrt(Tr))).^2;
    sm = 1+sqrt(2); ep = 1-sqrt(2); om = 0.0778; ps = 0.45724;
end
```

```
% compressibility factor (Z)
state = upper(state); beta = om*Pr./Tr; q = ps*al./(om*Tr); % beta and q
if strcmp(eos,'VR') % virial (VR) EOS: vapor phase
    B0 = 0.083 - 0.422./(Tr.^1.6); B1 = 0.139 - 0.172./(Tr.^4.2);
    B = R*Tc.*(B0 + w.*B1)./Pc; Z = 1 + B.*P./(R*T);
else
    fV = @(Z) 1+beta-q*beta.*(Z-beta)./((Z+ep*beta).*(Z+sm*beta)) - Z;
    fL = @(Z) beta+(Z+ep*beta).*(Z+sm*beta).*(1+beta-Z)./(q.*beta) - Z;
    switch state
        case 'V', Z = fzero(fV, 1);
         case 'L', Z = fzero(fL, beta);
    end
end
V = Z*R*T/P; % cm^3/mol
% departure function
a = ps*al*R^2*Tc.^2/Pc; b = om*R*Tc./Pc; % a, b
Ad = a*P/(R^2*T^2); Bd = b*P/(R*T); % A, B
if strcmp(eos,'VR') % virial EOS: vapor phase
    dH = -Pr*(1.0972/Tr^2.6-0.083/Tr+w*(0.8944/Tr^5.2-0.139/Tr))*R*T;
    dS = -Pr*(0.675/Tr^2.6 + w*0.722/Tr^5.2)*R;
elseif strcmp(eos,'RK') % Redlich-Kwong EOS
    dH = (Z - 1 - 1.5*ps/(om*Tr^1.5) * log(1 + b/V))*R*T;
    dS = (log(Z-Bd) - 0.5*ps/(om*Tr^1.5) * log(1 + b/V))*R;
elseif strcmp(eos,'VDW') % van der Waals EOS
    dH = (Z - 1 - 3.375*Bd/Tr/Z)*R*T;
    dS = R*log(Z-Bd);
elseif strcmp(eos,'SRK') % Soave-Redlich-Kwong EOS
    dH = (Z - 1 + (-kappa*sqrt(Tr)/(1 + kappa*(1-sqrt(Tr))) - 1)*(Ad/Bd)
    *log(1 + Bd/Z))*R*T;
    dS = (log(Z - Bd) - kappa*sqrt(Tr)/(1 + kappa*(1-sqrt(Tr)))*(Ad/Bd)
    *log(1 + Bd/Z))*R;
else % Peng-Robinson EOS
    dH = (Z-1 -(Ad/(Bd*sqrt(8)))*(1+ kappa*sqrt(Tr)/sqrt(al))
    *log((Z + sm*Bd)/(Z + ep*Bd)))*R*T;
    dS = (log(Z-Bd) - (Ad/(Bd*sqrt(8)))*(kappa*sqrt(Tr)/sqrt(al))
    *log((Z + sm*Bd)/(Z + ep*Bd)))*R;
end
% R = 83.14 cm^3*bar/mol/K = 8.314 J/mol/K
dH = dH/10; dS = dS/10; % dH:J/mol, dS:J/mol/K
end
```

The function *delHS* calculates changes in enthalpy and entropy of pure fluids due to phase changes. This function uses the function *deptfun* to evaluate the departure function. The simple syntax is

```
[dH dS] = delHS(state,eos,T1,P1,T2,P2,A,B,C,D,Tc,Pc,w)
```

where state denotes the state of the fluid (liquid: L, vapor: V); eos is the equation of state being used (VR, VDW, RK, SRK, PR); T1, T2 and P1, P2 are the temperature (K), and pressure (*bar*), at states 1 and 2, respectively; A, B, C, and D are coefficients of Cp equations; Tc and Pc are the critical temperature (K), and pressure (*bar*), respectively; w is the acentric factor; dH is the change in enthalpy (J/mol); and dS is the change in entropy ($J/(mol·K)$).

```
function [dH dS] = delHS(state,eos,T1,P1,T2,P2, A,B,C,D,Tc,Pc,w)
% delHS.m: calculates changes in H and S of pure fluids due to phase change
% inputs
```

```
%        state: fluid state (liquid: L, vapor: V)
%        eos: type of the equation of state (VR, VDW, RK, SRK, PR)
%        T1,P1: temperature (K) and pressure (bar) at state 1
%        T2,P2: temperature (K) and pressure (bar) at state 2
%        A, B, C, D: coefficients of Cp equation
% outputs:
%        dH: change in enthalpy
%        dS: change in entropy
% changes in enthalpy(H) and entropy(S)
[Z1 V1 dH1 dS1] = deptfun(state,eos,T1,P1,Tc,Pc,w);
[Z2 V2 dH2 dS2] = deptfun(state,eos,T2,P2,Tc,Pc,w);
% phase change in ideal gas
R = 8.314; fH = @(T) A + B*T + C*T.^2 + D*T.^3; fS = @(T) A./T + B + C*T + D*T.^2;
dHi = integral(fH,T1,T2); dSi = integral(fS,T1,T2); dSi = integral(fS,T1,T2) -
R*log(P2./P1);
% changes in H and S due to phase change
dH = dH2 + dHi - dH1; dS = dS2 + dSi - dS1;
end
```

Example 4.7 Enthalpy and Entropy Departure of n-Butane Gas[13,14]

Determine the enthalpy departure $H^R = H - H^{ig}$, and the entropy departure $S^R = S - S^{ig}$, for n-butane gas at $50bar$, and $500K$. For n-butane gas, $T_c = 425.1K$, $P_c = 37.96bar$, and $w = 0.2$.

Solution

Set state = 'v'. The script *resbutane* calls the function *deptfun* to calculate residual properties.

```
% resbutane.m: residual properties of n-butane
eosset = {'VR', 'VDW', 'RK', 'SRK', 'PR'};
Tc = 425.1; Pc = 37.96; w = 0.2; T = 500; P = 50; state = 'v';
for i = 1:length(eosset)
     eos = eosset{i}; [Z V dH dS] = deptfun(state,eos,T,P,Tc,Pc,w);
      fprintf('The equation of state=%s: Z=%g H^R=%g S^R=%g\n', eos,Z,dH,dS);
end
```

The solution can now be obtained by executing the script *resbutane*.

```
>> resbutane
The equation of state=VR: Z=0.740084 H^R=-3845.11 S^R=-5.52747
The equation of state=VDW: Z=0.660991 H^R=-3935.39 S^R=-5.42065
The equation of state=RK: Z=0.685188 H^R=-4502.78 S^R=-6.54207
The equation of state=SRK: Z=0.722389 H^R=-4821.29 S^R=-7.40783
The equation of state=PR: Z=0.690914 H^R=-4984.72 S^R=-7.42094
```

Example 4.8 Enthalpy and Entropy Departures of Propane Gas[15]

Propane gas undergoes a change of state from an initial condition of $5bar$, and $105\,°C$, (state 1) to $25bar$, and $190\,°C$, (state 2).

(1) Determine the enthalpy departure $H - H^{ig}$, and the entropy departure $S - S^{ig}$, at each state from the Peng-Robinson equation.

(2) Calculate changes in enthalpy and entropy for a change from state 1 to state 2. For propane gas, $T_c = 369.8K$, $P_c = 4\ 249MPa$, (= 42.49bar), and $w = 0.152$. The heat capacity coefficients are given by $A = -4.224$, $B = 0.3063$, $C = -1.586 \times 10^{-4}$, and $D = 3.215 \times 10^{-8}$.

Solution

(1) Set state = 'v' and eos = 'pr' (Peng-Robinson equation). The following commands produce results at each state:

```
>> Tc = 369.8; Pc = 42.49; w = 0.152; T = 378.15; P = 5; eos = 'pr'; state = 'v';
>> T1 = 378.15; P1 = 5; T2 = 463.15; P2 = 25;
>> [Z1 V1 dH1 dS1] = deptfun(state,eos,T1,P1,Tc,Pc,w) % state 1
Z1 =
  0.9574
V1 =
  6.0199e+03
dH1 =
-400.4959
dS1 =
 -0.7082
>> [Z2 V2 dH2 dS2] = deptfun(state,eos,T2,P2,Tc,Pc,w) % state 2
Z2 =
  0.8891
V2 =
  1.3694e+03
dH2 =
 -1.4898e+03
dS2 =
 -2.2923
```

Results are summarized in Table 4.5.

(2) The following commands produce the desired outputs:

```
>> A = -4.224; B = 0.3063; C = -1.586e-4; D = 3.215e-8; Tc = 369.8;
Pc = 42.49; w = 0.152;
>> eos = 'pr'; state = 'v'; T1 = 378.15; P1 = 5; T2 = 463.15; P2 = 25;
>> [dH dS] = delHS(state,eos,T1,P1,T2,P2, A,B,C,D,Tc,Pc,w)

dH =
 7.3155e+03
dS =
  5.0285
```

We can see that $\Delta H = 7,\ 315.5J/mol$, and $\Delta S = 5.0285J/(mol \cdot K)$.

TABLE 4.5

Enthalpy and Entropy Departures at Each State

	State 1		State 2	
	$H - H^{ig}$,(J/mol)	$S - S^{ig}$,($J/(mol \cdot K)$)	$H - H^{ig}$,(J/mol)	$S - S^{ig}$,($J/(mol \cdot K)$)
Reference[16]	−400.512	−0.7083	−1,489.9	−2.2925
Results	−400.496	−0.7082	−1,489.8	−2.2923

4.2.3 ENTHALPY OF MIXTURE

The mixing rules and the combining rule for the calculation of a, and b, are

$$a = \sum_i \sum_j x_i x_j a_{ij}, \; a_{ij} = \sqrt{a_i a_j}\,(1 - k_{ij}), \; b = \sum_i x_i b_i$$

where

k_{ij}, is a binary interaction parameter specific to an $i - j$, molecular pair
a_{ij}, and b_i, are the parameters for pure component i,
The enthalpy of a mixture is given by

$$H = H_{ig}^0 + \Delta H$$

where H_{ig}^0, can be obtained from the relation

$$H_{ig}^0 = \int_{T_0}^{T} C_{pV}^0 \, dT = \sum_{i=1}^{5} \frac{a_k}{k}(T^k - T_0^k)$$

Here, values of $a_i (i = 1, ..., 5)$, can be found elsewhere.[17]

The function *khmix* estimates the enthalpy of a mixture. In this function, RK (Redlich-Kwong), SRK (Soave-Redlich-Kwong), and PR (Peng-Robinson) equations can be used as the equation of state. This function has the syntax

```
[Z H] = khmix(x,P,T,state,eos,Pc,Tc,w,k,Afi)
```

where x is the mole fractions of all components (column vector); P is the pressure (Pa); T is the temperature (K); state denotes the state of the fluid ('L': liquid, 'V': vapor); eos is the equation of state being used ('RK', 'SRK', or 'PR'); mxp is a structure containing properties of all components; Z is the compressibility factor; H is the enthalpy of the mixture (J/mol); Pc is the critical pressure (Pa, column vector); Tc is the critical temperature (K, column vector); w is the acentric factors of all components (column vector); k is the binary interaction parameters matrix ($n \times n$, symmetric matrix); and Afi is the coefficient matrix of the ideal gas heat capacity relation ((number of components) $\times$, (number of constants) matrix) ($J/(mol \cdot K)$).

```
function [Z H] = khmix(x,P,T,state,eos, Pc,Tc,w,k,Afi)
% Estimation of enthalpy of mixture
% Inputs:
%     x: mole fractions of all components (column vector)
%     P, T: pressure(Pa) and temperature(K)
%     state: fluid state('L': liquid, 'V': vapor)
%     eos: equation of state('RK', 'SRK', or 'PR')
%     Pc, Tc: critical P(Pa) and T(K) (column vector)
%     w- acentric factors of all components (column vector)
%     k: binary interaction parameter matrix(n x n symmetric)
%     Afi: coefficients of ideal gas heat capacity relation(J/mol/K)
% Outputs:
%     Z: compressibility factor
%     H: enthalpy of mixture (J/mol) (1btu/lbmole = 2.326 J/mol)
R = 8.314; % gas constant: m^3 Pa/(mol K) = J/mol-K
Tr = T./Tc; Pr = P./Pc; nc = length(x); eos = upper(eos); state = upper(state);
switch eos
```

```
case{'RK'}
      ai = sqrt(0.4278./(Pc.*Tr.^2.5)); bi = 0.0867./(Pc.*Tr);
      A = sum(x.*ai); B = sum(x.*bi); Z = roots([1 -1 B*P*(A^2/B-B*P-1)
      -A^2*(B*P)^2/B]);
case{'SRK'}
      mx = 0.48+1.574*w-0.176*w.^2; al = (1+mx.*(1-sqrt(Tr))).^2;
      ai = 0.42747*al.*Pr./(Tr.^2); bi = 0.08664*Pr./Tr;
      am = sqrt(ai'*ai).*(1-k);
      A = x*am*x'; B = sum(x.*bi); Z = roots([1 -1 A-B-B^2 -A*B]);
case{'PR'}
      mx = 0.37464+1.54226*w-0.26992*w.^2; al = (1+mx.*(1-sqrt(Tr))).^2;
      ai = 0.45723553*al.*Pr./(Tr.^2); bi = 0.0777961*Pr./Tr;
      am = sqrt(ai'*ai).*(1-k);
      A = x*am*x'; B = sum(x.*bi); Z = roots([1 B-1 A-3*B^2-2*B B^3+B^2-A*B]);
end
iz = abs(imag(Z)); Z(and(iz>0,iz<=1e-6)) = real(Z(and(iz>0,iz<=1e-6)));
for i = 1:length(Z), zind(i) = isreal(Z(i)); end
Z = Z(zind);
if state == 'L', Z = min(Z); else, Z = max(Z); end
V = R*T*Z/P; % m^3/mol
% Hv0 = int(Cp)
Tf = (T-273.15)*1.8 + 32; % T: K->F
Hv0 = x*(Afi(:,1)*Tf + Afi(:,2)*Tf^2/2 + Afi(:,3)*Tf^3/3 + Afi(:,4)*Tf^4/4 + Afi
(:,5)*Tf^5/5);
Hv0 = Hv0*2.326; % Btu/lbmole -> J/mol
% compute enthalpy
switch eos
  case{'RK'}, H = Hv0 + R*T*(Z - 1 - 3*(A^2)*log(1 + B*P/Z)/(2*B));
  case{'SRK'}
    hsum = 0;
    for i = 1:nc
      for j = 1:nc
        hsum = hsum + x(i)*x(j)*am(i,j)*(1 - mx(i)*sqrt(Tr(i))/(2*sqrt(al(i))) -...
           mx(j)*sqrt(Tr(j))/(2*sqrt(al(j))));
      end
    end
    H = Hv0 + R*T*(Z - 1 - log((Z + B)/Z)*hsum/B);
  case{'PR'}
    hsum = 0;
    for i = 1:nc
      for j = 1:nc
        hsum = hsum + x(i)*x(j)*am(i,j)*(1 - mx(i)*sqrt(Tr(i))/(2*sqrt(al(i))) -...
           mx(j)*sqrt(Tr(j))/(2*sqrt(al(j))));
      end
    end
    H = Hv0 + R*T*(Z - 1 - log((Z + B)/Z)*hsum/B);
end
end
```

Example 4.9   Enthalpy of Mixture

Estimate the liquid-phase enthalpy of the mixture of methane(1)/ethane(2)/propane(3) at $-158K$, and 6.8947bar. The liquid-phase mole fractions of components are $x_1 = 0.419$, $x_2 = 0.3783$, $x_3 = 0.2027$, and the properties of each component are shown in Table 4.6 (ΔH_f, ΔG_f: J/mol). Table 4.7 shows coefficients of the ideal gas heat capacity relation for each component.

TABLE 4.6

Properties of Each Component of the Mixture[18]

Component	x	w	$T_c(K)$	$P_c(bar)$	ΔH_f	ΔG_f
Methane	0.4190	0.012	190.6	45.99	−74520	−50460
Ethane	0.3783	0.100	305.3	48.72	−83820	−31855
Propane	0.2027	0.152	369.8	42.48	−104680	−24290

TABLE 4.7

Coefficients of the Ideal Gas Heat Capacity Relation

Component	a_1	$a_2 \times 10^2$	$a_3 \times 10^5$	$a_4 \times 10^8$	$a_5 \times 10^{11}$
Methane	8.245223	0.380633	0.8864745	−0.746115	0.182296
Ethane	11.51606	1.40309	0.854034	−1.106078	0.31622
Propane	15.58683	2.504953	1.404258	−3.52626	1.864467

Solution

The script *enthmix* specifies the required data and calls the function *khmix* to perform calculations.

```
% enthmix.m
state = 'L'; nx = [0.419 0.3783 0.2027]; P = 6.8947e5; T = 158; eos = 'rk';
Pc = [45.99 48.72 42.48]*1e5; Tc = [190.6 305.3 369.8]; w = [0.012 0.1 0.152];
k = zeros(3,3);
Afi = [8.245223 0.3806333e-2 0.8864745e-5 -0.7461153e-8 0.182296e-11;
    11.51606 0.140309e-1 0.854034e-5 -0.1106078e-7 0.31622e-11;
    15.58683 0.2504953e-1 0.1404258e-4 -0.352626e-7 0.1864467e-10];
state = upper(state); if state == 'L', x = nx; else x = ny; end
[Z H] = khmix(x,P,T,state,eos,Pc,Tc,w,k,Afi);
fprintf('Equation of state: %s, State: %s',upper(eos),upper(state));
fprintf('\nCompressibility factor = %g, Enthalpy = %g (J/mol)\n',Z,H);
```

The results of executing the script *enthmix* are
```
>> enthmix
Equation of state: RK, State: L
Compressibility factor = 0.027695, Enthalpy = -18474.5 (J/mol)
```

Table 4.8 summarizes the results for each equation of state.

4.3 FUGACITY COEFFICIENT

4.3.1 Fugacity Coefficients of Pure Species

The fugacity of a pure component, f, is defined as

TABLE 4.8

Estimated Enthalpy and Z by Different Equations of State

Equation of State	Compressibility Factor	Enthalpy
RK	0.0276950	−18474.5
SRK[1]	0.0276512	−10664.5
PR	0.0245126	−12089.2

1 RK = Redlich-Kwong; SRK = Soave-Redlich-Kwong; PR = Peng-Robinson.

$$f = \phi P$$

The fugacity coefficient ϕ, is determined first, followed by multiplication by pressure to give f. The equation of state can be used to find ϕ. Table 4.9 summarizes descriptions for the fugacity coefficient provided by various equations of state.

The function *phigas* estimates the fugacity coefficient using the virial and cubic equations of state. The basic syntax is

```
[phig f] = phigas(state,eos,T,P,Tc,Pc,w)
```

where state denotes the state of the fluid (liquid: L, vapor: V); eos is the equation of state being used (VR, VDW, RK, SRK, PR); T and P are the temperature (K), and pressure (*bar*), respectively; Tc and Pc are the critical temperature (K), and pressure (*bar*), respectively; w is the acentric factor; phig is the fugacity factor; and f is the fugacity (*bar*).

```
function [phig f] = phigas(state,eos,T,P,Tc,Pc,w)
% Estimates the fugacity coefficient using the virial and cubic EOS.
% input
%     state: fluid state('L': liquid, 'V': vapor)
%     eos: equation of state (VR, VDW, RK, SRK, PR)
%     P,T: pressure(bar) and temperature(K)
%     Tc,Pc: critical T(K) and P(bar)
%     w- acentric factor
% output:
```

TABLE 4.9

Fugacity Coefficient Model Based on Equation of State

Equation of State	Fugacity Coefficient Equation
Virial	$\ln\phi = \frac{BP}{RT} = \frac{P_r}{T_r}(B^0 + wB^1)$
van der Waals	$\ln\phi = Z - 1 - \ln\left\{Z\left(1 - \frac{b}{V}\right)\right\} - \frac{a}{RTV}$
Redlich-Kwong	$\ln\phi = Z - 1 - \ln\left\{Z\left(1 - \frac{b}{V}\right)\right\} - \frac{a}{bRT}\ln\left(1 + \frac{b}{V}\right)$
Soave-Redlich-Kwong Peng-Robinson	$\ln\phi = Z - 1 - \ln(Z - B) - \frac{a(T)}{bRT(\sigma - \varepsilon)}\ln\left[\frac{Z + \sigma B}{Z + \varepsilon B}\right]$
Experimental data	$\ln\phi = \int_0^P \frac{Z-1}{P}dP$

```
%      phig: fugacity coefficient
%      f: fugacity (bar)

% Tr and Pr (reduced T and P)
Tr = T/Tc; Pr = P/Pc; R = 83.14; % cm^3*bar/mol/K
eos = upper(eos);
switch eos
    case 'VDW', al = 1; sm = 0; ep = 0; om = 0.125; ps = 0.42188; kappa = 0;
    case 'RK', al = 1./sqrt(Tr); sm = 1; ep = 0; om = 0.08664; ps = 0.42748;
    case 'SRK'
        kappa = 0.480 + 1.574*w - 0.176*w^2;
        al = (1 + kappa*(1-sqrt(Tr))).^2; sm = 1; ep = 0; om = 0.08664; ps = 0.42748;
    otherwise % PR or VR
        kappa = 0.37464 + 1.54226*w - 0.26992*w^2;
        al = (1 + kappa*(1-sqrt(Tr))).^2; sm = 1+sqrt(2); ep = 1-sqrt(2);
        om = 0.0778; ps = 0.45724;
end
% compressibility factor (Z)
state = upper(state); beta = om*Pr./Tr; q = ps*al./(om*Tr); % beta and q
if strcmp(eos,'VR') % virial EOS: vapor phase
    B0 = 0.083 - 0.422./(Tr.^1.6); B1 = 0.139 - 0.172./(Tr.^4.2);
    B = R*Tc.*(B0 + w.*B1)./Pc; Z = 1 + B.*P./(R*T);
else
    fV = @(Z) 1+beta-q*beta.*(Z-beta)./((Z+ep*beta).*(Z+sm*beta)) - Z;
    fL = @(Z) beta+(Z+ep*beta).*(Z+sm*beta).*(1+beta-Z)./(q.*beta) - Z;
    switch state
            case 'V', Z = fzero(fV, 1);
            case 'L', Z = fzero(fL, beta);
        end
end
V = Z*R*T/P; % cm^3/mol
% fugacity coefficient
a = ps*al*R^2*Tc.^2/Pc; b = om*R*Tc./Pc; % a, b
qi = a/(b*R*T); Bd = b*P/(R*T); % A, B
if strcmp(eos,'VR'), phig = exp(Pr*(B0 + w*B1)./Tr); % virial EOS: vapor phase
elseif strcmp(eos,'VDW'), phig = exp(Z - 1 - log(Z.*(1 - b/V)) - a./(R*T*V));
elseif strcmp(eos,'RK'), phig = exp(Z-1 - log(Z.*(1 - b/V)) - (a/(b*R*T)) * log
(1+b/V));
else % SRK or PR
    phig = exp(Z-1 - log(Z-Bd) - (qi/(sm-ep))*log((Z + sm*Bd)/(Z + ep*Bd)));
end
f = phig*P; % fugacity (bar)
end
```

Example 4.10 Fugacity of Acetylene Gas[19]

Find the fugacity of acetylene at $250K$, and $10bar$. For acetylene, $T_c = 308.3K$, $P_c = 6.139MPa$, (= $61.39bar$), and $w = 0.187$.

Solution
Using the Soave-Redlich-Kwong (SRK) equation, we obtain the following results:
```
>> Tc = 308.3; Pc = 61.39; w = 0.187; T = 250; P = 10; eos = 'srk'; state = 'v';
>> [phig f] = phigas(state,eos,T,P,Tc,Pc,w)
phig =
```

```
      0.8956
f =
      8.9559
```

Table 4.10 shows the results obtained using the virial and cubic equations of state.

4.3.2 Fugacity Coefficient of a Species in a Mixture

The fugacity coefficient of species i, in a mixture, $\hat{\phi}_i$, is defined as

$$\hat{\phi}_i = \frac{\hat{f}_i}{y_i P}$$

4.3.2.1 Fugacity Coefficient from the Virial Equation of State

From the virial equation of state, the fugacity coefficient of component i, in a mixture, $\hat{\phi}_i = \hat{f}_i/(y_i P)$, is given by[20]

$$\ln\hat{\phi}_i = \frac{P}{RT}\left(2\sum_j y_j B_{ji} - B\right), \ B = \sum_k \sum_j y_k y_j B_{kj}$$

For a binary mixture, we have

$$\ln\hat{\phi}_1 = \frac{P}{RT}(2y_1 B_{11} + 2y_2 B_{12} - B), \ \ln\hat{\phi}_2 = \frac{P}{RT}(2y_1 B_{12} + 2y_2 B_{22} - B)$$

4.3.2.2 Fugacity Coefficient from the Cubic Equations of State

For the i, component in a mixture, let

$$a_i(T) = \Psi\frac{\alpha(T_{ri})R^2 T_{ci}^2}{P_{ci}}, \ b_i = \Omega\frac{RT_{ci}}{P_{ci}}, \ \beta_i = \frac{b_i P}{RT}, \ q_i = \frac{a_i(T)}{b_i RT}$$

and use the mixing rule to obtain the parameters a, and b. The fugacity coefficient of component i, in a mixture is expressed as[21]

$$\ln\hat{\phi}_i = \frac{b_i}{b}(Z - 1) - \ln(Z - \beta) - \bar{q}_i I$$

where

$$I = \frac{1}{\sigma - \epsilon}\ln\left(\frac{Z + \sigma\beta}{Z + \epsilon\beta}\right), \ q = \frac{a}{bRT}, \ \bar{q}_i = q\left(1 + \frac{\bar{a}_i}{a} - \frac{b_i}{b}\right), \ \bar{a}_i = \frac{\partial(na)}{\partial n_i}\bigg|_{T,n_j}$$

TABLE 4.10

Estimated Fugacities and Fugacity Coefficients

Equation of State	VR	VDW	RK	SRK	PR
Fugacity coefficient	0.8938	0.9208	0.9007	0.8956	0.8889
Fugacity (*bar*),	8.9383	9.2075	9.0071	8.9559	8.8893

If we use the Peng-Robinson equation, the equation for $\ln\hat{\phi}_i$, is written as

$$\ln\hat{\phi}_i = -\ln(Z - \beta) + \frac{\beta_i}{\beta}(Z - 1) - \frac{q}{\sqrt{8}}\left(\frac{2}{a}\sum_j x_j a_{ij} - \frac{\beta_i}{\beta}\right)\ln\left(\frac{Z + (1 + \sqrt{2})\beta}{Z + (1 - \sqrt{2})\beta}\right)$$

4.3.2.3 Fugacity Coefficient from the van der Waals Equation of State

From the van der Waals equation of state, we have the following description for the fugacity coefficient of component i, in a mixture:

$$\ln\hat{\phi}_i = -\ln(Z - \beta) + \frac{\beta_i}{Z - \beta} - \frac{2}{Z}\sum_j x_j A_{ij}, A_{ij} = \frac{a_{ij}P}{R^2 T^2} = \frac{a_{ij}\beta}{bRT}$$

The function *phimix* estimates the fugacity coefficients of all components in a mixture from cubic equations of state. This function has the syntax

```
[Z,V,phi] = phimix(ni,P,T,Pc,Tc,w,k,state,eos)
```

where ni is the number of moles (or mole fractions) of each component in the mixture (vector); P is the pressure (Pa); T is the temperature (K); w is the acentric factor; k is a matrix of binary interaction parameters; state denotes the state of the fluid ('L': liquid, 'V': vapor); eos is the equation of state being used ('RK', 'SRK', or 'PR'); Z is the compressibility factor; V is the molar volume; and phi is the fugacity coefficients of all components (vector).

```
function [Z,V,phi] = phimix(ni,P,T,Pc,Tc,w,k,state,eos)
% Estimates fugacity coefficients of all components in a mixture using
% the cubic equation of state
% Inputs:
%     ni: number of moles (or mole fractions) of each component (vector)
%     P, T: pressure(Pa) and temperature(K)
%     Pc, Tc: critical P(Pa) and T(K) of all components (vector)
%     w- acentric factors of all components (vector)
%     k: symmetric matrix of binary interaction parameters (n x n)
%     state: fluid state('L': liquid, 'V': vapor)
%     eos: equation of state('RK', 'SRK', or 'PR')
% Outputs:
%     V: molar volume (m3/mol)
%     Z: compressibility factor
%     phi: fugacity coefficient vector
ni = ni(:); Pc = Pc(:); Tc = Tc(:); w = w(:); % column vector
x = ni/sum(ni); % mole fraction
R = 8.314; % gas constant: m^3 Pa/(mol K) = J/mol-K
Tref = 298.15; % reference T(K)
Pref = 1e5; % reference P(Pa)
Tr = T./Tc; eos = upper(eos); state = upper(state);
switch eos
  case{'RK'}
    ep = 0; sm = 1; om = 0.08664; ps = 0.42748; al = 1./sqrt(Tr);
  case{'SRK'}
    ep = 0; sm = 1; om = 0.08664; ps = 0.42748;
    al = (1+(0.48+1.574*w-0.176*w.^2).*(1-sqrt(Tr))).^2;
  case{'PR'}
    ep = 1 - sqrt(2); sm = 1 + sqrt(2); om = 0.07780; ps = 0.45724;
```

```
      al = (1+(0.37464+1.54226*w-0.26992*w.^2).*(1-sqrt(Tr))).^2;
end
ai = ps*(R^2).*al.*(Tc.^2)./Pc; am = sqrt(ai*ai').*(1 - k); % nxn matrix
a = x'*am*x; bi = om*R*Tc./Pc; b = x'*bi; beta = b*P/R/T; q = a/(b*R*T);
% compressibility factor(Z) and molar volume (V)
state = upper(state);
c(1) = 1; c(2) = (sm +ep)*beta - (1+beta);
c(3) = beta*(q + ep*sm*beta -(1+beta)*(sm+ep));
c(4) = -beta^2*(q +(1+beta)*ep*sm);
% Roots
Z = roots(c); iz = abs(imag(Z)); Z(and(iz>0,iz<=1e-6)) =
real(Z(and(iz>0,iz<=1e-6)));
for i = 1:length(Z), zind(i) = isreal(Z(i)); end
Z = Z(zind);
if state == 'L', Z = min(Z); else, Z = max(Z); end
V = R*T*Z/P;
% fugacity coefficients
bara = (2*ni'*am - a*ones(1,length(ni)))'; barb = bi;
phi = exp((Z - 1)*barb/b - log((V - b)*Z/V) + (a/(b*R*T))/(ep - sm)*...
   log((V + sm*b)/(V + ep*b))*(1 + bara/a - barb/b));
end
```

Example 4.11   Fugacity Coefficients in a Mixture[22]

Determine the fugacity coefficients of all components in a nitrogen(1)/methane(2) mixture by the Peng-Robinson equation at $100\,K$, and $0.4119\,MPa\,(4.119bar)$. In this mixture, the mole fraction of nitrogen is $y_1 = 0.958$. For nitrogen, $T_{c1} = 126.1\,K$, $P_{c1} = 3.394\,MPa\,(33.94bar)$, and $w_1 = 0.04$; and for methane, $T_{c2} = 190.6\,K$, $P_{c2} = 4.604\,MPa\,(46.04bar)$, and $w_2 = 0.011$.

Solution

The following commands produce a vector of fugacity coefficients:

```
>> state = 'v'; k = zeros(2,2); eos='pr'; T = 100; P = 4.119e5; ni = [0.958 0.042];
>> Tc = [126.1 190.6]; Pc = [33.94 46.04]*1e5; w = [0.04 0.011];
>> [Z,V,phi] = phimix(ni,P,T,Pc,Tc,w,k,state,eos)
Z =
   0.9059
V =
   0.0018
phi =
   0.9162
   0.8473
```

4.4 ACTIVITY COEFFICIENT

The Gibbs-Duhem equation may be used as the basis for estimating the vapor-phase composition and the activity coefficients. For a binary mixture, this equation can be expressed as

$$x_1\left(\frac{\partial \ln\gamma_1}{\partial x_1}\right)_{P,T} + x_2\left(\frac{\partial \ln\gamma_2}{\partial x_1}\right)_{P,T} = 0$$

where

x_i, is the mole fraction of component i, in the vapor phase

γ_i, is the liquid-phase activity coefficient of component i,

The activity coefficient of component i, is defined as

$$\gamma_i = \frac{y_i P}{x_i P_i^v}$$

where

y_i, is the mole fraction of component i, in the vapor phase
P, is the total pressure
P_i^v, is the vapor pressure of component i,

The combination of these two relations, followed by some manipulations, yields

$$\frac{dy_1}{dx_1} = \frac{y_1(1-y_1)}{y_1-x_1}\frac{d\ln P}{dx_1}$$

If $P - x_1$, data are available, P, can be expressed as a polynomial function of x_1, and the derivative $d\ln P/dx_1$, can be written as a function of x_1. Integration of this relation from $x_1 = 0$, to $x_1 = 1$, gives y_1. At the initial condition ($x_1 = 0$, $y_1 = 0$), the denominator of the derivative term becomes zero and l'Hopital's rule should be employed to give the following relation to be used in the integration:

$$\frac{dy_1}{dx_1} \approx \frac{\Delta y_1}{\Delta x_1} \approx \frac{\epsilon_{y_1}}{\epsilon_{x_1}} \approx (1-2y_1)(d\ln P/dx_1)\Big|_{x_1=0,y_1=0}$$

From this relation, $\epsilon_{y_1} = (d\ln P/dx_1)|_{x_1=0}\cdot\epsilon_{x_1}$,is evaluated and used as the initial value for y_1. As for ϵ_{x_1}, a very small value (e.g., 0.00001) may be used.

Example 4.12 Vapor-Phase Composition of Benzene/Acetic Acid System[23]

Data of liquid-phase composition versus total pressure for a benzene(1)/acetic acid(2) system at 50 °C, are presented in Table 4.11. Use the Gibbs-Duhem equation to estimate the composition of the vapor phase and the activity coefficients.

Solution

From the data, we can see that the vapor pressure of benzene is $P_1^v = 271 mmHg$, and that of acetic acid is $P_2^v = 57.52 mmHg$. The built-in function *polyfit* may be used in the regression of the total pressure data. From the regression, the 4th-order polynomial turns out to represent the data adequately:

TABLE 4.11

Total Pressure for a Benzene(1)/Acetic Acid(2) System at $50°C$,[24]

x_1	$P\,(mmHg)$	x_1	$P\,(mmHg)$
0.0	57.52	0.8286	250.20
0.0069	58.2	0.8862	259.00
0.1565	126.00	0.9165	261.11
0.3396	175.30	0.9561	264.45
0.4666	189.50	0.9840	266.53
0.6004	224.30	1.0	271.00
0.7021	236.00		

$$P(x_1) = -365.7643x_1^4 + 915.1596x_1^3 - 894.0699x_1^2 + 556.8470x_1 + 56.2570$$

Differentiation of this equation yields

$$P'(x_1) = -1463.0572x_1^3 + 2745.4788x_1^2 - 1788.1398x_1 + 556.8470$$

The derivative term is represented as

$$\frac{d\ln P}{dx_1} = \frac{1}{P}\frac{dP}{dx_1} = \frac{-1463.0572x_1^3 + 2745.4788x_1^2 - 1788.1398x_1 + 556.8470}{-365.7643x_1^4 + 915.1596x_1^3 - 894.0699x_1^2 + 556.8470x_1 + 56.2570}$$

Thus, we have

$$\frac{dy_1}{dx_1} = \frac{y_1(1-y_1)}{y_1-x_1}\frac{-1463.0572x_1^3 + 2745.4788x_1^2 - 1788.1398x_1 + 556.8470}{-365.7643x_1^4 + 915.1596x_1^3 - 894.0699x_1^2 + 556.8470x_1 + 56.2570}$$

This differential equation is defined by the function *bzacfun*.

```
function dy = bzacfun(x,y)
% dy/dx for benzene/acetic acid system
dy = (y*(1-y)/(y-x)) * (-1463.0572*x^3 + 2745.4788*x^2 - 1788.1398*x ...
  + 556.8470)/(-365.7643*x^4 + 915.1596*x^3 - 894.0699*x^2 + 556.8470*x + 56.2570);
end
```

The script file *bzacP* performs the data regression and calculates the activity coefficients. This script generates plots of P, vapor-phase composition, and activity coefficients as a function of the liquid-phase composition. The initial conditions are set to $\epsilon_{x_1} = 10^{-5}$, and $y_1(0) = \epsilon_{y_1}$.

```
% bzacP.m: vapor phase composition and activity coefficient
% for benzene/acetic acid system
x1 = [0.0 0.0069 0.1565 0.3396 0.4666 0.6004 0.7021 0.8286 0.8862 0.9165 0.9561
0.9840 1.0];
P = [57.52 58.2 126.00 175.30 189.50 224.30 236.00 250.20 259.00 261.11 264.45
266.53 271.00];
c = polyfit(x1,P,4) % regression polynomial (4th order)
dc = polyder(c) % differentiation of the polynomial
x10 = 1e-5; xinv = [x10 1]; y10 = x10*dc(end)/c(end);
[x,y] = ode45(@bzacfun,xinv,y10); % solve the differential eqn.
P1v = P(end); P2v = P(1); Px = polyval(c,x);
gam1 = y.*Px./x/P1v; gam2 = (1-y).*Px./(1-x)/P2v;
subplot(1,2,1), plot(x,y), xlabel('x_1'), ylabel('y_1') % x1-y1 graph
subplot(1,2,2), plot(x,gam1,x,gam2,'.-'), xlabel('x_1'), ylabel('\gamma')
legend('\gamma_1','\gamma_2','Location','best'), axis([0 1 0 4])
```

The script *bzacP* produces the following results and generates the plots shown in Figure 4.3:

```
>> bzacP
c =
-365.7643 915.1596 -894.0699 556.8470 56.2570
dc =
  1.0e+03 *
  -1.4631 2.7455 -1.7881 0.5568
```

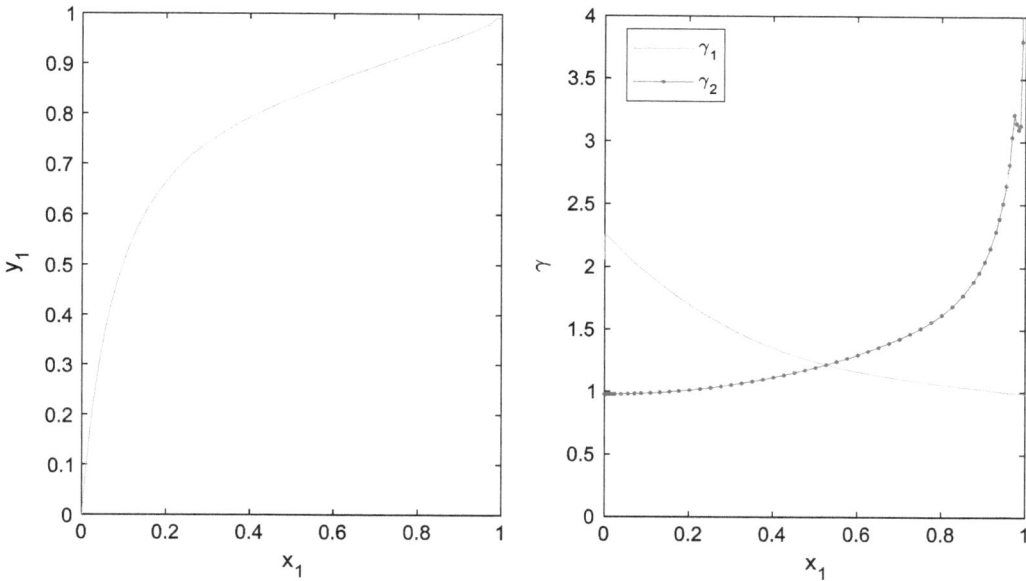

FIGURE 4.3 Vapor-phase composition (left) and activity coefficients (right) for benzene(1)/acetic acid(2) system.

4.4.1 ACTIVITY COEFFICIENT MODELS

4.4.1.1 Wilson equation

The Wilson equation may be used to estimate activity coefficients for miscible nonideal systems. For a binary mixture, the activity coefficients γ_1, and γ_2, are obtained from

$$\ln\gamma_1 = -\ln\theta_1 + \frac{(g_{12}\theta_2 - g_{21}\theta_1)x_2}{\theta_1\theta_2},\ \ln\gamma_2 = -\ln\theta_2 + \frac{(g_{12}\theta_2 - g_{21}\theta_1)x_1}{\theta_1\theta_2}$$

where

x_i, is the mole fraction

$\theta_1 = x_1 + g_{12}x_2$, $\theta_2 = x_2 + g_{21}x_1$,

g_{12}, and g_{21}, can be determined using the experimental data

The parameters of the Wilson equation can be determined by using the azeotropic data. At a low-pressure azeotropic point, the liquid-phase activity coefficients can be obtained from the relation $\gamma_i = P/P_i$.[25] At the azeotropic point T_a, the vapor pressure P_i, may be estimated by using a simple relation such as the Antoine equation. Thus, the parameters g_{ij}, can be determined if the composition data are available. For a binary system, we can get two nonlinear equations:

$$f_1 = \ln\gamma_1 + \ln\theta_1 - \frac{(g_{12}\theta_2 - g_{21}\theta_1)x_2}{\theta_1\theta_2},\ f_2 = \ln\gamma_2 + \ln\theta_2 + \frac{(g_{12}\theta_2 - g_{21}\theta_1)x_1}{\theta_1\theta_2}$$

4.4.1.2 van Laar equation

The van Laar equation is frequently used to correlate activity coefficient data for non-ideal systems. The van Laar equations for the correlation of binary activity coefficients are given by[26]

$$\log\gamma_1 = \frac{Ax_2^2}{\left(\frac{A}{B}x_1 + x_2\right)^2}, \ \log\gamma_2 = \frac{Bx_1^2}{\left(x_1 + \frac{B}{A}x_2\right)^2}$$

where parameters A, and B, constants for a particular binary mixture, may be determined using experimental data.

Example 4.13   Estimation of Parameters of the Wilson Equation

The temperature of the azeotrope for an ethanol(1)/n-octane(2) mixture at $P = 760mmHg$, is $77\,°C$, and the composition at the azeotrope is 78% ethanol and 22% n-octane (% by weight). At the temperature of the azeotrope T, the vapor pressures of ethanol and n-octane may be obtained by the Antoine equation $\log P_i = A_i - B_i/(T + C_i)$, where T, is the azeotrope temperature ($°C$), and P_i, is the vapor pressure ($mmHg$). For ethanol, $A_1 = 8.04494$, $B_1 = 1554.3$, and $C_1 = 222.65$; and for n-octane, $A_2 = 6.92374$, $B_2 = 1355.126$, and $C_2 = 209.517$. The molecular weights of ethanol and n-octane are 46.07 and 114, respectively. Determine the Wilson equation coefficients for this system.

Solution

The Wilson equation parameters can be estimated using the vapor pressures obtained by the Antoine equation. The function *wilact* defines the system of nonlinear equations in terms of mole fractions. This function takes x_1, x_2, γ_1, and γ_2, as inputs. The script *wilsonpar* uses the built-in function *fsolve* to solve the nonlinear system.

```
% wilsonpar.m: Wilson equation parameters for a binary system
w1 = 78; mw1 = 46.07; mw2 = 114; Ta = 77; P = 760;
A1 = 8.04494; B1 = 1554.3; C1 = 222.65; A2 = 6.92374; B2 = 1355.126; C2 = 209.517;
x1 = (w1/mw1)/(w1/mw1 + (100-w1)/mw2); x2 = 1 - x1; % mole fraction
P1 = 10^(A1 - B1/(Ta + C1)); P2 = 10^(A2 - B2/(Ta + C2)); % vapor pressure
gam1 = P/P1; gam2 = P/P2; % activity coefficients
% solve the nonlinear system (g(1)=g12, g(2)=g21)
g0 = [0.1 0.1]; g = fsolve(@wilact,g0,[],x1,x2,gam1,gam2);
g12 = g(1), g21 = g(2)

function f = wilact(g,x1,x2,gam1,gam2)
t1 = x1 + g(1)*x2; t2 = x2 + g(2)*x1;
f = [log(gam1) + log(t1) - (g(1)*t2 - g(2)*t1)*x2/(t1*t2);
     log(gam2) + log(t2) + (g(1)*t2 - g(2)*t1)*x1/(t1*t2)];
end
```

The script *wilsonpar* produces the following results:

```
>> wilsonpar
g12 =
  0.5713
g21 =
  0.0998
```

Example 4.14   Estimation of Activity Coefficients by Margules Equation

In a binary liquid mixture of chloroform (1) and 1,4-dioxane (2) at $50\,°C$, the activity coefficients of chloroform (γ_1), and 1,4-dioxane (γ_2), can be estimated from Margules equations given by

$$\gamma_1 = \exp[x_2^2\{A_{12} + 2(A_{21} - A_{12})x_1\}], \ \gamma_2 = \exp[x_1^2\{A_{21} + 2(A_{12} - A_{21})x_2\}]$$

where x_1, is the mole fraction of chloroform, x_2, is the mole fraction of 1,4-dioxane, and A_{12}, and A_{21}, are the Margules parameters for the binary system. The following relation can be applied to estimate A_{12}, and A_{21}:

$$\frac{G^E}{RTx_1 x_2} = A_{21}x_1 + A_{12}x_2$$

where G^E/RT, is the dimensionless excess Gibbs free energy. Table 4.12 shows data for G^E/RT, obtained from the vapor-liquid equilibrium experiment. Estimate A_{12}, and A_{21}, and determine the value of x_1, such that $\gamma_1 = \gamma_2$.

Solution

The Gibbs free energy equation can be rearranged as follows:

$$\frac{G^E}{RT} = A_{21}x_1^2 (1 - x_1) + A_{12}x_1 (1 - x_1)^2$$

The Margules parameters for the binary system can be obtained by solving the following set of linear equations:

$$Y = \begin{bmatrix} (G^E/RT)_1 \\ (G^E/RT)_2 \\ \vdots \\ (G^E/RT)_n \end{bmatrix} = \begin{bmatrix} x_{1,1}^2(1 - x_{1,1}) & x_{1,1}(1 - x_{1,1})^2 \\ x_{1,2}^2(1 - x_{1,2}) & x_{1,2}(1 - x_{1,2})^2 \\ \vdots & \vdots \\ x_{1,n}^2(1 - x_{1,n}) & x_{1,n}(1 - x_{1,n})^2 \end{bmatrix} \begin{bmatrix} A_{21} \\ A_{12} \end{bmatrix} = XA \Rightarrow A = \begin{bmatrix} A_{21} \\ A_{12} \end{bmatrix} = (X^T X)^{-1} X^T Y$$

The value of x_1, at which $\gamma_1 = \gamma_2$, can be found by solving the following nonlinear equation:

$$f(x_1) = \exp[(1 - x_1)^2 \{A_{12} + 2(A_{21} - A_{12})x_1\}] - \exp[x_1^2 \{A_{21} + 2(A_{12} - A_{21})(1 - x_1)\}] = 0$$

The script *mgpar* calculates the Margules parameters and the value of x_1, at which $\gamma_1 = \gamma_2$.

```
% mgpar.m
x1 = [0.0932 0.1248 0.1757 0.2000 0.2626 0.3615 0.4750 0.5555 0.6718];
Ge = [-0.064 -0.086 -0.120 -0.133 -0.171 -0.212 -0.248 -0.252 -0.245];
Y = Ge'; X = [x1.^2.*(1-x1); x1.*(1-x1).^2]';
A = inv(X'*X)*X'*Y; A12 = A(2); A21 = A(1);  % Margules parameters
fprintf('A12 = %g, A21 = %g\n', A12, A21);
f = @(x) exp((1-x)^2*(A12+2*(A21-A12)*x)) - exp(x^2*(A21+2*(A12-A21)*(1-x)));
x0 = 0.5; x = fsolve(f,x0)  % find x1 such that gamma1=gamma2

>> mgpar
```

TABLE 4.12

Experimental Data for G^E/RT, in a Chloroform (1)/1,4-Dioxane (2) System

x_1,	0.0932	0.1248	0.1757	0.2000	0.2626	0.3615	0.4750	0.5555	0.6718
G^E/RT,	−0.064	−0.086	−0.120	−0.133	−0.171	−0.212	−0.248	−0.252	−0.245

```
A12 = -0.721339, A21 = -1.28591
x =
  0.5666
```

We can see that the Margules parameters for the binary system are $A_{12} = -0.721339$, and $A_{21} = -1.28591$, and that $\gamma_1 = \gamma_2$, when $x_1 = 0.5666$.

4.4.2 Activity Coefficients by the Group Contribution Method

A molecule can be considered to be a combination of functional groups. Among activity models based on the contribution of functional groups, the UNIQUAC (universal quasi-chemical) and UNIFAC (UNIQUAC functional group activity coefficient) methods are widely used.

4.4.2.1 UNIQUAC Method

The activity coefficient can be represented as a sum of a combinatorial term, to account for molecular size and shape differences, and a residual term to account for molecular interactions[27]:

$$\ln\gamma_i = \ln\gamma_i^C + \ln\gamma_i^R$$

where

$$\ln\gamma_i^C = 1 - J_i + \ln J_i - 5q_i\left(1 - \frac{J_i}{L_i} + \ln\frac{J_i}{L_i}\right)$$

$$\ln\gamma_i^R = q_i\left(1 - \ln s_i - \sum_j \frac{\theta_j \tau_{ij}}{s_j}\right)$$

$$\tau_{ji} = \exp\left(-\frac{u_{ji} - u_{ii}}{RT}\right), J_i = \frac{r_i}{\sum_j r_j x_j}, L_i = \frac{q_i}{\sum_j q_j x_j}, s_i = \tau_{ki}\sum_k \theta_k, \theta_i = \frac{x_i q_i}{\sum_j x_j q_j}$$

$$r_i = \sum_k v_k^{(i)} R_k, q_i = \sum_k v_k^{(i)} Q_k$$

In these equations, r_i, is a parameter representing a relative molecular volume and q_i, is a parameter representing a relative molecular surface area, given by the sum of R_k, and Q_k, parameters of functional groups comprising the component, respectively. Values of the parameters R_k, and Q_k, can be found elsewhere.[28,29]

4.4.2.2 UNIFAC Method

The UNIFAC method is based on the UNIQUAC method, and the assumptions regarding coordination numbers and so forth are similar to those in the UNIQUAC method. The activity coefficient can be represented as the sum of a combinatorial term and a residual term[30]:

$$\ln\gamma_i = \ln\gamma_i^C + \ln\gamma_i^R$$

where

$$\ln\gamma_i^C = 1 - J_i + \ln J_i - 5q_i\left(1 - \frac{J_i}{L_i} + \ln\frac{J_i}{L_i}\right)$$

$$\ln\gamma_i^R = q_i\left(1 - \sum_k\left(\frac{\theta_k\beta_{ik}}{s_k} - e_{ki}\ln\frac{\beta_{ik}}{s_k}\right)\right)$$

$$e_{ki} = \frac{\nu_k^{(i)}Q_k}{q_i}, \; \beta_{ik} = \sum_m e_{mi}\tau_{mk}, \; \tau_{ij} = \exp\left(-\frac{a_{ij}}{T}\right), \; J_i = \frac{r_i}{\sum_j r_j x_j}, \; L_i = \frac{q_i}{\sum_j q_j x_j}, \; \theta_k = \frac{x_i q_i e_{ki}}{\sum_j x_j q_j}$$

Here, a_{ij}, are group interaction parameters.[31,32]

The function *unifgam* calculates activity coefficients using the UNIFAC method. The basic syntax is

```
gam = unifgam(k,R,Q,nu,amn,Nc,x,T)
```

where k is the number of functional groups; R and Q are the vectors of volumes and surface areas for each functional group, respectively; nu is the number of functional groups contained in each component; amn is the matrix of group interaction parameters; Nc is the number of components; x is the vector of liquid-phase mole fraction; T is the temperature (K); and gam is the vector of activity coefficients for each component.

```
function gam = unifgam(k,R,Q,nu,amn,Nc,x,T)
% Estimation of activity coefficients using the UNIFAC method
% input:
%      R,Q: vectors of volumes and surface areas for each functional group
%      nu: number of functional groups contained in each component
%      (row: functional groups k, column: components i)
%      amn: matrix of group interaction parameters
%      Nc: number of components
%      x: vector of liquid-phase mole fraction
%      T: temperature (K)
% output:
%      gam: vector of activity coefficients for each component
r = zeros(1,Nc); q = zeros(1,Nc); tau = zeros(k,k); ek = zeros(k,Nc);
theta = zeros(1,k); beta = zeros(Nc,k); s = zeros(1,k); z = 10;
for j = 1:Nc, r(j) = sum(R.*nu(:,j)'); q(j) = sum(Q.*nu(:,j)');
end % row vector(r,q)
for i = 1:k, ek(i,:) = nu(i,:)*Q(i)./q; end
tau = exp(-amn/T);
for i = 1:Nc
    for j = 1:k, beta(i,j) = sum(ek(:,i).*tau(:,j)); end
end
for i = 1:k, theta(i) = sum(x.*q.*ek(i,:))/sum(x.*q); end
for i = 1:k, s(i) = sum(theta'.*tau(:,i)); end
J = r/sum(r.*x); L = q/sum(q.*x); gamc = 1 - J + log(J) - 5*q.*(1 - J./L + log(J./
L)); gamr = [];
for i = 1:Nc
  sumb = 0;
   for j = 1:k
     gamval = theta(j)*beta(i,j)/s(j) - ek(j,i)*log(beta(i,j)/s(j));
     sumb = sumb + gamval;
   end
  gamr = [gamr q(i)*(1 - sumb)];
end
gam = exp(gamc + gamr);
end
```

Example 4.15   Activity Coefficients by the UNIFAC Method[32]

Determine γ_1, and γ_2, for the binary system of diethylamine(1)/n-heptane(2) at $T = 308.15\,K$, when $x_1 = 0.4$, and $x_2 = 0.6$. The subgroups involved are indicated by the chemical formulas as
   Diethylamine(1): $CH_3 - CH_2CH - CH_2 - CH_3$,
   n-Heptane(2): $CH_3 - (CH_2)_5 - CH_3$,

Solution

We can see that there are three functional groups: CH_3, (k = 1), CH_2, (k = 2), and CH_2NH, (k = 3). Table 4.13 shows the subgroups, their identification numbers k, values of parameters R_k, and Q_k, and $v_j^{(i)}$, which represents the number of j, group in component i.[28]

Since CH_3, and CH_2, belong to main group 1 and CH_2NH, belongs to main group 15, the a_{mn}, matrix is given by

The script *actunifgam* sets the required data and calls the function *unifgam* to perform the UNIFAC calculation.

```
% actunifgam.m
Nc = 2; k = 3; % numbers of components(Nc) and functional groups(k)
R = [0.9011 0.6744 1.2070]; % vector of volumes for each functional group
Q = [0.8480 0.5400 0.9360]; % vector of surface areas for each functional group
nu = [2 2; 1 5; 1 0]; % number of functional groups (row: functional groups k,
column: components i)
amn = [0 0 255.7; 0 0 255.7; 65.33 65.33 0]; % matrix of group interaction
parameters
T = 308.15; x = [0.4 0.6]; % mole fraction of each component (T: K)
gam = unifgam(k,R,Q,nu,amn,Nc,x,T)
```

The script *actunifgam* generates the following results:
```
>> actunifgam
gam =
  1.1330   1.0470
```

TABLE 4.13
Parameters of Subgroups

Functional Group	j	R_k	Q_k	$v_j^{(1)}$	$v_j^{(2)}$
CH_3,	1	0.9011	0.8480	2	2
CH_2,	2	0.6744	0.5400	1	5
CH_2NH,	3	1.2070	0.9360	1	0

		Main Group 1		Main Group 15
		j = 1	j = 2	j = 3
Main Group 1	j = 1	0	0	255.7
	j = 2	0	0	255.7
Main Group 15	j = 3	65.33	65.33	0

Example 4.16 Activity Coefficients for a Four-Component Mixture by the UNIFAC Method

Estimate activity coefficients of all components for a system n-hexane(1)/ethanol(2)/methylcyclopentane(3)/benzene(4). The given pressure P, is 1 atm, the temperature T, is 334.82 K, and the liquid-phase mole fractions of the components are $x_1 = 0.162$, $x_2 = 0.068$, $x_3 = 0.656$, and $x_4 = 0.114$. Each component consists of the following functional groups:

n-Hexane(1): $2CH_3 - 4CH_2$,
Ethanol(2): $CH_3 - CH_2 - OH$,
Methylcyclopentane(3): $3CH_3 - CH_2 - CH - C$,
Benzene(4): $6ACH$,

Solution

We can see that there are six functional groups: CH_3, ($j = 1$), CH_2, ($j = 2$), CH, ($j = 3$), and C, ($j = 4$), ACH, ($j = 5$), OH, ($j = 6$). Table 4.14 shows the subgroups, their identification numbers k, values of the parameters R_k, and Q_k, and $\nu_j^{(i)}$, which represents the number of group j, in component i.[28]

CH_3, CH_2, CH, and C, belong to main group 1, ACH, belongs to main group 3, and OH, belongs to main group 5. Thus, we can construct the a_{mn}, parameter matrix shown in Table 4.15.[31]

The script *actmixunifgam* sets the required data and calls the function *unifgam* to perform the UNIFAC calculation.

```
% actmixunifgam.m
Nc = 4; k = 6; % numbers of components(Nc) and functional groups(k)
R = [0.9011 0.6744 0.4469 0.2195 0.5313 1.0000]; % vector of volumes for each
functional group
Q = [0.848 0.540 0.228 0.000 0.400 1.200]; % vector of surface areas for each
functional group
% number of functional groups contained in each component
% (row: functional groups k, column: components i)
nu = [2 1 3 0; 4 1 1 0; 0 0 1 0; 0 0 1 0; 0 0 0 6; 0 1 0 0];
amn = [0 0 0 0 61.13 986.5; 0 0 0 0 61.13 986.5; 0 0 0 0 61.13 986.5;...
       0 0 0 0 61.13 986.5; -11.12 -11.12 -11.12 -11.12 0 636.1;...
       156.4 156.4 156.4 156.4 89.60 0]; % matrix of group interaction parameters
T = 334.82; x = [0.162 0.068 0.656 0.114]; % % mole fraction of each component (T: K)
gam = unifgam(k,R,Q,nu,amn,Nc,x,T)
```

TABLE 4.14
Parameters of Subgroups

Functional Group	j	R_k	Q_k	$\nu_j^{(1)}$	$\nu_j^{(2)}$	$\nu_j^{(3)}$	$\nu_j^{(4)}$
CH_3,	1	0.9011	0.848	2	1	3	0
CH_2,	2	0.6744	0.540	4	1	1	0
CH,	3	0.4469	0.228	0	0	1	0
C,	4	0.2195	0.000	0	0	1	0
ACH,	5	0.5313	0.400	0	0	0	6
OH,	6	1.0000	1.200	0	1	0	0

Source: Poling, B. E. et al., *The Properties of Gases and Liquids*, 5th ed., McGraw-Hill, New York, NY, 2001, pp. 8.78–8.81.

The script *actmixunifgam* generates the following results:
```
>> actmixunifgam
gam =
  1.0463   8.1079   1.0392   1.3088
```

Table 4.16 shows typical activity coefficient models available in commercial process simulators, and Figure 4.4 shows a decision tree to help in the choice of thermodynamic property model.[33]

4.5 VAPOR-LIQUID EQUILIBRIUM

4.5.1 VAPOR-LIQUID EQUILIBRIUM BY RAOULT'S LAW

If we assume that the vapor phase is an ideal gas and the liquid phase is an ideal solution, the equilibrium criterion becomes

$$y_i P = x_i P_i^{sat} \, (i = 1, 2, ..., N)$$

The ratio of the vapor-phase mole fraction to the liquid-phase mole fraction, K_i, is defined as

$$K_i = \frac{y_i}{x_i} = \frac{P_i^{sat}}{P}$$

K_i, can be estimated approximately by the relation[34]

$$K_i \approx \frac{P_{c,i}}{P} 10^{\frac{7}{3}(1+w)\left(1-\frac{1}{T_{r,i}}\right)}$$

which is sometimes called the shortcut K,-ratio.

From the expression to Raoult's law, the vapor-phase mole fraction is given by

$$y_j = \frac{x_j P_j^{sat}}{P}$$

where

x_j, is the liquid-phase mole fraction of species j,
P, is the total pressure
P_j^{sat}, is the vapor pressure of pure species j, at the temperature of the system
$\Sigma_j y_j = 1$, and this equation may be summed over all species to yield

---

TABLE 4.15
Interaction Parameter Matrix for a Four-Component System

		Main Group 1				Main Group 3	Main Group 5
		j = 1	j = 2	j = 3	j = 4	j = 5	j = 6
Main Group 1	j = 1	0	0	0	0	61.13	986.5
	j = 2	0	0	0	0	61.13	986.5
	j = 3	0	0	0	0	61.13	986.5
	j = 4	0	0	0	0	61.13	986.5
Main Group 3	j = 5	−11.12	−11.12	−11.12	−11.12	0	636.1
Main Group 5	j = 6	156.4	156.4	156.4	156.4	89.60	0

TABLE 4.16
Typical Activity Coefficient Models[11]

Type	Name	Phase(s)
BIP	NRTL	L,L1-L2
	Wilson	L
	van Laar	L
	Scatchard-Hilderbrand	L
	UNIQUAC	L,L1-L2
Group contribution	UNIFAC	L,L1-L2
	PSRK	V,L
Electrolyte models	Electrolyte NRTL	L,L1-L2
	Pitzer	L
	Bromley-Pitzer	L

Source: Mariano M. Martin (Editor), *Introduction to software for chemical engineers*, CRC Press, Taylor & Francis Group, Boca Raton, FL, 2015, pp. 302–303.

$$\sum_j y_j = 1 = \sum_j K_j x_j = \frac{1}{P} \sum_j x_j P_j^{sat}$$

where $K_j = y_j/x_j = P_j^{sat}/P$. For dew point calculation where liquid-phase compositions are not known, we use

$$\sum_j x_j = 1 = \sum_j \frac{y_j}{K_j} = P \sum_j \frac{y_j}{P_j^{sat}}$$

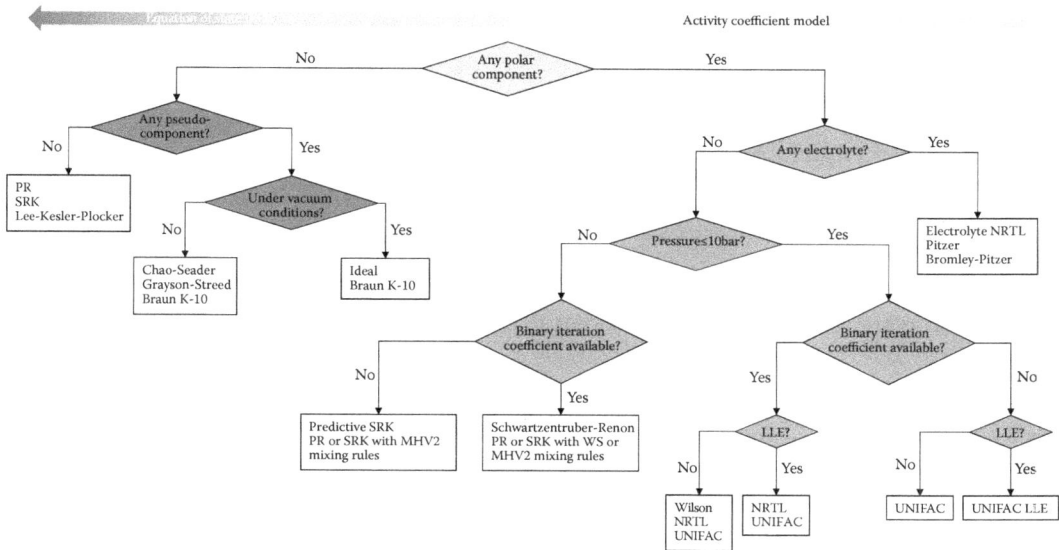

FIGURE 4.4 Decision tree for the selection of thermodynamic property models. (Modified from Mariano M. Martin (Editor), *Introduction to Software for Chemical Engineers*, CRC Press, Taylor & Francis Group, Boca Raton, FL, 2015, p. 305.)

In flash calculations, let F, V, and L, be flow rates (*lbmol/hr*), of multicomponent feed, vapor, and liquid phases, respectively, and let z_j, $y_{j,}$, and x_j, be the mole fractions of species j, in the feed, vapor, and liquid phases, respectively. Combination and rearrangement of the total mass balance and the component balance $F = V + L$, $Fz_j = Vy_j + Lx_j (j = 1, ..., N)$, yield

$$x_j = \frac{z_j}{1 + \alpha(K_j - 1)}, \; y_j = K_j x_j = \frac{K_j z_j}{1 + \alpha(K_j - 1)}$$

where N, is the number of components and $\alpha = V/F$. Using the relation $\sum_{j=1}^{N} x_j = \sum_{j=1}^{N} y_j = 1$, (i.e., $\sum_{j=1}^{N}(x_j - y_j) = 0$), we have the following nonlinear equation:

$$f(\alpha) = \sum_{j=1}^{N}(x_j - y_j) = \sum_{j=1}^{N} \frac{(1 - K_j)z_j}{1 + \alpha(K_j - 1)} = 0$$

Example 4.17 Estimation of Pressure by Raoult's Law

A liquid mixture containing 60*mol%*, of n-pentane(1) and 40*mol%*, n-heptane(2) enters a flash drum at a low pressure. The vapor and liquid streams from the drum are in equilibrium. Both streams are assumed to be ideal, and Raoult's law can be applied. The vapor pressure of each component, $P_{sat,i}$,(*kPa*), at temperature T,(°C), can be obtained using the Antoine equation given by

$$\ln P_{sat,i} = A_i - \frac{B_i}{T(°C) + C_i}(i = 1, 2)$$

where $A_1 = 13.8183$, $A_2 = 13.8587$, $B_1 = 2477.07$, $B_2 = 2991.32$, $C_1 = 233.21$, $C_2 = 216.64$.

(1) Determine the operating pressure P,(*kPa*), when the temperature of the flash drum is $T = 60°C$, and 65% of the feed is vaporized. What is the composition of each product stream at this operating pressure?

(2) Plot the operating pressure as a function of the fraction vaporized at 60°C.

Solution

(1) Let the mole fraction of the feed, the product liquid stream, and the product vapor stream be z, x, and y, respectively, and the molar fraction of the feed that is vaporized be v. From the component mass balance (basis: 1 mole feed mixture) and Raoult's law, we have

$$z_1 = y_1 v + x_1(1 - v), \; y_1 P = x_1 P_{sat,1}, \; (1 - y_1)P = (1 - x_1)P_{sat,2}$$

Data: $z_1 = 0.6$, $v = 0.65$,

The set of nonlinear equations to be solved is defined in the function *binflash* as follows:

```
function f = binflash(x,v,T,z,A,B,C)
% v- fraction of feed vaporized, t: temp(deg.C), z: feed composition
% A, B, C: parameters of Antoine equation
% x(1): x1, x(2): y1, x(3): P
for k = 1:2, Ps(k) = exp(A(k) - B(k)/(T + C(k))); end
f = [x(1)*(1-v) + x(2)*v - z; x(1)*Ps(1) - x(2)*x(3);
(1-x(1))*Ps(2) - (1-x(2))*x(3)];
end
```

As initial conditions, we set $x_{10} = 0.1$, $y_{10} = 0.6$, and $P_0 = 50$. Then the initial value of the vector x, in the program will be $x = [x(1)x(2)x(3)] = [0.10.650]$.

```
>> v = 0.65; T = 60; z = 0.6; A = [13.8183 13.8587]; B = [2477.07 2991.32];
C = [233.21 216.64];
>> x0 = [0.1 0.6 50]; x = fsolve(@binflash, x0, [], v, T, z, A, B, C)
x =
  0.2606   0.7827   71.5515
```

We can see that P = 71.5515, $x_1 = 0.2606$, and $y_1 = 0.7827$.

(2) The script *pvplot* uses the built-in nonlinear equation solver *fsolve* and generates the required plot shown in Figure 4.5 using the function *binflash*.

```
% pvplot.m
A = [13.8183 13.8587]; B = [2477.07 2991.32]; C = [233.21 216.64];
T = 60; z = 0.6; v = 0:0.01:1; n = length(v); x0 = [0.1 0.6 50];
for k = 1:n, x = fsolve(@binflash,x0,[],v(k),T,z,A,B,C); P(k) = x(3); end
plot(v,P), xlabel('v(vaporized frac.)'), ylabel('P(kPa)'), grid
```

Example 4.18 Bubble Point Estimation

Determine the bubble point temperature for a mixture of 32mol%,n-hexane, 31mol%,n-heptane, 25mol%,n-octane, and 12mol%,n-nonane at 1.5bar, total pressure. The vapor pressure of the pure species j, is given by the Antoine equation

$$\log P_j^{sat} = A - \frac{B}{C + T - 273.15} \quad (T : K, P_j^{sat} : bar)$$

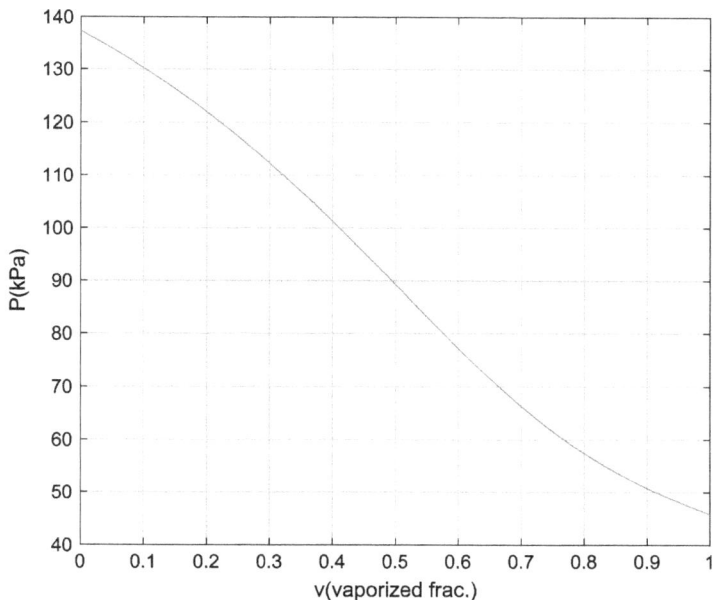

FIGURE 4.5 Plot of P versus v.

TABLE 4.17
Antoine Constants

Component	A	B	C
n-Hexane	4.00139	1170.875	224.317
n-Heptane	4.02023	1263.909	216.432
n-Octane	4.05075	1356.360	209.635
n-Nonane	4.07356	1438.03	202.694

where the Antoine constants for each component are shown in Table 4.17.

Solution

We use the relation

$$\sum_{j=1}^{n} y_j = \sum_{j=1}^{n} \frac{x_j P_j^{sat}}{P} = 1$$

where the vapor pressure of the pure species j, P_j^{sat}, is given by Antoine equation. The bubble point temperature is given by solving the following nonlinear equation:

$$\sum_{j=1}^{n} \frac{x_j}{P} 10^{\left(A_j - \frac{B_j}{C_j+T-273.15}\right)} - 1 = 0$$

The script *btraoult* defines the nonlinear equation and uses the built-in function solver *fsolve* to determine the bubble point temperature.

```
% btraoult.m: bubble point temperature by Raoult's law
% mixture: 32% n-hexane(1), 31% n-heptane(2), 25% n-octane(3), 12% n-nonane(4)
A = [4.00139, 4.02023, 4.05075, 4.07356]; % Antoine constants
B = [1170.875, 1263.909, 1356.360, 1438.03];
C = [224.317, 216.432, 209.635, 202.694];
x = [0.32 0.31 0.25 0.12]; % Compositions
P = 1.5; % total pressure (bar)
f = @(T) sum(x.*10.^(A-B./(C+T-273.15)))/P - 1; % define the nonlinear equation
(T: K)
T0 = 400; T = fsolve(f,T0); T = T - 273.15 % T0: initial guess

>> btraoult
T =
 106.5252
```

Example 4.19   P and T Plots by Raoult's Law[35]

A binary system of acetonitrile(1)/nitromethane(2) conforms closely to Raoult's law. Vapor pressures for the pure species are given by the following Antoine equations:

$$\ln P_1 = A_1 - \frac{B_1}{T+C_1} = 14.2724 - \frac{2945.47}{T+224.0}, \ln P_2 = A_2 - \frac{B_2}{T+C_2} = 14.2043 - \frac{2972.64}{T+209.0}$$

where T, is the temperature ($°C$).

(1) Generate a graph showing the total pressure P, versus x_1, and y_1, for a temperature of 75 °C.

(2) Generate a graph showing T, versus x_1, and y_1, for a pressure of $P = 70 kPa$.

Solution

(1) For a given T,

$$P = \sum_i x_i P_i^{sat} = x_1 P_1^{sat} + x_2 P_2^{sat} = P_2^{sat} + (P_1^{sat} - P_2^{sat})x_1$$

$$= \exp\left(A_2 - \frac{B_2}{T + C_2}\right) + \left\{\exp\left(A_1 - \frac{B_1}{T + C_1}\right) - \exp\left(A_2 - \frac{B_2}{T + C_2}\right)\right\}x_1$$

$$y_1 = \frac{x_1 P_1^{sat}}{P} = \frac{x_1}{P}\exp\left(A_1 - \frac{B_1}{T + C_1}\right)$$

The script *bubbp* plots P, versus x_1, and y_1. Implementation of the script *bubbp* produces Figure 4.6.

```
% bubbp.m: calculation of total P and T
clear all;
T = 75; A1 = 14.2724; B1 = 2945.47; C1 = 224; A2 = 14.2043; B2 = 2972.64; C2 = 209;
fp = @(x) exp(A2-B2/(T+C2)) + x*(exp(A1-B1/(T+C1)) - exp(A2-B2/(T+C2)));
% total pressure
x1 = 0:0.01:1; y1 = []; P = [];
for k = 1:length(x1)
   Ps = feval(fp,x1(k)); fy = @(x) x*(exp(A1-B1/(T+C1)))/Ps;
   y = feval(fy,x1(k)); y1 = [y1 y]; P = [P Ps];
end
plot(x1,P,y1,P,'.-'), xlabel('x_1,y_1'), ylabel('P')
legend('P-x_1','P-y_1','location','best')
```

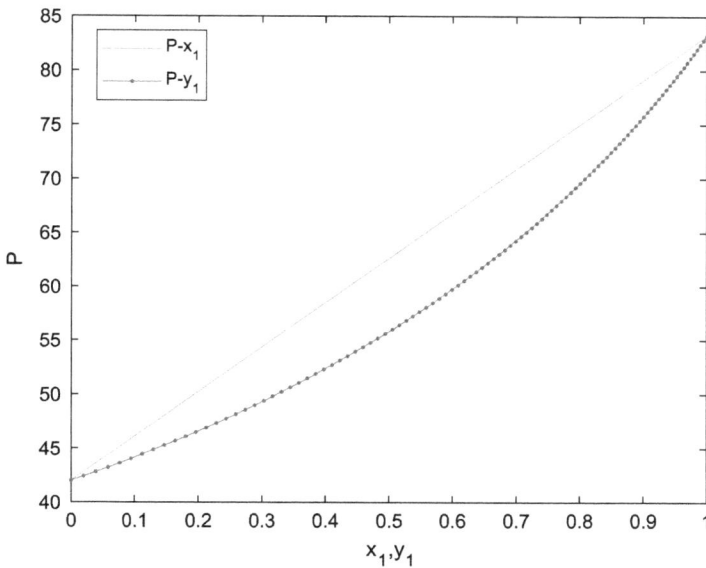

FIGURE 4.6 Vapor pressure of acetonitrile(1)/nitromethane(2) system.

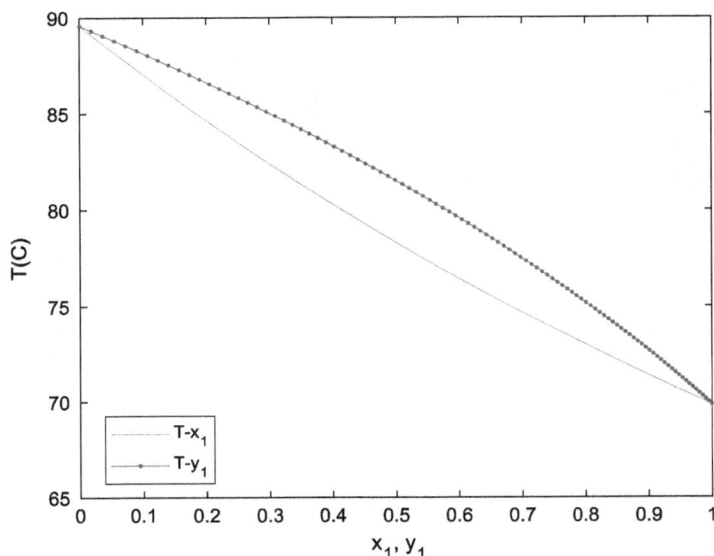

FIGURE 4.7   Dew point of an acetonitrile(1)/nitromethane(2) system.

(2) For a given pressure P, T, can be found by solving the nonlinear equation

$$\exp\left(A_2 - \frac{B_2}{T + C_2}\right) + \left\{\exp\left(A_1 - \frac{B_1}{T + C_1}\right) - \exp\left(A_2 - \frac{B_2}{T + C_2}\right)\right\}x_1 - P = 0$$

Once T, is known, the vapor-phase composition can be determined from the relation

$$y_1 = \frac{x_1 P_1^{sat}}{P} = \frac{x_1}{P}\exp\left(A_1 - \frac{B_1}{T + C_1}\right)$$

The script *dewpt* uses the built-in solver *fzero* to solve the nonlinear equation. Implementation of the script *dewpt* produces the plot of T, versus x_1, and y_1, shown in Figure 4.7.

```
% dewpt.m: calculation of dew point
clear all;
A1 = 14.2724; B1 = 2945.47; C1 = 224; A2 = 14.2043; B2 = 2972.64; C2 = 209;
T0 = 60; P = 70; x = 0:0.01:1; T = []; y = [];
for k = 1:length(x)
  x1 = x(k);
  fp = @(T) exp(A2-B2/(T+C2)) + x1*(exp(A1-B1/(T+C1)) - exp(A2-B2/(T+C2)))-P;
  T1 = fzero(fp,T0); y1 = x1*exp(A1-B1/(T1+C1))/P; T = [T T1]; y = [y y1];
end
plot(x,T,y,T,'.-'), xlabel('x_1, y_1'), ylabel('T(C)')
legend('T-x_1','T-y_1','location','best')
```

4.5.2 VAPOR-LIQUID EQUILIBRIUM BY MODIFIED RAOULT'S LAW

4.5.2.1 Dew Point and Bubble Point Calculations

Insertion of an activity coefficient, γ_i, into Raoult's law yields the modified Raoult's law:

$$y_i P = x_i \gamma_i P_i^{sat} \, (i = 1, 2, \ldots, N)$$

The equilibrium ratio (or K,-value), K_i, is defined as

$$K_i = \frac{y_i}{x_i} = \frac{\gamma_i P_i^{sat}}{P}$$

According to this equation, $y_i = K_i x_i$, and $x_i = y_i/K_i$. Summation relations give

$$\sum_{i=1}^{N} y_i = \sum_{i=1}^{N} K_i x_i = 1, \; \sum_{i=1}^{N} x_i = \sum_{i=1}^{N} \frac{y_i}{K_i} = 1$$

In vapor-liquid equilibrium calculations, the following relations are commonly used:

$$P = \sum_{i=1}^{N} x_i \gamma_i P_i^{sat}, \; y_i = \frac{x_i \gamma_i P_i^{sat}}{P} = K_i x_i \, \text{or} \, x_i = \frac{y_i P}{\gamma_i P_i^{sat}} = \frac{y_i}{K_i}, \; \sum_{i=1}^{N} y_i = 1, \; \sum_{i=1}^{N} x_i = 1$$

Bubble point pressure calculation (BubbleP): For a given temperature T, and liquid-phase compositions x_i, the total pressure P, and vapor-phase compositions y_i, are calculated. The values of the vapor pressure and the activity coefficient of species i, (P_i^{sat}, and γ_i), are estimated first. Then P, and y_i, can be obtained from the relations

$$P = \sum_{i=1}^{N} x_i \gamma_i P_i^{sat}, \; y_i = \frac{x_i \gamma_i P_i^{sat}}{P}$$

Dew point pressure calculation (DewP): For a given temperature T, and vapor-phase compositions y_i, the total pressure P, and liquid-phase compositions x_i, are calculated. The vapor pressure of species i, (P_i^{sat}), is determined first from the given T, followed by the estimation of liquid-phase compositions x_i, using the nonlinear equation

$$P = \sum_{i=1}^{N} x_i \gamma_i P_i^{sat}, \; f(x) = 1 - \sum_{i=1}^{N} x_i = 1 - \sum_{i=1}^{N} \frac{y_i P}{\gamma_i P_i^{sat}} = 0$$

Bubble point temperature calculation (BubbleT): For a given total pressure P, and liquid-phase compositions x_i, temperature T, and vapor-phase compositions y_i, are calculated. The solution of the following nonlinear equation yields the value of T:

$$f(T) = P - \sum_{i=1}^{N} x_i \gamma_i P_i^{sat} = 0$$

Dew point temperature calculation (DewT): For a given total pressure P, and vapor-phase compositions y_i, temperature T, and liquid-phase compositions x_i, are calculated. We first assume the value of T, and calculate the vapor pressure. Then we assume an initial value of the liquid-phase mole fraction x_i, (usually Raoult's law is used: $x_i = P y_i/P_i^{sat}$). We calculate the activity coefficient γ_i, and solve the following nonlinear equation to obtain the value of T:

$$f(T) = 1 - \sum_{i=1}^{N} \frac{y_i P}{\gamma_i P_i^{sat}} = 0$$

Using the resultant T, the relation $x_i = y_i P/(\gamma_i P_i^{sat})$, is evaluated and compared with the initial guess. This procedure is repeated until the value of x_i, does not show any changes.

Inputs and outputs for each calculation are summarized in Table 4.18.

TABLE 4.18

Inputs and Outputs for Vapor-Liquid Equilibrium Calculations

Calculation Type	Inputs	Outputs
Bubble point pressure (Bubble P),	$T, x_i,$	$P, y_i,$
Dew point pressure (Dew P),	$T, y_i,$	$P, x_i,$
Bubble point temperature (Bubble T),	$P, x_i,$	$T, y_i,$
Dew point temperature (Dew T),	$P, y_i,$	$T, x_i,$

Example 4.20 Equilibrium Calculations Using the Modified Raoult's Law[36]

For a methanol(1)/methyl acetate(2) binary system, reasonable correlations for the activity coefficients are given by

$$\ln\gamma_1 = Ax_2^2, \ \ln\gamma_2 = Ax_1^2, \ A = 2.771 - 0.00523T$$

The Antoine equation provides expressions for vapor pressures (kPa):

$$\ln P_1^{sat} = 16.59158 - \frac{3643.31}{T - 33.324}, \ \ln P_2^{sat} = 14.25326 - \frac{2665.54}{T - 53.424}$$

where T, is the temperature (K).

(1) Estimate P, and y_i, for $T = 318.15K$, and $x_1 = 0.25$.
(2) Estimate P, and x_i, for $T = 318.15K$, and $y_1 = 0.60$.
(3) Estimate T, and y_i, for $P = 101.33kPa$, and $x_1 = 0.85$.
(4) Estimate T, and x_i, for $P = 101.33kPa$, and $y_1 = 0.40$.

Solution

(1) For the given temperature and liquid-phase composition, P_i^{sat}, and γ_i, are calculated. These values are used to determine P, and y_i, from the relations $P = \sum_{i=1}^{N} x_i\gamma_i P_i^{sat}$, and $y_i = x_i\gamma_i P_i^{sat}/P$.

(2) P_i^{sat}, is determined from the given temperature, and the composition is found by solving the nonlinear equation

$$P = \sum_{i=1}^{N} x_i\gamma_i P_i^{sat}, f(x) = 1 - \sum_{i=1}^{N} x_i = 1 - \sum_{i=1}^{N} \frac{y_i P}{\gamma_i P_i^{sat}} = 0$$

(3) The value of T, is obtained from the solution of the nonlinear equation

$$f(T) = P - \sum_{i=1}^{N} x_i\gamma_i P_i^{sat} = 0$$

(4) The vapor pressure is estimated using the assumed T, and the initial value of the liquid-phase mole fraction x_i, is assumed. Usually Raoult's law is used to assume x_i: $x_i = Py_i/P_i^{sat}$. The activity coefficient γ_i, is calculated and T, is determined from the nonlinear equation

$$f(T) = 1 - \sum_{i=1}^{N} \frac{y_i P}{\gamma_i P_i^{sat}} = 0$$

The resultant T, is used to evaluate $x_i = y_i P/(\gamma_i P_i^{sat})$, which is compared with the initial guess. This procedure is repeated until x_i, converges on a fixed value.

The script *binVLE* implements these calculation procedures. Using given Antoine equation parameters, this script performs vapor-liquid equilibrium calculations.

```
% binVLE.m: VLE calculation for binary systems using modified Raoult's law
% Antoine vapor pressure equation parameters
A = [16.59158 14.25326]; B = [3643.31 2665.54]; C = [33.424 53.424];
% (1) Bubble P calculation
T = 318.15; x1 = 0.25; x = [x1 1-x1]; % data
gamma = gamma12(x1,T); Psat = vp12(A,B,C,T);
P = sum(x.*gamma.*Psat); y = (x.*gamma.*Psat)/P;
fprintf('(1) Bubble P: P = %g, y1 = %g, y2 = %g\n',P,y(1),y(2));
fprintf('         gamma1 = %g, gamma2 = %g\n',gamma(1),gamma(2));
% (2) Dew P calculation
T = 318.15; y1 = 0.60; y = [y1 1-y1]; Psat = vp12(A,B,C,T); % data
x10 = 0.7; % initial guess for mole fraction x
[x1, fval] = fzero(@dewpf,x10,[],y,T,Psat); x = [x1 1-x1];
gamma = gamma12(x1,T); P = sum(x.*gamma.*Psat);
fprintf('(2) Dew P: P = %g, x1 = %g, x2 = %g\n',P,x(1),x(2));
fprintf('         gamma1 = %g, gamma2 = %g\n',gamma(1),gamma(2));
% (3) Bubble T calculation
P = 101.33; x1 = 0.85; x = [x1 1-x1]; % data
T0 = 300; % initial guess for temperature
[T, fval] = fzero(@bubtf,T0,[],x1,A,B,C,P);
gamma = gamma12(x1,T); Psat = vp12(A,B,C,T); y = (x.*gamma.*Psat)/P;
fprintf('(3) Bubble T: T = %g, y1 = %g, y2 = %g\n',T,y(1),y(2));
fprintf('         gamma1 = %g, gamma2 = %g\n',gamma(1),gamma(2));
% (4) Dew T calculation
P = 101.33; y1 = 0.40; y = [y1 1-y1]; % data
T0 = 330; % initial guess for temperature
Psat0 = vp12(A,B,C,T0); x1 = y1*P/Psat0(1); crx = 1; cx = 1e-6;
while crx > cx
    [T, fval] = fzero(@dewfun,T0,[],y1,x1,A,B,C,P);
    gamma = gamma12(x1,T); Psat = vp12(A,B,C,T);
  x = y*P./(gamma.*Psat); crx = abs(x - x1); x1 = x(1);
end
fprintf('(4) Dew T: T = %g, x1 = %g, x2 = %g\n',T,x(1),x(2));
fprintf('         gamma1 = %g, gamma2 = %g\n',gamma(1),gamma(2));

function gamma = gamma12(x1,T) % Calculation of activity coefficients
A = 2.771 - 0.00523*T; % T(K)
gamma(1) = exp(A*(1-x1)^2); gamma(2) = exp(A*x1^2);
end
function Psat = vp12(A,B,C,T) % Vapor pressure by Antoine equation (T: K)
Psat = exp(A - B./(T - C));
end
function f = dewpf(x1,y,T,Psat) % Define equation for (2) dew P calculation
x = [x1 1-x1]; gamma = gamma12(x1,T); P = sum(x.*gamma.*Psat);
f = sum(y*P./(gamma.*Psat)) - 1;
end
function f = bubtf(T,x1,A,B,C,P) % Define equation for (3) Bubble T calculation
x = [x1 1-x1]; gamma = gamma12(x1,T); Psat = vp12(A,B,C,T);
f = sum(x.*gamma.*Psat) - P;
end
```

```
function fp = dewfun(T,y1,x1,A,B,C,P) % Define equation for (4) Dew T calculation
y = [y1 1-y1]; Psat = vp12(A,B,C,T); gamma = gamma12(x1,T);
fp = sum(y*P./(gamma.*Psat)) - 1;
end
```

Implementation of the script *binVLE* produces following outputs:

```
>> binVLE
(1) Bubble P: P = 73.5003, y1 = 0.282205, y2 = 0.717795
            gamma1 = 1.86401, gamma2 = 1.07164
(2) Dew P: P = 62.8945, x1 = 0.816927, x2 = 0.183073
            gamma1 = 1.03780, gamma2 = 2.09348
(3) Bubble T: T = 331.201, y1 = 0.66967, y2 = 0.33033
            gamma1 = 1.02365, gamma2 = 2.11816
(4) Dew T: T = 326.697, x1 = 0.460197, x2 = 0.539803
            gamma1 = 1.36283, gamma2 = 1.25231
```

The outputs can be summarized as shown in Table 4.19.

4.5.2.2 Flash Calculation by the Modified Raoult's Law

The dew point and bubble point calculation methods can be used in the estimation of the quantities and compositions for the flash evaporation of an ideal multi-component mixture. Figure 4.8 shows a flash evaporator.

In Figure 4.8, F, V, and L, are the total feed, vapor, and liquid flow rates, respectively, and z_j, y_j,, and x_j, are the mole fractions of component j, in the feed, vapor, and liquid streams, respectively. The total number of components is assumed to be N. From the overall mass and component balances $F = V + L$, $Fz_j = Vy_j + Lx_j$ $(j = 1, ..., N)$, and using $y_j = K_j x_j$, we have

$$x_j = \frac{z_j}{1 + \alpha(K_j - 1)}, \, y_j = K_j x_j = \frac{K_j z_j}{1 + \alpha(K_j - 1)}$$

where $\alpha = \frac{V}{F}$, $K_j = \gamma_j P_j^{sat}/P$, P_j^{sat}, is the vapor pressure of species j, and P, is the total pressure. Substitution of these equations into the summing relations $\sum_{j=1}^{N} x_j = \sum_{j=1}^{N} y_j = 1$, (i.e., $\sum_{j=1}^{N}(x_j - y_j) = 0$), yields the following nonlinear equation:

$$f(\alpha) = \sum_{j=1}^{N}(x_j - y_j) = \sum_{j=1}^{N}\frac{(1 - K_j)z_j}{1 + \alpha(K_j - 1)} = 0$$

At the dew point, $\alpha = 1$, (all the feed is evaporated) and

---

TABLE 4.19
Results of VLE Calculations

Problem Type	P(kPa)	T(K)	x_1	y_1	γ_1	γ_2
Bubble P,	73.5003	*318.15*	*0.25*	0.28221	1.86401	1.07164
Dew P,	62.8945	*318.15*	0.81693	*0.60*	1.03780	2.09348
Bubble T,	*101.33*	331.201	*0.85*	0.66967	1.02365	2.11816
Dew T,	*101.33*	326.696	0.46019	*0.40*	1.36283	1.25231

Note: Values in bold *italics* are inputs.

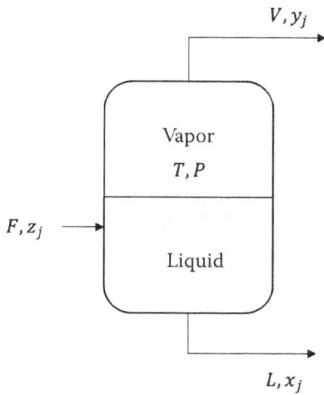

V, y_j

Vapor
T, P

F, z_j

Liquid

L, x_j **FIGURE 4.8**   Flash evaporator.

$$f(T) = \sum_{j=1}^{N} \frac{z_j}{K_j} - 1 = 0$$

At the bubble point, $\alpha = 0$, (evaporation of the feed is initiated) and we have

$$f(T) = \sum_{j=1}^{N} z_j K_j - 1 = 0$$

P_j^{sat}, is given by the Antoine vapor pressure equation:

$$\log P_j = A_j - \frac{B_j}{C_j + T} \quad (P_j^{sat} : mmHg, \ T : {}^\circ C)$$

Example 4.21   Flash Evaporator[37]

A feed stream of an ideal four-component mixture is fed into a flash evaporator. The composition of the feed stream is given in Table 4.20 with the Antoine equation constants. The flash drum operates under high pressure, between 15 and 25 atm, with a feed stream at 50 °C. Estimate the percentage of the total feed at 50 °C, that is evaporated, α,(= V/F), and the corresponding mole fractions in the liquid and vapor streams fbubble point temperature,or $P = 16$, 18, 20, and 24 atm. Calculate the dew point and bubble point temperatures of the feed stream.

Solution
The function *sumf* defines the summing relations. The equation for dew point temperature, $f(T) = \sum_{j=1}^{N}(z_j/K_j) - 1 = 0$, and bubble point temperature, $f(T) = \sum_{j=1}^{N} z_j K_j - 1 = 0$, are defined by the functions *zdk* and *zmk*, respectively. The script *flashdrum* calls the subfunctions *sumf*, *zdk*, and *zmk* and uses the built-in function *fzero* to solve the nonlinear equations.

```
% flashdrum.m: flash calculation using modified Raoult's law
zf = [0.1 0.25 0.5 0.15]; % feed composition
A = [6.64380 6.82915 6.80338 6.80776]; % Antoine constants
B = [395.74 663.72 804.00 935.77];
C = [266.681 256.681 247.040 238.789];
P = [16 18 20 24]*760; % pressure (mmHg)
% compute alpha
```

TABLE 4.20

Feed Composition and Antoine Equation Constants

Component	Mole Fraction(z_j)	Antoine Equation Constants		
		A	B	C
Ethylene	0.10	6.64380	395.74	266.681
Ethane	0.25	6.82915	663.72	256.681
Propane	0.50	6.80338	804.00	247.040
n-Butane	0.15	6.80776	935.77	238.789

```
T = 50; Pv = 10.^(A - B./(C + T)); % vapor pressure (mmHg) (T: deg.C)
x = []; y = []; alpha = []; n = length(P);
for i = 1:n
     k = Pv/P(i); ax = fzero(@sumf,0.5,[],zf,k); xv = zf./(1 + ax*(k - 1));
     yv = k.*xv; x = [x xv']; y = [y yv']; alpha = [alpha ax];
end
Patm = P/760;
% dew-point: alpha = 1
Tdew = [];
for i = 1:n, T = fzero(@zdk,50,[],zf,P(i),A,B,C); Tdew = [Tdew T]; end
% bubble-point: alpha = 0
Tbub = [];
for i = 1:n, T = fzero(@zmk,50,[],zf,P(i),A,B,C); Tbub = [Tbub T]; end
Patm, x, y, alpha, Tdew, Tbub

function f = sumf(alp,zf,k)
f = sum(zf.*(1-k)./(1 + alp*(k-1)));
end
function f = zdk(T,zf,P,A,B,C) % dew-point calculation: alpha=1
k = 10.^(A - B./(C + T)) / P; f = sum(zf./k) - 1;
end
function f = zmk(T,zf,P,A,B,C) % bubble-point calculation: alpha=0
k = 10.^(A - B./(C + T)) / P; f = sum(zf.*k) - 1;
end
```

The script *flashdrum* generates the following outputs:

```
>> flashdrum
Patm =
   16     18     20     24
x =
   0.0052   0.0067   0.0086   0.0136
   0.0685   0.0851   0.1032   0.1409
   0.4874   0.5380   0.5708   0.5945
   0.4389   0.3702   0.3174   0.2510
y =
   0.1055   0.1212   0.1398   0.1848
   0.2605   0.2875   0.3139   0.3571
   0.5007   0.4914   0.4692   0.4072
   0.1333   0.1000   0.0771   0.0508
alpha =
   0.9454   0.8148   0.6967   0.5045
```

```
Tdew =
 52.1955  57.5182  62.4550  71.4161
Tbub =
 -29.4195 -23.3346 -17.6971  -7.4850
```

The results may be summarized as shown in Table 4.21.

4.5.3 Vapor-Liquid Equilibrium Using Ratio of Fugacity Coefficients

4.5.3.1 Dew Point and Bubble Point Calculations[38]

Introduction of the vapor-phase fugacity coefficient to the modified Raoult's law yields, at equilibrium,

$$y_i \hat{\phi}_i P = x_i \gamma_i f_i = x_i \gamma_i \phi_i^{sat} P_i^{sat}$$

This can be rewritten as

$$y_i \frac{\hat{\phi}_i}{\phi_i^{sat}} P = y_i \Phi_i P = x_i \gamma_i P_i^{sat}$$

where $\Phi_i = \hat{\phi}_i / \phi_i^{sat}$, is the ratio of fugacity coefficients. The application of virial expansion gives

$$\Phi_i = \exp\left(\left[B_{ii}(P - P_i^{sat}) + \frac{1}{2} P \sum_j \sum_k y_j y_k (2\delta_{ji} - \delta_{jk})\right]\frac{1}{RT}\right)$$

$$\delta_{ji} = 2B_{ji} - B_{jj} - B_{ii}, \ \delta_{jk} = 2B_{jk} - B_{jj} - B_{kk}, \ \delta_{ii} = \delta_{jj} = 0, \ \delta_{ij} = \delta_{ji}$$

In dew point and bubble point calculations, the following relations are used:

$$y_i = \frac{x_i \gamma_i P_i^{sat}}{\Phi_i P}, \ x_i = \frac{y_i \Phi_i P}{\gamma_i P_i^{sat}}, \ \sum_i y_i = \sum_i x_i = 1, \ P = \sum_i \frac{x_i \gamma_i P_i^{sat}}{\Phi_i}, \ P = \frac{1}{\sum_i \left(\frac{y_i \Phi_i}{\gamma_i P_i^{sat}}\right)}$$

TABLE 4.21
Results of Flash Calculations

Component		$P(atm)$, 16	18	20	24
$\alpha = V/F$,		0.9454	0.8148	0.6967	0.5045
Dew point $T(°C)$,		52.1955	57.5182	62.4550	71.4161
Bubble point $T(°C)$,		−29.4195	−23.3346	−17.6971	−7.4850
x_j,	Ethylene	0.0052	0.0067	0.0086	0.0136
	Ethane	0.0685	0.0851	0.1032	0.1409
	Propane	0.4874	0.5380	0.5708	0.5945
	n-Butane	0.4389	0.3702	0.3174	0.2510
y_j,	Ethylene	0.1055	0.1212	0.1398	0.1848
	Ethane	0.2605	0.2875	0.3139	0.3571
	Propane	0.5007	0.4914	0.4692	0.4072
	n-Butane	0.1333	0.1000	0.0771	0.0508

Bubble point pressure calculations (BubbleP): Total pressure P, and vapor-phase composition y_i, are calculated for a given temperature T, and liquid-phase composition x_i.

1) Compute P_i^{sat}, and γ_i, set $\Phi_i = 1$, and calculate $P_{old} = \sum_i x_i \gamma_i P_i^{sat}/\Phi_i$.
2) Determine $y_i = x_i \gamma_i P_i^{sat}/(\Phi_i P_{old})$, and Φ_i.
3) Find $P_{new} = \sum_i x_i \gamma_i P_i^{sat}/\Phi_i$. If the criterion $|P_{old} - P_{new}| < \epsilon$, is not satisfied, set $P_{old} = P_{new}$, and go to 2).

Dew point pressure calculations (DewP): For the given values of T, and y_i, P, and x_i, are estimated.

1) Calculate P_i^{sat}, set $\Phi_i = 1$, and $\gamma_i = 1$, and compute $P_{old} = 1/\sum_i (y_i \Phi_i/(\gamma_i P_i^{sat}))$. Find $x_i = y_i \Phi_i P_{old}/(\gamma_i P_i^{sat})$, and γ_i, and estimate P_{old}, again.
2) Determine Φ_i.
3) Calculate $x_i = y_i \Phi_i P_{old}/(\gamma_i P_i^{sat})$, normalize x_i, and find γ_i. Repeat this procedure until the change in γ_i, from one iteration to the next is less than some tolerance.
4) Calculate $P_{new} = 1/\sum_i (y_i \Phi_i/(\gamma_i P_i^{sat}))$. If the criterion $|P_{old} - P_{new}| < \epsilon$, is not satisfied, set $P_{old} = P_{new}$, and go to 2).

Bubble point temperature calculations (BubbleT): For given values of P, and x_i, T, and y_i, are estimated.

1) Set $\Phi_i = 1$, and compute $T_i^{sat} = B_i/(A_i - \ln P) - C_i$, and $T = \sum_i x_i T_i^{sat}$.
2) Calculate P_i^{sat}, and γ_i. For a component j, calculate $P_j^{sat} = P/[\sum_i (x_i \gamma_i/\Phi_i)(P_i^{sat}/P_j^{sat})]$, and $T = B_j/(A_j - \ln P_j^{sat}) - C_j$.
3) Determine P_i^{sat}, and find $y_i = x_i \gamma_i P_i^{sat}/(\Phi_i P)$. Calculate Φ_i, and γ_i, and compute $P_j^{sat} = P/[\sum_i (x_i \gamma_i/\Phi_i)(P_i^{sat}/P_j^{sat})]$, and $T = B_j/(A_j - \ln P_j^{sat}) - C_j$. Repeat this procedure until the change in T, from one iteration to the next is less than some tolerance.

Dew point temperature calculations (DewT): For given values of P, and y_i, T, and x_i, are estimated.

1) Set $\Phi_i = 1$, and $\gamma_i = 1$, and compute $T_i^{sat} = B_i/(A_i - \ln P) - C_i$, and $T = \sum_i x_i T_i^{sat}$.
2) Find P_i^{sat}. For component j, calculate $P_j^{sat} = P \sum_i (y_i \Phi_i/\gamma_i)(P_j^{sat}/P_i^{sat})$, and $T = B_j/(A_j - \ln P_j^{sat}) - C_j$.
3) Determine P_i^{sat}, and Φ_i, and calculate $x_i = y_i \Phi_i P/(\gamma_i P_i^{sat})$, and γ_i.
4) Calculate $P_j^{sat} = P \sum_i (y_i \Phi_i/\gamma_i)(P_j^{sat}/P_i^{sat})$, and $T = B_j/(A_j - \ln P_j^{sat}) - C_j$.
5) Find P_i^{sat}, and Φ_i.
6) Calculate $x_i = y_i \Phi_i P/(\gamma_i P_i^{sat})$, followed by normalization, and determine γ_i. Repeat this procedure until the change in γ_i, from one iteration to the next is less than some tolerance.
7) Calculate $P_j^{sat} = P \sum_i (y_i \Phi_i/\gamma_i)(P_j^{sat}/P_i^{sat})$, and $T_{new} = B_j/(A_j - \ln P_j^{sat}) - C_j$. If the criterion $|T - T_{new}| < \epsilon$, is not satisfied, set $T = T_{new}$, and go to 5).

Example 4.22 Bubble Point P for a Two-Component System

A liquid mixture contains chloroform (1) and ethanol (2) at 60 °C. At equilibrium,

$$y_1 \hat{\phi}_1 P = x_1 \gamma_1 P_1{}^{sat}, \ (1 - y_1)\hat{\phi}_2 P = (1 - x_1)\gamma_2 P_2{}^{sat}$$

where x_1, and y_1, are the mole fractions of chloroform in the liquid and vapor phases, respectively; $P_i{}^{sat}$, is the vapor pressure of component i; γ_i, is the activity coefficient of component i; and $\hat{\phi}_i$, is the fugacity coefficient of component i. Generate profiles of the bubble point pressure (P), as a function of x_1, and y_1, using the data given below. The activity coefficient γ_i, is given by

$$\ln\gamma_1 = (1 - x_1)^2\{A_{12} + 2(A_{21} - A_{12})x_1\}, \ \ln\gamma_2 = x_1^2\{A_{21} + 2(A_{12} - A_{21})(1 - x_1)\}$$

and $\hat{\phi}_i$, can be obtained from

$$\ln\hat{\phi}_1 = \frac{1}{RT}\{B_{11}(P - P_1{}^{sat}) + P(1 - y_1)^2\delta_{12}\}, \ \ln\hat{\phi}_2 = \frac{1}{RT}\{B_{22}(P - P_2{}^{sat}) + Py_1^2\delta_{12}\}$$

where $\delta_{12} = 2B_{12} - B_{11} - B_{22}$, R, is the gas constant, and $T(K)$, is the temperature.

 Data: $P_1{}^{sat} = 83.25kPa$, $P_2{}^{sat} = 37.97kPa$, $A_{12} = 0.59$, $A_{21} = 1.42$, $B_{11} = -963cm^3/mol$, $B_{22} = -1523cm^3/mol$, $B_{12} = 52cm^3/mol$.

Solution

 For a given x_1, the solution of the system of nonlinear equations

$$x_1\gamma_1P_1{}^{sat} - y_1\hat{\phi}_1P = 0, \ (1 - x_1)\gamma_2P_2{}^{sat} - (1 - y_1)\hat{\phi}_2P = 0$$

yields y_1, and P. The script *bpbin* solves the system of nonlinear equations using the built-in solver *fsolve* and produces profiles of the bubble point pressure (P), as a function of x_1, and y_1, as shown in Figure 4.9.

```
% bpbin.m: bubble point P for binary system
T = 60+273.15; R = 8.314; % t=60 deg.C
P1s = 83.25e3; P2s = 37.97e3; A12 = 0.59; A21 = 1.42;
B11 = -963e-6; B22 = -1523e-6; B12 = 52e-6; % (m^3/mol)
x = [0:0.01:1]; n = length(x); % x1
gam1 = exp((1-x).^2 .*(A12 + 2*(A21 - A12)*x)); % activity coefficient gamma1
gam2 = exp(x.^2 .*(A21 + 2*(A12 - A21)*(1-x))); % activity coefficient gamma2
```

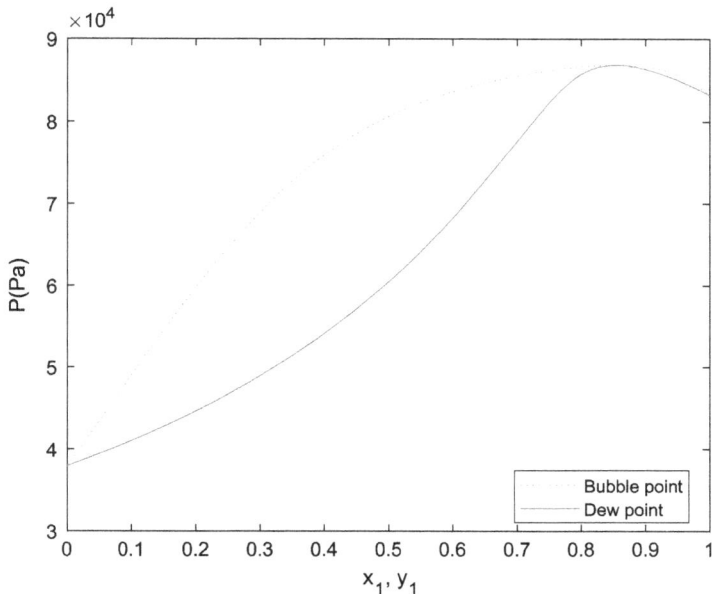

FIGURE 4.9 Pressure profile as a function of liquid and vapor compositions.

```
d12 = 2*B12 - B11 - B22;
for k = 1:n % z(1)=y1, z(2)=P
f = @(z) [x(k)*gam1(k)*P1s-z(1)*z(2)*exp((B11*(z(2)-P1s)+z(2)*d12*(1-z(1))
    ^2)/R/T);...
    (1-x(k))*gam2(k)*P2s-(1-z(1))*z(2)*exp((B22*(z(2)-P2s)+z(2)*d12*(z(1))
^2)/R/T)];
    z0 = [0.5, 5e4]; z = fsolve(f,z0); y(k) = z(1); P(k) = z(2);
end
plot(x,P,':',y,P), xlabel('x_1, y_1'), ylabel('P(Pa)')
legend('Bubble point','Dew point','location','best')
```

Example 4.23 Bubble T Calculations for a Four-Component System[39]

Determine the bubble point temperature and vapor-phase mole fractions for an n-hexane(1)/ethanol(2)/methylcyclopentane(3)/benzene(4) system. The given pressure is $1atm$, and the given liquid-phase mole fractions are $x_1 = 0.162$, $x_2 = 0.068$, $x_3 = 0.656$, and $x_4 = 0.114$. Vapor pressure can be estimated using the Antoine equation $\ln P_i^{sat} = A_i - B_i/(T + C_i)$, ($T$:K, P:atm). The parameters for the Antoine equation are

$$A_1 = 9.2033, \ A_2 = 12.2786, \ A_3 = 9.1690, \ A_4 = 9.2675$$

$$B_1 = 2697.55, \ B_2 = 3803.98, \ B_3 = 2731.00, \ B_4 = 2788.51$$

$$C_1 = -48.78, \ C_2 = -41.68, \ C_3 = -47.11, \ C_4 = -52.36$$

The virial coefficients (cm^3/mol), for each component are

$$B_{11} = -1360.1, \ B_{12} = -657.0, \ B_{13} = -1274.2, \ B_{14} = -1218.8$$

$$B_{22} = -1174.7, \ B_{23} = -621.8, \ B_{24} = -589.7, \ B_{33} = -1191.9$$

$$B_{34} = -1137.9, \ B_{44} = -1086.9$$

Solution

This problem belongs to the bubble point temperature calculations. Thus, the computational procedure begins by setting $\Phi_i = 1$, and calculating $T_i^{sat} = B_i/(A_i - \ln P) - C_i$, and $T = \Sigma_i x_i T_i^{sat}$. The gas constant $R = 82.06 cm^3 \cdot atm/(mol \cdot K)$, is used in the calculation of Φ_i. For the calculation of activity coefficients, the UNIFAC method is employed. The script *bubbleT* implements each computational step for bubble point temperature calculations. This script calls the function *unifgam* to estimate activity coefficients.

```
% bubbleT.m : bubble temperature of a mixture
clear all;
% Data
n = 4; P = 1; % n: number of components, P: total pressure (atm)
Rg = 82.06; % gas constant (cm^3 atm/(mol K))
x = [0.162 0.068 0.656 0.114]; % liquid-phase mole fraction
% Antoine equation constants
A = [9.2033 12.2786 9.1690 9.2675]; B = [2697.55 3803.98 2731.00 2788.51];
C = [-48.78 -41.68 -47.11 -52.36];
% Virial coefficients
Bij = zeros(n,n); Bij(1,1) = -1360.1; Bij(1,2) -657.0; Bij(1,3) = -1274.2; Bij
(1,4) = -1218.8;
Bij(2,2) = -1174.7; Bij(2,3) = -621.8; Bij(2,4) = -589.7; Bij(3,3) = -1191.9;
Bij(3,4) = -1137.9;
```

```
Bij(4,4) = -1086.9;
% Data for calculation of activity coefficients using UNIFAC method
k = 6; % number of functional groups (k)
R = [0.9011 0.6744 0.4469 0.2195 0.5313 1.0000]; % volume vector for each func-
tional group
Q = [0.848 0.540 0.228 0.000 0.400 1.200]; % surface area vector for each func-
tional group
nu = [2 1 3 0; 4 1 1 0; 0 0 1 0; 0 0 1 0; 0 0 0 6; 0 1 0 0]; % number of functional groups
amn = [0 0 0 0 61.13 986.5; 0 0 0 0 61.13 986.5; 0 0 0 0 61.13 986.5;...
       0 0 0 0 61.13 986.5; -11.12 -11.12 -11.12 -11.12 0 636.1;...
       156.4 156.4 156.4 156.4 89.60 0]; % interaction parameter matrix
Terr = 1; Tcrit = 1e-6; % Initialization
% delta(i,j)
for i = 1:n-1, Bij(:,i) = Bij(i,:); end
delta = zeros(n,n);
for i = 1:n
  for j = 1:n
    delta(j,i) = 2*Bij(j,i) - Bij(i,i) - Bij(j,j);
    if j == i, delta(i,j) = 0; end
  end
end
for i = 1:n-1, delta(:,i) = delta(i,:); end
% Step 1)
PHIi = ones(1,n); % PHI_i = 1 for each component
Tsat = B./(A - log(P)) - C; %Ti^sat
T = sum(x.*Tsat);
% Step 2)
Psat = exp(A - B./(T + C)); % Pi^sat
gamma = unifgam(k,R,Q,nu,amn,n,x,T);
P1sat = P/(sum(x.*gamma.*Psat./PHIi/Psat(1))); % j=1
T = B(1)/(A(1) - log(P1sat)) - C(1);
% Step 3)
Told = T;
while Terr > Tcrit
  Psat = exp(A - B./(Told + C)); y = x.*gamma.*Psat./PHIi/P;
  for i = 1:n % PHI
    sumy = 0;
    for j = 1:n
      for ki = 1:n, sumy = sumy + y(j)*y(ki)*(2*delta(j,i) - delta(j,ki)); end
    end
    PHIi(i) = exp((Bij(i,i)*(P - Psat(i)) + sumy*P/2)/(Rg*Told));
  end
  gamma = unifgam(k,R,Q,nu,amn,n,x,Told);
  P1sat = P/(sum(x.*gamma.*Psat./PHIi/Psat(1))); % j=1
  T = B(1)/(A(1) - log(P1sat)) - C(1); Terr = abs(Told - T); Told = T;
end
T, y, Psat,PHIi, gamma
```

The script *bubbleT* generates the following outputs:
```
>> bubbleT
T =
 335.1492
y =
```

TABLE 4.22
Results of Bubble T Calculations

Component	x_i(Data)	$y_{i,}$(Exp)[40]	Results of Calculations			
			y_i	P_i^{sat}	Φ_i	γ_i
n-Hexane(1)	0.162	0.140	0.1357	0.8053	1.0059	1.0462
Ethanol(2)	0.068	0.274	0.2765	0.5048	1.0052	8.0957
Methylcyclopentane(3)	0.656	0.503	0.5040	0.7317	0.9897	1.0391
Benzene(4)	0.114	0.083	0.0838	0.5525	0.9834	1.3084
$T(K)$,	Experimental: 334.85 K, Calculated: 335.1492 K,					

```
   0.1357   0.2765   0.5040   0.0838
Psat =
   0.8053   0.5048   0.7317   0.5525
PHIi =
   1.0059   1.0052   0.9897   0.9834
gamma =
   1.0462   8.0957   1.0391   1.3084
```

The results can be summarized as shown in Table 4.22.

4.5.3.2 Flash Calculations Using Fugacity Coefficients

Flash calculations based on the ratio of fugacity coefficients are somewhat more complex than calculations based on Raoult's law and K,-value correlation. But the primary equations used in the previous flash calculation by the modified Raoult's law are unchanged except $K_j = y_j/x_j = \gamma_j P_j^{sat}/\Phi_j P$ ($j = 1, 2, ...,N$).

The procedure for flash calculations can be summarized as follows:

1) Input temperature, pressure, feed compositions (z_j), and constants. For dew point P, calculations, set $y_j = z_j$, and for bubble point P, calculations, set $x_j = z_j$, and check $P_{dew} < P < P_{bubble}$. If P, is out of this range, stop the calculation procedure. Initialize values of γ_j, Φ_j, α.
2) Calculate $K_j = \gamma_j P_j^{sat}/(\Phi_j P)$, and find new α, from $f(\alpha) = \sum_{j=1}^{N} (1 - K_j)z_j/(1 + \alpha(K_j - 1)) = 0$.
3) Calculate $x_j = z_j/(1 + \alpha(K_j - 1))$, $y_j = K_j x_j$, γ_j, and Φ_j. Repeat steps 2) and 3) until changes in α, x_j, and y_j, from one iteration to the next are less than some tolerances.

Example 4.24   Flash Calculations for a Four-Component System[41]

Perform flash calculations for a four-component system consisting of n-hexane(1)/ethanol(2)/methylcyclopentane(3)/benzene(4). The given pressure and temperature are $1 atm$, and $334.15K$, and the given mole fractions of the feed stream are $z_1 = 0.25$, $z_2 = 0.40$, $z_3 = 0.20$, and $z_4 = 0.15$. Vapor pressure can be estimated using the Antoine equation $\ln P_i^{sat} = A_i - B_i/(T + C_i)$, ($T: K$, $P: atm$). The parameters for the Antoine equation are

$$A_1 = 9.2033, A_2 = 12.2786, A_3 = 9.1690, A_4 = 9.2675$$

$$B_1 = 2697.55, B_2 = 3803.98, B_3 = 2731.00, B_4 = 2788.51$$

$$C_1 = -48.78, C_2 = -41.68, C_3 = -47.11, C_4 = -52.36$$

The virial coefficients $(cm^3/mol,)$ for the components are

$$B_{11} = -1360.1, B_{12} = -657.0, B_{13} = -1274.2, B_{14} = -1218.8$$

$$B_{22} = -1174.7, B_{23} = -621.8, B_{24} = -589.7, B_{33} = -1191.9$$

$$B_{34} = -1137.9, B_{44} = -1086.9$$

Solution

The UNIFAC method is used to calculate activity coefficients. The bulit-in function *fzero* is used to solve the nonlinear equation to find α. The script *flashbubP* implements each computational step for flash calculations (bubble point pressure calculations). This script calls the function *unifgam* to estimate activity coefficients.

```
% flashbubP.m : flash calculations for multicomponent system (Bubble P)
clear all;
% Data
n = 4; P = 1; T = 334.15; % n: number of components, P: atm, T: K
Rg = 82.06; % gas constant (cm^3 atm/(mol K))
z = [0.25 0.40 0.20 0.15]; % feed stream mole fraction
% Antoine equation constants
A = [9.2033 12.2786 9.1690 9.2675]; B = [2697.55 3803.98 2731.00 2788.51];
C = [-48.78 -41.68 -47.11 -52.36];
% Virial coefficients
Bij = zeros(n,n);
Bij(1,1) = -1360.1; Bij(1,2) -657.0; Bij(1,3) = -1274.2; Bij(1,4) = -1218.8;
Bij(2,2) = -1174.7; Bij(2,3) = -621.8; Bij(2,4) = -589.7;
Bij(3,3) = -1191.9; Bij(3,4) = -1137.9; Bij(4,4) = -1086.9;
% Data for calculation of activity coefficients using UNIFAC method
k = 6; % number of functional groups (k)
R = [0.9011 0.6744 0.4469 0.2195 0.5313 1.0000]; % volume vector for each func-
tional group
Q = [0.848 0.540 0.228 0.000 0.400 1.200]; % surface area vector for each func-
tional group
nu = [2 1 3 0; 4 1 1 0; 0 0 1 0; 0 0 1 0; 0 0 0 6; 0 1 0 0]; % number of functicnal groups
amn = [0 0 0 0 61.13 986.5; 0 0 0 0 61.13 986.5; 0 0 0 0 61.13 986.5;...
    0 0 0 0 61.13 986.5; -11.12 -11.12 -11.12 -11.12 0 636.1;...
    156.4 156.4 156.4 156.4 89.60 0]; % interaction parameter matrix
Aerr = 1; Acrit = 1e-6; % Initialization
% delta(i,j)
for i = 1:n-1, Bij(:,i) = Bij(i,:); end
delta = zeros(n,n);
for i = 1:n
  for j = 1:n
    delta(j,i) = 2*Bij(j,i) - Bij(i,i) - Bij(j,j);
    if j == i, delta(i,j) = 0; end
  end
end
for i = 1:n-1, delta(:,i) = delta(i,:); end
% Step 1)
```

```
x = z; % Bubble P
Psat = exp(A - B./(T + C)); % Pi^sat
gamma = unifgam(k,R,Q,nu,amn,n,x,T);
PHIi = ones(1,n); % initial guess
y = x.*gamma.*Psat./PHIi/P;
for i = 1:n % PHI
  sumy = 0;
  for j = 1:n
    for m = 1:n, sumy = sumy + y(j)*y(m)*(2*delta(j,i) - delta(j,m)); end
  end
  PHIi(i) = exp((Bij(i,i)*(P - Psat(i)) + sumy*P/2)/(Rg*T));
end
% Step 2)
K = gamma.*Psat./(PHIi*P); alpha0 = 0.5; alphaold = alpha0;
% Step 3)
while Aerr > Acrit
  falpha = @(alpha) sum((1-K).*z./(1 + alpha*(K - 1)));
  alpha = fzero(falpha, alphaold); x = z./(1 + alpha*(K - 1)); y = K.*x;
  gamma = unifgam(k,R,Q,nu,amn,n,x,T);
  for i = 1:n % PHI
    sumy = 0;
    for j = 1:n
      for m = 1:n, sumy = sumy + y(j)*y(m)*(2*delta(j,i) - delta(j,m)); end
    end
    PHIi(i) = exp((Bij(i,i)*(P - Psat(i)) + sumy*P/2)/(Rg*T));
  end
  K = gamma.*Psat./(PHIi*P); Aerr = abs(alpha - alphaold); alphaold = alpha;
end
alpha, x, y, K, gamma
```

Implementation of the script *flashbubP* produces the following results:

```
>> flashbubP
alpha =
  0.8363
x =
  0.1496   0.5770   0.1316   0.1418
y =
  0.2697   0.3654   0.2134   0.1516
K =
```

TABLE 4.23
Results of Flash Calculations

Component	z_i(Given)	Results of Calculations			
		x_i	y_i	K_i	γ_i
n-Hexane(1)	0.25	0.1496	0.2697	1.8028	2.3332
Ethanol(2)	0.40	0.5770	0.3654	0.6332	1.3150
Methylcyclopentane(3)	0.20	0.1316	0.2134	1.6212	2.2622
Benzene(4)	0.15	0.1418	0.1516	1.0690	1.9669

Note: T : 334. 15K, P : 1atm, calculated value of $\alpha = V/F$: 0.8363

```
 1.8028    0.6332    1.6212    1.0690
gamma =
 2.3332    1.3150    2.2622    1.9669
```

The results can be summarized as shown in Table 4.23.

4.6 VAPOR-LIQUID-LIQUID EQUILIBRIUM

As a mixture of water (80 *mol%*), and isobutanol (20 *mol%*), is heated at a constant pressure of 1 *atm*, two liquid phases coexist. One of these liquid phases is composed principally of water and the other one is composed principally of isobutanol. When the bubble point temperature is reached at 88.54 °C, the two liquid phases are in equilibrium with the vapor phase.[42] When the temperature is raised above the bubble point, only a single liquid phase and the vapor phase are present until the dew point temperature is reached. Figure 4.10 shows the phase separation curve for the water-isobutanol system.

Figure 4.11 shows the three-phase flash drum.[42] The temperature and pressure in the flash drum are kept constant at T, and P, respectively, and the number of components in the system is n_c. The overall mass balance is given by

$$F = L_1 + L_2 + V$$

The component mass balance can be written as

$$F z_i = L_1 x_{i,1} + L_2 x_{i,2} + V y_i$$

The phase equilibrium relations are

$$y_i = K_{i,1} x_{i,1}, \; y_i = K_{i,2} x_{i,2}, \; K_{i,j} = \frac{\gamma_{i,j} P_i}{P}$$

where Raoult's law is used, $\gamma_{i,j}$, is the activity coefficient of species i, in phase j, and P_i, is the vapor pressure of species i. Sums of mole fractions give

$$\sum_{i=1}^{n_c} y_i = 1, \; \sum_{i=1}^{n_c} x_{i,1} = 1, \; \sum_{i=1}^{n_c} x_{i,2} = 1 \Longrightarrow \sum_{i=1}^{n_c} y_i - \sum_{i=1}^{n_c} x_{i,1} = 0, \; \sum_{i=1}^{n_c} x_{i,1} - \sum_{i=1}^{n_c} x_{i,2} = 0$$

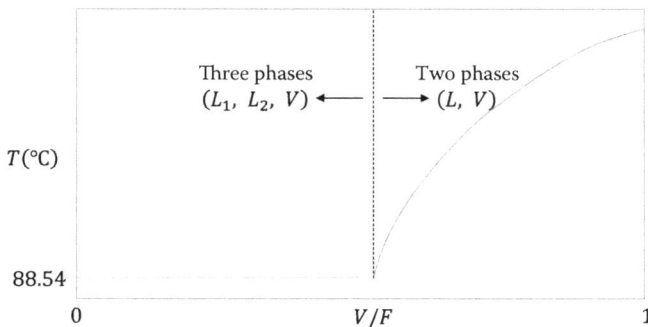

FIGURE 4.10 Phase separation curve for a water-isobutanol system ($P = 1 atm$). (Modified from Cutlip, M. B. and M. Shacham, *Problem Solving in Chemical and Biochemical Engineering with POLYMATH, Excel, and MATLAB*, 2nd ed., Prentice-Hall, Boston, MA, 2008, p. 528.)

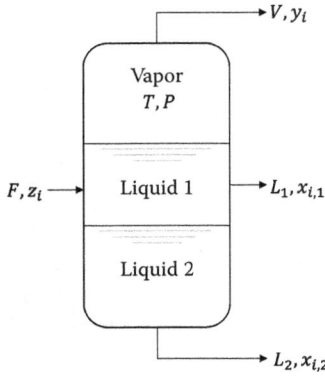

FIGURE 4.11 Three-phase flash drum. (Modified from Cutlip, M. B. and M. Shacham, *Problem Solving in Chemical and Biochemical Engineering with POLYMATH, Excel, and MATLAB*, 2nd ed., Prentice-Hall, Boston, MA, 2008, p. 528.)

From the overall mass balance, we can see that $L_1 = \xi(F - V)$, and $L_2 = (1 - \xi)(F - V)$, where $\alpha = V/F$, and $\xi = L_1/(L_1 + L_2)(0 \le \alpha, \xi \le 1)$. Substitution of these equations into the component mass balance yields

$$z_i = \left[\frac{\xi(1 - \alpha)}{K_{i,1}} + \frac{(1 - \xi)(1 - \alpha)}{K_{i,2}} + \alpha \right] y_i \text{ or } y_i = \frac{z_i}{\left[\frac{\xi(1 - \alpha)}{K_{i,1}} + \frac{(1 - \xi)(1 - \alpha)}{K_{i,2}} + \alpha \right]}$$

Then, from the summing relations, we have

$$\sum_{i=1}^{n_c} y_i - \sum_{i=1}^{n_c} x_{i,1} = \sum_{i=1}^{n_c} \frac{(K_{i,1} - 1)z_i}{\xi(1 - \alpha) + (1 - \xi)(1 - \alpha)\frac{K_{i,1}}{K_{i,2}} + K_{i,1}\alpha} = 0$$

and

$$\sum_{i=1}^{n_c} x_{i,1} - \sum_{i=1}^{n_c} x_{i,2} = \sum_{i=1}^{n_c} \frac{(K_{i,1}/K_{i,2} - 1)z_i}{\xi(1 - \alpha) + (1 - \xi)(1 - \alpha)\frac{K_{i,1}}{K_{i,2}} + K_{i,1}\alpha} = 0$$

At the bubble point, $V = 0$, $F = V + L_1 + L_2 = L_1 + L_2$, and $y_i = K_{i,1}x_{i,1} = K_{i,2}x_{i,2}$. Combining the component mass balance and summing relations yields

$$z_i = x_{i,1}\left[\xi + (1 - \xi)\frac{K_{i,1}}{K_{i,2}} \right], \sum_{i=1}^{n_c} x_{i,1} - \sum_{i=1}^{n_c} y_i = 0, \sum_{i=1}^{n_c} x_{i,1} - \sum_{i=1}^{n_c} x_{i,2} = 0$$

At the dew point, $L_1 = L_2 = 0$, $V = F$, $y_i = z_i$, and we have

$$y_i = K_i x_i, \sum_{i=1}^{n_c} x_i = 1$$

where $x_i = x_{i,1}$.

In the two-phase region, the isothermal flash calculation can be performed by introduction of $L = F - V$, and $\alpha = V/F$, into the component mass balance as

$$x_i = \frac{z_i}{(1 - \alpha) + \alpha K_i}$$

Example 4.25 Water-Isobutanol Equilibrium Calculations[43]

A mixture of isobutanol(1) (20 *mol%*), and water(2) (80 *mol%*), is heated to the bubble point at a constant pressure of 1*atm*. We assume that Raoult's law, $K_{i,j} = \gamma_{i,j} P_i / P$, can be applied. The vapor pressure of species i, P_i, for each component is given by the Antoine equation:

$$\log(P_1) = 7.62231 - \frac{1417.9}{191.15 + T} \text{ and } \log(P_2) = 8.10765 - \frac{1750.29}{235 + T} (T : {}^\circ C)$$

Activity coefficients of isobutanol(1) and water(2) are given by

$$\log \gamma_{1,j} = \frac{1.7 x_{2,j}^2}{\left(2.43 x_{1,j} + x_{2,j}\right)^2} \text{ and } \log \gamma_{2,j} = \frac{0.7 x_{1,j}^2}{\left(x_{1,j} + 0.412 x_{2,j}\right)^2} \text{ (j: phase)}$$

Determine the bubble point and dew point temperatures at 1*atm*. Plot the fraction evaporated (α), as a function of the boiling temperature between these two temperatures.

Solution

Bubble point temperature: From the relation $z_i = x_{i,1}[\xi + (1 - \xi) K_{i,1} / K_{i,2}]$, two nonlinear equations are obtained:

$$f(x_{i,1}) = x_{i,1} - \frac{z_i}{[\xi + (1 - \xi)\frac{K_{i,1}}{K_{i,2}}]} = 0 (i = 1, 2)$$

From $K_{i,1} x_{i,1} = K_{i,2} x_{i,2}$, another two equations are obtained:

$$f(x_{i,2}) = x_{i,2} - x_{i,1} \frac{K_{i,1}}{K_{i,2}} = 0 (i = 1, 2)$$

The sum of mole fractions gives

$$f(\xi) = \sum_{i=1}^{2} (x_{i,1} - x_{i,2}) = 0$$

Using $y_i = K_{i,1} x_{i,1}$, and $\sum_{i=1}^{n_c} x_{i,1} - \sum_{i=1}^{n_c} y_i = 0$, we have

$$f(T) = \sum_{i=1}^{2} x_{i,1}(1 - K_{i,1}) = 0$$

Additional equations required are

$$P_1 = 10^{7.62231 - \frac{1417.9}{191.15+T}}, P_2 = 10^{8.10765 - \frac{1750.29}{235+T}}$$

$$\gamma_{1,j} = 10^{\frac{1.7 x_{2,j}^2}{(2.43 x_{1,j} + x_{2,j})^2}}, \gamma_{2,j} = 10^{\frac{0.7 x_{1,j}^2}{(x_{1,j} + 0.412 x_{2,j})^2}}, K_{i,j} = \frac{\gamma_{i,j} P_i}{P}$$

Initial guesses for solutions are set to

$$x_{1,1}(0) = x_{2,2}(0) = 0, x_{2,1}(0) = x_{1,2}(0) = 1, T(0) = 100, \beta(0) = 0.8$$

The function *beq* defines the set of nonlinear equations.

```
function f = beq(t,z,P)
% t(1)=x11, t(2)=x21, t(3)=x12, t(4)=x22, t(5)=T, t(6)=beta
% Required relations
P1 = 10^(7.62231 - 1417.9/(191.15 + t(5))); P2 = 10^(8.10765 - 1750.29/(235 +
t(5)));
gam11 = 10^(1.7*t(2)^2/((2.43*t(1) + t(2))^2)); gam12 = 10^(1.7*t(4)^2/
((2.43*t(3) + t(4))^2));
gam21 = 10^(0.7*t(1)^2/((t(1) + 0.412*t(2))^2)); gam22 = 10^(0.7*t(3)^2/((t
(3) + 0.412*t(4))^2));
k11 = gam11*P1/P; k12 = gam12*P1/P; k21 = gam21*P2/P; k22 = gam22*P2/P;
% Nonlinear equations
f(1,1) = t(1) - z(1)/(t(6) + (1-t(6))*k11/k12); % i=1
f(2,1) = t(2) - z(2)/(t(6) + (1-t(6))*k21/k22); % i=2
f(3,1) = t(3) - t(1)*k11/k12; f(4,1) = t(4) - t(2)*k21/k22; f(5,1) = t(1) +t(2) -
t(3) - t(4);
f(6,1) = t(1)*(1-k11) + t(2)*(1-k21);
end
```

The following commands set the initial conditions and initial guesses for the solutions and calls the function *beq* to solve the nonlinear equations using the built-in function *fsolve*.

```
>> x110 = 0; x220 = 0; x210 = 1; x120 = 1; T0 = 100; beta0 = 0.8;
>> z = [0.2 0.8]; P = 760; t0 = [x110 x210 x120 x220 T0 beta0];
>> [x,fval] = fsolve(@beq, t0, [], z, P); x
x =
    0.0227   0.9773   0.6866   0.3134   88.5396   0.7329
```

We can see that

$$x_{11} = 0.0227, \ x_{21} = 0.9773, \ x_{12} = 0.6866, \ x_{22} = 0.3134, \ T = 88.5396 \,°\text{C}, \ \beta = 0.7329$$

The vector *fval* in the results, whose elements converge to zeros, depicts function values of nonlinear equations.

Dew point temperature: From $y_i = K_i x_i$, we have $f(x_i) = x_i - y_i/K_i = 0 (i = 1, 2)$. From $\sum_{i=1}^{2} x_i = 1$, we have $f(T) = x_1 + x_2 - 1 = 0$.

Accompanying relations are

$$P_1 = 10^{7.62231 - \frac{1417.9}{191.15+T}}, \ P_2 = 10^{8.10765 - \frac{1750.29}{235+T}}$$

$$\gamma_1 = 10^{\frac{1.7x_2^2}{(2.43x_1+x_2)^2}}, \ \gamma_2 = 10^{\frac{0.7x_1^2}{(x_1+0.412x_2)^2}}, \ K_i = \frac{\gamma_i P_i}{P} (i = 1, \ 2)$$

Initial guesses are set to $x_1(0) = 0.2, x_2(0) = 0.8, T(0) = 100$. The set of nonlinear equations is defined by the function *dpf*.

```
function f = dpf(t,y,P)
% t(1)=x1, t(2)=x2, t(3)=T
% Required relations
P1 = 10^(7.62231 - 1417.9/(191.15 + t(3))); P2 = 10^(8.10765 - 1750.29/(235 +
t(3)));
gam1 = 10^(1.7*t(2)^2/((2.43*t(1) + t(2))^2)); gam2 = 10^(0.7*t(1)^2/((t(1) +
0.412*t(2))^2));
```

```
k1 = gam1*P1/P; k2 = gam2*P2/P;
% Nonlinear equations
f = [t(1) - y(1)/k1; t(2) - y(2)/k2; t(1) + t(2) - 1];
end
```

The following commands set initial conditions and initial guesses for solutions and call the function *dpf* to solve the nonlinear equations using the built-in function *fsolve*.

```
>> x10 = 0.2; x20 = 0.8; T0 = 100; y = [0.2 0.8]; P = 760; t0 = [x10 x20 T0];
>> x = fsolve(@dpf, t0, [], y, P)
x =
  0.0079   0.9921   93.9671
```

We can see that $x_1 = 0.0079$, $x_2 = 0.9921$, and $T = 93.9671\,°C$.

PROBLEMS

4.1 For ethane, the critical temperature and pressure are $305.3K$, and $48.08\,atm$, respectively, and the acentric factor is 0.1. Use the virial equation of state to plot the compressibility factor Z, versus the pressure in the range of $24.04 \leq P \leq 480.8(atm)$, (that is, $0.5 \leq P_r \leq 10$), for $T = 457.95, 610.6$, and $3053K$, (i.e., $T_r = 1.5, 2, 10$).

4.2 Estimate the compressibility factor Z, for isopropanol vapor at $200\,°C$, and $10bar$, by the Lee-Kesler equation.[44] The critical temperature and pressure of isopropanol are $508.3K$, and $47.62bar$, respectively, and the acentric factor is 0.668.

4.3 Calculate the compressibility factor Z, for carbon dioxide in the range of $0.5 \leq P_r \leq 1$, for $T_r = 1, 1.4, 1.8$, and 2 by the van der Waals equation of state. Plot Z, versus P_r, for each value of $T_r = 1, 1.4, 1.8$, and 2. For carbon dioxide, the critical temperature and pressure are $304.2\,K$, and $72.9\,atm$, respectively.

4.4 Use the Peng-Robinson equation of state to calculate the specific volume (cm^3/g), of CO_2, at $310\,K$, and at $P = 8$, and $75\,bar$. Compare the results with the experimental values of 70.58 and 3.90, respectively. For CO_2, the molecular weight is 44 and $w = 0.228$, $T_c = 304.2K$, $P_c = 73.82bar$.[45]

4.5 Estimate the enthalpy departure $H^R = H - H^{ig}$, and entropy departure $S^R = S - S^{ig}$, for 1-butene vapor at $70\,bar$, and $473.15\,K$. Use the virial state of equation. For 1-butene, $T_c = 420K$, $P_c = 40.43bar$, and $w = 0.191$.[46]

4.6 Table P4.6 shows experimental values for the enthalpy departure of isobutane at $175\,°C$.[47] Calculate the theoretical values of enthalpy departure by the Peng-Robinson equation and compare the results with the experimental values shown in Table P4.6. For isobutane, the critical temperature and pressure are $408.1K$, and $36.48bar$, respectively, and the acentric factor is 0.181.

TABLE P4.6

Experimental Values for the Enthalpy Departure of Isobutane

P (atm),	10	20	35	70
$H - H_{ig}$ (J/g),	−15.4	−32.8	−64.72	−177.5

Source: Elliott, J. R. and Lira, C. T., *Introductory Chemical Engineering Thermodynamics*, 2nd ed., Prentice-Hall, 2012, p.329.

TABLE P4.7(1)
Properties for Each Component

Component	x (mole fraction)	w	$T_c(K)$	$P_c(bar)$	ΔH_f	ΔG_f
Methane	0.9852	0.012	190.6	45.99	−74520	−50460
Ethane	0.01449	0.100	305.3	48.72	−83820	−31855
Propane	0.000312	0.152	369.8	42.48	−104680	−24290

Source: Smith, J. M. et al., *Introduction to Chemical Engineering Thermodynamics*, 7th ed., McGraw-Hill, 2005, p.680–686.

4.7 Calculate the enthalpy for a methane(1)/ethane(2)/propane(3) system at $158K$, and $6.8947\ bar$, by cubic equations of state (Redlich-Kwong, Soave-Redlich-Kwong, and Peng-Robinson). Vapor-phase mole fractions and properties for each component are shown in Table P4.7(1), where ΔH_f, and ΔG_f, are in J/mol.[48] Table P4.7(2) shows constants for the heat capacity equation (C_p:$Btu/(lbmol\cdot°F)$, T:$°F$).[49]

$$C_p = a_1 + a_2 T + a_3 T^2 + a_4 T^3 + a_5 T^4$$

4.8 Table P4.8 shows experimental measurements on compressibility factors at various pressures for ammonia at $100\ °C$.[50] Determine fugacity coefficients from the data given in Table P4.8 at each pressure using a numerical integration scheme. In this case, use the built-in function *cumtrapz*. Calculate the fugacity coefficients at each pressure from cubic equations of state, and compare the results with those obtained from the numerical integration by plotting both results versus pressure.

4.9 The temperature of the azeotrope for an ethanol(1)/benzene(2) system at $P = 760mmHg$, is 68.2 °C, and the composition at the azeotrope is 32.4% ethanol and 67.6% benzene (% by weight). The vapor pressures of ethanol and benzene may be obtained by the Antoine equation $logP_i = A_i − B_i/(T + C_i)$, where T, is the azeotrope temperature (°C), and P_i, is the vapor pressure ($mmHg$). For ethanol, $A_1 = 8.04494$, $B_1 = 1554.3$, and $C_1 = 222.65$; and for benzene, $A_2 = 6.90565$, $B_2 = 1211.033$, and $C_2 = 220.79$. The molecular weights of ethanol and benzene are 46.07 and 78, respectively. Determine the Wilson equation coefficients for this system.

4.10 Estimate the activity coefficients γ_1, and γ_2, for a 2-propanol(1)/water(2) system by the UNIFAC method at $T = 353.52\ K$. The liquid-phase compositions are $x_1 = 0.6854$, and $x_2 = 0.3146$. Functional groups for each component can be represented as

TABLE P4.7(2)
Constants for the Heat Capacity Equation

Component	a_1	$a_2 \times 10^2$	$a_3 \times 10^5$	$a_4 \times 10^8$	$a_5 \times 10^{11}$
Methane	8.245223	0.380633	0.8864745	−0.746115	0.182296
Ethane	11.51606	1.40309	0.854034	−1.106078	0.31622
Propane	15.58683	2.504953	1.404258	−3.52626	1.864467

Source: Henley, E. J. and Seader, J. D., *Equilibrium-stage Separation Operations in Chemical Engineering*, John Wiley & Sons, 1981, pp. 720–724.

TABLE P4.8
Experimental Z Data for Ammonia

P (atm)	Z	P (atm)	Z	P (atm)	Z
1.374	0.0028	30.47	0.8471	300	0.3212
3.537	0.9828	33.21	0.831	400	0.4145
5.832	0.9728	36.47	0.8111	500	0.506
8.632	0.9599	40.41	0.7864	600	0.5955
11.352	0.9468	45.19	0.7538	700	0.6828
14.567	0.9315	51.09	0.7102	800	0.7684
19.109	0.9085	58.28	0.6481	900	0.8507
22.84	0.889	100	0.1158	1000	0.9333
26.12	0.8714	200	0.2221	1100	1.014

Source: Cutlip, M. B. and Shacham, M., Problem Solving in Chemical and Biochemical Engineering with POLYMATH, Excel, and MATLAB, 2nd ed., Prentice-Hall, Boston, MA, 2008, p.265.

$$(CH_3)_2 - CH - OH$$

2-propanol(1): $(CH_3)_2 - CH - OH$, water(2): H_2O

Values of R_k, Q_k, $v_j^{(i)}$ ($i = 1,2$), for each functional group can be found elsewhere.[51]

4.11 Determine the dew point temperature for a mixture of 32 mol% n-hexane, 31% n-heptane, 25% n-octane, and 12% n-nonane at 1.5 bar, total pressure.[52] The vapor pressure of the pure species j, is given by the Antoine equation

$$\log P_j^{sat} = A - \frac{B}{C + T - 273.15} (T : K, P_j^{sat} : bar)$$

where the Antoine constants for each component are shown in Table P4.11.

4.12 A mixture containing 40 mol% n-pentane (1) and 60 mol% n-heptane (2) is introduced into a flash drum at a pressure P, and temperature T. It is assumed that Raoult's law can be applied, and the vapor pressure $P^{sat}(kPa)$, of each component can be obtained from the Antoine equation

$$\ln P^{sat} = A - \frac{B}{T(°C) + C}$$

TABLE P4.11
Antoine Constants

Component	A	B	C
n-Hexane	4.00139	1170.875	224.317
n-Heptane	4.02023	1263.909	216.432
n-Octane	4.05075	1356.360	209.635
n-Nonane	4.07356	1438.03	202.694

Generate plots of the operating pressure P, x_1, and y_1, as a function of the fraction of the feed vaporized at $T = 350K$. The Antoine constants for the components are

$$A_1 = 13.8183, B_1 = 2477.07, C_1 = 233.21$$

$$A_2 = 13.8587, B_2 = 2991.32, C_2 = 216.64$$

4.13 A ternary mixture containing 30 mol% n-pentane (1), 50 mol% n-hexane (2) and 20 mol% n-heptane (3) at 90 °C, and 2 atm, splits into liquid and vapor phases in a flash drum. The vapor pressure of component i, P_i^{sat} ($mmHg$), is given by the Antoine equation

$$\log P_i^{sat} = A - \frac{B}{T(°C) + C}$$

and the equilibrium constant K_i, of component i, is given by

$$K_i = \frac{P_i^{sat}}{P}$$

Determine the fraction of the feed mixture that is vaporized (α), and the compositions of the liquid and vapor phases (x_i, and y_i; $i = 1,2,3$). The Antoine coefficients are

$$n\text{-pentane}(1): A_1 = 6.85221, B_1 = 1064.630, C_1 = 232.0$$

$$n\text{-hexane}(2): A_2 = 6.87776, B_2 = 1171.530, C_2 = 224.366$$

$$n\text{-heptane}(3): A_3 = 6.90240, B_3 = 1268.115, C_3 = 216.9$$

4.14 A mixture of isobutanol(1) (20 $mol\%$), and water(2) (80 $mol\%$), is heated to the bubble point at $P = 1\,atm$. The bubble temperature for this system at $P = 1\,atm$, is known to be $T_b = 88.5396\,°C$. Assume that this system follows Raoult's law, $k_i = \gamma_i P_i / P$ ($i = 1, 2$). The vapor pressure of species i, P_i, for each component is given by the Antoine equation

$$\log(P_1) = 7.62231 - \frac{1417.9}{191.15 + T}, \log(P_2) = 8.10765 - \frac{1750.29}{235 + T}(T:°C)$$

Activity coefficients of isobutanol(1) and water(2) are given by

$$\log(\gamma_{1,j}) = \frac{1.7x_{2,j}^2}{(2.43x_{1,j} + x_{2,j})^2}, \log\gamma_{2,j} = \frac{0.7x_{1,j}^2}{(x_{1,j} + 0.412x_{2,j})^2}(j\text{:phase})$$

TABLE P4.15

Constants for the Antoine Equation

Component	Feed Composition	Constants		
		A	B	C
Methane	0.08	6.64380	395.74	266.681
Ethane	0.21	6.82915	663.72	256.681
Propane	0.38	6.80338	804.00	247.040
n-Butane	0.20	6.80776	935.77	238.789
n-Pentane	0.13	6.85296	1064.84	232.012

Determine the fraction evaporated ($\alpha = V/F$), and the mole fraction of each component at the boiling temperature.

4.15 Perform flash calculations for a five-component system consisting of methane/ethane/propane/n-butane/n-pentane. The feed stream is fed into the flash drum at $40\,°C$. The vapor pressure can be estimated using the Antoine equation $\log P_i^{sat} = A_i - B_i/(T + C_i)$, ($P$:$mmHg$, T: $°C$). Feed compositions and parameters for the Antoine equation are given in Table P4.15. Determine the dew point and bubble point temperatures of the feed stream. Calculate the mole fractions of effluent vapor and liquid streams and the fraction evaporated ($\alpha = V/F$), for $P = 18$, 22, and 26 atm.

4.16 A feed consisting of acetone(1)/acetonitrile(2)/nitromethane(3) enters a flash drum operating at $80\,°C$, and 110 kPa. The feed composition is $z_1 = 0.45$, $z_2 = 0.35$, $z_3 = 0.20$. Assuming that Raoult's law is appropriate to this system, calculate V, L, x_i, and y_i. The vapor pressures (kPa), of the pure species at $80\,°C$, are $P_1^{sat} = 195.75$, $P_2^{sat} = 97.84$, $P_3^{sat} = 50.32$.[53]

REFERENCES

1. Lim, K. H., *Thermodynamics for Engineers and Scientists*, Jayoo Academi, Busan, Korea, p. 338, 2011.
2. Smith, J. M., H. C. Van Ness, and M. M. Abbott, *Introduction to Chemical Engineering Thermodynamics*, 7th ed., McGraw-Hill, New York, NY, p. 102, 2005.
3. Lim, K. H., *Thermodynamics for Engineers and Scientists*, Jayoo Academi, Busan, Korea, p. 340, 2011.
4. Adidharma, H. and V. Temyanko, *Mathcad for Chemical Engineers*, Trafford publishing, Victoria, BC, Canada, p. 35, 2007.
5. Lee, B. I. and M. G. Kesler, A generalized thermodynamic correlation based on three-parameter corresponding states, *AIChE Journal*, 21(3), p. 510, 1975.
6. Raman, R., *Chemical Process Computations*, Elsevier Applied Science Publishers, Barking, Essex, UK, p. 23, 1985.
7. Smith, J. M., H. C. Van Ness, and M. M. Abbott, *Introduction to Chemical Engineering Thermodynamics*, 8th ed., McGraw-Hill, New York, NY, p. 100, 2018.
8. Smith, J. M., H. C. Van Ness, and M. M. Abbott, *Introduction to Chemical Engineering Thermodynamics*, 8th ed., McGraw-Hill, New York, NY, pp. 100–101, 2018.
9. Smith, J. M., H. C. Van Ness, and M. M. Abbott, *Introduction to Chemical Engineering Thermodynamics*, 8th ed., McGraw-Hill, New York, NY, p. 101, 2018.
10. Adidharma, H. and V. Temyanko, *Mathcad for Chemical Engineers*, Trafford publishing, Victoria, BC, Canada, p. 85, 2007.
11. Mariano M. Martin (Editor), *Introduction to software for chemical engineers*, CRC Press, Taylor & Francis Group, Boca Raton, FL, pp. 302–303, 2015.
12. Smith, J. M., H. C. Van Ness, and M. M. Abbott, *Introduction to Chemical Engineering Thermodynamics*, 8th ed., McGraw-Hill, New York, NY, p. 138, 2018.
13. Smith, J. M., H. C. Van Ness, and M. M. Abbott, *Introduction to Chemical Engineering Thermodynamics*, 7th ed., McGraw-Hill, New York, NY, pp. 219–220, 2005.
14. Lim, K. H., *Thermodynamics for Engineers and Scientists*, Jayoo Academi, Busan, Korea, pp. 389–393, 2011.
15. Elliott, J. R. and C. T. Lira, *Introductory Chemical Engineering Thermodynamics*, 2nd ed., Prentice-Hall, Boston, MA, pp. 314–315, 2012.
16. Elliott, J. R. and C. T. Lira, *Introductory Chemical Engineering Thermodynamics*, 2nd ed., Prentice-Hall, Boston, MA, p. 314, 2012.
17. Henley, E. J. and J. D. Seader, *Equilibrium-Stage Separation Operations in Chemical Engineering*, John Wiley & Sons, Inc., Hoboken, NJ, pp. 720–724, 1981.
18. Smith, J. M., H. C. Van Ness, and M. M. Abbott, *Introduction to Chemical Engineering Thermodynamics*, 7th ed., McGraw-Hill, New York, NY, pp. 680–686, 2005.
19. Elliott, J. R. and C. T. Lira, *Introductory Chemical Engineering Thermodynamics*, 2nd ed., Prentice-Hall, Boston, MA, p. 352, 2012.
20. Elliott, J. R. and C. T. Lira, *Introductory Chemical Engineering Thermodynamics*, 2nd ed., Prentice-Hall, Boston, MA, p. 591, 2012.
21. Smith, J. M., H. C. Van Ness, and M. M. Abbott, *Introduction to Chemical Engineering Thermodynamics*, 8th ed., McGraw-Hill, New York, NY, p. 497, 2018.

22. Elliott, J. R. and C. T. Lira, *Introductory Chemical Engineering Thermodynamics*, 2nd ed., Prentice-Hall, Boston, MA, pp. 587–588, 2012.

23. Cutlip, M. B. and M. Shacham, Problem Solving in Chemical and Biochemical Engineering with POLYMATH, Excel, and MATLAB, 2nd ed., Prentice-Hall, Boston, MA, p. 274, 2008.

24. Washburn, Edward W. (ed.), *International Critical Tables*, 1st ed., Vol. III, McGraw-Hill, New York, p. 287, 1928.

25. 26. Sandler, S. I., *Chemical, Biochemical and Engineering Thermodynamics*, 4th ed., John Wiley & Sons, Inc., Hoboken, NJ, p. 521, 2006.

26. Sandler, S. I., *Chemical, Biochemical and Engineering Thermodynamics*, 4th ed., John Wiley & Sons, Inc., Hoboken, NJ, p. 433, 2006.

27. Smith, J. M., H. C. Van Ness, and M. M. Abbott, *Introduction to Chemical Engineering Thermodynamics*, 8th ed., McGraw-Hill, New York, NY, p. 731, 2018.

28. Poling, B. E., J. M. Prausnitz, and J. P. O'Connell, *The Properties of Gases and Liquids*, 5th ed., McGraw-Hill, New York, NY, pp. 8.78–8.81, 2001.

29. Lim, K. H., *Thermodynamics for Engineers and Scientists*, Jayoo Academi, Busan, Korea, pp. 832–833, 2011.

30. Smith, J. M., H. C. Van Ness, and M. M. Abbott, *Introduction to Chemical Engineering Thermodynamics*, 8th ed., McGraw-Hill, New York, NY, pp. 732–734, 2018.

31. Poling, B. E., J. M. Prausnitz, and J. P. O'Connell, *The Properties of Gases and Liquids*, 5th ed., McGraw-Hill, New York, NY, pp. 8.82–8.93, 2001.

32. Smith, J. M., H. C. Van Ness, and M. M. Abbott, *Introduction to Chemical Engineering Thermodynamics*, 8th ed., McGraw-Hill, New York, NY, pp. 734–736, 2018.

33. Mariano M. Martin (Editor), *Introduction to software for chemical engineers*, CRC Press, Taylor & Francis Group, Boca Raton, FL, p. 305, 2015.

34. Elliott, J. R. and C. T. Lira, *Introductory Chemical Engineering Thermodynamics*, 2nd ed., Prentice-Hall, Boston, MA, p. 376, 2012.

35. Smith, J. M., H. C. Van Ness, and M. M. Abbott, *Introduction to Chemical Engineering Thermodynamics*, 8th ed., McGraw-Hill, New York, NY, pp. 455–456, 2018.

36. Smith, J. M., H. C. Van Ness, and M. M. Abbott, *Introduction to Chemical Engineering Thermodynamics*, 8th ed., McGraw-Hill, New York, NY, pp. 457–459, 2018.

37. Cutlip, M. B. and M. Shacham, Problem Solving in Chemical and Biochemical Engineering with POLYMATH, Excel, and MATLAB, 2nd ed., Prentice-Hall, Boston, MA, p. 268, 2008.

38. Elliott, J. R. and C. T. Lira, *Introductory Chemical Engineering Thermodynamics*, 2nd ed., Prentice-Hall, Boston, MA, pp. 835–837, 2012.

39. Smith, J. M., H. C. Van Ness, and M. M. Abbott, *Introduction to Chemical Engineering Thermodynamics*, 7th ed., McGraw-Hill, New York, NY, p. 550, 2005.

40. Smith, J. M., H. C. Van Ness, and M. M. Abbott, *Introduction to Chemical Engineering Thermodynamics*, 7th ed., McGraw-Hill, New York, NY, p. 551, 2005.

41. Smith, J. M., H. C. Van Ness, and M. M. Abbott, *Introduction to Chemical Engineering Thermodynamics*, 7th ed., McGraw-Hill, New York, NY, p. 554, 2005.

42. Cutlip, M. B. and M. Shacham, Problem Solving in Chemical and Biochemical Engineering with POLYMATH, Excel, and MATLAB, 2nd ed., Prentice-Hall, Boston, MA, p. 528, 2008.

43. Cutlip, M. B. and M. Shacham, *Problem Solving in Chemical and Biochemical Engineering with POLYMATH, Excel, and MATLAB*, 2nd ed., Prentice-Hall, Boston, MA, pp. 527–534, 2008.

44. Smith, J. M., H. C. Van Ness, and M. M. Abbott, *Introduction to Chemical Engineering Thermodynamics*, 7th ed., McGraw-Hill, New York, NY, pp. 89–90, 2005.

45. Elliott, J. R. and C. T. Lira, *Introductory Chemical Engineering Thermodynamics*, 2nd ed., Prentice-Hall, Boston, MA, p. 268, 2012.

46. Smith, J. M., H. C. Van Ness, and M. M. Abbott, *Introduction to Chemical Engineering Thermodynamics*, 7th ed., McGraw-Hill, New York, NY, p. 235, 2005.

47. Elliott, J. R. and C. T. Lira, *Introductory Chemical Engineering Thermodynamics*, 2nd ed., Prentice-Hall, Boston, MA, p. 329, 2012.

48. Smith, J. M., H. C. Van Ness, and M. M. Abbott, *Introduction to Chemical Engineering Thermodynamics*, 7th ed., McGraw-Hill, New York, NY, pp. 680–686, 2005.

49. Henley, Ernest J. and J. D. Seader, *Equilibrium-stage Separation Operations in Chemical Engineering*, John Wiley & Sons, pp. 720–724, 1981.

50. Cutlip, M. B. and M. Shacham, Problem Solving in Chemical and Biochemical Engineering with POLYMATH, Excel, and MATLAB, 2nd ed., Prentice-Hall, Boston, MA, p. 265, 2008.
51. Bruce E. Poling, John M. Prausnitz and John P. O'Connell, *The Properties of Gases and Liquids*, 5th ed., McGraw-Hill, New York, NY, pp. 8.78–8.81, 2001.
52. Nayef Ghasem, *Computer Methods in Chemical Engineering*, CRC Press, Taylor & Francis Group, Boca Raton, FL, pp. 3–4, 2012.
53. Smith, J. M., H. C. Van Ness, and M. M. Abbott, *Introduction to Chemical Engineering Thermodynamics*, 7th ed., McGraw-Hill, New York, NY, pp. 368–369, 2005.

5 Fluid Mechanics

5.1 LAMINAR FLOW

5.1.1 REYNOLDS NUMBER

The Reynolds number is defined by

$$N_{Re} = \frac{Dv\rho}{\mu}$$

where
  D is the inside pipe diameter
  v is the velocity
  ρ is the density of the fluid
  μ is the viscosity of the fluid
  If the volumetric flow rate is given, N_{Re} can be represented as

$$N_{Re} = \frac{\rho v D}{\mu} = 506\frac{Q\rho}{ud} = 6.31\frac{W}{ud}$$

where
  $D\,(ft)$ and $d\,(in.)$ are the inside pipe diameters
  $\mu\,(cp)$ is the viscosity of the fluid
  $Q\,(gpm)$ and $W\,(lb/hr)$ are the volumetric flow rate and mass flow rate, respectively

5.1.2 FLOW IN A HORIZONTAL PIPE

Consider an incompressible Newtonian fluid flowing at steady state inside a horizontal circular pipe at constant temperature. Figure 5.1 shows a schematic of this laminar flow in the x direction, where the inside pipe radius is R.
  The shear stress at the radius r, τ_{rx}, and the velocity, v_x, are given by[1]

$$\tau_{rx} = \left(\frac{\Delta P}{2L}\right)r, \; v_x = \frac{R^2\Delta P}{4uL}\left[1 - \left(\frac{r}{R}\right)^2\right]$$

where
  $\Delta P = P_0 - P_L$
  R is the inside pipe radius
  L is the length of the pipe
  The average velocity $v_{x,avg}$ is obtained from

$$v_{x,avg} = \frac{1}{\pi R^2}\int_0^R 2\pi r v_x dr = \frac{R^2\Delta P}{8\mu L}$$

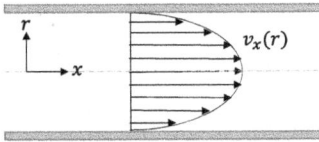

FIGURE 5.1   Laminar flow of a Newtonian fluid in a horizontal pipe.

The function *hzpipe* calculates the average velocity of the laminar flow inside a horizontal circular pipe and plots velocity distribution inside the pipe. The basic syntax is

```
v = hzpipe(L,R,mu,delP)
```

where L (*m*) and R (*m*) are the length and the radius of the pipe, respectively; mu (*kg/(m·sec)*) is the viscosity of the fluid; delP (Pa) is the pressure drop; and v (*m/sec*) is the average velocity.

```
function v = hzpipe(L,R,mu,delP)
% Average velocity and velocity distribution of the laminar flow inside a hor-
izontal circular pipe
% inputs
%  L: length of pipe (m), R: radius of pipe (m)
%  mu: viscosity (kg/m/sec), delP: pressure drop (Pa)
% output
%  v: average velocity (m/s)
r = linspace(-R,R,100);
vr = delP*R^2.*(1 - (r/R).^2)/(4*mu*L); v = delP*R.^2/(8*mu*L);
plot(vr,r), grid, axis([0 1.5 -R,R]), xlabel('v_x(r)'), ylabel('r'), axis([0 1
-R R])
end
```

Example 5.1 Newtonian Fluid Flow inside a Horizontal Pipe

Water flows inside a horizontal circular pipe at 25°C. Determine the average velocity and plot the velocity distribution inside the pipe. The length and the inside radius of the pipe are 15 and 0.009 *m*, respectively, the viscosity of water is $8.937 \times 10^{-4} kg/(m·sec)$, and the pressure drop is $\Delta P = 520 Pa$.

Solution

The following commands produce the desired outputs and illustrate the velocity distribution as shown in Figure 5.2:

```
>> L = 15; R = 0.009; mu = 8.937e-4; delP = 520; Vavg = hzpipe(L,R,mu,delP)
Vavg =
   0.3927
```

5.1.3 LAMINAR FLOW IN A HORIZONTAL ANNULUS

An incompressible fluid is flowing within an annulus between two concentric horizontal pipes as shown in Figure 5.3.[2] For a Newtonian fluid, the velocity distribution v_x and average velocity v_{avg} can be expressed as[3]

$$v_x = \frac{\Delta P}{4\mu L}\left[R_2^2 - r^2 + \frac{R_1^2}{\ln(R_2/R_1)}\left(\ln\frac{r}{R_2}\right)\right], \quad v_{avg} = \frac{\Delta P}{8\mu L}\left[R_1^2 + R_2^2 - \frac{R_2^2 - R_1^2}{\ln(R_2/R_1)}\right]$$

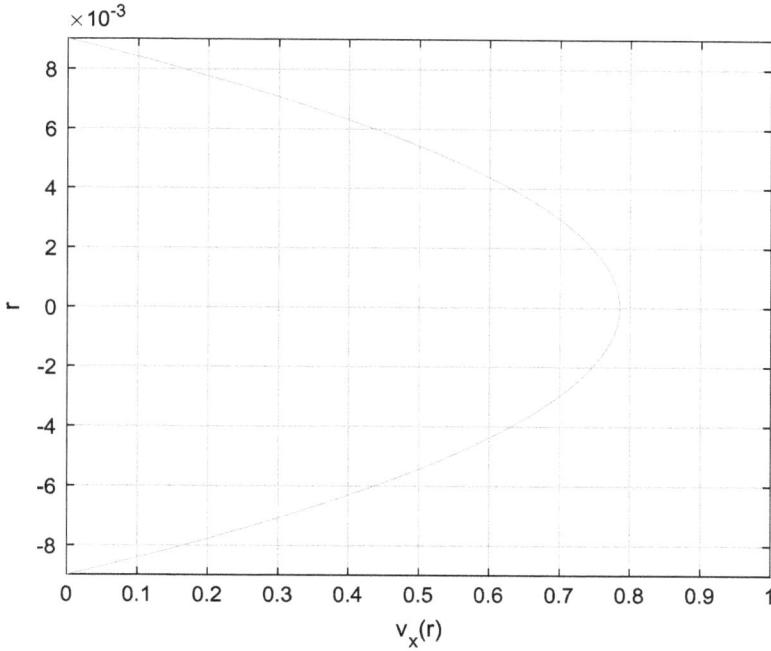

FIGURE 5.2 Velocity distribution inside a circular pipe.

where

$\Delta P = P_0 - P_L$

R_1 and R_2 are the inside radius of inner and outer pipe, respectively

For a Newtonian fluid, the shear stress τ_{rx} is given by

$$\tau_{rx} = -\mu \frac{dv_x}{dr}$$

For some non-Newtonian fluids, the shear stress τ_{rx} can be represented as

$$\tau_{rx} = -K \left| \frac{dv_x}{dr} \right|^{n-1} \frac{dv_x}{dr}$$

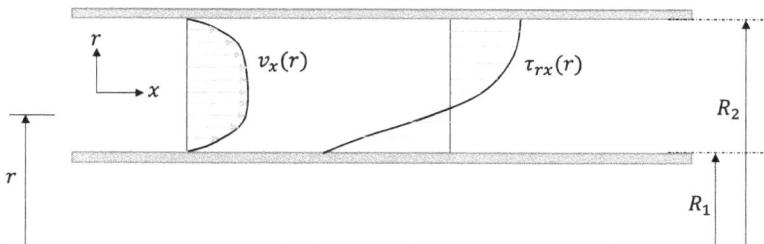

FIGURE 5.3 Laminar flow in a horizontal annulus. (Modified from Cutlip, M.B. and Shacham, M., *Problem Solving in Chemical and Biochemical Engineering with POLYMATH, Excel, and MATLAB*, 2nd ed., Prentice-Hall, Boston, MA, 2008, p. 294.)

For a Newtonian laminar flow in a horizontal annulus, the velocity distribution and average velocity are calculated by the function *hzannpipe*. The syntax is

```
avgv = hzannpipe(L,R1,R2,mu,delP)
```

This function generates plots showing distributions of the velocity and the shear stress.

```
function avgv = hzannpipe(L,R1,R2,mu,delP)
% Calculates the average velocity and plots velocity and shear stress
% distribution for Newtonian laminar flow in a horizontal annulus.
% input
%   L: length of pipe (m)
%   R1,R2: inside radius of inner and outer pipe (m)
%   mu: viscosity (kg/m/s), delP: pressure drop (Pa)
% output
%   avgv: average velocity (m/s)
n = 100; r = linspace(R1,R2,n); h = (R2-R1)/n;
v = delP*(R2^2 -r.^2 + (R2^2 -R1^2)/(log(R2/R1)).*log(r/R2))./(4*mu*L);
avgv = delP*(R1.^2 + R2.^2 - (R2.^2 -R1.^2)./(log(R2./R1)))./(8*mu*L);
dv = diff(v); dv = [dv dv(end)]; taurx = -mu.*dv./h;
subplot(1,2,1), plot(v,r), grid, axis([0 0.2 R1 R2]), xlabel('v_x(r)'),
ylabel('r')
subplot(1,2,2), plot(taurx,r), grid, axis tight, xlabel('\tau_{rx}(r)'),
ylabel('r')
end
```

Example 5.2 Laminar Flow of Water in a Horizontal Annulus

Water flows inside a horizontal annulus at 25°C. Determine the average velocity and plot the velocity and shear stress distributions inside the annulus. The length and the inside radii of the inner and outer pipes are 12, 0.025, and 0.036 m, respectively. The viscosity of water is $8.937 \times 10^{-4} kg/(m \cdot sec)$, and the pressure drop is $\Delta P = 110 Pa$.

Solution

The following commands calls the function *hzannpipe* to produce the plots of velocity and shear stress distributions shown in Figure 5.4.

```
>> mu = 8.937e-4; delP = 110; L = 12; R1 = 0.025; R2 = 0.036; avgv = hzannpipe
(L,R1,R2,mu,delP)
avgv =
  0.1037
```

5.1.4 Vertical Laminar Flow of a Falling Film

Consider a liquid flowing down a vertical surface. The liquid is a fully established film, where the velocity profile does not vary in the direction of flow, as shown in Figure 5.5.

From the z-momentum balance, the differential equation for the momentum flux is given by[4]

$$\frac{d\tau_{xz}}{dz} = \rho g, \ \tau_{xz}(x = 0) = 0$$

For a Newtonian fluid, the momentum flux (shear stress) is related to the velocity gradient by

(a)

(b)

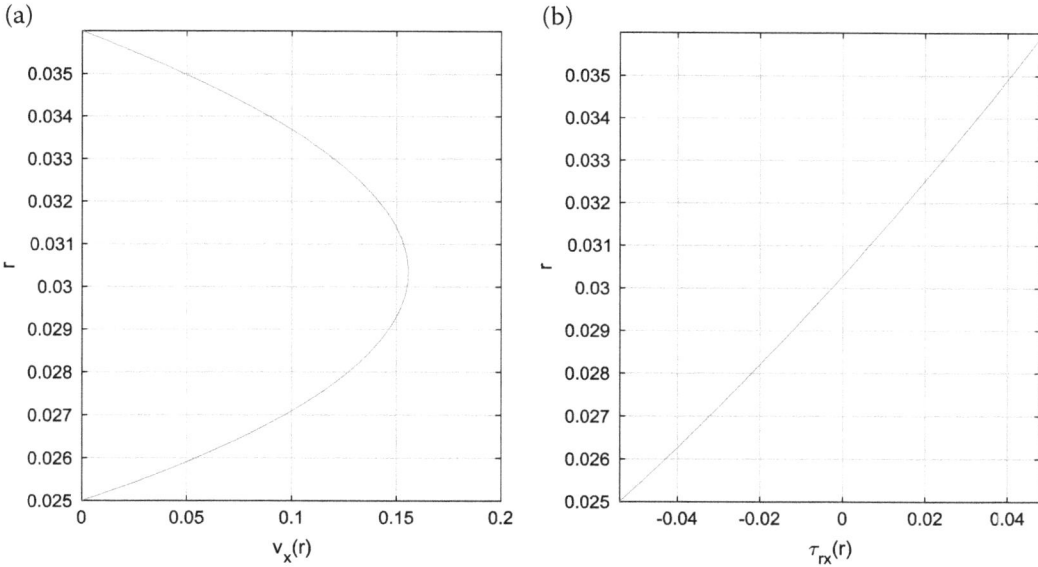

FIGURE 5.4 (a) Velocity ($v_x(r)$) and (b) shear stress ($\tau_{rx}(r)$) distributions for a laminar flow in a horizontal annulus.

$$\tau_{xz} = -\mu \frac{dv_z}{dx}(x = \delta \text{ to } v_z = 0)$$

Integration of the differential equation followed by substitution of boundary conditions yields the velocity distribution v_z and the average velocity $v_{z,avg}$ as

$$v_z = \frac{\rho g \delta^2}{2\mu}\left[1 - \left(\frac{x}{\delta}\right)^2 \right], \ v_{z,avg} = \frac{\rho g \delta^2}{3\mu}$$

The function *vtwall* calculates the velocity distribution and average velocity for a Newtonian fluid. The syntax is

```
[avgv] = vtwall(rho,mu,delta)
```

where rho and mu are the density and viscosity of the fluid, respectively, and delta is the film thickness.

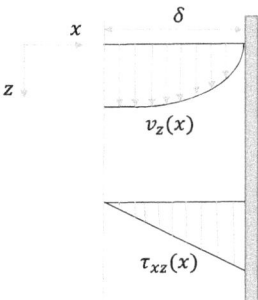

FIGURE 5.5   Vertical laminar flow of a liquid film.

```
function [avgv] = vtwall(rho,mu,delta)
% Calculates velocity distribution and average velocity for vertical laminar
flow of a falling film
% input
%   rho: density (kg/m^3), mu: viscosity (kg/m/s)
%   delta: film thickness (m)
% output
%   avgv: average velocity (m/s)
g = 9.8; h = delta/100; x = [0:h:delta]; Rg = rho*g*delta^2/2/mu; v = Rg.*(1 - (x/
delta).^2); avgv = Rg*2/3;
plot(x,v), grid, ylabel('v_z(x)'), xlabel('x'), axis([0 delta 0 max(v)])
end
```

Example 5.3 Velocity Distribution of a Falling Film

Plot the velocity distribution $v_z(x)$ and calculate the average velocity for a falling film of a Newtonian fluid with $\rho = 780 kg/m^3$ and $\mu = 0.172 kg/(m{\cdot}sec)$. The film thickness is $\delta = 0.0021 m$.

Solution

The function *vtwall* produces the desired plot, as shown in Figure 5.6:

```
>> rho = 780; mu = 0.172; delta = 0.0021; [avgv] = vtwall(rho,mu,delta)
avgv =   0.0653
```

5.1.5 FALLING PARTICLES

The terminal velocity v_t of a falling spherical particle is given by

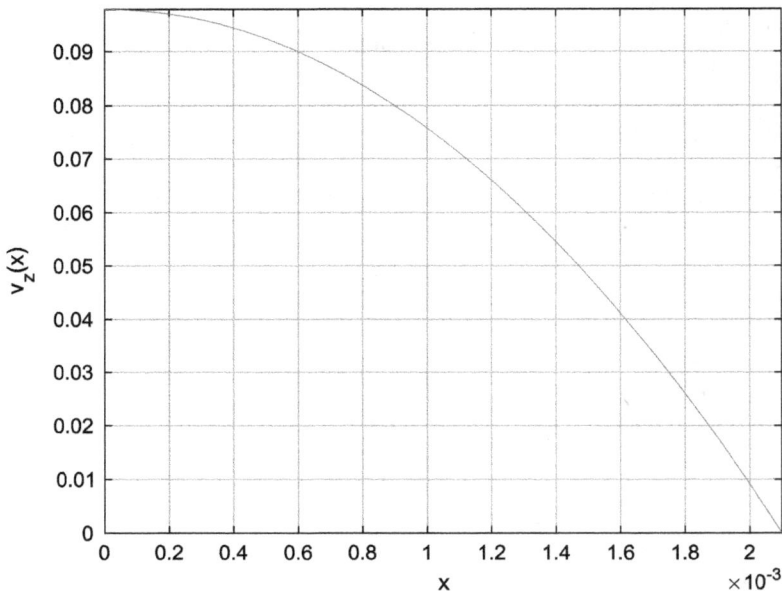

FIGURE 5.6 Velocity distribution for a falling film of a Newtonian fluid.

$$v_t = \sqrt{\frac{4g\left(\rho_p - \rho\right)D_p}{3C_D\rho}}$$

where

v_t is the terminal velocity (m/sec)
g is the acceleration of gravity (= 9. $8m/sec^2$)
ρ_p is the density of the particle (kg/m^3)
ρ is the fluid density (kg/m^3)
D_p is the diameter of the spherical particle (m)
C_D is a dimensionless drag coefficient
Rearrangement of the terminal velocity equation gives

$$3C_D\rho v_t^2 - 4g\left(\rho_p - \rho\right)D_p = 0$$

At the terminal velocity, the drag coefficient can be expressed as a function of the Reynolds number N_{Re}[5]:

$$C_D = \begin{cases} \dfrac{24}{N_{Re}} & : N_{Re} < 0.1 \\ \dfrac{24}{N_{Re}}(1 + 0.14N_{Re}^{0.7}) & : 0.1 \leq N_{Re} \leq 1000 \\ 0.44 & : 1,000 < N_{Re} \leq 350,000 \\ 0.19 - \dfrac{8 \times 10^4}{N_{Re}} & : 350,000 < N_{Re} \end{cases}$$

where $N_{Re} = D_p v_t \rho / \mu$ and μ is the fluid viscosity ($kg/(m \cdot sec)$ or $Pa \cdot sec$).
The nonlinear equation to be solved is defined by the function *vtfun*. The syntax is

```
fvt = vtfun(x, rp, ro, mu, dp)
```

where rp and ro are the densities of the particle and the fluid, respectively, mu is the viscosity of the fluid, and dp is the diameter of the particle.

```
function fvt = vtfun(x, rp, ro, mu, dp)
% x: unknown terminal velocity
g = 9.8; Nre = dp*ro*x/mu; % Reynolds number
if Nre < 0.1, Cd = 24./Nre;
elseif Nre <1000, Cd = (1 + 0.14*Nre.^0.7)*24./Nre;
elseif Nre < 3.5e5, Cd = 0.44;
else, Cd = 0.19 - 8e4./Nre; end
fvt = 3*Cd*ro*x.^2 - 4*g*(rp - ro)*dp; % nonlinear equation
end
```

Example 5.4 Terminal Velocity of a Falling Particle

Estimate the terminal velocity of a spherical particle falling in water at 25°C. For the particle, $\rho_p = 1780kg/m^3$ and $D_p = 0.2mm$. The viscosity and density of water are $\mu = 8.391 \times 10^{-4}kg/(m \cdot sec)$ and $994.6kg/m^3$, respectively.

Solution

The nonlinear equation defined by the function *vtfun* can be solved by using the built-in function *fzero*. vt0 is an initial guess for the terminal velocity. The following commands produce the terminal velocity (*m/sec*):

```
>> rp = 1780; ro = 994.6; dp = 2e-4; mu = 8.931e-4; vt0 = 1e-3; vt = fzero(@
vtfun,vt0,[], rp, ro, mu, dp)
vt = 0.0145
```

5.2 FRICTION FACTOR

The Fanning friction factor f is related to the Reynolds number N_{Re} through a set of correlations, and depends on whether the flow regime is laminar, transitional, or turbulent. A simple relation between f and N_{Re} can be expressed as

$$f = aN_{Re}^b$$

where a and b are constants. The friction factor is affected by the roughness of the surface at a high Reynolds number ($N_{Re} \geq 2000$).

For laminar flow ($N_{Re} \leq 2000$), the Fanning friction factor f and Darcy friction factor f_D are given by

$$f = \frac{16}{N_{Re}}, f_D = \frac{64}{N_{Re}} = 4f$$

The Darcy friction factor is four times the Fanning friction factor (i.e., $f_D = 4f$).

For a fully developed turbulent flow regime ($N_{Re} > 3000$) in rough pipes, the correlations can be expressed in terms of the surface roughness parameter ϵ/D. The Shacham equation[6] (SC) is an explicit form given by

$$f = \frac{1}{16}\left[\log\left\{\frac{\epsilon/D}{3.7} - \frac{5.02}{N_{Re}}\log\left(\frac{\epsilon/D}{3.7} + \frac{14.5}{N_{Re}}\right)\right\}\right]^{-2}$$

The equation given by Colebrook[7] (CB) has received wide acceptance:

$$\frac{1}{\sqrt{f}} = -1.7372\ln\left[\frac{\epsilon}{3.7D} + \frac{1.255}{N_{Re}\sqrt{f}}\right]$$

Another widely used implicit equation is the Colebrook-White equation[8] (CBW), given by

$$\frac{1}{\sqrt{f}} = -4.0\log\left(\frac{\epsilon}{D} + \frac{4.67}{N_{Re}\sqrt{f}}\right) + 2.28$$

The Haaland equation[9] (HA) is an explicit equation given by

$$f = \left[-3.6\log\left\{\frac{6.9}{N_{Re}} + \left(\frac{\epsilon/D}{3.7}\right)^{\frac{10}{9}}\right\}\right]^{-2} \quad (4 \times 10^4 < N_{Re} < 10^8, 0 < \epsilon/D < 0.05)$$

The Chen equation[10] (CH) is explicit and easier to use than the Colebrook equation. This can be expressed as

TABLE 5.1
Values of Absolute Pipe Roughness

Pipe Material	$\epsilon(ft)$
Riveted steel	0.003~0.03
Concrete	0.001~0.01
Wood stave	0.0006~0.003
Cast iron	0.00085
Galvanized iron	0.0005
Asphalted cast iron	0.0005
Commercial steel or wrought iron	0.00015
Drawn tubing	0.000005

Source: de Nevers, N., *Fluid Mechanics for Chemical Engineers*, 3rd ed., McGraw-Hill, New York, NY, 2005, p. 187.

$$\frac{1}{\sqrt{f}} = -4\log\left[\frac{\epsilon}{3.7D} - \frac{5.02}{N_{Re}}\log A\right], A = \frac{\epsilon}{3.7D} + \left(\frac{6.7}{N_{Re}}\right)^{0.9}$$

For commercial steel pipes, the surface roughness is about $\epsilon = 0.00015 ft$. Table 5.1 shows the values of ϵ for various pipe materials.[11]

As a single equation spanning all fluid-flow regimes and surface roughness parameter ϵ/D, the Churchill equation[12] is used:

$$f = 2\left[\left(\frac{8}{N_{Re}}\right)^{12} + \frac{1}{(A + B)^{1.5}}\right]^{1/12}$$

$$A = (2.457\ln C)^{16}, B = \left(\frac{37530}{N_{Re}}\right)^{16}, C = \left(\frac{7}{N_{Re}}\right)^{0.9} + 0.27\frac{\epsilon}{D}$$

For turbulent flow through smooth pipes where $\epsilon/D = 0$, the Nikuradse equation[13] (NK)

$$\frac{1}{\sqrt{f}} = 4.0\log\left(N_{Re}\sqrt{f}\right) - 0.4$$

or the Blasius equation (BS)

$$f = 0.0791 N_{Re}^{-1/4}$$

can be used.

The function *frfactor* calculates the friction factor for the given values of ϵ/D and N_{Re}.

```
function frfactor(eD, Nre)
% Friction factor correlations
% inputs:
%  eD: surface roughness parameter
%  Nre: Reynolds number
f0 = 5e-3; % initial guess
fSC = 1./(log10(eD/3.7 - (5.02./Nre).*log10(eD/3.7+14.5/Nre))).^2 /16; %
Shacham eqn.
```

```
funCB  =  @(f)   (1/sqrt(f)  +  1.7372*log(eD/3.7  +  1.255/Nre/sqrt(f)));  %
Colebrook eqn.
fCB = fzero(funCB, f0);
funCBW = @(f) (1/sqrt(f) + 4*log10(eD + 4.67/Nre/sqrt(f)) - 2.28); % Colebrook-
White eqn.
fCBW = fzero(funCBW, f0);
fHA = 1./(-3.6*log10(6.9/Nre + (eD/3.7)^(10/9)))^2; % Haaland eqn.
Av = eD/3.7 + (6.7/Nre)^0.9; % Chen eqn.
fCH = 1./(-4*log10(eD/3.7 - 5.02*log10(Av)/Nre)).^2;
funNK = @(f) (1./sqrt(f) - 4*log10(Nre*sqrt(f)) + 0.4); % Nikuradse eqn.
fNK = fzero(funNK, f0);
fBS = 0.0791*Nre^(-1/4); % Blasius eqn.
fprintf('Shacham eqn.: f = %g\n', fSC);
fprintf('Colebrook eqn.: f = %g\n', fCB);
fprintf('Colebrook-White eqn.: f = %g\n', fCBW);
fprintf('Haaland eqn.: f = %g\n', fHA);
fprintf('Chen eqn.: f = %g\n', fCH);
fprintf('Nikuradse eqn.: f = %g\n', fNK);
fprintf('Blasius eqn.: f = %g\n', fBS);
end
```

Example 5.5 Friction Factor

Determine the friction factors using various correlation equations (i.e., SC, CB, CBW, HA, CH, NK, and BS) at $N_{Re} = 2 \times 10^4$ and $N_{Re} = 3.2 \times 10^7$ for smooth pipes, where $\varepsilon/D = 0$.

Solution

$$N_{Re} = 2 \times 10^4$$

```
>> eD = 0; Nre = 2e4; frfactor(eD, Nre)
Shacham eqn.: f = 0.00648922
Colebrook eqn.: f = 0.00647063
Colebrook-White eqn.: f = 0.00647323
Haaland eqn.: f = 0.00643718
Chen eqn.: f = 0.00648215
Nikuradse eqn.: f = 0.00647573
Blasius eqn.: f = 0.00665149
```

$$N_{Re} = 3.2 \times 10^7$$

```
>> eD = 0; Nre = 3.2e7; frfactor(eD, Nre)
Shacham eqn.: f = 0.0017349
Colebrook eqn.: f = 0.00172196
Colebrook-White eqn.: f = 0.00172236
Haaland eqn.: f = 0.0017363
Chen eqn.: f = 0.00172145
Nikuradse eqn.: f = 0.00172273
Blasius eqn.: f = 0.00105169
```

Example 5.6 Friction Factor Correlation

Determine an empirical correlation to represent the relationship between the Reynolds number and the Fanning friction factor for a smooth pipe using the data given in Table 5.2.[14]

TABLE 5.2
Relation between Reynolds Number (N_{Re}) and Friction Factor (f)

N_{Re}	f
8500	0.008
20000	0.0065
30000	0.006
60000	0.005
700000	0.003
1000000	0.0028
10000000	0.002

Source: Kapuno, R.R.A., *Programming for Chemical Engineers*, Infinity Science Press, Hingham, MA, 2008, p. 105.

Solution

Assume a simple relation $f = aN_{Re}^b$. Taking the logarithm on both sides, we have a linear equation:

$$\log(f) = \log(a) + b\log(N_{Re})$$

Constants a and b can be determined from simple linear regression. The script *ffcor* uses the built-in function *polyfit* to perform the linear regression calculation.

```
% ffcor.m: friction factor correlation
Nre = [8500 20000 30000 60000 700000 1000000 10000000]; % Reynolds number
f = [0.008 0.0065 0.006 0.005 0.003 0.0028 0.002]; % friction factor
x = log10(Nre); y = log10(f); p = polyfit(x,y,1); % linear regression by polyfit
a = 10^p(2), b = p(1) % identified parameters
Nrex = linspace(min(Nre),max(Nre),100); fvx = a*Nrex.^b; % compute f by the
correlation
plot(Nre,f,'o',Nrex,fvx), legend('Data','Correlation')
xlabel('N_{Re}(Reynolds number)'), ylabel('f(Friction factor)')

>> ffcor
a =
  0.0468
b =
 -0.2001
```

The correlation is given by $f = aN_{Re}^b = 0.0468N_{Re}^{-0.2}$. The graph of this correlation and the data points is shown in Figure 5.7.

5.3 FLOW OF FLUIDS IN PIPES

Application of the Bernoulli equation between two points in a pipeline where a fluid flows yields

$$\frac{1}{\rho}\int_1^2 dP + \frac{1}{g_c}\int_1^2 vdv + \int_1^2 dz = 0$$

Integration of this equation gives

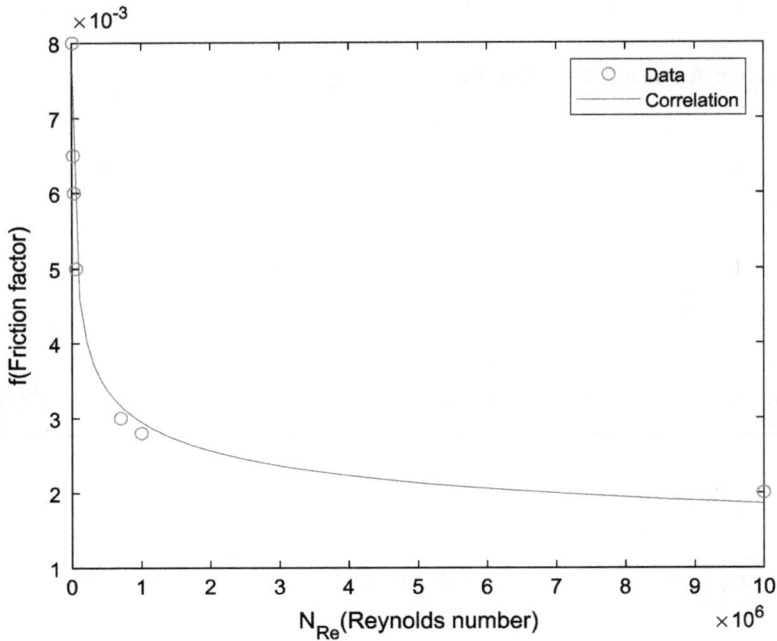

FIGURE 5.7 Reynolds number versus friction factor.

$$\frac{1}{\rho}(P_2 - P_1) + \frac{v_2^2 - v_1^2}{2g_c} + (z_2 - z_1) = 0$$

or

$$\frac{g_c \Delta P}{\rho} + \frac{v_2^2 - v_1^2}{2} + g_c \Delta z = 0$$

where
  P is the pressure of the fluid (lb_f/ft^2)
  ρ is the density of the fluid (lb_m/ft^3)
  v is the velocity of the fluid (ft/sec)
  g_c is the constant($32. 174(lb_m/lb_f)(ft/sec^2)$)
  z is the elevation of the fluid (ft)
  subscript 1 and 2 denote the conditions at the initial and final points, respectively
  Incorporating the head loss due to friction, h_L, with constant pipe diameter (i.e., $v_1 = v_2$), this relation becomes

$$\frac{\Delta P}{\rho} + \Delta z = h_L$$

5.3.1 FRICTION LOSS

The friction factor can be used to estimate the pressure drop, ΔP_f, due to friction loss from

$$\Delta P_f = 2f\rho \frac{Lv^2}{Dg_c}$$

where

f is the dimensionless friction factor

L is the pipe length

The friction loss, F_f, for an isothermal liquid flow in uniform circular pipes can be estimated by the relation

$$F_f = \frac{\Delta P_f}{\rho} = 2f\frac{Lv^2}{Dg_c} = f_D\frac{Lv^2}{2Dg_c}$$

The energy balance for an incompressible flow in a pipe with diameter $D\,(ft)$ and length $L\,(ft)$ can be written as

$$-\frac{1}{2}v^2 + g\Delta z + \frac{g_c\Delta P}{\rho} + g_c F_f = -\frac{1}{2}v^2 + g\Delta z + \frac{g_c\Delta P}{\rho} + 2f\frac{Lv^2}{D} = 0$$

From this equation, the fluid velocity is given by

$$v = \sqrt{\frac{g\Delta z + g_c\Delta P/\rho}{1/2 - 2fL/D}}$$

If the fluid performs mechanical work, the energy balance gives

$$\frac{\Delta v^2}{2} + g\Delta z + \frac{\Delta P}{\rho} + F_f = -\frac{\dot{W}_s}{\dot{m}}$$

Example 5.7 Determination of Pipe Diameter

Water flows inside a smooth pipeline at 21°C with a flow rate of 0.0085 m^3/sec. The length of the pipeline is 250 m, and the elevation of the upstream point is high enough to overcome the friction loss $F_f = 27.5J/kg$. The Fanning friction factor is given by $f = 0.0468N_{R_z}^{-0.2}$, and the density and viscosity of water at 21°C are $\rho = 997.91kg/m^3$ and $u = 0.000982kg/(m\cdot sec)$, respectively. Determine the diameter of the pipeline.

Solution

The flow rate equation $Q = Av = (\pi D^2/4)v$ can be solved for v to give $v = 4Q/(\pi D^2)$. Substitution of this velocity into the Fanning equation gives

$$f = 0.0468N_{Re}^{-0.2} = 0.0468\left(\frac{Dv\rho}{\mu}\right)^{-0.2} = 0.0468\left(\frac{4\rho Q}{\pi\mu D}\right)^{-0.2}$$

Substitution of this relation into the equation for friction loss yields

$$F_f = \frac{\Delta P_f}{\rho} = 2f\frac{Lv^2}{Dg_c} = (2)(0.0468N_{Re}^{-0.2})\frac{Lv^2}{Dg_c} = 0.0936\left(\frac{4\rho Q}{\pi\mu D}\right)^{-0.2}\left(\frac{L}{Dg_c}\right)\left(\frac{4Q}{\pi D^2}\right)^2$$

Thus, the pipe diameter $D\,(m)$ can be obtained from the solution of the nonlinear equation

$$g(D) = 0.0936\left(\frac{4\rho Q}{\pi\mu D}\right)^{-0.2}\left(\frac{L}{Dg_c}\right)\left(\frac{4Q}{\pi D^2}\right)^2 - F_f = 0$$

The script *pipedia* uses the built-in function *fzero* to solve the equation as follows:

```
% pipedia.m
rho = 997.92; Q = 0.0085; mu = 0.000982; L = 250; gc = 1; Ff = 27.5; % data
g = @(D) 0.0936*(4*rho*Q/(pi*mu*D))^(-0.2)*(L/D/gc)*(4*Q/pi/D^2)^2 - Ff; %
define eqn
D0 = 1; D = fzero(g,D0) % calls fzero to solve the eqn.

>> pipedia
D =
  0.0995
```

Example 5.8 Flow in a Pipe[15]

Figure 5.8 shows a pipeline that delivers water at a constant temperature from point 1 to point 2. At point 1, the pressure and elevation are $p_1 = 150psig$ and $z_1 = 0ft$, respectively, and at point 2, the pressure p_2 is atmospheric and the elevation is $z_2 = 300ft$. The density $\rho(lbm/ft^3)$ and viscosity $\mu(lbm/(ft{\cdot}sec))$ of the water can be estimated from the following equations:

$$\rho = 62.122 + 0.0122T - 1.54 \times 10^{-4}T^2 + 2.65 \times 10^{-7}T^3 - 2.24 \times 10^{-10}T^4$$

$$\ln\mu = -11.0318 + \frac{1057.51}{T + 214.624}$$

where T is in °F.

(1) Determine the volumetric flow rate (gal/min) at $T = 60°F$ for a pipeline with an effective length of $L = 1000ft$ and made of nominal 8 $in.$ schedule 40 commercial steel pipe ($\epsilon = 0.00015ft$).

(2) Estimate the flow velocities v in ft/sec and flow rates q in gal/min for pipelines at $T = 60°F$ with effective lengths of $L = 500, 1000, 1500, \cdots, 10,000$ ft and made of nominal 4, 5, 6, and 8 $in.$ schedule 40 commercial steel pipe. Prepare plots of velocity v versus D and L, and flow rate q versus D and L.

(3) Repeat part (1) at $T = 40, 60,$ and $100°F$ and display the results in a table showing temperature, density, viscosity, and flow rates.

Solution

(1) The general mechanical energy balance on an incompressible liquid applied to this problem yields

$$-\frac{1}{2}v^2 + g\Delta z + \frac{g_c\Delta P}{\rho} + 2f\frac{Lv^2}{D} = 0$$

The dimensionless friction factor f is a function of the Reynolds number N_{Re}, and N_{Re} is a function of the flow velocity v. This equation can be solved for v to yield the nonlinear equation

$$f(v) = v - \sqrt{\frac{g\Delta z + g_c\Delta P/\rho}{1/2 - 2fL/D}} = 0$$

This equation can be solved using the built-in function *fzero*. Input data to be supplied to the function *fzero* consist of temperature T(°F), pipe length L(ft), inside diameter D($in.$), pipe roughness rf(ϵ, ft), elevation difference dz(ft), and pressure difference dP(psi). The inside

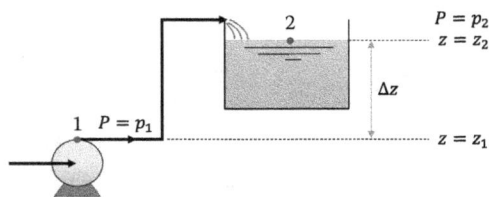

FIGURE 5.8 Water flow in a pipeline. (From Cutlip, M.B. and Shacham, M., *Problem Solving in Chemical and Biochemical Engineering with POLYMATH, Excel, and MATLAB*, 2nd ed., Prentice-Hall, Boston, MA, 2008, p. 110.)

diameter of the 8 *in.* schedule 40 steel pipe is 7.981 *in.*, rf= 0.00015*ft*, dz = $z_2 - z_1$ = 300*ft*, and dP= $p_2 - p_1$ = −150*psi*. The function *vwfun* defines the nonlinear equation to be solved.

```
function fv = vwfun(v,T,L,D,rf,dz,dP)
g = 32.174; gc = 32.174; D = D/12; dP = 144*dP; % psi->lbf/ft^2
eD = rf./D; % roughness factor
% density(rho; lbm/ft^3) and viscosity(mu; lbm/ft/s) at T
rho = 62.122+0.0122*T-(1.54e-4)*T.^2+(2.65e-7)*T.^3-(2.24e-10)*T.^4;
mu = exp(-11.0318 + 1057.51./(T+214.624));
Nre = D.*v.*rho./mu; % Reynolds number
if Nre < 2100, f = 16./Nre;
else, den = 16*(log10(eD/3.7 - 5.02*log10(eD/3.7+14.5./Nre)./Nre)).^2; f = 1./
den; end
fv = v - sqrt((g*dz + gc*dP./rho)./(0.5 - 2*f.*L./D));
end
```

The following commands define the data and call the function *fzero* to produce the desired outputs:

```
>> clear all; T=60; L=1000; D=7.981; rf=0.00015; dz=300; dP=-150;
>> v0 = 10; v = fzero(@vwfun, v0, [], T,L,D,rf,dz,dP), q = (7.481*60)*(pi*v.*(D/
12).^2)/4 % gpm
v =
  11.6133
q =
  1.8110e+03
```

We can see that v = 11.6133*ft/sec* and q = 1, 811*gpm*.

(2) The inside diameters of 4, 5, 6, and 8 *in.* schedule 40 steel pipe are 4.026, 5.047, 6.065, and 7.981 *in.*, respectively. Thus, the row vector D representing the inside diameters can be set as D= [4.026 5.047 6.065 7.981]. The script *vwfig* calculates flow velocities and flow rates for each D and L and generates plots showing v and q versus D and L.

```
% vwfig.m: calculates flow velocities and flow rates
clear all; T = 60; rf = 0.00015; dz = 300; dP = -150;
D = [4.026 5.047 6.065 7.981]; nD = length(D);
L = 500:500:10000; nL = length(L); vr = []; v0 = 10; qr = [];
for i = 1:nD
  for j = 1:nL
     v(j) = fzero(@vwfun, v0, [], T,L(j),D(i),rf,dz,dP);
     q(j) = (7.481*60)*(pi*v(j).*(D(i)/12).^2)/4; % gpm
  end
  vr = [vr v']; qr = [qr q'];
end
figure(1), plot(L,vr(:,1),'o',L,vr(:,2),'*',L,vr(:,3),'+',L,vr(:,4),'d')
legend('D=4in','D=5in','D=6in','D=8in'), axis([500 10000 0 20]), xlabel('L
(ft)'), ylabel('Velocity, v(ft/s)')
figure(2), plot(L,qr(:,1),'o',L,qr(:,2),'*',L,qr(:,3),'+',L,qr(:,4),'d')
legend('D=4in','D=5in','D=6in','D=8in'),   xlabel('L(ft)'),   ylabel('Flow
rate, q(gpm)')
```

Figure 5.9 shows the plots of flow velocity and flow rate versus pipe length and diameter produced by the script *vwfig*.

(3) Set T=[40 60 100] and use the built-in function *fzero* to solve the nonlinear equation. The script *qrmT* performs the calculation procedures.

(a)

(b)

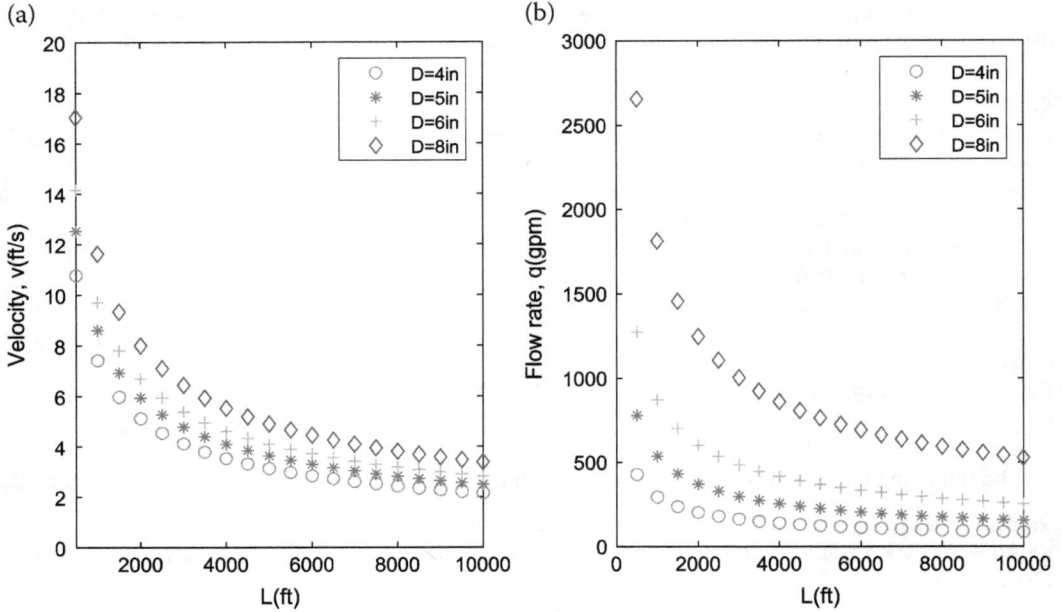

FIGURE 5.9   (a) Flow velocity and (b) flow rate versus pipe length and diameter.

```
% qrmT.m: calculates properties at various T
clear all; T=[40 60 100]; nT = length(T); L=1000; D=7.981; rf=0.00015; dz=300;
dP=-150; v0 = 10;
rhor = []; mur = []; qr = [];
for i = 1:nT
  v = fzero(@vwfun, v0, [], T(i),L,D,rf,dz,dP); % calls fzero to solve the eqn
  q = (7.481*60)*(pi*v.*(D/12).^2)/4;
    rho = 62.122+0.0122*T(i)-(1.54e-4)*T(i).^2+(2.65e-7)*T(i).^3-(2.24e-
10)*T(i).^4;
  mu = exp(-11.0318 + 1057.51./(T(i)+214.624));
  rhor = [rhor rho]; mur = [mur mu]; qr = [qr q];
end
qr, rhor, mur
```

The script *qrmT* produces the following results, which are summarized in Table 5.3.

```
>> qrmT
qr =
 1.0e+03 *
  1.7843   1.8110   1.8752
rhor =
 62.3800  62.3539  62.0446
mur =
  0.0010   0.0008   0.0005
```

Example 5.9 Liquid Flow in a Pipeline

Water is drained from a reservoir through a pipeline from point 1, where the pressure is $P_1 = 205kPa$, to point 2, where the pressure is $P_2 = 125kPa$ (see Figure 5.10). Calculate the mass flow rate in kg/sec for water that is required to generate $1.2MW$ from the turbine installed between

TABLE 5.3

Calculation Results from Property Equations

Temperature(°F)	Density(lbm/ft^3)	Viscosity($lbm/(ft\cdot sec)$)	Flow Rate(gpm)
40	62.3800	0.0010	1,784.3
60	62.3539	0.0008	1,811.0
100	62.0446	0.0005	1,875.2

point 1 and point 2. The distance from the reservoir (point 1) to the turbine is 120m, and the distance from the turbine to point 2 is 5m. Assume that the temperature and fluid velocity are constant and the friction loss is negligible.

Solution

From assumptions of constant velocity and negligible friction loss, the momentum balance becomes

$$\frac{\Delta v^2}{2} + g\Delta z + \frac{\Delta P}{\rho} + F_f = g\Delta z + \frac{\Delta P}{\rho} = -\frac{\dot{W_s}}{\dot{m}}$$

This equation can be solved for the mass flow rate to give

$$\dot{m} = -\frac{\dot{W_s}}{g\Delta z + \frac{\Delta P}{\rho}}$$

The following commands produce the desired result:

```
>> P1=205; P2=125; dP=P2-P1; dz=-125; rho=1000; g=9.81; Ws=1.2e6; mdot = -Ws/
(dP/rho + g*dz)
mdot =
   978.5294
```

Example 5.10 Steam Flow in a Pipeline

Superheated steam flows through a pipe with inner diameter (d_i) of 0.75 cm and length (z) of 12 m. The inlet temperature and pressure (P_1) of the steam are 145°C and 510 kPa, respectively, and the viscosity of the steam is $13.8 \times 10^{-6} N\cdot sec/m^2$. The outlet pressure ($P_2$) can be determined by solving the following equation:

$$P_1^2 = P_2^2 + \frac{G^2 RT}{M}\left\{\frac{fz}{d_i} + 2\ln\left(\frac{P_1}{P_2}\right)\right\}$$

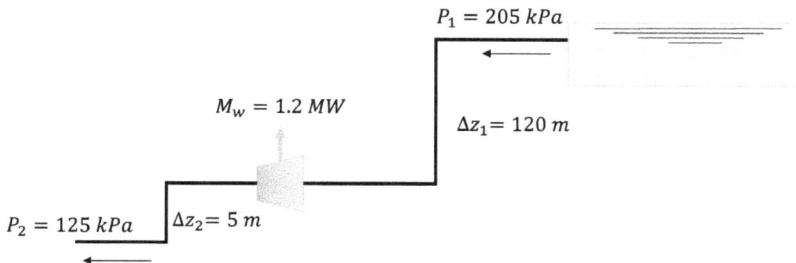

$P_1 = 205\ kPa$

$M_w = 1.2\ MW$

$\Delta z_1 = 120\ m$

$P_2 = 125\ kPa$ $\Delta z_2 = 5\ m$

FIGURE 5.10 Flow in a pipeline with a turbine.

where M is the molecular weight of water ($= 18kg/kmol$), G is the mass flow rate of the steam, and the friction factor f is given by $f = \{0.79\ln(Re) - 1.64\}^{-2}$ (*Re*: Reynolds number). Plot the pressure drop versus G when G varies from 20 to 50 $kg/(sec \cdot m^2)$.

Solution

The script *pressdrop* calls the built-in solver *fsolve* to solve the nonlinear equation and generates the required plot as shown in Figure 5.11.

```
% pressdrop.m
di = 0.75e-2; z = 12; T = 418.15; P1 = 510; mu = 13.8e-6; M = 18; R = 8.314e-3; x0
= 400;
G = 20:0.1:50; n = length(G);
for k = 1:n
  Re = di*G(k)/mu; f = 1/(0.79*log(Re) - 0.64)^2;
  fP = @(x) x.^2 + (G(k)^2*R*T/M)*(f*z/di + 2*log(P1./x)) - P1^2; % define eqn.
  P2(k) = fsolve(fP,x0); % use the built-in solver fsolve to solve the eqn.
end
plot(G, P2), grid, xlabel('G(kg/sec/m^2)'), ylabel('P_1-P_2 (kPa)')
```

Example 5.11 Flow Rate and Pressure Drop

Water is flowing through a horizontal 6 *in*. Schedule 40 mild steel pipe at $1atm$ and 25°C. The inside diameter and the length of the pipe are $0.154m$ and $1500m$, respectively, and the roughness factor is $0.0000457m$. The inlet and outlet conditions are $P_1 = 20atm$, $z_1 = 0m$, $P_2 = 2atm$, and $z_2 = 100m$. The density of water is $1000kg/m^3$ and the viscosity is $0.001kg/(m \cdot sec)$. Calculate the inlet volumetric flow rate and the pressure drop $\Delta P_f = 2f\rho Lv^2/D$. The friction factor f is given by the Colebrook equation as follows:

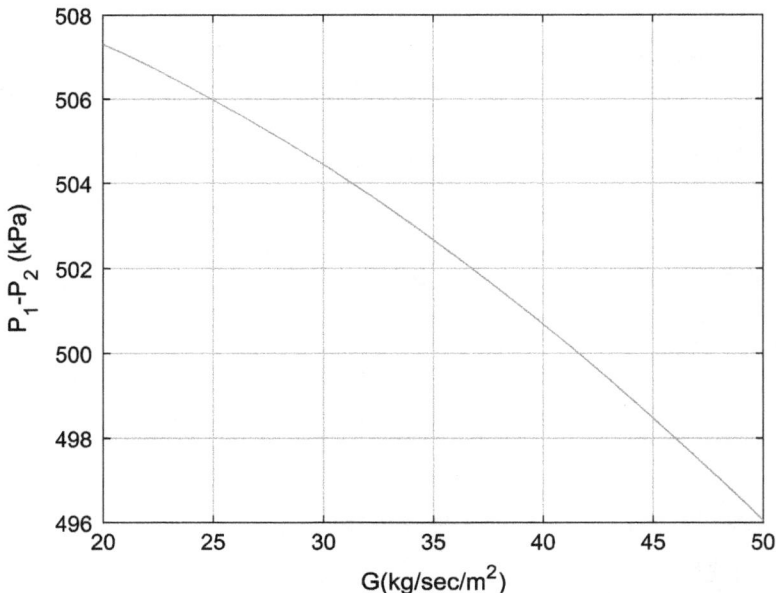

FIGURE 5.11 Pressure drop vs. mass flow rate.

$$\frac{1}{\sqrt{f}} = -1.7372\ln\left(\frac{\epsilon}{3.7D} + \frac{1.255}{N_{Re}\sqrt{f}}\right)$$

Solution

The energy balance for the incompressible flow in a pipe is given by

$$-\frac{1}{2}v^2 + g\Delta z + \frac{\Delta P}{\rho} + 2f\frac{Lv^2}{D} = 0$$

We have two nonlinear equations with two unknown variables. The equations are defined by the function *qdpfun* in the script *qdP*. The script *qdP* defines the data and uses the built-in function *fsolve* to find roots of the nonlinear equations.

```
% qdP.m: plot of volumetric flow rate vs. dP
clear all;
dP = -18*1.01325e5; dz = 100; rho = 1e3; mu = 1e-3; L = 1500; D = 0.154; rh = 4.57e-
5; % data
x0 = [5 0.001]; % initial guess (x(1)=v, x(2)=f)
x = fsolve(@qdpfun,x0,[],dP,dz,rho,mu,L,D,rh);
v = x(1); f = x(2); Q = x(1)*pi*D^2/4; % volumetric flow rate
dPf = 2*f*rho*L*v^2/D; % pressure drop due to friction loss
fprintf('Volumetric flow rate of water = %g m^3/sec\n', Q);
fprintf('Pressure drop due to friction loss = %g kPa\n', dPf/1000);
function fun = qdpfun(x,dP,dz,rho,mu,L,D,rh)
v = x(1); f = x(2); g = 9.8; Nre = D*v*rho/mu; % Reynolds number
fun = [-v^2/2 + g*dz + dP/rho + 2*f*L*v^2/D; 1/sqrt(f) + 1.7372*log(rh/3.7/D +
1.255/Nre/sqrt(f))];
end
```

Execution of the script *qdP* yields the following results:

```
>> qdP
Volumetric flow rate of water = 0.0610367 m^3/sec
Pressure drop due to friction loss = 849.219 kPa
```

Example 5.12 Unsteady-State Flow in a Pipe[16]

The momentum balance for the fluid flow through a long pipe with constant density (ρ) and viscosity (μ) is given by

$$\rho\frac{\partial v_z}{\partial t} = \frac{P_0 - P_L}{L} + \mu\frac{1}{r}\frac{\partial}{\partial r}\left(r\frac{\partial v_z}{\partial r}\right)$$

The initial and boundary conditions are given by

$$t = 0 : v_z = 0, r = 0 : v_z = \text{finite}, r = R : v_z = 0$$

The momentum equation can be expressed in dimensionless form as

$$\frac{\partial\phi}{\partial\tau} = 4 + \frac{1}{\xi}\frac{\partial}{\partial\xi}\left(\xi\frac{\partial\phi}{\partial\xi}\right), \phi(0,\xi) = 0, \phi(\tau,1) = 0, \frac{\partial\phi(\tau,0)}{\partial\xi} = 0$$

where

$$\phi = \frac{v_z}{(P_0 - P_L)R^2/(4\mu L)} = \frac{v_z}{v_{max}}, \, \xi = \frac{r}{R}, \, \tau = \frac{vt}{R^2}$$

The differential equation can be solved by using the built-in PDE solver *pdepe*. The standard form provided by the MATLAB to be used in *pdepe* is given by

$$g\left(x, t, u, \frac{\partial u}{\partial x}\right)\frac{\partial u}{\partial t} = x^{-m}\frac{\partial}{\partial x}\left(x^m f\left(x, t, u, \frac{\partial u}{\partial x}\right)\right) + r\left(x, t, u, \frac{\partial u}{\partial x}\right)$$

By direct comparison between the dimensionless momentum equation and the MATLAB standard form, we have

$$m = 1, \, g\left(x, t, u, \frac{\partial u}{\partial x}\right) = 1, \, r\left(x, t, u, \frac{\partial u}{\partial x}\right) = 4, f\left(x, t, u, \frac{\partial u}{\partial x}\right) = \frac{\partial \phi}{\partial \xi}$$

The standard forms of the initial and boundary conditions provided by MATLAB are given by

$$t = t_0 : u(x, t_0) = u_0(x)$$

$$x = a, \, x = b : p(x, t, u) + q(x, t)f\left(x, t, u, \frac{\partial u}{\partial x}\right) = 0$$

By comparison between the dimensionless initial and boundary conditions and MATLAB standard forms, we have

$$\phi(0, \xi) = 0 \rightarrow u_0(x) = 0$$

$$\phi(\tau, 1) = 0 \rightarrow p = ur, q = 0$$

$$\frac{\partial \phi(\tau, 0)}{\partial \xi} = 0 \rightarrow p = 0, q = 1$$

In order to use the built-in function *pdepe*, the differential equation and the initial and boundary conditions should be defined as functions. The plot of ϕ as functions of τ and ξ and the profiles of τ as a function of ξ can be produced as follows:

1. Define the differential equation to be solved: the function *fleqn* defines the PDE.
2. Define the initial and boundary conditions: the function *itcon* defines the initial condition and the function *bncon* sets up the boundary conditions.
3. Define the ranges of x and t and execute the built-in function *pdepe*: the script *unsflow* defines the range of x and t and calls *pdepe* to solve the PDE.

```
% unsflow.m: unsteady flow through pipe
clear all;
x = linspace(0,1,100); % 100 values from 0 to 1
t = linspace(0,1,60); % time span: 60 sec
m = 1; res = pdepe(m,@fleqn,@itcon,@bncon,x,t); u = res(:,:,1);
subplot(1,2,1), surf(x,t,u), colormap(gray), xlabel('\xi'), ylabel('\tau'),
zlabel('\phi')
shading interp, hold on % surface plot
surf(-x,t,u), colormap(gray), xlabel('\xi'), ylabel('\tau'), zlabel('\phi')
shading interp, hold off % axisymmetric plot
subplot(1,2,2)
% plot tau as a function of xi
```

```
for k = 1:length(t), plot(x,u(k,:),'k'), xlabel('\xi'), ylabel('\tau'), hold
on; end
for k = 1:length(t), plot(-x,u(k,:),'k'), xlabel('\xi'), ylabel('\tau'), hold
on; end
hold off
function [c,f,s] = fleqn(x,t,u,DuDx)
c = (1); f = DuDx; s = 4;
end
function u0 = itcon(x)
u0 = 0;
end
function [pl,ql,pr,qr] = bncon(xl,ul,xr,ur,t)
pl = 0; ql = 1; % 1st boundary
pr = ur; qr = 0; % 2nd boundary
end
```

Execution of the script *unsflow* yields the desired plots, as shown in Figure 5.12.

5.3.2 EQUIVALENT LENGTH OF VARIOUS FITTINGS AND VALVES

The equivalent pipe length is a convenient method for determining the overall pressure drop ΔP in a pipe. The equivalent length is added to the length of actual straight pipe, L_{st}, to give the total length of pipe, L_{Total}:

$$L_{Total} = L_{st} + L_{eq}$$

The equations for pressure drop ΔP and head loss ΔH can be expressed as

$$\Delta P = \left(f_D \frac{L}{D} + \sum K \right) \frac{\rho v^2}{2g_c}, \ \Delta H = \left(f_D \frac{L}{D} + \sum K \right) \frac{v^2}{2g_c}$$

(a)   (b)

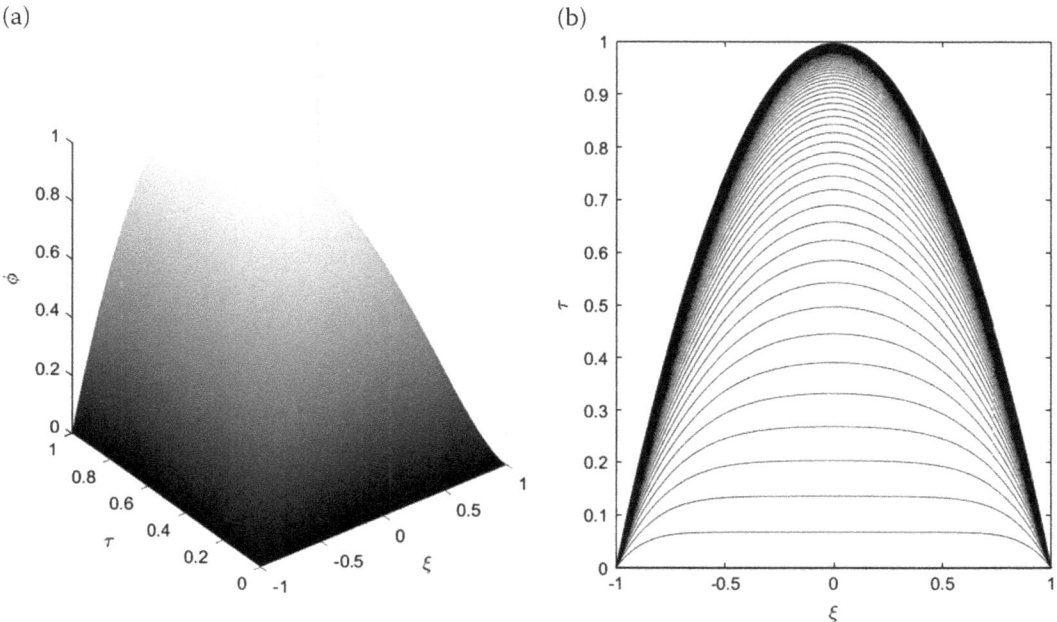

FIGURE 5.12   (a) Plot of ϕ and (b) the profiles of τ.

where

  D is the internal diameter of the pipe (ft)
  f_D is the Darcy friction factor
  K is the excess head loss
  L is the pipe length (ft)

Example 5.13 Pump Power[17]

Figure 5.13 shows the layout of a three-tank system. We need to feed tank 2 and tank 3 from storage tank 1. The flow rate from tank 1 is $0.042 m^3/sec$, and the density and viscosity of the liquid are $1000 kg/m^3$ and $0.001\ kg/(m{\cdot}sec)$, respectively. The tanks are open to atmosphere, and the characteristics of the pipes are given in Table 5.4. Determine the power of the pump to be bought for this transport operation and the flow rates to each tank.

Solution

Application of the Bernoulli equation yields

$$\frac{P_1}{\rho} + \frac{v_1^2}{2g} + \hat{W}_s + Z_1 = \frac{P_B}{\rho} + \frac{v_1^2}{2g} + \Delta H_1 + Z_B$$

$$\frac{P_B}{\rho} + \frac{v_2^2}{2g} + Z_B = \frac{P_2}{\rho} + \frac{v_2^2}{2g} + \Delta H_2 + Z_2, \quad \frac{P_B}{\rho} + \frac{v_2^2}{2g} + Z_B = \frac{P_3}{\rho} + \frac{v_3^2}{2g} + \Delta H_3 + Z_3$$

where P_i is the pressure, v_i is the velocity, $\hat{W}_s$ is the power of the pump, Z_i is the potential energy, and ΔH_i represents the friction loss. ΔH_i is given by

$$\Delta H_i = \frac{f_i L_i v_i^2}{2 g d_i}$$

where f_i is the friction factor and L_i is the equivalent length of the pipe. f_i is given by the Colebrook correlation as

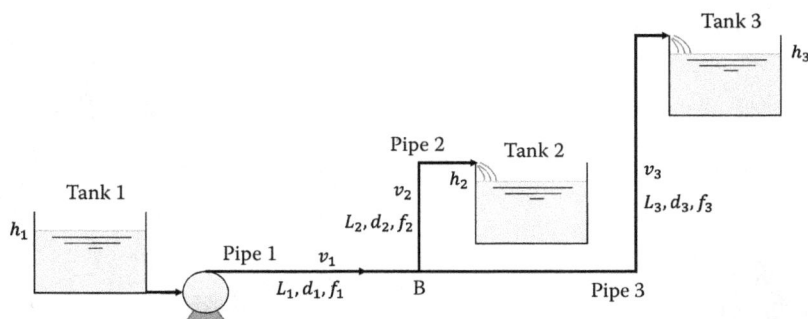

FIGURE 5.13   Layout of three-tank system.

TABLE 5.4
Specifications of the Pipe Layout

	Pipe 1	Pipe 2	Pipe 3
$h(m)$	8	35	50
$d(m)$	0.1524	0.0625	0.1016
$L(m)$	600	200	100
ε	5×10^{-6}	5×10^{-6}	5×10^{-6}

$$\frac{1}{\sqrt{f_i}} = -4\log\left(\frac{\varepsilon/d_i}{3.7} + \frac{1.256}{Re_i\sqrt{f_i}}\right), \; Re_i = \frac{d_i v_i \rho}{\mu}$$

Since the tanks are open to atmosphere, $P_1 = P_2 = P_3$. For incompressible fluids, $Q_1\rho = Q_2\rho + Q_3\rho$, where the volumetric flow rate Q_i is given by $Q_i = \pi d_i^2 v_i/4$. After some rearrangement, the problem can be formulated as follows:

$$\Delta H_i = \frac{f_i L_i v_i^2}{2 g d_i}, \; Re_i = \frac{d_i v_i \rho}{\mu} (i = 1, 2, 3)$$

$$\hat{W}_s + Z_1 - Z_2 - \Delta H_1 - \Delta H_2 = 0, \; \hat{W}_s + Z_1 - Z_3 - \Delta H_1 - \Delta H_3 = 0$$

$$d_1^2 v_1 - d_2^2 v_2 - d_3^2 v_3 = 0$$

$$\frac{1}{\sqrt{f_i}} + 4\log\left(\frac{\varepsilon/d_i}{3.7} + \frac{1.256}{Re_i\sqrt{f_i}}\right) = 0 (i = 1, 2, 3)$$

where $Z_1 = 8m$, $Z_1 = 35m$, and $Z_1 = 50m$. The function *pipsys* defines the set of nonlinear equations to be solved.

```
function fun = pipsys(x,L,d,rhg,Z,rho,mu,g,Q)
% x(1)=Ws; x(2)=v2, x(3)=v3, x(4)=f1, x(5)=f2, x(6)=f3
v1 = 4*Q/(pi*d(1)^2); Ws = x(1); f = [x(4) x(5) x(6)]; v = [v1 x(2) x(3)];
dH = f.*L.*v.^2 ./(2*g*d); Re = d.*v.*rho/mu;
fun(1,1) = Ws + Z(1) - Z(2) - dH(1) - dH(2);
fun(2,1) = Ws + Z(1) - Z(3) - dH(1) - dH(3);
fun(3,1) = v(1)*d(1)^2 - v(2)*d(2)^2 - v(3)*d(3)^2;
fun(4,1) = 1/sqrt(f(1)) + 4*log10(rhg(1)/d(1)/3.7 + 1.256/Re(1)/sqrt(f(1)));
fun(5,1) = 1/sqrt(f(2)) + 4*log10(rhg(2)/d(2)/3.7 + 1.256/Re(2)/sqrt(f(2)));
fun(6,1) = 1/sqrt(f(3)) + 4*log10(rhg(3)/d(3)/3.7 + 1.256/Re(3)/sqrt(f(3)));
end
```

The script *sol3tank* uses the built-in solver *fsolve* to solve the nonlinear equations defined by the function *pipsys*.

```
% sol3tank.m
clear all;
% Data
L = 100*[6 2 1]; d = [0.1524 0.0625 0.1016];
ep = 5e-6*ones(1,3); Z = [8 35 50]; rho = 1e3; mu = 1e-3; g = 9.8; Q1 = 0.042;
% x(1)=Ws; x(2)=v2, x(3)=v3, x(4)=f1, x(5)=f2, x(6)=f3
x0 = [10 5 5 1e-3 1e-3 1e-3]; % initial condition
x = fsolve(@pipsys,x0,[],L,d,ep,Z,rho,mu,g,Q1); % Solve nonlinear equation system
Ws = x(1); v2 = x(2); v3 = x(3); Q2 = pi*v2*d(2)^2 / 4; Q3 = pi*v3*d(3)^2 / 4;
fprintf('Power of the pump = %g m of column of water\n', Ws);
fprintf('Volumetric flow rate through pipe 1 = %g m^3/sec\n', Q1)
fprintf('Volumetric flow rate through pipe 2 = %g m^3/sec\n', Q2)
fprintf('Volumetric flow rate through pipe 3 = %g m^3/sec\n', Q3)
```

Execution of the script *sol3tank* yields the following results:
```
>> sol3tank
Power of the pump = 47.7323 m of column of water
Volumetric flow rate through pipe 1 = 0.042 m^3/sec
```

Volumetric flow rate through pipe 2 = 0.0160875 m^3/sec
Volumetric flow rate through pipe 3 = 0.0259125 m^3/sec

5.3.3 EXCESS HEAD LOSS

K is a dimensionless factor defined as the excess head loss in a pipe fitting and expressed in velocity heads. As a correlation for K, the 2-K method is known to be suitable for any pipe size.[18] The 2-K method can be expressed as

$$K = \frac{K_1}{N_{Re}} + K_\infty \left(1 + \frac{1}{d} \right)$$

where
  d is the internal diameter of the attached pipe (in.)
  K_1 is K for a fitting at $N_{Re} = 1$
  K_∞ is K for a large fitting at $N_{Re} = \infty$
  The conversion between the equivalent pipe length and the resistance coefficient K can be represented as

$$K = f \frac{L_{eq}}{R_h} = 4f \frac{L_{eq}}{D} = f_D \frac{L_{eq}}{D}$$

Table 5.5 shows the values of K_1 and K_∞ for the 2-K method.[19]

5.3.4 PIPE REDUCTION AND ENLARGEMENT

When the pipe size changes, the velocity head $v^2/(2g_c)$ also changes. The velocity is inversely proportional to the flow area and thus to the diameter squared. Therefore, K is inversely proportional to the velocity squared, and we have the relation

$$K_2 = K_1 \left(\frac{D_2}{D_1} \right)^4$$

Table 5.6 shows how K varies with changes in pipe size.[20]

5.3.5 OVERALL PRESSURE DROP

For steady-state flow, the Bernoulli equation yields[21]

$$\int_{P_1}^{P_2} \frac{dP}{\rho} + \frac{V_2^2 - V_1^2}{2g_c} + \frac{g}{g_c}(Z_2 - Z_1) + \Delta H = \hat{W}_s$$

where

$$\Delta H = \sum_{\text{all sections of straight pipe}} \frac{v^2}{2g_c} \left(4f \frac{L}{D} \right) + \sum_{\text{fitting, values, etc.}} K \frac{v^2}{2g_c}$$

and $\hat{W}_s$ is the power required in the pump.

Sizing of process pipes often involves trial and error. Usually, a pipe size is selected first and then the Reynolds number, friction factor, and coefficient of resistance are calculated. Then the

TABLE 5.5

Velocity Head Factors of Pipe Fittings

	Type of Fitting			K_1	K_∞
Elbows	90°	Standard (R/D=1), screwed		800	0.40
		Standard (R/D=1), flanged/welded		800	0.25
		Long-radius (R/D=1.5), all types		800	0.20
		Mitered elbows (R/D=1.5)	1 weld (90°)	1000	1.15
			2 welds (45°)	800	0.35
			3 welds (30°)	800	0.30
			4 welds (22.5°)	800	0.27
			5 welds (18°)	800	0.25
	45°	Standard (R/D=1), all types		500	0.20
		Long-radius(R/D=1.5), all types		500	0.15
		Mitered, 1 weld, 45°		500	0.25
		Mitered, 2 weld, 22.5°		500	0.15
	180°	Standard (R/D=1), screwed		1000	0.60
		Standard (R/D=1), flanged/welded		1000	0.35
		Long-radius (R/D=1.5), all types		1000	0.30
Tees	Used as elbow	Standard, screwed		500	0.70
		Long-radius, screwed		800	0.40
		Standard, flanged or welded		800	0.80
		Stub-in-type branch		1000	1.00
	Run-through tee	Screwed		200	0.10
		Flanged or welded		150	0.05
		Stub-in-type branch		100	0.00
Valves	Gate ball, plug	Full line size, β=1.0		300	0.10
		Reduced trim, β=0.9		500	0.15
		Reduced trim, β=0.8		1000	0.25
	Globe, standard			1500	4.00
	Globe, angle or Y-type			1000	2.00
	Diaphragm, dam type			1000	2.00
	Butterfly			800	0.25
	Check	Lift		2000	10.00
		Swing		1500	1.50
		Tilting-disk		1000	0.50

Source: Coker, A.K., *Chemical Process Design, Analysis and Simulation*, Gulf Publishing Company, Houston, TX, 1995, p. 155.

pressure drop per 100 *ft* of pipe, ΔP_{100}, is computed. For a given volumetric flow rate and physical properties of a single-phase fluid, ΔP_{100} for laminar and turbulent flows is given by

Laminiar flow: $\Delta P_{100} = 0.0273\mu Q/d^4 (psi/100ft)$

Turbulent flow: $\Delta P_{100} = 0.0216 f_D \rho Q^2/d^5 (psi/100ft)$

where d is the internal diameter of the pipe (*in.*) and ρ is the density of the fluid (lb/ft^3). If a mass flow rate and physical properties of a single-phase fluid are given, ΔP_{100} for laminar and turbulent flows is given by

Laminiar flow: $\Delta P_{100} = 0.0034\mu W/(d^4\rho)\,(psi/100ft)$

Turbulent flow: $\Delta P_{100} = 0.000336 f_D\,W^2/(d^5\rho)\,(psi/100ft)$

Multiplying ΔP_{100} by the total pipe length between two points and adding the pipe elevation yield the overall pressure drop ΔP (*psi*):

TABLE 5.6

Correlation of K for Fitting Caused by a Change in Pipe Size

Fitting Type		Inlet N_{Re}	K Based on Inlet Velocity Head
	1. Square reduction	$N_{Re} \leq 2500$	$K = \left[1.2 + \frac{160}{N_{Re}} \right]\left[\left(\frac{D_1}{D_2}\right)^4 - 1 \right]$
		$N_{Re} > 2500$	$K = \left(0.6 + 0.48 f_D \right)\left(\frac{D_1}{D_2}\right)^2 \left[\left(\frac{D_1}{D_2}\right)^2 - 1 \right]$
	2. Tapered reduction	All	Multiply K from Type 1 by $\sqrt{\sin(\theta/2)}$ for $45° < \theta < 180°$ or $1.6\sin(\theta/2)$ for $0° < \theta < 45°$
	3. Thin, sharp orifice	$N_{Re} \leq 2500$	$K = \left[2.7 + \left(\frac{D_2}{D_1}\right)^2 \left(\frac{120}{N_{Re}} - 1\right) \right]\left[1 - \left(\frac{D_2}{D_1}\right)^2 \right]\left[\left(\frac{D_1}{D_2}\right)^4 - 1 \right]$
		$N_{Re} > 2500$	$K = \left[2.7 - \left(\frac{D_2}{D_1}\right)^2 \left(\frac{4000}{N_{Re}}\right) \right]\left[1 - \left(\frac{D_2}{D_1}\right)^2 \right]\left[\left(\frac{D_1}{D_2}\right)^4 - 1 \right]$
	4. Square expansion	$N_{Re} \leq 4000$	$K = 2\left[1 - \left(\frac{D_1}{D_2}\right)^4 \right]$
		$N_{Re} > 4000$	$K = \left(1 + 0.8 f_D \right)\left[1 - \left(\frac{D_1}{D_2}\right)^2 \right]^2$
	5. Tapered expansion	All	If $\theta > 45°$, use K from Type 4; otherwise multiply K from Type 4 by $2.6\sin(\theta/2)$

Source: Coker, A.K., *Chemical Process Design, Analysis and Simulation*, Gulf Publishing Company, Houston, TX, 1995, p. 157.

$$\Delta P = \Delta P_{100} \cdot \frac{L_{total}}{100} + \frac{\rho \Delta z}{144} \ (psi)$$

The pipe size for a given flow rate is often selected on the assumption that ΔP is close to or less than the available pressure difference between two points in the pipeline. If ΔP and mass flow rate W are given, the pipe diameter D can be determined from the relation

$$D = \left[\frac{2f\,(L + L_e)}{\rho \Delta P} \left(\frac{4W}{\pi} \right)^2 \right]^{1/5}$$

Example 5.14 Power Requirement from a Pump[22]

Water is to be delivered to the upper tank shown in Figure 5.14. Determine the required power output from the pump at steady state. All of the piping is 4 *in.* internal-diameter smooth circular pipe. The Shacham equation can be used to estimate the friction factor in the turbulent region.
  Data: $\rho = 62.4 lb_m/ft^3$, $\mu = 1cp = 0.0006905 lb_m/(ft\cdot sec)$, $Q = 12ft^3/min = 0.2ft^3/sec$, K(sudden contraction)= 0.45, K(sudden expansion)= 1, K(90° elbow)= 0.5, $\epsilon = 0.0005$, $g_c = 32.174 lb_m ft/(lb_f \cdot sec^2)$.

Solution
  The values of the pressure and the velocity at point 1 are equal to those at point 2. Thus, from the steady-state mechanical energy balance, we get

$$W_s = \frac{g}{g_c}(Z_2 - Z_1) + \Delta H, \ \Delta H = \sum_{\substack{\text{all sections of straight pipe}}} \frac{v^2}{2g_c}\left(4f\frac{L}{D}\right) + \sum_{\substack{\text{fitting, values, etc.}}} K\frac{v^2}{2g_c}$$

The script *powfun* performs the required calculations.

```
% powfun.m
rho = 62.4; mu = 6.905e-4; D = 1/3; Q = 0.2; K = 0.45+3*0.5+1; gc = 32.174;
L = 5+300+100+120+20; eD = 5e-5;
v = 4*Q/(pi*D^2); m = Q*rho; Nre = D*v*rho/mu; % Reynolds 수
```

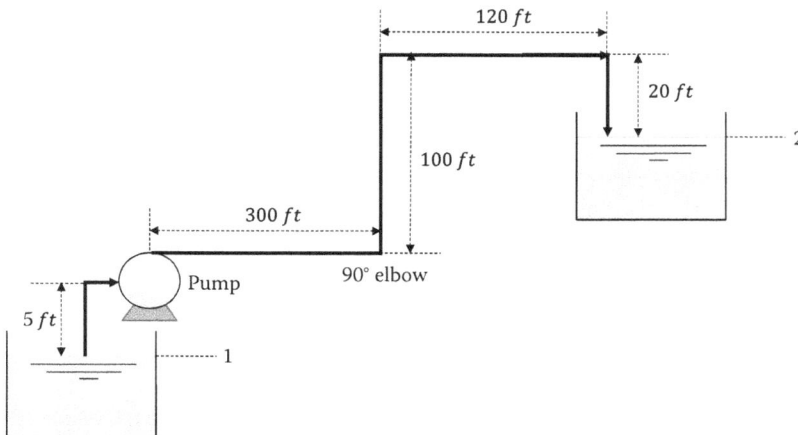

FIGURE 5.14 Pipeline flow with friction loss due to fittings. (From Bird, R.B. et al., *Transport Phenomena*, 2nd ed., John Wiley & Sons, Inc., Hoboken, NJ, 2002, pp. 207–208.)

```
if Nre < 2100
  f = 16./Nre;
else % Shacham eqn
  den = 16*(log10(eD/3.7 - 5.02*log10(eD/3.7+14.5./Nre)./Nre)).^2; f = 1./den;
end
dH = (4*f*L/D + K)*v^2/(2*gc); Ws = m*(dH + (105-20));
v,f,Ws
```

The script *powfun* produces the following outputs. From the results, we can see that $W_S = 1096.8 ft·lb_f/sec$.

```
>> powfun
v =
  2.2918
f =
  0.0049
Ws =
  1.0968e+03
```

Example 5.15 Internal Diameter of a Pipe

A process vapor is to be condensed with cooling water flowing inside the tubes of a shell-and-tube heat exchanger. The mass flow rate of cooling water is $0.36 kg/sec$, and the allowable pressure drop is $10 kPa$. Determine the diameter of the tube to be used in the condenser. The friction factor in the turbulent region may be given by the Shacham equation.
   Data: = 85. 9kg/m^3, $\mu = 4.4 \times 10^{-4} N·sec/m^2 (kg/m/sec)$, $L = 4m$, $\epsilon = 5 \times 10^{-6}m$, $K = 0.3$.

Solution
   Find the Reynolds number from the relation $v = 4W/(\pi D^2 \rho)$. The resulting Reynolds number is used to calculate the friction factor. The solution of the nonlinear equation

$$f(D) = D - \left[\frac{2f(L + L_e)}{\rho \Delta P} \left(\frac{4W}{\pi} \right)^2 \right]^{\frac{1}{5}} = 0$$

yields the diameter $D(m)$. In this equation, L_e is defined by $L_e = KD/(4f)$. The function *dwfun* defines the nonlinear equation.

```
function fD = dwfun(D,W,rho,mu,rf,dP,L,K)
eD = rf./D; % roughness factor
v = 4*W./(pi*rho*D.^2); % m/s
Nre = D.*v.*rho./mu; % Reynolds number
if Nre < 2100
  f = 16./Nre;
else % Shacham eqn
  den = 16*(log10(eD/3.7 - 5.02*log10(eD/3.7+14.5./Nre)./Nre)).^2; f = 1./den;
end
Le = K*D/(4*f); fD = D - ((2*f*(L + Le))./(rho*dP) * (4*W/pi)^2).^0.2;
end
```

The following commands set the data and call the built-in function *fzero* to solve the nonlinear equation defined by the function *dwfun*:

```
>> W=0.36; rho=85.9; mu=4.4e-4; rf=5e-6; dP=1e4; L=4; K=0.3; D0=0.1;
>> D = fzero(@dwfun,D0,[],W,rho,mu,rf,dP,L,K)
D =  0.0261
```

Example 5.16 Pipeline Flow[23]

A pipeline isometric (Figure 5.15) shows a 6 *in.* schedule 40 steel pipeline with six 90° LR (long-radius) elbows and two flow-through tees. The actual length of the pipe is 78 *ft*. The mass flow rate of the fluid is 75,000 *lb/hr*, the viscosity and density of the fluid are 1.25 *cP* and 64.30 *lb/ft*^{3}, respectively, and the pipe roughness is ϵ = 0.00015*ft*. Calculate the equivalent length of pipe fittings and the total length of the pipe.

Solution

The script *eqplen* performs the required calculations. In the script, the matrix *cfdat* represents the values of K_1 and K_∞ for pipes with large radius.

```
% eqplen.m: calculate the equivalent length of pipe fittings and total length of
the pipe
cfdat = zeros(14,3); cfdat(2,1) = 6; cfdat(11,1) = 2;
cfdat(:,2) = 100*[10 8 5 3 5 10 15 10 8 15 1.5 1 1.6 0]';
cfdat(:,3) = 0.1*[3 2 1.5 1 1.5 2.5 40 20 2.5 15 0.5 0 5 10]';
g = 32.2; d = 6.065; D = d/12; pr = 1.5e-4; w = 75000; mu = 1.25; rho = 64.3; % data
Lst = 78; z = 8; Area = pi*D^2/4; Nre = 6.31*w/(d*mu); v = 0.0509*w/(rhc*d^2); Q =
w/(8.02*rho);
if Nre <= 2100 % laminar flow
  f = 64/Nre; % Darcy friction factor
else % turbulent: friction factor by Chen eqn.
   Av = pr/3.7/D + (6.7/Nre)^0.9; f = 4./(-4*log10(pr/D/3.7 - 5.02*log10
(Av)/Nre)).^2;
end
Ksum = sum(cfdat(1:12,1).*cfdat(1:12,2)); Klsum = sum(cfdat(1:12,1).*cfdat
(1:12,3));
```

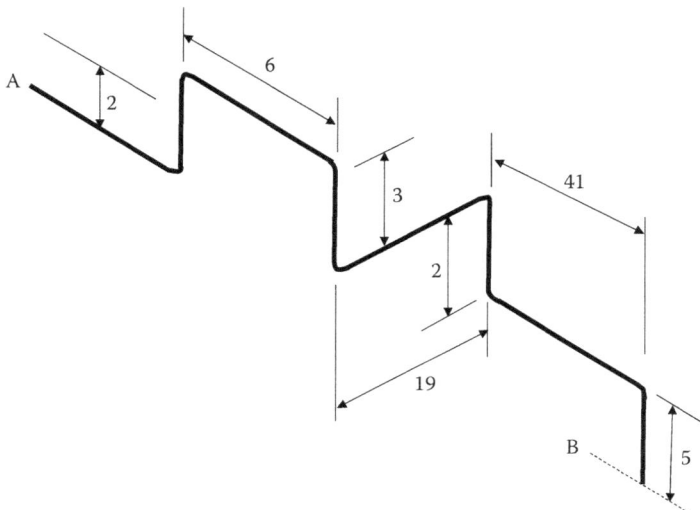

FIGURE 5.15 Pipeline isometric layout.

```
Kt = Ksum/Nre + Klsum*(1 + 1/d);
Kee1 = sum(cfdat(13:14,1).*cfdat(13:14,2)); Kee2 = sum(cfdat
(13:14,3));
K = Kee1/Nre + Kee2 + Kt; Kt = Kt + K + f*Lst/D; Leq = K*D/f, L = Lst + Leq
if Nre <= 2100, delP = 0.0034*mu*w/(d^4*rho);
else, delP = 0.000336*f*w^2/(d^5*rho); end
delPsi = delP*L/100 + z*rho/144; delH = 0.000483*f*L*w^2/(d^5*rho^2) + z;
```

The script *eqplen* produces the following results:
```
>> eqplen
Leq =
  38.3099
L =
  116.3099
```

The equivalent length of pipe fittings, L_{eq}, is 38.31 *ft*, and the total length of the pipe is $L = 116.31 ft$.

5.3.6 PIPELINE NETWORK

In the simulation of steady-state flows of a single fluid in a network of pipes, a large set of simultaneous equations has to be solved, and often the calculations are complex and iterative. The fundamental equation to be satisfied is mass flow conservation, which states that the flow into and out of each node should be equal. For the simple pipeline network shown in Figure 5.16, the pressure drop from node i to node j is given by

$$\Delta p_{ij} = \frac{32 f \rho \Delta L_{i,j}}{\pi^2 D^5} q_{ij}^2$$

where $q_{i,j}$ represents the volumetric flow rate between node i and node j. The pressure drop equation states that the sum of pressure drops around any loop k should be zero:

$$\sum \xi_{ik} \Delta p_{ik} = 0$$

where ξ_{ik} is the direction of flow relative to flow loop k. Flows in a clockwise direction (assumed flow direction) are treated as positive, and those in a counterclockwise direction are treated as negative.

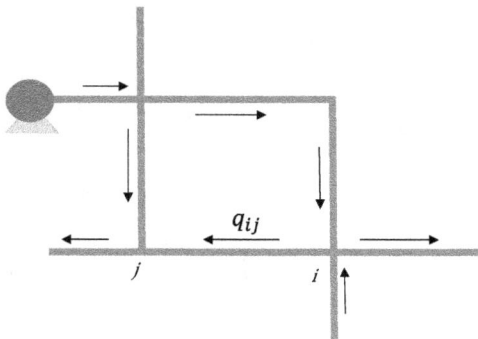

FIGURE 5.16 A simple pipeline network.

Example 5.17 Flow in a Pipeline Network[24]

Water at 25°C is flowing in the pipeline network shown in Figure 5.17. The pressure at the exit of the pump is $P_0 = 15bar$ (gauge pressure), and the water is discharged at atmospheric pressure (P_5) at the end of the pipeline. All the pipes are $6in.$ schedule 40 steel with an inside diameter of 0.154 m. The equivalent lengths of the pipes connecting different nodes are $L_{01} = 100m$, $L_{12} = L_{23} = L_{45} = 300m$, and $L_{13} = L_{24} = L_{34} = 1200m$. Determine all the flow rates q_{ij} and pressures at nodes 1, 2, 3, and 4 for the pipeline network. The pipe roughness is $\epsilon = 4.6 \times 10^{-5}m$, and the friction factor f may be given by the Shacham equation. At 25°C, the viscosity of water is $8.931 \times 10^{-4} kg/(m{\cdot}sec)$.

Solution

Application of the conservation of mass balance yields

$$\text{Node ①: } q_{01} - q_{12} - q_{13} = 0$$

$$\text{Node ②: } q_{12} - q_{23} - q_{24} = 0$$

$$\text{Node ③: } q_{13} - q_{23} - q_{34} = 0$$

$$\text{Node ④: } q_{24} - q_{34} - q_{45} = 0$$

The summation of pressure drops around various loops gives

$$\text{Loop I: } \Delta p_{01} + \Delta p_{12} + \Delta p_{24} + \Delta p_{45} + \Delta p_{pump} = 0$$

$$\text{Loop II: } \Delta p_{13} - \Delta p_{12} - \Delta p_{23} = 0$$

$$\text{Loop III: } \Delta p_{23} - \Delta p_{24} - \Delta p_{34} = 0$$

where $\Delta p_{pump} = 0 - 15 = -15bar = -15 \times 10^5 Pa$. The function *pnetq* defines the system of nonlinear equations. In this function, x(k) denotes q_{ij}.

```
function fun = pnetq(x,D,rho,mu,dP0,L)
% x1=q01, x2=q12, x3=q13, x4=q23, x5=q24, x6=q34, x7=q45
% L1=L01, L2=L12, L3=L13, L4=L23, L5=L24, L6=L34, L7=L45
A = pi*D.^2/4; eD = 4.6e-5./D;
for k = 1:7
  Nre = D*x(k)*rho/mu/A; % Shacham eqn.
  f = 1./(log10(eD/3.7 - (5.02./Nre).*log10(eD/3.7+14.5/Nre))).^2 /16;
  dP(k) = 32*f*rho*L(k)*(x(k))^2/(pi^2*D^5);
end
fun = [x(1) - x(2) - x(3); x(2) - x(4) - x(5); x(3) + x(4) - x(6); x(5) + x(6) -
x(7);
```

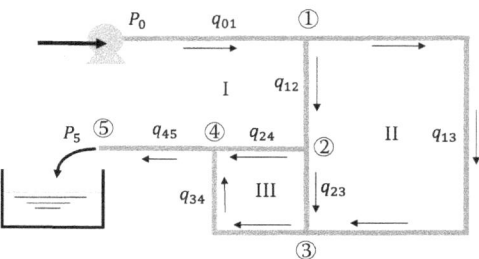

FIGURE 5.17 Pipeline network. (From Cutlip, M.B. and Shacham, M., *Problem Solving in Chemical and Biochemical Engineering with POLYMATH, Excel, and MATLAB*, 2nd ed., Prentice-Hall, Boston, MA, 2008, p. 309.)

```
    dP(1) + dP(2) + dP(5) + dP(7) + dP0; -dP(2) + dP(3) - dP(4); dP(4) - dP(5) +
dP(6)];
end
```

The initial values for x are all set to 0.1. The system of nonlinear equations defined by the function *pnetq* may be solved by using the built-in function *fsolve*.

```
>> L = 100*[1 3 12 3 12 12 3]; dP0 = -15e5; rho = 997.08; mu = 8.931e-4; D = 0.154; x0
= 0.1*ones(1,7);
>> x = fsolve(@pnetq,x0,[],D,rho,mu,dP0,L)
x =
    0.1098    0.0730    0.0368    0.0177    0.0553    0.0545    0.1098
```

We can see that

$q_{12} = 0.073$, $q_{13} = 0.0368$, $q_{23} = 0.0177$, $q_{24} = 0.0553$, $q_{34} = 0.0545$, $q_{45} = 0.1098(m^3/sec)$.

The Lang-Miller method[25] to solve pipe network problems uses the Churchill generalized friction factor correlation that spans the full range of the Reynolds numbers. The pressure drop is expressed as[26]

$$\Delta P = cf\dot{m}^2, \quad c = \frac{32L}{\rho D_i^5 \pi^2}$$

where

c is a constant for a given fluid and pipe

the subscript i denotes a loop in the network

L and D are the length and internal diameter of pipelines in the corresponding loop, respectively

The basic assumption in the Lang-Miller method is that the friction factor can be empirically represented as

$$f = a\dot{m}_i^{-b}$$

where a and b are both assumed constants over the range of the Reynolds numbers for one single iteration. Using this relation, we have

$$\Delta P_i = cf\dot{m}^2 = ac\dot{m}_i^{2-b} = \phi\dot{m}_i^{2-b}$$

For each loop in the pipeline network, we can get the following equations:

$$f = 2\left[\left(\frac{8}{N_{Re}}\right)^{12} + \frac{1}{(A+B)^{1.5}}\right]^{1/12}, \quad A = (2.457\ln C)^{16}$$

$$B = \left(\frac{37530}{N_{Re}}\right)^{16}, \quad C = \left(\frac{7}{N_{Re}}\right)^{0.9} + 0.27\frac{\epsilon}{D}$$

$$b = \frac{N_{Re}}{8}\left[\left[\frac{8}{N_{Re}}\right]^{13} + \frac{1}{(A+B)^{2.5}}\left(\frac{dA}{dN_{Re}} + \frac{dB}{dN_{Re}}\right)\right]\frac{1}{(f/2)^{12}}$$

$$\frac{dA}{dN_{Re}} = \frac{39.312}{C}(2.457\ln C)^{15}\frac{dC}{dN_{Re}}, \quad \frac{dB}{dN_{Re}} = -\frac{16B}{N_{Re}}, \quad \frac{dC}{dN_{Re}} = -0.9\left(\frac{7}{N_{Re}}\right)^{0.9}\frac{1}{N_{Re}}, \quad a = f\dot{m}^b$$

Substitution of the equation for pressure drop into the summation relation yields

$$\sum_i \xi_{ij} \phi_i \dot{m}_i^{2-b_i} = 0 (j = 1, 2, \cdots, n)$$

where n is the number of loops.

Example 5.18 Four-Loop Network[27]

A pipeline network with four flow loops is illustrated in Figure 5.18. The lengths and internal diameters of the pipes are shown in Table 5.7. Calculate the mass flow rate in each pipe that satisfies the pressure drop equation. A pipe roughness of $\epsilon = 5 \times 10^{-6} m$ is assumed uniformly for all pipes. The density and viscosity of the fluid are $881 kg/m^3$ and $5 \times 10^{-4} Ns/m^2$, respectively, and the inlet flow rate is $\dot{m}_0 = 334 kg/sec$.

Solution

Application of the conservation of mass balance at each junction yields the following nine equations:

$$m_1 + m_4 - 334 = 0, \ m_1 - m_2 - m_5 = 0, \ m_3 - m_4 + m_8 = 0,$$

$$m_2 + m_3 - m_7 - m_9 = 0, \ m_5 - m_6 - 42 = 0, \ m_6 + m_7 - m_{12} - 108 = 0,$$

$$m_{11} + m_{12} - 9 = 0, \ m_9 + m_{10} - m_{11} - 88 = 0, \ m_8 - m_{10} - 87 = 0$$

From the fact that the sum of pressure drops at each loop is equal to zero, we can get three equations:

$$\phi_1 m_1^{2-b_1} + \phi_2 m_2^{2-b_2} - \phi_3 m_3^{2-b_3} - \phi_4 m_4^{2-b_4} = 0$$

$$- \phi_2 m_2^{2-b_2} + \phi_5 m_5^{2-b_5} + \phi_6 m_6^{2-b_6} - \phi_7 m_7^{2-b_7} = 0$$

$$\phi_3 m_3^{2-b_3} - \phi_8 m_8^{2-b_8} + \phi_9 m_9^{2-b_9} - \phi_{10} m_{10}^{2-b_{10}} = 0$$

Now we have 12 nonlinear equations to be solved. The nonlinear equation system is defined by the function *pipnet*.

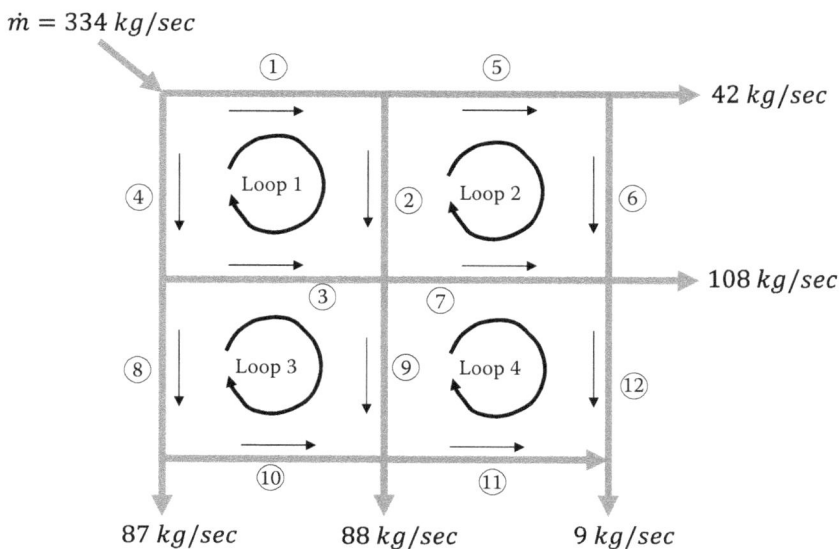

FIGURE 5.18 Four-loop pipeline network. (From Raman, R., *Chemical Process Computations*, Elsevier Applied Science Publishers, Barking, Essex, UK, 1985, pp. 234–235.)

TABLE 5.7

Lengths and Internal Diameters of the Pipes

Flow(i)	Diameter (D_i, m)	Length (L_i, m)	Flow(i)	Diameter (D_i, m)	Length (L_i, m)
1	0.508	915	7	0.305	915
2	0.406	1220	8	0.406	1220
3	0.406	915	9	0.305	1220
4	0.610	1220	10	0.406	915
5	0.508	915	11	0.305	915
6	0.406	1220	12	0.305	1220

Source: Raman, R., *Chemical Process Computations*, Elsevier Applied Science Publishers, Barking, Essex, UK, 1985, pp. 234–235.

```
function fm = pipnet(m,rho,mu,D,L,rf)
v = 4*m./(pi*D.^2*rho); Nre = D.*v*rho/mu; C = (7./Nre).^0.9 + 0.27*rf./D;
A = (2.457*log(C)).^16; B = (37530./Nre).^16; f = 2*((8./Nre).^12 + 1./
((A+B).^1.5)).^(1/12);
dC = -(0.9./Nre).*(7./Nre).^0.9; dA = (39.312./C).*dC.*(2.457*log(C)).^15; dB
= -16*B./Nre;
b = (Nre/8)./((f/2).^12).*((8./Nre).^13 + (dA+dB)./((A+B).^2.5)); c = 32.*L./
(rho*pi^2*D.^5);
a = f.*m.^b; phi = a.*c;
fm = [m(1) + m(4) - 334; m(1) - m(2) - m(5); m(3) - m(4) + m(8);
    m(2) + m(3) - m(7) - m(9); m(5) - m(6) - 42; m(6) + m(7) - m(12) - 108;
    m(11) + m(12) - 9; m(9) + m(10) - m(11) - 88; m(8) - m(10) - 87;
    phi(1)*m(1)^(2-b(1)) + phi(2)*m(2)^(2-b(2)) - phi(3)*m(3)^(2-b(3)) - phi
(4)*m(4)^(2-b(4));
    -phi(2)*m(2)^(2-b(2)) + phi(5)*m(5)^(2-b(5)) + phi(6)*m(6)^(2-b(6)) - phi
(7)*m(7)^(2-b(7));
    phi(3)*m(3)^(2-b(3)) - phi(8)*m(8)^(2-b(8)) + phi(9)*m(9)^(2-b(9)) - phi
(10)*m(10)^(2-b(10))];
end
```

The script file *pipnetwk* defines the data, calls the function *pipnet*, and solves the nonlinear system using the built-in function *fsolve*.

```
% pipnetwk.m
rho = 881; mu = 5e-4;
D = [0.508 0.406 0.406 0.610 0.508 0.406 0.305 0.406 0.305 0.406 0.305 0.305];
L = [915 1220 915 1220 915 1220 915 1220 1220 915 915 1220];
rf = 5e-6*ones(1,length(D)); m0 = 20*ones(1,length(D));
m = fsolve(@pipnet,m0,[],rho,mu,D,L,rf)
```

The script *pipnetwk* produces the following outputs:
```
>> pipnetwk
m =
 144.8934   41.8913   55.6231  189.1066  103.0021   61.0021   36.2300  133.4834
 61.2845   46.4834
 19.7680  -10.7680
```

The negative value implies that the flow direction is opposite to the assumed direction indicated in Figure 5.18.

5.4 FLOW THROUGH A TANK

5.4.1 OPEN TANK

Figure 5.19 shows an open tank flow system.[28] The change of the liquid level with respect to time is given by

$$A\frac{dz}{dt} = F_1 - F_2$$

where A is the cross-sectional area of the tank. Each flow rate is a function of pressure difference and is given by

$$F_1 = C_{v1}\sqrt{P_1 - P_2}, \; F_2 = C_{v2}\sqrt{P_2 - P_3}, \; P_2 = P_0 + \rho g z$$

Now we introduce dimensionless variables and parameters:

$$z^* = \frac{z}{z_0}, \; t^* = \frac{t}{\tau}, \; F^* = \frac{F}{F_C}, \; P^* = \frac{P}{P_{10}}, \; \tau = \frac{Az_0}{F_C}, \; F_C = C_{v1}\sqrt{P_{10} - P_{20}}$$

Then, because $dz = z_0 dz^*$, $dt = \tau dt^*$, and $F_i = F_c F^* (i = 1, 2)$, we get the following dimensionless balance equations:

$$\frac{dz^*}{dt^*} = F_1^* - F_2^*, \; F_1^* = \sqrt{\frac{P_{10}}{P_{10} - P_{20}}}\sqrt{P_1^* - P_2^*}$$

$$F_2^* = \left(\frac{C_{v2}}{C_{v1}}\right)\sqrt{\frac{P_{10}}{P_{10} - P_{20}}}\sqrt{P_2^* - P_3^*}, \; P_2^* = P_0^* + \left(\frac{\rho g z_0}{P_{10}}\right)z^*$$

5.4.2 ENCLOSED TANK

For an enclosed tank as shown in Figure 5.20, the pressure $P_0 = P_G$ above the liquid surface is not constant any more.[28] The movement of the liquid surface up and down will compress and expand the gas and cause the pressure to change. Assuming an ideal gas, we can use the equation of state

$$P_G = \frac{nRT_G}{V_G}$$

where V_G and T_G are the gas volume and temperature, respectively. Under the assumption of adiabatic conditions, $c_V dT_G = -(P_G/m)dV_G$, and we can get

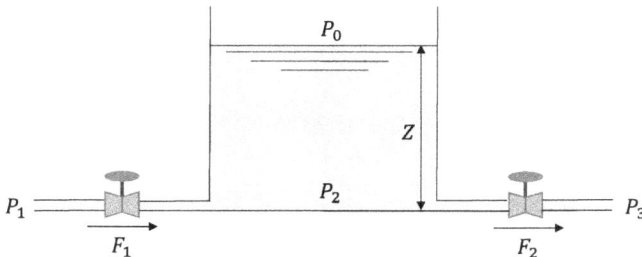

FIGURE 5.19 Open tank flow system. (From Ramirez, W.F., *Computational Methods for Process Simulation*, 2nd ed., Butterworth Heinemann, Oxford, UK, 1997, p. 145.)

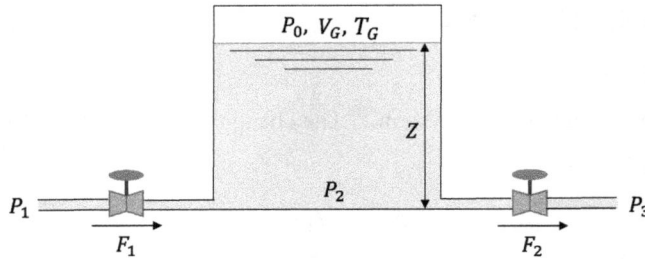

FIGURE 5.20 Enclosed tank. (From Ramirez, W.F., *Computational Methods for Process Simulation*, 2nd ed., Butterworth Heinemann, Oxford, UK, 1997, p. 145.)

$$\frac{T_G}{T_{G0}} = \left(\frac{V_{G0}}{V_G}\right)^{R/(c_v M_w)}$$

where M_w is the molecular weight of gas (air). The gas volume is given by

$$V_G = V_0 - Az$$

We define dimensionless variables and parameters as

$$z^* = \frac{z}{z_0}, \; t^* = \frac{t}{\tau}, \; F^* = \frac{F}{F_C}, \; P^* = \frac{P}{P_{10}}, \; V_G^* = \frac{V_G}{V_{G0}}$$

$$T_G^* = \frac{T_G}{T_{G0}}, \; \tau = \frac{Az_0}{F_C}, \; F_C = C_{v1}\sqrt{P_{10} - P_{20}}$$

Then, from the relations $dz = z_0 dz^*$, $dt = \tau dt^*$, $F_i = F_C F^* (i = 1, 2)$, $V_G = V_{G0} V_G^*$, and $T_G = T_{G0} T_G^*$, we have the following equations:

$$\frac{dz^*}{dt^*} = F_1^* - F_2^*, \; F_1^* = \sqrt{\frac{P_{10}}{P_{10} - P_{20}}}\sqrt{P_1^* - P_2^*}, \; F_2^* = \left(\frac{C_{v2}}{C_{v1}}\right)\sqrt{\frac{P_{10}}{P_{10} - P_{20}}}\sqrt{P_2^* - P_3^*}$$

$$P_G^* = \left(\frac{P_{G0}}{P_{10}}\right)\frac{T_G^*}{V_G^*}, \; P_2^* = P_G^* + \left(\frac{\rho g z_0}{P_{10}}\right)z^*, \; V_G^* = \frac{V_0}{V_{G0}} - \frac{Az_0}{V_{G0}}z^*, \; T_G^* = \left(\frac{1}{V_G^*}\right)^{R/(c_v M_w)}$$

Example 5.19 Flow through an Enclosed Tank[29]

For a flow involving an enclosed tank, plot the dimensionless level (z^*), flow rates (F_1^*, F_2^*), and pressure (P_2^*) versus the dimensionless time (t^*) within the range of $0 \leq t^* \leq 0.4$. Assume that the molecular weight of the gas is 29 (air) and $P_{G0} = 0.735 P_{10}$.

Data: $\rho = 1000 kg/m^3$, $A = 0.465 m^2$, $z_0 = 3.05m$, $g = 9.8 m/sec^2$
$C_{v1} = C_{v2} = 0.012 m^2/(N^{\frac{1}{2}} \cdot sec)$, $P_{10} = P_1 = 1.38 \times 10^5 N/m^2$, $P_{30} = P_3 = 1.08 \times 10^5 N/m^2$
$P_0 = 1.014 \times 10^5 N/m^2$, $V_0 = 2.83 m^3$, $V_{G0} = 1.415 m^3$, $T_{G0} = 338.6K$
$c_V = 0.2 cal/(g \cdot K)$, $R = 1.987 cal/(gmol \cdot K)$

Solution

Differential equations are defined by the function *cltank*. The script *encdtank* calls the function *cltank* and generates the graph.

```
% encdtank.m
Cv1 = 1.2e-2; Cv2 = Cv1; P0 = 1.014e5; P10 = 1.38e5; P30 = 1.08e5;
P1 = P10; P3 = P30; z0 = 3.05; rho = 1e3; A = 0.465; g = 9.8; P20 = P0 + rho*g*z0;
cv = 0.2; R = 1.987; Mw = 29; V0 = 2.83; Vg0 = 1.415;
k1 = sqrt(P10/(P10-P20)); k2 = k1*Cv2/Cv1;
k3 = rho*g*z0/P10; k4 = V0/Vg0; k5 = 0.735; k6 = R/(cv*Mw);
tspan    =    [0    0.4];    z0    =    1;    [t    z]    =    ode45(@cltank,tspan,z0,
[],k1,k2,k3,k4,k5,k6,P10,P1,P3);
Vg = k4 - z; Tg = (1./Vg).^k6; Pg = k5*Tg./Vg; P1s = P1/P10; P3s = P3/P10; P2s = Pg
+ k3*z;
F1s = k1*sqrt(P1s - P2s); F2s = k2*sqrt(P2s - P3s);
plot(t,z,t,F1s,'--',t,F2s,'.-',t,P2s,':'),  legend('z','F_1','F_2','P_2'),
xlabel('t(dimensionless)')
function dz = cltank(t,z,k1,k2,k3,k4,k5,k6,P10,P1,P3)
P1s = P1/P10; P3s = P3/P10; Vg = k4 - z; Tg = (1/Vg)^k6; Pg = k5*Tg/Vg;
P2s = Pg + k3*z; F1s = k1*sqrt(P1s - P2s); F2s = k2*sqrt(P2s - P3s); dz = F1s - F2s;
end
```

The script *encdtank* produces the plot shown in Figure 5.21.

5.5 FLOW MEASUREMENT: ORIFICE AND VENTURI METER

Applying the Bernoulli equation to an incompressible fluid flow in a horizontal installation between points 1 (upstream) and 2 (downstream), we have

$$\frac{g_c \Delta P}{\rho} + \frac{v_2^2 - v_1^2}{2} = 0$$

where $\Delta P = p_2 - p_1$. Since $v_1 A_1 = v_2 A_2$, v_1 can be represented as

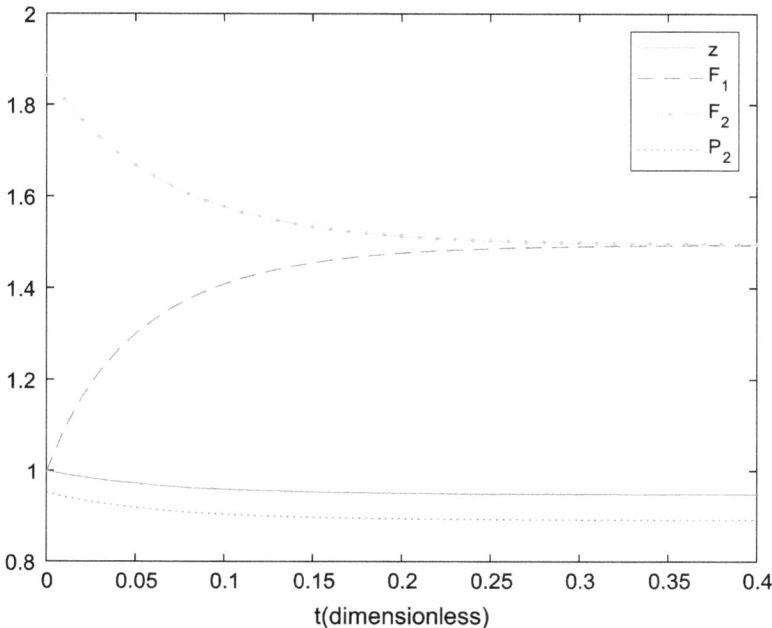

FIGURE 5.21 Dimensionless level, flow rate, and pressure versus time.

$$v_1 = v_2 \left(\frac{A_2}{A_1} \right) = v_2 \left(\frac{d_2^2}{d_1^2} \right) = \beta^2 v_2$$

where $\beta = d_2/d_1$ is the ratio of orifice (d_2) and pipe (d_1) diameters. From these two equations, v_2 is given by

$$v_2 = \sqrt{\frac{2g_c(p_1 - p_2)}{\rho(1 - \beta^4)}}$$

In general, the cross-sectional area at the point of pressure measurement (at vena contracta) is less than that of the orifice. Usually, the area of vena contracta is given by

$$A_2 = C_D A_o$$

where
    C_D is the coefficient of discharge
    A_o is the orifice area
Then the mass flow rate can be represented as

$$G = \rho v_2 A_2 = \rho C_D A_o \sqrt{\frac{2g_c(p_1 - p_2)}{\rho(1 - \beta^4)}}$$

The flow through orifice and venturi meters also depends on the expansion factor, Y, and the thermal expansion factor, F_a. Incorporation of Y and F_a into the mass flow rate equation yields

$$G = \rho C_D Y F_a A_o \sqrt{\frac{2g_c(p_1 - p_2)}{\rho(1 - \beta^4)}}$$

The value of the coefficient of discharge (C_D) is normally determined experimentally. In general, C_D is a function of the Reynolds number (N_{Re}), diameter ratio (β), and type of the measuring device. Table 5.8 shows calculation methods of C_D for various types of the meter.[30]
    The expansion factor (Y) can be estimated as follows:

$$\text{Orifice meter:} Y = 1 - (0.41 + 0.35\beta^4)\left(\frac{\Delta p}{\gamma p_1} \right)$$

$$\text{Venturimeter and nozzle:} Y = \sqrt{\left(\frac{\gamma r^{2/\gamma}}{\gamma - 1} \right)\left(\frac{1 - \beta^4}{1 - \beta^4 r^{2/\gamma}} \right)\left(\frac{1 - r^{(\gamma-1)/\gamma}}{1 - r} \right)}$$

where
    γ is the ratio of specific heats
    $r = p_2/p_1$ is the pressure ratio
    The pressure drop, Δp, does not represent the correct value. In fact, the actual pressure drop, Δp_a, is the nonrecoverable and is less than Δp. Table 5.9 shows equations for Δp_a for various meter types.[30]
    As the pressure ratio $r = p_2/p_1$ decreases, the flow through the orifice or Venturi meter increases until the velocity in the throat reaches the velocity of sound. The value of p_2/p_1 at which the throat velocity is sonic is called the critical pressure ratio, r_c, and the flow at such a condition is referred to as a critical or choked flow. In this case, the flow through the orifice, G, depends only upon the upstream pressure. r_c and G can be represented as follows:

TABLE 5.8
Calculation Methods of C_D (d: Pipe Diameter)

Type of Meter	Range of $d\,(m)$	C_D
Orifice: corner taps		$C_D = 0.5959 + 0.0312\beta^{2.1} - 0.184\beta^8 + \frac{91.71\beta^{2.5}}{N_{Re}^{0.75}}$
Orifice: flange taps	$0.0584 \le d$	$C_D = 0.5959 + 0.0312\beta^{2.1} - 0.184\beta^8 + \frac{0.09}{39.37d}\left(\frac{\beta^4}{1-\beta^4}\right) - \frac{0.0337\beta^3}{39.37d} + \frac{91.71\beta^{2.5}}{N_{Re}^{0.75}}$
	$0.0508 < d < 0.0584$	$C_D = 0.5959 + 0.0312\beta^{2.1} - 0.184\beta^8 + \frac{0.039}{39.37d}\left(\frac{\beta^4}{1-\beta^4}\right) - \frac{0.0337\beta^3}{39.37d} + \frac{91.71\beta^{2.5}}{N_{Re}^{0.75}}$
	Other diameters	$C_D = K_{SB}\sqrt{1-\beta^4}$
		$K_{SB} = 0.598 + 0.468(\beta^4 + 10\beta^{12}) + \frac{(0.87+8.1\beta^4)}{\sqrt{N_{Re}}}$
Orifice: $1d$–$1/2d$ taps		$C_D = 0.5959 + 0.0312\beta^{2.1} - 0.184\beta^8 + 0.039\left(\frac{\beta^4}{1-\beta^4}\right) - 0.0158\beta^3 + \frac{91.71\beta^{2.5}}{N_{Re}^{0.75}}$
Restriction orifice		$C_D = 0.6274 - 0.2354\beta + 0.7858\beta^2\,(0.2 < C_D \le 0.9)$
		$C_D = 0.61\,(\beta \le 0.2)$
		$C_D = 0.73$ (critical flow)
Venturi meter		$0.98 \le C_D \le 0.99$

Source: Arun Datta, *Process engineering and design using visual BASIC*, CRC Press, Taylor & Francis Group, New York, NY, 2008, pp. 120–127.

TABLE 5.9
Equations for Δp_a

Type of Meter	Δp_a
Orifice	$\Delta p_a = \left(\dfrac{\sqrt{1-\beta^4} - C_D\beta^2}{\sqrt{1-\beta^4} + C_D\beta^2}\right)\Delta p$
Venturi (15° divergent angle)	$\Delta p_a = (0.436 - 0.86\beta + 0.59\beta^2)\Delta p$
Venturi (7° divergent angle)	$\Delta p_a = (0.218 - 0.42\beta + 0.38\beta^2)\Delta p$

Source: Arun Datta, *Process engineering and design using visual BASIC*, CRC Press, Taylor & Francis Group, New York, NY, 2008, pp. 120–127.

$$r_c = \left(\frac{2}{1+\gamma}\right)^{\gamma/(\gamma+1)}, \quad G = C_D Y F_a A_o \sqrt{2\rho p_1 \gamma \left(\frac{2}{1+\gamma}\right)^{(\gamma+1)/(\gamma-1)}}$$

Example 5.20 Liquid Flow through an Orifice[31]

A gas flows through a 35 *mm* restriction orifice in a 100 NB (nominal bore) schedule 40 pipe. Estimate the pressure drop due to the orifice using the given data.

Data: mass flow rate $G = 3960\,kg/hr$, pipe inside diameter $d = 0.10226m$, gas density $\rho = 10.25\,kg/m^3$, gas viscosity $\mu = 0.013\,cP$, operating pressure $p_1 = 1200\,kPa$(abs), temperature $T_1 = 20°C$, $\gamma = 1.3$, compressibility $Z = 1$, $F_a = 1$.

Solution

The pressure drop Δp can be obtained from

$$Y = 1 - (0.41 + 0.35\beta^4)\left(\frac{\Delta p}{\gamma p_1}\right) = \frac{G}{\rho C_D F_a A_o}\sqrt{\frac{\rho(1-\beta^4)}{2g_c(p_1-p_2)}}$$

The script *dpgasflow* computes Δp using the given data.

```
% dpgasflow.m: pressure drop for gas flow
G = 3900/3600; d1 = 0.10226; d2 = 0.035; p1 = 1.2e6; Fa = 1; rho = 10.25; mu = 1.3e-
5; gam = 1.3; % data
beta = d2/d1; A1 = pi*d1^2/4; q1 = G/rho; v1 = q1/A1; % upstream velocity (m/s)
Nre = d1*v1*rho/mu; % Reynolds number
Cd = 0.6274 - 0.2354*beta + 0.7858*beta^2; % restriction orifice
Ao = pi*d2^2/4; % orifice area
fun  = @(x)  1-(0.41+0.35*beta^4)*x/gam/p1  -  sqrt(rho*(1-beta^4)/2/x)*G/
(rho*Ao*Cd*Fa);
x0 = 100; dp = fsolve(fun,x0);
Y = 1-(0.41+0.35*beta^4)*dp/gam/p1;
fprintf('Expansion factor = %g\n', Y); fprintf('Pressure drop = %g kPa\n',
dp/1000)
```

```
>> dpgasflow
Expansion factor = 0.969132
Pressure drop = 116.089 kPa
```

5.6 FLOW OF NON-NEWTONIAN FLUIDS

5.6.1 VELOCITY PROFILE

For a law fluid, the constitutive equation fornon-Newtonian fluid with power law characteristics, the shear stress can be described by

$$\tau_{rx} = -k\left|\frac{dv_x}{dr}\right|^{n-1}\frac{dv_x}{dr}$$

where
  $k(N\cdot sec^n/m^2)$ is a parameter
  n is the flow behavior index
  When $n = 1$, the fluid is Newtonian. For $n > 1$, the fluid is dilatant, and for $n < 1$ the fluid is pseudoplastic. The velocity profile for a non-Newtonian fluid flowing in a horizontal circular pipe is given by

$$v_x = \frac{n}{n+1}\left(\frac{\Delta P}{2kL}\right)^{1/n} R^{(n+1)/n}\left[1 - \left(\frac{r}{R}\right)^{(n+1)/n}\right]$$

Integration of the velocity profile followed by division by cross-sectional area yields the average velocity as

$$v_{x,avg} = \frac{1}{\pi R^2}\int_0^R 2\pi r v_x dr = \frac{n}{3n+1}\left(\frac{\Delta P}{2kL}\right)^{1/n} R^{(n+1)/n}$$

The function *nnhzpipe* calculates the average velocity and the shear stress and generates a plot of velocity profile for laminar flow of a non-Newtonian fluid in a horizontal pipe.

```
function [avgv taurx] = nnhzpipe(L,R,K,n,delP)
% average velocity and velocity profile for a non-Newtonian fluid flowing in a
horizontal pipe
% input
%   L: pipe length (m),  R: pipe diameter (m),  K: parameter (N*s^n/m^2)
%   n: flow index,  delP: pressure drop (Pa)
% output
%   avgv: avg. velocity (m/s)
h = 2*R/100; r = [-R:h:R]; % h: step size
Rn = R.^((n+1)/n); Pn = (delP./(2*K*L)).^(1/n);
v = (1 - (abs(r)/R).^((n+1)/n)) .* Pn .* Rn * n/(n+1);
avgv = Rn .* Pn .* n/(3*n+1); dvr = diff(v)/h; dvr = [dvr dvr(end)]; taurx =
-K.*abs(dvr).^(n-1).*dvr;
subplot(1,2,1), plot(v,r), grid, xlabel('v_x(r)'), ylabel('r'), axis tight
subplot(1,2,2), plot(taurx,r), grid, xlabel('\tau_{rx}(r)'), ylabel('r'),
axis tight
end
```

Example 5.21 Flow of a Non-Newtonian Fluid in a Horizontal Pipe

A non-Newtonian fluid is flowing in a horizontal pipe where $L = 15m$, $r = 0.009m$, $n = 2$, $k = 1.0 \times 10^{-6} N \cdot sec^n /m^2$, and $\Delta P = 110 Pa$. Calculate the average velocity, and plot the velocity profile and shear stress versus radius.

Solution

The function *nnhzpipe* calculates the average velocity and generates the plots shown in Figure 5.22:

```
>> L=15; R=0.009; K=1e-6; n=2; delP=110; avgv = nnhzpipe(L,R,K,n,delP)
avgv =
  0.4671
```

5.6.2 Reynolds Number

The constitutive equation for pipe flow of a Bingham plastic fluid is given by

$$\tau_{rz} = \tau_y + \eta\left(-\frac{du}{dr}\right)$$

where
u is the velocity
r and z are radial and axial coordinates, respectively
The modified Reynolds number is defined as

$$N_{ReB} = \frac{Dup}{\eta}$$

(a)

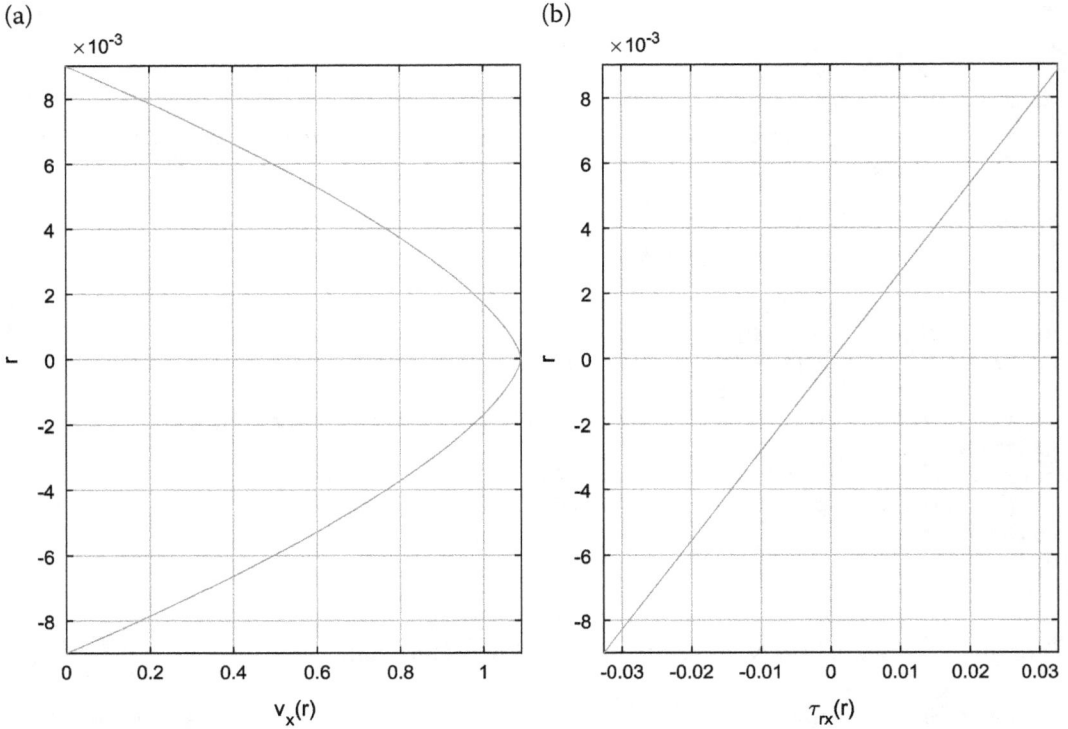

(b)

FIGURE 5.22 (a) Velocity profile and (b) shear stress for a non-Newtonian fluid flowing in a horizontal pipe.

The Hedstrom number is defined as

$$N_{He} = \frac{\tau_y D^2 \rho}{\eta^2}$$

Using N_{He}, the modified Reynolds number can be expressed as

$$N_{ReB} = \frac{N_{He}}{8\xi}\left(1 - \frac{4}{3}\xi + \frac{1}{3}\xi^4\right)$$

where $\xi = \tau_y/\tau_w$.

For a power law fluid, the constitutive equation for one-dimensional pipe flow of a Bingham plastic fluid is given by

$$\tau_{rx} = -k\left|\frac{du}{dr}\right|^{n-1}\frac{du}{dr}$$

The Reynolds number for pipe flow of power law fluids (density: ρ) is given by[32]

$$N_{Re} = \frac{D^n u^{2-n}\rho}{8^{n-1}k\left(\frac{3n+1}{4n}\right)^n}$$

where
   D is the pipe diameter
   u is the velocity
   The generalized Reynolds number can be defined as

$$N_{Reg} = \frac{D^{n'} u^{2-n'} \rho}{8^{n'-1} k'}$$

For Bingham plastic fluids,

$$n' = \frac{\xi^4 - 4\xi + 3}{3(1 - \xi^4)}, \; k' = \frac{\tau_y}{\xi} \left[\frac{3\eta\xi}{\tau_y(\xi^4 - 4\xi + 3)} \right]^{n'}$$

For power law fluids,

$$n' = n, \; k' = k\left(\frac{3n + 1}{4n} \right)^n$$

5.6.3 Friction Factor

For a Bingham plastic fluid, the relationship between the Fanning friction factor and the Reynolds number can be expressed as[33]
   Laminar region ($N_{ReB} \leq 2100$): $\frac{1}{N_{ReB}} = \frac{f}{16} - \frac{1}{6}\frac{N_{He}}{N_{ReB}^2} + \frac{\frac{1}{3}N_{He}^4}{f^3 N_{ReB}^8}$

   Turbulent region[34]: $\frac{1}{\sqrt{f}} = 4\log(N_{ReB}\phi(\xi)\sqrt{f}) - 0.4$
where N_{He} is the Hedstrom number and ϕ is defined as

$$\phi(\xi) = (1 - \xi)\left(1 - \frac{4}{3}\xi + \frac{1}{3}\xi^4 \right)$$

$\xi = \tau_y/\tau_w$ is obtained from the solution of the equation

$$N_{ReB} - \frac{N_{He}}{8\xi}\left(1 - \frac{4}{3}\xi + \frac{1}{3}\xi^4 \right) = 0$$

For power law fluids, the relationship between the Fanning friction factor and the Reynolds number can be expressed as
   Laminar region ($N_{Re} \leq 2100$): $f = 16/N_{Re}$
   Turbulent region[35]: $\frac{1}{\sqrt{f}} = \frac{4}{n^{0.75}}\log(N_{Re}f^{1-n/2}) - \frac{0.4}{n^{1.2}}$
   If we use the generalized Reynolds number N_{Reg}, the relationship between the Fanning friction factor and the Reynolds number can be expressed as[36]
   Laminar region ($N_{Reg} \leq 2100$): $f = \frac{16}{N_{Reg}}$

   Turbulent region: $\frac{1}{\sqrt{f}} = \frac{4}{n'^{0.75}}\log\left(N_{Reg}f^{1-n'^2/2} \right) - \frac{0.4}{n'^{1.2}}$

Example 5.22 Pipe Diameter for Non-Newtonian Flow
   Calculate the pipe diameter required for the horizontal flow of a power law fluid in a pipe.[37]
   Data: density $\rho = 961 kg/m^3$, mass flow rate $\dot{m} = 6.67 kg/sec$, pipe length $L = 10 m$, pipe roughness $\epsilon = 5 \times 10^{-6} m$, pressure drop $\Delta P = 15 kPa$, $K = 1.8$, $n = 0.64$, $k = 1.48 N \cdot sec^{2-n}/m^2$.

Solution

The Reynolds number and the friction factor are calculated first, and the diameter D is determined by solving the nonlinear equation

$$f(D) = D - \left[\frac{2f(L + L_e)}{\rho \Delta P} \left(\frac{4W}{\pi} \right)^2 \right]^{1/5} = 0$$

where $L_e = kD/(4f)$. The calculation procedure is performed by the function *nnDfun*.

```
function fD = nnDfun(D,W,rho,dP,rf,L,k,n,K)
eD = rf./D; % roughness
v = 4*W./(pi*rho*D.^2); % m/s
Nre = D^n*v^(2-n)*rho/(8^(n-1)*k*((3*n+1)/(4*n))^n); % Reynolds number
if Nre < 2100, f = 16/Nre;
else
  f0 = 16/Nre; fe = @(x) 4*log10(Nre*x^(1-n/2))/n^0.75 - 0.4*n^(-1.2) - 1/sqrt(x);
  f = fzero(fe,f0);
end
Le = K*D/(4*f); fD = D - ((2*f*(L + Le))./(rho*dP) * (4*W/pi)^2).^0.2;
end
```

The following commands produces the desired result. As an initial guess for D, set D0 = 0.1.

```
>> W=6.67; rho=961; dP=1.5e4; rf=5e-6; L=10; k=1.48; n=0.64; K=1.8;
>> D0 = 0.1; D = fzero(@nnDfun,D0, [],W,rho,dP,rf,L,k,n,K)
D =
   0.0888
```

5.7 COMPRESSIBLE FLUID FLOW IN PIPES

5.7.1 CRITICAL FLOW AND THE MACH NUMBER

The flow rate of a compressible fluid in a pipe with a given upstream pressure approaches a certain maximum value that it cannot exceed even with reduced downstream pressure. The maximum velocity that a compressible fluid can attain in a pipe is known as the sonic velocity, V_s, and can be expressed as

$$V_s = 223 \sqrt{\frac{kT}{M_w}} = 68.1 \sqrt{\frac{kP_1}{\rho_1}} \ (ft/sec)$$

where the subscript 1 denotes upstream. At the sonic velocity, a critical pressure, P_c, is attained. If P_c is less than the terminal pressure P_2, the flow is subcritical. If P_c is greater than P_2, the flow is critical. The critical pressure can be found from the Crocker equation:

$$P_c = \frac{G}{11400d^2} \sqrt{\frac{RT}{k(k+1)}} \ (psia)$$

where R is the molar gas constant given by $R = 1544S_g/29$ and the specific gravity S_g is defined by $S_g = (molecular\ weight\ of\ the\ gas) / (molecular\ weight\ of\ air)$. The upstream fluid velocity is determined from the relation $V = 0.0509G/(d^2\rho_1)(ft/sec)$. A recommended compressible fluid velocity for trouble-free operation is $V \le 0.6V_s$.[38] Table 5.10 shows the design criteria for carbon steel vapor lines.[39]

TABLE 5.10

Recommended Velocity and Maximum ΔP for Carbon Steel Vapor Lines

Type		Recommended Velocity (ft/sec)	Maximum ΔP (psi/100ft)
Fluid pressure (psig)	Subatmospheric		0.18
	$0 < P \leq 50$		0.15
	$50 < P \leq 150$		0.30
	$150 < P \leq 200$		0.35
	$200 < P \leq 500$		1.0
	$P > 500$		2.0
Tower overhead (psia)	$P > 50$	40–50	0.2–0.5
	Atmospheric	60–100	
	Vacuum ($P < 10$)		0.05–0.1
Compressor	Piping suction	75–200	0.5
	Piping discharge	100–250	1.0
	Gas lines with battery limit		0.5
	Refrigerant suction lines	15–35	
	Refrigerant discharge lines	35–60	
Steam lines	Saturated	200	
	Superheated	250	
	Steam pressure (psig) 0–50	167	0.25
	Steam pressure (psig) 50–150	117	0.40
	Steam pressure (psig) 150–300		1.0
	Steam pressure (psig) > 300		1.5
High-pressure steam lines	short (L < 600 ft)		1.0
	Long (L > 600 ft)		0.5
Exhaust steam lines	Exhaust steam lines		0.5
	P > atmosphere		0.5
	Leads to exhaust header		1.5
Relief valve	Discharge	$0.5\,V_s$	
	Entry point at silencer	V_s	

Source: Coker, A.K., *Chemical Process Design, Analysis and Simulation*, Gulf Publishing Company, Houston, TX, 1995, p. 164.

The Mach number, M, is the velocity of the gas divided by the velocity of sound in the gas, and can be represented as $M = V/V_s$. The exit Mach number for a compressible isothermal fluid has been shown to be $1/\sqrt{k}$, where k is the ratio of the fluid specific heat capacity at constant pressure to that at constant volume (i.e., $k = C_p/C_v$). The flow is subsonic for $M < 1/\sqrt{k}$, sonic for $M = 1/\sqrt{k}$, and supersonic for $M > 1/\sqrt{k}$. Table 5.11 shows the k values for some common gases.

5.7.2 COMPRESSIBLE ISOTHERMAL FLOW

Estimation of the maximum flow rate and pressure drop of a compressible isothermal flow is based on the following assumptions:

1. The fluid is an isothermal compressible fluid.
2. The gas obeys the ideal gas law.

TABLE 5.11

k **Values for Some Common Gases**

Gas	Molecular Weight	$k = C_p/C_v$	Gas	Molecular Weight	$k = C_p/C_v$
Acetylene	26.0	1.30	Hydrogen	2.0	1.41
Air	29.0	1.40	Methane	16.0	1.32
Ammonia	17.0	1.32	Methyl chloride	50.5	1.20
Argon	39.9	1.67	Natural gas	19.5	1.27
Butane	58.1	1.11	Nitric oxide	30.0	1.40
Carbon dioxide	44.0	1.30	Nitrogen	28.0	1.41
Carbon monoxide	28.0	1.40	Nitrous oxide	44.0	1.31
Chlorine	70.9	1.33	Oxygen	32.0	1.40
Ethane	30.0	1.22	Propane	44.1	1.15
Ethylene	28.0	1.22	Propylene	42.1	1.14
Helium	4.0	1.66	Sulfur dioxide	64.1	1.26
Hydrogen chloride	36.5	1.41			

3. The friction factor is constant along the pipe.
4. No mechanical work is done on or by the system.
5. Steady-state flow is maintained.

Figure 5.23 shows the distribution of fluid energy with work done by the pump and heat added to the system.[40] The Bernoulli equation for the steady flow of a fluid can be written as

$$\int_1^2 \frac{dP}{\rho} + \frac{V^2}{2g_c\alpha} + \frac{g}{g_c}\Delta Z + h_L + \delta W_s = 0$$

where α is the dimensionless velocity distribution and $\alpha = 1$ for turbulent or plug flow. Since no mechanical work is done on the system, $\delta W_s = 0$, and $Z_2 = Z_1$ for horizontal pipe.

The velocity head $h_L = KV^2/(2g_c)$ and the mass flow rate $G = \rho VA$ are constant. Then the maximum flow rate through the pipe can be expressed as

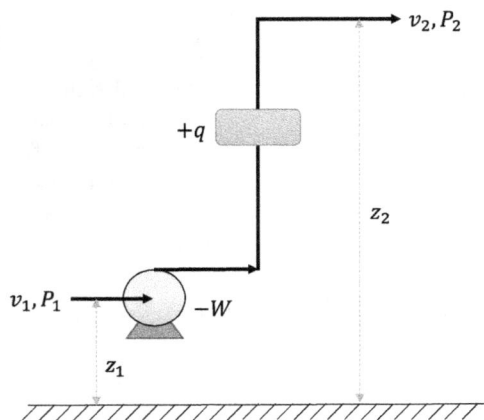

FIGURE 5.23 A single-stream piping system. (From Coker, A.K., *Chemical Process Design, Analysis and Simulation*, Gulf Publishing Company, Houston, TX, 1995, p. 166.)

$$G = 1335.6d^2 \sqrt{\left(\frac{\rho_1}{K_{total} + 2\ln\frac{P_1}{P_2}}\right)\left(\frac{P_1^2 - P_2^2}{P_1}\right)} \; (lb/hr)$$

where K_{total} is the total velocity head due to friction, fittings, and valves and is given by

$$K_{total} = f_D \frac{L}{D} + \Sigma K_f$$

If the pressure drop due to velocity acceleration is relatively small compared with the pressure drop due to friction, the term $\ln(P_1/P_2)$ can be neglected. Then the pressure drop ΔP can be expressed as

$$\Delta P \cong P_1 - \sqrt{P_1^2 - \frac{P_1 G^2 K_{total}}{\rho_1(1335.6d^2)^2}} \; (psi)$$

Table 5.12 shows the friction factors for clean commercial steel pipes with complete turbulent flow.[41]

Example 5.23 Compressible Fluid Flow[42]

Calculate the maximum flow rate of natural gas through a ruptured exchanger tube.

Data: Exchanger tube = 3/4 *in.*, Schedule 160, inside diameter d = 0.614 *in.*, tube length = 20 *ft*, pipe friction factor (complete turbulence) = 0.026, compressibility factor Z = 0.9, gas temperature = 100°F, molecular weight = 19.5, pressure in tubes (P_1) = 1110 *psig*, relief valve set pressure on the shell side (P_2) = 400 *psig*, ratio of specific heat capacities k = 1.27, fluid viscosity = 0.012 *cP*, resistance coefficient due to fittings and valves K = 2.026.

Solution

The solution procedure is as follows:

1. Calculate the cross-sectional area A and density ρ.
2. Calculate $K_{total} = f_D(L/D) + \Sigma K_f$ where $\Sigma K_f = K$.
3. Calculate $= 1335.6d^2 \sqrt{\left(\frac{\rho_1}{K_{total} + 2\ln\frac{P_1}{P_2}}\right)\left(\frac{P_1^2 - P_2^2}{P_1}\right)}$.
4. Calculate $N_{Re} = \frac{6.31G}{\mu D}$, $V = \frac{0.0509G}{d^2\rho_1}$, $V_s = 223\sqrt{\frac{kT}{M_w}} = 68.1\sqrt{\frac{kP_1}{\rho_1}}$
5. Calculate $M = \frac{V}{V_s}$, $R = \frac{1544S_g}{29}$, $P_c = \frac{G}{11400d^2}\sqrt{\frac{RT}{k(k+1)}}$.

The script *compflow* performs the calculation procedure.

```
% compflow.m: compressible fluid flow
d = 0.614; Lst = 20; f = 0.026; K = 2.026; Z = 0.9; T = 100; Mw = 19.5;
P1 = 1124.7; P2 = 414.7; k = 1.27; mu = 0.012; T = T+460; % T(R)
D = d/12; Area = pi*D^2/4; % cross sectional area (ft^2)
rho = P1*Mw/(10.72*Z*T); % density (lb/ft^3)
delP = P1 - P2; Kp = f*Lst/D; Kt = K + Kp; %Ktotal
G = 1335.6*d^2 * sqrt(rho*(P1^2 - P2^2)/(Kt + 2*log(P1/P2))/P1);
Nre = 6.31*G/(D*mu); Vg = 0.0509*G/(rho*d^2);
Vs = 223*sqrt(k*T/Mw); Mach = Vg/Vs; Sg = Mw/29; R = 1544/(29*Sg);
Pc = (G/(11400*d^2))*sqrt(R*T/(k*(k+1)));
G, Vg, Mach, Pc
```

TABLE 5.12
Friction Factors for Complete Turbulent Flow in Commercial Steel Pipes

Nominal Size (*in.*)	Friction Factor (*f*)	Nominal Size (*in.*)	Friction Factor (*f*)
1	0.023	10	0.0136
1.5	0.0205	12	0.0132
2	0.0195	14	0.0125
3	0.0178	16	0.0122
4	0.0165	18	0.12
5	0.016		0.0118
6	0.0152	24	0.0116
8	0.0142		

Source: Coker, A.K., *Chemical Process Design, Analysis and Simulation*, Gulf Publishing Company, Houston, TX, 1995, p. 167.

The script *compflow* produces the following outputs:

```
>> compflow
G =
  8.3969e+03
Vg =
 279.2890
Mach =
   0.2074
Pc =
 242.3052
```

We can see that the maximum flow rate is 8396.9 *lb/hr* and the fluid velocity is 279.289 *ft/sec*. Since the Mach number at the inlet is 0.2074, the flow is subsonic.

5.7.3 Choked Flow

When a compressible fluid is stored in a vessel at high pressure, it can escape the vessel at very high velocity to a low-downstream-pressure environment. If the pressure difference between the vessel and the downstream environment is high, the fluid will escape at sonic velocity, which is called a choked flow.[43] The flow of a compressible fluid flow can be expressed as follows:

$$Q = 53.64 Y d^2 \sqrt{\frac{p_1 \Delta p}{K T s_g}}$$

where
Q is the volumetric flow rate (m^3/sec)
Y is the expansion factor
d is the diameter (m)
p_1 is the vessel pressure (kPa)
Δp is the pressure difference (kPa)
K is the flow resistance factor of pipes and fittings
T is the temperature (K)
s_g is the specific gravity (s_g of air $= 1$)

In a choked flow, Δp depends on the sonic velocity limiting factor, which in turn depends on K and the ratio of specific heats κ. A typical value of κ for oil and gas applications is 1.3. When $\kappa = 1.3$, the limiting value of $\Delta p/p_1$ can be represented as a function of K as follows:

$$\frac{\Delta p}{p_1} = \left(0.9953 + \frac{0.9054}{\sqrt{K}} + \frac{0.1173}{K} - \frac{0.0195}{K^{1.5}}\right)^{-1}$$

For $\kappa = 1.3$, the expansion factor, Y, can be represented as

$$Y = 0.0415\ln K + 0.6097$$

If the flow is not choked, Y is defined as

$$Y = 1 - m\left(\frac{\Delta p}{p_1}\right)$$

where m is defined as

$$m = \frac{1 - Y_{choked}}{\left(\Delta p/p_1\right)_{choked}}$$

Example 5.24 Gas Release from a Cylinder[44]

A gas being stored in a cylinder at 60°C is released to the atmosphere through a 6 m 80 NB (schedule 80) pipe (ID = 0.0779m). The specific gravity of the gas is 0.42. Assume that the Darcy friction factor is $f_D = 0.0175$, the ratio of specific heats is 1.3, and the gas in the cylinder is not limiting. Calculate the flow rates of the gas when the pressure in the cylinder is 800kPa(gauge) and 150kPa(gauge).

Solution
The calculation procedure is as follows:

1. Determine K_{total}: For pipe, $K_{pipe} = f_D L_{eq}/D = (0.0175)(6/0.0779) = 1.348$, K at the entrance is $K_{entrance} = 0.5$, and K at the exit is $K_{exit} = 1.0$. Thus $K_{total} = K_{pipe} + K_{entrance} + K_{exit} = 2.848$.
2. Estimate $\Delta p/p_1$ and find the limiting $\Delta p/p_1$. If the limiting $\Delta p/p_1$ is less than the estimated value, the flow is sonic.
3. Calculate the limiting value of Y and the flow rate Q.

This procedure is performed by the script *gasrelease*.

```
% gasrelease.m: gas release from a cylinder
T = 60; sg = 0.42; d = 0.0779; K = 2.848; % total K
p0 = 101.3; p1 = p0 + [800 150]; n = length(p1); % pressure (kPa)
dp1c = 1/(0.9953 + 0.9054/sqrt(K) + 0.1173/K - 0.0195/K^1.5); % limiting dp/p1
dp1e = (p1-p0)./p1; % estimated dp/p1
Y = 0.0415*log(K) + 0.6097; % limiting expansion factor
for k = 1:n
  if dp1c < dp1e(k) % flow is sonic
    dp = dp1c*p1(k); % limiting pressure differnece (kPa)
```

```
      Q = 3600*53.64*Y*d^2 * sqrt(dp*p1(k)/(K*(T+273.15)*sg)); % flow rate
(m^3/hr)
         fprintf('Flow rate (cylinder pressure: %g kPaG) = %g m^3/sec
\n',p1(k)-p0,Q);
   else % flow is subsonic
      m = (1-Y)/dp1c; % slope
      Y = 1 - m*dp1e(k); % unchoked flow
      dp = dp1c*p1(k); % limiting pressure differnece (kPa)
      Q = 3600*53.64*Y*d^2 * sqrt(dp*p1(k)/(K*(T+273.15)*sg)); % flow rate
(m^3/hr)
         fprintf('Flow rate (cylinder pressure: %g kPaG) = %g m^3/sec
\n',p1(k)-p0,Q);
   end
end

>> gasrelease
Flow rate (cylinder pressure: 800 kPaG) = 27588.1 m^3/sec
Flow rate (cylinder pressure: 150 kPaG) = 7951.55 m^3/sec
```

Example 5.24 Pressure Drop due to the Orifice[45]

A gas flows through a 10 *mm* restriction orifice in a 100 NB schedule 40 pipe. The flow may be considered a choked (or critical) flow. Estimate the expansion factor Y and pressure drop due to the orifice using the given data.

Data: mass flow rate $G = 582 kg/hr$, pipe inside diameter $d = 0.10226m$, gas density $\rho = 10.25 kg/m^3$, gas viscosity $\mu = 0.013 cP$, operating pressure $p_1 = 1200 kPa$(abs), temperature $T_1 = 20°C$, $\gamma = 1.3, K = 1, F_a = 1$.

Solution

For choked flow,

$$G = C_D Y F_a A_o \sqrt{2\rho p_1 \gamma \left(\frac{2}{1+\gamma}\right)^{(\gamma+1)/(\gamma-1)}}, \quad \frac{\Delta p}{p_1} = \left(0.9953 + \frac{0.9054}{\sqrt{K}} + \frac{0.1173}{K} - \frac{0.0195}{K^{1.5}}\right)^{-1}$$

We use $C_D = 0.73$ for choked (critical) flow. The script *dpgasflowc* computes Y and Δp using the given data.

```
% dpgasflowc.m: pressure drop for critical flow
clear all;
G = 582/3600; d1 = 0.10226; d2 = 0.01; p1 = 1.2e6; Fa = 1; rho = 10.25; mu = 1.3e-5;
gam = 1.3; K = 1;
beta = d2/d1; A1 = pi*d1^2/4; q1 = G/rho; v1 = q1/A1; % upstream velocity (m/s)
Ao = pi*d2^2/4; % orifice area
Nre = d1*v1*rho/mu; % Reynolds number
Cd = 0.73; % critical flow
Y = G/(Cd*Fa*Ao)/sqrt(2*rho*p1*gam*(2/(1+gam))^((gam+1)/(gam-1)));
dp = p1/(0.9953 + 0.9054/sqrt(K) + 0.1173/K - 0.0195/K^1.5);
fprintf('Expansion factor = %g\n', Y); fprintf('Pressure drop = %g kPa\n',
dp/1000)

>> dpgasflowc
Expansion factor = 0.852008
Pressure drop = 600.45 kPa
```

5.8 TWO-PHASE FLOW IN PIPES

5.8.1 FLOW PATTERNS

Flow regimes in horizontal two-phase flow can be classified into seven types: stratified, wavy, annular, plug, slug, bubble or froth, and spray or mist. The Baker diagram shown in Figure 5.24 can be referred to determine the type of flow in a process pipeline.[46] Establishing the flow regime involves determining the Baker parameters B_x and B_y from the two-phase system's characteristics and physical properties. These parameters can be expressed as[47]

$$B_x = 531\left(\frac{W_L}{W_G}\right)\left(\frac{\sqrt{\rho_L\rho_G}}{\rho_L^{2/3}}\right)\left(\frac{\mu_L^{1/3}}{\sigma_L}\right), \, B_y = 2.16\left(\frac{W_G}{A}\right)\frac{1}{\sqrt{\rho_L\rho_G}}$$

where W_L and W_G are flow rates (*lb/hr*) in the liquid and vapor phases, respectively. Table 5.13 shows the characteristic linear velocities of the gas and liquid phases in each flow regime.[48]

The equations representing the boundaries of the flow regimes shown in Figure 5.24 are

$$C_1 : \ln B_y = 9.774459 - 0.6548(\ln B_x)$$

$$C_2 : \ln B_y = 8.67964 - 0.1901(\ln B_x)$$

$$C_3 : \ln B_y = 11.3976 - 0.6084(\ln B_x) + 0.0779(\ln B_x)^2$$

$$C_4 : \ln B_y = 10.7448 - 1.6265(\ln B_x) + 0.2839(\ln B_x)^2$$

$$C_5 : \ln B_y = 14.569802 - 1.0173(\ln B_x)$$

$$C_6 : \ln B_y = 7.8206 - 0.2189(\ln B_x)$$

The Baker parameter B_y for two-phase flow regimes is calculated by the function *twophreg*. This function computes B_y for a given B_x, which can be a vector.

```
function By = twophreg(Bx)
```

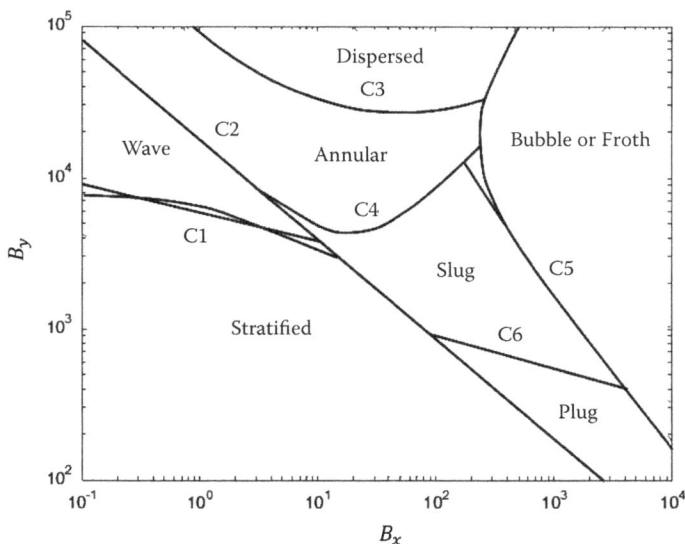

FIGURE 5.24 Baker parameters for two-phase flow regimes. (From Coker, A.K., *Chemical Process Design, Analysis and Simulation*, Gulf Publishing Company, Houston, TX, 1995, p. 172.)

```
% 2-phase flow regimes
% Bx, By: Baker parameters
By = [];
C1 = exp(9.774459 - 0.6548*log(Bx));
C2 = exp(8.67964 - 0.1901*log(Bx));
C3 = exp(11.3976 - 0.6084*log(Bx) + 0.0779*log(Bx).^2);
C4 = exp(10.7448 - 1.6265*log(Bx) + 0.2839*log(Bx).^2);
C5 = exp(14.569802 - 1.0173*log(Bx));
C6 = exp(7.8206 - 0.2189*log(Bx));
By = [C1' C2' C3' C4' C5' C6'];
end
```

5.8.2 Pressure Drop

The two-phase pressure drop in horizontal pipes, ΔP_T, can be expressed as

$$\frac{\Delta P_T}{100ft} = \frac{\Delta P_G}{100ft} \cdot Y_G \, (psi/100ft)$$

where
$\Delta P_G/100ft(psi/100ft)$ is the pressure drop of gas if flowing alone in the pipe
Y_G is the two-phase flow modulus
Y_G is a function of the Lockhart-Martinelli two-phase flow modulus X, which is defined as

$$X = \sqrt{\frac{\Delta P_L}{\Delta P_G}} \, , \, Y_G = f(X)$$

where $\Delta P_L/100ft$ (psi) is the pressure drop of liquid if flowing alone in the pipe. Equations of Y_G for different flow regimes are shown in Table 5.14.[49]

For wave flow, the Huntington friction factor F_H is used to determine the two-phase pressure loss:

$$H_x = \left(\frac{W_L}{W_G}\right)\left(\frac{\mu_L}{\mu_G}\right), \, \ln F_H = 0.2111\ln(H_x) - 0.3993, \, \frac{\Delta P_T}{100ft} = \frac{0.000366 F_H W_G^2}{d^5 \rho_G}$$

TABLE 5.13

Characteristic Linear Velocities of Two-Phase Flow Regimes

Flow Regime	Liquid-Phase Velocity (ft/sec)	Vapo- Phase Velocity (ft/sec)
Bubble or froth	5~15	0.5~2
Plug	2	< 4
Stratified	< 0.5	0.5~10
Wave	< 1.0	> 15
Slug	15	3~50
Annular	< 0.5	> 20
Dispersed, spray, or mist	Close to vapor velocity	> 200

Source: Coker, A.K., *Chemical Process Design, Analysis and Simulation*, Gulf Publishing Company, Houston, TX, 1995, p. 174.

---

where

W_L and W_G are the flow rates of the liquid and vapor phases (*lb/hr*), respectively
F_H is the Huntington friction factor
A is the inside cross-sectional area of the pipe (*ft²*)
ρ_G is the density of the vapor (*lb/ft²*)
μ_L and μ_G are the viscosities of liquid and vapor phases (*cP*), respectively
The pressure drop of a liquid or vapor flowing alone in a straight pipe can be expressed as

$$\frac{\Delta P_T}{100ft} = \frac{0.000366 f_D\, W_x^2}{d^5 \rho_G}\, (psi/100ft)$$

where W_x is the flow rate of the liquid or vapor (*lb/hr*). Modification of this relation yields the relationship representing the overall pressure drop of the two-phase flow for the total length of pipe plus fittings based on the gas-phase pressure drop:

$$\Delta P_{T_{overall}} = \frac{0.000366 \cdot f_D \cdot W_G^2 \cdot Y_G \cdot L}{100 d^5 \rho_G}\, (psi)$$

The velocity of the two-phase fluid can be given by

$$V = \frac{0.0509}{d^2}\left(\frac{W_G}{\rho_G} + \frac{W_L}{\rho_L} \right)$$

5.8.3 CORROSION AND EROSION

High velocities in a two-phase flow system lead to corrosion and erosion. An index based on velocity head can indicate whether corrosion or erosion may become significant at a particular

---

TABLE 5.14
Equations of Y_G for Two-Phase Flow Regimes

Flow Regime	Y_G
Bubble/froth	$Y_G = \left[\frac{14.2 X^{0.75}}{(W_L/A)^{0.1}} \right]^2$
Plug	$Y_G = \left[\frac{27.315 X^{0.855}}{(W_L/A)^{0.17}} \right]^2$
Stratified	$Y_G = \left[\frac{15400 X}{(W_L/A)^{0.8}} \right]^2$
Slug	$Y_G = \left[\frac{1190 X^{0.815}}{\sqrt{W_L/A}} \right]^2$
Annular	$Y_G = (aX^b)^2,\ a = 4.8 - 0.3125d,\ b = 0.343 - 0.021d$ d = pipe inside diameter (in) (if d>12, set d=10)
Dispersed/spray	$Y_G = [\exp(C_0 + C_1(\ln X) + C_2(\ln X)^2 + C_3(\ln X)^3)]^2$ $C_0 = 1.4659,\ C_1 = 0.49138,\ C_2 = 0.04887,\ C_3 = -0.000349$

Source: Coker, A.K., *Chemical Process Design, Analysis and Simulation*, Gulf Publishing Company, Houston, TX, 1995, pp. 177–178.

velocity.[50] This index can be used to determine the range of mixture densities and velocities below which corrosion/erosion should not occur. This index is

$$\rho_m U_M^2 \leq 10{,}000$$

The mixture density ρ_M and mixture velocity U_M are given by

$$\rho_M = \frac{W_L + W_G}{\left(\frac{W_L}{\rho_L} + \frac{W_G}{\rho_G}\right)}, \quad U_M = U_G + U_L = \frac{1}{3600A}\left(\frac{W_G}{\rho_G} + \frac{W_L}{\rho_L}\right)$$

where U_L and U_G are the velocities of liquid and vapor (ft/sec), respectively.

Example 5.25 Pressure Drop in a Two-Phase Flow[51]

Determine the pressure drop of a 5 mile (26,400 ft) length in a 6 in. schedule 40 (ID = 6.065in.) pipe, for a 5,000 bpd (77,956 lb/hr) rate of salt water ($\gamma_w = 1.07$), having a vapor flowing at 6 mmscfd (12,434 lb/hr) with 0.65 specific gravity at a temperature of 110°F. Physical properties of the fluid are shown in Table 5.15.

Solution

The script *twophdP* calculates the pressure drop. This script calls the function *twophmod* to determine the two-phase flow regime and two-phase flow modulus Y_G. The function *twophmod* includes subfunctions to calculate the value of Y_G in each flow regime.

```
% twophdP.m: pressure drop in two-phase flow
clear all;
rhol = 66.7; rhog = 2.98; mul = 1; mug = 0.02; sigl = 70; % data
L = 26400; d = 6.065; Wl = 77956; Wg = 12434; D = d/12; eD = 0.00015/D;
% liquid/vapor phase friction factor
Nrel = 6.31*Wl/(d*mul); Nreg = 6.31*Wg/(d*mug);
if Nrel <= 2100, fL = 64/Nrel;
else, Av = eD/3.7 + (6.7/Nrel)^0.9; fL = 4./(-4*log10(eD/3.7 - 5.02*log10(Av)/
Nrel)).^2; end
if Nreg <= 2100, fG = 64/Nreg;
else, Av = eD/3.7 + (6.7/Nreg)^0.9; fG = 4./(-4*log10(eD/3.7 - 5.02*log10(Av)/
Nreg)).^2; end
% Determine two-phase flow modulus (Yg) and flow regimes
fr   = {'stratified','wave','plug','slug','bubble','annular','dispersed'};
rind = 7;
delPl = 3.66e-4 * fL * Wl^2/(d^5 * rhol); delPg = 3.66e-4 * fG * Wg^2/(d^5 * rhog);
[Yg rind] = twophmod(rhol,rhog,mul,sigl,Wl,Wg,delPl,delPg,d);
if rind == 2
```

TABLE 5.15
Physical Properties

Physical Properties	Liquid	Vapor
Flow rate (lb/hr)	77,956	12,434
Density (lb/ft^3)	66.7	2.98
Viscosity (cP)	1.0	0.02
Surface tension ($dyne/cm$)	70.0	

```
fH = exp(0.2111*log(Wl*mul/(Wg*mug)) - 3.993);
  delPt = 3.66e-4 * fH * Wg^2 * L/(d^5 * rhog * 100);
else, delPt = Yg*delPg*L/100; end
fprintf('Flow regime: %s\n', fr{rind}); fprintf('Two-phase flow modulus (Yg): %
g\n', Yg);
fprintf('Total pressure drop(psi): %g\n', delPt);
function [Yg rind] = twophmod(rhol,rhog,mul,sigl,Wl,Wg,delPl,delPg,d)
% Calculation of two-phase flow modulus (Yg)
% input:
%  rhol,rhog: density of liquid and vapor (lb/ft^3), mul: liquid viscosity (cP)
%  sigl: liquid surface tension (dyne/cm), Wl,Wg: flow rates of liquid and vapor
(lb/h)
%  delPl,delPg: pressure drop for liquid-only/vapor-only flow, d: pipe inside
diameter (in)
% output:
%  Yg: two-phase flow modulus,  rind: flow regime index
% Calculate Baker parameter Bx and By
D = d/12; A = pi*D^2/4; Bx = 531*(Wl/Wg)*(sqrt(rhol*rhog)/(rhol^(2/3)))*(mul^
(1/3)/sigl);
By = 2.16*(Wg/A)/sqrt(rhol*rhog); C = twophreg(Bx);
% Classify flow regimes and calculate Yg for each flow regime
x = sqrt(delPl/delPg); Wa = Wl/A;
if By <= C(1)
  if By <= C(2), [Yg rind] = strat(x,Wa); % By < C1,C2
  else, [Yg rind] = wave(x,Wa); end % C2 < By < C1
  else % C1 < By
    if By < C(5) % C1 < By < C5
      if By < C(6), [Yg rind] = plug(x,Wa);  % C1 < By < C5,C6
      else % C1,C6 < By < C5
        if By < C(4), [Yg rind] = slug(x,Wa); % C1,C6 < By < C4,C5
        else % C1,C4,C6 < By < C5
          if By <= C(3), [Yg rind] = annul(x,d); % C1,C4,C6 < By < C3,C5
          else, [Yg rind] = dispr(x); % C1,C3,C4,C6 < By < C5
          end
        end
      end
    else % C1,C5 < By
      if Bx > 150, [Yg rind] = bubb(x,Wa);
      else % Bx <= 150
        if By <= C(3), [Yg rind] = annul(x,d); % C1,C5 < By < C3
        else, [Yg rind] = dispr(x); % C1,C3,C5 < By
        end
      end
    end
  end
end

function [Yg rind] = strat(x,Wa)
rind = 1; Yg = (15400*x/(Wa^0.8))^2;
end

function [Yg rind] = wave(x,Wa)
rind = 2; Yg = 0;
end
```

```
function [Yg rind] = plug(x,Wa)
rind = 3; Yg = (27.315*x^0.855 / Wa^0.17)^2;
end

function [Yg rind] = slug(x,Wa)
rind = 4; Yg = (1190*x^0.815/sqrt(Wa))^2;
end

function [Yg rind] = bubb(x,Wa)
rind = 5; Yg = (14.2*x^0.75 / Wa^0.1)^2;
end

function [Yg rind] = annul(x,d)
rind = 6; dx = d; if d > 12, dx = 10; end;
Yg = ((4.8 - 0.3125*dx)*x^(0.343-0.021*dx))^2;
end

function [Yg rind] = dispr(x)
rind = 7; a0 = 1.4659; a1 = 0.49138; a2 = 0.04887; a3 = -0.000349;
Yg = (exp(a0 + a1*log(x) + a2*(log(x))^2 + a3*(log(x))^3))^2;
end
```

The script *twophdP* produces the following outputs:
```
>> twophdP
Flow regime: annular
Two-phase flow modulus (Yg): 10.0125
Total pressure drop(psi): 97.7391
```

5.8.4 Vapor-Liquid Two-Phase Vertical Downflow

In a vertical downflow, large vapor bubbles, known as slug flow, are formed in the liquid stream. With bubbles greater than 1 *in.* in diameter and a liquid viscosity less than 100 *cP*, slug flow can be represented by the Froude numbers $N_{Fr,L}$(for liquid) and $N_{Fr,G}$(for vapor). These numbers are defined as

$$N_{Fr,L} = \frac{V_L}{\sqrt{gD}} \sqrt{\frac{\rho_L}{\rho_L - \rho_G}}, \; N_{Fr,G} = \frac{V_G}{\sqrt{gD}} \sqrt{\frac{\rho_L}{\rho_L - \rho_G}}$$

where D is the inside diameter of the pipe (*ft*). V_L and V_G are the superficial velocities based on the total pipe cross section and are given by

$$V_L = \frac{W_L}{3600\rho_L A}, \; V_G = \frac{W_G}{3600\rho_G A} \; (ft/sec)$$

where $A = \pi D^2/4$ (*ft²*). If $N_{Fr,L} < 0.31$, the vertical pipe is self-venting and the bubbles rise. If the Froude number is in the range of $0.3 < N_{Fr,L} < 1.0$, the flow is pulse flow, and pressure pulsation and vibration are produced. If $N_{Fr,L} > 1.0$, the friction force offsets the effect of gravity and thus requires no pressure gradient in the vertical downflow liquid.[52]

5.8.5 Pressure Drop in Flashing Steam Condensate Flow

When a liquid is flowing near its saturation point in a pipeline, decreased pressure will cause vaporization. The higher the pressure difference, the greater the vaporization, resulting in flashing of the liquid. The pressure drop for a flashed condensate mixture can be determined by the following calculation procedure:

1. Calculate the flashed steam rate W_G (*lb/hr*) and flashed condensate rate W_L (*lb/hr*):

$$W_{FL} = B(\ln P_C)^2 - A, \; A = 0.00671(\ln P_h)^{2.27}, \; B = 0.0088 + 10^{-4}e^X$$

$$X = 6.122 - \left(\frac{16.919}{\ln P_h}\right), \; W_G = W_{FL} \cdot W, \; W_L = W - W_G$$

where
   W_{FL} is the weight fraction of condensate flashed to vapor
   P_C is the steam condensate pressure before flashing (*psia*)
   P_h is the flashed condensate header pressure (*psia*)
   The temperature of flashed condensate, T_{FL}(°F), is given by

$$T_{FL} = 115.68(P_h)^{0.226}$$

2. Determine the density of flashed vapor ρ_G(*lb/ft³*) and flashed condensate ρ_L(*lb/ft³*) and the density of the flashed condensate/vapor mixture ρ_M(*lb/ft³*):

$$\rho_G = 0.0029P_h^{0.938}, \; \rho_L = 60.827 - 0.078P_h + 0.00048P_h^2 - 0.0000013P_h^3, \; \rho_M = \frac{W_G + W_L}{\left(\frac{W_G}{\rho_G} + \frac{W_L}{\rho_L}\right)}$$

3. Calculate the pressure drop ΔP_T(*psi/100ft*). For turbulent flow,

$$f = -\frac{0.25}{\left[-\log\left(\frac{0.000486}{d}\right)\right]^2} \; \text{(d: internal pipe diameter (\textit{in.}))}$$

in.

$$V = \frac{3.054}{d^2}\left[\frac{W_G}{\rho_g} + \frac{W_L}{\rho_L}\right], \; \Delta P_T = \frac{0.000336fW^2}{d^5\rho_m}$$

where V is the velocity of flashed condensate mixture (*ft/min*). If $V \geq 5,000ft/min$, the condensate mixture may cause some problems to the piping system.

5.9 FLOW THROUGH PACKED BEDS

The pressure drop, ΔP_T(*lb/in.²*), in the flow of fluids through packed beds of granular particles can be obtained from the Ergun equation[53]:

$$\Delta P_T = \frac{B_L}{144} \cdot \frac{1 - \epsilon}{\epsilon^3} \cdot \frac{G^2}{D_p g_c \rho}\left[\frac{362.85\mu(1 - \epsilon)}{D_p G} + 1.75\right]$$

$$\epsilon = \frac{\rho_c - \rho_b}{\rho_c}, \; \rho = \frac{M_W P}{10.73ZT}, \; D_p = \frac{6(1 - \epsilon)}{S}, \; S = \frac{1 - \epsilon}{V_p} \cdot A_p$$

$$V_p = \frac{1}{1728}\left(P_D^2 \cdot \frac{\pi}{4} \cdot P_L\right), \; A_p = \frac{1}{144}\left(\frac{\pi P_D^2}{2} + \pi P_D P_L\right), \; N_{Re} = \frac{D_p G}{2.419\mu(1 - \epsilon)}$$

where

B_L is the length of the packed bed (ft)

ϵ is the void fraction of the packed bed

ρ_c is the catalyst density (lb/ft^3)

ρ_b is the density of the packed bed (lb/ft^3)

ρ is the fluid density (lb/ft^3)

D_p is the effective particle diameter (ft)

$psia$ is the surface area of the packed bed (ft bed)

P is the molecular weight of the fluid

P is the fluid pressure (M_W)

T is the fluid temperature ($°R$)

A_P is the particle area (ft^2)

V_p is the particle volume (ft^3)

P_D is the particle diameter ($in.$)

P_L is the particle length ($in.$)

Z is the compressibility factor

G is the superficial fluid mass velocity ($lb/(hr \cdot ft^2)$) and is given by $G = W/A$

W is the mass flow rate of the fluid (lb/hr)

A is the cross-sectional area of the packed bed (ft^2)

Example 5.26 Pressure Drop for Flow through a Packed Bed[54]

Estimate the pressure drop in a 60 ft length of 1 1/2 $in.$ (schedule 40, inside diameter= 1.61$in.$ (= 0.134ft)) pipe packed with catalyst pellets 1/4 $in.$ diameter when 104.4 lb/hr of gas is passing through the bed. The temperature is constant along the length of the pipe at 260°C. The void fraction is 45% and the properties of the gas are similar to those of air at this temperature (density=0.413 lb/ft^3, viscosity=0.0278 cP). The entering pressure is 10 atm.

Solution

The function *dpcatbed* calculates the pressure drop.

```
function dPt = dpcatbed(W,Bd,Bl,Pd,Pl,ep,mu,rho)
% Calculates the pressure drop for flow through a packed bed
gc = 4.17e8; A = pi*Bd^2/4; G = W/A; % superficial mass flow rate (lb/h/ft^2)
Ap = (pi*Pd^2/2 + pi*Pd*Pl)/144; Vp = pi*Pl*Pd^2/(4*1728);
S = Ap*(1-ep)/Vp; ePd = 6*(1-ep)/S; Nre = G*ePd/(2.419*mu*(1 - ep));
if Nre <1, fP = 150/Nre;
elseif Nre < 1e4, fP = 150/Nre + 1.75;
else, fP = 1.75; end
dPt  =  Bl*(1-ep)*G^2*(150*2.419*mu*(1-ep)/ePd/G  +  1.75)/(144*ep^3*ePd*
gc*rho);
end
```

The following commands produces desired outputs:

```
>> Bd = 0.134; Bl = 60; Pd = 0.25; Pl = 0.25; ep = 0.45; W = 104.4; mu = 0.0278; rho
= 0.413;
>> dPt = dpcatbed(W,Bd,Bl,Pd,Pl,ep,mu,rho)
```

```
dPt =
  68.6034
```

We can see that the total pressure drop is 68.6034 $lb/in.^2$.

PROBLEMS

5.1 Water flows inside a horizontal circular pipe at 25°C. Determine the average velocity and plot the velocity distribution inside the pipe. The length and the inside radius of the pipe are 12 and 0.01 m, respectively, the viscosity of water is $8.937 \times 10^{-4} kg/(m \cdot sec)$, and the pressure drop is $\Delta P = 520 Pa$.

5.2 A non-Newtonian fluid is flowing in a horizontal pipe where $L = 10m$, $r = 0.009m$, n (flow behavior index)= 0.46, $k = 0.012 N \cdot sec^n /m^2$, and $\Delta P = 100 Pa$. Calculate the average velocity and plot the velocity profile versus radius.

5.3 Water flows down an inclined flat plate of depth $L = 0.004m$ and unit width, as shown in Figure P5.3. For water, the density is 1000 kg/m^3 and the viscosity is $9.93 \times 10^{-4} Pa \cdot sec$. Plot v_x versus z when $\theta = 30°$.

5.4 Water flows inside a smooth horizontal circular pipe at 21°C. The inside diameter of the pipe is $0.1m$ and the velocity is $0.5 m/sec$. Calculate the pressure drop when the pipe length is $200m$. At 21°C, the density of water is $\rho = 997.92 kg/m^3$ and the viscosity is $\mu = 0.000982 kg/(m \cdot sec)$. Assume that the friction factor is given by $f = 0.0468 N_{Re}^{-0.2}$.

5.5 Determine the friction factors using various correlation equations (i.e., SC, CB, CBW, HA, CH, NK, and BS) at $N_{Re} = 2 \times 10^4$ and $N_{Re} = 3.2 \times 10^7$ for pipes where $\epsilon/D = 0.0001$.

5.6 Water flows inside a smooth pipeline at 25°C with a flow rate of 3 $liter/sec$. The length of the pipeline is $120m$ and the maximum possible pressure drop is $\Delta P = 100 kPa$. The density and viscosity of water at 25°C are 994.6 kg/m^3 and $\mu = 8.931 \times 10^{-4} kg/(m \cdot sec)$, respectively. The Fanning friction factor is given by $f = 16/N_{Re}$ for laminar flow and (Nikuradse equation) for turbulent flow. Determine the diameter of the pipeline.

5.7 Water is flowing in a horizontal 1 $in.$ schedule 40 smooth pipe at $5 m/sec$ and 25°C. The inside diameter and the length of the pipe are $2.66cm$ and 15 m, respectively. The density of water is $1000 kg/m^3$ and the viscosity is $0.001 kg/(m \cdot sec)$. The inlet pressure is 1 atm. Determine the pressure drop in the pipe.

5.8 Water is flowing through a horizontal 1 $in.$ schedule 40 steel pipe at $1 atm$ and 25°C. The inside diameter and the length of the pipe are $0.0266m$ and $20m$, respectively, and the roughness factor is $0.00005m$. The pressure drop in the pipe is equal to $118 kPa$. The density of water is $1000 kg/m^3$ and the viscosity is $0.001 kg/(m \cdot sec)$. Determine the inlet volumetric flow rate of water. The friction factor can be obtained from the Colebrook equation.

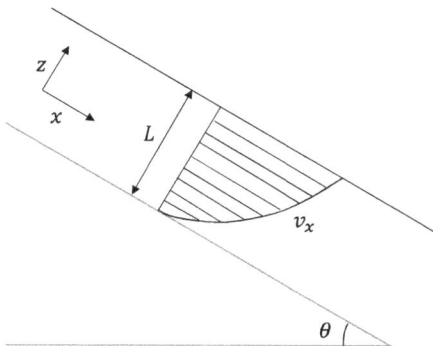

FIGURE P5.3 Flow down an inclined plate.

5.9 Figure P5.9 shows a pipeline that delivers water at constant temperature T from point 1, where the pressure is p_1 and the elevation is z_1, to point 2, where the pressure is p_2 and the elevation is z_2.[55] The density ρ (lbm/ft^3) and viscosity μ ($lbm/(ft \cdot sec)$) of water at T are given by

$$\rho = 62.122 + 0.0122T - 1.54 \times 10^{-4}T^2 + 2.65 \times 10^{-7}T^3 - 2.24 \times 10^{-10}T^4$$

$$\ln\mu = -11.0318 + \frac{1057.51}{T + 214.624}$$

At point 1, $p_1 = 150psig$ and $z_1 = 0ft$, and at point 2, $p_2 = 0psig$ and $z_2 = 300ft$. Assume that $T = 60°F$.

1. Calculate the flow rate q (gal/min) of water when the water is delivered through 8 *in.* schedule 40 pipe with $\epsilon = 0.00015ft$. The effective length of the pipe is $L = 5000ft$.
2. Plot the calculated flow rates as a function of pressure difference $\Delta p = p_2 - p_1$ when p_1 increases up to $200psig$ ($p_1 \le 200psig$). What is the minimal value of p_1 needed to start flow?

5.10 Natural gas is pumped through a horizontal 6 *in.* schedule 40 cast-iron pipe at $420kg/hr$ and $25°C$. The inside diameter and the length of the pipe are $0.154m$ and $24km$, respectively, and the roughness factor is $0.00026m$. The gas consists of 85 mol% CH_4 and 15 mol% CO_2. The density of the gas is $2.879kg/m^3$ and the viscosity is $1.2 \times 10^{-5}kg/(m \cdot sec)$. The inlet pressure is $3.45bar$. Determine the pressure drop in the pipe. The friction factor can be obtained from the Colebrook equation.

5.11 Water is being pumped from a feed tank at $20°C$ to a storage tank $15m$ high at a rate of $Q = 18m^3/hr$ through 4 *in.* schedule 40 steel pipes (inside diameter $= 0.1023m$), as shown in Figure P5.11.[56] The efficiency of the pump is $0.000046m$. The roughness factor is $0.000046m$, the density of water is $998kg/m^3$, and the viscosity is $0.001kg/(m \cdot sec)$. Determine the power needed for the pump to overcome the pressure loss in the pipeline. Let the contraction loss factor $K_c = 0.55$, the 90°elbow loss factor $K_e = 0.75$, and the expansion loss factor $K_{ex} = 1$. The Fanning friction factor can be estimated by the Colebrook equation.

5.12 Figure P5.12 shows a pipeline isometric of a 6 *in.* schedule 40 steel pipeline with six 90° LR (long radius) elbows and two flow-through tees.[57] The actual length of the pipe is $78ft$. Kerosene flows through the pipeline with a flow rate of $1026 gpm$ at $321°F$. The density of kerosene at $60°F$ is $51 lb/ft^3$, and the specific gravity at $60°F$ and $321°F$ is 0.82 and 0.72, respectively. The viscosity of kerosene at $321°F$ is $0.3 cP$. Determine the pressure drop between point A and point B. The friction factor can be calculated by the Chen equation.

5.13 A pipeline network with two flow loops is illustrated in Figure P5.13. The lengths and internal diameters of the pipes are shown in Table P5.13. Calculate the mass flow rate in each flow loop. A pipe roughness of $\epsilon = 5 \times 10^{-6}m$ is assumed uniformly for all pipes. The density and viscosity of the fluid are $991 kg/m^3$ and $1.2 \times 10^{-4}N \cdot sec/m^2$, respectively, and the inlet flow rate to the network is $\dot{m} = 50kg/sec$.

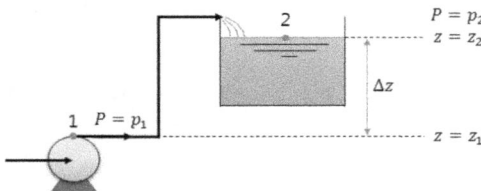

FIGURE P5.9 Water flow in a pipeline. (From Cutlip, M. B. and M. Shacham, *Problem Solving in Chemical and Biochemical Engineering with POLYMATH, Excel, and MATLAB*, 2nd ed., Prentice-Hall, Boston, MA, 2008, p. 307.)

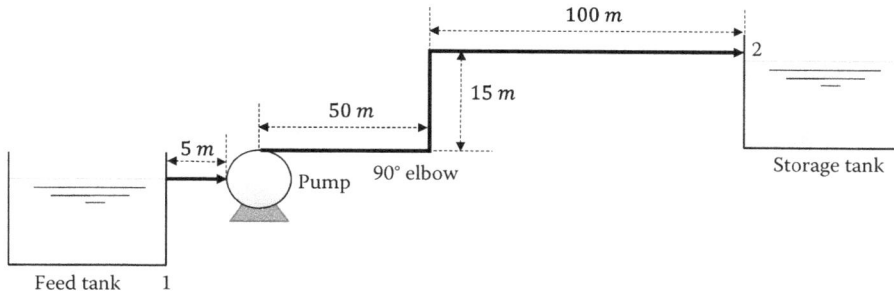

FIGURE P5.11 Pipeline flow with friction loss. (From Nayef Ghasem, *Computer Methods in Chemical Engineering*, CRC Press, Taylor & Francis Group, New York, NY, 2012, pp. 68–71.)

5.14 Calculate the Froude number and the flow condition for the two-phase flow in a 6 *in.* (schedule 40) vertical pipe. The flow rates of liquid and vapor are $W_L = 6930 lb/hr$ and $W_G = 1444 lb/hr$, respectively, and the densities of liquid and vapor are $\rho_L = 61.8 lb/ft^3$ and $\rho_G = 0.135 lb/ft^3$, respectively.

5.15 In an open tank flow system, $\rho = 1000 kg/m^3$, $A = 0.465 m^2$, $z_0 = 3.05 m$, $g = 9.8 m/sec^2$, $C_{v1} = C_{v2} = 0.012 m^2/(N^{1/2} \cdot sec)$, $P_{10} = P_1 = 1.38 \times 10^5 N/m^2$, $P_{30} = P_3 = 1.08 \times 10^5 N/m^2$, $P_0 = 1.014 \times 10^5 N/m^2$. Plot the dimensionless level z^*, the dimensionless flow rates F_1^* and F_2^*, and the dimensionless pressure P_2^* versus dimensionless time t^*.[58]

5.16 A liquid flows through an orifice (diameter: 6 *mm*) with flange taps in a 50 NB schedule 40 pipe. Estimate the maximum nonrecoverable pressure drop Δp_a using the given data.

Data: mass flow rate $G = 2500 kg/hr$, pipe inside diameter $d = 0.05248 m$, liquid density $\rho = 1000 kg/m^3$, liquid viscosity $\mu = 0.7 cP$, operating pressure $p_1 = 1100 kPa$(abs), temperature $T_1 = 20°C$, $Y = F_a = 1$.

5.17 Calculate the pressure drop for each 6 *in.* (schedule 40) condensate header when the flow rate = 10, 000 *lb/hr*, steam condensate pressure (P_c)= 114.7 *psia*, and header pressure(P_h)= 14.7 *psia*.

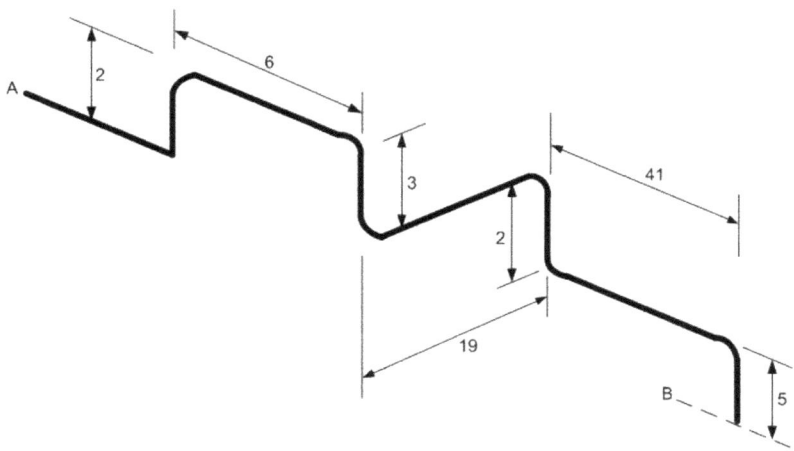

FIGURE P5.12 Kerosene flow in a pipeline. (From Coker, A. K., *Chemical Process Design, Analysis and Simulation*, Gulf Publishing Company, Houston, TX, 1995, p. 195.)

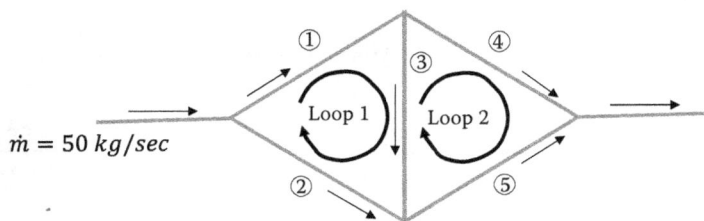

FIGURE P5.13 Two-loop network.

TABLE P5.13
Lengths and Internal Diameters of the Pipes

Flow(i)	Diameter(D_i, m)	Length(L_i, m)
1	0.051	17
2	0.061	17
3	0.040	20
4	0.061	19
5	0.051	19

REFERENCES

1. Geankoplis, C. J., *Transport Processes and Separation Process Principles*, 4th ed., Prentice-Hall, Boston, MA, p. 85, 2003.
2. Cutlip, M. B. and M. Shacham, Problem Solving in Chemical and Biochemical Engineering with POLYMATH, Excel, and MATLAB, 2nd ed., Prentice-Hall, Boston, MA, p. 294, 2008.
3. Cutlip, M. B. and M. Shacham, Problem Solving in Chemical and Biochemical Engineering with POLYMATH, Excel, and MATLAB, 2nd ed., Prentice-Hall, Boston, MA, p. 295, 2008.
4. Bird, R. B., W. E. Stewart, and E. N. Lighfoot, *Transport Phenomena*, 2nd ed., John Wiley & Sons, Hoboken, NJ, p. 44, 2002.
5. de Nevers, N., *Fluid Mechanics for Chemical Engineers*, 3rd ed., McGraw-Hill, New York, NY, p. 211, 2005.
6. Shacham, M., A review of non-iterative friction factor correlations for the calculation of pressure drop in pipes, *Industrial & Engineering Chemistry Fundamentals*, 19, pp. 228–229, 1980.
7. Colebrook, C. F., Turbulent flow in pipes with particular references to the transition region between the smooth and rough pipe laws, *Journal of the Institution of Civil Engineers (London)*, 11, p. 133, 1938–1939.
8. Colebrook, C. F. and C. M. White, Characteristics of flow in the transition region between the smooth and rough pipe, *Journal of the Institution of Civil Engineers (London)*, 10, pp. 99–108, 1938–1939.
9. Haaland, S. E., Simple and explicit formulas for the friction factor in turbulent flow, *Transactions of the ASME, Journal of Fluids Engineering*, 105, p. 89, 1983.
10. Coker, A. K., *Chemical Process Design, Analysis and Simulation*, Gulf Publishing Company, Houston, TX, p. 158, 1995.
11. de Nevers, N., *Fluid Mechanics for Chemical Engineers*, 3rd ed., McGraw-Hill, New York, NY, p. 187, 2005.
12. Churchill, S. W., Friction factor equations spans all fluid-flow regimes, *Chemical Engineering*, 84(24), p. 91, 1977.
13. Nikuradse, J., New development in pipe flow optimization modeling, *VDI-Forschungsheft*, 65(4), p. 356, 1932.
14. Kapuno, R. R. A., *Programming for Chemical Engineers*, Infinity Science Press, Hingham, MA, p. 105, 2008.

15. Cutlip, M. B. and M. Shacham, Problem Solving in Chemical and Biochemical Engineering with POLYMATH, Excel, and MATLAB, 2nd ed., Prentice-Hall, Boston, MA, p. 110, 2008.
16. Mariano M. Martin (Editor), *Introduction to software for chemical engineers*, CRC Press, Taylor & Francis Group, Boca Raton, FL, pp. 104–107, 2015.
17. Mariano M. Martin (editor), *Introduction to Software for Chemical Engineers*, CRC Press, Taylor & Francis Group, New York, NY, pp. 39–43, 2015.
18. Hooper, W. B., The two-K method predicts head loss in pipe fittings, *Chemical Engineering*, August 24, pp. 96–100, 1981.
19. Coker, A. K., *Chemical Process Design, Analysis and Simulation*, Gulf Publishing Company, Houston, TX, p. 155, 1995.
20. Coker, A. K., *Chemical Process Design, Analysis and Simulation*, Gulf Publishing Company, Houston, TX, p. 157, 1995.
21. Bird, R. B., W. E. Stewart, and E. N. Lightfoot, *Transport Phenomena*, 2nd ed., John Wiley & Sons, Inc., Hoboken, NJ, p. 207, 2002.
22. Bird, R. B., W. E. Stewart and E. N. Lightfoot, *Transport Phenomena*, 2nd ed., John Wiley & Sons, Inc., Hoboken, NJ, pp. 207–208, 2002.
23. Coker, A. K., *Chemical Process Design, Analysis and Simulation*, Gulf Publishing Company, Houston, TX, pp. 194–195, 1995.
24. Cutlip, M. B. and M. Shacham, Problem Solving in Chemical and Biochemical Engineering with POLYMATH, Excel, and MATLAB, 2nd ed., Prentice-Hall, Boston, MA, p. 309, 2008.
25. Lang, F. D. and B. L. Miller, Pipe network analysis based on the generalized friction factor correlation of Churchill, *Chemical Engineering*, 88(13), p. 95, 1981.
26. Raman, R., *Chemical Process Computations*, Elsevier Applied Science Publishers, Barking, Essex, UK, pp. 231–232, 1985.
27. Raman, R., *Chemical Process Computations*, Elsevier Applied Science Publishers, Barking, Essex, UK, pp. 234–235, 1985.
28. Ramirez, W.F., *Computational Methods for Process Simulation*, 2nd ed., Butterworth Heinemann, Oxford, UK, p. 145., 1997.
29. Ramirez, W. F., *Computational Methods for Process Simulation*, 2nd ed., Butterworth Heinemann, Oxford, UK, pp. 155–157, 1997.
30. Arun Datta, *Process engineering and design using visual BASIC*, CRC Press, Taylor & Francis Group, New York, NY, pp. 120–127, 2008.
31. Arun Datta, *Process engineering and design using visual BASIC*, CRC Press, Taylor & Francis Group, New York, NY, p. 174, 2008.
32. Raman, R., *Chemical Process Computations*, Elsevier Applied Science Publishers, Barking, Essex, UK, p. 226, 1985.
33. Raman, R., *Chemical Process Computations*, Elsevier Applied Science Publishers, Barking, Essex, UK, p. 225, 1985.
34. Tomita, Y., A study of non-Newtonian flow in pipe lines, *Bulletin of the Japan Society of Mechanical Engineers*, 2, p. 10, 1959.
35. Dodge, D. W. and A. B. Metzner, Flow non-Newtonian fluids correlation of laminar, transition and turbulent flow regions, *AIChE Journal*, 5, p. 189, 1959.
36. Raman, R., *Chemical Process Computations*, Elsevier Applied Science Publishers, Barking, Essex, UK, p. 227, 1985.
37. Raman, R., *Chemical Process Computations*, Elsevier Applied Science Publishers, Barking, Essex, UK, pp. 229–230, 1985.
38. Coker, A. K., *Chemical Process Design, Analysis and Simulation*, Gulf Publishing Company, Houston, TX, p. 163, 1995.
39. Coker, A. K., *Chemical Process Design, Analysis and Simulation*, Gulf Publishing Company, Houston, TX, p. 164, 1995.
40. Coker, A. K., *Chemical Process Design, Analysis and Simulation*, Gulf Publishing Company, Houston, TX, p. 166, 1995.
41. Coker, A. K., *Chemical Process Design, Analysis and Simulation*, Gulf Publishing Company, Houston, TX, p. 167, 1995.
42. Coker, A. K., *Chemical Process Design, Analysis and Simulation*, Gulf Publishing Company, Houston, TX, p. 198, 1995.
43. Arun Datta, *Process engineering and design using visual BASIC*, CRC Press, Taylor & Francis Group, New York, NY, pp. 150–151, 2008.

44. Arun Datta, *Process engineering and design using visual BASIC*, CRC Press, Taylor & Francis Group, New York, NY, pp. 151–152, 2008.

45. Arun Datta, *Process engineering and design using visual BASIC*, CRC Press, Taylor & Francis Group, New York, NY, p. 175, 2008.

46. Coker, A. K., *Chemical Process Design, Analysis and Simulation*, Gulf Publishing Company, Houston, TX, p. 172, 1995.

47. Coker, A. K., *Chemical Process Design, Analysis and Simulation*, Gulf Publishing Company, Houston, TX, p. 175, 1995.

48. Coker, A. K., *Chemical Process Design, Analysis and Simulation*, Gulf Publishing Company, Houston, TX, p. 174, 1995.

49. Coker, A. K., *Chemical Process Design, Analysis and Simulation*, Gulf Publishing Company, Houston, TX, pp. 177–178, 1995.

50. Coulson, J. M. and J. F. Richardson, *Chemical Engineering*, Vol. 1, 3rd ed., Pergamon Press, Oxford, UK, pp. 91–92, 1978.

51. Coker, A. K., *Chemical Process Design, Analysis and Simulation*, Gulf Publishing Company, Houston, TX, pp. 200–201, 1995.

52. Coker, A. K., *Chemical Process Design, Analysis and Simulation*, Gulf Publishing Company, Houston, TX, p. 182, 1995.

53. Ergun, S., Fluid flow through packed columns, *Chemical Engineering Progress*, 48(2), pp. 89–92, 1952.

54. Fogler, H. S., *Essentials of Chemical Reaction Engineering*, Pearson Education International, Boston, MA, pp. 175–176, 2011.

55. Cutlip, M. B. and M. Shacham, Problem Solving in Chemical and Biochemical Engineering with POLYMATH, Excel, and MATLAB, 2nd ed., Prentice-Hall, Boston, MA, p. 307, 2008.

56. Nayef Ghasem, *Computer methods in chemical engineering*, CRC Press, Taylor & Francis Group, New York, NY, pp. 68–71, 2012.

57. Coker, A. K., *Chemical Process Design, Analysis and Simulation*, Gulf Publishing Company, Houston, TX, p. 195, 1995.

58. Ramirez, W. F., *Computational Methods for Process Simulation*, 2nd ed., Butterworth Heinemann, Oxford, UK, pp. 150–151, 1997.

6 Chemical Reaction Engineering

6.1 CHARACTERISTICS OF REACTION RATES

6.1.1 ESTIMATION OF REACTION RATE CONSTANT AND REACTION ORDER

The Arrhenius equation for the reaction rate constant is given by

$$k_A = Ae^{-E/(RT)} \text{ or } \ln k_A = \ln A - \frac{E}{R}\left(\frac{1}{T}\right)$$

where
   A is the frequency factor
   E is the activation energy (J/mol)
   R is the gas constant (= $8.314 J/(mol \cdot K)$)
   T is the absolute temperature (K)
A and E can be determined by using experimental data for reaction rates at different temperatures.
   Consider a reaction whose rate is given by

$$\frac{dC_A}{dt} = -k'C_A^{\alpha}$$

Integration of this equation yields

$$t = \frac{1}{k'(1-\alpha)}[C_{A0}^{1-\alpha} - C_A^{1-\alpha}]$$

Using concentration (C_A)-time data obtained in batch experiments, k' and α can be calculated from the solution of the corresponding nonlinear equation system.

Example 6.1 Estimation of the Activation Energy[1]
   Estimate the activation energy for the decomposition of benzene diazonium chloride to produce chlorobenzene and nitrogen using the information given in Table 6.1 for this 1st-order reaction.

Solution
   We start by recalling the relation

$$\ln k = \ln A - \frac{E}{R}\left(\frac{1}{T}\right)$$

The plot of $\ln k$ versus $1/T$ yields a straight line. Thus, A and E can be obtained from the equation of the straight line. Parameters of a 1st-order line can be calculated by the built-in function *polyfit*, which performs regression analysis. The script *rateconst* generates a plot of experimental data versus temperature and the resulting straight line.

```
% rateconst.m
T = [313 319 323 328 333]; k = 1e-3*[0.43 1.03 1.80 3.55 7.17]; % data
x = 1./T; y = log(k); c = polyfit(x,y,1); A = exp(c(2)), E = -c(1)*8.314
xv = linspace(min(x),max(x),100); yv = polyval(c,xv); plot(xv,yv,x,y,'o'),
xlabel('1/T(K^-1)'), ylabel('ln(k)')
```

TABLE 6.1

Reaction Rate Constant for the Decomposition Reaction of Benzene Diazonium Chloride

$T(K)$	313.0	319.0	323.0	328.0	333.0
$k(sec^{-1})$	0.00043	0.00103	0.00180	0.00355	0.00717

Execution of the script *rateconst* gives A and E and yields the plot of the line and the data points shown in Figure 6.1:

```
>> rateconst
A =
  8.0303e+16
E =
  1.2148e+05
```

We can see that $A = 8.0303 \times 10^{16}\ sec^{-1}$ and $E = 1.2148 \times 10^5\ J/mol = 121.48\ kJ/mol$.

Example 6.2 Estimation of Rate Constant and Reaction Order[2]

Concentration (C_A)-time data were obtained in batch experiments for the liquid-phase reaction $A + B \rightarrow C$ as shown in Table 6.2. For this reaction, the rate can be represented as $-r_A = k' C_A^\alpha$. The concentration of B is assumed to be constant. Estimate the reaction order α and the rate constant k'.

Solution

At the point (t_i, C_{Ai}), the nonlinear equation

$$f(k', \alpha) = t_i k' (1 - \alpha) - [C_{A0}^{1-\alpha} - C_{Ai}^{1-\alpha}] = 0$$

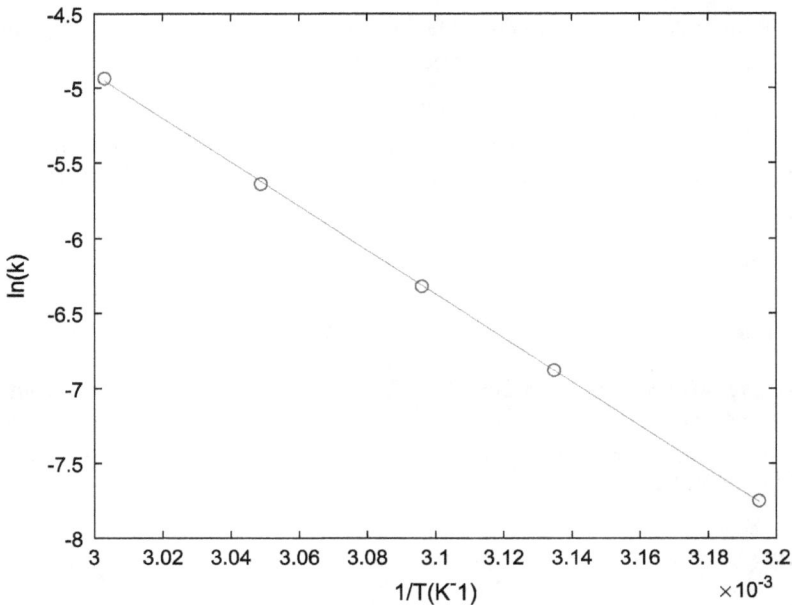

FIGURE 6.1   Reaction rate constant versus temperature.

TABLE 6.2

Concentration-Time Data for a Batch Reaction

$t \, (min)$	0	50	100	150	200	250	300
$C_A \, (mol/dm^3)$	0.05	0.038	0.0306	0.0256	0.0222	0.0195	0.0174

is to be solved. If the number of data points is n, then we have to solve the nonlinear equation system consisting of n equations. The nonlinear equation system is defined by the subfunction *frx* in the script *regorder*. The built-in function *fsolve* is used to solve the nonlinear equation system.

```
% regorder.m
t = [0 50 100 150 200 250 300]; Ca = [0.05 0.038 0.0306 0.0256 0.0222 0.0195
0.0174]; % data
Ca0 = 0.05; x = fsolve(@frx,[0.2 2],[],t,Ca0,Ca) % initial value: k=0.2, alpha=2

function f = frx(x,t,Ca0,Ca)
% x(1)=k', x(2)=alpha
n = length(t);
for i = 1:n, f(i,1) = x(1)*(1-x(2))*t(i) - Ca0^(1-x(2)) + Ca(i)^(1-x(2)); end
end

>> regorder
x =
  0.1449   2.0412
```

We can see that $k' = 0.1449 \, dm^3/(mol \cdot min)$ and $\alpha = 2.0412$.

Example 6.3 Estimation of Reaction Order

A pure substance A is decomposed into B according to the elementary reaction $A \rightarrow B$. The reaction rate must be found by using the experimental data shown in Table 6.3. Determine the most probable order (0, 1, or 2) of the reaction and the corresponding rate constant.

Solution

The reaction rate is expressed as $- r_A = -dC_A/dt = kC_A{}^n$, and we have to determine k and n.

$n = 0$ (0th-order reaction): $dC_A/dt = -k \Rightarrow C_A - C_{A0} = -kt \Rightarrow C_A = -kt + C_{A0}$

$n = 1$ (1st-order reaction): $dC_A/dt = -kC_A \Rightarrow \ln C_A - \ln C_{A0} = -kt \Rightarrow \ln C_A = -kt + \ln C_{A0}$

$n = 2$ (2nd-order reaction): $dC_A/dt = -kC_A{}^2 \Rightarrow 1/C_{A0} - 1/C_A = -kt \Rightarrow 1/C_A = kt + 1/C_{A0}$

The relationship between t and C_A is of the form $y = at + b$, where $y = C_A$, $a = -k$, $b = C_{A0}$ for $n = 0$; $y = \ln C_A$, $a = -k$, $b = \ln C_{A0}$ for $n = 1$; and $y = \frac{1}{C_A}$, $a = k$, $b = \frac{1}{C_{A0}}$ for $n = 2$. a and b can be calculated by using the pseudo-inverse relation as follows:

$$X = (M^T M)^{-1} M^T Y$$

TABLE 6.3

Change of Concentration of Reactant (C_A) with Time (t)

$t \, (sec)$	0	2	4	7	9	18
$C_A \, (mol/liter)$	1.48	1.01	0.67	0.58	0.51	0.32

where

$$M = \begin{bmatrix} t_1 & 1 \\ t_2 & 1 \\ \vdots & \vdots \\ t_n & 1 \end{bmatrix}, \quad X = \begin{bmatrix} a \\ b \end{bmatrix}, \quad Y = \begin{bmatrix} y_1 \\ y_2 \\ \vdots \\ y_n \end{bmatrix}$$

The script *rxnrate* computes the vector X for each reaction order and generates plots of C_A versus t obtained from each reaction rate.

```
% rxnrate.m: determination of reaction order
t = [0 2 4 7 9 18]; Ca = [1.48 1.01 0.67 0.58 0.51 0.32]; % data
M = [t' ones(length(t),1)]; Mi = inv(M'*M)*M';
Y0 = Ca'; Y1 = (log(Ca))'; Y2 = (1./Ca)'; X0 = Mi*Y0; X1 = Mi*Y1; X2 = Mi*Y2;
k0 = -X0(1); Ca0 = X0(2); k1 = -X1(1); Ca1 = exp(X1(2)); k2 = X2(1); Ca2 = 1/X2(2);
fprintf('0th order: k = %g, Ca0 = %g\n1st order: k = %g, Ca0 = %g
\n',k0,Ca0,k1,Ca1)
fprintf('2nd order: k = %g, Ca0 = %g\n',k2,Ca2)
% plot of each reaction rate and data
tv = 0:0.1:20; % time interval for plotting
Caz = -k0*tv + Ca0; Caf = exp(-k1*tv + log(Ca1)); Cas = 1./(k2*tv + 1./Ca2);
plot(tv,Caz,':',tv,Caf,'--',tv,Cas,t,Ca,'o'),   xlabel('t(sec)'),   ylabel
('C_A(mol/l)')
legend('0th order','1st order','2nd order','Data','location','best')

>> rxnrate
0th order: k = 0.0547749, Ca0 = 1.12683
1st order: k = 0.0784168, Ca0 = 1.13896
2nd order: k = 0.132451, Ca0 = 1.28474
```

Figure 6.2 shows the resultant plot produced by the script *rxnrate*. We can see that the data are fitted best by the 2nd–order reaction rate given by

$$\frac{dC_A}{dt} = -0.13245 C_A{}^2$$

6.1.2 Reaction Equilibrium

The constant representing chemical equilibrium is defined in terms of the activities of the species. The equilibrium constant for the reaction $aA + bB \leftrightarrow cC + dD$ is expressed as

$$K = \frac{a_C^c a_D^d}{a_A^a a_B^b}$$

Since the activity is the fugacity divided by the fugacity of the standard state, it is the fugacity in the gas-phase reaction. When the pressure is not high (usually less than 10 *atm*), the gas can be considered an ideal gas and the equilibrium constant can be represented in terms of the mole fractions (y_i) of the species:

$$K = \frac{y_C^c y_D^d}{y_A^a y_B^b} P^{c+d-a-b}$$

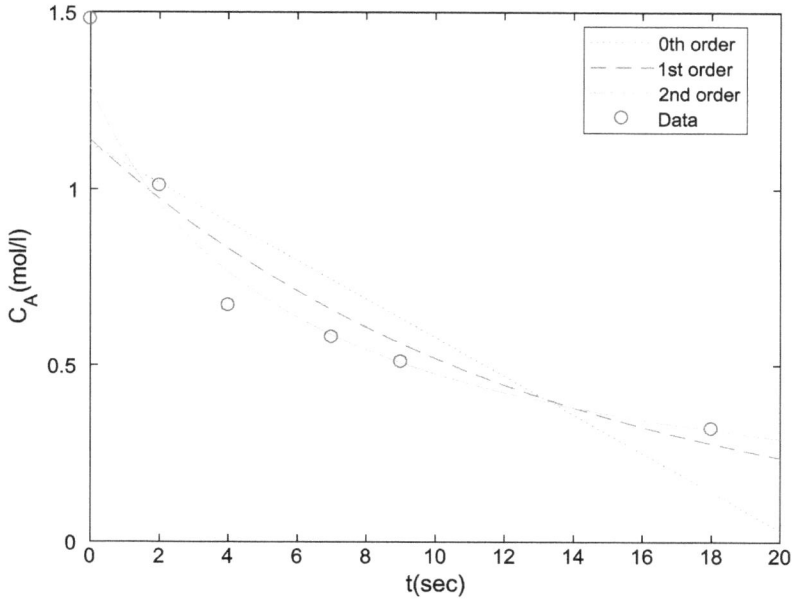

FIGURE 6.2 Comparison of reaction orders.

Example 6.4 Equilibrium of Water-Gas Shift Reaction[3]

Consider the water-gas shift reaction to make hydrogen for fuel cell applications:

$$CO + H_2O \leftrightarrow CO_2 + H_2$$

The equilibrium constant has been found to be $K = 148.4$ at $500\ K$. The reaction feed consists of 1 *mol* of CO and 1 *mol* of H_2O. Determine the compositions at equilibrium.

Solution

$$K = 148.4 = \frac{y_{CO_2} y_{H_2}}{y_{CO} y_{H_2O}}$$

Let x be the number of moles consumed in the reaction. Then we have

$$K = 148.4 = \frac{(x/2)(x/2)}{[(1-x)/2][(1-x)/2]} = \frac{x^2}{(1-x)^2}$$

x is obtained from the solution of the nonlinear equation

$$f(x) = 148.4 - \frac{x^2}{(1-x)^2} = 0$$

```
>> f = @(x) 148.4-x^2/(1-x)^2; x0 = 0.5; x = fzero(f,x0)  % initial guess = 0.5.
Uses the built-in solver fzero.
x =
   0.9241
```

6.1.3 REACTION CONVERSION

For batch reaction systems, the relation between the reaction conversion and the concentration is given by

$$C_A = \frac{N_A}{V} = \frac{N_{A0}(1 - X)}{V}$$

For a constant reactor volume ($V = V_0$),

$$C_A = \frac{N_{A0}(1 - X)}{V_0} = C_{A0}(1 - X)$$

For flow reaction systems, the concentration is represented as

$$C_A = \frac{F_A}{v} = \frac{F_{A0}(1 - X)}{v}$$

$v = v_0$ for a liquid-phase reaction, and we have

$$C_A = \frac{F_{A0}(1 - X)}{v_0} = C_{A0}(1 - X)$$

For a gas-phase reaction, v can be represented as

$$v = v_0(1 + \epsilon X)\frac{P_0}{P}\frac{T}{T_0}$$

and we have

$$C_A = \frac{F_{A0}(1 - X)}{v} = \frac{F_{A0}(1 - X)}{v_0(1 + \epsilon X)}\frac{P}{P_0}\frac{T_0}{T} = C_{A0}\left(\frac{1 - X}{1 + \epsilon X}\right)\frac{P}{P_0}\frac{T_0}{T}$$

The order of the reaction and corresponding reaction rate constant can be estimated using measured data for concentration and conversion. For a decomposition reaction of reactant A carried out in a continuous-stirred tank reactor (CSTR), the relation between the flow rate of the reactant, v_0, and the conversion of A, X_A, may be represented as

$$v_0 = \frac{kVC_{A0}^{n-1}(1 + X_A)^n}{X_A(1 + \epsilon X_e)^n}$$

where
   k is the reaction rate constant
   C_{A0} is the initial concentration of A
   V is the reactor volume
   This equation can be rewritten as

$$\ln\left(\frac{v_0 C_{A0} X_A}{V}\right) = \ln(k) + n\ln\left(\frac{C_{A0}(1 + X_A)}{1 + \epsilon X_A}\right)$$

This is a typical linear relation of the form $z = \beta x + \gamma y$.

Example 6.5 Equilibrium Conversion[4]

The reversible gas-phase decomposition of nitrogen tetroxide (N_2O_4) to nitrogen dioxide, $N_2O_4 \leftrightarrow 2NO_2$, is to be carried out at a constant temperature of 340 K. The feed consists of pure N_2O_4 at 202.6 kPa (2 atm). The concentration equilibrium constant, K_C, at 340 K is 0.1 mol/dm^3, and the rate constant is $k_{N_2O_4} = 0.5\ min^{-1}$. Calculate the equilibrium conversion of N_2O_4 in a flow reactor. The equilibrium constant is given by

$$K_C = \frac{C_{Be}^2}{C_{Ae}}$$

Solution

For flow system,

$$K_C = \frac{C_{Be}^2}{C_{Ae}} = \frac{[2C_{A0}X_e/(1 + \epsilon X_e)]^2}{C_{A0}(1 - X_e)/(1 + \epsilon X_e)} = \frac{4C_{A0}X_2^2}{(1 - X_e)/(1 + \epsilon X_e)}$$

Rearrangement of this equation yields a nonlinear equation

$$f(X_e) = 4C_{A0}X_e^2 - K_C(1 - X_e)(1 + \epsilon X_e) = 0$$

where $C_{A0} = y_{A0}P_0/(RT_0)$, $y_{A0} = 1$, $P_0 = 2\ atm$, $R = 0.082\ atm \cdot dm^3/(mol \cdot K)$, $T_0 = 340\ K$, and $\epsilon = y_{A0}$
$\delta = 1(2 - 1) = 1$. The equilibrium conversion can be determined by using the built-in function fzero:

```
>> P = 2; T = 340; R = 0.082; Kc = 0.1; ya0 = 1; epsilon = 1; Ca0 = ya0*P/(R*T); x0
= 0.5;
>> fF = @(x) 4*Ca0*x^2 - Kc*(1-x)*(1+epsilon*x); Xe = fzero(fF,x0)
Xe =
   0.5084
```

6.1.4 SERIES REACTIONS

Consider the reaction sequence in which species B is the desired product:

$$A \xrightarrow{k_1} B \xrightarrow{k_2} C$$

Reaction rates can be represented as

$$\frac{dC_A}{dt} = -k_1 C_A, \quad \frac{dC_B}{dt} = k_1 C_A - k_2 C_B, \quad \frac{dC_C}{dt} = k_2 C_C$$

A is fed into a batch reactor, and the initial concentration of A is C_{A0}. Part of A is converted to the desired product B. If the reaction is allowed to proceed for a long time in the batch reactor, part of B will be converted to the undesired product C. Integration of the rate equations with the initial conditions $C_A = C_{A0}$, $C_B = C_C = 0$ at $t = 0$ gives the concentration of each species as

$$C_A = C_{A0}e^{-k_1 t}, \quad C_B = C_{A0}\frac{k_1}{k_2 - k_1}\left(e^{-k_1 t} - e^{-k_2 t}\right), \quad C_C = C_{A0} - C_A - C_B$$

For series reactions carried out in a batch or a plug-flow reactor, the maximum possible concentration of B is given by

$$C_{Bmax} = C_{A0}\left(\frac{k_1}{k_2}\right)^{k_2/(k_2 - k_1)}$$

C_{Bmax} is reached at

$$t_{max} = \frac{\ln(k_2/k_1)}{k_2 - k_1}$$

The reactor volume is given by

$$V = vt$$

where v is the volumetric flow rate.

Example 6.6 Series Reactions

An elementary liquid-phase series reaction is carried out in a batch reactor:

$$A \xrightarrow{k_1} B \xrightarrow{k_2} C$$

In this reaction, A is decomposed to the desired product B and the flow rate of the feed containing A is 25 *liter/min*. Determine the maximum concentration of B and the time when the concentration of B reaches the maximum value. The initial concentrations of A, B, and C are $C_{A0} = 2.5$ *mol/liter*, $C_B = C_C = 0$. The reaction rate constants are $k_1 = 0.2\ min^{-1}$, and $k_2 = 0.1\ min^{-1}$.

Solution

The script *serrb* integrates differential equations using the built-in function *ode45* and plots concentrations versus time.

```
% serrb.m
v = 25; k1 = 0.2; k2 = 0.1; Ca0 = 2.5; Cb0 = 0; Cc0 = 0; tspan = 0:0.01:15; C0 = [Ca0
Cb0 Cc0];
dCdt = @(t,C) [-k1*C(1); k1*C(1)-k2*C(2); k2*C(2)]; [t,C]=ode45(dCdt,
tspan, C0);
plot(t,C(:,1),'-',t,C(:,2),':',t,C(:,3),'--');
xlabel('Time(min)'), ylabel('Concentration(mol/liter)'), legend('A','B','C');
Cmax = max(C(:,2)), tmax = t(find(C(:,2) == Cmax)), Vol = v*tmax
```

The script *serrb* produces the following results and generates the concentration profiles shown in Figure 6.3:

```
>> serrb
Cmax =
  1.2500
```

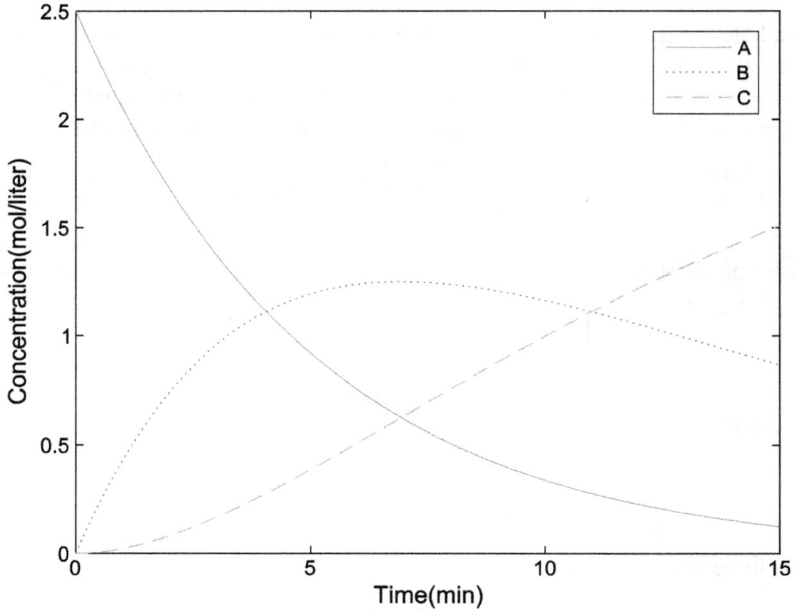

FIGURE 6.3   Concentration profiles in series reactions.

```
tmax =
  6.9300
Vol =
  173.2500
```

6.2 CONTINUOUS-STIRRED TANK REACTORS (CSTRS)

6.2.1 CONCENTRATION CHANGES WITH TIME

Suppose that the elementary irreversible liquid-phase reaction

$$A + B \rightarrow C, \ -r_A = kC_A C_B$$

is carried out in a series of three CSTRs as shown in Figure 6.4. The volume of each reactor is $V_i (i = 1, 2, 3)$, the volumetric flow rate of each stream is $v_i (i = 1, 2, 3)(dm^3/min)$, and the concentration of species i in the reactor j is $C_{ij} (i = A, B; j = 1, 2, 3)(gmol/dm^3)$.

For a liquid-phase reaction, the volume change with reaction may be neglected. The material balances on each reactor yield the following ordinary differential equations:

$$\frac{dC_{A1}}{dt} = \frac{1}{V_1}(v_{0A} C_{A0} - v_1 C_{A1} - kV_1 C_{A1} C_{B1}), \ \frac{dC_{B1}}{dt} = \frac{1}{V_1}(v_{0B} C_{B0} - v_1 C_{B1} - kV_1 C_{A1} C_{B1})$$

$$\frac{dC_{A2}}{dt} = \frac{1}{V_2}(v_1 C_{A1} - v_2 C_{A2} - kV_2 C_{A2} C_{B2}), \ \frac{dC_{B2}}{dt} = \frac{1}{V_2}(v_1 C_{B1} - v_2 C_{B2} - kV_2 C_{A2} C_{B2})$$

$$\frac{dC_{A3}}{dt} = \frac{1}{V_3}(v_2 C_{A2} - v_3 C_{A3} - kV_3 C_{A3} C_{B3}), \ \frac{dC_{B3}}{dt} = \frac{1}{V_3}(v_2 C_{B2} - v_3 C_{B3} - kV_3 C_{A3} C_{B3})$$

Example 6.7 CSTRs in Series[5]

The elementary irreversible liquid-phase reaction

$$A + B \rightarrow C, \ r_A = kC_A C_B$$

is carried out in a series of three identical CSTRs. The initial concentrations of A and B are $C_{A0} = C_{B0} = 2 \ gmol/dm^3$, the inlet flow rates of A and B are $v_{0A} = v_{0B} = 6 \ dm^3/min$, the reaction rate constant is $k = 0.5$, the volume of each reactor is $V_1 = V_2 = V_3 = 200 \ dm^3$, and the flow rate of each stream is $v_1 = v_2 = v_3 = 12 \ dm^3/min$. Plot the concentration of A exiting each reactor, $C_{Ai} (i = 1, 2, 3)$, during start-up to the final time $t = 20(0 \le t \le 20)$.

Solution

The initial concentrations of A and B are C_{A0} and C_{B0}, respectively, in the first reactor only, and are zero in the remaining reactors. In the system of differential equations defined in the script

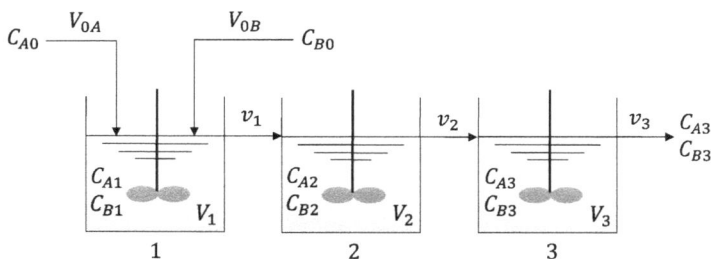

FIGURE 6.4　CSTRs in series.

ser3v, $C(1) = C_{A1}$, $C(3) = C_{A2}$, $C(5) = C_{A3}$, $C(2) = C_{B1}$, $C(4) = C_{B2}$, and $C(6) = C_{B3}$. The script *ser3v* uses the built-in function ode45 to calculate concentrations.

```
% ser3v.m
v0 = 6; v = 12; V = 200; Ca0 = 2; Cb0 = 2; k = 0.5; tspan = 0:0.01:20; C0 = [Ca0 Cb0 0 0
0 0];
dCdt = @(t,C) [(v0*Ca0-v*C(1)-k*V*C(1)*C(2))/V; (v0*Cb0-v*C(2)-k*V*C(1)*
C(2))/V;
           (v*C(1)-v*C(3)-k*V*C(3)*C(4))/V; (v*C(2)-v*C(4)-k*V*C(3)*C(4))/V;
           (v*C(3)-v*C(5)-k*V*C(5)*C(6))/V; (v*C(4)-v*C(6)-k*V*C(5)*C(6))/V];
[t,C]=ode45(dCdt, tspan, C0); plot(t,C(:,1),'-',t,C(:,3),':',t,C(:,5),'--');
xlabel('Time(min)'),    ylabel('Concentration(gmol/dm^3)'),    legend('C_
{A1}','C_{A2}','C_{A3}');
```

The script *ser3v* produces the concentration profiles shown in Figure 6.5.

Example 6.8 pH Neutralization Reaction[6]

Figure 6.6 shows a CSTR where an acidic solution is neutralized with an alkaline solution. The neutralization reaction takes place between a strong acid (HA) and a strong base (BOH) in the presence of a buffer agent (BX).

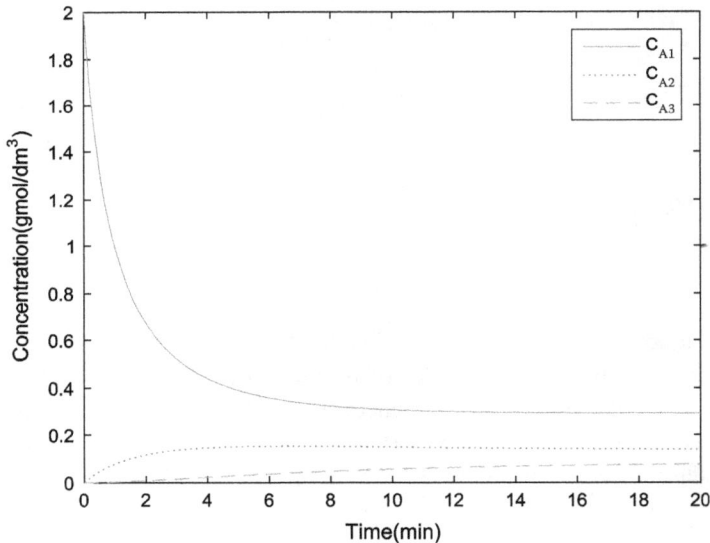

FIGURE 6.5 Concentration profiles in the train of three CSTRs.

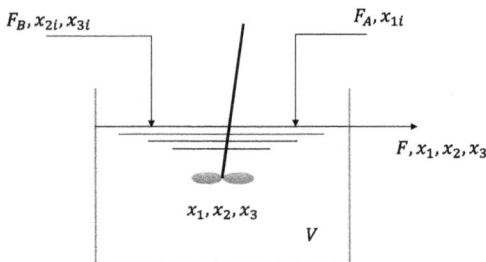

FIGURE 6.6 A pH neutralization reactor.

The following reactions occur in the reactor:

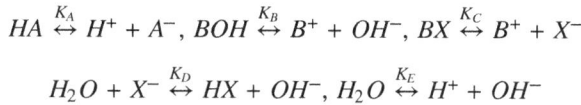

$$HA \overset{K_A}{\leftrightarrow} H^+ + A^-, \ BOH \overset{K_B}{\leftrightarrow} B^+ + OH^-, \ BX \overset{K_C}{\leftrightarrow} B^+ + X^-$$

$$H_2O + X^- \overset{K_D}{\leftrightarrow} HX + OH^-, \ H_2O \overset{K_E}{\leftrightarrow} H^+ + OH^-$$

The equilibrium constants are given by

$$K_A = \frac{[H^+][A^-]}{[HA]}, \ K_B = \frac{[B^+][OH^-]}{[BOH]}, \ K_C = \frac{[B^+][X^-]}{[BX]}, \ K_D = \frac{[HX][OH^-]}{[X^-]}, \ K_E = [H^+][OH^-]$$

The base and buffer agents are both highly soluble, and $[HA] \approx 0$, $[BOH] \approx 0$, and $[BX] \approx 0$. Thus we can assume that $K_A \to \infty$, $K_B \to \infty$, and $K_C \to \infty$. The invariant species are given by

$$x_1 = [HA] + [A^-] \approx [A^-], \ x_2 = [BOH] + [BX] + [B^+] \approx [B^+]$$

$$x_3 = [BX] + [HX] + [X^-] \approx [HX] + [X^-]$$

From the material and charge balances, we have

$$\frac{dx_1}{dt} = \frac{F_A}{V}(x_{1i} - x_1) - \frac{F_B}{V}x_1, \ \frac{dx_2}{dt} = \frac{F_B}{V}(x_{2i} - x_2) - \frac{F_A}{V}x_2, \ \frac{dx_3}{dt} = \frac{F_B}{V}(x_{3i} - x_3) - \frac{F_A}{V}x_3$$

$$[H^+] + x_2 + x_3 - x_1 - \frac{K_E}{[H^+]} - \frac{x_3}{1 + \frac{K_D[H^+]}{K_E}} = 0, \ pH = -\log_{10}[H^+]$$

Using the given data, generate plots displaying the concentration of each species and pH as a function of time.

Data: $x_{1i} = 0.0012 \ mol/liter \ (HCl)$, $x_{2i} = 0.002 \ mol/liter \ (NaOH)$, $x_{3i} = 0.0025 \ mol/liter$ $(NaHCO_3)$, $K_D = 10^{-7} \ mol/liter$, $K_E = 10^{-14} \ mol^2/liter^2$, $F_A = 0.01667 \ liter/sec$, $F_B = 0.002333$ $liter/sec$, $V = 2.5 \ liter$.

Solution

The script *pHcstr* solves the set of ODEs by using the built-in function *ode45* and produces the plots displaying concentrations of each species and pH as a function of time shown in Figure 6.7.

```
% pHcstr.m: a neutralization cstr
clear all;
x1i = 1.2e-3; x2i = 2e-3; x3i = 2.5e-3; Kd = 1e-7; Ke = 1e-14; % data
Fa = 0.01667; Fb = 2.333e-3; V = 2.5; av = Fa/V; bv = Fb/V; % data
xi = [x1i x2i x3i];
% define ODE system
dx = @(t,x) [av*(xi(1) - x(1)) - bv*x(1); bv*(xi(2) - x(2)) - av*x(2); bv*(xi(3)
- x(3)) - av*x(3)];
tspan = [0 400]; [t x] = ode45(dx,tspan,xi); % Solve ODEs using ode45
x1 = x(:,1); x2 = x(:,2); x3 = x(:,3);
% Determine [H+]
Ken = Ke*ones(length(t),1); Kdn = Kd*ones(length(t),1);
f = @(h) h+x2+x3-x1-Ken./h-x3./(1+Kdn.*h./Ken);
H0 = 1e-3*ones(length(t),1); % Initial guesses
Hp = fsolve(f,H0); pH = -log10(Hp);
% Plot concentrations and pH
subplot(1,2,1), plot(t,x1,t,x2,':',t,x3,'--'), grid, xlabel('t(sec)'),
ylabel('x_i(mol/liter)'), legend('x_1','x_2','x_3')
subplot(1,2,2), plot(t,pH), grid, xlabel('t(sec)'), ylabel('pH')
```

(a)

(b)

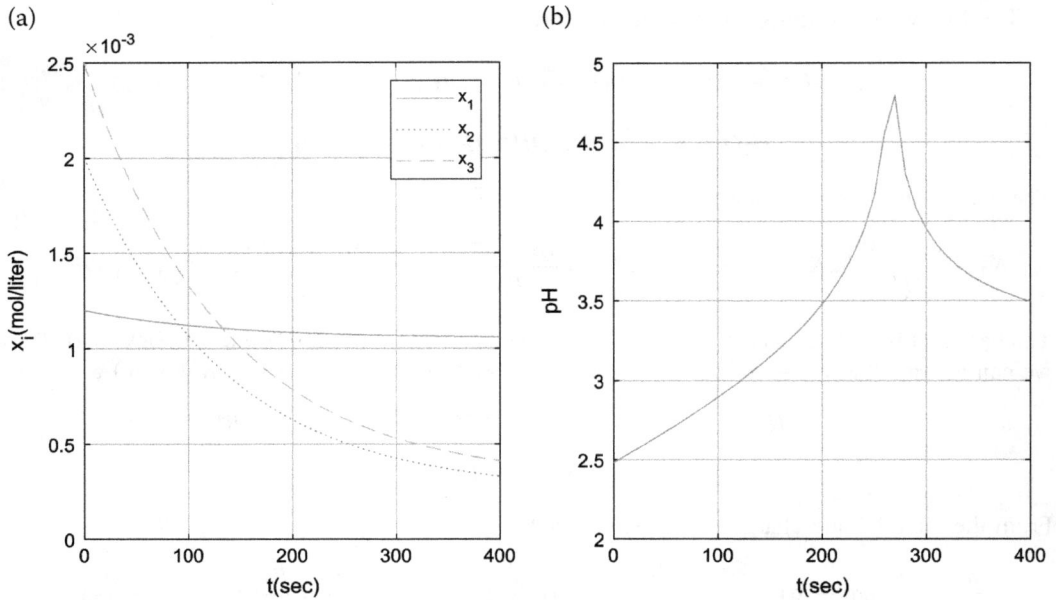

FIGURE 6.7   Profiles of (a) concentration of each species and (b) pH.

6.2.2 Nonisothermal Reaction

Figure 6.8 shows a well-mixed, constant-volume CSTR. A single 1st-order exothermic reaction $A \rightarrow B$ is carried out in the reactor. To remove the heat of reaction, the reactor is surrounded by a jacket through which a cooling liquid flows. Let's assume that the heat loss to the surroundings is negligible and that the densities and heat capacities of the reactants and products are both equal and constant.

The reaction rate is given by

$$- r_A = kC_A, \ k = \alpha e^{-E/RT}$$

The rate of heat transferred from the reactor to the cooling jacket is given by

$$Q = UA(T - T_j)$$

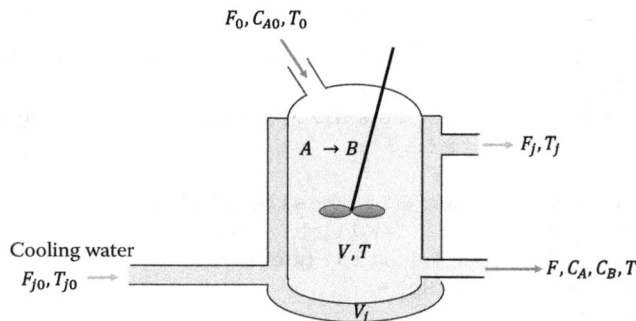

FIGURE 6.8   A CSTR with a cooling jacket.

where

U is the overall heat transfer coefficient

A is the heat transfer area

T is the temperature of the reactor

T_j is the temperature of the cooling jacket

The mass balance for species A yields

$$F_0 C_{A0} - FC_A - kVC_A = 0 \text{ or } C_A = \frac{F_0 C_{A0}}{F + kV}$$

From the energy balance on the reactor, we have

$$\rho C_p (F_0 T_0 - FT) - \lambda kVC_A - UA(T - T_j) = 0$$

The energy balance on the cooling jacket gives

$$\rho_j C_j F_j (T_{j0} - T_j) + UA(T - T_j) = 0 \text{ or } T_j = \frac{\rho_j C_j F_j T_{j0} + UAT}{\rho_j C_j F_j + UA}$$

where

λ is the heat of reaction

ρ and ρ_j are the densities of the reactants and cooling liquid, respectively

C_p and C_j are the heat capacities of the reactants and cooling liquid, respectively

Example 6.9 Exothermic CSTR

A 1st-order irreversible liquid-phase reaction is to be carried out in a CSTR. The reaction rate is given by

$$A \rightarrow B, \ -r_A = k_0 e^{-E/T}$$

where $T(K)$ is the temperature of the fluid in the reactor. The reaction is exothermic, and a cooling medium with a temperature of $T_c(K)$ will be used. From the mass and energy balances, we have

$$\tau \frac{dC_A}{dt} = (C_{A0} - C_A) - (-r_A \tau), \ \tau \frac{dT}{dt} = \left(\frac{-\Delta H_r}{C_p} \right) \left(\frac{-r_A}{C_{A0}} \right) \tau - (1 + \kappa)(T - T_c)$$

where

$\tau(hr)$ is the residence time of the fluid in the reactor

κ is the heat transfer parameter

C_p is the heat capacity of the solution in the reactor

$\Delta H_r (J/mol)$ is the heat of the reaction

C_{A0} (mol A/cm^3) is the initial concentration of A in the feed

Plot C_A and T as a function of the reaction time $t(hr)$ using the given data. What are the steady-state values of C_A and T?

Data: $k_0 = 460 \ hr^{-1}$, $E = 1380 \ K$, $C_{A0} = 0.4 \ mol/cm^3$, $\tau = 0.18 \ hr$, $T_c = 298.15 \ K$, $\kappa = 78$, $C_p = 32 \ J/(mol.K)$, $\Delta H_r = -151,080 + 2(T - 298.15) \ J/mol$

Solution

The script *plcstr* solves the system of differential equations using the built-in function ode45 and generates the required plots, shown in Figure 6.9.

```
% plcstr.m
Ca0 = 0.4; k0 = 460; E = 1380; tau = 0.18; Tc = 298.15; kappa = 78; Cp = 32; % data
```

(a)

(b)

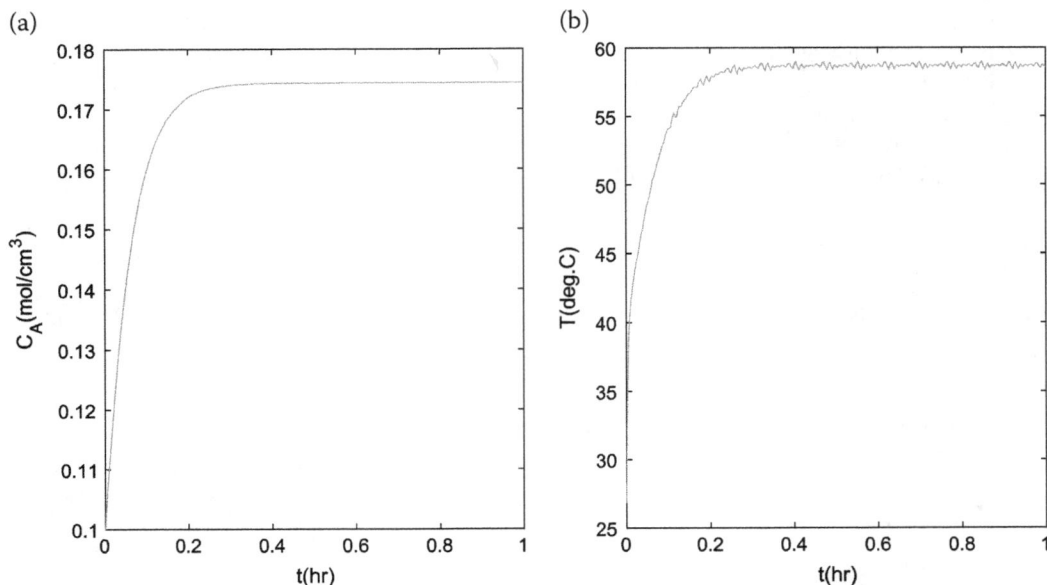

FIGURE 6.9   Plot of (a) C_A and (b) T as a function of the reaction time t.

```
dy = @(t,x) [(Ca0 - x(1))/tau - (-(-k0*exp(-E/x(2))*x(1))); % define differ-
ential equations
     (-(-151080 + 2*(x(2) - 298.15))/Cp)*(-(-k0*exp(-E/x(2))*x(1))/Ca0) -
(1+kappa)*(x(2) - Tc)/tau ];
tspan = [0 1]; y0 = [0.1 300]; [t y] = ode45(dy, tspan, y0); % solve ODE by ode45
y(:,2) = y(:,2) - 273.15; % K -> deg.C
subplot(1,2,1), plot(t,y(:,1)), xlabel('t(hr)'), ylabel('C_A(mol/cm^3)')
subplot(1,2,2), plot(t,y(:,2)), xlabel('t(hr)'), ylabel('T(deg.C)')
```

At steady state, the left-hand sides of the differential equations are zero:

$$0 = (C_{A0} - C_A) - (-r_A\tau), \quad 0 = \left(\frac{-\Delta H_r}{C_p}\right)\left(\frac{-r_A}{C_{A0}}\right)\tau - (1 + \kappa)(T - T_c)$$

The resultant system of nonlinear equations is defined by the subfunction *cstrst*. The script *stcstr* defines the data and solves the system of equations using the built-in nonlinear solver *fsolve*. As the initial conditions, we set $C_{A0} = 0.1$ and $T_0 = 300$. Then the initial value of the vector x in the program will be $x_0 = [x(1)x(2)] = [C_{A0}T_0] = [0.1300]$.

```
% stcstr.m
Ca0 = 0.4; k0 = 460; E = 1380; tau = 0.18; Tc = 298.15; kappa = 78; Cp = 32;
x0 = [0.1 300]; x = fsolve(@cstrst, x0, [], Ca0, k0, E, tau, Tc, kappa, Cp);
x(2) = x(2) - 273.15 % K -> deg.C

function y = cstrst(x,Ca0, k0, E, tau, Tc, kappa, Cp)
% x(1) = Ca, x(2) = T
ra = -k0*exp(-E/x(2))*x(1); dHr = -151080 + 2*(x(2) - 298.15);
y = [(Ca0 - x(1))/tau - (-ra); (-dHr/Cp)*(-ra/Ca0) - (1+kappa)*(x(2) - Tc)/
tau ];
end
```

Execution of the script *stcstr* yields the following results:

```
>> solstcstr
x =
  0.1744  58.6983
```

We can see that, at steady state, $C_A = C_{As} = 0.1744 \ mol/cm^3$ and $T = T_s = 58.6983°C$.

Example 6.10 Exothermic Irreversible Reaction[7]

The 1st-order exothermic irreversible reaction $A \xrightarrow{k} B$ takes place in the CSTR shown in Figure 6.10. The reactant A is supplied continuously to the reactor with a flow rate $F_i \ (m^3/hr)$, a concentration $C_{Af} \ (kmol/m^3)$, and a temperature $T_f \ (°C)$. A cooling jacket surrounds the reactor to remove the exothermic heat. A coolant with a flow rate $F_j \ (m^3/hr)$ and an inlet temperature $T_{j0} \ (°C)$ takes out the heat to maintain the desired reaction temperature.

It is assumed that the exit flow rate $F \ (m^3/hr)$ is proportional to $\sqrt{h}$ and can be represented as $F = \sqrt{10 S_A h}$, where $S_A \ (m^2)$ is the cross-sectional area of the reactor and $h \ (m)$ is the liquid level in the reactor. From the material balances for the reactor and for species A, we obtain

$$\frac{dh}{dt} = \frac{F_i}{S_A} - \sqrt{\frac{10h}{S_A}}, \ \frac{dC_A}{dt} = \frac{F_i}{S_A h}(C_{Af} - C_A) - A_1 e^{-E/(RT)} C_A$$

The energy balance for the reactor gives

$$\frac{dT}{dt} = \frac{F_i}{S_A h}(T_f - T) + \left(\frac{-\Delta H}{\rho C_p}\right) A_1 e^{-E/(RT)} C_A - \frac{U_i S_h}{\rho C_p S_A h}(T - T_j)$$

where $S_h \ (m^2)$ is the heat transfer area given by $S_h = (\pi/4)D^2 + \pi D h = S_A + \pi D h$. The coolant temperature T_j in the cooling jacket is assumed to be constant $(T_j = T_{j0})$.

(1) Using the given data, generate profiles of h, C_A, and T as a function of reaction time $t \ (0 \le t \le 20 \ hr)$.

(2) Determine steady-state operating points and generate the heat profile as a function of the reactor temperature and the reactor temperature profile as a function of the jacket temperature at steady state.

Data: $C_{Af} = 10.0 \ kmol/m^3$, $D = 2.335 \ m$, $F_i = 10 \ m^3/h$, $T_f = T_j = T_{j0} = 25°C$,
$(-\Delta H) = 5960 \ kcal/kmol$, $A_1 = 3.49308 \times 10^7 \ hr^{-1}$, $E = 11843 \ kcal/kmol$,
$\rho C_p = 500 \ kcal/(m^3 \cdot °C)$, $U_i = 70 \ kcal/(m^2 \cdot °C \cdot hr)$, $R = 1.987 \ kcal/(kmol \cdot K)$.

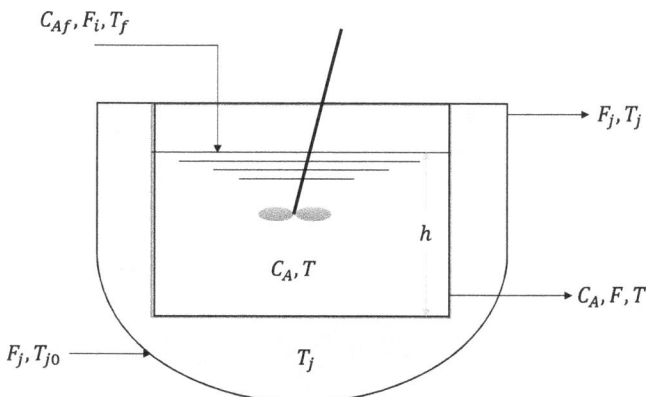

FIGURE 6.10 Schematic representation of a CSTR.

Solution

(1) The script *cstrun* sets the initial conditions and uses the built-in function *ode45* to solve the differential equation system defined by the subfunction *cstrxn*.

```
% cstrun.m
% data for CSTR
Caf = 10; D = 2.335; Fi = 10; Tf = 25; Tj = 25; dH = 5960; A1 = 3.49308e7;
E = 11843; rCp = 500; Ui = 70; R = 1.987;
% initial guess and time span
X0 = [1 5 20]; tspan = [0 20];
[t X] = ode45(@cstrxn,tspan,X0,[],Caf,D,Fi,Tf,Tj,dH,A1,E,rCp,Ui,R);
% plot results
subplot(2,2,1), plot(t,X(:,1)), xlabel('t(hr)'), ylabel('h(m)'), grid
subplot(2,2,2),   plot(t,X(:,2)),   xlabel('t(hr)'),   ylabel('C_A(kmol/
m^3)'), grid
subplot(2,2,3), plot(t,X(:,3)), xlabel('t(hr)'), ylabel('T(deg.C)'), grid

function dX = cstrxn(t,X,Caf,D,Fi,Tf,Tj,dH,A1,E,rCp,Ui,R)
% X(1) = h, X(2) = Ca, X(3) = T(deg.C)
Sa = (pi/4)*D^2; Sh = Sa + pi*D*X(1);
a1 = Fi/Sa/X(1); k1 = A1*exp(-E/R/(X(3)+273.15));
b1 = dH/rCp; c1 = Ui*Sh/(rCp*Sa*X(1));
dX = [Fi/Sa - sqrt(10*X(1)/Sa); a1*(Caf - X(2)) - k1*X(2);
  a1*(Tf - X(3)) + b1*k1*X(2) - c1*(X(3) - Tj)];
end
```

The resultant profiles of h, C_A, and T are shown in Figure 6.11.

(2) At steady state,

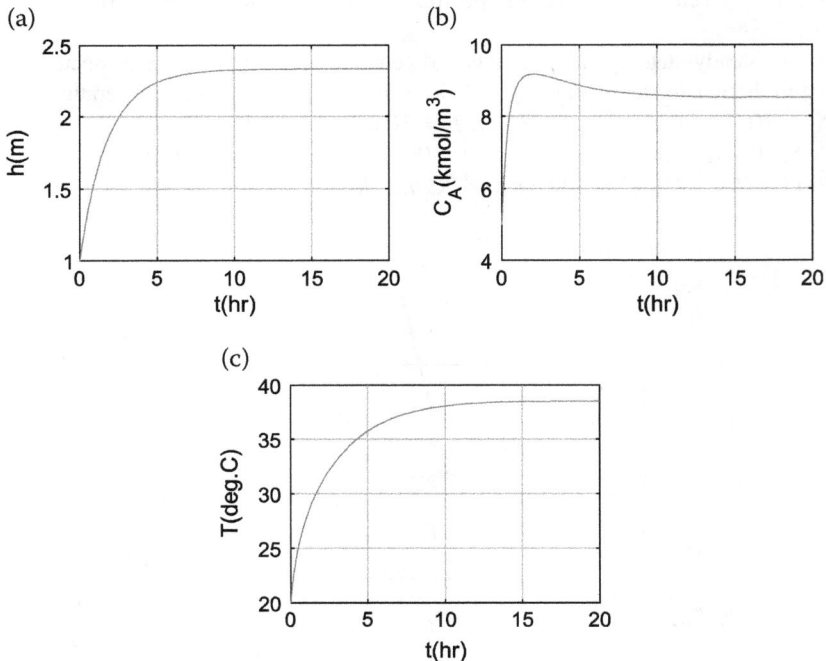

FIGURE 6.11 Profiles of (a) h, (b) C_A, and (c) T.

$$\frac{dh_s}{dt} = 0 = \frac{F_i}{S_A} - \sqrt{\frac{10h_s}{S_A}}, \quad \frac{dC_{As}}{dt} = 0 = \frac{F_i}{S_A h_s}(C_{Af} - C_{As}) - A_1 e^{-E/(RT_s)} C_{As}$$

$$\frac{dT_s}{dt} = 0 = \frac{F_i}{S_A h_s}(T_f - T_s) + \left(\frac{-\Delta H}{\rho C_p}\right) A_1 e^{-E/(RT_s)} C_{As} - \frac{U_i S_{hs}}{\rho C_p S_A h_s}(T_s - T_j)$$

where $S_{hs} = (\pi/4)D^2 + \pi D h_s = S_A + \pi D h_s$ and h_s is given by $h_s = F_i^2/(10 S_A)$. The steady-state conditions are determined by solving the following set of nonlinear equations:

$$0 = \frac{F_i}{S_A h_s}(C_{Af} - x(1)) - A_1 e^{-E/(Rx(2))} x(1)$$

$$0 = \frac{F_i}{S_A h_s}(T_f - x(2)) + \left(\frac{-\Delta H}{\rho C_p}\right) A_1 e^{-E/(Rx(2))} x(1) - \frac{U_i S_{hs}}{\rho C_p S_A h_s}(x(2) - T_j)$$

where $x(1) = C_{As}$ and $x(2) = T_s$. Depending on the initial guesses for C_{As} and T_s, three sets of solutions are obtained. The script *sscstr* solves the set of nonlinear equations for three different initial guesses. This script also generates the plots showing three different steady states. The energy equation at steady state can be rearranged as $Q_g = Q_r$ where

$$Q_r = U_i S_{hs}(T_s - T_j) + F_i \rho C_p (T_s - T_f), \quad Q_g = (-\Delta H) S_A h_s A_1 e^{-E/(RT_s)} C_{As}$$

C_{As} is given by

$$C_{As} = \frac{F_i C_{Af}}{F_i + S_A h_s A_1 e^{-E/(RT_s)}}$$

The steady-state jacket temperature is given by

$$T_{js} = T_s + \frac{1}{U_i S_{hs}} \left\{ \rho C_p F_i (T_s - T_f) - (-\Delta H) S_A h_s A_1 e^{-E/(RT_s)} C_{As} \right\}$$

Plots of Q_g and Q_r as a function of T_s and T as a function of T_j exhibit multiple steady states, as shown in Figure 6.12.

```
% sscstr.m: steady-state solution for a CSTR
clear all;
Caf = 10; D = 2.335; Fi = 10; Tf = 25; Tj = 25; dH = 5960; A1 = 3.49308e7;
E = 11843; rCp = 500; Ui = 70; R = 1.987;
Sa = (pi/4)*D^2; hs = Fi^2 / (10*Sa); Sh = Sa + pi*D*hs;
a1 = Fi/Sa/hs; b1 = dH/rCp; c1 = Ui*Sh/(rCp*Sa*hs);
fs = @(X) [a1*(Caf-X(1))-A1*exp(-E/R/(X(2)+273.15))*X(1); % X(1) = Cas, X(2)
= Ts
  a1*(Tf - X(2)) + b1*A1*exp(-E/R/(X(2)+273.15))*X(1) - c1*(X(2) - Tj)];
% Solve nonlinear equations for three different initial guesses.
x01 = [8 20]; x1 = fsolve(fs,x01); x02 = [5 70]; x2 = fsolve(fs,x02); x03 = [2
120]; x3 = fsolve(fs,x03);
fprintf('Initial guess: Cas = %g, Ts = %g',x01(1), x01(2));
fprintf(' Steady-state values: Cas = %g, Ts = %g\n',x1(1), x1(2));
fprintf('Initial guess: Cas = %g, Ts = %g',x02(1), x02(2));
fprintf(' Steady-state values: Cas = %g, Ts = %g\n',x2(1), x2(2));
fprintf('Initial guess: Cas = %g, Ts = %g',x03(1), x03(2));
```

(a)

(b)

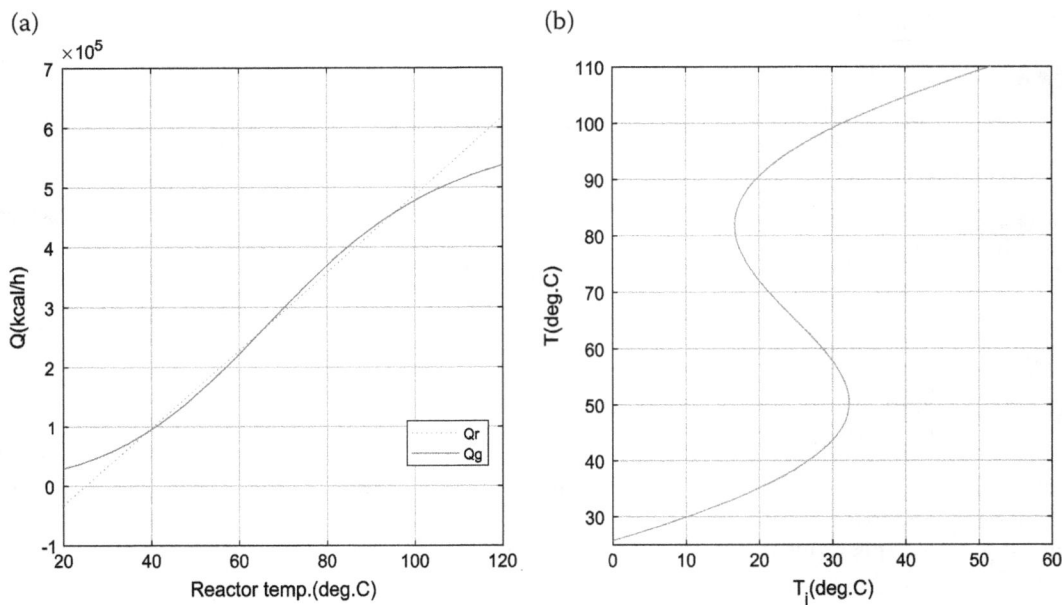

FIGURE 6.12   Multiple steady states: (a) heat profile and (b) reactor temperature profile.

```
fprintf(' Steady-state values: Cas = %g, Ts = %g\n',x3(1), x3(2));
% Generate heat and temperature profiles.
Ts = [20:0.1:120]; k1s = A1*exp(-E/R./(Ts+273.15)); Cas = a1*Caf./(k1s+a1);
Qr = Ui*Sh*(Ts-Tj) + Fi*rCp*(Ts-Tf); Qg = dH*Sa*hs*Cas.*k1s;
Tjs = Ts + (rCp*Fi*(Ts-Tf) - dH*Sa*hs*Cas.*k1s)/(Ui*Sh);
subplot(1,2,1) % Heat profile
plot(Ts,Qr,':',Ts,Qg), xlabel('Reactor temp.(deg.C)'), ylabel('Q(kcal/h)')
legend('Qr','Qg','location','best'), grid
subplot(1,2,2) % Temperature profile
plot(Tjs,Ts), axis([0 60 25 110]), grid, xlabel('T_j(deg.C)'), ylabel('T(deg.C)')
```

Execution of the script *sscstr* yields multiple steady states as following:

```
>> sscstr
Initial guess: Cas = 8, Ts = 20  Steady-state values: Cas = 8.52464, Ts = 38.5302
Initial guess: Cas = 5, Ts = 70  Steady-state values: Cas = 5.63247, Ts = 65.0537
Initial guess: Cas = 2, Ts = 120 Steady-state values: Cas = 2.31324, Ts = 95.4936
```

Example 6.11 Nonisothermal CSTR

The liquid-phase 1st-order irreversible exothermic reaction $A \rightarrow B$ is carried out in a CSTR. The reaction rate is given by $-r_A = kC_A$, where $k = \alpha e^{-E/RT}$.

(1) Calculate the steady-state values of C_A, T, and T_j.

(2) This exothermic reaction system may exhibit multiple steady states. The energy balance on the reactor yields

$$\rho C_p(F_0 T_0 - FT) - \lambda k V C_A - UA(T - T_j) = 0$$

Substituting $k = \alpha e^{-E/RT}$, $C_A = F_0 C_{A0}/(F + kV)$, and $T_j = (\rho_j C_j F_j T_{j0} + UAT)/(\rho_j C_j F_j + UA)$ into this equation gives

$$f(T) = \rho C_p(F_0 T_0 - FT) - \frac{F_0 C_{A0} V \lambda \alpha e^{-E/RT}}{F + \alpha e^{-E/RT} V} - UA \rho_j C_j F_j \left(\frac{T - T_{j0}}{\rho_j C_j F_j + UA} \right) = 0$$

Solve the nonlinear equation $f(T) = 0$ and plot $f(T)$ versus T to verify multiple steady states. Required data are given as follows:

Flow rate of the reactant feed	$F_0 = 40\,ft^3/hr$
Flow rate of the product stream	$F = 40\,ft^3/hr$
Initial concentration of species A	$C_{A0} = 0.55\,lbmol/ft^3$
Reactor volume	$V = 48\,ft^3$
Flow rate of the cooling water	$F_{j0} = F_j = 49.9\,ft^3/hr$
Heat capacity of the reactant	$C_p = 0.75\,btu/(lb_m \cdot {}^\circ R)$
Heat capacity of the cooling water	$C_j = 1\,btu/(lb_m \cdot {}^\circ R)$
Rate constantDifferent steady states	$\alpha = 7.08 \times 10^{10}\,hr^{-1}$
Activation energy	$E = 30{,}000\,btu/lbmol$
Density of the reactant	$\rho = 50\,lb_m/ft^3$
Density of the cooling water	$\rho_j = 62.3\,lb_m/ft^3$
Overall heat transfer coefficient	$U = 150\,btu/(hr \cdot ft^2 \cdot {}^\circ R)$
Heat transfer area	$A = 250\,ft^2$
Inlet temperature of the cooling water	$T_{j0} = 530^\circ R$
Inlet temperature of the reactant	$T_0 = 530^\circ R$
Heat of the reaction	$\lambda = -30{,}000\,btu/lbmol$
Volume of the cooling jacket	$V_j = 12\,ft^3$
Gas constant	$R = 1.9872\,btu/(lbmol \cdot {}^\circ R)$

Solution

(1) We can formulate the mass and energy balances that apply to the reactor and the cooling jacket. Required steady-state values can be obtained from the solution of the system of nonlinear equations consisting of these balances. Rearrangement of the mass and energy balances yields the following nonlinear equations ($x_1 = C_A$, $x_2 = T$, $x_3 = T_j$):

Mass balance on the reactor: $f_1 = F_0 C_{A0} - Fx_1 - \alpha V e^{-E/Rx_2} x_1 = 0$
Energy balance on the reactor: $f_2 = \rho C_p(F_0 T_0 - Fx_2) - \lambda \alpha V e^{-E/Rx_2} x_1 - UA(x_2 - x_3) = 0$
Energy balance on the cooling jacket: $f_3 = \rho_j C_j F_j(T_{j0} - x_3) + UA(x_2 - x_3) = 0$

Set the initial estimates of the final solution as $x_0 = [C_{A0}\ \ T_0\ \ T_{j0}]$. The script *exocstr* uses the built-in function *fsolve* to give the desired results.

```
% exocstr.m
% Data:
F0 = 40; F = 40; Fj = 49.9;Ca0=0.55; V = 48; rho = 50; rhoj = 62.3; Cp = 0.75; Cj = 1; A
= 250; U = 150;
T0 = 530; Tj0 = 530; alp=7.08e10; lam = -3e4; E = 3e4; R = 1.9872;
% Define nonlinear equations
fun = @(x) [F0*Ca0 - F*x(1) - alp*V*x(1)*exp(-E/R/x(2));
    rho*Cp*(F0*T0-F*x(2))-lam*alp*V*x(1)*exp(-E/R/x(2))-U*A*(x(2) - x(3));
    rhoj*Cj*Fj*(Tj0 - x(3)) + U*A*(x(2) - x(3))];
```

```
% Solution of nonlinear equation system
x0 = [Ca0 T0 Tj0]; % initial guess
x = fsolve(fun, x0); Ca = x(1), T = x(2), Tj = x(3)
```

The script *exocstr* produces the following outputs:

```
>> exocstr
Ca =
    0.5214
T =
    537.8548
Tj =
    537.2534
```

(2) The nonlinear equation

$$f(T) = \rho C_p (F_0 T_0 - FT) - \frac{F_0 C_{A0} V \lambda \alpha e^{-E/RT}}{F + \alpha e^{-E/RT} V} - UA \rho_j C_j F_j \left(\frac{T - T_{j0}}{\rho_j C_j F_j + UA} \right) = 0$$

is solved by the script *multst*, which also generates a plot $f(T)$ versus T.

```
% multst.m
% Data:
F0 = 40; F = 40; Fj = 49.9; Ca0=0.55; V = 48; rho = 50; rhoj = 62.3; Cp = 0.75; Cj = 1; A
= 250; U = 150;
T0 = 530; Tj0 = 530; alp=7.08e10; lam = -3e4; E = 3e4; R = 1.9872;
% Define nonlinear functions
fT = @(T) [rho*Cp*(F0*T0 - F*T) - (F0*Ca0*V*lam*alp*exp(-E/R./T))./...
    (F+alp*V*exp(-E/R./T)) - U*A*rhoj*Cj*Fj*(T-Tj0)./(rhoj*Cj*Fj + U*A)];
% Solve nonlinear equations by using the built-in solver fzero
T0 = T0+150; % initial guess
x = fzero(fT, T0)
% Plot of f(T) versus T
Tv = 500:0.1:700; Fv = fT(Tv); Fv0 = Fv*0; plot(Tv,Fv,Tv,Fv0), xlabel('T(R)'),
ylabel('f(T)')
```

Different steady states can be determined by solving the nonlinear equation with different initial estimates of the final solution. The script *multst* computes T for three different initial estimates of $T_0 = 530, 580, 680$ (°R) and generates the plot of $f(T)$ versus T shown in Figure 6.13.

```
Initial temperature(T0) = 530, Calculated temperature(T) = 537.8548
Initial temperature(T0) = 580, Calculated temperature(T) = 590.3498
Initial temperature(T0) = 680, Calculated temperature(T) = 671.278
```

We can see that the reaction system exhibits three different steady states of $T_s = 537.8548, 590.3498, 671.2783$ (°R).

6.2.3 MULTIPLE REACTIONS IN A CSTR

For q liquid-phase reactions occurring where N different species are present, the mass balance on each reactor yields[8]

$$F_{10} - F_1 = -r_1 V = V \sum_{i=1}^{q} (-r_{i1}) = V f_1(C_1, \cdots, C_N)$$

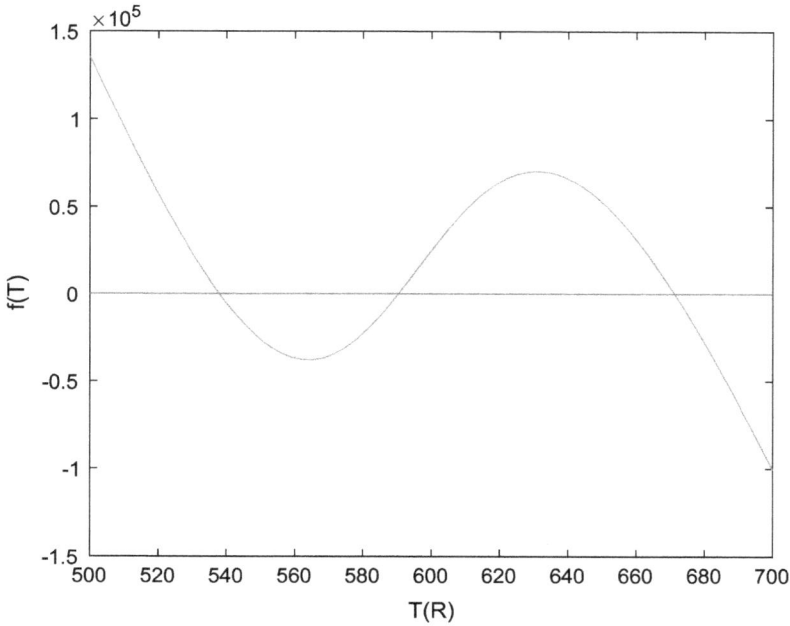

FIGURE 6.13 Multiple steady-state solutions for the exothermic reaction system.

$$F_{20} - F_2 = -r_2 V = V \sum_{i=1}^{q} (-r_{i2}) = V f_2 (C_1, \cdots, C_N)$$

$$\vdots$$

$$F_{j0} - F_j = -r_j V = V \sum_{i=1}^{q} (-r_{ij}) = V f_j (C_1, \cdots, C_N)$$

$$\vdots$$

$$F_{N0} - F_N = -r_N V = V \sum_{i=1}^{q} (-r_{iN}) = V f_N (C_1, \cdots, C_N)$$

Example 6.12 Multiple Reactions in a Liquid-Phase CSTR[9]

The following liquid-phase reactions take place in a 2500 dm^3 CSTR:

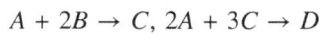

$$A + 2B \rightarrow C, \quad 2A + 3C \rightarrow D$$

The reaction rate for each species is expressed as

$$r_A = -k_a C_A C_B^2 - \frac{2}{3} k_c C_A^2 C_C^3, \quad r_B = -2k_a C_A C_B^2, \quad r_C = k_a C_A C_B^2 - k_c C_A^2 C_C^3, \quad r_D = \frac{1}{3} k_c C_A^2 C_C^3$$

Determine the concentrations of A, B, C, and D exiting the reactor, along with the exiting selectivity, which is defined as $S_{C/D} = C_C/C_D$.

Data:
$k_a = 10$ $(dm^3/mol)^2/min$ $k_c = 15$ $(dm^3/mol)^4/min$, $v_0 = 100$ dm^3/min $C_{A0} = C_{B0} = 2$ mol/dm^3

Solution

The mole balance for each species gives

$$f(C_A) = v_0 C_{A0} - v_0 C_A + r_A V = 0, f(C_B) = v_0 C_{B0} - v_0 C_B + r_B V = 0,$$

$$f(C_C) = -v_0 C_C + r_C V = 0, f(C_D) = -v_0 C_D + r_D V = 0$$

The system of nonlinear equations is defined by the function *cstrmult*. The dependent variable vector C is defined as $C^T = [C_A \quad C_B \quad C_C \quad C_D]$.

```
function fC = cstrmult(C,ka,kc,Ca0,Cb0,v0,V)
% C(1)=Ca, C(2)=Cb, C(3)=Cb, C(4)=Cd
ra = -ka*C(1)*C(2)^2 - 2*kc*C(1)^2*C(3)^3/3; rb = -2*ka*C(1)*C(2)^2;
rc = ka*C(1)*C(2)^2 - kc*C(1)^2*C(3)^3; rd = kc*C(1)^2*C(3)^3/3;
fC = [v0*Ca0 - v0*C(1) + ra*V; v0*Cb0 - v0*C(2) + rb*V; -v0*C(3) + rc*V; -v0*C(4)
+ rd*V];
end
```

The script *multcstr* calls the function *cstrmult* and employs the built-in function *fsolve* to solve the system of nonlinear equations. In the calculation of the selectivity $S_{C/D}$, the value of the selectivity is set to zero when C_D is very small (remember that the initial value of C_D is 0).

```
% multcstr.m
clear all;
ka = 10; kc = 15; V = 2500; v0 = 100; Ca0 = 2; Cb0 = 2; C0 = [Ca0 Cb0 0 0];
[C fval] = fsolve(@cstrmult,C0,[],ka,kc,Ca0,Cb0,v0,V);
Ca = C(:,1); Cb = C(:,2); Cc = C(:,3); Cd = C(:,4);
n = length(Cd); Scd = zeros(1,n);
for i = 1:n
  if Cd(i) <= 1e-4, Scd(i) = 0; else, Scd(i) = Cc(i)/Cd(i); end
end
fprintf('The final concentration of each species: \n');
fprintf('Caf=%g, Cbf=%g, Ccf=%g, Cdf=%g\n',Ca(end),Cb(end),Cc(end),Cd(end));
fprintf('The final selectivity: Scdf = %g\n',Scd(end));
```

The script *multcstr* generates the following outputs:

```
>> multcstr
The final concentration of each species:
Caf = 0.532653, Cbf = 0.0848008, Ccf = 0.192978, Cdf = 0.25487
The final selectivity: Scdf = 0.757153
```

Example 6.13 van de Vusse Reaction[10]

The following van de Vusse reaction is carried out in a CSTR:

$$A \rightarrow B \rightarrow C, 2A \rightarrow D$$

The component material balance yields

$$\frac{dC_A}{dt} = -k_1 C_A - k_3 C_A^2 + \frac{F}{V}(C_{Af} - C_A), \frac{dC_B}{dt} = k_1 C_A - k_2 C_B - \frac{F}{V}C_B$$

where $V(liter)$ is the reactor volume, $F(liter/hr)$ is the feed flow rate, $k_i(i = 1,2,3)$ are kinetic constants, $C_{Af}(mol/liter)$ is the feed concentration of reactant A, and C_A and C_B are the concentrations of A and B, respectively.

(1) Determine the steady-state values of C_A and C_B.

(2) Plot C_A and C_B as a function of reaction time t $(0 \leq t \leq 0.06)$.

Data: $V = 1$ *liter*, $F = 25$ *liter*/hr, $C_{Af} = 10$ *mol/liter*, $k_1 = 50$ hr^{-1}, $k_2 = 100$ hr^{-1}, $k_3 = 10$ *liter*$/(mol{\cdot}hr)$

Solution

(1) At the steady state,

$$\frac{dC_{As}}{dt} = 0 = -k_1 C_{As} - k_3 C_{As}{}^2 + \frac{F}{V}(C_{Af} - C_{As}), \frac{dC_{Bs}}{dt} = 0 = k_1 C_{As} - k_2 C_{Bs} - \frac{F}{V} C_{Bs}$$

```
>> V = 1; F = 25; Caf = 10; k1 = 50; k2 = 100; k3 = 10; % data
>> f = @(Cs) [-k1*Cs(1) - k3*Cs(1)^2 + F*(Caf - Cs(1))/V; k1*Cs(1) - k2*Cs(2) -
F*Cs(2)/V];
>> Cs0 = [5 5]; Cs = fsolve(f,Cs0)
Cs =
  2.5000   1.0000
```

We can see that $C_{As} = 2.5$ *mol/liter* and $C_{Bs} = 1$ *mol/liter*.

(2) The concentration profiles shown in Figure 6.14 can be generated by the following script *vandrxn*:

```
% vandrxn.m
V = 1; F = 25; Caf = 10; k1 = 50; k2 = 100; k3 = 10; % data
df = @(t,C) [-k1*C(1) - k3*C(1)^2 + F*(Caf - C(1))/V; k1*C(1) - k2*C(2) -
F*C(2)/V];
tspan = [0 0.06]; C0 = [10 0]; [t C] = ode45(df,tspan,C0); % solve ODEs by ode45
plot(t,C(:,1),t,C(:,2),':'), legend('C_A','C_B'), xlabel('t'), ylabel('C(t)')
```

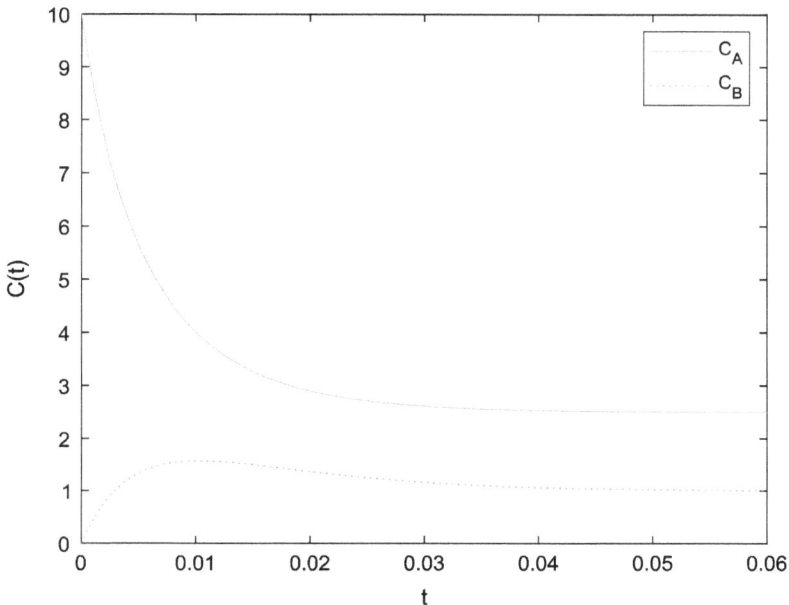

FIGURE 6.14 Profiles of $C_A(t)$ and $C_B(t)$.

6.3 BATCH REACTORS

6.3.1 ESTIMATION OF BATCH REACTION PARAMETERS

A conversion reaction of A is carried out in a batch reactor. A mass balance on the batch reactor yields

$$\frac{dC_A}{dt} = -kC_A^n$$

If the measured data on the concentration of A (C_A) with respect to time t are available, the reaction order and rate constant can be determined by applying the differential method. Taking the logarithm of the rate equation yields

$$\ln\left(-\frac{dC_A}{dt}\right) = \ln(k) + n\ln(C_A)$$

Thus, determination of the reaction order n and the rate constant k becomes a typical linear regression problem.

Example 6.14 Reaction Parameters in a Batch Reactor[11]

The liquid-phase bromination of xylene at 17°C is carried out in a batch reactor. Iodine is used as a catalyst, and small quantities of the reactant bromine are introduced into the reactor containing the reactant xylene in considerable excess. The concentrations of the reactant xylene and the catalyst iodine are approximately constant during the reaction. A mass balance on the batch reactor yields

$$\frac{dC_{Br_2}}{dt} = -kC_{Br_2}^n$$

where

C_{Br_2} is the concentration of bromine ($gmol/dm^3$)
k is a pseudo rate constant that depends on the iodine and xylene concentrations
n is the reaction order

Data on the concentration of bromine (C_{Br_2}) are shown in Table 6.4. Estimate the rate constant k and the reaction order n.

TABLE 6.4
Concentration of Bromine versus Time

$t\,(min)$	$C_{Br_2}\,(gmol/dm^3)$	$t\,(min)$	$C_{Br_2}\,(gmol/dm^3)$
0	0.3335	19.60	0.1429
2.25	0.2965	27.00	0.1160
4.50	0.2660	30.00	0.1053
6.33	0.2450	38.00	0.0830
8.00	0.2255	41.00	0.0767
10.25	0.2050	45.00	0.0705
12.00	0.1910	47.00	0.0678
13.50	0.1794	57.00	0.0553
15.60	0.1632	63.00	0.0482
17.85	0.1500		

Solution

From the rate equation $dC_{Br_2}/dt = -kC_{Br_2}^n$ (C_{Br_2}: concentration of bromine), we have

$$\ln\left(-\frac{dC_{Br_2}}{dt}\right) = \ln(k) + n\ln\left(C_{Br_2}\right)$$

The built-in function *diff* can be used to execute the numerical differentiation of the data. Then the linear regression method is employed to calculate n and k. The relation $x = (A^T A)^{-1} A^T b$ can be used in the linear regression, where $x(1) = n$ and $x(2) = \ln(k)$. The script *rnk* performs the calculation procedure.

```
% rnk.m
t = [0 2.25 4.50 6.33 8.00 10.25 12.00 13.50 15.60 17.85 ...
   19.60 27.00 30.00 38.00 41.00 45.00 47.00 57.00 63.00];
Cb = [0.3335 0.2965 0.2660 0.2450 0.2255 0.2050 0.1910 0.1794 0.1632 ...
   0.1500 0.1429 0.1160 0.1053 0.0830 0.0767 0.0705 0.0678 0.0553 0.0482];
dCb = diff(Cb)./diff(t); dCb = [dCb dCb(end)]; % numerical differentiation
n = length(t); A = [(log(Cb))' ones(n,1)]; x = (A'*A)^-1*A'*(log(-dCb))', n = x
(1), k = exp(x(2))
```

The script *rnk* produces the following outputs:

```
>> rnk
x =
   1.5128
  -2.4454
n =
   1.5128
k =
   0.0867
```

Example 6.15 Nonisothermal Batch Reactor

The following exothermic consecutive reactions are carried out in a batch reactor fitted with a cooling coil through which cooling water is passed to remove the exothermic heat:

$$A \xrightarrow{k_1} B \xrightarrow{k_2} C$$

From the material balances for species A and B, we obtain

$$\frac{dC_A}{dt} = -k_1 C_A^2, \quad \frac{dC_B}{dt} = k_1 C_A^2 - k_2 C_B$$

where
k_1 and k_2 are the reaction rate constants
C_A and C_B are the concentrations of species A and B, respectively
k_1 and k_2 are represented as

$$k_1 = A_1 e^{-E_1/(RT)}, \quad k_2 = A_2 e^{-E_2/(RT)}$$

The energy balance for the batch reactor gives

$$\frac{dT}{dt} = \frac{(-\Delta H_1)}{\rho C_p} k_1 C_A^2 + \frac{(-\Delta H_2)}{\rho C_p} k_2 C_B + \frac{U_j A_j}{\rho C_p V}(T_s - T) - \frac{U_c A_c}{\rho C_p V}(T - T_c)$$

where

$(-\Delta H_1)$ is the heat of reaction for $A \rightarrow B$

$(-\Delta H_2)$ is the heat of reaction for $B \rightarrow C$

T_s and T_c are the steam and coolant temperatures, respectively

U_j and U_c are the overall heat transfer coefficients of the jacket and coolant, respectively

Plot C_A, C_B, and T as a function of reaction time t ($0 \leq t \leq 6000\ sec$).

Data: $C_{A0} = 1.5\ kmol/m^3$, $C_{B0} = 0.0\ kmol/m^3$, $A_1 = 1.2\ m^3/(kmol\cdot sec)$, $A_2 = 180.0\ sec^{-1}$, $E_1 = 2.1 \times 10^4\ kJ/kmol$, $E_2 = 4.3 \times 10^4\ kJ/kmol$, $(-\Delta H_1) = 4.09 \times 10^4\ kJ/kmol$, $(-\Delta H_2) = 8.24 \times 10^4\ kJ/kmol$, $\rho = 1000\ kg/m^3$, $T_c = 20°C$, $U_j = 1.2\ kJ/(m^2\cdot°C\cdot sec)$, $U_c = 3.0\ kJ/(m^2\cdot°C\cdot sec)$, $T_s = 110°C$, $R = 8.314\ kJ/(kmol\cdot K)$, $A_c/V = 18.6\ m^2/m^3$, $A_j/V = 31.5 m^2/m^3$, $C_p = 1.0\ kJ/(kg\cdot°C)$.

Solution

The given data are defined in the script *batdat*:

```
% batdat.m: data for a batch reactor
A1 = 1.2; A2 = 180; E1 = 2.1e4; E2 = 4.3e4; R = 8.314;
dH1 = 4.09e4; dH2 = 8.24e4; rho = 1000; Cp = 1;
Tc = 20; Ts = 110; Uj = 1.2; Uc = 3.0; AcV = 18.6; AjV = 31.5;
```

The script *batdat* is called by the subfunction *batrxn*, which defines the differential equations. The script *batrun* sets the initial conditions and uses the built-in function ode45 to solve the differential equation system.

```
% batrun.m
X0 = [1.5 0 50]; tspan = [0 6000];
[t X] = ode45(@batrxn,tspan,X0);
subplot(1,2,1), plot(t,X(:,1),t,X(:,2),'--')
xlabel('t(s)'), ylabel('C(kmol/m^3)'), legend('C_A(t)','C_B(t)')
subplot(1,2,2), plot(t,X(:,3)), xlabel('t(s)'), ylabel('T(deg.C)')
function dX = batrxn(t,X)
% X(1) = Ca, X(2) = Cb, X(3) = T(deg.C)
batdat; % supply data
a1 = dH1/(rho*Cp); a2 = dH2/(rho*Cp); a3 = Uj*AjV/(rho*Cp); a4 = Uc*AcV/(rho*Cp);
k1 = A1*exp(-E1/R/(273.15+X(3))); k2 = A2*exp(-E2/R/(273.15+X(3)));
dX = [-k1*X(1)^2; k1*X(1)^2 - k2*X(2); a1*k1*X(1)^2 + a2*k2*X(2) + a3*(Ts-X(3))
- a4*(X(3)-Tc)];
end
```

The resultant concentration and temperature profiles are shown in Figure 6.15.

6.3.2 Semibatch Reactors

One of the best reasons to use semibatch reactors is to enhance selectivity in liquid-phase reactions.[12] For example, consider the following two simultaneous reactions being carried out in a semibatch reactor, where the desired product is D:

$$A + B \xrightarrow{k_D} D,\ r_D = k_D C_A^2 C_B$$

$$A + B \xrightarrow{k_U} U,\ r_U = k_U C_A C_B^2$$

The instantaneous selectivity $S_{D/U}$ is the ratio of these two rates:

(a)

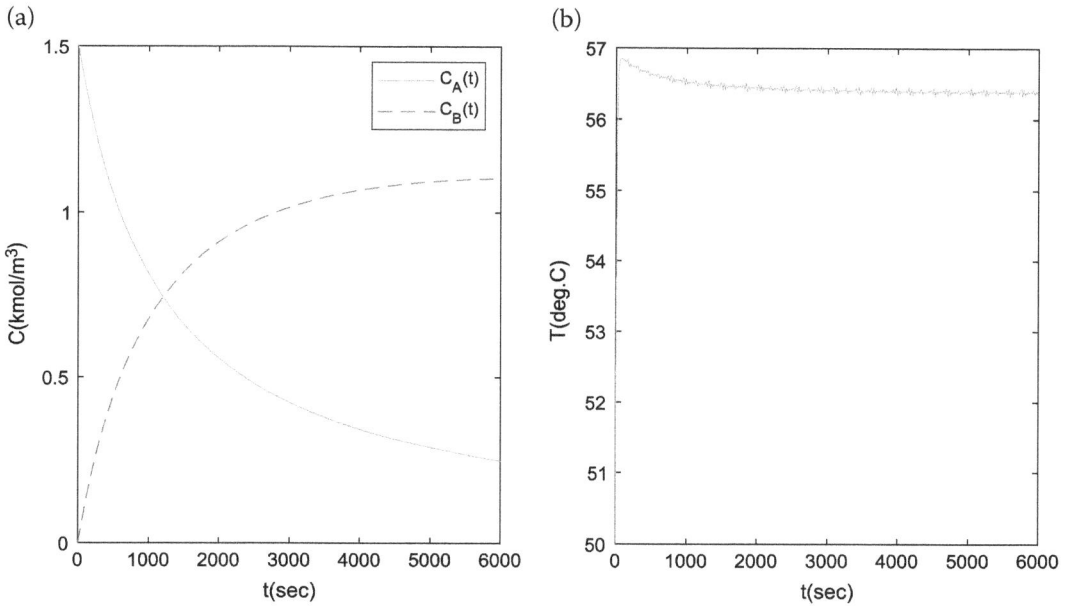

(b)

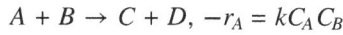

FIGURE 6.15 (a) Concentration profile and (b) temperature profile.

$$S_{D/U} = \frac{r_D}{r_U} = \frac{k_D C_A^2 C_B}{k_U C_A C_B^2} = \frac{k_D}{k_U} \frac{C_A}{C_B}$$

Consider a semibatch reactor that is charged with pure A and to which B is fed slowly to A as shown in Figure 6.16. The elementary liquid-phase reaction is carried out in the reactor[13]:

$$A + B \rightarrow C + D, \quad -r_A = k C_A C_B$$

The mole balance on each species yields

$$\frac{dC_A}{dt} = r_A - \frac{v_0 C_A}{V}, \quad \frac{dC_B}{dt} = \frac{v_0 (C_{B0} - C_B)}{V} + r_B, \quad \frac{dC_C}{dt} = r_C - \frac{v_0 C_C}{V}, \quad \frac{dC_D}{dt} = r_D - \frac{v_0 C_D}{V}$$

$$- r_A = -r_B = r_C = r_D$$

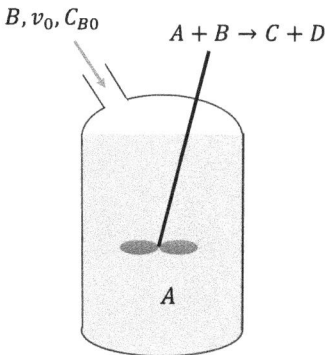

FIGURE 6.16 Semibatch reactor.

The semibatch reactor volume can be expressed as a function of time, and the conversion of species A is represented in terms of the reactor volume V as

$$V = V_0 + v_0 t, \quad X = \frac{C_{A0} V_0 - C_A V}{C_{A0} V_0}$$

Example 6.16 Semibatch Reactor[14]

Methyl bromide is produced by the irreversible liquid-phase reaction

$$CNBr\,(A) + CH_3NH_2\,(B) \rightarrow CH_3Br\,(C) + NCNH_2\,(D)$$

The reaction is carried out isothermally in a semibatch reactor. The reaction rate is given by $-r_A = kC_A C_B$, where $k = 2.2\ dm^3/(sec \cdot mol)$. The initial volume of liquid in the reactor is $V_0 = 5\ dm^3$. An aqueous solution of methyl amine (B) at a concentration of $C_{B0} = 0.025\ mol/dm^3$ is to be fed at a volumetric flow rate of $v_0 = 0.05\ dm^3/sec$ to an aqueous solution of bromine cyanide (A) contained in the reactor. The initial concentration of bromine cyanide is $C_{A0} = 0.05\ mol/dm^3$. Solve for the concentration of each species (A and B) and the rate of reaction as a function of time. Plot the results versus time ($0 \leq t \leq 500\ sec$).

Solution

Since $-r_A = -r_B = r_C = r_D$, we get the following differential equations:

$$\frac{dC_A}{dt} = r_A - \frac{v_0 C_A}{V}, \quad \frac{dC_B}{dt} = \frac{v_0(C_{B0} - C_B)}{V} + r_A, \quad \frac{dC_C}{dt} = -r_A - \frac{v_0 C_C}{V}, \quad \frac{dC_D}{dt} = -r_A - \frac{v_0 C_D}{V}$$

$$V = V_0 + v_0 t, \quad -r_A = kC_A C_B$$

The subfunction *semibrx* defines these differential equations, and the script *rxncon* calls the function *semibrx* and uses the built-in function *ode45* to integrate the system of differential equations. The script *rxncon* calculates the conversion of A as a function of time and generates the plots shown in Figure 6.17.

```
% rxncon.m
clear all;
k = 2.2; v0 = 0.05; V0 = 5; Cb0 = 0.025; Ca0 = 0.05; % data
x0 = [Ca0 0 0 0]; tspan = [0 500]; % initial guess and time interval
[t x] = ode45(@semibrx,tspan,x0,[],k,v0,V0,Cb0); % solve ODE system
Ca = x(:,1); Cb = x(:,2); Cc = x(:,3); Cd = x(:,4);
V = V0 + v0*t; Xa = (Ca0*V0 - Ca.*V)/(Ca0*V0); rA = k*Ca.*Cb;
subplot(1,2,1), plot(t,Ca, t,Cb,':', t,Cc,'.-', t,Cd,'--')
xlabel('t(sec)'),   ylabel('Concentration(mol/dm^3)'),   legend('C_A','C_B',
'C_C','C_D')
subplot(1,2,2), plot(t,rA), xlabel('t(sec)'), ylabel('Reaction rate(mol/
dm^3sec)')

function dxdt = semibrx(t,x,k,v0,V0,Cb0)
% x(1)=Ca, x(2)=Cb, x(3)=Cc, x(4)=Cd
rA = -k*x(1)*x(2); V = V0 + v0*t;
dxdt = [rA - v0*x(1)/V; rA + (Cb0 - x(2))*v0/V; -rA - v0*x(3)/V; -rA - v0*x(4)/V];
end
```

(a)   (b)

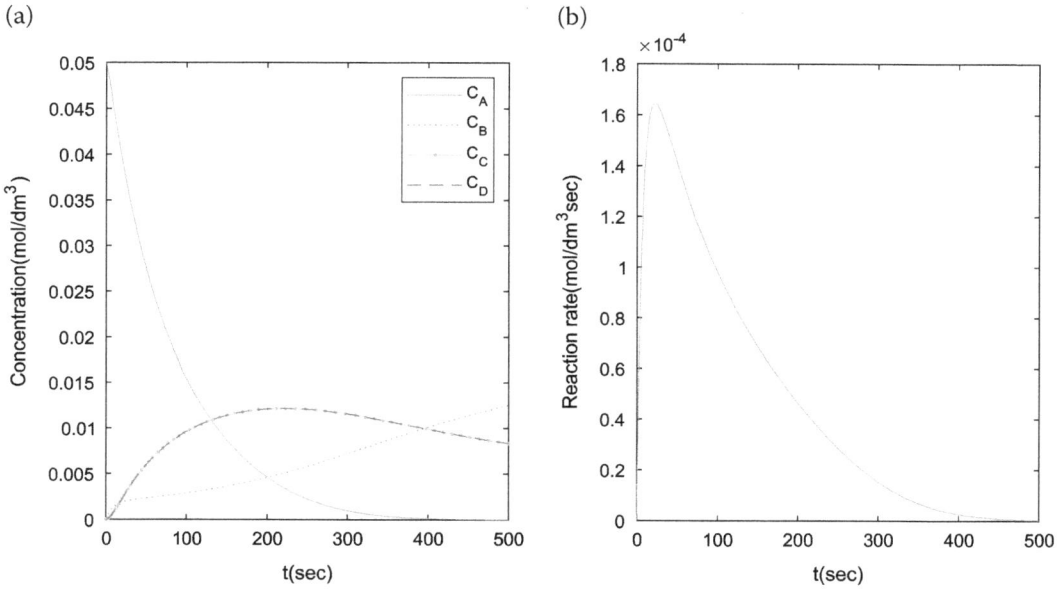

FIGURE 6.17   (a) Concentration and (b) reaction rate versus time.

6.4 PLUG-FLOW REACTORS

6.4.1 Isothermal Plug-Flow Reactor

The mass balance for a plug-flow reactor can be expressed as

$$\frac{dX}{dV} = \frac{-r_A}{F_{A0}} \text{ or } v_0 \frac{dC_A}{dV} = r_A$$

If the reaction is carried out in isothermal isobaric conditions and the rate of reaction is taken as 1st order, $F_{A0} = C_{A0}v_0$, $\epsilon = y_0\delta$, and

$$-r_A = kC_A = kC_{A0}\left(\frac{1-X}{1+\epsilon X}\right)\frac{P}{P_0}\frac{T_0}{T} = kC_{A0}\left(\frac{1-X}{1+\epsilon X}\right)$$

Thus, the mass balance equation can be expressed as

$$\frac{dX}{dV} = \frac{k(1-X)}{v_0(1+\epsilon X)}, \ X|_{V=0} = 0$$

If the reactant consists of pure A, ϵ is given by $\epsilon = y_0\delta = \delta$. But if the reactant contains $z\%$ of A, ϵ is given by $\epsilon = y_0\delta = z\delta/100$.

Example 6.17 Isothermal Plug-Flow Reactor[15]

Components A and C are fed to a plug-flow reactor in equimolar amounts, and the reaction $2A \rightarrow B$ takes place in the reactor. The mass balance on each species yields

$$v_0 \frac{dC_A}{dV} = -2kC_A^2, \ v_0 \frac{dC_B}{dV} = kC_A^2, \ v_0 \frac{dC_C}{dV} = 0$$

where $v_0 = 0.5$ m/sec and $k = 0.3$ $m^3/(kmol \cdot sec)$. The initial concentrations of the species are $C_{A0} = 2$ $kmol/m^3$, $C_{B0} = 0$ and $C_{C0} = 2$ $kmol/m^3$, and the volume of the reactor represented as the total reactor length is $V_f = 2.4$ m. Plot the concentration change of each species as a function of the reactor volume (represented in length) V.

Solution

The differential equations are defined first, and the built-in function *ode45* is employed to solve the differential equations. The script *isotplr* produces the plot shown in Figure 6.18.

```
% isotplr.m
v0 = 0.5; k = 0.3; Vf = 2.4; Vi = [0 Vf]; C0 = [2 0 2]; % data
dC = @(V,C) [-2*k*C(1)^2/v0; k*C(1)^2/v0; 0]; [V C] = ode45(dC,Vi,C0);
plot(V,C(:,1),V,C(:,2),'.-',V,C(:,3),'--'), legend('C_A','C_B','C_C')
xlabel('V(length, m)'), ylabel('Concentration(kmol/m^3)')
```

Example 6.18 HDA Reaction in a Plug-Flow Reactor[16]

m-Xylene is produced by the hydrodealkylation (HDA) of mesitylene in a plug-flow reactor. The HDA reaction is to be carried out isothermally at 1500 R and 35 atm. Two reactions occur in the reactor:

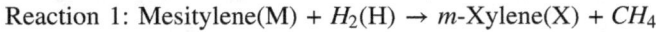

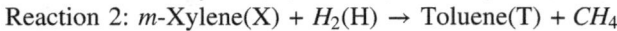

Reaction 1: Mesitylene(M) + H_2(H) → *m*-Xylene(X) + CH_4

Reaction 2: *m*-Xylene(X) + H_2(H) → Toluene(T) + CH_4

Reaction 2 is not desirable because it consumes the desired product *m*-xylene to produce toluene. Mass balance yields the following set of differential equations:

$$\frac{dC_H}{d\tau} = -k_1 C_H^{\frac{1}{2}} C_M - k_2 C_X C_H^{\frac{1}{2}}, \; \frac{dC_M}{d\tau} = -k_1 C_H^{\frac{1}{2}} C_M, \; \frac{dC_X}{d\tau} = k_1 C_H^{1/2} C_M - k_2 C_X C_H^{1/2}$$

where

τ is the residence time

k_1 is the reaction constant of Reaction 1

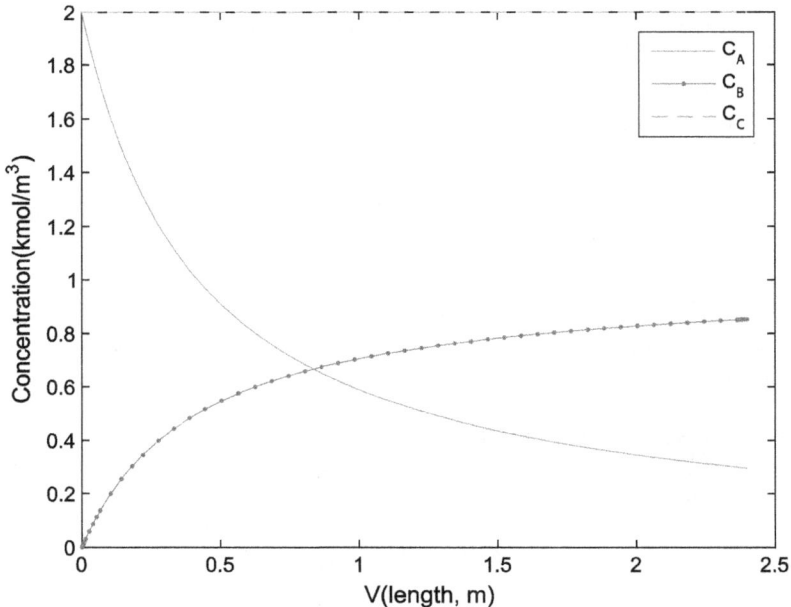

FIGURE 6.18   Concentration versus reactor length.

k_2 is the reaction constant of Reaction 2

C_H, C_M, and C_X are the concentration of H_2, mesitylene, and m-xylene, respectively

Plot the concentrations of H_2, mesitylene, and m-xylene as a function of τ ($0 \leq \tau \leq 0.5\ hr$). What is the optimum residence time of the reactor to give the maximum product concentration?

Data: $k_1 = 55.2\,(ft^3/lbmol)^{1/2}/hr$, $k_2 = 30.2\,(ft^3/lbmol)^{1/2}/hr$, $C_H(0) = 0.021$, $C_M(0) = 0.0105$, $C_X(0) = 0.0$.

Solution

The system of differential equations is defined by the subfunction *pfrmx*. The script *pfrmxplot* defines the required data, uses the built-in function ode45 to find the concentrations as a function of residence time, and determines the optimum residence time at which the concentration of the desired product (m-xylene) is at maximum.

```
% pfrmxplot.m
% Plot concentrations
k = [55.2 30.2]; C0 = [0.021 0.0105 0]; tf = 0.5; % data and initial conditions
[t C] = ode45(@pfrmx,[0 tf],C0,[],k,C0); Ch = C(:,1); Cm = C(:,2); Cx = C(:,3);
plot(t,Ch,'--',t,Cm,':',t,Cx),    xlabel('\tau(hr)'),    ylabel('C(lbmol/
ft^3)'), legend('C_H','C_M','C_X')
% Find optimum t
Cxm = max(Cx); ti = find(Cx == Cxm); opmt = t(ti);
fprintf('Optimum residence time = %g\n', opmt);
fprintf('Maximum concentration of m-xylene = %g\n', Cxm);
function dC = pfrmx(t,C,k,C0)
% C(1) = Ch, C(2) = Cm, C(3) = Cx
% k = [k1 k2], C0 = [Ch(0) Cm(0) Cx(0)]
dC = [-k(1)*C(1)^0.5*C(2) - k(2)*C(3)*C(1)^0.5; -k(1)*C(1)^0.5*C(2);
  k(1)*C(1)^0.5*C(2) - k(2)*C(3)*C(1)^0.5];
end
```

Execution of the script *pfrmxplot* produces the following results and the plot shown in Figure 6.19.

```
>> pfrmxplot
Optimum residence time = 0.201855
Maximum concentration of m-xylene = 0.00506664
```

Example 6.19 Multiple Reaction in a Plug-Flow Reactor[17]

The following four gas-phase reactions take place simultaneously on a metal oxide-supported catalyst in a plug-flow reactor (PFR):

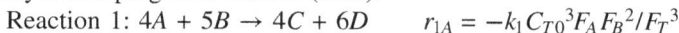

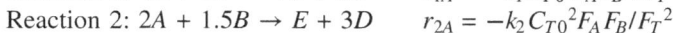

Reaction 1: $4A + 5B \rightarrow 4C + 6D$ $r_{1A} = -k_1 C_{T0}^3 F_A F_B^2/F_T^3$
Reaction 2: $2A + 1.5B \rightarrow E + 3D$ $r_{2A} = -k_2 C_{T0}^2 F_A F_B/F_T^2$
Reaction 3: $2C + B \rightarrow 2F$ $r_{3B} = -k_3 C_{T0}^3 F_B F_C^2/F_T^3$
Reaction 4: $4A + 6C \rightarrow 5E + 6D$ $r_{4C} = -k_4 C_{T0}^{5/3} F_C F_A^{2/3}/F_T^{5/3}$

where

C_{T0} is the total concentration at the entrance to the reactor

$F_i (i = A, B, C, D, E, F)$ is the molar flow rate of component i

F_T is the total molar flow rate, given by $F_T = \sum F_i$

In these reactions, $A = NH_3$, $B = O_2$, $C = NO$, $D = H_2O$, $E = N_2$, and $F = NO_2$. From mole balances we have

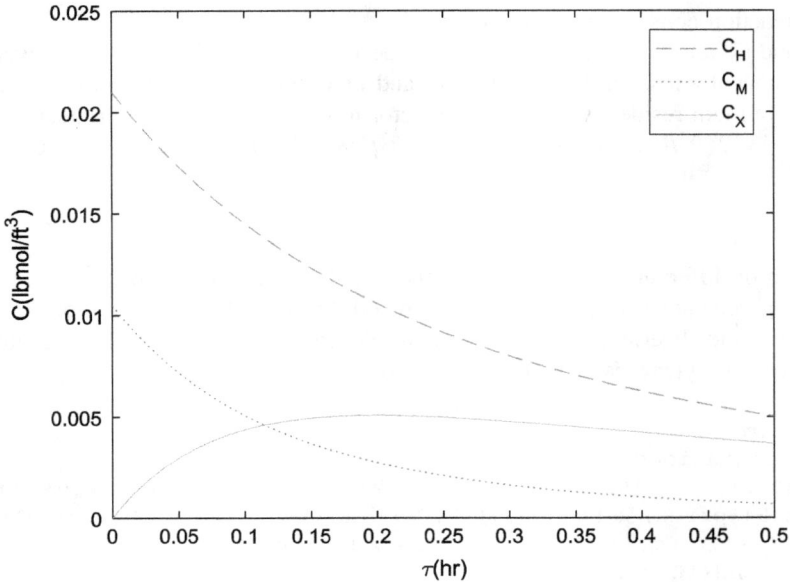

FIGURE 6.19 Concentration of each compound as a function of residence time.

$$\frac{dF_A}{dV} = r_{1A} + r_{2A} + \frac{2}{3}r_{4C}, \quad \frac{dF_B}{dV} = \frac{5}{4}r_{1A} + \frac{3}{4}r_{2A} + r_{3B}, \quad \frac{dF_C}{dV} = -r_{1A} + 2r_{3B} + r_{4C}$$

$$\frac{dF_D}{dV} = -\frac{3}{2}r_{1A} - \frac{3}{2}r_{2A} - r_{4C}, \quad \frac{dF_E}{dV} = -\frac{1}{2}r_{2A} - \frac{5}{6}r_{4C}$$

Plot the molar flow rate profiles as a function of position (V, volume) in a PFR ($0 \le V \le 10$ *liter*).
 Data: $k_1 = 5.0$ $(liter/mol)^2/min$, $k_2 = 2.0$ $liter/(mol \cdot min)$, $k_3 = 10.0$ $(liter/mol)^2/min$, $k_4 = 5.0$ $(liter/mol)^{2/3}/min$, $F_{A0} = F_{B0} = 10$ *mol/min*, $F_{C0} = F_{D0} = F_{E0} = F_{F0} = 0$, $C_{T0} = 2$ *mol/liter*.

Solution
 The system of differential equations is defined by the subfunction *pf*. The script *pfrgpr* defines the required data, uses the built-in function *ode45* to find the molar flow rates, and plots the molar flow rate profiles as a function of position in the reactor volume, as shown in Figure 6.20.

```
% pfrgpr.m
k = [5, 2, 10, 5]; F0 = [10, 10, 0, 0, 0, 0]; Vf = 10; Ct0 = 2;
[V F] = ode45(@pf, [0 Vf], F0, [], k, Ct0);
Fa = F(:,1); Fb = F(:,2); Fc = F(:,3); Fd = F(:,4); Fe = F(:,5); Ff = F(:,6);
plot(V, Fa, '--', V, Fb, ':', V, Fc, V, Fd, '.-',   V, Fe, V, Ff),   xlabel('V(liter)'),
ylabel('F_i(mol/min)')
legend('F_A', 'F_B', 'F_C', 'F_D', 'F_E', 'F_F')

function dF = pf(V,F,k,Ct0)
% F(1) = A, F(2) = B, F(3) = C, F(4) = D, F(5) = E, F(6) = F
% k = [k1 k2 k3 k4]
Ft = sum(F);
r1A = -k(1) * Ct0^3 * F(1)*F(2)^2 / (Ft^3); r2A = -k(2) * Ct0^2 * F(1)*F(2)
/ (Ft^2);
```

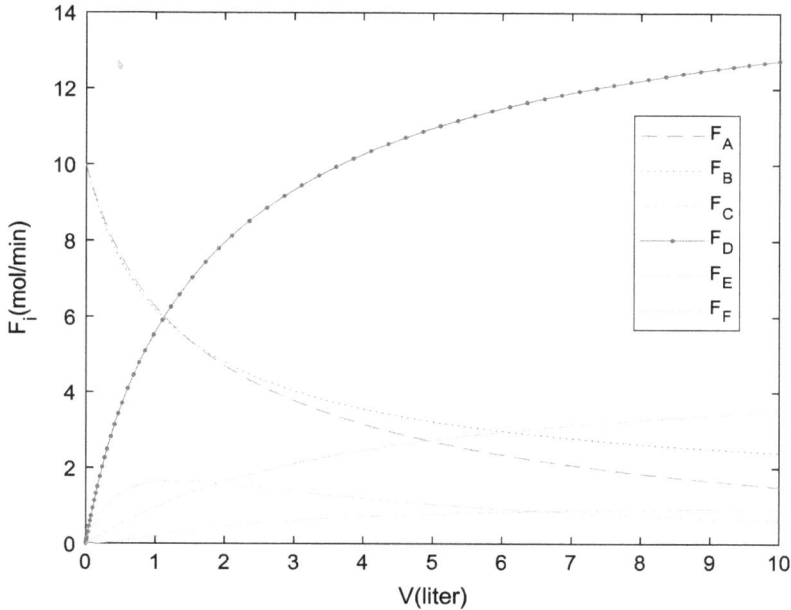

FIGURE 6.20 Molar flow rate profiles.

```
r3B = -k(3) * Ct0^3 * F(2)*F(3)^2 / (Ft^3); r4C = -k(4) * Ct0^(5/3) * F(3)*F(1)^
(2/3) / (Ft^(5/3));
dF = [r1A + r2A + (2/3)*r4C; (5/4)*r1A + (3/4)*r2A + r3B; -r1A + 2*r3B + r4C;
-(3/2)*r1A - (3/2)*r2A - r4C; -(1/2)*r2A - (5/6)*r4C; -2*r3B];
end
```

6.4.2 NONISOTHERMAL PLUG-FLOW REACTOR

The material balance on a plug-flow reactor can be expressed as a 1st-order hyperbolic partial differential equation as follows:

$$\frac{\partial C}{\partial t} + v\frac{\partial C}{\partial x} = r(C), \; C(0, x) = C_0, \; C(t, 0) = C_f$$

where v is the inlet velocity. Application of the method of lines with the central difference formula yields

$$\frac{dC_i}{dt} = -v_i\left(\frac{C_{i+1} - C_{i-1}}{2h}\right) - r(C_i)(i = 2, 3, \cdots, n - 1)$$

$$\frac{dC_i}{dt} = -v_i\left(\frac{C_{i+1} - C_f}{2h}\right) - r(C_i)(i = 1)$$

$$\frac{dC_i}{dt} = -v_i\left(\frac{C_i - C_{i-1}}{h}\right) - r(C_i)(i = n)$$

Example 6.20 Nonisothermal Plug-Flow Reactor[18]

A liquid-phase reaction is carried out in a nonisothermal plug-flow reactor. The model equations are given by

$$\frac{\partial C}{\partial t} + v\frac{\partial C}{\partial x} = r(C),\ C(0,x) = C_0,\ C(t,0) = C_f$$

$$\frac{\partial T}{\partial t} + v\frac{\partial T}{\partial x} = \frac{(-\Delta H_r)r(C)}{\rho C_p} = g(C),\ T(0,x) = T_0,\ T(t,0) = T_f$$

where v is assumed to be constant. The reaction rate is given by

$$r(C) = -\frac{kC}{\sqrt{1 + K_r C^2}},\ k = k_0 e^{-E/(RT)},\ K_r = K_{r0} e^{-\Delta E_r/(RT)}$$

Using the specified rate constants $k = k_1$ and $K_r = K_{r1}$ at $T = T_1$, k and K_r can be represented as

$$k = k_1 \exp\left\{-\frac{E}{R}\left(\frac{1}{T} - \frac{1}{T_1}\right)\right\},\ K_r = K_{r1}\exp\left\{-\frac{\Delta E_r}{R}\left(\frac{1}{T} - \frac{1}{T_1}\right)\right\}$$

Application of the method of lines yields ($\phi = C$ or T, $h = r$ or g)

$$\frac{d\phi_i}{dt} = -v\left(\frac{\phi_{i+1} - \phi_{i-1}}{2h}\right) - h(\phi_i)(i = 2, 3, \cdots, n-1)$$

$$\frac{d\phi_i}{dt} = -v\left(\frac{\phi_{i+1} - \phi_f}{2h}\right) - h(\phi_i)(i = 1)$$

$$\frac{d\phi_i}{dt} = -v\left(\frac{\phi_i - \phi_{i-1}}{h}\right) - h(\phi_i)(i = n)$$

Generate concentration and temperature profiles using the given data.

Data: $C_f = 1\ mol/m^3$, $T_f = T_1 = 450\ K$, $k_1 = 2$, $K_{r1} = 1$, $E = 60\ kJ/mol$, $\Delta E_r = -10\ kJ/mol$, $(-\Delta H_r) = 100\ kJ/mol$, $\rho C_p = 800\ J/(mol \cdot K)$, $L = 2\ m$, $v = 0.4\ m/min$, $n = 50$.

Solution

The subfunction *dfr* defines the system of difference equations to be integrated. The script *pfrnon* uses the built-in function *ode15s* to integrate the system of nonlinear equations defined by *dfr* and produces three-dimensional plots of concentration and temperature as shown in Figure 6.21 when $n = 50$.

```
% pfrnon.m: non-isothermal PFR using the method of line (MoL)
clear all;
% Data and parameters
n = 50; pf.L = 2; pf.v = 0.4; pf.n = n; pf.Cf = 1; pf.Tf = 450; pf.E = 6e4;
pf.dEr = -1e4; pf.dH = -1e5; pf.rCp = 800; pf.T1 = 450; pf.k1 = 0.2; pf.Kr1 = 1;
pf.R = 8.314;
% Initialization
Cf = pf.Cf; Tf = pf.Tf; h = pf.L/n; w = [0:n]*h; w = w(2:end);
Z0 = [ones(1,n)*Cf; ones(1,n)*Tf]; Z0 = reshape(Z0,2*n,1); tspan = [0:0.1:10];
% Employ stiff system solver
[t Z] = ode15s(@dfr,tspan,Z0,[],pf);
C = Z(:,1:2:end); T = Z(:,2:2:end); % Concentration and temperature
% Display results
subplot(1,2,1),   mesh(w,t,C);   xlabel('x(m)'),   ylabel('t(min)'),   zlabel
('C(mol/m^3)'), colormap(gray)
```

(a)   (b)

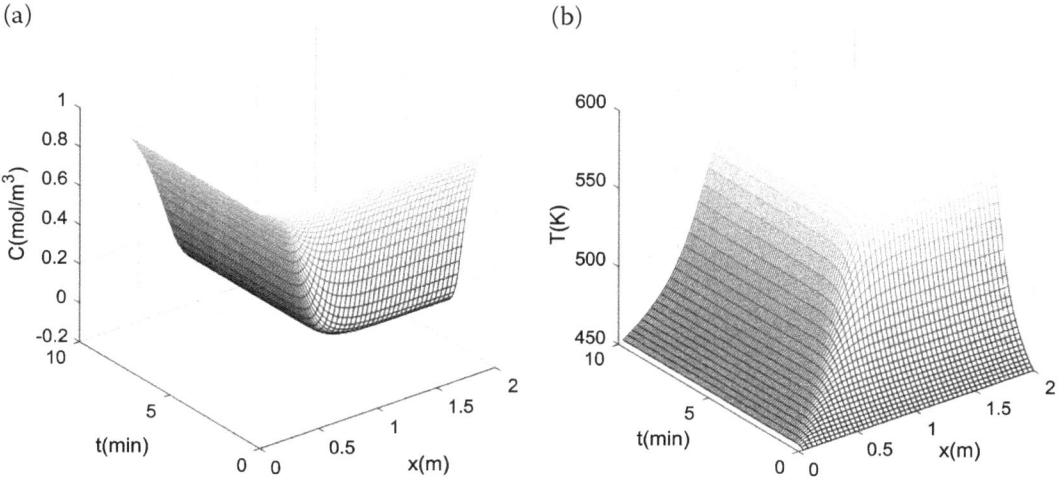

FIGURE 6.21 Profiles of (a) concentration and (b) temperature.

```
subplot(1,2,2), mesh(w,t,T); xlabel('x(m)'), ylabel('t(min)'), zlabel
('T(K)'), colormap(gray)

function dZ = dfr(t,Z,pf)
% Non-isothermal PFR
% Retrieve data
L = pf.L; v = pf.v; n = pf.n; Cf = pf.Cf; Tf = pf.Tf; E = pf.E; dEr = pf.dEr;
T1 = pf.T1; k1 = pf.k1; Kr1 = pf.Kr1; dH = pf.dH; R = pf.R; rCp = pf.rCp;
% Initialization
h = L/n; Z = reshape(Z,2,n); C = Z(1,:); T = Z(2,:); dC = zeros(n,1); dT =
zeros(n,1);
% Difference equations
for i = 1:n
  k = k1*exp(-(E/R)*(1/T(i) - 1/T1)); Kr = Kr1*exp(-(dEr/R)*(1/T(i) - 1/T1));
  rx = k*C(i)/sqrt(1 + Kr*C(i)^2);
  if i == 1, s = (v/h)*(C(i) - Cf); d = (v/h)*(T(i) - Tf);
  else, s = (v/h)*(C(i) - C(i-1)); d = (v/h)*(T(i) - T(i-1)); end
  dC(i) = - s - rx; dT(i) = - d + (-dH)*rx/rCp; % difference model equation
end
dZ = reshape([dC'; dT'],2*n,1);
end
```

6.4.3 ADIABATIC REACTION IN A PLUG-FLOW REACTOR

Suppose that the elementary gas-phase reversible reaction $A \leftrightarrow B$ is carried out in a plug-flow reactor in which pressure drop is neglected ($P = P_0$) and pure A enters the reactor. The mole balance gives

$$\frac{dX}{dV} = \frac{-r_A}{F_{A0}}$$

where F_{A0} is molar flow rate of component A fed to the reactor. The reaction rate and rate constants can be expressed by

$$-r_A = k\left(C_A - \frac{C_B}{K_C}\right), \quad k = k_1(T_1)\exp\left[\frac{E}{R}\left(\frac{1}{T_1} - \frac{1}{T}\right)\right], \quad K_C = K_{C2}(T_2)\exp\left[\frac{\Delta H_{R_x}^\circ}{R}\left(\frac{1}{T_2} - \frac{1}{T}\right)\right]$$

where $k_1(T_1)$ and $K_{C2}(T_2)$ are values of k and K_C at T_1 and T_2, respectively. For the gas-phase reaction, $\epsilon = 0$ and we have

$$C_A = C_{A0}(1 - X)\frac{T_0}{T}, \; C_B = C_{A0}X\frac{T_0}{T}$$

where T_0 is the temperature of the feed stream. Combination of these equations yields

$$-r_A = kC_{A0}\left[(1 - X) - \frac{X}{K_C}\right]\frac{T_0}{T}$$

From the energy balance on the reactor, T can be expressed as a function of conversion as

$$T = \frac{X\left[-\Delta H^\circ_{R_x}(T_R)\right] + \Sigma\, \Theta_i C_{pi} T_0 + X\Delta C_p T_R}{\Sigma\, \Theta_i C_{pi} + X\Delta C_p}$$

Example 6.21 Adiabatic Liquid-Phase Isomerization of n-Butane[19]

n-Butane (C_4H_{10}) is to be isomerized to isobutane in a plug-flow reactor:

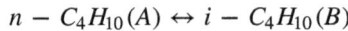

$$n - C_4H_{10}(A) \leftrightarrow i - C_4H_{10}(B)$$

The reaction is an elementary reversible reaction to be carried out adiabatically in the liquid phase under high pressure using essentially trace amounts of a liquid catalyst. The feed enters at $T_0 = 330\,K$. At $T_1 = 360\,K$ and $T_2 = 333\,K$, it is known that $k_1(T_1) = 31.1\,hr^{-1}$ and $K_{C2}(T_2) = 3.03\,hr^{-1}$. A mixture of 90 mol% n-butane and 10 mol% i-pentane, which is considered inert, is to be processed at 70% conversion. The molar flow rate of the mixture is 163 $kmol/hr$. Plot the conversion X, equilibrium conversion X_e, temperature T, and reaction rate $-r_A$ down the length of the reactor.

Data: $\Delta H^\circ_{R_x} = -6900\,J/mol$, activation energy $E = 65,700\,J/mol$, $C_{A0} = 9.3\,kmol/m^3$, $R = 8.314\,J/(mol\cdot K)$, $C_{p_{n-B}} = 141\,J/(mol\cdot K)$, $C_{p_{i-B}} = 141\,J/(mol\cdot K)$, $C_{p_{i-P}} = 161\,J/(mol\cdot K)$.

Solution

From the given data, we can see that

$$F_{A0} = 0.9F_{T0} = (0.9)(163) = 146.7\,kmol/hr, \; T_0 = 330\,K$$

$$\Sigma\, \Theta_i C_{pi} = C_{pA} + \Theta_I C_{pI} = \left(141 + \frac{0.1}{0.9}161\right) = 158.889\,J/(mol\cdot K)$$

Since $\Delta C_p = C_{pB} - C_{pA} = 141 - 141 = 0$, we have

$$T = \frac{X\left[-\Delta H^\circ_{R_x}(T_R)\right] + \Sigma\, \Theta_i C_{pi} T_0}{\Sigma\, \Theta_i C_{pi}} = T_0 + \frac{X\left[-\Delta H^\circ_{R_x}(T_R)\right]}{\Sigma\, \Theta_i C_{pi}} = 330 + \frac{-(-6900)}{158.889}X$$

$$= 330 + 43.4265X$$

Substitution of the concentrations $C_A = C_{A0}(1 - X)$ and $C_B = C_{A0}X$ into the reaction rate equation gives

$$-r_A = kC_{A0}\left[1 - \left(1 + \frac{1}{K_C}\right)X\right]$$

The change of conversion with respect to reactor volume is given by

$$\frac{dX}{dV} = \frac{-r_A}{F_{A0}}$$

At equilibrium, $- r_A = 0$ and the equilibrium conversion is given by

$$X_e = \frac{K_c}{1 + K_c}$$

The subfunction *fad* defines the differential equation. The script *adbpfr* sets the data and employs the built-in function *ode45* to solve the differential equation defined by the subfunction *fad*.

```
% adbpfr.m
clear all;
Ca0 = 9.3; Fa0 = 146.7; T1 = 360; T2 = 333; k1 = 31.1; K2 = 3.03;
E = 65700; R = 8.314; dH = -6900; Vspan = [0 5]; X0 = 0;
[V X] = ode45(@fad,Vspan,X0,[],Ca0,Fa0,T1,T2,k1,K2,E,R,dH);
T = 330 + 43.4265*X; k = k1*exp(E*(1/T1 - 1./T)/R); Kc = K2*exp(dH*(1/T2 - 1./
T)/R);
ra = -k*Ca0.*(1 - (1 + 1./Kc).*X); Xe = Kc./(1+Kc);
subplot(2,2,1), plot(V,T), xlabel('V'), ylabel('T(K)')
subplot(2,2,2), plot(V,-ra), xlabel('V'), ylabel('-r_A')
subplot(2,2,3), plot(V,X,V,Xe,'--'), xlabel('V'), ylabel('X,X_e'), le-
gend('X','Xe')
fprintf('Conversion (X) and equilibrium conversion (Xe): Xf = %g, Xef = %g\n',X
(end),Xe(end));
fprintf('Final temperature: Tf = %g\n',T(end));
fprintf('Final reaction rate: raf = %g\n',-ra(end));

function dX = fad (V,X,Ca0,Fa0,T1,T2,k1,K2,E,R,dH)
T = 330 + 43.4265*X; k = k1*exp(E*(1/T1 - 1/T)/R); Kc = K2*exp(dH*(1/T2 - 1/
T)/R);
ra = -k*Ca0*(1 - (1 + 1/Kc)*X); dX = -ra/Fa0;
end
```

The script *adbpfr* produces the following outputs and Figure 6.22:

```
>> adbpfr
Conversion (X) and equilibrium conversion (Xe): Xf = 0.714056, Xef = 0.714063
Final temperature: Tf = 361.009
Final reaction rate: raf = 0.00312054
```

6.5 CATALYTIC REACTORS

6.5.1 CHARACTERISTICS OF CATALYTIC REACTION

Steps in a catalytic reaction consist of diffusion from the bulk to the external surface of the catalyst, surface reaction, and desorption. For the hydrodemethylation reaction in which toluene (T) reacts with hydrogen (H) to produce benzene (B) and methane (M), the reaction rate can be expressed by

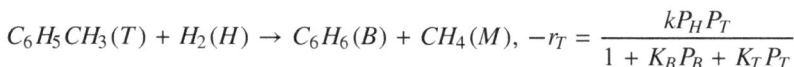

$$C_6H_5CH_3(T) + H_2(H) \rightarrow C_6H_6(B) + CH_4(M), \quad -r_T = \frac{kP_HP_T}{1 + K_BP_B + K_TP_T}$$

where $P_i(i = T, H, B, M)$ is the partial pressure of each species and k, K_B, K_T are parameters.

(a)

(b)

(c)

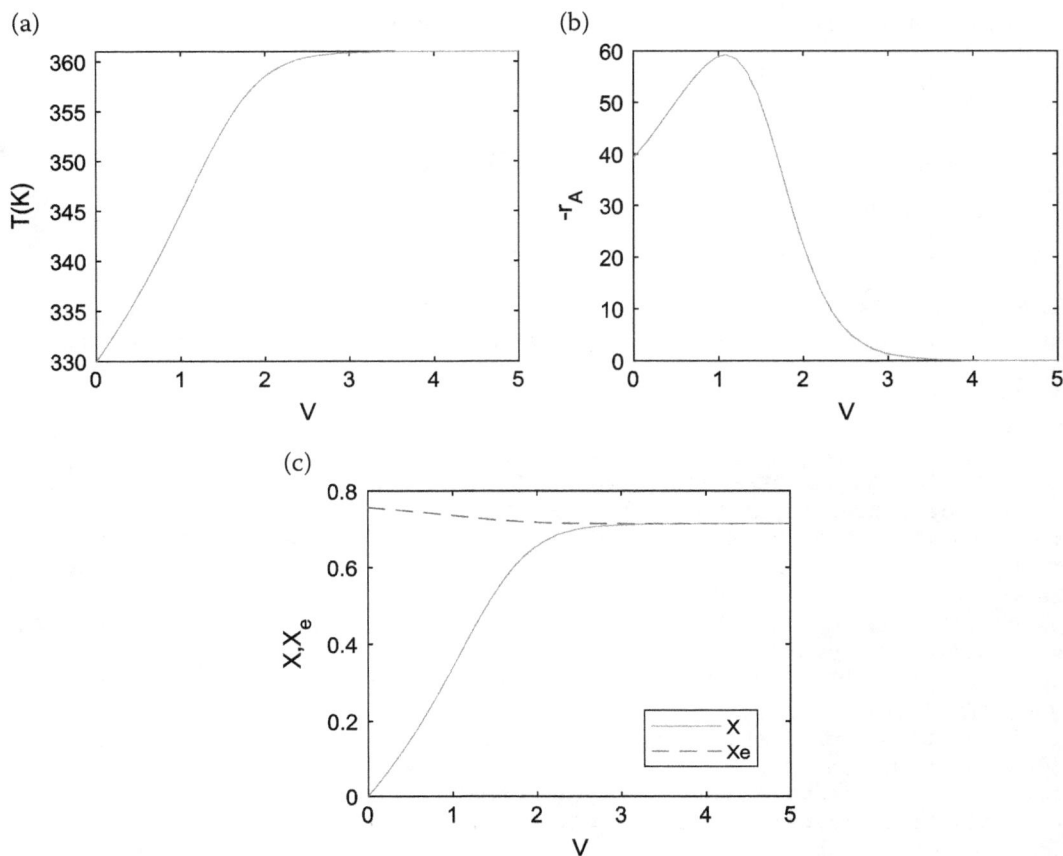

FIGURE 6.22 Profiles of (a) temperature, (b) reaction rate, and (c) conversion.

Consider an isothermal gas-phase reaction being carried out in a packed-bed catalytic reactor. The rate of change of conversion X with respect to the catalyst weight W can be expressed by

$$\frac{dX}{dW} = \frac{-r_A}{F_{A0}}$$

and the concentration of the key component (A) is given by

$$C_A = C_{A0}\left(\frac{1-X}{1+\epsilon X}\right)\frac{P}{P_0} = C_{A0}\left(\frac{1-X}{1+\epsilon X}\right)y$$

where $y = P/P_0$, $\epsilon = y_{A0}\delta = F_{A0}\delta/F_{T0}$, and δ is the difference in the stoichiometric ratio. When there is no change in the number of moles, $\epsilon = 0$. The pressure drop in a packed-bed reactor is given by[20]

$$\frac{dy}{dW} = -\frac{\alpha}{2}\left(\frac{1+\epsilon X}{y}\right)\frac{T}{T_0}$$

where α is the pressure drop parameter. The pressure drop along the reactor length z is expressed by

$$\frac{dP}{dz} = -\beta_0 \frac{P_0}{P}\left(\frac{T}{T_0}\right)\frac{F_T}{F_{T0}}$$

where

β_0 is a constant related to a packed bed

P_0 and T_0 are the initial pressure and temperature

F is the molar flow rate

The ratio of volumetric flow rates is given by

$$f = \frac{v}{v_0} = \frac{1 + \epsilon X}{y}$$

Example 6.22 Estimation of Catalytic Reaction Parameters[21]

Table 6.5 shows reaction rates and partial pressures for the hydrodemethylation reaction in which toluene (T) reacts with hydrogen (H) to produce benzene (B) and methane (M). The reaction rate can be expressed by

$$-r_T = \frac{kP_H P_T}{1 + K_B P_B + K_T P_T}$$

Use the regression method along with the data in Table 6.5 to find the best estimates of the rate law parameters k, K_B, and K_T.

Solution

Rearrangement of the given rate equation gives

TABLE 6.5
Reaction Rate and Partial Pressure

$-r_T \times 10^{10}$	Partial Pressure (atm)			
	Toluene (P_T)	Hydrogen (P_H)	Methane (P_M)	Benzene (P_B)
71.0	1	1	1	0
71.3	1	1	4	0
41.6	1	1	0	1
19.7	1	1	0	4
42.0	1	1	1	1
17.1	1	1	0	5
71.8	1	1	0	0
142.0	1	2	0	0
284.0	1	4	0	0
47.0	0.5	1	0	0
71.3	1	1	0	0
117.0	5	1	0	0
127.0	10	1	0	0
131.0	15	1	0	0
133.0	20	1	0	0
41.8	1	1	1	1

$$-\frac{P_H P_T}{r_T} = \frac{1}{k} + \left(\frac{K_B}{k}\right)P_B + \left(\frac{K_T}{k}\right)P_T$$

We can use the linear regression method to estimate parameters. The script *pbrpar* employs the linear regression method to estimate the parameters and generates a plot of both measured and estimated reaction rates.

```
% pbrpar.m
r = 1e-10*[71.0 71.3 41.6 19.7 42.0 17.1 71.8 142.0 284.0 47.0 71.3 117.0 127.0
131.0 133.0 41.8];
Pt = [1 1 1 1 1 1 1 1 1 0.5 1 5 10 15 20 1]; Ph = [1 1 1 1 1 1 1 2 4 1 1 1 1 1 1 1];
Pm = [1 4 0 0 1 0 0 0 0 0 0 0 0 0 0 1]; Pb = [0 0 1 4 1 5 0 0 0 0 0 0 0 0 0 1];
n = length(r); A = [ones(n,1) Pb' Pt']; b = Ph.*Pt./r;
x = inv(A'*A)*A'*b; k = 1/x(1), Kb = x(2)>k, Kt = x(3)*k
rc = k*Ph.*Pt./(1 + Kb*Pb + Kt*Pt); nc = 1:n;
plot(nc,r*1e10,'o',nc,rc*1e10,'*'), legend('Data','Estimated'), xlabel('Run
#'), ylabel('-10^{10}r')
```

The script *pbrpar* calculates the rate law parameters k, K_B, and K_T and generates the plot shown in Figure 6.23:

```
>> pbrpar
k =
 1.4047e-08
Kb =
 logical
 1
Kt =
 1.0058
```

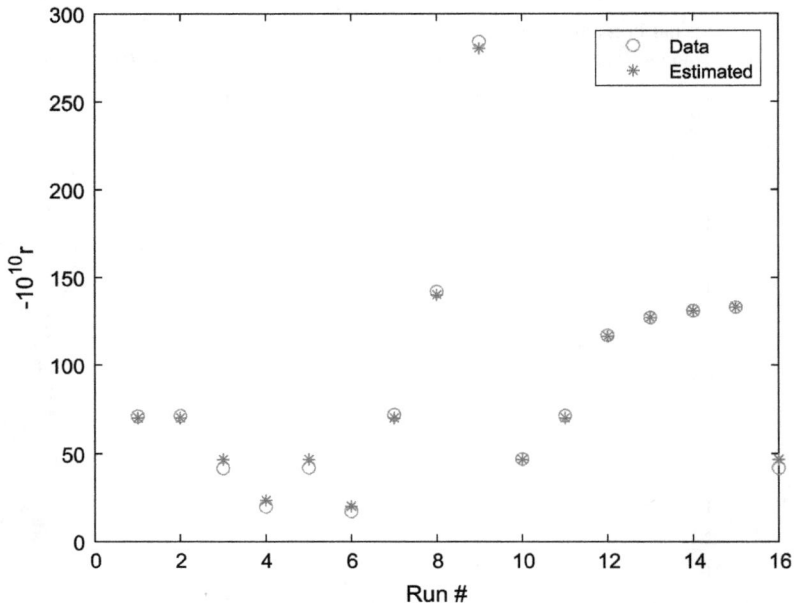

FIGURE 6.23 Comparison of reaction rates.

6.5.2 Diffusion and Reaction in a Catalyst Pellet

A reaction taking place in a spherical catalyst pellet of radius R can be expressed as

$$\frac{D_e}{r^2}\frac{d}{dr}\left(r^2\frac{dC_A}{dr}\right) = D_e\frac{d^2C_A}{dr^2} + \frac{2D}{r}\frac{dC_A}{dr} = r_A, \; C_A|_{r=R} = C_{A0}, \; \left.\left(\frac{2D}{r}\frac{dC_A}{dr} - r_A\right)\right|_{r=0} = 0$$

where
  D is the effective diffusivity
  C_A is the concentration of the reactant
  The reaction is represented by Langmuir-Hinshelwood reaction kinetics as

$$r_A = \frac{kC_A}{\sqrt{1 + K_r C_A{}^2}}$$

Application of the finite difference formula yields

$$\frac{D_e}{h^2}(C_{i+1} - 2C_i + C_{i-1}) + \frac{D_e}{r_i h}(C_{i+1} - C_{i-1}) - r_A(C_i) = 0 \, (i = 2, 3, \cdots, n)$$

$$\frac{2D_e}{h^2}(C_{i+1} - C_i) - r_A(C_i) = 0 \, (i = 1), \; C_i - C_{A0} = 0 \, (i = n + 1)$$

where n denotes the number of subintervals.

Example 6.23 Concentration Profile[22]

Generate the concentration profiles for various C_{A0} when $R = 0.2 \, cm$, $k = 100 \, sec^{-1}$, $D_e = 0.25 \, cm^6/mol$, and $K_r = 10^9 \, cm^6/mol^2$. Try four different values of C_{A0}: $C_{A0} = 5 \times 10^{-5}$, 10×10^{-5}, 15×10^{-5}, and $20 \times 10^{-5} \, mol/cm^3$.

Solution

The function *catLH* defines the system of difference equations to be solved. The script *rxnLH* uses the built-in function *fsolve* and produces the profiles of C_A shown in Figure 6.24.

```
function z = catLH(y,De,n,h,k,Kr,Ca0)
% Diffusion and reaction in a catalyst pellet by Langmuir-Hinshelwood kinetics
% y: dimensionless concentration, De: effective diffusivity
m = n+1;
for i = 1:m, x(i) = h*(i-1); end % radial distance
for i = 1:m
  rxn = k*y(i)/sqrt(1 + Kr*y(i)^2);
  if i == 1, z(i) = 2*De*(y(i+1) - y(i))/h^2 - rxn;
  elseif i == m, z(i) = y(i) - Ca0; % i = n+1
  else % i = 2,3,...,n
    z(i) = De*(y(i+1)-2*y(i)+y(i-1))/h^2 + De*(y(i+1)-y(i-1))/(h*x(i)) - rxn;
  end
end
% rxnLH.m: diffusion and reaction in a catalyst pellet by Langmuir-Hinshelwood
kinetics
clear all;
n = 50; R = 0.2; k = 100; De = 0.25; Kr = 1e9; % data
m = n+1; h = R/n; y0 = zeros(n+1,1); Ca0 = 5e-5*[1 2 3 4]; p = [];
for i = 1:m, x(i) = h*(i-1); end % radial distance
```

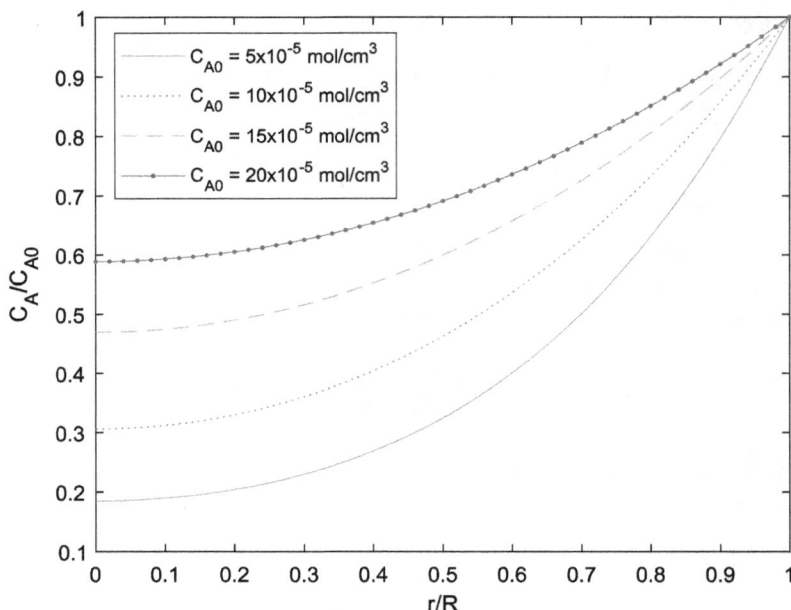

FIGURE 6.24   Effect of concentration change at the catalyst surface.

```
for i = 1:length(Ca0)
  y0 = Ca0(i)*ones(n+1,1); y = fsolve(@catLH,y0,[],De,n,h,k,Kr,Ca0(i)); p = [p
y/Ca0(i)];
end
x = x/R; plot(x,p(:,1),x,p(:,2),':',x,p(:,3),'--',x,p(:,4),'.-')
xlabel('r/R'), ylabel('C_A/C_{A0}')
legend('C_{A0} = 5x10^{-5} mol/cm^3','C_{A0} = 10x10^{-5} mol/cm^3',...
       'C_{A0}  =  15x10^{-5}  mol/cm^3','C_{A0}  =  20x10^{-5}  mol/cm^3
','location','best')
```

6.5.3 CATALYTIC REACTIONS IN A PACKED-BED REACTOR

Packed-bed reactors are among the most common reactors used in the chemical industry, due to their high conversion rate per catalyst weight compared to other catalytic reactors. Packed-bed reactors are very versatile and are used in many chemical processing applications, such as absorption, distillation, stripping, separation, and catalytic reactions. Typical packed-bed reactors consist of a chamber, such as a tube or channel that contains catalyst particles or pellets, and a liquid or gas that flows through the catalyst. The liquid or gas interacts with the catalyst across the length of the reactor tube, altering the chemical composition of the substance.

The packed catalyst in the reactor can be modeled as a porous structure, which leads to particle transport with different orders of magnitude, making the analysis of mass and energy transport a challenging task. Catalyst pellets are usually granular, and a catalyst particle's radius is typically in the order of magnitude of 1 *mm*. The space located between catalyst particles is described as the macroporous structure of the bed, while pores inside the catalyst form what is known as the microstructure. These catalyst particles can be loaded into the packed-bed reactor in several ways: as a single bed, as separate shells, or in tubes. Catalysts are typically made from nickel, copper, osmium, platinum, and rhodium. Some catalysts are made of precious metal on 3.175 *mm* ceramic beads and are used in electric catalytic oxidizers that treat air streams contaminated with volatile organic compounds.

When designing a packed-bed reactor, one must take into account the active life of the catalyst. This will affect the length of time a bed of catalyst may be used, and thus how long the reactor may be run before the catalyst needs to be regenerated. Another challenge when designing a packed-bed reactor lies in the pressure drop that occurs across the length of the reactor. The pressure drop can be reduced by using larger catalyst particles, but this causes lower intraparticle diffusion, making the reaction progress slower. The trade-off here is to find a particle size that is large enough to limit the pressure drop and small enough to allow the reaction to proceed at a fast enough rate.

Example 6.24 Gas-Phase Reaction in a Packed-Bed Reactor[23]

The irreversible gas-phase catalytic reaction

$$A + B \rightarrow C + D$$

is to be carried out in a packed-bed reactor with four different catalysts (Catalyst 1, Catalyst 2, Catalyst 3, Catalyst 4). For each catalyst, the rate expression has a different form:

Catalyst 1: $-r_{A1} = kC_A C_B/(1 + K_A C_A)$
Catalyst 2: $-r_{A2} = kC_A C_B/(1 + K_A C_A + K_C C_C)$
Catalyst 3: $-r_{A3} = kC_A C_B/(1 + K_A C_A + K_B C_B)^2$
Catalyst 4: $-r_{A4} = kC_A C_B/(1 + K_A C_A + K_B C_B + K_C C_C)^2$

The initial concentrations of the reactants are $C_{A0} = C_{B0} = 1.0 \ gmol/dm^3$ at the reactor inlet, and the molar feed flow rate of A is $F_{A0} = 1.5 \ gmol/min$. There is a total of $W_{max} = 2 \ kg$ of each catalyst used in the reactor. The reaction rate constant is $k = 10 \ dm^6/(kg \cdot min)$, and the various catalytic parameters K_A, K_B, and K_C are given by $K_A = 1 \ dm^3/gmol$, $K_B = 2 \ dm^3/gmol$, and $K_C = 20 \ dm^3/gmol$. Calculate and plot the conversion X versus the catalyst weight W for each of the catalytic rate expressions when the reactor operation is at a constant pressure and $\alpha = 0.4$.

Solution

The mass balance for each catalyst yields $dX/dW = -r_{Ai}/F_{A0} (i = 1, 2, 3, 4)$. Since there is no change in the number of moles, $\epsilon = 0$ and we have $C_A = C_B = C_{A0}(1 - X)$, $C_C = C_D = C_{A0}X$. The function *pbrmf* defines the differential equations.

```
function dxdw = pbrmf(w,x,k,fa0,ca0,ka,kb,kc)
kfc = k*ca0^2/fa0;
dxdw = [kfc*(1-x(1))^2/(1+ka*ca0*(1-x(1)));
    kfc*(1-x(2))^2/(1+ka*ca0*(1-x(2))+kc*ca0*x(2));
    kfc*(1-x(3))^2/(1+ka*ca0*(1-x(3))+kb*ca0*(1-x(3)))^2;
    kfc*(1-x(4))^2/(1+ka*ca0*(1-x(4))+kb*ca0*(1-x(4))+kc*ca0*x(4))^2];
end
```

The built-in function *ode45* can be used to integrate the differential equations. The script *catwt* generates the plot of the conversion versus the catalyst weight for each of the catalytic rate expressions (see Figure 6.25).

```
% catwt.m
ca0 = 1; fa0 = 1.5; k = 10; ka = 1; kb = 2; kc = 20; wf = 2;  % data
wint = [0 wf]; x0 = [0 0 0 0];
[w x] = ode45(@pbrmf,wint,x0,[],k,fa0,ca0,ka,kb,kc);
plot(w,x(:,1),w,x(:,2),':',w,x(:,3),'.-',w,x(:,4),'--'), xlabel('W(kg)'),
ylabel('X')
legend('X_1','X_2','X_3','X_4','location','best')
```

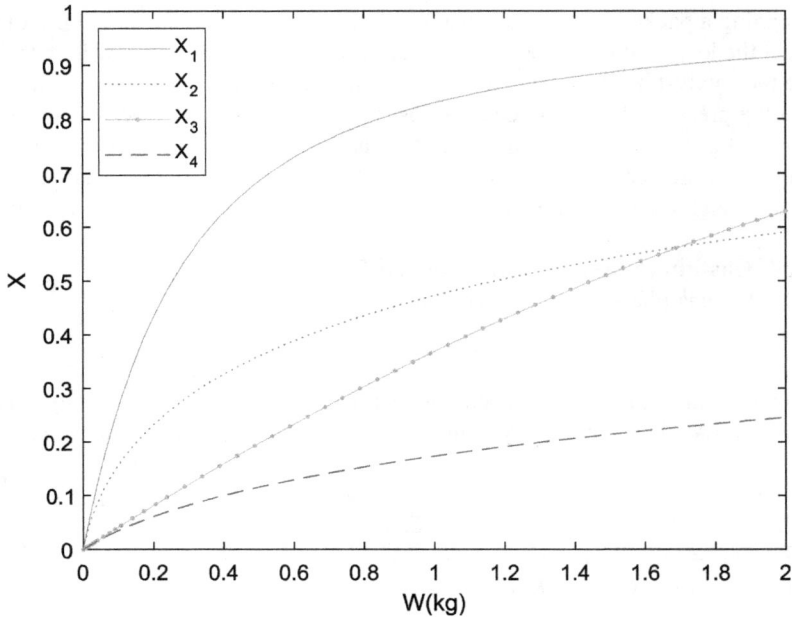

FIGURE 6.25 Conversion versus catalyst weight.

Example 6.25 Complex Reactions in a Packed-Bed Reactor[24]

The complex gas-phase reactions

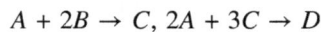

$$A + 2B \rightarrow C, \ 2A + 3C \rightarrow D$$

take place isothermally in a packed-bed reactor. These reactions follow elementary rate laws, and the reaction rates for the species can be written as

$$r_A = \frac{dF_A}{dW} = -k_a C_A C_B^2 - \frac{2}{3} k_c C_A^2 C_C^3, \ r_B = \frac{dF_B}{dW} = -2k_a C_A C_B^2$$

$$r_C = \frac{dF_C}{dW} = k_a C_A C_B^2 - k_c C_A^2 C_C^3, \ r_D = \frac{dF_D}{dW} = \frac{1}{3} k_c C_A^2 C_C^3$$

where

W is the catalyst weight (kg)

F_i is the molar flow rate of species i

The concentration of each species can be expressed by

$$C_i = C_{T0} \left(\frac{F_i}{F_T} \right) y \ (i = A, B, C, D)$$

where $F_T = F_A + F_B + F_C + F_D$ and the rate of change of y with respect to W is given by

$$\frac{dy}{dW} = -\frac{\alpha}{2y} \left(\frac{F_T}{F_{T0}} \right), \ y(0) = 1$$

The selectivity is defined by $S_{C/D} = F_C/F_D$. Plot the molar flow rate of each species and the selectivity as a function of catalyst weight, W ($0 \le W \le 1000 \ (kg)$).

Data: $C_{T0} = 0.2\ mol/dm^3$, $\alpha = 0.0019\ kg^{-1}$, $k_a = 100\ (dm^3/mol)^2/(min \cdot kg_{cat})$, $k_c = 1500\ dm^{15}/(mol^4 \cdot min \cdot kg_{cat})$, $F_{T0} = 20\ mol/min$ $(F_{A0} = F_{B0} = 10\ mol/min)$

Solution

The function *pbrmult* defines the differential equations. In the function, the dependent variable vector x is defined as $x^T = [\,F_A\quad F_B\quad F_C\quad F_D\quad y\,]$.

```
function frx = pbrmult(W,x,ka,kc,Ct0,Ft0,alpha)
% x(1)=Fa, x(2)=Fb, x(3)=Fb, x(4)=Fd, x(5)=y
Ft = x(1) + x(2) + x(3) + x(4);
for i = 1:4, C(i) = Ct0*x(i)*x(5)/Ft; end
frx = [-ka*C(1)*C(2)^2 - 2*kc*C(1)^2*C(3)^3/3; -2*ka*C(1)*C(2)^2;
      ka*C(1)*C(2)^2 - kc*C(1)^2*C(3)^3; kc*C(1)^2*C(3)^3/3; - alpha*Ft/
(2*x(5)*Ft0)];
end
```

The script *compflr* calls the function *pbrmult*, employs the built-in function *ode45* to solve the differential equations, and generates plots of molar flow rates and selectivity as a function of catalyst weight. To avoid division by zero, the value of the selectivity is set to zero when F_D is very small.

```
% compflr.m
clear all;
ka = 100; kc = 1500; Ct0 = 0.2; Ft0 = 20; alpha = 0.0019; % data
x0 = [10 10 0 0 1]; Wv = [0 1000]; [W x] = ode45(@pbrmult,Wv,x0,
[],ka,kc,Ct0,Ft0,alpha);
Fa = x(:,1); Fb = x(:,2); Fc = x(:,3); Fd = x(:,4); y = x(:,5);
n = length(W); Scd = zeros(1,n);
for i = 1:n
  if Fd(i) <= 1e-4, Scd(i) = 0; else, Scd(i) = Fc(i)/Fd(i); end
end
subplot(1,2,1), plot(W,Fa,W,Fb,':',W,Fc,'.-',W,Fd,'--',W,y,'.')
xlabel('W'), ylabel('F_i'), legend('F_A','F_B','F_C','F_D','y')
subplot(1,2,2), plot(W,Scd), xlabel('W'), ylabel('S_{C/D}')
fprintf('Final molar flow rate of each species: \n');
fprintf(' Faf=%g, Fbf=%g, Fcf=%g, Fdf=%g\n',Fa(end),Fb(end),Fc(end),Fd(end));
fprintf('Final value of y: yf = %g\n',y(end)); fprintf('Selectivity: Scdf = %g
\n',Scd(end));
```

The script *compflr* produces the following outputs and generates Figure 6.26:

```
>> compflr
Final molar flow rate of each species:
 Faf = 4.29365, Fbf = 0.340968, Fcf = 3.51426, Fdf = 0.438418
Final value of y: yf = 0.257788
Selectivity: Scdf = 8.01579
```

Example 6.26 Conversion in a Packed-Bed Reactor[25]

Ethylene oxide (C) can be produced by the vapor-phase catalytic oxidation of ethylene (A) with air (B):

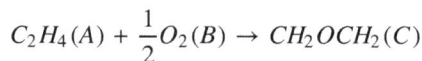

$$C_2H_4(A) + \frac{1}{2}O_2(B) \rightarrow CH_2OCH_2(C)$$

$$-r_A = kP_A^{1/3}P_B^{2/3} = kP_{A0}\left(\frac{1}{2}\right)^{2/3}\left(\frac{1-X}{1+\epsilon X}\right)y = k'\left(\frac{1-X}{1+\epsilon X}\right)y$$

(a)

(b)

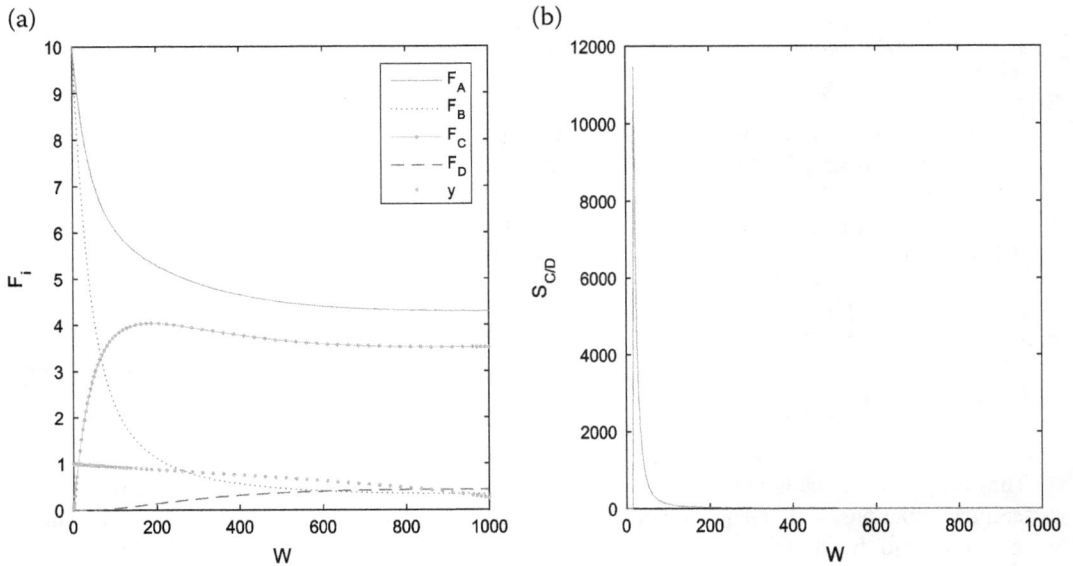

FIGURE 6.26   Profiles of (a) molar flow rates and (b) selectivity.

The ratio of the volumetric flow rate, f, is given by

$$f = \frac{v}{v_0} = \frac{1 + \epsilon X}{y}$$

Calculate the catalyst weight necessary to achieve 60% conversion and plot X, y, f, and the reaction rate as a function of catalyst weight.

   Data: $k = 0.00392 \; mol/(atm{\cdot}kg_{cat}{\cdot}sec)$, $F_{A0} = 0.1362 \; mol/sec$, $F_{B0} = 0.068 \; mol/sec$, $P_0 = 10 \; atm$, $\alpha = 0.0367 \; kg^{-1}$.

Solution
   Substitution of the reaction rate equation into the relation

$$\frac{dX}{dW} = \frac{-r_A}{F_{A0}}$$

gives

$$\frac{dX}{dW} = \frac{-r_A'}{F_{A0}} = \frac{k'}{F_{A0}}\left(\frac{1-X}{1+\epsilon X}\right)y, \; X(0) = 0$$

Since the temperature is constant, we have

$$\frac{dy}{dW} = -\frac{\alpha}{2}\left(\frac{1+\epsilon X}{y}\right)\frac{T}{T_0} = -\frac{\alpha}{2}\left(\frac{1+\epsilon X}{y}\right), \; y(0) = 1$$

where

$$F_I = F_{B0}\left(\frac{79}{21}\right), \; F_{T0} = F_{A0} + F_{B0} + F_I, \; y_{A0} = \frac{F_{A0}}{F_{T0}}, \; P_{A0} = y_{A0}P_0, \; k' = kP_{A0}\left(\frac{1}{2}\right)^{2/3},$$

$$\delta = 1 - \frac{1}{2} - 1,\ \epsilon = y_{A0}\delta$$

Differential equations are integrated over the interval $W_{span} = [0\ \ W_f]$. The goal is to determine W_f at which the conversion $X = 0.6$ is achieved. This problem can be solved by using the bisection method. The initial search interval can be set as $15 \leq W_f \leq 25$. We guess the initial search interval and reduce the interval iteratively using the bisection method. The subfunction *pbconv* defines differential equations. The script *catwtx* calls the subfunction *pbconv* and employs the built-in solver *ode45* to find solutions.

```
% catwtx.m
clear all;
k = 0.00392; Fa0 = 0.1362; Fb0 = 0.068; P0 = 10; alpha = 0.0367; % data
Fi = Fb0*(79/21); Ft0 = Fa0 + Fb0 + Fi; ya0 = Fa0/Ft0;
Pa0 = ya0*P0; kp = k*Pa0*(0.5)^(2/3); delta = -0.5; epsilon = ya0*delta;
Wfa = 15; Wfb = 25; % initial guess of catalyst weight
z0 = [0 1]; Xf = 0.6;
while (abs(Wfa-Wfb) >= 1e-3)
        Wfm  =  (Wfa+Wfb)/2;   [Wa   Za]   =   ode45(@pbconv,[0   Wfa],z0,
[],kp,Fa0,epsilon,alpha);
[Wm Zm] = ode45(@pbconv,[0 Wfm],z0,[],kp,Fa0,epsilon,alpha);
[Wb Zb] = ode45(@pbconv,[0 Wfb],z0,[],kp,Fa0,epsilon,alpha);
Xa = Za(end,1); Xm = Zm(end,1); Xb = Zb(end,1);
if (Xa-Xf)*(Xm-Xf) < 0, Wfb = Wfm;
else, Wfa = Wfm; end
end
X = Zm(:,1); y = Zm(:,2); W = Wm;
fprintf('Catalyst weight = %g, Conversion = %g\n',Wfm, X(end));
fm = (1 + epsilon*X)./y; % ratio of the volumetric flow rate
rp = -kp*(1 - X).*y./(1 + epsilon*X); % reaction rate
subplot(1,2,1), plot(Wm,X,Wm,y,':',Wm,fm,'--'), legend('X','y','f'), xlabel
('W(kg)'), grid on
subplot(1,2,2), plot(Wm,-rp), xlabel('W(kg)'), ylabel('-r_A'), grid on
function dz = pbconv(w,z,kp,Fa0,epsilon,alpha)
% x = z(1), y = z(2)
dz = [kp*(1-z(1))*z(2)/(Fa0*(1 + epsilon*z(1))); - alpha*(1 + epsilon*z(1))/
(2*z(2))];
end
```

The script *catwtx* produces the desired results and plots X, y, f, and the reaction rate as a function of catalyst weight (see Figure 6.27).

```
>> catwtx
Catalyst weight = 20.5634, Conversion = 0.599998
```

6.5.4 PACKED-BED REACTOR WITH AXIAL MIXING

A packed-bed reactor with axial mixing can be expressed by a parabolic partial differential equation as

$$\frac{\partial C}{\partial t} + v\frac{\partial C}{\partial x} = D\frac{\partial^2 C}{\partial x^2} - kC$$

(a)

(b)

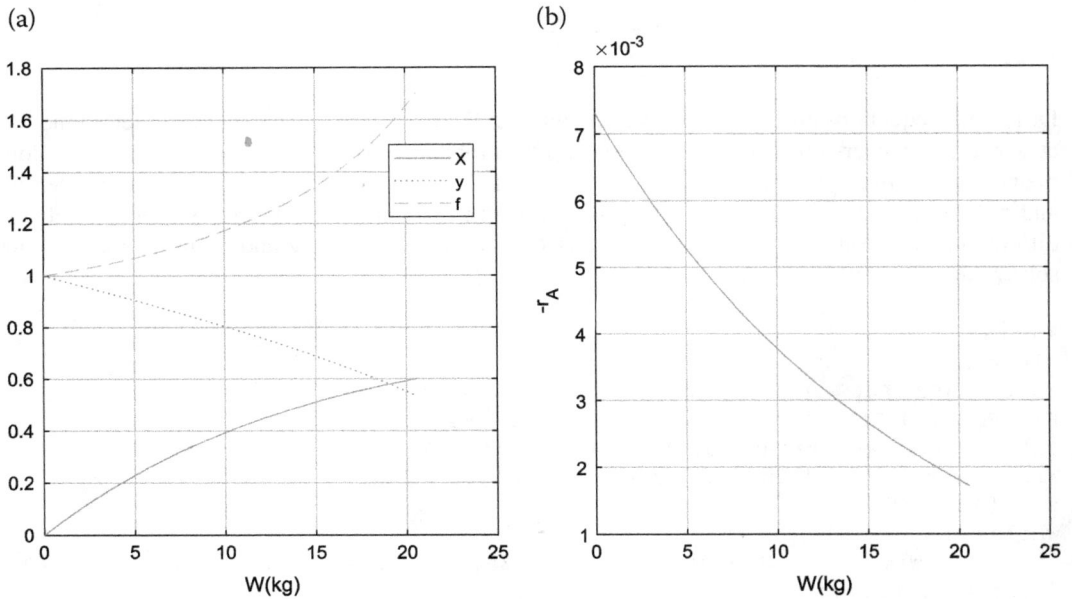

FIGURE 6.27   (a) Conversion and flow ratio, and (b) reaction rate versus catalyst weight.

where
  $C(t, x)$ is the concentration of the interested component
  D is the diffusion coefficient
  x is the distance to the direction of the flow
  Suppose that the initial and boundary conditions are given by

$$C(0, x) = C_0, \; C(t, 0) = C_f, \; \frac{\partial C}{\partial x}\bigg|_{t,x=L} = 0$$

where L is the length of the reactor. Introduction of some dimensionless variables and parameters, followed by some algebraic manipulation, yields

$$\frac{\partial \phi}{\partial \tau} + \frac{\partial \phi}{\partial \xi} = \frac{1}{Pe} \frac{\partial^2 \phi}{\partial \xi^2} - Da\phi$$

where $\phi = \frac{C}{C_f}$, $\tau = \frac{tv}{L}$, $\xi = \frac{x}{L}$, $Pe = \frac{vL}{D}$, $Da = \frac{kL}{v}$. The initial and boundary conditions are given by

$$\phi(0, \xi) = \phi_0, \; \phi|_{\xi=0} = 1, \; \frac{\partial \phi}{\partial \xi}\bigg|_{\xi=1} = 0$$

If the reactor length L is divided into n subintervals, the step size is given by $h = L/n$. Using the central difference formula, we have a set of n coupled ODEs:

$$\frac{d\phi_i}{d\tau} = -\left(\frac{\phi_{i+1} - \phi_{i-1}}{2h}\right) + \left(\frac{\phi_{i+1} - 2\phi_i + \phi_{i-1}}{Pe_i h^2}\right) - Da_i \phi_i \; (i = 2, 3, \cdots, n-1)$$

$$\frac{d\phi_i}{d\tau} = -\left(\frac{\phi_{i+1} - 1}{2h}\right) + \left(\frac{\phi_{i+1} - 2\phi_i + 1}{Pe_i h^2}\right) - Da_i \phi_i \; (i = 1)$$

$$\frac{d\phi_i}{d\tau} = \left(\frac{-2\phi_i + 2\phi_{i-1}}{Pe_i h^2}\right) - Da_i\phi_i \, (i = n)$$

These difference equations can be solved by using the method of lines. At steady state, $\partial\phi/\partial\tau = 0$ and we have

$$\frac{d^2\phi}{d\xi^2} - Pe\frac{d\phi}{d\xi} - PeDa\phi = 0$$

Example 6.27 Packed-Bed Reactor with Axial Dispersion

Produce concentration profiles $\phi(\tau, \xi)$ for a plug-flow reactor with axial dispersion using the method of lines when $Da = 2$ and $Pe = 5$ $(0 \leq \tau, \xi \leq 1)$.

Solution

The function *pfrdiff* defines the system of difference equations to be integrated.

```
function dC = pfrdiff(t,C,pf)
Pe = pf.Pe; Da = pf.Da; n = pf.n; % Retrieve data
h = 1/n; dC = zeros(n,1); % Initialization
% Difference equations
for k = 1:n
  if k == 1, s = (C(k+1) - 1)/(2*h); d = (C(k+1) - 2*C(k) + 1)/(Pe*h^2);
  elseif k == n, s = 0; d = (-2*C(k) + 2*C(k-1))/(Pe*h^2);
  else, s = (C(k+1) - C(k-1))/(2*h); d = (C(k+1) - 2*C(k) + C(k-1))/(Pe*h^2); end
  dC(k) = - s + d- Da*C(k);
end
end
```

The script *pfraxd* sets the data and parameters and employs the built-in function *ode45* to obtain the concentrations as a function of τ and ξ $(0 \leq \tau, \xi \leq 1)$. Figure 6.28 shows the concentration profile for $n = 50$.

```
% pfraxd.m: PFR with axial diffusion using the method of line (MoL)
clear all;
n = 50; pf.n = n; pf.Pe = 1; pf.Da = 2; % Data and parameters
h = 1/n; Z0 = ones(n,1); tspan = [0 1]; % initialize
[t,C] = ode45(@pfrdiff,tspan,Z0,[],pf); % solve the set of ODEs
Cs = [1, C(end,:)]; % steady-state by MoL
x   =  [h:h:1];  mesh(x,t,C);  xlabel('\xi'),  ylabel('\tau'),  zlabel('\phi
(\tau,\xi)')
```

6.5.5 Oxidation of SO_2 in a Packed-Bed Reactor[26,27]

Fixed-bed catalytic reactors with periodic flow-reversal operation are used to oxidize SO_2 (sulfur dioxide), especially that in the exhausted gas from the nonferrous metal industry. A one-dimensional two-phase unsteady-state model describing this reactor can be derived from the mass conservation of SO_2 and the heat for each phase. The Crank-Nicolson predictor-corrector method on a nonuniform spatial grid may be used in solving the derived partial differential equations. From the numerical simulation of SO_2 oxidation over vanadium catalysts in a small reactor, it is known that autothermal oxidation of low-concentration SO_2 is feasible.

Commercial catalysts usually comprise the active component V_2O_5 (vanadium pentoxide) and M_2O (alkali metal oxides). The contents of V_2O_5 are usually 3% to 10% by weight, and the contents

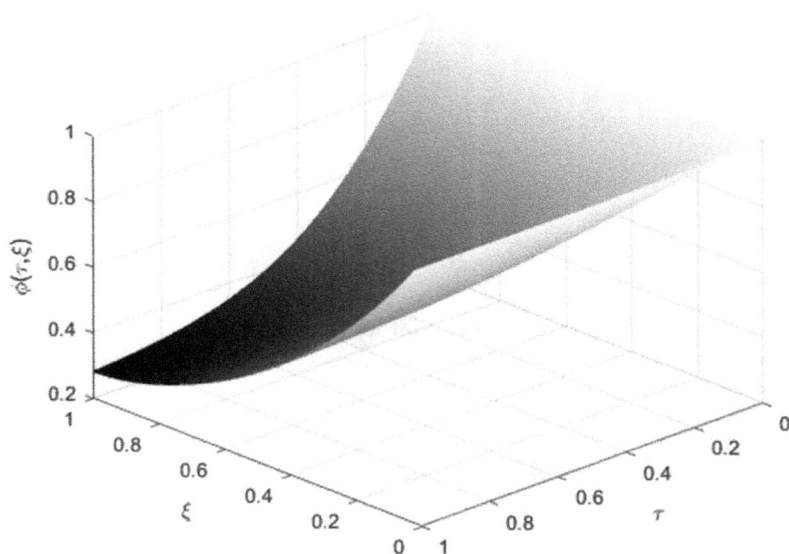

FIGURE 6.28 Concentration profile in the plug-flow reactor with axial dispersion.

of M_2O are, depending on the species used and the combination of various alkali metals, 6% to 26% by weight, with the molar ratio of alkali metal to vanadium usually being 2 to 5.5.

H_2SO_4(sulfuric acid) is nowadays obtained virtually exclusively by oxidation of SO_2 to SO_3 (sulfur trioxide) in the contact/double contact process with subsequent hydrolysis. In this process, SO_2 is oxidized to SO_3 by means of molecular oxygen over vanadium-containing catalysts in a plurality of adiabatic layers (beds) arranged in series. The SO_3 formed can be removed from the gas stream by intermediate absorption in order to achieve higher total conversion (double contact process). Depending on the bed, the reaction is carried out in a temperature range from 340°C to 680°C, with the maximum temperature decreasing with increasing bed number because of the decreasing SO_2 content.

For the formation of H_2SO_4, sulfur is used as raw material. SO_2 obtained by burning sulfur is oxidized in a fixed-bed reactor to yield H_2SO_4: $S + O_2 \leftrightarrow SO_3$. Assume that a packed-bed reactor using V_2O_5 in a pipe of 2 $in.$ 12 BWG (inside diameter = 0.0453 m) is used. The converters operate close to atmosphere, and we assume that the gas is fed into the reactor at 202 kPa. We want to generate profiles of conversion, temperature, and pressure as a function of the length of the tube (or the weight of catalyst).

The reactor kinetics can be expressed as

$$F_{A0}\frac{dx}{dw} = -r_A{}'$$

where x is the conversion and w is the weight of catalyst (kg). The reaction rate $-r_A{}' = -r_{SO_2}{}'$ is given by

$$-r'_{SO_2} = k\sqrt{\frac{P_{SO_2}}{P_{SO_3}}}\left\{P_{O_2} - \left(\frac{P_{SO_3}}{K_p P_{SO_2}}\right)^2\right\}$$

where K_p is the equilibrium constant. If the conversion x is below 5%, the reaction rate does not dependent on x, that is, $-r'_{SO_2} = -r'_{SO_2}(x = 0.05)$ for $x \leq 0.05$. K_p is given as

$$K_p = 3.1415 \times 10^{-3} \exp\left\{ \frac{42{,}311}{1.987(1.8(T - 273.15) + 491.67)} - 11.24 \right\} (K_p : Pa^{-1/2}, \ T: K)$$

and the rate constant $k \, (molSO_2/kgcat \cdot sec \cdot Pa)$ is given as

$$k = 9.8692 \times 10^{-3} \exp\left\{ \frac{-176{,}008}{(1.8(T - 273.15) + 491.67)} - 110.1 \ln(1.8(T - 273.15) + 491.67) \right.$$

$$\left. + 912.8 \right\} (T: K)$$

The stoichiometric of the reaction is

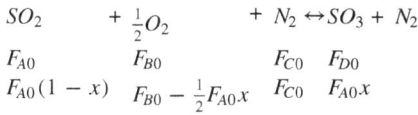

$$SO_2 \quad + \tfrac{1}{2}O_2 \quad + N_2 \leftrightarrow SO_3 + N_2$$

$$F_{A0} \quad\quad F_{B0} \quad\quad\quad F_{C0} \quad F_{D0}$$

$$F_{A0}(1 - x) \quad F_{B0} - \tfrac{1}{2}F_{A0}x \quad F_{C0} \quad F_{A0}x$$

Let $\Theta_i = F_{i0}/F_{A0}$. Then the total number of moles F_T is given by

$$F_T = F_{A0}\left\{ (1 - x) + \Theta_{O_2} - \frac{1}{2}x + \Theta_{N_2} + \Theta_{SO_3} + x \right\} = F_{A0}\left(1 + \Theta_{O_2} + \Theta_{N_2} + \Theta_{SO_3} \right) + F_{A0}\delta x$$

$$= F_{T0} + F_{A0}\delta x$$

where $\delta = -1/2$ and $F_{T0} = F_{A0}\left(1 + \Theta_{O_2} + \Theta_{N_2} + \Theta_{SO_3} \right)$. Assuming ideal gases, concentrations are represented by

$$C_T = \frac{F_T}{V} = \frac{P}{RT}, \ C_{T0} = \frac{F_{T0}}{V_0} = \frac{P_0}{RT_0}$$

from which we have

$$V = V_0 \frac{P_0}{P} \frac{T}{T_0} \frac{F_T}{F_{T0}} = V_0 \frac{P_0}{P} \frac{T}{T_0} \frac{F_{T0} + F_{A0}\delta x}{F_{T0}} = V_0 \frac{P_0}{P} \frac{T}{T_0}\left(1 + \frac{F_{A0}}{F_{T0}}\delta x \right)$$

Then the concentration of the species i, C_i, is given as

$$C_i = \frac{F_i}{V} = \frac{F_i}{V_0 \frac{P_0}{P} \frac{T}{T_0} \frac{F_T}{F_{T0}}} = C_{T0} \frac{F_i}{F_T} \frac{P}{P_0} \frac{T_0}{T} = C_{T0} \frac{F_{A0}(\Theta_i + \nu_i x)}{F_{T0} + F_{A0}\delta x} \frac{P}{P_0} \frac{T_0}{T} = C_{A0} \frac{\Theta_i + \nu_i x}{1 + \varepsilon x} \frac{P}{P_0} \frac{T_0}{T}$$

where $C_{A0} = C_{T0} F_{A0}/F_{T0}$ and $\varepsilon = (F_{A0}/F_{T0})\delta = y_{A0}\delta$. Thus the partial pressure of component i, P_i, is given as

$$P_i = C_i RT = C_{A0} \frac{T_0}{P_0}\left(\frac{\Theta_i + \nu_i x}{1 + \varepsilon x} \right)RP = P_{A0}\left(\frac{P}{P_0} \right)\left(\frac{\Theta_i + \nu_i x}{1 + \varepsilon x} \right)$$

where $P_{A0} = P_{SO_2,0} = C_{A0}RT_0$. Substitution of $P_i (i = SO_2, SO_3)$ into the reaction kinetics gives

$$\frac{dx}{dw} = -\frac{r_{SO_2}'}{F_{A0}} = \frac{k}{F_{A0}} \sqrt{\frac{P_{SO_2}}{P_{SO_3}}} \left\{ P_{O_2} - \left(\frac{P_{SO_3}}{K_p P_{SO_2}} \right)^2 \right\}$$

We can see that

$$\Theta_{SO_2} = 1, \ \Theta_{SO_3} = 0, \ \Theta_{O_2} = 0.11/0.1 = 1.1, \ \Theta_{N_2} = 0.79/0.1 = 7.9, \ \nu_{SO_2} = -1, \ \nu_{SO_3} = 1, \ \nu_{O_2}$$
$$= -1/2$$

Then the above equation can be rewritten as

$$\frac{dx}{dw} = \frac{k}{F_{A0}} \sqrt{\frac{1-x}{x}} \left\{ P_{SO_2,0} \left(\frac{1.1 - x/2}{1 + \varepsilon x} \right) \frac{P}{P_0} - \left(\frac{x}{K_p(1-x)} \right)^2 \right\}$$

For our example,

$$\varepsilon = \delta y_{SO_2,0} = \left(-\frac{1}{2} \right) \frac{0.1}{0.1 + 0.11 + 0.79} = -0.05, \ F_{A0} = 0.00237 \ mol/sec$$

Then we have

$$\frac{dx}{dw} = \frac{k}{0.00237} \sqrt{\frac{1-x}{x}} \left\{ 20265 \left(\frac{1.1 - x/2}{1 - 0.05x} \right) \frac{P}{P_0} - \left(\frac{x}{K_p(1-x)} \right)^2 \right\}$$

The energy balance can be expressed as follows:

$$\dot{Q} - W_s - F_{A0} \sum_{i=1}^{n} \int_{T_0}^{T} \Theta_i C_{pi} dT - \left\{ \Delta H_R(H_R) + \int_{T_0}^{T} \Delta C_p dT \right\} F_{A0} x = 0$$

Differentiating with respect to V, we have

$$\frac{d\dot{Q}}{dV} - F_{A0} \left(\sum_{i=1}^{n} \Theta_i C_{pi} + x\Delta C_p \right) \frac{dT}{dV} - \left\{ \Delta H_R(H_R) + \int_{T_0}^{T} \Delta C_p dT \right\} F_{A0} \frac{dx}{dV} = 0$$

Substituting $d\dot{Q}/dV = UA(T_a - T)$ and $F_{A0} dx/dV = -r_A$ into the energy balance, we obtain

$$UA(T_a - T) - F_{A0} \left(\sum_{i=1}^{n} \Theta_i C_{pi} + x\Delta C_p \right) \frac{dT}{dV} - (-r_A) \left\{ \Delta H_R(H_R) + \int_{T_0}^{T} \Delta C_p dT \right\} = 0$$

Since $A = \pi D/A_c = \pi D/(\pi D^2/4) = 4/D$ and $W = \rho_b V$, we have, after some rearrangement,

$$\frac{dT}{dW} = \frac{4U(T_a - T)/(\rho_b D) + (-r_A')\Delta H_R}{F_{A0}(\sum_{i=1}^{n} \Theta_i C_{pi} + x\Delta C_p)}$$

where

$$\Delta H_R = \Delta H_R(H_R) + \int_{T_0}^{T} \Delta C_p dT$$

The standard heat of reaction and heat capacity for each species $(C_p : J/(mol \cdot K), T : K)$ are:

$$\Delta H_R(298K) = -98,480 \ J/molSO_2$$

$$C_{p,SO_2} = 23.852 + 0.066989T - 4.961 \times 10^{-5}T^2 + 1.3281 \times 10^{-8}T^3$$

$$C_{p,O_2} = 28.106 - 3.68 \times 10^{-6}T + 1.7459 \times 10^{-5}T^2 - 1.065 \times 10^{-8}T^3$$

$$C_{p,SO_3} = 16.370 + 0.14591T - 1.12 \times 10^{-4}T^2 + 3.2324 \times 10^{-8}T^3$$

$$C_{p,N_2} = 31.150 - 0.01357T + 2.6796 \times 10^{-5}T^2 - 1.168 \times 10^{-8}T^3$$

From these relationships, we obtain

$$\Delta H_R = -98{,}480 - 21.535(T - 298) + 0.0395(T^2 - 298^2) - 2.371 \times 10^{-5}(T^3 - 298^3)$$
$$+ 6.11675 \times 10^{-9}(T^4 - 298^4)$$

$$\Delta C_p = -21.535 + 0.0789T - 7.112 \times 10^{-5}T^2 + 2.447 \times 10^{-8}T^3$$

$$\sum_{i=1}^{n} \Theta_i C_{pi} = 300.85 - 0.0402T + 1.8 \times 10^{-4}T^2 - 9.071 \times 10^{-8}T^3$$

The pressure drop along the tube reactor is given by the Ergun equation:

$$\frac{dP}{dW} = -\frac{G(1 - \phi)(1 + \varepsilon x)}{\rho_b A_c \rho_0 g_c D_p \phi^3} \left\{ \frac{150(1 - \phi)\mu}{D_p} + 1.752G \right\} \frac{P_0}{P} \frac{T}{T_0}$$

where $G\,(kg/m^2/sec)$ is the superficial mass velocity, ϕ is the porosity of the packed bed, $D_p\,(m)$ is the particle diameter, $\mu\,(Pa \cdot sec)$ is the viscosity of the gas, and $\rho_0\,(kg/m^3)$ is the gas density.

In short, we need to solve the following differential equations:

$$\frac{dx}{dV} = -\frac{r_{SO_2}'}{F_{SO_2,0}}, \quad \frac{dT}{dW} = \frac{4U(T_a - T)/(\rho_b D) + (-r_{SO_2}')\Delta H_R}{F_{SO_2,0}(\sum_{i=1}^{n} \Theta_i C_{pi} + x\Delta C_p)}$$

$$\frac{dP}{dW} = -\frac{G(1 - \phi)(1 + \varepsilon x)}{\rho_b A_c \rho_0 g_c D_p \phi^3} \left\{ \frac{150(1 - \phi)\mu}{D_p} + 1.752G \right\} \frac{P_0}{P} \frac{T}{T_0}$$

Required data are given by
$D = 0.0453\,m$, $U = 17\,J/(sec \cdot K \cdot m^2)$, $T_a = 700\,K$, $T_0 = 750\,K$, $\rho_0 = 0.866\,kg/m^3$, $\rho_b = 542$ kg/m^3, $D_p = 0.00457\,m$, $P_0 = 202{,}650\,Pa$, $\phi = 0.45$, $G = 0.433\,kg/(m^2 \cdot sec)$, $\mu = 3.72 \times 10^{-5}\,Pa \cdot sec$, $F_{T0} = 0.02153\,mol/sec$.

The function *so3de* defines the differential equation system. The data are supplied into the function by the structure *sa*.

```
function dz = so3de(w,z,sa)
% The reaction model is based on the following references:
%  [1] Harrer, T. S., Kirk Othmer encyclopedia of chemical technology,
%     2nd ed., Vol.19, Wiley-Interscience, New York, NY, p.470, 1969.
%  [2] Mariano M. Martin (Editor), Introduction to software for chemical
%     engineers, CRC Press, Taylor & Francis Group, Boca Raton, FL, p.136-
145, 2015.
% Retrieve data
Ta = sa.Ta; T0 = sa.T0; Pt0 = sa.Pt0; rho0 = sa.rho0; rhob = sa.rhob;
ya0 = sa.ya0; yb0 = sa.yb0; yc0 = sa.yc0; Ft0 = sa.Ft0; G = sa.G;
epn = sa.epn; phi = sa.phi; mu = sa.mu; D = sa.D; Dp = sa.Dp; U = sa.U;
% Assign variables
```

```
x = z(1); T = z(2); P = z(3);
% Rate and equilibrium constants
k = 9.8692e-3 * exp(-1.76008e5/(1.8*(T - 273.15) + 491.67) -...
   110.1*log((1.8*(T-273.15)+491.67)) + 912.8);
Kp = 3.1415e-3 * exp(42311/(1.987*(1.8*(T-273.17) + 491.67)) - 11.24);
% parameters (a: SO2, b: O2, c: N2, d: SO3)
phib = yb0/ya0; Pa0 = Pt0*ya0; Fa0 = Ft0*ya0; Ac = pi*D^2/4;
Cpsum = 300.85 - 0.0402*T + 1.8e-4*T^2 - 9.071e-8*T^3;
dCp = -21.535 + 0.0789*T - 7.112e-5*T^2 + 2.447e-8*T^3;
dHr = - 98480 - 21.535*(T-298) + 0.0395*(T^2-298^2) -...
   2.371e-5*(T^3-298^3) + 6.11675e-9*(T^4-298^4);
if x < 0.05
  xs = 0.05;
    r = k*sqrt((1-xs)/xs)*(Pa0*((1.1-xs/2)/(1+epn*xs))*P/Pt0 - (xs/(Kp*(1-
xs)))^2);
else
  r = k*sqrt((1-x)/x)*(Pa0*((1.1-x/2)/(1+epn*x))*P/Pt0 - (x/(Kp*(1-x)))^2);
end
dz(1,1) = r/Fa0;
dz(2,1) = (4*U*(Ta-T)/(rhob*D) + r*(-dHr))/(Fa0*(Cpsum + x*dCp));
dz(3,1)    =    -G*(1-phi)*(1+epn*x)*Pt0*T*(150*(1-phi)*mu/Dp    +    1.752*G)
/ (P*T0*...
      rhob*Ac*rho0*Dp*phi^3);
end
```

The script *so3rxn* defines the data and uses the built-in function *ode15s* to solve the differential equation system. The script produces the desired profiles shown in Figure 6.29.

```
% so3rxn.m: oxidation of SO2 using fixed-bed reactor
clear all;
% Data
sa.Ta = 700; sa.T0 = 750; sa.Pt0 = 202650; % initial temperature and pressure
sa.rho0 = 0.866; sa.rhob = 542; % density
sa.ya0 = 0.1; sa.yb0 = 0.11; sa.yc0 = 0.79; % a:SO2, b:O2, c:N2, d:SO3
sa.Ft0 = 0.02153; sa.G = 0.433; % feed rate(Ft0) and mass velocity(G)
sa.epn = -0.05; sa.phi = 0.45; sa.mu = 3.72e-5; % parameters and viscosity
sa.D = 0.0453; sa.Dp = 4.57e-3; % tube diameter(D) and particle diameter(Dp)
sa.U = 17; % overall heat transfer coefficient(J/s/m^2/K)
% Solve differential equations: x = z(1); T = z(2); P = z(3);
wspan = [0 4]; z0 = [0, sa.T0, sa.Pt0]; [w z] = ode15s(@so3de,wspan,z0,[],sa);
% Plot results
x = z(:,1); T = z(:,2); P = z(:,3);
subplot(2,2,1), plot(w,x), grid, xlabel('W(kg)'), ylabel('X')
subplot(2,2,2), plot(w,T), grid, xlabel('W(kg)'), ylabel('T(K)')
subplot(2,2,3), plot(w,P), grid, xlabel('W(kg)'), ylabel('P(Pa)')
```

6.5.6 STRAIGHT-THROUGH TRANSPORT REACTOR

A catalytic gas-phase reaction $A \rightarrow B$ is carried out in a straight-through transport reactor where the catalyst bed is moving through the reactor. When the catalyst bed moves upward from the bottom of the reactor ($z = 0$), the mass balance yields

(a)

(b)

(c)

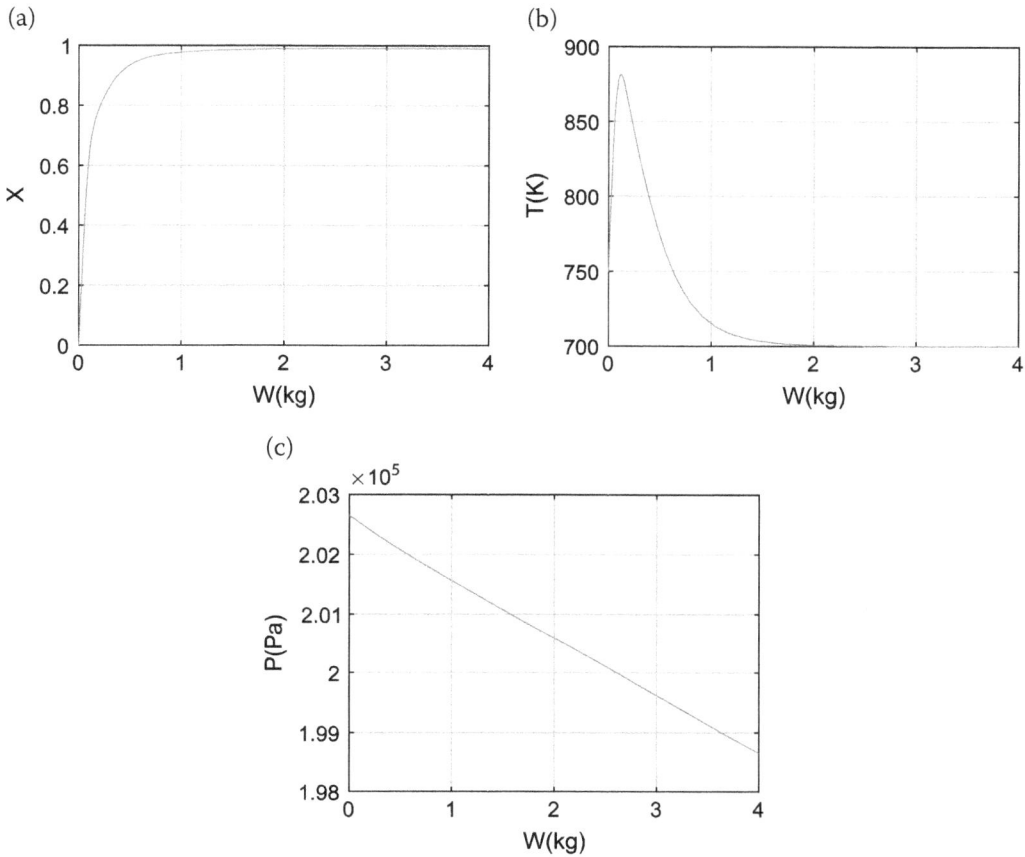

FIGURE 6.29 Profiles of (a) conversion, (b) temperature, and (c) pressure.

$$v_0 C_{A0} \frac{dx_A}{dz} = -r_A A_c, \ x_A|_{z=0} = 0$$

where A_c is the cross-sectional area of the reactor. The reaction rate per unit volume of catalyst bed can be represented as the particular activity multiplied by the catalytic rate expression:

$$-r_A = \frac{akC_A}{1 + K_A C_A}, \ C_A = C_{A0}(1 - x_A), \ C_{B0} = C_{A0} x_A$$

where a denotes the activity of the catalyst. If the deactivation of the catalyst is negligible, the mass balance yields

$$\frac{dx_A}{dz} = \frac{akC_A}{(1 + K_A C_A)u}, \ u = \frac{v_0}{A_c}, \ x_A|_{z=0} = 0$$

where u is the moving velocity of the catalyst bed.

When the deactivation of the catalyst is significant, three types of catalyst deactivation may be examined: coking, sintering, and poisoning. Let a_c, a_s, and a_p be the activity of the catalyst when deactivation is caused by coking, sintering, and poisoning, respectively. These activities are given by[28]

$$a_c = \frac{1}{1 + A'\sqrt{\frac{z}{u}}}, \quad \frac{da_s}{dz} = -\frac{k_{ds}a_s^2}{u}, \quad \frac{da_p}{dz} = -\frac{k_{dp}a_p C_B}{u}$$

Example 6.28 Gas-Phase Reaction of Gas Oil[29]

The rate of the gas-phase cracking reaction of a gas oil (A)

$$A \rightarrow B$$

can be represented by

$$-r_A = \frac{akC_A}{1 + K_A C_A}, \quad C_A = C_{A0}(1 - x_A), \quad C_{B0} = C_{A0}x_A$$

The catalyst particles are assumed to move upward with the mean gas velocity, given by $u = 8$ m/sec. The reaction is to be carried out at 750°F under constant temperature and pressure. The volume change with reaction, pressure drop, and temperature variation may be neglected.

Plot the conversion of A (X_A) and the catalyst activity versus the reactor length z for the three types of catalyst deactivation (coking, sintering, and poisoning). The height of the reactor is $z_f = 6$ m, and the initial activity of the catalyst is assumed to be 1.

Data:

$k = 30$ sec^{-1}, $K_A = 5$ $m^3/kgmol$, $C_{A0} = 0.2$ $kgmol/m^3$, $A' = 12$ $sec^{-1/2}$, $k_{ds} = 17.5$ sec^{-1}, k_{dp}

$= 140$ $dm^3/(mol·sec)$

Solution

Let $X_{Ai}(i = 1, 2, 3)$ be the conversion for each deactivation type (1: coking, 2: sintering, 3: poisoning), and let $a_i (i = 1, 2, 3)$ be the corresponding catalyst activities. If we define the variable vector y as $y(i) = x_{Ai} (i = 1, 2, 3)$ and $y(i) = a_{i-2} (i = 4, 5)$, we have the following differential equations:

$$\frac{dy(1)}{dz} = \frac{a_1 kC_{A0}(1 - y(1))}{(1 + K_A C_{A0}(1 - y(1)))u}, \quad \frac{dy(2)}{dz} = \frac{y(4)kC_{A0}(1 - y(2))}{(1 + K_A C_{A0}(1 - y(2)))u}, \quad a_1 = \frac{1}{1 + A'\sqrt{\frac{z}{u}}},$$

$$\frac{dy(3)}{dz} = \frac{y(5)kC_{A0}(1 - y(3))}{(1 + K_A C_{A0}(1 - y(3)))u}, \quad \frac{dy(4)}{dz} = -\frac{k_{ds}y(4)^2}{u}, \quad \frac{dy(5)}{dz} = -\frac{k_{dp}y(5)C_{A0}y(3)}{u}$$

The script *mcatb* defines these differential equations, uses the built-in solver *ode45* to find solutions, and produces the curves shown in Figure 6.30.

```
% mcatb.m
% y1=Xa1, y2=Xa2, y3=Xa3, y4=a2, y5=a3
clear all;
u=8; Ap=12; k=30; Ka=5; Ca0=0.2; kds=17.5; kdp=140;
zspan = [0 6]; y0 = [0 0 0 1 1];
dydz = @(z,y) [(1/(1+Ap*sqrt(z/u)))*k*Ca0*(1-y(1))/(1+Ka*Ca0*(1-y(1)))/u;
        y(4)*k*Ca0*(1-y(2))/(1+Ka*Ca0*(1-y(2)))/u;
        y(5)*k*Ca0*(1-y(3))/(1+Ka*Ca0*(1-y(3)))/u;
        -kds*y(4)^2/u;
        -kdp*Ca0*y(3)*y(5)/u];
[z,y] = ode45(dydz, zspan, y0);
subplot(1,2,1), plot(z,y(:,1),'-',z,y(:,2),':',z,y(:,3),'--');
xlabel('z(m)'), ylabel('X_A'), legend('X_{A1}','X_{A2}','X_{A3}');
subplot(1,2,2), plot(z,y(:,4),'-',z,y(:,5),':'), xlabel('z(m)'), ylabel
('a'), legend('a_s','a_p');
```

(a)

(b)

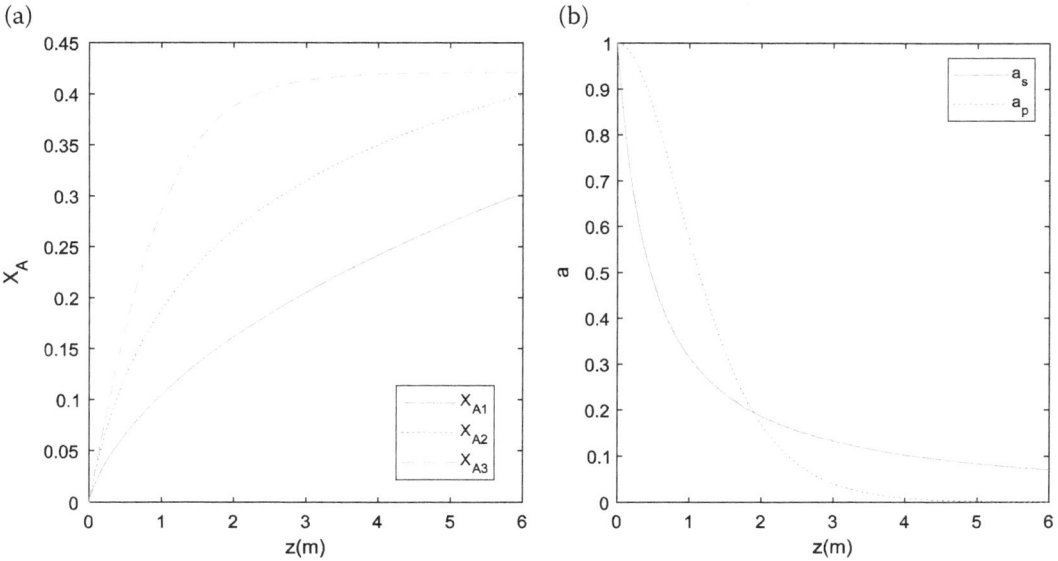

FIGURE 6.30　(a) Conversion and (b) catalyst activity versus reactor height.

6.5.7 Steady-State Nonisothermal Reactions

Consider the reversible gas-phase reaction

$$A + B \leftrightarrow 2C$$

to be carried out in a plug-flow reactor or a packed-bed reactor in which pressure drop is neglected $(P = P_0)$. The mole balance on each species yields

$$\frac{dX}{dV} = \frac{-r_A}{F_{A0}}, \frac{dF_A}{dV} = r_A, \frac{dF_B}{dV} = r_B, \frac{dF_C}{dV} = r_C$$

where the reaction rates can be expressed as

$$-r_A = k_1\left(C_A C_B - \frac{C_C^2}{K_C}\right), r_B = r_A, r_C = -2r_A$$

The reaction rate constants are given by

$$k = k_1(T_1)\exp\left[\frac{E}{R}\left(\frac{1}{T_1} - \frac{1}{T}\right)\right], K_C = K_{C2}(T_2)\exp\left[\frac{\Delta H_{Rx}^\circ}{R}\left(\frac{1}{T_2} - \frac{1}{T}\right)\right]$$

where $k_1(T_1)$ and $K_{C2}(T_2)$ are values of k and K_C at T_1 and T_2, respectively. For a gas-phase reaction, $\epsilon = 0$, and the concentration of each species can be expressed as

$$C_A = C_{A0}(1 - X)\frac{T_0}{T}, C_B = C_{A0}(\Theta_B - X)\frac{T_0}{T}, C_C = 2C_{A0}X\frac{T_0}{T}$$

or

$$C_A = C_{T0}\frac{F_A}{F_T}\frac{T_0}{T}, C_B = C_{T0}\frac{F_B}{F_T}\frac{T_0}{T}, C_C = C_{T0}\frac{F_C}{F_T}\frac{T_0}{T}$$

where T_0 is the feed temperature and $F_T = F_A + F_B + F_C$.

The energy balance yields

$$\frac{dT}{dV} = \frac{Ua(T_a - T) + (-r_A)(-\Delta H_{Rx})}{F_{A0}[C_{pA} + \Theta_B C_{pB} + X\Delta C_p]} = \frac{Ua(T_a - T) + (-r_A)(-\Delta H_{Rx})}{F_A C_{pA} + F_B C_{pB} + F_C C_{pC}}$$

If the heat transfer fluid (coolant) temperature, T_a, is not constant, the energy balance on the heat exchange fluid gives

Co − current flow: $\frac{dT_a}{dV} = \frac{Ua(T - T_a)}{\dot{m}_c C_{pc}}$

Counter − current flow: $\frac{dT_a}{dV} = \frac{Ua(T_a - T)}{\dot{m}_c C_{pc}}$

where

U is the overall heat transfer coefficient

$\dot{m}_c$ is the mass flow rate of the heat transfer fluid

C_{pc} is the heat capacity of the heat transfer fluid

Example 6.29 Isomerization of *n*-Butane[30]

The isomerization reaction of *n*-butane (C_4H_{10}) to isobutane,

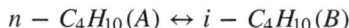

$$n - C_4H_{10}(A) \leftrightarrow i - C_4H_{10}(B)$$

is to be carried out in a bank of 10 tubular reactors; each reactor is $V = 5\ m^3$. The bank reactors are double-pipe heat exchangers with the reactants flowing in the inner pipe and $Ua = 5,000\ kJ/(m^3 \cdot hr \cdot K)$.

A mixture of 90 mol% *n*-butane and 10 mol% *i*-pentane, which is considered inert, is to be processed at 70% conversion. The molar flow rate of the mixture is $163\ kmol/hr$. The bank reactors can be considered as a countercurrent heat changer. The entering temperature of the reactants is $T_0 = 305\ K$, and the entering coolant temperature is $T_a = 310\ K$. For a countercurrent heat exchanger, this value is the entering coolant temperature $T_{a0}(=310\ K)$ at $V = V_{final} = 5\ m^3$. In order to find the coolant temperature at the outlet ($V = 0$), we guess T_a at $V = 0$ and see if it matches T_{a0} at $V = V_{final} = 5\ m^3$. If it doesn't match, we guess again.

The mass flow rate of the coolant is $\dot{m}_c = 500\ kg/hr$, and the heat capacity of the coolant is $C_{pc} = 28\ kJ/(kg \cdot K)$. The temperature in any one of the reactors cannot rise above 325 K. At $T_1 = 360\ K$ and $T_2 = 333\ K$, it is known that $k_1(T_1) = 31.1\ hr^{-1}$ and $K_{C2}(T_2) = 3.03\ hr^{-1}$. Plot the conversion X, equilibrium conversion X_e, temperature T, and reaction rate $-r_A$ down the length of the reactor.

Data: $\Delta H_{Rx}^{\circ} = -6900\ J/mol$, activation energy $E = 65,700\ J/mol$, $C_{A0} = 9.3\ kmol/m^3$, $R = 8.314\ J/(mol \cdot K)$, $C_{pA} = C_{pB} = 141\ J/(mol \cdot K)$, $C_{p_{i-P}} = 161\ J/(mol \cdot K)$.

Solution

For each reactor,

$$F_{A0} = 0.9F_{T0} = (0.9)(163) \times \frac{1}{10} = 14.67\ kmol A/hr,\ T_0 = 305\ K$$

$$C_{p0} = \sum \Theta_i C_{pi} = C_{pA} + \Theta_I C_{pI} = \left(141 + \frac{0.1}{0.9}161\right) = 158.889\ J/(mol \cdot K)$$

$$\Delta C_p = C_{pB} - C_{pA} = 141 - 141 = 0$$

The concentration of each species can be expressed as

$$C_A = C_{A0}(1 - X0),\ C_B = C_{A0}X$$

Substitution of these relations into the rate equation gives

$$-r_A = kC_{A0}\left[1 - \left(1 + \frac{1}{K_c}\right)X\right]$$

The rate of change of conversion with respect to the reactor volume is given by

$$\frac{dX}{dV} = \frac{-r_A}{F_{A0}}$$

At equilibrium, $- r_A = 0$, and the equilibrium conversion is given by

$$X_e = \frac{K_c}{1 + K_c}$$

The energy balance yields

$$\frac{dT}{dV} = \frac{Ua(T_a - T) + r_A\Delta H_{Rx}}{F_{A0}\sum \Theta_i C_{pi}} = \frac{r_A\Delta H_{Rx} - Ua(T - T_a)}{F_{A0}C_{p0}}$$

For a countercurrent heat exchanger,

$$\frac{dT_a}{dV} = \frac{Ua(T_a - T)}{\dot{m}_c C_{pc}}$$

We guess $T_a(V = 0) = T_{a0}$ and see if it matches $T_a = 310\ K$ at $V = V_{final} = 5\ m^3$.

The differential equations are defined by the function *exbco*. The input argument hx denotes the type of heat exchanger: for a countercurrent heat exchanger, hx = 'cn', for a cocurrent heat exchanger, hx = 'co', and for constant T_a, hx = 'ct'.

```
function dz = exbco(V,z,Ca0,Fa0,Cp0,Cpc,Ua,m,T1,T2,k1,K2,E,R,dH,hx)
% z(1)=Ta, z(2)=X, z(3)=T
k = k1*exp(E*(1/T1 - 1/z(3))/R); Kc = K2*exp(dH*(1/T2 - 1/z(3))/R); ra =
-k*Ca0*(1 - (1 + 1/Kc)*z(2));
if hx == 'co', dz(1) = Ua*(z(3) - z(1))/(m*Cpc);
elseif hx == 'cn', dz(1) = -Ua*(z(3) - z(1))/(m*Cpc);
else, dz(1) = 0; end
dz(2) = -ra/Fa0; dz(3) = (ra*dH - Ua*(z(3)-z(1)))/(Fa0*Cp0); dz = dz';
end
```

The script *isomnbt* sets data, calls the function *exbco*, and employs the built-in solver *ode45* to solve the system of differential equations. For a countercurrent heat exchanger, the initial value of T_a at $V = 0$ is unknown. T_a is updated iteratively until the temperature at $V = V_{final} = 5\ m^3$ is satisfied by using a numerical analysis method such as the bisection scheme. In this script, the initial search region of $305 \le T_a \le 320$ is used.

```
% isomnbt.m
clear all;
Ca0 = 9.3; Fa0 = 14.67; T1 = 360; T2 = 333; k1 = 31.1; K2 = 3.03; % data
E = 65700; R = 8.314; dH = -6900; Cp0 = 158.889; Cpc = 28; Ua = 5000; m = 500; hx =
'cn'; % data and conditions
criT = 1e-3; Taf = 310; errT = 10; Vi = [0 5];
if hx == 'cn' % guess Ta(z1)
```

```
Ta01 = 305; Ta02 = 320; Ta0m = (Ta01+Ta02)/2;
while errT >= criT
   z01 = [Ta01 0 305]; z02 = [Ta02 0 305]; z0m = [Ta0m 0 305];
                                [V      z1]    =      ode45(@exbco,Vi,z01,
[],Ca0,Fa0,Cp0,Cpc,Ua,m,T1,T2,k1,K2,E,R,dH,hx);
                                [V      zm]    =      ode45(@exbco,Vi,z0m,
[],Ca0,Fa0,Cp0,Cpc,Ua,m,T1,T2,k1,K2,E,R,dH,hx);
                                [V      z2]    =      ode45(@exbco,Vi,z02,
[],Ca0,Fa0,Cp0,Cpc,Ua,m,T1,T2,k1,K2,E,R,dH,hx);
   if (z1(end,1)-Taf)*(zm(end,1)-Taf) < 0, Ta02 = Ta0m; Ta0m = (Ta01+Ta02)/2;
   else, Ta01 = Ta0m; Ta0m = (Ta01+Ta02)/2; end
   errT = abs(zm(end,1) - Taf);
end
else
         z0     =     [310     0     305];     [V      zm]    =      ode45(@exbco,Vi,z0,
[],Ca0,Fa0,Cp0,Cpc,Ua,m,T1,T2,k1,K2,E,R,dH,hx);
end
Ta = zm(:,1); X = zm(:,2); T = zm(:,3); k = k1*exp(E*(1/T1 - 1./T)/R); Kc = K2*exp
(dH*(1/T2 - 1./T)/R);
ra = -k*Ca0.*(1 - (1 + 1./Kc).*X); Xe = Kc./(1+Kc);
subplot(2,2,1),   plot(V,T,V,Ta,'--'),   xlabel('V'),   ylabel('T(K)'),   le-
gend('T','T_a')
subplot(2,2,2),   plot(V,X,V,Xe,'--'),   xlabel('V'),   ylabel('X,X_e'),   le-
gend('X','Xe')
subplot(2,2,3), plot(V,-ra), xlabel('V'), ylabel('-r_A')
fprintf('Conversion (X) and equilibrium conversion (Xe): Xf = %g, Xef = %g\n',X
(end),Xe(end));
fprintf('Final T and Ta: Tf = %g, Taf = %g\n',T(end), Ta(end));
fprintf('Final reaction rate: raf = %g\n',-ra(end));
```

The script *isomnbt* produces the following outputs and the profiles shown in Figure 6.31:

```
>> isomnbt
Conversion (X) and equilibrium conversion (Xe): Xf = 0.779843, Xef = 0.784512
Final T and Ta: Tf = 310.153, Taf = 310.001
Final reaction rate: raf = 0.0505495
```

6.6 CRACKING AND POLYMERIZATION

6.6.1 CRACKING REACTION

The catalytic cracking of hydrocarbons is considered one of the most important processes in the oil refining industry. It is a chain reaction that is believed to follow the carbonium ion theory. This chain mechanism involves three elementary steps: initiation, propagation, and termination. The initiation step is represented by the attack of an active site on the reactant molecule to produce the activated complex that, in the gas phase or when using liquid superacids, corresponds to the formation of a carbocation. The chain propagation is represented by the transfer of a hydride ion from a reactant molecule to an adsorbed carbenium ion. The termination step corresponds to desorption of the adsorbed carbenium ion to give an olefin while restoring the initial active site.

In the thermal cracking of hydrocarbons, coke formation on the internal reactor wall limits the onstream time of a furnace. In an ethane cracking reactor, the inside surface of a cracking coil gradually fouls with carbonaceous deposits. The formation of coke has major consequences on the operation of the furnace. In the first place, the coke layer hampers the heat transfer to the process gas. To maintain the desired performance, the external tube metal temperature has to be raised. The

(a)

(b)

(c)

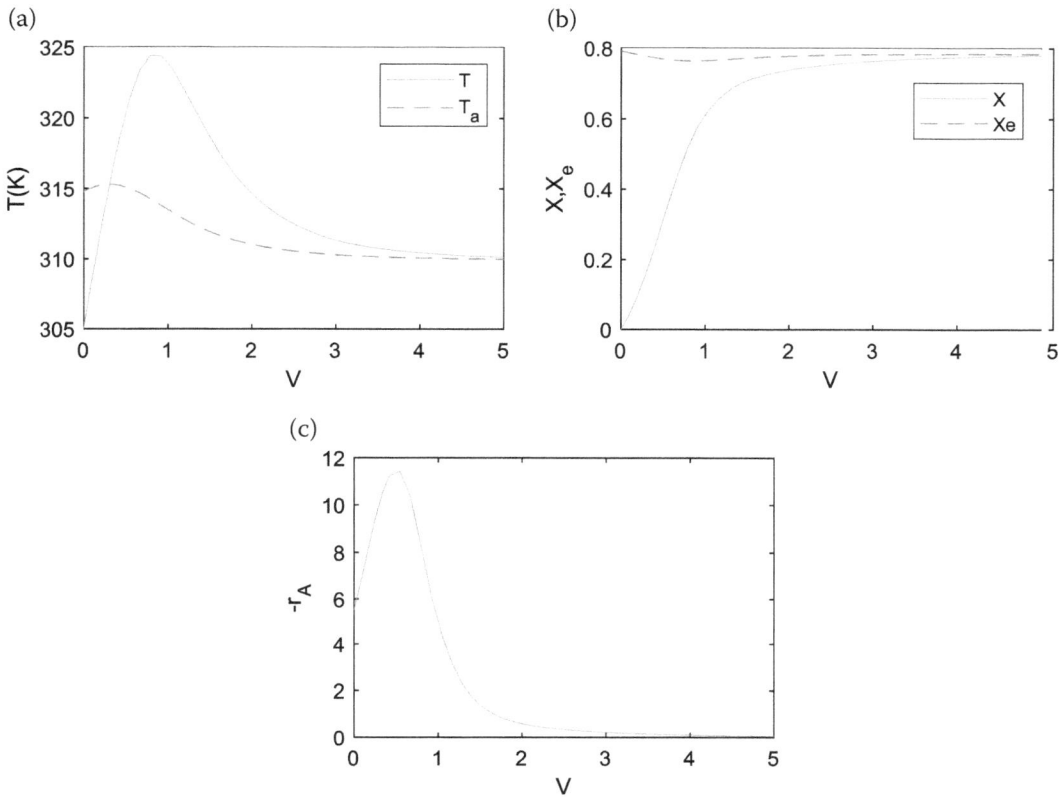

FIGURE 6.31 Profiles of (a) temperature, (b) conversion, and (c) reaction rate.

run has to be stopped when it reaches the limit imposed by the coil metallurgy. Further, the thermal efficiency of the radiation section of the firebox decreases. In addition, fouling leads to a reduction of the tube inside diameter, causing a higher pressure drop over the reactor. This is detrimental for the ethylene selectivity. Too high an inlet pressure is another reason for stopping the production and decoking by a controlled combustion with a mixture of steam and air. Which of the effects actually is limiting depends on the reactor geometry, operation conditions, and feedstock composition. Modeling a thermal cracking reactor requires the integration of a set of continuity equations for the process gas species, together with the energy and pressure drop equations.[31]

Ethylene can be produced using thermal cracking of light hydrocarbons such as ethane. In the steam cracking process to produce ethylene, the change in the molar flow of component j can be represented as a function of reactor length as follows:

$$\frac{dF_j}{dV} = \sum \alpha_{ij} r_i, \ r_i = k_i \prod_j C_j = k_i \prod_j \left(\frac{F_j P}{F_T RT}\right), \ k_i = A_i e^{-E_i/(RT)}$$

where
  k_i is the reaction rate constant
  C_j is the molar concentration of component j
  F_j is the molar flow rate of component j
  R is the gas constant
  P is the reactor pressure
  T is the absolute temperature

F_T is the total molar flow rate, which is given by

$$F_T = F_{steam} + \sum_j F_j$$

Example 6.30 Ethane Cracking Reaction[32]

An ethane cracking process uses 4680 $kgmol/hr$ ($F_{C_2H_6}$ = 1300 mol/sec) of ethane as a feed to produce ethylene. Reactions taking place in the steam cracking process are shown in Table 6.6. Steam is introduced at a rate of F_{steam} = 0.4 $F_{C_2H_6}$, and the reactor pressure and temperature are P = 3 atm and T = 1073.15 K. Generate concentration profiles of ethane and ethylene as a function of reactor length V ($0 \le V \le 20000$ $liter$).

Solution

The change in molar flow rate of each component j as a function of reactor length can be represented as follows:

$$\frac{dF_{C_2H_6}}{dV} = -r_1 - 2r_2 - r_5, \quad \frac{dF_{C_2H_4}}{dV} = r_1 - r_4 - r_5, \quad \frac{dF_{C_2H_2}}{dV} = r_3 - r_4, \quad \frac{dF_{C_3H_6}}{dV} = -r_3 + r_5$$

$$\frac{dF_{C_3H_8}}{dV} = r_2, \quad \frac{dF_{C_4H_6}}{dV} = r_4, \quad \frac{dF_{CH_4}}{dV} = r_2 + r_3 + r_5, \quad \frac{dF_{H_2}}{dV} = r_1$$

The function *C2H6fun* defines the differential equation system:

```
function dF = C2H6fun(V,F,P,T,Fs)
% 1: C2H6, 2: C2H4, 3: C2H2, 4: C3H6; 5: C3H8, 6: C4H6, 7: CH4, 8: H2
Re = 1.987; Ri = 0.08314; % gas constant
k1 = 4.65e13*exp(-65210/(Re*T)); k2 = 3.85e11*exp(-65210/(Re*T));
k3 = 9.81e8*exp(-36920/(Re*T));  k4 = 1.03e12*exp(-41260/(Re*T));
k5 = 7.08e13*exp(-60430/(Re*T));
Ft = sum(F) + Fs; C = F*P/(Ft*Ri*T);
r1 = k1*C(1); r2 = k2*C(1)^2; r3 = k3*C(4); r4 = k4*C(2)*C(3); r5 = k5*C(1)*C(2);
dF = [-r1 - 2*r2 - r5; r1 - r4 - r5; r3 - r4; -r3 + r5; r2; r4; r2 + r3 + r5; r1];
end
```

TABLE 6.6

Reactions Occurring in the Ethane Steam Cracking Process

Reaction	Reaction Rate ($mol/(liter \cdot sec)$)	Rate Constant (R = 1.987)
$C_2H_6 \xrightarrow{k_1} C_2H_4 + H_2$	$r_1 = k_1 C_{C_2H_6}$	$k_1 = 4.65 \times 10^{13} e^{-65210/(RT)}$
$2C_2H_6 \xrightarrow{k_2} C_3H_8 + CH_4$	$r_2 = k_2 C_{C_2H_6}{}^2$	$k_2 = 3.85 \times 10^{11} e^{-65210/(RT)}$
$C_3H_6 \xrightarrow{k_3} C_2H_2 + CH_4$	$r_3 = k_3 C_{C_3H_6}$	$k_3 = 9.81 \times 10^8 e^{-36920/(RT)}$
$C_2H_2 + C_2H_4 \xrightarrow{k_4} C_4H_6$	$r_4 = k_4 C_{C_2H_2} C_{C_2H_4}$	$k_4 = 1.03 \times 10^{12} e^{-41260/(RT)}$
$C_2H_4 + C_2H_6 \xrightarrow{k_5} C_3H_6 + CH_4$	$r_5 = k_5 C_{C_2H_4} C_{C_2H_6}$	$k_5 = 7.08 \times 10^{13} e^{-60430/(RT)}$

Source: Nayef Ghasem, *Computer Methods in Chemical Engineering*, CRC Press, Taylor & Francis Group, Boca Raton, FL, 2012, pp. 250–253.

The script *etcrack* defines the data and initial conditions required in the solution of the differential equations. The built-in function *ode45* is employed to solve the differential equation system. The fractional conversion of ethane and the final molar flow rate of each component at the end of the reactor volume are also calculated.

```
% etcrack.m: ethane cracking
% 1: C2H6, 2: C2H4, 3: C2H2, 4: C3H6; 5: C3H8, 6: C4H6, 7: CH4, 8: H2
% Solution of differential equation system
P = 3; T = 1073.15; Ri = 0.08314; F10 = 1300; Fs = 0.4*F10; % data
Vspan = [0 20000]; F0 = zeros(1,8); F0(1) = F10;
[V F] = ode45(@C2H6fun,Vspan,F0,[],P,T,Fs); n = length(V); C = [];
for k = 1:n, Ft = sum(F(k,:)); C = [C; F(k,:)*P/(Ft*Ri*T)]; end
% Concentration profiles
C1 = C(:,1); C2 = C(:,2);
plot(V,C1,V,C2,'--'), xlabel('V(liter)'), ylabel('C(mol/l)'), grid, legend
('C_2H_6','C_2H_4')
% Conversion and molal flow rate at the end of the reactor volume:
x = (F10 - F(end,1))/F10; % conversion
fprintf('Fractional conversion of ethane = %g\n', x);
fprintf('Flow rate of each component at the end of the reactor volume:\n');
fprintf('C2H6 = %g mol/s,  C2H4 = %g mol/s\n', F(end,1), F(end,2));
fprintf('C2H2 = %g mol/s,  C3H6 = %g mol/s\n', F(end,3), F(end,4));
fprintf('C3H8 = %g mol/s,  C4H6 = %g mol/s\n', F(end,5), F(end,6));
fprintf('CH4 = %g mol/s,   H2 = %g mol/s\n', F(end,7), F(end,8));
```

Execution of the script *etcrack* produces the following results and the concentration profiles shown in Figure 6.32.

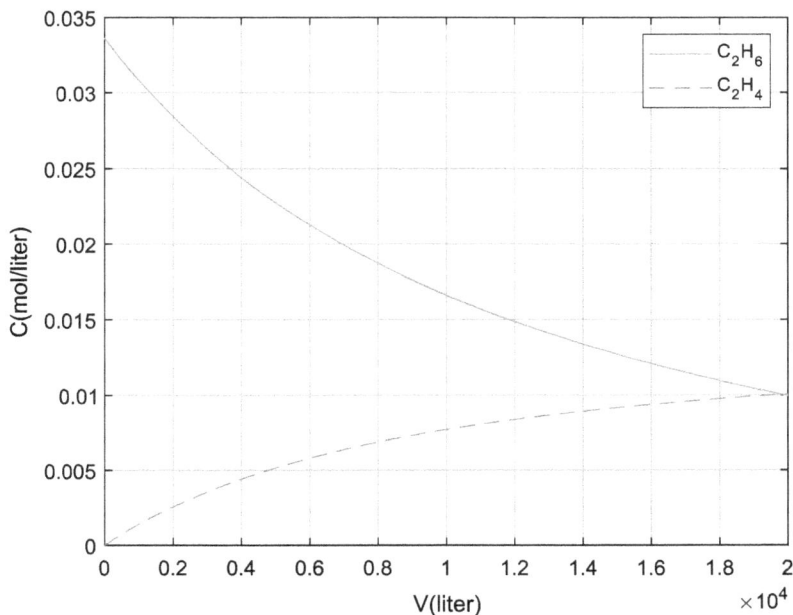

FIGURE 6.32 Concentration profiles for ethane and ethylene.

```
>> etcrack
Fractional conversion of ethane = 0.549538
Flow rate of each component at the end of the reactor volume:
C2H6 = 585.601 mol/sec,  C2H4 = 592.134 mol/sec
C2H2 = 5.56981 mol/sec,  C3H6 = 5.69562 mol/sec
C3H8 = 0.0843035 mol/sec, C4H6 = 33.1886 mol/sec
CH4 = 83.2968 mol/sec,    H2 = 669.777 mol/sec
```

Example 6.31 Cracking of Acetone in a Plug-Flow Reactor[33]

The irreversible vapor-phase cracking reaction of acetone (A) to ketene (B) and methane (C)

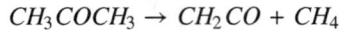

$$CH_3COCH_3 \rightarrow CH_2CO + CH_4$$

is carried out adiabatically in a plug-flow reactor. The reaction is 1st-order with respect to acetone, and the reaction rate is given by $-r_A = kC_A$ (C_A: concentration of acetone, $gmol/m^3$). From the mass balance equations for the plug-flow reactor, the rate of change of the molar flow rate of each species F_A, F_B, and F_C ($gmol/sec$) with respect to reactor volume V is given by

$$\frac{dF_A}{dV} = r_A, \ \frac{dF_B}{dV} = -r_A, \ \frac{dF_C}{dV} = -r_A$$

The rate constant $k\,(sec^{-1})$ can be expressed as a function of temperature $T\,(K)$:

$$\ln k = 34.34 - \frac{34222}{T}$$

For a gas-phase reactor, the concentration of acetone $C_A\,(gmol/m^3)$ can be represented as $C_A = 1000y_A P/(8.31T)$. The mole fraction of species i, y_i, is given by $y_i = F_i/(F_A + F_B + F_C)$ $(i = A, B, C)$ and the conversion of acetone can be calculated from $x_A = (F_{A0} - F_A)/F_{A0}$. An energy balance on a differential volume of the reactor yields

$$\frac{dT}{dV} = \frac{-r_A(-\Delta H)}{F_A C_{pA} + F_B C_{pB} + F_C C_{pC}}$$

where $\Delta H\,(J/gmol)$ is the heat of reaction at temperature T and $C_{pi}\,(i = A, B, C)$ are the molar heat capacities $(J/(gmol \cdot K))$ of acetone (A), ketene (B), and methane (C) and are given by[34]

$$\Delta H = 80770 + 6.8(T - 298) - 0.00575(T^2 - 298^2) - 1.27 \times 10^{-6}(T^3 - 298^3)$$

$$C_{pA} = 26.2 + 0.183T - 45.86 \times 10^{-6}T^2, \ C_{pB} = 20.04 + 0.0945T - 30.95 \times 10^{-6}T^2$$

$$C_{pC} = 13.39 + 0.077T - 18.91 \times 10^{-6}T^2$$

The acetone feed flow rate to the reactor is 8000 kg/hr (= 38.3 $gmol/sec$), the inlet temperature is $T = 1150\,K$, and the reactor operates at a constant pressure of $P = 162\,kPa$ (1.6 atm). The volume of the reactor is 4 m^3.

(1) Calculate the flow rate $(gmol/sec)$ and the mole fraction of each species at the reactor outlet.

(2) In order to increase the conversion of acetone, it is suggested to feed nitrogen along with the acetone. The total molar feed rate is maintained constant at 38.3 $gmol/sec$. Calculate the final conversions and temperatures for the cases where 28.3, 18.3, 8.3, 3.3, and 0.0 $gmol/sec$ nitrogen is fed into the reactor, and plot the results as a function of reactor volume. The heat capacity of nitrogen is given by

$$C_{pN_2} = 6.25 + 0.00878T - 2.1 \times 10^{-8}T^2.$$

(3) Calculate the final conversions and temperatures in the reactor operating at a pressure range of $1.6\ atm \le P \le 5\ atm$ for acetone feed rates of 10, 20, 30, 35, and 38.3 *gmol/sec*. The inlet temperature is $T = 1035\ K$, and nitrogen is fed to maintain the total feed rate at 38.3 *gmol/sec* in all cases. Prepare plots of final conversion versus P and F_{A0} and final temperature versus P and F_{A0}.[35]

Solution

(1) Application of mass and energy balances yields a system of differential equations with four unknown variables (F_A, F_B, F_C, T). The function *adfun* defines the differential equations:

```
function dX = adfun(V,X,pf)
% Differential equations for cracking reaction of acetone
% pf=[P FN2], X(1)== FA, X(2)=FB, X(3)=FC, X(4)=T
P  =  pf(1);  FN2  =  pf(2);  T  =  X(4);  CA  =  1000*(X(1)./(X(1)+X(2)+X(3)
+FN2))*P/(8.31*T);
k = exp(34.34 - 34222/T); dH = 80770 + 6.8*(T-298) - 5.75e-3*(T^2 -298^2) -
1.27e-6*(T^3 -298^3);
CpA = 26.2 + 0.183*T - 45.86e-6*T^2; CpB = 20.04 + 0.0945*T - 30.95e-6*T^2;
CpC = 13.39 + 0.077*T - 18.91e-6*T^2; CpN2 = 6.25 + 0.00878*T - 2.1e-8*T^2; rA
= -k.*CA;
dX = [rA; -rA; -rA; -rA*(-dH)./(X(1)*CpA + X(2)*CpB + X(3)*CpC + FN2*CpN2)];
end
```

The system of differential equations can be solved by the built-in function *ode45*. The following commands produces the desired outputs:

```
>> pf = [162 0]; Vspan = [0 4]; X0 = [38.3 0 0 1150]; [V X] = ode45(@adfun, Vspan,
X0, [], pf);
>> fprintf('\nFA = %g, FB = %g, FC = %g, T = %g\n', X(end,1), X(end,2), X(end,3),
X(end,4))
FA = 19.9777,  FB = 18.3223,  FC = 18.3223,  T = 920.878
```

(2) When nitrogen is added to the feed stream, the energy balance becomes

$$\frac{dT}{dV} = \frac{-r_A(-\Delta H)}{F_A C_{pA} + F_B C_{pB} + F_C C_{pC} + F_{N_2} C_{pN_2}}$$

and the mole fraction of each species, y_i, is given by

$$y_i = \frac{F_i}{F_A + F_B + F_C + F_{N_2}} (i = A,\ B,\ C)$$

These equations are contained in the function *adfun*, which defines the system of differential equations. The script *convtemp* computes the final conversions and temperatures corresponding to the flow rates of nitrogen.

```
% convtemp.m
P = 162; Vspan = [0 4]; FN2 = [28.3 18.3 8.3 3.3 0.0]; nF = length(FN2);
for i = 1:nF
  X0 = [38.3-FN2(i) 0 0 1150]; pf = [P FN2(i)];
  [V X] = ode45(@adfun,Vspan,X0,[],pf);
  xc(i) = (X0(1) - X(end,1))/X0(1); Tr(i) = X(end,4);
end
xc, Tr
```

```
>> convtemp
xc =
  0.5899    0.5183    0.4913    0.4829    0.4784
Tr =
926.3222 921.0767 920.4939 920.6749 920.8778
```

The results are summarized in Table 6.7.

The plots displaying the final conversions and temperatures for the case when the rate of nitrogen is 28.3 *gmol/sec* can be generated by the script *finxtemp* (see Figure 6.33):

```
% finxtemp.m
P = 162; Vspan = [0 4]; FN2 = 28.3; X0 = [38.3-FN2 0 0 1150]; pf = [P FN2];
[V X] = ode45(@adfun,Vspan,X0,[],pf);
xc = (X0(1) - X(:,1))/X0(1);
subplot(1,2,1),  plot(V,X(:,4)),  xlabel('Reactor  volume(m^3)'),  ylabel
('Temperature(K)'), grid
subplot(1,2,2),  plot(V,xc),  xlabel('Reactor  volume(m^3)'),  ylabel
('Conversion'), grid
```

TABLE 6.7

Final Conversions and Temperatures

Nitrogen rate(*gmol/sec*)	28.3	18.3	8.3	3.3	0.0
Final conversion	0.5899	0.5183	0.4913	0.4829	0.4784
Final temperature (K)	926.32	921.08	920.49	920.67	920.88

(a)   (b)

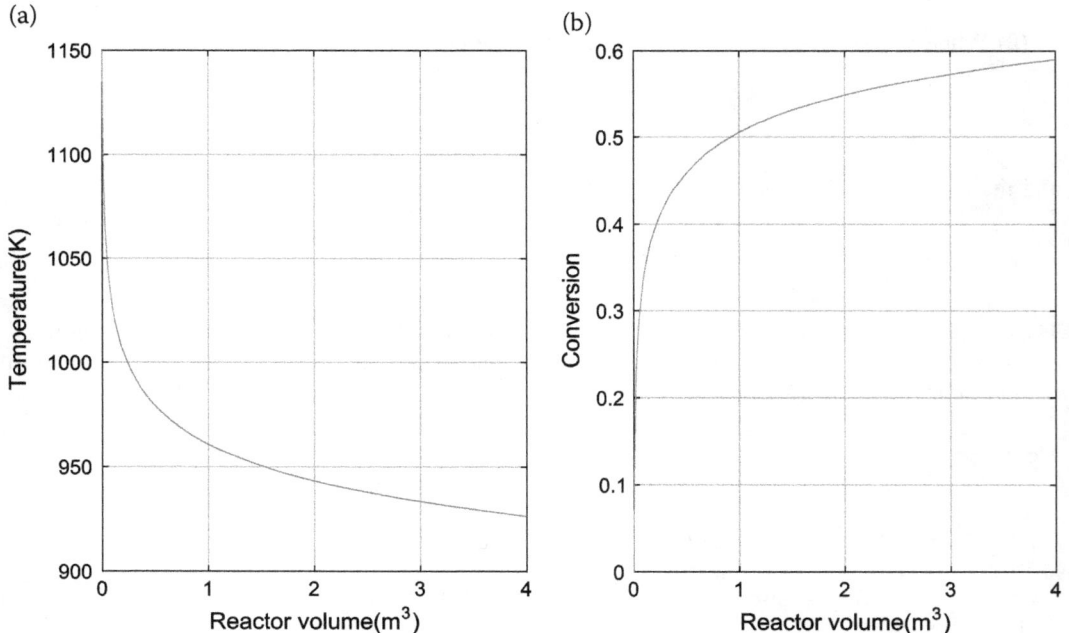

FIGURE 6.33   (a) Temperature and (b) conversion versus reactor volume.

(3) For acetone feed rates of $F_{A0} = 10, 20, 30, 35, 38.3$ *mol/sec*, the nitrogen flow rate is given by $F_{N_2} = 38.3 - F_{A0}$. The script *finconvt* calls the function *adfun*, calculates final conversions and temperatures using the built-in function *ode45*, and produces the plots shown in Figure 6.34.

```
% finconvt.m: calculates final conversions and temperatures
P = [1.6:0.2:5]*101.325; Vspan = [0 4]; FA0 = [10 20 30 35 38.3];
FN2 = 38.3-FA0; nP = length(P); nF = length(FA0);
xc = zeros(nF,nP); T = zeros(nF,nP);
for i = 1:nF
  for j = 1:nP
    X0 = [FA0(i) 0 0 1035]; pf = [P(j) FN2(i)];
    [V X] = ode45(@adfun, Vspan, X0, [], pf);
    xc(i,j) = (X0(1) - X(end,1))/X0(1); T(i,j) = X(end,4);
  end
end
P = P/101.325; % P: atm
subplot(1,2,1), plot(P,T(1,:),'o',P,T(2,:),'*',P,T(3,:),'x',P,T(4,:),'d',
P,T(5,:),'v'), grid
xlabel('P(atm)'), ylabel('T(K)'), legend('F_A0=10','F_A0=20','F_A0=30','
F_A0=35','F_A0=38.3')
subplot(1,2,2), plot(P,xc(1,:),'o',P,xc(2,:),'*',P,xc(3,:),'x',P,xc(4,:),
'd',P,xc(5,:),'v'), grid
xlabel('P(atm)'), ylabel('x_A'), legend('F_A0=10','F_A0=20','F_A0=30','
F_A0=35','F_A0=38.3')
```

6.6.2 Polymerization of Methyl Methacrylate (MMA)[36]

The polymerization of methyl methacrylate (MMA) is a radical-based polymerization that requires the use of initializers. The reaction mechanism can be expressed as follows:

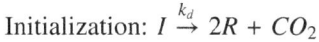

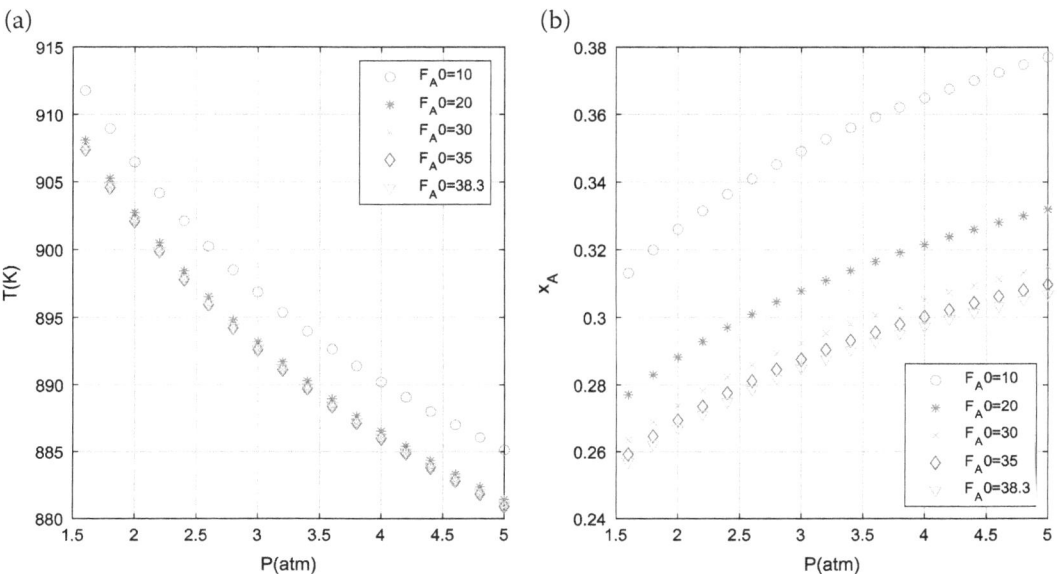

Initialization: $I \overset{k_d}{\rightarrow} 2R + CO_2$

(a)

(b)

FIGURE 6.34 (a) Temperature and (b) final conversion versus pressure and acetone flow rate.

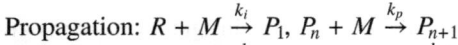

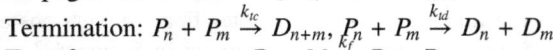

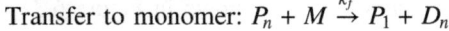

Propagation: $R + M \xrightarrow{k_i} P_1$, $P_n + M \xrightarrow{k_p} P_{n+1}$

Termination: $P_n + P_m \xrightarrow{k_{tc}} D_{n+m}$, $P_n + P_m \xrightarrow{k_{td}} D_n + D_m$

Transfer to monomer: $P_n + M \xrightarrow{k_f} P_1 + D_n$

where

    I is the initiator

    R is the active radicals

    M is the monomer

    P is the polymer

    D is the dead polymer

    $k_d\,(sec^{-1})$ is the initiator decomposition rate constant

    $k_i\,(m^3{\cdot}mol^{-1}{\cdot}sec^{-1})$ is the initial propagation rate constant

    $k_p\,(m^3{\cdot}mol^{-1}{\cdot}sec^{-1})$ is the propagation rate constant

    $k_{tc}\,(m^3{\cdot}mol^{-1}{\cdot}sec^{-1})$ is the termination rate constant

    $k_{td}\,(m^3{\cdot}mol^{-1}{\cdot}sec^{-1})$ is the disproportion rate constant

    $k_f\,(m^3{\cdot}mol^{-1}{\cdot}sec^{-1})$ is the monomer transfer rate constant

For a specific polymerization of MMA, the rate constants are represented as[37,38]

$$k_d = 1.69 \times 10^{14} \exp\left(-\frac{125{,}400}{8.314T}\right), \quad k_{p0} = 491.7 \exp\left(-\frac{18{,}220}{8.314T}\right)$$

$$k_{t0} = k_{td0} = 9.8 \times 10^4 \exp\left(-\frac{2937}{8.314T}\right), \quad k_i = k_p, \; k_i \cong 0, \; k_{tc} \cong 0$$

The volumetric relationships can be represented by using the parameter ψ, defined by

$$\psi = \frac{\gamma}{\rho_m \varphi_m V_m{}^* V_{fm} + \rho_p \varphi_p V_p{}^* V_{fp}} \left(\frac{\rho_m \varphi_m V_m{}^*}{\xi_{13}} + \rho_p \varphi_p V_p{}^*\right)$$

where

    $\rho_m\,(kg/m^3)$ is the monomer density

    $\rho_p\,(kg/m^3)$ is the polymer density

    $V_{fm}\,(m^3/kg)$ is the volume of gel, cage, and crystal effects for the monomer

    $V_{fp}\,(m^3/kg)$ is the corresponding volume for the polymer

    φ_m is the volumetric fraction of the monomer with respect to the reaction volume V

    φ_p is the corresponding volumetric fraction for the polymer with respect to V

    ξ_{i3} are the ratios of critical volumes of the monomer ($i = 1$) and the initiator ($i = I$) over the polymer

    $V_m{}^*$ and $V_p{}^*$ are the critical volumes of the monomer and polymer, respectively

    γ is usually equal to 1. These parameters can be defined as follows:

$$\rho_m = 966.5 - 1.1(T - 273.1), \quad \rho_p = 1200$$

$$V_{fm} = 0.149 + 2.9 \times 10^{-4}(T - 273.1), \quad V_{fp} = 0.0194 + 1.3 \times 10^{-4}(T - 273.1 - 105)$$

$$\varphi_m = \frac{M{\cdot}MW_m}{\rho_m}\left(\frac{1}{\dfrac{M \cdot MW_m}{\rho_m} + \dfrac{(M_0 - M)MW_m}{\rho_p}}\right), \quad \varphi_p = 1 - \varphi_m$$

$$\xi_{13} = \frac{V_m{}^*{\cdot}MW_m}{V_p{}^*{\cdot}M_{jp}}, \quad \xi_{I3} = \frac{V_I{}^*{\cdot}MW_I}{V_p{}^*{\cdot}M_{jp}}$$

$$V_I^* = 8.25 \times 10^{-4}, \; V_p^* = 7.7 \times 10^{-4}, \; V_m^* = 8.22 \times 10^{-4}$$

$$M_{jp} = 0.18781, \; MW_m = 0.10013, \; MW_I = 0.077$$

where MW_m and MW_I are molecular weights for the monomer and initiator, respectively. The initiator efficiency (f) is given by

$$f = \frac{f_0}{1 + \theta_f(T) \dfrac{M}{V} \dfrac{1}{\exp\{\xi_{I3}(-\psi + \psi_{ref})\}}}$$

where f_0 is the initial efficiency of the initiator (=1) and $\theta_f \, (m^3/mol)$ is the adjustable parameter for the initiator efficiency. ψ_{ref} is defined by $\psi_{ref} = \gamma/V_{fp}$. The termination rate constant k_t and the propagation rate constant k_p are given by

$$k_t = \frac{1}{\dfrac{1}{k_{t0}} + \theta_t(T) + \mu_n^2 \dfrac{\lambda_0}{V} \dfrac{1}{\exp(-\psi + \psi_{ref})}}, \; k_p = \frac{1}{\dfrac{1}{k_{p0}} + \theta_p(T) \dfrac{\lambda_0}{V} \dfrac{1}{\exp\{\xi_{I3}(-\psi + \psi_{ref})\}}}$$

where θ_t and θ_p are adjustable parameters for the disproportion termination rate constant and the propagation rate constant, respectively. These parameters are given by

$$\log_{10}\{10^3 \theta_f(T)\} = -40.86951 + 1.7179 \times 10^4 T^{-1}$$

$$\log_{10}\{\theta_t(T)\} = 124.1 - 1.0314 \times 10^5 T^{-1} + 2.2735 \times 10^7 T^{-2}$$

$$\log_{10}\{\theta_p(T)\} = 80.3 - 7.5 \times 10^4 T^{-1} + 1.765 \times 10^7 T^{-2}$$

The mixture volume V is calculated as follows:

$$V = \frac{M \cdot MW_m}{\rho_m} + \frac{(M_0 - M) MW_m}{\rho_p}$$

where M_0 is the initial moles of monomer. The equations describing the polymerization of MMA can be summarized as follows:

$$\frac{dI}{dt} = -k_d I + F$$

$$\frac{dM}{dt} = -(k_p + k_f) \frac{\lambda_0 M}{V} - k_i \frac{RM}{V}$$

$$\frac{dR}{dt} = 2 f k_d I - k_i \frac{RM}{V}$$

$$\frac{d\lambda_0}{dt} = k_i \frac{RM}{V} - k_t \frac{\lambda_0^2}{V}$$

$$\frac{d\lambda_1}{dt} = k_i \frac{RM}{V} + k_p \frac{\lambda_0 M}{V} - k_t \frac{\lambda_0 \lambda_1}{V} + k_f \frac{M(\lambda_0 - \lambda_1)}{V}$$

$$\frac{d\lambda_2}{dt} = k_i\frac{RM}{V} + k_p\frac{M(\lambda_0 + 2\lambda_1)}{V} - k_t\frac{\lambda_0\lambda_2}{V} + k_f\frac{M(\lambda_0 - \lambda_2)}{V}$$

$$\frac{d\mu_0}{dt} = k_f\frac{\lambda_0 M}{V} + \left(k_{td} + \frac{1}{2}k_{tc}\right)\frac{\lambda_0^2}{V}$$

$$\frac{d\mu_1}{dt} = k_f\frac{\lambda_1 M}{V} + k_t\frac{\lambda_0\lambda_1}{V}$$

$$\frac{d\mu_2}{dt} = k_f\frac{\lambda_2 M}{V} + k_t\frac{\lambda_0\lambda_2}{V} + k_{tc}\frac{\lambda_1^2}{V}$$

$$\frac{dQ}{dt} = -57,700\left\{-\left(k_p + k_f\right)\frac{\lambda_0 M}{V} - k_i\frac{RM}{V}\right\}$$

where

F is the feed rate

$Q(J)$ is the energy

λ_i is the momentum of the radical species i

μ_i is the momentum of the dead polymeric species i

λ_i and μ_i are given by

$$\lambda_i = \sum_{n=1}^{\infty} n^i R_n, \quad \mu_i = \sum_{n=1}^{\infty} n^i P_n$$

M_n and M_w can be given in terms of λ_i and μ_i as follows:

$$M_n = MW_m\left(\frac{\lambda_1 + \mu_1}{\lambda_0 + \mu_0}\right), \quad M_w = MW_m\left(\frac{\lambda_2 + \mu_2}{\lambda_1 + \mu_1}\right)$$

The conversion (X) of the reaction is defined by

$$X = \frac{M_0 - M}{M_0}$$

Example 6.32 Polymerization of MMA[37–39]

As an example, suppose that a reactor is fed with 1500 kg of monomer with a fixed temperature of 350 K. We assume that 1% of the monomer is fed into the reactor in 1 min at a constant rate. The conversion and energy generation profiles, monomer and initiator profiles, and average molecular weight profiles are desired.

Solution

The differential equations to be solved are defined by the function *pmmar*. The data are fed into the function via the structure *pm*.

```
function df = pmmar(t,x,pm)
% The reaction model is based on the following references:
% [1] Seth,V. and Gupta,S.K., J. of Polymer Eng, 15(3-4), pp. 283-323 (1995).
% [2] Ghosh,P., Gupta,S.K. and Saraf,D.N., Chemical Eng. J., 70, pp. 25-35 (1998).
% [3] Ray,A.B., Saraf,D.N. and Gupta,S.K., Polymer Eng. & Science, 35, pp. 1290-
1299 (1995).
% Retrieve data
M0 = pm.M0; MWm = pm.MWm; MWi = pm.MWi; Mjp = pm.Mjp; rhop = pm.rhop;
```

```
Vms = pm.Vms; Vps = pm.Vps; Vis = pm.Vis; T = pm.T; gam = 1; eff0 = 1;
% Assign variables
I = x(1); M = x(2); R = x(3); L0 = x(4); L1 = x(5); L2 = x(6);
N0 = x(7); N1 = x(8); N2 = x(9); Qg = x(10);
% Set values of temperature dependent parameters
rhom = 966.5 - 1.1*(T - 273.1); % monomer density (kg/m^3)
kd = 1.69e14 * exp(-125400/(8.314*T)); kp0 = 491.7 * exp(-18220/(8.314*T));
ktd0 = 9.8e4 * exp(-2937/(8.314*T));
Vm = 0.149 + 2.9e-4 * (T - 273.1); Vp = 0.0194 + 1.3e-4 * (T - 273.1 - 105);
% Ref. [3]
thet = 10^(124.1 - 1.0314e5 / T + 2.2735e7 / T^2); % theta_t
thep = 10^(80.3 - 7.5e4 / T + 1.765e7 / T^2); % theta_p
thef = 1e-3 * 10^(-40.86951 + 1.7179e4 / T); % theta_f
% Set parameters (Ref.[1])
gv = gam/Vp; eta13 = Vms*MWm/(Vps*Mjp); etai3 = Vis*MWi/(Vps*Mjp);
V = M*MWm/rhom + (M0 - M)*MWm/rhop; % volume of mixture (m^3)
psim = M*MWm/(rhom*V); psip = 1 - psim;
if L0+N0 == 0, rm = 0; else rm = (L1+N1)/(L0+N0); end % Ref. [1]
% Ref. [3]
ps  =  gam*(rhom*psim*Vms/eta13  +  rhop*psip*Vps)/(rhom*psim*Vms*Vm  +
rhop*psip*Vps*Vp);
eff = eff0/(1 + thef*(M/V)/exp(etai3*(-ps+gv)));
% Reaction rate constants
ktd = 1/(1/ktd0 + thet*rm^2*(L0/V)/exp(-ps+gv));
kp = 1/(1/kp0 + thep*(L0/V)/exp(eta13*(-ps+gv)));
% Ref. [2]
ki = kp; kf = 0; ktc = 0;
if t<60, fr = M0*0.01*100/(242*60);  % Ref. [1]
else fr = 0; end
% Define differential equations (Ref. [1])
df(1,1) = -kd*I + fr;
df(2,1) = -(kp+kf)*L0*M/V - ki*R*M/V;
df(3,1) = 2*eff*kd*I - ki*R*M/V;
df(4,1) = ki*R*M/V - ktd*L0^2/V;
df(5,1) = ki*R*M/V + kp*M*L0/V - ktd*L0*L1/V + kf*M*(L0-L1)/V;
df(6,1) = ki*R*M/V + kp*M*(L0+2*L1)/V - ktd*L0*L2/V + kf*M*(L0-L2)/V;
df(7,1) = kf*M*L0/V + (ktd+ktc/2)*L0^2/V;
df(8,1) = kf*M*L1/V + ktd*L0*L1/V;
df(9,1) = kf*M*L2/V + ktd*L0*L2/V + ktc*L1^2/V;
df(10,1) = -57700*(-(kp+kf)*L0*M/V - ki*R*M/V);
end
```

The script *pmmarxn* defines the data structure *pm* and uses the built-in function *ode15s* to solve the differential equations. The conversion (X), the number average molecular weight (PMn), and the weight average molecular weight (PMw) are calculated and all the required profiles are produced, as shown in Figure 6.35.

```
% pmmarxn.m: PMMA polymerization reaction
% Data structure
pm.M0 = 1.5e4; pm.MWm = 0.10013; pm.MWi = 0.077; pm.Mjp = 0.18781;
pm.rhop = 1200; pm.Vms = 8.22e-4; pm.Vps = 7.7e-4; pm.Vis = 8.25e-4; pm.T = 350;
tspan = [0 6000]; M0 = pm.M0; x0 = zeros(1,10); x0(2) = M0; % Initial values
[t x] = ode15s(@pmmar,tspan,x0,[],pm);
I = x(:,1); M = x(:,2); R = x(:,3); L0 = x(:,4); L1 = x(:,5); L2 = x(:,6);
```

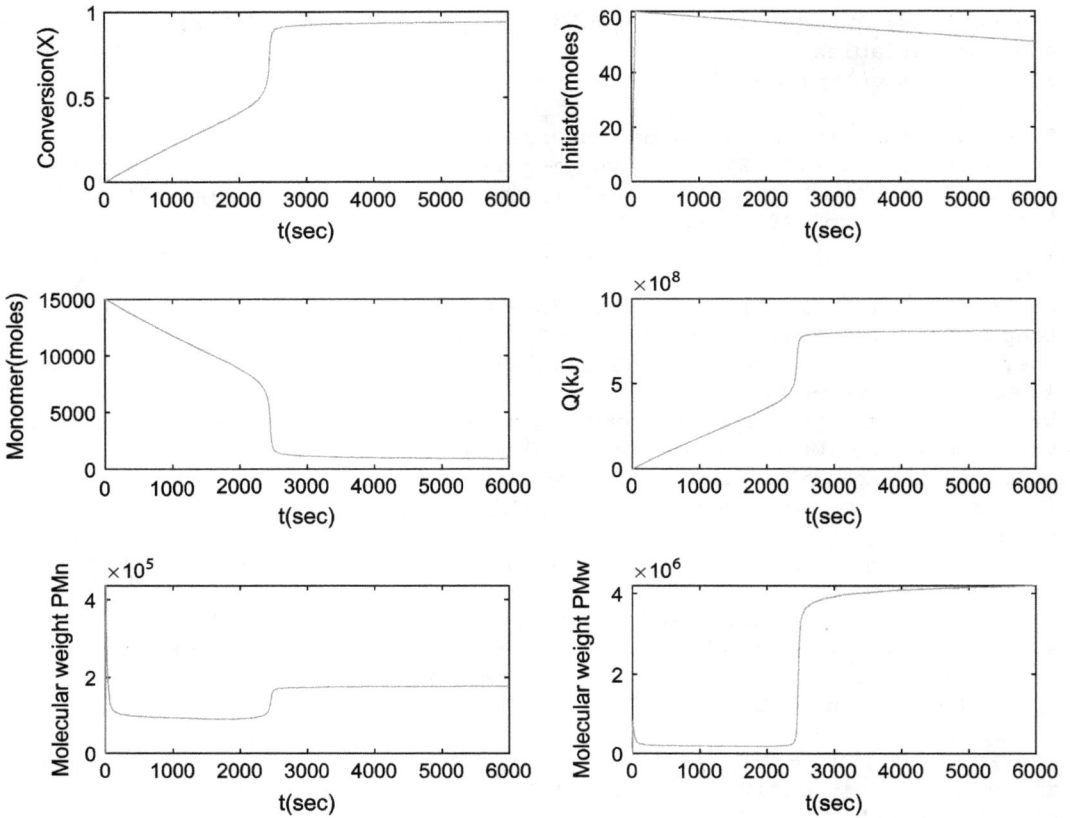

FIGURE 6.35 Profiles of the PMMA polymerization reactor performance.

```
N0 = x(:,7); N1 = x(:,8); N2 = x(:,9); Q = x(:,10);
X = (M0 - M)/M0; % conversion
mom0 = L0 + N0; mom1 = L1 + N1; mom2 = L2 + N2;
smom0 = size(mom0); n = smom0(1,1);
for k = 2:n % calculate molecular weight
  Mn(k,1) = mom1(k-1,1)/mom0(k-1,1); Mw(k,1) = mom2(k-1,1)/mom1(k-1,1);
end
for k = 1:n,  Pd(k,1) = Mw(k,1)/Mn(k,1); end
PMn = 100.13*Mn; PMw = 100.13*Mw;
% Plot results
subplot(3,2,1), plot(t,X), xlabel('t(s)'), ylabel('Conversion(X)')
subplot(3,2,2), plot(t,I), xlabel('t(s)'), ylabel('Initiator(moles)')
subplot(3,2,3), plot(t,M), xlabel('t(s)'), ylabel('Monomer(moles)')
subplot(3,2,4), plot(t,Q), xlabel('t(s)'), ylabel('Q(kJ)')
subplot(3,2,5), plot(t,PMn), xlabel('t(s)'), ylabel('Molecular weight PMn')
subplot(3,2,6), plot(t,PMw), xlabel('t(s)'), ylabel('Molecular weight PMw')
```

6.7 MICROREACTORS

Microreactors are characterized by high surface-to-volume ratios in their microstructured regions that contain tubes or channels.[40] Microreactors, with their high surface-to-volume ratios, are able to absorb heat created from a reaction much more efficiently than any batch reactor. A typical channel width is $100\,\mu m$, with a length of $2\,cm$. These small geometric dimensions result in intensified mass and heat transfer, which often leads to increased yield and selectivity compared to the classical batch approach. Mixing quality is crucial for many reactions where the molar ratio between reactants needs to be controlled precisely in order to suppress side reactions. In fact, the core part of any microreactor is the mixing regime. A microreactor that contains a number of different catalysts fixed in different compartments connected via a microfluidic network or reactor modules connected by microtubing may well be the optimal chemical production unit. Each compartment may operate simultaneously, which leads to efficient use of the microreactor.

Example 6.33 Decomposition Reaction in a Microreactor[41]

The gas-phase decomposition reaction

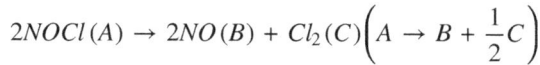

$$2NOCl\,(A) \rightarrow 2NO\,(B) + Cl_2\,(C)\left(A \rightarrow B + \frac{1}{2}C\right)$$

is carried out in a microreactor. The reaction is 2nd-order, and the reaction rate is given by

$$-r_A = kC_A^2 = kC_{T0}^2\left(\frac{F_A}{F_T}\right)^2$$

where $C_{T0} = P/(RT)$, $C_i = C_{T0}F_i/F_T$ ($i = A, B, C$), and $F_T = F_A + F_B + F_C$. Plot the molar flow rate of each species (F_A, F_B, F_C) as a function of reactor volume. The rate constant is given by

$$k = k_0 \exp\left(\frac{E}{1.987}\left(\frac{1}{500} - \frac{1}{T}\right)\right)$$

where $E = 24000$.

Data: $F_{A0} = 22.6\,\mu mol/sec = 2.26 \times 10^{-5}\,mol/sec$, $k_0 = 0.29\,dm^3/(mol\cdot sec)$, $R = 8.314\,kPa\cdot dm3/(mol\cdot K)$, $P = 1641\,kPa$, $T = 698\,K$

Solution

$$\frac{dF_i}{dV} = r_i\,(i = A, B, C)$$

Since $r_A/(-1) = r_B/(1) = r_C/(1/2)$, $r_B = -r_A$ and $r_C = -(1/2)r_A$. Thus

$$\frac{dF_A}{dV} = -kC_{T0}^2\left(\frac{F_A}{F_T}\right)^2, \quad \frac{dF_B}{dV} = kC_{T0}^2\left(\frac{F_A}{F_T}\right)^2, \quad \frac{dF_C}{dV} = \frac{k}{2}C_{T0}^2\left(\frac{F_A}{F_T}\right)^2$$

The initial conditions are $F_{A0} = 22.6$, $F_{B0} = 0$, $F_{C0} = 0$. Here the integration is performed in the range of $0 \leq V \leq 0.00001$. The script *microrx* defines the differential equations, calls the built-in solver *ode23s* to integrate them, and produces the curves shown in Figure 6.36. The system of differential equations is stiff, and the built-in solver *ode23s* is used here.

```
% microrx.m
Fa0 = 2.26e-5; R = 8.314; P0 = 1641; T0 = 698; E = 24000;
Ct0 = P0/(R*T0); Vf = 1e-5; x0 = [Fa0 0 0];
k = 0.29*exp(E/1.987*(1/500-1/T0));
dxdv = @(t,x) [-k*Ct0^2*(x(1)/ sum(x))^2; k*Ct0^2*(x(1)/ sum(x))^2; (k/
2)*Ct0^2*(x(1)/ sum(x))^2];
[V x] = ode23s(dxdv, [0 Vf],x0); Fa = x(:,1); Fb = x(:,2); Fc = x(:,3);
fprintf('Final values: Fa = %g, Fb = %g, Fc = %g\n', Fa(end), Fb(end), Fc(end));
```

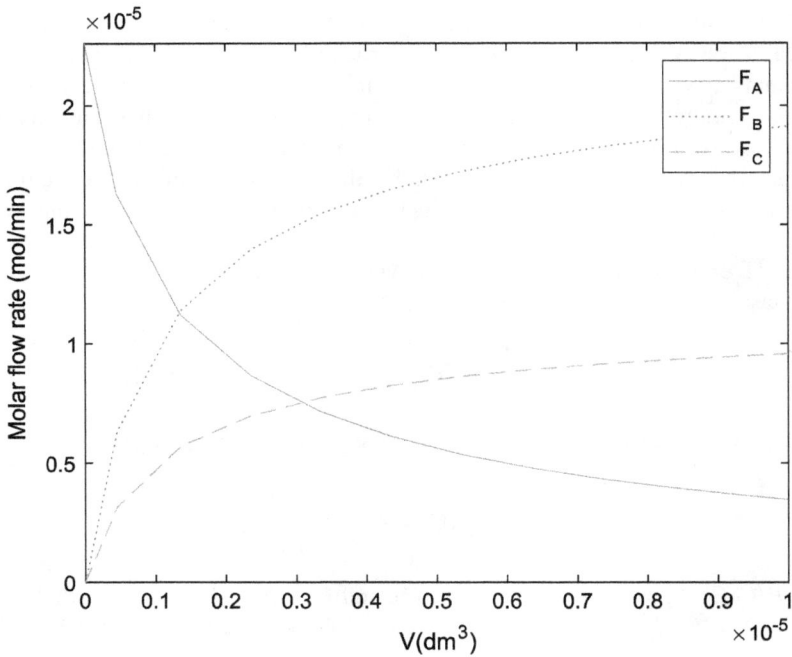

FIGURE 6.36 Profiles of microreactor molar flow rates.

```
plot(V,Fa,V,Fb,':',V,Fc,'--'), xlabel('V(dm^3)'), axis tight
ylabel('Molar flow rate (mol/min)'), legend('F_A','F_B','F_C')

>> microrx
Final values: Fa = 3.47619e-06, Fb = 1.91238e-05, Fc = 9.5619e-06
```

6.8 MEMBRANE REACTORS

A membrane can either act as a barrier to certain components while being permeable to others, prevent certain components such as particulates from contacting the catalyst, or contain reactive sites and be a catalyst in itself. Figure 6.37 shows a schematic of an inert membrane reactor with catalyst pellets on the feed side.[42]

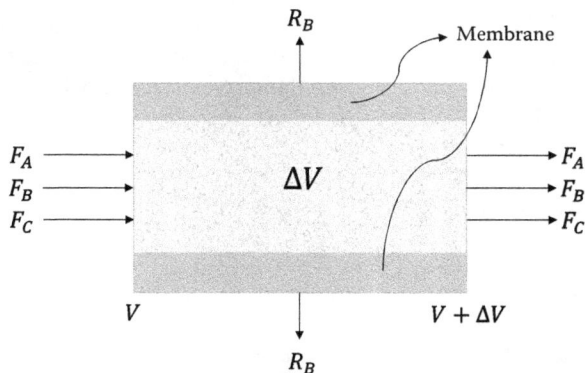

FIGURE 6.37 Schematic of a membrane reactor with catalyst pellets.

Consider a reversible reaction

$$A \leftrightarrow B + C$$

taking place on the catalyst side of a membrane reactor with catalyst. The mass balance for each component gives

$$\frac{dF_A}{dV} = r_A, \ \frac{dF_B}{dV} = r_B - R_B, \ \frac{dF_C}{dV} = r_C, \ R_B = k_c C_B$$

Example 6.34 Membrane Reactor[43]

A reversible reaction $A \leftrightarrow B + C$ takes place on the catalyst side of a membrane reactor with catalyst pellets. The reaction rate and the rate of diffusion of the product B out of the reactor, R_B, are given by

$$-r_A = k\left(C_A - \frac{C_B C_C}{K_C}\right), \ R_B = k_c C_B$$

(1) Plot the molar flow rates of each species as a function of reactor volume V ($0 \le V \le 500 \ (dm^3)$).
(2) Calculate the conversion of A at $V = 400 \ dm^3$.
   Data: $k = 0.7 \ min^{-1}$, $K_C = 0.05 \ mol/dm^3$, $k_c = 0.2 \ min^{-1}$, $F_{A0} = 10 \ mol/min$, $F_{B0} = F_{C0} = 0$, $P = 830.6 \ kPa$, $T = 500 \ K$, $R = 8.314 \ kPa \cdot dm^3/(mol \cdot K)$

Solution
   Relative reaction rates yield

$$\frac{r_A}{-1} = \frac{r_B}{1} = \frac{r_C}{1} \Rightarrow r_B = -r_A, \ r_C = -r_A$$

The concentration of each species is given by

$$C_i = C_{T0}\left(\frac{F_i}{F_T}\right)(i = A, B, C), \ C_{T0} = \frac{P}{RT}$$

The mass balance on each component gives

$$\frac{dF_A}{dV} = r_A, \ \frac{dF_B}{dV} = r_B - R_B = -r_A - k_c C_{T0}\left(\frac{F_B}{F_T}\right), \ \frac{dF_C}{dV} = r_C = -r_A$$

Combining the reaction rate equation and concentration equations yields

$$-r_A = k\left(C_A - \frac{C_B C_C}{K_C}\right) = kC_{T0}\left[\frac{F_A}{F_T} - \frac{C_{T0}}{K_C}\left(\frac{F_B}{F_T}\right)\left(\frac{F_C}{F_T}\right)\right], \ F_T = F_A + F_B + F_C$$

The subfunction *mrf* defines differential equations to be solved. The script *membrx* calls the subfunction *mrf* and employs the built-in function *ode45* to solve the system of differential equations.

```
% membrx.m
Fa0 = 10; k = 0.7; kc = 0.2; Kc = 0.05; R = 8.314; P = 830.6; T = 500;
Ct0 = P/(R*T); Vf = 500; Vc = 400; x0 = [Fa0 0 0];
[V x] = ode45(@mrf,[0 Vf],x0,[],k,kc,Kc,Ct0);
```

```
Fa = x(:,1); Fb = x(:,2); Fc = x(:,3);
fprintf('Final values: Fa = %g, Fb = %g, Fc = %g\n', Fa(end), Fb(end), Fc(end));
plot(V,Fa,V,Fb,':',V,Fc,'--'), xlabel('V(dm^3)'), axis tight
ylabel('Molar flow rate (mol/min)'), legend('F_A','F_B','F_C')
for i = 1:length(V)
  if V(i) >= Vc, iV = i; break; end
end
Fc = Fa(iV) + (Fa(iV+1)-Fa(iV))*(Vc - V(iV))/(V(iV+1)-V(iV)); Xa = (Fa0 -
Fc)/Fa0;
fprintf('At V = %g, Fa = %g and the conversion of A = %g\n', Vc, Fc, Xa);
function dxdv = mrf(v,x,k,kc,Kc,Ct0)
% x(1) = Fa, x(2) = Fb, x(3) = Fc
Ft = sum(x); rA = -k*Ct0*(x(1)/Ft - Ct0*x(2)*x(3)/(Kc*Ft^2));
dxdv = [rA; -rA - kc*Ct0*x(2)/Ft; -rA];
end
```

The script *membrx* produces the following results and Figure 6.38:

```
>> membrx
Final values: Fa = 3.99507, Fb = 1.83459, Fc = 6.00493
At V = 400, Fa = 4.32946 and the conversion of A = 0.567054
```

Example 6.35 Multiple Reactions in a Membrane Reactor[44]
The gas-phase reactions

$$A + B \longrightarrow D, A + B \longrightarrow U$$

take place in a membrane reactor. The reaction rate for each component is given by

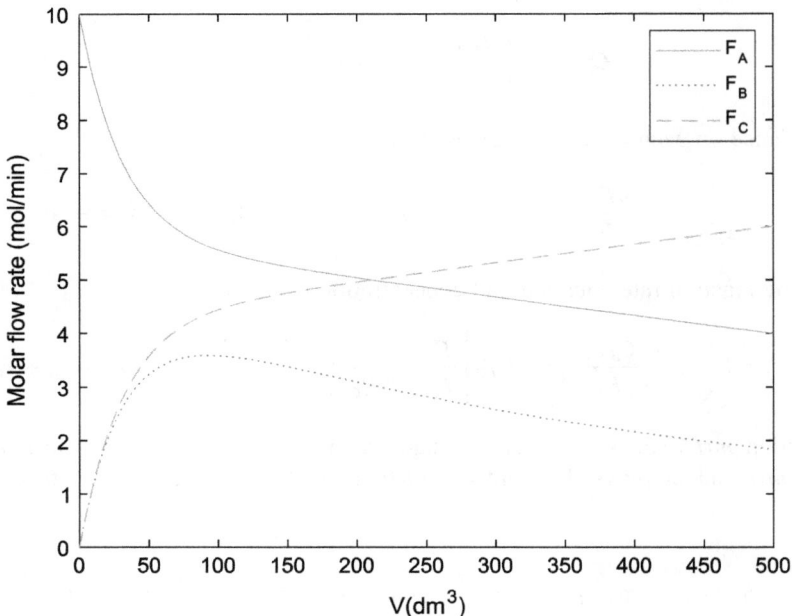

FIGURE 6.38 Molar flow rates versus reactor volume.

$$r_A = -k_1 C_A^2 C_B - k_2 C_A C_B^2, \; r_B = -k_1 C_A^2 C_B - k_2 C_A C_B^2, \; r_D = k_1 C_A^2 C_B, \; r_U = k_2 C_A C_B^2$$

and the concentration of each species is given by

$$C_i = C_{T0} \frac{F_i}{F_T} \; (i = A, B, D, U)$$

The component mass balance on each species yields

$$F_T = F_A + F_B + F_D + F_U, \; \frac{dF_A}{dV} = r_A, \; \frac{dF_B}{dV} = r_B + R_B, \; \frac{dF_D}{dV} = r_D, \; \frac{dF_U}{dV} = r_U$$

where $R_B = F_{B0}/V_t$ and V_t is the reactor volume (dm^3). The selectivity is defined by $S_{D/U} = F_D/F_U$.

Plot the molar flow rates and the overall selectivity as a function of reactor volume V ($0 \le V \le 50 \, dm^3$). The molar flow rate of A entering the reactor is 4 mol/sec, and that of B entering through the membrane, F_{B0}, is 4 mol/sec.

Data: $k_1 = 2 \, dm^6/(mol^2 \cdot sec)$, $k_2 = 3 \, dm^6/(mol^2 \cdot sec)$, $C_{T0} = 0.8 \, mol/dm^3$, $V_t = 50 \, dm^3$

Solution

The subfunction *mbf* defines differential equations, where the dependent variable vector x is defined by $x^T = [\, F_A \; F_B \; F_D \; F_U \,]$. The script *mbrmult* calls the subfunction *mbf* and uses the built-in function *ode45* to find solutions. The initial value of the vector x is set to x0 = [4 0 0 0], and that of F_U is set to zero. But when the value of F_U is very small, the value of the overall selectivity is set to zero to avoid division by zero.

```
% mbrmult.m
clear all;
k1 = 2; k2 = 3; Ct0 = 0.8; Fb0 = 4; Vt = 50; Fai = 4; % data
Rb = Fb0/Vt; x0 = [Fai 0 0 0]; Vspan = [0 Vt];
[V x] = ode45(@mbf,Vspan,x0,[],k1,k2,Ct0,Rb);
Fa = x(:,1); Fb = x(:,2); Fd = x(:,3); Fu = x(:,4); n = length(V); Sdu =
zeros(1,n);
for i = 1:n
  if Fu(i) <= 1e-6, Sdu(i) = 0; else, Sdu(i) = Fd(i)/Fu(i); end
end
subplot(1,2,1),     plot(V,Fa,V,Fb,':',V,Fd,'.-',V,Fu,'--'),     xlabel('V
(dm^3)'), ylabel('F_i(mol/s)')
legend('F_A','F_B','F_D','F_U','Location','best')
subplot(1,2,2), plot(V,Sdu), xlabel('V(dm^3)'), ylabel('S_{D/U}')
fprintf('Final molar flow rates: \n');
fprintf('Faf = %g,   Fbf = %g,   Fdf = %g,   Fuf = %g\n',Fa(end),Fb(end),Fd
(end),Fu(end));
fprintf('Overall selectivity: Sduf = %g\n',Sdu(end));
function frx = mbf (V,x,k1,k2,Ct0,Rb)
% x(1)=Fa, x(2)=Fb, x(3)=Fd, x(4)=Fu
C = Ct0*x/sum(x);
frx = [-k1*C(1)^2*C(2) - k2*C(1)*C(2)^2; -k1*C(1)^2*C(2) - k2*C(1)*C(2)^2 + Rb;
   k1*C(1)^2*C(2); k2*C(1)*C(2)^2];
end
```

The script *mbrmult* produces the following outputs and Figure 6.39:

```
>> mbrmult
Final molar flow rates:
Faf = 1.35139, Fbf = 1.35139, Fdf = 1.90998, Fuf = 0.738632
Overall selectivity: Sduf = 2.58584
```

(a)

(b)

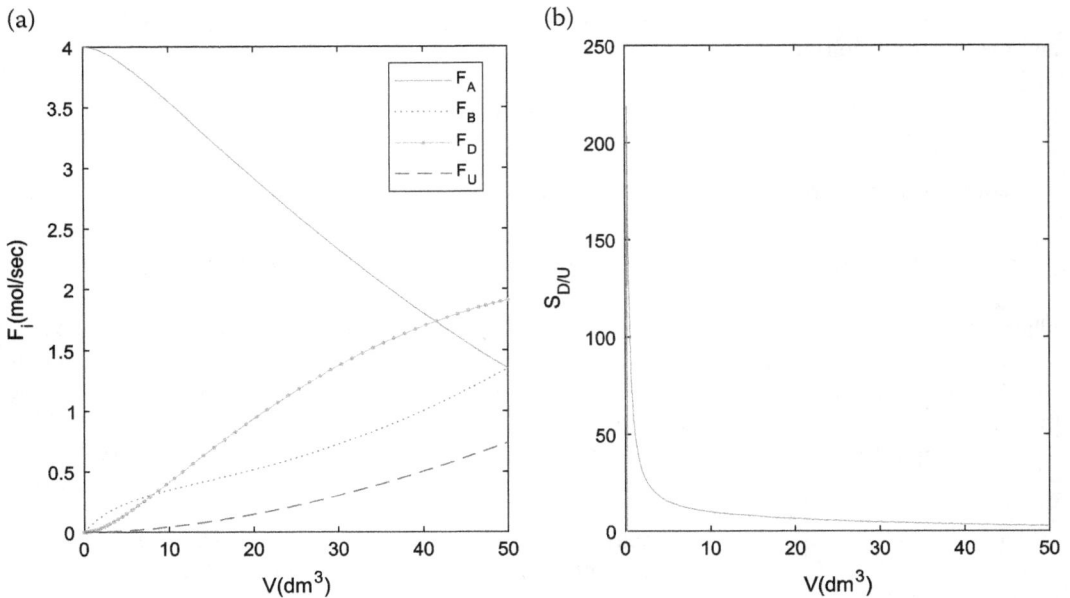

FIGURE 6.39 (a) Molar flow rates and (b) overall selectivity versus reactor volume.

6.9 BIOCHEMICAL REACTION: CELL GROWTH MODELS

Microorganisms grow under various physical, chemical, and nutritional conditions. Microorganisms convert the nutrients extracted from the medium into biological compounds. Microbial mass increases with time and may be represented by

$$\sum S + X \rightarrow \sum P + nX$$

where S is the substrate, X is cell mass, and P is the product. Cellular composition and biosynthetic capabilities are dependent upon growth conditions. Description of the relationship of the specific growth rate to substrate concentration is based on the saturation kinetics.

While many models exist for the growth rate of new cells, the most commonly used expression is the Monod equation:

$$r_g = \mu C_c$$

where
   r_g is the cell growth rate ($g/(dm^3 \cdot sec)$)
   C_c is the cell concentration (g/dm^3)
   μ denotes the specific growth rate (sec^{-1}), given by

$$\mu = \mu_{max} \frac{C_s}{K_s + C_s}$$

where
   μ_{max} is the maximum specific growth reaction rate (sec^{-1})
   K_s is the Monod constant (g/dm^3)
   C_s is the substrate concentration (g/dm^3)
   A combination of these equations yields the Monod equation for bacterial cell growth rate

$$r_g = \frac{\mu_{max} C_s C_c}{K_s + C_s}$$

The yield coefficients are defined as

$$Y_{c/s} = -\frac{\Delta C_c}{\Delta C_s},\ Y_{\frac{p}{c}} = \frac{\Delta C_p}{\Delta C_c},\ Y_{\frac{p}{s}} = -\frac{\Delta C_p}{\Delta C_s},\ Y_{\frac{s}{c}} = \frac{1}{Y_{\frac{c}{s}}},\ Y_{\frac{c}{p}} = \frac{1}{Y_{\frac{p}{c}}},\ Y_{s/p} = \frac{1}{Y_{p/s}}$$

where C_p is the concentration of the product (g/dm^3), and the subscripts c, p, and s represent cell, product, and substrate, respectively.

For a batch system, the cell mass balance gives

$$\frac{dC_c}{dt} = r_g - r_d$$

where $r_g - r_d$ denotes the net rate of formation of living cells. The rate of substrate consumption, r_{sm}, is given by

$$r_{sm} = mC_c$$

The mass balance on the substrate yields[45]

$$\frac{dC_s}{dt} = Y_{s/c}(-r_g) - mC_c$$

The rate of product formation, r_p, can be related to the net rate of substrate consumption, $-r_s$, through the following balance when $m = 0$:

$$V\frac{dC_p}{dt} = r_p V = Y_{p/s}(-r_s)V$$

Many models have been proposed to describe the substrate-limited growth phase. Table 6.8 shows typical models to represent the specific growth rate μ.[46]

Example 6.36 Bacteria Growth in a Batch Reactor[47]

Glucose-to-ethanol fermentation is to be carried out in a batch reactor using an organism. The cell growth rate r_g, cell consumption rate r_d, and substrate consumption rate r_{sm} are given by

$$r_g = \mu_{max}\left(1 - \frac{C_p}{C_p^*}\right)^{0.52}\frac{C_c C_s}{K_s + C_s},\ r_d = k_d C_c,\ r_{sm} = mC_c$$

respectively. Plot the concentrations of cells, substrate, and product, and the rates r_g, r_d, and r_{sm} as functions of time.

Data: the initial cell concentration = 1 g/dm^3, the substrate (glucose) concentration = 250 g/dm^3, $C_p^* = 93\ g/dm^3$, $Y_{c/s} = 0.08\ g/g$, $Y_{p/s} = 0.45\ g/g$, $Y_{p/c} = 5.6\ g/g$, $n = 52$, $\mu_{max} = 0.33\ hr^{-1}$, $K_s = 1.7\ g/dm^3$, $k_d = 0.01\ hr^{-1}$, $m = 0.03\ g/(g{\cdot}hr)$

Solution

Substituting rate equations into the mass balance equations

$$\frac{dC_c}{dt} = r_g - r_d,\ \frac{dC_s}{dt} = Y_{s/c}(-r_g) - r_{sm},\ \frac{dC_p}{dt} = Y_{p/s}r_g$$

TABLE 6.8
Fermentation Models

Model	Proposed by
$\mu(S) = \frac{\mu_m S}{K_m + S}$	Monod (1942)
$\mu(S) = \mu_m \left\{ 1 - \exp\left(-\frac{S}{K_m}\right) \right\}$	Tessier (1942)
$\mu(S) = \frac{\mu_m S^\lambda}{K_m + S^\lambda}, \ \lambda > 0$	Moser (1958)
$\mu(S, x) = \frac{\mu_m S}{K_c x + S}$	Contois (1959)
$\mu(S, P) = \mu_m \left(\frac{S}{K_m + S}\right)\left(\frac{1}{1 + P/K_p}\right)$	Aiba et al. (1965)
$\mu(S) = \frac{\mu_m}{2K_m}(K_m + S - \sqrt{K_m + S^2 - 4K_m S})$	Powell (1967)
$\mu(S) = \frac{\mu_m S}{K_m + S + (S^2/K_1)}$	Edwards (1970)
$\mu(S, A) = \mu_m{}^a \frac{S}{K_m + S}\left(\frac{A}{K_a + A} + \frac{1}{1 + K_b A}\right) + \mu_m{}^b, \ \mu_m = 2\mu_m{}^a + \mu_m{}^b$	Peringer et al. (1972)
$\mu(S, H^+) = \frac{\mu_m S}{\left(1 + \frac{H^+}{K_1} + \frac{K_2}{H^+}\right)(K_m + S + S^2/K_i(1 + K_3/H^+))}$	Jackson and Edwards (1975)
$\mu(S, A) = \mu_m \frac{SA}{(K_m + S)(K_a + A)}$	Olsson (1976)
$\mu(S, P) = \mu_m \left(\frac{S}{K_m + S + (S^2/K_1)}\right)\left(\frac{K_p}{K_p + P}\right)\left(1 - \frac{P}{P_f}\right)$	Dourado and Calvet (1983)
$\mu(S, A, P) = \left(\frac{K_1 S}{K_m + S} + \frac{K_2 P}{K_p + P}\right)\left(\frac{A}{K_a + A} + K_3 A - K_4\right)$	Williams et al. (1984)

S: substrate concentration, x: cell mass concentration, P: product concentration, P_f: inhibition constant,
A: dissolved oxygen concentration, H^+: hydrogen ion concentration, K_j ($j = 1,2,3,4, a, b, c, i, m, p$): constant, μ: specific growth
rate, μ_m: maximum specific growth rate.
Source: Jana, A. K., *Chemical Process Modelling and Computer Simul*ation, 2nd ed., PHI Learning Private, Ltd., Delhi, India,
2011, p. 101.

gives the following relations:

$$\frac{dC_c}{dt} = \mu_{max}\left(1 - \frac{C_p}{C_p^*}\right)^{0.52}\frac{C_c C_s}{K_s + C_s} - k_d C_c, \ \frac{dC_s}{dt} = -Y_{s/c}\mu_{max}\left(1 - \frac{C_p}{C_p^*}\right)^{0.52}\frac{C_c C_s}{K_s + C_s} - mC_c$$

The subfunction *gf* defines the system of differential equations, where the dependent variable C is defined by $C^T = [C_c \ C_s \ C_p]$.

The script *usegeferm* calls the function *geferm* and employs the built-in function *ode45* to solve the system of differential equations. The initial value of C is set to C0 = [1 250 0].

```
% geferm.m
clear all;
mumax=0.33; Cps=93; Ks=1.7; kd=0.01; m=0.03; Ysc=1/0.08; Ypc=5.6; % data
C0 = [1 250 0]; tspan = [0 12]; % initial concentration and time span
[t C] = ode45(@gf,tspan,C0,[],mumax,Cps,Ks,kd,m,Ysc,Ypc); Cc = C(:,1); Cs =
C(:,2); Cp = C(:,3);
```

```
rd = kd*Cc; rsm = m*Cc; rg = mumax*(1-Cp/Cps).^0.52 .* Cc.*Cs./(Ks+Cs);
subplot(2,2,1), plot(t,Cc), xlabel('t(hr)'), ylabel('C_c(g/dm^3)')
subplot(2,2,2), plot(t,Cs,t,Cp,'--'), xlabel('t(hr)'), ylabel('C(g/dm^3)'),
legend('C_s','C_p','Location','best')
subplot(2,2,3), plot(t,rg,t,rsm,'.-',t,rd,'--'), xlabel('t(hr)'), ylabel
('rates(g/dm^3/hr)')
legend('r_g','r_sm','r_d','Location','best')
fprintf('Final concentrations: Ccf = %g, Csf = %g, Cpf = %g\n',Cc(end),Cs
(end),Cp(end));
fprintf('Final reaction rates: rgf = %g, rsmf = %g, rdf = %g\n',rg(end),rsm
(end),rd(end));
function dC = gf (t,C,mumax,Cps,Ks,kd,m,Ysc,Ypc)
% C(1)=Cc, C(2)=Cs, C(3)=Cp
rd = kd*C(1); rsm = m*C(1); rg = mumax*(1 - C(3)/Cps)^0.52 * C(1)*C(2)/(Ks
+ C(2));
dC = [rg - rd; -rg*Ysc - rsm; Ypc*rg];
end
```

The script *geferm* produces the following outputs and Figure 6.40:

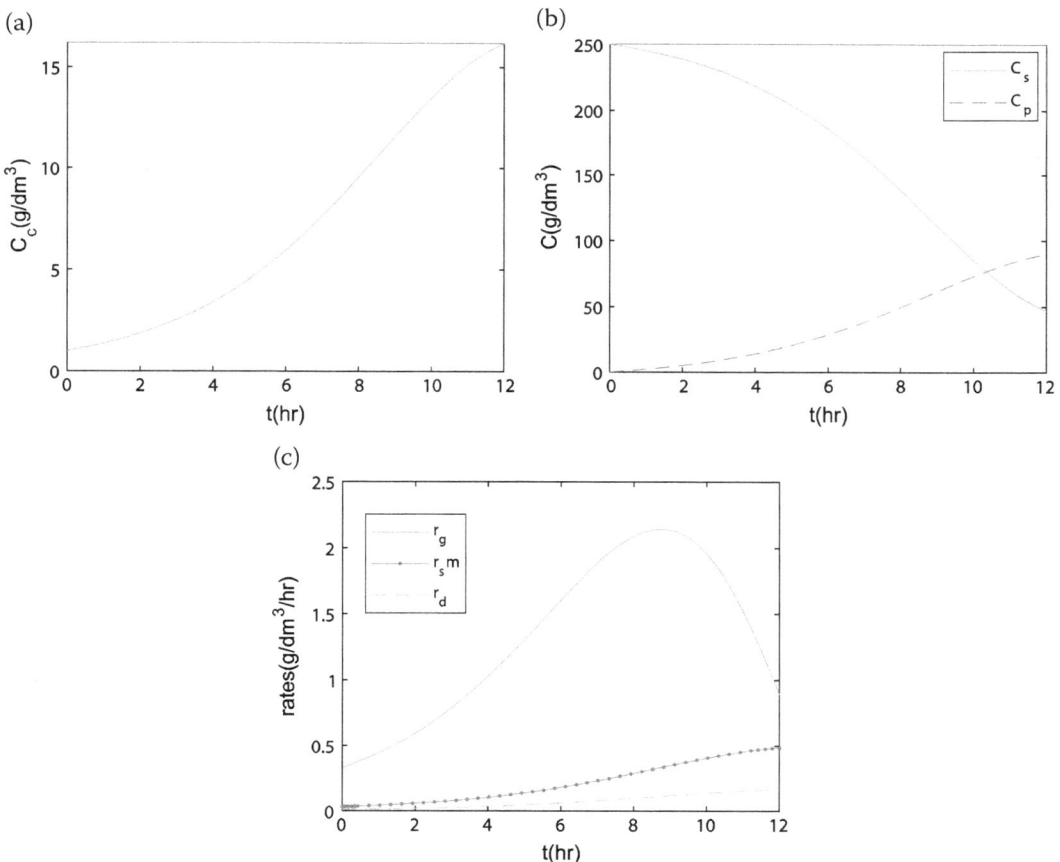

FIGURE 6.40   (a) Cell concentration, (b) substrate and product concentrations, and (c) reaction rates as a function of time.

```
>> geferm
Final concentrations: Ccf = 16.1841, Csf = 46.9352, Cpf = 89.8229
Final reaction rates: rgf = 0.890418, rsmf = 0.485522, rdf = 0.161841
```

Example 6.37 Chemostat

A simple model that describes the essential features of microorganism growth is given by the following equations:

$$\frac{dx}{dt} = -Dx + y_1\mu x, \quad \frac{dS}{dt} = -\mu x + D(S_f - S), \quad \frac{dP}{dt} = y_2\mu x - DP, \quad \mu = \frac{\mu_m S}{K_m + S}$$

where

x is the biomass concentration
S is the substrate concentration
D is the dilution rate
y_1 and y_2 are yield constants
S_f is the substrate concentration in the feed
μ_m denotes the biomass' maximum specific growth rate
P represents the concentration of the microorganism
K_m is the saturation constant
Plot x, S, and P as a function of t using the given data.
 Data: $x(0) = 0.03 \ g/ml$, $S_f = 5 \ g/ml$, $D = 0.1 \ hr^{-1}$, $S(0) = 5 \ g/ml$, $P(0) = 0.0 \ g/ml$, $\mu_m = 0.6 \ hr^{-1}$, $y_1 = 0.8$, $y_2 = 0.7$, $K_m = 0.28 \ g/ml$.

Solution

The script *chemostat* generates the profiles of x, S, and P shown in Figure 6.41.

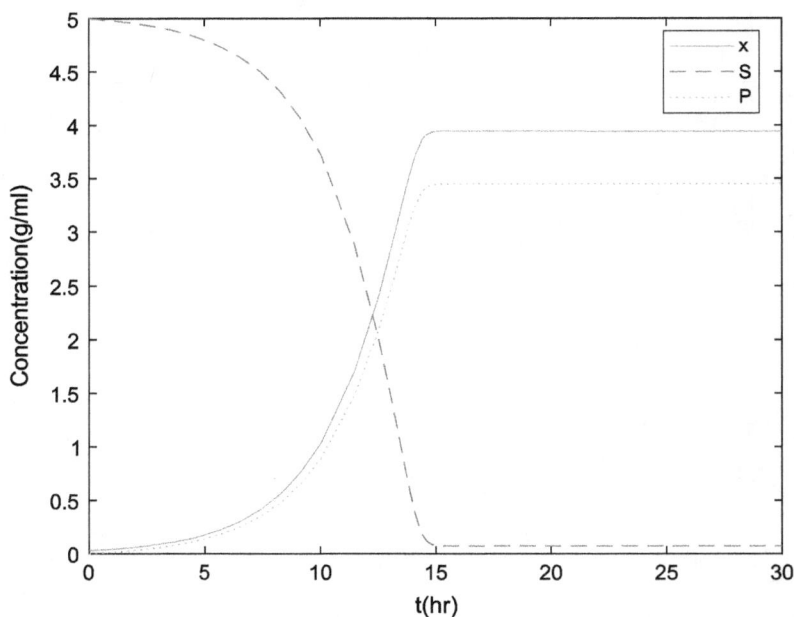

FIGURE 6.41   Transient response of a chemostat.

```
% chemostat.m: calculation of a chemostat (continuous stirred biological reactor)
clear all;
D = 0.1; Sf = 5; y1 = 0.8; y2 = 0.7; mum = 0.6; Km = 0.28; % data
X0 = 0.03; S0 = 5; P0 = 0; % initial conditions (z(1) = x, z(2) = S, z(3) = P)
dz = @(t,z) [-D*z(1) + y1*mum*z(2)*z(1)/(Km + z(2));
  - mum*z(2)*z(1)/(Km + z(2)) + D*(Sf - z(2));
  y2*mum*z(2)*z(1)/(Km + z(2)) - D*z(3)];
tint = [0 30]; z0 = [X0, S0, P0];
[t z] = ode15s(dz,tint,z0); % z(1) = x, z(2) = S, z(3) = P
x = z(:,1); S = z(:,2); P = z(:,3);
plot(t,x,t,S,'--',t,P,':'), xlabel('t(hr)')
ylabel('Concentration(g/ml)'), legend('x','S','P','location','best')
```

Example 6.38 Fermentation Batch Reaction[48]

Ethanol is produced in a baker's yeast fermenter operated in batch mode. The mass balance equations can be summarized as

$$\frac{dV}{dt} = F, \quad \frac{dx}{dt} = \left(\mu - \frac{F}{V}\right)x, \quad \frac{dS}{dt} = -\sigma x + \frac{F}{V}\left(S_f - S\right), \quad \frac{dP}{dt} = \pi x - \frac{F}{V}P$$

$$\mu = \frac{0.408S}{0.22 + S}e^{-0.028P}, \quad \sigma = 10\mu, \quad \pi = \frac{S}{0.44 + S}e^{-0.015P}$$

where

V is the reactor volume (*liter*)

F is the feed rate (*liter/hr*)

μ is the specific growth rate (hr^{-1})

σ is the substrate consumption rate

x is the biomass concentration (*g/liter*)

S is the substrate concentration (*g/liter*)

π is the product formation rate

P represents the ethanol concentration (*g/liter*)

Plot x, S, P, μ, and π as a function of t using the given data.

Data: $x(0) = 0.2 \ g/liter$, $S(0) = 100 \ g/liter$, $P(0) = 0.0 \ g/liter$, $V(0) = 1.0 \ liter$, $S_f = 100 \ g/liter$, $F = 1 \ liter/hr$.

Solution

The script *fembat* generates the profiles of x, S, P, μ, and π shown in Figure 6.42.

```
% fembat.m: batch bioreactor for ethanol production
F = 1; Sf = 100; % data
dx = @(t,x) [F;
  (0.408*x(3)*exp(-0.028*x(4))/(0.22+x(3)) - F/x(1))*x(2);
  -4.08*x(3)*x(2)*exp(-0.028*x(4))/(0.22+x(3)) + F*(Sf - x(3))/x(1);
  x(3)*x(2)*exp(-0.015*x(4))/(0.44+x(3)) - F*x(4)/x(1)];
tint = [0 20]; x0 = [1 0.2 100 0];
[t X] = ode45(dx,tint,x0); % x(1)=V, x(2)=x, x(3)=S, x(4)=P
x = X(:,2); S = X(:,3); P = X(:,4);
mu = 0.408*S.*exp(-0.028*P)./(0.22 + S);
phi = S.*exp(-0.015*P)./(0.44 + S);
subplot(1,2,1), plot(t,x,t,S,'--',t,P,':'), xlabel('t(hr)')
ylabel('Concentration(g/liter)'), grid, legend('x','S','P','location','best')
subplot(1,2,2), plot(t,mu,t,phi,'--'), xlabel('t(hr)')
ylabel('\mu and \pi'), grid, legend('\mu','\pi','location','best')
```

(a)

(b)

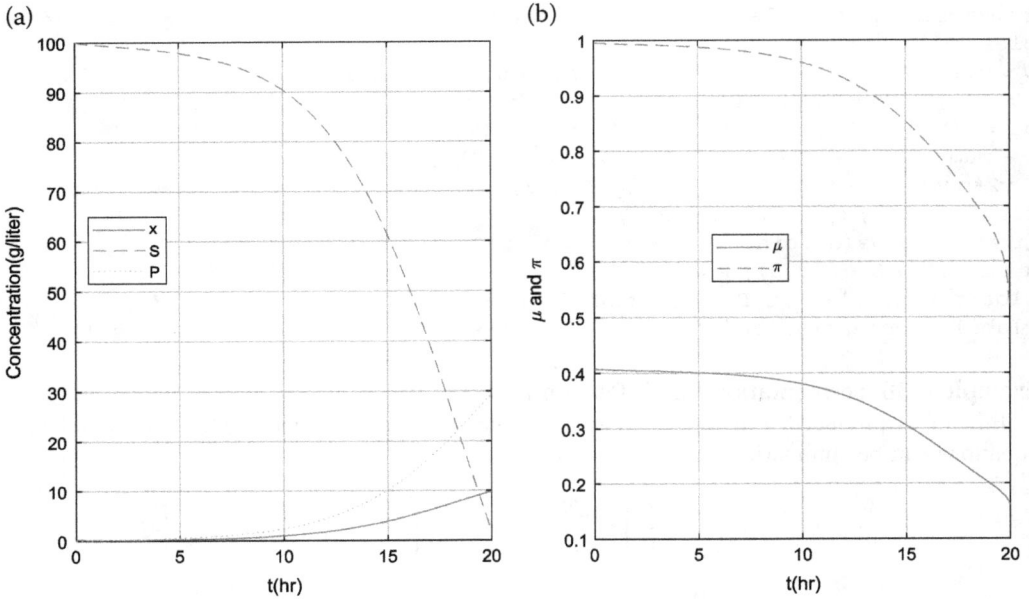

FIGURE 6.42   Profiles of (a) x, S, and P, and (b) μ and π.

Example 6.39 Biological Reactor[49]

In a biological reactor, a simple microbial culture involves a single biomass (x) growing on a single substrate (S) and yielding a single product (P). The reaction model consists of the following set of equations:

$$\frac{dx}{dt} = (\mu - D)x, \quad \frac{dS}{dt} = D(S_f - S) - \frac{\mu x}{Y}, \quad \frac{dP}{dt} = -DP + (\alpha_1\mu + \alpha_2)x,$$

$$\mu = \mu_m\left(\frac{S}{K_m + S + K_1 S^2}\right)\left(1 - \frac{P}{P_f}\right)$$

where

μ represents the specific growth rate

D is the dilution rate

S_f is the substrate concentration in the feed

Y is the yield coefficient

μ_m denotes the maximum specific growth rate

K_m is the saturation constants

K_1 and P_f are the inhibition constants

Generate profiles of x, S, and P as a function of time ($0 \leq t \leq 120$) using the given data. What are the steady-state values of x, S, and P?

Data: $x_0 = 1$ *g/liter,* $S_0 = 50$ *g/liter,* $P_0 = 0$ *g/liter,* $Y = 0.4$ *g/g,* $D = 0.202$ *hr^{-1},* $\alpha_1 = 2.2$ *g/g,* $\alpha_2 = 0.2$ *hr^{-1},* $P_f = 50$ *g/liter,* $\mu_m = 0.48$ *hr^{-1},* $K_1 = 0.04545$ *g/liter,* $K_m = 1.2$ *g/liter,* $S_f = 20$ *g/liter.*

Solution

The subfunction *mbrxn* defines the differential equations. At steady state, we have three nonlinear equations defined by the subfunction *mbsrxn*. The script *micbrx* defines the data structure (*mbdat*) and uses the built-in function *ode45* and *fsolve* to solve the set of differential equations and determine steady-state values. The initial values are set to $[x(0) S(0) P(0)] = [1500]$ to solve the ODEs, and the initial guesses are set to $[x_0 S_0 P_0] = [5510]$ to solve the steady-state nonlinear equations.

```
% micbrx.m
clear all;
% Define data structure (mbdat)
mbdat.Y = 0.4; mbdat.D = 0.202; mbdat.alp1 = 2.2; mbdat.alp2 = 0.2;
mbdat.Pf = 50; mbdat.mum = 0.48; mbdat.K1 = 0.04545; mbdat.Km = 1.2; mbdat.Sf = 20;
% Solve ODEs
x0 = 1; S0 = 50; P0 = 0; % Initial values
tint = [0 120]; z0 = [x0 S0 P0]; [t z] = ode45(@mbrxn,tint,z0,[],mbdat);
x = z(:,1); S = z(:,2); P = z(:,3);
subplot(1,2,1), plot(t,x,t,P,'--'), grid, xlabel('t(h)'), ylabel('x and P (g/
l)'), legend('x','P','location','best')
subplot(1,2,2), plot(t,S), grid, xlabel('t(h)'), ylabel('S (g/l)')
% Determine steady-state values
xs0 = 5; Ss0 = 5; Ps0 = 10; zs0 = [xs0 Ss0 Ps0]; zs = fsolve(@mbsrxn,zs0,[],mbdat);
fprintf('At steady-state, x = %g, S = %g, P = %g\n', zs(1), zs(2), zs(3));
function dz = mbrxn(t,z,mbdat)
% z(1)=x, z(2)=S, z(3)=P
Y = mbdat.Y; D = mbdat.D; alp1 = mbdat.alp1; alp2 = mbdat.alp2;
Pf = mbdat.Pf; mum = mbdat.mum; K1 = mbdat.K1; Km = mbdat.Km; Sf = mbdat.Sf;
mu = mum*z(2)*(1-z(3)/Pf)/(Km+z(2)+K1*(z(2))^2);
dz = [(mu - D)*z(1); D*(Sf - z(2)) - mu*z(1)/Y; -D*z(3) + (alp1*mu + alp2)*z(1)];
end
function f = mbsrxn(z,mbdat)
% z(1)=x, z(2)=S, z(3)=P
Y = mbdat.Y; D = mbdat.D; alp1 = mbdat.alp1; alp2 = mbdat.alp2;
Pf = mbdat.Pf; mum = mbdat.mum; K1 = mbdat.K1; Km = mbdat.Km; Sf = mbdat.Sf;
mu = mum*z(2)*(1-z(3)/Pf)/(Km+z(2)+K1*(z(2))^2);
f = [(mu - D)*z(1); D*(Sf - z(2)) - mu*z(1)/Y; -D*z(3) + (alp1*mu + alp2)*z(1)];
end
```

Execution of the script *micbrx* yields the desired profiles shown in Figure 6.43 and values of the steady state as follows:

```
>> micbrx
At steady-state, x = 5.99579, S = 5.01053, P = 19.1272
```

Example 6.40 *E. coli* Consumption by Amoebas[50]

The predictive Tsuchiya equations of *E. coli* consumption by amoebas are

$$\frac{dS}{dt} = \frac{F}{V}(S_0 - S) - \frac{c_1 \mu_{m1} N_1 S}{K_1 + S}, \quad \frac{dN_1}{dt} = -\frac{F}{V}N_1 + \frac{\mu_{m1} N_1 S}{K_1 + S} - \frac{c_2 \mu_{m2} N_1 N_2}{K_2 + N_1},$$

$$\frac{dN_2}{dt} = -\frac{F}{V}N_2 + \frac{\mu_{m2} N_1 N_2}{K_2 + N_1}$$

where
  S is the substrate
  N_1 is the number of bacteria
  N_2 is the number of amoebas

(a)   (b)

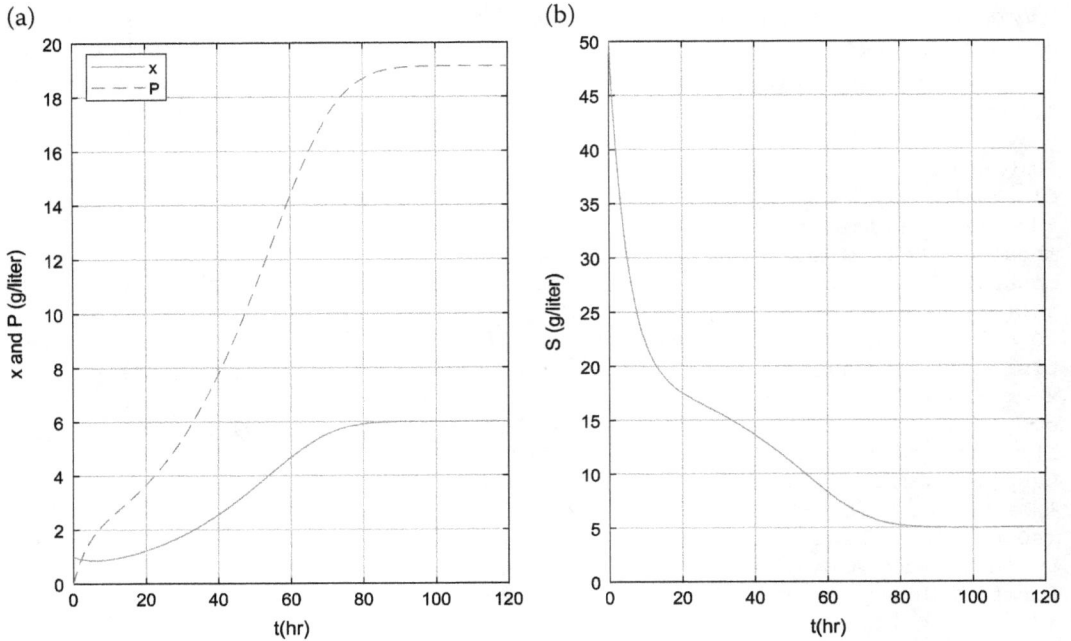

FIGURE 6.43   Profiles of (a) x and P and (b) S.

Generate time profiles of bacteria and amoebas using the given data ($0 \leq t \leq 1000\ hr$).
   Data: $F/V = 0.0625\ hr^{-1}$, $S_0 = 0.5\ mg/ml$, $\mu_{m1} = 0.25\ hr^{-1}$, $\mu_{m2} = 0.24\ hr^{-1}$,
$K_1 = 5 \times 10^{-4}\ mg/ml$, $K_2 = 4 \times 10^8\ ml^{-1}$, $c_1 = 3.3 \times 10^{-10}\ mg$, $c_2 = 1.4 \times 10^3$, $N_1(0) =$
$1.3 \times 10^9\ ml^{-1}$, $N_2(0) = 4 \times 10^5\ ml^{-1}$.

Solution
   The subfunction *pred* defines the system of differential equations. The script *predprey* uses the
given data to produce plots of $\log N_1(t)$ and $\log N_2(t)$ as a function of time, as shown in Figure 6.44.

```
% predprey.m: predator-prey model (E.Coli - Amoeba)
clear all;
pdat.Fv = 0.0625; pdat.S0 = 0.5; pdat.m1 = 0.25; pdat.m2 = 0.24; % data
pdat.K1 = 5e-4; pdat.K2 = 4e8; pdat.c1 = 3.3e-10; pdat.c2 = 1.4e3;
N10 = 1.3e9; N20 = 4e5; % initial conditions
x0 = [pdat.S0, N10, N20]; tspan = [0 1000];
[t x] = ode15s(@pred,tspan,x0,[],pdat);
plot(t,log10(x(:,2)),t,log10(x(:,3)),'--'), xlabel('t(hr)'), ylabel('log_
{10}(N_1) and log_{10}(N_2)')
legend('Bacteria(N_1)','Amoeba(N_2)','location','best')
function dx = pred(t,x,pdat)
S = x(1); N1 = x(2); N2 = x(3);
% Retrieve data
S0 = pdat.S0; Fv = pdat.Fv; c1 = pdat.c1; c2 = pdat.c2;
m1 = pdat.m1; m2 = pdat.m2; K1 = pdat.K1; K2 = pdat.K2;
% Define differential equations
dx = [Fv*(S0 - S) - (c1*m1*N1*S)/(K1 + S);
    -Fv*N1 + (m1*N1*S)/(K1 + S) - (c2*m2*N1*N2)/(K2 + N1);
    -Fv*N2 + (m2*N1*N2)/(K2 + N1)];
end
```

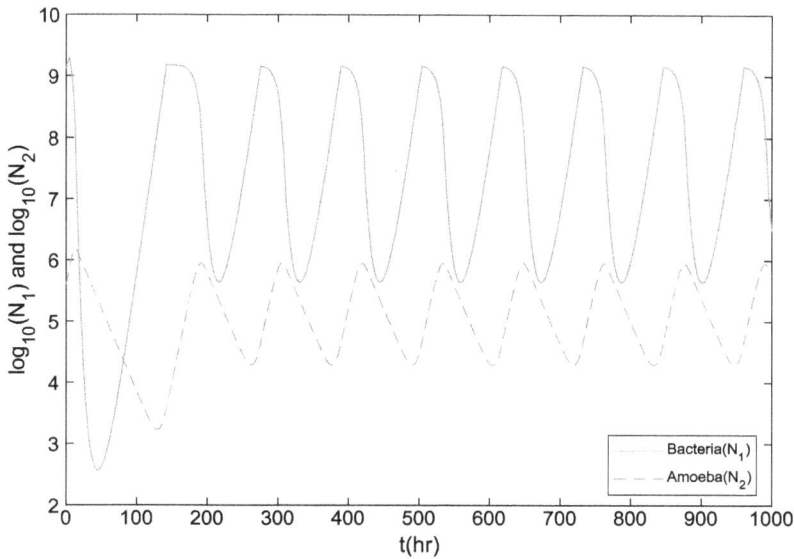

FIGURE 6.44 Time profiles of the number of bacteria (prey) and amoebas (predator).

PROBLEMS

6.1 A homogeneous irreversible gas-phase reaction whose stoichiometry is represented by

$$A \rightarrow B + 2C$$

is carried out in a CSTR with volume $V = 1 \ dm^3$ at 300°C and 0.9125 atm. The data for conversion, X_A, versus the feed flow rate $v_0 \ (dm^3/sec)$ at reactor conditions where the reactor feed consists of pure reactant A are shown in Table P6.1.[51] A mass balance on reactant A for this reactor yields

$$v_0 = \frac{kVC_{A0}^{n-1}(1 + X_A)^n}{X_A(1 + 2X_A)^n}$$

Estimate the reaction order n with respect to A and the corresponding value of the reaction rate coefficient k. The initial concentration of A is $C_{A0} = 0.1942 \ gmol/dm^3$.

6.2 The liquid-phase reaction $A + B \rightarrow C$ was carried out in a batch reactor. The concentration of reactant A was measured as a function of time and is shown in Table P6.2. The concentration of B is assumed to be constant. Use the integral method to confirm that the reaction is 2nd-order with regard to species A. Calculate the reaction rate constant.[52]

6.3 The liquid-phase reaction $A + B \rightarrow C$ takes place in a batch reactor. The reaction rate can be expressed by

$$-r_A = k'C_A^\alpha$$

TABLE P6.1
Conversion versus Feed Flow Rate

v_0	250	100	50	25	10	5	2.5	1	0.5
X_A	0.45	0.5562	0.6434	0.7073	0.7874	0.8587	0.8838	0.9125	0.95

TABLE P6.2
Concentration Data

$t\,(min)$	0	50	100	150	200	250	300
$C_A\,(mol/dm^3)$	0.05	0.038	0.0306	0.0256	0.0222	0.0195	0.0174

The concentration of reactant A is measured as a function of time and is shown in Table P6.3. The concentration of B is assumed to be constant. Use the differential method to calculate the reaction order α and the reaction rate constant k'.[53]

6.4 The gas-phase 1st-order decomposition reaction of dimethyl ether

$$CH_3OCH_3 \rightarrow CH_4 + CO + H_2$$

is carried out in a constant-volume batch reactor at 552°C. A mass balance on the batch reactor for the 1st-order reaction gives

$$\ln\left(\frac{3P_0 - P}{2P_0}\right) = -kt$$

where
P_0 is the initial pressure
P is the measured pressure
k is the rate constant

Table P6.4 shows pressure changes in the decomposition reaction. Determine the 1st-order rate constant from the data given in Table P6.4.[54]

6.5 The reversible gas-phase decomposition of nitrogen tetroxide, N_2O_4, to nitrogen dioxide, NO_2,

$$N_2O_4 \leftrightarrow 2NO_2$$

TABLE P6.3
Concentration Data

$t\,(min)$	2	48	95	152	196	245	300
$C_A\,(mol/dm^3)$	0.035	0.033	0.030	0.025	0.022	0.019	0.017

TABLE P6.4
Pressure Changes in Decomposition of Dimethyl Ether

$t\,(sec)$	$P\,(mmHg)$	$t\,(sec)$	$P\,(mmHg)$
0	420	182	891
57	584	219	954
85	662	261	1013
114	743	299	1054
145	815		

is to be carried out at constant temperature. The feed consists of pure N_2O_4 at 340 K and 202.6 kPa (2 atm). The concentration equilibrium constant, K_C, at 340 K is 0.1 mol/dm^3, and the rate constant is $k_{N_2O_4} = 0.5\ min^{-1}$. Calculate the equilibrium conversion of N_2O_4 in a constant-volume batch reactor.[55] The concentration equilibrium constant is given by $K_C = C_{Be}^2/C_{Ae}$.

6.6 Gas-phase reactions are taking place in a constant-volume batch reactor:

$$A + B \leftrightarrow C + D,\ B + C \leftrightarrow X + Y,\ A + X \leftrightarrow Z$$

The nonlinear equilibrium relationships are expressed as

$$K_1 = \frac{C_C C_D}{C_A C_B},\ K_2 = \frac{C_X C_Y}{C_B C_C},\ K_3 = \frac{C_Z}{C_A C_X}$$

$$C_A = C_{A0} - C_D - C_Z,\ C_B = C_{B0} - C_D - C_Y,\ C_C = C_D - C_Y,\ C_Y = C_X + C_Z$$

The initial concentrations of A and B are $C_{A0} = C_{B0} = 1.5\ mol/liter$, and the equilibrium constants are $K_1 = 1.06$, $K_2 = 2.63$, $K_3 = 5$. Calculate the concentration of each species at equilibrium when the initial estimates of D, X, and Z are $C_D = C_X = C_Z = 0\ mol/liter$.[56] Can we get feasible solutions for nonzero initial estimates such as $C_D = C_X = C_Z = 1$?

6.7 Pure butanol is to be fed into a semibatch reactor containing pure ethyl acetate to produce butyl acetate and ethanol. The reaction can be expressed as

$$CH_3COOC_2H_5\,(A) + C_4H_9OH\,(B) \leftrightarrow CH_3COOC_4H_9\,(C) + C_2H_5OH\,(D)$$

The mole balance, rate law, and stoichiometry equations are as follows:

Mole balance: $\frac{dN_A}{dt} = r_A V,\ \frac{dN_B}{dt} = r_A V + v_0 C_{B0},\ \frac{dN_C}{dt} = -r_A V,\ \frac{dN_D}{dt} = -r_A V,$

Rate law: $-r_A = k\left(C_A C_B - \frac{C_C C_D}{K_e}\right)$

Stoichiometry: $C_A = \frac{N_A}{V},\ C_B = \frac{N_B}{V},\ C_C = \frac{N_C}{V},\ C_D = \frac{N_D}{V}$

The overall mass balance and conversion are given by

$$\frac{dV}{dt} = v_0,\ x_A = \frac{N_{A0} - N_A}{N_{A0}}$$

respectively. The reaction is carried out at 300 K. At this temperature, the equilibrium constant based on concentrations is $K_e = 1.08$ and the reaction rate constant is $k = 9 \times 10^{-5}\ dm^3/gmol$. Initially, there is $V(0) = 200\ dm^3$ of ethyl acetate in the reactor, and butanol is fed at a rate of $v_0 = 0.05\ dm^3/sec$ for a period of 4000 sec from the start of reactor operation. The initial concentrations of ethyl acetate in the reactor and butanol in the feed stream are $C_{A0} = 7.72\ gmol/dm^3$ and $C_{B0} = 10.93\ gmol/dm^3$, respectively.[57] Plot the conversion x_A of ethyl acetate versus time for the first 5000 sec of reactor operation.

6.8 Complex liquid-phase irreversible reactions

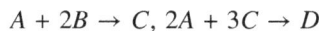

$$A + 2B \rightarrow C,\ 2A + 3C \rightarrow D$$

are taking place in a semibatch reactor. Initially, the reactor contains component B, whose initial concentration is $C_{B0} = 0.2\ mol/dm^3$. The reactant A is fed to B in the reactor at a rate of $F_{A0} = 3\ mol/min$. The volumetric flow rate of the reactant is 10 dm^3/min, and the initial reactor volume is $V = 1000\ dm^3$. The reaction rate for each species can be expressed as

$$r_A = -k_a C_A C_B^2 - \frac{2}{3}k_c C_A^2 C_C^3,\ r_B = -2k_a C_A C_B^2,\ r_C = k_a C_A C_B^2 - k_c C_A^2 C_C^3,\ r_D = \frac{1}{3}k_c C_A^2 C_C^3$$

where $C_i = N_i/V$ $(i = A, B, C, D)$ and $V = V_0 + v_0t$. The mole balances give

$$\frac{dN_A}{dt} = r_A V + F_{A0}, \quad \frac{dN_B}{dt} = r_B V, \quad \frac{dN_C}{dt} = r_C V, \quad \frac{dN_D}{dt} = r_D V$$

The selectivity is defined by $S_{C/D} = N_C/N_D$. Plot the number of moles of each species and the selectivity as functions of time for the first 100 minutes $(0 \le t \le 100min)$ of reactor operation.[58]

6.9 The elementary irreversible liquid-phase reaction

$$A + B \rightarrow C, \quad -r_A = kC_A C_B$$

is carried out in a series of three identical CSTRs. The initial concentrations of A and B are $C_{A0} = C_{B0} = 2$ $gmol/dm^3$, the inlet flow rates of A and B are $v_{0A} = v_{0B} = 6$ dm^3/min, the reaction rate constant is $k = 0.5$, the volume of each reactor is $V_1 = V_2 = V_3 = 200$ dm^3, and the flow rate of each stream is $v_1 = v_2 = v_3 = 12$ dm^3/min.

(1) Calculate the steady-state concentrations of A and B in each reactor.

(2) Plot the concentration of B exiting each reactor, C_{Bi} $(i = 1, 2, 3)$, during start-up to the final time $t = 20$ $(0 \le t \le 20)$.

6.10 A reaction $A + 2B \rightarrow C$ is carried out in a CSTR at the isothermal condition, as shown in Figure P6.10. The reaction rate is given by

$$-r_A = \frac{k_1 C_A C_B^2}{1 + k_2 C_A}$$

where k_1 and k_2 are reaction constants, and the concentrations of the components are given by $C_A = F_{A0}(1 - x_A)/q$ and $C_B = (F_{B0} - 2F_{A0}x_A)/q$. The reactor volume is obtained from $V = F_{A0}x_A/(-r_A)$, F_{A0} and F_{B0} denote molar flow rates of A and B in the feed stream, respectively, and q is the volumetric flow rate of the inlet stream. Assume that the outlet flow rate from the reactor is equal to the inlet flow rate. Plot the conversion of A, x_A, as a function of the reactor volume $(0 \le V \le 100)$.

Data: $q = 3$ $liter/min$, $F_{A0} = 10$ mol/min, $F_{B0} = 20$ mol/min, $k_1 = 0.012$ $liter^6/(mol^2 \cdot hr)$, $k_2 = 0.053$ $liter^3/mol$.

6.11 Propylene glycol (C) can be produced by the hydrolysis of propylene oxide (A) in an adiabatic CSTR. The reaction can be carried out at low temperature when sulfuric acid is used as the catalyst:

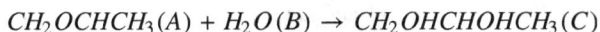

$$CH_2OCHCH_3(A) + H_2O(B) \rightarrow CH_2OHCHOHCH_3(C)$$

From the mole balance, the conversion x is given by

$$x = x_{MB} = \frac{\tau k}{1 + \tau k} = \frac{\tau A e^{-E/RT}}{1 + \tau A e^{-E/RT}}$$

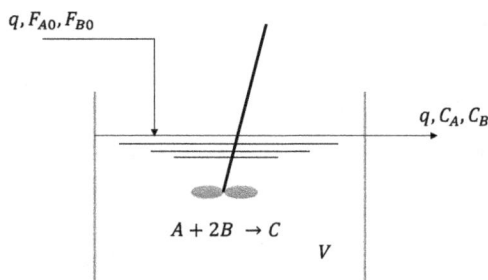

FIGURE P6.10   Isothermal CSTR.

where

$\tau = V/v_0$ (V: reactor volume, v_0: volumetric flow rate entering the reactor)

$k = Ae^{-E/RT}$ is the reaction rate constant

The energy balance gives[59]

$$x = x_{EB} = \frac{C_{pi}(T - T_i)}{-[\Delta H_R^0 + \Delta C_p(T - T_R)]}$$

where

ΔH_R^0 is the standard heat of the reaction

C_{pi} is the heat capacity of the mixture entering the reactor at T_i

ΔC_p is the overall change in heat capacity

(1) Determine T and x by solving these two nonlinear equations.

(2) Plot x_{MB} and x_{EB} as a function of T ($530 \le T \le 630$).

Data: $V = 40.1\,ft^3$, $v_0 = 326.3\,ft^3/hr$, $A = 16.96 \times 10^{12}\,hr^{-1}$, $E = 32,400\,Btu/lbmol$, $R = 1.987\,Btu/(lbmol{\cdot}R)$, $C_{pi} = 400\,Btu/(lbmol{\cdot}R)$, $\Delta C_p = -7\,Btu/(lbmol{\cdot}R)$, $T_i = 75°F$, $T_R = 68°F$, $\Delta H_R^0 = -36,400\,Btu/lbmol$.

6.12 The 1st-order exothermic irreversible reaction $A \xrightarrow{k} B$ takes place in the CSTR shown in Figure P6.12. Reactant A is supplied continuously to the reactor with flow rate $F\,(liter/sec)$, concentration $C_{Af}\,(mol/liter)$, and temperature $T_f\,(K)$. A cooling jacket surrounds the reactor to remove the exothermic heat. A coolant with flow rate $F_j\,(liter/sec)$ and inlet temperature $T_{j0}\,(K)$ takes out the heat to maintain the desired reaction temperature. From the material and energy balances for the reactor, we obtain

$$\frac{dC_A}{dt} = \frac{F}{V}(C_{Af} - C_A) - k_0 e^{-E/(RT)} C_A$$

$$\frac{dT}{dt} = \frac{F}{V}(T_f - T) + \left(\frac{-\Delta H}{\rho C_p}\right) k_0 e^{-E/(RT)} C_A - \frac{UA}{\rho C_p V}(T - T_j)$$

$$\frac{dT_j}{dt} = \frac{F_j}{V_j}(T_{j0} - T) + \frac{UA}{\rho_j C_{pj} V_j}(T - T_j)$$

Determine steady-state operating points and generate profiles of C_A, T, and T_j as a function of time using the given data. The steady state depends on the initial guess for T. Find three different steady states by changing the initial guess for T.

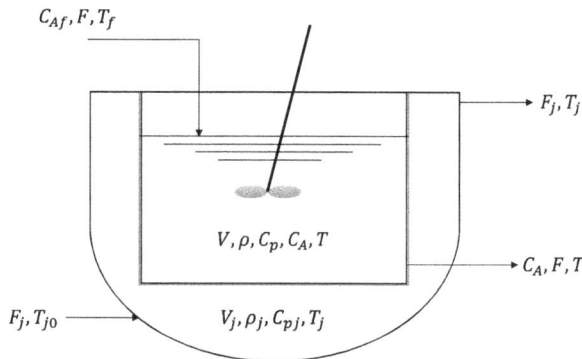

FIGURE P6.12   Schematic representation of a jacketed CSTR.

Data: $C_{Af} = 4.1\ mol/liter, k_0 = 798.5\ sec^{-1}, E/R = 4500\ K, (-\Delta H) = 2.516 \times 10^5\ J/mol, UA = 20.6\ kW/K,\ F = 25.2\ liter/sec,\ V = 255\ liter,\ \rho = 1000\ kg/m^3,\ C_p = 2498\ J/(kg\cdot K),\ T_f = 352\ K, F_j = 4.98\ liter/sec,\quad V_j = 42\ liter,\quad \rho_j = 810\ kg/m^3,\quad C_{pj} = 5100\ J/(kg\cdot K),\quad T_{j0} = 298.15\ K0 \le t \le 100sec.$

6.13 Methane and water are formed from carbon monoxide and hydrogen using a nickel catalyst. Table P6.13 shows calculated reaction rates. The rate equation can be expressed as

$$r'_{CH_4} = \frac{aP_{CO}P_{H_2}^{\beta_1}}{1 + bP_{H_2}^{\beta_2}}$$

Estimate the parameters a, b, β_1, and β_2.[60]

6.14 The irreversible 1st-order decomposition of di-tert-butyl peroxide is to be carried out in an isothermal plug-flow reactor in which there is no pressure drop. The reaction can be expressed as

$$A \rightarrow B + 2C$$

The reaction rate constant for this 1st-order reaction is $k = 0.08\ min^{-1}$. The reactor volume is $200\ dm^3$, and the entering volumetric flow rate is maintained constant at $10\ dm^3/min$. The feed stream consists of pure A at a concentration of $C_{A0} = 1.0\ gmol/dm^3$. Plot the conversion X as a function of reactor volume.[61]

6.15 The irreversible 1st-order decomposition of di-tert-butyl peroxide is to be carried out in an isothermal plug-flow reactor in which there is no pressure drop. The reactor volume is $200\ dm^3$, and the entering volumetric flow rate is maintained constant at $10\ dm^3/min$. The initial concentration of reactant A is $C_{A0} = 1.0\ gmol/dm^3$.[61]

(1) The reaction can be expressed as $A \rightarrow B + 2C$, and the reaction rate constant for this 1st-order reaction is $k = 0.08\ min^{-1}$. The feed consists of 5% A and 95% nitrogen gas as an inert component. Plot the conversion X as a function of reactor volume.

(2) The reaction can be expressed as $3A \rightarrow B$, and the reaction rate constant for this 1st-order reaction is $k = 0.08\ min^{-1}$. The feed consists of 5% A and 95% nitrogen gas as an inert component. Plot the conversion X as a function of reactor volume.

6.16 A irreversible liquid-phase reaction $A \rightarrow B$ is carried out in a plug-flow reactor. The reaction rate is expressed as

$$-r_A = kC_A^n$$

where n is the reaction order and k is the rate constant. The reactor volume is $V = 1.5\ dm^3$, and the entering volumetric flow rate is maintained constant at $v_0 = 0.9\ dm^3/min$. The initial concentration

---

TABLE P6.13

Calculated Reaction Rates for Methane Formation Reaction

P_{CO} (atm)	P_{H_2} (atm)	r'_{CH_4} (Reaction Rate)
1.0	1.0	5.20×10^{-3}
1.8	1.0	13.2×10^{-3}
4.08	1.0	30.0×10^{-3}
1.0	0.1	4.95×10^{-3}
1.0	0.5	7.42×10^{-3}
1.0	4.0	5.25×10^{-3}

of A is $C_{A0} = 1.0\ gmol/dm^3$, and the rate constant is $k = 1.1$. Plot the conversion X as a function of reactor volume for $n = 0, 1, 2,$ and 3.

6.17 A liquid-phase reaction is carried out in a nonisothermal plug-flow reactor. The model equations are

$$\frac{\partial C}{\partial t} + v\frac{\partial C}{\partial x} = r(C),\ C(0, x) = C_0,\ C(t, 0) = C_f$$

$$\frac{\partial T}{\partial t} + v\frac{\partial T}{\partial x} = \frac{(-\Delta H_r)r(C)}{\rho C_p} = g(C),\ T(0, x) = T_0,\ T(t, 0) = T_f$$

where v is assumed to be constant. The reaction rate is given by

$$r(C) = -\frac{kC}{\sqrt{1 + K_r C^2}},\ k = k_0 e^{-E/(RT)},\ K_r = K_{r0}e^{-\Delta E_r/(RT)}$$

Using the specified rate constants $k = k_1$ and $K_r = K_{r1}$ at $T = T_1$, k and K_r can be represented as

$$k = k_1 \exp\left\{-\frac{E}{R}\left(\frac{1}{T} - \frac{1}{T_1}\right)\right\},\ K_r = K_{r1}\exp\left\{-\frac{\Delta E_r}{R}\left(\frac{1}{T} - \frac{1}{T_1}\right)\right\}$$

Application of the method of lines yields ($\phi = C$ or T, $h = r$ or g)[18]

$$\frac{d\phi_i}{dt} = -v\left(\frac{\phi_{i+1} - \phi_{i-1}}{2h}\right) - h(\phi_i)(i = 2, 3, \cdots, n - 1)$$

$$\frac{d\phi_i}{dt} = -v\left(\frac{\phi_{i+1} - \phi_f}{2h}\right) - h(\phi_i)(i = 1)$$

$$\frac{d\phi_i}{dt} = -v\left(\frac{\phi_i - \phi_{i-1}}{h}\right) - h(\phi_i)(i = n)$$

The conversion is given by $X = (C_f - C)/C_f$. Generate plots of conversion as a function of x and temperature as a function of t using the given data.[18]

  Data: $C_f = 1\ mol/m^3$, $T_f = T_1 = 450\ K$, $k_1 = 2$, $K_{r1} = 1$, $E = 60\ kJ/mol$, $\Delta E_r = -10\ kJ/mol$, $(-\Delta H_r) = 100\ kJ/mol$, $\rho C_p = 800\ J/(mol\cdot K)$, $L = 2m$, $v = 0.4\ m/min$, $n = 50$.

6.18 The catalytic gas-phase 1st-order reaction

$$A \rightarrow B(-r_A = kC_A)$$

is to be carried out in a packed-bed reactor under isothermal operation. The reactant is pure A with an inlet concentration of $C_{A0} = 1.0\ gmol/dm^3$, the entering pressure is 25 *atm*, and the entering volumetric flow rate is $v_0 = 1\ dm^3/min$. The 1st-order reaction rate constant based on reactant A is $k = 1\ dm^3/(kg\ cat\cdot min)$. Plot both the conversion X and the relative pressure $y = P/P_0$ as a function of the weight of the packing W ($0 \le W \le W_{max} = 2\ kg$) for the pressure drop parameter $\alpha = 0.12\ kg^{-1}$.[62]

6.19 The irreversible gas-phase catalytic reaction

$$A + B \rightarrow C + D$$

is to be carried out in a packed-bed reactor with four different catalysts (Catalyst 1, Catalyst 2, Catalyst 3, Catalyst 4). For each catalyst, the rate expression has a different form:

Catalyst1 : $- r_{A1} = \dfrac{kC_A C_B}{1 + K_A C_A}$

Catalyst2 : $- r_{A2} = \dfrac{kC_A C_B}{1 + K_A C_A + K_C C_C}$

Catalyst3 : $- r_{A3} = \dfrac{kC_A C_B}{(1 + K_A C_A + K_B C_B)^2}$

Catalyst4 : $- r_{A4} = \dfrac{kC_A C_B}{(1 + K_A C_A + K_B C_B + K_C C_C)^2}$

The initial concentrations of the reactants are $C_{A0} = C_{B0} = 1.0 \ gmol/dm^3$ at the reactor inlet, and the molar feed flow rate of A is $F_{A0} = 1.5 \ gmol/min$. A total of $W_{max} = 2 \ kg$ of each catalyst is used in the reactor. The reaction rate constant is $k = 10 \ dm^6/(kg{\cdot}min)$, and the various catalytic parameters K_A, K_B, and K_C are given by $K_A = 1 \ dm^3/gmol$, $K_B = 2 \ dm^3/gmol$, and $K_C = 20 \ dm^3/gmol$. The pressure ratio within the reactor is given by[63]

$$\frac{dy}{dW} = -\frac{\alpha}{2y}$$

where $y = P/P_0$ and α is a constant ($\alpha = 0.4$). Calculate and plot the conversion X versus catalyst weight W for each of the catalytic rate expressions.

6.20 A gas-phase catalytic reaction $A \rightarrow B$ is carried out in a packed-bed reactor where the catalyst activity is decaying. The reaction with deactivation follows the rate expression

$$-r_A = kaC_A$$

where a is the catalyst activity. The packed-bed reactor can be approximated by three CSTRs in series, as shown in Figure P6.20.[64] Changes in the concentration of A and B in each of the three reactors can be represented as

$$\frac{dC_{Ai}}{dt} = \frac{v_0}{V}(C_{A(i-1)} - C_{A(i)}) + r_{Ai}, \quad \frac{dC_{Bi}}{dt} = \frac{v_0}{V}(C_{B(i-1)} - C_{B(i)}) - r_{Ai}$$

(1) It is assumed that the catalyst activity follows the deactivation kinetics given by

$$\frac{da_i}{dt} = -k_d a_i \ (i = 1, 2, 3)$$

Plot the concentration of A in each of the three reactors as a function of time to 60 minutes ($0 \le t \le 60$).

(2) It is assumed that the catalyst activity function is given by

$$\frac{da_i}{dt} = -k_d a_i C_{B(i)} \ (i = 1, 2, 3)$$

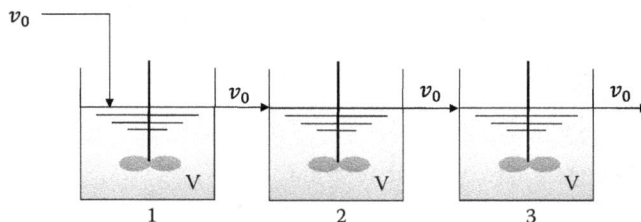

FIGURE P6.20 Train of three CSTRs.

Plot the concentration of A in each of the three reactors as a function of time to 60 minutes ($0 \leq t \leq 60$).

Data: $k_d = 0.01\ min^{-1}$, $k = 0.9\ dm^3/(dm^3(cat)\cdot min)$, $C_{A0} = 0.01\ gmol/cm^3$, $C_{B0} = 0$, $v_0 = 5\ dm^3/min$, $V = 10\ dm^3$, $a_i(0) = 1.0 (i = 1, 2, 3)$, $C_{A(i)}(0) = C_{B(i)}(0) = 0 (i = 1, 2, 3)$

6.21 A reaction taking place in a spherical catalyst pellet of radius R can be expressed as

$$\frac{D}{r^2}\frac{d}{dr}\left(r^2\frac{dC_A}{dr}\right) = D\frac{d^2C_A}{dr^2} + \frac{2D}{r}\frac{dC_A}{dr} = r_A,\ C_A|_{r=R} = C_{A0},\ \frac{dC_A}{dr}\bigg|_{r=0} = 0$$

where D is the effective diffusivity, C_A is the concentration of the reactant, and $r_A = kC_A$ denotes the rate of reaction. Introducing dimensionless variables $\phi = C_A/C_{A0}$, $\eta = r/R$, and $\zeta^2 = kR^2/D$, this equation can be rewritten as follows:

$$\frac{d^2\phi}{d\eta^2} + \frac{2}{\eta}\frac{d\phi}{d\eta} - \zeta^2\phi = 0,\ \phi|_{\eta=1} = 1,\ \frac{d\phi}{d\eta}\bigg|_{\eta=0} = 0$$

where ζ^2 is known as the Thiele modulus. Generate the concentration profile $\phi(\eta)$ for various values of ζ ($\zeta = 1, 2, 5, 10$) using the finite difference method.

6.22 The hydrodemethylation of toluene (T) is to be carried out in a packed-bed catalytic reactor. In the reactor, toluene is reacted with hydrogen (H) to produce benzene (B) and methane (M):

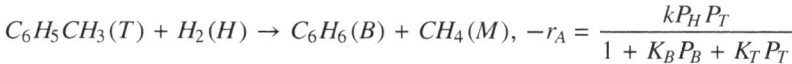

$$C_6H_5CH_3(T) + H_2(H) \rightarrow C_6H_6(B) + CH_4(M),\ -r_A = \frac{kP_H P_T}{1 + K_B P_B + K_T P_T}$$

The molar feed rate of toluene to the reactor is $F_{T0} = 50\ mol/min$, and the reactor inlet is at 40 atm and 640°C. The feed consists of 30% toluene, 45% hydrogen, and 25% inerts. The mole balance gives

$$\frac{dX}{dW} = \frac{-r_T}{F_{T0}}$$

where

X is the conversion
W is the catalyst weight
The partial pressure of each species is given by

$$P_T = P_{T0}(1 - X)y,\ P_H = P_{T0}(\Theta_{H_2} - X)y,\ P_B = P_{T0}Xy$$

where $\Theta_{H_2} = 0.45/0.30 = 1.5$, $P_{T0} = y_{T0}P_0 = (0.3)(40) = 12\ atm$, $y = P/P_0 = \sqrt{1 - \alpha W}$, and α is the pressure drop parameter. Plot the conversion, the pressure ratio $y = P/P_0$, and the partial pressure of each species as a function of catalyst weight.[65]

Data: $k = 8.7 \times 10^{-4}\ mol/(atm^2 \cdot kg_{cat} \cdot min)$, $K_B = 1.39\ atm^{-1}$, $K_T = 1.038\ atm^{-1}$, $\alpha = 9.8 \times 10^{-5}kg^{-1}$

6.23 SO_2 is oxidized to form SO_3 in a nonisothermal plug-flow reactor. The conversion can be represented as $X = C_{SO_2}(z)/C_{SO_2}(z = 0)$. The mass and energy balances on the reactor yield[66]

$$\frac{dX}{dz} = -50r,\ \frac{dT}{dz} = -4.1(T - T_a) + 1.02 \times 10^4 r,\ X(0) = 1,\ T(0) = 673.2$$

$$r = \frac{X\sqrt{1 - 0.167(1 - X)} - 2.2(1 - X)/K_e}{[k_1 + k_2(1 - X)]^2},\ K_e = -11.02 + \frac{11570}{T}$$

$$\ln k_1 = -14.96 + \frac{11070}{T}, \quad \ln k_2 = -1.331 + \frac{2331}{T}, \quad T_a = 673.2$$

Plot the profiles of X and T as a function of dimensionless reactor length (z).

6.24 The isomerization reaction of n-butane (C_4H_{10}) to isobutane

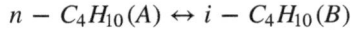

$$n - C_4H_{10}(A) \leftrightarrow i - C_4H_{10}(B)$$

is to be carried out in a bank of 10 tubular reactors; each reactor has $V = 5\ m^3$. The bank reactors are double-pipe heat exchangers with the reactants flowing in the inner pipe and $Ua = 5,000\ kJ/(m^3 \cdot hr \cdot K)$.

A mixture of 90 mol% n-butane and 10 mol% i-pentane, which is considered inert, is to be processed at 70% conversion. The molar flow rate of the mixture is 163 $kmol/hr$. The bank reactors can be considered as a countercurrent heat changer. The entering temperature of the reactants is $T_0 = 305\ K$, and the entering coolant temperature is $T_{a0} = 310\ K$.

The mass flow rate of the coolant is $m_c = 500\ kg/hr$ and the heat capacity of the coolant is $C_{pc} = 28\ kJ/(kg \cdot K)$. The temperature in any one of the reactors cannot rise above $325K$. At $T_1 = 360\ K$ and $T_2 = 333\ K$, it is known that $k_1(T_1) = 31.1\ hr^{-1}$ and $K_{C2}(T_2) = 3.03\ hr^{-1}$. Plot the conversion X, equilibrium conversion X_e, temperature T, coolant temperature T_a, and reaction rate $-r_A$ down the length of the reactor for each of the following cases[67]:

(1) Cocurrent heat exchange.
(2) Constant T_a.

Data: $\Delta H_{Rx}^\circ = -6900\ J/mol$, activation energy $E = 65,700\ J/mol$, $C_{A0} = 9.3\ kmol/m^3$, $R = 8.314\ J/(mol \cdot K)$, $C_{pA} = C_{pB} = 141\ J/(mol \cdot K)$, $C_{P_i-P} = 161\ J/(mol \cdot K)$

6.25 In the development of an artificial kidney, a reaction that converts urea to ammonia and CO_2 needs to be studied. Table P6.25 shows a set of experimental data using a certain amount of urease. In the Table P6.25, S is the concentration of urea and $-r$ is the rate of urea conversion. It is known that $-r$ can be represented by a Michaelis-Menten equation of the form

$$-r = \frac{V_m S}{K_m + S}$$

where V_m and K_m are parameters for a certain amount of urease. Determine the time t needed to reduce the concentration of urea from 0.5 to 0.2 $kmol/m^3$ using the same amount of urease as in Table P6.25.[68] The time t needed to reduce the concentration of urea from S_0 to S_f is given by

$$t = \int_{S_f}^{S_0} \frac{1}{-r} dS$$

6.26 The rate of production formation in the single substrate enzyme-catalyzed reaction $E + S \rightarrow E + P$ is given by

$$\frac{dP}{dt} = \frac{\mu_m S}{K_m + S}$$

TABLE P6.25

Reaction Rate of Urea Conversion

$S\ (kmol/m^3)$	0.2	0.02	0.01	0.005	0.002
$-r\ (mol/(m^3 \cdot sec))$	1.08	0.55	0.38	0.2	0.09

---

where

E is the enzyme

P is the product

S is the substrate

μ_m and K_m are parameters to be determined

Table P6.26 shows experimental data obtained for an enzyme-catalyzed reaction. Estimate μ_m and K_m.

6.27 Figure P6.27 shows a continuous-stirred tank bioreactor (CSTB) where a fermentation process is performed with a substrate by the action of microorganisms. In Figure P6.27, x is the biomass concentration, S is the substrate concentration, F represents the volumetric flow rate of the feed stream, and V is the volume of the reactor. The model of the bioreactor can be summarized as:

$$\frac{dx}{dt} = (\mu - D)x, \quad \frac{dS}{dt} = D(S_f - S) - \frac{\mu x}{Y}, \quad \mu = \frac{\mu_m S}{K_m + S}$$

where

μ is the specific growth rate (hr^{-1})

D represents the dilution rate (hr^{-1})

S_f is the substrate concentration in the feed stream $(g/liter)$

μ_m is the maximum specific growth rate

K_m is the limiting substrate concentration when the specific growth rate is equal to half the maximum specific growth rate

Y denotes the yield

Using the given data, plot x and S as a function of time. Profiles of x and S are dependent upon the initial conditions.[69] Here we set $x_0 = S_0 = 0.5$.

Data: $\mu_m = 0.53\ hr^{-1}$, $K_m = 0.12\ g/liter$, $D = 0.3h^{-1}$, $S_f = 4\ g/liter$, $Y = 0.4$.

6.28 In a biochemical reactor, the consumption of substrate (S) promotes the growth of biomass (x) and formation of product (P). The dynamic model consists of the following three differential equations[70]:

TABLE P6.26

Kinetic Data for an Enzyme-Catalyzed Reaction

dP/dt	1.138	0.872	0.695	0.592	0.498	0.446	0.389	0.351
$S\,(g/liter)$	20.1	10.5	6.71	5.02	4.03	3.29	2.92	2.53

F, x_f, S_f

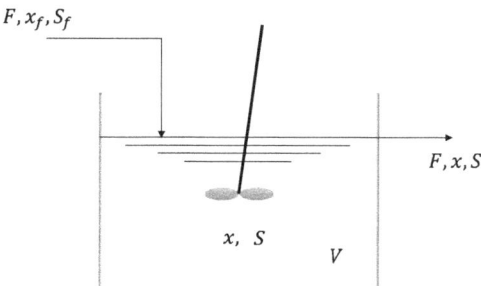

F, x, S

x, S

V

FIGURE P6.27   A continuous fermenter.

$$\frac{dx}{dt} = r_1 - Dx, \quad \frac{dS}{dt} = -y_1 r_1 - y_2 r_2 + D(S_f - S), \quad \frac{dP}{dt} = r_2 - DP, \quad r_1 = \mu x, \quad r_2 = \nu x$$

$$\mu = \mu_m \left(\frac{S}{K_{m1} + S + (S^2/K_1)} \right)\left(\frac{K_p}{K_p + P} \right)\left(1 - \frac{P}{P_f} \right), \quad \nu = \nu_m \left(\frac{S}{K_{m2} + S} \right)$$

where

r_1 and r_2 represent the growth and the biosynthesis reaction rate, respectively
y_1 and y_2 are yield constants
D is the dilution rate
S_f is the substrate concentration in the feed
μ_m denotes the biomass' maximum specific growth rate
ν_m represents the product's maximum specific synthesis rate
K_{m1} and K_{m2} are the saturation constants
K_1, P_f and K_p are the inhibition constants
Generate profiles of x, S, and P as a function of time ($0 \le t \le 150$) using the given data. What are the steady-state values of x, S, and P?

Data: $D = 0.15\ hr^{-1}$, $y_1 = 5.1$, $y_2 = 1.5$, $\mu_m = 0.3\ hr^{-1}$, $\nu_m = 0.11\ hr^{-1}$, $K_{m1} = 0.26\ g/liter$, $K_{m2} = 9.5\ g/liter$, $K_1 = 297\ (g/liter)^2$, $P_f = 85\ g/liter$, $K_p = 8\ g/liter$, $S_f = 95\ g/liter$.

6.29 The model of the baker's yeast production process in a batch reactor is

$$\frac{dx}{dt} = \mu(S)x - \frac{F}{V}x, \quad \frac{dS}{dt} = -\frac{\mu(S)x}{Y_x} - \frac{\nu x}{Y_P} + \frac{F}{V}(S_f - S), \quad \frac{dP}{dt} = \nu x - \frac{F}{V}P$$

$$\frac{dV}{dt} = F, \quad \mu(S) = \frac{\mu_m S}{K_m + S + S^2/K_1}$$

where

x is the biomass concentration ($g/liter$)
S is the substrate concentration ($g/liter$)
P is the product concentration ($g/liter$)
V is the reactor volume ($liter$)
F is the feed rate ($liter/hr$)
μ is the specific growth rate (hr^{-1})
S_f is the substrate concentration in the feed ($g/liter$)
Y_P and Y_x are the yield coefficients
μ_m, K_m, K_1, and ν are the kinetic parameters
Generate profiles of x, S, and P as a function of time ($0 \le t \le 150$) using the given data.[71]

Data: $x_0 = 1\ g/liter$, $S_0 = 0.5\ g/liter$, $P_0 = 0\ g/liter$, $V_0 = 150\ liter$, $Y_x = 0.5\ g/g$, $Y_P = 1.2\ g/g$, $\nu = 0.004\ liter/hr$, $\mu_m = 0.02\ liter/hr$, $K_m = 0.05\ g/liter$, $K_1 = 5\ g/liter$, $S_f = 200\ g/liter$, $F = 0.5\ liter/hr$.

REFERENCES

1. Fogler, H. S., *Essentials of Chemical Reaction Engineering*, Pearson Education International, Boston, MA, pp. 90–92, 2011.
2. Fogler, H. S., *Essentials of Chemical Reaction Engineering*, Pearson Education International, Boston, MA, pp. 262–263, 2011.
3. Bruce A. F., *Introduction to Chemical Engineering Computing*, John Wiley & Sons, Inc., Hoboken, NJ, pp. 43–47, 2006.
4. Fogler, H. S., *Essentials of Chemical Reaction Engineering*, Pearson Education International, Boston, MA, pp. 125–128, 2011.

5. Cutlip, M. B. and M. Shacham, *Problem Solving in Chemical and Biochemical Engineering with POLYMATH, Excel, and MATLAB*, 2nd ed., Prentice-Hall, Boston, MA, pp. 462–463, 2008.
6. Jana, A. K., *Chemical Process Modelling and Computer Simulation*, 2nd ed., PHI Learning Private, Ltd., Delhi, India, pp. 78–82, 2011.
7. Jana, A. K., *Chemical Process Modelling and Computer Simulation*, 2nd ed., PHI Learning Private, Ltd., Delhi, India, pp. 62–74, 2011.
8. Fogler, H. S., *Essentials of Chemical Reaction Engineering*, Pearson Education Inc., Boston, MA, pp. 311–312, 2011.
9. Fogler, H. S., *Essentials of Chemical Reaction Engineering*, Pearson Education Inc., Boston, MA, pp. 312–313, 2011.
10. Jana, A. K., *Chemical Process Modelling and Computer Simulation*, 2nd ed., PHI Learning Private, Ltd., Delhi, India, p. 40, 2011.
11. Cutlip, M. B. and M. Shacham, *Problem Solving in Chemical and Biochemical Engineering with POLYMATH, Excel, and MATLAB*, 2nd ed., Prentice-Hall, Boston, MA, pp. 465–467, 2008.
12. Fogler, H. S., *Essentials of Chemical Reaction Engineering*, Pearson Education International, Boston, MA, p. 226, 2011.
13. Fogler, H. S., *Essentials of Chemical Reaction Engineering*, Pearson Education International, Boston, MA, pp. 227–229, 2011.
14. Fogler, H. Scott, *Essentials of Chemical Reaction Engineering*, Pearson Education International, Boston, MA, pp. 230–232, 2011.
15. Finlayson, B. A., *Introduction to Chemical Engineering Computing*, John Wiley & Sons, Inc., Hoboken, NJ, pp. 118–120, 2006.
16. H. Scott Fogler, *Elements of Chemical Reaction Engineering*, 4th ed., Prentice Hall PTR, Upper Saddle River, Boston, NJ, pp. 340–342, 2006.
17. H. Scott Fogler, *Elements of Chemical Reaction Engineering*, 4th ed., Prentice Hall PTR, Upper Saddle River, Boston, NJ, pp. 352–354, 2006.
18. Niket S. Kaisare, *Computational techniques for process simulation and analysis using MATLAB*, CRC Press, Taylor & Francis Group, Boca Raton, FL, pp. 159–163, 2018.
19. Fogler, H. S., *Essentials of Chemical Reaction Engineering*, Pearson Education International, Boston, MA, pp. 496–500, 2011.
20. Cutlip, M. B. and M. Shacham, Problem Solving in Chemical and Biochemical Engineering with POLYMATH, Excel, and MATLAB, 2nd ed., Prentice-Hall, Boston, MA, p. 454, 2008.
21. Fogler, H. S., *Essentials of Chemical Reaction Engineering*, Pearson Education International, Boston, MA, pp. 451–452, 2011.
22. Niket S. Kaisare, *Computational Techniques for Process Simulation and Analysis using MATLAB*, CRC Press, Taylor & Francis Group, Boca Raton, FL, p. 308, 2018.
23. Cutlip, M. B. and M. Shacham, *Problem Solving in Chemical and Biochemical Engineering with POLYMATH, Excel, and MATLAB*, 2nd ed., Prentice-Hall, Boston, MA, pp. 485–486, 2008.
24. Fogler, H. S., *Essentials of Chemical Reaction Engineering*, Pearson Education International, Boston, MA, pp. 308–311, 2011.
25. Fogler, H. S., *Essentials of Chemical Reaction Engineering*, Pearson Education International, Boston, MA, pp. 183–187, 2011.
26. Mariano M. Martin (Editor), *Introduction to Software for Chemical Engineers*, CRC Press, Taylor & Francis Group, Boca Raton, FL, pp. 136–145, 2015.
27. Harrer, T. S., *Kirk Othmer Encyclopedia of Chemical Technology*, 2nd ed., Vol.19, Wiley-Interscience, New York, NY, p. 470, 1969.
28. Cutlip, M. B. and M. Shacham, *Problem Solving in Chemical and Biochemical Engineering with POLYMATH, Excel, and MATLAB*, 2nd ed., Prentice-Hall, Boston, MA, pp. 491–492, 2008.
29. Cutlip, M. B. and M. Shacham, *Problem Solving in Chemical and Biochemical Engineering with POLYMATH, Excel, and MATLAB*, 2nd ed., Prentice-Hall, Boston, MA, pp. 491–495, 2008.
30. Fogler, H. S., *Essentials of Chemical Reaction Engineering*, Pearson Education International, Boston, MA, pp. 530–535, 2011.
31. Patrick M. Plehiers, Geert C. Reyniers, and Gilbert F. Froment, Simulation of the run length of an ethane cracking furnace, *Ind. Eng. Chem. Res.*, 29, pp. 636–641, 1990.
32. Nayef Ghasem, *Computer Methods in Chemical Engineering*, CRC Press, Taylor & Francis Group, Boca Raton, FL, pp. 250–253, 2012.
33. Cutlip, M. B. and M. Shacham, *Problem Solving in Chemical and Biochemical Engineering with POLYMATH, Excel, and MATLAB*, 2nd ed., Prentice-Hall, Boston, MA, pp. 119–122, 2008.

34. Fogler, H. S., *Elements of Chemical Reaction Engineering*, 3rd ed., Prentice Hall, Boston, MA, p. 523, 1999.
35. Cutlip, M. B. and M. Shacham, Problem Solving in Chemical and Biochemical Engineering with POLYMATH, Excel, and MATLAB, 2nd ed., Prentice-Hall, Boston, MA, p. 173, 2008.
36. Mariano M. Martin (Editor), *Introduction to Software for Chemical Engineers*, CRC Press, Taylor & Francis Group, Boca Raton, FL, pp. 127–131, 2015.
37. Seth, V. and Gupta, S. K., Free radical polymerizations associated with the Trommsdorff effect under semibatch reactor conditions: an improvement model, *Journal of Polymer Engineering*, 15(3-4), pp. 283–323, 1995.
38. Ghosh, P., Gupta, S. K. and Saraf, D. N., An experimental study on bulk and solution polymerization of methyl methacrylate with responses to step changes in temperature, *Chemical Engineering Journal*, 70, pp. 25–35, 1998.
39. Ray, A. B., Saraf, D. N. and Gupta, S. K., Free radical polymerizations associated with the Trommsdorff effect under semibatch reactor conditions - I: Modelling, *Polymer Engineering & Science*, 35, pp. 1290–1299, 1995.
40. Fogler, H. S., *Essentials of Chemical Reaction Engineering*, Pearson Education International, Boston, MA, p. 212, 2011.
41. Fogler, H. S., *Essentials of Chemical Reaction Engineering*, Pearson Education International, Boston, MA, pp. 213–216, 2011.
42. Fogler, H. S., *Essentials of Chemical Reaction Engineering*, Pearson Education International, Boston, MA, p. 218, 2011.
43. Fogler, H. S., *Essentials of Chemical Reaction Engineering*, Pearson Education International, Boston, MA, pp. 220–224, 2011.
44. Fogler, H. S., *Essentials of Chemical Reaction Engineering*, Pearson Education International, Boston, MA, pp. 317–320, 2011.
45. Fogler, H. S., *Essentials of Chemical Reaction Engineering*, Pearson Education International, Boston, MA, p. 386, 2011.
46. Jana, A. K., *Chemical Process Modelling and Computer Simulation*, 2nd ed., PHI Learning Private, Ltd., Delhi, India, p. 101, 2011.
47. Fogler, H. S., *Essentials of Chemical Reaction Engineering*, Pearson Education International, Boston, MA, pp. 387–389, 2011.
48. Jana, A. K., *Chemical Process Modelling and Computer Simulation*, 2nd ed., PHI Learning Private, Ltd., Delhi, India, pp. 105–108, 2011.
49. Jana, A. K., *Chemical Process Modelling and Computer Simulation*, 2nd ed., PHI Learning Private, Ltd., Delhi, India, p. 111, 2011.
50. King, M. R. and N. A. Mody, *Numerical and Statistical Methods for Bioengineering*, Cambridge University Press, Cambridge, UK, p. 416, 2011.
51. Cutlip, M. B. and M. Shacham, Problem Solving in Chemical and Biochemical Engineering with POLYMATH, Excel, and MATLAB, 2nd ed., Prentice-Hall, Boston, MA, p. 476, 2008.
52. Fogler, H. S., *Essentials of Chemical Reaction Engineering*, Pearson Education International, Boston, MA, pp. 251–252, 2011.
53. Fogler, H. S., *Essentials of Chemical Reaction Engineering*, Pearson Education International, Boston, MA, pp. 255–258, 2011.
54. Cutlip, M. B. and M. Shacham, Problem Solving in Chemical and Biochemical Engineering with POLYMATH, Excel, and MATLAB, 2nd ed., Prentice-Hall, Boston, MA, p. 467, 2008.
55. Fogler, H. S., *Essentials of Chemical Reaction Engineering*, Pearson Education International, Boston, MA, pp. 125–128, 2011.
56. Cutlip, M. B. and M. Shacham, Problem Solving in Chemical and Biochemical Engineering with POLYMATH, Excel, and MATLAB, 2nd ed., Prentice-Hall, Boston, MA, p. 223, 2008.
57. Cutlip, M. B. and M. Shacham, *Problem Solving in Chemical and Biochemical Engineering with POLYMATH, Excel, and MATLAB*, 2nd ed., Prentice-Hall, Boston, MA, pp. 458–459, 2008.
58. Fogler, H. Scott, *Essentials of Chemical Reaction Engineering*, Pearson Education International, Boston, MA, pp. 314–315, 2011.
59. Fogler, H. Scott, *Essentials of Chemical Reaction Engineering*, Pearson Education International, Boston, MA, pp. 549–554, 2011.
60. Fogler, H. S., *Essentials of Chemical Reaction Engineering*, Pearson Education International, Boston, MA, pp. 267–270, 2011.

61. Cutlip, M. B. and M. Shacham, Problem Solving in Chemical and Biochemical Engineering with POLYMATH, Excel, and MATLAB, 2nd ed., Prentice-Hall, Boston, MA, p. 445, 2008.
62. Cutlip, M. B. and M. Shacham, Problem Solving in Chemical and Biochemical Engineering with POLYMATH, Excel, and MATLAB, 2nd ed., Prentice-Hall, Boston, MA, p. 453, 2008.
63. Cutlip, M. B. and M. Shacham, Problem Solving in Chemical and Biochemical Engineering with POLYMATH, Excel, and MATLAB, 2nd ed., Prentice-Hall, Boston, MA, p. 486, 2008.
64. Cutlip, M. B. and M. Shacham, *Problem Solving in Chemical and Biochemical Engineering with POLYMATH, Excel, and MATLAB*, 2nd ed., Prentice-Hall, Boston, MA, pp. 488–489, 2008.
65. Fogler, H. S., *Essentials of Chemical Reaction Engineering*, Pearson Education International, Boston, MA, pp. 453–455, 2011.
66. Finlayson, B. A., *Introduction to Chemical Engineering Computing*, John Wiley & Sons, Inc., Hoboken, NJ, pp. 121–123, 2006.
67. Fogler, H. S., *Essentials of Chemical Reaction Engineering*, Pearson Education International, Boston, MA, pp. 530–535, 2011.
68. Adidharma, H. and V. Temyanko, *Mathcad for Chemical Engineers*, Trafford publishing, Victoria, BC, Canada, p. 88, 2007.
69. Jana, A. K., *Chemical Process Modelling and Computer Simulation*, 2nd ed., PHI Learning Private, Ltd., Delhi, India, pp. 102–103, 2011.
70. Jana, A. K., *Chemical Process Modelling and Computer Simulation*, 2nd ed., PHI Learning Private, Ltd., Delhi, India, pp. 110–111, 2011.
71. Jana, A. K., *Chemical Process Modelling and Computer Simulation*, 2nd ed., PHI Learning Private, Ltd., Delhi, India, p. 112, 2011.

7 Mass Transfer

7.1 DIFFUSION

7.1.1 ONE-DIMENSIONAL DIFFUSION

Consider binary gas-phase diffusion of component A during evaporation of a pure liquid in a simple diffusion tube. Figure 7.1 shows a cylindrical tube where liquid A is evaporating into a gas mixture of A and B from a liquid layer of pure A near the bottom of the tube.[1]

Fick's law for the flux of A can be expressed as

$$\frac{dN_A}{dz} = 0, \quad \frac{dx_A}{dz} = -\frac{(1 - x_A)N_A}{D_{AB}C}$$

where

C is the total concentration $(kgmol/m^3)$
D_{AB} is the molecular diffusivity of A in B (m^2/sec)
N_A is the mole flux of A $(kgmol/(m^2 \cdot sec))$
$x_A = C_A/C$ is the mole fraction of A

At $z = z_1$, x_A is given by $x_A = x_{A1} = P_{A0}/P$. The gas mixture is assumed to be an ideal gas. Then at constant temperature and pressure, the total concentration $C (= P/RT)$ and D_{AB} can be considered constant and the concentration profile may be expressed as[2,3]

$$\frac{1 - x_A}{1 - x_{A1}} = \left(\frac{1 - x_{A2}}{1 - x_{A1}}\right)^{(z-z_1)/(z_2-z_1)}, \quad N_A|_{z=z_1} = D_{AB}C\frac{(x_{A1} - x_{A2})}{(z_2 - z_1)(x_B)_{lm}}$$

where

$$(x_B)_{lm} = \frac{x_{B2} - x_{B1}}{\ln (x_{B2}/x_{B1})} = \frac{x_{A1} - x_{A2}}{\ln\left(\frac{1 - x_{A2}}{1 - x_{A1}}\right)}$$

The binary diffusivities for gases are known to vary according to the absolute temperature raised to a power of 1.75.[4] If the diffusivity at T_1 is known, the diffusivity at T may be approximated by[5]

$$D_{AB} = D_{AB}|_{T_1}\left(\frac{T}{T_1}\right)^{1.75}$$

Example 7.1 Binary Diffusion[6]

Methanol (A) is evaporated into a stream of dry air (B) in a cylindrical tube at $328.5 \ K$. The distance from the tube inlet to the liquid surface is $z_2 - z_1 = 0.238 \ m$. At $T = 328.5 \ K$, the vapor pressure of methanol is $P_{A0} = 68.4 \ kPa$ and the total pressure is $P = 99.4 \ kPa$. The binary molecular diffusion coefficient of methanol in air under these conditions is $D_{AB} = 1.991 \times 10^{-5} \ m^2/sec$.

Calculate the constant molar flux of methanol within the tube at steady state and plot the mole fraction profile of methanol from the liquid surface to the flowing air stream. Compare the calculated molar flux with that obtained from the equation

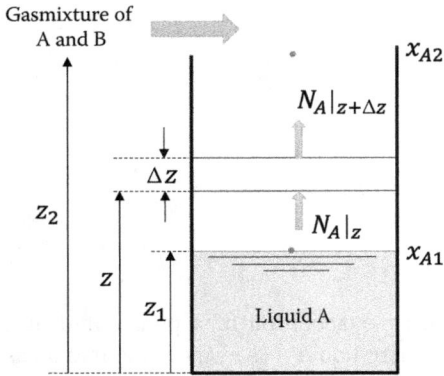

FIGURE 7.1   Gas-phase diffusion of A through gas mixture of A and B. (From Cutlip, M.B. and Shacham, M., *Problem Solving in Chemical and Biochemical Engineering with POLYMATH, Excel, and MATLAB*, 2nd ed., Prentice-Hall, Boston, MA, 2008, p. 384.)

$$N_{Az} = D_{AB} C \frac{x_0}{(z_2 - z_1)(x_B)_{lm}}, \quad (x_B)_{lm} = \frac{x_0}{\ln(1/(1-x_0))}$$

Solution

At $z = z_1$, $x_{A1} = P/P_{A0}$. We guess N_A, solve the differential equation $dx_A/dz = -(1 - x_A)N_A/(D_{AB}C)$, and check whether the condition $x_{A2} = 0$ is satisfied. If $x_{A2} \neq 0$, the calculation procedure is repeated using another guess of N_A. The bisection method can be effectively used in the iterative calculation procedure. The subfunction *dxz* defines the differential equation. The script *xaz* employs the built-in function *ode45* to solve the differential equation and determines N_A by using the bisection method.

```
% xaz.m
Dab = 1.991e-5; T = 328.5; R = 8.314; P = 99.4; Pa0 = 68.4; Tf = 295; z2 = 0.238; z1 =
0; zv = [z1 z2];
C = P/(R*T); x0 = Pa0/P; critN = 1e-11; errN = 1; Na1 = 3.5e-6; Na2 = 3.6e-6;
while errN > critN
    Nam = (Na1+Na2)/2;
    [z x1] = ode45(@xaz,zv,x0,[],Na1,Dab,C);
    [z x2] = ode45(@xaz,zv,x0,[],Na2,Dab,C);
    [z xm] = ode45(@xaz,zv,x0,[],Nam,Dab,C);
    if x1(end)*xm(end) < 0, Na2 = Nam;
    else, Na1 = Nam; end
    errN = abs(Na1 - Na2);
end
xblm = x0/(log(1/(1-x0))); Nanal = Dab*C*x0/((z2-z1)*xblm);
fprintf('Estimated Nab = %e, xA = %8.6f\n',Nam, xm(end));
fprintf('Analytic Nab = %e\n',Nanal);
plot(z,xm), xlabel('z (m)'), ylabel('x_A'), grid, axis tight
function dx = dxz(z,x,Na,Dab,C)
dx = - (1-x)*Na/Dab/C;
end
```

The script *xaz* produces the following outputs and the plot shown in Figure 7.2:
```
>> xaz
Estimated Nab = 3.547504e-06, xA = -0.000000
Analytic Nab = 3.547502e-06
```

We can see that $N_{AB} = 3.5475 \times 10^{-6} \ m^2/sec$ and $x_{A2} = -4.4158 \times 10^{-7} \approx 0$.

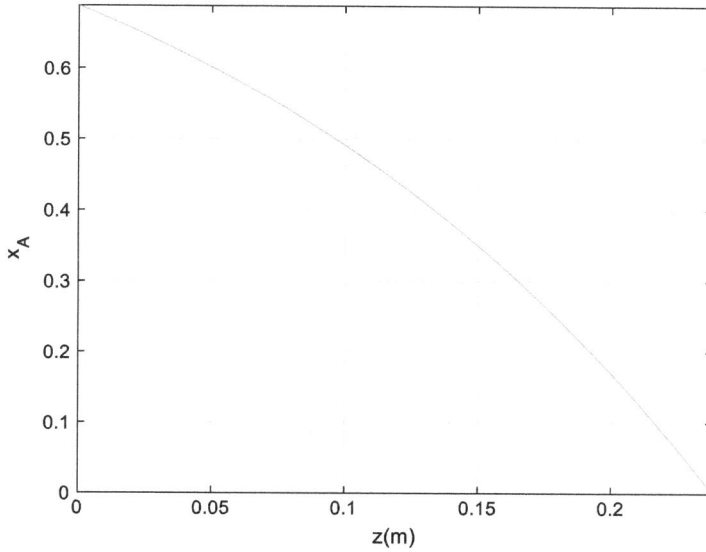

FIGURE 7.2 Mole fraction of methanol along the diffusion path.

7.1.2 MULTI-COMPONENT DIFFUSION IN GASES

For multi-component diffusion in gases at low density to the positive z direction, the Maxwell-Stefan equation[7] can be expressed as

$$\frac{dC_i}{dz} = \sum_{i=1}^{n} \frac{x_i N_i - x_j N_j}{D_{ij}}$$

where

$C_i(kgmol/m^3)$ is the concentration of species i
x_i is the mole fraction of species i
$N_i(kgmol/(m^2 \cdot sec))$ is the mole flux of species i
$D_{ij}(m^2/sec)$ is the molecular diffusivity of i in j
n is the number of components
For a gas mixture of three components (A, B, C), the Maxwell-Stefan equation yields

$$\frac{dC_A}{dz} = \frac{(x_A N_B - x_B N_A)}{D_{AB}} + \frac{(x_A N_C - x_C N_A)}{D_{AC}}$$

$$\frac{dC_B}{dz} = \frac{(x_B N_A - x_A N_B)}{D_{AB}} + \frac{(x_B N_C - x_C N_B)}{D_{BC}}$$

$$\frac{dC_C}{dz} = \frac{(x_C N_A - x_A N_C)}{D_{AC}} + \frac{(x_C N_B - x_B N_C)}{D_{BC}}$$

where $D_{ij} = D_{ji}$ ($i, j = A, B, C$).

Example 7.2 Multi-Component Diffusion of Gases[8]

Gases A and B are diffusing through stagnant gas C at a temperature of 55°C and a pressure of 0.2 *atm* from point 1 (z_1) to point 2 (z_2). The distance between these two points is 0.001 *m*. The molar flux of B is measured to be $N_B = -4.143 \times 10^{-4}$ $kgmol/(m^2 \cdot sec)$ (that is, gas B diffuses from z_2 to z_1). The gas mixture is assumed to be an ideal gas. Estimate the molar flux of A (N_A).

    Data:

$$C_{A1} = 2.229 \times 10^{-4}, \; C_{A2} = 0, \; C_{B0} = 0, \; C_{B2} = 2.701 \times 10^{-3}, \; C_{C1} = 7.208 \times 10^{-3}, \; C_{C2} = 4.730$$

$$\times 10^{-3}, \; C_{AB} = 1.47 \times 10^{-4}, \; C_{AC} = 1.075 \times 10^{-4}, \; D_{BC} = 1.245 \times 10^{-4}$$

Solution

    Since component C is stagnant, $N_C = 0$. A simple way to solve this problem is to assume a value of N_A, solve the differential equations, and check whether the concentration conditions at z_2 are satisfied. The initial guess for N_A can be determined by $N_A = -D_{AC}(C_{A2} - C_{A1})/L$, assuming the gas is a binary mixture of A and C. This procedure is repeated using updated N_A until the concentration conditions at z_2 are satisfied. The bisection method can be used to update N_A. The total concentration is given by $C_t = n/V = P/RT$ from the ideal gas law. The subfunction *mf* defines the differential equations. The script *mdif* employs the built-in function *ode45* to estimate N_A.

```
% mdif.m
Ca1=2.229e-4; Cb1=0; Cc1=7.208e-3; Ca2=0; Cb2=2.701e-3; Cc2=4.73e-3;
D1=1.075e-4; D2=1.245e-4; D3=1.47e-4; P=0.2;T=328; R=82.057e-3; Ct = P/(R*T);
L=0.001; Nb=-4.143e-4; Nc=0; Na=-D3*(Ca2-Ca1)/L;
zspan = [0 L]; c0 = [Ca1 Cb1 Cc1]; critN = 1e-10; errA = 1; Na1 = Na/2; Na2 = 2*Na;
iter = 1;
while errA > critN
    Nam = (Na1+Na2)/2;
    [z c1] = ode45(@mf,zspan,[Ca1,Cb1,Cc1],[],D1,D2,D3,Na1,Nb,Nc,Ct);
    [z c2] = ode45(@mf,zspan,[Ca1,Cb1,Cc1],[],D1,D2,D3,Na2,Nb,Nc,Ct);
    [z cm] = ode45(@mf,zspan,[Ca1,Cb1,Cc1],[],D1,D2,D3,Nam,Nb,Nc,Ct);
    if c1(end,1)*cm(end,1) < 0, Na2 = Nam; % check whether Ca(z2)=0 is satisfied
    else, Na1 = Nam; end
    errA = abs(Na1 - Na2); iter = iter+1;
end
c = cm; xa = c(:,1)/Ct; xb = c(:,2)/Ct; xc = c(:,3)/Ct;
plot(z,xa,z,xb,':',z,xc,'.-'), legend('x_A','x_B','x_C'), xlabel('Distance
z(m)'), ylabel('Mole fraction')
iter, Na = Nam

function dc = mf(z,c,D1,D2,D3,Na,Nb,Nc,Ct)
% c1=Ca, c2=Cb, c3=Cc, D1=Dab, D2=Dbc, D3=Dac
xa = c(1)/Ct; xb = c(2)/Ct; xc = c(3)/Ct;
dc = [(xa*Nb-xb*Na)/D1 + (xa*Nc-xc*Na)/D3;
        (xb*Na-xa*Nb)/D1 + (xb*Nc-xc*Nb)/D2;
        (xc*Na-xa*Nc)/D3 + (xc*Nb-xb*Nc)/D2];
end
```

    The script *mdif* produces the following results and the curves shown in Figure 7.3:

```
>> mdif
iter =
   20
Na =
   2.3471e-05
```

    We can see that the value of the molar flux of A converges to $N_A = 2.3471 \times 10^{-5}$ $kgmol/(m^2 \cdot sec)$ after 20 iterations.

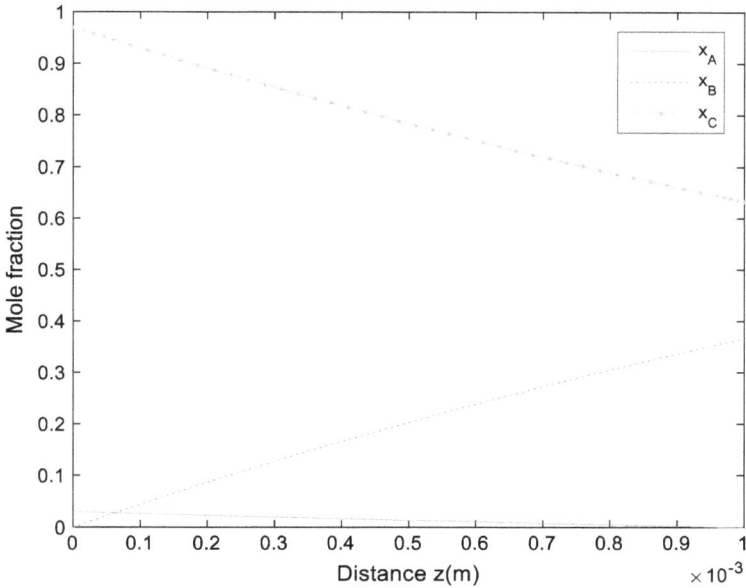

FIGURE 7.3 Mole fraction profiles for components A, B, and C.

7.1.3 DIFFUSION FROM A SPHERE

When component A diffuses through stagnant fluid B from a sphere to a surrounding medium (see Figure 7.4), the mass balance gives

$$\frac{d\left(N_A r^2\right)}{dr} = 0, \quad \frac{dp_A}{dr} = -\frac{RTN_A}{D_{AB}}\left(1 - \frac{p_A}{P}\right)$$

where

R is the gas constant ($=8314.34 \ m^3 \cdot Pa/(kgmol \cdot K)$)
T is the absolute temperature (K)
P is the total pressure (Pa)
Integration of this equation yields[9]

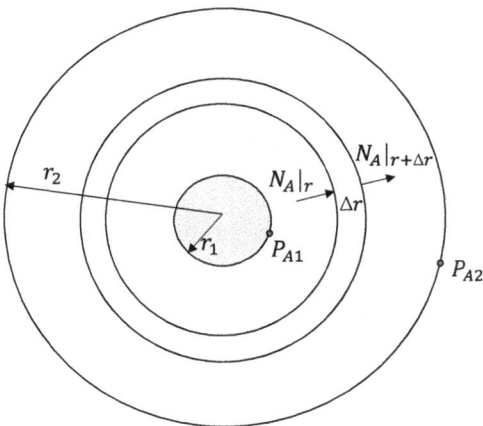

FIGURE 7.4 Diffusion from a sphere. (From Cutlip, M.B. and Shacham, M., *Problem Solving in Chemical and Biochemical Engineering with POLYMATH, Excel, and MATLAB*, 2nd ed., Prentice-Hall, Boston, MA, 2008, p. 392.)

$$N_{A1} = \frac{D_{AB}}{RTr_1} \frac{(p_{A1} - p_{A2})}{p_{BM}}, \quad p_{BM} = \frac{p_{A1} - p_{A2}}{\ln\left(\frac{P - p_{A2}}{P - p_{A1}}\right)}$$

Example 7.3 Diffusion from a Solid Sphere[10]

Dichlorobenzene (A), suspended in stagnant air (B), is sublimed at $25\,°C$ and atmospheric pressure. The sublimation is taking place at the surface of a sphere of solid dichlorobenzene with a radius of $3 \times 10^{-3}\ m$. The vapor pressure of A at $25\,°C$ is $1\ mmHg$, and the diffusivity in air is $7.39 \times 10^{-6}\ m^2/sec$. The density of A is $1458\ kg/m^3$ and the molecular weight is 147. Calculate the rate of sublimation (flux) and plot it as a function of the radius r.

Solution

The flux N_A can be determined from

$$\frac{d(N_A r^2)}{dr} = 0, \quad \frac{dp_A}{dr} = -\frac{RTN_A}{D_{AB}}\left(1 - \frac{p_A}{P}\right), \quad N_A = \frac{(N_A r^2)}{r^2}$$

Since $z_2 = p_A$ is the vapor pressure of dichlorobenzene at $r = r_0(p_A(r_0) = 1\ mmHg = 101,$ $325\ Pa/760 = 133.32\ Pa)$, $z_2(r_0) = 133.32\ Pa$. Because the initial condition $z_1(r_0)$ is not given, the initial condition for $z_1 = N_A r^2$ has to be determined such that $p_A = 0$ is satisfied at large r (for example, $r = 20\ m$). If x is defined as $r - r_0 = x$, $r = r_0 + x$ and $dr = dx$. The subfunction spf defines the differential equations. The script $sphdf$ uses the built-in function $ode45$ to calculate the flux.

```
% sphdf.m
Dab = 7.39e-6; T = 298.15; P = 1.01325e5; p0 = 133.32; r0 = 0.003;
xspan = [0 20]; critN = 1e-16; errN = 1; Na1 = 1e-12; Na2 = 2e-12; % initial guesses
while errN > critN % bisection method
    Nam = (Na1+Na2)/2;
    [x z1] = ode45(@spf,xspan,[Na1,p0],[],T,P,Dab,r0);
    [x z2] = ode45(@spf,xspan,[Na2,p0],[],T,P,Dab,r0);
    [x zm] = ode45(@spf,xspan,[Nam,p0],[],T,P,Dab,r0);
    if z1(end,2)*zm(end,2) < 0, Na2 = Nam;
    else, Na1 = Nam; end
    errN = abs(Na1 - Na2);
end
r = r0 + x; Na = zm(:,1)./r.^2; pf = zm(end,2);
fprintf('Flux at r=r0: %7.5e, partial pressure at r=inf: %7.5f\n', Na(1), pf);
plot(r(1:25),Na(1:25)), xlabel('r(m)'), ylabel('N_A')

function dzdx = spf(x,z,T,P,Dab,r0)
R = 8314.34;
dzdx = [0; -R*T*z(1)*(1-z(2)/P)/(Dab*(x+r0)^2)];
end
```

The script $sphdf$ generates the following results and the plot shown in Figure 7.5. We can see that the partial pressure at $r = r_\infty$ is $0.0002\ Pa$ and the molar flux is $N_A|_{r=0} = 1.32575 \times 10^{-7}\ kgmol/(m^2 \cdot sec)$.

```
>> sphdf
Flux at r=r0: 1.32575e-07, partial pressure at r=inf: 0.00021
```

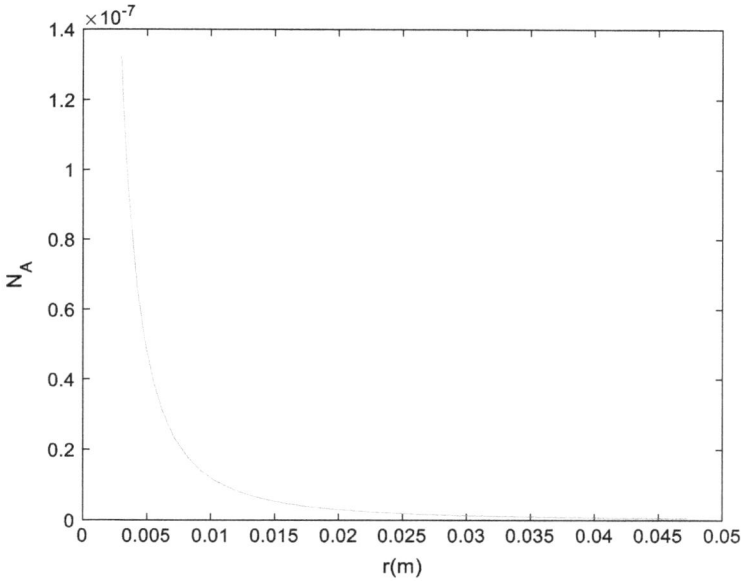

FIGURE 7.5 Molar flux versus radius.

7.1.4 MASS TRANSFER COEFFICIENT

When component A is transferred through stagnant medium B from the surface of a solid sphere, the mass transfer coefficient of A through B, k_c' (m/sec), can be represented in terms of the Sherwood number N_{sh}[11]:

$$N_{sh} = k_c' \frac{D_p}{D_{AB}} = 2 + 0.552 N_{re}^{0.53} N_{sc}^{1/3}$$

where
D_p (m) is diameter of the solid sphere
$N_{re} = D_p v \rho / \mu$ is the Reynolds number
$N_{sc} = \mu / (\rho D_{AB})$ is the Schmidt number
ρ (kg/m^3) is the gas density
The molar flux of A through B from a solid sphere, N_A, can be expressed in terms of k_c':

$$N_A = \frac{k_c' P}{RT} \frac{(p_{A1} - p_{A2})}{p_{BM}}$$

Example 7.4 Drug Delivery by Dissolution of Pill Coating[12]

The pills to deliver three particular drugs all have a solid spherical inner core of pure drug D surrounded by a spherical outer coating of A. The outer coating and the drug dissolve at different rates in the stomach due to their difference in solubility. A person takes all three different pills at the same time. Assume that the stomach is well mixed and that the pills remain in the stomach while they are dissolving.

Let the diameter of pill i be D_i, the diameter of pure drug D in pill i be D_{di}, and the mass transfer coefficient for pill i be k_{Li} ($i = 1, 2, 3$). If the concentration of coating in the stomach is C_{AS} (mg/cm^3), the concentration of drug in the stomach is C_{DS} (mg/cm^3), the concentration of drug in the body is $C_{DB} (mg/kg)$, and the solubilities of the outer pill layer and inner drug core at stomach conditions are S_A and $S_D (mg/cm^3)$, respectively, then the mass balances on volumes of pills yield $dD_i/dt = -(2k_{Li}/\rho)(S_A - C_{As})$, $D_1(0) = 0.5 cm$, $D_2(0) = 0.4\ cm$, $D_3(0) = 0.35\ cm$, and we have

$$\frac{dD_i}{dt} = \begin{cases} -\frac{2k_{Li}}{\rho}(S_D - C_{DS}) : 10^{-5} \le D_i \le D_{di} = 0.3cm \\ \\ 0 : D_i \le 10^{-5}cm \end{cases}$$

$$\frac{dC_{AS}}{dt} = \frac{1}{V}(S_A - C_{AS})\pi \sum_{i=1}^{3} S_{W_i}k_{Li}D_i^2 - \frac{C_{AS}}{\tau}, \quad \frac{dC_{DS}}{dt} = \frac{1}{V}(S_D - C_{DS})\pi \sum_{i=1}^{3} (1 - S_{W_i})k_{Li}D_i^2 - \frac{C_{DS}}{\tau}$$

$$k_{Li} = \frac{1.2}{D_i} (i = 1, 2, 3), \quad S_{W_i} = \begin{cases} 1 : D_i > 0.3 \\ 0 : D_i \le 0.3 \end{cases} (i = 1, 2, 3)$$

where

$V (liter)$ is the volume of fluid in the stomach

$\tau (hr)$ is the residence time in the stomach

Plot the diameters of pills (D_1, D_2, D_3) and C_{AS} and C_{DS} as a function of time for up to 150 minutes ($0 \le t \le 150\ min$) after the pills are taken.

Data: $V = 1200\ cm^3$, $\tau = 240\ min$, $S_A = 1\ mg/cm^3$, $S_D = 0.4\ mg/cm^3$, $\rho = 1414.7\ mg/cm^3$

Solution

The system of differential equations is defined by the subfunction *ruf*. The script *drugf* plots variations of diameter and concentration as a function of time, as shown in Figure 7.6.

```
% drugf.m
x10 = 0.5; x20 = 0.4; x30 = 0.35; x40 = 0; x50 = 0;
V = 1200; tau = 240; Sa = 1; Sd = 0.4; rho = 1414.7;
tspan = [0 150]; x0 = [x10 x20 x30 x40 x50];
[t x] = ode45(@ruf,tspan,x0,[],V,rho,tau,Sa,Sd);
D1 = x(:,1); D2 = x(:,2); D3 = x(:,3); Cas = x(:,4); Cds = x(:,5);
subplot(1,2,1), plot(t,D1,t,D2,':',t,D3,'.-'), xlabel('t(min)'), ylabel
('D(cm)')
legend('D_1','D_2','D_3'), axis tight
subplot(1,2,2), plot(t,Cas,t,Cds,':'), xlabel('t(min)'), ylabel('C(mg/cm^3)')
```

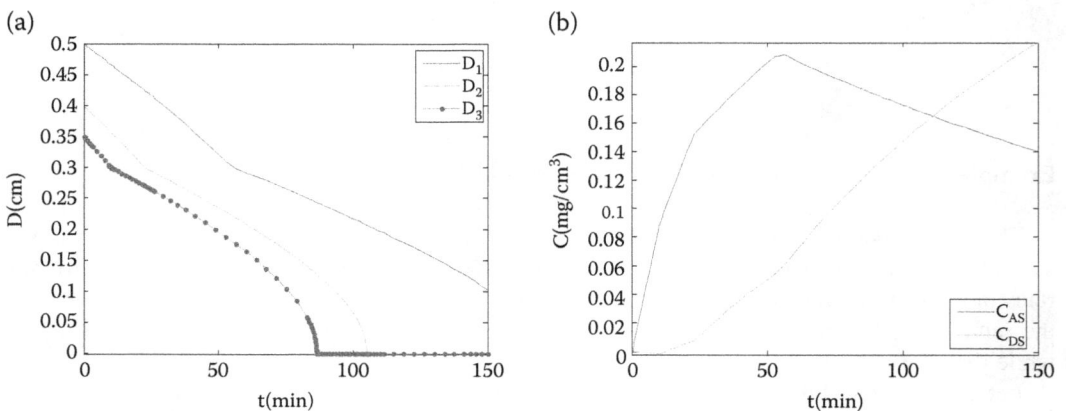

FIGURE 7.6 (a) Diameter variations for three pills; (b) concentration variations.

```
axis tight, legend('C_{AS}','C_{DS}','Location','best')
function dx = ruf(t,x,V,rho,tau,Sa,Sd)
% x(i)=D(i) (i=1,2,3), X(4)=C_AS, x(5)=C_Ds
sum0 = 0; sum1 = 0;
for i = 1:3
    kL(i) = 1.2/x(i); Sw(i) = 0;
    if x(i) > 0.3, fd(i) = -2*kL(i)*(Sa - x(4))/rho; Sw(i) = 1;
    elseif x(i) <= 0.3 && x(i) >= 1e-5, fd(i) = -2*kL(i)*(Sd - x(5))/rho;
    else, fd(i) = 0; end
    sum0 = sum0 + Sw(i)*kL(i)*x(i); sum1 = sum1 + (1-Sw(i))*kL(i)*x(i);
end
fd(4) = pi*(Sa - x(4))*sum0/V - x(4)/tau; fd(5) = pi*(Sd - x(5))*sum1/V - x
(4)/tau;
dx = [fd(1) fd(2) fd(3) fd(4) fd(5)]';
end
```

7.1.5 Diffusion in Isothermal Catalyst Particles

For a spherical catalyst particle, the mass balance on a differential volume within the particle yields

$$\frac{d}{dr}(N_A r^2) = -k_1 a C_A r^2, \ N_A = (N_A r^2)/r^2$$

where
N_A is the flux of reactant A
r is the radius of the spherical catalyst particle
k_1 is the 1st-order rate constant based on particle volume
C_A is the concentration of reactant
At $r = 0$, $d(N_A r^2)/dr = 0$. Fick's law for the diffusion of reactant A can be expressed as

$$\frac{dC_A}{dr} = -\frac{N_A}{D_e}$$

where D_e is the effective diffusivity for the diffusion of reactant A in the porous particle. The boundary condition for this equation is that the concentration of A at the particle surface is given by $C_A = C_{As}$ when $r = R$. The isothermal internal effectiveness factor, η, is defined as the ratio of the average reaction rate within the particle to the reaction rate at the concentration of the particle surface:

$$\eta = \frac{(-r_A)|_{avg}}{(-r_A)|_{surface}}$$

η can be calculated from

$$\eta = \frac{\int_0^R k_1 a C_A (4\pi r^2) dr}{k_1 a C_{As} (4\pi R^3/3)} = \frac{3}{C_{As} R^3} \int_0^R C_A r^2 dr$$

Differentiation of this equation with respect to r gives

$$\frac{d\eta}{dr} = \frac{3 C_A r^2}{C_{As} R^3}, \ \eta|_{r=0} = 0$$

The analytical solution to the diffusion is given by[13]

$$\eta = \frac{3}{\phi^2}\{\phi\coth(\phi) - 1\}, \; \phi = R\sqrt{\frac{k_1 a C_{As}^{n-1}}{D_e}}$$

where

ϕ is the Thiele modulus

n is the reaction order

For cylindrical catalyst particles,

$$\frac{d}{dr}(N_A r) = -k_1 a C_A^r, \; N_A = \frac{(N_A r)}{r}, \; \frac{dC_A}{dr} = -\frac{N_A}{D_e}$$

$$\eta = \frac{\int_0^R k_1 a C_A (2\pi r)\, dr}{k_1 a C_{As}(\pi R^2)} = \frac{2}{C_{As} R^2}\int_0^R C_A r\, dr, \; \frac{d\eta}{dr} = \frac{2 C_A r}{C_{As} R^2}, \; \eta|_{r=0} = 0$$

For a catalytic gas-phase reaction being carried out in a reactor packed with a porous catalyst layer, the mass balance on component A in the z direction from the top of the porous layer yields

$$\frac{dN_A}{dz} = r_A, \; N_A|_{z=L} = 0$$

where N_A is the molar flux of A. At $z = L$, $N_A = 0$. The rate of change of the concentration of A, C_A, with respect to z is expressed as

$$\frac{dC_A}{dz} = \frac{1}{D_e}[x_A(N_A + N_B) - N_A]$$

where the effective diffusivity of the catalyst layer, D_e, is given by

$$D_e = \frac{D_{AB}\epsilon_p \sigma}{\tilde{\tau}}$$

where

D_{AB} is the binary diffusivity

ϵ_p is the catalyst porosity

σ is the constriction factor

$\tilde{\tau}$ is the tortuosity

Example 7.5 Diffusion with Reaction in Catalyst Particles[14]

Calculate the concentration profile for C_A and determine the effectiveness factor η for a 1st-order irreversible reaction in a spherical particle, where $R = 0.5$ cm, $D_e = 0.1$ cm^2/sec, $C_{As} = 0.2$ $gmol/cm^3$, and $k_1 a = 6.4$ sec^{-1}.

Solution

Since the initial condition $C_A|_{r=0}$ is unknown, we first assume $C_A|_{r=0}$ and solve the system of differential equations. The initial condition $C_A|_{r=0}$ is updated iteratively until the condition $C_A|_{r=R} = C_{As}$ is satisfied. Once the initial condition is fixed, other values can be calculated. The subfunction *rxf* defines the system of differential equations. The script *rxnf* employs the built-in function *ode45* to integrate the system of differential equations and uses the bisection method to determine the initial condition $C_A|_{r=0}$.

```
% rxnf.m
```

```
Cas = 0.2; R = 0.5; De = 0.1; k1a = 6.4; rspan = [0 R]; critN = 1e-6; errN = 1; Ca1 =
1e-2; Ca2 = 3e-2;
while errN > critN
    Cam = (Ca1+Ca2)/2;
    [r x1] = ode45(@rxf,rspan,[0,0,Ca1],[],Cas,R,De,k1a);
    [r x2] = ode45(@rxf,rspan,[0,0,Ca2],[],Cas,R,De,k1a);
    [r xm] = ode45(@rxf,rspan,[0,0,Cam],[],Cas,R,De,k1a);
    if (x1(end,3)-Cas)*(xm(end,3)-Cas) < 0, Ca2 = Cam;
    else, Ca1 = Cam; end
    errN = abs(Ca1 - Ca2);
end
Na = xm(:,1)./r.^2; effc = xm(end,2); Ca = xm(:,3);
ephi = R*sqrt(k1a/De); effa = 3*(ephi*coth(ephi)-1)/ephi^2;
fprintf('Effectiveness factor at r=R (calculated): %7.5f\n', effc);
fprintf('Effectiveness factor at r=R (analytic): %7.5f\n', effa);
plot(r,Ca), xlabel('r(cm)'), ylabel('C_A(gmol/cm^3)'), grid

function dxdr = rxf(r,x,Cas,R,De,k1a)
% x1=Nar^2, x2=eta, x3=Ca
if r == 0, Na = 0; else Na = x(1)/r^2; end
dxdr = [-k1a*x(3)*r^2; 3*x(3)*r^2/(Cas*R^3); -Na/De];
end
```

The script *rxnf* produces the following outputs and the plot shown in Figure 7.7:

```
>> rxnf
Effectiveness factor at r=R (calculated): 0.56301
Effectiveness factor at r=R (analytic): 0.56300
```

Example 7.6 Reaction and Diffusion in a Porous Catalytic Layer[15]

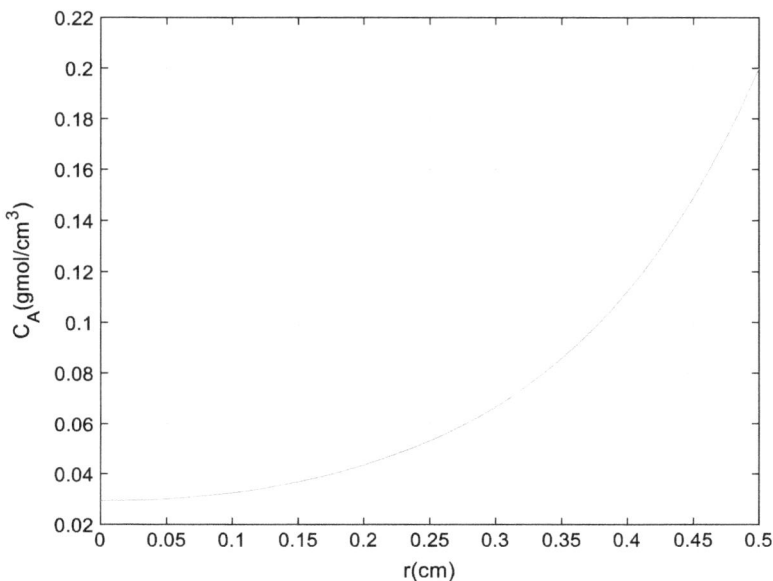

FIGURE 7.7   Concentration profile in a spherical catalyst.

A catalytic gas-phase reversible reaction between components A and B, $2A \leftrightarrow B$, is taking place in a porous catalyst layer in a reactor. The reaction rate for reactant A is given by

$$r_A = -k\left(C_A^2 - \frac{C_B}{K_c}\right) gmol/cm^3 \cdot sec$$

where the rate constant $k = 8 \cdot 10^4\ cm^3/(sec \cdot gmol)$ and the equilibrium constant $K_c = 6 \times 10^5\ cm^3/gmol$. The thickness of the catalytic layer is $L = 0.2\ cm$, the effective diffusivity of A in B for this layer is $D_e = 0.01\ cm^2/sec$, the total concentration of A and B is $C_t = 4 \times 10^{-5}\ gmol/cm^3$, and the concentrations of A and B at the surface of the catalytic layer are $C_{As} = 3 \times 10^{-5}\ gmol/cm^3$ and $C_{Bs} = 1 \times 10^{-5}\ gmol/cm^3$, respectively. Calculate the effectiveness factor for the given reaction (η) and plot C_A and N_A as a function of the depth of the catalytic layer (z).

Solution

Substitution of the relation $C_B = C_t - C_A$ into the rate equation yields

$$\frac{dN_A}{dz} = -k\left(C_A^2 - \frac{C_B}{K_c}\right) = -k\left(C_A^2 - \frac{C_t - C_A}{K_c}\right),\ N_A|_{z=L} = 0$$

From the reaction stoichiometry, the relation between the molar fluxes of A and B is given by

$$N_B = -\frac{1}{2}N_A$$

From Fick's law, we have[16]

$$\frac{dC_A}{dz} = \frac{1}{D_e}[x_A(N_A + N_B) - N_A]$$

Substitution of the definition of x_A in terms of concentrations, $x_A = C_A/C_t$, into this equation yields

$$\frac{dC_A}{dz} = \frac{N_A}{2D_e}\left(\frac{C_A}{C_t} - 2\right),\ C_A|_{z=0} = C_{As}$$

The effectiveness factor can be expressed as

$$\eta = \frac{(-r_A)|_{avg}}{(-r_A)|_{surface}} = \frac{N_{As}/L}{k\left(C_{As}^2 - \frac{C_t - C_{As}}{K_c}\right)}$$

Since the initial condition $N_{As} = N_A|_{z=0}$ is unknown, it should be assumed to solve the differential equations. The initial condition $N_A|_{z=0}$ that satisfies the boundary condition of zero flux at $z = L$, $N_A|_{z=L} = 0$, can be determined by iterative calculations. Once the initial condition is determined, other variables can be calculated. The subfunction crf defines the differential equations. The script crxf determines the initial condition, calculates the effectiveness factor, and generates the plots of C_A and N_A versus the depth of the catalytic layer shown in Figure 7.8.

```
% crxf.m
Cas = 3e-5; Ct=4e-5; L=0.2; De=0.01; k=8e4; Kc=6e5; zspan = [0 L]; critN = 1e-10;
errN = 1;
Na1 = 2e-6; Na2 = 5e-6;
while errN > critN
    Nam = (Na1+Na2)/2;
```

(a)

(b)

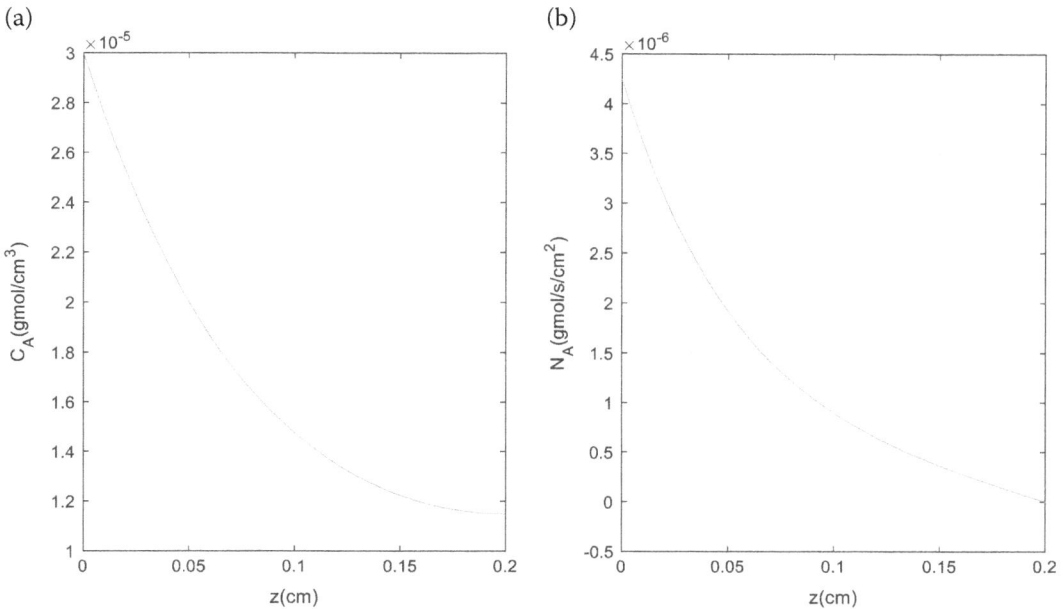

FIGURE 7.8   (a) Concentration profile and (b) molar flux profile.

```
    [z x1] = ode45(@crf,zspan,[Na1,Cas],[],Ct,k,Kc,De);
    [z x2] = ode45(@crf,zspan,[Na2,Cas],[],Ct,k,Kc,De);
    [z xm] = ode45(@crf,zspan,[Nam,Cas],[],Ct,k,Kc,De);
    if x1(end,1)*xm(end,1) < 0, Na2 = Nam;
    else, Na1 = Nam; end
    errN = abs(Na1 - Na2);
end
Na = xm(:,1); Ca = xm(:,2); ras = Cas^2 - (Ct-Cas)/Kc; effc = Na(1)/(L*k*ras);
fprintf('Effectiveness factor: %7.5f\n', effc);
fprintf('Molar flux of A at z=0: %12.8f\n', Na(1));
subplot(1,2,1), plot(z,Ca), xlabel('z(cm)'), ylabel('C_A(gmol/cm^3)'), grid
subplot(1,2,2), plot(z,Na), xlabel('z(cm)'), ylabel('N_A(gmol/s/cm^2)'), grid

function dx = crf(z,x,Ct,k,Kc,De)
% x1=Na, x2=Ca
dx = [-k*(x(2)^2 - (Ct-x(2))/Kc); x(1)*(x(2)/Ct - 2)/(2*De)];
end

>> crxf
Effectiveness factor: 0.30068
Molar flux of A at z = 0: 0.00000425
```

7.1.6 Unsteady-State Diffusion in a One-Dimensional Slab

Unsteady-state diffusion of component A in a one-dimensional slab can be expressed as[17]

$$\frac{\partial C_A}{\partial t} = D_{AB}\frac{\partial^2 C_A}{\partial x^2}, \ C_A|_{x=0} = \frac{C_{A0}}{K}, \ \frac{\partial C_A}{\partial x}\bigg|_{x=L} = 0$$

where

 K is the distribution coefficient

 D_{AB} is the binary diffusivity

Figure 7.9 shows a one-dimensional slab of width Δx. A central difference formula for the second derivative can be used to approximate the partial derivatives in this equation:

$$\frac{dC_{Ai}}{dt} = \frac{D_{AB}}{(\Delta x)^2}(C_{Ai+1} - 2C_{Ai} + C_{Ai-1})(2 \le i \le n - 1)$$

where Δx is the length of the subinterval. At $x = 0$,

$$k_c(C_{A0} - C_{A1}) = -D_{AB}\frac{\partial C_A}{\partial x}\bigg|_{x=0} \quad \text{and} \quad \frac{\partial C_A}{\partial x}\bigg|_{x=0} = \frac{-C_{A3} + 4C_{A2} - 3C_{A1}}{2\Delta x}$$

and we have

$$C_{A1} = \frac{2k_c C_{A0}\Delta x - D_{AB}C_{A3} + 4D_{AB}C_{A2}}{3D_{AB} + 2k_c K\Delta x}$$

where k_c is the external mass transfer coefficient (m/sec). At $x = L$,

$$\frac{\partial C_A}{\partial x}\bigg|_{x=L} = 0 \text{ and } \frac{\partial C_{An}}{\partial x} = \frac{3C_{An} - 4C_{An-1} + C_{An-2}}{2\Delta x} = 0$$

This equation can be solved for C_{An} to yield[18]

$$C_{An} = \frac{4C_{An-1} - C_{An-2}}{3}$$

Example 7.7 Unsteady-State Diffusion in a One-Dimensional Slab[19]

A slab with a thickness of 0.004 m has one surface suddenly exposed to a solution containing component A with $C_{A0} = 6 \times 10^{-3}$ $kgmol/m^3$. The other surface of the slab is supported by an insulated solid allowing no mass transport. The binary diffusivity is $D_{AB} = 1 \times 10^{-9}$ m^2/sec, the distribution coefficient is $K = 1.5$, and the external mass transfer coefficient is $k_c = 1 \times 10^{-6}$ m/sec. At $t = 0$, the concentration of component A at the surface of the solution side is

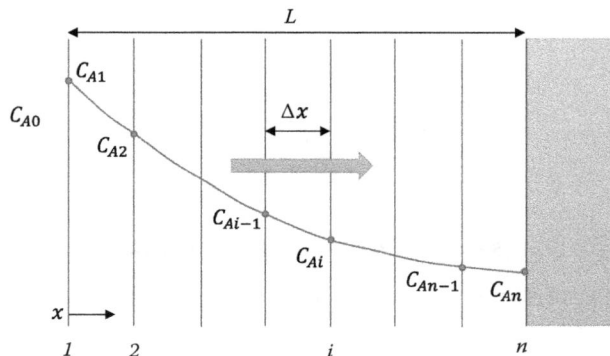

FIGURE 7.9   Unsteady-state diffusion in a one-dimensional slab.

$C_{A1} = 0.001\ kgmol/m^3$ and that at the solid side is $C_{An} = 0.002\ kgmol/m^3$. Calculate and plot the concentrations within the slab after 20,000 sec. The interior of the slab may be divided into several intervals of width 0.0005 m (that is, $\Delta x = 0.0005\ m$).

Solution

If $\Delta x = 0.0005\ m$, $0.0004/0.0005 = 8$ and the number of intervals is $n = 9$. The system of differential equations can be expressed as

$$\frac{dC_{Ai}}{dt} = \frac{D_{AB}}{(\Delta x)^2}(C_{Ai+1} - 2C_{Ai} + C_{Ai-1})(2 \le i \le 8)$$

When $t = 0$, $C_{A1} = 0.001\ kgmol/m^3$ and $C_{A9} = 0.002\ kgmol/m^3$; and when $t > 0$, $C_{A1} = C_{A0}/K$ and $C_{A9} = (4C_{A8} - C_{A7})/3$. The initial concentration profile within the slab is assumed to be linear. In this case, the initial concentration at the node i is given by

$$C_{Ai} = C_{A1} + \frac{(C_{A9} - C_{A1})}{8}(i - 1)(i = 1, 2, \cdots, 9)$$

The subfunction *slf* defines the system of differential equations. The script *sldif* calculates the concentrations and generates the curves shown in Figure 7.10.

```
% sldif.m
% x(1)=Ca2, x(2)=Ca3, ..., x(7)=Ca8
Dab = 1e-9; delx = 5e-4; ca10 = 1e-3; ca90 = 2e-3; ca0 = 6e-3; K = 1.5;
for k = 1:9, x0(k) = ca10 + (ca90 - ca10)*(k-1)/8; end
c0 = [x0(2) x0(3) x0(4) x0(5) x0(6) x0(7) x0(8)]; tf = 20000; tspan = [0 tf];
```

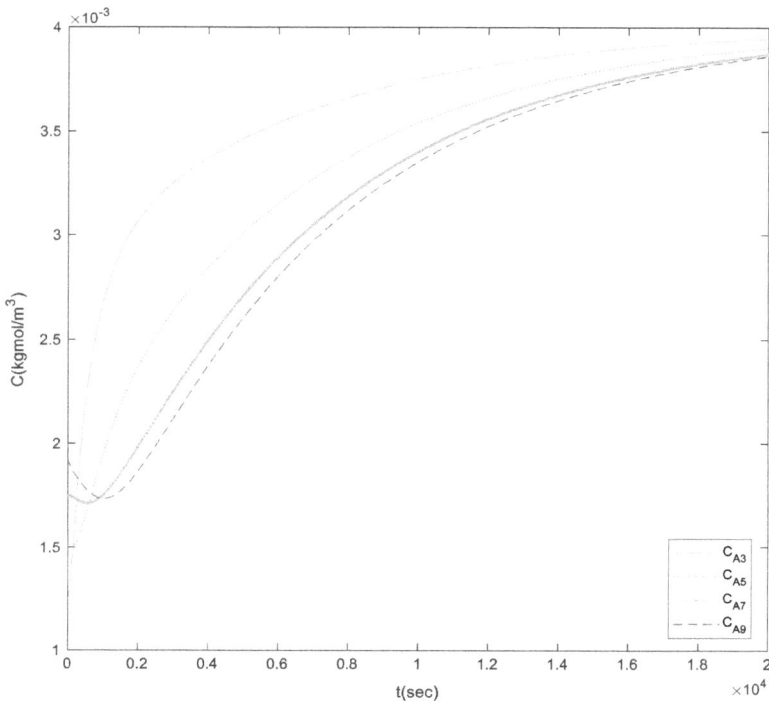

FIGURE 7.10 Concentration profile in a one-dimensional slab.

```
[t x] = ode45(@slf,tspan,c0,[],Dab,delx,ca0,ca10,ca90,K);
nt = length(t); xc1 = [ca10 ones(1,nt-1)*ca0/K]; % Ca1
xc9 = [ca90 ((4*x(2:end,7) - x(2:end,6))/3)']; % Ca9
xc = [];
for k = 1:7, xc = [xc x(:,k)]; end
c = [xc1' xc xc9'];
plot(t,c(:,3),t,c(:,5),':',t,c(:,7),'.-',t,c(:,9),'--'),  xlabel('t(sec)'),
 ylabel('C(kgmol/m^3)')
legend('C_{A3}','C_{A5}','C_{A7}','C_{A9}','location','best')

function dx = slf(t,x,Dab,delx,ca0,ca10,ca90,K)
% x(1)=Ca2, x(2)=Ca3, ..., x(7)=Ca8
if t == 0, ca1 = ca10; ca9 = ca90;
else, ca1 = ca0/K; ca9 = (4*x(7) - x(6))/3; end
dx = [Dab*(x(2)-2*x(1)+ca1)/(delx^2); Dab*(x(3)-2*x(2)+x(1))/(delx^2);
Dab*(x(4)-2*x(3)+x(2))/(delx^2); Dab*(x(5)-2*x(4)+x(3))/(delx^2);
Dab*(x(6)-2*x(5)+x(4))/(delx^2); Dab*(x(7)-2*x(6)+x(5))/(delx^2);
Dab*(ca9-2*x(7)+x(6))/(delx^2)];
end
```

7.1.7 DIFFUSION IN A FALLING LAMINAR FILM

CO_2 gas is absorbed into a falling liquid film of alkaline solution in which there is a 1st-order irreversible reaction, as shown in Figure 7.11. The resulting concentration of the dissolved CO_2 in the film is quite small, so that the viscosity of the liquid is not affected. The velocity distribution for the steady-state laminar flow down a vertical wall, v_z (m/sec), is given by[20]

$$v_z = \frac{\rho g \delta^2}{2\mu}\left[1 - \left(\frac{x}{\delta}\right)^2\right] = v_{z_{max}}\left[1 - \left(\frac{x}{\delta}\right)^2\right]$$

A steady-state mass balance on a differential volume within the liquid film yields the partial differential equation[21]

$$v_z\frac{\partial C_A}{\partial z} = D_{AB}\frac{\partial^2 C_A}{\partial x^2} - k'C_A$$

where
   C_A (kgmol/m³) is the concentration of dissolved CO_2
   D_{AB} (m²/sec) is the diffusivity of dissolved CO_2 in the alkaline solution
   k' (sec⁻¹) is a 1st-order reaction rate constant for the neutralization reaction

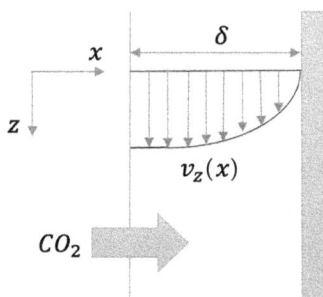

FIGURE 7.11   Mass transfer within a falling laminar film. (From Bird, R.B. et al., *Transport Phenomena*, John Wiley & Sons, Hoboken, NJ, 2002, p. 562.)

The boundary conditions are given by

$$C_A|_{z=0} = 0, \; C_A|_{x=0,z>0} = C_{As}, \; \frac{\partial C_A}{\partial x}\bigg|_{x=\delta,z\geq0} = 0$$

The partial differential equation can be solved numerically using the method of lines. The finite difference expressions for the partial differential equation and for the velocity equation can be combined to give

$$\frac{dC_{Ai}}{dz} = \frac{1}{v_{z_{max}}\left[1 - \left(\frac{(i-1)\Delta x}{\delta}\right)^2\right]}\left[\frac{D_{AB}}{(\Delta x)^2}(C_{Ai+1} - 2C_{Ai} + C_{Ai-1}) - k'C_{Ai}\right](2 \leq i \leq n-1)$$

$$C_{Ai}|_{z=0} = 0, \; C_{A1} = C_{As}, \; C_{An} = \frac{4C_{An-1} - C_{An-2}}{3}$$

The finite difference elements are shown in Figure 7.12.

Example 7.8 Mass Transfer in a Falling Laminar Film[22]

CO_2 gas is absorbed into a falling liquid film of alkaline solution in which there is no reaction. The film thickness is $\delta = 3 \times 10^{-4}$ m, the maximum velocity is $v_{z_{max}} = 0.6$ m/sec, and the diffusivity of dissolved CO_2 in the alkaline solution is $D_{AB} = 1.5 \times 10^{-9}$ m²/sec. Use the numerical method of lines with 10 intervals to calculate the concentration of dissolved CO_2 at $z = 1$ m (see Figure 7.12).

Solution

Since the liquid film is divided into 10 intervals, $n = 11$ and we have nine differential equations.

$$\frac{dC_{Ai}}{dz} = \frac{1}{v_{z_{max}}\left[1 - \left(\frac{(i-1)\Delta x}{\delta}\right)^2\right]}\left[\frac{D_{AB}}{(\Delta x)^2}(C_{Ai+1} - 2C_{Ai} + C_{Ai-1}) - k'C_{Ai}\right](2 \leq i \leq 10)$$

$$C_{Ai}|_{z=0} = 0, \; C_{A1} = C_{As}, \; C_{A11} = \begin{cases} (4C_{A10} - C_{A9})/3 \\ 0 : 4C_{A10} < C_{A9} \end{cases}$$

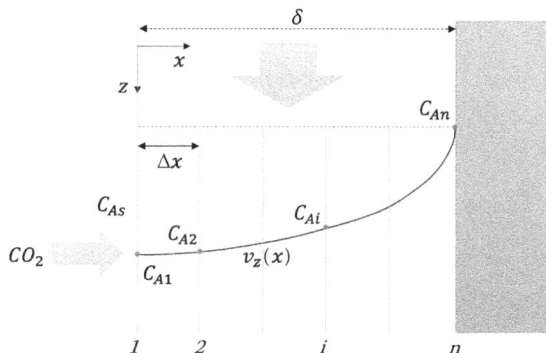

FIGURE 7.12 Mass transfer within a falling laminar film divided into small intervals. (From Cutlip, M.B. and Shacham, M., *Problem Solving in Chemical and Biochemical Engineering with POLYMATH, Excel, and MATLAB*, 2nd ed., Prentice-Hall, Boston, MA, 2008, p. 440.)

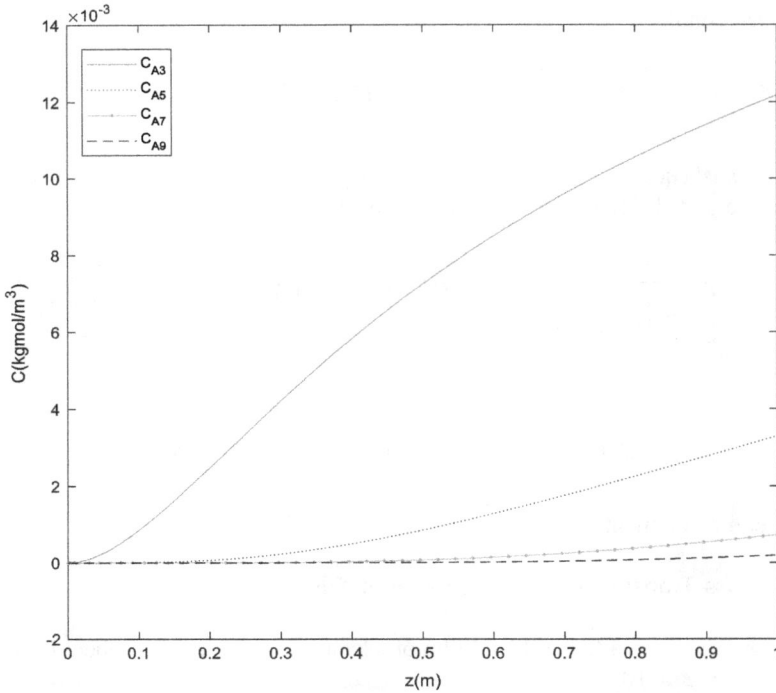

FIGURE 7.13   Concentration profile as a function of distance z.

The subfunction *bsf* defines the system of differential equations, and the script *absf* produces the curves shown in Figure 7.13.

```
% absf.m
% x(1)=Ca2, x(2)=Ca3, ..., x(9)=Ca10
Dab = 1.5e-9; delt = 3e-4; delx = delt/10; vm = 0.6; cas = 0.03; kp = 0;
c0 = zeros(1,9); % initial concentration profile
zf = 1; zspan = [0 zf]; [z x] = ode45(@bsf,zspan,c0,[],Dab,kp,delx,delt,vm,cas);
nz = length(z); xc1 = cas*ones(1,nz); % Ca1
for k = 1:nz
    if (4*x(k,9)<x(k,8)), xc11(k) = 0;
    else, xc11(k) = (4*x(k,9) - x(k,8))/3; end
end
xc = [];
for k = 1:9, xc = [xc x(:,k)]; end
c = [xc1' xc xc11']; plot(z,c(:,3),z,c(:,5),':',z,c(:,7),'.-',z,c(:,9),'--')
xlabel('z(m)'),    ylabel('C(kgmol/m^3)'),    legend('C_{A3}','C_{A5}','C_
{A7}','C_{A9}','location','best')

function dxdz = bsf(z,x,Dab,kp,delx,delt,vm,cas)
% x(1)=Ca2, x(2)=Ca3, ..., x(9)=Ca10
ca1 = cas;
if (4*x(9)<x(8)), ca11 = 0;
else, ca11 = (4*x(9) - x(8))/3; end
dxdz = [(Dab*(x(2)-2*x(1)+ca1)/(delx^2) - kp*x(1))/(vm*(1 - (1*delx/delt)^2));
(Dab*(x(3)-2*x(2)+x(1))/(delx^2) - kp*x(2))/(vm*(1 - (2*delx/delt)^2));
(Dab*(x(4)-2*x(3)+x(2))/(delx^2) - kp*x(3))/(vm*(1 - (3*delx/delt)^2));
```

```
(Dab*(x(5)-2*x(4)+x(3)))/(delx^2) - kp*x(4))/(vm*(1 - (4*delx/delt)^2));
(Dab*(x(6)-2*x(5)+x(4)))/(delx^2) - kp*x(5))/(vm*(1 - (5*delx/delt)^2));
(Dab*(x(7)-2*x(6)+x(5)))/(delx^2) - kp*x(6))/(vm*(1 - (6*delx/delt)^2));
(Dab*(x(8)-2*x(7)+x(6)))/(delx^2) - kp*x(7))/(vm*(1 - (7*delx/delt)^2));
(Dab*(x(9)-2*x(8)+x(7)))/(delx^2) - kp*x(8))/(vm*(1 - (8*delx/delt)^2));
(Dab*(call-2*x(9)+x(8)))/(delx^2) - kp*x(9))/(vm*(1 - (9*delx/delt)^2))];
end
```

7.2 EVAPORATION

7.2.1 SINGLE-EFFECT EVAPORATORS

In evaporation, the vapor from a boiling liquid solution is removed and a more concentrated solution remains. The properties of the solution being concentrated and of the vapor being removed bear greatly on the type of evaporator used and the pressure and temperature of the evaporation process. The factors to be considered in the operation include concentration in the liquid, foaming, solubility, scale deposition, and materials of construction. Figure 7.14 shows a typical single-effect evaporator.

In a single-effect evaporator, the latent heat of condensation of the steam is transferred through a heating surface to vaporize solvent from a boiling solution. Two enthalpy balances are needed: one for the steam side and one for the solution side. It is assumed that the superheat and the subcooling of the condensate are negligible. The difference between the enthalpy of the steam and that of the condensate is represented by the latent heat of condensation of the steam:

$$q_s = m_s(h_s - H_s) = -m_s \lambda_s$$

where
  h_s is the specific enthalpy of the condensate
  H_s is the specific enthalpy of the steam
  m_s is the flow rate of the steam
  q_s is the heat transfer rate through the heating surface from the steam
  λ_s is the latent heat of condensation of the steam

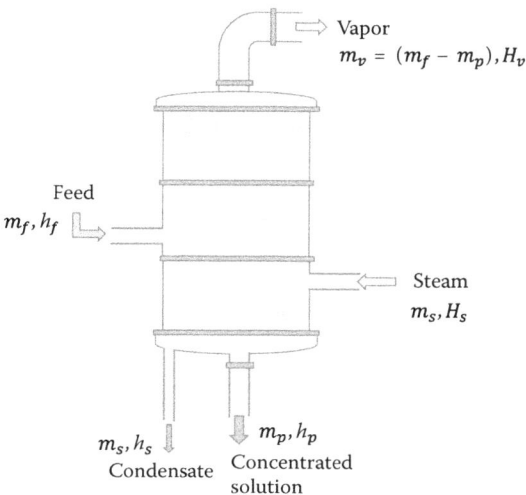

FIGURE 7.14 A single-effect evaporator.

The enthalpy balance for the solution side yields

$$q = (m_f - m_p)H_v - m_f h_f + m_p h_p$$

where

h_f is the specific enthalpy of the thin liquor

H_v is the specific enthalpy of the vapor

h_p is the specific enthalpy of the concentrated solution

q is the heat transfer rate from the heating surface to the liquid solution

In the absence of heat losses, the heat transferred to the heating surface from the steam equals that transferred from the heating surface to the liquid, and $q = -q_s$. Combination of the equations on these quantities gives

$$q = m_s \lambda_s = (m_f - m_p)H_v - m_f h_v + m_p h_p$$

h_f and h_p depend upon the temperature and concentration of the solution. From the enthalpy and mass balances, the heat transfer area A and first effect m_{ev} are given by

$$A = \frac{q}{U(T_s - T_v)}, \; m_{ev} = m_f - m_p$$

where T_s and T_v are the temperatures of the saturated steam and the saturated vapor, respectively.

The function *sinevap* calculates the single-effect evaporation process. The syntax is

```
res = sinevap(evdat)
```

The function requires the structure variable evdat, which consists of seven fields: evdat.mf is the feed flow rate (*lb/hr*), evdat.Tf is the feed temperature (°F), evdat.xf is the solid fraction in the feed solution, evdat.xp is the solid fraction in the concentrated solution, evdat.Ps is the inlet steam pressure (*psia*), evdat.Pv is the outlet vapor pressure (*psia*), and evdat.U is the overall heat transfer coefficient (*Btu/(ft²·hr·°F)*).

The function *sinevap* executes the following calculation procedure:

1) Converts the units of temperature and pressure to K and Pa, respectively.
2) Find the saturation temperatures of the steam and the solution using the equation

$$f(\tau) = a_1\tau + a_2\tau^{1.5} + a_3\tau^3 + a_4\tau^{3.5} + a_5\tau^4 + a_6\tau^{7.5} - (1-\tau)\ln\left(\frac{p}{p_c}\right) = 0$$

3) Use the function *satsteam* to calculate properties at the saturation temperature. When this function is called, the unit of temperature should be converted to °C.
4) Determine q and A from the mass balance.

```
function res = sinevap(evdat)
% Calculation of single-effect evaporator
% Data
Tc = 647.096; % critical temperature of H2O (K)
Pc = 22064000; % critical pressure of H2O (Pa)
T0 = 273.15; xf = evdat.xf; xp = evdat.xp; mf = evdat.mf;
Tf = (evdat.Tf-32)/1.8 + 273.15; % F->K
Ps = evdat.Ps*6894.757; Pv = evdat.Pv*6894.757; % psia->Pa
U = evdat.U; cf = 1/2326; % J/kg->Btu/lb
% Saturation temperature
```

```
a = [-7.85951783 1.84408259 -11.7866497 22.6807411 -15.9618719 1.80122502];
gs = @(x) a(1)*x + a(2)*x^1.5 + a(3)*x^3 + a(4)*x^3.5 + a(5)*x^4 +a(6)*x^7.5 -
(1-x)*log(Ps/Pc);
gv = @(x) a(1)*x + a(2)*x^1.5 + a(3)*x^3 + a(4)*x^3.5 + a(5)*x^4 +a(6)*x^7.5 -
(1-x)*log(Pv/Pc);
xs = fzero(gs,0.5); xv = fzero(gv,0.5); Ts = Tc*(1 - xs); Tv = Tc*(1 - xv);
% Enthalpy
sts = satsteam(Ts-T0); stv = satsteam(Tv-T0); stf = satsteam(Tf-T0);
Hs = sts.hV*cf; hs = sts.hL*cf; % steam
Hv = stv.hV*cf; hv = stv.hL*cf; % solution
hf = stf.hL*cf;
% Material balance
mp = xf*mf/xp; % Btu/hr
mev = mf - mp; q = mev*Hv - mf*hf + mp*hv; ms = q/(Hs - hs);
Ts = (Ts - 273.15)*1.8+32; % K->F
Tv = (Tv - 273.15)*1.8+32; dT = Ts - Tv; A = q/U/dT;
% Assign output structure (heating surface(A) and steam amount(ms))
res.ms = ms; res.A = A;
end


function stpr = satsteam(Ts)
% Properties of the saturated steam at a given temperature Ts(C)
% stpr: a structure containing results of calculations
% Critical properties
Tc = 647.096;  % critical temperature(K)
Pc = 22064000; % critical pressure(Pa)
rhoc = 322;    % critical density(kg/m^3)
% Definition of parameters and constants
Ts = Ts + 273.15; alpha0 = 1000; % alpha_0(J/kg)
phi0 = 1000/647.096;
a = [-7.85951783 1.84408259 -11.7866497 22.6807411 -15.9618719 1.80122502];
b = [1.99274064 1.09965342 -0.510839303 -1.75493479 -45.5170352 -674694.45];
c  =  [-2.03150240  -2.68302940  -5.38626492  -17.2991605  -44.7586581
-63.9201063];
d = [-5.65134998e-8 2690.66631 127.287297 -135.003439 0.981825814];
alphad = -1135.905627715; phid = 2319.5246; theta = Ts/Tc; tau = 1 - theta;
% saturated steam pressure
tw = (Tc/Ts)*(a(1)*tau + a(2)*tau^1.5 + a(3)*tau^3+a(4)*tau^3.5 + a(5)*tau^4 +
a(6)*tau^7.5);
Ps = Pc*exp(tw);
% density of saturated liquid
rhoL =rhoc*(1 + b(1)*tau^(1/3) + b(2)*tau^(2/3) + b(3)*tau^(5/3) +...
    b(4)*tau^(16/3) + b(5)*tau^(43/3) + b(6)*tau^(110/3));
% density of saturated steam
vw = c(1)*tau^(1/3) + c(2)*tau^(2/3) + c(3)*tau^(4/3) + c(4)*tau^3 + c(5)*tau^
(37/6) + c(6)*tau^(71/6);
rhoV = rhoc*exp(vw);
% specific volume
vL = 1/rhoL;  % saturated liquid
vV = 1/rhoV;  % saturated steam
% alpha
alpha = alpha0*(alphad + d(1)*theta^(-19) + d(2)*theta + d(3)*theta^4.5 + d
(4)*theta^5 + d(5)*theta^54.5);
% phi
phi = phi0*(phid + (19/20)*d(1)*theta^(-20) + d(2)*log(theta) +...
```

```
    (9/7)*d(3)*theta^3.5 + (5/4)*d(4)*theta^4 + (109/117)*d(5)*theta^53.5);
% dp/dT
tv = 7.5*a(6)*tau^6.5 + 4*a(5)*tau^3 + 3.5*a(4)*tau^2.5 +...
    3*a(3)*tau^2 + 1.5*a(2)*tau^0.5 + a(1) + log(Ps/Pc);
dpdT = (-Ps/Ts)*tv;
% enthalpy
hL = alpha + (Ts/rhoL)*dpdT; % saturated liquid
hV = alpha + (Ts/rhoV)*dpdT; % saturated steam
% entropy
sL = phi + (1/rhoL)*dpdT;  % saturated liquid
sV = phi + (1/rhoV)*dpdT; % saturated steam
% result structure
stpr.T = Ts; stpr.P = Ps; stpr.vL = vL; stpr.vV = vV; stpr.hL = hL;
stpr.hV = hV; stpr.sL = sL; stpr.sV = sV;
end
```

Example 7.9 Single-Effect Evaporator

A vertical-tube single-effect evaporator is used to concentrate 29, 000 *lb/hr* of a 25 % solution of organic colloid to 60 % solution. The feed temperature is 60 °F. Assume that the boiling point elevation is negligible, the physical properties of the solution are similar to those of water, and the specific enthalpy of the solution, h_p, is the same as that of the steam, h_v. The absolute pressure of the saturated steam being introduced is 25 *psia*. The absolute pressure in the vapor space is 1.69 *psia* and the overall heat transfer coefficient is 300 *Btu/(ft²·hr· °F)*. Calculate the heating surface required *(ft²)* and the amount of steam consumed *(lb/hr)*.

Solution

The following commands produce the desired outputs:

```
>> evdat.mf = 29000; evdat.Tf = 60; evdat.xf = 0.25;
>> evdat.xp = 0.6; evdat.Ps = 25; evdat.Pv = 1.69; evdat.U = 300;
>> res = sinevap(evdat)
res =
    ms: 2.0039e+04
    A: 529.3588
```

We can see that the amount of steam consumed is m_s = 20,039 *lb/hr* and the heating surface required is 529.3588 *ft²*.

7.2.2 VAPORIZERS

Figure 7.15 shows a typical vaporizer, in which a liquid enters at the bottom with a flow rate F_i *(liter/hr)* and temperature T_i (°C). Steam is used as a heating medium. The assumptions adopted for the vaporizer can be summarized as follows:

- Heat losses are negligible.
- Perfect mixing and vapor-liquid equilibrium are maintained in the vessel.
- Vapor holdup is negligible.
- The vapor pressure of the liquid is the same as the pressure in the vapor phase $(P = P_V)$.
- The density (ρ) and heat capacity (C_p) remain constant.
- The ideal gas law can be applied to the vapor phase.

The material and energy balances equations are as follows:

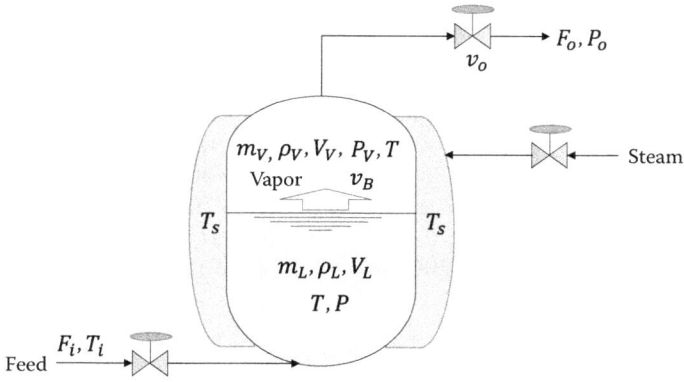

FIGURE 7.15　Schematic diagram of a vaporizer.

$$\frac{dm_L}{dt} = \frac{d\,(\rho_L V_L)}{dt} = F_i \rho_L - v_B, \frac{dm_V}{dt} = \frac{d\,(\rho_V V_V)}{dt} = v_B - F_o \rho_V = v_B - v_o$$

$$\frac{d\,(\rho_L V_L C_p T)}{dt} = F_i \rho_L C_p T_i - v_B \left(\lambda + C_p T\right) + Q = F_i \rho_L C_p T_i - v_B \left(\lambda + C_p T\right) + UA\,(T_s - T)$$

where
　$V_L\,(liter)$ and $V_V\,(=V - V_L)(liter)$ are the volumes of the liquid and vapor phases, respectively
　$F_o\,(liter/hr)$ is the volumetric vapor flow rate through the exit valve
　$m_L\,(kg)$ and $m_V\,(kg)$ are the masses of the liquid and vapor phases, respectively
　$\lambda\,(cal/g)$ is the latent heat of vaporization at $T\,(^\circ C)$
　$C_p\,(cal/(g\cdot {}^\circ C))$ is the heat capacity
　$V\,(liter)$ is the vessel volume
　$\rho\,(g/cm^3)$ is the density
　If the heat Q is supplied by the steam, it can be represented as

$$Q = UA\,(T_s - T)$$

where
　U is the overall heat transfer coefficient
　A is the heat transfer area
　The flow rate through the exit valve, $v_o\,(kg/hr)$, is given by

$$v_o = F_o \rho_V = K_V \sqrt{P_V\,(P_V - P_o)}$$

where $K_V\,(kg/(hr \cdot bar))$ is the valve constant and $P_o\,(bar)$ is the exit pressure. From the ideal gas law,

$$P_V V_V = PV_V = \frac{m_V RT}{M_W}$$

The vapor pressure is given by the Antoine equation

$$\ln P = \ln P_V = A - \frac{B}{T + C}$$

Example 7.10 Simple Boiler[23]

The simple boiler shown in Figure 7.15 is used to generate steam by receiving heat Q from an external source. The dynamics of the temperature T are assumed to be negligible. Plot the masses of the liquid and vapor phases (m_L and m_V) as a function of time using the given data. For water, the Antoine constants are $A = 18.3036$, $B = 3816.44$, $C = -46.13$ (T:K, P:mmHg).

Data: $Q = 6000\ kcal/hr$, $\lambda = 475\ cal/g$, $C_p = 1 cal/g/\,°C$, $T_i = 20\,°C$, $F_i = 1000\ liter/hr$, $\rho_L = 1 kg/liter$, $V = 5500\ liter$, $V_L(t = 0) = 2800\ liter$, $m_V(t = 0) = 9\ kg$, $K_V = 40\ kg/(hr·bar)$, $P_o = 10\ bar$, $M_W = 18\ g/mol$, $R = 0.08206\ (liter·atm)/(mol·K)$

Solution
The solution procedure can be summarized as follows:

1. $V_V = V - V_L = V - m_L/\rho_L$ (liter)
2. Calculate $T(K)$ from $\exp(A - B/(T + C)) - 760000\, m_V RT/(V_V M_W) = 0$ and $P = P_V$ from $P(bar) = 0.0012986 \times \exp(A - B/(T + C))$.
3. Calculate $v_B = (F_i \rho_L C_p T_i + Q)/(\lambda + C_p(T - 273.15))$ and $v_o = K_V \sqrt{P(P - P_o)}$.
4. $dm_V/dt = v_B - v_o$, $dm_L/dt = F_i \rho_L - v_B$

The script *vapdat* defines the data. The differential equations are defined by the subfunction *vapr*, and the script *vapcm* calls the subfunction vapr, uses the built-in solver *ode45* to solve the equation system, and produces the mass profiles shown in Figure 7.16.

```
% vapdat.m: define data
Q = 6000; lambda = 475; Cp = 1; Ti = 20; Fi = 1000; rhol = 1;
V = 5500; Kv = 40; P0 = 10; Mw = 18; R = 0.08206;
A = 18.3036; B = 3816.44; C = -46.13;
```

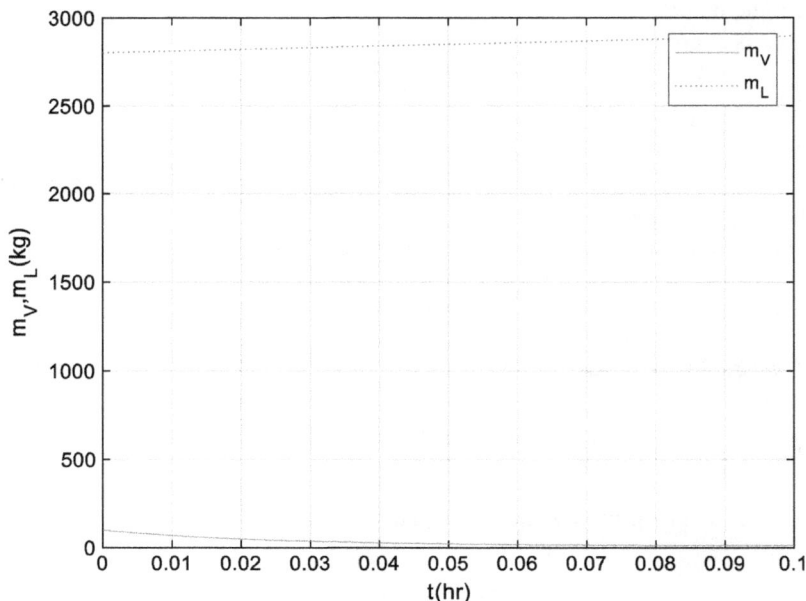

FIGURE 7.16   Masses of the liquid and vapor phases.

```
% vapcm.m
% differential equations are defined by the subfunction vapr
vapdat; % retrieve data
mL0 = 2800; mV0 = 100; z0 = [mV0 mL0]; tspan = [0 0.1]; [t z] = ode45(@
vapr,tspan,z0);
plot(t,z(:,1),t,z(:,2),':'), grid, xlabel('t(h)'), ylabel('m_V,m_L(kg)'),
legend('m_V','m_L')

function dz = vapr(t,z)
% define differential equations: z(1)=mV, z(2)=mL
vapdat; % retrieve data
mV = z(1); mL = z(2); Vv = V - mL/rhol;
% Determine T by using bisection method
f = @(T) exp(A-B/(T+C)) - 7.6e5*mV*R*T/(Vv*Mw); Ta = 273.15; Tb = Ta + 400;
if f(Ta)*f(Tb) > 0, display('No solution T'), return; end
Tm = (Ta + Tb)/2;
while abs(Ta-Tb) > 1e-2
    if f(Ta)*f(Tm) < 0, Tb = Tm; else, Ta = Tm; end; Tm = (Ta + Tb)/2;
end
T = Tm;
% Define differential eqns
P = exp(A - B/(T + C)) / 750.044; % P: bar
vB = (Fi*rhol*Cp*Ti + Q)/(lambda + Cp*(T - 273.15));
vO = Kv*sqrt(P*(P - P0));
dz = [vB - vO; Fi*rhol - vB];
end
```

7.2.3 MULTIPLE-EFFECT EVAPORATORS

There are several methods of feeding in the evaporation process. The usual method of feeding a multiple-effect evaporator is to introduce the thin liquid into the first effect and send it in turn through the other effects. This method of feeding is called forward-feed. Figure 7.17 shows a schematic of a forward-feed triple-effect evaporator. The fresh feed is added to the first effect and flows to the next in the same direction as the vapor flow. The concentration of the liquid increases from the first effect to the last effect. Connections are made so that the vapor from one effect serves as the heating medium for the next effect. Figure 7.18 shows a typical multiple-effect evaporator.

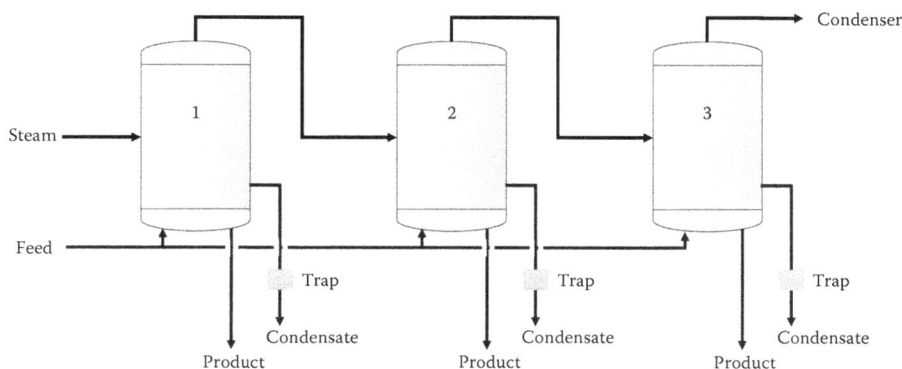

FIGURE 7.17 Schematic of a forward-feed triple-effect evaporator.

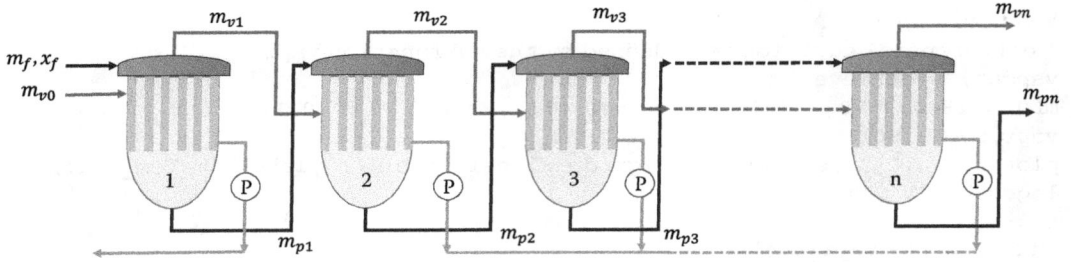

FIGURE 7.18   Simplified diagram of forward-feed m-effect evaporator.

Let the evaporation continue until the nth effect. Let the flow rate and the enthalpy of the liquor from the kth effect be m_{pk} and h_{pk}, respectively, and those of the vapor phase from the kth effect be m_{vk} and H_{vk}, respectively. From the overall material balance $m_f x_f = m_{pn} x_{pn}$, we have

$$m_{pn} = \frac{m_f x_f}{x_{pn}}$$

The mass and energy balances on the first effect yield

$$m_f = m_{p1} + m_{v1}, \; m_f h_f + m_{v0}(H_{v0} - h_{v0}) = m_{v1} H_{v1} + m_{p1} h_{p1}$$

The mass and energy balances on the kth effect yield

$$m_{p,k-1} = m_{pk} + m_{vk}, \; m_{p,k-1} h_{p,k-1} + m_{v,k-1}\left(H_{v,k-1} - h_{v,k-1}\right) = m_{pk} h_{pk} + m_{vk} H_{vk}$$

The flow rate of liquor from the kth effect is given by

$$m_{pk} = m_f - m_{v1} - m_{v2} - \cdots - m_{vk} = m_f - \sum_{i=1}^{k} m_v$$

Combination of these equations after rearrangement gives

Effect 1: $(H_{v0} - h_{v0})m_{v0} + (h_{p1} - H_{v1})m_{v1} = (h_{p1} - h_f)m_f$

Effect 2: $(H_{v1} - h_{v1} - h_{p1} + h_{p2})m_{v1} + (h_{p2} - H_{v2})m_{v2} = (h_{p2} - h_{p1})m_f$

Effect k: $(h_{pk} - h_{p,k-1})\sum_{i=1}^{k-2} m_{vi} + (H_{v,k-1} - h_{v,k-1} + h_{pk} - h_{p,k-1})m_{v,k-1}$
$\quad\quad + (h_{pk} - H_{vk})m_{vk} = (h_{pk} - h_{p,k-1})m_f \, (3 \le k \le n-1)$

Effect n: $(H_{vn} - h_{p,n-1})\sum_{i=1}^{n-2} m_{vi} + (H_{vn} + H_{v,n-1} - h_{v,n-1} - h_{p,n-1})m_{v,n-1}$

$$= \left\{ (h_{pn} - h_{p,n-1}) + (H_{vn} - h_{pn})\left(1 - \frac{x_f}{x_{pn}}\right) \right\} m_f$$

The last equation can be obtained by using

$$m_{pn} = m_f - \sum_{i=1}^{n} m_{vi} = \frac{x_f}{x_{pn}} m_f, \; m_{vn} = \left(1 - \frac{x_f}{x_{pn}}\right) m_f - \sum_{i=1}^{n-1} m_{vi}$$

The heat transfer area is considered to be the same in each effect: $A_1 = A_2 = \cdots = A_n$. Then the temperature drop between adjacent effects is given by

$$\Delta T_i = \frac{\dfrac{1}{U_i}}{\dfrac{1}{U_1} + \dfrac{1}{U_2} + \cdots + \dfrac{1}{U_n}} \cdot \Delta T_{total}$$

The heat transfer rate (q) at each effect can be calculated from the mass and energy balances. The heat transfer area (A) can be determined using the value of q. The calculation is repeated until the same heat transfer area is obtained for each effect. The calculation procedure can be summarized as follows:

1) Determine the saturation temperatures for the steam and vapor at a given pressure. Evaluate the enthalpy at the saturation temperature.
2) Calculate ΔT_i and determine the saturation temperature in each effect, T_i.
3) Calculate the enthalpy of the vapor and liquid phases at the temperature T_i.
4) Find the flow rate of the vapor from each effect using mass and energy balances on each effect.
5) Determine the heat transfer rate in each effect and calculate the heat transfer area A_i.
6) Compare values of A_i at each effect. If all A_i's are the same, stop the procedure. If not, calculate $\Delta T_i = q_i/(A_i U_i)$ and go to step 3).

The function *multievapSI* performs the calculation procedure for multiple-effect evaporation. The basic syntax is
```
res = multievapSI(evdat)
```

The function requires the structure variable evdat, which consists of seven fields: evdat.xp is the solid fraction in the concentrated solution at the final effect, evdat.xf is the solid fraction in the feed solution, evdat.Ps is the inlet steam pressure (Pa), evdat.Pn is the outlet vapor pressure from the last effect (Pa), evdat.mf is the feed flow rate (kg/hr), evdat.Tf is the feed temperature ($^\circ C$), and evdat.U is the overall heat transfer coefficient at each effect ($J/(m^2 \cdot hr \cdot K)$). The output variable res is a structure containing the results and consists of four fields: res.T is a vector containing the temperature at each effect ($^\circ C$), res.A is a vector of the heat transfer area (m^2), res.mv is a vector of the vapor flow rate from each effect (kg/hr), and res.iter denotes the iteration number.

```
function res = multievapSI(evdat)
% Multiple-effect evaporator calculation (SI unit system)
% Data
crit = 1e-3; Tc = 647.096; % critical temperature of H2O (K)
Pc = 22064000; % critical pressure of H2O (Pa)
T0 = 273.15; xf = evdat.xf; xpn = evdat.xp; mf = evdat.mf;
Tf = evdat.Tf + T0; % C -> K
Ps = evdat.Ps; % steam pressure (Pa)
Pn = evdat.Pn; % pressure of the final effect (Pa)
U = evdat.U; % overall heat transfer coefficient
n = length(U);
% Saturation temperature
a = [-7.85951783 1.84408259 -11.7866497 22.6807411 -15.9618719 1.80122502];
gs = @(x) a(1)*x + a(2)*x^1.5 + a(3)*x^3 + a(4)*x^3.5 + a(5)*x^4 + a(6)*x^7.5 -
(1-x)*log(Ps/Pc);
gn = @(x) a(1)*x + a(2)*x^1.5 + a(3)*x^3 + a(4)*x^3.5 + a(5)*x^4 + a(6)*x^7.5 -
(1-x)*log(Pn/Pc);
xs = fzero(gs,0.5); xn = fzero(gn,0.5);
Ts = Tc*(1 - xs); % steam temperature (K)
Tn = Tc*(1 - xn); % temperature of the final effect (K)
% Enthalpy
sts = satsteam(Ts-T0); % saturated steam
stv = satsteam(Tn-T0); % final effect
stf = satsteam(Tf-T0); % feed
Hv0 = sts.hV; hp0 = sts.hL; % J/kg
```

```
Hvn = stv.hV; hpn = stv.hL; hf = stf.hL;
% Mass balances and temperature drop
mpn = xf*mf/xpn; % kg/hr
dTtotal = Ts - Tn; % K
sumU = sum(1./U);
dT = (1./U)*dTtotal/sumU; % K
critA = 10; oldA = 10*ones(1,n); iter = 0;
while critA >= crit
    T(1) = Ts - dT(1);
    for j = 2:n, T(j) = T(j-1) - dT(j); end
    for j = 1:n
        sprop(j) = satsteam(T(j)-T0); Hv(j) = sprop(j).hV; hp(j) = sprop(j).hL;
    end
    % balance equations
    evM = [Hv0-hp0 hp(1)-Hv(1) 0; 0 Hv(1)+hp(2)-2*hp(1) hp(2)-Hv(2);...
        0 Hv(3)-hp(2) Hv(2)+Hv(3)-2*hp(2)];
    evb = [hp(1)-hf; hp(2)-hp(1); Hv(3)-hp(2)+(hp(3)-Hv(3))*xf/xpn]*mf;
    mv = evM\evb; q(1) = mv(1)*(Hv0-hp0);
    for j = 2:n, q(j) = mv(j)*(Hv(j)-hp(j)); end
    Area = q./(U.*dT); avgA = sum(Area)/n; critA = sum(abs(Area - oldA)); dT =
dT.*Area/avgA;
    if abs(sum(dT) - dTtotal) >= crit, dT = dT*dTtotal/sum(dT); end
    iter = iter + 1; oldA = Area;
end
% Results
res.T = T-T0; % vector of temperature of each effect (C)
res.A = Area; % vector of area of each effect (m^2)
res.mv = mv; % vector of vapor flow rate from each effect (kg/hr)
res.iter = iter; % number of iterations
end
```

Example 7.11 Multiple-Effect Evaporator

A triple-effect forward-feed evaporator is being used to evaporate an organic colloid solution containing 15% solids to a concentrated solution of 60%. Saturated steam at 205,603 Pa is being used, and the pressure in the vapor phase of the third effect is 8756 Pa. The feed rate is 20,412 kg/hr at 15.6 °C. The heat capacity of the liquid solutions is the same as that of water over the whole concentration range. The coefficients of heat transfer have been estimated as $U_1 = 1.084 \times 10^7$, $U_2 = 7.155 \times 10^6$, and $U_3 = 3.986 \times 10^6 \, J/(m^2 \cdot hr \cdot K)$. If the heat transfer areas of each of the three effects are to be equal, calculate the heat transfer area (m^2) and the amount of steam consumed (kg/hr). The boiling point elevation of the solutions is assumed to be negligible.

Solution

The enthalpy balance on each effect gives

Effect1 : $(H_{v0} - h_{v0})m_{v0} + (h_{p1} - H_{v1})m_{v1} = (h_{p1} - h_f)m_f$

Effect2 : $(H_{v1} - h_{v1} - h_{p1} + h_{p2})m_{v1} + (h_{p2} - H_{v2})m_{v2} = (h_{p2} - h_{p1})m_f$

Effect3 : $(H_{v3} - h_{p2})m_{v1} + (H_{v3} + H_{v2} - h_{v2} - h_{p2})m_{v2} = \left\{ (H_{v3} - h_{p2}) + (h_{p3} - H_{v3})\dfrac{x_f}{x_{p3}} \right\} m_f$

Assuming that the enthalpy of the condensate is the same as that of the solution (that is, $h_{vi} = h_{pi}$), we have

$$(H_{v0} - h_{p0})m_{v0} + (h_{p1} - H_{v1})m_{v1} = (h_{p1} - h_f)m_f$$

$$(H_{v1} - 2h_{p1} + h_{p2})m_{v1} + (h_{p2} - H_{v2})m_{v2} = (h_{p2} - h_{p1})m_f$$

$$(H_{v3} - h_{p2})m_{v1} + (H_{v3} + H_{v2} - 2h_{p2})m_{v2} = \left\{(H_{v3} - h_{p2}) + (h_{p3} - H_{v3})\frac{x_f}{x_{p3}}\right\}m_f$$

Rearrangement of these equations yields

$$\begin{bmatrix} H_{v0} - h_{p0} & h_{p1} - H_{v1} & 0 \\ 0 & H_{v1} - 2h_{p1} + h_{p2} & h_{p2} - H_{v2} \\ 0 & H_{v3} - h_{p2} & H_{v3} + H_{v2} - 2h_{p2} \end{bmatrix} \begin{bmatrix} m_{v0} \\ m_{v1} \\ m_{v2} \end{bmatrix} = \begin{bmatrix} (h_{p1} - h_f) \\ (h_{p2} - h_{p1}) \\ \left(H_{v3} - h_{p2}\right) + \left(h_{p3} - H_{v3}\right)\frac{x_f}{x_{pn}} \end{bmatrix} m_f$$

The following commands show the results of calculations by the function *multievapSI*:

```
>> evdat.xp = 0.60; evdat.xf = 0.15; evdat.Ps = 205602.9; evdat.Pn = 8756;
>> evdat.mf = 20412; evdat.Tf = 15.6; evdat.U = [10.834 7.155 3.986]*10^6;
>> results = multievapSI(evdat)
results =
        T: [100.4974 81.7076 43.2312]
        A: [79.4099 79.4099 79.4099]
       mv: [3x1 double]
     iter: 6
>> results.mv
ans =
     1.0e+03 *
     8.0540
     4.6341
     5.0781
```

We can see that the heat transfer area converges to $79.41m^2$ after six iterations.

7.3 ABSORPTION

7.3.1 ABSORPTION BY TRAY COLUMN

In the absorption process, a soluble vapor is absorbed, by means of a liquid in which the solute gas is more or less soluble, from its mixture with an inert gas. Usually, absorption is used to remove impurities, contaminants, and pollutants or to recover valuable chemicals. Figure 7.19 shows a typical gas absorption process.

For the absorber shown in Figure 7.19, L and V are the molar flow rates of the liquid and gas phases, respectively, and x_i and y_i are the mole fractions of the liquid and gas phases at the ith stage, respectively. Assuming constant L and V and thermodynamic equilibrium in each stage, the mass balance on the ith stage gives

$$\frac{d(Mx_i + Wy_i)}{dt} = Lx_{i-1} + Vy_{i+1} - Lx_i - Vy$$

where M and W are the molar holdups of the liquid and gas phases for the ith stage, respectively. W may be negligible compared to M, and M is assumed to be constant for each stage. The mass balance can be rewritten as

$$\frac{dx_i}{dt} = \frac{L}{M}x_{i-1} + \frac{V}{M}y_{i+1} - \frac{L}{M}x_i - \frac{V}{M}y_i$$

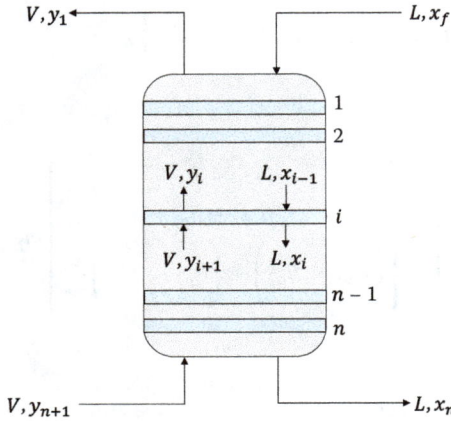

FIGURE 7.19   Typical gas absorption process.

The equilibrium relation in the gas-liquid may be represented as $y_i = a x_i$. Substitution of this equilibrium relation into the mass balance equation gives

$$\frac{dx_i}{dt} = \frac{L}{M} x_{i-1} - \frac{(L + Va)}{M} x_i + \frac{Va}{M} x_{i+1}$$

$$\frac{dx_1}{dt} = \frac{L}{M} x_f - \frac{(L + Va)}{M} x_1 + \frac{Va}{M} x_2$$

$$\frac{dx_n}{dt} = \frac{L}{M} x_{n-1} - \frac{(L + Va)}{M} x_n + \frac{V}{M} y_{n+1}$$

For $n = 5$, these equations can be represented as

$$
\begin{bmatrix} \dot{x}_1 \\ \dot{x}_2 \\ \dot{x}_3 \\ \dot{x}_4 \\ \dot{x}_5 \end{bmatrix}
=
\begin{bmatrix}
-\frac{(L+Va)}{M} & \frac{Va}{M} & 0 & 0 & 0 \\
\frac{L}{M} & -\frac{(L+Va)}{M} & \frac{Va}{M} & 0 & 0 \\
0 & \frac{L}{M} & -\frac{(L+Va)}{M} & \frac{Va}{M} & 0 \\
0 & 0 & \frac{L}{M} & -\frac{(L+Va)}{M} & \frac{Va}{M} \\
0 & 0 & 0 & \frac{L}{M} & -\frac{(L+Va)}{M}
\end{bmatrix}
\begin{bmatrix} x_1 \\ x_2 \\ x_3 \\ x_4 \\ x_5 \end{bmatrix}
+
\begin{bmatrix}
\frac{L}{M} & 0 \\
0 & 0 \\
0 & 0 \\
0 & 0 \\
0 & \frac{V}{M}
\end{bmatrix}
\begin{bmatrix} x_f \\ y_6 \end{bmatrix}
$$

or

$$\dot{x} = Ax + Bu$$

At steady state, the time derivatives are zero:

$$0 = Ax_s + Bu_s$$

Thus, the steady-state value of x is given by

$$x_s = -A^{-1} Bu_s$$

Introduction of a step change in feed composition, Δu, yields

$$\Delta \dot{x} = A \Delta x + B \Delta u, \quad \Delta y = C \Delta x + D \Delta u = \begin{bmatrix} \Delta x_5 \\ \Delta y_1 \end{bmatrix}, \quad C = \begin{bmatrix} 0 & 0 & 0 & 0 & 1 \\ a & 0 & 0 & 0 & 0 \end{bmatrix}, \quad D = \begin{bmatrix} 0 & 0 \\ 0 & 0 \end{bmatrix}$$

where

Δy is a response vector

Δx_5 and Δy_1 are the mole fractions of the liquid and gas phases from the absorption column

Example 7.12 Benzene Absorption Column

A five-stage absorption column is to be used to remove benzene from the process gas using an oil stream. The oil feed flow rate is 1.33 $kgmol/min$, and the gas feed rate and the mole fraction of benzene in the gas feed are 1.667 $kgmol$ air/min and 0.1, respectively. The liquid molar holdup for each stage is 6.667 $kgmol$ and the equilibrium relation is given by $y_i = 0.5x_i$. The oil feed does not contain any benzene ($x_f = 0$).

(1) Calculate the steady-state composition for each stage.

(2) The benzene composition of the gas stream entering the column (y_6) suddenly changes from 0.1 to 0.15. Plot the compositions of the liquid and vapor streams leaving the column as a function of time. Also plot the profile of the liquid-phase composition for each stage.

Solution

The numerical data yield the following matrices:

$$A = \begin{bmatrix} -0.325 & 0.125 & 0 & 0 & 0 \\ 0.2 & -0.325 & 0.125 & 0 & 0 \\ 0 & 0.2 & -0.325 & 0.125 & 0 \\ 0 & 0 & 0.2 & -0.325 & 0.125 \\ 0 & 0 & 0 & 0.2 & -0.325 \end{bmatrix}, B = \begin{bmatrix} 0.2 & 0 \\ 0 & 0 \\ 0 & 0 \\ 0 & 0 \\ 0 & 0.25 \end{bmatrix}, U_s = \begin{bmatrix} x_{fs} \\ y_{6s} \end{bmatrix} = \begin{bmatrix} 0.0 \\ 0.1 \end{bmatrix}$$

$$x_s = -A^{-1}BU_s$$

(1)
```
>> A = [-0.325 0.125 0 0 0;0.2 -0.325 0.125 0 0;0 0.2 -0.325 0.125 0;
          0 0 0.2 -0.325 0.125;0 0 0 0.2 -0.325];
>> B = [0.2 0;0 0;0 0;0 0;0 0.25]; Us = [0.0;0.1]; xs = -inv(A)*B*Us
xs =
    0.0076
    0.0198
    0.0392
    0.0704
    0.1202
```

Thus,

$$x_s = -A^{-1}BU_s = \begin{bmatrix} 0.0076 \\ 0.0198 \\ 0.0392 \\ 0.0794 \\ 0.1202 \end{bmatrix}$$

(2) The step change is $\Delta u = [0 \quad 0.05]^T$, which is equal to the unit step change multiplied by 0.05. Addition of the steady-state values to the step response yields the actual composition. The script *absbz* generates the desired profile plots shown in Figure 7.20.

(a)

(b)

(c)

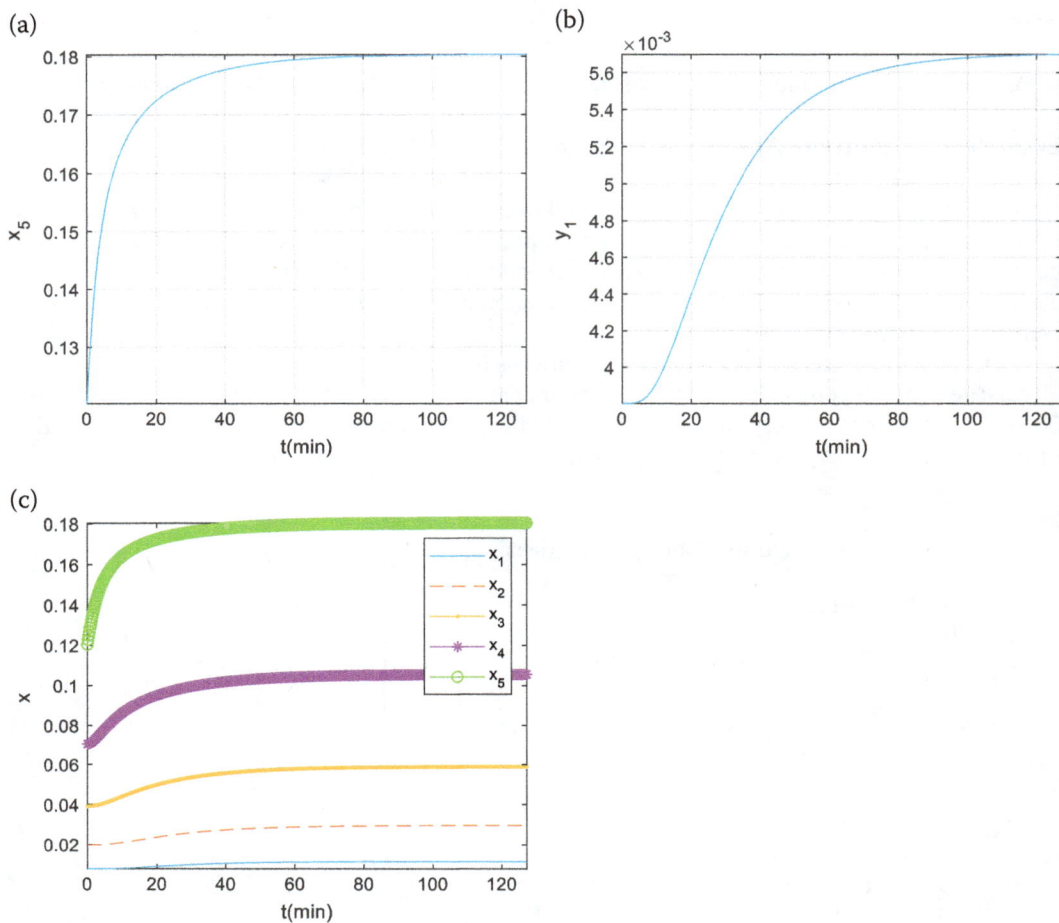

FIGURE 7.20 (a, b): Composition profiles of outlet streams and (c) liquid-phase mole fractions.

```
% absbz.m
A = [-0.325 0.125 0 0 0;0.2 -0.325 0.125 0 0;0 0.2 -0.325 0.125 0; 0 0 0.2 -0.325
0.125;0 0 0 0.2 -0.325];
B = [0.2 0;0 0;0 0;0 0;0 0.25]; C = [0 0 0 0 1;0.5 0 0 0 0]; D = [0 0;0 0];
Us = [0.0;0.1]; xs = -inv(A)*B*Us; ys = C*xs + D*Us; % Steady-state
[y,x,t] = step(A,B,C,D,2); y = 0.05*y; x = 0.05*x;
for k = 1:2, y(:,k) = y(:,k) + ys(k); end; % Y(t) = ys + y(t)
for k = 1:5, x(:,k) = x(:,k) + xs(k); end; % X(t) = xs + x(t)
subplot(2,2,1), plot(t,y(:,1)), xlabel('t(min)'), ylabel('x_5'), grid, axis
tight
subplot(2,2,2), plot(t,y(:,2)), xlabel('t(min)'), ylabel('y_1'), grid, axis
tight
subplot(2,2,3), plot(t,x(:,1),t,x(:,2),'--',t,x(:,3),'.-',t,x(:,4),'*-',t,x
(:,5),'o-')
xlabel('t(min)'), ylabel('x'), axis tight, legend('x_1','x_2','x_3','x_4',
'x_5')
```

Example 7.13 Absorption in a Tray Column[24]

Figure 7.21 shows a tray column used in an absorption operation. The operating curve of the absorption operation in the column is given by

$$
y = \frac{\left(\frac{L}{V}\right)\frac{x}{1-x} + \left\{\frac{y_p}{1-y_p} - \left(\frac{L}{V}\right)\frac{x_0}{1-x_0}\right\}}{1 + \left(\frac{L}{V}\right)\frac{x}{1-x} + \left\{\frac{y_p}{1-y_p} - \left(\frac{L}{V}\right)\frac{x_0}{1-x_0}\right\}}
$$

where

x and y are the mole fractions of the solute in the liquid and gas phases, respectively
L and V are the molar flow rates of the solute-free liquid and gas, respectively ($V = 85 \ kmol/hr$)
x_0 and y_p are the mole fractions of the solute in the liquid and gas phases at the top of the column, respectively
Equilibrium data are given in Table 7.1.
(1) Find a polynomial that fits the equilibrium data best. Plot the data and the fitting curve for $0 \le x \le 0.18$.
(2) On the same graph, plot operating curves for $L = 170$, 150, and 130 $kmol/hr$.
(3) The minimum value of L ($= L_{min}$) must be determined. Find L_{min} using the fact that the operating curve is tangent to the equilibrium curve at $L = L_{min}$ (i.e., these curves have a common tangent line at a point).
(4) Plot the equilibrium curve along with the operating curve when $L = L_{min}$.

Solution

(1) The number of data points is five, and a polynomial with order less than 5 would suffice to approximate the data. The order can be selected based on the root-mean-square error between the data and the fitting polynomial. It turns out that the best approximating curve, $P(x)$, is the 4th-order polynomial. Figure 7.22(a) shows the results of fitting by the 4th-order polynomial.

TABLE 7.1
Equilibrium Data

x	0.0	0.033	0.072	0.117	0.171
y^*	0.0	0.0396	0.0829	0.1127	0.136

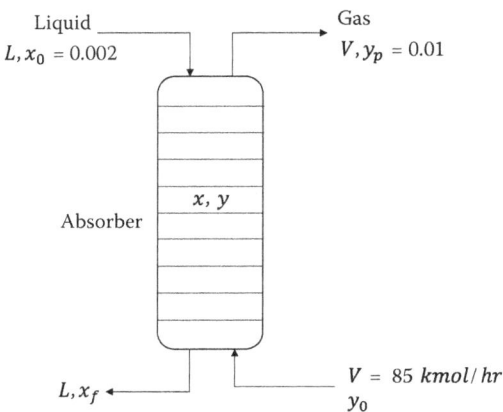

Liquid
$L, x_0 = 0.002$

Gas
$V, y_p = 0.01$

x, y
Absorber

L, x_f

$V = 85 \ kmol/hr$
y_0

FIGURE 7.21 A tray column used in the absorption operation.

(2) The operating curve of the absorption operation is evaluated for each value of L when $0 \leq x \leq 0.18$. The script *trayabs* produces a plot of the operating curve for each L.

(3) When the operating curve becomes tangent to the equilibrium curve at $x = x_{min}$ (and $L = L_{min}$), the difference between these two curves will be zero. The minimum value of L ($= L_{min}$) can be found by using the bisection method. As the initial search interval, we can choose $[L_a \ L_b] = [80 \ 120]$. L_{min} is found to be $L_{min} = 91.2307 \ kmol/hr$.

(4) The script *trayabs* generates a plot of the equilibrium curve along with the operating curve when $L = L_{min}$. We can determine the point $x = x_{min}$ at which the operating curve becomes tangent to the equilibrium curve when $L = L_{min}$ by solving the nonlinear equation

$$ f(x) = \frac{\left(\frac{L_{min}}{V}\right)\frac{x}{1-x} + \left\{\frac{y_p}{1-y_p} - \left(\frac{L_{min}}{V}\right)\frac{x_0}{1-x_0}\right\}}{1 + \left(\frac{L_{min}}{V}\right)\frac{x}{1-x} + \left\{\frac{y_p}{1-y_p} - \left(\frac{L_{min}}{V}\right)\frac{x_0}{1-x_0}\right\}} - P(x) = 0 $$

where $P(x)$ is the 4th-order polynomial approximating the equilibrium curve. The script *trayabs* uses the built-in function *fsolve* to yield $x_{min} = 0.059$.

```
% trayabs.m
clear all;
x = [0.0 0.033 0.072 0.117 0.171]; % equilibrium data (x)
```

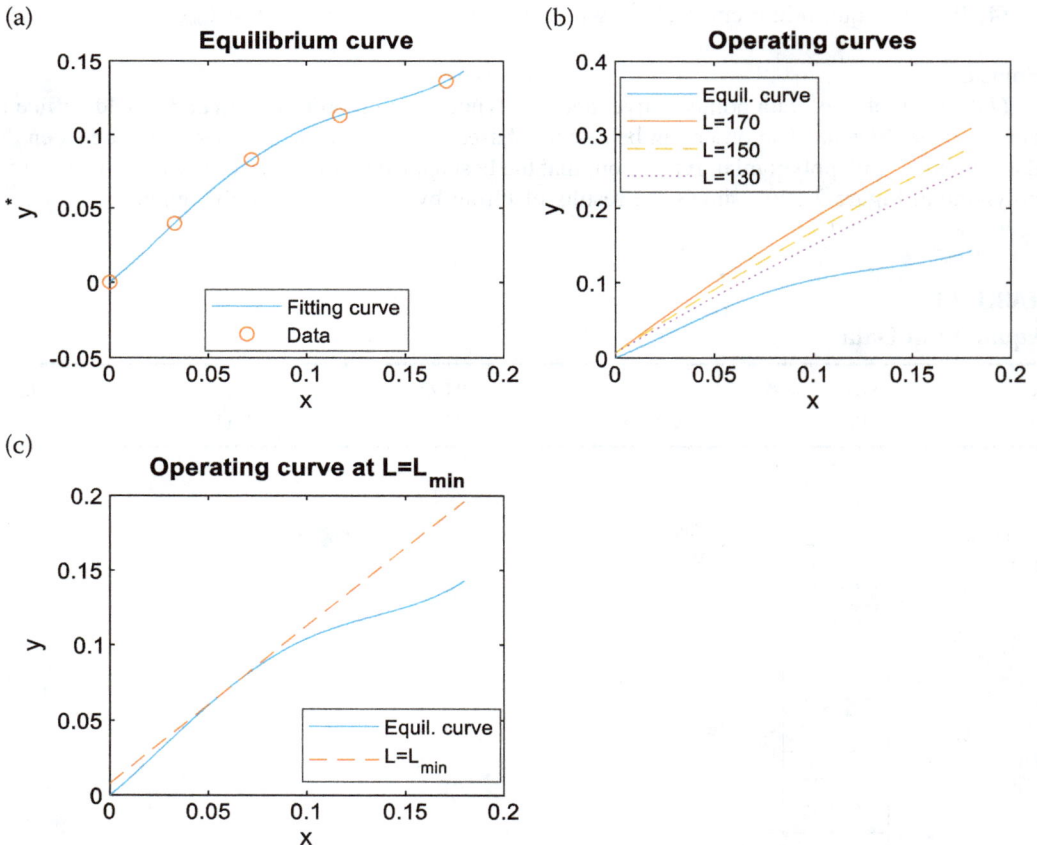

(a)

(b)

(c)

FIGURE 7.22 (a) Equilibrium curve, (b) operating curves, and (c) minimum liquid flow rate.

```
ye = [0.0 0.0396 0.0829 0.1127 0.136]; % equilibrium data (y*)
V = 85; % solute-free gas flow rate (kmol/hr)
x0 = 0.002; yp = 0.01; % mole fractions at the top of the column

% (1) Curve fitting on the equilibrium data (order <=4)
P3 = polyfit(x,ye,3);  % fit the data by 3rd-order polynomial
P4 = polyfit(x,ye,4);  % fit the data by 4th-order polynomial
y3 = polyval(P3,x); y4 = polyval(P4,x);
rmse3 = sqrt(sum((ye-y3).^2)); rmse4 = sqrt(sum((ye-y4).^2)); % find RMSE
n = 3; P = P3; if rmse3 > rmse4, n = 4; P = P4; end % choose order of the fitting curve
fprintf('Order of the fitting polynomial = %g\n', n);
xi = 0:0.001:0.18; yi = polyval(P,xi); % generate data for plot
subplot(2,2,1), plot(xi,yi,x,ye,'o'), xlabel('x'), ylabel('y^*')
title('Equilibrium curve'), legend('Fitting curve','Data','location','best')

% (2) plots of operating curves for L = 170, 150 and 130 110 kmol/hr
L = [170 150 130]; r = L/V; ypr = yp/(1-yp); x0r = x0/(1-x0);
y = [];
for k = 1:length(L), w = r(k)*xi./(1-xi) + ypr - r(k)*x0r; z = w./(1 + w); y = [y
z']; end
subplot(2,2,2), plot(xi,yi,xi,y(:,1),xi,y(:,2),'--',xi,y(:,3),':')
xlabel('x'), ylabel('y'), title('Operating curves')
legend('Equil. curve','L=170','L=150','L=130','location','best')

% (3) find the point at which the operating curve becomes tangent to the equili-
brium curve
% use bisection method
La = 80; Lb = 120; Lm = (La + Lb)/2; crit = abs(La - Lb);
r = La/V; w = r*xi./(1-xi) + ypr - r*x0r; opa = w./(1 + w);
r = Lb/V; w = r*xi./(1-xi) + ypr - r*x0r; opb = w./(1 + w);
r = Lm/V; w = r*xi./(1-xi) + ypr - r*x0r; opm = w./(1 + w);
Da = min(opa - yi); Db = min(opb - yi); Dm = min(opm - yi);
while crit > 1e-6
    if (Da*Dm) < 0, Lb = Lm; Db = Dm; Lm = (La + Lb)/2;
    else, La = Lm; Da = Dm; Lm = (La + Lb)/2; end
    r = Lm/V; w = r*xi./(1-xi) + ypr - r*x0r; opm = w./(1 + w);
    Dm = min(opm - yi); crit = abs(La - Lb);
end
Lmin = Lm; fprintf('Minimum liquid flow rate = %g kmol/h\n', Lmin);

% (4) plot of the operating line when L = Lmin
rm = Lmin/V; w = rm*xi./(1-xi) + ypr - rm*x0r; z = w./(1 + w);
subplot(2,2,3),, plot(xi,yi,xi,z,'--')
xlabel('x'), ylabel('y'), title('Operating curve at L=L_{min}')
legend('Equil. curve','L=L_{min}','location','best')
% determine x at which the operating curve becomes tangent to the equilibrium
curve
f = @(x) (rm*x./(1-x) + ypr - rm*x0r)/(1+rm*x./(1-x) + ypr - rm*x0r) -...
    (P(1)*x.^4 + P(2)*x.^3 + P(3)*x.^2 + P(4)*x + P(5));
x0 = 0.03; xmin = fsolve(f,x0);
fprintf('The operating curve becomes tangent to the equilibrium curve at x = %g.
\n', xmin);

>> trayabs
Order of the fitting polynomial = 4
```

```
Minimum liquid flow rate = 91.2307 kmol/h
The operating curve becomes tangent to the equilibrium curve at x = 0.0591771.
```

7.3.2 MOMENTUM AND MASS TRANSFER IN THE ABSORPTION COLUMN

In an absorption column where simultaneous mass and momentum transfer take place, a falling liquid film contacts with a gas phase. The mass transfer rates are low and the velocity of the falling liquid is not affected by the mass transfer. Thus, we can assume that the liquid properties are constant. The momentum balance yields

$$\frac{d\tau_{yx}}{dy} + \rho g = 0$$

Solving the momentum balance equation, we have

$$v_x = \frac{\rho g \delta^2}{\mu} \left\{ \frac{y}{\delta} - \frac{1}{2} \left(\frac{y}{\delta} \right)^2 \right\}$$

where
  ρ is the liquid density
  μ is the liquid viscosity
  δ is the thickness of the liquid film
  The mass balance can be represented as

$$\frac{\partial N_{ax}}{\partial x} + \frac{\partial N_{ay}}{\partial y} = 0$$

where

$$N_{ax} = -D_A \frac{\partial C_a}{\partial x} + x_a (N_{ax} + N_{bx}), \quad N_{ay} = -D_A \frac{\partial C_a}{\partial y} + x_a (N_{ay} + N_{by})$$

where D_A is the diffusion coefficient. For short contact times, we have

$$N_{ax} \cong x_a (N_{ax} + N_{bx}) = C_a v_x, \quad N_{ay} \cong -D_A \frac{\partial C_a}{\partial y}$$

Thus,

$$\frac{\partial N_{ax}}{\partial x} + \frac{\partial N_{ay}}{\partial y} = \frac{\partial (C_a v_x)}{\partial x} - D_A \frac{\partial}{\partial y} \left(\frac{\partial C_a}{\partial y} \right) = 0$$

Substitution of the equation for v_x and rearrangement give[25]

$$2 \frac{\rho g \delta^2}{\mu D_A} \left\{ \frac{y}{\delta} - \frac{1}{2} \left(\frac{y}{\delta} \right)^2 \right\} \frac{\partial C_a}{\partial x} = \frac{\partial}{\partial y} \left(\frac{\partial C_a}{\partial y} \right)$$

The initial and boundary conditions are

$$x = 0 : C_a(0, y) = 0, \quad y = 0 : \frac{\partial C_a}{\partial y} = 0, \quad y = \delta : C_a = C_0$$

The differential equation can be solved by using the built-in function *pdepe*. The standard form to be used in the *pdepe* is given by

$$g\left(x, t, u, \frac{\partial u}{\partial x}\right)\frac{\partial u}{\partial t} = x^{-m}\frac{\partial}{\partial x}\left(x^m f\left(x, t, u, \frac{\partial u}{\partial x}\right)\right) + r\left(x, t, u, \frac{\partial u}{\partial x}\right)$$

By direct comparison between the differential equation and the *pdepe* standard form, we have

$$m = 0,\ g\left(x, t, u, \frac{\partial u}{\partial x}\right) = 2\frac{\rho g \delta^2}{\mu D_A}\left\{\frac{y}{\delta} - \frac{1}{2}\left(\frac{y}{\delta}\right)^2\right\},\ r\left(x, t, u, \frac{\partial u}{\partial x}\right) = 0,\ f\left(x, t, u, \frac{\partial u}{\partial x}\right) = \frac{\partial C_a}{\partial y}$$

The standard forms of the initial and boundary conditions provided by *pdepe* are given by

$$t = t_0 : u(x, t_0) = u_0(x)$$

$$x = a,\ x = b : p(x, t, u) + q(x, t)f\left(x, t, u, \frac{\partial u}{\partial x}\right) = 0$$

By comparison between the initial and boundary conditions and the *pdepe* standard forms, we have

$$C_a(0, y) = 0 \rightarrow u_0(x) = 0$$

$$\frac{\partial C_a}{\partial y} = 0 \rightarrow p = 0,\ q = 1$$

$$C_a = C_0 \rightarrow p = ur - C_0,\ q = 0$$

In order to use the built-in function *pdepe*, the differential equation and initial and boundary conditions should be defined as functions.

Example 7.14 Concentration Profile[25]

Let's produce profiles of C_a as functions of L and δ when $C_0 = 0.1\ kgmol/m^3$, $\delta = 0.01\ m$, $\rho g \delta^2/D_A = 0.5$, and $\mu = 2.1 \times 10^{-5}\ Pa\cdot sec$.

Solution

The subfunction *abeqn* defines the differential equation to be solved, the subfunction *abitcon* defines the initial condition, and the subfunction *abbncon* defines the boundary conditions. The script *abscolm* defines the ranges of x and t and executes the built-in function *pdepe*. Execution of the script *abscolm* yields the desired plots, shown in Figure 7.23.

```
% abscolm.m: absorption column film mass transfer
clear all;
x = linspace(0,0.01,100); % 100 values from 0 to 0.01
t = linspace(0,1,100); % 100 values from 0 to 1
m = 0;
res = pdepe(m,@abeqn,@abitcon,@abbncon,x,t);
u = res(:,:,1);
subplot(1,2,1), surf(x,t,u), colormap(gray), xlabel('\delta(m)'), ylabel('L
(m)'), zlabel('C_a(kgmol/m^3)')
shading interp % surface plot
subplot(1,2,2)
% plot Ca as a function of delta(m)
```

(a)

(b)

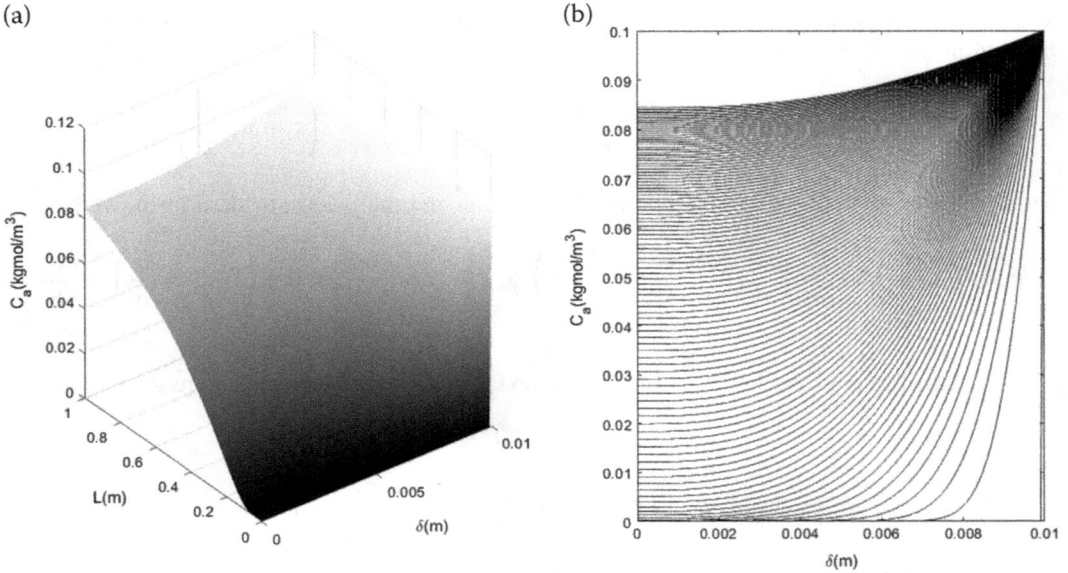

FIGURE 7.23   Profiles of C_a as a function of (a) L and (b) δ.

```
for k = 1:length(t), plot(x,u(k,:),'k'), xlabel('\delta(m)'), ylabel('C_a
(kgmol/m^3)'), hold on, end
hold off

function [c,f,s] = abeqn(x,t,u,DuDx)
delt = 0.01; mu = 2.1e-5; c = 2*(1/2)*(x/delt - (1/2)*(x/delt)^2)/mu; f = DuDx; s
= 0;
end
function u0 = abitcon(x)
u0 = 0;
end

function [pl,ql,pr,qr] = abbncon(xl,ul,xr,ur,t)
C0 = 0.1;
pl = 0; ql = 1; % y = 0
pr = ur - C0; qr = 0; % y = delta
end
```

7.3.3 Packed-Bed Absorber

The packed-bed absorber is typically used to reduce gaseous pollutants from process streams. The operation of the absorber is affected by contaminant solubility in the absorbing liquid, the liquid-to-gas ratio, pressure drop, stream flow rates, and configuration details including packing materials, liquid distributors, and entrainment separators. The solubility, which significantly affects the amount of a pollutant or solute to be absorbed, is a function of temperature and pressure. The equilibrium solubility of a gas-liquid system can be expressed by Henry's law:

$$y = \mathcal{H}x$$

where

y is the mole fraction of the gas in equilibrium with the liquid

$\mathcal{H}$ is Henry's constant

x is the mole fraction of the solute in equilibrium

In an absorber, the height of a transfer unit, H_t, is a function of the packing type, liquid and gas flow rates, liquid properties, pollutant solubility and concentration, and operating temperature. The packing depth required in the absorber, H_{pd}, is determined from H_t and the theoretical number of overall transfer units, N_t, as follows:

$$H_{pd} = N_t H_t$$

N_t can be estimated by

$$N_t = \frac{S_f}{S_f - 1} \ln \left\{ \left(1 - \frac{1}{S_f} \right) \left(\frac{y_i - mx_i}{y_o - mx_i} \right) + \frac{1}{S_f} \right\}$$

where m is the slope of the equilibrium line and S_f is the absorption factor defined by

$$S_f = \frac{L_m}{mG_m}$$

The slope m can be obtained from the following equation:

$$m = \frac{y_o{}^* - y_i{}^*}{x_o - x_i}$$

where $y_o{}^*$ and $y_i{}^*$ are the mole fractions of the pollutant in the vapor phase in equilibrium with those existing and entering the absorber in the liquid, x_o and x_i, respectively.

7.3.3.1 Packed-Bed Column Diameter

The size of a packed column is significantly affected by the gas velocity and the packing factor, $F_p\,(m^2/m^3)$, of the packing material used. An increase in gas velocity will eventually cause the liquid to completely fill the void spaces in the packing. This condition is referred to as flooding, and the gas velocity at which it occurs is the flooding velocity. A typical operating range for the gas velocity is 50%–75% of the flooding velocity. The packing size is inversely proportional to the packing factor: the column diameter decreases as the size of the packing material increases for the same gas flow rate. Table 7.2 shows typical values of the packing factor for various packing materials.[26]

The relation between the superficial gas flow rate entering the absorber, $G_f\,(kg/(sec\cdot m^2))$, and the L/G ratio can be expressed in the form

$$Y = \frac{G_f{}^2 \psi F_p \mu_L{}^{0.2}}{\rho_G \rho_L g} = 10^z, \; z = -1.668 - 1.085 \log X - 0.297 (\log X)^2 , \; X = \left(\frac{L}{G} \right) \sqrt{\frac{\rho_G}{\rho_L}}$$

where

ψ is the ratio of the density of the scrubbing liquid to water

$\mu_L\,(cP)$ is the solvent viscosity

ρ_G and $\rho_L\,(kg/m^3)$ are the densities of the gas and liquid, respectively

$$g = 9.82 \; m/sec^2$$

Solving this equation for G_f, we have

$$G_f = \sqrt{\frac{\rho_G \rho_L g Y}{\psi F_p \mu_L{}^{0.2}}}$$

TABLE 7.2
Values of the Packing Factor for Various Packing

Packing Material	Size in.	mm	Bulk Density (kg/m^3)	Surface Area $a(m^2/m^3)$	Packing Factor $F_p(m^{-1})$
Raschig rings ceramic	0.5	13	881	368	2100
	1.0	25	673	190	525
	1.5	38	689	128	310
	2.0	51	651	95	210
	3.0	76	561	69	120
Metal (density for carbon steel)	0.5	13	1201	417	980
	1.0	25	625	207	375
	1.5	38	785	141	270
	2.0	51	593	102	190
	3.0	76	400	72	105
Pall rings metal (density for carbon steel)	0.625	16	593	341	230
	1.0	25	481	210	160
	1.25	32	385	128	92
	2.0	51	353	102	66
	3.0	76	273	66	52
Plastics (density for polypropylene)	0.625	16	112	341	320
	1.0	25	88	207	170
	1.5	38	76	128	130
	2.0	51	68	102	82
	3.5	89	64	85	52
Intalox saddles ceramic	0.5	13	737	480	660
	1.0	25	673	253	300
	1.5	38	625	194	170
	2.0	51	609	108	130
	3.0	76	577		72

Source: Gavin Towler and Ray Sinnott, *Chemical Engineering Design*, Butterworth-Heinemann, Burlington, MA, 2008, p. 744.

The cross-sectional area of the column, $A(m^2)$, is given by

$$A = \frac{\pi D_t^2}{4} = \frac{G_m}{fG_f}$$

where f is the flooding factor. The column diameter $D_t(m)$ is given by

$$D_t = \sqrt{\frac{4A}{\pi}} = \sqrt{\frac{4G_m}{\pi fG_f}}$$

Example 7.15 Gas Absorber[27]

Pure water is used to remove 90% of the SO_2 from a gas stream in a gas absorber. The flow rate of the gas stream is 103 kg/min, and the stream contains 3 vol.% SO_2. The minimum liquid flow rate is found to be 2450 kg/min, and the operating liquid flow rate is 1.5 times the minimum value.

The operating temperature and pressure are 293 K and 101.32 kPa. The gas velocity should not be greater than 70% ($f = 0.7$) of the flooding velocity, and 2 $in.$ ceramic Intalox saddles are used as the packing material. Find the column diameter.

Data: $\rho_G = 1.17\ kg/m^3$, $\rho_L = 1000\ kg/m^3$, $F_p = 130\ (m^2/m^3)$, $\psi = 1$, $\mu_L = 0.8\ cP$, $L_m = 2450\ kg/min$, $G = 103\ kg/min$, $f = 0.7$.

Solution
The script *absdiam* defines the data and calculates the required diameter.

```
% absdiam.m: diameter of a SO2 absorber
% Data
rhog = 1.17; rhol = 1000; Fp = 130; phi = 1; mul = 0.8; Lm = 2450; G = 103; g = 9.82; f
= 0.7;
% Find diameter
L = 1.5*Lm; X = (L/G)*sqrt(rhog/rhol); z = -1.668 - 1.085*log(X) - 0.297*(log
(X))^2; Y = 10^z;
Gf = sqrt(rhog*rhol*g*Y/(phi*Fp*mul^0.2)); % superficial velocity
Gr = 0.7*Gf; A = (G/60)/Gr; Dt = sqrt(4*A/pi);
fprintf('Column diameter = %g m\n', Dt)

>> absdiam
Column diameter = 1.69744 m
```

7.3.3.2 Packed-Bed Column Height
The film mass transfer coefficients $k_G\ (kmol/(m^2\!\cdot\!sec\!\cdot\!atm))$ and $k_L\ (m/sec)$, and the effective wetted area of packing, a_w, can be obtained using the correlation proposed by Onda[28]:

$$k_G\left(\frac{RT}{a_p D_G}\right) = 5.23\left(\frac{G}{a_p \mu_G}\right)^{0.70} Sc_G^{1/3}(a_p d_p)^{-2}$$

$$k_L\left(\frac{\rho_L}{g\mu_L}\right)^{1/3} = 0.0051\left(\frac{L}{a_w \mu_L}\right)^{2/3} Sc_L^{-1/2}(a_p d_p)^{0.40}$$

$$a_w = a_p\left[1 - \exp\left\{-1.45\left(\frac{\sigma_c}{\sigma}\right)^{0.75}\left(\frac{L}{a_p \mu_L}\right)^{0.1}\left(\frac{L^2 a_p}{\rho_L^2 g}\right)^{-0.05}\left(\frac{L^2}{\rho_L \sigma a_p}\right)^{0.2}\right\}\right]$$

where
$a_p\ (m^2/m^3)$ is the total packing surface area per packed-bed volume
$D_G\ (m^2/sec)$ is the diffusivity in the gas phase
Sc_G is the gas-phase Schmidt number, given by $Sc_G = \mu_G/(\rho_G D_G)$
Sc_L is the liquid-phase Schmidt number, given by $Sc_L = \mu_L/(\rho_L D_L)$
$d_p\ (m)$ is the equivalent diameter of the packing, given by $d_p = 6(1 - \varepsilon)/a_p$
G and $L\ (kg/(m^2\!\cdot\!sec))$ are the superficial gas and liquid mass velocities, respectively
μ_G and $\mu_L\ (kg/(m\!\cdot\!sec))$ are the gas and liquid viscosities, respectively
$\rho_L\ (kg/m^3)$ is the liquid density
$\sigma\ (N/m)$ is the water surface tension

$$R = 0.08314\ m^3\!\cdot\!bar/(K\!\cdot\!kmol)$$

$$g = 9.81 \ m/sec^2$$

$\sigma_c = 61 \ dyne/cm$ for ceramic packing, $75 \ dyne/cm$ for steel packing, and $33 \ dyne/cm$ for plastic packing

The heights of the transfer units for the gas phase, H_G, and the liquid phase, $H_L \ (m)$, are given by

$$H_G = \frac{G_m}{k_G a_w P}, \ H_L = \frac{L_m}{k_L a_w C_T}$$

where

G_m and $L_m (kmol/(m^2{\cdot}sec))$ are the molar gas and liquid flow rates per cross-sectional area, respectively

$P \ (bar)$ is the column pressure

$C_T (kmol/m^3)$ is the total concentration

The overall height of transfer units (HTUs), H_t, is given by

$$H_t = H_G + \left(\frac{m G_m}{L_m} \right) H_L$$

Cornell et al. (1960) presented empirical equations for predicting the height of the gas and liquid film transfer units.[29,30] The heights of the transfer units for the gas film, H_G, and for the liquid film, H_L, are given by

$$H_G = \frac{0.011 \psi_h \sqrt{Sc_G} \left(\frac{D_c}{0.305} \right)^{1.11} \left(\frac{H_{pd}}{3.05} \right)^{0.33}}{\sqrt{L f_1 f_2 f_3}} \cong \frac{0.0253 \psi_h \sqrt{Sc_G} \left(\frac{H_{pd}}{3.05} \right)^{0.33}}{\sqrt{L f_1 f_2 f_3}}$$

$$H_L = 0.305 \phi_h \sqrt{Sc_L} \ K_3 \left(\frac{H_{pd}}{3.05} \right)^{0.15}$$

where

$D_c \ (m)$ is the column diameter

$H_{pd} \ (m)$ is the packing depth

ψ_h and ϕ_h are factors for H_G and H_L, respectively

$L (kg/(m^2{\cdot}sec))$ is the liquid flow rate per unit area column cross-sectional area

K_3 is the percentage flooding correction factor

f_1 is the liquid viscosity correction factor

f_2 is the liquid density correction factor

f_3 is the surface tension correction factor

For flooding less than 45%, $K_3 = 1$, and for a higher percentage of flooding F, K_3 is given by

$$K_3 = -0.\,014 \ F + 1.685$$

For Berl saddles packing (particle size = 1.5 in. (38 mm)), ψ_h and ϕ_h are given by

$$\psi_h = 0.834 + 5.29F - 0.12F^2 + 0.0009F^3, \ \phi_h = 0.034 L_f^{0.4}$$

where $L_f (kg/(m^2{\cdot}sec))$ is the liquid mass velocity. $f_i \ (i = 1,2,3)$ are given by

$$f_1 = \left(\frac{\mu_L}{\mu_w} \right)^{0.16}, f_2 = \left(\frac{\rho_w}{\rho_L} \right)^{1.25}, f_3 = \left(\frac{\sigma_w}{\sigma_L} \right)^{0.8}$$

where $\mu_w = 1.0\ cP$, $\rho_w = 1000\ kg/m^3$, and $\sigma_w = 72.8\ mN/m$. Again, the overall height of transfer units (HTUs), H_t, is given by

$$H_t = H_G + \left(\frac{mG_m}{L_m}\right)H_L$$

and

$$H_{pd} = N_t H_t = N_t\left\{H_G + \left(\frac{mG_m}{L_m}\right)H_L\right\}$$

The total height of the column is obtained from

$$Z = 14H_{pd} + 1.02D + 2.81\,(2 \le D \le 12ft,\ 4 \le Z \le 12ft)$$

The surface area, S, of the absorber is given by

$$S = \pi D\left(Z + \frac{D}{2}\right)$$

Example 7.16 Gas Absorption[31]

For the SO_2/H_2O system shown in Figure 7.24, the overall height of transfer units, H_t, is found to be 0.6 m. Determine the total height of packing, H_{pd}, required to achieve 90% reduction in the inlet concentration. The packing is 2 in Raschig rings ceramic, and the operating conditions are $P = 1\ atm$ and $T = 20\,°C$.

Data: $\mathcal{H} = 26$, $G_m = 206\ kmol/hr$, $L_m = 12{,}240\ kmol/hr$, $x_2 = 0.0$, $y_1 = 0.03$, $y_2 = 0.003$.

Solution

The number of theoretical transfer units, N_t, is given by

$$N_t = \frac{S_f}{S_f - 1}\ln\left\{\left(1 - \frac{1}{S_f}\right)\left(\frac{y_1 - mx_2}{y_2 - mx_2}\right) + \frac{1}{S_f}\right\},\ S_f = \frac{L_m}{\mathcal{H}G_m}$$

The total packing height is given by

$$H_{pd} = N_t H_t$$

The script *SO2abs* calculates the required packing height.

```
% SO2abs.m: absorption of SO2 by water
Ht = 0.6; H = 26; Gm = 206; Lm = 12240; x2 = 0.0; y1 = 0.03; y2 = 0.003; % data
Sf = Lm/(H*Gm); % absorption factor
Nt = Sf/(Sf-1)*log((1 - 1/Sf)*((y1 - H*x2)/(y2 - H*x2)) + 1/Sf); % number of
transfer units
Hpd = Nt*Ht; % total packing height
fprintf('Number of theoretical transfer units = %g\n', Nt);
fprintf('Total packing height = %g m\n', Hpd);

>> SO2abs
Number of theoretical transfer units = 3.20402
Total packing height = 1.92241 m
```

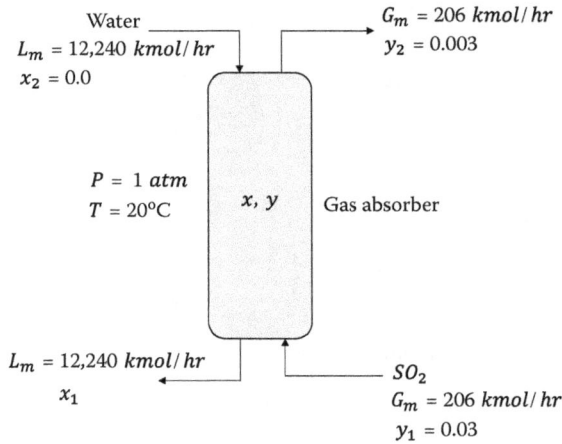

Water $\quad\quad$ $G_m = 206\ kmol/hr$
$L_m = 12{,}240\ kmol/hr$ $\quad\quad$ $y_2 = 0.003$
$x_2 = 0.0$

$P = 1\ atm$
$T = 20°C$ $\quad$ x, y $\quad$ Gas absorber

$L_m = 12{,}240\ kmol/hr$
x_1 $\quad\quad\quad\quad$ SO_2
$\quad\quad\quad\quad\quad\quad$ $G_m = 206\ kmol/hr$
$\quad\quad\quad\quad\quad\quad$ $y_1 = 0.03$

FIGURE 7.24 $\quad SO_2/H_2O$ system.

7.4 BINARY DISTILLATION

7.4.1 McCabe-Thiele Method[32]

Figure 7.25 shows a simple distillation column with a total condenser and a reboiler. The feed typically enters close to the middle of the column. Vapor flows from stage to stage up the column, while liquid flows from stage to stage down the column. The vapor from the top tray is condensed to liquid in the condenser and a portion of that liquid, L, is returned as reflux. The rest of that vapor is withdrawn as the overhead product stream, D. The reflux ratio is defined as

$$R = \frac{L}{D}$$

A portion of the liquid at the bottom of the column is withdrawn as a bottoms product, while the rest is vaporized in the reboiler and returned to the column.

The operating line for the rectifying section (top section of column, above the feed stage n_F) can be represented as

$$y = \left(\frac{R}{R+1}\right)x + \left(\frac{1}{R+1}\right)x_D$$

where

x and y are the mole fractions of the key (light key) component in the liquid and vapor, respectively

x_D is the mole fraction of the key component in the distillate

The operating line for the stripping section (bottom section of the column, below the feed stage) can be expressed as

$$y = \left(\frac{V_B + 1}{V_B}\right)x - \left(\frac{1}{V_B}\right)x_B$$

The buildup ratio, V_B, is the ratio of the flow rate of vapor leaving the reboiler, $\bar{V}$, to the bottoms product rate, B, and is given by $V_B = \bar{V}/B$. x_B is the mole fraction of the key component in the bottoms product. The condition of the feed stream fed into the feed stage can be characterized by

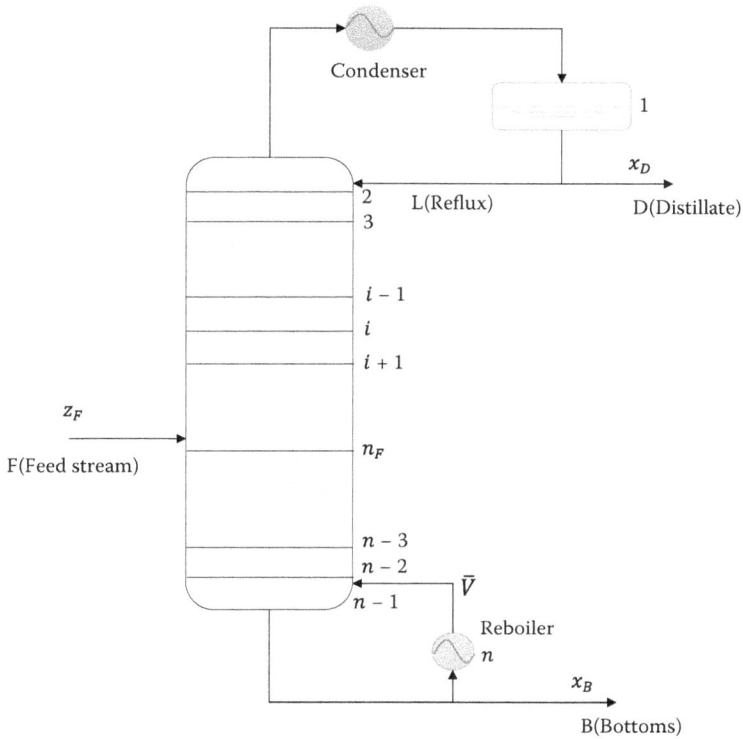

FIGURE 7.25 Schematic diagram of a distillation column. (From Henley, E.J. et al., *Separation Process Principles*, 3rd ed., John Wiley & Sons, Inc., Hoboken, NJ, 2011, pp. 285–290.)

the parameter q, which is defined as the ratio of the increase in molar reflux rate across the feed stage to the molar feed rate:

$$q = \frac{\bar{L} - L}{F}$$

Alternatively, q can be obtained from the mass balance around the feed stage as

$$q = 1 + \frac{\bar{V} - V}{F}$$

For a subcooled liquid feed, $q > 1$; for a bubble point liquid feed, $q = 1$; for a partially vaporized feed, $0 < q < 1$; for a dew point vapor feed, $q = 0$; and for a superheated vapor feed, $q < 0$. q-line equation is represented as

$$y = \left(\frac{q}{q-1}\right)x - \left(\frac{z_F}{q-1}\right)$$

From the rectifying operating line and the q-line equation, we have

$$x_q = \frac{(R+1)z_F + (q-1)x_D}{R+q}, \quad y_q = \frac{Rz_F + qx_D}{R+q}$$

If these relations are substituted into the stripping operating line to eliminate V_B, the stripping operating line can be rewritten as

$$y = \left(\frac{y_q - x_B}{x_q - x_B}\right)x - \left(\frac{y_q - x_q}{x_q - x_B}\right)x_B$$

The relative volatility, α, represents a numerical measure of separation. For a binary mixture, the relative volatility can be expressed in terms of equilibrium vapor and liquid mole fractions from the K-value:

$$\alpha = \frac{y/x}{(1 - y)/(1 - x)}$$

Solving this equation for y,

$$y = \frac{\alpha x}{1 + (\alpha - 1)x}$$

The function *bindistMT* determines the number of stages and the operating lines for a binary distillation using the McCabe-Thiele method. The basic syntax is
```
[feedn, totaln] = bindistMT(alpha,q,zf,xd,xb,R)
```

where alpha is the relative volatility, q is the feed condition parameter, zf is the mole fraction of the key component in the feed stream, xd is the mole fraction of the key component in the distillate, xb is the mole fraction of the key component in the bottoms product, and R is the reflux ratio. In the output vector, feedn is the location of the feed stage and totaln is the number of total stages.

```
function [feedn, totaln] = bindistMT(alpha,q,zf,xd,xb,R)
% Calculation of binary distillation using McCabe/Thiele method
% input
%  alpha: relative volatility
%  q: feed condition parameter
%  zf,xd,xb: mole fractions of feed, distillate and bottoms
%  R: reflux ratio
% output
%  feedn: feed stage
%  totaln: number of total stages
%  Initialization and calculation of equilibrium curve
y = 0:0.1:1; ye = y; xe = ye./(alpha + (1-alpha)*ye);
xq = ((R+1)*zf + (q-1)*xd)/(R + q); yq = (R*zf + q*xd)/(R + q);
plot(xe,ye,'r'); hold on, axis([0 1 0 1]); set(line([0 1],[0 1]),'Color',[0 0 0]);
set(line([xd   xq],[xd   yq]),'Color',[1   0   1]);   set(line([zf   xq],[zf
yq]),'Color',[1 0 1]);
set(line([xb xq],[xb yq]),'Color',[1 0 1]);
%Rectifying section
i = 1; xop(1) = xd; yop(1) = xd; y = xd;
while (xop(i) > xq)
     xop(i+1) = y./(alpha + (1-alpha)*y);
     yop(i+1)=R*xop(i+1)/(R+1) + xd/(R+1); % rectifying operating line
     y = yop(i+1);
     set(line([xop(i) xop(i+1)],[yop(i) yop(i)]),'Color',[0 0 1]);
  if (xop(i+1) > xq), set(line([xop(i+1) xop(i+1)],[yop(i) yop(i+1)]),'Color',
[0 0 1]); end
   i = i+1;
end
```

```
feedn = i-1;
% Stripping section
c1 = (yq - xb)/(xq - xb); c2 = (yq - xq)/(xq - xb); yop(i) = c1*xop(i) - c2*xb; y =
yop(i);
set(line([xop(i) xop(i)],[yop(i-1) yop(i)]),'Color',[0 0 1]);
while (xop(i)>xb)
    xop(i+1) = y/(alpha + (1-alpha)*y); yop(i+1) = c1*xop(i+1) - c2*xb; y = yop
(i+1);
    set(line([xop(i) xop(i+1)],[yop(i) yop(i)]),'Color',[0 0 1]);
   if (xop(i+1) > xb), set(line([xop(i+1) xop(i+1)],[yop(i) yop(i+1)]),'
Color',[0 0 1]); end
  i=i+1;
end
set(line([xop(i) xop(i)],[yop(i-1) yop(i)]),'Color',[0 0 1]); hold off,
xlabel('x'), ylabel('y'), totaln = i-1;
fprintf('Feed stage = %g\n', feedn); fprintf('Number of stages = %g\n',
totaln);
end
```

Example 7.17 McCabe-Thiele Operating Lines

A binary mixture is to be distilled in a distillation column to give a distillate of $x_D = 0.9$ and a bottoms composition of $x_B = 0.1$. The feed composition is $z_F = 0.5$ and the reflux ratio is $R = 1.5$. The feed is partially vaporized and $q = 0.8$. The equilibrium equation is given by $y = \alpha x/(1 + (1 - \alpha)x)$ with $\alpha = 2.45$. Determine the location of the feed stage and the number of total stages, and plot the equilibrium curve, q-line, and operating lines as a function of the liquid-phase mole fraction.

Solution

The following commands produce the desired outputs and the plot shown in Figure 7.26:
```
>> alpha = 2.45; q = 0.80; zf = 0.5; xd = 0.9; xb = 0.1; R = 1.5;
>> [feedn, totaln] = bindistMT(alpha,q,zf,xd,xb,R);
Feed stage = 5
Number of stages = 10
```

7.4.2 IDEAL BINARY DISTILLATION

In a staged column, a number of stages can be lumped to form an equivalent stage and yield a compartment system. This method helps generate a reduced-order model of separation processes without linearization. The distillation column shown in Figure 7.27(a) includes one feed stream, which is a saturated liquid mixture containing two components. Figure 7.27(b) and (c) show the feed stage and the ith stage for a binary distillation column, respectively. The following assumptions are introduced to construct the compartmental distillation model:

- The molar flow rates of the vapor and liquid are constant.
- The vapor holdup in each compartment is negligible.
- The liquid holdup in each compartment is constant.
- Perfect mixing is accomplished in each section.
- The heat losses are negligible.
- Equilibrium is achieved in each tray.

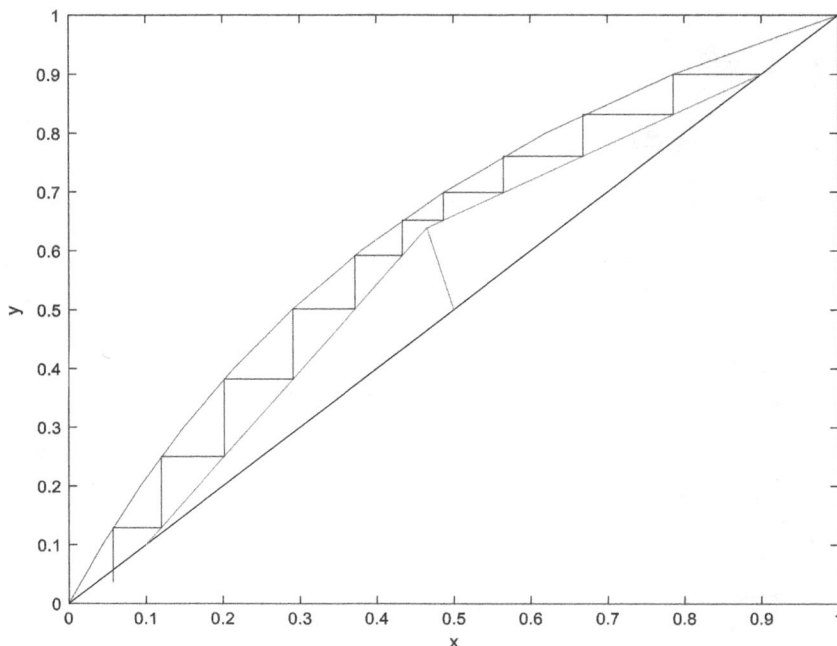

FIGURE 7.26 Determination of the number of stages by the McCabe-Thiele method.

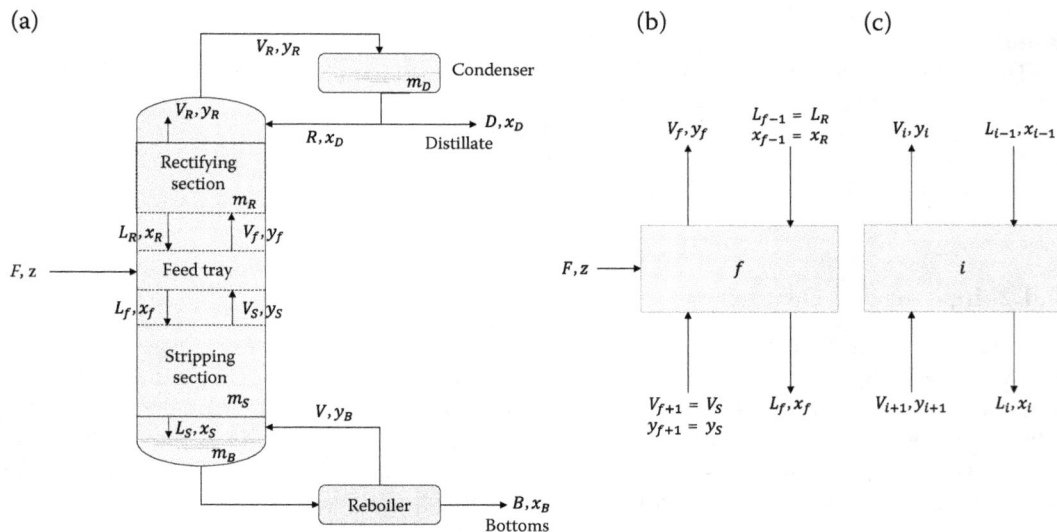

FIGURE 7.27 (a) Binary distillation column, (b) feed stage, and (c) ith stage.

- The relative volatility remains constant throughout the column.
- The dynamics of the condenser and the reboiler are negligible.

The vapor-liquid equilibrium curve is given by

$$y = \frac{\alpha x}{1 + (\alpha - 1)x}$$

The mass balance around the ith stage gives

$$\frac{d(M_i x_i)}{dt} = L_{i-1}x_{i-1} + V_{i+1}y_{i+1} - L_i x_i - V_i y_i$$

where M_i is the molar liquid holdup in the ith stage. The modeling equations can be summarized as follows:

Reboiler: $\frac{dx_B}{dt} = \frac{1}{m_B}(L_S x_S - B x_B - V y_B)$

Stripping section: $\frac{dx_S}{dt} = \frac{1}{m_S}\{L_S(x_f - x_S) + V(y_B - y_S)\}$

Feed plate: $\frac{dx_f}{dt} = \frac{1}{m_f}\{L_R(x_R - x_f) + F(z - x_f) + V(y_S - y_f)\}$

Rectifying section: $\frac{dx_R}{dt} = \frac{1}{m_R}\{L_R(x_D - x_R) + V(y_f - y_R)\}$

Condenser: $\frac{dx_D}{dt} = \frac{V}{m_D}(y_R - x_D)$

Equilibrium: $y_B = \frac{\alpha x_B}{1+(\alpha-1)x_B}$, $y_S = \frac{\alpha x_S}{1+(\alpha-1)x_S}$, $y_f = \frac{\alpha x_f}{1+(\alpha-1)x_f}$, $y_R = \frac{\alpha x_R}{1+(\alpha-1)x_R}$

From assumptions, we can see that $L_R = R$, $L_S = R + F$, and $B = L_S - V$.

The function *dyndist* defines the system of differential equations representing the dynamics of a binary distillation process. The basic syntax is

dx = dyndist(t,x,dpar,dels)

The input argument dpar is a structure variable consisting of parameter fields: dpar.alpha = α (relative volatility), dpar.n = n (number of total stages), dpar.nf = n_F (feed stage), dpar.F = F (feed flow rate), dpar.zf = z_F (feed composition), dpar.q = q (feed condition), dpar.R = R (reflux rate), dpar.Vs = V_s (reboiler vapor rate), dpar.md = M_D (condenser holdup), dpar.mb = M_B (reboiler holdup), and dpar.mt = M_T (liquid holdup in each stage). The input argument dels is a structure variable that defines step changes in operating variables.

```
function dx = dyndist(t,x,dpar,dels)
% System of differential equations for a binary distillation column
% Equilibrium relation: y=alpha*x/(1+(alpha-1)*x)
% Solver: [t x] = ode45(@dyndist,[t0 tf],x0,[],dpar,dels)
% Parameters
alpha = dpar.alpha; n = dpar.n; nf = dpar.nf; Fi = dpar.F;
zfi = dpar.zf; q = dpar.q; Ri = dpar.R; Vsi = dpar.Vs;
md = dpar.md; mb = dpar.mb; mt = dpar.mt;
delR = dels.delR; delRt = dels.delRt; delV = dels.delV; delVt = dels.delVt;
delz = dels.delz; delzt = dels.delzt; delF = dels.delF; delFt = dels.delFt;
% Changes in operating conditions
if t < delRt, R = Ri; else R = Ri + delR; end
if t < delVt, Vs = Vsi; else Vs = Vsi + delV; end
if t < delzt, zf = zfi; else zf = zfi + delz; end
if t < delFt, F = Fi; else F = Fi + delF; end
% Floe rates
Lr = R; Ls = R + F*q; B = Ls - Vs; D = F - B; Vr = Vs + F*(1-q);
% Initialization and phase equilibrium
dx = zeros(n,1); y = alpha*x./(1 + (alpha-1)*x);
% Condenser
dx(1) = Vr*(y(2)-x(1))/md;
```

```
% Rectifying section
for i = 2:nf-1, dx(i) = (Lr*x(i-1)+Vr*y(i+1)-Lr*x(i)-Vr*y(i))/mt; end
% Feed stage
dx(nf) = (Lr*x(nf-1)+Vs*y(nf+1)+F*zf-Ls*x(nf)-Vr*y(nf))/mt;
% Stripping section
for i = nf+1:n-1, dx(i) = (Ls*x(i-1)+Vs*y(i+1)-Ls*x(i)-Vs*y(i))/mt; end
% Reboiler
dx(n) = (Ls*x(n-1)-B*x(n)-Vs*y(n))/mb;
end
```

The composition profile may be sensitive to the initial conditions. It is common practice to use the steady-state values as the initial conditions. At steady state, the compositions do not change with time, and the derivative terms in the balance equations all become zero. The function *ssdist* calculates the steady-state liquid compositions for all stages. The syntax is
`f = ssdist(x,dpar)`

where dpar is the parameter structure, whose fields are dpar.alpha = α, dpar.n = n, dpar.nf = n_F, dpar.F = F, dpar.zf = z_F, dpar.q = q, dpar.R = R, and dpar.D = D.

```
function f = ssdist(x,dpar)
alpha = dpar.alpha; n = dpar.n; nf = dpar.nf; F = dpar.F;
zf = dpar.zf; q = dpar.q; R = dpar.R; D = dpar.D;
Lr = R; B = F - D; Ls = R + F*q; Vs = Ls - B; Vr = Vs + F*(1-q);
y = alpha*x./(1 + (alpha-1)*x);
f(1) = (Vr*y(2) - (D + R)*x(1)); % condenser
for i = 2:nf-1, f(i) = Lr*x(i-1) + Vr*y(i+1) - Lr*x(i) - Vr*y(i); end
f(nf) = Lr*x(nf-1) + Vs*y(nf+1) - Ls*x(nf) - Vr*y(nf) + F*zf;
for i = nf+1:n-1, f(i) = Ls*x(i-1) + Vs*y(i+1) - Ls*x(i) - Vs*y(i); end
f(n) = (Ls*x(n-1) - B*x(n) - Vs*y(n)); % reboiler
end
```

The system of nonlinear equations defined by the function *ssdist* can be solved using the built-in solver *fsolve*. The results obtained from the solution are used as initial values for the compositions.

Example 7.18 Simple Binary Distillation[33]

A saturated liquid mixture containing 1-propanol and ethanol is fed into the column shown in Figure 7.27(a). Using the given data, plot the composition profiles as a function of time ($0 \le t \le 12$ *min*).
Data: feed flow rate $F = 100$ *gmol/min*, feed composition $z = 0.5$, reflux rate $R = 128.01$ *gmol/min*, vapor rate $V = 178.01$ *gmol/min*, tray holdup $m_R = m_f = m_S = 10$ *gmol*, reflux drum holdup $m_D = 100$ *gmol*, bottom holdup $m_B = 100$ *gmol*, relative volatility $\alpha = 2.0$.

Solution
The subfunction *cmeqn* defines the set of differential equations. The script *compdist* defines the data structure cmdat and uses the built-in function *ode45* to solve the system of differential equations. The resultant composition profiles are shown in Figure 7.28.

```
% compdist.m
clear all;
% Define data structure (cmdat)
cmdat.F = 100; cmdat.z = 0.5; cmdat.R = 128.01; cmdat.V = 178.01; cmdat.mr = 10;
cmdat.mf = 10;
```

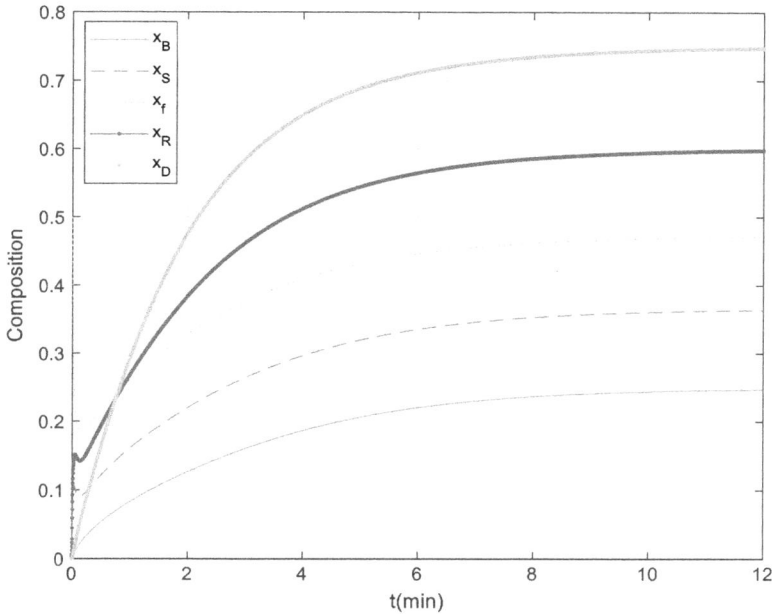

FIGURE 7.28 Composition profiles.

```
cmdat.ms = 10; cmdat.md = 100; cmdat.mb = 100; cmdat.alpa = 2;
% Solve ODEs: x(1)=xb, x(2)=xs, x(3)=xf, x(4)=xr, x(5)=xd
xb0 = 0; xs0 = 0; xf0 = cmdat.z; xr0 = 0; xd0 = 0; % Initial values
tint = [0 12]; x0 = [xb0 xs0 xf0 xr0 xd0];
[t x] = ode45(@cmeqn,tint,x0,[],cmdat); % solve differential eqn system
xb = x(:,1); xs = x(:,2); xf = x(:,3); xr = x(:,4); xd = x(:,5);
plot(t,xb,t,xs,'--',t,xf,':',t,xr,'.-',t,xd,'.'), grid, xlabel('t(min)')
ylabel('Composition'), legend('x_B','x_S','x_f','x_R','x_D','location','best')

function dx = cmeqn(t,x,cmdat)
% x(1)=xb, x(2)=xs, x(3)=xf, x(4)=xr, x(5)=xd
F = cmdat.F; z = cmdat.z; R = cmdat.R; V = cmdat.V; alpa = cmdat.alpa;
mr = cmdat.mr; mf = cmdat.mf; ms = cmdat.ms; md = cmdat.md; mb = cmdat.mb;
yb = alpa*x(1)/(1 + (alpa-1)*x(1)); ys = alpa*x(2)/(1 + (alpa-1)*x(2));
yf = alpa*x(3)/(1 + (alpa-1)*x(3)); yr = alpa*x(4)/(1 + (alpa-1)*x(4));
Lr = R; Ls = R + F; B = Ls - V;
dx = [(Ls*x(2) - B*x(1) - V*yb)/mb;
     (Ls*(x(3) - x(2)) + V*(yb - ys))/ms;
     (Lr*(x(4) - x(3)) + F*(z - x(3)) + V*(ys - yf))/mf;
     (Lr*(x(5) - x(4)) + V*(yf - yr))/mr;
     V*(yr - x(5))/md];
end
```

Example 7.19 Dynamics of a Binary Distillation Column

A 30-stage column with the overhead condenser as stage 1, the feed stage as stage 15, and the reboiler as stage 30 is used to distill a binary mixture. The relative volatility, α, is 1.5. The feed rate is $F = 1$ *mol/min*, the feed composition is $z_F = 0.5$, and the feed condition is $q = 1$ (bubble point liquid). The reflux flow rate is $R = 2.7$ *mol/min* and the vapor rate leaving the reboiler is

3.2 *mol/min*. The holdups in the condenser and the reboiler are both 5 *mol*, and the holdup in each stage is maintained constant at 0.5 *mol*.

At $t = 10$ *min*, there is a 1% step change in the reflux flow rate. Plot the liquid compositions of the distillate and the bottoms for $0 \leq t \leq 400$. Also plot the liquid composition profile along the stages at $t = 400$ *min*.

Solution

The script *compdyndist* sets the field values of the structures dpar and dels for the given operating conditions. Then the function *ssdist* is called to calculate the steady-state compositions, which are used as initial values. Finally, the function *dyndist* is called and the built-in solver *ode45* is employed to solve the system of differential equations.

```
% compdyndist.m
dpar.alpha = 1.5; dpar.n = 30; dpar.nf = 15; dpar.F = 1; dpar.zf = 0.5; dpar.q = 1;
dpar.R = 2.7;
dpar.Vs = 3.2; dpar.D = 0.5; dpar.md = 5; dpar.mb = 5; dpar.mt = 0.5; dels.delR =
0.01*dpar.R;
dels.delRt = 10; dels.delV = 0; dels.delVt = 0; dels.delz = 0; dels.delzt = 0;
dels.delF = 0;
dels.delFt = 0; t0 = 0; tf = 400; x0 = 0.5*ones(1,dpar.n); nv = 1:dpar.n;
x0 = fsolve(@ssdist,x0,[],dpar); % steady-state to be used as initial
conditions
for i = 1:length(x0), if x0(i) <= 0, x0(i) = -x0(i); end; end
[t x] = ode45(@dyndist,[t0 tf],x0,[],dpar,dels);
subplot(1,2,1), plot(t,x(:,1),t,x(:,end),'--'), xlabel('t(min)'), ylabel
('x'),legend('x_D','x_B')
subplot(1,2,2), plot(nv,x(end,:)), xlabel('n'), ylabel('x_i'), axis tight
```

In the script, "dels.delR = 0.01*dpar.R;" denotes a 1% step change in the reflux flow at "dels.delRt = 10;" (that is, $t = 10$ *min*). The script *compdyndist* generates the composition profiles shown in Figure 7.29.

7.5 MULTI-COMPONENT DISTILLATION: SHORTCUT CALCULATION

7.5.1 FENSKE-UNDERWOOD-GILLILAND METHOD

7.5.1.1 Fenske Equation: The Minimum Number of Stages

The minimum number of equilibrium stages, N_{min}, can be calcuthe set of stage temperatureslated by using the Fenske equation

$$N_{min} = \frac{\log\left[\left(\frac{x_{LK}}{x_{HK}}\right)_D \left(\frac{x_{HK}}{x_{LK}}\right)_B\right]}{\log[\alpha_{(LK/HK)_{avg}}]}$$

where

N_{min} is the minimum number of equilibrium stages, including the reboiler and the condenser
x_{LK} is the mole fraction of the light key component
x_{HK} is the mole fraction of the heavy key component
$\alpha_j = k_j/k_{HK}$ is the relative volatility of the component j
$\alpha_{(LK/HK)_{avg}}$ is the geometric average volatility of the light key component, given by $\alpha_{(LK/HK)_{avg}} = \sqrt{\alpha_{TD}\alpha_{TB}}$

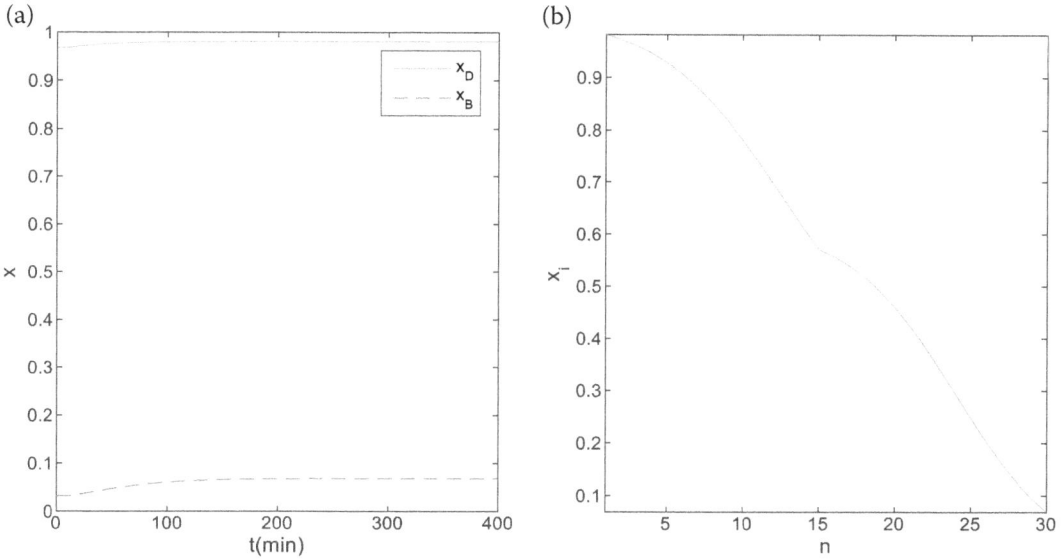

FIGURE 7.29 (a) Top and bottom and (b) stagewise composition profiles.

In the calculation of $\alpha_{(LK/HK)_{avg}}$, the temperature at the top of the column, T_D(dew point temperature of the distillate) and the temperature at the bottom of the column, T_B(bubble point temperature of the bottoms), are determined, followed by the calculation of α_{TD} at T_D and α_{TB} at T_B.

If we use the component molar flow rates d and b, the minimum number of stages is given by

$$N_{min} = \frac{\log\left[\left(\frac{d_{LK}}{d_{HK}}\right)\left(\frac{b_{HK}}{b_{LK}}\right)\right]}{\log\sqrt{(\alpha_{i,j})_D(\alpha_{i,j})_B}}$$

7.5.1.2 Underwood Equation: The Minimum Reflux

The Underwood equation is used to calculate the minimum reflux. First, the following equation is solved for the parameter θ:

$$1 - q = \sum_{j=1}^{n_c} \frac{\alpha_{jF}x_{jF}}{(\alpha_{jF} - \theta)}$$

where
   q is the feed condition parameter
   x_{jF} is the mole fraction of component j in the feed stream
   α_{jF} is the relative volatility of component j in the feed
   n_c is the total number of components
   The minimum reflux ratio, R_{min}, is calculated from the equation

$$R_{min} + 1 = \sum_{j=1}^{n_c} \frac{\alpha_{jD}x_{jD}}{(\alpha_{iD} - \theta)}$$

where
   α_{jD} is the relative volatility of component j in the distillate
   x_{jD} is the mole fraction of component j in the distillate
   $\alpha_{j,HK}$ may be used instead of α_{jF} and α_{jD}.

7.5.1.3 Gilliland's Correlation[34]

The number of theoretical equilibrium stages required for a given separation at a given reflux ratio is often determined by empirical correlations in terms of the abscissa, X, and the ordinate, Y:

$$X = \frac{R - R_{min}}{R + 1}, \; Y = \frac{N - N_{min}}{N + 1}$$

The Gilliland's correlation is an experimental curve that gives the relation between X and Y. Many equations have been proposed to describe the Gilliland curve for multi-component distillation. Some of these equations are as follows:

Hengstebeck (1961) correlation[35]:

$$\log Y = -1.3640187 - 3.0920489Z - 3.407344729Z^2 - 1.74673876Z^3 - 0.33268897Z^4$$

$$Z = \log X$$

Liddle (1968) correlation[36]:

$$0.0 \le X \le 0.01 : Y = 1.0 - 18.5715X$$

$$0.01 \le X \le 0.90 : Y = 0.545827 - 0.591422X + \frac{0.002743}{X}$$

$$0.90 \le X \le 1.0 : Y = 0.16595 - 0.16595X$$

Van Winkle and Todd (1971) correlation[37]:

$$0.0078 < X < 0.125 : Y = 0.5039 - 0.5968X - 0.0908(\log X)$$

$$0.125 < X \text{ to } X = 1.0 : Y = 0.6257 - 0.9868X + 0.516X^2 - 0.1738X^3$$

Molokanov (1972) correlation[38]:

$$Y = 1 - \exp\left[\left(\frac{1 + 54.4X}{11 + 117.2X}\right)\left(\frac{X - 1}{\sqrt{X}}\right)\right](0 \le X, Y \le 1.0)$$

Hohmann and Lockhart (1972) correlation[39]:

$$X = 0, Y = 0.65; \; X = 1.0, Y = 0.067; Y = \frac{0.65 - 0.5X}{1 + 1.25X}$$

Eduljee (1975) correlation[40]:

$$X = 0, Y = 1.0; \; X = 1.0, Y = 0; Y = 0.75(1 - X^{0.5668})$$

Chang (1985) correlation[41]:

$$X = 0, Y = 1.0; \; X = 1.0, Y = 0; Y = 1 - \exp\left(1.49 + 0.315X - \frac{1.805}{X^{0.1}}\right)$$

Harg (1985) correlation[42]:

$$X = 0, Y = 1.0; \; X = 1.0, Y = 0; Y = 1 - X^{1/3}$$

McCormick (1988) correlation[43]:

$$X = 0, \; Y = 1.0; \; X = 1.0, \; Y = 0; \; Y = 1 - X^B, \; B = 0.105\,(logX) + 0.44$$

McCormick's correlation gives a good agreement in the normal operating range.

The ratio of the number of stages above the feed stage to the number below the feed stage can be obtained using the Kirkbride equation[44]:

$$\log\left(\frac{m}{p}\right) = 0.206\log\left\{\left(\frac{B}{D}\right)\left(\frac{x_{HK}}{x_{LK}}\right)_F\left[\frac{(x_{LK})_B}{(x_{HK})_D}\right]^2\right\}$$

where

m is the number of theoretical stages above the feed stage, including the condenser
p is the number of theoretical stages below the feed stage, including the reboiler
D and B are the molar flow rates of the distillate and bottoms, respectively

7.5.1.4 Feed Stage Location

The location of the optimal feed stage can be estimated by using the Kirkbride equation:

$$\frac{N_R}{N_S} = \left[\left(\frac{z_{F,HK}}{z_{F,LK}}\right)\left(\frac{x_{B,LK}}{x_{D,HK}}\right)^2\left(\frac{B}{D}\right)\right]^{0.206}$$

where

the subscript R denotes the rectifying section
S denotes the stripping section

7.5.1.5 Determination of the Number of Stages by the Smoker Equation

Combination of the vapor-liquid equilibrium relationship

$$y = \frac{\alpha x}{1 + (\alpha - 1)x}$$

and the operating line

$$y = mx + b$$

gives

$$m(\alpha - 1)x^2 + [m + (\alpha - 1)b - \alpha]x + b = 0$$

Let k be the real root between 0 and 1. Then k satisfies the equation

$$m(\alpha - 1)k^2 + [m + (\alpha - 1)b - \alpha]k + b = 0$$

k is the value of the x-ordinate at the point where the extended operating lines intersect the vapor-liquid equilibrium curve. The number of stages required, N, is given by

$$N = \frac{\log\left[\frac{x_0^*(1 - \beta x_n^*)}{x_n^*(1 - \beta x_0^*)}\right]}{\log\left(\frac{\alpha}{mc^2}\right)}$$

where $\beta = mc(\alpha - 1)/(\alpha - mc^2)$.

In the rectifying section, m and b can be expressed in terms of the reflux ratio R:

$$m = \frac{R}{R + 1}, \, b = \frac{x_D}{R + 1}$$

m is the slope of the operating line between x_n^* and x_0^*, which are given by

$$x_0 = x_D, \, x_0^* = x_D - k, \, x_n = z_F, \, x_n^* = z_F - k$$

where x_D and z_F are the mole fractions in the distillate and the feed stream, respectively.

In the stripping section, m and b can be represented as

$$m = \frac{Rz_F + x_D - (R + 1)x_B}{(R + 1)(z_F - x_B)}, \, b = \frac{(z_F - x_D)x_B}{(R + 1)(z_F - x_B)}, \, x_0^* = z_F - k, \, x_n^* = x_n - k$$

If the feed stream is not at its bubble point, z_F is replaced by the value of x at the intersection of operating lines given by[45]

$$z_F^* = \frac{b + \frac{z_F}{q - 1}}{\frac{q}{q - 1} - m}$$

For distillation at total reflux, the number of stages required is given by

$$N = \frac{\log\left[\frac{x_0(1 - x_n)}{x_n(1 - x_0)}\right]}{\log \alpha}$$

Example 7.20 Simple Distillation Calculation[46]

A distillation column is to be used to separate i-butane from a mixture of lighter compounds consisting of ethane, propane, i-butane, and n-butane. The feed stream enters the column as liquid at its bubble point, and the column pressure is 7 bar. It is required that 95% of the i-butane in the feed be recovered in the bottoms, and the bottoms stream must contain no more than 0.1% propane.

(1) Calculate the minimum number of stages required to achieve the desired separation at total reflux using the Fenske equation.

(2) Determine the minimum reflux ratio required to achieve the desired separation with an infinite number of stages using the Underwood equation.

(3) Estimate the number of theoretical stages required if the actual reflux ratio is given by $R = 1.5 \, R_m$ using the Gilliland correlation. Assume that the relation between X and Y is represented by the Eduljee correlation. Determine the location of the optimal feed stage using the Kirkbride equation. Table 7.3 shows the feed compositions and the coefficients of the Antoine equation.

Solution

(1) The first step of the solution involves calculating the distillate composition and bottoms composition. The vapor pressure of each component, P_i, is determined by the Antoine equation. The dew point can be obtained using the relationship $\sum x_i = 1 = \sum y_i/k_i$, and the bubble point can be obtained using the relationship $\sum y_i = 1 = \sum k_i x_i$, $k_i = P_i/P$.

(2) To calculate the minimum reflux ratio using the Underwood equation, determine the temperature of the feed stream and calculate k_{jF} at this temperature. In the evaluation of the equation

$$1 - q = \sum_{j=1}^{n_c} \frac{\alpha_{jF} x_{jF}}{(\alpha_{jF} - \theta)}$$

TABLE 7.3

Feed Composition and Antoine Equation Coefficients

| | Component | Feed Composition (Mole Fraction) | Antoine Equation Coefficients | | |
			A	B	C
1	Ethane	0.15	3.93835	659.739	−16.719
2	Propane (light key)	0.18	4.53678	1149.36	24.906
3	*i*-Butane (heavy key)	0.18	4.3281	1132.108	0.918
4	*n*-Butane	0.49	4.35576	1175.581	−2.071

set $\alpha_{jF} = k_{jF}/k_{HK,F}$ and $q = 1$ because the feed is saturated liquid.
(3) We use

$$R = 1.5R_{min}, \; X = \frac{R - R_{min}}{R + 1}, \; Y = \frac{N - N_{min}}{N + 1}, \; Y = 0.75(1 - X^{0.5668})$$

The value of $h = m/p$ can be obtained using the Kirkbride equation. Then we
have $N = m + p = (1 + h)p$, $p = N/(1 + h)$, $m = ph$

The script *fugdist* performs the calculation procedure described above.

```
% fugdist.m
% 1: ethane, 2: propane(LK), 3: i-butane(HK), 4: n-butane
xF = [0.15 0.18 0.18 0.49]; % feed composition (mole fraction)
A = [3.93835 4.53678 4.3281 4.35576]; B = [659.739 1149.36 1132.108 1175.581];
C = [-16.719 24.906 0.918 -2.071]; P = 7;
% Compositions of D and B (D: distillate, B: bottom)
Br(1) = 0; Br(3) = 0.95*xF(3); Br(4) = xF(4); Br(2) = 0.001*(Br(3) + Br(4))/0.999;
Dr(1) = xF(1); Dr(2) = xF(2) - Br(2); Dr(3) = 0.05*xF(3); Dr(4) = 0;
xD = Dr/sum(Dr); xB = Br/sum(Br);
% Dew point(D) and bubble point(B)
Td0 = 270; Tb0 = 300;
fD = @(Td) sum(P*xD./(10.^(A - B./(Td + C)))) - 1;
fB = @(Tb) sum(xB.*10.^(A - B./(Tb + C))/P) - 1;
Td = fzero(fD, Td0); Tb = fzero(fB, Tb0);
% Fenske equation
kD = (10.^(A - B./(Td + C)))/P; kB = (10.^(A - B./(Tb + C)))/P;
num = log((xD(2)/xD(3))*(xB(3)/xB(2))); den = log(sqrt((kD(2)/kD(3)*(kB(2)/
kB(3))))) ;
Nm = num/den;
% Temperature of feed stream
Tf0 = 300; fF = @(Tf) sum(xF.*10.^(A - B./(Tf + C))/P) - 1;
Tf = fzero(fF, Tf0); kF = (10.^(A - B./(Tf + C)))/P;
% Underwood equation
th0 = 1.5;
fTh = @(th) sum(xF.*10.^(A - B./(Tf + C))/P./(10.^(A - B./(Tf + C))/P - th*10^
(A(3) - B(3)/(Tf+C(3)))/P));
theta = fzero(fTh, th0); alpD = kD/kD(3); Rm = sum(alpD.*xD./(alpD - theta)) - 1;
% Gilliland equation
R = 1.5*Rm; X = (1.5 - 1)*Rm/(R + 1); Y = 0.75*(1 - X^0.5658); N = (Y+Nm)/(1-Y);
% Kirkbride equation
c1 = sum(Br)/sum(Dr); c2 = xF(3)/xF(2); c3 = (xB(2)/xD(3))^2;
h = (c1*c2*c3)^0.206; p = N/(1+h); m = p*h;
```

```
% Print results
fprintf('Bubble point(B) = %8.4f, Dew point(D) = %8.4f', Tb, Td);
fprintf('\nFeed temp.(F) = %8.4f', Tf);
fprintf('\ntheta = %8.4f, min. number of stages = %7.4f', theta, Nm);
fprintf('\nMin. reflux ratio = %7.4f, actual number of stages = %7.4f',N, Rm);
fprintf('\nStages above the feed stage = %7.4f', m);
fprintf('\nStages below the feed stage = %7.4f\n', p);
```

The script produces the following outputs:

```
>> fugdist
Bubble point(B) = 333.1795, Dew point(D) = 274.0267
Feed temp.(F) = 285.4040
theta = 1.6731, min. number of stages = 8.5438
Min. reflux ratio = 17.3676, actual number of stages = 0.6452
Stages above the feed stage = 3.9784
Stages below the feed stage = 13.3892
```

Example 7.21 Minimum Reflux Ratio[47]

The debutanizer shown in Figure 7.30 is to be used to treat a mixture consisting of eight components listed in Table 7.4. The minimum reflux ratio R_{min} (= L_{min}/D) can be determined by solving the following Underwood equations:

$$\sum_i \frac{\alpha_{i,3}z_{Fi}}{\alpha_{i,3} - \theta_1} = 1 - q, \ \sum_i \frac{\alpha_{i,3}z_{Fi}}{\alpha_{i,3} - \theta_2} = 1 - q, \ \sum_i \frac{\alpha_{i,3}d_i}{\alpha_{i,3} - \theta_1} = D(1 + R_{min})$$

$$\sum_i \frac{\alpha_{i,3}d_i}{\alpha_{i,3} - \theta_2} = D(1 + R_{min}), \ \sum_i d_i = D$$

where
 z_{Fi} is the mole fraction of each species in the feed
 $\alpha_{i,3}$ is the relative volatility between components i and 3
 q is the feed quality
 d_i is the molar flow rate of each component in the distillate
 D is the molar flow rate of the distillate withdrawn from the debutanizer
 L_{min} is the molar flow rate of the liquid returned to the debutanizer at the minimum reflux
 The quantities θ_1 and θ_2 have to satisfy $\alpha_{4,3} < \theta_2 < \alpha_{3,3} < \theta_1 < \alpha_{2,3}$. Calculate the minimum reflux ratio using the data shown in Table 7.4. Let $q = 0.87$.

Solution

The equations can be rearranged as follows:

$$f_1 : \sum_i d_i - D = 0, f_2 : \sum_i \frac{\alpha_{i,3}z_{Fi}}{\alpha_{i,3} - \theta_1} - (1 - q) = 0, f_3 : \sum_i \frac{\alpha_{i,3}z_{Fi}}{\alpha_{i,3} - \theta_2} - (1 - q) = 0$$

$$f_4 : \sum_i \frac{\alpha_{i,3}d_i}{\alpha_{i,3} - \theta_1} - D(1 + R_{min}) = 0, f_5 : \sum_i \frac{\alpha_{i,3}d_i}{\alpha_{i,3} - \theta_2} - D(1 + R_{min}) = 0$$

The vector of the unknown variables is defined as $x = [x(1)x(2)x(3)x(4)x(5)] = [\theta_1\theta_2d_4DR_{min}]$. The subfunction *debutanizer* defines the system of equations. As initial conditions, we set = $[x(1)x(2)x(3)x(4)x(5)] = [1.50.835001.5]$. The script *rmin* uses the built-in solver *fsolve* to get the desired results.

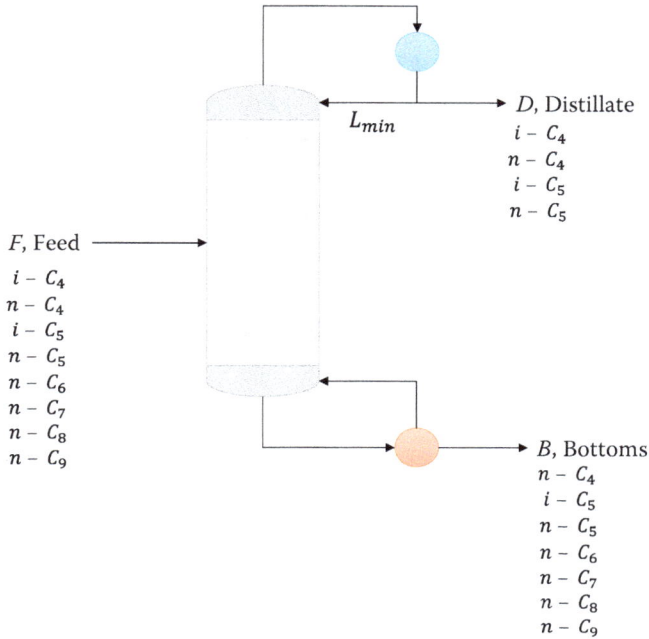

FIGURE 7.30   A debutanizer.

TABLE 7.4

Data for a Debutanizer

i	Component	z_{Fi}	$\alpha_{i,3}$	$d_i\,(lbmol/hr)$
1	$i - C_4$	0.0137	2.43	12
2	$n - C_4$	0.5113	1.93	442
3	$i - C_5$	0.0411	1.00	13
4	$n - C_5$	0.0171	0.765	d_4
5	$n - C_6$	0.0262	0.362	0
6	$n - C_7$	0.0446	0.164	0
7	$n - C_8$	0.3106	0.0720	0
8	$n - C_9$	0.0354	0.0362	0

```
% rmin.m
zf = [0.0137 0.5113 0.0411 0.0171 0.0262 0.0446 0.3106 0.0354];
relvol = [2.43 1.93 1.00 0.765 0.362 0.164 0.0720 0.0362];
d = [12 442 13 0 0 0 0 0]; q = 0.87; x0 = [1.5 0.8 3 500 1.5];
x = fsolve(@debutanizer, x0, [], zf, q, relvol, d)
function f = debutanizer(x, zf, q, relvol, d)
% zf: feed composition, q: feed quality, relvol: relative volatility, d: com-
ponent rate in D
% x(1): theta1, x(2): theta2, x(3): d(4), x(4): D, x(5): Rmin
d(4) = x(3);
```

```
f = [sum(d) - x(4);
    sum(relvol.*zf./(relvol - x(1))) - (1-q);
    sum(relvol.*zf./(relvol - x(2))) - (1-q);
    sum(relvol.*d./(relvol - x(1))) - x(4)*(1 + x(5));
    sum(relvol.*d./(relvol - x(2))) - x(4)*(1 + x(5))];
end

>> rmin
x =
  1.0449   0.7801   2.5724 469.5724   0.4655
```

We can see that $\theta_1 = 1.0449$, $\theta_2 = 0.7801$, $d_4 = 2.5724$ *lbmol/hr*, $D = 469.5724$ *lbmol/hr*, and $R_{min} = 0.4655$. The flow rate of liquid at the minimum reflux, L_{min}, is given by $L_{min} = R_{min}D = 218.586$ *lbmol/hr*.

Example 7.22 Shortcut Distillation Calculation[48]

A mixture of 33% n-hexane, 37% n-heptane and 30% n-octane is to be separated in a distillation column. The feed stream is 60% vapor at 105 °C. The distillate should contain 0.01 mole fraction n-heptane and the bottom product should contain 0.01 mole fraction n-hexane. The feed molar flow rate is 100 *mol/hr* and the operating pressure is 1.2 *atm*. Table 7.5 shows the parameters of the Antoine equation given by

$$\log P = A - \frac{B}{T + C} \quad (P : mmHg, \ T : °C)$$

(1) Determine the minimum number of trays at infinite reflux.
(2) Find the number of ideal trays required for separation if the reflux ratio is $1.5R_{min}$.
(3) Determine the optimum feed tray.

Solution
Let LK = n-hexane and HK = n-heptane. The script *mshortcut* produces the desired results.

```
% mshortcut.m: multicomponent distillation by shortcut method
% 1: n-hexane(C6H14), 2: n-heptane(C7H16), 3: n-octane(C8H18)
% LK: n-hexane(1), HK: n-heptane(2), HNK: n-octane(3)
clear all;
% Data
xf = [0.33, 0.37, 0.3]; xd = [0.99, 0.01, 0]; xb(1) = 0.01; % mole fraction
F = 100; Tf = 105; P = 1.2; % F: mol/h, T: deg.C, P: atm
q = 0.4; % feed stream is 60% vapor
% Antoine parameters
A = [6.87024, 6.89385, 6.90940]; B = [1168.720, 1264.370, 1349.820];
```

TABLE 7.5

Parameters of Antoine Equation for Each Component (*P:mmHg, T:°C*)

Component	A	B	C
n-hexane, C_6H_{14}	6.87024	1168.720	224.210
n-heptane, C_7H_{16}	6.89385	1264.370	216.636
n-octane, C_8H_{18}	6.90940	1349.820	209.385

```
C = [224.210, 216.636, 209.385];
% Determine product flow rates and composition
% z(1)=D, z(2)=xb(2), z(3)=xb(3)
mf   = @(z)  [F*xf(1)-z(1)*xd(1)-(F-z(1))*xb(1);   F*xf(2)-z(1)*xd(2)-(F-z
(1))*z(2);
      F*xf(3)-z(1)*xd(3)-(F-z(1))*z(3)];
z0 = [5, 0.5, 0.5]; z = fsolve(mf,z0); Dist = z(1); xb(2) = z(2); xb(3) = z(3);
Bott = F - Dist; % bottom product rate
% LK and HK compositions
xlkf = xf(1); xhkf = xf(2); xlkd = xd(1); xhkd = xd(2); xlkb = xb(1); xhkb = xb(2);
% Determine boiling points of top stream (D) and bottom stream (B)
Df = @(Td) (760*P)*sum(xd./10.^(A - B./(Td + C))) - 1;
Bf = @(Tb) sum(xb.*10.^(A - B./(Tb + C)))/(760*P) - 1;
Td0 = 100; Tb0 = 100;
Td = fzero(Df,Td0); % boiling point of distillate (D)
Tb = fzero(Bf,Tb0); % boiling point of bottom product (B)
% Find distribution coefficients Ki = yi/xi = Pivp/P
Kf = 10.^(A - B./(Tf + C))/(760*P); % feed stream
Kd = 10.^(A - B./(Td + C))/(760*P); % distillate
Kb = 10.^(A - B./(Tb + C))/(760*P); % bottom product stream
% Relative volatility (alpa_LK/HK = alpa12) and average alpa
alpaf12 = Kf(1)/Kf(2); % feed stream
alpad12 = Kd(1)/Kd(2); % distillate
alpab12 = Kb(1)/Kb(2); % bottom product stream
avgalpa = (alpaf12*alpad12*alpab12)^(1/3); % average relative volatility
% (1) Minimum number of trays
Nmin = log((xlkd/xhkd)*(xhkb/xlkb))/log(avgalpa); % Fenske eqn.
% Minimum reflux ratio: Underwood eqn.
alpaf = Kf/Kf(2); % relative volatility of feed stream
alpad = Kd/Kd(2); % relative volatility of distillate
funf = @(theta) 1 - q - sum(alpaf.*xf./(alpaf - theta)); theta0 = 1.5;
theta = fzero(funf,theta0);
Rmin = sum(alpad.*xd./(alpad - theta)) - 1; % minimum reflux ratio
Ract = 1.5*Rmin; % R = 1.5*Rmin
% (2) ideal number of trays if R=1.5Rmin: Gilliland correlation
X = (Ract - Rmin)/(Ract + 1); Y = 0.75*(1 - X^0.5668); % Eduljee correlation
N = ceil((Nmin + Y)/(1 - Y)); % number of trays (integer)
% (3) feed tray location
Nrs = (xhkf*xlkb^2*Bott/(xlkf*xhkd^2*Dist))^0.206; % Kirkbride eqn.
Ns = ceil(N/(Nrs + 1)); % number of stripping section trays
Nr = N - Ns; % number of rectifying section trays
Ftray = Nr; % feed tray location from top
% Results
fprintf('Dew point(D) = %g (deg.C), bubble point(B) = %g (deg.C)\n', Td, Tb);
fprintf('(1) the minimum number of trays = %g\n', Nmin);
fprintf('(2) the number of ideal trays (R=1.5*Rmin) = %g\n', N);
fprintf('(3) feed tray location = %g\n', Ftray);

>> mshortcut
Dew point(D) = 75.186 (deg.C), bubble point(B) = 114.174 (deg.C)
(1) the minimum number of trays = 10.1088
(2) the number of ideal trays (R=1.5*Rmin) = 18
(3) feed tray location = 9
```

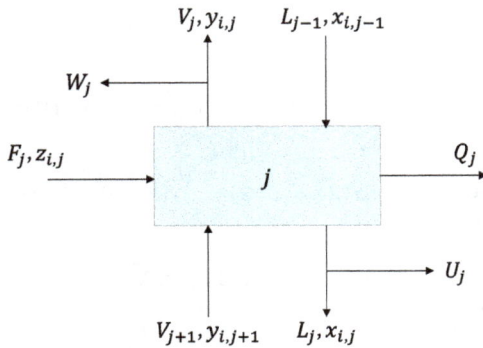

FIGURE 7.31 Equilibrium stage. (From Cutlip, M.B. and Shacham, M., *Problem Solving in Chemical and Biochemical Engineering with POLYMATH, Excel and MATLAB*, 2nd ed., Prentice-Hall, Boston, MA, 2008, p. 552.)

7.6 RIGOROUS STEADY-STATE DISTILLATION CALCULATIONS

7.6.1 EQUILIBRIUM STAGE

In rigorous steady-state distillation calculations, the mass balance, energy balance, and phase equilibrium equations for each stage are used. Steady-state distillation calculations consist of solving the system of nonlinear equations for each stage. Figure 7.31 shows a general equilibrium stage.[49]

In Figure 7.31, V and L are the vapor and liquid molar flow rates (*lbmol/hr*), respectively; F is the molar feed rate (*lbmol/hr*); U and W are the liquid and vapor side cuts (*lbmol/hr*), respectively; H and h are the vapor and liquid enthalpies (*Btu/lbmol*), respectively; Q is the heat transfer rate (*Btu/hr*); x and y are the mole fractions of liquid and vapor, respectively; z is the feed composition; the subscript i denotes component; and j denotes stage (numbered from the top to the bottom).

Let the number of total stages be n and the number of components be n_c. Balances around the jth stage are as follows:

Component mass balance: $L_{j-1}x_{i,j-1} + V_{j+1}y_{i,j+1} + F_j z_{i,j} = (L_j + U_j)x_{i,j} + (V_j + W_j)y_{i,j}$

Overall mass balance: $L_{j-1} + V_{j+1} + F_j = L_j + U_j + V_j + W_j$

Energy balance: $L_{j-1}h_{j-1} + V_{j+1}H_{j+1} + F_j h_{Fj} = (L_j + U_j)h_j + (V_j + W_j)H_j$

Vapor-liquid equilibrium: $y_{i,j} = k_{i,j}x_{i,j}$

Summation of mole fractions: $\sum_{i=1}^{n_c} y_{i,j} = 1$, $\sum_{i=1}^{n_c} x_{i,j} = 1$

Example 7.23 Three-Stage Distillation Column[50]

A simple distillation column with three theoretical stages—shown in Figure 7.32—is to be used to separate a mixture n-butane(1) and n-pentane(2).[51] A feed stream consisting of 0.23 *lbmol/hr* n-butane and 0.77 *lbmol/hr* n-pentane enters the column as liquid at its bubble point on stage 2 ($j = 2$). The operating pressure of the column is 120 *psia*, and a total condenser is used. The amount of heat added to the reboiler is 10,000 *Btu/hr*. Distillate and bottoms are removed at the rate of $D = 0.25$ and $B = 0.75$ *lbmol/hr*, respectively.

Calculate the temperatures of all stages and of the condenser, the flow rates and compositions of the vapor and liquid flows in the column, and the compositions of the distillate and the bottoms. The vapor pressures can be estimated by the Antoine equation, and the molar enthalpies of the pure compounds can be estimated by the quadratic equations of T. Table 7.6 shows the Antoine equation coefficients and the enthalpy correlation equations.

Solution

The built-in function *fsolve* can be used to solve the system of nonlinear equations consisting of the component mass balances, the energy balance, and the summations of mole fractions for each stage. There are 14 unknown variables and 14 nonlinear equations:

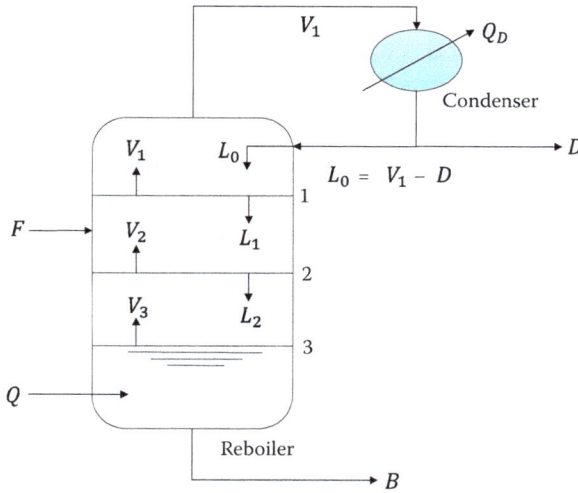

FIGURE 7.32 Three-stage distillation column. (From Cutlip, M.B. and Shacham, M., *Problem Solving in Chemical and Biochemical Engineering with POLYMATH, Excel and MATLAB*, 2nd ed., Prentice-Hall, Boston, MA, 2008, p. 553.)

TABLE 7.6

Antoine Equation Coefficients and Enthalpy Equations

Component		*n*-Butane (1)	*n*-Pentane (2)
Antoine equation coefficients (T:°C)	A	6.80776	6.85296
	B	935.77	1064.84
	C	238.789	232.012
Enthalpy equation (*Btu/lbmol*) (T:°F)	Liquid	$h_1 = 0.04T^2 + 29.6T$	$h_2 = 0.025T^2 + 38.5T$
	Vapor	$H_1 = -0.04T^2 + 43.8T + 8003$	$H_2 = 0.007T^2 + 31.7T + 12004$

$$\sum_{i=1}^{n_c} k_{i,j}x_{i,j} = 1 \, (j = 0, 1, F, 2, 3)$$

$$-\left((V_1 - L_0)k_{i,1} + L_1\right)x_{i,1} + V_2 k_{i,2}x_{i,2} = 0 \, (i = 1, 2)$$

$$L_1 x_{i,1} - \left(V_2 k_{i,2} + L_2\right)x_{i,2} + V_3 k_{i,3}x_{i,3} + Fz_i = 0 \, (i = 1, 2)$$

$$-V_1 H_1 + V_2 H_2 - L_1 h_1 + L_0 h_0 = 0$$

$$-V_2 H_2 + V_3 H_3 + Fh_{Ff} + L_1 h_1 - L_2 h_2 = 0$$

$$L_2 x_{i,2} - \left(V_3 k_{i,3} + B\right)x_{i,3} = 0 \, (i = 1, 2)$$

$$-V_3 H_3 + Q + L_2 h_2 - L_3 h_3 = 0$$

The subfunction *fd* defines the nonlinear equations for each stage. The variable vector s contains 14 unknown variables as its elements: $s(1) = T_0$, $s(2) = T_1$, $s(3) = T_f$, $s(4) = T_2$, $s(5) = T_3$, $s(6) = x_{11}$, $s(7) = x_{21}$, $s(8) = x_{12}$, $s(9) = x_{22}$, $s(10) = x_{13}$, $s(11) = x_{23}$, $s(12) = V_1$, $s(13) = V_2 V_2$, and $s(14) = V_3$. The script *rdist* calls the function *rdist* and employs the function *fsolve* to solve the nonlinear equations.

```
% rdist.m
% s1: T0, s2: T1, s3: Tf, s4: T2, s5: T3, s6: x11, s7: x21
% s8: x12, s9: x22, s10: x13, s11: x23, s12: V1, s13: V2, s14: V3
F = 1; D = 0.25; z = [0.23 0.77]; P = 760*120/14.7; Q = 1e4;
T0 = 200; T1 = 145; Tf = 200; T2 = 190; T3 = 210; x11 = 0.65; x21 = 0.35;
x12 = 0.43; x22 = 0.57; x13 = 0.33; x23 = 0.76; V1 = 1.1; V2 = 1; V3 = 1.1;
s0 = [T0 T1 Tf T2 T3 x11 x21 x12 x22 x13 x23 V1 V2 V3];
s = fsolve(@fd,s0,[],F,D,z,P,Q);
T0 = s(1); T1 = s(2); Tf = s(3); T2 = s(4); T3 = s(5);
x11 = s(6); x21 = s(7); x12 = s(8); x22 = s(9); x13 = s(10); x23 = s(11); V1 = s(12);
V2 = s(13); V3 = s(14);
fprintf('T0=%8.4f,      T1=%8.4f,      Tf=%8.4f,      T2=%8.4f,      T3=%8.4f
\n',T0,T1,Tf,T2,T3);
fprintf('x11=%6.4f, x12=%6.4f, x13=%6.4f\n', x11, x12, x13);
fprintf('x21=%6.4f, x22=%6.4f, x23=%6.4f\n', x21, x22, x23);
fprintf('V1=%6.4f, V2=%6.4f, V3=%6.4f\n', V1, V2, V3);
function f = fd(s,F,D,z,P,Q)
% i=1: n-butane, i=2:n-Pentane)
% s1: T0, s2: T1, s3: Tf, s4: T2, s5: T3, s6: x11, s7: x21
% s8: x12, s9: x22, s10: x13, s11: x23, s12: V1, s13: V2, s14: V3
% Antoine coefficients and parameters
A = [6.80776 6.85296]; B = [935.77 1064.84]; C = [238.789 232.012];
h1c = [0.04 29.6]; h2c = [0.025 38.5]; H1c = [-0.04 43.8 8003]; H2c = [0.007 31.7
12004];
% Equilibrium constants
k0 = 10.^(A - B./((s(1)-32)*5/9 + C)) / P; k1 = 10.^(A - B./((s(2)-32)*5/9 + C))
/ P;
kf = 10.^(A - B./((s(3)-32)*5/9 + C)) / P; k2 = 10.^(A - B./((s(4)-32)*5/9 + C))
/ P;
k3 = 10.^(A - B./((s(5)-32)*5/9 + C)) / P;
% Composition vector
x1 = [s(6) s(7)]; x2 = [s(8) s(9)]; x3 = [s(10) s(11)]; x0 = k1.*x1;
% Enthalpy
hL0 = sum(([s(1)^2 s(1)]*[h1c' h2c']).*x0); hL1 = sum(([s(2)^2 s(2)]*[h1c'
h2c']).*x1);
hLf = sum(([s(3)^2 s(3)]*[h1c' h2c']).*z); hL2 = sum(([s(4)^2 s(4)]*[h1c'
h2c']).*x2);
hL3 = sum(([s(5)^2 s(5)]*[h1c' h2c']).*x3); hV1 = sum(([s(2)^2 s(2) 1]*[H1c'
H2c']).*k1.*x1);
hV2 = sum(([s(4)^2 s(4) 1]*[H1c' H2c']).*k2.*x2); hV3 = sum(([s(5)^2 s(5)
1]*[H1c' H2c']).*k3.*x3);
% Mass balance
B = F - D; L0 = s(12) - D; L1 = s(13) - D; L2 = s(14) + B; L3 = B;
% Definition of equations
f(1,1) = sum(k0.*x0) - 1; f(2,1) = sum(k1.*x1) - 1; f(3,1) = sum(kf.*z) - 1; f
(4,1) = sum(k2.*x2) - 1;
f(5,1) = sum(k3.*x3) - 1; f(6,1) = -((s(12) - L0)*k1(1) + L1)*s(6) + s
(13)*k2(1)*s(8);
f(7,1) = -((s(12) - L0)*k1(2) + L1)*s(7) + s(13)*k2(2)*s(9);
f(8,1) = -s(12)*hV1 + s(13)*hV2 - L1*hL1 + L0*hL0;
```

```
f(9,1) = L1*s(6) - (s(13)*k2(1)+L2)*s(8) + s(14)*k3(1)*s(10) + F*z(1);
f(10,1) = L1*s(7) - (s(13)*k2(2)+L2)*s(9) + s(14)*k3(2)*s(11) + F*z(2);
f(11,1) = -s(13)*hV2 + s(14)*hV3 + hLf + L1*hL1 - L2*hL2;
f(12,1) = L2*s(8) - (s(14)*k3(1) + B)*s(10);
f(13,1) = L2*s(9) - (s(14)*k3(2) + B)*s(11);
f(14,1) = -s(14)*hV3 + Q + L2*hL2 - L3*hL3;
end
```

The script *rdist* produces the following outputs:

```
>> rdist
T0=188.2848, T1=206.4285, Tf=215.4895, T2=218.5461, T3=226.5515
x11=0.3272, x12=0.1991, x13=0.1225
x21=0.6728, x22=0.8009, x23=0.8775
V1=1.0537, V2=1.0388, V3=1.0417
```

The results can be summarized as shown in Table 7.7.

7.6.2 RIGOROUS DISTILLATION MODEL: MESH EQUATIONS

For a steady-state stage, we assume that phase equilibrium is achieved, no chemical reactions occur, and entrainment of liquid drops in vapor and occlusion of vapor bubbles in liquid are negligible. Figure 7.33 represents a general equilibrium stage.[52] There are vapor and liquid side streams, and heat can be transferred at a rate of Q_j from (+) or to (-) stage j.

Entering stage j is a feed stream of molar flow rate F_j with composition in mole fraction $z_{i,j}$ and molar enthalpy h_F. The feed pressure is equal to stage pressure P_j. Interstage liquid from stage $j - 1$ of molar flow rate L_{j-1} is also entering stage j, as well as interstage vapor from stage $j + 1$ of molar flow rate V_{j+1}. Leaving stage j is vapor of properties $y_{i,j}$, h_{V_j}, T_j, and P_j. The vapor stream may be divided into a vapor side stream of molar flow rate W_j and an interstage stream of molar flow rate V_j to be sent to stage $j - 1$. Also leaving stage j is a liquid of properties $x_{i,j}$, h_{L_j}, T_j, and P_j in equilibrium with the vapor. This liquid may be divided into a side stream of molar flow rate U_j and an interstage stream of molar flow rate L_j to be sent to stage $j + 1$.

Figure 7.34 shows a distillation column with N equilibrium stages that are numbered from the top down.[52] There are N stages and the number of components in the feed mixture is C. Stage 1 ($j = 1$) is the partial condenser and stage N ($j = N$) is the reboiler.

Let the amount of component i in the jth stage be $X_{i,j}$. Using the relationship $y_{i,j} = v_{i,j}/V_j$ and $x_{i,j} = l_{i,j}/L_j$, the mass balances for the jth stage yield

$$M_{i,j} = V_{j+1}y_{i,j+1} + L_{j-1}x_{i,j-1} + F_jz_{i,j} - (V_j + W_j)y_{i,j} - (L_j + U_j)x_{i,j} = 0$$

TABLE 7.7
Calculation Results for a Three-Stage Column

Variable	Stage 1	Feed Stage	Stage 2	Stage 3
Temperature(°F)	206.4285	215.4895	218.5461	226.5515
Vapor flow rate(*lbmol/hr*)	1.0537	-	1.0388	1.0417
Composition of *n*-butane	0.3272	0.23	0.1991	0.1225
Composition of *n*-pentane	0.6728	0.77	0.8009	0.8775

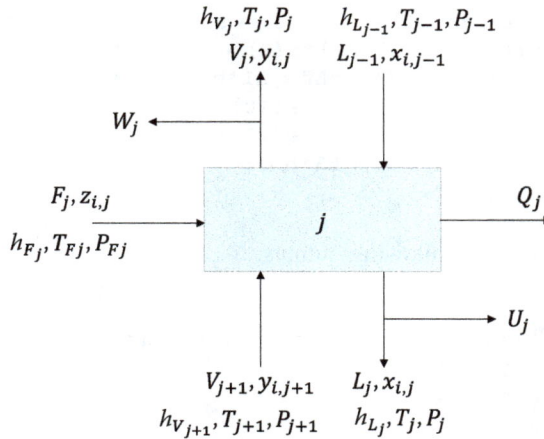

FIGURE 7.33 A steady-state equilibrium stage model. (From Henley, E.J., Seader, et al., *Separation Process Principles*, 3rd ed., John Wiley & Sons, Hoboken, NJ, 2011, p. 411.)

$$M_{i,j} = v_{i,j+1} + l_{i,j-1} + f_{i,j} - \left(1 + S_{V_j}\right)v_{i,j} - \left(1 + S_{L_j}\right)l_{i,j} = 0$$

where $S_{V_j} = W_j/V_j$ and $S_{L_j} = U_j/L_j$.

The phase equilibrium equation for each component gives

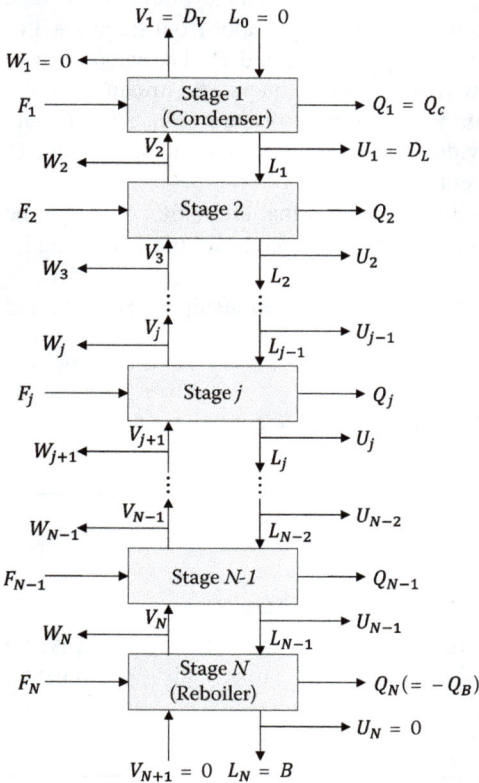

FIGURE 7.34 A distillation column with N stages. (From Henley, E.J., et al., *Separation Process Principles*, 3rd ed., John Wiley & Sons, Hoboken, NJ, 2011, p. 411.)

$$E_{i,j} = y_{i,j} - K_{i,j}x_{i,j} = \frac{v_{i,j}}{V_j} - K_{i,j}\frac{l_{i,j}}{L_j} = 0$$

The summation of mole fractions for each stage should be equal to 1:

$$S_{y,j} = \sum_{i=1}^{C} y_{i,j} - 1 = \sum_{i=1}^{C} \frac{v_{i,j}}{V_j} - 1 = 0, \ S_{x,j} = \sum_{i=1}^{C} x_{i,j} - 1 = \sum_{i=1}^{C} \frac{l_{i,j}}{L_j} - 1 = 0$$

The energy balance for each stage gives

$$H_j = L_{j-1}h_{L_{j-1}} + V_{j+1}h_{V_{j+1}} + F_j h_{F_j} - (L_j + U_j)h_{L_j} - (V_j + W_j)h_{V_j} - Q_j = 0$$

Using the summation of component flow relations $V_j = \sum_{i=1}^{C} v_{i,j}$ and $L_j = \sum_{i=1}^{C} l_{i,j}$ and $F_j = \sum_{i=1}^{C} f_{i,j}$, this equation becomes

$$H_j = h_{V_{i+1}} \sum_{i=1}^{C} v_{i,j+1} + h_{L_{j-1}} \sum_{i=1}^{C} l_{i,j-1} + h_{F_j} \sum_{i=1}^{C} f_{i,j}$$

$$- \left(1 + S_{V_j}\right)h_{V_j} \sum_{i=1}^{C} v_{i,j} - \left(1 + S_{L_j}\right)h_{L_j} \sum_{i=1}^{C} l_{i,j} - Q_j = 0$$

For each equilibrium stage, there are $2C + 3$ MESH (mass, equilibrium, summations, and enthalpy) equations.

The total mass balance can be obtained from the summation of component mass balance equations from stage 1 to stage j:

$$\sum_{k=1}^{j} \sum_{i=1}^{C} M_{i,k} = L_0 - L_j + V_{j+1} - V_1 + \sum_{k=1}^{j} (F_k - U_k - W_k) = 0$$

Since $L_0 = 0$, this equation can be rewritten as

$$L_j = V_{j+1} - V_1 + \sum_{k=1}^{j} (F_k - U_k - W_k)(1 \leq j \leq N - 1)$$

The summation of component mass balances from stage j to stage N gives

$$\sum_{k=j}^{N} \sum_{i=1}^{C} M_{i,k} = V_{N+1} - V_j + L_{j-1} - L_N + \sum_{k=j}^{N} (F_k - U_k - W_k) = 0$$

Since $V_{N+1} = 0$, this equation can be expressed as

$$V_j = L_{j-1} - L_N + \sum_{k=j}^{N} (F_k - U_k - W_k)$$

7.6.3 Tridiagonal Matrix Method

The tridiagonal matrix method introduced by Wang and Henke[53] is a fast and accurate technique for calculating the component and total flow rates. In the tridiagonal matrix method, the calculation for a stage is based on the calculation for the previous stage, and the accumulation of calculation errors can be significant when the number of stages is large. In order to avoid the accumulation of calculation errors, the revised Thomas algorithm can be used. In the Thomas algorithm, the calculation begins from the upper left corner of the matrix (i.e., the first stage). Similar calculations are repeated along the diagonal direction of the matrix.

Calculation of the stage liquid composition $x_{i,j}$: The mass balances and equilibrium equations give

$$\left\{ V_j - V_1 + \sum_{k=1}^{j-1} (F_k - U_k - W_k) \right\} x_{i,j-1}$$

$$- \left\{ \left(V_j + W_j \right) K_{i,j} + V_{j+1} - V_1 + \sum_{k=1}^{j} (F_k - U_k - W_k) + U_j \right\} x_{i,j} + V_{j+1} K_{i,j+1} x_{i,j+1} = -F_j z_{i,j}$$

This equation can be rewritten as

$$P_j x_{i,j-1} + Q_j x_{i,j} + R_j x_{i,j+1} = S_j$$

where

$$P_j = V_j - V_1 + \sum_{k=1}^{j-1} (F_k - U_k - W_k)(2 \le j \le N)$$

$$Q_j = V_1 - \left(V_j + W_j \right) K_{i,j} - U_j - V_{j+1} - \sum_{k=1}^{j} (F_k - U_k - W_k)(1 \le j \le N)$$

$$R_j = V_{j+1} K_{i,j+1}(1 \le j \le N - 1)$$

$$S_j = -F_j z_{i,j}(1 \le j \le N)$$

$$x_{i,N+1} = x_{i,0} = 0, \ V_{N+1} = W_1 = U_N = 0$$

Rearrangement of the equations for all stages ($j = 1, 2, \cdots, N$) yields

$$\begin{bmatrix} 1 & r_1 & 0 & \cdots & 0 \\ 0 & 1 & r_2 & \cdots & \vdots \\ \vdots & \vdots & \ddots & \ddots & 0 \\ \vdots & \vdots & \vdots & 1 & r_{N-1} \\ 0 & \cdots & \cdots & 0 & 1 \end{bmatrix} \begin{bmatrix} x_{i,1} \\ x_{i,2} \\ \vdots \\ x_{i,N-1} \\ x_{i,N} \end{bmatrix} = \begin{bmatrix} s_1 \\ s_2 \\ \vdots \\ s_{N-1} \\ s_N \end{bmatrix}$$

where

$$r_j = \frac{R_j}{Q_j - r_{j-1} P_j}(2 \le j \le N - 1), \ r_1 = \frac{R_1}{Q_1}$$

$$s_j = \frac{S_j - s_{j-1} P_j}{Q_j - r_{j-1} P_j}(2 \le j \le N), \ s_1 = \frac{S_1}{Q_1}$$

The liquid composition $x_{i,j}$ for stage j is given by

$$r_0 = s_0 = 0, \ x_{i,N} = s_N, \ x_{i,j} = s_j - r_j x_{i,j+1}(1 \le j \le N - 1)$$

Calculation of the stage vapor component flow rate $v_{i,j}$: The calculation procedure for $v_{i,j}$ is similar to the previous one. The absorption factor $A_{i,j}$ can be defined as $A_{i,j} = L_j/(K_{i,j} V_j)$. Then, since $l_{i,j} = L_j/(K_{i,j} V_j) = A_{i,j} v_{i,j}$, the component mass balance gives

$$M_{i,j} = v_{i,j+1} + A_{i,j-1}v_{i,j-1} + f_{i,j} - (1 + S_{V_j})v_{i,j} - (1 + S_{L_j})A_{i,j}v_{i,j} = 0$$

This equation can be rewritten as

$$A_{i,j-1}v_{i,j-1} + B_j v_{i,j} + v_{i,j+1} = D_j$$

where

$$B_j = -1 - S_{V_j} - (1 + S_{L_j})A_{i,j}, \quad D_j = -f_{i,j}$$

Collecting equations for all stages yields

$$\begin{bmatrix} 1 & p_1 & 0 & \cdots & 0 \\ 0 & 1 & p_2 & \cdots & \vdots \\ \vdots & \vdots & \ddots & \ddots & 0 \\ \vdots & \vdots & \vdots & 1 & p_{N-1} \\ 0 & \cdots & \cdots & 0 & 1 \end{bmatrix} \begin{bmatrix} v_{i,1} \\ v_{i,2} \\ \vdots \\ v_{i,N-1} \\ v_{i,N} \end{bmatrix} = \begin{bmatrix} q_1 \\ q_2 \\ \vdots \\ q_{N-1} \\ q_N \end{bmatrix}$$

where

$$p_j = \frac{1}{B_i - p_{i-1}A_{i,j-1}}(1 \le j \le N - 1), \quad q_j = \frac{D_j - q_{j-1}A_{i,j-1}}{B_j - p_{j-1}A_{i,j-1}}(1 \le j \le N)$$

and $A_{i,0} = p_0 = q_0 = 0$. From these equations, the component vapor flow rate is given by

$$v_{i,N} = q_N, \quad v_{i,j} = q_j - p_j v_{i,j+1}(1 \le j \le N - 1)$$

Calculation of the stage liquid component flow rate $l_{i,j}$: The calculation procedure for $l_{i,j}$ is similar to the previous one for $v_{i,j}$. The relative volatility and the stripping factor $S_{i,j}$ can be expressed as

$$\alpha_{i,j} = \frac{K_{i,j}}{K_{b,j}}, \quad S_{b,j} = \frac{K_{b,j}V_j}{L_j}, \quad S_{i,j} = \frac{K_{i,j}V_j}{L_j} = \alpha_{i,j}S_{b,j}$$

where the subscript b denotes the reference component. If we define R_L and R_V as

$$R_{L_j} = 1 + \frac{U_j}{L_j} = 1 + S_{L_j}, \quad R_{V_j} = 1 + \frac{W_j}{V_j} = 1 + S_{V_j}$$

the component vapor flow rate can be represented as

$$v_{i,j} = \frac{K_{i,j}V_j}{L_j}l_{i,j} = \frac{K_{i,j}}{K_{b,j}}\frac{K_{b,j}V_j}{L_j}l_{i,j} = \alpha_{i,j}S_{b,j}l_{i,j}$$

Combination of these equations yields

$$l_{i,j-1} + W_j l_{i,j} + Z_j l_{i,j+1} = G$$

where

$$W_j = -R_{L_j} - \alpha_{i,j}S_{b,j}R_{V_j}, \quad Z_j = \alpha_{i,j+1}S_{b,j+1}, \quad G_j = -f_{i,j}$$

Collecting the balance equations for all stages gives

$$
\begin{bmatrix}
1 & p_1 & 0 & \cdots & 0 \\
0 & 1 & p_2 & \cdots & \vdots \\
\vdots & \vdots & \ddots & \ddots & 0 \\
\vdots & \vdots & \vdots & 1 & p_{N-1} \\
0 & \cdots & \cdots & 0 & 1
\end{bmatrix}
\begin{bmatrix}
l_{i,1} \\
l_{i,2} \\
\vdots \\
l_{i,N-1} \\
l_{i,N}
\end{bmatrix}
=
\begin{bmatrix}
q_1 \\
q_2 \\
\vdots \\
q_{N-1} \\
q_N
\end{bmatrix}
$$

where

$$
p_j = \frac{Z_j}{W_j - p_{j-1}} (1 \leq j \leq N - 1), \; q_j = \frac{G_j - q_{j-1}}{W_j - p_{j-1}} (1 \leq j \leq N), \; p_0 = q_0 = 0
$$

Consequently, the component liquid flow rates are given by

$$
l_{i,N} = q_N, \; l_{i,j} = q_j - p_j l_{i,j+1} (1 \leq j \leq N - 1)
$$

7.6.4 Bubble Point (BP) Method

In the BP method, a form of the equilibrium equation and summation equations are used to calculate the stage temperature. A new set of stage temperatures is calculated during each iteration from the bubble point equations. The BP method generally works best for narrow-boiling, ideal, or nearly ideal systems, where composition has a greater effect on temperature than the latent heat of vaporization.[54]

The condenser load Q_1 and reboiler load Q_N can be expressed as

$$
Q_1 = (L_1 + V_1 + U_1 + W_1 - F_1)h_{V2} - V_1 h_{V1} - (L_1 + U_1)h_{L1} + F_1 h_{F1}
$$

$$
Q_N = \sum_{j=1}^{N} (F_j h_{Fj} - U_j h_{Lj} - W_j h_{Vj}) - \sum_{i=1}^{N-1} Q_j - V_1 h_{V1} - L_N h_{LN}
$$

The liquid flow rate L_j and vapor flow rate V_j at the jth stage are given by

$$
L_j = V_{j+1} - V_1 + \sum_{k=1}^{j} (F_k - U_k - W_k)(1 \leq j \leq N - 1), \; L_0 = 0
$$

$$
V_{j+1} = \frac{d_j - a_j V_j}{b_j} (2 \leq N \leq j - 1), \; V_2 = L_1 + V_1 + U_1 + W_1 - F_1
$$

$$
a_j = h_{L_{j-1}} - h_{V_j}, \; b_j = h_{V_{j+1}} - h_{L_j}
$$

$$
d_j = \left(h_{L_j} - h_{L_{j-1}}\right) \sum_{k=1}^{j-1} (F_k - U_k - W_k) + F_j\left(h_{L_j} - h_{F_j}\right) + W_j\left(h_{V_j} - h_{L_j}\right) + Q_j
$$

In the calculation procedure of the BP method, the liquid composition $x_{i,j}$ and vapor composition $y_{i,j}$ are normalized as

$$
x_{i,j} = \frac{x_{i,j}}{\sum_i^C x_{i,j}}, \; y_{i,j} = \frac{y_{i,j}}{\sum_{i=1}^C y_{i,j}}
$$

The solution is considered to converge when sets of stage temperatures at the kth iteration are within some prescribed tolerance of the corresponding sets of stage temperatures at the $k - 1$th iteration. A simple criterion based on successive sets of stage temperatures can be used for the convergence test:

$$J_T = \sum_{j=1}^{N} \left\{ T_j^{(k)} - T_j^{(k-1)} \right\}^2 \leq \epsilon N \, (0 < \epsilon < 1)$$

The BP algorithm can be summarized as follows:

1) Initialize the input variables (F_j, $z_{i,j}$, T_F, P_F, h_F, P_j, U_j, W_j, Q_j(excluding Q_1 and Q_N), N, L_1(reflux flow rate), V_1). Set $k = 1$.
2) Set the initial values of the tear variables T_j and V_j.
3) Calculate $x_{i,j}$ using $x_{i,N} = S_N$, $x_{i,j} = s_j - r_j x_{i,j+1}(1 \leq j \leq N - 1)$, $r_0 = s_0 = 0$, and normalize the resulting $x_{i,j}$ by $x_{i,j} = x_{i,j}/\sum_{i=1}^{C} x_{i,j}$. This procedure has to be done for all components.
4) Use the bubble point calculation method to find a new set of T_j and $y_{i,j}$. The equilibrium equation $E_{i,j} = y_{i,j} - K_{i,j} x_{i,j} = 0$ is used in the calculation.
5) Calculate the condenser load Q_1 and the reboiler load Q_N:

$$Q_1 = (L_1 + V_1 + U_1 + W_1 - F_1)h_{V2} - V_1 h_{V1} - (L_1 + U_1)h_{L1} + F_1 h_{F1}$$

$$Q_N = \sum_{j=1}^{N} (F_j h_{Fj} - U_j h_{Lj} - W_j h_{Vj}) - \sum_{j=1}^{N-1} Q_j - V_1 h_{V1} - L_N h_{LN}$$

6) Calculate V_j and L_j sequentially:

$$V_{j+1} = \frac{d_j - a_j V_j}{b_j}(2 \leq N \leq j - 1), \quad V_2 = L_1 + V_1 + U_1 + W_1 - F_1$$

$$L_j = V_{j+1} - V_1 + \sum_{k=1}^{j} (F_k - U_k - W_k)(1 \leq j \leq N - 1), \quad L_0 = 0$$

7) Check the convergence using the criterion $J_T = \sum_{j=1}^{N} \{T_j^{(k)} - T_j^{(k-1)}\}^2 \leq \epsilon N$. If the set of stage temperatures does not converge, adjust the tear variables again, set $k = k + 1$, and go to step 3).

The function *distBPeu* performs this calculation procedure (in the function name, eu denotes use of the English unit system). The basic syntax is

```
[x,y,T,L,V,iter] = distBPeu(opdat,mxdat)
```

The structure opdat contains operating conditions as fields, and the structure mxdat contains physical properties and parameters as its fields. This function calls the function *phiHeu* to calculate the enthalpy of the mixture and the fugacity of each component.

```
function [x,y,T,L,V,iter] = distBPeu(opdat,mxdat)
% Calculation of multicomponent distillation using BP method (English unit)
% Set data
eos = opdat.eos; N = opdat.N; nc = opdat.nc; F = opdat.F; Tf = opdat.Tf; Pf =
opdat.Pf; hF = opdat.hF;
P = opdat.P; T = opdat.T; V = opdat.V; L = opdat.L; criv = opdat.criv;
U = opdat.U; W = opdat.W; Q = opdat.Q; T0 = opdat.T0; nf = opdat.nf;
hV = opdat.hV; hL = opdat.hL; fstate = upper(opdat.fstate);
x = opdat.x; y = opdat.y; z = opdat.z; Pc = mxdat.Pc; Ant = mxdat.Ant;
```

```
% Initialization
a = zeros(1,N); b = zeros(1,N); d = zeros(1,N); V(2) = L(1) + V(1) + U(1) + W(1) -
F(1);
% Feed enthalpy
for j = 1:length(nf)
      [Z H phi] = phiHeu(z(:,nf(j))',Pf(nf(j)),Tf(nf(j)),fstate,eos,opdat,
mxdat); hF(nf(j)) = H; % Btu/mol
end
T = T0; Told = T; criT = 10; iter = 1;
% Initialization of K(i,j): K(i,j) at T0
for j = 1:N
      Pv0(:,j) = Pc(:).*exp(Ant(:,1) - Ant(:,2)./(T(j) + Ant(:,3))); K(:,j) =
Pv0(:,j)./P(j); % Raoult's law
end
while criT > criv
   % Calculate x(i,j)
   for i = 1:nc
     for j = 2:N-1
           Pj(j) = V(j) - V(1) + sum(F(1:j-1) - U(1:j-1) - W(1:j-1));
           Qj(j) = V(1) - (V(j)+W(j))*K(i,j) - U(j) - V(j+1) - sum(F(1:j) - U(1:j)
- W(1:j));
           Rj(j) = V(j+1)*K(i,j+1); Sj(j) = -F(j)*z(i,j);
     end
     Pj(N) = V(N) - V(1) + sum(F(1:N-1) - U(1:N-1) - W(1:N-1));
     Qj(1) = V(1) - (V(1)+W(1))*K(i,1) - U(1) - V(2) - (F(1) - U(1) - W(1));
     Qj(N) = V(1) - (V(N)+W(N))*K(i,N) - U(N) - sum(F(1:N) - U(1:N) - W(1:N));
     Rj(1) = V(2)*K(i,2); Sj(1) = -F(1)*z(i,1); Sj(N) = -F(N)*z(i,N);
     r(1) = Rj(1)/Qj(1); s(1) = Sj(1)/Qj(1);
     for j = 2:N-1
           r(j) = Rj(j)/(Qj(j) - r(j-1)*Pj(j)); s(j) = (Sj(j) - s(j-1)*Pj(j))/(Qj
(j) - r(j-1)*Pj(j));
     end
     s(N) = (Sj(N) - s(N-1)*Pj(N))/(Qj(N) - r(N-1)*Pj(N)); x(i,N) = s(N);
     for j = N-1:-1:1, x(i,j) = s(j) - r(j)*x(i,j+1); end
   end
   for j = 1:N, x(:,j) = x(:,j)/sum(x(:,j)); end
   % Calculate new T(j) and y(i,j) using BP method
   for j = 1:N % calculate y(i,j)
         y(:,j) = K(:,j).*x(:,j); y(:,j) = y(:,j)/sum(y(:,j));
   end
   % Antoine eqn. (T: deg.F)
   for j = 1:N % calculate T(j) (T in F)
         f = @(Tv) sum(Pc'.*exp(Ant(:,1) - Ant(:,2)./(Tv +Ant(:,3))).*x(:,j)/P
(j)) - 1; T(j) = fzero(f,T(j));
   end
   for j = 1:N
         lstate = 'L';
         [Z H phi] = phiHeu(x(:,j)',P(j),T(j),lstate,eos,opdat,mxdat);
         phiL = phi; hL(j) = H; % Btu/lbmol
         vstate = 'V'; [Z H phi] = phiHeu(y(:,j)',P(j),T(j),vstate,eos,opdat,
mxdat);
      phiV = phi; hV(j) = H; % Btu/lbmol
      K(:,j) = phiL./phiV;
   end
```

```
% Calculate loads for the condenser and reboiler
 Q(1) = (L(1) + V(1) + U(1) + W(1) - F(1))*hV(2) - V(1)*hV(1) - (L(1) + U(1))*hL
(1) + F(1)*hF(1);
 Q(N) = sum(F.*hF - U.*hL - W.*hV) - sum(Q(1:N-1)) - V(1)*hV(1) - L(N)*hL(N);
 % Calculate V(j) and L(j) sequentially
 for j = 2:N
      a(j) = hL(j-1) - hV(j); b(j-1) = hV(j) - hL(j-1);
      d(j) = (hL(j)-hL(j-1))*sum(F(1:j-1)-U(1:j-1)-W(1:j-1)) + ...
           F(j)*(hL(j)-hF(j)) + W(j)*(hV(j)-hL(j)) + Q(j);
 end
 a(1) = -hV(1); b(N) = -hL(N); d(1) = F(1)*(hL(1)-hF(1)) + W(1)*(hV(1)-hL(1)) +
Q(1);
 for j = 2:N-1
    V(j+1) = (d(j) - a(j)*V(j))/b(j); L(j) = V(j+1) - V(1) + sum(F(1:j) - U(1:j) -
W(1:j));
 end
 % Check convergence
 criT = sum(abs(Told - T)); Told = T; iter = iter + 1;
end
end
```

The function *phiHeu* calculates the enthalpy of the mixture and the fugacity coefficient of each component using the English unit system. The basic syntax is
```
[Z H phi] = phiHeu(x,P,T,state,eos,opdat,mxdat)
```

where x is the mole fraction vector, P and T are the pressure (psia) and temperature (°F), respectively, state denotes the fluid state ('L': liquid, 'V': vapor), and eos is the equation of state being used ('RK', 'SRK', or 'PR'). opdat is a structure containing operating conditions as its fields. The structure mxdat includes physical properties and parameters as its fields: mxp.Pc is a critical pressure (psia) vector, mxp.Tc is a critical temperature (R) vector, mxp.w is a vector of acentric factors, mxp.k is a matrix of binary interaction parameters ($n \times n$ diagonal matrix), and mxp.Afiis a matrix of heat capacity integration coefficients (dimension: number of components×number of parameters). In the output arguments, Z is the compressibility factor, H is the mixture enthalpy (Btu/lbmol), and phi is a vector denoting the fugacity coefficients of each component.

```
function [Z H phi] = phiHeu(x,P,T,state,eos,opdat,mxdat)
% Calculation of mixture enthalpy and fugacity coefficient of component
% Inputs:
% x: mole fraction vector
% P, T: pressure (Psia) and temperature (F)
% state: fluid state ('L': liquid, 'V': vapor)
% eos: equation of state ('RK', 'SRK', or 'PR')
% mxp: physical property structure
%   mxp.Pc, mxp.Tc: critical pressure (Psia) and temperature (R) vector
%   mxp.w: acentric factor for each component (vector)
%   mxp.k: nxn diagonal matrix of binary interaction parameters
%   mxp.Afi: heat capacity integration coefficients for ideal gases
% Outputs:
% Z: compressibility factor
% H: mixture enthalpy (Btu/lbmol)
% phi: fugacity coefficient
w = mxdat.w; k = mxdat.k; Tc = mxdat.Tc/1.8; Pc = 6894.8*mxdat.Pc; % T and P in K
and Pa
```

```
T = (T-32)/1.8 + 273.15; P = 6894.8*P; Afi = mxdat.Afi; % Btu/lbmole(1Btu/lbmole =
2.326 J/mol)
nc = opdat.nc; R = 8.314; % gas constant: m^3 Pa/(mol K) = J/mol-K
Tr = T./Tc; Pr = P./Pc; nc = opdat.nc; eos = upper(eos); state = upper(state);
switch eos
case{'RK'}
ai = sqrt(0.4278./(Pc.*Tr.^2.5)); bi = 0.0867./(Pc.*Tr); A = sum(x.*ai); B =
sum(x.*bi);
Z = roots([1 -1 B*P*(A^2/B-B*P-1) -A^2*(B*P)^2/B]);
case{'SRK'}
mx = 0.48+1.574*w-0.176*w.^2; al = (1+mx.*(1-sqrt(Tr))).^2;
ai = 0.42747*al.*Pr./(Tr.^2); bi = 0.08664*Pr./Tr;
am = sqrt(ai'*ai).*(1-k); A = x*am*x'; B = sum(x.*bi); Z = roots([1 -1 A-B-
B^2 -A*B]);
case{'PR'}
mx = 0.37464+1.54226*w-0.26992*w.^2; al = (1+mx.*(1-sqrt(Tr))).^2;
ai = 0.45723553*al.*Pr./(Tr.^2); bi = 0.0777961*Pr./Tr;
am = sqrt(ai'*ai).*(1-k); A = x*am*x'; B = sum(x.*bi);
Z = roots([1 B-1 A-3*B^2-2*B B^3+B^2-A*B]);
end
iz = abs(imag(Z)); Z(and(iz>0,iz<=1e-6)) = real(Z(and(iz>0,iz<=1e-6)));
for i = 1:length(Z), zind(i) = isreal(Z(i)); end
Z = Z(zind);
if state == 'L', Z = min(Z);
else, Z = max(Z); end
V = R*T*Z/P; % m^3/mol
% Enthalpy and fugacity coefficients
% Hv0 = int(Cp) Btu/lbmol, Afi uses T in F
Te = (T-273.15)*1.8 + 32; % T: K->F
Hv0 = x*(Afi(:,1)*Te + Afi(:,2)*Te^2/2 + Afi(:,3)*Te^3/3 +Afi(:,4)*Te^4/4 + Afi
(:,5)*Te^5/5);
Hv0 = Hv0*2.326; % 1Btu/lbmole=2.326J/mol
switch eos
  case{'RK'}
    H = Hv0 + R*T*(Z - 1 - 3*(A^2)*log(1 + B*P/Z)/(2*B));
    phi = exp((Z-1)*bi/B - log(Z-B*P) - (A^2/B)*(2*ai/A - bi/B)*log(1 + B*P/Z));
  case{'SRK'}
    hsum = 0;
    for i = 1:nc
      for j = 1:nc
            hsum = hsum + x(i)*x(j)*am(i,j)*(1 - mx(i)*sqrt(Tr(i))/(2*sqrt(al
(i))) -...
                    mx(j)*sqrt(Tr(j))/(2*sqrt(al(j))));
      end
    end
    H = Hv0 + R*T*(Z - 1 - log((Z + B)/Z)*hsum/B);
    phi = exp((Z-1)*bi/B - log(Z-B) - (A/B)*(2*sqrt(ai)/sqrt(A) - bi/B)*log((Z
+B)/Z));
case{'PR'}
    hsum = 0;
    for i = 1:nc
        for j = 1:nc
                hsum = hsum + x(i)*x(j)*am(i,j)*(1 - mx(i)*sqrt(Tr(i))/
(2*sqrt(al(i))) -...
            mx(j)*sqrt(Tr(j))/(2*sqrt(al(j))));
```

```
      end
   end
   H = Hv0 + R*T*(Z - 1 - log((Z + B)/Z)*hsum/B);
   phi = exp((Z-1)*bi/B - log(Z-B) - (A/B/sqrt(8))*(2*x(1:end)*am(1:end,:)/
A -...
            bi/B)*log((Z+(1+sqrt(2))*B)/(Z+(1-sqrt(2))*B)));
end
H = H/2.326; % J/mol -> Btu/lbmol
end
```

Example 7.24 Distillation of a Five-Component Mixture

Figure 7.35 shows a distillation column where a feed consisting of ethane(1)/propane(2)/ n-butane(3)/n-pentane(4)/n-hexane(5) is distilled.[52] The number of total stages including the condenser and the reboiler is 16. Feed stream 1 enters the first feed stage (sixth stage from the top) at 170°F and 300 $psia$ with a flow rate of $F_6 = 41$ $lbmol/hr$ and a composition of $z_{1,6} = 0.061$, $z_{2,6} = 0.342$, $z_{3,6} = 0.463$, $z_{4,6} = 0.122$, and $z_{5,6} = 0.012$. Feed stream 2 enters the second feed stage (ninth stage from the top) at 230°F and 275 $psia$ with a flow rate of $F_9 = 59$ $lbmol/hr$ and a composition of $z_{1,9} = 0.0085$, $z_{2,9} = 0.1017$, $z_{3,9} = 0.3051$, $z_{4,9} = 0.5085$ and $z_{5,9} = 0.0762$. The pressure in the column is kept constant as 240 $psia$; the flow rate of the vapor stream to the first stage (V_1) is 15 $lbmol/hr$ and that of the liquid stream to the first stage (L_1) is 150 $lbmol/hr$. The top product is obtained as liquid from the first stage and the flow rate of the liquid side cut is $U_1 = 5$ $lbmol/hr$. An intercooler is implemented in the third stage and the cooling rate is $Q_3 = 200,000$ Btu/hr. The flow rate of the liquid side cut from the third stage is $U_3 = 3$ $lbmol/hr$, and that of the vapor side cut from the 13th stage is $W_{13} = 37$ $lbmol/hr$.

The specific heat C_{pv}^0 ($Btu/lbmol$) and the vapor pressure (P_i^S) are given by

$$C_{pv}^0 = a_1 + a_2 T + a_3 T^2 + a_4 T^3 + a_5 T^4 \quad (T : °F)$$

$$\ln\frac{P_i^s}{P_c} = A_1 - \frac{A_2}{T + A_3} \quad (T : °F, \, P_c : \text{critical pressure})$$

Table 7.8 shows the physical properties and Table 7.9[55] the parameters of specific heat equations for each component. Calculate the stage temperatures and vapor and liquid compositions, and plot the results versus the stage number.[56]

Solution

The script c5distBPeu specifies parameters for each component and operating conditions, and calls the function distBPeu to perform the BP calculations. The K-value for component i is obtained from $K_i = \phi_{iL}/\phi_{iV}$, and the enthalpy and the fugacity coefficients for the vapor and liquid phases in each stage are calculated by the function phiHeu.

```
% c5distBPeu.m: calculation of 5-component distillation using BP method
% (English unit)
% Operating conditions and parameters
clear all;
opdat.N = 16; opdat.nc = 5; % N: total stages, nc: number of components
intv = zeros(1,opdat.N); intc = zeros(opdat.nc,opdat.N);
opdat.F = intv; opdat.Tf = intv; opdat.Pf = intv; opdat.hF = intv;
opdat.P = intv; opdat.T = intv; opdat.V = intv; opdat.L = intv;
opdat.U = intv; opdat.W = intv; opdat.Q = intv; opdat.eos = 'pr';
opdat.hV = intv; opdat.hL = intv; opdat.nf = 0; opdat.x = intc; opdat.y = intc;
opdat.z = intc;
```

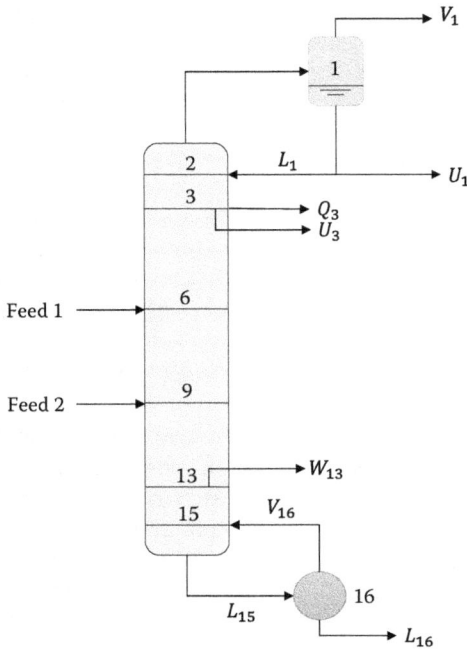

FIGURE 7.35 A simple 16-stage distillation column. (From Henley, E.J. et al., *Separation Process Principles*, 3rd ed., John Wiley & Sons, Hoboken, NJ, 2011, p. 411.)

```
% Physical properties and parameters for each component
mxdat.Pc = [709.8 617.4 550.7 489.5 440.0]; % critical pressure (psia)
mxdat.Tc = [550.0 665.9 765.3 845.9 914.2]; % critical temperature (R)
mxdat.k = zeros(opdat.nc,opdat.nc); % matrix of binary interaction parameters
mxdat.w = [0.1064 0.1538 0.1954 0.2387 0.2972]; % acentric factors
% Antoine equation parameters
mxdat.Ant = [5.38389 2847.921 434.898; 5.35342 3371.084 414.488;
5.74162  4126.385  409.5179;  5.853654  4598.287  394.4148;  6.03924  5085.758
382.794];
% Specific heat parameters (a(i))
mxdat.Afi = [11.51606 0.140309e-1 0.0854034e-4 -0.110608e-7 0.316220e-11;
15.58683 0.2504953e-1 0.1404258e-4 -0.352626e-7 1.864467e-11;
20.79783 0.314329e-1 0.192851e-4 -0.458865e-7 2.380972e-11;
25.64627 0.389176e-1 0.239729e-4 -0.584262e-7 3.079918e-11
30.17847 0.519926e-1 0.030488e-4 -0.27640e-7 1.346731e-11];
% Operating conditions (P: psia, T: F, flow: lbmol/hr)
```

TABLE 7.8
Physical Properties of Feed Mixture

Component	P_c (psia)	T_c (R)	u	A_1	A_2	A_3
Ethane C_2H_6 (1)	709.8	550.0	0.1064	5.38389	2847.921	434.898
Propane C_3H_8 (2)	617.4	665.9	0.1538	5.35342	3371.084	414.488
n-Butane C_4H_{10} (3)	550.7	765.3	0.1954	5.74162	4126.385	409.5179
n-Pentane C_5H_{12} (4)	489.5	845.9	0.2387	5.85365	4598.287	394.4148
n-Hexane C_5H_{12} (5)	440.0	914.2	0.2972	6.03924	5085.758	382.794

TABLE 7.9

Parameters of the Specific Heat Equation for Each Component

Component	a_1	$a_2 \times 10$	$a_3 \times 10^4$	$a_4 \times 10^7$	$a_5 \times 10^{11}$
Ethane C_2H_6(1)	11.51606	0.140309	0.085404	−0.110608	0.316220
Propane C_3H_8(2)	15.58683	0.250495	0.140426	−0.352626	1.864467
n-Butane C_4H_{10}(3)	20.79783	0.314329	0.192851	−0.458865	2.380972
n-Pentane C_4H_{10}(4)	25.64627	0.389176	0.239729	−0.584262	3.079918
n-Hexane C_5H_{12}(5)	30.17847	0.519926	0.030488	−0.27640	1.346731

```
opdat.nf = [6 9]; % feed stage
opdat.fstate = 'V'; % feed state
opdat.F(opdat.nf) = [41 59]; opdat.V(1) = 15; opdat.L(1) = 150;
opdat.U(1) = 5; opdat.U(3) = 3; opdat.W(13) = 37;
opdat.L(opdat.N) = sum(opdat.F(opdat.nf)) - opdat.V(1) - opdat.U(1) - opdat.U
(3) - opdat.W(13);
opdat.Pf(opdat.nf) = [300 275]; opdat.Tf(opdat.nf) = [170 230];
opdat.P = 240*ones(1,opdat.N); opdat.Q(3) = 2e5;
opdat.z(:,opdat.nf(1)) = [0.061 0.342 0.463 0.122 0.012];
opdat.z(:,opdat.nf(2)) = [0.0085 0.1017 0.3051 0.5085 0.0762];
% Initialization and guess
opdat.T0 = (300/opdat.N)*[1:opdat.N]; % guess stage temperatures (F)
opdat.V(2:opdat.N-1) = 170; opdat.criv = 1e-3; % convergence criterion
% BP calculation
[x,y,T,L,V,iter] = distBPeu(opdat,mxdat);
fprintf('Number of iterations (convergence criterion: %g): %d\n', opdat.-
criv, iter)
disp('Temperature: '), T
Nx = 1:opdat.N;
x1 = x(1,:); x2 = x(2,:); x3 = x(3,:); x4 = x(4,:); x5 = x(5,:);
y1 = y(1,:); y2 = y(2,:); y3 = y(3,:); y4 = y(4,:); y5 = y(5,:);
subplot(2,2,1), plot(Nx,T), xlabel('Stage'),ylabel('T(F)'), axis tight
subplot(2,2,2), plot(x1,Nx,x2,Nx,'--',x3,Nx,'.-',x4,Nx,':',x5,Nx,'*-')
xlabel('x'),ylabel('Stage'), legend('x_1','x_2','x_3','x_4','x_5'), axis
([0 1 1 16])
subplot(2,2,3), plot(y1,Nx,y2,Nx,'--',y3,Nx,'.-',y4,Nx,':',y5,Nx,'*-')
xlabel('y'),ylabel('Stage'), legend('y_1','y_2','y_3','y_4','y_5'), axis
([0 1 1 16])
```

The script *c5distBPeu* calculates the number of iterations and temperatures, and generates the temperature and composition profiles shown in Figure 7.36:

```
>> c5distBPeu
Number of iterations (convergence criterion: 0.001): 40
Temperature:
T =
102.0861 112.6251 118.4383 124.3199 132.5619 146.0313 157.7010 173.5853
195.1363 205.4129 215.8198 226.3022 237.5966 249.2418 262.1954 276.7688
```

(a)

(b)

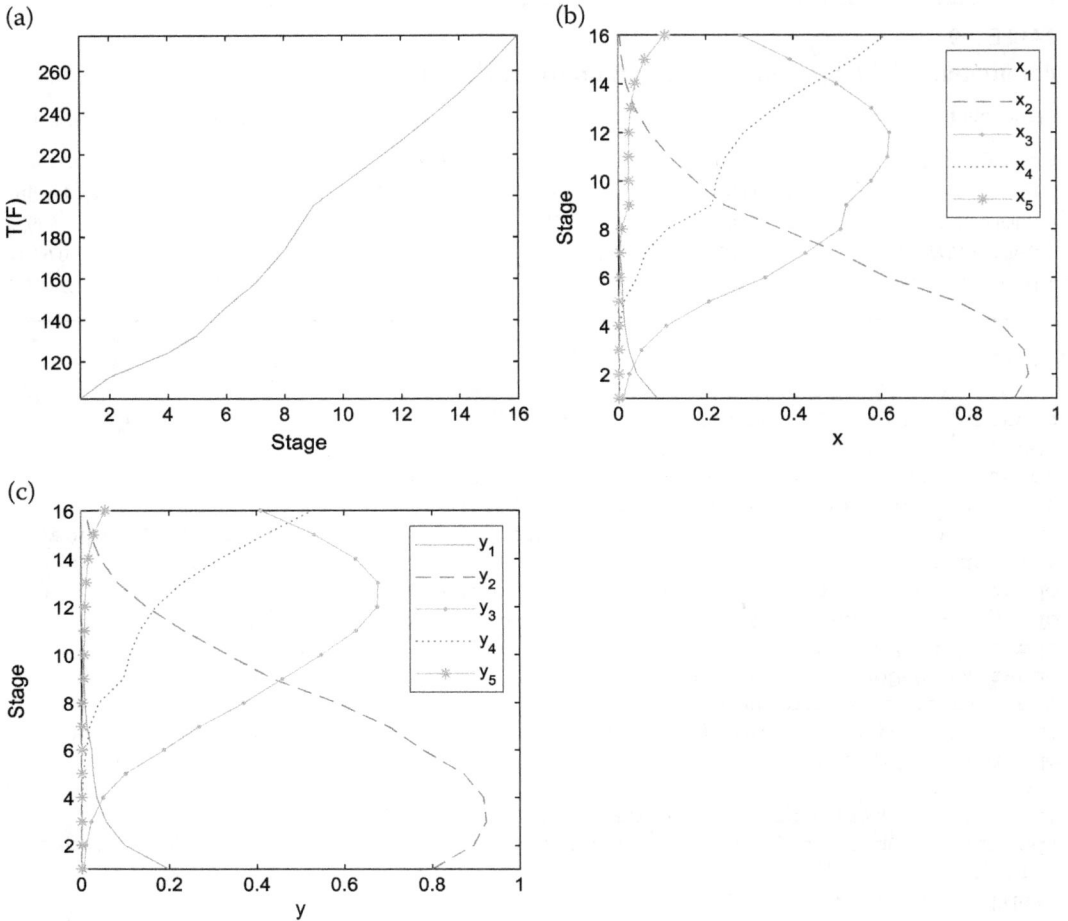

(c)

FIGURE 7.36 Profiles of (a) temperature, (b) liquid composition, and (c) vapor composition.

7.7 DIFFERENTIAL DISTILLATION

In simple differential distillation, the liquid is heated slowly to the boiling point and the vapor is collected and condensed. Figure 7.37 shows a simple single-stage batch distillation apparatus. At the start, the vapor condensed contains the richest fraction of the more volatile component. But as the vaporization continues, its concentration decreases.

The concentration of the more volatile component, x, in the liquid solution (L) changes through time. Thus, we can formulate the equation using the mass balance based on the more volatile component. Let dx be the change of the concentration of the more volatile component in the liquid solution, dL be the fractional amount of the liquid evaporated, and y be the concentration of the more volatile component in the vapor (V). The balance equation gives

$$xL = (x - dx)(L - dL) + ydL = xL - Ldx - xdL + dxdL + ydL$$

The magnitude of the term $dxdL$ is negligible compared to other terms in this equation. Then we have

$$xL = xL - Ldx - xdL + ydL$$

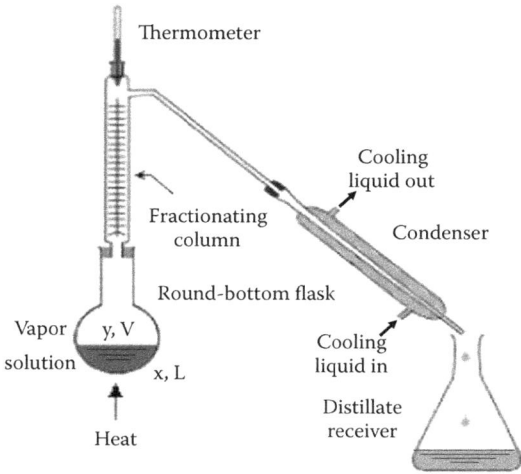

FIGURE 7.37   A simple single-stage batch distillation apparatus.

Rearrangement of this equation yields

$$-\frac{dL}{L} = \frac{dx}{x-y} \text{ or } \frac{dL}{dx} = \frac{L}{y-x}$$

Integration of both sides of this equation gives

$$-\int_{L_1}^{L_2} \frac{dL}{L} = \int_{x_1}^{x_2} \frac{dx}{x-y} \Rightarrow -\ln\frac{L_2}{L_1} = \int_{x_1}^{x_2} \frac{dx}{x-y}$$

where
   L_1 is the initial number of moles of the liquid solution
   L_2 is the number of moles of liquid after evaporation takes place

Example 7.25 Differential Distillation

A solution containing 70 moles of benzene and 50 moles of toluene is distilled using the simple differential distillation apparatus at 1 *atm* until only 50 moles of liquid is left. Calculate the composition of the liquid after the distillation process. Table 7.10 shows the phase equilibrium data for the benzene-toluene mixture.t

Solution

$$x_1 = \frac{70}{70+50} = 0.5833, \ \ln\frac{L_2}{L_1} = -\ln\frac{50}{120} = 0.8755$$

Thus, the value of x_2 satisfying

$$-\ln\frac{L_2}{L_1} = 0.8755 = \int_{0.5833}^{x_2} \frac{dx}{x-y} = \int_{0.5833}^{x_2} \frac{dx}{f(x)}$$

is to be determined. The built-in functions *quad* or *integral* can be used for the integration. The given equilibrium data can be regressed using a 4th-order polynomial. The script *ddbztol* performs the calculation procedure.

TABLE 7.10

Phase Equilibrium Data for Benzene-Toluene Mixture

x	0.0	0.1	0.2	0.3	0.4	0.5	0.6	0.7	0.8	0.9	1.0
y	0.000	0.211	0.378	0.512	0.623	0.714	0.791	0.856	0.911	0.959	1.000

```
% ddbztol.m
x = [0.0 0.1 0.2 0.3 0.4 0.5 0.6 0.7 0.8 0.9 1.0];
y = [0.000 0.211 0.378 0.512 0.623 0.714 0.791 0.856 0.911 0.959 1.000];
p = polyfit(x,y,4);
f = @(x) 1./(x-(p(1)*x.^4+p(2)*x.^3+p(3)*x.^2+p(4)*x+p(5)));
x1 = 0.5833; vfix = 0.8755; crit = 1e-2;
for x2 = x1:-1e-3:0.0, if abs(integral(f,x1,x2)-vfix) <= crit, break, end
end
xf = x2

>> ddbztol
xf =
   0.3983
```

We can see that the benzene composition in the residue solution after the distillation operation is $x_f = 0.3983$.

Example 7.26 Single-Stage Batch Distillation[57]

A single-stage batch distillation apparatus is used for separation of ethanol (2) from water (1). A liquid mixture of 40 *mol%* water ($x_1 = 0.4$) and 60 *mol%* ethanol ($x_2 = 0.6$) is charged initially to the still pot. The amount of the initial charge is 100 *kgmol* ($L_1 = 100$). The distillation is carried out at 1 *atm* total pressure and is continued until the water mole fraction reaches 0.8 ($x_1 = 0.8$). Plot the temperature profile and the amount of liquid remaining in the still as a function of the water composition. Calculate the temperature and the composition of the liquid remaining in the still after the distillation process. The vapor-liquid equilibrium follows Raoult's law: $k_j = \gamma_j P_j/P$. The vapor pressure P_j is given by the Antoine equation, and the activity coefficient of component j, γ_j, is given by

$$\log\gamma_1 = (1 - x_1)^2\{c_1 + 2x_1(c_2 - c_1)\}, \quad \log\gamma_2 = (1 - x_2)^2\{c_2 + 2x_2(c_1 - c_2)\}$$

where $c_1 = 0.3781$ and $c_2 = 0.6848$. The temperature change with respect to the liquid composition may be represented by

$$\frac{dT}{dx_1} = K(1 - k_1x_1 - K_2x_2) \quad (K\text{:constant})$$

The constant K may take a large value, such as 500,000. The Antoine equation parameters for water and ethanol are given in Table 7.11.

Solution

The subfunction *bf* defines the system of algebraic differential equations. The script *btdist* calls the function *bf* and employs the built-in function *ode45* to find solutions.

```
% btdist.m
%% z(1) = T, z(2) = L
A = [7.96681 8.04494]; B = [1668.21 1554.3]; C = [228 222.65];
```

TABLE 7.11

Antoine Equation Parameters for Water and Ethanol ($T{:}°C, P{:}mmHg$)

	Component	Composition (Mole Fraction)	Antoine Equation Parameters		
			A	B	C
1	Water	0.4	7.96681	1668.21	228
2	Ethanol	0.6	8.04494	1554.3	222.65

```
c = [0.3781 0.6848]; Pt = 760; K = 5e5; x1v = [0.4 0.8]; z0 = [79 100];
[x1 z] = ode45(@bf,x1v,z0,[],Pt,A,B,C,c,K);
T = z(:,1); L = z(:,2); fprintf('Tfinal = %g, Lfinal = %g\n',T(end),L(end))
subplot(1,2,1), plot(x1,T), xlabel('x_1'), ylabel('T(C)')
subplot(1,2,2), plot(x1,L), xlabel('x_1'), ylabel('L(kgmole)')

function dz = bf(x1,z,Pt,A,B,C,c,K)
% z(1) = T, z(2) = L
x2 = 1-x1; x = [x1 x2]; Pj = 10.^(A - B./(z(1) + C));
gam(1) = 10^((1-x1)^2*(c(1) + 2*x1*(c(2)-c(1)))); gam(2) = 10^((1-x2)^2*(c(2)
+ 2*x2*(c(1)-c(2)))));
k = gam.*Pj/Pt;
dz(1,1) = K*(1 - k(1)*x1 - k(2)*x2); dz(2,1) = z(2)/(x1*(k(1)-1));
end
```

The script btdist generates the following results and the plots shown in Figure 7.38:

```
>> btdist
Tfinal = 82.7562, Lfinal = 13.0123
```

We can see that the final temperature is 82.76 °C and the amount of the liquid remaining in the still is 13.01 *kgmol*.

7.8 FILTRATION

Filtration is used to remove solid particles from a fluid by passing the fluid through a medium. Fluid flows through the medium due to a pressure difference across the medium. The filter medium must remove the solids to be filtered from the slurry and give a clear filtrate, and allow the filter cake to be removed easily. The pressure drop of fluid through filter cake, for a laminar flow, can be estimated by the Carman-Kozeny relation as

$$\frac{\Delta p_c}{L} = -\frac{k_1 \mu v (1 - \varepsilon)^2 S_0^2}{\varepsilon^3}$$

where
  $\Delta p_c (N/m^2)$ is the pressure drop in the cake
  $L (m)$ is the thickness of the cake
  k_1 is a constant
  $\mu (Pa{\cdot}sec)$ is the viscosity
  $v (m/sec) = (dV/dt)/A$ is the velocity
  $A (m^2)$ is the filter area
  $V (m^3)$ is the volume of filtrate collected up to time $t (sec)$
  ε is the void fraction (porosity) of the cake

(a)

(b)

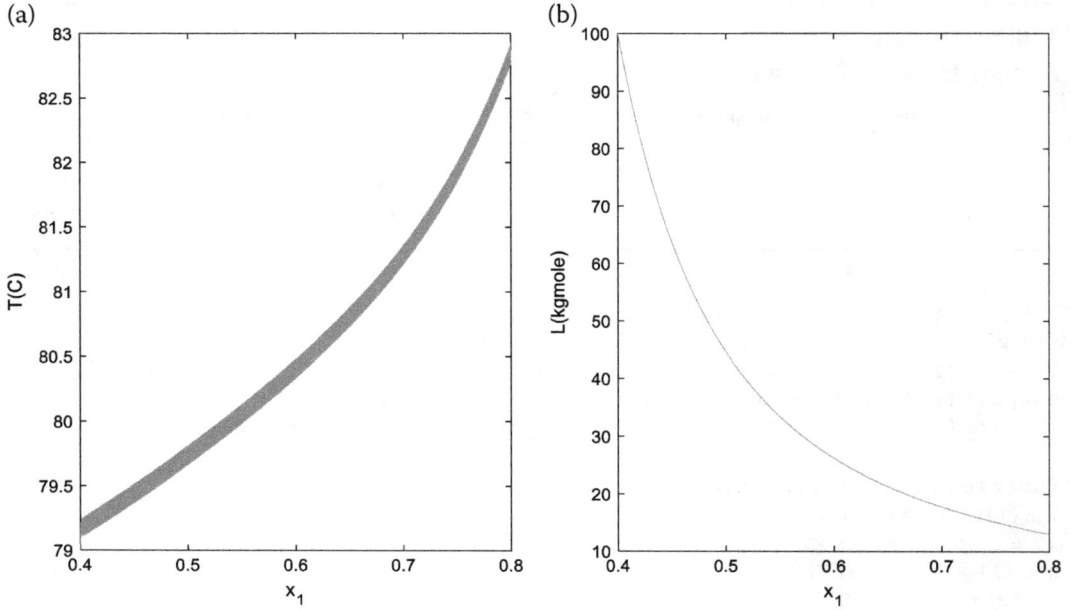

FIGURE 7.38   (a) Temperature and (b) liquid profiles in the batch distillation process.

$S_0 (m^2/m^3)$ is the specific surface area of the particle
Applying the definition of v and material balance, we have[58]

$$\frac{dV}{Adt} = \frac{-\Delta p_c \rho_p \varepsilon^3 A}{\mu c_s V k_1 (1-\varepsilon) S_0{}^2} = \frac{-\Delta p_c A}{\alpha \mu c_s V}, \ \alpha = \frac{k_1(1-\varepsilon)S_0{}^2}{\rho_p \varepsilon^3}$$

where
  $\rho_p (kg/m^3)$ is the density of solid particles in the cake
  $c_s (kg/m^3)$ is the solid mass per unit volume of filtrate
  For the filter medium resistance, we have

$$\frac{dV}{Adt} = \frac{-\Delta p_f}{\mu R_m}$$

where
  $\Delta p_f (N/m^2)$ is the pressure drop in the medium
  $R_m (m^{-1})$ is the resistance of the medium to filtrate flow
  Combination of these two equations yield

$$\frac{dV}{Adt} = \frac{-\Delta p}{\mu \left(\frac{\alpha c_s V}{A} + R_m \right)} = \frac{-\Delta p}{\frac{\mu \alpha c_s}{A}(V+V_e)} = \frac{-(\Delta p_c + \Delta p_f)}{\frac{\mu \alpha c_s}{A}(V+V_e)}$$

where $V_e = AR_m/(\alpha c_s)$ is the filtrate volume necessary to build up a fictitious filter cake whose resistance is equal to R_m.V can be related to the amount of accumulated dry cake, $W (kg)$, as follows:

$$W = c_s V = \left(\frac{\rho c_x}{1 - m c_x} \right) V$$

where

$\rho \, (kg/m^3)$ is the filtrate density

c_x is the mass fraction of solid in the slurry

m is the mass ratio of wet cake to dry cake

Usually α depends on Δp and an empirical equation $\alpha = \alpha_0 \, (-\Delta p)^s$ or $\alpha = \alpha_0' \, \{1 + \beta \, (-\Delta p)^{s'}\}$ is used where α_0, α_0', s and s' are constants.

For constant pressure filtration, we have[59]

$$\frac{dt}{dV} = \frac{\alpha \mu c_s}{A^2 \, (-\Delta p)} V + \frac{\mu R_m}{A \, (-\Delta p)} = K_p V + B, \; K_p = \frac{\alpha \mu c_s}{A^2 \, (-\Delta p)}, \; B = \frac{\mu R_m}{A \, (-\Delta p)}$$

If α is constant and the cake is incompressible, this equation can be integrated to give, after some rearrangement

$$\frac{t}{V} = \frac{K_p V}{2} + B$$

A filtration cycle is followed by washing of the filter cake. For constant-pressure filtration, the rate of washing, $(dV/dt)_f \, (m^3/sec)$, is given by

$$\left(\frac{dV}{dt}\right)_f = \frac{1}{K_p V_f + B}$$

where $V_f \, (m^3)$ is the filtrate volume at the end of filtration. For a plate-frame filter press, the washing rate is one-fourth of the final filtration rate, or $(1/4)(dV/dt)_f$.

In continuous filtration, the resistance of the filter medium is negligible compared with the cake resistance. Thus we set $B = 0$ and integration gives

$$t = \frac{K_p V^2}{2}$$

In a rotary-drum filter, t is less than the total cycle time t_c and can be represented as $t = f t_c$, where f is the fraction of the cycle used for cake formation. In this case, the flow rate $v \, (m/sec)$ is given by

$$v = \frac{V}{t_c A} = \sqrt{\frac{2f \, (-\Delta p)}{\alpha \mu c_s t_c}}$$

If cycle times are short in continuous filtration, t can be represented as

$$t = f t_c = \frac{K_p V^2}{2} + BV$$

and the flow rate is given by[60]

$$\frac{V}{t_c A} = \frac{1}{\alpha c_s} \left\{ \sqrt{\frac{R_m^2}{t_c^2} + \frac{2\alpha c_s f \, (-\Delta p)}{\mu t_c}} - \frac{R_m}{t_c} \right\}$$

For constant rate filtration, the pressure drop $\Delta p \, (N/m^2)$ is given by

$$-\Delta p = \left(\frac{\alpha \mu c_s}{A^2} \frac{dV}{dt}\right) V + \left(\frac{\mu R_m}{A} \frac{dV}{dt}\right) = K_v V + C, \; K_v = \left(\frac{\alpha \mu c_s}{A^2} \frac{dV}{dt}\right), \; C = \left(\frac{\mu R_m}{A} \frac{dV}{dt}\right)$$

Substituting $V = tdV/dt$ into this equation, we have[61]

$$-\Delta p = \left\{ \frac{\alpha \mu c_s}{A^2} \left(\frac{dV}{dt} \right)^2 \right\} t + \left(\frac{\mu R_m}{A} \frac{dV}{dt} \right)$$

Example 7.27 Constant-Pressure Filtration[62]

Table 7.12 shows experimental data for filtration of $CaCO_3$ slurry in water at 298.2 K at a constant pressure of $(-\Delta p) = 338 \, kN/m^2$ using a plate-frame press. The filter area is $A = 0.0439 \, m^2$ and the slurry concentration is $c_s = 23.47 \, kg/m^3$. At 298.2 K, the viscosity of water is $\mu = 8.937 \times 10^{-4} \, kg/(m \cdot sec)$. Evaluate the constants α and R_m. Assume that α is constant and the cake is incompressible.

Solution

Since α is constant and the cake is incompressible, we use the relationship $t/V = K_p V/2 + B$ and the required constants can be obtained from K_p and B as $\alpha = K_p A^2 (-\Delta p)/(\mu c_s)$ and $R_m = A(-\Delta p)B/\mu$. K_p and B can be estimated as follows:

$$\begin{bmatrix} (t/V)_1 \\ \vdots \\ (t/V)_n \end{bmatrix} = \begin{bmatrix} (V/2)_1 & 1 \\ \vdots & \vdots \\ (V/2)_n & 1 \end{bmatrix} \begin{bmatrix} K_p \\ B \end{bmatrix} \Rightarrow Y = CX \Rightarrow X = (C^T C)^{-1} C^T Y$$

```
% filtconsP.m: constant pressure filtration
clear all;
t = [4.4, 9.5, 16.3, 24.6, 34.7, 46.1, 59.0, 73.6, 89.4, 107.3]'; % t (s)
V = 1e-3*[0.498, 1.0, 1.501, 2.0, 2.498, 3.002, 3.506, 4.004, 4.502, 5.009]'; %
V(m^3)
dP = 338e3; A = 0.0439; cs = 23.47; mu = 8.937e-4;
n = length(t); Y = t./V; C = [V/2 ones(n,1)]; X = inv(C'*C)*C'*Y;
Kp = X(1); B = X(2); alpa = Kp*A^2*dP/(mu*cs); Rm = A*dP*B/mu;
fprintf('Alpha = %g m/kg\n', alpa); fprintf('Rm = %g m^(-1)\n', Rm);

>> filtconsP
Alpha = 1.79188e+11 m/kg
Rm = 1.12631e+11 m^(-1)
```

7.9 MEMBRANE SEPARATION

In membrane processes for gas separation, a gas phase is present on both sides of the membrane, which is usually a polymer such as rubber, polyamide, and so on. The solute gas first dissolves in

TABLE 7.12
Filtrate Volume

t (sec)	V (m^3)	t (sec)	V (m^3)
4.4	0.498×10^{-3}	46.1	3.002×10^{-3}
9.5	1.000×10^{-3}	59.0	3.506×10^{-3}
16.3	1.501×10^{-3}	73.6	4.004×10^{-3}
24.6	2.000×10^{-3}	89.4	4.502×10^{-3}
34.7	2.498×10^{-3}	107.3	5.009×10^{-3}

the membrane and then diffuses in the solid to the other gas phase. High-pressure feed gas is supplied to one side of the membrane and permeates normal to the membrane. The permeate leaves in a direction normal to the membrane, accumulating on the low-pressure side. Because of the very high diffusion coefficient in gases, concentration gradients in the gas phase in the direction normal to the surface of the membrane are very small. Hence, gas film resistances compared to the membrane resistance may be neglected.

There are several cases in the operation of a membrane module. Separate theoretical models can be derived for different types of operation. In deriving models for gas separation by membranes, isothermal conditions and negligible pressure drop in the feed stream are assumed. It is also assumed that pressure drop in the gas stream can be calculated by using the Hagen-Poiseuille equation and that the permeability of each component is constant (i.e., no interactions between different components).

7.9.1 COMPLETE-MIXING MODEL FOR GAS SEPARATION

In a complete-mixing model, changes in compositions of gas streams are assumed to be negligible. Figure 7.39 shows a flow diagram for a complete-mixing model.

The overall material balance is given by

$$L_f = L_R + V_p$$

where

L_f is the molar flow rate of the feed stream

L_R is the molar flow rate of the reject stream

V_p is the molar flow rate of the permeate stream

The cut or fraction of the feed permeated, θ, is expressed as

$$\theta = \frac{V_p}{L_f}$$

For a binary feed consisting of components A and B, the rate of permeation of component A is given by

$$\frac{q_A}{A_m} = \frac{V_p y_p}{A_m} = \left(\frac{P_A}{t}\right)(p_h x_R - p_l y_p)$$

where

P_A is the permeability of A in the membrane

q_A is the flow rate of A in the permeate

A_m is the membrane area

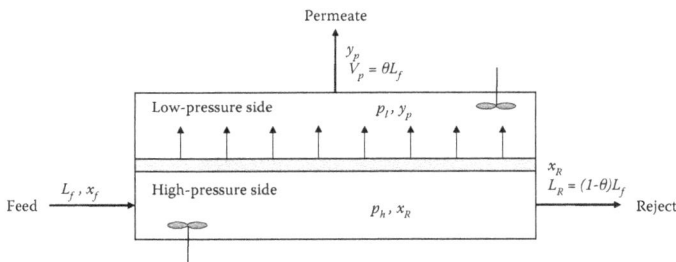

FIGURE 7.39 Complete-mixing flow pattern.

p_h is the total pressure in the high-pressure side
p_l is the total pressure in the low-pressure side
x_R is the mole fraction of A in the reject
y_p is the mole fraction of A in the permeate
t is the membrane thickness
The rate of permeation of component B is given by

$$\frac{q_B}{A_m} = \frac{V_p(1 - y_p)}{A_m} = \left(\frac{P_B}{t}\right)[p_h(1 - x_R) - p_l(1 - y_p)]$$

where P_B is the permeability of B in the membrane. From these two equations, we have

$$y_p\left\{(1 - x_R) - r\left(1 - y_p\right)\right\} = \alpha(1 - y_p)(x_R - ry_p)$$

where $\alpha = P_A/P_B = P_{CO_2}/P_{N_2}$ and $r = p_l/p_h$. Solving this equation for y_p yields

$$y_p = \frac{-b + \sqrt{b^2 - 4ac}}{2a}$$

where $a = 1 - \alpha$, $b = -1 + \alpha + (1/r) + (\alpha - 1)x_R/r$, and $c = -\alpha x_R/r$. The concentrations of the reject and permeate streams as well as the cut are given by

$$x_R = \frac{x_f - \theta y_p}{(1 - \theta)}, \; y_p = \frac{x_f - (1 - \theta)x_R}{\theta}, \; \theta = \frac{x_f - x_R}{y_p - x_R}$$

The membrane area can be expressed as

$$A_m = \frac{\theta L_f y_p}{(P_A/t)(p_h x_R - p_l y_p)}$$

The recovery, r_c, defined as the ratio of the amount of A permeated to the amount of A supplied, is given by

$$r_c = \frac{V_p y_p}{L_f x_f} \times 100$$

If all the feed supplied is permeated (that is, $V_p = L_f$), $\theta = 1$ and $y_p = x_f$. Hence,

$$x_R = \frac{x_f\{1 + r(\alpha - 1)(1 - x_f)\}}{x_f(1 - \alpha) + \alpha}$$

This equation represents the minimum reject composition. In calculations, two cases can be considered: (1) x_f, x_R, α, r are given and y_p, θ, A_m are to be determined; (2) x_f, θ, α, r are given and y_p, x_R, A_m are to be determined.

The function *cpm1ex* calculates the membrane separation process based on the complete-mixing model for a binary feed. The syntax is

```
res = cpm1ex
```

This function calls the script *cpmdata* that specifies membrane data and operating conditions.

```
function res = cpm1ex
```

```
cpmdata;
Pa = Pm(1); a = 1 - alpa; b = -1 + alpa + 1./r + (alpa-1)*xr./r;
c = -alpa*xr./r; yp = (-b+sqrt(b.^2 - 4*a.*c))./(2*a); % permeate mole fraction
theta = (xf-xr)./(yp-xr); % stage-cut
Am = (theta.*qf.*yp)./((Pa./t).*(ph.*xr - pl.*yp)); % membrane area
rc = qf*theta*yp/(qf*xf); % recovery ratio
xom = (xf*(1+r*(alpa-1)*(1-xf)))/(xf*(1-alpa)+alpa);
% Results: res = [yp, xr, Am, theta, rc]
res.yp = yp; res.theta = theta; res.Am = Am; res.rc = rc; res.xr = xr;
end
```

Example 7.28 Complete-Mixing Model[63]

A membrane is to be used to separate a gaseous mixture of A and B. The feed flow rate is $L_f = 1 \times 10^4 \ cm^3/sec$ and the feed composition of A is $x_f = 0.5$ (mole fraction). The desired composition of the reject is $x_R = 0.25$. Calculate the permeate composition y_p, the stage-cut θ, and the membrane area A_m.

Data: membrane thickness $t = 2.54 \times 10^{-3} \ cm$, feed pressure $p_h = 80 \ cmHg$, permeate-side pressure $p_l = 20 \ cmHg$

Permeability of A: $P_A = 50 \times 10^{-10} \ cm^3 \cdot cm/(sec \cdot cm^2 \cdot cmHg)$,
Permeability of B: $P_B = 5 \times 10^{-10} \ cm^3 \cdot cm/(sec \cdot cm^2 \cdot cmHg)$

Solution
The script *cpmdata* specifies the membrane data and operating conditions.

```
% cpmdata.m: gas mixture and membrane data
t = 0.00254; % membrane thickness (cm)
Pm = [50 5]*1e-10; % permeability (cm^3*cm/(sec*cm^2*cmHg)
alpa = Pm(1)/Pm(2);
ph = 80; % feed-side pressure (cmHg)
pl = 20; % permeate-side pressure (cmHg)
r = pl/ph; % pressure ratio (Plow/Phigh)
qf = 1e4; % feed flow rate (cm^3/sec(STP))
xf = 0.5; % feed composition (mole fraction)
xr = 0.25; % desired reject composition (mole fraction)
```

The function *cpm1ex* produces the following results:

```
>> cpm1ex
ans =
yp: 0.6036
theta: 0.7071
Am: 2.7344e+08
rc: 0.8535
xr: 0.2500
```

7.9.2 CROSS-FLOW MODEL FOR GAS SEPARATION

In a cross-flow pattern, the velocity of the high-pressure gas stream is large enough that this stream is in plug flow and flows parallel to the membrane. The flow of the permeate stream in the low-pressure side is essentially perpendicular to the membrane. It is assumed that there is no mixing in either the low-pressure or the high-pressure side. Thus, the permeate composition at any point along the membrane is determined by the relative rates of permeation of the feed components at that point. Figure 7.40 shows a flow diagram for a cross-flow model.

Rearranging the equations representing the local permeation rate over a differential membrane area dA_m at any point gives

$$\frac{y}{1-y} = \frac{\alpha(x-ry)}{(1-x)-r(1-y)}$$

where $\alpha = P_A/P_B = P_{CO_2}/P_{N_2}$ and $r = p_l/p_h$. The analytical solution was given by Weller and Steiner as[64]

$$\frac{(1-\theta^*)(1-x)}{(1-x_f)} = \left(\frac{u_f - E/D}{u - E/D}\right)^R \left(\frac{u_f - \alpha + F}{u - \alpha + F}\right)^S \left(\frac{u_f - F}{u - F}\right)^T$$

where

$$\theta^* = \frac{L_f - L}{L_f} = 1 - \frac{L}{L_f}, \; i = \frac{x}{1-x}\left(orx = \frac{i}{i+1}\right), \; u = -Di + \sqrt{D^2i^2 + 2Ei + F^2}$$

$$D = \frac{1}{2}[(1-\alpha)r + \alpha], \; E = \frac{\alpha}{2} - DF, \; F = -\frac{1}{2}[(1-\alpha)r - 1], \; R = \frac{1}{2D-1}$$

$$S = \frac{\alpha(D-1)+F}{(2D-1)(\alpha/2 - F)}, \; T = \frac{1}{1 - D - (E/F)}$$

u_f denotes the value of u at $i = i_f = x_f/(1-x_f)$ and θ^* represents the fraction permeated up to x. At the outlet, $x = x_R$ and θ^* is equal to the total permeation fraction θ. The total membrane area, A_m, may be obtained from the equation

$$A_m = \frac{L_f}{p_h(P_B/t)} \int_{i_R}^{i_f} \frac{\left(1-x_f\right)\left(\frac{u_f - E/D}{u - E/D}\right)^R \left(\frac{u_f - \alpha + F}{u - \alpha + F}\right)^S \left(\frac{u_f - F}{u - F}\right)^T}{(f_i - i)\left(\frac{1}{1+i} - \frac{r}{1+f_i}\right)} di$$

where $i_f = \frac{x_f}{1-x_f}$, $i_R = \frac{x_R}{1-x_R}$.

The function *crflex* calculates the membrane separation process based on the cross-flow model for a binary feed mixture. The basic syntax is

```
res = crflex
```

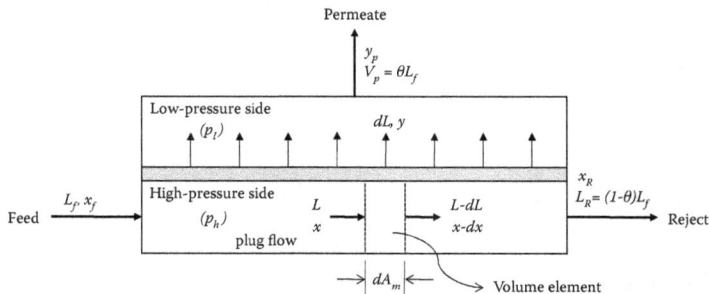

FIGURE 7.40 Cross-flow pattern.

This function calls the script *crfdata*, which specifies the membrane data and operating conditions. The subfunction *mArea* defines the function to be integrated to give the membrane area.

```
function res = crf1ex
% Membrane separation process for binary feed using cross-flow model
clear all;
crfdata;
Pa = Pm(1); Pb = Pm(2); D1 = ((1-alpa)*r + alpa)/2;
F1 = -((1-alpa)*r - 1)/2; E1 = alpa/2 - D1.*F1;
R1 = 1./(2*D1-1); S1 = (alpa.*(D1-1) + F1)./((2*D1 - 1).*(alpa/2 - F1));
T1 = 1./(1 - D1 - E1./F1);
i0 = xr./(1-xr); i2 = xf./(1-xf);
ur = -D1.*i0 + sqrt((D1.^2).*i0.^2 + 2*E1.*i0 + F1.^2);
uf = -D1.*i2 + sqrt((D1.^2).*i2.^2 + 2*E1.*i2 + F1.^2);
theta = 1 - ((1-xf)./(1-xr)).*((uf-E1./D1)./(ur-E1./D1)).^R1...
.*((uf-alpa+F1)./(ur-alpa+F1)).^S1 .*((uf-F1)./(ur-F1)).^T1;
yp = (xf - (1-theta)*xr)/theta;
Ami = quad(@(x) mArea(x,D1,E1,F1,R1,S1,T1,alpa,r,xf,uf), i0, i2);
Am = Ami*qf*t/(ph*Pb); rc = theta*yp./xf; % recovery ratio
% Calculated variables: yp(permeate mole fraction), Am(area)
% theta(stage-cut) or xr(reject composition)
% Results: results = [yp, xr, Am, theta, rc]
res.yp = yp; res.theta = theta; res.Am = Am; res.rc = rc; res.xr = xr;
end


function y = mArea(x,D,E,F,R,S,T,alpa,r,xf,uf)
u = -D.*x + sqrt((D.^2).*(x.^2) + 2*E.*x + F.^2);
fi = (D.*x - F) + sqrt((D.^2)*(x.^2) + 2*E.*x + F.^2);
ud = (1-xf).*((uf-E/D)./(u-E/D)).^R.*((uf-alpa+F)./(u-alpa+F)).^S...
.*((uf-F)./(u-F)).^T;
y = ud./((fi-x).*(1./(1+x) - r./(1+fi)));
end
```

Example 7.29 Cross-Flow Model[65]

A membrane is to be used to separate a gaseous mixture of *A* and *B*. The feed flow rate is $L_f = 1 \times 10^4$ cm^3/sec and the feed composition of *A* is $x_f = 0.5$ (mole fraction). The desired composition of the reject is $x_R = 0.25$. Use the cross-flow model to determine the permeate composition y_p, the stage-cut θ, and the membrane area A_m.

Data: membrane thickness $t = 2.54 \times 10^{-3}$ cm, feed pressure $p_h = 80 cmHg$, permeate-side pressure $p_l = 20$ $cmHg$
   Permeability of *A*: $P_A = 50 \times 10^{-10}$ $cm^3 \cdot cm/(sec \cdot cm^2 \cdot cmHg)$,
   Permeability of *B*: $P_B = 5 \times 10^{-10}$ $cm^3 \cdot cm/(sec \cdot cm^2 \cdot cmHg)$

Solution
The script *crfdata* specifies the membrane data and operating conditions.

```
% crfdata.m: cross-flow calculation data
t = 0.00254; % membrane thickness (cm)
Pm = [50 5]*1e-10; % permeability(cm^3*cm/(s*cm^2*cmHg)
alpa = Pm(1)/Pm(2); % ratio of permeabilities
ph = 80; % feed side pressure(cmHg)
pl = 20; % permeate side pressure(cmHg)
```

```
r = pl/ph; % pressure ratio (Plow/Phigh)
qf = 1e4; % feed rate(cm^3/s(STP))
xf = 0.5; % Feed composition (mole fraction)
theta = []; % stage-cut
xr = 0.25; % desired reject composition (mole fraction)
```

The function *crflex* generates the following outputs:

```
>> crflex
ans =
yp: 0.7670
theta: 0.4835
Am: 1.2677e+08
rc: 0.7418
xr: 0.2500
```

PROBLEMS

7.1 Methanol (A) is evaporated into a stream of dry air (B) in a cylindrical tube at 328.5 K. The distance from the tube inlet to the liquid surface is $z_2 - z_1 = 0.238$ m. At $T = 328.5$ K, the vapor pressure of methanol is $P_{A0} = 68.4$ kPa and the total pressure is $P = 99.4$ kPa. The binary molecular diffusion coefficient of methanol in air under these conditions is $D_{AB} = 1.991 \times 10^{-5}$ m^2/sec. The temperature profile in the tube exhibits a linear characteristic from the liquid surface ($T = 328.5$ K) to the tube inlet ($T = 295$ K). Calculate the molar flux of methanol within the tube and plot the mole fraction profile of methanol from the liquid surface to the flowing air stream.

7.2 Gases A and B are diffusing through stagnant gas C at a temperature of 55 °C and a pressure of 0.2 atmospheres from point 1 (z_1) to point 2 (z_2). The distance between these two points is 0.001 m. The molar flux of A is measured to be $N_A = 2.115 \times 10^{-5}$ $kgmol/(m^2 \cdot sec)$. The gas mixture is assumed to be an ideal gas. Estimate the molar flux of B (N_B) and plot the mole fraction profile for each component as a function of z.

 Data: $C_{A1} = 2.229 \times 10^{-4}$, $C_{A2} = 0$, $C_{B0} = 0$, $C_{B2} = 2.701 \times 10^{-3}$, $C_{C1} = 7.208 \times 10^{-3}$, $C_{C2} = 4.730 \times 10^{-3}$, $D_{AB} = 1.47 \times 10^{-4}$, $D_{AC} = 1.075 \times 10^{-4}$, $D_{BC} = 1.245 \times 10^{-4}$

7.3 Dichlorobenzene (A), suspended in stagnant air (B), is sublimed at 25 °C and atmospheric pressure. The sublimation is taking place at the surface of a sphere of solid dichlorobenzene with a radius of 3×10^{-3} m. The material balance in the gas phase gives

$$(N_A 4\pi r^2)|_r \Delta t = \left(\frac{V p_A}{RT}\right)\Bigg|_{t+\Delta t} - \left(\frac{V p_A}{RT}\right)\Bigg|_t \Rightarrow \frac{dp_A}{dt} = \frac{4\pi r^2 k_c' P}{V} \frac{(p_A - p_{A2})}{p_{BM}}$$

The material balance in the spherical particle gives

$$0 = (N_A 4\pi r^2)|_r \Delta t + \left(\frac{4}{3}\pi r^3 \frac{\rho_A}{M_A}\right)\Bigg|_{t+\Delta t} - \left(\frac{4}{3}\pi r^3 \frac{\rho_A}{M_A}\right)\Bigg|_t \Rightarrow \frac{dr}{dt} = -\frac{N_A M_A}{\rho_A}$$

where $N_A = k_c' P(p_A - p_{A2})/(RT p_{BM})$, $p_{BM} = (p_A - p_{A2})/\ln((P - p_{A2})/(P - p_A))$, and $k_c' = D_{AB}/r$. The vapor pressure of dichlorobenzene at 25 °C is 1 $mmHg$, and the diffusivity in air is 7.39×10^{-6} m^2/sec. The density of dichlorobenzene is 1458 kg/m^3 and the molecular weight is 147.

 Calculate the time necessary for the complete sublimation of a single particle of dichlorobenzene if the particle is enclosed in a volume of $V = 0.05$ m^3. Plot the partial pressure p_A as a function of the radius r.

7.4 Calculate the concentration profile for C_A and determine the effectiveness factor η for a 1st-order irreversible reaction in a cylindrical catalyst particle, where $R = 0.5$ cm, $D_e = 0.1$ cm^2/sec, $C_{As} = 0.2$ gmol/cm^3, and $k_1 a = 6.4$ sec^{-1}.[66]

7.5 A triple-effect forward-feed evaporator is being used to evaporate an organic colloid solution containing 15% solids to a concentrated solution of 60% solids. Saturated steam at 29.82 psia is being used, and the pressure in the vapor phase of the third effect is 1.27 psia. The feed rate is 45,000 lb/hr at 60 °F. The heat capacity of the liquid solutions is the same as that of water over the whole concentration range. The overall heat transfer coefficients have been estimated as $U_1 = 530$, $U_2 = 350$, and $U_3 = 195$ Btu/(ft^2·hr·°F). If the heat transfer areas of each of the three effects are to be equal, calculate the heat transfer area (ft^2) and the amount of steam consumed (lb/hr). The boiling point elevation of the solutions is assumed to be negligible.

7.6 SO_2 in air is absorbed by pure water at $P = 1$ atm and $T = 20$ °C in the gas absorber shown in Figure P7.6. The gas feed stream is 1.389 kg/sec containing 8 mol% SO_2 in air, and 95% recovery of SO_2 is required.[67] The flow rate of inlet solvent water is 29.5 kg/sec. The solubility of SO_2 in water is shown in Table P7.6.[67] The packing material is $1\frac{1}{2}$in. ceramic Intalox saddles. Determine the number of transfer units N_t, the heights of transfer units for the gas film, H_G, and the liquid film, H_L, and the packing height of the column, H_{pd}, required to fulfill the requirement using Cornell's method.

Data: $D_L = 1.7 \times 10^{-9}$ m^2/sec, $D_G = 1.45 \times 10^{-5}$ m^2/sec, $\mu_L = 1 \times 10^{-3}$ N·sec/m$^2 = 1cP$, $\mu_G = 1.8 \times 10^{-5}$ N·sec/m^2, $F_p = 170$ m^{-1}, $\psi = 1$, $f = 0.66$, $f_1 = f_2 = f_3 = 1$, $\psi_h = 80$, $\phi_h = 0.1$, $K_3 = 0.85$.

7.7 Figure P7.7 shows a packed column being used to scrub a gaseous feed consisting of NH_3 and inert gas with water. The mole fraction of NH_3 in the feed stream is $y_f = 0.4$, and the flow rate of the inert gas is $V_n = 900$ lbmol/hr. The desired concentration of NH_3 in the outlet gas stream is $y_d = 0.032$. The estimated value of the gas film mass transfer coefficient, k_{ya}, is constant at 30 lbmol/(hr·ft^3). The diameter of the column is $D = 4ft$. The required height of the packed column, z, which can be obtained from

$$z = \frac{V_n}{k_{ya}S}\int_{y_d}^{y_f}\frac{1}{(1-y)^2\ln\left(\frac{1-y_i}{1-y}\right)}dy = \frac{V_n}{k_{ya}S}I_y$$

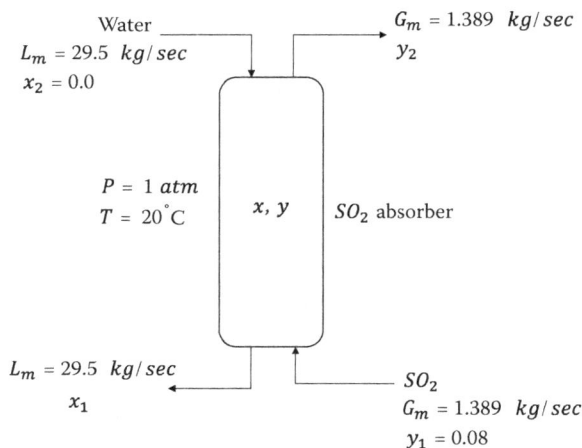

FIGURE P7.6 SO_2 absorber. (Modified from Nayef Ghasem, *Computer Methods in Chemical Engineering*, CRC Press, Taylor & Francis Group, Boca Raton, FL, 2012, p. 356.)

TABLE P7.6
Solubility of SO_2 in Water

Mole Fraction of SO_2 in Water (x)	Mole Fraction of SO_2 in Air (y)
0.00014	0.002
0.00042	0.008
0.00085	0.018
0.00141	0.034
0.00198	0.051
0.00283	0.077

Source: Nayef Ghasem, *Computer Methods in Chemical Engineering*, CRC Press, Taylor & Francis Group, Boca Raton, FL, 2012, p. 356.

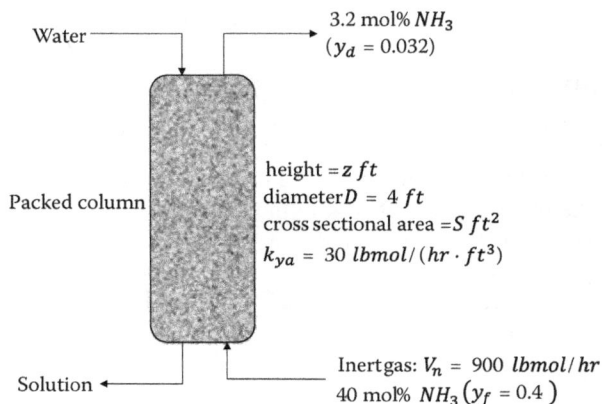

FIGURE P7.7 A packed column to scrub a gaseous feed containing NH_3.

where
  S is the cross-sectional area of the packed column
  y is the mole fraction of NH_3 in the bulk phase
  y_i is the mole fraction of NH_3 at the water-gas interface
  Table P7.7 shows the dependency of y_i on y. Determine z and plot z as a function of y_d $(0.032 \le y_d \le 0.20)$.[68]

7.8 A binary mixture is to be distilled in a distillation column to give a distillate of $x_D = 0.98$ and a bottoms composition of $x_B = 0.01$. The feed composition is $z_F = 0.5$ and the reflux ratio is $R = 2.7$. The feed is a mixture of vapor and liquid and $q = 1$. The equilibrium equation is given by $y = \alpha x/(1 + (1 - \alpha)x)$, with $\alpha = 2.5$. Determine the location of the feed stage and the number of total stages, and plot the equilibrium curve, q-line, and operating lines as a function of the liquid-phase mole fraction.)

7.9 A binary mixture of 40 *mol%* benzene and 60 *mol%* toluene is to be distilled in a distillation column. The feed rate is 10*kmol/hr*, and the desired results are 99.2 *mol%* of benzene in the distillate and 98.6 *mol%* toluene in the bottom product. The relativity volatility, α_{BT} (benzene/toluene), is 2.354. The reflux is returned as a saturated liquid, and the distillation column has a total condenser and a partial reboiler. Determine the following[69]:

TABLE P7.7

y_i **as a Function of** y

y	0.032	0.080	0.125	0.166	0.206	0.243	0.278	0.311	0.342	0.372	0.400
y_i	0.004	0.029	0.056	0.084	0.115	0.148	0.183	0.221	0.261	0.304	0.349

1) Minimum number of trays
2) Optimum feed stage
3) Minimum reflux ratio and actual reflux ratio
4) Actual number of trays and number of rectifying section trays (use the Eduljee correlation)
5) Rectifying liquid and vapor, stripping the liquid and vapor flow rates

7.10 A mixture of 30 *mol%* benzene, 25 *mol%* toluene, and 45 *mol%* ethyl benzene is to be separated in a distillation column at atmospheric pressure. 98% of the benzene and only 1% of the toluene is to be recovered in the distillate stream.[70] The reflux ratio is 2, and all ethylbenzene is assumed to go to the bottoms. The molar flow rate of the saturated liquid feed is 100 *kmol/hr*. Table P7.10 shows the parameters of the Antoine equation given by[71]

$$\ln P = A - \frac{B}{T + C} (P : kPa, \ T : °C)$$

1) Determine the minimum number of ideal trays.
2) Calculate the approximate composition of the product.
3) Find the number of actual trays.
4) Determine the minimum number of trays if the distillation is carried out at 0.2 *atm* (boiling point = 55 °C).

7.11 A 41-stage column with the overhead condenser as stage 1, the feed stage as stage 21, and the reboiler as stage 41 is used to distill a binary mixture. The relative volatility, α, is 2.5. The feed rate is $F = 1$ *mol/min*, the feed composition is $z_F = 0.5$, and the feed condition is $q = 1$ (bubble point liquid). The flow rate of the distillate leaving the column is 0.5 *mol/min*. Plot the steady-state liquid composition profile along with the stages for reflux flow rates of $R = 2.4$, 2.7, and 3.0 *mol/min*.

7.12 A 20-stage column with the overhead condenser as stage 1, the feed stage as stage 10, and the reboiler as stage 20 is used to distill a binary mixture. The relative volatility, α, is 2.5. The feed rate is $F = 1$ *mol/min*, the feed composition is $z_F = 0.5$, and the feed condition is $q = 1$ (bubble point liquid). The reflux flow rate is $R = 3$ *mol/min*, and the distillate leaving the column is 0.5 *mol/min*.

TABLE P7.10

Parameters of Antoine Equation for Each Component (*P:kPa, T:°C*)

Component	A	B	C
Benzene, C_6H_6	13.7819	2726.81	217.572
Toluene, C_7H_8	13.9320	3056.96	217.625
Ethylbenzene, C_8H_{10}	13.9726	3259.93	212.300

Source: Smith, J. M. et al., *Introduction to Chemical Engineering Thermodynamics*, 8th ed., McGraw-Hill Education, New York, NY, 2018, p. 653.

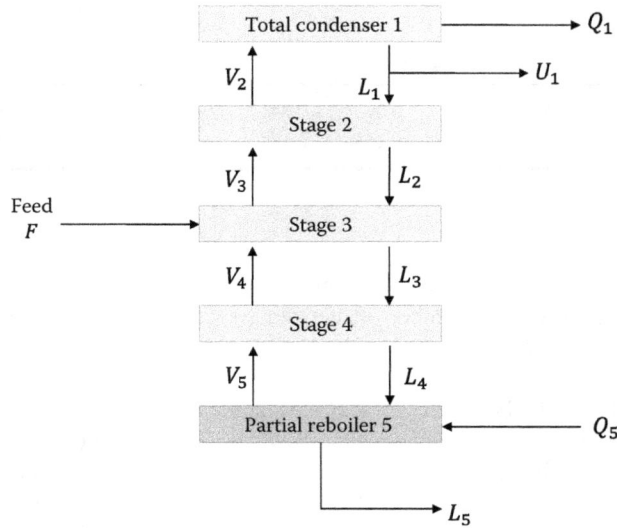

FIGURE P7.13 Five-stage distillation column. (From Henley, E. J. et al., *Separation Process Principles*, 3rd ed., John Wiley & Sons, Inc., Hoboken, NJ, 2011, p. 417.)

The holdups in the condenser and the reboiler are both 5 *mol*, and the holdup in each stage is maintained constant at 0.5 *mol*. At $t = 2$ *min*, there is a 0.2 *mol/min* step change in the reflux flow rate. Plot the liquid compositions of the distillate and the bottoms for $0 \leq t \leq 30$. Also plot the liquid composition profile along the stages at $t = 30$ *min*.

7.13 A three-component feed mixture of propane(1)/*n*-butane(2)/-pentane(3) is to be distilled in the five-stage distillation column shown in Figure P7.13. The feed composition is $z_{1,3} = 0.3$, $z_{2,3} = 0.3$, $z_{3,3} = 0.4$ (mole fraction), and the feed is supplied to the third stage with a flow rate of $F_j = F_3 = 100$ *lbmol/hr*. The feed is saturated liquid at 122.33°F and 100 *psia*. The pressure at each stage is maintained constant at 100 *psia*. The flow rate of the liquid side cut is $U_1 = L_1/2$, and the flow rate of the bottoms is $L_5 = 50$ *lbmol/hr*. The specific heat C_{pv}^0 (*Btu/lbmol*) and the vapor pressure for each component (P_i^s) are given by

$$C_{pv}^0 = a_1 + a_2 T + a_3 T^2 + a_4 T^3 + a_5 T^4 \, (T : °F)$$

$$\ln \frac{P_i^s}{P_c} = A_1 - \frac{A_2}{T + A_3} \, (T : °F, P_c : \text{critical pressure})$$

The physical properties and parameters for the specific heat equation are shown in Table P7.13(1) and Table P7.13(2).[72] Plot the temperature profile along the stages. Also plot the profiles of liquid and vapor compositions along the stages.[73]

7.14 Figure P7.14 shows a distillation column where a feed mixture consisting of methane(1)/ ethane(2)/propane(3)/*n*-butane(4)/*n*-pentane(5) is distilled. The number of total stages including the condenser and the reboiler is 13. The feed stream enters the feed stage (seventh stage from the top) at 105 °F and 400 *psia*, with a flow rate of $F_j = F_7 = 800$ *lbmol/hr* and a composition of $z_{1,7} = 0.2$, $z_{2,7} = 0.4625$, $z_{3,7} = 0.3$, $z_{4,7} = 0.03125$, and $z_{5,7} = 0.00625$. The pressure in the column is kept constant at 400 *psia*; and the flow rate of the vapor stream to the first stage (V_1) is 530 *lbmol/hr* and that of the liquid stream to the first stage (L_1) is 1000 *lbmol/hr*. The top product is obtained as vapor from the first stage, and the flow rate of the liquid side cut is $U_1 = 0$ *lbmol/hr*.

The specific heat C_{pv}^0 (*Btu/lbmol*) and the vapor pressure (P_i^s) are given by

TABLE P7.13(1)
Physical Properties of Each Component

Component	P_c(psia)	T_c(R)	ω	A_1	A_2	A_3
Propane C_3H_8(1)	617.4	665.9	0.1538	5.35342	3371.084	414.488
n-Butane C_4H_{10}(2)	550.7	765.3	0.1954	5.74162	4126.385	409.5179
n-Pentane C_5H_{12}(3)	489.5	845.9	0.2387	5.85365	4598.287	394.4148

TABLE P7.13(2)
Parameters for Specific Heat Equations for Each Component

Component	a_1	$a_2 \times 10$	$a_3 \times 10^4$	$a_4 \times 10^7$	$a_5 \times 10^{11}$
Propane C_3H_8(1)	15.58683	0.250495	0.140426	−0.352626	1.864467
n-Butane C_4H_{10}(2)	20.79783	0.314329	0.192851	−0.458865	2.380972
n-Pentane C_5H_{12}(3)	25.64627	0.389176	0.239729	−0.584262	3.079918

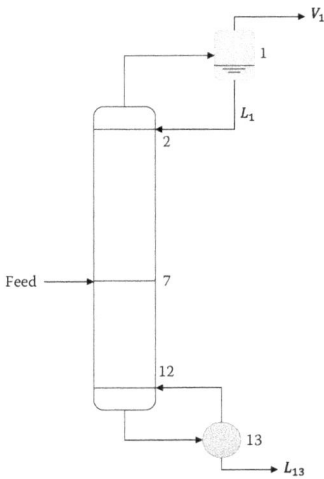

FIGURE P7.14 13-stage distillation column. (From Henley, E. J. et al., *Separation Process Principles*, 3rd ed., John Wiley & Sons, Inc., Hoboken, NJ, 2011, p. 419.)

$$C_{pv}^0 = a_1 + a_2 T + a_3 T^2 + a_4 T^3 + a_5 T^4 \ (T : °F)$$

$$\ln\frac{P_i^s}{P_c} = A_1 - \frac{A_2}{T + A_3} \ (T : °F, \ P_c:\text{critical pressure})$$

Table P7.14(1) shows the physical properties and Table P7.14(2) shows the parameters of specific heat equations for each component.[72] Calculate the stage temperatures and vapor and liquid compositions, and plot the results versus the stage number.[74]

7.15 A mixture of 120 moles of n-pentane and 80 moles of n-heptane is distilled in a differential distillation apparatus at 1 *bar* until 80 moles of the liquid is left. Determine the composition of the liquid left in the apparatus. Table P7.15 shows the equilibrium data for the mixture of n-pentane

TABLE P7.14(1)
Physical Properties for Each Component

Component	$P_c(psia)$	$T_c(R)$	ω	A_1	A_2	A_3
Methane $CH_4(1)$	673.1	343.9	0.0	5.14135	1742.638	452.974
Ethane $C_2H_6(2)$	709.8	550.0	0.1064	5.38389	2847.921	434.898
Propane $C_3H_8(3)$	617.4	665.9	0.1538	5.35342	3371.084	414.488
n-Butane $C_4H_{10}(4)$	550.7	765.3	0.1954	5.74162	4126.385	409.5179
n-Pentane $C_5H_{12}(5)$	489.5	845.9	0.2387	5.853654	4598.287	394.4148

TABLE P7.14(2)
Parameters of Specific Heat Equations for Each Component

Component	a_1	$a_2 \times 10$	$a_3 \times 10^4$	$a_4 \times 10^7$	$a_5 \times 10^{11}$
Methane $CH_4(1)$	8.24522	0.038063	0.0886474	−0.074612	0.182296
Ethane $C_2H_6(2)$	11.51606	0.140309	0.0854034	−0.110608	0.316220
Propane $C_3H_8(3)$	15.58683	0.250495	0.140426	−0.352626	1.864467
n-Butane $C_4H_{10}(4)$	20.79783	0.314329	0.192851	−0.458865	2.380972
n-Pentane $C_5H_{12}(5)$	25.64627	0.389176	0.239729	−0.584262	3.079918

TABLE P7.15
Equilibrium Data for the Mixture of n-Pentane and n-Heptane

x	1.0	0.87	0.59	0.40	0.25	0.15	0.06	0.0
y	1.0	0.98	0.93	0.84	0.70	0.52	0.27	0.0

and n-heptane, where x and y are the mole fractions of n-pentane in the liquid and vapor phases, respectively.

7.16 $CaCO_3$ slurry in water is to be filtered at 298.2 K with a pressure drop of $(-\Delta p) = 67$ kPa using a rotary drum filter having a 33% submergence ($f = 0.33$) of the drum in the slurry. The solids concentration in the slurry is $c_x = 0.191$ $kg\,(solid)/kg\,(slurry)$. The mass ratio of wet cake to dry cake is $m = 2$. At 298.2 K, the viscosity of the filtrate is $\mu = 8.937 \times 10^{-4}$ $kg/(m\cdot sec)$ and the density is $\rho = 996.9$ kg/m^3. The filter cycle time is $t_c = 250sec$ and the specific cake resistance is represented by $\alpha = 4.37 \times 10^9(-\Delta p)^{0.3}$, where $(-\Delta p)$ is in Pa and α in m/kg. Determine the filter area required to filter $\dot{m} = 0.778$ $kg\,(slurry)/sec$.[75]

7.17 A membrane is to be used to separate air (oxygen(A) + nitrogen(B)). The feed flow rate is $L_f = 1 \times 10^6$ cm^3/sec and the feed composition of A is $x_f = 0.209$ (mole fraction). The stage-cut is $\theta = 0.2$ and the ratio of the permeability of oxygen to that of nitrogen is $\alpha = 10$. Use the complete-mixing model to calculate the permeate composition y_p, the reject composition x_R, and the membrane area A_m.[76]

Data: membrane thickness $t = 2.54 \times 10^{-3}$ cm, feed pressure $p_h = 190$ $cmHg$, permeate-side pressure $p_l = 19$ $cmHg$, permeability of A: $P_A = 500 \times 10^{-10}$ $cm^3\cdot cm/(sec\cdot cm^2\cdot cmHg)$

7.18 A membrane is to be used to separate air (oxygen(A) + nitrogen(B)). The feed flow rate is $L_f = 1 \times 10^6 \ cm^3/sec$ and the feed composition of A (oxygen) is $x_f = 0.209$ (mole fraction). The stage-cut is $\theta = 0.2$ and the ratio of the permeability of oxygen to that of nitrogen is $\alpha = 10$. Use the cross-flow model to calculate the permeate composition y_p, the reject composition x_R, and the membrane area A_m.[65]

Data: membrane thickness $t = 2.54 \times 10^{-3} \ cm$, feed pressure $p_h = 190 \ cmHg$, permeate-side pressure $p_l = 19 \ cmHg$,

   permeability of A: $P_A = 500 \times 10^{-10} \ cm^3 \cdot cm/(sec \cdot cm^2 \cdot cmHg)$,

   permeability of B: $P_B = 50 \times 10^{-10} \ cm^3 \cdot cm/(sec \cdot cm^2 \cdot cmHg)$

REFERENCES

1. Cutlip, M. B. and M. Shacham, Problem Solving in Chemical and Biochemical Engineering with POLYMATH, Excel, and MATLAB, 2nd ed., Prentice-Hall, Boston, MA, p. 384, 2008.
2. Geankoplis, C. J., *Transport Processes and Separation Process Principles*, 4th ed., Prentice Hall, Boston, MA, p. 418, 2003.
3. Bird, R. B., W. E. Stewart, and E. N. Lightfoot, *Transport Phenomena*, 2nd ed., John Wiley & Sons, Hoboken, NJ, p. 551, 2002.
4. Geankoplis, C. J., *Transport Processes and Separation Process Principles*, 4th ed., Prentice Hall, Boston. MA, p. 426, 2003.
5. Cutlip, M. B. and M. Shacham, Problem Solving in Chemical and Biochemical Engineering with POLYMATH, Excel, and MATLAB, 2nd ed., Prentice-Hall, Boston, MA, p. 386, 2008.
6. Cutlip, M. B. and M. Shacham, *Problem Solving in Chemical and Biochemical Engineering with POLYMATH, Excel, and MATLAB*, 2nd ed., Prentice-Hall, Boston, MA, pp. 385–386, 2008.
7. Bird, R. B., W. E. Stewart, and E. N. Lightfoot, *Transport Phenomena*, 2nd ed., John Wiley & Sons, Hoboken, NJ, p. 538, 2002.
8. Cutlip, M. B. and Shacham, M., Problem Solving in Chemical and Biochemical Engineering with POLYMATH, Excel, and MATLAB, 2nd ed., Prentice-Hall, Boston, MA, p. 413, 2008.
9. Geankoplis, C. J., *Transport Processes and Separation Process Principles*, 4th ed., Prentice-Hall, Boston, MA, p. 421, 2003.
10. Cutlip, M. B. and M. Shacham, Problem Solving in Chemical and Biochemical Engineering with POLYMATH, Excel, and MATLAB, 2nd ed., Prentice-Hall, Boston, MA, p. 391, 2008.
11. Geankoplis, C. J., *Transport Processes and Separation Process Principles*, 4th ed., Prentice-Hall, Boston, MA, p. 482, 2003.
12. Cutlip, M. B. and M. Shacham, *Problem Solving in Chemical and Biochemical Engineering with POLYMATH, Excel, and MATLAB*, 2nd ed., Prentice-Hall, Boston, MA, pp. 396–398, 2008.
13. Bird, R. B., W. E. Stewart, and E. N. Lightfoot, *Transport Phenomena*, 2nd ed., John Wiley & Sons, Hoboken, NJ, p. 566, 2002.
14. Cutlip, M. B. and M. Shacham, Problem Solving in Chemical and Biochemical Engineering with POLYMATH, Excel, and MATLAB, 2nd ed., Prentice-Hall, Boston, MA, p. 402, 2008.
15. Cutlip, M. B. and M. Shacham, *Problem Solving in Chemical and Biochemical Engineering with POLYMATH, Excel, and MATLAB*, 2nd ed., Prentice-Hall, Boston, MA, pp. 406–408, 2008.
16. Geankoplis, C. J., *Transport Processes and Separation Process Principles*, 4th ed., Prentice-Hall, Boston, MA, p. 417, 2003.
17. Geankoplis, C. J., *Transport Processes and Separation Process Principles*, 4th ed., Prentice-Hall, Boston, MA, p. 506, 2003.
18. Cutlip, M. B. and M. Shacham, *Problem Solving in Chemical and Biochemical Engineering with POLYMATH, Excel, and MATLAB*, 2nd ed., Prentice-Hall, Boston, MA, pp. 430–431, 2008.
19. Cutlip, M. B. and M. Shacham, *Problem Solving in Chemical and Biochemical Engineering with POLYMATH, Excel, and MATLAB*, 2nd ed., Prentice-Hall, Boston, MA, pp. 428–429, 2008.
20. Bird, R. B., W. E. Stewart, and E. N. Lightfoot, *Transport Phenomena*, 2nd ed., John Wiley & Sons, Hoboken, NJ, p. 562, 2002.
21. Cutlip, M. B. and M. Shacham, Problem Solving in Chemical and Biochemical Engineering with POLYMATH, Excel, and MATLAB, 2nd ed., Prentice-Hall, Boston, MA, p. 438, 2008.
22. Cutlip, M. B. and M. Shacham, Problem Solving in Chemical and Biochemical Engineering with POLYMATH, Excel, and MATLAB, 2nd ed., Prentice-Hall, Boston, MA, p. 439, 2008.

23. Jana A. K., *Chemical Process Modelling and Computer Simulation*, 2nd ed., PHI Learning Private, Delhi, India, p. 340, 2011.

24. Adidharma, H. and V. Temyanko, *Mathcad for Chemical Engineers*, Trafford publishing, Victoria, BC, Canada, pp. 89–90, 2007.

25. Martin M. M. (Editor), *Introduction to Software for Chemical Engineers*, CRC Press, Taylor & Francis Group, Boca Raton, FL, pp.107–111, 2015.

26. Towler, G. and R. Sinnott, *Chemical Engineering Design*, Butterworth-Heinemann, Burlington, MA, p. 744, 2008.

27. Ghasem, N., *Computer Methods in Chemical Engineering*, CRC Press, Taylor & Francis Group, Boca Raton, FL, p. 342, 2012.

28. Onda K., Takeushi H. and Y. Okumoto, Mass transfer coefficients between gas and liquid in packed columns, *J. Chem. Eng. Japan*, 1(1), pp. 56–62, 1968.

29. Cornell, D., W. G. Knapp, and J. R. Fair, Mass transfer efficiency in packed columns, *Chem. Eng. Prog.*, 56(July, p. 120) and 68(August, p. 172), 1960.

30. Towler G. and Sinnott R., *Chemical Engineering Design*, Butterworth-Heinemann, Burlington, MA, p.753 (2008).

31. Nayef G., *Computer Methods in Chemical Engineering*, CRC Press, Taylor & Francis Group, Boca Raton, FL, pp. 352–353, 2012.

32. Henley, E. J., J. D. Seader, and D. K. Roper, *Separation Process Principles*, 3rd ed., John Wiley & Sons, Inc., Hoboken, NJ, pp. 285–290, 2011.

33. Jana A. K., *Chemical Process Modelling and Computer Simulation*, 2nd ed., PHI Learning Private, Ltd., Delhi, India, p. 127, 2011.

34. Gilliland, E. R., Estimate of the number of theoretical plates as a function of reflux ratio, *Industrial and Engineering Chemistry*, 32, p. 1220, 1940.

35. Hengstebeck, R. J., *Distillation*, Reinhold, New York, NY, p. 234, 1961.

36. Liddle, C. J., Improved short-cut method for distillation calculations, *Chemical Engineering*, 75, p. 137, October 21, 1968.

37. Van Winkle, M. and W. G. Todd, Optimum fractionation design by simple graphical methods, *Chemical Engineering*, 78, p. 136, September 20, 1971.

38. Molkavov, Y. K., T. P. Koralina, N. I. Mazurina, and G. A. Nikiforov, An approximation method for calculating the basic parameters of multicomponent fractionation, *International Chemical Engineering*, 12(2), pp. 29–212, 1972.

39. Hohman, E. C. and F. J. Lockhart, Fractional distillation of multicomponent mixtures, *Chemical Technology*, 2, p. 614, 1972.

40. Eduljee, H. E., Equations replace Gilliland plot, *Hydrocarbon Processing*, 54(9), p. 120, 1975.

41. Chang, H. Y., Gilliland plot in one equation, *Hydrocarbon Processing*, 64(3), p. 48, 1985.

42. Harg, K., Equation proposed, *Hydrocarbon Processing*, 64(3), p. 49, 1985.

43. McCormick, J. E., A correlation for distillation stages and reflux, *Chemical Engineering*, 95, p. 75, September 26, 1988.

44. Kirkbride, C. G., Process design procedure for multicomponent fractionators, *Petroleum Refinery*, 23(9), p. 87, 1944.

45. Coker, A. K., *Chemical Process Design, Analysis and Simulation*, Gulf Publishing Company, Houston, TX, p. 516, 1995.

46. Cutlip, M. B. and M. Shacham, *Problem Solving in Chemical and Biochemical Engineering with POLYMATH, Excel and MATLAB*, 2nd ed., Prentice-Hall, Boston, MA, pp. 544–545, 2008.

47. Adidharma, H. and V. Temyanko, *Mathcad for Chemical Engineers*, Trafford publishing, Victoria, BC, Canada, pp. 56–57, 2007.

48. Ghasem N., *Computer Methods in Chemical Engineering*, CRC Press, Taylor & Francis Group, Boca Raton, FL, pp. 299–301, 2012.

49. Cutlip, M.B. and Shacham, M., Problem Solving in Chemical and Biochemical Engineering with POLYMATH, Excel and MATLAB, 2nd ed., Prentice-Hall, Boston, MA, p. 552, 2008.

50. Cutlip, M. B. and M. Shacham, *Problem Solving in Chemical and Biochemical Engineering with POLYMATH, Excel and MATLAB*, 2nd ed., Prentice-Hall, Boston, MA, pp. 551–552, 2008.

51. Cutlip, M.B. and Shacham, M., Problem Solving in Chemical and Biochemical Engineering with POLYMATH, Excel and MATLAB, 2nd ed., Prentice-Hall, Boston, MA, p. 553, 2008.

52. Henley, E.J., Seader, J.D., and Roper, D.K., *Separation Process Principles*, 3rd ed., John Wiley & Sons, Hoboken, NJ, p. 411, 2011.

53. Wang, J. C. and G. E. Henke, Tridiagonal matrix for distillation, *Hydrocarbon Processing*, 45(8), p. 155, 1966.
54. Kister, H. Z., *Distillation: Design*, McGraw-Hill, Inc., New York, NY, p. 153, 1992.
55. Henley, E. J. and J. D. Seader, *Equilibrium-Stage Separation Operations in Chemical Engineering*, John Wiley & Sons, Inc., Hoboken, NJ, pp. 716–722, 1981.
56. Henley, E. J. and J. D. Seader, *Equilibrium-Stage Separation Operations in Chemical Engineering*, John Wiley & Sons, Inc., Hoboken, NJ, p. 568, 1981.
57. Cutlip, M. B. and M. Shacham, *Problem Solving in Chemical and Biochemical Engineering with POLYMATH, Excel and MATLAB*, 2nd ed., Prentice-Hall, Boston, MA, pp. 559–560, 2008.
58. Geankoplis, C. J., *Transport Processes and Separation Process Principles*, 4th ed., Prentice Hall, Upper Saddle River, NJ, pp. 910–913, 2003.
59. McCabe, W. L., J. C. Smith, and P. Harriott, *Unit Operations of Chemical Engineering*, 7th ed., McGraw-Hill, New York, NY, p. 1024, 2005.
60. Geankoplis, C. J., *Transport Processes and Separation Process Principles*, 4th ed., Prentice Hall, Upper Saddle River, NJ, p. 918, 2003.
61. Geankoplis, C. J., *Transport Processes and Separation Process Principles*, 4th ed., Prentice Hall, Upper Saddle River, NJ, p. 919, 2003.
62. Geankoplis, C. J., *Transport Processes and Separation Process Principles*, 4th ed., Prentice Hall, Upper Saddle River, NJ, pp. 913–915, 2003.
63. Geankoplis, C. J., *Transport Processes and Separation Process Principles*, 4th ed., Prentice Hall, Boston, MA, p. 853, 2003.
64. Geankoplis, C. J., *Transport Processes and Separation Process Principles*, 4th ed., Prentice Hall, Boston, MA, pp. 859–860, 2003.
65. Geankoplis, C. J., *Transport Processes and Separation Process Principles*, 4th ed., Prentice Hall, Boston, MA, pp. 861–863, 2003.
66. Cutlip, M. B. and M. Shacham, *Problem Solving in Chemical and Biochemical Engineering with POLYMATH, Excel and MATLAB*, 2nd ed., Prentice-Hall, Boston, MA, p. 402, Prentice-Hall, 2008.
67. Ghasem, N., *Computer Methods in Chemical Engineering*, CRC Press, Taylor & Francis Group, Boca Raton, FL, p. 356, 2012.
68. Adidharma, H. and V. Temyanko, *Mathcad for Chemical Engineers*, Trafford publishing, Victoria, BC, Canada, pp. 88–89, 2007.
69. Ghasem, N., *Computer Methods in Chemical Engineering*, CRC Press, Taylor & Francis Group, Boca Raton, FL, p. 274, 2012.
70. Ghasem, N., *Computer Methods in Chemical Engineering*, CRC Press, Taylor & Francis Group, Boca Raton, FL, pp. 309–310, 2012.
71. Smith, J. M., H. C. Van Ness, M. M. Abbott, and M. T. Swihart, *Introduction to Chemical Engineering Thermodynamics*, 8th ed., McGraw-Hill Education, New York, NY, p. 653, 2018.
72. Henley, E. J. and J. D. Seader, *Equilibrium-Stage Separation Operations in Chemical Engineering*, John Wiley & Sons, Hoboken, NJ, pp. 716–722, 1981.
73. Henley, E. J., J. D. Seader, and D. K. Roper, *Separation Process Principles*, 3rd ed., John Wiley & Sons, Inc., Hoboken, NJ, pp. 417–419, 2011.
74. Henley, E. J., J. D. Seader, and D. K. Roper, *Separation Process Principles*, 3rd ed., John Wiley & Sons, Inc., Hoboken, NJ, pp. 418–421, 2011.
75. Geankoplis, C. J., *Transport Processes and Separation Process Principles*, 4th ed., Prentice Hall, Boston, MA, pp. 918–919, 2003.
76. Geankoplis, C. J., *Transport Processes and Separation Process Principles*, 4th ed., Prentice Hall, Boston, MA, p. 857, 2003.

8 Heat Transfer

8.1 ONE-DIMENSIONAL HEAT TRANSFER

8.1.1 Heat Transfer in a One-Dimensional Slab

Heat transfer in a one-dimensional slab involves conduction, convection, and radiation to the surroundings. Steady-state heat conduction occurring within a one-dimensional slab without heat generation follows Fourier's law, given by

$$\frac{q_x}{A} = -k\frac{dT}{dx} \approx -k\frac{\Delta T}{\Delta x}$$

where
q_x (W or J/sec) is the heat transfer rate in the x direction
A (m^2) is the cross-sectional area normal to the direction of heat conduction
k ($W/(m{\cdot}K)$) is the thermal conductivity of the solid medium
The convective heat transfer between solids and fluid can be described by

$$\frac{q_x}{A} = h(T_w - T_f)$$

where
h ($W/(m^2{\cdot}K)$) is the heat transfer coefficient
T_w (K) is the surface temperature of the solid
T_f (K) is the fluid temperature
The heat flux at the interface between a solid and a fluid can be expressed as

$$\left.\frac{q_x}{A}\right|_S = \left.-k\frac{dT}{dx}\right|_S = h(T_w - T_f)$$

where the subscript S denotes the solid surface.

Example 8.1 Heat Transfer in a One-Dimensional Slab[1]

Figure 8.1 shows a one-dimensional slab with heat conduction and radiation. One surface of the slab is maintained at temperature T_1, and the other surface at temperature T_2 has radiative heat transfer with the surroundings that act as a black body at temperature T_a. The radiation from the slab surface can be represented by the Stefan-Boltzmann law:

$$\left.\frac{q_x}{A}\right|_{x=\Delta x} = \sigma(T_2^4 - T_a^4)|_{x=\Delta x}\,(\sigma = 5.676 \times 10^{-8}W/(m^2{\cdot}K^4))$$

Calculate and plot the temperature profile within the slab. What is the corresponding value of T_2? The thermal conductivity of the solid slab, k, is dependent upon temperature and is given by $k = 30(1 + 0.002T)$. Assume that the convective heat transfer between the slab and the surroundings is negligible.
   Data: $T_1 = 290\ K$, $T_a = 1273\ K$, $\Delta x = 0.2\ m$

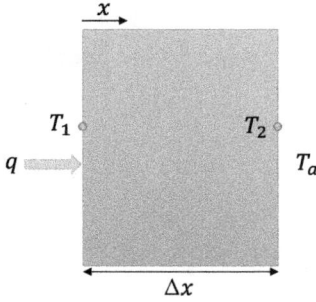

FIGURE 8.1 One-dimensional heat transfer. (From Cutlip, M.B. and Shacham, M., *Problem Solving in Chemical and Biochemical Engineering with POLYMATH, Excel, and MATLAB*, 2nd ed., Prentice-Hall, Boston, MA, 2008, pp. 209–210.)

Solution

$$q_x = -kA\frac{dT}{dx} = -30(1 + 0.002T)A\frac{dT}{dx} = \sigma(T_2^4 - T_a^4) \Longrightarrow \frac{dT}{dx} = -\frac{\sigma(T_2^4 - T_a^4)}{30(1 + 0.002T)A}$$

First set an initial estimate for T_2 is provided and solve the differential equation to find a new T_2. The iterations are continued until the value of T_2 converges. The subfunction *slab1T* defines the differential equation to be solved. The script *ht1D* performs the iterative calculations. In the script, k is the number of iterations carried out until the convergence is achieved, and T(end) denotes an updated value of T_2 at each iteration.

```
% ht1D.m: one-dimensional heat transfer (conduction and radiation)
A = 1; T2 = 700; Ta = 1273; sigma = 5.676e-8; xspan = [0 0.2]; T0 = 290; delT = 10;
criT = 1e-3; k = 1;
while delT > criT
  [x T] = ode45(@slab1T, xspan, T0, [], A, T2, Ta, sigma);
  delT = abs(T2 - T(end)); T2 = T(end); k = k+1;
end
plot(x,T), xlabel('x(m)'), ylabel('T(K)'), grid
k, T(end)
function dT = slab1T(x,T,A,T2,Ta,sigma)
% heat transfer within a one-dimensional slab
dT = - sigma*(T2^4 - Ta^4)/(30*(1 + 0.002*T)*A);
end
```

The script *ht1D* generates the following outputs and the plot shown in Figure 8.2:
```
>> ht1D
k = 10,  T2 = 729.167
```

Example 8.2 One-Dimensional Heat Conduction[2]

A flat plate of infinite length is cooled using airflow over both sides of the plate. The initial uniform temperature of the plate is $T_i = 340\,°C$ and the ambient temperature is $T_a = 25\,°C$. The mid − plane temperature of the plate (T) at time t is given by

$$T = T_a + (T_i - T_a)\cdot\frac{4\sin(\gamma)}{2\gamma + \sin(2\gamma)}\cdot\exp\left(-\frac{4\gamma^2\alpha t}{D^2}\right)$$

where

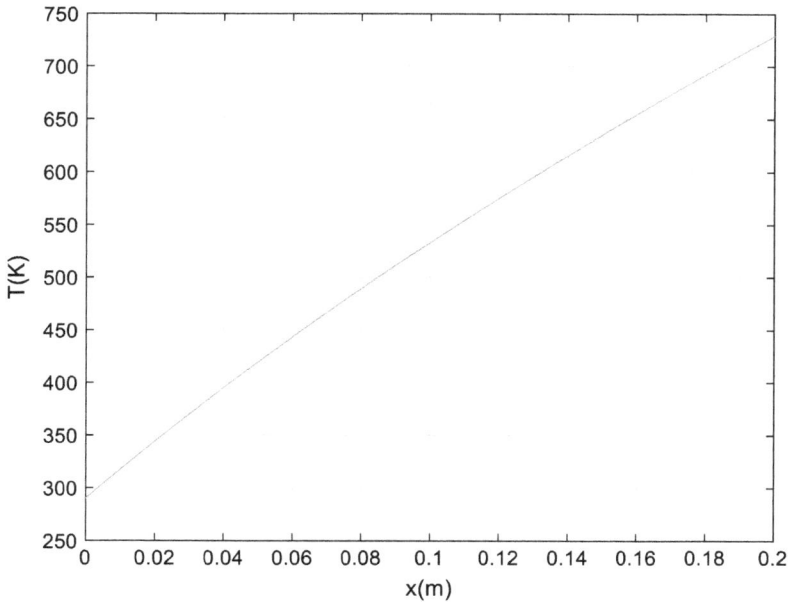

FIGURE 8.2   Heat transfer within a one-dimensional slab.

D is the thickness of the plate $(=0.05\ m)$
α is the thermal diffusivity of the plate $(=4.97 \times 10^{-7}\ m^2/sec)$
γ is the first positive root of the equation

$$\gamma \tan(\gamma) - \frac{hD}{2k} = 0$$

where
  k is the thermal conductivity of the plate $(=1.18\ W/(m{\cdot}K))$
  h is the convection heat transfer coefficient $(=98.6\ W/(m^2{\cdot}K))$
  Plot the mid-plane temperature of the plate (T) as a function of t $(0 \le t \le 15\,(min))$.

Solution
   The script *htcond* generates the following output and the plot shown in Figure 8.3:
```
% htcond.m
D = 0.05; h = 98.6; k = 1.18; alpa = 4.97e-7; Ta = 25; Ti = 340; t = linspace
(0,15*60,2000);
f = @(x) x*tan(x) - h*D/k/2; x0 = 0.1; gam = fzero(f,x0);
T = Ta + (Ti-Ta)*4*sin(gam).*exp(-4*gam^2*alpa*t/D^2)/(2*gam+sin(2*gam));
fprintf('gamma = %g\n', gam);
plot(t/60,T), grid,xlabel('t(min)'), ylabel('T(deg.C)')

>> htcond
gamma = 1.08991
```

Example 8.3 Heat Transfer by Radiation[3]

A coating on a curved surface is cured by exposing it to an infrared heater. The system is located in a large room, and the heat transfer is assumed to be entirely due to radiation. The

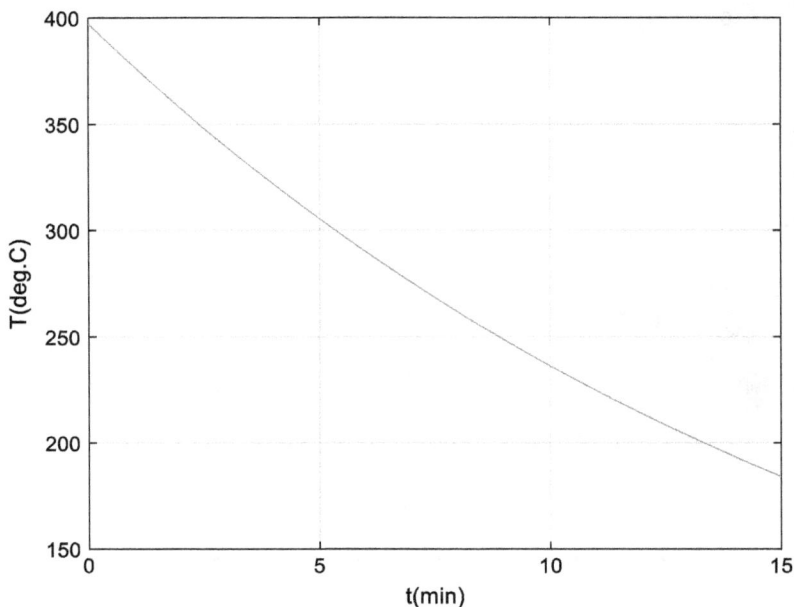

FIGURE 8.3 Plot of the mid-plane temperature of the plate versus t.

surface radiosity of the heater and the view factors can be determined by solving the following equations:

$$q_1 = \frac{\varepsilon_1 A_1}{1 - \varepsilon_1}(\sigma T_1^4 - J_1) = F_{12}A_1(J_1 - J_2) + F_{13}A_1(J_1 - \sigma T_3^4)$$

$$-q_2 = \frac{\varepsilon_2 A_2}{1 - \varepsilon_2}(\sigma T_2^4 - J_2) = F_{12}A_1(J_2 - J_1) + F_{23}A_2(J_2 - \sigma T_3^4)$$

$$F_{12} = \frac{2}{\pi xy}\left[\ln\sqrt{\frac{(1 + x^2)(1 + y^2)}{1 + x^2 + y^2}} + x\sqrt{1 + y^2}\,\tan^{-1}\left(\frac{x}{\sqrt{1 + y^2}}\right) + y\sqrt{1 + x^2}\,\tan^{-1}\left(\frac{y}{\sqrt{1 + x^2}}\right)\right.$$

$$\left. - x\tan^{-1}x - y\tan^{-1}y\right]$$

$$F_{13} = 1 - F_{12}, \ F_{23} = \frac{F_{13}A_1}{A_2}, \ x = \frac{W}{H}, \ y = \frac{L}{H}$$

where
    ε_1 and ε_2 are the emissivities of the heater and the surface, respectively
    q_1 is the heater power requirement
    q_2 is the heat transfer rate to the surface
    A_1 and A_2 are the areas of the heater and the surface, respectively
    σ is the Stefan-Boltzmann constant
    J_1 and J_2 are the surface radiosities of the heater and the surface, respectively
    F_{ij} is the view factor between surface$i$ and surface$j$
    T_1 is the temperature of the heater
    T_2 is the temperature of the surface

T_3 is the wall temperature
W is the width of the heater
H is the distance between the heater and the surface
L is the length of both the heater and the surface
Using the following data, calculate q_1.

Data: $A_1 = 10$ m^2, $A_2 = 15$ m^2, $W = H = 1$ m, $L = 10$ m, $T_1 = 1000$ K, $T_3 = 300$ K, $\varepsilon_1 = 0.9$, $\varepsilon_2 = 0.5$, $q_2 = 77.1$ kW, $\sigma = 5.67 \times 10^{-8}$ $W/(m^2 \cdot K^4)$

Solution

The unknown variables are J_1, J_2, and T_2. The equations can be rearranged as follows:

$$f_1 : \frac{\varepsilon_1 A_1}{1 - \varepsilon_1}(\sigma T_1^4 - J_1) - F_{12}A_1(J_1 - J_2) - F_{13}A_1(J_1 - \sigma T_3^4) = 0$$

$$f_2 : \frac{\varepsilon_2 A_2}{1 - \varepsilon_2}(\sigma T_2^4 - J_2) - F_{12}A_1(J_2 - J_1) - F_{23}A_2(J_2 - \sigma T_3^4) = 0$$

$$f_3 : q_2 + \frac{\varepsilon_2 A_2}{1 - \varepsilon_2}(\sigma T_2^4 - J_2) = 0$$

The subfunction *frad* defines the system of nonlinear equations. The data are fed into the function via the structure *radt*. The script *sufrad* assigns the data and calls the built-in solver *fsolve* to find the roots of the nonlinear equations.

```
% sufrad.m
radt.A1 = 10; radt.A2 = 15; radt.W = 1; radt.H = 1; radt.L = 10;
radt.sig = 5.67e-8; radt.T1 = 1000; radt.T3 = 300; radt.ep1 = 0.9;
radt.ep2 = 0.5; radt.q2 = 77100;
% x(1) = J1, x(2) = J2, x(3) = T2
x0 = [500 500 500]; % initial guess
x = fsolve(@frad, x0, [], radt)
q1 = radt.ep1*radt.A1*(radt.sig*(radt.T1)^4 - x(1))/(1-radt.ep1)
q2 = -radt.ep2*radt.A2*(radt.sig*x(3)^4 - x(2))/(1-radt.ep2)

function f = frad(x,radt)
% x(1) = J1, x(2) = J2, x(3) = T2
A1 = radt.A1; A2 = radt.A2; W = radt.W; H = radt.H; L = radt.L; sig = radt.sig;
T1 = radt.T1; T3 = radt.T3; ep1 = radt.ep1; ep2 = radt.ep2; q2 = radt.q2;
X = W/H; Y = L/H;
F12 = (2/(pi*X*Y))*(log(sqrt((1+X^2)*(1+X^2)/(1+X^2+Y^2))) +...
X*sqrt(1+Y^2)*atan(X/sqrt(1+Y^2))  +  Y*sqrt(1+X^2)*atan(Y/sqrt(1+X^2))  -
X*atan(X) - Y*atan(Y));
F13 = 1 - F12; F23 = F13*A1/A2;
f = [ep1*A1*(sig*T1^4 - x(1))/(1-ep1) - F12*A1*(x(1)-x(2)) - F13*A1*(x(1)-
sig*T3^4);
ep2*A2*(sig*x(3)^4 - x(2))/(1-ep2) - F12*A1*(x(2)-x(1)) - F23*A2*(x(2)-sig*T3^4);
  q2 + ep2*A2*(sig*x(3)^4 - x(2))/(1-ep2)];
end
```

Execution of the script *sufrad* generates the desired results:

```
>> sufrad
x =
 1.0e+04 *
  5.1222   0.6026   0.0354
```

```
q1 =
  4.9306e+05
q2 =
  77100
```

We can see that $J_1 = 5.1222 \times 10^4$ W/m^2, $J_2 = 0.6026 \times 10^4$ W/m^2, $T_2 = 354$ K, and $q_1 = 4.9306 \times 10^5$ W.

Example 8.4 Heat Transfer by Conduction and Convection[4]

A pin made of pure aluminum is used to conduct heat away from an electronic device. Generate the temperature profile of the pin as a function of time and calculate the steady-state temperature for each of the following cases:

    1 The pin is considered as a single lump (Figure 8.4(a)).
    2 The pin is assumed to consist of five equal lumps (Figure 8.4(b)).

The ambient temperature $T_a = 25°C$, the convective heat transfer coefficient $h_c = 20$ $W/(m^2 \cdot °C)$, the length of the pin is $L = 0.01$ m, the diameter of the pin is $D = 0.002$ m, the circular area of the base of the pin through which the heat conduction take places is $A_k = 3.14 \times 10^{-6}$ m^2, and the surface area of the pin exposed to the air is $A_c = 6.28 \times 10^{-5}$ m^2. At $t = 0$, the pin temperature $T = 25°C$ and the base temperature $T_b = 100°C$. Thermal properties of pure aluminum are: density $\rho = 2707$ kg/m^3, thermal conductivity $k = 220$ $W/(m \cdot °C)$, specific heat $C_p = 896$ $J/(kg \cdot °C)$.

Solution
    (1) The energy balance gives

$$\rho C_p V \frac{dT}{dt} = \frac{kA_k}{L}(T_b - T) - h_c A_c (T - T_a)$$

which can be rearranged as

$$\frac{dT}{dt} = \left(\frac{1}{C_h R_c}\right)T_a + \left(\frac{1}{C_h R_k}\right)T_b - \left(\frac{1}{C_h R_c} + \frac{1}{C_h R_k}\right)T$$

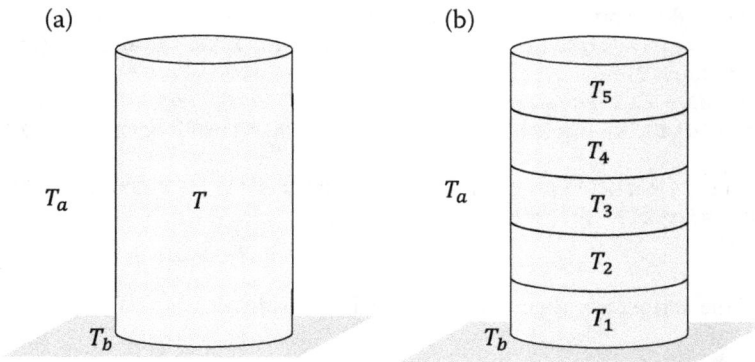

FIGURE 8.4 Pin model: (a) single lump, (b) five equal-sized lumps.

where $C_h = \rho C_p V$, $R_c = 1/h_c A_c$, $R_k = L/kA_k$, and the volume V is given by $V = A_k L$. The script *pintemp* produces the desired plot, shown in Figure 8.5. We can see that at steady state, the pin temperature reaches $T = 98.66$ °C.

```
% pintemp.m: pin temperature profile
Ta = 25; L = 0.01; Ac = 6.28e-5; Ak = 3.14e-6; hc = 20; % data
rho = 2707; k = 220; Cp = 896; % properties of pure aluminum
V = Ak*L; Ch = rho*Cp*V; Rc = 1/(hc*Ac); Rk = L/(k*Ak);
T0 = Ta; Tb = 100; % initial pin and base temperature at t = 0
tspan = [0 8]; dT = @(t,T) (1/(Ch*Rc))*Ta + (1/(Ch*Rk))*Tb - (1/(Ch*Rc) + 1/
(Ch*Rk))*T;
[t,T] = ode45(dT,tspan,T0); plot(t,T), xlabel('t(s)'), ylabel('T(deg.C)')
Ts = ((1/(Ch*Rc))*Ta + (1/(Ch*Rk))*Tb)/(1/(Ch*Rc) + 1/(Ch*Rk)); % st-st temp.
fprintf('At steady-state, T(pin) = %g deg.C\n',Ts);

>> pintemp
At steady-state, T(pin) = 98.6607 deg.C
```

(2) If the pin is divided into five equal pieces, the height of each element is $L = L_0/5 = 0.002\ m$ and the surface area of each element exposed to the air is $A_c = \pi DL$. From the energy balance for each element of the five-lump model, we obtain

$$\frac{dT_1}{dt} = \frac{1}{C_h}\left\{-\left(\frac{1}{R_{k0}} + \frac{1}{R_k} + \frac{1}{R_c}\right)T_1 + \left(\frac{1}{R_k}\right)T_2 + \left(\frac{1}{R_c}\right)T_a + \left(\frac{1}{R_{k0}}\right)T_b\right\}$$

$$\frac{dT_2}{dt} = \frac{1}{C_h}\left\{\left(\frac{1}{R_k}\right)T_1 - \left(\frac{2}{R_k} + \frac{1}{R_c}\right)T_2 + \left(\frac{1}{R_k}\right)T_3 + \left(\frac{1}{R_c}\right)T_a\right\}$$

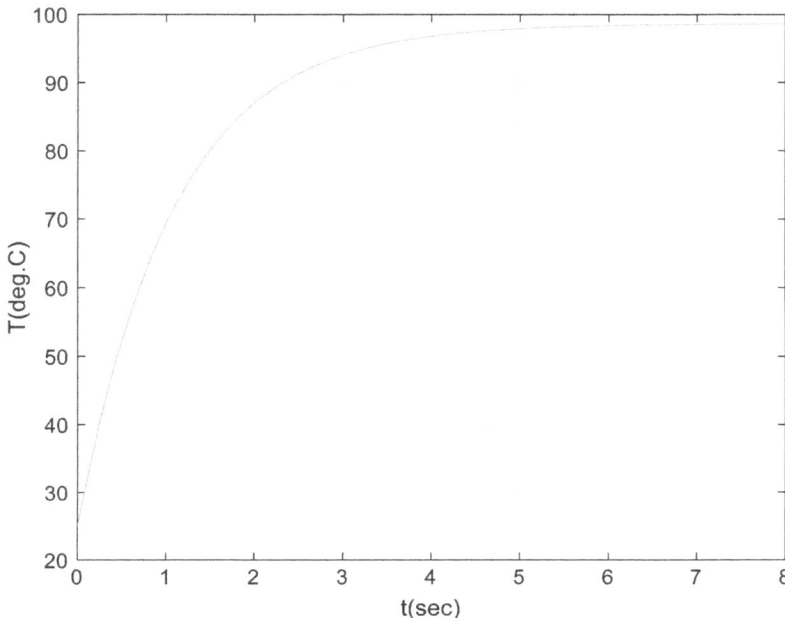

FIGURE 8.5 Profile of the pin temperature.

$$\frac{dT_3}{dt} = \frac{1}{C_h}\left\{\left(\frac{1}{R_k}\right)T_2 - \left(\frac{2}{R_k} + \frac{1}{R_c}\right)T_3 + \left(\frac{1}{R_k}\right)T_4 + \left(\frac{1}{R_c}\right)T_a\right\}$$

$$\frac{dT_4}{dt} = \frac{1}{C_h}\left\{\left(\frac{1}{R_k}\right)T_3 - \left(\frac{2}{R_k} + \frac{1}{R_c}\right)T_4 + \left(\frac{1}{R_k}\right)T_5 + \left(\frac{1}{R_c}\right)T_a\right\}$$

$$\frac{dT_5}{dt} = \frac{1}{C_h}\left\{\left(\frac{1}{R_k}\right)T_4 - \left(\frac{1}{R_k} + \frac{1}{R_c} + \frac{1}{R_{ce}}\right)T_5 + \left(\frac{1}{R_c} + \frac{1}{R_{ce}}\right)T_a\right\}$$

In these equations, the parameters are defined as $C_h = \rho C_p V_e$, $V_e = A_k L$, $R_{k0} = L/(2kA_k)$, $R_k = L/(kA_k)$, $R_c = 1/(h_c A_c)$, and $R_{ce} = 1/(h_c A_k)$. The script *pinelemtemp* produces the desired plot, shown in Figure 8.6, and computes the pin element temperatures at steady state.

```
% pinelemtemp.m: temperature profiles of pin elements
Ta = 25; L0 = 0.01; D = 0.002; L = L0/5; Ac = pi*D*L; Ak = 3.14e-6; hc = 20; % data
rho = 2707; k = 220; Cp = 896; % properties of pure aluminum
V = Ak*L; Ch = rho*Cp*V; Rc = 1/(hc*Ac); Rce = 1/(hc*Ak); Rk = L/(k*Ak); Rk0 =
L/(2*k*Ak);
T0 = Ta*ones(1,5); Tb = 100; % initial pin and base temperature at t = 0
tspan = [0 3];
dT = @(t,T) [(-(1/Rk0+1/Rk+1/Rc)*T(1)+T(2)/Rk+Ta/Rc+Tb/Rk0)/Ch;
        (T(1)/Rk-(2/Rk+1/Rc)*T(2)+T(3)/Rk+Ta/Rc)/Ch;
        (T(2)/Rk-(2/Rk+1/Rc)*T(3)+T(4)/Rk+Ta/Rc)/Ch;
        (T(3)/Rk-(2/Rk+1/Rc)*T(4)+T(5)/Rk+Ta/Rc)/Ch;
        (T(4)/Rk-(1/Rk+1/Rc+1/Rce)*T(5)+(1/Rc+1/Rce)*Ta)/Ch];
[t,T] = ode45(dT,tspan,T0);
T1 = T(:,1); T2 = T(:,2); T3 = T(:,3); T4 = T(:,4); T5 = T(:,5);
```

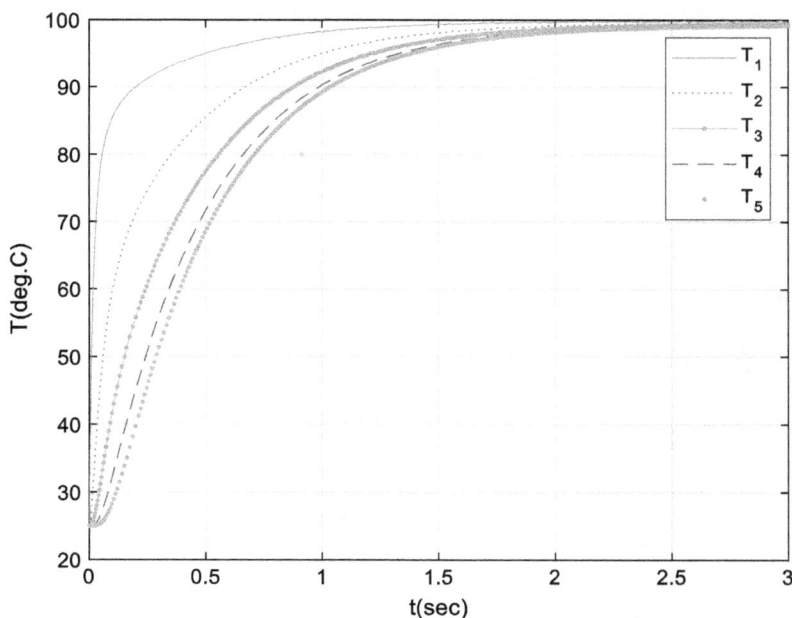

FIGURE 8.6 Temperature profiles of pin elements.

```
plot(t,T1,t,T2,':',t,T3,'.-',t,T4,'--',t,T5,'.'),  xlabel('t(s)'),  ylabel
('T(deg.C)'), grid
legend('T_1','T_2','T_3','T_4','T_5')
% Steady-state temperature
Tst = @(x) [-(1/Rk0+1/Rk+1/Rc)*x(1)+x(2)/Rk+Ta/Rc+Tb/Rk0;
       x(1)/Rk-(2/Rk+1/Rc)*x(2)+x(3)/Rk+Ta/Rc;
       x(2)/Rk-(2/Rk+1/Rc)*x(3)+x(4)/Rk+Ta/Rc;
       x(3)/Rk-(2/Rk+1/Rc)*x(4)+x(5)/Rk+Ta/Rc;
       x(4)/Rk-(1/Rk+1/Rc+1/Rce)*x(5)+(1/Rc+1/Rce)*Ta];
x0 = Ta*ones(1,5); Ts = fsolve(Tst,x0);
fprintf('Steady-state temperature:\nPin element 1 = %g deg.C\n',Ts(1));
fprintf('Pin element 2 = %g deg.C\n',Ts(2)); fprintf('Pin element 3 = %g
deg.C\n',Ts(3));
fprintf('Pin element 4 = %g deg.C\n',Ts(4)); fprintf('Pin element 5 = %g
deg.C\n',Ts(5));

>> pinelemtemp
Steady-state temperature:
Pin element 1 = 99.8577 deg.C
Pin element 2 = 99.6276 deg.C
Pin element 3 = 99.4518 deg.C
Pin element 4 = 99.3302 deg.C
Pin element 5 = 99.2627 deg.C
```

8.1.2 HEAT TRANSFER THROUGH MULTILAYERS OF SLABS AND CYLINDERS

Figure 8.7 shows a multilayer slab of more than one material.[5] At interfaces between two different solids, the temperature is continuous. The heat flux continuity can be described by

$$T = T_1|_{x=intf} = T_2|_{x=intf}, \frac{q_x}{A} = -k_1\frac{dT_1}{dx}\bigg|_{intf} = -k_2\frac{dT_2}{dx}\bigg|_{intf}$$

where the subscript *intf* represents the interface between two different solids.

Since the heat flow q is the same in each layer, the Fourier equation for each layer (A, B, and C) can be written as

$$q = \frac{k_A A}{\Delta x_A}(T_1 - T_2) = \frac{k_B A}{\Delta x_B}(T_2 - T_3) = \frac{k_C A}{\Delta x_C}(T_3 - T_4)$$

where A is the heat transfer area. Rearrangement of this equation gives

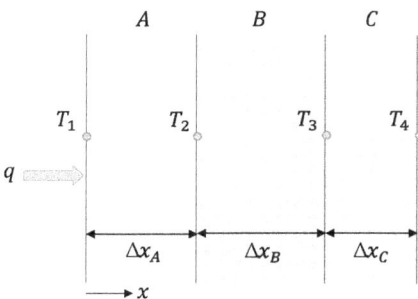

FIGURE 8.7 One-dimensional heat transfer through a multilayer solid slab. (From Cutlip, M. B. and Shacham, M., *Problem Solving in Chemical and Biochemical Engineering with POLYMATH, Excel, and MATLAB*, 2nd ed., Prentice-Hall, Boston, MA, 2008, p. 334.)

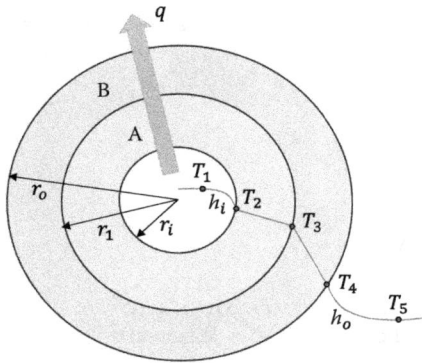

FIGURE 8.8 Heat transfer through a multilayer cylinders. (From Cutlip, M.B. and Shacham, M., *Problem Solving in Chemical and Biochemical Engineering with POLYMATH, Excel, and MATLAB*, 2nd ed., Prentice-Hall, Boston, MA, 2008, p. 334.)

$$\frac{q}{A} = \frac{T_1 - T_4}{\Delta x_A / k_A + \Delta x_B / k_B + \Delta x_C / k_C}$$

Heat transfer often occurs in multilayer cylinders in the process industries. At steady state, the heat transfer rate q is the same for each layer. From Fourier's law, the heat transfer rate is given by[6]

$$q = \frac{T_1 - T_5}{\frac{1}{h_i A_i} + \frac{r_1 - r_i}{k_A A_{Alm}} + \frac{r_0 - r_1}{k_B A_{Blm}} + \frac{1}{h_o A_o}}$$

where

$r_j\,(j = o,\ 1,\ i)$ represents the radius of each cylinder

h_i and h_o ($btu/(hr{\cdot}ft^2{\cdot}°F)$) are the heat transfer coefficients inside and outside the cylinder, respectively

k_A and k_B ($btu/(hr{\cdot}ft{\cdot}°F)$) are the heat conductivities

The inner and outer surface areas (ft^2) are given by $A_i = 2\pi r_i L$ and $A_o = 2\pi r_o L$, respectively, and the area between the cylinders (ft^2) is given by $A_1 = 2\pi r_1 L$. The log-mean area (ft^2) is defined by

$$A_{Alm} = \frac{A_1 - A_i}{\ln(A_1/A_i)}, \quad A_{Blm} = \frac{A_o - A_1}{\ln(A_o/A_1)}$$

Figure 8.8 shows two concentric hollow cylinders[5], that is, a pipe with insulation around it.

Example 8.5 Heat Transfer through a Multilayer Slab[7]

Figure 8.9 shows a multilayer slab through which heat is transferred.[5] The thickness of each layer of the slab is $L_A = 0.015$ m, $L_B = 0.1$ m, and $L_C = 0.075$ m, and the thermal conductivity of each slab is $k_A = 0.0151$, $k_B = 0.0433$, and $k_C = 0.762$ ($W/(m{\cdot}K)$).

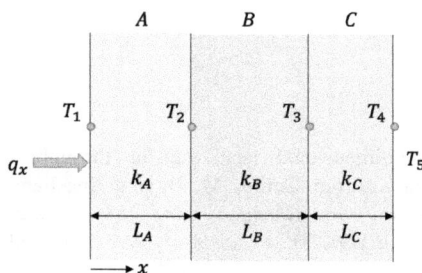

FIGURE 8.9 Heat transfer through a multilayer slab. (From Cutlip, M.B. and Shacham, M., *Problem Solving in Chemical and Biochemical Engineering with POLYMATH, Excel, and MATLAB*, 2nd ed., Prentice-Hall, Boston, MA, 2008, p. 334.)

1 Calculate the heat flux through the slab if the interior surface is at $T_1 = 255\ K$ and the exterior surface is at $T_4 = 298\ K$.

2 It is proposed to reduce the heat loss by 50% by increasing the thickness of slab B, L_B. What value of L_B is required?

3 A new slab is to be used instead of the slab B. The thermal conductivity of the new slab is given by $k = 2.5e^{-1225/T}$ $(T:K)$, and the thickness of the new slab is the same as that of the slab B. Calculate the heat flux through the new slab if the interior surface is at $T_1 = 255\ K$ and the exterior surface is at $T_4 = 298\ K$, and plot the temperature profile within the slab. Assume that $q_x/A = -15\ W/m^2$.

Solution

(1) $q_x/A = (T_1 - T_4)/(\Delta L_A/k_A + \Delta L_B/k_B + \Delta L_C/k_C)$. The following commands calculate the value of q_x/A.

```
>> T1 = 255; T4 = 298; L = [0.015 0.1 0.075]; k = [0.151 0.043 0.762]; qx = (T1-T4)/
sum(L./k)
qx =
 -17.0409
```

We can see that $q_x/A = -17.041 W/m^2$.

(2) The heat loss per unit area is $q_x/A = -17.0409\ W/m^2$. Hence the value of, ΔL_B is to be determined such that $(-17.0409)(0.5) = -8.5204\ W/m^2$. ΔL_B can be directly calculated by

$$\Delta L_B = k_B \left\{ \frac{T_1 - T_4}{q_x/A} - (\Delta L_A/k_A + \Delta L_C/k_C) \right\}$$

```
>> T1 = 255; T4 = 298; L = [0.015 0.075]; k = [0.151 0.762]; kB = 0.043;
>> dLB = kB*((T1-T4)/(-8.5204) - sum(L./k))
dLB =
 0.2085
```

(3) According to Fourier's law, one-dimensional heat transfer by conduction for each layer gives

$$\frac{dT}{dx} = \begin{cases} -\dfrac{q_x/A}{k_A} : 0 \leq x \leq L_A \\[2ex] -\dfrac{q_x/A}{2.5e^{-1225/T}} : L_A \leq x \leq L_A + L_B \\[2ex] -\dfrac{q_x/A}{k_C} : L_A + L_B \leq x \leq L_A + L_B + L_C \end{cases}$$

The subfunction *slabmT* defines these differential equations. The total thickness of the multilayer slab is $L_t = L_A + L_B + L_C = 0.19\ m$. Thus, the integration interval is $0 \leq x \leq L_t$. The script *slabtemp* employs the built-in function *ode45* to solve the system of differential equations defined by the subfunction *slabmT* and generates the plot shown in Figure 8.10.

```
% slabtemp.m
LA = 0.015; LB = 0.1; LC = 0.075; Lt = LA+LB+LC; kA = 0.151; kC = 0.762;
qx = -15; T0 = 255; xspan = [0 Lt];
[x T] = ode45(@slabmT, xspan, T0, [], LA,LB,kA,kC,qx);
plot(x,T), grid, axis([0 Lt 250 310]), xlabel('x(m)'), ylabel('T(K)')

function dT = slabmT(x,T,LA,LB,kA,kC,qx)
```

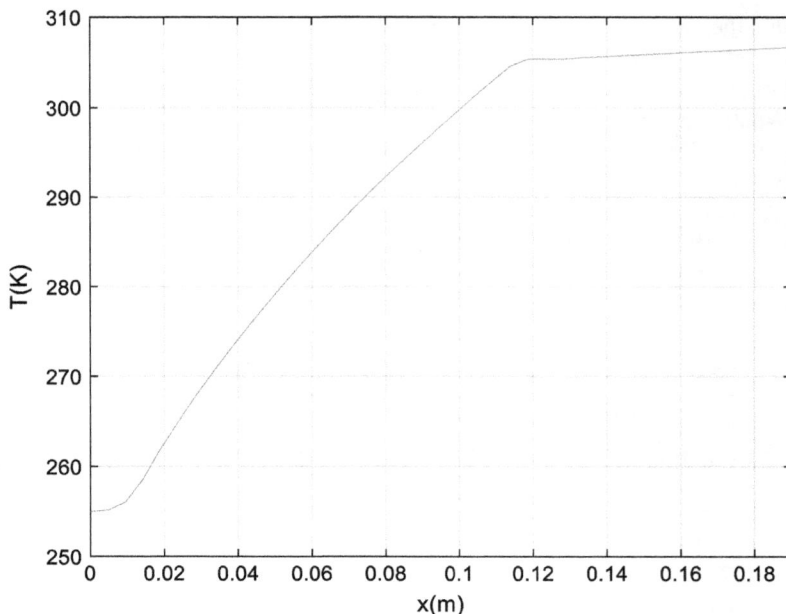

FIGURE 8.10 Temperature profile through a multilayer slab.

```
% slabmT.m: heat transfer through multilayer slab
if x <= LA, dT = -qx/kA;
elseif x <= LA+LB, dT = -qx/(2.5*exp(-1225/T));
else, dT = -qx/kC; end
end
```

8.1.3 HEAT TRANSFER IN A WIRE

The passage of current in an insulated wire generates heat. Figure 8.11 shows the differential volume of a cylindrical wire of radius R and length L.[8]

The energy balance on a cylinder shell of thickness Δr yields

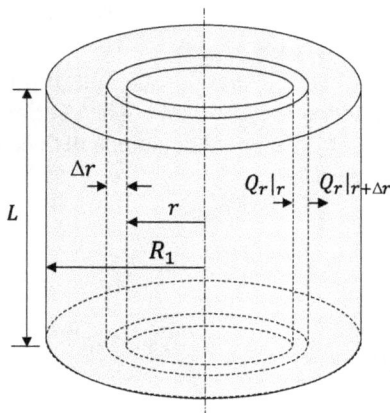

FIGURE 8.11 Differential volume of a wire. (From Cutlip, M.B. and Shacham, M., *Problem Solving in Chemical and Biochemical Engineering with POLYMATH, Excel, and MATLAB*, 2nd ed., Prentice-Hall, Boston, MA, 2008, p. 339.)

$$Q_r = -k\frac{dT}{dr}, \quad \frac{d}{dr}(rQ_r) = \frac{I^2}{k_e}r$$

where

Q_r (W/m^2) is the heat flux
k $(W/(m{\cdot}K))$ is the heat conductivity
I $(amps/m^2)$ is the current
k_e $(\Omega^{-1}m^{-1})$ is the electrical conductivity

Example 8.6 Heat Transfer in a Wire[8]

An insulated wire is carrying an electrical current. The wire surface is maintained at $T_1 = 15°C$, and the electrical and thermal conductivities are given by $k = 5$ $W/(m{\cdot}K)$ and $k_e = 1.4 \times 10^5$ $e^{0.0035T}$ $\Omega^{-1}m^{-1}$. The wire radius is $R_1 = 0.004$ m, and the total current is maintained at $I_t = 400$ $amps$. Calculate and plot the temperature and heat flux within the wire.

Solution

$$Q_r = -k\frac{dT}{dr}, \quad \frac{d}{dr}(rQ_r) = \frac{I^2}{k_e}r, \quad Q_r = \frac{(rQ_r)}{r}, \quad I = \frac{I_t}{\pi R_1^2/4}$$

The energy balance gives

$$\frac{dT}{dr} = -\frac{U}{kr}, \quad \frac{dU}{dr} = \frac{I^2}{k_e}r, \quad T(R_1) = T_1, \quad U|_{r=0} = 0$$

where $U = rQ_r$. The subfunction *funQT* defines these differential equations. The temperature at $r = 0$ is unknown. Thus, an initial guess of T at $r = 0$ is used to solve the system of differential equations to give $T(r = R_1)$. The value of T at $r = 0$ is updated at each iteration until the condition for $T(r = R_1)$ is satisfied. The bisection method may be used in the iterations; the script *bisecQT* performs the iterative calculations.

```
% bisecQT.m: calculation of heat flux and temperature in a wire
clear all;
T1 = 288.15; R1 = 0.004; k = 5; Ir = 400/(pi*R1^2);
T0a = 300; T0b = 400; % guess initial temperature
Teps = 1e-3; Terr = 1e3; iter = 1;
while Terr > Teps
      T0m = (T0a+T0b)/2;
      [r,za] = ode45(@funQT,[0 R1],[T0a 0],[],k,R1,Ir);
      [r,zb] = ode45(@funQT,[0 R1],[T0b 0],[],k,R1,Ir);
      [r,zm] = ode45(@funQT,[0 R1],[T0m 0],[],k,R1,Ir);
      a = za(end,1); b = zb(end,1); m = zm(end,1);
      if (m-T1) > 0, T0b = T0m;
      else, T0a = T0m; end
      Terr = abs(T0b - T0a); iter = iter+1;
end
Qr = zm(:,2)./r; T = zm(:,1);
subplot(1,2,1), plot(r,Qr), xlabel('r(m)'), ylabel('Qr(W/m^2)'), grid;
subplot(1,2,2), plot(r,T), xlabel('r(m)'), ylabel('T(K)'), grid;
fprintf('Number of iterations: %d, Qr at r=R1 (W/m^2): %g, ', iter, Qr(end))
fprintf('T at r=0 (K): %g\n', T(1))
```

(a)

(b)

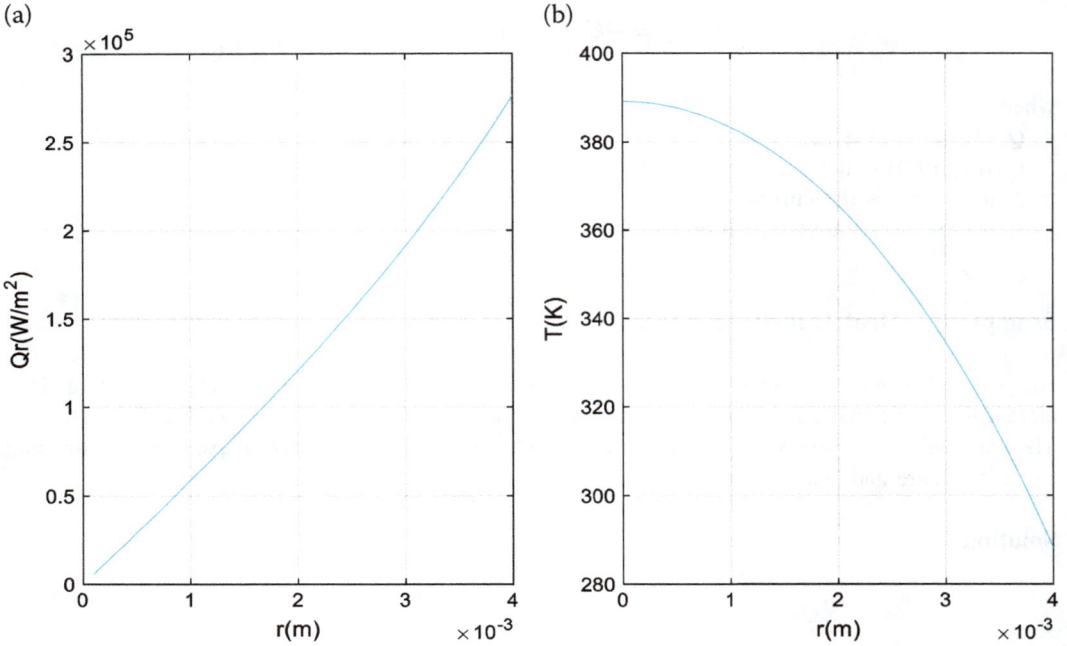

FIGURE 8.12   (a) Heat flux and (b) temperature profile in a wire.

```
function dz = funQT(r,z,k,R1,Ir)
if r > 0, Qr = z(2)/r;
else, Qr = 0; end
dz(1) = -Qr/k; dz(2) = Ir^2*r/(1.4e5*exp(0.0035*z(1))); dz = dz';
end
```

The script *bisecQT* produces the following results and the curves shown in Figure 8.12:

```
>> bisecQT
Number of iterations: 18, Qr at r=R1(W/m^2): 276472, T at r=0(K): 389.245
```

8.1.4 Heat Loss through Pipe Flanges[9]

Pipes may be connected by bolting together two identical flanges on the pipe ends. Figure 8.13 shows a cross section of pipe flange. The two flanges are symmetrical, and we can consider just one flange with heat loss from one exposed circular face and exposed rim, as shown in Figure 8.13. The heat transfer rate can be represented as a function of the radius r. At the inner surface where $r = R_1$, the temperature is maintained as the fluid temperature T_0. The temperature at $r = R_2$ is determined by heat convection from the surface to the ambient temperature T_a.

The heat balance on the different elements gives

$$\frac{dT}{dr} = -\frac{q}{k}, \quad \frac{d}{dr}(rq) = -\frac{hr}{B}(T - T_a), \quad q = \frac{(rq)}{r}$$

The initial and boundary conditions are

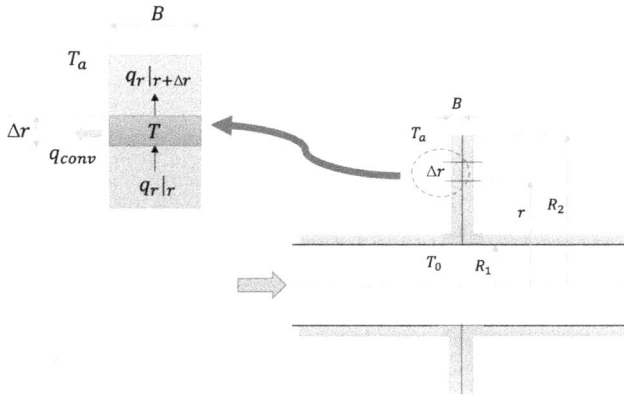

FIGURE 8.13 Cross section of a pipe flange. (From Cutlip, M.B. and Shacham, M., *Problem Solving in Chemical and Biochemical Engineering with POLYMATH, Excel, and MATLAB*, 2nd ed., Prentice-Hall, Boston, MA, 2008, pp. 347–349.)

$$T|_{r=R_1} = T_0, \; (rq)|_{r=R_2} = R_2 h (T - T_a)|_{r=R_2}$$

The heat flux q_r is given by $q_r = 2\pi r B q$. In order to determine the temperature profile and the heat loss flux q, the initial condition for rq should be determined so that the boundary condition is satisfied. This can be accomplished by defining a nonlinear equation

$$f = (rq)|_{r=R_2} - R_2 h (T - T_a)|_{r=R_2} = 0$$

The initial condition should satisfy this equation.

Example 8.7 Heat Loss through Pipe Flanges[10]

Consider the union formed by two aluminum flanges. Calculate the total heat loss flux q from a single flange when the average ambient temperature $T_a = 60°F$. Plot the heat transfer rate q_r versus the radius r. The thermal conductivity of the aluminum pipe and the flange is 133 $Btu/(hr \cdot ft \cdot °F)$, the thickness of the flange is 1 in., and $R_1 = 0.0833$ ft and $R_2 = 0.25$ ft. The fluid in the pipe is at $T_0 = 260°F$ and the heat transfer coefficient to the surroundings is constant at $h = 3$ $Btu/(hr \cdot ft^2 \cdot °F)$.

Solution
The subfunction *funQL* defines the differential equations. The value of the product rq at $r = R_1$ is unknown. Thus, we have to provide an initial guess for $rq|_{r=R_1}$ to solve the differential equations. This procedure continues iteratively until the equation

$$f = (rq)|_{r=R_2} - R_2 h (T - T_a)|_{r=R_2} = 0$$

is satisfied. The bisection method can be used effectively to update the guess. The script *bisecQL* performs these calculations.

```
% bisecQL.m: calculation of heat loss from pipe flange
clear all;
T0 = 260; Ta = 60; R1 = 0.0833; R2 = 0.25; B = 0.5/12; k = 133; h = 3;
qa = 200; qb = 800; % assume initial range for heat loss flux
feps = 1e-3; ferr = 1e3; iter = 1;
while ferr > feps
```

(a)

(b)

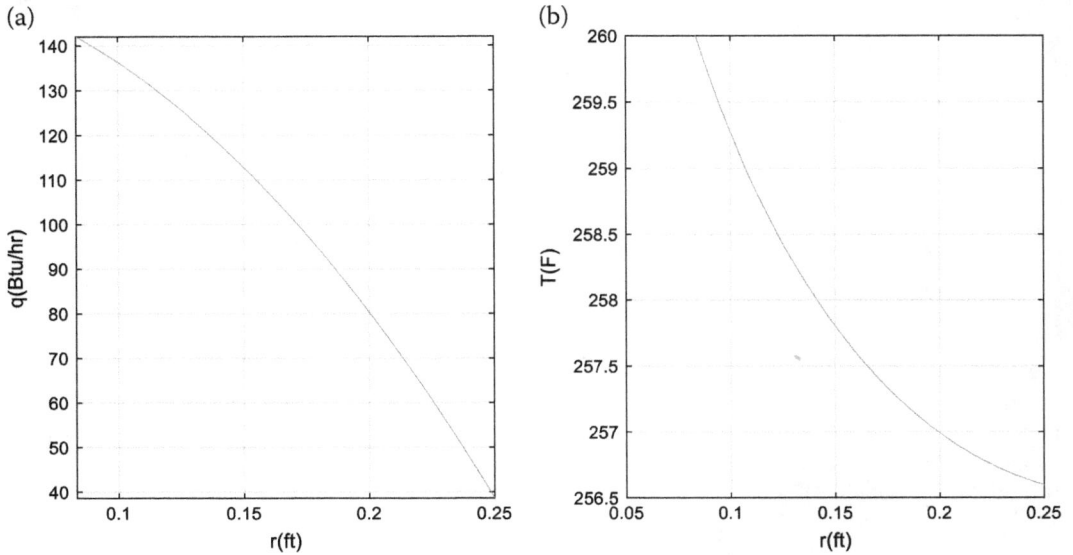

FIGURE 8.14 Profiles of (a) heat transfer rate and (b) temperature.

```
qm = (qa+qb)/2; % midpoint of the assumed temperature range
[r,ya] = ode45(@funQL,[R1 R2],[T0 qa],[],Ta,k,h,B);
[r,yb] = ode45(@funQL,[R1 R2],[T0 qb],[],Ta,k,h,B);
[r,ym] = ode45(@funQL,[R1 R2],[T0 qm],[],Ta,k,h,B);
fa = ya(end,2) - R2*h*(ya(end,1)-Ta); fb = yb(end,2) - R2*h*(yb(end,1)-Ta);
fm = ym(end,2) - R2*h*(ym(end,1)-Ta);
if (fa*fm) < 0, qb = qm;
else, qa = qm; end
ferr = abs(qb - qa); iter = iter+1;
end
Qr = ym(:,2)./r; T = ym(:,1); qrate = 2*pi*ym(:,2)*B;
subplot(1,2,1), plot(r,qrate), xlabel('r(ft)'), ylabel('q(Btu/hr)'), grid,
axis tight;
subplot(1,2,2), plot(r,T), xlabel('r(ft)'), ylabel('T(F)'), grid;
fprintf('Number of iterations: %d, heat transfer rate at r=R2 (Btu/h): %g\n',
iter, qrate(end))
fprintf('T at r=R2 (K): %g\n', T(end))
function dy = funQL(r,y,Ta,k,h,B)
q = y(2)/r; dy(1) = -q/k; dy(2) = -h*r*(y(1)-Ta)/B; dy = dy';
end
```

The script *bisecQL* produces the following outputs and the curves shown in Figure 8.14:
```
>> bisecQL
Number of iterations: 21, heat transfer rate at r=R2 (Btu/h): 38.602
T at r=R2 (K): 256.599
```

8.1.5 HEAT TRANSFER IN A LAMINAR FLOW THROUGH A CYLINDER[11]

Consider the flow of a liquid in a cylinder with radius R. For laminar flow, the velocity profile is given by

$$v_z = \frac{(P_0 - P_L)R^2}{4\mu L}\left\{1 - \left(\frac{r}{R}\right)^2\right\} = v_{max}\left\{1 - \left(\frac{r}{R}\right)^2\right\}$$

where $v_{max} = (P_0 - P_L)R^2/(4\mu L)$. From the energy balance in cylindrical coordinates due to convection and conduction, we have

$$\rho C_p \frac{v_z}{k}\frac{\partial T}{\partial z} = \rho C_p \frac{v_{max}}{k}\left\{1 - \left(\frac{r}{R}\right)^2\right\}\frac{\partial T}{\partial z} = \frac{1}{r}\frac{\partial}{\partial r}\left(r\frac{\partial T}{\partial r}\right)$$

Assume that the initial and boundary conditions are given as follows:

$$z = 0 : T(0, r) = T_0, \ r = 0 : \frac{\partial T(z, 0)}{\partial r} = 0, \ r = R : -k\left(\frac{\partial T}{\partial r}\right) = Q$$

The partial differential equation can be solved by using the built-in function *pdepe*. The standard form to be used in *pdepe* is given by

$$g\left(x, t, u, \frac{\partial u}{\partial x}\right)\frac{\partial u}{\partial t} = x^{-m}\frac{\partial}{\partial x}\left(x^m f\left(x, t, u, \frac{\partial u}{\partial x}\right)\right) + r\left(x, t, u, \frac{\partial u}{\partial x}\right)$$

By direct comparison between the differential equation and the standard form, we have $(T \to u, z \to t, r \to x)$

$$m = 1, \ g\left(x, t, u, \frac{\partial u}{\partial x}\right) = \rho C_p \frac{v_{max}}{k}\left\{1 - \left(\frac{r}{R}\right)^2\right\}, \ r\left(x, t, u, \frac{\partial u}{\partial x}\right) = 0, \ f\left(x, t, u, \frac{\partial u}{\partial x}\right) = \frac{\partial T}{\partial r}$$

The standard forms of the initial and boundary conditions to be used in the built-in function *pdepe* are given by

$$t = t_0 : u(x, t_0) = u_0(x)$$

$$x = a, \ x = b : p(x, t, u) + q(x, t)f\left(x, t, u, \frac{\partial u}{\partial x}\right) = 0$$

By comparison between the initial and boundary conditions and the standard forms, we have

$$T(0, r) = T_0 \to u_0(x) = T_0$$

$$r = 0 : \frac{\partial T(z, 0)}{\partial r} = 0 \to p = 0, \ q = 1$$

$$r = R : -k\left(\frac{\partial T}{\partial r}\right) = Q \to p = -Q, \ q = -k$$

In order to use the built-in function *pdepe*, the differential equation and initial and boundary conditions should be defined as functions.

Example 8.8 Heat Transfer in a Cylindrical Laminar Flow[11]

For the flow of a liquid in a cylinder with radius R, produce profiles of T as functions of L and r when $R=0.03$ m, $Q=100$ $kJ/(sec \cdot m^2)$, $T_0 = 500$ K, $\rho = 1000$ kg/m^3, $C_p = 4.2$ $kJ/(kg \cdot K)$, $k = 0.1$ $kJ/(sec \cdot m \cdot K)$, and $v_{max}/k = 5$.

Solution

The subfunction *hteqn* defines the differential equation to be solved, the subfunction *htitcon* defines the initial condition, and the subfunction *htbncon* defines the boundary condition. The script *htcylinder* defines ranges of *x* and *t* and executes the built-in function *pdepe*. Execution of the script *htcylinder* yields the desired plots, shown in Figure 8.15.

```
% htcylinder.m: heat transfer in cylindrical laminar flow
clear all;
R = 0.03; tmax = 0.5; m = 1;
x = linspace(0,R,40); % subdivision from 0 to R
t = linspace(0,tmax,20); % subdivision from 0 to tmax
res = pdepe(m,@hteqn,@htitcon,@htbncon,x,t);
u = res(:,:,1);
subplot(1,2,1), mesh(x,t,u), colormap(jet), xlabel('r(m)'), ylabel('L(m)'),
zlabel('T(K)')
shading interp % surface plot
subplot(1,2,2)
% plot T as a function of r(m)
for k = 1:length(t), plot(x,u(k,:),'k'), xlabel('r(m)'), ylabel('TK)'), hold
on, end
grid, hold off
function [c,f,s] = hteqn(x,t,u,DuDx)
Cp = 4.2; rho = 1e3; vk = 5; R = 0.03; c = rho*Cp*vk*(1 - (x/R)^2); f = DuDx; s = 0;
end
function u0 = htitcon(x)
T0 = 500; u0 = T0;
end

function [pl,ql,pr,qr] = htbncon(xl,ul,xr,ur,t)
pl = 0; ql = 1; % r = 0
k = 0.1; Q = 100; pr = -Q; qr = -k; % r = R
end
```

(a)   (b)

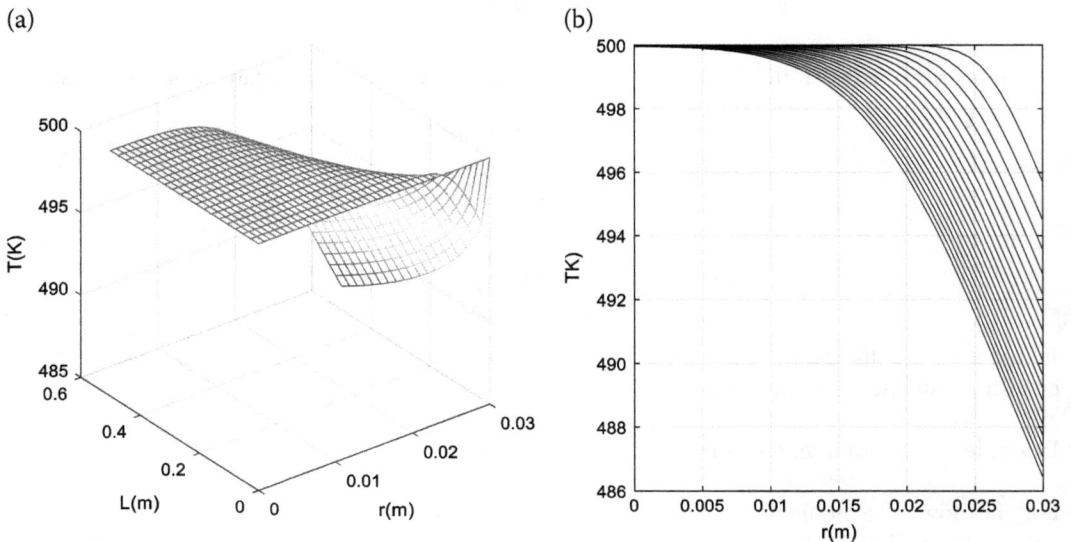

FIGURE 8.15 Profiles of T as a function of (a) L and r and (b) r.

Example 8.9 Heat Transfer in a Laminar Flow through a Cylinder[12]

Consider the flow of a liquid in a cylinder with radius R and length L. For laminar flow, the velocity profile is given by

$$v_z = \frac{(P_0 - P_L)R^2}{4\mu L}\left\{1 - \left(\frac{r}{R}\right)^2\right\} = v_{max}\left\{1 - \left(\frac{r}{R}\right)^2\right\}$$

where $v_{max} = (P_0 - P_L)R^2/(4\mu L)$. From the energy balance in cylindrical coordinates due to convection and conduction, we have

$$v_z \frac{\partial T}{\partial z} = \frac{\alpha}{r}\frac{\partial}{\partial r}\left(r\frac{\partial T}{\partial r}\right)$$

where $\alpha = k/(\rho C_p)$. Assume that the initial and boundary conditions are given as follows:

$$z = 0 : T(0, r) = T_0, \quad r = 0 : \frac{\partial T(z, 0)}{\partial r} = 0, \quad r = R : T(z, R) = T_b$$

Generate the temperature profile using the method of lines when $R = 0.05$ m, $T_0 = 300$ K, $T_b = 400$ K, $L = 2$ m, $\alpha = 10^{-4}$ m^2/sec, and $v_{max} = 0.5$ m/sec. Application of the difference formula for radian nodes 1 to n yields the following differential equations in the axial direction:

$$\frac{dT}{dz} = \frac{\alpha}{v_i}\left\{\frac{T_{i+1} - 2T_i + T_{i-1}}{h^2} + \frac{1}{r_i}\left(\frac{T_{i+1} - T_{i-1}}{2h}\right)\right\}(i = 2, 3, \cdots, n-1)$$

$$\frac{dT}{dz} = \frac{\alpha}{v_i}\left\{\frac{T_b - 2T_i + T_{i-1}}{h^2} + \frac{1}{r_i}\left(\frac{T_b - T_{i-1}}{2h}\right)\right\}(i = n)$$

$$\frac{dT}{dz} = \frac{2\alpha}{v_i}\left(\frac{T_{i+1} - T_i}{h^2}\right)(i = 1)$$

In this problem, let $n = 20$.

Solution

The function *pflowht* defines the differential equations to be integrated:

```
function dT = pflowht(x,T,pf)
% heat transfer in fluid flowing through a pipe
% Retrieve data
r = pf.r; v = pf.v; n = pf.n; h = pf.h; alpa = pf.alpa; Tb = pf.Tb; dT = zeros(n,1);
% Difference model equations
for k = 1:n
  if k == 1, s = 2*(T(k+1)-T(k))/h^2; d = 0;
  elseif k == n, s = (Tb- 2*T(k)+T(k-1))/h^2; d = (Tb-T(k-1))/(2*h*r(k+1));
  else, s = (T(k+1)-2*T(k)+T(k-1))/h^2; d = (T(k+1)-T(k-1))/(2*h*r(k)); end
  dT(k) = (alpa/v(k))*(s + d);
end
end
```

The script *httube* sets the data and uses the built-in function *ode15s* to generate the 3D profile of the temperature as functions of z and r as shown in Figure 8.16.

(a)

(b)

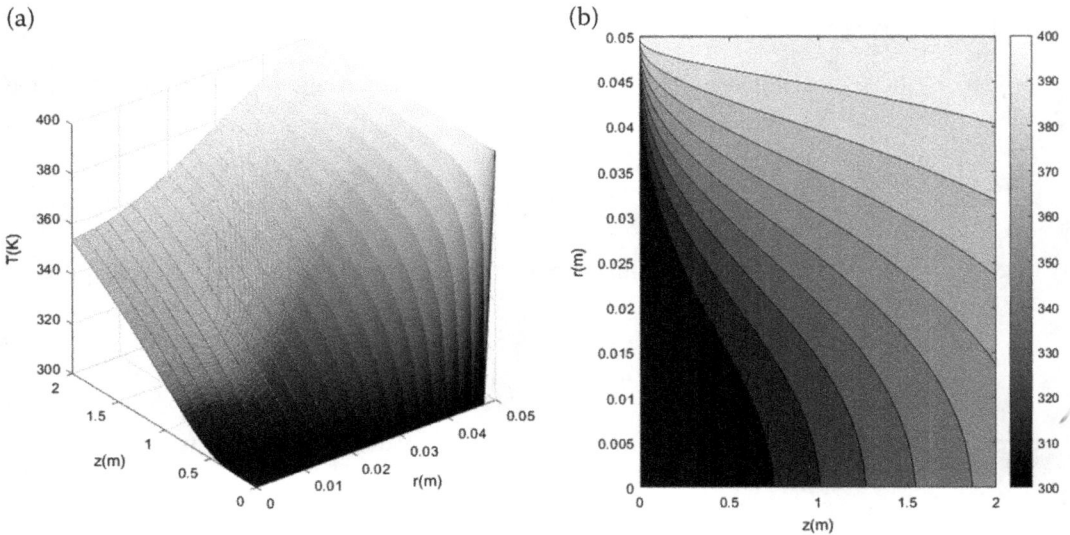

FIGURE 8.16   Profiles of T in (a) 3D and (b) 2D images.

```
% httube.m: heat transfer in fluid flowing through a pipe
clear all;
% Data and parameters
L = 2; R = 0.05; % pipe length (L) and radius (R)
n = 20; h = R/n; r = [0:n]'*h; alpa = 1e-4; Ti = 300; Tb = 400; vmax = 0.5;
v = vmax*(1-(r/R).^2); % parabolic velocity profile
pf.r = r; pf.v = v; pf.n = n; pf.h = h; pf.alpa = alpa; pf.Tb = Tb;
% Solve PDE
T0 = ones(n,1)*Ti; opn = odeset('relTol',1e-12,'absTol',1e-10);
[x,T] = ode15s(@pflowht, [0,L],T0,opn,pf);
% Display results
na = length(x); Tw = Tb*ones(na,1); T = [T,Tw];
subplot(1,2,1), mesh(r,x,T), xlabel('r(m)'), ylabel('x(m)'),
zlabel('T(K)'), colormap(gray)
subplot(1,2,2), contourf(x,r,T'), xlabel('x(m)'), ylabel('r(m)'),
colormap(gray)
```

8.2 MULTIDIMENSIONAL HEAT CONDUCTION

8.2.1 UNSTEADY-STATE HEAT CONDUCTION

For heat conduction in solids, the velocity terms are zero, and the temperature profile can be represented as a three-dimensional unsteady-state heat conduction equation

$$\rho C_p \frac{\partial T}{\partial t} = k\left(\frac{\partial^2 T}{\partial x^2} + \frac{\partial^2 T}{\partial y^2} + \frac{\partial^2 T}{\partial z^2}\right)$$

where C_p is the heat capacity at constant pressure and the heat conductivity k is assumed to be constant within the solid. The two-dimensional unsteady-state heat conduction equation can be expressed as

$$\frac{\partial T}{\partial t} = \alpha \left(\frac{\partial^2 T}{\partial x^2} + \frac{\partial^2 T}{\partial y^2} \right)$$

where $\alpha = k/\rho C_p$. The initial and boundary conditions associated with the partial differential equations can be classified into three categories.[13]

Dirichlet conditions (first kind): The values of the dependent variable are given at fixed values of the independent variable. For example,

$$T = f(x)(t = 0, 0 \le x \le 1)$$

or

$$T = T_0(t = 0, 0 \le x \le 1)$$

Boundary conditions can be written as

$$T = f(t)(x = 0, t > 0), T = T_1(x = 1, t > 0)$$

Neumann conditions (second kind): The derivative of the dependent variable is given as a constant or as a function of the independent variable. For example,

$$\frac{\partial T}{\partial x} = 0(x = 1, t \ge 0)$$

Cauchy conditions: Combined Dirichlet and Neumann conditions are classified as Cauchy conditions.

Robbins conditions (third kind): The derivative of the dependent variable is given as a function of the dependent variable itself. For example,

$$k\frac{\partial T}{\partial x} = h\left(T - T_f \right)(x = 0, t \ge 0)$$

8.2.2 Method of Lines[14]

In the method of lines, only the spatial derivatives are discretized using finite differences, not the time derivatives. For the one-dimensional parabolic equation $\partial u/\partial t = \alpha \partial^2 u/\partial x^2$, the discretization of only the spatial derivative term yields

$$\frac{du_i}{dt} = \frac{\alpha}{\Delta x^2}(u_{i+1} - 2u_i + u_{i-1})$$

The complete set of ordinary differential equations for $0 \le i \le N$ can be written as

$$\frac{du_1}{dt} = \frac{\alpha}{\Delta x^2}(u_2 - 2u_1 + u_0)$$

$$\vdots$$

$$\frac{du_i}{dt} = \frac{\alpha}{\Delta x^2}(u_{i+1} - 2u_i + u_{i-1})$$

$$\vdots$$

$$\frac{du_{N-1}}{dt} = \frac{\alpha}{\Delta x^2}(u_N - 2u_{N-1} + u_{N-2})$$

Chemical Engineering Computation with MATLAB®

The equations at the boundary,

$$\frac{du_0}{dt} = \frac{\alpha}{\Delta x^2}(u_1 - 2u_0 + u_{-1}), \frac{du_N}{dt} = \frac{\alpha}{\Delta x^2}(u_{N+1} - 2u_N + u_{N-1})$$

are specified according to the boundary conditions. For example, if a Dirichlet condition is given, $u_0 = \beta$ (constant) and

$$\frac{du_0}{dt} = 0, \; u_0(0) = \beta$$

If a Neumann condition is given at this boundary, that is, $du/dx|_{0,t} = 0$ at $x = 0$, the partial derivative is replaced by a central difference approximation $du/dx|_{0,t} = (u_1 - u_{-1})/(2\Delta x) = 0$ and we have

$$\frac{du_0}{dt} = \frac{\alpha}{\Delta x^2}(2u_1 - 2u_0)$$

The function *parab1D* solves the unsteady-state one-dimensional parabolic partial differential equation. The function *parab2D* solves the unsteady-state two-dimensional parabolic partial differential equation. When a Dirichlet condition is given at the boundary for a one-dimensional parabolic partial differential equation, the function *parabDbc* can also be used to solve the problem. The basic syntax of the functions *parab1D*, *parab2D*, and *parabDbc* is as follows:

```
function [x,t,u] = parab1D(nx, nt, dx, dt, alpha, u0, bc, func, varargin)
function [u r] = parabDbc(nx, nt, dx, dt, alpha, u0, bci, bcf)
function [x,y,t,u] = parab2D(nx,ny,nt,dx,dy,dt,alpha,u0,bc,func,varargin)
```

where u is the matrix of dependent variables (u(x,t)); nx is the number of divisions in the x direction; ny is the number of divisions in the y direction; nt is the number of divisions in the t direction; dx and dy are the distances between grid points in x and y directions, respectively; dt is the time step; and u0 is the distribution vector for u at t = 0.

In the function *parab1D*, bc is the matrix containing the types and values of boundary conditions in the x direction as its elements (2 × 2 or 2 × 3 matrix). The first and second rows of bc represent the conditions at lower x and upper x, respectively. The first column of bc specifies the type of the boundary conditions (1: Dirichlet condition, values of u specified in the second column; 2: Neumann condition, values of u' specified in the second column; 3: Robbins condition, constants and coefficients for u specified in the second and third columns). bci and bcf are the values of the boundary conditions at each end of x.

In the function *parab2D*, bc is the matrix containing the types and values of boundary conditions in the x and y directions as its elements (4 × 2 or 4 × 3 matrix). The order of the boundary conditions is lower x, upper x, lower y, and upper y (these are stored in the first four rows of the matrix bc). The first column of bc specifies the type of the boundary conditions (1: Dirichlet condition, values of u specified in the second column; 2: Neumann condition, values of u' specified in the second column; 3: Robbins condition, constants and coefficients for u specified in the second and third columns).

```
function [x,t,u] = parab1D(nx,nt,dx,dt,alpha,u0,bc,func,varargin)
% Solve 1-dimensional parabolic PDE
% Outputs:
%    x, t; vectors of x and t values
%    u: matrix of dependent variables [u(x,t)]
% Inputs:
```

```
%     nx, nt: number of divisions in x and t direction
%     dx, dy: x and t increment
%     alpha: coefficient of equation
%     u0: vector of u distribution at t=0
%     bc: a matrix containing types and values of boundary conditions
%         in x direction (2x2 or 2x3 matrix)
%         row 1: conditions at lower x, row 2: conditions at upper x
%         1st column: type of condition
%         (1: Dirichlet condition, values of u in 2nd column
%          2: Neumann condition, values of u' in 2nd column
%          3: Robbins condition, 2-3 columns contain constant and coef. of u)

% Initialization
if nargin < 7, error('Insufficient number of inputs.'); end
nx = fix(nx); x = [0:nx]*dx; nt = fix(nt); t = [0:nt]*dt; r = alpha*dt/dx^2;
u0 = (u0(:).')'; % confirm column vector
if length(u0) ~= nx+1, error('Incorrect length of the initial condition
vector.'); end
[a,b] = size(bc);
if a ~= 2, error('Invalid number of boundary conditions.'); end
if b < 2 | b > 3, error('Invalid boundary condition.'); end
if b == 2 & max(bc(:,1)) <= 2, bc = [bc zeros(2,1)]; end
u(:,1) = u0; c = zeros(nx+1,1);
% Iteration according to t
for n = 2:nt+1 % lower x boundary condition
  switch bc(1,1)
    case 1, A(1,1) = 1; c(1) = bc(1,2);
    case {2, 3}
       A(1,1) = -3/(2*dx) - bc(1,3); A(1,2) = 2/dx; A(1,3) = -1/(2*dx); c(1) =
bc(1,2);
  end
  % Interior points
  for i = 2:nx
    A(i,i-1) = -r; A(i,i) = 2*(1+r); A(i,i+1) = -r;
    c(i) = r*u(i-1,n-1) + 2*(1-r)*u(i,n-1) + r*u(i+1,n-1);
    if nargin >= 8 % Nonhomogeneous equation
       intercept = feval(func,0,x(i),t(n),varargin{:});
      slope = feval(func,1,x(i),t(n),varargin{:}) - intercept; A(i,i) = A(i,i)
- dt*slope;
         c(i) = c(i) + dt*feval(func,u(i,n-1),x(i),t(n-1),varargin{:}) +
dt*intercept;
    end
  end
  switch bc(2,1) % upper x boundary condition
    case 1, A(nx+1,nx+1) = 1; c(nx+1) = bc(2,2);
    case {2, 3}
      A(nx+1,nx+1) = 3/(2*dx) - bc(2,3);
      A(nx+1,nx) = -2/dx; A(nx+1,nx-1) = 1/(2*dx); c(nx+1) = bc(2,2);
  end
  u(:,n) = inv(A)*c;
end
end
function [x,y,t,u] = parab2D(nx,ny,nt,dx,dy,dt,alpha,u0,bc,func,varargin)
% Solve 2-dimensional parabolic PDE: use explicit formula
% Outputs:
```

```
%    x, y, t: vectors of x, y and t values
%    u: 3D array of dependent variables [u(x,y,t)]
% Inputs:
%    nx, ny, nt: number of divisions in x, y and t direction
%    dx, dy: x and y increments
%    dt: t increments (leave empty to use the default values)
%    alpha: coefficient of equation
%    u0: matrix of u distribution at t=0 [u0(x,y)]
%    bc: a matrix containing types and values of boundary conditions
%        in x and y directions (4x2 or 4x3 matrix)
%     order of appearing: lower x, upper x, lower y, upper y
%     in rows 1 to 4 of the matric bc
%     1st column: type of condition
%     (1: Dirichlet condition, values of u in 2nd column
%      2: Neumann condition, values of u' in 2nd column
%      3: Robbins condition, 2-3 columns contain constant and coef. of u)

% Initialization
if nargin < 9, error(' Invalid number of inputs.'); end
nx = fix(nx); x = [0:nx]*dx; ny = fix(ny); y = [0:ny]*dy;
% Check dt for stability
tmax = dt*nt;
if isempty(dt) | dt > (dx^2+dy^2)/(16*alpha)
  dt = (dx^2+dy^2)/(16*alpha); nt = tmax/dt+1;
  fprintf('\ndt is adjusted to %6.2e (nt=%3d)\n',dt,fix(nt))
end
nt = fix(nt); t = [0:nt]*dt; rx = alpha*dt/dx^2; ry = alpha*dt/dy^2;
[r0,c0] = size(u0);
if r0 ~= nx+1 | c0 ~= ny+1
  error('Incorrect size of the initial condition matrix.')
end
[a,b] = size(bc);
if a ~= 4, error('Invalid number of boundary conditions.'); end
if b < 2 | b > 3, error('Invalid boundary condition.'); end
if b == 2 & max(bc(:,1)) <= 2, bc = [bc zeros(4,1)]; end
% Solution of PDE
u(:,:,1) = u0;
for n = 1:nt
  for i = 2:nx
    for j = 2:ny
      u(i,j,n+1) = rx*(u(i+1,j,n)+u(i-1,j,n))+ ry*(u(i,j+1,n) + u(i,j-1,n)) +
(1-2*rx-2*ry)*u(i,j,n);
        if nargin >= 10
                u(i,j,n+1) = u(i,j,n+1) + dt*feval(func,u(i,j,n),x(i),y(j),t
(n),varargin{:});
        end
      end
    end
  end
  % Lower x boundary condition
  switch bc(1,1)
    case 1, u(1,2:ny,n+1) = bc(1,2) * ones(1,ny-1,1);
    case {2, 3}
      u(1,2:ny,n+1) = (-2*bc(1,2)*dx + 4*u(2,2:ny,n+1) - u(3,2:ny,n+1)) / (2*bc
(1,3)*dx + 3);
  end
```

```
  % Upper x boundary condition
  switch bc(2,1)
    case 1, u(nx+1,2:ny,n+1) = bc(2,2) * ones(1,ny-1,1);
    case {2, 3}
     u(nx+1,2:ny,n+1) = (-2*bc(2,2)*dx - 4*u(nx,2:ny,n+1) + u(nx-1,2:ny,n+1))
/ (2*bc(2,3)*dx - 3);
  end
  % Lower y boundary condition
  switch bc(3,1)
    case 1, u(2:nx,1,n+1) = bc(3,2) * ones(nx-1,1,1);
    case {2, 3}
     u(2:nx,1,n+1) = (-2*bc(3,2)*dy + 4*u(2:nx,2,n+1) - u(2:nx,3,n+1)) / (2*bc
(3,3)*dy + 3);
  end
  % Upper y boundary conditionCorner nodes
  switch bc(4,1)
    case 1, u(2:nx,ny+1,n+1) = bc(4,2) * ones(nx-1,1,1);
    case {2, 3}
     u(2:nx,ny+1,n+1) = (-2*bc(4,2)*dy - 4*u(2:nx,ny,n+1) + u(2:nx,ny-1,n+1))
/ (2*bc(4,3)*dy - 3);
  end
end
% Corner nodes
u(1,1,:) = (u(1,2,:) + u(2,1,:)) / 2;
u(nx+1,1,:) = (u(nx+1,2,:) + u(nx,1,:)) / 2;
u(1,ny+1,:) = (u(1,ny,:) + u(2,ny+1,:)) / 2;
u(nx+1,ny+1,:) = (u(nx+1,ny,:) + u(nx,ny+1,:)) / 2;
end
function [u r] = parabDbc(nx,nt,dx,dt,alpha,u0,bci,bcf)
% Solve 1-dimensional parabolic PDE (constant Dirichlet boundary conditions)
% Outputs:
%    u: matrix of dependent variables [u(x,t)]
% Inputs:
%    nx, nt: number of divisions in x and t direction
%    dx, dy: x and t increments
%    alpha: coefficient of equation
%    u0: row vector of u distribution at t=0
%    bci, bcf: boundary conditions at both ends of x (scalar)

r = alpha*dt/dx^2; A = zeros(nx-1,nx-1); u = zeros(nt+1,nx+1); u(:,1) = bci*ones
(nt+1,1);
u(:,nx+1) = bcf*ones(nt+1,1); u(1,:) = u0; A(1,1) = 1 + 2*r; A(1,2) = -r;
for i = 2:nx-2, A(i,i) = 1 + 2*r; A(i,i-1) = -r; A(i,i+1) = -r; end
A(nx-1,nx-2) = -r; A(nx-1,nx-1) = 1 + 2*r; b(1,1) = u0(2) + u0(1)*r;
for i = 2:nx-2, b(i,1) = u0(i+1); end
b(nx-1,1) = u0(nx) + u0(nx+1)*r; [L U] = lu(A);
for j = 2:nt+1
  y = L\b; x = U\y; u(j,2:nx) = x'; b = x;
  b(1,1) = b(1,1) + bci*r; b(nx-1,1) = b(nx-1,1) + bcf*r;
end
end
```

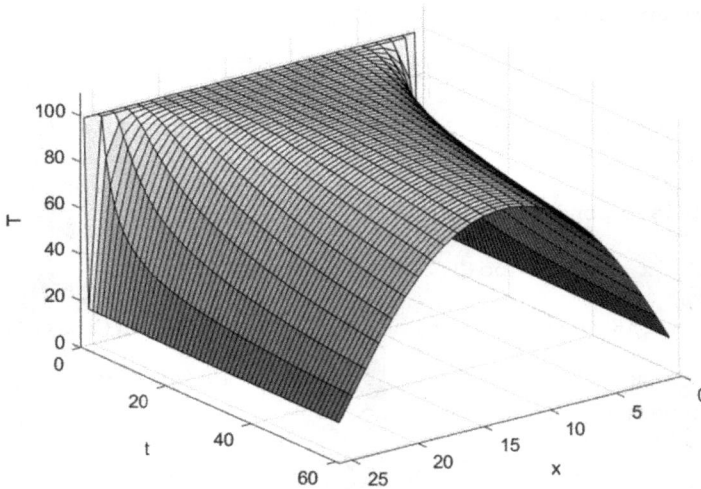

FIGURE 8.17   Temperature profile within a brick wall.

Example 8.10 One-Dimensional Parabolic PDE for Heat Transfer

A wall made of brick is 0.5 m thick, and the temperature of the wall is 100°C at $t = 0$. The thermal diffusivity of the brick is $\alpha = 4.52 \times 10^{-7}$ m/sec^2. The temperature on the both sides of the wall is suddenly dropped to 18°C. Calculate and plot the temperature profile within the brick wall during 5 hr (18,000 sec) with the interval of 300 sec. Assume that the number of nodes in the x direction is 25.

Solution

Since the temperature at both sides of the wall is given as fixed values, the boundary conditions are Dirichlet conditions and the function *parabDbc* can be used to solve the problem. The script *htdiri* produces desired outputs and the temperature profile shown in Figure 8.17:

```
% htdiri.m
alpha = 4.52e-7; nx = 25; dx = 0.5/nx; dt = 300; nt = 18000/dt;
u0 = 100*ones(1,nx+1); bci = 18; bcf = 18;
[u r] = parabDbc(nx, nt, dx, dt, alpha, u0, bci, bcf);
surf(u), axis([0 nx+1 0 nt+1 0 110]), view([-217 30]), xlabel('x'), ylabel
('t'), zlabel('T'), r

>> htdiri
r =
  0.3390
```

8.2.3 STEADY-STATE HEAT CONDUCTION

Steady-state heat conduction within a three-dimensional solid can be expressed by the elliptic partial differential equation

$$\frac{\partial^2 T}{\partial x^2} + \frac{\partial^2 T}{\partial y^2} + \frac{\partial^2 T}{\partial z^2} = 0$$

The two-dimensional steady-state heat conduction problem can be described by

$$\nabla^2 T = 0 \text{ (Laplace equation)}$$

$$\nabla^2 T = f(x, y) \text{ (Poisson equation)}$$

$$\nabla^2 T + g(x, y)T = f(x, y) \text{ (Helmholtz equation)}$$

where $\nabla^2 u = (\partial^2 u/\partial x^2) + (\partial^2 u/\partial y^2)$. Introducing the central difference approximation to the Laplace equation $\nabla^2 u = (\partial^2 u/\partial x^2) + (\partial^2 u/\partial y^2) = 0$, we have

$$\frac{1}{\Delta x^2}\left(u_{i+1,j} - 2u_{i,j} + u_{i-1,j}\right) + \frac{1}{\Delta y^2}\left(u_{i,j+1} - 2u_{i,j} + u_{i,j-1}\right) = 0$$

Rearrangement of this equation yields

$$-2\left(\frac{1}{\Delta x^2} + \frac{1}{\Delta y^2}\right)u_{i,j} + \left(\frac{1}{\Delta x^2}\right)u_{i+1,j} + \left(\frac{1}{\Delta x^2}\right)u_{i-1,j} + \left(\frac{1}{\Delta y^2}\right)u_{i,j+1} + \left(\frac{1}{\Delta y^2}\right)u_{i,j-1} = 0$$

The function *ellipticPDE* solves the elliptic partial differential equation using this equation.

```
function [x,y,U] = ellipticPDE(nx,ny,dx,dy,bc,f)
% Solve 2-dimensional elliptic PDE
% output:
%   x, y: vectors of x and y values
%   U: matrix of dependent variables (U(x,y))
% input:
%   nx, ny: number of divisions in x and y direction
%   dx, dy: x and y increments
%   f: constant for Poisson equation
%   bc: a matrix containing types and values of boundary conditions
%      in x and y directions (4x2 or 4x3 matrix)
%      order of appearing: lower x, upper x, lower y, upper y
%      in rows 1 to 4 of the matric bc
%      1st column: type of condition:
%      (1: Dirichlet condition, values of u in 2nd column
%       2: Neumann condition, values of u' in 2nd column
%       3: Robbins condition, 2-3 columns contain constant and coef. of u)

% Initialization
if nargin < 5, error('Invalid number of inputs.'); end
[a,b] = size(bc);
if a ~= 4, error('Invalid number of boundary conditions.'); end
if b < 2 | b > 3, error('Invalid boundary condition.'); end
if b == 2 & max(bc(:,1)) <= 2, bc = [bc zeros(4,1)]; end
if nargin < 6 | isempty(f), f = 0; end
nx = fix(nx); x = [0:nx]*dx; ny = fix(ny); y = [0:ny]*dy; dx2 = 1/dx^2; dy2 = 1/dy^2;
% Coefficient matrix and constant vector
n = (nx+1)*(ny+1); A = zeros(n); c = zeros(n,1);
onex = diag(diag(ones(nx-1))); oney = diag(diag(ones(ny-1)));
% Interior nodes
```

```
i = [2:nx];
for j = 2:ny
   ind = (j-1)*(nx+1)+i; A(ind,ind) = -2*(dx2+dy2)*onex;
     A(ind,ind+1) = A(ind,ind+1) + dx2*onex; A(ind,ind-1) = A(ind,ind-1) +
dx2*onex;
   A(ind,ind+nx+1) = A(ind,ind+nx+1) + dy2*onex;
   A(ind,ind-nx-1) = A(ind,ind-nx-1) + dy2*onex; c(ind) = f*ones(nx-1,1);
end
% Lower x boundary condition
switch bc(1,1)
   case 1
       ind = ([2:ny]-1)*(nx+1)+1; A(ind,ind) = A(ind,ind) + oney; c(ind) = bc
(1,2)*ones(ny-1,1);
   case {2, 3}
        ind = ([2:ny]-1)*(nx+1)+1; A(ind,ind) = A(ind,ind) - (3/(2*dx) + bc
(1,3))*oney;
     A(ind,ind+1) = A(ind,ind+1) + 2/dx*oney; A(ind,ind+2) = A(ind,ind+2) - 1/
(2*dx)*oney;
     c(ind) = bc(1,2)*ones(ny-1,1);
end
% Upper x boundary condition
switch bc(2,1)
   case 1
     ind = [2:ny]*(nx+1); A(ind,ind) = A(ind,ind) + oney; c(ind) = bc(2,2)*ones
(ny-1,1);
   case {2, 3}
     ind = [2:ny]*(nx+1); A(ind,ind) = A(ind,ind) + (3/(2*dx) - bc(2,3))*oney;
     A(ind,ind-1) = A(ind,ind-1) - 2/dx*oney;
       A(ind,ind-2) = A(ind,ind-2) + 1/(2*dx)*oney; c(ind) = bc(2,2)*ones
(ny-1,1);
end
% Lower y boundary condition
switch bc(3,1)
   case 1
   ind = [2:nx]; A(ind,ind) = A(ind,ind) + onex; c(ind) = bc(3,2)*ones(nx-1,1);
   case {2, 3}
     ind = [2:nx]; A(ind,ind) = A(ind,ind) - (3/(2*dy) + bc(3,3))*onex;
     A(ind,ind+nx+1) = 2/dy*onex; A(ind,ind+2*(nx+1)) = -1/(2*dy)*onex;
     c(ind) = bc(3,2)*ones(nx-1,1);
end
% Upper y boundary condition
switch bc(4,1)
   case 1
       ind = ny*(nx+1)+[2:nx]; A(ind,ind) = A(ind,ind) + onex; c(ind) = bc
(4,2)*ones(nx-1,1);
   case {2, 3}
    ind = ny*(nx+1)+[2:nx]; A(ind,ind) = A(ind,ind) + (3/(2*dy) - bc(4,3))*onex;
     A(ind,ind-(nx+1)) = A(ind,ind-(nx+1)) - 2/dy*onex;
       A(ind,ind-2*(nx+1)) = A(ind,ind-2*(nx+1)) + 1/(2*dy)*onex; c(ind) = bc
(4,2)*ones(nx-1,1);
end
% Corner nodes
A(1,1) = 1; A(1,2) = -1/2; A(1,nx+2) = -1/2; c(1) = 0;
A(nx+1,nx+1) = 1; A(nx+1,nx) = -1/2; A(nx+1,2*(nx+1)) = -1/2; c(nx+1) = 0;
A(ny*(nx+1)+1,ny*(nx+1)+1) = 1; A(ny*(nx+1)+1,ny*(nx+1)+2) = -1/2;
```

```
A(ny*(nx+1)+1, (ny-1)*(nx+1)+1) = -1/2; c(ny*(nx+1)+1) = 0;
A(n,n) = 1; A(n,n-1) = -1/2; A(n,n-(nx+1)) = -1/2; c(n) = 0;
u = inv(A)*c; % solve equation
% Rearrange results in matrix form
for k = 1:ny+1, U(k,1:nx+1) = u((k-1)*(nx+1)+1:k*(nx+1))'; end
end
```

Example 8.11 Two-Dimensional Elliptic Equations for Heat Transfer[15]

The temperature profile in a two-dimensional thin metal plate can be described by the two-dimensional elliptic partial differential equation

$$\frac{\partial^2 T}{\partial x^2} + \frac{\partial^2 T}{\partial y^2} = f$$

where f is assumed to be constant. The metal plate is made of an alloy that has a melting point of 800 °C and a thermal conductivity of 16 $W/(m\cdot K)$. The plate is subject to an electric current that creates a uniform heat source within the plate. The amount of heat generated is $Q' = 100\ kW/m^3$. All four edges of the plate are in contact with a fluid at 25 °C. The set of Robbins boundary conditions is

$$\left.\frac{\partial T}{\partial x}\right|_{0,y} = 5\{T(0,y) - 25\}, \left.\frac{\partial T}{\partial x}\right|_{1,y} = 5\{25 - T(1,y)\},$$

$$\left.\frac{\partial T}{\partial y}\right|_{x,0} = 5\{T(x,0) - 25\}, \left.\frac{\partial T}{\partial y}\right|_{x,1} = 5\{25 - T(x,1)\}$$

Plot the temperature profiles within the plate.

In order to solve the set of equations, all the values of the dependent variables have to be rearranged as a column vector and numbered. The finite difference approximation for this problem has the form

$$-2\left(\frac{1}{\Delta x^2} + \frac{1}{\Delta y^2}\right)u_n + \left(\frac{1}{\Delta x^2}\right)u_{n+1} + \left(\frac{1}{\Delta x^2}\right)u_{n-1} + \left(\frac{1}{\Delta y^2}\right)u_{n+p+1} + \left(\frac{1}{\Delta y^2}\right)u_{n-p-1} = f$$

When the Laplace equation is being solved, $f = 0$. For the Poisson equation, the value of f is assumed constant throughout the plate.

If the boundary condition is of the Dirichlet type, u_N=(constant). However, if the boundary condition is of the Neumann or Robbins type, forward or backward difference is used to evaluate the 1st-order derivative at the boundaries.

$x = 0$: forward difference (N is a node on the line $x = 0$)

$$\left.\frac{\partial u}{\partial x}\right|_{x=0} = \frac{1}{2\Delta x}(-3u_N + 4u_{N+1} - u_{N+2})$$

$x = L$: backward difference (N is a node on the line $x = L$)

$$\left.\frac{\partial u}{\partial x}\right|_{x=L} = \frac{1}{2\Delta x}(3u_N - 4u_{N-1} + u_{N-2})$$

$y = 0$: forward difference (N is a node on the line $y = 0$)

$$\left.\frac{\partial u}{\partial y}\right|_{y=0} = \frac{1}{2\Delta y}\left(-3u_N + 4u_{N+p+1} - u_{N+2p+2}\right)$$

$y = L$: backward difference (N is a node on the line $y = L$)

$$\left.\frac{\partial u}{\partial y}\right|_{y=L} = \frac{1}{2\Delta y}\left(3u_N - 4u_{N-p-1} + u_{N-2p-2}\right)$$

Solution

The script *tempPDE* specifies the dimensions of the plate and the boundary conditions. This script calls the function *ellipticPDE* to solve the partial differential equation and produces the temperature profile shown in Figure 8.18.

```
% tempPDE.m
% Solve Laplace/Poisson equations using finite difference method
clear all;
distx = 1; % length of the plate in x direction (m)
disty = 1; % width of the plate in y direction (m)
ndx = 20; % number of divisions in x direction
ndy = 20; % number of divisions in y direction
rhf = -100e3/16; % right-hand side of the equation
% Boundary conditions: 1) Dirichlet, 2) Neumann, 3) Robbins
bc(1,1) = 3;   % Lower x boundary condition: Robbins
bc(1,2) = -5*25; % constant (beta)
bc(1,3) = 5; % coefficient (gamma)
bc(2,1) = 3; % Upper x boundary condition: Robbins
bc(2,2) = 5*25; % constant (beta)
bc(2,3) = -5; % coefficient (gamma)
bc(3,1) = 3; % Lower y boundary condition: Robbins
```

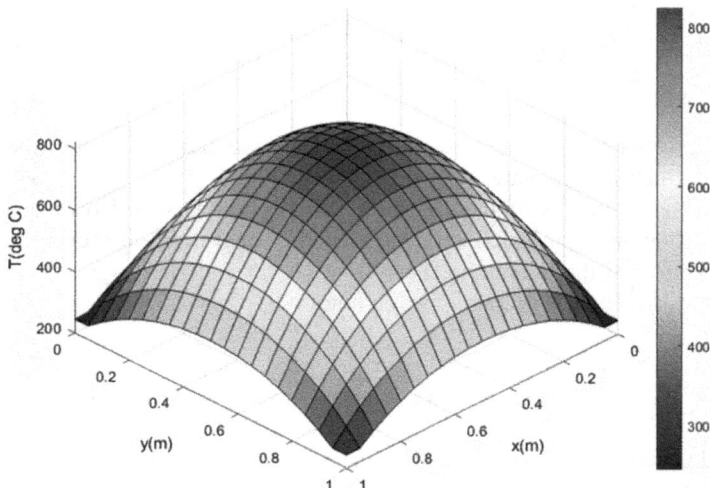

FIGURE 8.18 Temperature profile within a metal plate.

```
bc(3,2) = -5*25; % constant (beta)
bc(3,3) = 5; % coefficient (gamma)
bc(4,1) = 3; % Upper y boundary condition
bc(4,2) = 5*25; % constant (beta
bc(4,3) = -5; % coefficient (gamma)
[x,y,T] = ellipticPDE(ndx,ndy,distx/ndx,disty/ndy,bc,rhf); colormap(jet)
surf(y,x,T), xlabel('x(m)'), ylabel('y(m)'), zlabel('T(deg C)'), colorbar,
view(135,45)
```

8.3 HEAT EXCHANGERS

8.3.1 Log-Mean Temperature Difference

In designing a countercurrent shell-and-tube heat exchanger, it is important to minimize the number of shells. For the simple countercurrent shell-and-tube heat exchanger shown in Figure 8.19, the minimum number of shells and the log-mean temperature difference can be given as follows[16]:

$$\Delta t_1 = T_1 - t_2, \ \Delta t_2 = T_2 - t_1, \ P = \frac{t_2 - t_1}{T_1 - t_1}, \ R = \frac{T_1 - T_2}{t_2 - t_1}$$

i) $R \neq 1$:

$$P' = \frac{1 - \left(\frac{PR-1}{P-1}\right)^{1/N}}{R - \left(\frac{PR-1}{P-1}\right)^{1/N}}, \ F = \frac{\frac{\sqrt{R^2+1}}{R-1}\log\left(\frac{1-P'}{1-RP'}\right)}{\log A}, \ A = \frac{\frac{2}{P'} - 1 - R + \sqrt{R^2+1}}{\frac{2}{P'} - 1 - R - \sqrt{R^2+1}}$$

ii) $R = 1$:

$$P'' = \frac{P}{N - P(N-1)}, \ F = \frac{\left(\frac{\sqrt{R^2+1}\cdot P''}{\ln 10(1-P'')}\right)}{\log B}, \ B = \frac{\frac{2}{P''} - 2 + \sqrt{2}}{\frac{2}{P''} - 2 - \sqrt{2}}$$

$$\Delta T_{lm} = \frac{\Delta t_1 - \Delta t_2}{\ln\left(\frac{\Delta t_1}{\Delta t_2}\right)}$$

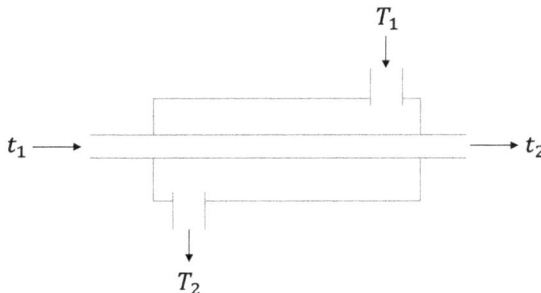

FIGURE 8.19 A simple shell-and-tube heat exchanger.

where

    Δt_1 is the larger terminal temperature difference

    Δt_2 is the smaller terminal temperature difference

    N is the number of shells required

    ΔT_{lm} is the log-mean temperature difference

    The updated log-mean temperature difference is given by

$$\Delta T_{lm} = F \cdot \Delta T_{lm}$$

where F is the correction factor.

The updated log-mean temperature difference is evaluated using the following procedure:

1)   Input required data and calculate P and R.
2)   Set $N = 1$ and compare the value of R with 1.
3)   If $R \neq 1$, calculate P' and A:
      i) If $A > 0$, calculate F and go to step 5).
      ii) If $A < 0$, set $N = N + 1$ and go to step 3).

4)   If $R = 1$, calculate P'' and B:
      i) If $B > 0$, calculate F and go to step 5).
      ii) If $B < 0$, set $N = N + 1$ and go to step 3).

5)   If $F \leq 0.75$, set $N = N + 1$ and go to step 3).
6)   If $\Delta t_1 = \Delta t_2$, $\Delta T_{lm} = \Delta t_1$.
7)   If $\Delta t_1 \neq \Delta t_2$, calculate ΔT_{lm}.
8)   Update ΔT_{lm}.

This procedure is executed by the script *shellLMTD*.

```
% shellLMTD.m : number of required shells and LMTD
clear all;
% Data
T1 = input('Hot fluid inlet temperature (deg.F): ');
T2 = input('Hot fluid outlet temperature (deg.F): ');
t1 = input('Cold fluid inlet temperature (deg.F): ');
t2 = input('Cold fluid outlet temperature (deg.F): ');
N = 1; Dt1 = T2 - t1; Dt2 = T1 - t2;
P = (t2 - t1)/(T1 - t1); R = (T1 - T2)/(t2 - t1); % Compute P and R
A = -1; B = -1; F = 0.1; % Initial F
while (F <= 0.75) % Test F and R
  if ( R > 1 || R < 1)
     while (A < 0)
        Pp = (1 - ((P*R-1)/(P-1))^(1/N)) / (R-((P*R-1)/(P-1))^(1/N));
        A = (2/Pp - 1 - R + sqrt(R^2 +1)) / (2/Pp - 1 - R - sqrt(R^2 +1));
        if (A < 0), N = N + 1; end
     end
     F = ( sqrt(R^2 +1) * log10((1-Pp)/(1-Pp*R)) / (R-1) ) / (log10(A));
  else % R = 1
     while (B < 0)
        Ppp = P / (N-P*(N-1)); B = (2/Ppp - 2 + sqrt(2)) / (2/Ppp - 2 - sqrt(2));
        if (B < 0), N = N + 1; end
     end
     F = ( sqrt(R^2 +1) * Ppp / (log(10*(1-Ppp))) ) / (log10(B));
  end
```

```
  if (F <= 0.75), N = N + 1; end
end
if (Dt1 == Dt2), LMTD = Dt1; else, LMTD = (Dt1-Dt2)/(log(Dt1/Dt2)); end %
Compute LMTD
cLMTD = F*LMTD; % Corrected LMTD
% Output
fprintf('\nNumber of shells = %3d\n', N); fprintf('F factor = %9.4f\n', F);
fprintf('Corrected LMTD = %9.4f\n', cLMTD);
```

Example 8.12 Number of Shells and Log-Mean Temperature Difference

A hot fluid is cooled by cooling water in a countercurrent shell-and-tube heat exchanger. Determine the number of shells required, the correction factor F, and the updated log-mean temperature difference (LMTD).

  Data: hot fluid inlet temperature $(T_1) = 250°F$, hot fluid outlet temperature $(T_2) = 100°F$, cooling water inlet temperature $(t_1) = 80°F$, cooling water outlet temperature $(t_2) = 120°F$.

Solution

  The script *shellLMTD* produces the following results:
```
>> shellLMTD
Hot fluid inlet temperature (deg.F): 250
Hot fluid outlet temperature (deg.F): 100
Cold fluid inlet temperature (deg.F): 80
Cold fluid outlet temperature (deg.F): 120
Number of shells = 2
F factor = 0.9189
Corrected LMTD = 54.0029
```

8.3.2 SIMPLIFIED HEAT EXCHANGER CALCULATION[17,18]

Shell-and-tube heat exchangers are the common type in chemical plants. Figure 8.20 shows a typical heat exchanger with one shell pass and two tube passes. The tube side is for high-temperature and high-pressure fluids, and the shell side is for more viscous, lower-flow-rate, evaporating and condensing fluids. In the heat exchanger, heat is transferred from the hot fluid to the cold fluid through the tube walls. Heat transfer coefficients, pressure drops, and the heat transfer area that need to be determined are dependent upon the geometric configuration of the heat exchanger. Designing a shell-and-tube heat exchanger follows the following steps[17]:

1) Determine the required heat duty for the heat exchanger to meet.
2) Choose the streams that will be placed on the tube side and the shell side.
3) Compute the required heat transfer area using the assumed overall heat transfer coefficient.
4) Select suitable tube specifications.
5) Compute the tube cross-sectional area.
6) Estimate the number of tubes and tube passes.
7) Estimate outside and inside film heat transfer coefficients.
8) Calculate the overall heat transfer coefficient and compare it with the assumed value used in 3).
9) Repeat the calculation by changing the baffle spacing and re-estimating shell-side film heat transfer coefficients until the difference between the assumed U and estimated U is confined within a small range.

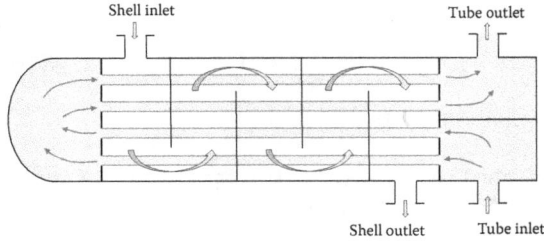

FIGURE 8.20 Schematic illustration of a one shell pass and two tube passes (1–2) heat exchanger. (Modified from Nayef Ghasem, *Computer Methods in Chemical Engineering*, CRC Press, Taylor & Francis Group, Boca Raton, FL, 2012, p. 135.)

8.3.2.1 Heat Duty

The required heat duty Q can be represented as

$$Q = \dot{m}_h C_{p,h} \left(T_{h,in} - T_{h,out} \right) = \dot{m}_c C_{p,c} (T_{c,out} - T_{c,in})$$

where

$\dot{m}$ is the mass flow rate

C_p is the heat capacity

The subscripts denote h: hot stream, c: cold stream, in: inlet, and out: outlet. The basic design equation is given by[18]

$$Q = U_i A_i F \Delta T_{lm}$$

where

U_i is the overall heat transfer coefficient

A_i is the inside heat transfer area

F is the correction factor that is used with the log-mean temperature difference for a countercurrent heat exchanger ΔT_{lm}

ΔT_{lm} is given by

$$\Delta T_{lm} = \frac{(T_1 - t_2) - (T_2 - t_1)}{\ln \left(\frac{T_1 - t_2}{T_2 - t_1} \right)}$$

where

T_1 and t_1 are the hot and cold stream inlet temperatures, respectively

T_2 and t_2 are the corresponding outlet temperatures

The ratio of the temperature difference of the hot stream to that of the cold stream is defined as R:

$$R = \frac{T_{hot,in} - T_{hot,out}}{t_{cold,out} - t_{cold,in}} = \frac{T_1 - T_2}{t_2 - t_1}$$

The ratio of the temperature difference of the cold stream to the maximum temperature difference is defined as S:

$$S = \frac{t_{cold,out} - t_{cold,in}}{T_{hot,in} - t_{cold,in}} = \frac{t_2 - t_1}{T_1 - t_1}$$

The value of F depends on the arrangement of the streams within the heat exchanger. For one shell pass and two, four (or any multiple of two) tube passes, the value of F, F_{12}, is given by

$$F_{12} = \frac{\sqrt{R^2 + 1}\ln\left(\frac{1-S}{1-RS}\right)}{(R-1)\ln\left\{\frac{2-S(R+1-\sqrt{R^2+1})}{2-S(R+1+\sqrt{R^2+1})}\right\}}$$

For two shell passes and four, eight (or any multiple of four) tube passes, the value of F, F_{24}, is given by

$$F_{24} = \frac{\sqrt{R^2 + 1}\ln\left(\frac{1-S}{1-RS}\right)}{2(R-1)\ln\left\{\frac{2+2\sqrt{(1-S)(1-RS)}-S(R+1-\sqrt{R^2+1})}{2+2\sqrt{(1-S)(1-RS)}-S(R+1+\sqrt{R^2+1})}\right\}}$$

8.3.2.2 Overall Heat Transfer Coefficient

The overall heat transfer coefficient can be defined based on the outside or inside surface area of the tubes. The overall heat transfer coefficient based on the outside surface area of the tubes, U_o, is given by

$$\frac{1}{U_o} = \frac{1}{h_o} + \frac{\Delta x}{k_w}\left(\frac{A_o}{A_{lm}}\right) + \frac{1}{h_i}\left(\frac{A_o}{A_i}\right) + R_{fi}\left(\frac{A_o}{A_i}\right) + R_{fo}$$

where

h_o and h_i are the outside and inside film heat transfer coefficients, respectively
Δx is the tube wall thickness
k_w is the tube metal thermal conductivity
R_{fi} and R_{fo} are the inside and outside fouling resistances, respectively
A_o and A_i are the outside and inside areas of the tube, respectively
A_{lm} is the log-mean of A_o and A_i
Δx and A_{lm} can be represented as

$$\Delta x = \frac{D_o - D_i}{2}, \quad A_{lm} = \frac{A_o - A_i}{\ln\left(\frac{A_o}{A_i}\right)}$$

where D_o and D_i are the outside and inside diameters of the tube, respectively.

The overall heat transfer coefficient based on the inside surface area of the tubes, U_i, is given by

$$\frac{1}{U_i} = \left(\frac{D_i}{D_o}\right)\frac{1}{h_o} + \frac{D_i\Delta x}{D_{lm}k_w} + \frac{1}{h_i} + R_{fi} + \left(\frac{D_i}{D_o}\right)R_{fo}$$

where the log-mean diameter D_{lm} is given by

$$D_{lm} = \frac{D_o - D_i}{\ln\left(\frac{D_o}{D_i}\right)}$$

8.3.2.3 Tube-Side Heat Transfer Coefficient[17]

The inside film heat transfer coefficient, h_i, can be obtained using the Sieder-Tate equation. For a laminar flow,

$$Nu_i = \frac{h_i D_i}{k_i} = 1.86 \left\{ N_{Rei} N_{Pri} \left(\frac{D_i}{L} \right) \right\}^{1/3} \left(\frac{\mu_i}{\mu_w} \right)^{0.14}$$

and for a turbulent flow,

$$Nu_i = \frac{h_i D_i}{k_i} = 0.027 N_{Re,i}^{0.8} N_{Pr,i}^{1/3} \left(\frac{\mu_i}{\mu_w} \right)^{0.14}$$

where

$$N_{Re,i} = \frac{D_i u_i \rho_i}{\mu_i}, \quad N_{Pr,i} = \frac{C_{pi} \mu_i}{k_i}$$

and u_i, ρ_i, μ_i, and k_i are the velocity, density, viscosity, and thermal conductivity of the tube inside the fluid, respectively.

8.3.2.4 Shell-Side Heat Transfer Coefficient[17]

The Kern method or the Donohue equation can be used to calculate the shell-side heat transfer coefficient h_o. For a turbulent flow, the Kern method may be used:

$$Nu_o = \frac{h_o D_e}{k_o} = 0.36 N_{Re,o}^{0.55} N_{Pr,o}^{1/3} \left(\frac{\mu_o}{\mu_w} \right)^{0.14}$$

where
   D_e is the hydraulic effective diameter defined by 4(free area)/(wetted perimeter)
   μ_w is the fluid viscosity evaluated at the average wall temperature
   D_e and $N_{Re,o}$ can be expressed as follows:

$$D_e = \frac{4}{\pi D_o} \left(P_t^2 - \frac{\pi D_o^2}{4} \right), \quad N_{Re,o} = \frac{D_e v_{max} \rho_o}{\mu_o}$$

where
   P_t is the tube pitch
   v_{max} is the maximum fluid velocity flowing through the tube bank
   v_{max} can be obtained from dividing the shell-side volumetric flow rate by the shell-side cross-flow area.

Alternatively, the Donohue equation is based on the weighted average of the mass velocity of the shell-side fluid flowing parallel to the tubes, G_b, and flowing across the tubes, G_c, as follows:

$$\frac{h_o D_o}{k_o} = 0.2 \left(\frac{D_o G_e}{\mu_o} \right)^{0.6} \left(\frac{C_{po} \mu_o}{k_o} \right)^{0.33} \left(\frac{\mu_o}{\mu_w} \right)^{0.14}, \quad G_e = \sqrt{G_b G_c}$$

G_b and G_c are given by

$$G_b = \frac{\dot{m}}{S_b}, \quad G_c = \frac{\dot{m}}{S_c}, \quad S_b = \frac{\pi}{4} \left(f_b D_s^2 - N_b D_o^2 \right), \quad S_c = b D_s \left(1 - \frac{D_o}{P_t} \right)$$

where
  f_b is the fraction of the shell cross section occupied by the baffle window
  D_s is the shell inside diameter
  N_b is the number of tubes in the baffle window (= $f_b \times$ (number of tubes))
  b is the baffle spacing

8.3.2.5 Pressure Drop in the Tube Side[17]

The pressure drop for fluid flow without phase change through the tubes can be determined as follows:

$$- \Delta P_i = P_{in} - P_{out} = \frac{0.6 N_p f_D G_i^2 L}{g_c \rho_i D_i \phi}$$

where
  N_p denotes the number of tube passes
  $G_i = \rho_i u_i$ is the tube-side mass velocity
  $\phi = 1.02 \mu / \mu_w$ is the correction factor for the nonisothermal turbulent flow
  The Darcy friction factor f_D is given by

$$f_D = (1.82 \log N_{Rei} - 1.64)^{-2}$$

Alternatively, ΔP_i can be determined using the following equation:

$$\Delta P_i = \frac{f L N_p}{D} \left(\frac{1}{2} \rho_i v^2 \right)$$

where v is the average fluid velocity through a single tube. The Fanning friction factor f is given by $f = f_D / 4$

8.3.2.6 Pressure Drop in the Shell Side[17]

The pressure drop for fluid flow without phase change across the tubes in the shell side can be determined as follows:

$$- \Delta P_o = P_{in} - P_{out} = \frac{2 K_s N_R f' G_o^2}{g_c \rho_c \phi}$$

where K_s is the correction factor given by

$$K_s = 1.1 \left(\frac{\text{tube length}}{\text{baffle spacing}} \right) = 1.1 \left(\frac{L}{b} \right)$$

N_R, the number of tube rows across which the shell fluid flows, is given by

$$N_R = \frac{1}{2} (\text{number of tubes at center line}) \cong \frac{1}{2} \left(\frac{D_s}{P_t} \right)$$

and the modification friction factor, f', is given by

$$f' = \left\{ 0.044 + \frac{0.08 x_L}{(x_T - 1)^{0.43 + 1.13 / x_L}} \right\} \left(\frac{D_o G_o}{\mu_o} \right)^{-0.15}$$

where

x_T is the ratio of the pitch transverse to the flow of the outside tube
x_L is the ratio of the pitch parallel to the outside tube
For square pitch,

$$x_T = x_L = \frac{P_t}{D_o}$$

Alternatively, ΔP_o can be determined using the equation

$$\Delta P_o = \frac{2 f G_s^2 D_s (N_B + 1)}{\rho D_e (\mu/\mu_s)}$$

where

N_B is the number of baffles
ρ is the density of the shell-side fluid
G_s is the mass velocity on the shell side, given by
$G_s = \frac{\dot{m}}{S_m} = \frac{(\text{mass flow rate})}{D_s b (\text{clearance / pitch})}$
D_e is given by

$$D_e = \frac{4}{\pi D_o} \left(c S_n^2 - \frac{\pi D_o^2}{4} \right)$$

where S_n is the pitch. The constant c is $c = 1$ for a square pitch and $c = 0.86$ for a triangular pitch.

Example 8.13 Design of a Condenser[19]

Consider a 1–4 shell-and-tube heat exchanger to cool 62,000 *lb/hr* of diethanolamine (DEA) so-lution (0.2 mass fractions DEA/0.8 water) from 150°F to 120°F by using water at 75°F heated to 100°F as shown in Figure 8.21. Assume that the tube-inside fouling resistance is given by $R_{fi} = 0.004$ *ft²·hr·*°F/*Btu* and that the shell-side fouling resistance is negligible. Physical properties are given in Table 8.1. Calculate the baffle spacing, shell-side and tube-side overall heat transfer coefficients, tube inside and outside heat transfer areas, and shell-side and tube-side pressure drops using the given data. As an initial guess for U_i, use $U_i = 160$ *Btu/(ft²·hr·*°F).

Data: operating conditions: $\dot{m}_h = 62{,}000$ *lb/hr*, $T_1 = 150$ °F, $T_2 = 120$ °F, $t_1 = 75$ °F, $t_2 = 100$ °F, $u_i = 5$ *ft/sec*. Shell and tube; $L = 15$ *ft*, $D_o = 0.0625$ *ft*, $D_i = 0.04017$ *ft*, $D_s = 1.4375$ *ft*, $P_t = 1/12$ *ft*, $k_w = 30$ *Btu/(hr·ft·*°F), $R_{fi} = 0.004$ *ft²·hr·*°F/*Btu*, $R_{fo} = 0$

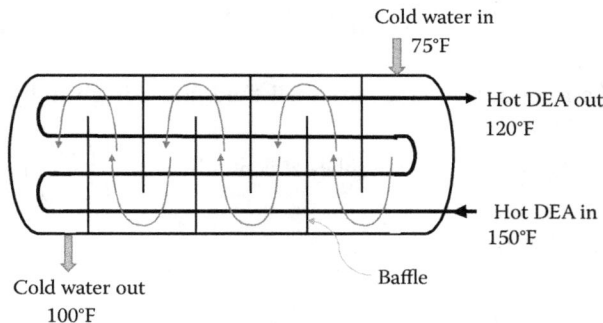

Cold water in
75°F

Hot DEA out
120°F

Hot DEA in
150°F

Cold water out
100°F

Baffle

FIGURE 8.21   Schematic of a 1–4 shell-and-tube heat exchanger.

TABLE 8.1
Physical Properties

Property	Tube-Side (Hot) DEA Solution (135°F)	Shell-Side (Cold) Water (87. 5°F)
$\rho\,(lb/ft^3)$	59.76	59.87
$C_p\,(Btu/(lb\cdot°F))$	0.92	1.0
$\mu\,(cP)$	0.75	0.77
$k\,(Btu/(hr\cdot ft\cdot°F))$	0.30	0.36

Solution

The calculation procedure can be summarized as follows:

1) Calculate the required heat duty Q_r and mass flow rate of the cold stream.
2) Assume U_i and find A_i and total cross − sectional area(tube − side) per pass A_{ci}.
3) Find the number of tubes per pass and heat transfer area per tube.
4) Find the number of tube passes N_p and adjust A_i based on N_p.
5) Calculate U_i using new A_i and D_e.
6) Calculate the cross-sectional area between baffles and shell axis and the shell-side mass velocity.
7) Determine the shell-side Reynolds number, Prandtl number, Nusselt number, and h_o.
8) Determine the tube-side Reynolds number, Prandtl number, Nusselt number, and h_i.
9) Determine the tube thickness and the log-mean diameter.
10) Determine U_i and $U_o = U_i(D_i/D_o)$.
11) Calculate f' and the shell-side and tube-side pressure drop.

The required heat duty Q_r should be provided by the heat exchanger. The heat provided by the heat exchanger, Q_d, can be considered as a function of the baffle spacing b. If b is regarded as unknown variable, the problem can be formulated as a nonlinear equation. In this case, we guess an initial value of b (for example, $b = 0$) and solve the nonlinear equation to get final value of b. The script *coolhxdat* defines the data and parameters.

```
% coolhxdat.m
% Physical properties
Cph = 0.92; Cpc = 1; rhoh = 59.76; rhoc = 59.87; kh = 0.3; kc = 0.36;
kw = 30; muh = 0.75; muc = 0.77; muwc = muc; muwh = muh;
% Operating conditions
mh = 6.2e4; ui = 5; T1 = 150; T2 = 120; t1 = 75; t2 = 100;
% Shell and tube
L = 15; Do = 0.0625; Di = 0.04017; Ds = 1.4375; Pt = 1/12; cl = 0.02083; Rfi = 0.004;
Rfo = 0;
% Parameters and basic properties
gc = 32.2; dTlm = ((T1-t2) - (T2-t1))/log((T1-t2)/(T2-t1)); % log-mean temp.
R = (T1-T2)/(t2-t1); S = (t2-t1)/(T1-t1);
F12den = (R-1)*log((2 - S*(R+1-sqrt(R^2+1)))/(2 - S*(R+1+sqrt(R^2+1))));
F12 = sqrt(R^2+1)*log((1-S)/(1-R*S))/F12den; % correction factor
Q = mh*Cph*(T1 - T2); mc = Q/(Cpc*(t2 - t1)); % Heat load and cold stream rate
Aci = mh/(rhoh*ui*3600); % tube-side total cross-sectional area
```

```
Nt = ceil(4*Aci/(pi*Di^2)); % number of tubes per pass
At = pi*Di*L; % heat transfer area per tube
dx = (Do - Di)/2; % tube thickness
Dlm = (Do - Di)/log(Do/Di); % log-mean diameter
De = (4/(pi*Do))*(Pt^2 - pi*Do^2/4); % hydraulic effective diameter
Ui = 150; % Assume Ui
```

The function *coolhxf* defines the equation $f(b) = Q_r - Q_d = 0$.

```
function fun = coolhxf(b)
% coolhxn.m: design of condenser
coolhxdat; % retireve data
Ai = Q/(Ui*F12*dTlm); % inside heat transfer area
Np = ceil(Ai/(At*Nt)); % number of tube passes
Ai = Np*Nt*pi*Di*L; % adjust Ai based on Np
De = (4/(pi*Do))*(Pt^2 - pi*Do^2/4); % hydraulic effective diameter
Acf = Ds*cl*b/Pt; % cross-sectional area between baffles and shell axis
Go = mc/Acf; % shell-side mass velocity
Nreo = De*Go/(muc*3600/1488); % shell-side Reynolds number
Npro = Cpc*(muc*3600/1488)/kc; % shell-side Prandtl number
Nuo = 0.36*Nreo^0.55*Npro^(1/3)*(muc/muwc)^0.14; % shell-side Nusselt number
ho = Nuo*kc/De; % shell-side heat transfer coefficient
Nrei = Di*rhoh*ui/(muh/1488); % tube-side Reynolds number
Npri = Cph*(muh*3600/1488)/kh; % tube-side Prandtl number
Nui = 0.027*Nrei^(0.8)*Npri^(1/3)*(muh/muwh)^0.14; % tube-side Nusselt number
hi = Nui*kh/Di; % tube-side heat transfer coefficient
Ui = 1/((Di/Do)/ho + (Di*dx)/(Dlm*kw) + 1/hi + Rfi + (Di/Do)*Rfo);
Qd = Ai*Ui*F12*dTlm; fun = Q - Qd;
end
```

The script *coolcal* uses the built-in function *fsolve* to determine the baffle spacing that satisfies $Q_r = Q_d$. We set $b = 0$ as an initial guess.

```
% coolcal.m: calculation of heat exchanger
coolhxdat; b0 = 0.0; % guess initial b
b = fsolve(@coolhxf,b0);
Acf = Ds*cl*b/Pt; % cross-sectional area between baffles and shell axis
Go = mc/Acf; % shell side mass velocity
Nreo = De*Go/(muc*3600/1488); % shell-side Reynolds number
Npro = Cpc*(muc*3600/1488)/kc; % shell-side Prandtl number
Nuo = 0.36*Nreo^0.55*Npro^(1/3)*(muc/muwc)^0.14; % shell-side Nusselt number
ho = Nuo*kc/De; % shell-side heat transfer coefficient
Nrei = Di*rhoh*ui/(muh/1488); % tube-side Reynolds number
Npri = Cph*(muh*3600/1488)/kh; % tube-side Prandtl number
Nui = 0.027*Nrei^(0.8)*Npri^(1/3)*(muh/muwh)^0.14; % tube-side Nusselt number
hi = Nui*kh/Di; % tube-side heat transfer coefficient
Ui = 1/((Di/Do)/ho + (Di*dx)/(Dlm*kw) + 1/hi + Rfi + (Di/Do)*Rfo); % new overall
heat transfer coefficients
Uo = Ui*Di/Do; xt = Pt/Do; xl = xt; Ks = 1.1*L/b; Nr = ceil(Ds/Pt/2);
fp = (0.044 + 0.08*xl/((xt - 1)^(0.43+1.13/xl)))*((Do*Go)/muc)^(-0.15);
dPo = 2*Ks*Nr*fp*(Go/3600)^2/(gc*rhoc*144); % shell-side pressure drop (psi)
fD = 1/(1.82*log10(Nrei) - 1.64)^2; Gi = rhoh*ui;
Ai = Q/(Ui*F12*dTlm); Ao = Q/(Uo*F12*dTlm);
Np = ceil(Ai/(At*Nt)); % number of tube passes
```

```
dPi = 0.6*Np*fD*Gi^2*L/(gc*rhoh*Di*144); % tube-side pressure drop (psi)
fprintf('Baffle spacing = %g in\n',b*12);
fprintf('Shell-side overall heat transfer coefficient = %g Btu/
(ft^2*h*F)\n',Uo);
fprintf('Tube-side overall heat transfer coefficient = %g Btu/
(ft^2*h*F)\n',Ui);
fprintf('Tube inside heat transfer area = %g ft^2\n',Ai);
fprintf('Tube outside heat transfer area = %g ft^2\n',Ao);
fprintf('Shell-side pressure drop = %g psi\n',dPo);
fprintf('Tube-side pressure drop = %g psi\n',dPi);


>> coolcal
Baffle spacing = 12.5176 in
Shell-side overall heat transfer coefficient = 94.2043 Btu/(ft^2*h*F)
Tube-side overall heat transfer coefficient = 146.571 Btu/(ft^2*h*F)
Tube inside heat transfer area = 261.229 ft^2
Tube outside heat transfer area = 406.444 ft^2
Shell-side pressure drop = 0.298752 psi
Tube-side pressure drop = 5.41203 psi
```

Example 8.14 Design of a Heater[20]

Design a 1–2 shell-and-tube heat exchanger to be used to heat raw water at 75°F to 80°F using 150,000 *lb/hr* of demineralized water that enters the exchanger at 95°F and exits at 85°F. Assume that the tube-side fouling resistance is $R_{fi} = 0.001$ $hr{\cdot}ft^2{\cdot}°F/Btu$ and that the shell-side fouling resistance, R_{fo}, is negligible. The schematic diagram of the exchanger is shown in Figure 8.22, and physical properties are given in Table 8.2. As an initial guess for U_i, use $U_i = 400$ $Btu/(ft^2{\cdot}hr{\cdot}°F)$.

Data: operating conditions: $\dot{m}_h = 150,000$ *lb/hr*, $T_1 = 95$ °F, $T_2 = 85$ °F, $t_1 = 75$ °F, $t_2 = 80$ °F, $u_i = 5$ *ft/sec*. Shell and tube; $L = 10$ *ft*, $D_o = 0.0625$ *ft*, $D_i = 0.05167$ *ft*, $D_s = 1.77083$ *ft*, $P_t = 1/12$ *ft*, $k_w = 30$ *Btu/h/ft/°F*, $cl = 0.02083$ *ft*, $R_{fi} = 0.001$ $ft^2{\cdot}h{\cdot}°F/Btu$, $R_{fo} = 0$

Solution
The calculation procedure is very similar to that of the previous example. The required heat duty Q_r should be provided by the heat exchanger. The heat provided by the heat exchanger, Q_d, can be considered as a function of the baffle spacing b. If b is regarded as unknown variable, the problem can be formulated as a nonlinear equation. In this case, we guess an initial value of b (for example, $b = 0$) and solve the nonlinear equation to get final value of b. The script *heatH2Odat* defines the data and parameters:

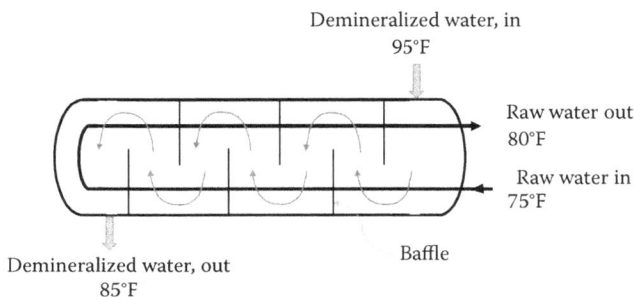

FIGURE 8.22 Schematic diagram of 1–2 raw-water heat exchanger.

TABLE 8.2
Physical Properties

Property	Shell-Side (Hot) Demineralized Water (90°F)	Tube-Side (Cold) Raw Water (78°F)
$\rho\,(lb/ft^3)$	62.4	62.4
$C_p\,(Btu/(lb/°F))$	1.0	1.01
$\mu\,(cP)$	0.81	0.92
$k\,(Btu/(hr\cdot ft\cdot °F))$	0.36	0.363

```
% heatH2Odat.m
% Physical properties (hot(shell): demineralized water, cold(tube): raw water)
Cph = 1.0; Cpc = 1.01; rhoh = 62.4; rhoc = 62.4; kh = 0.36; kc = 0.363;
muh = 0.81; muc = 0.92; muwc = muc; muwh = muh;
% Operating conditions
mh = 1.5e5; ui = 5; T1 = 95; T2 = 85; t1 = 75; t2 = 80;
% Shell and tube
L = 10; Do = 0.0625; Di = 0.05167; Ds = 1.77083; Pt = 1/12;
cl = 0.02083; Rfi = 0.001; Rfo = 0; kw = 30;
% Parameters and basic properties
gc = 32.2; Qr = mh*Cph*(T1 - T2); % heat load (Btu/h)
mc = Qr/(Cpc*(t2 - t1)); % cold stream rate (lb/h)
dTlm = ((T1-t2) - (T2-t1))/log((T1-t2)/(T2-t1)); % log-mean temp.
R = (T1-T2)/(t2-t1); S = (t2-t1)/(T1-t1);
F12den = (R-1)*log((2 - S*(R+1-sqrt(R^2+1)))/(2 - S*(R+1+sqrt(R^2+1))));
F12 = sqrt(R^2+1)*log((1-S)/(1-R*S))/F12den; % correction factor
Aci = mc/(rhoc*ui*3600); % tube-side total cross-sectional area (ft^2/pass)
Nt = ceil(4*Aci/(pi*Di^2)); % number of tubes per pass
At = pi*Di*L; % heat transfer area per tube
dx = (Do - Di)/2; % tube thickness
Dlm = (Do - Di)/log(Do/Di); % log-mean diameter
De = (4/(pi*Do))*(Pt^2 - pi*Do^2/4); % hydraulic effective diameter
Ui = 400; % Assume Ui
```

The function *heatH2O* defines the equation $f(b) = Q_r - Q_d = 0$.

```
function fun = heatH2O(b)
% heatH2O.m: design of H2O heater
heatH2Odat;
Ai = Qr/(Ui*F12*dTlm); % inside heat transfer area
Np = ceil(Ai/(At*Nt)); % number of tube passes
Ai = Np*Nt*pi*Di*L; % adjust Ai based on Np
Ui = Qr/(Ai*F12*dTlm); % adjust Ui
Acf = Ds*cl*b/Pt; % cross-sectional area between baffles and shell axis
Go = mh/Acf; % shell side mass velocity
Nreo = De*Go/(muh*3600/1488); % shell-side Reynolds number
Npro = Cph*(muh*3600/1488)/kh; %  shell-side Prandtl number
Nuo = 0.36*Nreo^0.55*Npro^(1/3)*(muh/muwh)^0.14; % shell-side Nusselt number
ho = Nuo*kh/De; % shell-side heat transfer coefficient
```

```
Nrei = Di*rhoc*ui/(muc/1488); % tube-side Reynolds number
Npri = Cpc*(muc*3600/1488)/kc; % tube-side Prandtl number
Nui = 0.027*Nrei^(0.8)*Npri^(1/3)*(muc/muwc)^0.14; % tube-side Nusselt number
hi = Nui*kc/Di; % tube-side heat transfer coefficient
Ui = 1/((Di/Do)/ho + (Di*dx)/(Dlm*kw) + 1/hi + Rfi + (Di/Do)*Rfo);
Qd = Ai*Ui*F12*dTlm;
fun = Qr - Qd;
end
```

The script *heatH2Ocal* uses the built-in function *fsolve* to determine the baffle spacing that satisfies $Q_r = Q_d$. We set $b = 0$ as an initial guess.

```
% heatH2Ocal.m: calculation of H2O heater
clear all; heatH2Odat; b0 = 0.0;
b = fsolve(@heatH2O,b0);
Ai = Qr/(Ui*F12*dTlm); % inside heat transfer area
Np = ceil(Ai/(At*Nt)); % number of tube passes
Acf = Ds*cl*b/Pt; % cross-sectional area between baffles and shell axis
Go = mh/Acf; % shell side mass velocity
Nreo = De*Go/(muh*3600/1488); % shell-side Reynolds number
Npro = Cph*(muh*3600/1488)/kh; % shell-side Prandtl number
Nuo = 0.36*Nreo^0.55*Npro^(1/3)*(muh/muwh)^0.14; % shell-side Nusselt number
ho = Nuo*kh/De; % shell-side heat transfer coefficient
Nrei = Di*rhoc*ui/(muc/1488); % tube-side Reynolds number
Npri = Cpc*(muc*3600/1488)/kc; % tube-side Prandtl number
Nui = 0.027*Nrei^(0.8)*Npri^(1/3)*(muc/muwc)^0.14; % tube-side Nusselt number
hi = Nui*kc/Di; % tube-side heat transfer coefficient
Ui = 1/((Di/Do)/ho + (Di*dx)/(Dlm*kw) + 1/hi + Rfi + (Di/Do)*Rfo);
Uo = Ui*Di/Do;
xt = Pt/Do; xl = xt; Ks = 1.1*L/b; Nr = ceil(Ds/Pt/2);
fp = (0.044 + 0.08*xl/((xt - 1)^(0.43+1.13/xl)))*((Do*Go)/muh)^(-0.15);
dPo = 2*Ks*Nr*fp*(Go/3600)^2/(gc*rhoh*144); % shell-side pressure drop (psi)
fD = 1/(1.82*log10(Nrei) - 1.64)^2; Gi = rhoc*ui;
dPi = 0.6*Np*fD*Gi^2*L/(gc*rhoc*Di*144); % tube-side pressure drop (psi)
Ai = Qr/(Ui*F12*dTlm); % inside heat transfer area
Ao = Qr/(Uo*F12*dTlm); % outside heat transfer area
fprintf('Number of tube passes = %g\n',Np);
fprintf('Baffle spacing = %g in\n',b*12);
fprintf('Shell-side overall heat transfer coefficient = %g Btu/(ft^2*h*F)\n',
Uo);
fprintf('Tube-side overall heat transfer coefficient = %g Btu/(ft^2*h*F)\n',Ui);
fprintf('Tube inside heat transfer area = %g ft^2\n',Ai);
fprintf('Tube outside heat transfer area = %g ft^2\n',Ao);
fprintf('Shell-side pressure drop = %g psi\n',dPo);
fprintf('Tube-side pressure drop = %g psi\n',dPi);

>> heatH2Ocal
Number of tube passes = 2
Baffle spacing = 7.64324 in
Shell-side overall heat transfer coefficient = 258.904 Btu/(ft^2*h*F)
Tube-side overall heat transfer coefficient = 313.17 Btu/(ft^2*h*F)
Tube inside heat transfer area = 412.308 ft^2
Tube outside heat transfer area = 498.728 ft^2
Shell-side pressure drop = 2.78669 psi
Tube-side pressure drop = 1.9091 psi
```

8.3.3 RIGOROUS HEAT EXCHANGER CALCULATION

8.3.3.1 Tube-Side Heat Transfer Coefficient

The tube-side Reynolds number is defined as

$$N_{Rei} = \frac{d_i G_i}{\mu_i}$$

where the mass flow rate G_i is obtained from

$$G_i = \frac{4W_i}{n\pi d_i^2}$$

where

W_i is the total mass flow rate of the tube-side fluid
n is the number of tubes per pass
The Graetz number G_{zi} for the tube-side fluid can be defined as

$$G_{zi} = \frac{d_i^2 G_i C_{pi}}{k_i L}$$

The Nusselt number, $N_{ui} = h_i d_i / k_i$, is specified according to the tube-side Reynolds number as

$$N_{Re} \le 2100 \,(\text{laminar flow}) : N_{ui} = \begin{cases} 3.66 + \frac{0.085 G_{zi}}{1 + 0.047 G_{zi}^{2/3}} \left(\frac{\mu_i}{\mu_\omega}\right)^{0.14} & : G_{zi} \le 100 \\[2mm] 1.86 G_{zi}^{1/3} \left(\frac{\mu_i}{\mu_\omega}\right)^{0.14} & : G_{zi} > 100 \end{cases}$$

$$2100 < N_{Rei} < 10^4 : N_{ui} = 0.116 (N_{Rei}^{2/3} - 125) P_{ri} \left[1 + \left(\frac{d_i}{L}\right)^{2/3} \right] \left(\frac{\mu_i}{\mu_\omega}\right)^{0.14}$$

$$N_{Rei} \ge 10^4 : N_{ui} = 0.023 N_{Rei}^{0.8} P_{ri}^{0.33} \left(\frac{\mu_i}{\mu_\omega}\right)^{0.14}$$

where the Prandtl number P_{ri} is defined as

$$P_{ri} = \frac{C_{pi} \mu_i}{k_i}$$

8.3.3.2 Shell-Side Heat Transfer Coefficient[21]

The local heat transfer coefficient on the shell side is calculated using the following procedure:
(1) Calculate the shell-side Reynolds number

$$N_{Res} = \frac{d_0 W_0}{\mu_0 S_m}$$

where S_m is the cross-flow area at or near the center line for one cross-flow section, given by[22]

$$S_m \begin{cases} l_b\left[D_s - d_{otl} + \dfrac{d_{otl} - d_0}{p_n}(p_T - d_0)\right] : \text{ square layouts} \\ l_b\left[D_s - d_{otl} + \dfrac{d_{otl} - d_0}{p_T}(p_T - d_0)\right] : \text{ triangular layouts} \end{cases}$$

where

l_b is the baffle spacing

D_s is the shell inside diameter

p_T is the tube pitch

d_{otl} is the shell outer tube limit

The tube pitch normal to flow, p_n, is defined as p_T for an in-line square layout, $p_T/\sqrt{2}$ for a rotated square layout, and $p_T/2$ for a triangular layout.

(2) Calculate the heat transfer coefficient for an ideal tube bank from[23]

$$\frac{h_o}{C_{po}G_o} = j_{H,s}\, P_{ro}^{-2/3}\left(\frac{\mu_o}{\mu_{0\omega}}\right)^{0.14}, \quad j_{H,s} = a_1 N_{Res}^{a2}\left(\frac{1.33 d_o}{p_T}\right)^a, \quad a = \frac{a_3}{1 + 0.14 N_{Res}^{a4}}$$

Table 8.3 shows the correlational coefficients $a_i\,(i = 1, 2, 3, 4)$ for various flow configurations.[24]

(3) Determine the correction factor ϕ_c for baffle configuration effects from

TABLE 8.3
Correlational Coefficients for Various Flow Configurations

Tube Arrangement	Range of N_{Res}	a_1	a_2	a_3	a_4
Triangular	$10^5 - 10^4$	0.321	−0.388	1.450	0.519
	$10^4 - 10^3$	0.321	−0.388		
	$10^3 - 10^2$	0.593	−0.477		
	$10^2 - 10$	1.360	−0.657		
	< 10	1.400	−0.667		
Square (in-line)	$10^5 - 10^4$	0.370	−0.395	1.187	0.370
	$10^4 - 10^3$	0.107	−0.266		
	$10^3 - 10^2$	0.408	−0.460		
	$10^2 - 10$	0.900	−0.631		
	< 10	0.970	−0.667		
Square (rotated)	$10^5 - 10^4$	0.370	−0.396	1.930	0.500
	$10^4 - 10^3$	0.370	−0.396		
	$10^3 - 10^2$	0.730	−0.500		
	$10^2 - 10$	0.498	−0.656		
	< 10	1.550	−0.667		

Source: Serth, R.W., *Process Heat Transfer,* Academic *Press,* Elsevier, Burlington, MA, 2007, p. 249.

$$\phi_c = F_{tc} + 0.54(1 - F_{tc})^{0.345}$$

where F_{tc} is the fraction of total tubes in cross-flow and is given by

$$F_{tc} = \frac{1}{\pi}\left[\pi + 2\left(\frac{D_s - 2l_c}{d_{otl}}\right)\sin\left(\cos^{-1}\frac{D_s - 2l_c}{d_{otl}}\right) - 2\cos^{-1}\frac{D_s - 2l_c}{d_{otl}}\right]$$

where l_c is the baffle cut in dimensions of length.

(4) Calculate the correction factor ϕ_l for baffle leakage effects using

$$\phi_l = \alpha + (1 - \alpha)\exp\left(-2.2\frac{S_{tb} + S_{sb}}{S_m}\right)$$

$$\alpha = 0.44\left(1 - \frac{S_{sb}}{S_{tb} + S_{sb}}\right), \quad S_{tb} = 15.81 d_o N_T (1 + F_{tc}), \quad S_{sb} = \frac{D_s \delta_{sb}}{2}\left[\pi - \cos^{-1}\left(1 - \frac{2l_c}{D_s}\right)\right]$$

where N_T is the total number of tubes in the bundle.

(5) Determine the correction factor ϕ_b for bundle bypassing effects from

$$\phi_b = \exp\left[-C_{bh}F_{bp}\left\{1 - 2\left(\frac{N_{ss}}{N_c}\right)^{1/3}\right\}\right]$$

$$F_{bp} = \frac{D_s - d_{otl}}{S_m}l_b, \quad N_c = \frac{D_s}{P_p}\left(1 - 2\frac{l_c}{D_s}\right), \quad C_{bh} = \begin{cases} 1.35 : N_{Res} \leq 100 \\ 1.25 : N_{Res} > 100 \end{cases}$$

where the tube pitch parallel to flow, p_p, is defined as p_T for in-line square layout, $p_T/\sqrt{2}$ for rotated square layout, and $\sqrt{3}\,p_T/2$ for triangular layout. $\phi_b = 1$ when $N_{sc}/N_c \geq 0.5$.

(6) Calculate the correction factor ϕ_r for adverse temperature gradient buildup at low Reynolds numbers by using

$$\phi_r = \begin{cases} 1 & : N_{Res} > 100 \\ 1 - \left(1 - \phi_r^*\right)(1.24 - 0.0124 N_{Res}) & : 20 \leq N_{Res} \leq 100 \\ \phi_r^* & : N_{Res} < 20 \end{cases}$$

$$\phi_r^* = 3.76[\ln(N_c + N_{c\omega})]^{-0.15 N_b^{-0.174}}, \quad N_{c\omega} = \frac{0.81 l_c}{P_p}, \quad N_b = \frac{L}{l_b} - 1$$

(7) Calculate the correction factor due to unequal baffle spacing at inlet and outlet:

$$\phi_s = \frac{(N_b - 1) + l_{in}^{*\,(1-n)} + l_{out}^{*\,(1-n)}}{(N_b - 1) + l_{in}^* + l_{out}^*}, \quad l^* = \frac{l_c}{l_b}$$

The exponent n takes on values of 0.333 for $N_{Res} \leq 100$ and 0.6 otherwise.

(8) The shell-side heat transfer coefficient is given by

$$h_o = h_{o,ideal}\phi_c\phi_l\phi_b\phi_r\phi_s$$

8.3.3.3 Pressure Drop in the Tube Side

The pressure drop for the tube-side fluid is calculated by

$$\Delta P_i = \frac{2 f_i\, u_i^2\, \rho_i L N_{tp}}{d_i\,(\mu_i/\mu_\omega)^{0.14}} + K\left(\frac{u_i^2}{2g}\right)\rho_i N_{tp}$$

The pressure losses in inlet and exit nozzles can be estimated from

$$\Delta P_{nozzel} = 1.5\rho_i\left(\frac{u_{nozzle}^2}{2}\right)$$

8.3.3.4 Pressure Drop in the Shell Side[23]

(1) Calculate the ideal tube bank friction factor from

$$f_s = b_1 N_{Res}^{b2}\left(\frac{1.33 d_o}{p_T}\right)^b,\; b = \frac{b_3}{1 + 0.14 N_{Res}^{b4}}$$

The correlation constants $b_i (i = 1, 2, 3, 4)$ are shown in Table 8.4 for various tube configurations.[24]

(2) Calculate the ideal tube bank pressure drop for one cross-flow path:

$$\Delta P_b = 2 f_s\, N_c\, \frac{G_o^2}{\rho_o}\left(\frac{\mu_o}{\mu_\omega}\right)^{-0.14}$$

TABLE 8.4
Correlation Constants for Various Tube Configurations

Tube Arrangement	Range of N_{Res}	b_1	b_2	b_3	b_4
Triangular	$10^5 - 10^4$	0.372	−0.123	7.00	0.500
	$10^4 - 10^3$	0.486	−0.152		
	$10^3 - 10^2$	4.570	−0.476		
	$10^2 - 10$	45.100	−0.973		
	< 10	48.000	−1.000		
Square (in-line)	$10^5 - 10^4$	0.391	−0.148	6.30	0.378
	$10^4 - 10^3$	0.0815	0.022		
	$10^3 - 10^2$	6.090	−0.602		
	$10^2 - 10$	32.100	−0.963		
	< 10	35.000	−1.000		
Square (rotated)	$10^5 - 10^4$	0.303	−0.126	6.59	0.520
	$10^4 - 10^3$	0.333	−0.136		
	$10^3 - 10^2$	3.500	−0.476		
	$10^2 - 10$	26.200	−0.913		
	< 10	32.000	−1.000		

Source: Serth, R.W., *Process Heat Transfer*, Academic Press, Elsevier, Burlington, MA, 2007, p. 249.

(3) Calculate the pressure drop for an ideal window section from

$$\Delta P_\omega = \begin{cases} (2 + 0.6N_{c\omega})\dfrac{W_o^2}{2\sqrt{S_m S_\omega}\rho_o} & : N_{Res} > 100 \\[3mm] 0.026\dfrac{W_o^2 \mu_0}{\sqrt{S_m S_\omega}\rho_o}\left(\dfrac{N_{c\omega}}{p_T - d_o} + \dfrac{l_b}{d_\omega^2}\right) + \dfrac{W_o^2}{\sqrt{S_m S_\omega}\rho_o} & : N_{Res} \leq 100 \end{cases}$$

$$S_\omega = \frac{D_s^2}{4}\left[\cos^{-1}\left(1 - \frac{2l_c}{D_s}\right) - \left(1 - \frac{2l_c}{D_s}\right)\sqrt{1 - \left(1 - \frac{2l_c}{D_s}\right)^2}\right] - \frac{N_T}{8}(1 - F_{tc})\pi d_o^2$$

$$d_\omega = \frac{4S_\omega}{(\pi/2)N_T(1 - F_{tc})d_o + 2D_s\cos^{-1}(1 - 2l_c/D_s)}$$

(4) Determine the correction factor R_l for the effect of baffle leakage on pressure drop[25]:

$$R_l = \exp\left[-1.33b\left(\frac{S_{tb} + S_{sb}}{S_m}\right)^m\right], \quad m = -0.15b + 0.8, \quad b = 1 + \frac{S_{sb}}{S_{tb} + S_{sb}}$$

(5) Calculate the correction factor R_b for the bundle bypassing effect on the pressure drop:

$$R_b = \exp\left[-C_{bp}F_{bp}\left\{1 - 2\left(\frac{N_{ss}}{N_c}\right)^{1/3}\right\}\right], \quad C_{bp} = \begin{cases} 4.5 : N_{Res} \leq 100 \\ 3.7 : N_{Res} > 100 \end{cases}$$

(6) Determine the correction factor for unequal baffle spacing at the inlet and outlet by

$$R_s = l_{in}^{*\,(2-n)} + l_{out}^{*\,(2-n)}, \quad n = \begin{cases} 1 : N_{Res} \leq 100 \\ 0.2 : N_{Res} > 100 \end{cases}$$

(7) The shell-side pressure drop is given by

$$\Delta P_o = R_b \Delta P_b\left[(N_b - 1)R_l + \left(1 + \frac{N_{c\omega}}{N_c}\right)R_s\right] + N_b R_l \Delta P_\omega$$

8.3.3.5 Heat Exchanger Calculation Procedure

The rigorous calculation procedure for a shell-and-tube heat exchanger can be summarized as follows:

1) Input required data: shell inside diameter (D_s), baffle spacing (l_b), tube outer diameter (d_o), tube inner diameter (d_i), tube layout (triangular, square, rotated square), tube pitch (p_T), tube length (L), thickness of the tube sheet (l_s), number of tube passes (N_{Tp}), total number of tubes (N_T), baffle cut (l_c), number of sealing strips (N_{ss}).
2) Specify two reference temperatures (T_{ref1}, T_{ref2}) and physical properties at these temperatures: the densities, viscosities, thermal conductivities, and heat capacities of shell- and tube-side fluids and the thermal conductivity of the tube material (k_ω).
3) Input the operating conditions: the inlet and outlet temperatures of the shell-side and tube-side fluids, flow rates of shell-side and tube-side flows, and fouling coefficients.

4) If a temperature or a flow rate is unknown, determine the unknown quantity from an energy balance equation. If there are two unknown quantities, assume one unknown+quantity and calculate the other one by energy balances.

5) Calculate the tube-side heat transfer coefficient using an appropriate correlation. For a start, assume that the viscosity correction factor, $(\mu_i/\mu_{iw})^{0.14}$, is equal to 1.

6) Calculate the shell-side heat transfer coefficient.

7) Estimate the tube wall temperature: $t_w = t_b + h_o(T_b - t_b)/(h_i + h_o)$

8) Calculate the viscosity correction factor $(\mu/\mu_w)^{0.14}$.

9) Repeat steps 5)-8) until two successive values of t_w agree within a preset tolerance.

10) Calculate the tube wall coefficient: $h_w = 2k_w/(d_o - d_i)$.

11) Calculate the overall heat transfer coefficient.

12) Calculate the log-mean temperature difference ΔT_{lm} and the correction factor $F = F_T$ for multiple tube passes.

13) If the heat duty is known, calculate the required heat transfer area $A_{req} = Q/(UF_T\Delta T_{lm})$. Find the difference between the actual heat transfer area and the calculated value and go to step 15).

14) If the heat duty is unknown, estimate the heat duty from $Q = U\Delta T_{lm}F_TA$. Using the estimated Q, iteratively calculate the unknown quantity assumed in step 4).

15) Determine the tube-side and shell-side pressure drop and stop the calculation procedure.

Various MATLAB functions perform these calculation steps. In the calculation of physical properties, two reference temperatures in the operation range are specified, and physical properties at these temperatures are used to find physical properties at a certain temperature T. For example,

liquid viscosity: $\mu = ae^{b/T}$, vapor viscosity: $\rho = \rho_{ref}T_{ref}/T$, $y = a + bT$ ($y = \mu, k, C_p, \rho$)

$$\rho = \rho_{ref}T_{ref}/T, \; y = a + bT \; (y = \mu, k, C_p, \rho):$$

The constants a and b are determined from two reference values. The function *hxvis* calculates the tube-side and shell-side viscosities. The function *hxthc* calculates heat conductivities, the function *hxcp* calculates heat capacities, and the function *hxrho* finds densities.

```
function mu = hxvis(muref,Tref,T,fstate)
% Calculate viscosity at T using two reference temperatures and viscosities
% input:
%  muref: viscosity vector at reference temperature vector Tref
%  Tref: reference temperature vector
%  T: temperature (K) at which viscosity is to be determined
%  fstate: state of fluid (1: liquid(mu = A*exp(B/T), 2: vapor(mu = A+BT))
% output:
%  mu: viscosity (Ns/m^2)
if fstate == 1 % liquid
  mu = muref(1)*exp(Tref(2)*(Tref(1) - T)/(T*Tref(1) - Tref(2))*log(muref(2)/
muref(1)));
else % vapor
  mu = (muref(2)*Tref(1) - muref(1)*Tref(2) + T*(muref(1)-muref(2)))/(Tref(1)
- Tref(2));
end
end
function xk = hxthc(xkref,Tref,T)
% Calculate heat conductivity at T using two ref. temperatures and
% conductivities
```

```
% input:
% xkref: heat conductivity vector at ref. temperature vector Tref
% Tref: reference temperature vector (K)
% T: temperature (K) at which heat conductivity is to be determined
% output:
% xk: heat conductivity (W/m/K) (xk = A+B*T)
xk = (xkref(2)*Tref(1) - xkref(1)*Tref(2) + T*(xkref(1)-xkref(2)))/(Tref(1) -
Tref(2));
end
function Cp = hxcp(cpref,Tref,T)
% Calculate heat capacity at T using two ref. temperatures and heat capacities
% input:
% cpref: heat capacity vector (J/kg/K) at ref. temperature vector Tref
% Tref: reference temperature vector (K)
% T: temperature (K) at which heat capacity is to be determined
% output:
% Cp: heat capacity (J/kg/K) (Cp = A+B*T)
Cp = (cpref(2)*Tref(1) - cpref(1)*Tref(2) + T*(cpref(1)-cpref(2)))/(Tref(1) -
Tref(2));
end
function rho = hxrho(rhoref,Tref,T,fstate)
% Calculate density at T using two ref. temperatures and densities
% input:
% rhoref: density vector (kg/m^3) at ref. temperature vector Tref
% Tref: reference temperature vector (K)
% T: temperature (K) at which density is to be determined
% fstate: state of fluid (1: liquid(rho = A+B*T, 2: vapor(rho = A/T))
% output:
% rho: density (kg/m^3)
if fstate == 1 % liquid
   rho = (rhoref(2)*Tref(1) - rhoref(1)*Tref(2) + T*(rhoref(1)-rhoref(2)))/
(Tref(1) - Tref(2));
else % vapor
  rho = rhoref(1)*Tref(1)/T;
end
end
```

The heat transfer coefficients for the tube side and shell side are calculated by the functions *htctube* and *htcshell*, respectively.

```
function Htube = htctube(Nre,Pr,D,L,Xk,Phi)
% Calculate convective heat transfer coefficient within tube
% input:
% Nre: Reynolds number       Pr: Prandtl number
% D: inside diameter of tube (mm)   L: tube length (m)
% Xk: thermal conductivity of fluid (W/m/K)   Phi: viscosity correction factor
% output:
% Htube: convective heat transfer coefficient within tube
Dm = D*1e-3; % mm->m
if Nre <= 2100
  Gw = Nre*Pr*Dm/L;
  if Gw > 100, Nu = 1.86*Phi*Gw^0.333;
  else, Nu = 3.66 + 0.085*Gw*Phi/(1 + 0.047*Gw^0.6667); end
elseif Nre < 1e4, Nu = 0.116*(Nre^0.6667 - 125)*Pr^0.333 * (1+(Dm/L)
^0.6667)*Phi;
```

```
else, Nu = 0.023*Phi*(Nre^0.8)*(Pr^0.333);
end
Htube = Nu*Xk/Dm;
end
function Hshell = htcshell(D,Ds,L,Lbc,Lbin,Lbout,Lc,Dotl,Dsb,Pt,Nt,
Nss,w,Visc,Cp,Xk,Phi,Layout)
% Calculate shell-side heat transfer coefficient
% input:
% D: outside diameter of tube (mm)    Ds: inside diameter of shell (mm)
% L: tube length (m)       Lbc: central baffle spacing (mm)
% Lbin: inlet baffle spacing (mm)     Lbout: outlet baffle spacing (mm)
% Lc: baffle cut (mm)        Dotl: shell outer tube limit(mm)
% Dsb: shell-baffle clearance (mm)    Pt: tube pitch (mm)
% Nt: total number of tubes in the bundle Nss: number of pairs of sealing strips
% w- flow rate in shell (kg/s)    Visc: shell-side fluid viscosity (Ns/m^2)
% Cf: heat capacity of shell-side fluid (J/kg/K)  Xk: heat conductivity of shell-
side fluid (W/m/K)
% Phi: viscosity correction factor
% Layout: tube layout (1:triangular, 2:in-line square, 3:rotated square)
% output:
%  Hshell: shell-side heat transfer coefficient
%  Correlational coefficients for tube arrangement
a1 = [0.321 0.321 0.593 1.360 1.400; 0.370 0.107 0.408 0.900 0.970; 0.370 0.370
0.730 0.498 1.550];
a2 = -[0.388 0.388 0.477 0.657 0.667; 0.395 0.266 0.460 0.631 0.667; 0.396 0.396
0.500 0.656 0.667];
a3 = [1.450 1.187 1.930]; a4 = [0.519 0.370 0.500];
switch Layout
  case 1, Pp = 0.866*Pt; Pn = Pt/2; Pd = Pt;
  case 2, Pp = Pt; Pn = Pt; Pd = Pn;
  otherwise, Pp = 0.7071*Pt; Pn = Pp; Pd = Pn;
end
Nc = Ds*(1 - 2*Lc./Ds)/Pp; Sm = Lbc*(Ds - Dotl + (Pt-D)*(Dotl-D)/Pd);
Nres = D*w*1e3/(Visc*Sm); % Shelläø Reynolds ¼ö
Pr = Cp*Visc/Xk; Ptd = Pt/D;
% Heat transfer coefficients for ideal tube bank
% (Nres: Reynolds number of shell-side fluid)
if Nres >= 1e4, J = 1; c1 = 1.25;
elseif Nres >= 1e3, J = 2; c1 = 1.25;
elseif Nres >= 100, J = 3; c1 = 1.25;
elseif Nres >= 10, J = 4; c1 = 1.35;
else J = 5; c1 = 1.35; end
a = a3(Layout)/(1 + 0.14*Nres^a4(Layout)); Hj = a1(Layout,J) * (1.33/Ptd)^a *
Nres^a2(Layout,J);
Hsi = Hj*Cp*Phi*w*Pr^(-0.6667) / (Sm*1e-6);
% Correction factor for baffle configuration effects
adm = (Ds - 2*Lc)/Dotl; adm1 = acos(adm); Ftc = (pi + 2*adm*sin(adm1) -
2*adm1)/pi;
Phic = Ftc + 0.54*(1 - Ftc)^0.345;
% Correction factor for baffle leakage effects
Stb = 0.6223*D*(1 + Ftc)*Nt; Ssb = Ds*Dsb*0.5*(pi - acos(1 - 2*Lc/Ds));
R1 = (Stb + Ssb)/Sm; R2 = Ssb/(Ssb + Stb); Phil = 0.44*(1-R2) + (1-0.44*(1-
R2))*exp(-2.2*R1);
% Correction factor for bundle bypassing
Fbp = (Ds - Dotl)*Lbc/Sm; Nsc = Nss/Nc;
```

```
if Nsc >= 0.5, Phib = 1;
else, if Nss == 0, c2 = 0; else c2 = (2*Nsc)^0.3333; end
  Phib = exp(-c1*c2*Fbp);
end
Nb = 1e3*L/Lbc + 1;
% Correction factor for adverse temperature gradient buildup
if Nres >= 100, Phir = 1;
else
  Ncw = 0.8*Lc/Pp; Phs = 1.51/((Nc+Ncw)*(Nb+1))^0.18;
  if Nres <= 20, Phir = Phs;
  elseif Nres <= 100, Phir = Phs - (1-Phs)*(0.25-0.0125*Nres); end
  if Phir <= 0.4, Phir = Phs; end
end
% Correction factor due to unequal baffle spacing at inlet and outlet
if Nres >= 100, An = 0.6; else An = 0.333; end
Phis  =  (Nb-1  +  (Lbin/Lbc)^(1-An)  +  (Lbout/Lbc)^(1-An))/(Nb-1  +  (Lbin
+Lbout)/Lbc);
% Shell-side heat transfer coefficient
Hshell = Hsi*Phic*Phil*Phib*Phir*Phis;
end
```

The pressure drops for the tube side and shell side are calculated by the functions *dptube* and *dpshell*, respectively.

```
function DPtube = dptube(D,L,rf,v,rho,Visc,Phi,Npass)
% Calculate tube-side pressure drop
% input:
% D: tube inside diameter (mm)    L: tube length (m)
% rf: tube roughness (mm)         v- tube-side flow rate (m/s)
% rho: fluid density (kg/m^3)     Visc: fluid viscosity (Ns/m^2)
% Phi: viscosity correction factor    Npass: number of tube passes in bundle
% output:
% DPtube: tube-side pressure drop (N/m^2)
g = 9.81; rfD = rf/D; Dm = 1e-3*D; Nre = Dm*v*rho/Visc;
if Nre <= 2100 % Laminar flow
  f = 16/Nre;
elseif Nre <= 4000 % Zigrang & Sylvester friction factor correlation eqn.
  t1 = rfD/3.7; t2 = 5.02/Nre; ftm = log10(t1 - t2*log10(t1 + 13/Nre));
  f = 1/(4*log10(t1) + t2*ftm)^2;
else % Round friction factor correlation eqn.
  f = 1/(3.6*log10(Nre/(0.135*(Nre*rfD + 6.5))))^2;
end
pD = 2*f*rho*v^2*L/Dm; dp1 = pD/Phi; dp2 = 4*rho*L*v^2/(2*g); DPtube = (dp1 +
dp2)*Npass;
end
function DPs = dpshell(D,Ds,L,Lbc,Lbin,Lbout,Lc,Dotl,Dsb,
Pt,Nt,Nss,w,rho,Visc,Phi,Layout)
% Calculate shell-side pressure drop
% input:
% D: tube outside diameter (mm)       Ds: shell inside diameter(mm)
% L: tube length (m)        Lbc: central baffle spacing (mm)
% Lbin: inlet baffle spacing (mm)     Lbout: outlet baffle spacing (mm)
% Lc: baffle cut (mm)       Dotl: shell outer tube limit (mm)
% Dsb: shell-baffle clearance (mm)    Pt: tube pitch (mm)
% Nt: total number of tubes in the bundle  Nss: number pairs of sealing strips
```

```
% w- shell-side flow rate (kg/s)      rho: fluid density (kg/m^3)
% Visc: shell-side fluid viscosity (Ns/m^2)   Phi: viscosity correction factor
% Layout: tube layout (1:triangular, 2:in-line square, 3:rotated square)
% output:
% DPs: shell-side pressure drop
% Correlational coefficients for tube arrangement
b1 = [0.372 0.486 4.570 45.100 48.000; 0.391 0.0815 6.090 32.100 35.000;
  0.303 0.333 3.500 26.200 32.000];
b2 = -[0.123 0.152 0.476 0.973 1.000; 0.148 -0.022 0.602 0.963 1.000; 0.126 0.136
0.476 0.913 1.000];
b3 = [7.00 6.30 6.59]; b4 = [0.500 0.378 0.520];
switch Layout
  case 1, Pp = 0.866*Pt; Pn = Pt/2; Pd = Pt;
  case 2, Pp = Pt; Pn = Pt; Pd = Pn;
  otherwise, Pp = 0.7071*Pt; Pn = Pp; Pd = Pn;
end
% Friction factor for ideal tube-bank
Nc = Ds*(1 - 2*Lc./Ds)/Pp; Ptd = Pt/D; Sm = Lbc*(Ds - Dotl + (Pt-D)*(Dotl-D)/Pd);
Nres = D*w*1e3/(Visc*Sm); % Shell-side Reynolds number
if Nres >= 1e4, J = 1; c1 = 3.7; elseif Nres >= 1e3, J = 2; c1 = 3.7;
elseif Nres >= 100, J = 3; c1 = 3.7; elseif Nres >= 10, J = 4; c1 = 4.5; else J = 5; c1
= 4.5; end
b = b3(Layout)/(1 + 0.14*Nres^b4(Layout)); Fj = b1(Layout,J) * (1.33/Ptd)^b *
Nres^b2(Layout,J);
Dpbi = 2*Fj*Nc*(w/(Sm*1e-6))^2 / (rho*Phi);
% Pressure drop for ideal window section
bdm = (Ds - 2*Lc)/Dotl; bd1 = acos(bdm); bd2 = 1 - 2*Lc/Ds;
Ftc = (pi + 2*bdm*sin(bd1) - 2*bd1)/pi;
Sw = (Ds^2/4)*(acos(bd2)-bd2*sqrt(1-bd2^2)) - (Nt/8)*(1-Ftc)*pi*D^2;
Dw = 4*Sw/(1.5708*Nt*(1-Ftc)*D + 2*Ds*bd2); Ncw = 0.81*Lc/Pp; Gw = 1e6*w/
sqrt(Sm*Sw);
if Nres >= 100, Dpw = 0.5*Gw^2 * (2+0.6*Ncw)/rho;
else, Dpw = (2.6e4*Visc*(Ncw/(Pt-D) + Lbc/Dw^2) + Gw)*Gw/rho; end
% Correction factor for baffle leakage effects
Stb = 0.6223*D*(1 + Ftc)*Nt; Ssb = Ds*Dsb*0.5*(pi - acos(bd2));
R1 = (Stb + Ssb)/Sm; R2 = Ssb/(Ssb + Stb); Pv = -0.15*(1+R2) + 0.8;
Rl = exp(-1.33*(1 + R2)*R1^Pv);
% Correction factor for bundle bypassing
Fbp = (Ds - Dotl)*Lbc/Sm; Nsc = Nss/Nc;
if Nsc >= 0.5, Rb = 1;
else
  if Nss == 0, c2 = 0; else c2 = (2*Nsc)^0.3333; end
  Rb = exp(-c1*c2*Fbp);
end
% Correction factor due to unequal baffle spacing at inlet and outlet
Nb = 1e3*L/Lbc + 1;
if Nres >= 100, An = 0.2; else An = 1; end
Rs = (Lbin/Lbc)^(2-An) + (Lbout/Lbc)^(2-An);
% Shell-side pressure drop
DPs = Rb*Dpbi*((Nb-1)*Rl + (1+Ncw/Nc)*Rs) + Nb*Rl*Dpw;
end
```

There are six critical variables to be specified: T_1 and T_2 (the inlet and outlet temperatures of the hot fluid), t_1 and t_2 (the inlet and outlet temperatures of the cold fluid), and W_i and W_o (the inlet and outlet mass flow rates of the fluid). If five variables are known, the sixth variable should be

TABLE 8.5

Eight Possible Combinations for Two Unknown Variables

Unknown Variables	Problem Type
All variables known	0
t_1, T_1	1
t_1, T_2	2
t_1, W_o	3
t_2, T_1	4
t_2, T_2	5
t_2, W_o	6
W_i, T_1	7
W_i, T_2	8

Source: Raman, R., *Chemical Process Computations*, Elsevier Applied Science Publishers, Barking, Essex, UK, 1985, p. 295.

determined beforehand and specified in the input. If two variables are unknown, then there are eight possible combinations.[26] A flag is specified to denote the type of combination, as shown in Table 8.5.

The input data required in the heat exchanger calculations consist of the following:

Heat exchanger geometry: N_T, N_{TP}, N_{ss}.

Tube arrangement: D_s, l_{bc}, l_{in}, l_{out}, l_c, l_s, δ_{sb}, δ_{otl}, L, d_o, d_i, p_T.

Physical properties: two values of reference temperatures, densities, viscosities, thermal conductivities, and heat capacities at the two reference temperatures for tube and shell fluids (liquid or vapor).

Process data: fouling resistances for the tube and shell sides, values for six critical variables (up to two trial values are specified if only four variables are known), and a flag for problem type (Table 8.5).

The script *hxndat* defines the input data. The main script *hxnst* calls the script *hxndat* to specify the input data and calls related functions to perform the calculations.

```
% hxnst.m: Shell-and-tube heat exchanger
clear all;
hxndat; % input data
% Problem type
if ptype > 0
  if ptype <= 3, Nvar = 1;    % calculate tube-side inlet temp.(Ti1)
  elseif ptype <= 6, Nvar = 2;  % calculate tube-side outlet temp.(Ti2)
  else Nvar = 3; end       % calculate tube-side flow rate(Wi)
end
Rw = 1e-3*(Do - Di)/(2*Xkw);  % Xkw: inverse of tube-wall heat conductivity
(1/hw)
A = pi*1e-3*Do*(L - 2e-3*Ls)*Nt;
if ptype == 0
  Cpim = hxcp(cpreft,Trt,(Ti1+Ti2)/2); % tube side
  Cpsm = hxcp(cprefs,Trs,(Ts1+Ts2)/2); % shell side
  Qi = Wi*Cpim*(Ti2 - Ti1); Qs = Ws*Cpsm*(Ts2 - Ts1);
  Qr = abs(Qi/Qs);
```

```
  if (abs(1-Qr) > 0.1)
     display('Tube-side and shell-side heat duty differ by more than 10%.');
     display('Tube-side heat duty is used as total duty.');
  end
  Q = Qi;
end
% Iterations until differences in successive values of key variables converge
to zero.
varC = 10; iter = 0;
while varC > 1e-3
  if ptype > 0
    % Calculation of unknown variables using energy balances
    Cpim = hxcp(cpreft,Trt,(Ti1+Ti2)/2); % tube-side
    Cpsm = hxcp(cprefs,Trs,(Ts1+Ts2)/2); % shell-side
    Qs = Ws*Cpsm*(Ts2 - Ts1); Qi = -Qs;
    iter1 = 0;
    switch Nvar
      case 1 % calculates Ti1(tube-side inlet temperature)
        Ti1new =  Ti2 - Qi/(Wi*Cpim); crT = 10;
        while crT >= 1e-3
          Ti1 = Ti1new; Cpim = hxcp(cpreft,Trt,(Ti1+Ti2)/2);
          Ti1new = Ti2 - Qi/(Wi*Cpim); crT = abs((Ti1new - Ti1)/Ti1new);
          iter1 = iter1 + 1;
        end
      case 2 %  calculates Ti2(tube-side outlet temperature)
        Ti2new = Ti1 + Qi/(Wi*Cpim); crT = 10;
        while crT >= 1e-3
          Ti2 = Ti2new; Cpim = hxcp(cpreft,Trt,(Ti1+Ti2)/2);
          Ti2new = Ti1 + Qi/(Wi*Cpim); crT = abs((Ti2new - Ti2)/Ti2new);
          iter1 = iter1 + 1;
        end
      case 3  % calculates Wi(tube-side flow rate)
        Wi = abs(Qs/(Cpim*(Ti2 - Ti1)));
    end
  end
  Tib = (Ti1 + Ti2)/2;
  % Shell-side heat transfer coefficient
  Tsb = (Ts1 + Ts2)/2;
  mus = hxvis(murefs,Trs,Tsb,fsS); % shell-side viscosity
  rhos = hxrho(rhorefs,Trs,Tsb,fsS); % shell-side density
  Cps = hxcp(cprefs,Trs,Tsb); % shell-side heat capacity
  xks = hxthc(xkrefs,Trs,Tsb); % shell-side heat conductivity
  Tw = (Tsb + Tib)/2; Twnew = Tw; crT = 10;
  while crT >= 1e-3
    if fsS == 1 % liquid
      Phis = (mus/hxvis(murefs,Trs,Tw,1))^0.14;
    elseif fsS == 2 % gas
      Phis = (Tsb/Tw)^0.25;
    end
    Hs = htcshell(Do,Ds,L-2e-3*Ls,Lbc,Lbin,Lbout,Lc,Dotl,Dsb,Pt,...
      Nt,Nss,Ws,mus,Cps,xks,Phis,Layout);
    % tube-side heat conductivity
    mut = hxvis(mureft,Trt,Tib,fsT); % tube-side viscosity
    rhot = hxrho(rhoreft,Trt,Tib,fsT); % tube-side density
    Cpt = hxcp(cpreft,Trt,Tib); % tube-side heat capacity
```

```
    xkt = hxthc(xkreft,Trt,Tib); % tube-side heat conductivity
    Ui = 4e6*Wi*Npass/(pi*rhot*Di^2*Nt); Rei = 1e-3*Di*Ui*rhot/mut;
    Pri = Cpt*mut/xkt;
    if fsT == 1, Phit = (mut/hxvis(mureft,Trt,Tw,1))^0.14;
    elseif fsT == 2, Phit = (Tw/Tib)^0.25; end
    Ht = htctube(Rei,Pri,Di,L,xkt,Phit);
    % Estimate tube wall temperature
    Tw = Tib + Hs/(Hs + Ht) * (Tsb - Tib); crT = abs((Tw - Twnew)/Tw); Twnew = Tw;
  end % end while

  % Calculate heat duty from heat transfer equations
  U = 1/(Do/(Di*Ht) + 1/Hs + Rw + Rds + Rdt); Dt1 = Ts1 - Ti2; Dt2 = Ts2 - Ti1;
  if (Dt1 <= 0 || Dt2 <= 0) break; end
  Delt = (Dt1 - Dt2)/log(Dt1/Dt2); Ft = 1;
  if Npass > 1
    R = (Ts1 - Ts2)/(Ti2 - Ti1); P = (Ti2 - Ti1)/(Ts1 - Ti1); tm = sqrt(R^2 + 1);
    Ft = tm*log((1-P)/(1-R*P))/((R-1)*log((2-P*(R+1-tm))/(2-P*(R+1+tm))));
  end
  Deltm = Delt*Ft;
  if ptype == 0, Areq = Qi/(U*Deltm); Da = (A - Areq)/A*100; break;
  end
  Q = U*A*Deltm;
  % Calculate assumed unknown variable
  Sgn = 1; if Qs < 0, Sgn = -1; end
  switch ptype
    case {1,4,7}
      Ts1new = Ts2 - Sgn*Q/(Ws*Cps); varC = abs((Ts1 - Ts1new)/Ts1new); Ts1 = Ts1new;
    case {2,5,8}
      Ts2new = Ts1 + Sgn*Q/(Ws*Cps); varC = abs((Ts2 - Ts2new)/Ts2new); Ts2 = Ts2new;
    case {3,6}
      Wsnew = abs(Q/((Ts2 - Ts1)*Cps)); varC = abs((Ws - Wsnew)/Wsnew); Ws = Wsnew;
  end
  iter = iter + 1;
end
% Pressure drop
DPs = dpshell(Do,Ds,L-2e-3*Ls,Lbc,Lbin,Lbout,Lc,Dotl,Dsb,Pt,Nt,Nss,Ws,
rhos,mus,Phis,Layout);
DPt = dptube(Di,L,rf,Ui,rhot,mut,Phit,Npass);
% Print results
fprintf('Overall heat transfer coefficient: U = %g(W/m^2/K)\n', U);
fprintf('Heat transfer coefficient: tube-side = %g(W/m^2/K), shell-side = %g(W/
m^2/K)\n',Ht,Hs);
fprintf('Heat duty: Q = %g(W)\n', Q);
fprintf('Pressure drop: tube-side = %g(Pa), shell-side = %g(Pa)\n', DPt,DPs);
fprintf('Tube-side: Ti1 = %g(K), Ti2 = %g(K), flow rate = %g(kg/sec)
\n',Ti1,Ti2,Wi);
fprintf('Shell-side: Ts1 = %g(K), Ts2 = %g(K), flow rate = %g(kg/sec)
\n',Ts1,Ts2,Ws);
```

Example 8.15 Shell-and-Tube Countercurrent Heat Exchanger[27]

Liquid benzene flowing at a rate of 4.8 *kg/sec* is cooled from 353 *K* in a four-pass shell-and-tube heat exchanger with cooling water flowing in the tubes. Inlet and outlet temperatures of cooling

TABLE 8.6
Physical Properties at Reference Temperatures

	Tube Side (Water)		Shell Side (Benzene)	
Reference temperature (K)	323	283	375	289
Density (kg/m^3)	988.1	999.7	798	885
Viscosity (mNs/m^2)	0.6	1.26	0.258	0.679
Heat conductivity ($W/(m \cdot K)$)	0.64	0.603	0.126	0.163
Heat capacity ($J/(kg \cdot K)$)	4183	4195	1980	1675

water are measured to be 298 and 303 K, respectively. The heat exchanger geometry is as follows: $d_o = 25.4$ mm, $d_i = 19.86$ mm, $N_T = 86$ (mounted on equilateral triangular pitch with a pitch ratio of 1.25), tube roughness = 0.025 mm, $L = 2$ m, $l_s = 27$ mm, $D_s = 305$ mm, $d_{otl} = 294$ mm, $\delta_{sb} = 4.45$ mm, $N_{ss} = 0$, $l_{bc} = 450$ mm, $l_{in}, l_{out} = 165$ mm, baffle cut for segmental baffles = 25%. Thermal conductivity of wall material is 45 $W/(m \cdot K)$. The fouling resistances for tube side and shell side are 0.00036 and 0.00018 m^2 K/W, respectively. The physical properties at reference temperatures are shown in Table 8.6.

Determine the exit temperature of benzene, the heat duty, the flow rate of cooling water, and the pressure drops on the shell and tube sides.

Solution

The script *hxndat* defines the input data as follows:

```
% hxndat.m: Shell-and-tube heat exchanger data
% Two values of reference temperatures and physical properties at the two
% reference temperatures for tube and shell fluids
Trt = [323 283]; Trs = [375 289]; % reference temperatures (tube ans shell)(K)
rhoreft = [988.1 999.7]; rhorefs = [798 885]; % densities (kg/m^3)
mureft = [0.6 1.26]; murefs = [0.258 0.679]; % viscosities (mNs/m^2)
xkreft = [0.64 0.603]; xkrefs = [0.126 0.163]; % thermal conductivities
cpreft = [4183 4195]; cprefs = [1980 1675]; % heat capacities (J/kg/K)
% Heat exchanger geometry
Do = 25.4; Di = 19.86; % tube outside diameter (Do,mm) and inside dia-
meter (Di,mm)
Xkw = 45; % heat conductivity of tube wall (W/m/K)
L = 2; % tube length (m)
Ls = 27; % tube sheet thickness (mm)
rf = 0.025; % tube roughness(mm)
Nt = 86; % total number of tubes in tube bundle
Layout = 1; % tube layout (1:triangular, 2:in-line square, 3:rotated square)
Pt = 1.25*Do; % tube pitch (mm)
Nss = 0; % number of pairs of sealing strips
Npass = 4; % number of passes
Rdt = 0.00036; Rds = 0.00018; % fouling resistance (m^2*K/W)
Ds = 305; % shell inside diameter(mm)
Dotl = 294; % shell outside tube limit(mm)
Dsb = 4.45; % shell-baffle clearance(mm)
Lbin = 165; Lbout = 165; % inlet and outlet baffle spacing(mm)
Lbc = 450; % central baffle spacing(mm)
Lc = 0.25*Ds; % baffle cut(mm)
```

```
% Specification of key variables
Ti1 = 298; Ti2 = 303; % inlet and outlet temperatures for tube-side(K)
Ts1 = 353; % Shell-side(hot stream) inlet temperature(K)
Ws = 4.8; % Shell-side(hot stream) mass flow rate(kg/sec)
fsT = 1; fsS = 1; % state of tube-side and shell-side fluids(1:liquid, 2:vapor)
ptype = 8; % problem type
Ts2 = 338; % guess outlet temperature of shell-side fluid
Wi = 5; % guess tube-side flow rate
```

The main script *hxnst* produces the following outputs:
```
>> hxnst
Overall heat transfer coefficient: U = 125.291(W/m^2/K)
Heat transfer coefficient: tube-side = 529.371(W/m^2/K), shell-side =
201.457(W/m^2/K)
Heat duty: Q = 80243(W)
Pressure drop: tube-side = 232564(Pa), shell-side = 26610.9(Pa)
Tube-side: Ti1 = 298(K), Ti2 = 303(K), flow rate = 3.77462(kg/sec)
Shell-side: Ts1 = 353(K), Ts2 = 344.138(K), flow rate = 4.8(kg/sec)
```

8.3.4 DOUBLE-PIPE HEAT EXCHANGER

A double-pipe finned-tube heat exchanger may be used when the heat duty is moderate (that is, $UA < 100,000$), when one stream is viscous liquid, or when flow rates are small. Heat transfer coefficients and film coefficients for vapor-liquid streams in the tube side and shell side can be determined using simple correlations.[28]

(1) Tube-side calculation

For liquid:

$$h_i = 0.023 Re^{0.8} Pr^{0.333} \frac{K}{D_{eq}} \left(\frac{\mu}{\mu_\omega} \right)^{0.14} \text{(where } \mu/\mu_\omega \approx 1), \ D_eq = D_ti, \ Re = (GD_eq)/\mu$$

If $Re < 10,000$,

$$h_i = J \cdot Pr^{0.333} \frac{K}{D_{eq}}$$

For vapor:

$$h_i = \frac{0.0144 C_p G^{0.8}}{D_{eq}^{0.2}}, \qquad \frac{1}{h_{if}} = \frac{1}{h_i} + F_f$$

where F_f accounts for the fouling correction.

(2) Shell-side calculation (bare tube): Use $D_{eq} = D_{si} - D_{to}$ in the tube-side calculation.

(3) Overall heat transfer coefficient

$$\frac{1}{U_o} = \frac{D_{to} - D_{ti}}{K} + h_{if} \frac{A_i}{A_o} \bigg|_{tube} \ h_{if} \bigg|_{shell}, \ \frac{A_i}{A_o} = \frac{D_{ti}}{D_{to}}$$

(4) Shell-side calculation (finned tube): In the tube-side calculation, set

$$D_{eq} = \frac{4N_f}{\pi(D_{si} + D_{to}) - N\theta + A_f}, \ N_f = C_s - \frac{\theta}{2} A_f, \ C_s = \frac{\pi}{4}(D_{si}^2 - D_{to}^2), A_f = 2HN$$

$$A_o = \pi D_{to} + A_f, \; X = H\sqrt{\frac{h_{if}|_{shell}}{6K\theta}}, \; \omega = \frac{\tanh X}{X} = \frac{1}{X}\left(\frac{e^X - e^{-X}}{e^X + e^{-X}}\right), \; \omega' = \omega\left(\frac{A_f}{A_o}\right) + \left(1 - \frac{A_f}{A_o}\right),$$

$$h_{ifd} = h_{if}\omega', \; \frac{1}{U_0} = \left.\frac{D_{to} - D_{ti}}{K}\right|_{tubewall} + \left.\frac{h_{if}A_i}{A_o}\right|_{tube} + h_{ifd}\Big|_{shell}, \; \frac{A_i}{A_o} = \frac{\pi D_{ti}}{\pi D_{to} + A_f}$$

where

h_i is the heat transfer film coefficient $(Btu/(ft^2 \cdot {}^\circ F \cdot hr))$
K is the thermal conductivity $(Btu/(ft \cdot {}^\circ F \cdot hr))$
Re is the Reynolds number
Pr is the Prandtl number
D_{eq} is the equivalent diameter (ft)
μ, μ_ω is the fluid viscosity (cP)
J is the heat transfer factor
G is the mass flow rate $(lb/(hr \cdot ft^2))$
C_p is the fluid heat capacity $(Btu/(lb \cdot {}^\circ F))$
A is the cross-sectional area (ft^2)
D_{ti} is the tube inside diameter $(in.)$
F_f is the fouling factor
h_{if} is the heat transfer film coefficient corrected for fouling $(Btu/(ft^2 \cdot {}^\circ F \cdot hr))$
D_{to} is the tube outside diameter $(in.)$
D_{si} is the shell inside diameter $(in.)$
A_i is the inside heat transfer area for the tube (ft^2/ft)
A_o is the outside heat transfer area for the tube (ft^2/ft)
U_o is the overall heat transfer coefficient $(Btu/(ft^2 \cdot {}^\circ F \cdot hr))$
N_f is the shell-side net free cross-sectional area (ft^2)
N is the number of fins
θ is the fin thickness (normally 0.035 $in.$)
H is the fin height $(in.)$
C_s is the cross-sectional area for the shell side without fins (ft^2)
A_f is the finned transfer area (ft^2)
X is the parameter used in the fin efficiency calculation
ω is the fin efficiency
ω' is the effective surface efficiency for the fins
h_{ifd} is the corrected film heat transfer rate $(Btu/(ft^2 \cdot {}^\circ F \cdot hr))$

Example 8.16 Finned-Tube Heat Transfer Coefficient

A double-pipe finned-tube heat exchanger is to be used to cool 30°API oil. Determine the tube-side and shell-side heat transfer coefficients and the overall heat transfer coefficient. The operating conditions and data are as follows:

Tube-side: flow rate of cooling water = 26,740 *lb/hr*, T_{in} = 85 °F, T_{out} = 120 °F,

$$F_f = 0.002, \; K = 0.366 \; Btu/(ft \cdot {}^\circ F \cdot hr), \; \mu = 0.72 \; cP, \; C_p = 1 \; Btu/(lb \cdot {}^\circ F)$$

$$= 18{,}000 \; lb/hr, \; T_{in} = 250 \,{}^\circ F, \; T_{out} = 150 \,{}^\circ F, \; F_f = 0.002,$$

$$K = 0.074 \; Btu/(ft \cdot {}^\circ F \cdot hr), \; \mu = 2.45 cP, \; C_p = 0.518 \; Btu/(lb \cdot {}^\circ F)$$

Heat exchanger geometry:
Shell: 3 *in.* (3.068*in.* ID, 3.5*in.* OD)

Tube: 1.5 *in*. (1.61 *in*. ID, 1.9 *in*. OD)
Nnumber of fins: 24, height 0.5 *in*. , width 0.035 *in*.

$$K = 25 \ Btu/(ft \cdot {}^\circ F \cdot hr)$$

Solution

The script *htcoef* calculates the heat transfer coefficients.

```
% htcoef.m : heat transfer coefficients (double-pipe finned-tube heat exchanger)
clear all;
% Data
Fft = 0.002; Kt = 0.366; mut = 0.72; Cpt = 1; Jt = 9.75; % Tube
Ffs = 0.002; Ks = 0.074; mus = 2.45; Cps = 0.518; % Shell
Dis = 3.068/12; Dos = 3.5/12; % shell size
Dit = 1.61/12; Dot = 1.9/12; % tube size
N = 24; Hf = 0.5/12; w = 0.035/12; % fin size
Gt = 26740; Gs = 18000; % flow rates
% Tube-side :
Deq = Dit; Pr = Cpt*(2.4191*mut)/Kt; % 1 cP = 2.4191 lb/(ft-hr)
Re = Gt*Deq/(2.4191*mut);
if (Re < 10000), hit = Jt* Pr^0.333 * Kt/Deq;
else, hit = 0.023 * Re^0.8 * Pr^0.333 * Kt / Deq; end
hift = 1 / (1/hit + Fft);
% Shell-side (bare tube) :
Deq = Dis - Dot; Pr = Cps*(2.4191*mus)/Ks; % 1 cP = 2.4191 lb/(ft-hr)
Re = Gs*Deq/(2.4191*mus);
if (Re < 10000), his = Jt* Pr^0.333 * Ks/Deq;
else, his = 0.023 * Re^0.8 * Pr^0.333 * Ks / Deq; end
hifs = 1 / (1/his + Ffs);
% Shell-side (finned-tube) :
Af = 2*Hf*N; Cs = pi*(Dis^2 - Dot^2)/4; Nf = Cs - w*Af/2;
Deq = 4*Nf/(pi*(Dis+Dot) - N*w + Af); Ao = pi*Dot + Af; X = Hf*sqrt
(hifs/(6*Kt*w));
e = (exp(X) - exp(-X))/(exp(X) + exp(-X))/X; ep = e*Af/Ao + 1 - Af/Ao; hifd =
hifs*ep;
Ar = pi*Dit/(pi*Dot + Af); Uo = 1/((Dot-Dit)/Kt + hift*Ar + hifd);
% Output
fprintf('Shell-side heat transfer coefficient: %10.7f\n', hifd);
fprintf('Tube-side heat transfer coefficient: %10.7f\n', hift);
fprintf('Fin efficiency: %10.7f\n', ep);
fprintf('Overall heat transfer coefficient: %10.7f\n', Uo);
```

The script *htcoef* generates the following results:

```
>> htcoef
Shell-side heat transfer coefficient: 12.3624311
Tube-side heat transfer coefficient: 41.0441459
Fin efficiency:  0.5070751
Overall heat transfer coefficient: 0.0516646
```

PROBLEMS

8.1 A hot metal cylinder D cm in diameter and L cm in length is cooled to a desired temperature, T_f, in a vacuum chamber. The wall temperature of the chamber, T_w, is kept constant. The time $t\,(sec)$ needed for the cooling process can be obtained from the following equation:

$$t = -\frac{\rho C_p V}{\varepsilon \sigma A} \int_{T_i}^{T_f} \frac{1}{T^4 - T_w^{\,4}} dT$$

where

ρ is the density of the cylinder
C_p is the specific heat of the cylinder
A and V are the surface area and the volume of the cylinder, respectively
ε is the emissivity
σ is the Boltzmann constant
T_i is the initial temperature of the cylinder
Determine t using the following data.
    Data: $D = 4\,cm$, $L = 12\,cm$, $T_i = 1100\,K$, $T_f = 580\,K$, $T_w = 298.15\,K$, $\rho = 8860\,kg/m^3$, $C_p = 532\,J/(kg{\cdot}K)$, $\varepsilon = 0.48$, $\sigma = 5.67 \times 10^{-8}\,J/(sec{\cdot}m^2{\cdot}K^4)$.

8.2 Convectional heat transfer occurs in a smooth tube of constant wall temperature (T_s), as shown in Figure P8.2.[29] Water at a temperature of $T_i(K)$ enters the tube at a mass rate of $\dot{m}\,(kg/sec)$. The water properties are given as a function of temperature ($273.15 \le T \le 373.15\,K$) as[29]:

$$k(T) = -0.496665 + 5.945727 \times 10^{-3}T - 7.482665 \times 10^{-6}T^2$$

$$\rho(T) = 246.631798 + 6.658227T - 0.018487T^2 + 1.541764 \times 10^{-5}T^3$$

$$\mu(T) = 0.40337 - 4.638597 \times 10^{-3}T + 2.01165 \times 10^{-5}T^2 - 3.890376 \times 10^{-8}T^3 + 2.827235$$
$$\times 10^{-11}T^4$$

$$C_p(T) = 1.8704258 \times 10^5 - 2.7097467 \times 10^3T + 16.0500521T^2 - 0.0474905T^3$$
$$+ 7.0174807 \times 10^{-5}T^4 - 4.1406453 \times 10^{-8}T^5$$

where

$k\,(W{\cdot}m^{-1}{\cdot}K^{-1})$ is the thermal conductivity
$\rho\,(kg{\cdot}m^{-3})$ is the density
$\mu\,(N{\cdot}sec{\cdot}m^{-2})$ is the viscosity
$C_p\,(J{\cdot}kg^{-1}{\cdot}K^{-1})$ is the specific heat
    The inside diameter of the tube is $D = 0.04\,m$, the wall temperature is $T_s = 342.15\,K$, and the inlet temperature is $T_i = 305.15\,K$.

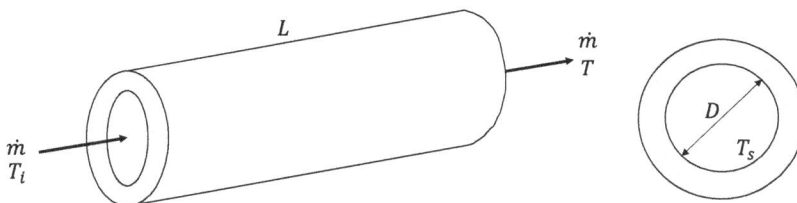

FIGURE P8.2 Convectional heat transfer in a tube.

  1 Calculate the exit water temperature $T(K)$ and the heat transfer rate $q(J/sec)$ when $\dot{m} = 0.25\ kg/sec$ and $L = 4\ m$.

  2 Plot T as a function of $\dot{m}$ $(0.01 \le \dot{m} \le 0.5\ kg/sec)$ when $L = 4\ m$.

  3 Plot T as a function of L $(2 \le L \le 8\ m)$ when $\dot{m} = 0.25\ kg/sec$.

8.3 A hot flue gas is separated from its surroundings by a brick wall whose thickness (D) is $0.22\ m$. The ambient temperature (T_a) is $298.15\ K$. From the energy balance, we obtain

$$k\left(\frac{T_i - T_0}{D}\right) + h(T_a - T_0) = \varepsilon\sigma(T_0^4 - T_a^4)$$

where

  k is the thermal conductivity of the brick wall

  h is the convectional heat transfer coefficient

  ε is the brick wall's surface emissivity

  σ is the Boltzmann constant

  T_0 is the temperature of the outer surface of the brick wall

  Generate a profile of T_0 as a function of the temperature T_i of the inner surface of the brick wall $(500 \le T_i \le 1000\ K)$ using the given data.

    Data: $k = 1.18\ W/(m{\cdot}K)$, $h = 16.2\ W/(m^2{\cdot}K)$, $\varepsilon = 0.71$, $\sigma = 5.67 \times 10^{-8}\ W/(m^2{\cdot}K^4)$, $D = 0.22\ m$, $T_a = 298.15\ K$.

8.4 A low-pressure saturated steam flows in an insulated 2 *in.* schedule 40 steel pipe with a thermal conductivity of 26 *Btu*/(*hr·ft·* °F) at 60 *psia* (292.73 °F), as shown in Figure P8.4. The pipe insulation has a thermal conductivity of 0.05 *Btu*/(*hr·ft·* °F). The steam-side and air-side heat transfer coefficients are $h_i = 2000$ *Btu*/(*hr·ft*$^2{\cdot}$°F) and $h_0 = 4$ *Btu*/(*hr·ft*^{2} °F), respectively. The air temperature is 70 °F.[30]

  1 Determine the heat flux q to the outside per foot if the insulation is 1 *in.* thick. Assume that the length of the pipe is $L = 1\ ft$.

  2 Determine the thickness of insulation (*in.*) required to keep the heat flux q at 50 *Btu/hr*.

8.5 A horizontal steel pipe with an external diameter of 0.033 *m* carries steam at $T_s = 450\ K$, as shown in Figure P8.5. The pipe is to be insulated with a material whose thermal conductivity $k(W/(m{\cdot}K))$ is given as a function of temperature T (K) by[31]

$$k = 6.6667 \times 10^{-10}T^3 - 1.54 \times 10^{-6}T^2 + 1.08976 \times 10^{-3}T - 0.14457$$

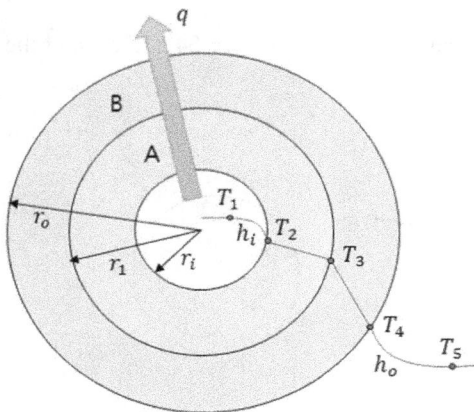

FIGURE P8.4 Heat transfer through multilayer cylinders. (From Cutlip, M. B. and M. Shacham, *Problem Solving in Chemical and Biochemical Engineering with POLYMATH, Excel, and MATLAB*, 2nd ed., Prentice-Hall, Boston, MA, 2008, p. 344.)

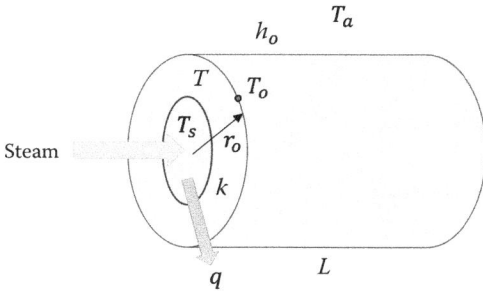

FIGURE P8.5   Heat loss from insulated pipe.

The external heat transfer coefficient from the surface of a cylinder due to natural convection in air is approximated by[32]

$$h = 1.32 \left(\frac{|\Delta T|}{D} \right)^{1/4}$$

where

$\Delta T \, (K)$ is the temperature difference between the surface and the air
$D \, (m)$ is the insulated cylinder diameter
The ambient temperature is $T_a = 300 \, K$. The pipe can be assumed to be at the same temperature as the steam due to the high thermal conductivity of steel. Assume that the length of the pipe is $L = 1 \, m$.

   1 Determine the heat loss per meter from the pipe without insulation.
   2 Determine the heat loss per meter from the pipe when the insulation thickness is 0.04 m.

8.6 The temperature distribution within a two-dimensional metal plate can be expressed by a two-dimensional elliptic partial differential equation of the general form

$$\frac{\partial^2 T}{\partial x^2} + \frac{\partial^2 T}{\partial y^2} = f$$

where f is a constant. A thin square metal plate of dimensions 1 m × 1 m is subject to four heat sources, which maintain the temperature on its four edges at $T(0, y) = 250\,°C$, $T(1, y) = 100\,°C$, $T(x, 0) = 500\,°C$, $T(x, 1) = 25\,°C$ (Dirichlet boundary conditions). The flat sides of the plate are insulated so that no heat is transferred through these sides. Determine the temperature profiles within the plate.[33]

8.7 The temperature distribution within a two-dimensional metal plate may be given by the two-dimensional elliptic partial differential equation:

$$\frac{\partial^2 T}{\partial x^2} + \frac{\partial^2 T}{\partial y^2} = f$$

where f is a constant. The plate is a thin square metal plate of dimensions 1 m × 1 m. Perfect insulation is installed on two edges (top and right), and the other two edges are maintained at constant temperatures. The set of Dirichlet and Neumann boundary conditions is given by

$$T(0, y) = 250\,°C, \quad \frac{\partial T}{\partial x}\bigg|_{1,y} = 0, \quad T(x, 0) = 500\,°C, \quad \frac{\partial T}{\partial x}\bigg|_{x,1} = 0$$

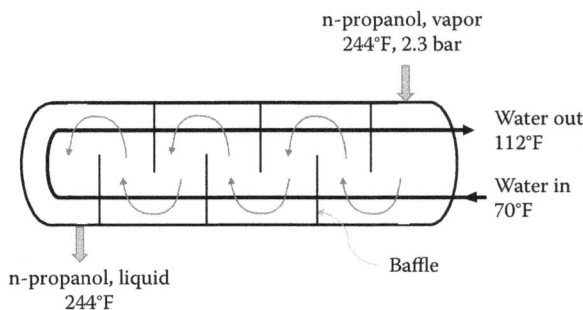

FIGURE P8.8   Schematic diagram of 1–2 *n*-propanol-water exchanger.

Determine the temperature profiles within the plate.[33]

8.8 The saturated vapor of *n*-propanol flowing at 158, 000 *lb/hr* is to be cooled in an existing 1–2 shell-and-tube heat exchanger. The exchanger contains 900 steel tubes (18 BWG, 14 *ft* long, 0.652 *in*. ID, 0.75 *in*. OD on a 15/16 triangular pitch in a 37 *in*. ID shell). The *n*-propanol vapor flows in the shell side and cooling water flows in the tube side, as shown in Figure P8.8. If the cooling water enters at 70 °F and exits at 112 °F, determine whether the exchanger will work or not.[34]

    Data: boiling point of *n*-propanol at 2.3 bar = 244 °F, latent heat of vaporization = 257 *Btu/lb*, thermal conductivity of the tubes = 25 *Btu/(hr·ft· °F)*, shell-side heat transfer coefficient = 1160 *Btu/(hr·ft² · °F)*, tube-side heat transfer coefficient = 195 *Btu/(hr·ft² · °F)*, fouling resistance for water = 0. 0038 *hr·ft² · °F/kJ*

8.9 Design a 1–2 shell-and-tube heat exchanger to cool ethylene glycol, flowing at 100, 000 *lb/hr* and 250 °F, to 130 °F. In the heat exchanger, water enters at 90 °F and exits at 120 °F.[35] Assume that the tube-side fouling resistance is R_{fi} = 0. 004 *hr·ft² · °F/Btu* and that the shell-side fouling resistance, R_{fo}, is negligible. As an initial guess for U_i, use U_i = 100 *Btu/(ft² ·hr· °F)*. The schematic diagram of the exchanger is shown in Figure P8.9, and physical properties are given in Table P8.9.

    Data: operating conditions: $\dot{m}_h$ = 100,000 *lb/hr*, T_1 = 250 °F, T_2 = 130 °F, t_1 = 90 °F, t_2 = 120 °F, u_i = 5 *ft/sec*. Shell and tube: L = 18*ft*, D_o = 0.0625 *ft*, D_i = 0.05167 *ft*, D_s = 2.4167 *ft*, P_t = 1/12 *ft*, k_w = 30 *Btu/(hr·ft· °F)*, R_{fi} = 0. 004 *ft²·hr· °F/Btu*, R_{fo} = 0.

8.10 Design a 1–2 shell-and-tube heat exchanger to heat a cold water stream.[36] The cold water enters the exchanger at 125 °F and 30*psi* and exits at 150 °F. A hot water stream is fed to the shell side at 250 °F and 100,000 *lb/hr*, and exits the exchanger at 190 °F. Assume that both the tube-side and shell-side fouling resistances are R_{fi} = R_{fo} = 0.002 *hr·ft² · °F/Btu*. The schematic diagram of the exchanger is shown in Figure P8.10, and physical properties at average temperatures are given in Table P8.10.[36] As an initial guess for U_i, use U_i = 200 *Btu/(ft² ·hr· °F)*.

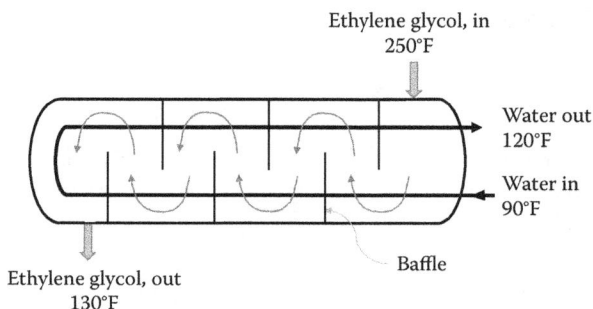

FIGURE P8.9   Schematic diagram of 1–2 ethylene glycol-water heat exchanger.

TABLE P8.9
Physical Properties

Property	Shell-Side (Hot) Ethylene Glycol (190°F)	Tube-Side (Cold) Water (105°F)
$\rho\,(lb/ft^3)$	68.6	62.4
$C_p\,(Btu/(lb\cdot°F))$	0.65	1.01
$\mu\,(cP)$	3.50	0.67
$k\,(Btu/(hr\cdot ft\cdot°F))$	0.16	0.363

TABLE P8.10
Physical Properties

Property	Shell-Side (Hot) Hot Water (220°F)	Tube-Side (Cold) Cold Water (138°F)
$\rho\,(lb/ft^3)$	60.0	61.4
$C_p\,(Btu/(lb\cdot°F))$	1.0	1.01
$\mu\,(cP)$	0.27	0.47
$k\,(Btu/(hr\cdot ft\cdot°F))$	0.39	0.38

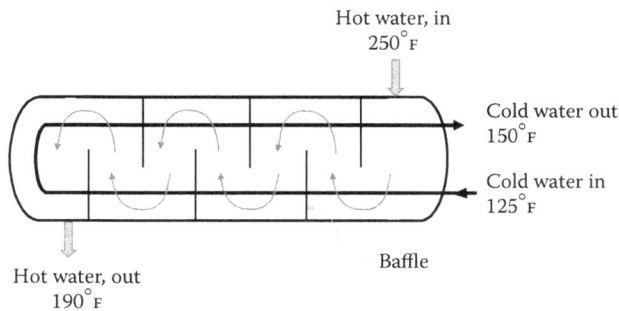

FIGURE P8.10 Schematic diagram of 1–2 raw-water heat exchanger.

Data: Operating conditions: $\dot{m}_h = 100{,}000\ lb/hr$, $T_1 = 250\,°F$, $T_2 = 190\,°F$, $t_1 = 125\,°F$, $t_2 = 150\,°F$, $u_i = 5\ ft/sec$, Shell and tube:

$L = 12\ ft$, $D_o = 0.0625\ ft$, $D_i = 0.05167\ ft$, $D_s = 1.60417\ ft$, $P_t = 1/12\ ft$, $k_w = 30\ Btu/(hr\cdot ft\cdot°F)$, $cl = 0.02083\ ft$, $R_{fi} = 0.002\ ft^2\cdot hr\cdot°F/Btu$, $R_{fo} = 0.002\ ft^2\cdot hr\cdot°F/Btu$.

8.11 Liquid benzene is to be cooled from 353 to 303 K in a four-pass shell-and-tube heat exchanger, with cooling water flowing in the tubes. The inlet temperature of the cooling water is measured to be 298 K, and the flow rate of the cooling water is 12 kg/sec. The heat exchanger

TABLE P8.11

Physical Properties at Reference Temperatures

	Tube Side (Water)		Shell Side (Benzene)	
Reference temperature (K)	323	283	375	289
Density (kg/m^3))	988.1	999.7	798	885
Viscosity $(m \cdot N \cdot sec/m^2)$	0.6	1.26	0.258	0.679
Heat conductivity $(W/(m \cdot K))$	0.64	0.603	0.126	0.163
Heat capacity $(J/(kg \cdot K))$	4183	4195	1980	1675

geometry is as follows: $d_o = 25.4$ *mm*, $d_i = 19.86$ *mm*, $N_T = 86$ (mounted on equilateral triangular pitch with a pitch ratio of 1.25), tube roughness $= 0.025$ *mm*, $L = 2$ *m*, $l_s = 27$ *mm*, $D_s = 305$ *mm*, $d_{otl} = 294$ *mm*, $\delta_{sb} = 4.45$ *mm*, $N_{ss} = 0$, $l_{bc} = 450$ *mm*, l_{in}, $l_{out} = 165$ *mm*, baffle cut for segmental baffles $= 25\%$. The thermal conductivity of the wall material is 45 45 $W/(m \cdot K)$. The fouling resistances for the tube side and shell side are 0.00036 and 0.00018 $m^2 \cdot K/W$, respectively. The physical properties at reference temperatures are shown in Table P8.11.

Determine the exit temperature of cooling water, the heat duty, the flow rate of benzene, and the pressure drops on the shell and tube sides.

8.12 A packed countercurrent cooling tower operating at $P = 1$ *atm* is designed to cool the process water. The height of the tower can be obtained from

$$z = \frac{G}{k_G MW_{air} P} \int_{H_1}^{H_2} \frac{1}{H_i - H} dH$$

where
    $G\,(kg \cdot sec^{-1} \cdot m^{-2})$ is the dry air flow rate
    $k_G\,(kgmol \cdot sec^{-1} \cdot m^{-3} \cdot Pa^{-1})$ is the volumetric mass transfer coefficient in the gas phase
    MW_{air} is the molecular weight of air ($= 29.0$ *kg/kgmol*)
    H_1 and $H_2\,(J/kg)$ are the enthalpy of the air entering and leaving the tower, respectively
    $H_i\,(J/kg)$ is the enthalpy of the air at the gas interface
    $H\,(J/kg)$ is the enthalpy of the air
    Table P8.12 shows values of H_i as a function of H. Determine z using the given data.[37]
    Data: $G = 1.356$ *kg/(sec·m²)*, $k_G = 1.207 \times 10^{-7}$ *kgmol/(sec·m³·Pa)*, $H_1 = 71.7 \times 10^3$ *J/kg*, $H_2 = 129.9 \times 10^3$ *J/kg*

TABLE P8.12

H_i as a Function of H

$H\,(10^3 J/kg)$	71.7	83.5	94.9	106.5	118.4	129.9
$H_i\,(10^3 J/kg)$	94.4	108.4	124.4	141.8	162.1	184.7

REFERENCES

1. Cutlip, M. B. and M. Shacham, *Problem Solving in Chemical and Biochemical Engineering with POLYMATH, Excel, and MATLAB*, 2nd ed., Prentice-Hall, Boston, MA, pp. 209–210, 2008.
2. Adidharma, H. and V. Temyanko, *Mathcad for Chemical Engineers*, Trafford publishing, Victoria, BC, Canada, pp. 36–37, 2007.
3. Adidharma, H. and V. Temyanko, *Mathcad for Chemical Engineers*, Trafford publishing, Victoria, BC, Canada, p. 59, 2007.
4. Kulakowski, B. T., J. F. Gardner, and J. Lowen Shearer, *Dynamic Modeling and Control of Engineering Systems*, Cambridge University Press, New York, NY, pp. 204–208, 2007.
5. Cutlip, M. B. and M. Shacham, Problem Solving in Chemical and Biochemical Engineering with POLYMATH, Excel, and MATLAB, 2nd ed., Prentice-Hall, Boston, MA, p. 334, 2008.
6. Geankoplis, C. J., *Transport Processes and Separation Process Principles*, 4th ed., Prentice-Hall, Boston, MA, pp. 246–247, 2003.
7. Cutlip, M. B. and M. Shacham, *Problem Solving in Chemical and Biochemical Engineering with POLYMATH, Excel, and MATLAB*, 2nd ed., Prentice-Hall, Boston, MA, pp. 334–335, 2008.
8. Cutlip, M. B. and M. Shacham, Problem Solving in Chemical and Biochemical Engineering with POLYMATH, Excel, and MATLAB, 2nd ed., Prentice-Hall, Boston, MA, p. 339, 2008.
9. Cutlip, M. B. and M. Shacham, Problem Solving in Chemical and Biochemical Engineering with POLYMATH, Excel, and MATLAB, 2nd ed., Prentice-Hall, Boston, MA, pp. 347–349, 2008.
10. Cutlip, M. B. and M. Shacham, Problem Solving in Chemical and Biochemical Engineering with POLYMATH, Excel, and MATLAB, 2nd ed., Prentice-Hall, Boston, MA, p. 347, 2008.
11. Mariano M. Martin (Editor), *Introduction to Software for Chemical Engineers*, CRC Press, Taylor & Francis Group, Boca Raton, FL, pp. 112–114, 2015.
12. Niket S. Kaisare, *Computational Techniques for Process Simulation and Analysis Using MATLAB*, CRC Press, Taylor & Francis Group, Boca Raton, FL, pp. 172–173, 2018.
13. Constantinides, A. and N. Mostoufi, *Numerical Methods for Chemical Engineers with MATLAB Applications*, Prentice-Hall, Boston, MA, pp. 370–372, 1999.
14. Constantinides, A. and N. Mostoufi, *Numerical Methods for Chemical Engineers with MATLAB Applications*, Prentice-Hall, Boston, MA, pp. 401–402, 1999.
15. Constantinides, A. and N. Mostoufi, *Numerical Methods for Chemical Engineers with MATLAB Applications*, Prentice-Hall, Boston, MA, pp. 382–385, 1999.
16. Coker, A. K., *Chemical Process Design, Analysis, and Simulation*, Gulf Publishing Company, Houston, TX, pp. 596–604, 1995.
17. Ghasem, N., *Computer Methods in Chemical Engineering*, CRC Press, Taylor & Francis Group, Boca Raton, FL, pp. 135–145, 2012.
18. Serth, R. W., *Process Heat Transfer*, Academic Press, Burlington, MA, p. 99, 2007.
19. Ghasem, N., *Computer Methods in Chemical Engineering*, CRC Press, Taylor & Francis Group, Boca Raton, FL, pp. 146–154, 2012.
20. Ghasem, N., *Computer Methods in Chemical Engineering*, CRC Press, Taylor & Francis Group, Boca Raton, FL, pp. 186–194, 2012.
21. Raman, R., *Chemical Process Computations*, Elsevier Applied Science Publishers, Barking, Essex, UK, pp. 287–290, 1985.
22. Serth, R. W., *Process Heat Transfer*, Academic Press, Elsevier, Burlington, MA, p. 254, 2007.
23. Serth, R. W., *Process Heat Transfer*, Academic Press, Elsevier, Burlington, MA, p. 248, 2007.
24. Serth, R. W., *Process Heat Transfer*, Academic Press, Elsevier, Burlington, MA, p. 249, 2007.
25. Serth, R. W., *Process Heat Transfer*, Academic Press, Elsevier, Burlington, MA, p. 259, 2007.
26. Raman, R., *Chemical Process Computations*, Elsevier Applied Science Publishers, Barking, Essex, UK, p. 295, 1985.
27. Raman, R., *Chemical Process Computations*, Elsevier Applied Science Publishers, Barking, Essex, UK, p. 298, 1985.
28. Coker, A. K., *Chemical Process Design, Analysis and Simulation*, Gulf Publishing Company, Houston, TX, pp. 632–634, 1995.
29. Adidharma, H. and V. Temyanko, *Mathcad for Chemical Engineers*, Trafford publishing, Victoria, BC, Canada, pp. 154–156, 2007.
30. Cutlip, M. B. and M. Shacham, Problem Solving in Chemical and Biochemical Engineering with POLYMATH, Excel, and MATLAB, 2nd ed., Prentice-Hall, Boston, MA, p. 344, 2008.

31. Cutlip, M. B. and M. Shacham, Problem Solving in Chemical and Biochemical Engineering with POLYMATH, Excel, and MATLAB, 2nd ed., Prentice-Hall, Boston, MA, p. 346, 2008.
32. Geankoplis, C. J., *Transport Processes and Separation Process Principles*, 4th ed., Prentice-Hall, Boston, MA, p. 305, 2003.
33. Constantinides, A. and N. Mostoufi, *Numerical Methods for Chemical Engineers with MATLAB Applications*, Prentice-Hall, Boston, MA, p. 382, 1999.
34. Ghasem, N., *Computer Methods in Chemical Engineering*, CRC Press, Taylor & Francis Group, Boca Raton, FL, pp. 163–165, 2012.
35. Ghasem, N., *Computer Methods in Chemical Engineering*, CRC Press, Taylor & Francis Group, Boca Raton, FL, pp. 170–177, 2012.
36. Ghasem, N., *Computer Methods in Chemical Engineering*, CRC Press, Taylor & Francis Group, Boca Raton, FL, pp. 196–207, 2012.
37. Adidharma, H. and V. Temyanko, *Mathcad for Chemical Engineers*, Trafford publishing, Victoria, BC, Canada, p. 86, 2007.

9 Process Control

9.1 LAPLACE TRANSFORM AND TRANSFER FUNCTION

9.1.1 LAPLACE TRANSFORM AND INVERSE LAPLACE TRANSFORM

The Laplace transform of a function $f(t)$ is defined as

$$F(s) = \mathscr{L}\{f(t)\} = \int_0^\infty f(t)e^{-st}dt$$

The inverse Laplace transform to find $f(t)$ from the given Laplace transform $F(s)$ is expressed as

$$f(t) = \mathscr{L}^{-1}[F(s)]$$

In the partial fraction expansion method used to find $f(t)$, the polynomial in the denominator of $F(s)$ is broken up into simpler partial fractions. $F(s)$ can be represented as

$$F(s) = \frac{b_m s^m + b_{m-1}s^{m-1} + \cdots + b_1 s + b_0}{s^n + a_{n-1}s^{n-1} + \cdots + a_1 s + a_0} + R(s)$$

where $n > m$.

Symbolic variables are used in the Laplace transform of a time-domain function using MATLAB. The built-in functions *laplace* and *ilaplace* are used in the Laplace and inverse Laplace transforms, respectively.

Example 9.1 Laplace Transform

Find the Laplace transform of $f(t) = 1 + t + t^2 + \sin at - t\cos bt$.

Solution

First create the symbolic variables t, a, and b using the built-in function *syms* and call the built-in function *laplace*. The following commands produce the Laplace transform of $f(t)$.

```
>> syms t a b;
f = 1 + t + t^2 + sin(a*t) - t*cos(b*t); Lf = laplace(f)
Lf =
(s + 1)/s^2 + a/(a^2 + s^2) + 1/(b^2 + s^2) - (2*s^2)/(b^2 + s^2)^2 + 2/s^3
```

Example 9.2 Inverse Laplace Transform

Find the inverse Laplace transform of $F(s) = 2s/(s^2 + 4s + 1)$.

Solution

Create the symbolic variable s and use the built-in function *ilaplace* to get $f(t)$. The following commands produce the inverse Laplace transform $f(t)$ of $F(s)$.

```
>> syms s;
>> F = 2*s / (s^2 + 4*s + 1); f = ilaplace(F)
```

```
f =
2*exp(-2*t)*(cosh(3^(1/2)*t) - (2*3^(1/2)*sinh(3^(1/2)*t))/3)
```

9.1.2 PARTIAL FRACTION EXPANSION

In MATLAB, the partial fraction expansion is performed by the built-in function *residue*. If the function $F(s)$ can be expanded into a sum of partial fractions and repeated factors, $F(s)$ can be written as

$$F(s) = \frac{N(s)}{D(s)} = \frac{b_m s^m + b_{m-1} s^{m-1} + \cdots + b_1 s + b_0}{s^n + a_{n-1} s^{n-1} + \cdots + a_1 s + a_0}$$

$$= \frac{r_1}{s - p_1} + \cdots + \frac{r_k}{s - p_k} + \frac{r_{q1}}{(s - p_q)} + \frac{r_{q2}}{(s - p_q)^2} + \cdots + K(s)$$

For partial fraction expansion by MATLAB, the built-in function *residue* can be used as

```
[R P K] = residue(N, D)
```

where $R = [r_1, r_2, \cdots]$, $p = [P_1, P_2, \cdots]$, and K is a row vector containing the coefficients of $K(s)$.

Example 9.3 Partial Fraction Expansion

$F(s) = (s^3 + 5s^2 + 9s + 7)/(s^2 + 3s + 2)$ can be expanded into $F(s) = -1/(s + 2) + 2/(s + 1) + (s + 2)$. Verify this using MATLAB.

Solution

The built-in function *residue* can be used to verify this.

```
>> N = [1 5 9 7]; D = [1 3 2]; [R P K] = residue(N, D)
R =
      -1
       2
P =
      -2
      -1
K =
       1    2
```

9.1.3 REPRESENTATION OF THE TRANSFER FUNCTION

In MATLAB, the Laplace transform can be represented by row vectors of coefficients of the numerator and denominator of the Laplace transform $F(s)$ in descending order. Data are inputted using brackets [], with numbers separated by a blank space or a comma (,). If the denominator is factorised, we can use the built-in function *conv* to expand the multiplication of these factors.

Once the coefficient vectors for the denominator and the numerator of a transfer function are specified, the MATLAB built-in function *tf* can be used to get the transfer function in the form of Laplace transform. In this case, other MATLAB commands can easily be used by storing the output of *tf* in a variable.

Example 9.4 Representation of the Transfer Function

Use MATLAB to display $G(s) = (2s + 1)/(s^2 + 3s + 2) = (2s + 1)/((s + 1)(s + 2))$.

Solution

The built-in function *conv* can be used to expand the product of two terms. The following commands represent $G(s)$ in expanded form:

```
>> num = [2 1]; den = conv([1 1], [1 2]); G = tf(num, den)
G =
    2 s + 1
  -------------
  s^2 + 3 s + 2
Continuous-time transfer function.
```

Example 9.5 Use of the Built-In Function *tf*

Find the transfer function G when the numerator is given by $2s + 1$ and the denominator is given by $s^2 + 3s + 2$.

Solution

The following commands produce the transfer function G:

```
>> num = [2 1]; den = conv([1 1], [1 2]); G = tf(num, den)
G =
    2 s + 1
  -------------
  s^2 + 3 s + 2
Continuous-time transfer function.
```

9.2 BLOCK DIAGRAMS

Block diagrams consist of unidirectional operational blocks that represent the transfer function of the variables of interest. The components of a chemical process can be represented as blocks. The interconnected blocks show clearly the flows of information and signals between adjacent process components. Figure 9.1 shows a simple feedback control system represented by a block diagram.

In Figure 9.1, each block represents a process element whose characteristics can be expressed in terms of algebraic or differential equations. Identification of mathematical expressions for each block is one of the important tasks a process engineer has to perform.

The MATLAB built-in functions *series*, *parallel*, and *feedback* can be used to synthesise block diagrams. The function *series* is used when the blocks are connected in series, the function *parallel*

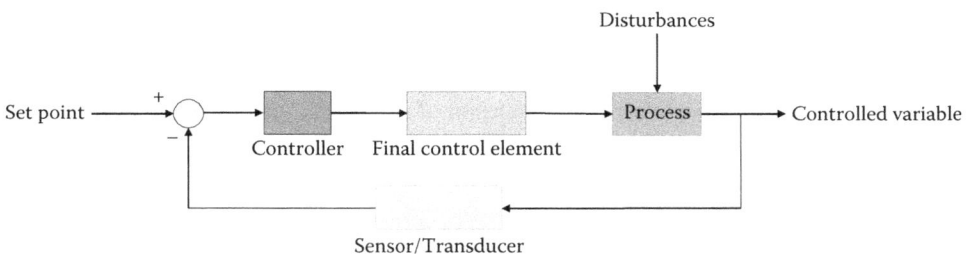

FIGURE 9.1   A block diagram representation of a simple feedback control system.

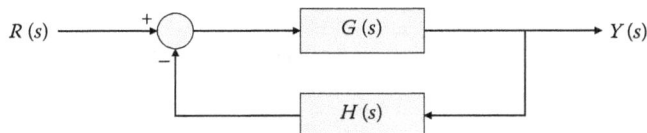

FIGURE 9.2 A simple feedback loop.

is used when the blocks are connected in parallel, and the function *feedback* is used to synthesise a block diagram when the output is used as a feedback signal.

Example 9.6 Overall Transfer Function

Find the overall transfer function for the simple feedback loop shown in Figure 9.2. In Figure 9.2, $G = (2s + 1)/(s^2 + 3s + 2)$, $H = 1/(s + 1)$.

Solution

The built-in function *feedback* produces the desired overall transfer function:

```
>> ng = [2 1]; dg = [1 3 2]; nh = 1; dh = [1 1]; [nt, dt] = feedback(ng, dg, nh,
dh, -1);
>> Gcl = tf(nt, dt)
Gcl =
     2 s^2 + 3 s + 1
  -------------------------
  s^3 + 4 s^2 + 7 s + 3
Continuous-time transfer function
```

The numerator and denominator of G are stored in ng and dg, respectively, and those of H are stored in nh and dh, respectively. The numerator and denominator of the overall transfer function are nt and dt, which are stored in Gcl. When the function *feedback* is used, negative or positive feedback should be specified by −1 (negative feedback) and +1 (positive feedback).

9.3 STATE-SPACE REPRESENTATION

A set of differential equations can be put in standard vector-matrix form:

$$\dot{x}(t) = Ax(t) + Bu(t), \; y(t) = Cx(t) + Du(t)$$

where
 x is the state variable vector
 u is the input vector
 y is the output vector

The time derivative is denoted by the overdot. In addition, A is the process matrix, B is the input matrix, C is the output matrix, and D is the feed-forward matrix. If the dimensions of the vectors x, u, and y are n, m, and r, respectively, the dimension of A is $n \times n$, that of B is $n \times m$, that of C is $r \times n$, and that of D is $r \times m$.

The resolvent matrix $\Phi(s)$ is defined as

$$\Phi(s) = (sI - A)^{-1}$$

This gives the time-domain representation $\Phi(t) = e^{At}$, and we have

$$\underline{x}(t) = \Phi(t)\underline{x}(0) + \int_0^t \Phi(t-\tau)B\underline{u}(\tau)d\tau = e^{At}\underline{x}(0) + \int_0^t e^{A(t-\tau)}B\underline{u}(\tau)d\tau$$

The Laplace transformation of the state-space model yields

$$Y(s) = CX(s) = C\Phi(s)B\cdot U(s) = G(s)U(s)$$

where

$$G(s) = C\Phi(s)B = C(sI - A)^{-1}B$$

is the transfer function matrix.

Once the matrices A, B, and C are known, the state-space model can be readily entered in MATLAB by defining the three matrices and using the *ss* command. The command *tf2ss* is used to create a state-space model from a given transfer function model. The command *ss2tf* generates a transfer model from the given state-space model.

Example 9.7 From Transfer Function to State-Space Model

Find a state-space model for the process whose transfer function is given by

$$G = \frac{2s + 1}{S^3 + 3s^2 + 2s + 1}.$$

Solution
The command *tf2ss* generates the matrices A, B, C, and D.

```
>> n = [2 1]; d = [1 3 2 1]; [A B C D] = tf2ss(n,d)
A =
    -3   -2   -1
     1    0    0
     0    1    0
B =
     1
     0
     0
C =
     0    2    1
D =
     0
```

Example 9.8 State-Space Representation

Find a state-space representation for the system given by

$$\frac{d^2y}{dt^2} + 1.5\frac{dy}{dt} + y = \frac{du}{dt} + 2u, \ y(0) = \frac{dy}{dt}(0) = u(0) = 0$$

Solution
To obtain the matrices A, B, C, and D, we first take the Laplace transform of the differential equation and supply the coefficient vectors of the numerator and denominator in descending order to the command *tf2ss*.

```
>> [A B C D] = tf2ss([1 2],[1 1.5 1])
A =
    -1.5000  -1.0000
     1.0000        0
B =
     1
     0
C =
     1    2
D =
     0
```

9.4 PROCESS DYNAMICS

9.4.1 DYNAMICS OF 1ST-ORDER PROCESSES

A 1st-order process is represented by the 1st-order differential equation

$$\tau \frac{dy}{dt} + y = Kx(t)$$

where

    y is the output
    x is the input
    τ is the time constant of the process
    K is the gain of the process
    Assume that the initial condition is given by $y(0) = 0$. The Laplace transform of this model yields

$$\frac{Y(s)}{X(s)} = G(s) = \frac{K}{\tau s + 1}$$

The output $Y(s)$ depends on the input $X(s)$. For a step input of magnitude A, the time-domain response is given by

$$Y(t) = KA(1 - e^{-t/\tau})$$

For a block pulse input of magnitude H and duration of T, the output $Y(s)$ is given by

$$Y(s) = \frac{KH}{s(\tau s + 1)}(1 - e^{-Ts})$$

and the time-domain response can be expressed as

$$Y(t) = KH[1 - e^{-t/\tau} - \{1 - e^{-(t-T)/\tau}\}u(t - T)]$$

The Laplace transform of a ramp input with a slope of A is $X(s) = A/s^2$. Thus, the output is given by

$$Y(s) = G(s)X(s) = \frac{KA}{s^2(\tau s + 1)} = \frac{KA\tau^2}{\tau s + 1} - \frac{KA\tau}{s} + \frac{KA}{s^2}$$

and the time-domain response is given by

$$Y(t) = KA(t + \tau e^{-t/\tau} - \tau)$$

A sinusoidal input with an amplitude of A and frequency of ω can be represented as $X(t) = A\sin(\omega t)u(t)$. The Laplace transform of this function is $X(s) = A\omega/(s^2 + \omega^2)$, and the output is given by

$$Y(s) = G(s)X(s) = \frac{KA\omega}{(\tau s + 1)(s^2 + \omega^2)} = \frac{KA}{1 + \tau^2\omega^2}\left(\frac{\tau^2\omega}{\tau s + 1} - \frac{\tau\omega s}{s^2 + \omega^2} + \frac{\omega}{s^2 + \omega^2}\right)$$

We can see that the time-domain response can be represented as

$$Y(t) = \frac{KA\omega\tau}{1 + \tau^2\omega^2}e^{-t/\tau} + \frac{KA}{\sqrt{1 + \tau^2\omega^2}}\sin(\omega t + \phi)$$

where ϕ is the phase angle given by $\phi = \tan^{-1}(-\tau\omega)$. As $t \to \infty$, the exponential term goes to zero and the time-domain output $Y(t)$ becomes a sine curve with constant amplitude and frequency:

$$Y(\infty) = \frac{KA}{\sqrt{1 + \tau^2\omega^2}}\sin(\omega t + \phi)$$

In this oscillation, the period is given by $T = 2\pi/\omega$. The frequency of the sinusoidal oscillation is represented either in hertz (number of cycles per second), Hz, or in radian per unit time, ω.

The step response of a 1st-order process can be obtained by the built-in function *step* in MATLAB. The *step* function plots the unit step response for each input-output pair of the system, assuming that the initial conditions are zero. The *step* function requires the transfer function and response time as input arguments.

The built-in function *impulse* calculates the impulse response of a 1st-order process. The *impulse* function plots the unit impulse response for each input-output pair of the system, assuming that the initial conditions are zero. This function also needs the transfer function and response time as input arguments.

The built-in function *lsim* calculates the response for a linear input. The *lsim* function plots the response of the system to an arbitrary input. This function needs the transfer function, the input form, and the response time as input arguments. The *lsim* function can also be used to calculate the response for a sinusoidal input.

Example 9.9 Step Response of a 1st-Order Process

Plot the unit step response of a 1st-order process during the time interval [0, 10]. The transfer function of the process is given by $3/(2s + 1)$.

Solution

The script *simstep* generates the desired step response curve.

```
% simstep.m
num = 3; den = [2 1]; t = [0:0.1:10];
y = step(num, den, t);
plot(t, y), grid, title('Step response of a 1st-order process'), xlabel('t
(sec)'), ylabel('Response y(t)')
```

Figure 9.3 shows the step response curve of the given 1st-order process.

The step response can also be obtained by Simulink®. To start the simulation, perform the following steps:

1)   Click on the "Simulink" icon in the HOME menu bar and select "Blank Model."
2)   A new editor window appears. In the menu bar of the new editor window, select the "Library Browser" icon (⛁).

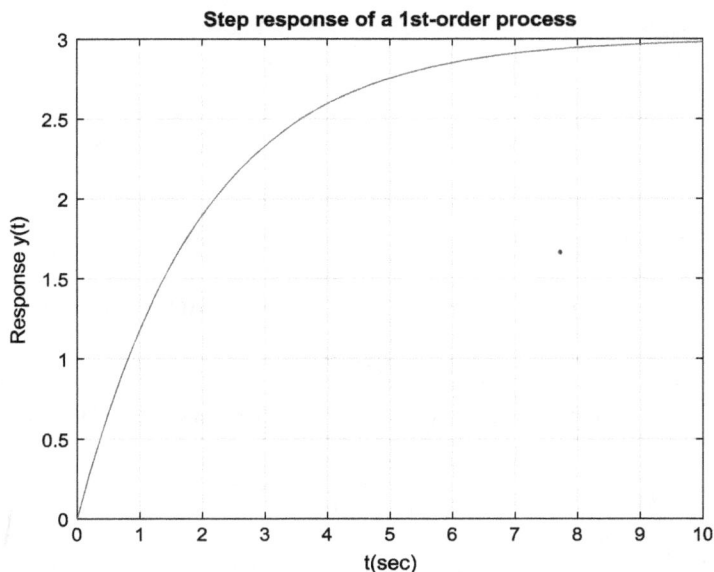

FIGURE 9.3 Step response of a 1st-order process.

3) Select and place in the editor window the "Step" block from the "Sources" library located in the "Simulink" submenu. This can be done by selecting the "Step" block while pressing the left mouse button and dragging it to the editor window.
4) Select and place in the editor window the "Transfer Fcn" (transfer function) block from the "Continuous" library and the "Scope" block from the "Sinks" library.
5) Construct a Simulink block diagram for the 1st-order process as shown in Figure 9.4. In order to connect the blocks in the editor editor window, simply connect the arrows in the adjacent blocks by dragging the mouse while pressing the left mouse button.
6) Specification of the numerator and denominator of the transfer function can be done in the editor window, which appears by double-clicking on the "Transfer Fcn" block.
7) In the editor window obtained by double-clicking on the "Step" block, set "Step time" to 0, select the "Apply" button, and press the "OK" button.
8) In the "Simulation" menu of the model editor window, select "Model Configuration Parameters", set the "Stop time" to 10(sec), and select "Run" in the "Simulation" menu.
9) The response curve can be seen by double-clicking the "Scope" block. Figure 9.5 shows the response curve displayed in the "Scope" block.

Example 9.10 Sinusoidal Response of a 1st-Order Process

Plot the response curve of the 1st-order system $3/(2s + 1)$ to the sinusoidal input $u = \sin(3t)$ during the time interval $[0, 10]$.

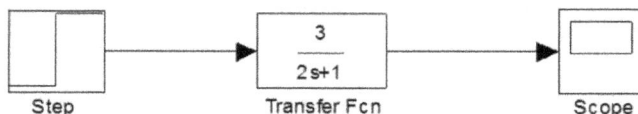

FIGURE 9.4 Simulink® block diagram of a 1st-order process.

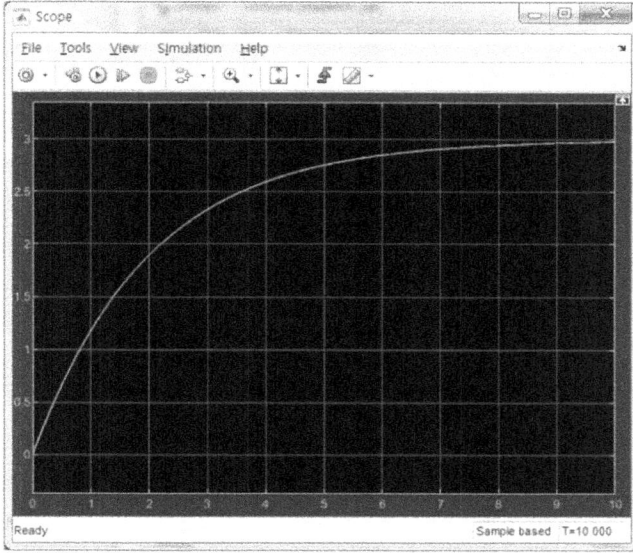

FIGURE 9.5 Step response curve in the Simulation Scope.

Solution

The script *sin1st* generates the response curve shown in Figure 9.6.

```
% sin1st.m: sinusoidal response of the 1st-order process
num = 3; den = [2 1]; G = tf(num,den);
t = [0:0.1:10]; u = sin(3*t); z = t*0; y = lsim(G,u,t);
plot(t,y,t,u,':'), legend('Output y(t)','Input u(t)'), hold on
plot(t,z), hold off, xlabel('t(sec)'), ylabel('Response y(t)')
title('Sinusoidal response of 1st-order process')
```

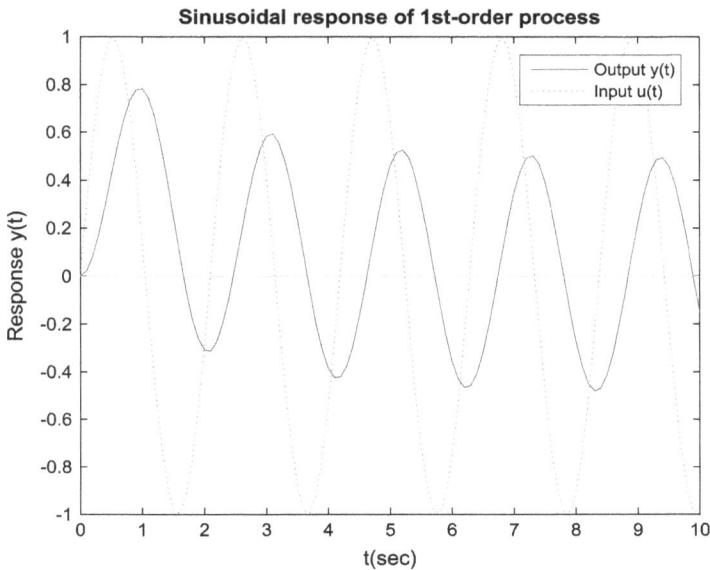

FIGURE 9.6 Sinusoidal response of a 1st-order process.

9.4.2 Dynamics of 2nd-Order Processes

A 2nd-order model is represented by a 2nd-order differential equation. The corresponding Laplace transform yields a transfer function whose denominator is a 2nd-order polynomial of s. The response of a 2nd-order process to various inputs can be found by similar methods to those used in 1st-order processes.

The general form of the transfer function of a 2nd-order process can be expressed as

$$G(s) = \frac{Y_s(s)}{X(s)} = \frac{K}{\tau^2 s^2 + 2\tau\zeta s + 1}$$

The response $Y(s)$ to the unit step input $X(s) = 1/s$ is given by

$$Y(s) = \frac{K}{s(\tau^2 s^2 + 2\tau\zeta s + 1)} = \frac{K}{\tau^2 s(s - r_1)(s - r_2)} = \frac{K}{\tau^2}\left\{\frac{A}{s} + \frac{B}{s - r_1} + \frac{C}{s - r_2}\right\}$$

where

$$r_1 = \frac{-\zeta + \sqrt{\zeta^2 - 1}}{\tau}, \ r_2 = \frac{-\zeta - \sqrt{\zeta^2 - 1}}{\tau}, \ A = \frac{1}{r_1 r_2}, \ B = \frac{1}{r_1(r_1 - r_2)}, \ C = \frac{1}{r_2(r_2 - r_1)}$$

The characteristics of the roots r_1 and r_2 depend on the decay ratio ζ:

1 $\zeta < 1$: The roots r_1 and r_2 are two complex conjugate poles. This situation is considered underdamped. The time-domain response $Y(t)$ is given by

$$Y(t) = K\left\{1 - \frac{1}{\sqrt{1 - \zeta^2}}e^{-\zeta t/\tau}\sin\left(\frac{\sqrt{1 - \zeta^2}}{\tau}t + \phi\right)\right\}, \ \phi = \tan^{-1}\left(\frac{\sqrt{1 - \zeta^2}}{\zeta}\right)$$

2 $\zeta = 1$: The roots $r_1 = r_2$ are two repeating poles and the time-domain response $Y(t)$ is given by

$$Y(t) = K\left\{1 - \left(1 + \frac{t}{\tau}\right)e^{-t/\tau}\right\}$$

3 $\zeta > 1$: In this case, the roots r_1 and r_2 are two distinct real poles, and the time-domain response $Y(t)$ is given by

$$Y(t) = K\left[1 - \frac{1}{2}e^{-\zeta t/\tau}\left\{\left(1 + \frac{\zeta}{\sqrt{\zeta^2 - 1}}\right)e^{\frac{\sqrt{\zeta^2-1}}{\tau}t} + \left(1 - \frac{\zeta}{\sqrt{\zeta^2 - 1}}\right)e^{-\frac{\sqrt{\zeta^2-1}}{\tau}t}\right\}\right]$$

For a sinusoidal input $X(t) = A\sin\omega t$, with $X(s) = A\omega/(s^2 + \omega^2)$, the output $Y(s)$ is given by

$$Y(s) = \frac{KA\omega}{(s^2 + \omega^2)(\tau^2 s^2 + 2\tau\zeta s + 1)}$$

The time-domain response $Y(t)$ can be expressed as

$$Y(t) = e^{-\zeta t/\tau}\left(C_1\cos\frac{\sqrt{1 - \zeta^2}}{\tau}t + C_2\sin\frac{\sqrt{1 - \zeta^2}}{\tau}t\right) + \frac{KA}{\sqrt{(1 - \tau^2\omega^2)^2 + (2\tau\zeta\omega)^2}}\sin(\omega t + \phi)$$

where

C_1 and C_2 are constants

$\phi = -\tan^{-1}(2\tau\omega\zeta/(1 - \tau^2\omega^2))$

As $t \to \infty$, the exponential term goes to zero and the output $Y(t)$ becomes a sinusoidal curve with constant frequency:

$$\lim_{t \to \infty} Y(t) = \frac{KA}{\sqrt{(1 - \tau^2\omega^2)^2 + (2\tau\zeta\omega)^2}} \sin(\omega t + \phi)$$

The step response of a 2nd-order process can be obtained by using the function step as in a 1st-order process. The response of a 2nd-order process to impulse, ramp, or other inputs can be found by similar methods to those used in a 1st-order process.

Example 9.11 Dynamics of Two Heated Tanks in Series[1]

Figure 9.7 shows two heated tanks in series. The two tanks are connected with pipes at the bottom and at a certain height H. If the liquid level of the first tank h_1 is greater than H, the liquid flows through both connecting pipes. If $h_1 < H$, the liquid flows through the bottom pipe only. Heats Q_1 and Q_2 are introduced into tanks from external heat sources. The cross-sectional areas of tank 1 and tank 2 are A_1 and A_2, respectively. From the material and energy balances for this system, we obtain

$$A_1 \frac{dh_1}{dt} = F_0 - F_{1t} - F_{1b}, \quad A_2 \frac{dh_2}{dt} = F_{1t} + F_{1b} - F_2$$

$$\frac{dT_1}{dt} = \frac{F_0}{A_1 h_1}(T_0 - T_1) + \frac{Q_1}{\rho C_p A_1 h_1}, \quad \frac{dT_2}{dt} = (F_{1t} + F_{1b})(T_1 - T_2) + \frac{Q_2}{\rho C_p A_2 h_2}$$

The flow rates in these equations are given by

$$F_{1b} = c_1\sqrt{h_1 - h_2}, \quad F_2 = c_2\sqrt{h_2}, \quad F_{1t} = \begin{cases} 0 & : h_1 \leq H \\ c_1\sqrt{h_1 - H} & : h_1 > H, h_2 \leq H \\ c_1\sqrt{h_1 - h_2} & : h_2 > H \end{cases}$$

where c_1 and c_2 are valve coefficients. Initially, both liquid tanks contain 100 l of water at 25°C. Generate time profiles of temperatures (T_1, T_2) and liquid levels (h_1, h_2) for $0 \leq t \leq 10$ (min) using the given data.

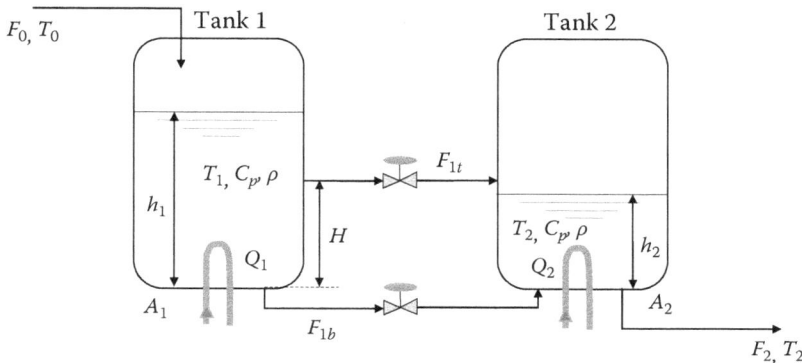

FIGURE 9.7   Schematic of two heated tanks in series.

Data: $A_1 = A_2 = 0.25 m^2$, $F_0 = 0.4 m^3/min$, $T_0 = 25°C$, $c_1 = c_2 = 0.6 m^{2.5}/min$, $\rho C_p = 4180 kJ/kg$, $Q_1 = Q_2 = 6000 kJ/min$, $H = 0.5 m$.

Solution

The function *LTmodel* defines the system of differential equations. The structure *ht* delivers the data required in the differential equations.

```
function dz = LTmodel(t,z,ht)
% z(1) = h1, z(2) = h2, z(3) = T1, z(4) = T2
% Retrieve data
A1 = ht.A1; A2 = ht.A2; F0 = ht.F0; T0 = ht.T0; H = ht.H;
c1 = ht.c1; c2 = ht.c2; rCp = ht.rCp; Q1 = ht.Q1; Q2 = ht.Q2;
h1 = z(1); h2 = z(2); T1 = z(3); T2 = z(4);
% Flow rates
F1b = c1*sqrt(h1 - h2); F2 = c2*sqrt(h2); dh = max(H,h1) - max(H,h2); F1t =
c1*sqrt(dh);
% Heat input
qi1 = Q1/(rCp*A1*h1); qi2 = Q2/(rCp*A2*h2);
% Differential equations
dz = [(F0 - F1t - F1b)/A1; (F1t + F1b - F2)/A2;
      (F0/A1/h1)*(T0 - T1) + qi1; (F1t + F1b)*(T1 - T2) + qi2];
end
```

The script *hybridtanks* defines the data and uses the built-in function *ode45* to solve the system of differential equations given by the function *LTmodel*. Time profiles of liquid levels and temperatures are shown in Figure 9.8.

```
% hybridtanks.m: hybrid two-tank heater system
% Data structure
ht.A1 = 0.25; ht.A2 = 0.25; ht.F0 = 0.4; ht.T0 = 25; ht.H = 0.5;
ht.c1 = 0.6; ht.c2 = 0.6; ht.rCp = 4180; ht.Q1 = 6000; ht.Q2 = 6000;
% Initial guesses
h10 = 0.4; h20 = 0.35; z0 = [h10; h20; 25; 25]; tspan = [0 10];
```

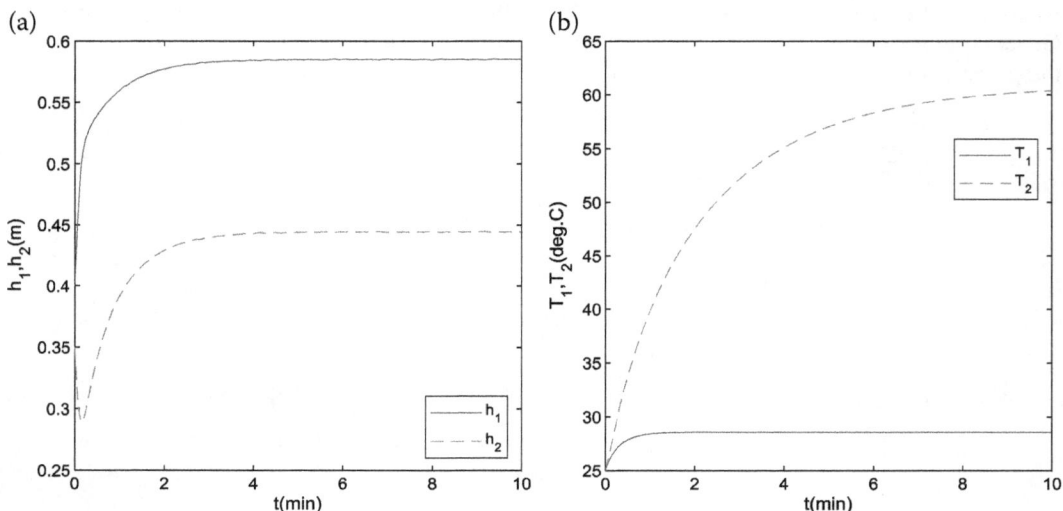

FIGURE 9.8 Profiles of (a) levels and (b) temperatures in the two-tank system.

```
% Solve DE system
[t,z] = ode45(@LTmodel,tspan,z0,[],ht); h1 = z(:,1); h2 = z(:,2); T1 = z(:,3);
T2 = z(:,4);
% Plot results
subplot(1,2,1), plot(t,h1,t,h2,'--'),xlabel('t(min)'), ylabel('h_1,h_2(m)')
legend('h_1','h_2','location','best')
subplot(1,2,2),  plot(t,T1,t,T2,'--'),xlabel('t(min)'),  ylabel('T_1,T_2
(deg.C)'), legend('T_1','T_2','location','best')
```

Example 9.12 Step Response of a 2nd-Order Process

In the 2nd-order process represented by $1/(\tau^2 s^2 + 2\tau\zeta s + 1)$, the value of the time constant is $\tau = 0.5$. Plot the step response curves when ζ is 0.5, 1.0, and 1.5 during the time interval [0, 10].

Solution

The script *step2ndpro* generates the desired curves as shown in Figure 9.9.

```
% step2ndpro.m
tau = 0.5; z1 = 0.5; z2 = 1; z3 = 1.5; num = 1; t = [0:0.1:10];
den1 = [tau*tau 2*tau*z1 1]; den2 = [tau*tau 2*tau*z2 1]; den3 = [tau*tau
2*tau*z3 1];
y1 = step(num, den1, t); y2 = step(num, den2, t); y3 = step(num, den3, t);
plot(t, y1, ':', t, y2, '--', t, y3), grid, xlabel('Time t(sec)'), ylabel
('Output y(t)')
title('Step responses of a 2nd-order process');
legend('Underdamped', 'Critically damped', 'Overdamped');
```

Simulink can be used to obtain the step response of a 2nd-order process. Consider the case when $\zeta = 0.3$. Since $\tau = 0.5$, the transfer function is given by $G(s) = 1/(0.25s^2 + 0.3s + 1)$. Connect blocks in the new editor window created by the MATLAB Simulink as shown in Figure 9.10.

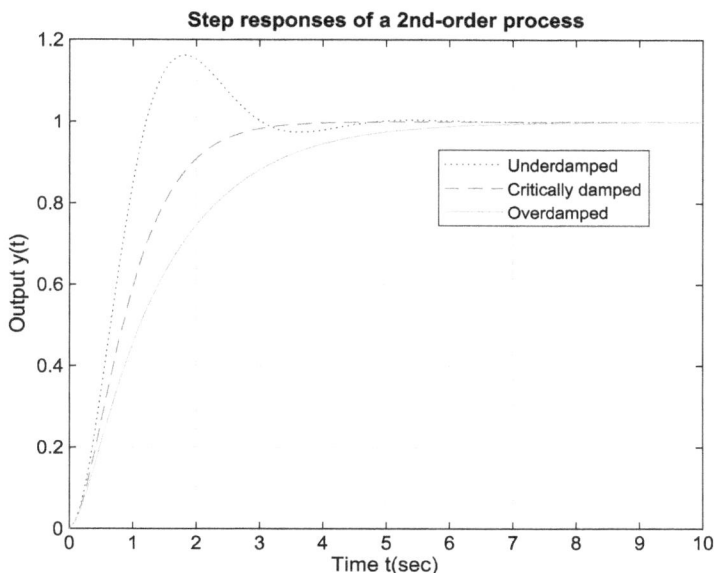

FIGURE 9.9 Step responses of a 2nd-order process.

Chemical Engineering Computation with MATLAB®

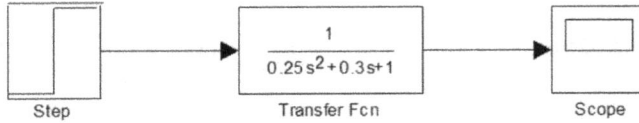

FIGURE 9.10 Simulink block diagram of a 2nd-order process.

Double-click on the "Transfer Fcn" block to set the transfer function, and double-click on the "Step" block to set the "Step time" to 0. In the "Simulation" in the editor window, select "Model Configuration Parameters" and set the calculation time to 10(sec). Select "Run" in the "Simulation" menu or click the "Run" icon (⊙) in the menu bar to get the step response curve. The result shown in Figure 9.11 can be obtained by double-clicking the "Scope" block.

9.4.3 DYNAMICS OF COMPLEX PROCESSES

9.4.3.1 Higher-Order Processes

Higher-order or complex processes are processes with interconnected 1st-order or 2nd-order processes, with time delays, or with complicated behaviour such as inverse responses. The general form of a higher-order process can be written as

$$G(s) = \frac{Y(s)}{X(s)} = \frac{K \prod_{j=1}^{m} (\tau_{dj} s + 1)}{\prod_{i=1}^{n} (\tau_{gi} s + 1)} (n > m)$$

The unit step $((X(s) = 1/s))$ response of a higher-order process can be represented as

$$Y(t) = K \left\{ 1 - \sum_{i=1}^{n} \frac{\prod_{j=1}^{m} (\tau_{gi} - \tau_{dj}) \tau_{gi}^{n-m-1}}{\prod_{j=1 (j \neq i)}^{m} (\tau_{gi} - \tau_{gj})} e^{-t/\tau_{gi}} \right\}$$

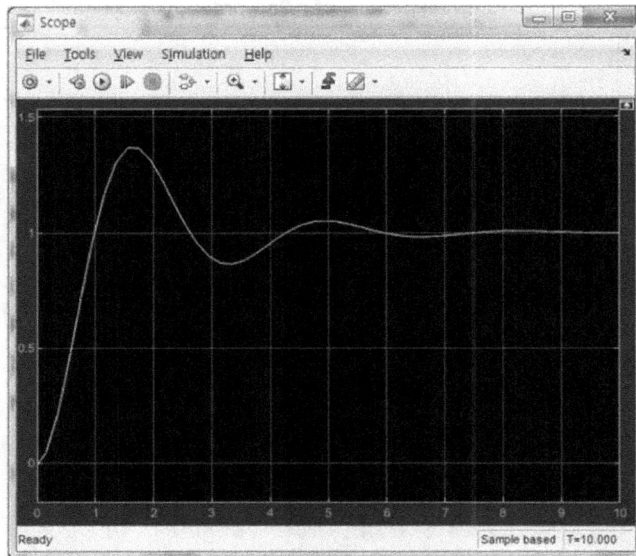

FIGURE 9.11 Step response displayed in the Scope window.

As with 1st- and 2nd-order processes, the step function can be used to get the step response of a higher-order process.

9.4.3.2 Lead/Lag

When $m = n = 1$, the transfer function of a higher-order process becomes

$$G(s) = \frac{Y(s)}{X(s)} = \frac{\tau_d s + 1}{\tau_g s + 1}$$

The term $1/(\tau_g s + 1)$ is called 1st-order lag, and the term $\tau_d s + 1$ is called 1st-order lead. The transfer function $G(s)$ is sometimes called lead/lag. The time-domain unit step response of a lead/lag is given by

$$Y(t) = 1 + \left(\frac{\tau_d}{\tau_g} - 1\right) e^{-t/\tau_g}$$

The built-in function *step* can be used to get the step response of a lead/lag.

9.4.3.3 Time Delay

Most chemical processes involve the movement of mass or energy, and there is a time delay (or dead time) associated with the movement. The transfer function of a time delay of magnitude θ is $e^{-\theta s}$. Thus, the transfer function of a 1st-order process with time delay is given by

$$Y(s) = \frac{Ke^{-\theta s}}{\tau s + 1}$$

and the transfer function of a 2nd-order process with time delay is given by

$$Y(s) = \frac{Ke^{-\theta s}}{(\tau_1 s + 1)(\tau_2 s + 1)}$$

In MATLAB, the time delay can be represented by using the 'iodelay' option. The basic syntax of this option is

```
delay = 1.6; set(G1, 'iodelay', delay);
```

Example 9.13 Step Response of a Higher-Order Process

Plot the unit step response curves for the higher-order process represented by $1/(2s + 1)^2$, $1/(2s + 1)^4$, $1/(2s + 1)^5$ during the time interval [0, 20].

Solution

The script *highresp* produces the step response curves of a higher-order process.

```
% highresp.m
num = 1; den1 = conv([2 1],[2 1]); den2 = conv(den1,den1); den3 = conv(den2,[2
1]); t = [0:0.1:20];
y1 = step(num, den1, t); y2 = step(num, den2, t); y3 = step(num, den3, t);
plot(t, y1, ':', t, y2, '--', t, y3), grid, xlabel('Time t(sec)'), ylabel
('Response y(t)')
title('Step response of higher-order process (n=2, 4, 5)'), legend('n = 2', 'n =
4', 'n = 5', 'location','best')
```

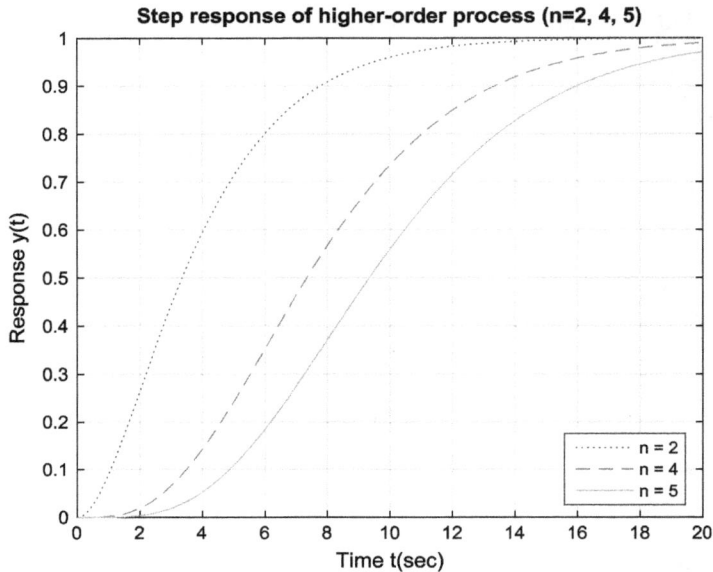

FIGURE 9.12 Step response curves of a higher-order process.

Figure 9.12 shows the step response curves of the higher-order process.

Example 9.14 Step Response of a 1st-Order Plus Time Delay Process

Plot the unit step response curve of a 1st-order process with time delay, the transfer function of which is given by

$$G_1(s) = \frac{Y(s)}{X(s)} = \frac{3e^{-1.6s}}{3s + 1}$$

Compare the result with the step response of the 4th-order process given by

$$G_2(s) = \frac{Y(s)}{X(s)} = \frac{3}{(0.1s + 1)(0.5s + 1)(s + 1)(3s + 1)}$$

Solution

The script *tdelay* uses the 'iodelay' option to generate the required plots, shown in Figure 9.13.

```
% tdelay.m
G1 = tf(3, [3 1]); delay = 1.6; set(G1, 'iodelay', delay);
G2 = tf(3, conv(conv([0.1 1], [0.5 1]), conv([1 1], [3 1])));
step(G1); hold on; step(G2, ':'); legend('G1', 'G2'), grid, hold off
```

The step response of a higher-order process $G_2(s)$ without time delay is very similar to that of a 1st-order process with time delay. This means that a higher-order process can be represented by a 1st-order process with time delay. In fact, many chemical processes can be effectively represented by a 1st-order-plus-time-delay model, which can be used in the computer control operation.

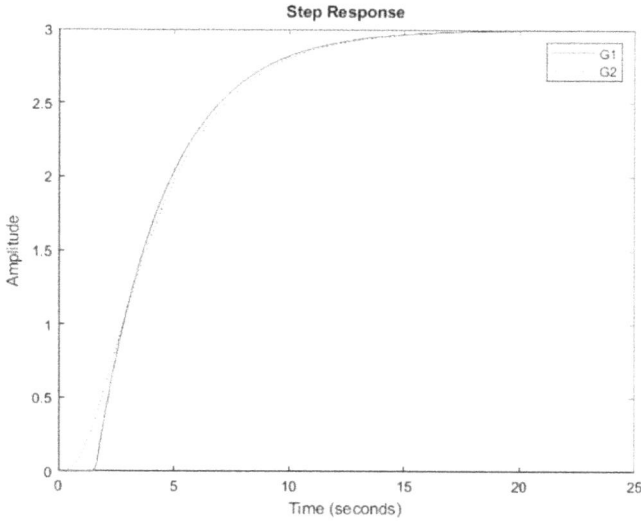

FIGURE 9.13　Step response of a process with time delay.

9.5　DYNAMICS OF FEEDBACK CONTROL LOOPS

9.5.1　SIMPLE FEEDBACK CONTROL LOOPS

Figure 9.14 shows a block diagram of a simple closed-loop feedback control system. Each variable is the Laplace transform of a deviation variable, and each block contains the transfer function. The overall transfer function can be written as

$$C = \frac{G_R G_c G_v G_T G_p}{1 + G_c G_v G_T G_p G_m} R + \frac{G_L}{1 + G_c G_v G_T G_p G_m} L$$

9.5.1.1　Servo Problem

The servo problem refers to closed-loop system behaviour for set-point changes only. In this case, no disturbance change occurs and $L = 0$. From Figure 9.14, it follows that

$$\frac{C}{R} = \frac{G_R G_c G_v G_T G_p}{1 + G_{OL}}, \quad G_{OL} = G_c G_v G_T G_p G_m$$

9.5.1.2　Regulator Problem

The regulator problem refers to closed-loop system behaviour for disturbance changes only. In this case, no set-point change occurs. Since $R = 0$, the overall closed-loop transfer function can be

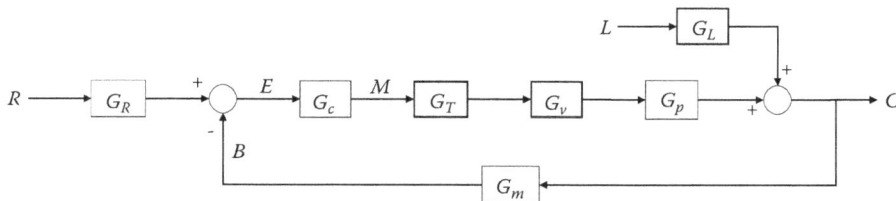

FIGURE 9.14　Block diagram of a closed-loop feedback control system.

rearranged to give the closed-loop transfer function for disturbance changes only. From Figure 9.14, it follows that

$$\frac{C}{L} = \frac{G_L}{1 + G_{OL}}$$

The step response of a closed-loop feedback control system can easily be obtained by using the built-in function *step,* especially when the process is 1st or 2nd order and a proportional controller is used.

Example 9.15 Step Response of a Feedback Control System

The overall closed-loop transfer function of a feedback control system is given by

$$C = \frac{K_c}{5s + 1 + K_c}R$$

Calculate and plot the closed-loop response to a unit step change in the set point for three values of the proportional controller gain: $K_c = 5$, 20, and 50.

Solution

The script *levelcon* calculates and plots the closed-loop response curves.

```
% levelcon.m
Kc1 = 5; Kc2 = 20; Kc3 = 50; t = [0:0.1:10]; num1 = Kc1; num2 = Kc2; num3 = Kc3;
den1 = [5 1+Kc1]; den2 = [5 1+Kc2]; den3 = [5 1+Kc3];
y1 = step(num1, den1, t); y2 = step(num2, den2, t); y3 = step(num3, den3, t);
plot(t, y1, ':', t, y2, '--', t, y3), xlabel('Time t(sec)'), ylabel('output y
(t)'), legend('Kc = 5', 'Kc = 20', 'Kc = 50')
```

Figure 9.15 shows the step response curves. We can see that the offset decreases as the controller gain increases.

Example 9.16 Step Responses for Proportional Control of a 2nd-Order Process

A proportional controller is to be used to control a 2nd-order process given by $G = 0.5/(s(0.5s + 1))$. Calculate and plot the closed-loop response to a unit step change in the set point for three values of the proportional controller gain: $K_c = 0.5$, 1, and 2. Assume that $K_v = K_m = 1$.

Solution

The closed-loop transfer function is given by

$$\frac{C}{R} = \frac{0.5K_c}{0.5s^2 + s + 0.5K_c}$$

The script *secpro* asks for the proportional controller gain and plots the closed-loop response curves.

```
% secpro.m : feedback control of 2nd-order process using P-controller
% Input controller gain
Kc = input('Controller gain = '); num = 0.5*Kc; den = [0.5 1 0.5*Kc];
% Unit step response
```

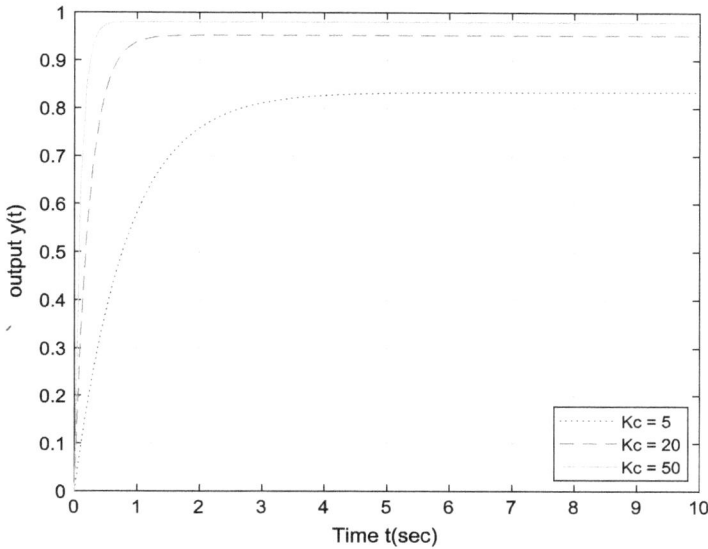

FIGURE 9.15   Step responses for a feedback control system.

```
t = [0:0.1:10]; y = step(num, den, t);
plot(t, y), grid, xlabel('Time t(sec)'), ylabel('Output C(t)');
```

The script *plotsecpro* calls the script *secpro* to input the controller gains and plot the step response curves.

```
% plotsecpro.m: calls secpro.m to plot step response curve
Secpro, hold on, secpro, secpro, hold off
text(2.1, 1.1, 'Kc=0.5'), text(4.1, 0.86, 'Kc=1'), text(5.1, 0.68, 'Kc=2')
```

The script *plotsecpro* generates the step response curves shown in Figure 9.16.

```
>> plotsecpro
Controller gain = 0.5
Controller gain = 1
Controller gain = 2
```

Figure 9.16 indicates that increasing K_c tends to speed up the response. But the closed-loop response becomes more oscillatory as K_c is increased. This means that the decay ratio becomes less than 1 when K_c is greater than a certain value.

Example 9.17 Step Response of a Feedback Control System Using a Proportional-Integral Controller

A proportional-integral (PI) controller is used in a feedback control system. The process transfer function is given by $G_p(s) = 5/((s + 1)(2s + 1))$, the gain of the control valve is $K_v = 0.01$, and the gain of the sensor/transducer is $K_m = 20$. Use Simulink to find the step response curve to a unit step change in the set point. The transfer function of the PI controller is given by $G_c(s) = 2(1 + 1/(5s))$.

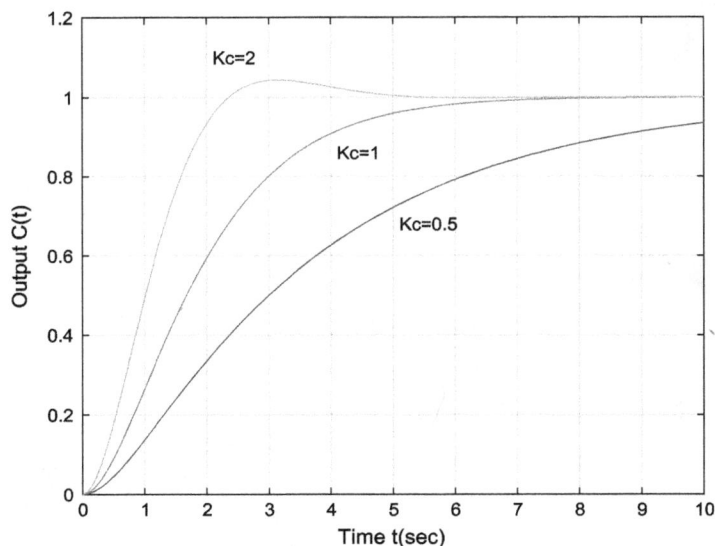

FIGURE 9.16   Step responses for feedback control of a 2nd-order process.

Solution

The Simulink block diagram is shown in Figure 9.17. In the "Simulation" menu of the model editor window, select "Model Configuration Parameters", and set the "Stop time" to 30(sec). The scope output is shown in Figure 9.18.

Example 9.18 Proportional Integral Control of a Batch Reactor[2]

The following exothermic consecutive reactions are carried out in a batch reactor fitted with a cooling coil through which cooling water is passed to remove the exothermic heat, as shown in Figure 9.19:

$$A \xrightarrow{k_1} B \xrightarrow{k_2} C$$

From the material balances for species A and B, we obtain

$$\frac{dC_A}{dt} = -k_1 C_A^2, \frac{dC_B}{dt} = k_1 C_A^2 - k_2 C_B$$

where
k_1 and k_2 are the reaction rate constants

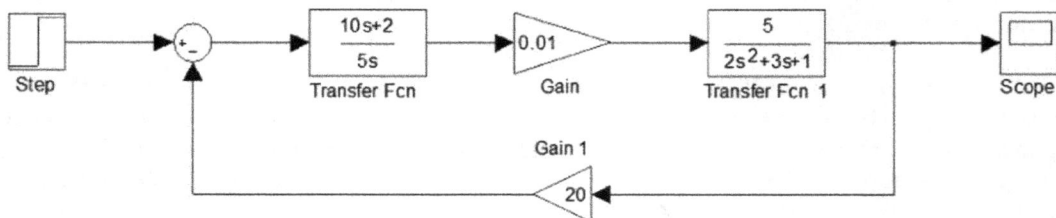

FIGURE 9.17   The Simulink block diagram for a feedback system using a PI controller.

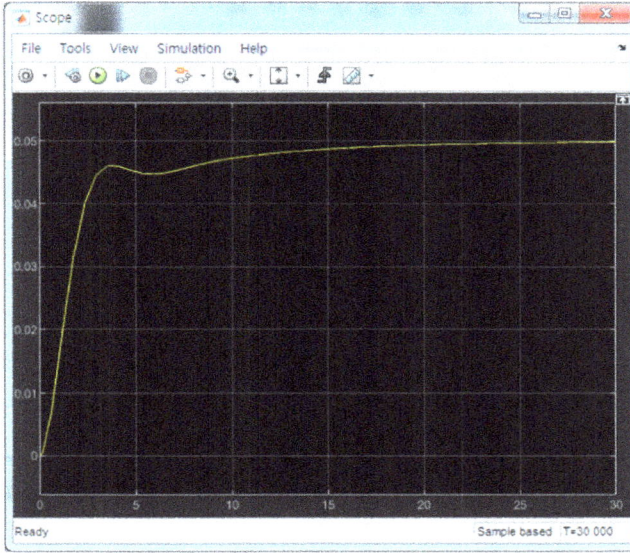

FIGURE 9.18 The scope output for a feedback system using a PI controller.

C_A and C_B are the concentrations of species A and B, respectively
k_1 and k_2 are represented as

$$k_1 = A_1 e^{-E_1/(RT)}, \; k_2 = A_2 e^{-E_2/(RT)}$$

The energy balance for the batch reactor gives

$$\frac{dT}{dt} = \frac{(-\Delta H_1)}{\rho C_p} k_1 C_A{}^2 + \frac{(-\Delta H_2)}{\rho C_p} k_2 C_B + \frac{U_j A_j}{\rho C_p V}(T_s - T) - \frac{U_c A_c}{\rho C_p V}(T - T_c)$$

where
$(-\Delta H_1)$ is the heat of reaction for $A \rightarrow B$
$(-\Delta H_2)$ is the heat of reaction for $B \rightarrow C$
T_s and T_c are the steam and coolant temperatures, respectively
U_j and U_c are the overall heat transfer coefficients of the jacket and coolant, respectively

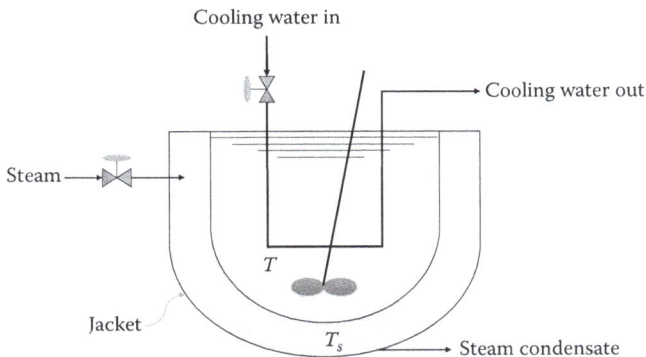

FIGURE 9.19 Schematic representation of a batch reactor control configuration.

U_c is assumed to be a function of the coolant flow rate F_c as

$$\frac{1}{U_c} = \frac{1}{4550 F_c^{0.8}} + \frac{1}{10.8}$$

For the present case, the reactor temperature should precisely followed the desired trajectory given by

$$T_d(t) = 54 + 71e^{-0.0025t}$$

One way to control $T(t)$ is to introduce a parameter u defined by

$$T_s = (T_{s,max} - T_{s,min})u + T_{s,min}, \quad U_c = (U_{c,min} - U_{c,max})u + U_{c,max}$$

$u = 0$ represents the maximum cooling and $u = 1$ the maximum heating of the system. Substituting these relations into the energy balance, followed by rearrangement, yields

$$\frac{dT}{dt} = \gamma_1 k_1 C_A^2 + \gamma_2 k_2 C_B + (a_1 + a_2 T) + (b_1 + b_2 T)u$$

where

$$\gamma_1 = \frac{(-\Delta H_1)}{\rho C_p}, \; \gamma_2 = \frac{(-\Delta H_2)}{\rho C_p}, \; a_1 = \frac{U_j A_j T_{s,min} + U_{c,max} A_c T_c}{\rho C_p V}, \; a_2 = -\frac{U_j A_j + U_{c,max} A_c}{\rho C_p V}$$

$$b_1 = \frac{U_j A_j (T_{s,max} - T_{s,min}) - (U_{c,max} - U_{c,min}) A_c T_c}{\rho C_p V}, \; b_2 = \frac{(U_{c,max} - U_{c,min}) A_c}{\rho C_p V}$$

A simple PI controller is used in the control, and u is determined by

$$u(t) = u_s + K_c \left\{ e(t) + \frac{1}{\tau_I} \int_0^t e(t)\,dt \right\}$$

where the control error $e(t)$ is given by $e(t) = T_d(t) - T(t)$ and the controller gain is arbitrarily chosen as $K_c = 0.1(°C^{-1})$ and $\tau_I = 360(sec)$. Plot C_A, C_B, F_c, T_d, T_s, and T as a function of reaction time t ($0 \le t \le 4000sec$).

Data: $C_{A0} = 1.0\,kmol/m^3$, $C_{B0} = 0.0\,kmol/m^3$, $A_1 = 1.1\,m^3/(kmol \cdot sec)$, $A_2 = 172.2sec^{-1}$, $E_1 = 2.09 \times 10^4\,kJ/kmol$, $E_2 = 4.18 \times 10^4\,kJ/kmol$, $(-\Delta H_1) = 4.18 \times 10^4\,kJ/kmol$, $(-\Delta H_2) = 8.36 \times 10^4\,kJ/kmol$, $\rho = 1000\,kg/m^3$, $T_c = 25°C$, $U_j = 1.16kJ/(m^2 \cdot °C \cdot sec)$, $U_{c,max} = 4.42\,kJ/(m^2 \cdot °C \cdot sec)$, $U_{c,min} = 1.39\,kJ/(m^2 \cdot °C \cdot sec)$, $T_{s,max} = 150°C$, $T_{s,min} = 70°C$, $R = 8.314\,kJ/(kmol \cdot K)$, $A_c/V = 17\,m^2/m^3$, $A_j/V = 30\,m^2/m^3$, $C_p = 1.0\,kJ/(kg \cdot °C)$, $u_s = 1.0$, $K_c = 0.1°C^{-1}$, $\tau_I = 360\,sec$.

Solution

The script *batPIcon* defines the data, sets the initial conditions, and uses the 4th-order Runge-Kutta method to solve the differential equation system.

```
% batPIcon.m: PI control for batch reactor
% Data
clear all;
A1 = 1.1; A2 = 172.2; E1 = 2.09e4; E2 = 4.18e4; R = 8.314;
dH1 = 4.18e4; dH2 = 8.36e4; rho = 1000; Cp = 1; Tc = 25;
Tsmax = 150; Tsmin = 70; Uj = 1.16; Ucmax = 4.42; Ucmin = 1.39;
AcV = 17; AjV = 30; Kc = 0.1; tauI = 360; us = 1; Tmax = 125;
dt = 0.1; tmax = 4000;
```

```
% Parameters
gam1 = dH1/(rho*Cp); gam2 = dH2/(rho*Cp);
a1 = (Uj*Tsmin*AjV + Ucmax*Tc*AcV)/(rho*Cp); a2 = -(Uj*AjV + Ucmax*AcV)/
(rho*Cp);
b1 = (Uj*AjV*(Tsmax - Tsmin) - (Ucmax - Ucmin)*AcV*Tc)/(rho*Cp);
b2 = (Ucmax - Ucmin)*AcV/(rho*Cp);
% Initialization
t = 0:dt:tmax+dt; n = length(t); T(1) = 25; Ca(1) = 1; Cb(1) = 0;
Td = 54 + 71*exp(-0.0025*t); % desired temperature trajectory
%Td(n+1) = 54 + 71*exp(-0.0025*(t(end)+dt));
er(1) = 100; erc(1) = 0; us = 1; u(1) = us;
Ts(1) = (Tsmax - Tsmin)*u(1) + Tsmin; Uc(1) = (Ucmin - Ucmax)*u(1) + Ucmax;
Fc(1) = (4550*(1/Uc(1) - 1/10.8))^(-1.25);
for k = 1:n-1
    % Begin 4th-order RK method
    T0 = T(k); Ca0 = Ca(k); Cb0 = Cb(k);
    k1 = gam1*A1*exp(-E1/R/(273.15+T0))*Ca0^2 +...
      gam2*A2*exp(-E2/R/(273.15+T0))*Cb0 + (a1+a2*T0) + (b1+b2*T0)*u(k);
    k11 = -A1*exp(-E1/R/(273.15+T0))*Ca0^2;
    k12 = A1*exp(-E1/R/(273.15+T0))*Ca0^2 - A2*exp(-E2/R/(273.15+T0))*Cb0;
    T1 = T(k) + k1*dt/2; Ca1 = Ca(k) + k11*dt/2; Cb1 = Cb(k) + k12*dt/2;
    k2 = gam1*A1*exp(-E1/R/(273.15+T1))*Ca1^2 +...
      gam2*A2*exp(-E2/R/(273.15+T1))*Cb1 + (a1+a2*T1) + (b1+b2*T1)*u(k);
    k21 = -A1*exp(-E1/R/(273.15+T1))*Ca1^2;
    k22 = A1*exp(-E1/R/(273.15+T1))*Ca1^2 - A2*exp(-E2/R/(273.15+T1))*Cb1;
    T2 = T(k) + k2*dt/2; Ca2 = Ca(k) + k21*dt/2; Cb2 = Cb(k) + k22*dt/2;
    k3 = gam1*A1*exp(-E1/R/(273.15+T2))*Ca2^2 +...
      gam2*A2*exp(-E2/R/(273.15+T2))*Cb2 + (a1+a2*T2) + (b1+b2*T2)*u(k);
    k31 = -A1*exp(-E1/R/(273.15+T2))*Ca2^2;
    k32 = A1*exp(-E1/R/(273.15+T2))*Ca2^2 - A2*exp(-E2/R/(273.15+T2))*Cb2;
    T3 = T(k) + k3*dt/2; Ca3 = Ca(k) + k31*dt/2; Cb3 = Cb(k) + k32*dt/2;
    k4 = gam1*A1*exp(-E1/R/(273.15+T3))*Ca3^2 +...
  gam2*A2*exp(-E2/R/(273.15+T3))*Cb3 + (a1+a2*T3) + (b1+b2*T3)*u(k);
    k41 = -A1*exp(-E1/R/(273.15+T3))*Ca3^2;
    k42 = A1*exp(-E1/R/(273.15+T3))*Ca3^2 - A2*exp(-E2/R/(273.15+T3))*Cb3;
    T(k+1) = T(k) + dt*(k1/6 + k2/3 + k3/3 + k4/6);
    Ca(k+1) = Ca(k) + dt*(k11/6 + k21/3 + k31/3 + k41/6);
    Cb(k+1) = Cb(k) + dt*(k12/6 + k22/3 + k32/3 + k42/6);
    % End of 4th-order RK method
    % PI control
    erc(k+1) = erc(k) + er(k)*dt; er(k+1) = Td(k+1) - T(k+1);
    u(k+1) = us + Kc*(er(k+1) + erc(k+1)/tauI);
    if u(k+1) >= 1, u(k+1) = 1; elseif u(k+1) <= 0, u(k+1) = 0; end
    % Calculate Ts, Uc and Fc
    Ts(k+1) = (Tsmax - Tsmin)*u(k+1) + Tsmin;
    Uc(k+1) = (Ucmin - Ucmax)*u(k+1) + Ucmax;
    Fc(k+1) = (4550*(1/Uc(k+1) - 1/10.8))^(-1.25);
end
subplot(1,2,1), plot(t,Ca,'--',t,Cb), xlabel('t(sec)'), ylabel('C(kmol/m^3)')
legend('C_A(t)','C_B(t)'), axis([0 4000 0 1])
subplot(1,2,2), plot(t,Td,'--',t,T,t,Ts,'.-'), xlabel('t(sec)'), ylabel
('Temp(deg.C)')
legend('Desired temp.','Reactor temp.','Steam temp.'), axis([0 4000 20 160])
```

The resultant concentration and temperature profiles are shown in Figure 9.20.

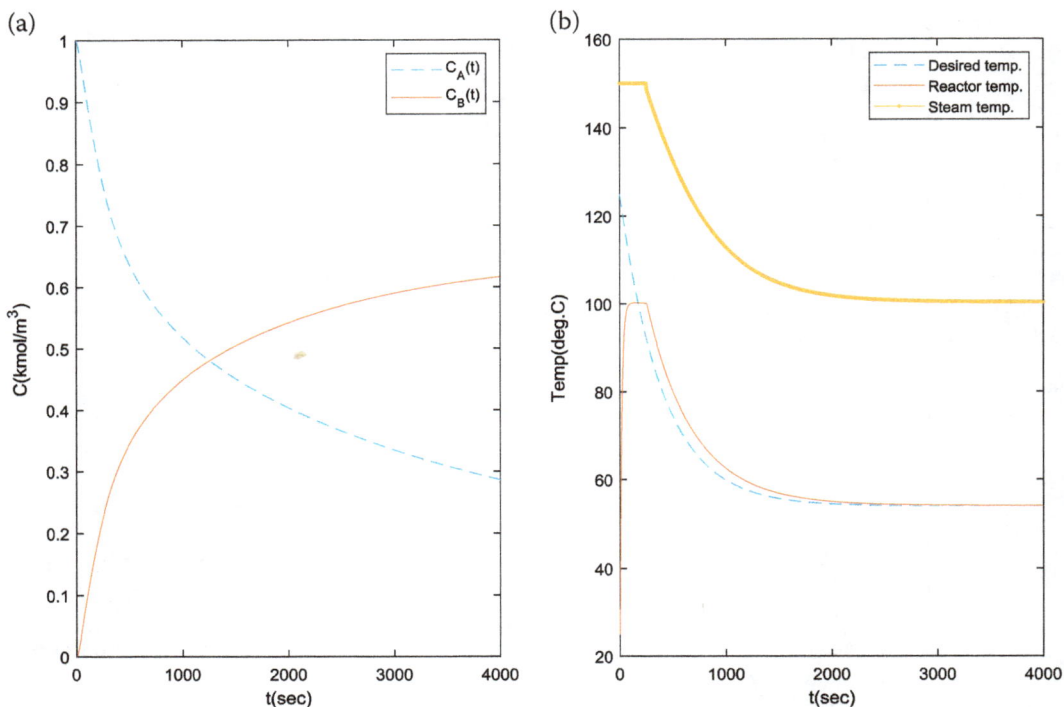

FIGURE 9.20 Profiles of (a) concentration and (b) temperature.

Example 9.19 Step Response of a Feedback Control System Using a PID Controller

Figure 9.21 shows a simple feedback control system where the process transfer function is $G_p(s) = 3/(4s^2 + s + 1)$ and a proportional-integral-derivative (PID) controller is used. The controller transfer function is $G_c(s) = K_c(1 + 1/(\tau_I s) + \tau_D s)$. Calculate and plot the closed-loop response to a unit step change in the set point for various values of τ_I and τ_D while the value of K_c is kept constant.

Solution

The script *tunpid* generates step response curves for different values of τ_D and τ_I, as shown in Figure 9.22.

```
% tunpid.m
tau = 2; zeta = 0.25; Kp = 3; Kc = 5; h(1,:) = '- '; h(2,:) = ': '; h(3,:) = '-.'; h
(4,:) = '--';
t0 = 0; delt = 0.1; fint = 20; ms = 1;
% Constant reset time (tauI)
tauI = 1; tauD = [0.5 1 5 10]; % try 4 different tauD
```

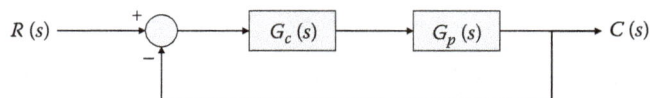

FIGURE 9.21 Simple feedback system.

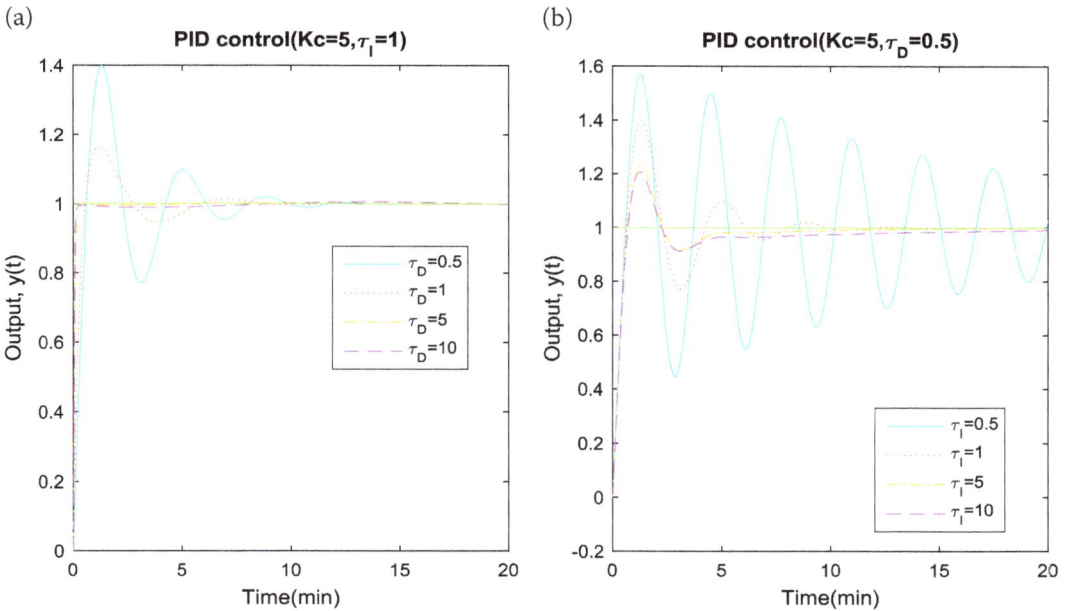

FIGURE 9.22   Effects of (a) derivative time and (b) reset time.

```
subplot(1,2,1)
for i = 1:length(tauD)
    num = Kc*Kp*[tauI*tauD(i) tauI 1]; d1 = tauI*tau^2;
    d2 = 2*tauI*tau*zeta+Kc*Kp*tauI*tauD(i); d3 = tauI*(1+Kc*Kp); d4 = Kc*Kp;
den = [d1 d2 d3 d4];
    [y,t] = stepnp(num, den, t0, delt, fint, ms); plot(t,y,h(i,:)), hold on
end
st = 1+0*t; plot(t,st), hold off, legend('\tau_D=0.5','\tau_D=1','\tau_D=5','
\tau_D=10','location','best')
xlabel('Time(min)'),  ylabel('Output,  y(t)'),  title('PID  control(Kc=5,
\tau_I=1)')
% Constant derivative time (tauD)
tauD = 0.5; tauI = [0.5 1 5 10]; % try 4 different tauI
subplot(1,2,2)
for i = 1:length(tauI)
    num = Kc*Kp*[tauI(i)*tauD tauI(i) 1]; d1 = tauI(i)*tau^2;
    d2 = 2*tauI(i)*tau*zeta+Kc*Kp*tauI(i)*tauD;
    d3 = tauI(i)*(1+Kc*Kp); d4 = Kc*Kp; den = [d1 d2 d3 d4];
    [y,t] = stepnp(num, den, t0, delt, fint, ms); plot(t,y,h(i,:)), hold on
end
st = 1+0*t; plot(t,st), hold off, legend('\tau_I=0.5','\tau_I=1','\tau_I=5','
\tau_I=10','location','best')
xlabel('Time(min)'),  ylabel('Output,  y(t)'),  title('PID  control(Kc=5,
\tau_D=0.5)')
```

The script *tunpid* calls the function *stepnp* to get the step responses. The function *stepnp* calculates step responses of proper single-input, single-output systems.

```
function [y,t] = stepnp(num, den, t0, delt, fint, ms);
```

```
% Calculate step response of a proper SISO system
% num : numerator of transfer function
% den : denominator of transfer function
% t0 : time at which unit step input is introduced
% delt : time step
% fint : final response time
% y : step response
% ms: step size
% Partial fraction of (transfer function)*(step input)
% (r: residue vector, p: pole vector, k: constant vector)
[r, p, k] = residue(num, conv(den, [1 0]));
% Set calculation time interval
t = t0 : delt : fint;
% Identify pole multiplicity
for j = 1 : size(p)
    n = 1;
    for i = 1 : size(p)
        if p(j) == p(i), if (i ~= j), n = n+1; end; end
  end
  mult(:, j) = n;
end
% Step response: use inverse Laplace transform
y = zeros(size(t));
j = 1;
while j <= size(p, 1)
    for i = 1 : mult(:, j), y = y + r(j+i-1)*((t-t0).^(i-1)).*exp(p(j)*(t-t0))/
factorial(i-1); end
    j = j + i;
end
y = ms*y;
end
```

9.5.2 CONTROL OF A CONTINUOUS-STIRRED TANK HEATER

Figure 9.23 shows a continuous-stirred tank heating process. The energy balance for the continuous-stirred tank heater with a constant flow rate can be expressed as

$$\rho C_p V \frac{dT}{dt} = w C_p (T_i - T) + Q$$

where
  ρ is the fluid density (kg/m^3)
  C_p is the fluid heat capacity ($kJ/(kg \cdot °C)$)
  V is the volume of the heating tank (m^3)
  w is the mass flow rate of the fluid (kg/min)
  Q is the heat flux from the heating source (kJ/min)
  Rearrangement of the energy balance equation yields

$$\frac{dT}{dt} = \frac{1}{\tau}(T_i - T) + \frac{K}{\tau}Q$$

where $\tau = \rho V/w$ and $K = 1/(wC_p)$. At steady state,

$$Q_s = \frac{1}{K}(T_s - T_{is}) = \frac{1}{K}(T_R - T_{is})$$

where T_R is the set point. It is assumed that $T_s = T_R$ at steady state.

Since the thermocouple is located at the outflow pipe a short distance downstream, the thermocouple will exhibit a time delay, θ, which is the time required for the output flow to reach the measurement point. Thus, the temperature sensed by the thermocouple can be represented as

$$T_1(t) = T(t - \theta)$$

To handle the time delay, we can use the 1st-order Padé approximation:

$$T_1(s) = e^{-\theta s}T(s) \approx \frac{1 - \theta s/2}{1 + \theta s/2}T(s)$$

The inverse Laplace transform on $T_1(s)$ gives

$$\frac{dT_1}{dt} = \frac{2}{\theta}\left(T - T_1 - \frac{\theta}{2}\frac{dT}{dt}\right)$$

If we assume that the thermocouple exhibits 1st-order dynamics with time constant τ_t and unity gain, the dynamics of the temperature T_t measured by the thermocouple can be represented as

$$\frac{T_t(s)}{T_1(s)} = \frac{1}{\tau_t s + 1} \quad \text{or} \quad \frac{dT_t}{dt} = \frac{1}{\tau_t}(T_1 - T_t)$$

If a PI controller is used, the controller output can be represented in terms of the heat rate Q supplied from the heater:

$$Q(t) = Q_s + K_c(T_R - T_t) + \frac{K_c}{\tau_I}\int_0^t (T_R - T_t)dt$$

where

$K_c(kJ/(min \cdot °C))$ is the proportional gain of the controller

$\tau_I(min)$ is the reset time

The accumulated control error is given by

$$C_e(t) = \int_0^t (T_R - T_t)dt$$

Differentiation of this equation gives

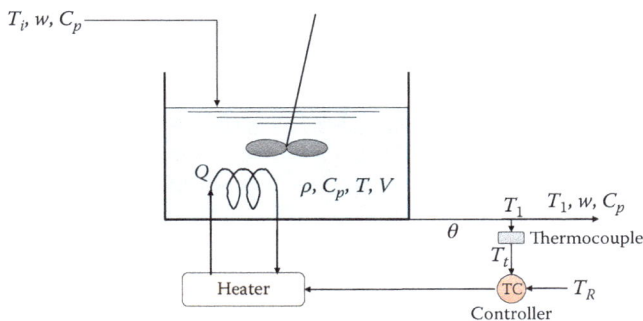

FIGURE 9.23 A continuous-stirred tank heating process.

$$\frac{dC_e(t)}{dt} = T_R - T_t$$

Substitution of these two equations into the controller output equation yields

$$Q(t) = Q_s + \frac{K_c}{\tau_I}C_e(t) + K_c\frac{dC_e(t)}{dt}$$

If a proportional controller is used, the controller output is given by

$$Q(t) = Q_s + K_c(T_R - T_t)$$

Example 9.20 Control of a Stirred Tank Heating Process

A continuous-stirred tank heating process consists of a stirred tank, heater, and PI controller. The liquid feed with density $\rho = 980\ kg/m^3$ and heat capacity of $C_p = 1.6\ kJ/(kg \cdot °C)$ flows into the heated tank at a constant mass flow rate of $w = 250\ kg/min$ and temperature $T_i = 50°C$. The volume of the tank is $V = 2.5\ m^3$. This stream is to be heated to a higher set-point temperature $T_R = 75°C$. The outlet temperature is measured by a thermocouple as T_t, and the heat flux $Q\ (kJ/min)$ supplied by the heater is adjusted by a PI controller with $K_c = 60\ kJ/(min·°C)$ and $\tau_I = 1.8\ min$. The thermocouple exhibits 1st-order dynamics with time constant $\tau_t = 3\ min$ and unity gain. The time delay between the heating tank and the thermocouple is $\theta = 1.2\ min$.

(1) The system is initially operating at steady state at a temperature of 75°C (set point). The inlet temperature T_i is suddenly changed to 30°C at time $t = 10\ min$. Assuming no control actions (open loop, $K_c = 0$), plot the profiles of the fluid temperature in the tank T, the measured temperature T_t, and the fluid outlet temperature T_1.

(2) If the controller is engaged (closed loop), plot the profiles of T, T_t, and T_1.

(3) If the PI controller is replaced by a proportional (P) controller with $K_c = 60\ kJ/(min·°C)$,
plot
   the profiles of T, T_t, and T_1.

Solution

The following system of differential equations is to be solved:

$$\frac{dT}{dt} = \frac{1}{\tau}(T_i - T) + \frac{K}{\tau}Q, \quad \frac{dT_1}{dt} = \frac{2}{\theta}\left(T - T_1 - \frac{\theta}{2}\frac{dT}{dt}\right)$$

$$\frac{dT_t}{dt} = \frac{1}{\tau_t}(T_1 - T_t), \quad \frac{dC_e(t)}{dt} = T_R - T_t, \quad Q(t) = Q_s + \frac{K_c}{\tau_I}C_e(t) + K_c\frac{dC_e(t)}{dt}$$

(1) The change in the inlet temperature can be represented as

$$T_i(t) = \begin{cases} T_{i1} : t < 10 \\ T_{i2} : t \geq 0 \end{cases}$$

where $T_{i1} = 0°C$ and $T_{i2} = 30°C$. The subfunction *htf* defines the system of differential equations. The script *csthcont* employs the built-in function *ode45* to solve the system of differential equations and produces the outputs shown in Figure 9.24.

```
% csthcont.m
% z(1) = T, z(2) = T1, z(3) = Tt, z(4) = Ce
rho = 980; V = 2.5; Cp = 1.6; w = 250; taut = 3.6; theta = 1.2; Tr = 75;
```

```
Ti = 50; tauI = 1.8; K = 1/(Cp*w); tau = rho*V/w; Qs = (Tr-Ti)/K;
z0 = [Tr Tr Tr 0]; % Integral time interval and initial conditions
tv = [0 60]; Kc = 0;
[t z] = ode45(@htf,tv,z0,[],tau,taut,tauI,K,Kc,theta,Qs,Ti,Tr);
T = z(:,1); T1 = z(:,2); Tt = z(:,3);
plot(t,T,t,Tt,':',t,T1,'.-'), xlabel('t(min)'), ylabel('T(C)')
legend('Tank','Thermocouple','Outlet','location','best')
function dz = htf(t,z,tau,taut,tauI,K,Kc,theta,Qs,Ti,Tr)
% z(1) = T, z(2) = T1, z(3) = Tt, z(4) = Ce
if t < 10, Ti = 50; else Ti = 30; end
Q = Qs + (Kc/tauI)*z(4) + Kc*(Tr - z(3));
dz(1,1) = (Ti - z(1))/tau + K*Q/tau;
dz(2,1) = 2*(z(1) - z(2) - (theta/2)*dz(1,1))/theta;
dz(3,1) = (z(2) - z(3))/taut; dz(4,1) = Tr - z(3);
end
```

(2) In the script *csthcont*, set values of Kc and tv as Kc = 60; tv = [0 200]; The script produces the output shown in Figure 9.25.

(3) When a proportional controller is used, the heat flux is given by

$$Q(t) = Q_s + K_c(T_R - T_t)$$

The subfunction *hpf* defines the system of differential equations. The script *csthcontp* uses the built-in function *ode45* to solve the system of differential equations and generate the curves shown in Figure 9.26. As can be seen, the closed responses exhibit oscillatory behaviour for very large K_c.

```
% csthcontp.m
% z(1) = T, z(2) = T1, z(3) = Tt
rho = 980; V = 2.5; Cp = 1.6; w = 250; taut = 3.6; theta = 1.2; Tr = 75;
Ti = 50; Kc = 60; K = 1/(Cp*w); tau = rho*V/w; Qs = (Tr-Ti)/K;
```

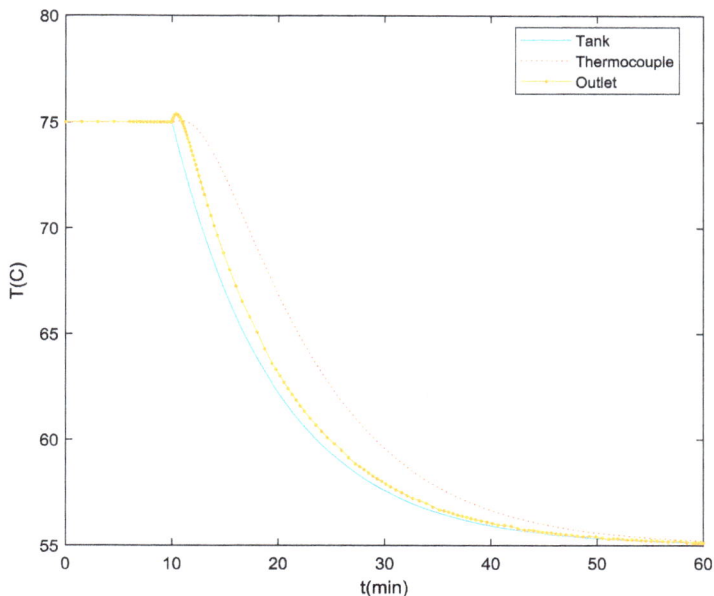

FIGURE 9.24 Temperature profiles.

```
tv = [0 60]; z0 = [Tr Tr Tr]; [t z] = ode45(@hpf,tv,z0,[],tau,taut,K,Kc,theta,
Qs,Ti,Tr);
T = z(:,1); T1 = z(:,2); Tt = z(:,3);
plot(t,T,t,Tt,':',t,T1,'.-'),  xlabel('t(min)'),  ylabel('T(C)'),  legend
('Tank','Thermocouple','Outlet')
function dz = hpf (t,z,tau,taut,K,Kc,theta,Qs,Ti,Tr)
% z(1) = T, z(2) = T1, z(3) = Tt
if t < 10, Ti = 50; else Ti = 30; end
Q = Qs + Kc*(Tr - z(3));
dz(1,1) = (Ti - z(1))/tau + K*Q/tau;
dz(2,1) = 2*(z(1) - z(2) - (theta/2)*dz(1,1))/theta;
dz(3,1) = (z(2) - z(3))/taut;
end
```

9.6 STABILITY OF FEEDBACK CONTROL SYSTEMS

A system is stable if the output response is bounded for any bounded input. The stability of a feedback control system can be analysed by using the characteristic equation. The overall transfer function of a closed-loop feedback control system is given by

$$C = \frac{G_R G_c G_v G_T G_p}{1 + G_c G_v G_T G_p G_m}R + \frac{G_L}{1 + G_c G_v G_T G_p G_m}L$$

The characteristic equation is

$$1 + G_c G_v G_T G_p G_m = 1 + G_{OL} = 0$$

The roots of the characteristic equation are called the poles of the closed-loop system. The closed-loop control system is stable if all the roots of the characteristic equation have negative real parts. In other words, the closed-loop system is stable if all the poles of the closed-loop transfer function

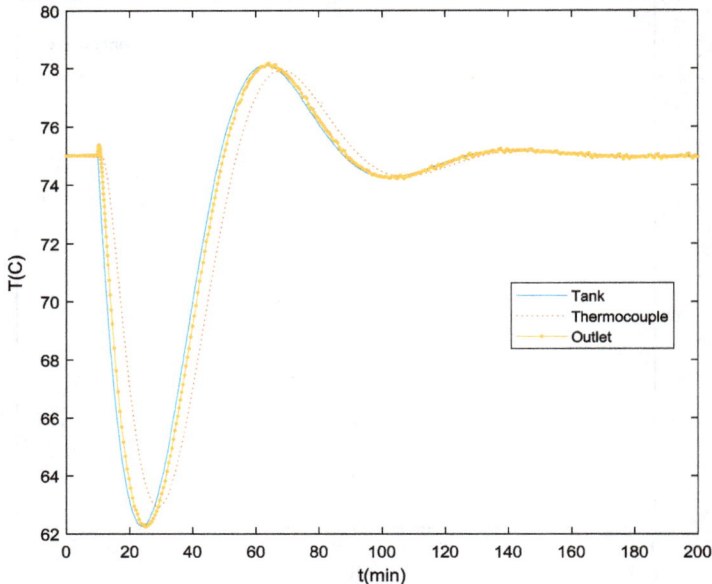

FIGURE 9.25   Response of a stirred tank heating process (PI controller).

(a)

(b)

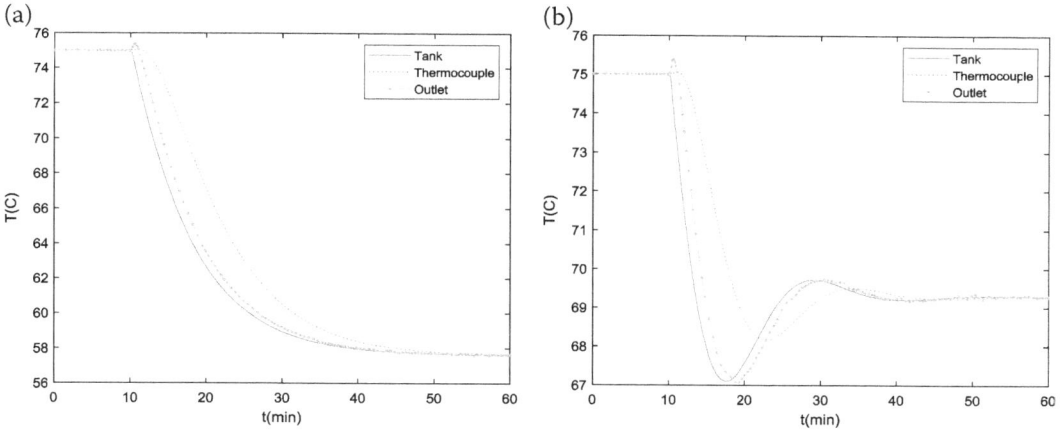

FIGURE 9.26   Response of a stirred tank heating process (P controller): (a) $K_c = 60$, (b) $K_c = 1000$.

lie in the left half-plane. If some roots of the characteristic equation lie on the imaginary axis, the real parts are zero and these roots can be represented as $r_{1,2} = \pm i\omega_u$. The response corresponding to these roots can be expressed as

$$C(s) = \frac{As + B}{s^2 + \omega_u^2} + (\cdots) = M\sin(\omega_u t + \phi) + (\cdots)$$

The response exhibits sinusoidal oscillations with amplitude M and frequency ω_u. The frequency ω_u is called the ultimate frequency, and the corresponding period $T_u = 2\pi/\omega_u$ is called the ultimate period.

Since the stability of a closed-loop feedback control system depends on the poles of the system, the stable region of the system can be represented graphically by displaying the poles on the complex plane. All of the poles must lie to the left of the imaginary axis in the complex plane for a system to be stable. A root locus diagram is a trajectory of poles and shows how the poles change when a parameter such as controller gain changes. By representing the poles for various values of the controller gain K_c on the complex plane, we can see the behaviour of the poles as a function of K_c.

A root locus diagram can easily be constructed by using the built-in function *rlocus*. The numerator and denominator of the transfer function should be supplied as input arguments to the *rlocus* function.

Example 9.21 Root Locus (1)

Plot the trajectory of poles when the controller gain K_c changes from 1 to 40 for the system with the characteristic equation given by

$$s^3 + 6s^2 + 11s + 6 + 2K_c = 0$$

Solution
The script *rootloc* plots the root locus diagram.

```
% rootloc.m
Kc = [0:1:40];
for i = 1:length(Kc)
    pol = [1 6 11 6+2*Kc(i)]; sol = roots(pol);
```

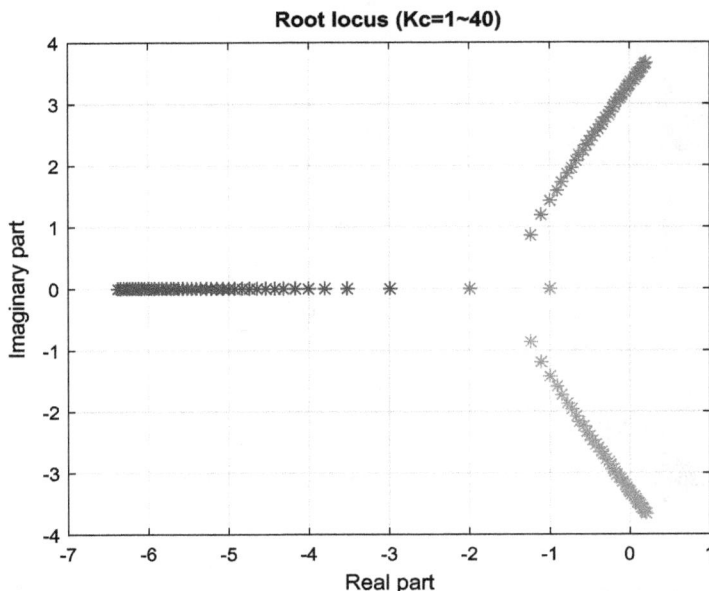

FIGURE 9.27   Locus of poles when the controller gain changes.

```
    for j = 1:length(sol), r(i,j) = sol(j); end
end
plot(r,'*'), grid, ylabel('Imaginary part'), xlabel('Real part'), title('Root
locus (Kc=1~40)')
```

When the controller gain is greater than 30, some of the poles lie to the right of the imaginary axis in the complex plane and the system becomes unstable. The script *rootloc* generates the locus of poles shown in Figure 9.27.

Example 9.22 Root Locus (2)

Plot the root locus diagram for a system with the characteristic equation given by

$$1 + \frac{K(s+3)}{s^2 + 2s} = 0$$

Solution

The numerator and denominator of the open-loop transfer function are $s + 3$ and $s^2 + 2s$, respectively. These are supplied to the built-in function *rlocus* as input arguments. The function sets the range of K automatically to construct the root locus diagram shown in Figure 9.28.

```
>> num = [1 3]; den = [1 2 0]; rlocus(num, den);
>> grid, title('Root locus'), xlabel('Real axis'), ylabel('Imaginary axis')
```

9.7 FREQUENCY RESPONSE ANALYSIS

The response of a stable system at large times is characterised by its amplitude and phase shift when the input is a sinusoidal wave. Consider a stable process whose transfer function is given by

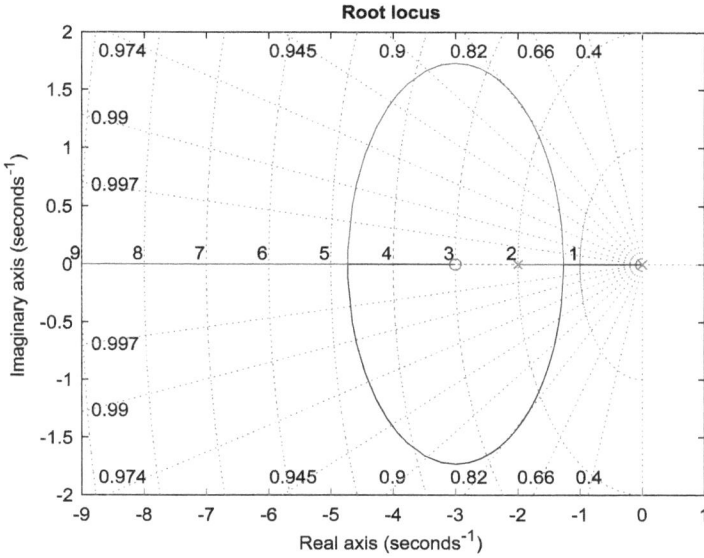

FIGURE 9.28 Root locus.

$$G(s) = \frac{Y(s)}{X(s)} = \frac{N(s)}{(s - r_1)(s - r_2) \cdots (s - r_n)}$$

The response to a general sinusoidal input, $x(t) = A\sin(\omega t)$, or $X(s) = A\omega/(s^2 + \omega^2)$, can be expressed as

$$Y(t) = a_1 e^{r_1 t} + a_2 e^{r_2 t} + \cdots + a_n e^{r_n t} + C\cos\omega t + \frac{D}{\omega}\sin\omega t$$

where C and D are constants. Since the system is stable, the real parts of all the roots $r_1, r_2, \cdots, r_n$ are negative. Thus, if the sinusoidal input is continued for a long time, the exponential terms will decay away and the time response will become a purely sinusoidal function represented by cosine and sine terms. The remaining cosine and sine terms can be combined to yield

$$Y(t) = AI\cos\omega t + \frac{AR\omega}{\omega}\sin\omega t = A\sqrt{R^2 + I^2}\sin(\omega t + \phi) = \hat{A}\sin(\omega t + \phi)$$

where $\hat{A} = A\sqrt{R^2 + I^2}$ and $\phi = \tan^{-1}(I/R)$. The amplitude ratio (AR) is given by

$$AR = \frac{\hat{A}}{A} = \frac{A\sqrt{R^2 + I^2}}{A} = \sqrt{R^2 + I^2} = |G(i\omega)|$$

We can see that the amplitude ratio is the magnitude of the complex number obtained by setting $s = i\omega$ in $G(s)$.

9.7.1 Bode Diagram

The Bode diagram is a graphical representation of the frequency response characteristics of a transfer function model. The Bode diagram consists of a log-log plot of the AR versus ω (radians/time) and a semilog plot of the phase angle ϕ versus ω. The built-in function *bode* constructs the Bode plot and calculates the amplitude ratio and the phase angle.

Example 9.23 Bode Diagram of a 2nd-Order Process

Plot the Bode diagram of a 2nd-order process $G(s) = 1/(2.25s^2 + 3\zeta s + 1)$ when $\zeta = 0, 0.25,$ 0.5, 0.75, and 1.

Solution

The script *bodezeta* generates the Bode plot for each value of ζ, as shown in Figure 9.29:

```
% bodezeta.m
w = logspace(-2, 1, 300); zeta = [0:0.25:1]; num = 1;
for i = 1:length(zeta),
    den = [2.25 3*zeta(i) 1]; [ar, phase] = bode(num, den, w);
    lar = 20*log10(ar);
    subplot(2,1,1), loglog(w,ar), xlabel('w(rad/min)'), ylabel('AR'), title
('Response of 2nd-order process')
    text(0.8, 20, 'zeta=0'), text(0.4, 0.2,'zeta=1'), grid, hold on
    subplot(2,1,2), semilogx(w, phase), grid, xlabel('w(rad/min)'), ylabel
('\phi'), hold on
end
hold off
```

Example 9.24 Bode Diagram of a 3rd-Order Process

Plot the Bode diagram of a 3rd-order process with time delay:

$$G(s) = \frac{(0.4s + 1)e^{-0.2s}}{(0.3s + 1)(s + 1)^2}$$

Solution

The script *delaybode* generates the Bode plot for the 3rd-order process with time delay. In the script, the option 'iodelay' is used to handle the time delay. The script *delaybode* produces the Bode plot shown in Figure 9.30.

```
% delaybode.m
num = [0.4 1]; den = conv([0.3 1], conv([1 1], [1 1]));
G = tf(num, den, 'iodelay', 0.2); [mag, phase, w] = bode(G);
subplot(2,1,1), loglog(w, squeeze(mag)), grid, ylabel('Amplitude'), xlabel
('Frequency(rad/time)')
subplot(2,1,2), semilogx(w, squeeze(phase)), grid, ylabel('Phase(deg)'),
xlabel('Frequency(rad/time)')
```

9.7.2 NYQUIST DIAGRAM

A Nyquist diagram, an alternative representation of frequency response information, is a graphical representation of the real and imaginary parts of $G(i\omega)$ on the s plane, with ω as the parameter. Because a complex number can be put in polar coordinates, the Nyquist diagram is also referred as the polar plot of $G(i\omega)$. The Bode diagram is easier to interpret, whereas the Nyquist diagram is more widely used in multiloop or multivariable analysis. The built-in functions *nyquist* and *polar* can both be used to construct the Nyquist diagram.

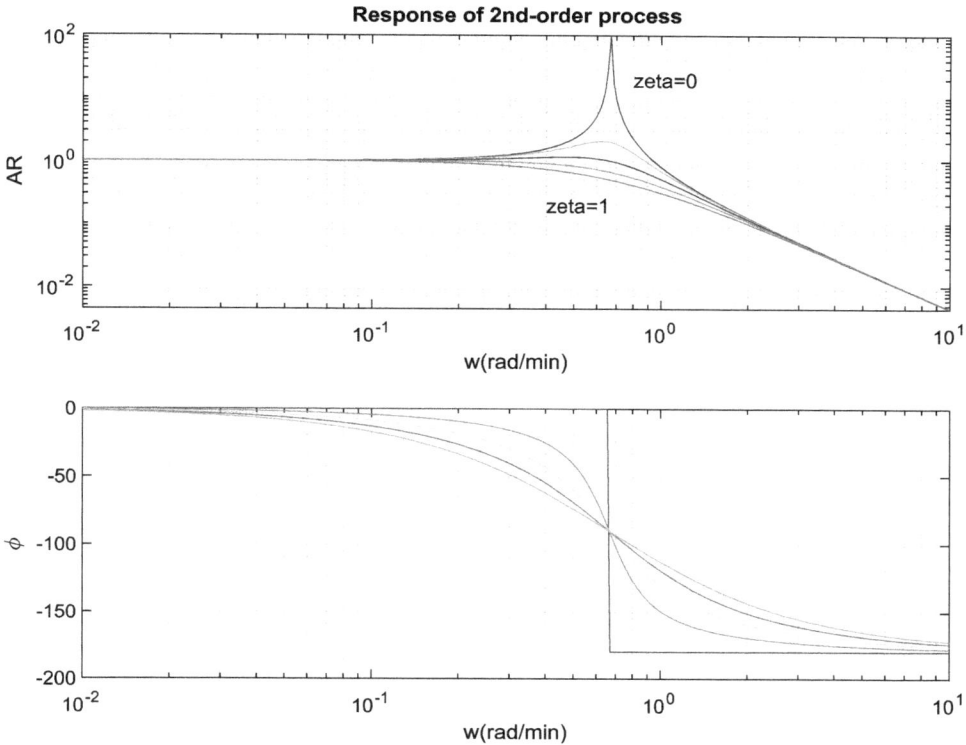

FIGURE 9.29 Bode diagram for a 2nd-order process.

Example 9.25 Nyquist Diagram of a 2nd-Order Process

Plot the Nyquist diagram for a 2nd-order process whose transfer function is

$$G(s) = \frac{2}{(10s + 1)(2.5s + 1)}$$

Solution

Both the built-in functions *polar* and *nyquist* can be used. The script *plotnyquist* uses the function *polar* and *nyquist* to generate the Nyquist diagrams shown in Figure 9.31. The Nyquist diagram can also be constructed by plotting the outputs obtained from the function *nyquist*.

```
% plotnyquist
w = logspace(-2, 1, 300); num = 2; den = conv([10 1], [2.5 1]); [x, p] = bode(num,
den, w);
subplot(1,2,1), polar((pi/180)*p, x), title('polar plot of a 2nd-order
process')
subplot(1,2,2), nyquist(num, den)
```

Example 9.26 Nyquist Diagram of a Time Delay

Plot the Nyquist diagram for a 1st-order process with a time delay whose transfer function is

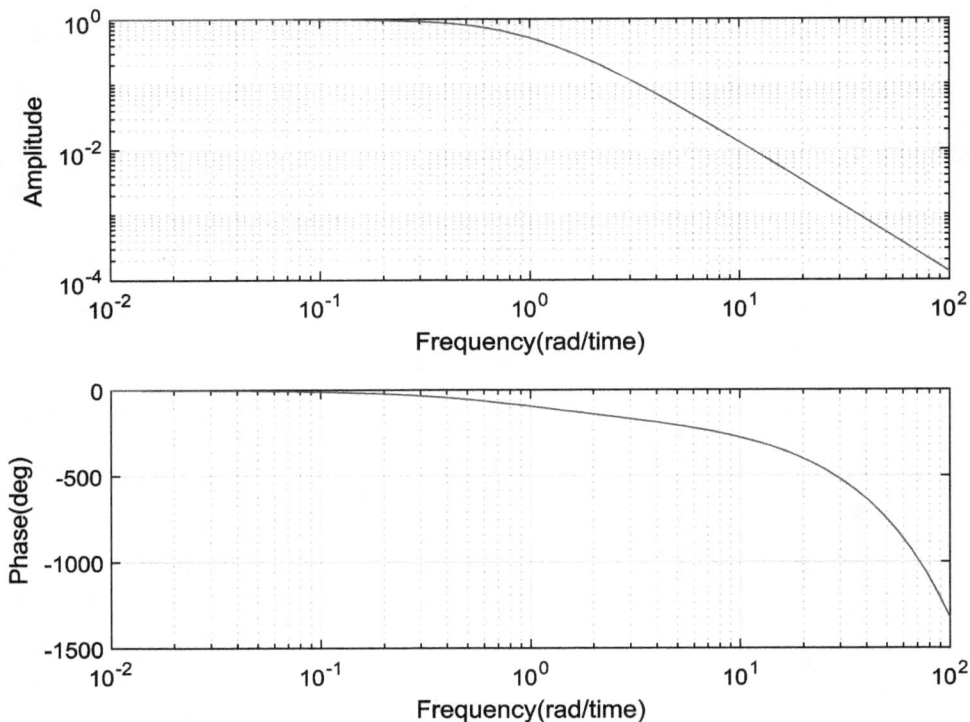

FIGURE 9.30 Bode diagram of a 3rd-order process with time delay.

$$G(s) = \frac{12.76e^{-s}}{5s + 1}$$

Solution

The following commands generate the desired Nyquist diagram, shown in Figure 9.32.

```
>> G = tf(12.76,[5 1]); set(G,'iodelay',1); nyquist(G), xlabel('Real axis'),
ylabel('Imaginary axis')
>> title('Nyquist diagram of a time delay')
```

9.7.3 NICHOLS CHART

A Nichols chart is a plot of the logarithmic magnitude against the phase lag. In MATLAB, a Nichols chart can be constructed by the built-in function *nichols*.

Example 9.27 Nichols Chart

Plot the Nichols chart for a 2nd-order process whose transfer function is

$$G(s) = \frac{2}{(10s + 1)(2.5s + 1)}$$

(a)

(b)

Polar plot of a 2nd-order process

Nyquist Diagram

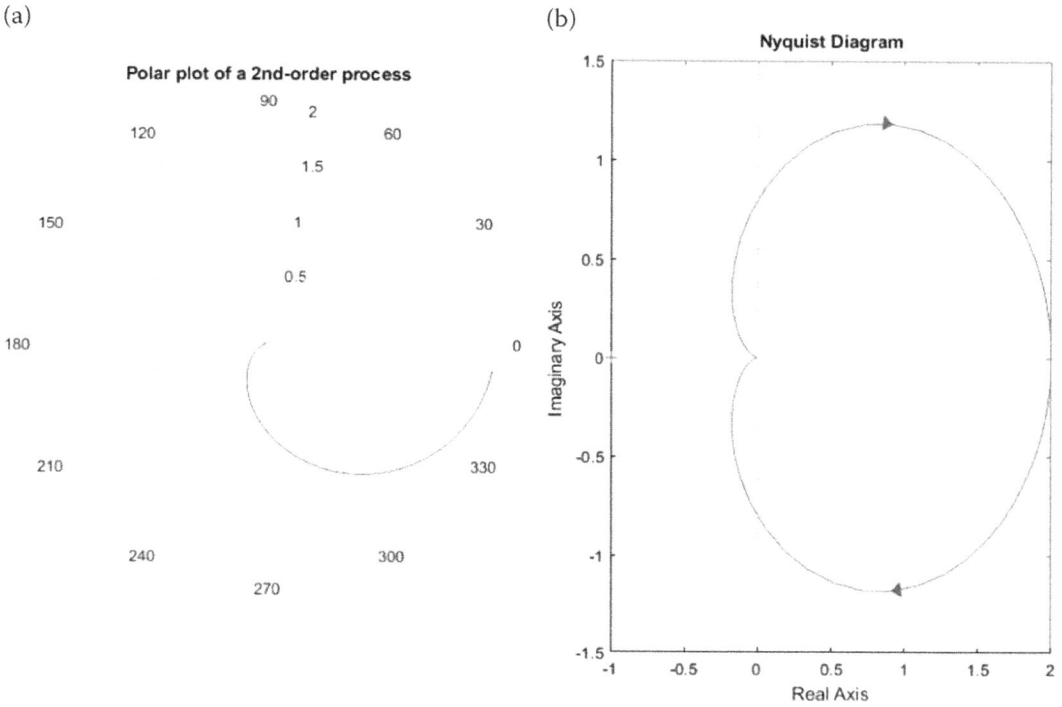

FIGURE 9.31 Nyquist plots of a 2nd-order process by (a) *polar* function and (b) *nyquist* function.

Solution

The following commands generate the desired Nichols chart, shown in Figure 9.33.

```
>> num = 2; den = conv([10 1], [2.5 1]); w = logspace(-2, 1, 300);
```

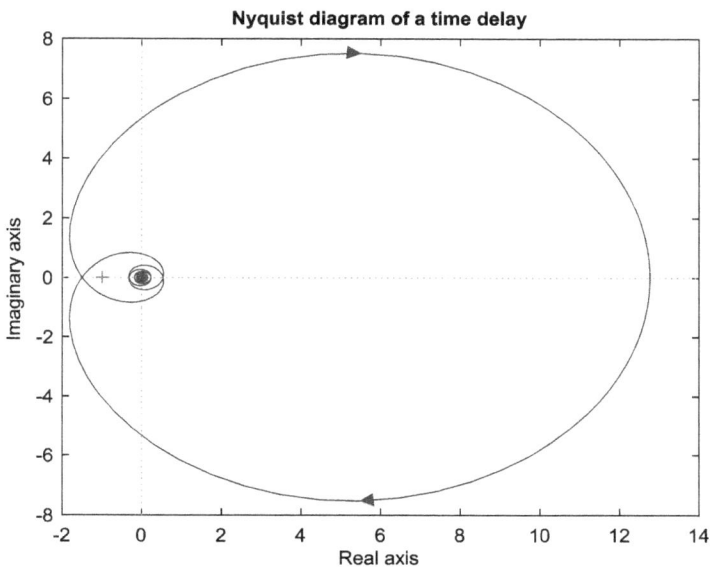

FIGURE 9.32 Nyquist diagram of a process with time delay.

```
>> nichols(num, den, w); ngrid
```

9.7.4 Gain and Phase Margins

Gain margin (GM) and phase margin (PM) provide quantitative measures of relative stability that indicate how close the system is to becoming unstable. The ultimate controller gain K_{cu} is the gain at which the system is at marginal stability. Thus, when the controller gain K_c is adjusted close to K_{cu}, the control system is close to an unstable state. Let AR_c be the value of the open-loop amplitude ratio at the ultimate frequency ω_c when $\phi = -180°$, and let ϕ_g be the phase angle at frequency ω_g when $AR = 1$. The GM and PM are defined as, respectively,

$$GM = \frac{1}{AR_c}, \quad PM = 180 + \phi_g$$

In practise, a controller is tuned so that a GM lies between 1.7 and 4.0 and a PM lies between 30° and 45°. The smaller the values of GM and PM, the closer the system is to the point of marginal stability, and the closed-loop system exhibits very oscillatory dynamics. The controller gain can be set by the ultimate gain K_{cu} and GM as

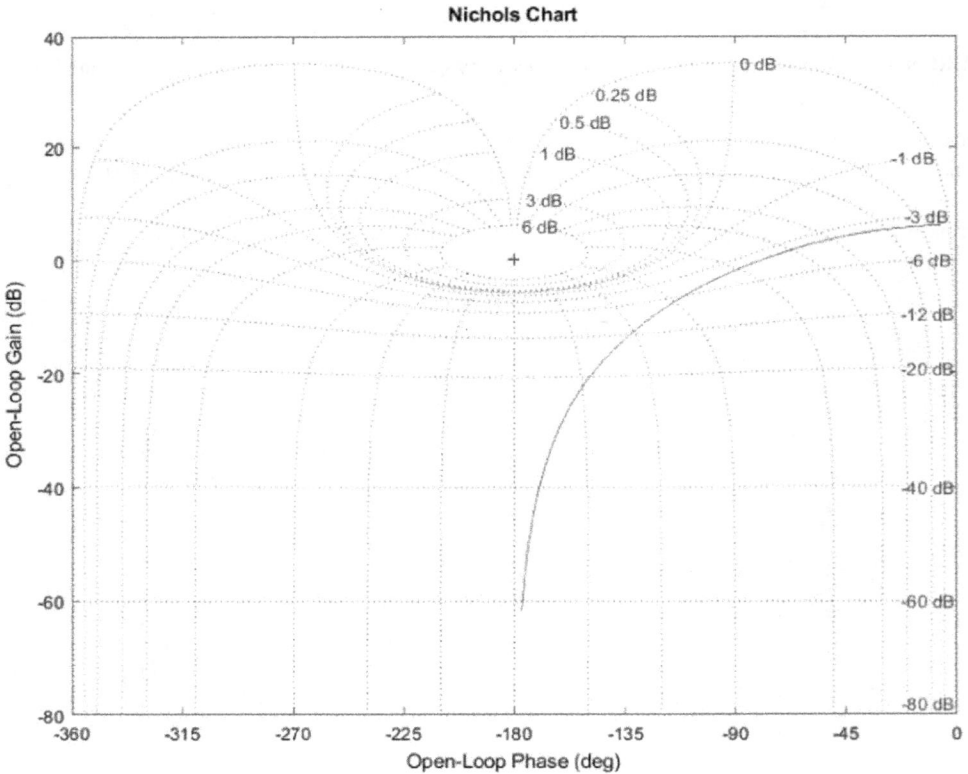

FIGURE 9.33   Nichols chart of a 2nd-order process.

$$K_c = \frac{K_{cu}}{\text{GM}}$$

In MATLAB, GM and PM are obtained from the built-in function *margin*.

Example 9.28 Ultimate Gain

The characteristic equation of a feedback control system using a proportional controller is given by

$$1 + K_c \frac{0.8e^{-2s}}{5s + 1} = 0$$

where K_c is the controller gain. Find the ultimate gain.

Solution

The following commands produce the desired outputs.

```
>> G = tf(0.8, [5 1]); delay = 2; [m, p, w] = bode(G); Mag = m(1, :);
>> Phase = p(1, :) - ((180/pi)*delay*w'); [Gm, Pm, Wcg, Wcp] = margin(Mag,
Phase, w)
Gm =
   5.7221
Pm =
   Inf
Wcg =
    0.8932
Wcp =
  NaN
```

We can see that $K_{cu} = 5.7221$ and $\omega_u = 0.8932$. The following commands generate plots for GM and PM, as shown in Figure 9.34:

```
>> G = tf(0.8, [5 1]); theta = 2; set(G, 'iodelay', theta); margin(G)
```

PROBLEMS

9.1 Expand $F(s) = (s^2 + 2s + 3)/(s^3 + 3s^2 + 3s + 1)$ into partial fractions.
9.2 Two blocks connected in series can be represented as $C = G_1G_2A$. Synthesise these two blocks. The transfer functions of the blocks are given by $G_1 = 2/(s + 1)$ and $G_2 = (2s + 1)/(s^2 + 3s + 2)$.
9.3 Two blocks connected in parallel can be represented as $C = (G_1 + G_2)A$. Synthesise these two blocks. The transfer functions of the blocks are given by $G_1 = 2/(s + 1)$ and $G_2 = (2s + 1)/(s^2 + 3s + 2)$.
9.4 The state-space representation of a process is given by

$$\frac{d}{dt}\begin{bmatrix} x_1 \\ x_2 \\ x_3 \end{bmatrix} = \begin{bmatrix} 0 & 1 & 0 \\ 0 & -1 & -3 \\ 1 & 0 & -7 \end{bmatrix}\begin{bmatrix} x_1 \\ x_2 \\ x_3 \end{bmatrix} + \begin{bmatrix} 0 \\ 1 \\ 1 \end{bmatrix}u, \; y = [100]\begin{bmatrix} x_1 \\ x_2 \\ x_3 \end{bmatrix}$$

Bode Diagram
Gm = 15.2 dB (at 0.895 rad/s), Pm = Inf

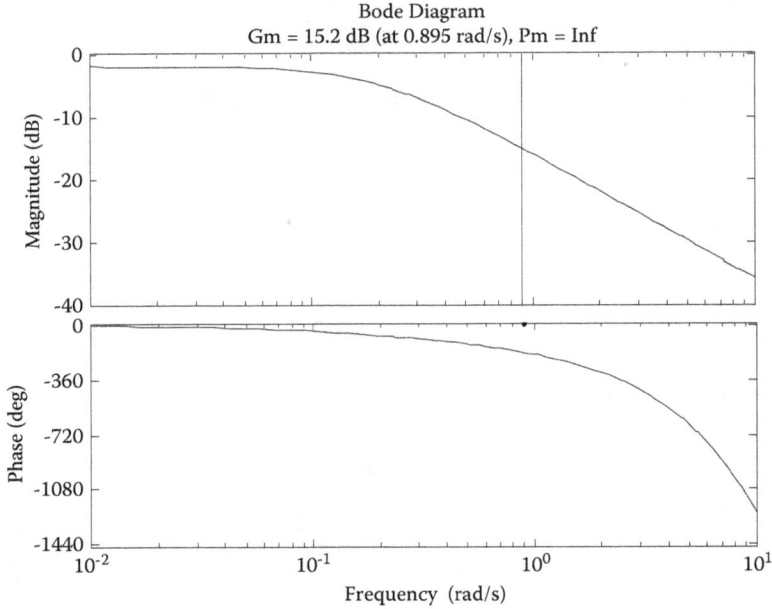

FIGURE 9.34 GM and PM of a process with time delay.

Find the transfer function for the process.

9.5 Find the impulse response of a 1st-order process with the transfer function given by $3/(2s + 1)$ during the time interval $[0, 10]$.

9.6 The transfer function of a 1st-order process is given by $3/(2s + 1)$. Plot the response curve of the process for the linear input $x = 1.2t$ during the time interval $[0, 10]$.

9.7 Figure P9.7 shows two level tanks. The outlet flow rates q_1 and q_2 are related to the liquid levels h_1 and h_2 in the tanks through the following equations:

$$q_1 = \frac{h_1 - h_2}{R_1}, \ q_2 = \frac{h_2}{R_2}$$

where R_1 and R_1 are the resistances of the valves. The levels h_1 and h_2 are maintained at $h_1 = h_{1s}$ and $h_2 = h_{2s}$, respectively, valves are closed ($q_1 = q_2 = 0$), and $q_i = 0$. At $t = 0$, a step change of magnitude w is introduced to the inlet flow rate q_i (m^3/min), and the valve below each tank is opened. Using the given data, plot h_1, h_2, q_1, and q_2 as a function of time ($0 \leq t \leq 50min$). The liquid density is assumed to be constant.

Data: $h_{1s} = 2 \ m$, $h_{2s} = 1.5 \ m$, $R_1 = R_2 = 6 \ min/ \ m^2$, $A_1 = 3 \ m^2$, $A_2 = 2 \ m^2$, $w = 0.15 \ m^3/min$.

9.8 In a 2nd-order process, the time constant and the process gain are given by $\tau = 2$ and $k = 3$, respectively. A step change of magnitude $\Delta m = 4$ is introduced to the process input. Plot the dimensionless response $y(t)/(k\Delta m)$ as a function of time t ($0 \leq t \leq 40$) for each value of the decay ratio $\zeta = 0.1, 0.4, 0.7$, and 1.

9.9 In a 2nd-order process, the time constant and the process gain are given by $\tau = 2$ and $k = 3$, respectively. A step change of magnitude $\Delta m = 4$ is introduced to the process input. Plot the

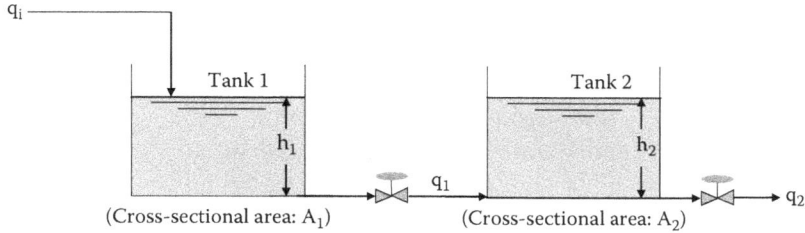

FIGURE P9.7   Two level tanks.

dimensionless response $y(t)/(k\Delta m)$ as a function of time t ($0 \le t \le 40$) for each value of the decay ratio $\zeta = 1$, 1.5, 2.0, and 2.5.

9.10 The transfer function of a 2nd-order process is given by $1/(\tau^2 s^2 + 2\tau\zeta s + 1)$, where $\tau = 0.5$. Plot the impulse response curve for each value of the decay ratio $\zeta = 0.5$, 1.0, and 1.5 during the time interval [0, 10].

9.11 The transfer function of a lead/lag element is given by $(\tau s + 1)/(s + 1)$. Plot the unit step response curve for each value of $\tau = 0$, 0.5, 1.0, 1.5, and 2 during the time interval [0, 10].

9.12 A proportional (P) controller is used in a feedback control system. The process transfer function is given by $G_p(s) = 5/(5s + 1)$, the gain of the control valve is $K_v = 0.01$, and the gain of the sensor/transducer is $K_m = 20$. Use Simulink to find the step response curve to a unit step change in the set point. The controller gain K_c is 10.

9.13 The following exothermic consecutive reactions are carried out in a batch reactor fitted with a cooling coil through which cooling water is passed to remove the exothermic heat, as shown in Figure P9.13[3]:

$$A \xrightarrow{k_1} B \xrightarrow{k_2} C$$

From the material balances for species A and B, we obtain

$$\frac{dC_A}{dt} = -k_1 C_A{}^2, \quad \frac{dC_B}{dt} = k_1 C_A{}^2 - k_2 C_B$$

where
k_1 and k_2 are the reaction rate constants
C_A and C_B are the concentrations of species A and B, respectively
k_1 and k_2 are represented as

$$k_1 = A_1 e^{-E_1/(RT)}, \quad k_2 = A_2 e^{-E_2/(RT)}$$

The energy balance for the batch reactor gives[3]

$$\frac{dT}{dt} = \frac{(-\Delta H_1)}{\rho C_p} k_1 C_A{}^2 + \frac{(-\Delta H_2)}{\rho C_p} k_2 C_B + \frac{U_j A_j}{\rho C_p V}(T_s - T) - \frac{U_c A_c}{\rho C_p V}(T - T_c)$$

where
$(-\Delta H_1)$ is the heat of reaction for $A \rightarrow B$
$(-\Delta H_2)$ is the heat of reaction for $B \rightarrow C$
T_s and T_c are the steam and coolant temperatures, respectively

U_j and U_c are the overall heat transfer coefficients of the jacket and coolant, respectively
For the present case, the reactor temperature should precisely follow the desired trajectory given by

$$T_d(t) = 54 + 71e^{-0.0025t}$$

One way to control $T(t)$ is to introduce a parameter u defined by

$$T_s = (T_{s,max} - T_{s,min})u + T_{s,min}, \quad U_c = (U_{c,min} - U_{c,max})u + U_{c,max}$$

$u = 0$ represents the maximum cooling and $u = 1$ represents the maximum heating of the system. Substitution of these relations into the energy balance, followed by rearrangement, yields

$$\frac{dT}{dt} = \gamma_1 k_1 C_A^2 + \gamma_2 k_2 C_B + (a_1 + a_2 T) + (b_1 + b_2 T)u$$

where

$$\gamma_1 = \frac{(-\Delta H_1)}{\rho C_p}, \quad \gamma_2 = \frac{(-\Delta H_2)}{\rho C_p}, \quad a_1 = \frac{U_j A_j T_{s,min} + U_{c,max} A_c T_c}{\rho C_p V}, \quad a_2 = -\frac{U_j A_j + U_{c,max} A_c}{\rho C_p V}$$

$$b_1 = \frac{U_j A_j (T_{s,max} - T_{s,min}) - (U_{c,max} - U_{c,min}) A_c T_c}{\rho C_p V}, \quad b_2 = \frac{(U_{c,max} - U_{c,min}) A_c}{\rho C_p V}$$

A simple proportional controller is used in the control, and u is determined by

$$u(t) = u_s + K_c e(t)$$

where the control error $e(t)$ is given by $e(t) = T_d(t) - T(t)$ and the controller gain is arbitrarily chosen as $K_c = 1.2(°C^{-1})$. Plot C_A, C_B, and T as a function of reaction time t ($0 \le t \le 6000\ sec$).

  Data: $C_{A0} = 1.0\ kmol/m^3$, $C_{B0} = 0.0\ kmol/m^3$, $A_1 = 1.1\ m^3/(kmol·sec)$, $A_2 = 172.2\ sec^{-1}$, $E_1 = 2.09 \times 10^4\ kJ/kmol$, $E_2 = 4.18 \times 10^4\ kJ/kmol$, $(-\Delta H_1) = 4.18 \times 10^4\ kJ/kmol$, $(-\Delta H_2) = 8.36 \times 10^4\ kJ/kmol$, $\rho = 1000\ kg/m^3$, $T_c = 25°C$, $U_j = 1.16\ kJ/(m^2·°C·sec)$, $U_{c,max} = 4.42\ kJ/(m^2·°C·sec)$, $U_{c,min} = 1.39\ kJ/(m^2·°C·sec)$, $T_{s,max} = 150°C$, $T_{s,min} = 70°C$, $R = 8.314\ kJ/(kmol·K)$, $A_c/V = 17\ m^2/m^3$, $A_j/V = 30\ m^2/m^3$, $C_p = 1.0\ kJ/(kg·°C)$, $u_s = 1.0$, $K_c = 1.2(°C^{-1})$.

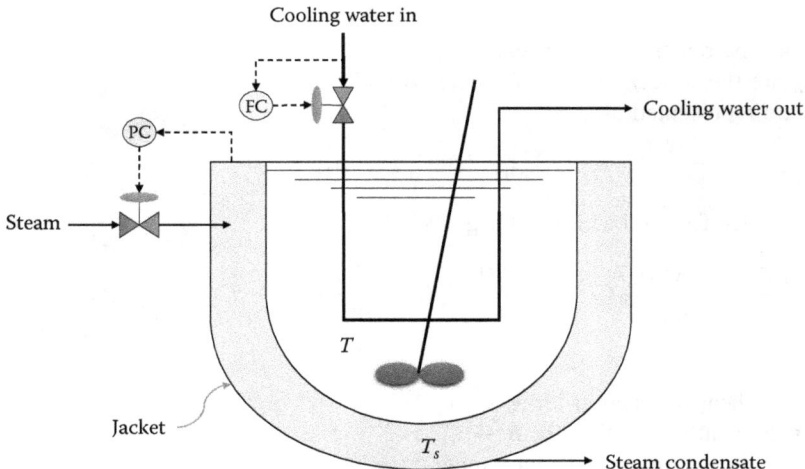

FIGURE P9.13 Schematic representation of a batch reactor control configuration.

9.14 A PI controller is used to control a 1st-order process $G = 0.8/(5s + 1)$. The transfer function of the controller is $G_c(s) = K_c(1 + 1/(\tau_I s))$, where $K_c = 1$ and $\tau_I = 10$. Plot the step response curve to a unit step change in the set-point.

9.15 Figure P9.15 shows a level control process. The liquid level of a tank, $h(m)$, is to be controlled at a desired steady-state value h_s when a step change is introduced to the inlet flow rate $q_i(m^3/min)$. A proportional-integral (PI) controller is used to maintain the liquid level at h_s. Let the outlet flow rate $q(m^3/min)$ be represented by

$$Q = -K_c\left(H + \frac{1}{\tau_I}\int_0^t H dt\right)$$

where Q and H are deviation variables defined by $Q = q - q_s(m^3/min)$ and $H = h - h_s(m)$, and $K_c(m^2/min)$ and $\tau_I(min)$ are the proportional gain and the integral time constant of the PI controller, respectively. The dynamic behaviour of H is to be observed for certain range of K_c and τ_I when q_i undergoes a sudden step change of magnitude 3. Plot H as a function of time t ($0 \le t \le 20$) for $K_c = 1, 2, 4, 8$ when $\tau_I = 0.2$, and for $\tau_I = 0.1, 0.2, 0.5, 1.0$ when $K_c = 2$. The cross-sectional area of the liquid tank (A) is $4\ m^2$, the liquid density is assumed to be constant, and the step change of magnitude 3 can be expressed as $dH(0)/dt = 3$.

9.16 A proportional-integral (PI) controller is used in a liquid-level control system. The process transfer function is $G(s) = 2/(4.8s + 1)$, the gain of the control valve is 0.08, and the reset time of the PI controller is $\tau_I = 2$. Use Simulink to create the step response curve to a unit step change in the inlet flow rate for each value of the controller gain $K_c = 3, 10, 20, 60$. To construct the PI controller block in Simulink, select the PID block from the Simulink $\rightarrow$ Continuous submenu in the Simulink Library Browser and drag the block to the block diagram in the editor window. Double-click the PID block to set the controller parameters. Set both the filter coefficient N and the derivative time to zeros.

9.17 Figure P9.17 shows two liquid-level tanks connected in an interacting fashion. Assuming constant fluid density, the material balance for each tank gives

$$A_1\frac{dh_1}{dt} = q_i - C_1\sqrt{h_1 - h_2}, \quad A_2\frac{dh_2}{dt} = C_1\sqrt{h_1 - h_2} - C_2\sqrt{h_2}$$

where
    A_1 and A_2 are cross-sectional areas of tank 1 and tank 2, respectively
    C_1 and C_2 are parameters

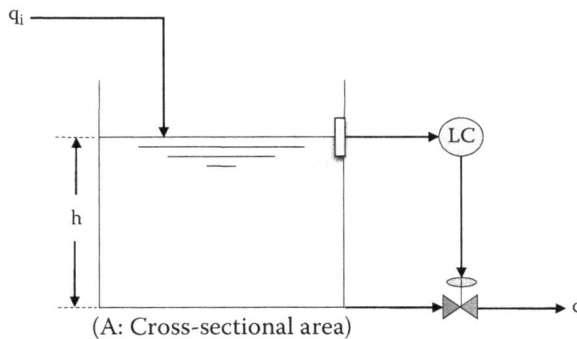

(A: Cross-sectional area)

FIGURE P9.15 Level control process.

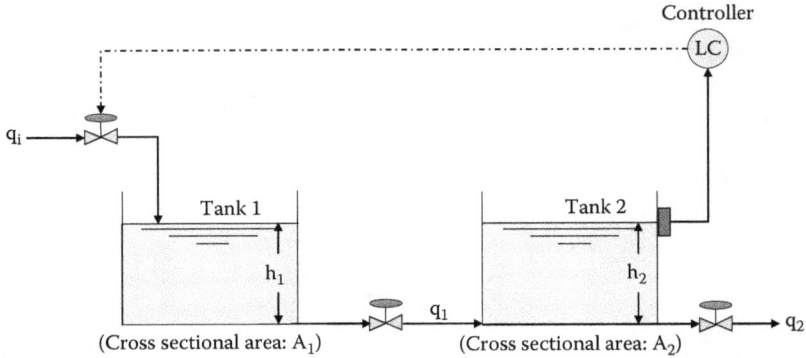

FIGURE P9.17 Two tanks connected in an interacting fashion.

The manipulated variable is the flow rate q_i (ft^3/min) to the first tank. When a PI controller is used, the inlet flow rate to the first tank is given by

$$\frac{q_i}{q_m} = r_m + K_c e(t) + \frac{K_c}{\tau_I} \int_0^t e(t)dt, \; e(t) = h_R - h_2$$

where

$K_c (ft^{-1})$ is the controller gain
$\tau_I (min)$ is the reset time
q_m is the maximum possible flow rate (that is, $0 \le q_i \le q_m$)
r_m is the fractional flow rate when there is no control action (that is, $q_i = rq_m$)
h_R is the set point for the second tank level
The accumulated control error can be represented as

$$C_e(t) = \int_0^t e(t)dt = \int_0^t (h_R - h_2)dt$$

Differentiation of this equation yields

$$\frac{dC_e(t)}{dt} = e(t) = h_R - h_2$$

Thus, the controller equation can be expressed as

$$\frac{q_i}{q_m} = r_m + \frac{K_c}{\tau_I} C_e(t) + K_c \frac{dC_e(t)}{dt}$$

If a P controller is used, the controller equation will be

$$\frac{q_i}{q_m} = r_m + K_c (h_R + h_2)$$

The following information and data apply: $A_1 = A_2$, $C_1 = 61.8 \, ft^{2.5}/min$, $C_2 = 30.9 \, ft^{2.5}/min$, $h_R = 6 \, ft$, $r_m = 0.4$, $q_m = 480 \, ft^3/min$, $d = 12 \, ft$ (d: tank diameter).

1 Plot the liquid level of each tank versus time when there is no control action.

2 A P controller is used to control the liquid level. Plot the liquid level of the second tank versus time when the controller gain is $K_c = 2$ and $K_c = 40$.

3 A PI controller is used to control the liquid level. Plot the liquid level of each tank versus time when the controller gain is $K_c = 0.2$ and the reset time is $\tau_I = 20\ min$.

9.18 Determine the stability of a system whose characteristic equation is given by

$$s^4 + 5s^3 + 3s^2 + 1 = 0$$

9.19 The transfer function of a PID controller can be expressed as

$$G_c(s) = \frac{M(s)}{E(s)} = K_P + \frac{K_I}{s} + K_D s = \frac{K_D s^2 + K_P s + K_I}{s}$$

where $K_c = K_P$, $K_c \tau_D = K_D$, and $K_c/\tau_I = K_I$. Plot the root locus diagram for arbitrary tuning parameters when the process transfer function is given by $G_p(s) = 1/s^2$.

9.20 Plot the Bode diagram for a 2nd-order process whose transfer function is given by

$$G(s) = \frac{2}{(10s + 1)(2.5s + 1)}$$

9.21 Plot the Nyquist diagram for a 3rd-order system whose transfer function is given by $G(s) = 1/(s^3 + 2s^2 + 2s + 1)$.

9.22 A P controller is used in a feedback control system whose characteristic equation is given by

$$1 + \frac{K_c}{(s + 1)(s + 2)(s + 3)} = 0$$

Find the ultimate gain.

REFERENCES

1. Kaisare, N. S., *Computational Techniques for Process Simulation and Analysis Using MATLAB*, CRC Press, Taylor & Francis Group, Boca Raton, FL, pp. 117–118, 2018.
2. Jana, A. K., *Chemical Process Modelling and Computer Simulation*, 2nd ed., PHI Learning Private, Ltd., Delhi, India, pp. 62–65, 2011.
3. Jana, A. K., *Chemical Process Modelling and Computer Simulation*, 2nd ed., PHI Learning Private, Ltd., Delhi, India, pp. 63–66, 2011.

10 Optimization

10.1 UNCONSTRAINED OPTIMIZATION

10.1.1 FIBONACCI METHOD

The Fibonacci method is most convenient when the interval containing the extremum (minimum or maximum) point or the interval of uncertainty is to be reduced to a given value in the least number of trials or function evaluations. If the Fibonacci numbers are represented as $F_0, F_1, F_2, \cdots, F_n, \cdots$, the sequence can be generated using

$$F_0 = 1, \ F_1 = 1, \ F_i = F_{i-1} + F_{i-2} \ (i = 2, \cdots, n)$$

Figure 10.1 shows the interval reduction procedure, which can be described as

$$I_1 = I_2 + I_3, \ I_2 = I_3 + I_4, \ \cdots, \ I_k = I_{k+1} + I_{k+2}, \ \cdots, \ I_{n-2} = I_{n-1} + I_n, \ I_{n-1} = 2I_n$$

Using the definition of the Fibonacci number, we have

$$I_1 = F_n I_n, \ I_2 = F_{n-1} I_n, \ I_{n-k} = F_{k+1} I_n$$

The minimization algorithm based on the Fibonacci method is as follows:

1) Specify two points x_1 and x_4 representing the initial interval $I_1 = |x_4 - x_1|$.
2) Specify the number of interval reductions n. If a desired accuracy ε is given, n can be defined as the smallest value such that $1/F_n < \varepsilon$.
3) Calculate $v = (\sqrt{5} - 1)/2$, $w = (1 - \sqrt{5})/(1 + \sqrt{5})$, $\alpha = v(1 - w^n)/(1 - w^{n+1})$.
4) Determine intermediate points $x_3 = \alpha x_4 + (1 - \alpha)x_1$ and calculate $f_3 = f(x_3)$.
5) for i = 1:n-1

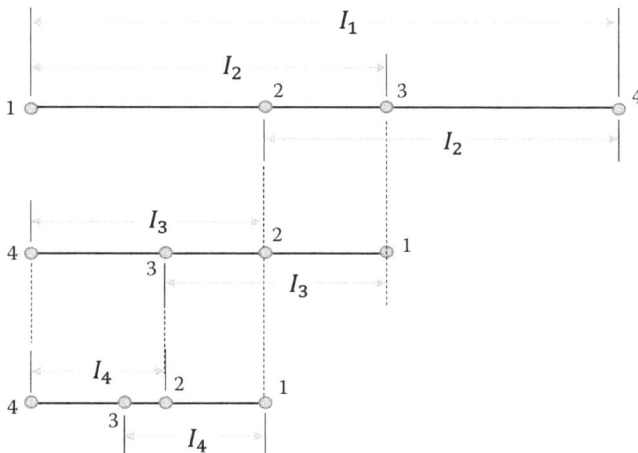

FIGURE 10.1 Interval reduction.

if i = n-1, $x_2 = 0.01x_1 + 0.99x_3$
   else $x_2 = \alpha x_1 + (1 - \alpha)x_4$
   endif
   calculate $f_2 = f(x_2)$.
   if $f_2 < f_3$, $x_4 = x_3$, $x_3 = x_2$, $f_3 = f_2$
   else $x_1 = x_4$, $x_4 = x_2$
   endif

$$\alpha = \frac{v(1 - w^{n-i})}{1 - w^{n-i+1}}$$

end for

The MATLAB function *fbnopt* implements the Fibonacci search algorithm. The basic syntax is
[x,f,fint] = fbnopt(objfun, a, b, n)

where *objfun* is the objective function, a and b are the initial interval points, n is the number of reductions, x is the resultant optimum point, f is the function value at the optimum point, and fint is the length of the final interval.

```
function [x, f, fint] = fbnopt(objfun, a, b, n)
% fbnopt.m: 1-dimensional Fibonacci search method
% inputs:
%    objfun: objective function
%    a, b: initial interval points
%    n: number of reduction
% output:
%    fint: size of final interval
%    x: optimum point
%    f: function value at the optimum point
% Sample run:
%       a = 0; b = 50; n = 20; fobj= @(x) -870*x + 102*x^2 - 5*x^3;
%       [x, f, fint] = fbnopt(fobj, a, b, n);

v = (sqrt(5)-1)/2; w = (1-sqrt(5))/(1+sqrt(5)); x1 = a; x4 = b; alpha = v*(1-
w^n)/(1-w^(n+1));
x3 = alpha*x4 + (1-alpha)*x1; f3 = objfun(x3);
for k = 1: n - 1
    if k == n - 1, x2 = 0.01*x1 + 0.99*x3;
    else, x2 = alpha*x1 + (1-alpha)*x4; end
    f2 = objfun(x2);
    if (f2 < f3), x4 = x3; x3 = x2; f3 = f2;
    else, x1 = x4; x4 = x2; f4 = f2; end
    alpha = v*(1- w^(n - k)) / (1-w^(n - k + 1));
end
x = x3; f = f3; fint = abs(x1 - x4);
end
```

Example 10.1 Fibonacci Search

Use the Fibonacci search method to find the minimum of $f(x) = 2x^2 \sin(1.5x) - 4x^2 + 3x - 1$. The initial interval can be specified as [−8,8]. Let the number of interval reduction be $n = 20$.

Solution

```
>> a = -8; b = 8; n = 20; fobj = @(x) 2*x^2*sin(1.5*x) - 4*x^2 + 3*x-1;
>> [x, f, fint] = fbnopt(fobj, a, b, n)
x =
  -5.7256
f =
-197.9705
fint =
  0.0015
```

10.1.2 GOLDEN SECTION METHOD

From the interval reduction procedure shown in Figure 10.1,

$$I_1 = I_2 + I_3, \ I_2 = I_3 + I_4, \ \cdots$$

As the number of iterations is increased, the ratio of the Fibonacci numbers F_{n-1}/F_n reaches the limit $\tau = (\sqrt{5} - 1)/2 = 0.61803$. The limit τ is called the golden ratio. Now we wish to maintain
From these relations, we have

$$I_2 = \tau I_1, \ I_3 = \tau I_2 = \tau^2 I_1$$

Substituting these equations into $I_1 = I_2 + I_3$, we have

$$1 = \tau + \tau^2$$

The positive root of this equation is given by

$$\tau = \frac{\sqrt{5} - 1}{2} = 0.61803$$

which is the golden ratio.

The interval reduction scheme by the golden section method follows the procedure used in the Fibonacci algorithm. The golden section algorithm is as follows:

1) Specify the initial point x_1 and the step size h. Evaluate $f_1 = f(x_1)$.
2) Let $x_2 = x_1 + h$ and evaluate $f_2 = f(x_2)$.
3) If $f_2 > f_1$, interchange points 1 and 2. Set $h = -h$.
4) Set $h = h/\tau$, $x_4 = x_2 + h$ and evaluate $f_4 = f(x_4)$.
5) If $f_4 > f_2$, go to step 7).
6) Let $x_1 = x_2$, $x_2 = x_4$ and go to step 4).
7) Set $x_3 = \tau x_4 + (1 - \tau)x_1$ and evaluate $f_3 = f(x_3)$.
8) If $f_2 < f_3$, set $x_4 = x_1$ and $x_1 = x_3$. Otherwise set $x_1 = x_2$, $x_2 = x_3$, and $f_2 = f_3$.
9) If the stopping criterion is not satisfied, go to step 7).

The MATLAB function *gsopt* implements the golden section search algorithm. The basic syntax is
`[x, f, n] = gsopt(objfun, x1, h, crit)`

where objfun is the objective function, x1 is the initial point, h is the step size, crit is the stopping criterion, x is the resultant optimum point, f is the function value at the optimum point, and n is the length of the final interval.

```
function [x, f, n] = gsopt(objfun, x1, h, crit)
% gsopt.m: golden section search method
% input:
%       objfun: objective function
%       x1: initial point
%       h: step size
%       crit: stopping criterion
% output:
%       x: optimal point
%       f: function value at the optimal point
%       n: number of function evaluations
% sample run:
%       crit=1e-6;,x1=1; h=0.1;
%       fun = @(x) 85*(1-x^3)^2 + (1-x^2) + 3*(1-x)^2;
%       [x, f, n] = gsopt(fun, x1, h, crit)

% Initialization
tau = (sqrt(5) - 1)/2; n = 0; f1 = objfun(x1); n = n+1;
x2 = x1 + h; f2 = objfun(x2); n = n+1;
% Golden section search
if f2 > f1, temp = x1; x1 = x2; x2 = temp; temp = f1; f1 = f2; f2 = temp; h = -h; end
while x1 < 1e50
      h = h/tau; x4 = x2 + h; f4 = objfun(x4); n = n+1;
      if f4 > f2, break; end
      f1 = f2; x1 = x2; f2 = f4; x2 = x4;
end
fold = (f1 + f2 + f4) / 3; ind = 0;
while x1 < 1e50
      if abs(x4-x1) < crit, break; end
      x3 = tau*x4 + (1-tau)*x1; f3 = objfun(x3); n = n+1;
      if f2 < f3, x4 = x1; x1 = x3; f4 = f1; f1 = f3;
      else, x1 = x2; x2 = x3; f1 = f2; f2 = f3; end
      fpr = (f1 + f2 + f4) / 3;
      if abs(fpr - fold) < crit, ind = ind + 1; if ind == 2, break; end
      else, ind = 0; end
      fold = fpr;
end
x = x2; f = f2;
end
```

Example 10.2 Golden Section Method

Use the golden section method to find the minimum of $f(x) = 2x^2 \sin(1.5x) - 4x^2 + 3x - 1$. The initial point can be specified as $x_1 = -5$. Let the step size be $h = 0.1$ and the stopping criterion be 1×10^{-6}.

Solution

```
>> crit = 1e-6; x1 = -5; h = 0.1; fobj = @(x) 2*x^2*sin(1.5*x) - 4*x^2 + 3*x-1;
>> [x, f, n] = gsopt(fobj, x1, h, crit)
x =
 -5.7248
f =
-197.9705
n =
  25
```

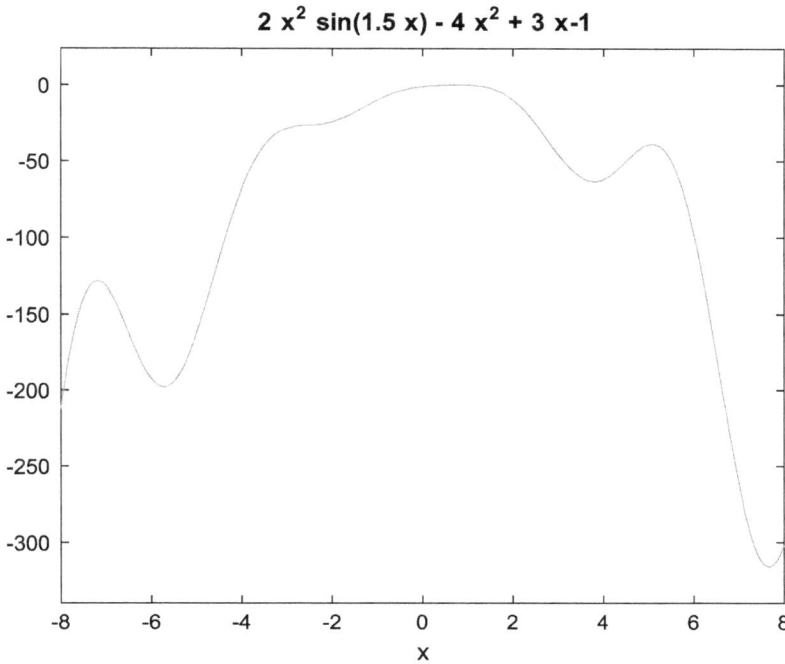

FIGURE 10.2   Plot of $f(x) = 2x^2\sin(1.5x) - 4x^2 + 3x - 1$.

Note that the given objective function has many local minimum values, as shown in Figure 10.2. Thus, different initial points give different results. For example, $x_1 = 1$ gives $x_{min} = 3.7939$ and $f_{min} = -63.265$, while $x_1 = 6$ gives $x_{min} = 7.6665$ and $f_{min} = -316.0254$.

```
>> x1 = 1; [x, f, n] = gsopt(fobj, x1, h, crit)
x =
  3.7939
f =
 -63.2650
n =
  28
>> x1 = 6; [x, f, n] = gsopt(fobj, x1, h, crit)
x =
  7.6665
f =
-316.0254
n =
  28
```

10.1.3 Brent's Quadratic Fit Method

Suppose that initial three data points are given as (x_1, f_1), (x_2, f_2), (x_3, f_3), where $x_1 < x_2 < x_3$ and $f_2 \leq \min(f_1, f_3)$. A quadratic function of a form $p(x) = a + bx + cx^2$ can be fitted through these points. Using the Lagrange polynomial, the function $p(x)$ can be expressed as

$$p(x) = f_1 \frac{(x - x_2)(x - x_3)}{(x_1 - x_2)(x_1 - x_3)} + f_2 \frac{(x - x_1)(x - x_3)}{(x_2 - x_1)(x_2 - x_3)} + f_3 \frac{(x - x_1)(x - x_2)}{(x_3 - x_1)(x_3 - x_2)}$$

The minimum point $x^* = x_4$ can be found by setting $dp(x)/dx = 0$, which gives

$$x^* = x_4 = \frac{BCf_1(x_2 + x_3) + ACf_2(x_1 + x_3) + ABf_3(x_1 + x_2)}{2(BCf_1 + ACf_2 + ABf_3)}$$

$$A = (x_1 - x_2)(x_1 - x_3), \quad B = (x_2 - x_1)(x_2 - x_3), \quad C = (x_3 - x_1)(x_3 - x_2)$$

A new interval of uncertainty $(x_{1new}, x_{2new}, x_{3new})$ is determined by comparing f_2 with $f_4 = f(x_4)$:

$$(x_{1new}, \ x_{2new}, \ x_{3new}) = \begin{cases} (x_1, \ x_2, \ x_4): x_4 > x_2, \ f_4 \geq f_2 \\ (x_2, \ x_4, \ x_3): x_4 > x_2, \ f_4 < f_2 \\ (x_4, \ x_2, \ x_3): x_4 < x_2, \ f_4 \geq f_2 \\ (x_1, \ x_4, \ x_2): x_4 < x_2, \ f_4 < f_2 \end{cases}$$

Brent's quadratic fit method starts with bracketing an interval that contains the minimum. At any step, five points (a, b, x, v, w) are considered. Points a and b bracket the minimum, x is the point which gives the minimum function value, w is the point which gives the second lowest function value, and v v is the previous value of w. Quadratic fitting is tried for x, v, and w, and the quadratic minimum point s is given as

$$s = x - \frac{1}{2} \frac{(x - w)^2 \{f(x) - f(v)\} - (x - v)^2 \{f(x) - f(w)\}}{(x - w)\{f(x) - f(v)\} - (x - v)\{f(x) - f(w)\}}$$

Suppose that the estimated points are equidistant, with an equal distance h. Then the estimated points w, x, v and their corresponding function values can be expressed as

$$w = x_0, \ x = x_1 = w + h = x_0 + h, \ v = x_2 = x + h = x_0 + 2h$$

$$f_0 = f(w) = f(x_0), f_1 = f(x) = f(x_1), f_2 = f(v) = f(x_2)$$

Substitution of these relations to the above equation yields

$$s = x_0 + \frac{h}{2} \left(\frac{4f_1 - 3f_0 - f_2}{2f_1 - f_0 - f_2} \right)$$

Brent's quadratic fit algorithm for the minimum is as follows[1]:

1) Specify three points a, b (form the interval), and x (gives the least function value).
2) Initialize w and v at x.
3) If x, w, and v are all distinct, go to step 5).
4) Find point u using the golden section search method for the larger of the two intervals $x - a$ or $x - b$. Go to step 7).
5) Perform quadratic fitting for x, w, and v, and find the minimum point u.
6) Adjust u into the larger interval of $x - a$ or $x - b$.
7) Calculate $f_u = f(u)$.
8) Find new values for a, b, x, w, and v.
9) Check whether the larger of the intervals $x - a$ or $x - b$ is smaller than the specified convergence criterion. If convergence has not been achieved, go to step 3).

The MATLAB function *brentopt* implements the Brent search algorithm. The basic syntax is
```
[xopt, fopt, nf] = brentopt(objfun, x1, h, crit)
```

where objfun is the objective function, x1 is the initial point, h is the step size, crit is the stopping criterion, xopt is the resultant optimum point, fopt is the function value at the optimum point, and nf is the number of function evaluations.

```
function [xopt, fopt, nf] = brentopt(objfun, x1, h, crit)
% brentopt.m: minimization(1-dimensional search) by Brent's algorithm
% input:
%      objfun: objective function
%      x1: initial point
%      h: step size
%      crit: convergence criterion
% output:
%      x: optimum point
%      f: function value at the optimum point
%      hf: number of function evaluations for search
% Sample run:
%      x1 = 0.01; h = 0.2; crit = 1e-6;
%      fun = @(x) exp(x) -3*x + 0.02/x - 0.00004/x;
%      [xopt, fopt, nf] = brentopt(fun, x1, h, crit)

% Initialization
clarge = 1e40; tau = (sqrt(5) - 1)/2; nf = 1; f1 = objfun(x1);
% Determine initial 3-points
x2 = x1 + h; nf = nf+1; f2 = objfun(x2);
if f2 > f1, temp = x1; x1 = x2; x2 = temp; temp = f1; f1 = f2; f2 = temp; h = -h; end
while x1 < clarge
    h = h/tau; x4 = x2 + h; nf = nf+1; f4 = objfun(x4);
    if f4 > f2, break; end; f1 = f2; x1 = x2; f2 = f4; x2 = x4;
end
x3 = x4; f3 = f4; a = x1; b = x3; fa = f1; fb = f3; ha = a; hb = b;
% Check whether ha < hb
if (b < a), ha = b; hb = a; end
x = x2; fx = fb; w = x; v = w; ev = 0; fx = f2; fw = fx; fv = fw;
while 1
    hm = (ha + hb)/2;
    % Check interval convergence
    if (abs(x - hm) <= crit - (hb - ha)/2), x2 = x; f2 = fx; return; end
    etemp = 0; p = 0; q = 0;
    if (abs(ev) > crit)
        r = (x - w)*(fx - fv); q = (x - v)*(fx - fw); p = (x - v)*q - (x - w)*r; q = 2*(q - r);
        if (q > 0), p = -p; else, q = -q; end;
        etemp = ev; ev = d; % length of the larger interva
    end
   ind1 = 1; if ((q * (ha - x) - p) < 0), ind1 = -1; end
   ind2 = 1; if ((q * (hb - x) - p) < 0), ind2 = -1; end
   if ((abs(p) >= abs(q*etemp/2)) | (ind1 == ind2))
      if (x < hm), ev = hb - x; else, ev = ha - x; end; d = (1 - tau)*ev;
   else
      d = p/q; u = x + d;
     if (((u - ha) < crit) | ((hb - u) < crit)), if (x < hm), d = crit; else, d = -crit;
end; end
   end
   end
```

```
if (abs(d) >= crit), u = x + d;
else, if (d > 0), u = x + crit; else, u = x - crit; end; end
nf = nf+1; fu = objfun(u);
% Set a, b, x, u, v, w for the next iteration
if (fu <= fx)
    if (u < x), hb = x; fb = fx;
    else, ha = x; fa = fx; end
    v = w; fv = fw; w = x; fw = fx; x = u; fx = fu;
else
    if (u < x), ha = u; fa = fu;
    else, hb = u; fb = fu; end
    if ((fu <= fw) | (w == x)), v = w; fv = fw; w = u; fw = fu;
    elseif ((fu <= fv) | (v == x) | (v == w)), v = u; fv = fu; end
end
% Check function convergence
if ((fa - fx) + (fb - fx) < crit), x2 = x; f2 = fx; return; end
xopt = x2; fopt = f2;
end
end
```

Example 10.3 Brent's Algorithm

Use Brent's search method to find the minimum of

$$f(x) = \exp(x) - 3x + \frac{0.02}{x} - \frac{0.00004}{x}$$

The initial point can be specified as $x_1 = 0.01$. Let the step size be $h = 0.2$ and the stopping criterion be 1×10^{-6} $x_1 = 0.01$. Let the step size be $h = 0.2$ and the stopping criterion be 1×10^{-6}.

Solution

```
>> x1 = 0.01; h = 0.2; crit = 1e-6; fun = @(x) exp(x) -3*x + 0.02/x - 0.00004/x;
>> [xopt, fopt, nf] = brentopt(fun, x1, h, crit)
xopt =
  1.0572
fopt =
 -0.2744
nf =
  12
```

10.1.4 SHUBERT-PIYAVSKII METHOD[2]

Consider a simple maximization problem of the form

Maximize $f(x)$

Subject to $a \le x \le b$

The function $f(x)$ is assumed to be Lipschitz continuous: there is a constant C whose value is assumed to be known, where $|f(x) - f(y)| \le C|x - y|$ for any $x, y \in [a, b]$. A sequence of points $x_0, x_1, x_2, \cdots$, converging to the global maximum is generated by sequential construction of a sawtooth cover over the function. The Shubert-Piyavskii algorithm is as follows:

1) Choose a first sample point x_0 at the midpoint of the interval $[a, b]$ as $x_0 = (a + b)/2$.
2) Draw a pair of lines with slope C at x_0: $y(x) = y_0 + C|x - x_0|$. The intersection of this pair

of lines with the endpoints of the interval $[a, b]$ gives an initial sawtooth consisting of two points, given by $[(t_1, z_1), (t_2, z_2)] = [(b, y_0 + c(b - a)/2), (a, y_0 + c(b - a)/2)]$.

3) Construct a sequence of the sawtooth vertices $[(t_1, z_1), (t_2, z_2), \cdots, (t_n, z_n)](z_1 \le z_{21} \le \cdots \le z_n)$.
4) Determine the new sawtooth cover $[(t_1, z_1), (t_2, z_2), \cdots, (t_{n-1}, z_{n-1}), (t_l, z_l), (t_r, z_r)]$, where $z_l = z_r = (z_n + y_n)/2$, $t_l = t_n - (z_n - y_n)/(2C)$, $t_r = t_n + (z_n - y_n)/(2C)$.
5) Stop the procedure when $|z_n - y_n| < \varepsilon$.

The MATLAB function *shubertopt* implements the Shubert-Piyavskii algorithm. The basic syntax is

```
[xopt, fopt, nf] = shubertopt(objfun, a, b, C, crit, nfmax)
```

where objfun is the objective function, x1 is the initial point, a and b are the endpoints of the initial interval, C is the Lipschitz constant, crit is the stopping criterion, nfmax is the maximum number of function evaluations, xopt is the resultant optimum point, fopt is the function value at the optimum point, and nf is the number of function evaluations.

```
function [xopt, fopt, nf] = shubertopt(objfun, a, b, C, crit, nfmax)
% shubertopt.m: optimization by Shubert-Piyavskii algorithm
%             1-D maximization of Lipschitz functions
% inputs:
%     objfun: objective function
%     a,b: end points of initial interval
%     C: Lipschitz constant
%     crit: tolerance
%     nfmax: maximum number of function evaluations
% outputs:
%     xopt: optimum point
%     fopt: function value at the optimum point
%     nf: number of function evaluations
% Sample run:
%     C = 8; a = -3; b = 8; crit = 1e-6; nfmax = 2000;
%     fun = @(x) -sin(x)-sin(3.5*x);
%     [xopt,fopt,nf] = shubertopt(fun, a, b, C, crit, nfmax)

nf = 0; x0 = (a + b)/2; nf = nf + 1; y0 = objfun(x0); ymax = y0; xmax = x0; fmax = y0 +
C*(b - a)/2;
T(1) = b; Z(1) = y0 + C*(b - a)/2; T(2) = a; Z(2) = y0 + C*(b - a)/2; n = 2;
while ((fmax - ymax) > crit & nf <= nfmax)
    tn = T(n); zn = Z(n); nf = nf + 1; yn = objfun(tn);
    if (yn > ymax), ymax = yn; xmax = tn; end
    zL = (zn + yn)/2; zR = zL; tL = tn - (zn - yn)/2/C; tR = tn + (zn - yn)/2/C;
    % Replace T(n) and Z(n) by tl,zl and tr,zr
    ind1 = 0; ind2 = 0;
    if (tL >= a & tL <= b), ind1 = 1; end
    if (tR >= a & tR <= b), ind2 = 1; end
    if (ind1 == 1 & ind2 == 0), T(n) = tL; Z(n) = zL;
    elseif (ind1 == 0 & ind2 == 1), T(n) = tR; Z(n) = zR;
    elseif (ind1 == 1 & ind2 == 1), T(n) = tL; Z(n) = zL; T(n+1) = tR; Z(n+1) = zR;
    n = n+1; end
    [Z, Indx] = sort(Z); T = T(Indx); fmax = Z(n);
end
xopt = xmax; fopt = ymax;
end
```

Example 10.4 Shubert-Piyavskii Algorithm

Use the Shubert-Piyavskii method to find the maximum of $f(x) = -\sin(1.2x)-\sin(3.5x)$. The initial interval can be specified as $[-3,8]$. Let the Lipschitz constant be $C = 8$, the stopping criterion be 1×10^{-6}, and the maximum number of function evaluations be 2000.

Solution

```
>> C = 8; a = -3; b = 8; crit = 1e-6; nfmax = 2000; fun = @(x) -sin(1.2*x)-
sin(3.5*x);
>> [xopt,fopt,nf] = shubertopt(fun, a, b, C, crit, nfmax)
xopt =
  3.2162
fopt =
  1.6239
nf =
    2000
```

10.1.5 Steepest Descent Method

Consider a minimization problem of the form

Minimize $f(x)$

where $x = [x_1, x_2, \cdots, x_n]^T$ is a column vector of n real-valued variables. Let x_k be the point at the kth iteration. Let's choose a downhill direction d and a step size $\alpha > 0$ such that the new point $x_k + \alpha d$ yields an improved function value: $f(x_k + \alpha d) < f(x_k)$. To select d, we can use the Taylor expansion on $f(x)$ about x_k:

$$f(x_k + \alpha d) = f(x_k) + \alpha \nabla f(x_k)^T d + O(\alpha^2)$$

where $\nabla f(x)^T = [\partial f(x)/\partial x_1, \partial f(x)/\partial x_2, \cdots, \partial f(x)/\partial x_n]$. We can see that δf, the change in f, is given as

$$\delta f = f(x_k + \alpha d) - f(x_k) = \alpha \nabla f(x_k)^T d + O(\alpha^2) \cong \alpha \nabla f(x_k)^T d$$

For $\delta f < 0$ (i.e., reduction in f), d should be a descent direction that satisfies

$$\nabla f(x_k)^T d < 0$$

In the steepest descent method, the direction $d = d_k$ at the kth iteration is chosen as

$$d_k = -\nabla f(x_k)$$

This choice of $d = d_k$ satisfies $\nabla f(x_k)^T d < 0$, since

$$\nabla f(x_k)^T d = \nabla f(x_k)^T d_k = -\nabla f(x_k)^T \nabla f(x_k) = -\|\nabla f(x_k)\|^2 < 0$$

Next we have to determine the step size $\alpha = \alpha_k$ along the direction d_k. As we move along d_k, the objective function f and the design variable depend only on $\alpha = \alpha_k$:

$$f(x_k + \alpha d_k) \equiv f(\alpha), \ x_k + \alpha d \equiv x(\alpha)$$

The directional derivative of the function f along the direction d_k can be expressed as

$$\frac{df\,(\alpha^*)}{d\alpha} = \nabla f\,(x_k + \alpha^* d_k)^T d_k$$

We may choose $\alpha = \alpha^*$ so as to minimize $f\,(\alpha) \equiv f\,(x_k + \alpha d_k)$. The steepest descent algorithm is given as follows:

1) Specify a starting point x_0 and stopping criterion ε. Set $k = 0$.
2) Calculate $\nabla f\,(x_k)$. Stop if $\|\nabla f\,(x_k)\| \le \varepsilon$. Otherwise, calculate a normalized direction vector $d_k = -\nabla f\,(x_k)/\|\nabla f\,(x_k)\|$.
3) Determine $\alpha_k = \alpha^*$ using an approximate line search method. Update the design variable: $x_{k+1} = x_k + \alpha_k d_k$.
4) Calculate the function value at the new point $f\,(x_{k+1})$. Stop if $|f\,(x_{k+1}) - f\,(x_k)| \le \varepsilon$. Otherwise, set $k = k + 1$, $x_k = x_{k+1}$ and go to step 2).

The MATLAB function *sdopt* implements the steepest descent algorithm. The basic syntax is
`[xopt,fopt, iter] = sdopt(fun, delfun, x0, alpha0, crit, kmax)`

where fun is the objective function, delfun is the gradient of the objective function, x0 is the initial point vector, alpha0 is the initial step size, crit is the stopping criterion, kmax is the maximum number of iterations, xopt is the resultant optimum point, fopt is the function value at the optimum point, and iter is the number of iterations.

```
function [xopt,fopt, iter] = sdopt(fun, delfun, x0, alpha0, crit, kmax)
% sdopt.m: steepest descent method
% inputs:
%      fun: objective function
%      delfun: gradient of fun
%      x0: starting point
%      alpha0: initial step size
%      crit: stopping criterion
%      kmax: maximum iterations
% outputs:
%      xopt: optimal point
%      fopt: function value at the optimal point (=f(xopt))
%      iter: number of iterations
% Example:
%      fun = @(x) 100*(x(2)-x(1)^2)^2 + (1-x(1))^2;
%      delfun = @(x) [-400*x(1)*(x(2)-x(1)^2)+2*(x(1)-1); 200*(x(2)-x(1)^2)];
%      x0 = [-1.2 1]; crit = 1e-6; alpha0 = 5; kmax = 1e3;
%      [xopt,fopt, iter] = sdopt(fun, delfun, x0, alpha0, crit, kmax)

nfmax = 10; % maximum function calls during a line search
ni = 2; % indicate poor interval reductions before sectioning
h = alpha0; beta = 0.9; xz = x0; f = fun(x0); x1 = 0; f1 = f; fs = f; k = 0; nc = 0; fold
= f;
while (1)
    k = k + 1;
    if k > kmax, disp('Number of maximum iterations exceeded.'); break; end
    x = xz; delf = delfun(x);
    if abs(norm(delf)) <= crit, break; end
    d = -delf'; % steepest descent direction vector
    d = d / norm(d); % row vector
    [xs, fs] = quadappx(fun, xz, x, d, x1, f1, h, nfmax, ni, crit);
```

```
          x1 = 0; f1 = fs; xz = xz + xs*d; h = beta * xs; fpr = fs;
          if abs(fpr - fold) < crit
              nc = nc + 1;
              if nc == ni, break; end; else, nc = 0; end
      fold = fpr;
  end
  iter = k; xopt = x; fopt = fs;
  end

  function [x2, f2] = quadappx(fun, xz, x, d, x1, f1, h, nfmax, ni, crit)
  % Quadratic approximation method for line search
  tau = (sqrt(5)-1)/2; x2 = x1 + h; f2 = fun(xz + x2*d);
  if f2 < f1
    while (1)
          h = h / tau; x3 = x2 + h; f3 = fun(xz + x3*d);
          if f3 > f2, break;
          else, f1 = f2; x1 = x2; f2 = f3; x2 = x3; end
    end
  else
    x3 = x2; f3 = f2;
    while (1)
          x2 = (1 - tau) * x1 + tau * x3; f2 = fun(xz + x2*d);
          if f2 <= f1, break; else, x3 = x2; f3 = f2; end
    end
  end
  sf = 0.05; % 0 < sf < 0.5
  if (x1 >= x2 | x2 >= x3), disp('Incorrect interval.'); return;
  elseif (f1 <= f2 | f2 >= f3), disp('Not 3-point pattern.'); return;
  end
  vs = 0; vc = 0; wc = 0; j = 1;
  while j <= nfmax
      sold = abs(x3-x1); fmold = (f1+f2+f3)/3.;
      if vs == 0
          A = (x1-x2)*(x1-x3); B=(x2-x1)*(x2-x3); C=(x3-x1)*(x3-x2);
          x4 = (f1*(x2+x3)/A + f2*(x1+x3)/B + f3*(x1+x2)/C)/(f1/A+f2/B+f3/C)/2;
      else
          if x2 <= (x1+x3)/2, x4 = x2 + (1-tau)*(x3-x2);
          else, x4 = x3 - (1-tau)*(x2-x1); end
          vs = 0;
      end
      %safeguard against coincident points
      dxs = sf*min(abs(x2-x1), abs(x3-x2));
      if abs(x4-x1) < dxs, x4 = x1+dxs;
      elseif abs(x4-x3) < dxs, x4 = x3-dxs;
      elseif abs(x4-x2) < dxs
          if x2 > (x1+x3)/2, x4 = x2-dxs;
          else, x4 = x2+dxs; end
      end
      f4 = fun(xz + x4*d);
      if (x4 > x2)
          if (f4 >= f2), x3 = x4; f3 = f4;
          else, x1 = x2; f1 = f2; x2 = x4; f2 = f4; end
      else
          if (f4 >= f2), x1 = x4; f1 = f4;
          else, x3 = x2; f3 = f2; x2 = x4; f2 = f4; end
```

```
  end
  snew = abs(x3-x1); fmnew = (f1+f2+f3)/3.;
  if abs(x3-x1) <= crit, break; end
  if abs(fmnew-fmold) <= crit
        wc = wc + 1;
        if wc == 2, break; end; else, wc = 0; end
  if snew/sold > tau, vc = vc + 1;
        if vc == ni, vc = 0; vs = 1; end
  else, vc = 0; vs = 0; end
end
end
```

Example 10.5 Steepest Descent Method

Use the steepest descent method to find the minimum of the Rosenbrock's function given by

$$f(x) = 100(x_2 - x_1^2)^2 + (1 - x_1)^2$$

The initial point can be specified as [-1.2, 1]. Use the initial step size of $\alpha_0 = 5$. The stopping criterion and the maximum number of function evaluations can be set as 1×10^{-6} and 10,000, respectively.

Solution

The gradient of the given function is

$$\nabla f(x) = \begin{bmatrix} 400x_1(x_1^2 - x_2) + 2(x_1 - 1) \\ 200(x_2 - x_1^2) \end{bmatrix}$$

```
>> fun = @(x) 100*(x(2)-x(1)^2)^2 + (1-x(1))^2;
>> delfun = @(x) [400*x(1)*(x(1)^2-x(2))+2*(x(1)-1); 200*(x(2)-x(1)^2)];
>> x0 = [-1.2 1]; crit = 1e-6; alpha0 = 5; kmax = 1e4;
>> [xopt,fopt, iter] = sdopt(fun, delfun, x0, alpha0, crit, kmax)
xopt =
   0.9761   0.9526
fopt =
  5.7226e-04
iter =
     2352
```

10.1.6 Newton's Method

Newton's method uses second derivatives, which is called the Hessian matrix. The basic idea is to construct a quadratic approximation to the objective function $f(x)$ and minimize the quadratic approximation. At a point x_k, the quadratic approximation can be constructed as

$$s(x) = f(x_k) + \nabla f(x_k)^T(x - x_k) + \frac{1}{2}(x - x_k)^T \nabla^2 f(x_k)(x - x_k)$$

The minimum of $s(x)$ is found by setting $\nabla s(x) = 0$, which gives

$$[\nabla^2 f(x_k)]d_k = -\nabla f(x_k)$$

where $d_k = x_{k+1} - x_k$. Thus, a new point is given by

$$x_{k+1} = x_k + d_k = x_k - [\nabla^2 f(x_k)]^{-1} \nabla f(x_k)$$

At the new point, we again calculate the gradient and the Hessian matrix and update the point. Newton's method is as follows:

1) Choose a starting point x_0. Set the stopping criterion ε and iteration index $k = 0$.
2) Calculate $\nabla f(x_k)$ and $\nabla^2 f(x_k)$. Stop if $\|\nabla f(x_k)\| \leq \varepsilon$. Otherwise, find the direction vector d_k.
3) Update the current point by $x_{k+1} = x_k + d_k$.
4) Calculate $f(x_{k+1})$. Stop if $|f(x_{k+1}) - f(x_k)| \leq \varepsilon$ is satisfied for two successive iterations. Otherwise, set $k = k + 1$, $x_k = x_{k+1}$ and go to step 2).

The MATLAB function *newtonopt* implements Newton's method. The basic syntax is
`[xopt,fopt, iter] = newtonopt(fun, x0, crit, kmax)`

where fun is the objective function, x0 is the initial point vector, crit is the stopping criterion, kmax is the maximum number of iterations, xopt is the resultant optimum point, fopt is the function value at the optimum point, and iter is the number of iterations. The Hessian matrix is calculated by the function *jacob*.

```
function [xopt,fopt, iter] = newtonopt(fun, x0, crit, kmax)
% newtonopt.m: minimization by Newton's method
% inputs:
%       fun: objective function
%       x0: starting point
%       crit: stopping criterion
%       kmax: maximum iterations
% outputs:
%       xopt: optimal point
%       fopt: function value at the optimal point (=f(xopt))
%       iter: number of iterations
% Example:
%       fun = @(x) 100*(x(2)-x(1)^2)^2 + (1-x(1))^2;
%       x0 = [-1.2 1]; crit = 1e-6; kmax = 1e3;
%       [xopt,fopt, iter] = newtonopt(fun, x0, crit, kmax)

h = 1e-4; fx = fun(x0); nf = length(fx); nx = length(x0);
xs(1,:) = x0(:)'; % initial row solution vector
for k = 1:kmax
    dx = -jacob(fun, xs(k,:), h)\fx(:); % -[df]^(-1)*fx
    xs(k+1,:) = xs(k,:) + dx'; fx = fun(xs(k+1,:));
    if norm(fx) < crit | norm(dx) < crit, break; end
end
x = xs(k+1,:); xopt = x; fopt = fx; iter = k;
end

function Hs = jacob(fun, x, h)
% Jacobian of f(x)
hd = 2*h; n = length(x); x = x(:)'; M = eye(n);
for k = 1:n, Hs(:,k) = (fun(x + M(k,:)*h) - fun(x-M(k,:)*h))'/hd; end
end
```

Example 10.6 Newton's Method

Use Newton's method to find the minimum of the Rosenbrock's function given by

$$f(x) = 100(x_2 - x_1^2)^2 + (1 - x_1)^2$$

The initial point can be specified as [-1.2, 1]. The stopping criterion and the maximum number of function evaluations can be set as 1×10^{-6} and 1000, respectively.

Solution

```
>> fun = @(x) 100*(x(2)-x(1)^2)^2 + (1-x(1))^2; x0 = [-1.2 1]; crit = 1e-6; kmax
= 1e3;
>> [xopt,fopt, iter] = newtonopt(fun, x0, crit, kmax)
xopt =
  0.9996   0.9993
fopt =
  9.6530e-07
iter =
  646
```

10.1.7 CONJUGATE GRADIENT METHOD

If $f(x)$ is quadratic and is minimized in each search direction, it converges in at most n iterations, because its search directions are conjugate. The conjugate gradient method can find the minimum of a quadratic objective function of n variables in n iterations. This method combines current information about the gradient vector with information on gradient vectors from previous iterations to get the new search direction. The new direction is calculated by a linear combination of the current gradient and the previous search direction. This method is also powerful on general nonquadratic functions. Consider the minimization of a quadratic function given by

$$s(x) = \frac{1}{2}x^T A x + c^T x$$

where
  $x = [x_1, x_2, \cdots, x_n]^T$ is a column vector of n real-valued variables
  A is a positive definite symmetric matrix
  c is a constant column vector
  The gradient of $s(x)$ at the point x_k can be expressed as

$$\nabla s(x_k) = g_k = A x_k + c$$

The search direction d is chosen as the steepest descent direction, and the first direction d_0 is chosen as $-g_0$. A new point x_{k+1} is determined by minimizing $s(x)$ along the direction d_k as follows:

$$x_{k+1} = x_k + \alpha_k d_k$$

The value of α_k is obtained from the minimization of $f(\alpha) = s(x_k + \alpha d_k)$. From $df(\alpha)/d\alpha = 0$, we have

$$\alpha_k = -\frac{d_k^T g_k}{d_k^T A d_k}, \; d_k^T g_{k+1} = 0$$

The new direction d_{k+1} is given by

$$d_{k+1} = -g_{k+1} + \beta_k d_k, \quad \beta_k = \frac{g_{k+1}^T(g_{k+1} - g_k)}{g_k^T g_k}$$

This update of d_{k+1} is referred to as the Polak-Ribière algorithm. Considering the fact that $g_{k+1}^T g_k = 0$, β_k can be calculated by

$$\beta_k = \frac{g_{k+1}^T g_{k+1}}{g_k^T g_k}$$

which is referred to as the Fletcher-Reeves algorithm.[3] Figure 10.3 illustrates the update procedure of d_k for a two-variable quadratic function.[4]

The conjugate gradient method is as follows:

1) Choose an initial point x_0. Set $k = 0$ and calculate $d_0 = -\nabla s(x_0)$.
2) Determine α_k and calculate x_{k+1}.
3) Calculate β_k and the next direction d_{k+1}.
4) Set $k = k + 1$ and go to step 2).

The MATLAB function *cgopt* implements the conjugate gradient algorithm. The basic syntax is
`[xopt,fopt, iter] = cgopt(fun, delfun, x0, alpha0, crit, kmax)`

where fun is the objective function, delfun is the gradient of the objective function, x0 is the initial point vector, alpha0 is the initial step size, crit is the stopping criterion, kmax is the maximum number of iterations, xopt is the resultant optimum point, fopt is the function value at the optimum point, and iter is the number of iterations.

```
function [xopt,fopt, iter] = cgopt(fun, delfun, x0, alpha0, crit, kmax)
% cgopt.m: conjugate gradient method (Fletcher-Reeves algorithm)
% inputs:
%      fun: objective function
%      delfun: gradient of fun
```

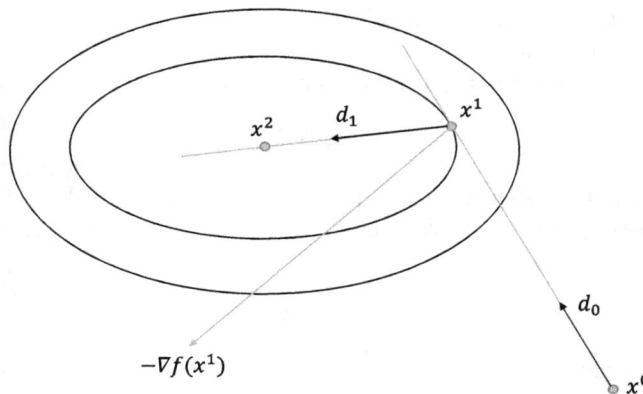

FIGURE 10.3 Conjugate gradient for a two-variable quadratic function. (From Belegundu, A.D. and Chandrupatla, T.R., *Optimization Concepts and Applications in Engineering*, 2nd ed., Cambridge University Press, New York, NY, 2011, p. 108.)

```
%     x0: starting point
%     alpha0: initial step size (line search)
%     crit: stopping criterion
%     kmax: maximum iterations
% outputs:
%     xopt: optimal point
%     fopt: function value at the optimal point (=f(xopt))
%     iter: number of iterations
% Example:
%     fun = @(x) 100*(x(2)-x(1)^2)^2 + (1-x(1))^2;
%     delfun = @(x) [-400*x(1)*(x(2)-x(1)^2)+2*(x(1)-1); 200*(x(2)-x(1)^2}];
%     x0 = [-1.2 1]; crit = 1e-6; alpha0 = 1; kmax = 1e3;
%     [xopt,fopt, iter] = cgopt(fun, delfun, x0, alpha0, crit, kmax)

n = length(x0); % n: number of variables
nfmax = 10; % maximum function calls during a line search
ni = 2; % indicate poor interval reductions before sectioning
h = alpha0; beta = 0.9; xz = x0; f = fun(x0); x1 = 0; f1 = f; fs = f; k = 0; nc = 0; nd =
0; fold = f;
while (1)
    k = k + 1;
    if k > kmax, disp('Maximum possible iterations exceeded.'); break; end
    nd = nd + 1; x = xz; delf = delfun(x); normd = norm(delf);
    if abs(normd) <= crit, break; end
    if nd == 1, d = -delf'; % row vector
    else, beta = (normd/normd0)^2; d = -delf' + beta*d0;
    end
    d0 = d; d = d/norm(d);
    [xs, fs] = quadappx(fun, xz, x, d, x1, f1, h, nfmax, ni, crit);
    x1 = 0; f1 = fs; xz = xz + xs*d; h = beta * xs; fpr = fs;
    if abs(fpr - fold) < crit
      nc = nc + 1; if nc == ni, break; end
    else, nc = 0;
    end
    fold = fpr; normd0 = normd;
    if nd == n+1, nd = 0; end
end
iter = k; xopt = x; fopt = fs;
end


function [x2, f2] = quadappx(fun, xz, x, d, x1, f1, h, nfmax, ni, crit)
% Quadratic approximation method for line search
tau = (sqrt(5)-1)/2;
x2 = x1 + h; f2 = fun(xz + x2*d);
if f2 < f1
    while (1)
      h = h / tau; x3 = x2 + h; f3 = fun(xz + x3*d);
      if f3 > f2, break; else, f1 = f2; x1 = x2; f2 = f3; x2 = x3; end
    end
else
    x3 = x2; f3 = f2;
    while (1)
      x2 = (1 - tau) * x1 + tau * x3; f2 = fun(xz + x2*d);
      if f2 <= f1, break; else, x3 = x2; f3 = f2; end
    end
```

```
end
sf = 0.05; % 0 < sf < 0.5
if (x1 >= x2 | x2 >= x3), disp('Incorrect interval.'); return;
elseif (f1 <= f2 | f2 >= f3), disp('Not 3-point pattern.'); return; end
vs = 0; vc = 0; wc = 0; j = 0;
while j <= nfmax
    sold = abs(x3-x1); fmold = (f1+f2+f3)/3.;
    if vs == 0
        A = (x1-x2)*(x1-x3); B=(x2-x1)*(x2-x3); C=(x3-x1)*(x3-x2);
        x4 = (f1*(x2+x3)/A + f2*(x1+x3)/B + f3*(x1+x2)/C)/(f1/A+f2/B+f3/C)/2;
    else
        if x2 <= (x1+x3)/2, x4 = x2 + (1-tau)*(x3-x2);
        else, x4 = x3 - (1-tau)*(x2-x1);
        end
        vs = 0;
    end
    dxs = sf*min(abs(x2-x1), abs(x3-x2));
    if abs(x4-x1) < dxs, x4 = x1+dxs;
    elseif abs(x4-x3) < dxs, x4 = x3-dxs;
    elseif abs(x4-x2) < dxs
        if x2 > (x1+x3)/2, x4 = x2-dxs; else, x4 = x2+dxs; end
    end
    f4 = fun(xz + x4*d);
    if (x4 > x2)
        if (f4 >= f2), x3 = x4; f3 = f4; else, x1 = x2; f1 = f2; x2 = x4; f2 = f4; end
    else
        if (f4 >= f2), x1 = x4; f1 = f4; else, x3 = x2; f3 = f2; x2 = x4; f2 = f4; end
    end
    snew = abs(x3-x1); fmnew = (f1+f2+f3)/3.;
    if abs(x3-x1) <= crit, break; end
    if abs(fmnew-fmold) <= crit
        wc = wc + 1; if wc == 2, break; end
    else, wc = 0;
    end
    if snew/sold > tau
        vc = vc + 1; if vc == ni, vc = 0; vs = 1; end
    else, vc = 0; vs = 0;
    end
end
end
```

Example 10.7 Conjugate Gradient Method

Use the conjugate gradient method to find the minimum of the Rosenbrock's function given by

$$f(x) = 100(x_2 - x_1^2)^2 + (1 - x_1)^2$$

The initial point can be specified as [-1.2, 1]. Use the initial step size of $\alpha_0 = 1$. The stopping criterion and the maximum number of function evaluations can be set as 1×10^{-6} and 10,000, respectively.

Solution

The gradient of the given function is given by

$$\nabla f(x) = \begin{bmatrix} 400x_1(x_1^2 - x_2) + 2(x_1 - 1) \\ 200(x_2 - x_1^2) \end{bmatrix}$$

```
>> fun = @(x) 100*(x(2)-x(1)^2)^2 + (1-x(1))^2;
>> delfun = @(x) [-400*x(1)*(x(2)-x(1)^2)+2*(x(1)-1); 200*(x(2)-x(1)^2)];
>> x0 = [-1.2 1]; crit = 1e-6; alpha0 = 1; kmax = 1e3;
>> [xopt,fopt, iter] = cgopt(fun, delfun, x0, alpha0, crit, kmax)
xopt =
  1.0000   1.0000
fopt =
  1.0380e-10
iter =
  28
```

10.1.8 QUASI-NEWTON METHOD

The search direction of the steepest descent method can be interpreted as being orthogonal to a linear approximation of the objective function at the current point x_k. Newton's method makes use of the 2nd-order approximation of $f(x)$ at x_k to determine a search direction.[5] The current point x_k can be updated by introducing a step-size parameter α_k:

$$x_{k+1} = x_k + \alpha_k d_k$$

where α_k is obtained from the minimization of $f(\alpha) = f(x_k + \alpha d_k)$. The direction vector d_k should be a descent direction to the function $f(x)$ at $x = x_k$:

$$\nabla f(x_k)^T d_k < 0 \quad \text{or} \quad -\nabla f(x_k)^T [\nabla^2 f(x_k)]^{-1} \nabla f(x_k) < 0$$

The Hessian can be replaced with a symmetric positive definite matrix F_k defined by

$$F_k = \nabla^2 f(x_k) + \gamma I$$

The parameter γ is chosen such that all eigenvalues of F_k are greater than a scalar $\delta > 0$.[6] The direction vector d_k can be obtained from

$$d_k = -[F_k]^{-1} \nabla f(x_k) = -H_k \nabla f(x_k)$$

where $H_k = [F_k]^{-1}$. Using this direction vector, the update formula can be written as

$$x_{k+1} = x_k + \alpha_k d_k = x_k - \alpha_k [F_k]^{-1} \nabla f(x_k) = x_k - \alpha_k H_k \nabla f(x_k)$$

The basic concept behind the quasi-Newton method is to start with a symmetric positive definite H and update it so that it contains the Hessian information. From the Taylor expansion, we have

$$\nabla f(x_{k+1}) = \nabla f(x_k) + \nabla^2 f(x_k)\delta_k$$

where $\delta_k = x_{k+1} - x_k$. Rearrangement of this equation gives

$$\nabla^2 f(x_k)\delta_k = \gamma_k = \nabla f(x_{k+1}) - \nabla f(x_k)$$

The matrix H_{k+1} should satisfy the quasi-Newton condition $H_{k+1}\gamma_k = \delta_k$. The matrix H_{k+1} can be updated according to

$$H_{k+1} = H_k + auu^T + bvv^T, \; H_0 = I$$

If we choose $u = \delta_k$ and $v = H_k\gamma_k$, the update formula can be expressed as

$$H_{k+1} = H_k - \frac{H_k\gamma_k\gamma_k^T H_k}{\gamma_k^T H_k\gamma_k} + \frac{\delta_k\delta_k^T}{\delta_k^T\gamma_k}$$

which is referred to as the Davidon-Fletcher-Powell (DFP) method.[3] Another well-known update formula is given by

$$H_{k+1} = H_k - \frac{\delta_k\gamma_k^T H_k + H_k\gamma_k\delta_k^T}{\delta_k^T\gamma_k} + \left(1 + \frac{\gamma_k^T H_k\gamma_k}{\delta_k^T\gamma_k}\right)\frac{\delta_k\delta_k^T}{\delta_k^T\gamma_k}$$

which is referred to as the Broyden-Fletcher-Goldfarb-Shanno (BFGS) method.

The MATLAB function *dfpopt* implements the David-Fletcher-Powell (DFP) method. The basic syntax is

```
[xopt,fopt, iter] = dfpopt(fun, delfun, x0, alpha0, crit, kmax)
```

where fun is the objective function, delfun is the gradient of the objective function, x0 is the initial point vector, alpha0 is the initial step size, crit is the stopping criterion, kmax is the maximum number of iterations, xopt is the resultant optimum point, fopt is the function value at the optimum point, and iter is the number of iterations.

```
function [xopt,fopt, iter] = dfpopt(fun, delfun, x0, alpha0, crit, kmax)
% dfpopt.m: quasi-Newton method (Davidon-Fletcher-Powell(DFP) algorithm)
% inputs:
%     fun: objective function
%     delfun: gradient of fun
%     x0: starting point
%     alpha0: initial step size (line search)
%     crit: stopping criterion
%     kmax: maximum iterations
% outputs:
%     xopt: optimal point
%     fopt: function value at the optimal point (=f(xopt))
%     iter: number of iterations
% Example:
%     fun = @(x) 100*(x(2)-x(1)^2)^2 + (1-x(1))^2;
%     delfun = @(x) [-400*x(1)*(x(2)-x(1)^2)+2*(x(1)-1); 200*(x(2)-x(1)^2)];
%     x0 = [-1.2 1]; crit = 1e-6; alpha0 = 1; kmax = 1e3;
%     [xopt,fopt, iter] = dfpopt(fun, delfun, x0, alpha0, crit, kmax)

n = length(x0); % n: number of variables
nfmax = 10; % maximum function calls during a line search
ni = 2; % indicate poor interval reductions before sectioning
h = alpha0; beta = 0.9; x = x0'; xz = x; f = fun(x); x1 = 0; f1 = f; fs = f;
k = 0; nc = 0; nd = 0; fold = f; H = eye(n);
while (1)
    k = k + 1;
    if k > kmax, disp('Maximum possible iterations exceeded.'); break; end
    x = xz; delf = delfun(x);
```

```
       if abs(norm(delf)) <= crit, break; end
       d = -H*delf; d = d/norm(d);
       [xs, fs] = quadappx(fun, xz, x, d, x1, f1, h, nfmax, ni, crit);
       x1 = 0; f1 = fs; xz = xz + xs*d; delf0 = delfun(xz); delf0 = delf0 - delf;
       d0 = H*delf0; Qh = delf0'*d0;
       if abs(Qh) > 1e-15, H = H - d0*d0'/Qh; end
       d0 = xs*d; Pq = d0'*delf0;
       if abs(Pq) > 1e-15, H = H + d0*d0'/Pq; end
       h = beta * xs; fpr = fs;
       if abs(fpr - fold) < crit
         nc = nc + 1; if nc == ni, break; end
       else, nc = 0;
       end
       fold = fpr; if nd == n+1, nd = 0; H = eye(n); end
   end
   iter = k; xopt = x; fopt = fs;
   end


   function [x2, f2] = quadappx(fun, xz, x, d, x1, f1, h, nfmax, ni, crit)
   % Quadratic approximation method for line search
   tau = (sqrt(5)-1)/2; x2 = x1 + h; f2 = fun(xz + x2*d);
   if f2 < f1
       while (1)
         h = h / tau; x3 = x2 + h; f3 = fun(xz + x3*d);
         if f3 > f2, break; else, f1 = f2; x1 = x2; f2 = f3; x2 = x3; end
       end
   else
       x3 = x2; f3 = f2;
       while (1)
         x2 = (1 - tau) * x1 + tau * x3; f2 = fun(xz + x2*d);
         if f2 <= f1, break; else, x3 = x2; f3 = f2; end
     end
   end
   sf = 0.05; % 0 < sf < 0.5
   if (x1 >= x2 | x2 >= x3), disp('Incorrect interval.'); return;
   elseif (f1 <= f2 | f2 >= f3), disp('Not 3-point pattern.'); return; end
   vs = 0; vc = 0; wc = 0; j = 0;
   while j <= nfmax
         sold = abs(x3-x1); fmold = (f1+f2+f3)/3.;
         if vs == 0
           A = (x1-x2)*(x1-x3); B=(x2-x1)*(x2-x3); C=(x3-x1)*(x3-x2);
           x4 = (f1*(x2+x3)/A + f2*(x1+x3)/B + f3*(x1+x2)/C)/(f1/A+f2/B+f3/C)/2;
       else
           if x2 <= (x1+x3)/2, x4 = x2 + (1-tau)*(x3-x2);
           else, x4 = x3 - (1-tau)*(x2-x1);
           end
           vs = 0;
       end
     end
     dxs = sf*min(abs(x2-x1), abs(x3-x2));
     if abs(x4-x1) < dxs, x4 = x1+dxs;
     elseif abs(x4-x3) < dxs, x4 = x3-dxs;
     elseif abs(x4-x2) < dxs
         if x2 > (x1+x3)/2, x4 = x2-dxs; else, x4 = x2+dxs; end
     end
     f4 = fun(xz + x4*d);
```

```
if (x4 > x2)
     if (f4 >= f2), x3 = x4; f3 = f4; else, x1 = x2; f1 = f2; x2 = x4; f2 = f4; end
else
     if (f4 >= f2), x1 = x4; f1 = f4; else, x3 = x2; f3 = f2; x2 = x4; f2 = f4; end
end
snew = abs(x3-x1); fmnew = (f1+f2+f3)/3.;
if abs(x3-x1) <= crit, break; end
if abs(fmnew-fmold) <= crit
     wc = wc + 1; if wc == 2, break; end
else, wc = 0;
end
if snew/sold > tau
     vc = vc + 1; if vc == ni, vc = 0; vs = 1; end
else, vc = 0; vs = 0;
end
end
end
```

Example 10.8 Quasi-Newton Method

Wood's function is given by

$$f(x) = 100(x_2 - x_1^2)^2 + (1 - x_1)^2 + 90(x_4 - x_3^2)^2 + (1 - x_3)^2$$
$$+ 10.1[(x_2 - 1)^2 + (x_4 - 1)^2] + 19.8(x_2 - 1)(x_4 - 1)$$

Use the Davidon-Fletcher-Powell (DFP) method to find the minimum of Wood's function. The initial point can be specified as [-3, -1, -3, -1]. Use an initial step size of $\alpha_0 = 2$. The stopping criterion and the maximum number of function evaluations can be set as 1×10^{-6} and 1000, respectively.

Solution

The gradient of the given function is given by

$$\nabla f(x) = \begin{bmatrix} 400x_1(x_1^2 - x_2) + 2(x_1 - 1) \\ 200(x_2 - x_1^2) + 20.2(x_2 - 1) + 19.8(x_4 - 1) \\ 360x_3(x_3^2 - x_4) + 2(x_3 - 1) \\ 180(x_4 - x_3^2) + 20.2(x_4 - 1) + 19.8(x_2 - 1) \end{bmatrix}$$

```
>> fun = @(x) 100*(x(2)-x(1)^2)^2+(1-x(1))^2+90*(x(4)-x(3)^2)^2+(1-x(3))
^2+10.1*((x(2)-1)^2+(x(4)-1)^2)+19.8*(x(2)-1)*(x(4)-1);
>> delfun = @(x) [400*x(1)*(x(1)^2-x(2))+2*(x(1)-1); 200*(x(2)-x(1)^2)
+20.2*(x(2)-1)+19.8*(x(4)-1); 360*x(3)*(x(3)^2-x(4))+2*(x(3)-1); 180*(x
(4)-x(3)^2)+20.2*(x(4)-1)+19.8*(x(2)-1)];
>> x0 = [-3 -1 -3 -1]; crit = 1e-6; alpha0 = 2; kmax = 1e3;
>> [xopt,fopt, iter] = dfpopt(fun, delfun, x0, alpha0, crit, kmax)
xopt =
     1.0000
     1.0000
     1.0000
     1.0000
```

```
fopt =
    1.1890e-12
iter =
    39
```

10.2 LINEAR PROGRAMMING

10.2.1 FORMULATION OF LINEAR PROGRAMMING PROBLEMS

The general form of a linear programming (LP) problem consists of an objective function to be minimized or maximized and a set of constraints. It can be stated as

$$\text{Minimize } f(x) = c^T x = c_1 x_1 + c_2 x_2 + \cdots + c_n x_n$$

$$\text{Subject to } Ax = b, \; x \geq 0$$

where $x = [x_1, x_2, \cdots, x_n]^T$ is a column vector of n real-valued variables. The constraints $Ax = b$ can be expressed as

$$a_{11} x_1 + a_{12} x_2 + \cdots + a_{1n} x_n = b_1$$

$$a_{21} x_1 + a_{22} x_2 + \cdots + a_{2n} x_n = b_2$$

$$\vdots$$

$$a_{m1} x_1 + a_{m2} x_2 + \cdots + a_{mn} x_n = b_m$$

The constraints can take the form of inequality relations. The inequality constraints can be converted into equality forms by introducing slack variables. If a constraint appears in the form of ≤ (less than or equal to) as

$$a_{j1} x_1 + a_{j2} x_2 + \cdots + a_{jn} x_n \leq b_j$$

it can be converted into an equality constraint by adding a nonnegative slack variable x_{n+1} as follows:

$$a_{j1} x_1 + a_{j2} x_2 + \cdots + a_{jn} x_n + x_{n+1} = b_j$$

Similarly, if a constraint appears in the form of ≥ (greater than or equal to) as

$$a_{j1} x_1 + a_{j2} x_2 + \cdots + a_{jn} x_n \geq b_j$$

it can be converted into an equality constraint by subtracting a non-negative x_{n+1} as

$$a_{j1} x_1 + a_{j2} x_2 + \cdots + a_{jn} x_n - x_{n+1} = b_j$$

where x_{n+1} is known as a surplus variable.

A feasible region, $\mathscr{F}$, is the set of all feasible solutions:

$$\mathscr{F} = \{x \mid Ax = b, \; x \geq 0\}$$

An optimal solution $x = x^*$ is an element of the feasible region: $x^* \in \mathscr{F}$. The value of the objective function at the optimal point, $f(x^*)$, is called the optimal value. If the LP solution set contains only one element, the LP problem is said to have a unique optimum. If the optimal value, $f(x^*)$, approaches positive or negative infinity, the LP problem is said to have an unbounded optimum. A pivot operation used in the solution of LP problems is a sequence of elementary row operations that reduce the coefficients of a specified variable to unity in one of the equations and zero elsewhere.

10.2.2 SIMPLEX METHOD

The simplex method starts with an initial basic feasible solution in canonical form. The solution is improved by finding a new one. When a particular basic feasible solution is found and cannot be improved by finding new basic feasible solution, it is said that optimality is reached. If the minimum of the objective function in the feasible region is finite, then a vertex minimizer exists. Suppose that x_0 is a vertex and is not a minimizer. The simplex method generates an adjacent vertex x_1 with $f(x_1) < f(x_0)$. This procedure is continued until a vertex minimizer is reached.

Let A_α be the matrix whose rows are the rows of A that are associated with the constraints that are active at x. The matrix A_α is referred to as the active constraint matrix.

Denote A_{a_k} and the index set J as

$$A_{a_k} = \begin{bmatrix} a_{j1}^T \\ a_{j2}^T \\ \vdots \\ a_{jn}^T \end{bmatrix}, \; J_{k+1} = [j_1, j_2, \cdots, j_{l-1}, i^*, j_{l+1}, \cdots, j_n]$$

where i^* is the index that achieves the minimum. Given a vertex x_k, the vertex x_{k+1} is adjacent to x_k if $A_{a_{k+1}}$ differs from A_{a_k} by one row. The simplex method can be summarized as follows:

1) Specify x_0 and form A_{a_0} and J_0. Set $k = 0$.
2) Solve $A_{a_k}^T \mu_k = c$ for μ_k.
3) Solve $A_{a_k} d_k = e_l$ for d_k.
4) Calculate the residual vector $r_k = Ax_k - b$. If the index set is empty, stop. Otherwise, compute $\alpha_k = \min(-r_i/(a_i^T d_k))$.
5) Set $x_{k+1} = x_k + \alpha_k d_k$. Update $A_{a_{k+1}}$ and J_{k+1}. Set $k = k + 1$ and go to step 2).

10.2.3 TWO-PHASE SIMPLEX METHOD

The two-phase simplex method used to solve LP problems consists of phases I and II. In phase I, the simplex algorithm is used to determine whether the LP problem has a feasible solution. If a feasible solution exists, it provides a basic feasible solution in canonical form to start phase II. Phase II uses the simplex algorithm to determine whether the problem has a bounded optimum. If a bounded optimum exists, phase II finds the basic feasible solution that is optimal. The two-phase simplex method consists of the following steps:[7]

1) Rearrange the given system so that all constant terms b_i are positive or zero by changing the signs on both sides of any of the constraint equations.
2) Rewrite the objection function as $c_1x_1 + c_2x_2 + \cdots + c_nx_n + (-f) = 0$. Introduce to this system a set of artificial variables (which serve as basic variables in phase I) $y_1, y_2, \cdots, y_m$ ($y_i \geq 0$) so that it becomes $a_{j1}x_1 + a_{j2}x_2 + \cdots + a_{jn}x_n + y_j = b_j$ ($b_j \geq 0, j = 1, \cdots, m$).
3) Phase I. Define w as the sum of the artificial variables: $w = y_1 + y_2 + \cdots + y_m$. Use the simplex algorithm to find $x_i \geq 0 (i = 1,2, \cdots, n)$ and $y_i \geq 0 (i = 1,2, \cdots, m)$. The canonical form can be rearranged as follows:

$$a_{j1}x_1 + a_{j2}x_2 + \cdots + a_{jn}x_n + y_j = b_j \, (b_j \geq 0, j = 1, \cdots, m)$$

$$c_1x_1 + c_2x_2 + \cdots + c_nx_n + (-f) = 0$$

$$d_1x_1 + d_2x_2 + \cdots + d_nx_n + (-w) = -w_0, \, d_i = -(a_{1i}$$
$$\cdots + a_{mi}), \, -w_0 = -(b_1\cdots + b_m)$$

4) If the minimum of $w > 0$, no feasible solution exists and the procedure is terminated. If the minimum of $w = 0$, start phase II by eliminating the w equation and the columns corresponding to each of the artificial variables from the array.

5) Phase II. Apply the simplex method to the adjusted canonical form at the end of phase I to obtain a solution that optimizes the objective function $f(x)$.

The MATLAB function *simplexlp* implements the two-phase simplex method. The basic syntax is
[xopt, fopt] = simplexlp(A, b, c, constr)

where A is the coefficient matrix of constraints, b is a column vector denoting the right-hand side of the constraints, c is the row vector representing the coefficients of the objective function, constr is a string denoting the type of constraints ('<' or '>' for inequalities and '=' for equalities), xopt is the resultant optimum point, and fopt is the function value at the optimum point.

```
function [xopt, fopt] = simplexlp(A, b, c, constr)
% simplexlp.m: 2-phase LP minimization problem
% Problem: Minimize f(x) = c*x subject to Ax constr b, x >= 0
% inputs:
%     A,b: coefficients of constraint equations (and inequalities)
%     c: coefficients of the objective function
%     constr: type of constraints (example: constr = '==<>>=')
% outputs:
%     xopt: optimum point
%     fopt: value of the objective function at x=xopt
% Example: Minimize f(x)=2x1 + 3x2 + 2x2 - x4 + x5
%     subject to 3x1-3x2+4x3+2x4-x5=0, x1+x2+x3+3x4+x5=2
%     A = [3 -3 4 2 -1;1 1 1 3 1]; b = [0; 2]; c = [2 3 2 -1 1]; constr = '==';
%     [xopt, fopt] = simplexlp(A, b, c, constr)

b = b(:); c = c(:)'; [m, n] = size(A); n1 = n; nleq = 0; neq = 0; ncomv = 0;
if length(c) < n, c = [c zeros(1,n-length(c))]; end
for j = 1:m
    temx = zeros(m,1); temx(j) = 1;
    if(constr(j) == '<') % <=: less than or equal to
      A = [A temx]; nleq = nleq + 1;
    elseif(constr(j) == '>') % >=: greater than or equal to
      A = [A -temx];
    else % =: equality constraints
      neq = neq + 1;
    end
end
lenA = length(A);
if nleq == m
    c = [c zeros(1,lenA-length(c))]; A = [A;c]; A = [A [b;0]];
    [maux, A, z] = compsim(A, n1+1:lenA, 1, 1);
```

```
else
    A = [A eye(m) b];
    if m > 1, w = -sum(A(1:m,1:lenA)); else, w = -A(1,1:lenA); end
    c = [c zeros(1,length(A)-length(c))]; A = [A;c]; A = [A;[w zeros(1,m)
-sum(b)]];
    maux = lenA+1:lenA+m; mv = maux; [maux, A, z] = compsim(A, maux, 2, 1);
    nc = lenA + m + 1; x = zeros(nc-1, 1); x(maux) = A(1:m, nc); xm = x(mv);
    incomv = intersect(maux, mv);
    if (any(xm) ~= 0), disp(sprintf('\n\n Empty feasible region\n')); return
    else, if ~isempty(incomv), ncomv = 1; end; end
    A = A(1:m+1,1:nc); A =[A(1:m+1,1:lenA) A(1:m+1,nc)];
    [maux, A, z] = compsim(A, maux, 1, 2);
end
if (z == inf | z == -inf), return; end
[m, n] = size(A); x = zeros(n,1); x(maux) = A(1:m-1,n);
x = x(1:n1); z = -A(m,n); t = find(A(m,1:n-1) == 0);
if length(t) > m-1, disp('There are infinite solutions'); end
if ncomv == 1, disp('Redundant constraint(s).'); end
xopt = x; fopt = z;
end

function [maux, A, z]= compsim(A, maux, k, ph)
% Main loop of the simplex primal algorithm.
% Bland's rule to prevente cycling is used.
[m, n] = size(A); [mi, col] = Bland(A(m,1:n-1));
while ~isempty(mi) & mi < 0 & abs(mi) > eps
    t = A(1:m-k,col);
    if all(t <= 0)
     z = -inf; disp(sprintf('\n Unbounded optimal solution with z=%s\n',z));
return
    end
    [row, small] = minrtest(A(1:m-k,n),A(1:m-k,col));
    if ~isempty(row)
    if abs(small) <= 100*eps & k == 1, [s,col] = Bland(A(m,1:n-1)); end
    A(row,:) = A(row,:)/A(row,col); maux(row) = col;
    for i = 1:m, if i ~= row, A(i,:) = A(i,:)-A(i,col)*A(row,:); end; end
    [mi, col] = Bland(A(m,1:n-1));
   end
end
z = A(m,n);
end

function [m, j] = Bland(D)
% Apply the Bland's rule to the array D.
% m: first negative number in D, j: index of the entry m.
ind = find(D < 0);
if ~isempty(ind), j = ind(1); m = D(j);
else, m = []; j = []; end
end

function [row, mi] = minrtest(a, b)
% Minimum ratio test on vector a and vector b.
% row: index of the pivot row, mi: value of the minimum ratio.
m = length(a); c = 1:m; a = a(:); b = b(:); l = c(b > 0);
[mi, row] = min(a(l)./b(l)); row = l(row);
```

end

Example 10.9 Two-Phase Simplex Method (1)

Find the solution of the following LP using the two-phase method:

$$\text{Minimize } f(x) = 2x_1 + x_2$$

$$-x_1 + x_2 \leq 1, \; 2x_1 + x_2 \leq 2, \; x_1 \geq 0, \; x_2 \geq 0$$

Solution

$$A = \begin{bmatrix} 2 & 5 \\ 1 & 1 \\ 3 & 1 \end{bmatrix}, \; b = \begin{bmatrix} 20 \\ 6 \\ 9 \end{bmatrix}, \; c = [2 \quad 1]$$

Since all the constraints are inequality relations (>=, greater than or equal to), constr can be set as '> > >'.

```
>> A = [2 5; 1 1; 3 1]; b = [20; 6; 9]; c = [2 1]; constr = '> > >';
>> [xopt, fopt] = simplexlp(A, b, c, constr)
xopt =
  1.5000
  4.5000
fopt =
  7.5000
```

We can see that the optimum point is $x_1 = 1.5$, $x_2 = 4.5$ and the optimal value is $f(x_{opt}) = 7.5$.

Example 10.10 Two-Phase Simplex Method (2)[8]

In the platform support system shown in Figure 10.4, cable 1 can support 120, *lb*, cable 2 can support 160 *lb*, and cable 3 and 4 can support 100 *lb* each. A weight of × acting at *a* from the left support and *b* from the right support causes reactions of *bx*/(*a*+*b*) and *ax*/(*a*+*b*) respectively. Determine the maximum load that the system can support.

Solution
The problem can be formulated as

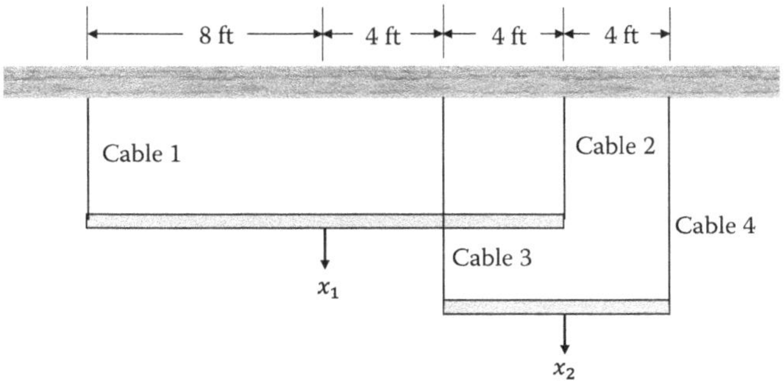

FIGURE 10.4 Platform support system.

$$\text{Maximize } f(x) = x_1 + x_2 \text{ or minimize } f(x) = -x_1 - x_2$$

$$\text{Subject to } x_2 \le 200, \ 4x_1 + 3x_2 \le 1280, \ 4x_1 + x_2 \le 960, \ x_1 \ge 0, \ x_2 \ge 0$$

```
>> A = [0 1;4 3;4 1]; b = [200; 1280; 960]; c = [-1 -1]; constr = '< < <';
>> [xopt, fopt] = simplexlp(A, b, c, constr)
xopt =
   170
   200
fopt =
  -370
```

10.2.4 Interior Point Method

Consider a linear programming problem rearranged in the standard form as

$$\text{Maximize } f(x) = c^T x$$

$$\text{Subject to } Ax = b, \ x \ge 0$$

where $x = [x_1, x_2, \cdots, x_n]^T$ is a column vector of n real-valued variables. In the interior point method, we start from a strictly interior point x^0 such that all the elements of x^0 are positive. From this interior point, we proceed in some direction to increase the objective function for maximization. This direction is obtained from the coefficient vector c and the coefficient matrix of constraints A. We use a scaling scheme and then a projection idea to find a reasonable direction. Various scaling schemes may be used; Karmarkar uses a nonlinear scaling scheme and applies it to an equivalent problem.[9] The interior point algorithm transforms the linear programming problem to a more convenient form and then searches through the interior of the feasible region using a direction of search toward its surface. The interior point method is described in the following steps[10]:

1) Let the number of variables be n and set $a(i, n+1) = b(i) - \sum_j a(i,j)$ and $c(n+1) = 1 \times 10^4$. Specify the initial point $x^0 = [11, \cdots, 1]$. Set $k = 0$.
2) Set $D^k = diag(x^k)$ and calculate an improved point using $x^{k+1} = x^k - s(D^k)^2(c - A^T\lambda^k)/norm(D^k(c - A^T\lambda^k))$ where $\lambda^k = (A(D^k)^2A^T)^{-1}A(D^k)^2c$ and the step s is chosen such that $s = \min\{norm(D^k(c - A^T\lambda^k))/(x_j^k(c_j - A_j^T\lambda^k))\} - \alpha$ where A_j is the jth column of the matrix A.
3) Stop if the primal and dual values of the objective functions are approximately equal. Otherwise set $k = k + 1$ and go to step 2).

The MATLAB function *barnslp* implements the interior point method by Barnes' algorithm. The basic syntax is

```
[xopt,fopt, basic] = barnslp(A,b,c,tol)
```

where A is the coefficient matrix of constraints, b is a column vector denoting the right-hand side of the constraints, c is a row vector representing the coefficients of the objective function, tol is the stopping criterion, xopt is the resultant optimum point, fopt is the function value at the optimum point, and basic is the list of basic variables.

```
function [xopt,fopt, basic] = barnslp(A,b,c,tol)
% barnslp.m: solution of LP by using Barnes' interior point method
% Problem: Minimize c'x subject to Ax = b (assumed to be non-degenerate)
```

```
% Inputs:
%      A,b: coefficients of constraint equations
%      c: coefficients of the objective function
% Outputs:
%      xopt: solution vector
%      fopt: function value at xsol
%      basic: list of basic variables
% Example: minimize f(x)=2x1 + 3x2 + 2x2 - x4 + x5
%      subject to 3x1-3x2+4x3+2x4-x5=0, x1+x2+x3+3x4+x5=2
%      A = [3 -3 4 2 -1;1 1 1 3 1]; b = [0; 2]; c = [2 3 2 -1 1]; tol = 1e-6;
%      [xopt,fopt, basic] = barnslp(A,b,c,tol)

% Initialization
x2 = [ ]; x = [ ]; [m n] = size(A); ctemp = zeros(1,n); clen = length(c);
for j = 1:clen, ctemp(j) = c(j); end
c = ctemp; aplus1 = b - sum(A(1:m,:)')'; cplus1 = 1000000;
A = [A aplus1]; c = [c cplus1]; B = [ ]; n = n+1; x0 = ones(1,n)'; x = x0;
alpha = 0.0001; lambda = zeros(1,m)'; iter = 0;
% Main step
while abs(c*x - lambda'*b) > tol
    x2 = x.*x; D = diag(x); D2 = diag(x2); AD2 = A*D2;
    lambda = (AD2*A')\(AD2*c'); dualres = c' - A'*lambda; normres =
    norm(D*dualres);
     for i = 1:n
     if dualres(i)>0, ratio(i) = normres/(x(i)*(c(i)-A(:,i)'*lambda));
     else, ratio(i) = inf; end
  end
    R = min(ratio) - alpha; x1 = x - R*D2*dualres/normres; x = x1; basiscount = 0;
    B = [ ]; basic = [ ]; cb = [ ];
    for k = 1:n
    if x(k)>tol, basiscount = basiscount+1; basic = [basic k]; end
  end
  % Non-degenerate problem
  if basiscount == m
    for k = basic, B = [B A(:,k)]; cb = [cb c(k)]; end
    primalsol = b'/B'; xopt = primalsol; break;
  end
  iter = iter + 1;
end
xopt = x(basic); fopt = c*x;
end
```

Example 10.11 Interior Point Method

Find the solution of the following LP using the interior point method:

$$\text{Minimize } f(x) = -2x_1 - x_2 - 4x_3$$

Solution

$$A = \begin{bmatrix} 1 & 1 & 1 & 1 & 0 \\ 1 & 2 & 3 & 0 & 1 \end{bmatrix}, b = \begin{bmatrix} 7 \\ 12 \end{bmatrix}, c = \begin{bmatrix} -2 & -1 & -4 \end{bmatrix}$$

```
>> A = [1 1 1 1 0; 1 2 3 0 1]; b = [7; 12]; c = [-2 -1 -4]; tol = 1e-6;
```

```
>> [xopt,fopt, basic] = barnslp(A,b,c,tol)
xopt =
      4.5000
      2.5000
fopt =
    -19.0000
basic =
     1    3
```

Example 10.12 Maximum Profit[11]

Five crude oils of different grades are processed in a refinery to produce gasoline, heating oil, jet fuel, and lube oil. Determine the amount of each crude oil to be processed per week in order to achieve the maximum profit. What is the profit per week? Table 10.1 shows the fractions of products that can be obtained from the crude oils, and Table 10.2 shows cost data for each crude oil. Price and demand data for products are shown in Table 10.3.

Solution
Let the amount of crude i be x_i ($i = 1,2,3,4,5$). Then

$$0 \le x_1 \le 80000, \ 0 \le x_2 \le 100000, \ 0 \le x_3 \le 100000, \ 0 \le x_4 \le 100000, \ 0 \le x_5 \le 60000$$

The maximum product demand (bbl/week) for each product is given by

Gasoline: $0.6x_1 + 0.5x_2 + 0.3x_3 + 0.4x_4 + 0.4x_5 \le 170{,}000$
Heating oil: $0.2x_1 + 0.2x_2 + 0.3x_3 + 0.3x_4 + 0.1x_5 \le 85{,}000$
Jet fuel: $0.1x_1 + 0.2x_2 + 0.3x_3 + 0.2x_4 + 0.2x_5 \le 75{,}000$
Lube oil: $0.2x_5 \le 30{,}000$

TABLE 10.1
Fractions of Products That Can Be Obtained from Crude Oils (bbl Product/bbl Crude)

Crude Oil	Gasoline	Heating Oil	Jet Fuel	Lube Oil
Crude 1	0.6	0.2	0.1	0
Crude 2	0.5	0.2	0.2	0
Crude 3	0.3	0.3	0.3	0
Crude 4	0.4	0.3	0.2	0
Crude 5	0.4	0.1	0.2	0.2

TABLE 10.2
Cost Data of Each Crude Oil

Crude Oil	Cost, $/bbl	Operating Cost, $/bbl	Availability, bbl/week
Crude 1	45	12.0	80,000
Crude 2	43	20.0	100,000
Crude 3	40	17.0	100,000
Crude 4	52	7.5	100,000
Crude 5	65	6.5	60,000

TABLE 10.3
Price and Demand for Each Product

Product	Price, $/bbl	Maximum Demand, bbl/week
Gasoline	105	170,000
Heating oil	95	85,000
Jet fuel	61	75,000
Lube oil	140	30,000

The profit $p(x)$ can be represented as

$$p(x) = 105(0.6x_1 + 0.5x_2 + 0.3x_3 + 0.4x_4 + 0.4x_5) + 95(0.2x_1 + 0.2x_2 + 0.3x_3 + 0.3x_4 + 0.1x_5)$$

$$+ 61(0.1x_1 + 0.2x_2 + 0.3x_3 + 0.2x_4 + 0.2x_5) + 140(0.2x_5)$$

$$- (45 + 12)x_1 - (43 + 20)x_2 - (40 + 17)x_3 - (52 + 7.5)x_4 - (65 + 6.5)x_5$$

$$= 31.1x_1 + 20.7x_2 + 21.3x_3 + 23.2x_4 + 20.2x_5$$

By introducing slack variables $(x_6, x_7, \cdots, x_{13})$, the optimization problem can be formulated as following:

$$\text{Minimize } p(x) = -31.1x_1 - 20.7x_2 - 21.3x_3 - 23.2x_4 - 20.2x_5$$

Subject to

$$0.6x_1 + 0.5x_2 + 0.3x_3 + 0.4x_4 + 0.4x_5 + x_6 = 170{,}000$$

$$0.2x_1 + 0.2x_2 + 0.3x_3 + 0.3x_4 + 0.1x_5 + x_7 = 85{,}000$$

$$0.1x_1 + 0.2x_2 + 0.3x_3 + 0.2x_4 + 0.2x_5 + x_8 = 75{,}000$$

$$x_1 + x_9 = 80000, \; x_2 + x_{10} = 100000, \; x_3 + x_{11} = 100000$$

$$x_4 + x_{12} = 100000, \; x_5 + x_{13} = 60000, \; x_i \geq 0 (i = 1, 2, \cdots, 13)$$

Then we have

$$A = \begin{bmatrix} 0.6 & 0.5 & 0.3 & 0.4 & 0.4 & 1 & 0 & 0 & 0 & 0 & 0 & 0 & 0 \\ 0.2 & 0.2 & 0.3 & 0.3 & 0.1 & 0 & 1 & 0 & 0 & 0 & 0 & 0 & 0 \\ 0.1 & 0.2 & 0.3 & 0.2 & 0.2 & 0 & 0 & 1 & 0 & 0 & 0 & 0 & 0 \\ 1 & 0 & 0 & 0 & 0 & 0 & 0 & 0 & 1 & 0 & 0 & 0 & 0 \\ 0 & 1 & 0 & 0 & 0 & 0 & 0 & 0 & 0 & 1 & 0 & 0 & 0 \\ 0 & 0 & 1 & 0 & 0 & 0 & 0 & 0 & 0 & 0 & 1 & 0 & 0 \\ 0 & 0 & 0 & 1 & 0 & 0 & 0 & 0 & 0 & 0 & 0 & 1 & 0 \\ 0 & 0 & 0 & 0 & 1 & 0 & 0 & 0 & 0 & 0 & 0 & 0 & 1 \end{bmatrix}, b = \begin{bmatrix} 170000 \\ 85000 \\ 75000 \\ 80000 \\ 100000 \\ 100000 \\ 100000 \\ 60000 \end{bmatrix}$$

The script *optcrude* uses the function *barnslp* to find the optimal solution. The function *barnslp* implements the interior point method by Barnes' algorithm.

```
% optcrude.m: LP by interior point method
A = zeros(8,13);
A(1:3,1:5) = [0.6 0.5 0.3 0.4 0.4;0.2 0.2 0.3 0.3 0.1;0.1 0.2 0.3 0.2 0.2];
A(1:3,6:8) = eye(3); A(4:8,1:5) = eye(5); A(4:8,9:13) = eye(5);
b = 1e4*[17 8.5 7.5 8 10 10 10 6]'; c = [-31.1 -20.7 -21.3 -23.2 -20.2]; tol = 1e-6;
[xopt,fopt,basic] = barnslp(A,b,c,tol)
```

Execution of optcrude produces the following results:
```
>> optcrude
xopt =
      1.0e+04 *
      8.0001
      8.9997
      6.9997
      8.0009
      6.0000
      1.0004
      3.0003
      1.9992
fopt =
  -8.9101e+06
basic =
    1    2    3    4    5   10   11   12
```

We can see that $x_1 = 80001$, $x_2 = 89993$, $x_3 = 70004$, $x_4 = 80004$, $x_5 = 60000$, $x_{10} = 10008$, $x_{11} = 29997$, $x_{12} = 19997$, and values of other variables are all zero. The desired solution is $[(x_1 \ x_2 \ x_3 \ x_4 \ x_5)] = [80001 \ 89993 \ 70004 \ 80004 \ 60000]$ and the maximum profit is $p(x) = 8,910,000$ ($/week).

10.3 CONSTRAINED OPTIMIZATION

10.3.1 ROSEN'S GRADIENT PROJECTION METHOD

Consider a simple minimization problem of the form
   Minimize $f(x)$
   Subject to $a^i x - b_i \leq 0 (i = 1, \cdots, m)$, $a^i x - b_i = 0 (i = m + 1, \cdots, m + l)$
where $x = [x_1, x_2, \cdots, x_n]^T$ is a column vector of n real-valued variables. Now, suppose that x^k is the current design point. It is required that x^k satisfy the given constraints. We have to determine a direction vector d, followed by a step length along this direction, which will give us a new and improved design point. Let q be the number of active constraints. We can introduce the tangent plane, which can be described by $n - q$ independent parameters. Define a matrix $B \in \mathbb{R}^{q \times n}$ that consists of rows of the gradient vectors of the active constraints:

$$B = [a^{1T}|a^{2T}|\cdots | a^q]^T$$

We find a direction vector d that satisfies $By = 0$ and makes $|-\nabla f(x^k) - d|$ a minimum. The direction d can be determined as the solution of the following minimization problem:
   Minimize $h(d) = (-\nabla f(x^k) - d)^T (-\nabla f(x^k) - d)$

Subject to $Bd = 0$

We can define the Lagrangian $L(d) = (\nabla f(x^k) + d)^T (\nabla f(x^k) + d) + \beta^T Bd$ and apply the optimality condition as follows:

$$\frac{\partial L^T}{\partial d} = (\nabla f(x^k)^T + d) + B^T \beta = 0$$

Using the constraint equation $Bd = 0$, we have

$$BB^T \beta = -B \nabla f(x^k)$$

Solving this equation, we have

$$\beta = -(BB^T)^{-1} B \nabla f(x^k)$$

Thus, the direction vector d is given by

$$d = -\nabla f(x^k) - B^T \beta = -\nabla f(x^k) + B^T (BB^T)^{-1} B \nabla f(x^k) = W(-\nabla f(x^k))$$

where

$$W = I - B^T (BB^T)^{-1} B$$

Rosen's gradient projection method for linear constraints consists of the following steps[12]:

1) Choose a feasible starting point x^0.
2) Determine the active set and construct the matrix B.
3) Calculate β and d.
4) If $d \neq 0$, determine α_k and update the current design point by $x^{k+1} = x^k + \alpha_k d$ and go to step 2).
5) If $\beta_j \geq 0$ for all j corresponding to active inequalities, stop the procedure. Otherwise, delete the row from B corresponding to the most negative component of β and go to step 3).

The MATLAB function *rosencopt* implements Rosen's gradient projection method. The basic syntax is
`[xopt,fopt,iter] = rosencopt(fun,delfun,x0,A,b,ne,m,crit)`

where fun is the objective function, delfun is the gradient of the objective function, x0 is the initial point vector, A is the coefficient matrix of constraints, b is a column vector denoting the right-hand side of the constraints, ne is the number of equality constraints, m is the number of inequality constraints, crit is the stopping criterion, xopt is the resultant optimum point, fopt is the function value at the optimum point, and iter is the number of iterations.

```
function [xopt,fopt,iter] = rosencopt(fun,delfun,x0,A,b,ne,m,crit)
% rosencopt.m: Rosen's gradient projection method
% Problem type:
%      Minimize f(x)
%      subject to Aj*x = b (j=1,...,ne), Aj*x <= b (j=ne+1,...,m)
%      Starting point should satisfy all constraints.
% inputs:
%      fun: objective function
%      delfun: gradient of fun
%      x0: starting point (must satisfy all constraints)
```

```
%       A: constraints coefficient matrix
%       b: right-hand side of constraints (column vector)
%       ne: number of equality constraints
%       m: number of inequality constraints
%       crit: stopping criterion
% outputs:
%       xopt: optimal point
%       fopt: function value at the optimal point (=f(xopt))
%       iter: number of iterations
% Example:
%       fun = @(x) -x(1)*x(2)*x(3);
%       delfun = @(x) [-x(2)*x(3); -x(1)*x(3); -x(1)*x(2)];
%       x0 = 10*[1 1 1]; crit = 1e-4; ne = 0; m = 8;
%       A = [-1 0 0;0 -1 0;0 0 -1;1 0 0;0 1 0;0 0 1;-1 -2 -2;1 2 2];
%       b = [0 0 0 42 42 42 0 72]';
%       [xopt,fopt,iter] = rosencopt(fun,delfun,x0,A,b,ne,m,crit,kmax)

x = x0; n = length(x0); % n: number of variables
f = fun(x); iter = 0;
while (1)
    iter = iter + 1; [nc, nca] = actcont(n,m,ne,crit,x,A,b); % active constraints
    [dv,d,Bm] = dirvec(delfun,n,nc,ne,crit,x,nca,A); % direction (row) vector
    if (abs(dv) < crit), break; end
    alphak = maxstep(delfun,n,m,nc,x,x0,d,A,b,nca); % maximum step size
    x = x0 + alphak*d; fp = delfun(x)'*d';
    if (fp > 0), x = bisec(delfun,n,alphak,x0,d); end
    f = fun(x); x0 = x;
end
xopt = x; fopt = f;
end


function [nc, nca] = actcont(n,m,ne,crit,x,A,b)
g = A*x' - b;
for k = 1:m, nca(k) = k; end
nc = ne;
for j = ne+1:m
  if (abs(g(j)) < crit), nc = nc + 1; ntemp = nca(j); nca(j) = nca(nc); nca(nc) =
ntemp; end
end
end


function x = bisec(delfun,n,alphak,x0,d)
mcrit = 1e-6; a1 = 0; a2 = alphak; aw = a2 - a1;
while (a2 - a1) > mcrit*aw
    am = (a1 + a2)/2; x = x0 + am*d; fp = delfun(x)'*d';
    if (fp < 0), a1 = am; elseif (fp > 0), a2 = am; else, break; end
end
x = x0 + a1*d;
end


function [dn,d,Bm] = dirvec(delfun,n,nc,ne,crit,x,nca,A)
df = delfun(x)'; % row vector
while (1)
    if (nc == 0)
        d = -df; dn = norm(d); if (dn < crit), break; end; d = d/dn; Bm(1) = 0; return;
```

```
       else
         for j = 1: nc, ic = nca(j);
            for jk = 1: nc
               jc = nca(jk); Am(j,jk) = 0.;
                 for k = 1: n, Am(j,jk) = Am(j,jk) + A(ic,k) * A(jc,k); end
         end
       end
       for j = 1: nc
         ic = nca(j); Bm(j) = 0.; for k = 1: n, Bm(j) = Bm(j) - A(ic,k) * df(k); end
       end
       Am = Am(1:nc,1:nc); Bm = Bm(1:nc)'; Bm = inv(Am)*Bm;
       for j = 1: n
            d(j) = -df(j); for k = 1: nc, kn = nca(k); d(j) = d(j) - A(kn,j) * Bm(k); end
       end
   end
   dn = norm(d);
   if (nc == ne & dn <= crit), break; end
   if (dn <= crit)
      Bmin = Bm(ne+1); imin = ne + 1;
      for j = ne + 1: nc, if (Bmin > Bm(j)), Bmin = Bm(j); imin = j; end; end
      if (Bmin >= 0.), break;
    else, ntemp = nca(imin); nca(imin) = nca(nc); nca(nc) = ntemp; nc = nc - 1; end
   else, d = d/dn; break;
   end
end
end

function alphak = maxstep(delfun,n,m,nc,x,x0,d,A,b,nca)
nq = 0;
for k = nc + 1: m
  cq = nca(k); c = -b(cq); aq = 0.;
  for j = 1:n, c = c + A(cq,j)*x(j); aq = aq + A(cq,j)*d(j); end
  if (aq ~= 0)
     am = -c/aq;
     if (am > 0.), nq = nq + 1;
        if (nq == 1), alphak = am;
        else, if (alphak > am), alphak = am; end
        end
     end
  end
end
if (nq == 0)
  alphak = 1;
  while (2)
     x = x0 + alphak*d; fp = delfun(x)'*d'; if (fp > 0.), break; end; alphak =
2*alphak;
  end
end
end
```

Example 10.13 Rosen's Gradient Projection Method

Find the solution of the following constrained minimization problem using Rosen's gradient projection method:

$$\text{Minimize } f(x) = 0.02x_1^2 + 1.2x_2^2 - 80$$

$$\text{Subject to } 3 - x_1 \leq 0,\ 12 - 10x_1 + x_2 \leq 0,\ -40 \leq x_1,\ x_2 \leq 40$$

As a starting point, use $x^0 = [4, 5]$. Note that this point satisfies all constraints.

Solution

The constraints can be rearranged as follows:

$$-x_1 \leq -3,\ 10x_1 + x_2 \leq -12,\ x_1 \leq 40,\ x_2 \leq 40,\ x_1 \leq 40,\ x_2 \leq 40$$

From these inequalities we have

$$A = \begin{bmatrix} -1 & 0 \\ -10 & 1 \\ -1 & 0 \\ 0 & -1 \\ 1 & 0 \\ 0 & 1 \end{bmatrix},\ b = \begin{bmatrix} -3 \\ -12 \\ 40 \\ 40 \\ 40 \\ 40 \end{bmatrix}$$

```
>> fun = @(x) 0.02*x(1)^2+1.2*x(2)^2-80; delfun = @(x) [0.04*x(1); 2.4*x(2)];
>> ne = 0; m = 6; crit = 1e-4; x0 = [4 5]; A = [-1 0;-10 1;-1 0;0 -1;1 0;0 1]; b = [-3
-12 40 40 40 40]';
>> [xopt,fopt,iter] = rosencopt(fun,delfun,x0,A,b,ne,m,crit)
xopt =
  3.0000   0.0000
fopt =
 -79.8200
iter =
  4
```

10.3.2 ZOUTENDIJK'S FEASIBLE DIRECTION METHOD

Consider a minimization problem with nonlinear constraints of the form

$$\text{Minimize } f(x)$$

$$\text{Subject to } g_i(x) \leq 0 (i = 1, \cdots, m)$$

where $x = [x_1, x_2, \cdots, x_n]^T$ is a column vector of n real-valued variables. We first define the active set H as

$$H = \{g_j(x^k) + \varepsilon \geq 0,\ i = 1, \cdots, m\}$$

Next we introduce an artificial variable α, which is given by

$$\alpha = \max \{\nabla f(x^k)^T d,\ \nabla g_j(x^k)^T d,\ j \in H\}$$

To obtain a descent feasible direction, we have to reduce α until it becomes a negative number. To do this, we can formulate the following subproblem:

$$\text{Minimize } \alpha$$

$$\text{Subject to } \nabla f(x^k)^T d \leq \alpha,\ \nabla g_j(x^k)^T d \leq \alpha,\ j \in H,\ -1 \leq d_i \leq 1\, (i = 1, \cdots, n)$$

The step-size problem is a constrained one-dimensional search and can be described as follows:

$$\text{Minimize } f(\alpha) = f(x^k + \alpha d)$$

$$\text{Subject to } g_j(\alpha) = g_j(x^k + \alpha d) \leq 0\, (i = 1, \cdots, m)$$

This problem can be rewritten as

$$\text{Minimize } -\beta = \alpha$$

$$\text{Subject to } \nabla f(x^k)^T s + \beta \leq c_0,\ \nabla g_j(x^k)^T s + \theta_j \beta \leq c_j\, (j \epsilon H),\ 0 \leq s_i \leq 2\, (i = 1, \cdots, n),\ \beta \geq 0$$

where $c_0 = \sum_{i=1}^n \partial f / \partial x_i$, $c_j = \sum_{i=1}^n \partial g_j / \partial x_i$. Zoutendijk's feasible direction method can be summarized as follows[13]:

1) Choose a feasible starting point x^0 which satisfies the constraints.
2) Determine the set H.
3) Solve the minimization problem (minimize $-\beta = \alpha$) to find β^* and d^*.
4) If $\beta^* = 0$, stop the procedure.
5) Otherwise, find the optimum step size α_k and update the current point as $x_{k+1} = x_k + \alpha_k d^*$ and go to step 2).

The MATLAB function *zoutopt* implements Zoutendijk's feasible direction method. The basic syntax is

`[xopt,fopt,iter] = zoutopt(funz,delf,delg,x0,xl,xu,nc,ncs,crit,kmax)`

where funz is the objective function and constraints, delf is the gradient of the objective function, delg is the gradient of constraints, x0 is the starting point vector, xl and xu are the lower and upper bounds on x, nc is the number of active constraints, ncs is the number of constraints, crit is the stopping criterion, kmax is the number of maximum possible iterations, xopt is the resultant optimum point, fopt is the function value at the optimum point, and iter is the number of iterations.

```
function [xopt,fopt,iter] = zoutopt(funz,delf,delg,x0,xl,xu,nc,ncs,crit,kmax)
% zoutopt.m: minimization by Zoutendijk's feasible direction method
% Problem type: min. f(x)
%          subject to gj(x) <= 0 (j=1,...,nca), xl <= x <= xu
% Inputs:
%     funz: objective function and constraints
%     delf: gradient of f
%     delg: gradient of g
%     x0: starting point (must satisfy all constraints)
%     xl,xu: lower and upper limit of x
%     nc: number of active constraints
%     ncs: number of constraints
%     crit: stopping criterion
```

```
%       kmax: maximum possible iterations
% Outputs:
%       xopt: optimal point
%       fopt: function value at the optimal point (=f(xopt))
%       iter: number of iterations
% Example:
%       delf = @(x) [-1; -2]; delg = @(x) [2*x(1); 12*x(2)];
%       x0=[1 0]; xl=[0 0]; xu=[10 10]; nc=0; ncs=1; crit=1e-4; kmax=1e3;
%       [xopt,fopt,iter] = zoutopt(funz,delf,delg,x0,xl,xu,nc,ncs,crit,kmax)

mcrit = 2e-3; gcrit = 1e-4; maxac = 20; maxac = 20; maxs = 30;
nv = length(x0); x = x0; fg = funz(x); f = fg(1); g = fg(2);
ic = 0; fold = f; iter = 0; f0 = f;
while (1)
     iter = iter + 1;
     [nc, na] = actcont(ncs, crit, g); % active constraints
     if (iter > kmax), disp('Maximim possible iterations exceeded.'); break; end
     [d, dn, beta] = dirvec(delf,delg,crit,nv,x,nc,xl,xu,mcrit); % direction
     vector
     if (abs(dn) < crit | abs(beta) < crit), break; end
     d = d/dn;
     alpha  = linsr(funz,delf,nv,ncs,x,nc,na,xl,xu,d,x0,maxs,gcrit,crit); %
calculate step size
  x = x0 + alpha*d; fg = funz(x); f = fg(1); x0 = x;
  if (abs(f - fold) < crit)
    ic = ic + 1; if (ic == 2), break; end
  else, ic = 0;
  end
  fold = f;
end
fg = funz(x); f = fg(1); g = fg(2); [nc,na] = actcont(ncs, crit, g); xopt = x; fopt
= f;
end

function [nc, na] = actcont(ncs, mcrit, g)
for i = 1:ncs, na(i) = i; end
nc = 0;
for k = 1:ncs
   if ( g(k) > -mcrit), nc = nc + 1; ntemp = na(k); na(k) = na(nc); na(nc) =
ntemp; end
end
end

function [d, dn, beta] = dirvec(delf,delg,crit,nv,x,nc,xl,xu,mcrit)
df = delf(x)'; % row vector
if (nc > 0), A = delg(x)'; end
df = df/norm(df); % normalization
if (nc > 0), for j = 1:nc, A(j,:) = A(j,:)/sqrt(A(j,:)*A(j,:)'); end; end
% active bounds
for k = 1:nv, if (xl(k) - x(k) + mcrit >= 0), nc = nc + 1; A(nc,1:nv) = 0; A(nc,k) =
-1; end; end
for k = 1:nv, if (x(k) - xu(k) + mcrit >= 0), nc = nc + 1; A(nc,1:nv) = 0; A(nc,k) =
1; end; end
if (nc == 0), beta = 1; d = -df; dn = norm(d); return; end
[beta, d] = simpx(nc,nv,df,A); dn = sqrt(d*d'); % feasible direction
```

```
end

function alpha = linsr(funz,delf,nv,ncs,x,nc,na,xl,xu,d,x0,maxs,gcrit,crit)
% Determine maximum step size using line search method
nlarge = 1e40; c = max(abs(xu - xl));
for k = 1:nv
  if (abs(d(k))*nlarge > c)
    if (d(k) < 0)
      cn = (xl(k) - x(k)) / d(k); if (cn < nlarge), nlarge = cn; end
    else
      cn = (xu(k) - x(k)) / d(k); if (cn < nlarge), nlarge = cn; end
    end
  end
end
abet = nlarge; x = x0 + abet * d; fg = funz(x); f = fg(1); g = fg(2); gmax = max(g);
inda = 1;
if (gmax <= 0), inda = 0; end
if (inda == 0), amax = abet;
else, xl = 0; [xm, fm] = nears(funz,xl,abet,x,d,x0,maxs,crit); amax = xm;
end
x = x0 + amax*d; df = delf(x)'; sdr = df*d';
if (sdr <= 0), alpha = amax; return; end
a1 = 0; a2 = amax; adif = a2 - a1;
while ((a2 - a1) > crit * adif)
  am = (a1 + a2)/2; x = x0 + am*d; df = delf(x)'; sdr = df*d';
  if (sdr == 0), break; end
  if (sdr < 0), a1 = am; elseif (sdr > 0), a2 = am; end
end
alpha = a1;
end


function [xm, fm] = nears(funz,xa,xb,x,d,x0,maxs,crit)
miter = 0;
while (1)
    xm = (xa + xb)/2; miter = miter + 1;
    if (miter > maxs), xm = xa; fm = fa; return; end
    x = x0 + xm*d; fg = funz(x); f = fg(1); g = fg(2); gmax = max(g); fm = f;
    if (gmax <= 0 & gmax >= -crit), return; end
    if (gmax < 0), xa = xm; fa = fm; else, xb = xm; end
end
end
function [beta, d] = simpx(nc,nv,df,A)
% Find search direction using the linear programming (simplex) method
Bg = 1e2; nrow = nc + nv + 2; nm = nc + nv + 1; Bm(1:nrow) = 0; Bm(1) = sum(df);
for j = 1:nc
  Bm(j+1) = 0; for k = 1: nv, Bm(j+1) = Bm(j+1) + A(j,k); end
end
for k = nc + 2:nrow - 1, Bm(k) = 2; end
for k = 1: nm, Bs(k) = nv + k + 1; end
ncol = nv + nm + 1;
for k = 1:nm, if (Bm(k) < 0), ncol = ncol + 1; Bs(k) = -ncol; end; end
Am(1:nrow,1:ncol) = 0; Am(1,1:nv) = df; Am(1,nv+1) = 1;
for k = 1:nc, for j = 1: nv, Am(k+1,j) = A(k,j); end; Am(k+1,nv+1) = 1; end
mi = 0;
for k = nc+2: nrow-1, mi = mi + 1; Am(k,mi) = 1; end
```

```
Am(nrow,nv+1) = -1; for k = 1: nm, Am(k,nv+k+1) = 1; end
nt = nv + nm + 1;
for k = 1: nm
  if (Bm(k) < 0)
     nt = nt + 1; Bm(k) = -Bm(k); for j = 1:ncol, Am(k,j) = -Am(k,j); end
     Am(k,nt) = 1; Am(nrow,nt) = Bg;
  end
end
for k = 1:nm
  if (Bs(k) < 0)
     Bs(k) = -Bs(k); for j = 1: ncol, Am(nrow,j) = Am(nrow,j) - Bg*Am(k,j); end
     Bm(nrow) = Bm(nrow) - Bg*Bm(k);
  end
end
while (1)
  nf0 = 0;
  for k = 1:ncol, if (Am(nrow, k) < 0), nf0 = 1; break; end; end
  if (nf0 == 0), break; end
  c = Bg;
  for j = 1:ncol, if (Am(nrow, j) < c), c = Am(nrow, j); iv = j; end; end
  ik = 0; jk = 0;
  for k = 1:nrow - 1
     if (Am(k,iv) > 0)
        jk = jk + 1; c1 = Bm(k)/(Am(k,iv) + 1e-10);
        if (jk == 1), c = c1; jp = k; else, if (c1 < c), c = c1; jp = k; end; end
        ik = 1;
     end
  end
  Bs(jp) = iv;
  if (ik == 0), disp('Unbounded objective function.'); break; end
  c1 = 1/Am(jp,iv); Bm(jp) = c1*Bm(jp);
  for j = 1:ncol, Am(jp,j) = c1*Am(jp,j); end
  for k = 1: nrow
     if (k ~= jp)
        c2 = Am(k,iv); for j = 1: ncol, Am(k,j) = Am(k,j) - c2*Am(jp,j); end
        Bm(k) = Bm(k) - c2*Bm(jp);
     end
  end
end
d(1:nv) = -1;
for k = 1: nm
  for j = 1: nv, if (j == Bs(k)), d(j) = Bm(k) - 1; end; end
end
beta = Bm(nrow);
end
```

Example 10.14 Zoutendijk's Feasible Direction Method

Find the solution of the following constrained minimization problem using Zoutendijk's feasible direction method:

$$\text{Minimize} f(x) = -(1.2x_1 + 3x_2)$$

$$f(x) = -(1.2x_1 + 3x_2)$$

$$\text{Subject to } g(x) = x_1^2 + 6x_2^2 - 1 \leq 0, \ 0 \leq x_1, \ x_2 \leq 10$$

As a starting point, use $x^0 = [1, 0]$. There are no active constraints (nc=0). Note that this point satisfies all constraints.

Solution

The gradients of $f(x)$ and $g(x)$ are given by

$$\frac{\partial f}{\partial x_1} = -1.2, \ \frac{\partial f}{\partial x_2} = -3, \ \frac{\partial g}{\partial x_1} = 2x_1, \ \frac{\partial g}{\partial x_2} = 12x_2$$

The lower and upper bounds on x can be written as $x^l = [00]$, $x^u = [1010]$. The objective function and the constraint can be defined as a subfunction *funz*, and the script *optzout* computes the desired results.

```
% optzout
delf = @(x) [-1.2; -3]; delg = @(x) [2*x(1); 12*x(2)];
x0 = [1 0]; xl = [0 0]; xu = [10 10]; nc = 0; ncs = 1; crit = 1e-4; kmax = 1e3;
[xopt,fopt,iter] = zoutopt(@funz,delf,delg,x0,xl,xu,nc,ncs,crit,kmax)

function fg = funz(x)
fg(1) = -(1.2*x(1) + 3*x(2));
fg(2) = x(1)^2 + 6*x(2)^2 - 1;
end

>> optzout
xopt =
  0.6998   0.2916
fopt =
 -1.7146
iter =
  5
```

10.3.3 Generalized Reduced Gradient (GRG) Method

The generalized reduced gradient (GRG) method handles nonlinear equality constraints. Consider a minimization problem with nonlinear constraints of the form

$$\text{Minimize } f(x)$$

$$\text{Subject to } h_i(x) \leq 0 (i = 1, \cdots, m), \ l_k(x) = 0 (k = 1, \cdots, l), \ x_j^L \leq x_j \leq x_j^U \ (j = 1, \cdots, n)$$

where $x = [x_1, x_2, \cdots, x_n]^T$ is a column vector of n real-valued variables. By adding a nonnegative slack variable to each of the inequality constraints, the problem can be rewritten as

$$\text{Minimize } f(x)$$

$$\text{Subject to } g_i(x) = 0 (i = 1, \cdots, m + l), \ x_j^L \leq x_j \leq x_j^U \ (j = 1, \cdots, n + m)$$

The GRG method is based on the elimination of variables using the equality constraints. Usually $n + m$ design variables are divided into two sets—independent variables (Y) and dependent variables (Z)—as

$$X = \begin{bmatrix} Y \\ Z \end{bmatrix}, \; Y = \begin{bmatrix} y_1 \\ \vdots \\ y_{n-l} \end{bmatrix}, \; Z = \begin{bmatrix} z_1 \\ \vdots \\ z_{m+l} \end{bmatrix}$$

The first derivatives of the objective and constraint functions are given by[14]

$$df(x) = \sum_{i=1}^{n-l} \frac{\partial f}{\partial y_i} dy_i + \sum_{i=1}^{m+l} \frac{\partial f}{\partial z_i} dz_i = \nabla_Y^T f dY + \nabla_Z^T f dZ$$

$$dg_i(x) = \sum_{i=1}^{n-l} \frac{\partial g_i}{\partial y_i} dy_i + \sum_{i=1}^{m+l} \frac{\partial g_i}{\partial z_i} dz_i$$

The last relation can be expressed as $dg = CdY + BdZ$. When Y is held fixed, $g(x) + dg(x) = 0$. Substitution of dg into this equation gives

$$dZ = B^{-1}\{-g(x) - CdY\}$$

If the constraints are satisfied at the vector x, $g(x) = 0$ and $dZ = -B^{-1}CdY$. This equation can be rewritten as

$$d_z = -B^{-1}Cd_y$$

The direction vector d is defined as $d = [d_z d_y]^T$. Substitution of d_z into the derivative of the objective function yields

$$df(x) = \nabla_Y^T f dY + \nabla_Z^T f dZ = (\nabla_Y^T f - \nabla_Z^T f B^{-1}C)d_y$$

The coefficient matrix is referred to as the generalized reduced gradient G_R:

$$G_R = \nabla_Y f - (B^{-1}C)^T \nabla_Z f$$

The new point is obtained as

$$x^{k+1} = x^k + \alpha_k d$$

The generalized reduced gradient method can be summarized as follows[15]:

1) Select a starting point x^0 that satisfies $g(x^0) = 0$ with $x^L \leq x^0 \leq x^U$.
2) Determine the basic and nonbasic variables and calculate the reduced gradient G_R.
3) Find the direction vector d. If $\|d\| \leq \varepsilon$, stop the procedure (ε: very small number).
4) Calculate the current step size α_k and determine x^{k+1}.
5) Check feasibility of the new point x^{k+1}: if this point is feasible, set $x^k = x^{k+1}$ and go to step 2). Otherwise, compute a corrected point x_c^{k+1}. If $f(x_c^{k+1}) < f(x^k)$, set $x^k = x_c^{k+1}$ and go to step 2). If $f(x_c^{k+1}) > f(x^k)$, set $\alpha_k = \alpha_k/2$, calculate x^{k+1}, and repeat step 5).

The MATLAB function *grgopt* implements the generalized reduced gradient method. The basic syntax is

```
[xopt,fopt,iter] = grgopt(grgfun,delgrgf,delgrgg,x0,xl,xu,kmax,crit)
```

where grgfun is the objective function and constraints, delgrgf is the gradient of the objective function, delgrgg is the gradient of constraints, x0 is the starting point vector, xl and xu are the lower and upper bounds on x, kmax is the number of maximum possible iterations, crit is the stopping criterion, xopt is the resultant optimum point, fopt is the function value at the optimum point, and iter is the number of iterations.

```
function [xopt,fopt,iter] = grgopt(grgfun,delgrgf,delgrgg,x0,xl,xu,kmax,crit)
% grgopt.m: minimization by the generalized reduced gradient (GRG) method
% Problem type: min. f(x)
%           subject to gj(x) = 0 (j=1,...,nca), xl <= x <= xu
% Inputs:
%     grgfun: objective function and constraints
%     delgrgf: gradient of f
%     delgrgg: gradient of g
%     x0: starting point (must satisfy all constraints)
%     xl,xu: lower and upper limit of x
%     kmax: maximum possible iterations
%     crit: stopping criterion
% Outputs:
%     xopt: optimal point
%     fopt: function value at the optimal point (=f(xopt))
%     iter: number of iterations
% Example:
%     (objective function and constraints are defined by function fg1)
%     df1 = @(x) [-1.2 -3 0]; dg1 = @(x) [2*x(1) 12*x(2) 1];
%     x0 = [0 0 1]; xl = [0 0 0]; xu = [10 10 10]; crit = 1e-4; kmax = 1e3;
%     [xopt,fopt,iter] = grgopt(@fg1,df1,dg1,x0,xl,xu,kmax,crit)

dcrit = crit/10; fcrit = crit*1e-2; x = x0; [f g] = grgfun(x); [ncs nv] = size
(delgrgg(x));
for k = 1: ncs, if (abs(g(k)) > crit), disp('Infeasible starting point.');
break; end; end
indc = 0; fold = f;
for iter = 1:kmax
    df = delgrgf(x); dg = delgrgg(x); for k = 1:nv, nbr(k) = k; end
    [nbr,df,dg] = redgr(nbr,nv,ncs,x,xl,xu,df,dg, crit); % basic variables
    dm = 0.;
  for k = ncs+1:nv
      nk = nbr(k); d(nk) = -df(nk);
      if (x(nk) < xl(nk)+crit & df(nk) > 0), d(nk) = 0; end
    if (x(nk) > xu(nk)-crit & df(nk) < 0), d(nk) = 0; end
    if (dm < abs(d(nk))) dm = abs(d(nk)); end
  end
  if (dm < crit), break; end
  for k = 1: ncs
      nk = nbr(k); d(nk) = 0;
      for j = ncs+1:nv, nj = nbr(j); d(nk) = d(nk) - dg(k,nj)*d(nj); end
  end
  x0 = x; alpha = findstep(nv,x,xl,xu,d); x = x0;
  for k = 1: nv, xtemp(k) = x(k) + alpha*d(k); end
  [ftemp g] = grgfun(xtemp); df = delgrgf(xtemp); dg = delgrgg(xtemp);
  fup = 0;
  for k = 1:nv, fup = fup + df(k)*d(k); end
  if (fup > 0 | ftemp > f)
      c = 0; x0 = x; [alpha,ftemp] = goldsec(nv,ncs,c,alpha,dcrit,d,x); x = x0;
      for k = 1:nv, xtemp(k) = x(k) + alpha*d(k); end
  end
  [c g] = grgfun(xtemp); ag = abs(g(1));
```

```
for k = 1:ncs, if (ag < abs(g(k))), ag = abs(g(k)); end; end
% If infeasible, apply Newton's method
if (ag > crit)
   blim = 20;
   for bi0 = 1:blim
        bnew = 0; bi2 = 12;
        for bi1 = 1:bi2
           bnew = bnew + 1; df = delgrgf(xtemp); dg = delgrgg(xtemp);
           [g] = rednewton (dg, g, nbr, ncs); bf = 0;
           for k = 1: ncs
              nk = nbr(k); xtemp(nk) = xtemp(nk) - g(k);
              if ((xtemp(nk) < xl(nk)) | (xtemp(nk) > xu(nk))), bf = 1; break; end
           end
           if (bf == 1), break; end; [c g] = grgfun(xtemp); ag1 = abs(g(1));
           for k = 1: ncs
                if (ag1 < abs(g(k))), ag1 = abs(g(k)); end
           end
           if (bf == 0 & ag1 < crit), break; end
           bf = 1;
           if (bnew > 3 | ag1 > ag), break; end
           ag = ag1;
        end
        if (bf == 0)
             [ftemp, g] = grgfun(xtemp); if (ftemp < f), break; end
        end
        alpha = alpha/2;
        for k = 1:nv, xtemp(k) = x(k) + alpha*d(k); end
        [c g] = grgfun(xtemp); ag = abs(g(1));
        for k = 1: ncs, if (ag < abs(g(k))), ag = abs(g(k)); end; end
     end
  end
  if (bf==0 & ftemp<f)
       for k = 1: nv, x(k) = xtemp(k); end; f = ftemp;
  end
  fcrit = abs(f)*fcrit + fcrit;
  if (abs(f - fold) < fcrit)
       indc = indc + 1; if (indc > 1), break; end
  else, indc = 0; end
  fold = f;
end
xopt = x; fopt = f;
end

function [x4,ft] = goldsec(nv,ncs,x1,x4,dcrit,d,x)
tau = (sqrt(5)-1)/2; x2 = tau*x1 + (1-tau)*x4;
for k = 1:nv, xtemp(k) = x(k) + x2*d(k); end
[f2, g] = grgfun(xtemp);
for count = 1:100
    x3 = tau*x4 + (1-tau)*x1;
    for k = 1:nv, xtemp(k) = x(k) + x3*d(k); end
    [f3, g] = grgfun(xtemp);
    if (f2 < f3), x4 = x1; x1 = x3;
    else, x1 = x2; x2 = x3; f2 = f3; end
    if (abs(x4 - x1) <= dcrit), break; end
end
```

```
x4 = x2; ft = f2;
end

function [g] = rednewton(dgr, g, nbr, ncs)
for k = 1: ncs-1
    nk = nbr(k);
    for jk = k+1:ncs
       c = dgr(jk,nk)/dgr(k,nk);
       for j = k + 1: ncs, nj = nbr(j); dgr(jk,nj) = dgr(jk,nj) - c*dgr(k,nj); end
       g(jk) = g(jk) - c*g(k);
  end
end
g(ncs) = g(ncs)/dgr(ncs,nbr(ncs));
for jm = 1:ncs-1
    jk = ncs - jm; im = nbr(jk); c = 1/dgr(jk,im); g(jk) = c*g(jk);
    for k = jk + 1: ncs, nk = nbr(k); g(jk) = g(jk) - c*dgr(jk,nk)*g(k); end
end
end

function [nbr,df,dg] = redgr(nbr,nv,ncs,x,xl,xu,df,dg,xcrit)
for k = 1: ncs
    nk = nbr(k); kcount = 0;
    for j = k: nv
      nj = nbr(j);
      if (x(nj) > xl(nj)+xcrit & x(nj) < xu(nj)-xcrit)
        kcount = kcount + 1;
        if (kcount == 1), pivot = dg(k,nj); jpv = j;
        else
            if (abs(pivot) < abs(dg(k,nj))), pivot = dg(k,nj); jpv = j; end
      end
    end
  end
  nbr(k) = nbr(jpv); nbr(jpv) = nk; nk = nbr(k);
  if (k == ncs), break; end
  for jk = k+1: ncs
      dgratio = dg(jk,nk)/dg(k,nk);
      for j = k+1:nv, nj = nbr(j); dg(jk,nj) = dg(jk,nj) - dgratio*dg(k,nj); end
  end
end
nbs = nbr(ncs);
for j = ncs + 1: nv
    nj = nbr(j); dg(ncs,nj) = dg(ncs,nj)/dg(ncs,nbs);
    for jm = 1:ncs-1
       jk = ncs-jm; km = nbr(jk); dgratio = 1/dg(jk,km); dg(jk,nj) = dgratio*dg
(jk, nj);
    for k = jk+1: ncs, nk = nbr(k); dg(jk,nj) = dg(jk,nj) - dgratio*dg(jk,nk)*dg
(k,nj); end
  end
end
% Calculate reduced gradient
for jk = ncs+1:nv
     km = nbr(jk); for j = 1: ncs, nj = nbr(j); df(km) = df(km) - dg(j,km)*df
(nj); end
end
end
```

```
function [alpha] = findstep(nv, x, xl, xu, d)
% Calculate maximum step size
kcount = 0;
for j = 1:nv
    au = xu(j) - x(j); % check upper bound
    if (d(j) > 1E-30*(au+1))
       kcount = kcount + 1; ad = au/d(j);
       if (kcount == 1), alpha = ad; end
       if (kcount > 1), if (alpha > ad), alpha = ad; end; end
  end
  al = x(j) - xl(j); % check lower bound
  if (-d(j) > 1E-30*(al + 1))
       kcount = kcount + 1; ad = -al/d(j);
       if (kcount == 1), alpha = ad; end
       if (kcount > 1), if (alpha > ad), alpha = ad; end; end
  end
end
end
```

Example 10.15 Generalized Reduced Gradient Method

Find the solution of the following constrained minimization problem using the generalized reduced gradient method:

$$\text{Minimize } f(x) = x_1^2 + x_2^2 + 2.3x_3^2 - 1.2x_4^2 - 4x_1 - 6x_2 - 20x_3 + 6x_4 + 100$$

Subject to

$$h_1(x) = x_1^2 + x_2^2 + x_3^2 + x_4^2 + x_1 - x_2 + x_3 - x_4 \leq 7$$

$$h_2(x) = x_1^2 + 2x_2^2 + x_3^2 + 2x_4^2 - x_1 - x_4 \leq 11$$

$$h_3(x) = 2x_1^2 + x_2^2 + x_3^2 + 2x_1 - x_2 - x_4 \leq 6, \quad -100 \leq x_1, x_2, x_3, x_4 \leq 100$$

The inequality constraints can be converted to equalities by introducing slack variables as follows:

$$g_1(x) = x_1^2 + x_2^2 + x_3^2 + x_4^2 + x_1 - x_2 + x_3 - x_4 + x_5 - 7 = 0$$

$$g_2(x) = x_1^2 + 2x_2^2 + x_3^2 + 2x_4^2 - x_1 - x_4 + x_6 - 11 = 0$$

$$g_3(x) = 2x_1^2 + x_2^2 + x_3^2 + 2x_1 - x_2 - x_4 + x_7 - 6 = 0$$

As a starting point, we can use $x^0 = [00007116]$. Note that this point satisfies all equality constraints. Set $k_{max} = 1000$ and crit $= 1 \times 10^{-4}$.

Solution

The objective function and the equality constraints can be defined as a function as follows:

```
function [f g] = grgfun(x)
x1 = x(1); x2 = x(2); x3 = x(3); x4 = x(4);
x5 = x(5); x6 = x(6); x7 = x(7); % slack variables
f = x1^2 + x2^2 + 2.3*x3^2 - 1.2*x4^2 - 4*x1 - 6*x2 - 20*x3 + 6*x4 + 100;
```

```
g(1) = x1^2 + x2^2 + x3^2 + x4^2 + x1 - x2 + x3 - x4 + x5 - 7;
g(2) = x1^2 + 2*x2^2 + x3^2 + 2*x4^2 - x1 - x4 + x6 - 11;
g(3) = 2*x1^2 + x2^2 + x3^2 + 2*x1 - x2 - x4 + x7 - 6;
end
```

The gradients of f(x) and $g(x)$ are given by

$$\frac{\partial f}{\partial x_1} = 2.3x_1 - 4, \quad \frac{\partial f}{\partial x_2} = 2x_2 - 6, \quad \frac{\partial f}{\partial x_3} = 4.6x_3 - 20, \quad \frac{\partial f}{\partial x_4} = -2.4x_1 + 6$$

$$\nabla g = \begin{bmatrix} \nabla g_1 \\ \nabla g_2 \\ \nabla g_3 \end{bmatrix} = \begin{bmatrix} 2x_1 + 1 & 2x_2 - 1 & 3x_3 + 1 & 2x_4 - 1 & 1 & 0 & 0 \\ 2x_1 - 1 & 4x_2 & 2x_3 & 4x_4 - 1 & 0 & 1 & 0 \\ 4x_1 + 2 & 2x_2 - 1 & 2x_3 & -1 & 0 & 0 & 1 \end{bmatrix}$$

The subfunctions *delgrgf* and *delgrgg* define the gradients of $f(x)$ and $g(x)$, respectively. The script *grgoptex* executes the function *grgopt* to find the solution.

```
% grgoptex.m
x0 = [0 0 0 0 7 11 6]; xl = [-100 -100 -100 -100 0 0 0]; xu = 100*ones(1,7); kmax =
1e3; crit = 1e-4;
[xopt,fopt,iter] = grgopt(@grgfun,@delgrgf,@delgrgg,x0,xl,xu,kmax,crit)
function df = delgrgf(x)
x1 = x(1); x2 = x(2); x3 = x(3); x4 = x(4); df = zeros(1,length(x));
df(1) = 2.3*x1-4; df(2) = 2*x2-6; df(3) = 4.6*x3-20; df(4) = -2.4*x4+6;
end
function dg = delgrgg(x)
x1 = x(1); x2 = x(2); x3 = x(3); x4 = x(4);
dg = [2*x1+1 2*x2-1 3*x3+1 2*x4-1 1 0 0;
      2*x1-1 4*x2   2*x3   4*x4-1 0 1 0;
      4*x1+2 2*x2-1 2*x3   -1     0 0 1];
end

>> grgoptex
xopt =
      0.0081   0.4922   0.6609   0.2043   6.3066   10.2076   6.0011
fopt =
      86.2196
iter =
      5
```

10.3.4 SEQUENTIAL QUADRATIC PROGRAMMING (SQP) METHOD

Consider a constrained minimization

$$\text{Minimize } f(x)$$

$$\text{Subject to } v_i(x) = 0 (i = 1, \cdots, p), \ w_k(x) \ge 0 (k = 1, \cdots, q)$$

where $x = [x_1, x_2, \cdots, x_n]^T$ is a column vector of n real-valued variables. Let $\{x^k, \lambda^k, \mu^k\}$ be the kth iterates and $\{\delta_x, \delta_\lambda, \delta_\mu\}$ be a set of increment vectors such that the following Karush-Kuhn-Tucker (KKT) conditions are satisfied at the next iterate $\{x^{k+1}, \lambda^{k+1}, \mu^{k+1}\} = \{x^k + \delta_x, \lambda^k + \delta_\lambda, \mu^k + \delta_\mu\}$:

$$\nabla_x \mathscr{L}(x, \lambda, \mu) = 0, \ v_i(x) = 0 (i = 1, \cdots, p), \ w_k(x) \ge 0 (k = 1, \cdots, q)$$

$$\mu \ge 0, \ \mu^k w_k(x) = 0 (k = 1, \cdots, q)$$

where $\mathscr{L}(x, \lambda, \mu)$ is the Lagrangian defined as

$$\mathscr{L}(x, \lambda, \mu) = f(x) - \sum_{i=1}^{p} \lambda_i v_i(x) - \sum_{k=1}^{q} \mu^k w_k(x)$$

We obtain the approximate KKT conditions as[16]

$$Z^k \delta_x + g^k - A_{ek}^T \lambda^{k+1} - A_{ik}^T \mu^{k+1} = 0, \; A_{ek}\delta_x = -v_k, \; A_{ik}\delta_x \geq -w_k, \; \mu^{k+1} \geq 0$$

$$(\mu^{k+1})_j(A_{ik}\delta_x + w_k)_j = 0 (j = 1, 2, \cdots, q)$$

where

$$Z^k = \nabla_x^2 f(x^k) - \sum_{i=1}^{p} (\lambda^k)_i \nabla_x^2 v_i(x^k) - \sum_{j=1}^{q} (\mu^k)_j \nabla_x^2 w_i(x^k), g^k = \nabla_x f(x^k),$$

$$A_{ek} = \begin{bmatrix} \nabla_x^T v_1(x^k) \\ \vdots \\ \nabla_x^T v_p(x^k) \end{bmatrix}, \; A_{ik} = \begin{bmatrix} \nabla_x^T w_1(x^k) \\ \vdots \\ \nabla_x^T w_q(x^k) \end{bmatrix}$$

Given $\{x^k, \lambda^k, \mu^k\}$, the approximate KKT conditions may be interpreted as the exact KKT conditions of the quadratic programming problem given by

$$\text{Minimize } h(x) = \frac{1}{2}\delta^T Z^k \delta + \delta^T g^k$$

$$\text{Subject to } A_{ek}\delta = -v_k, \; A_{ik}\delta \geq -w_k$$

If δ_x is a regular solution of this quadratic programming problem, the approximate KKT condition can be written as

$$Z^k \delta_x + g^k - A_{ek}^T \lambda^{k+1} - A_{aik}^T \hat{\mu}^{k+1} = 0$$

where the matrix A_{aik} consists of those rows of A_{ik} that satisfy the equality $(A_{ik}\delta_x + w_k)_j = 0$ and $\hat{\mu}^{k+1}$ is the associated Lagrange multiplier. If δ_x is the solution of this quadratic programming problem, the current point x^k can be updated by

$$x^{k+1} = x^k + \alpha_k \delta_x$$

The value of α_k is calculated as[17]

$$\alpha_k = 0.95 \min\{\alpha_1^*, \alpha_2^*\}, \; \alpha_1^* = \arg[\min_{0 \leq \alpha \leq 1}\varphi(\alpha)], \; \alpha_2^* = \max\{\alpha : w_j(x^k + \alpha\delta_x) \geq 0, j \in J\},$$

$$\varphi(\alpha) = f(x^k + \alpha\delta_x) + \beta \sum_{k=1}^{p} v_i^2(x^k + \alpha\delta_x) - \sum_{j=1}^{q} (\mu^{k+1})_j w_j(x^k + \alpha\delta_x)$$

where $J = \{j : (\mu^{k+1})_j = 0\}$. The Hessian Z^k can be updated by using the BFGS formula as

$$Z^{k+1} = Z^k + \frac{\eta^k(\eta^k)^T}{\delta_x^T \eta^k} - \frac{Z^k \delta_x \delta_x^T Z^k}{\delta_x^T Z^k \delta_x}, \; \eta^k = \theta\gamma^k + (1 - \theta)Z^k\delta_x$$

$$\gamma^k = (g^{k+1} - g^k) - (A_{e,k+1} - A_{e,k})^T \lambda^{k+1} - (A_{i,k+1} - A_{i,k})^T \mu^{k+1}$$

$$\theta = \begin{cases} 1 : \delta_x^T \gamma^k \geq 0.2 \delta_x^T Z^k \delta_x \\ \dfrac{0.8 \delta_x^T Z^k \delta_x}{\delta_x^T Z^k \delta_x - \delta_x^T \gamma^k} ; \text{otherwise} \end{cases}$$

The sequential quadratic programming algorithm is as follows:

1) Select x^0 and μ^0 such that $w_k(x^0) \geq 0 (k = 1, \cdots, q)$ and $\mu^0 \geq 0$. Set $k = 0$ and $Z^0 = I_n$.
2) Calculate g^k, A_{ek}, A_{ik}, v^k, and w^k.
3) Solve the quadratic programming problem for δ_x, and calculate the Lagrange multiplier λ^{k+1}, $\hat{\mu}^{k+1}$, and α_k.
4) Set $\delta_x = \alpha_k \delta_x$ and $x^{k+1} = x^k + \delta_x$.
5) If $\|\delta_x\| \leq \varepsilon$, stop the calculation procedure.
6) Calculate γ^k, θ, η^k, and Z^{k+1}. Set $k = k + 1$ and go to step 2).

The MATLAB function *sqpopt* implements the sequential quadratic programming method. The basic syntax is
`[xopt,fopt,iter] = sqpopt(fun,dfun,x0,lam0,mu0,crit)`

where fun is the objective function and constraints, dfun is the gradient of these functions, x0 is the starting point vector, lam0 and mu0 are the initial values of Lagrangian multipliers, crit is the stopping criterion, xopt is the resultant optimum point, fopt is the function value at the optimum point, and iter is the number of iterations.

```
function [xopt,fopt,iter] = sqpopt(fun,dfun,x0,lam0,mu0,crit)
% sqpopt.m: minimization by using SQP algorithm
% Problem type: minimize f(x) subject to h(x) = 0, g(x) >= b
% Inputs:
%          fun: objective and constraint functions
%          dfun: gradients of the objective and constraint functions
%          x0: starting point
%          lam0,mu0: initial Lagrange multipliers
%          crit: stopping criterion
% Outputs:
%          xopt: optimal point
%          fopt: objective function value at x=xopt
%          iter: number of iterations
% Example:
%    x0 = [3 3]'; lam0 = 1; mu0 = 4; crit = 1e-6;
%    [xopt,fopt,iter] = sqpopt(@fun,@dfun,x0,lam0,mu0,crit)

x = x0(:); n = length(x); % number of variables
lam1 = lam0 + 1; Hj = eye(n); fv = fun(x); aj = fv(2:lam1); cj = fv((lam1+1):
(lam1+mu0));
Gj = dfun(x); gj = Gj(:,1); Aej = Gj(:,2:lam1)'; Aij = Gj(:,(lam1+1):
(lam1+mu0))'; iter = 0; d = 1;
while d >= crit
    delx = quadpr(Hj,gj,Aej,-aj,Aij,-cj,zeros(n,1),crit);
    ad = Aij*(x+delx) + cj; k = find(ad <= crit); nk = length(k); muj = zeros(mu0,1);
    if nk == 0, lamj = inv(Aej*Aej')*Aej*(Hj*delx+gj);
    else
        Aaik = Aij(k,:); Aaj = [Aej; Aaik]; mun = inv(Aaj*Aaj')*Aaj*(Hj*delx+gj);
```

```
                lamj = mun(1:lam0); mujh = mun(lam1:end); muj(k) = mujh;
    end
    alpha = linsearch(fun,x,delx,lam1,muj,crit); delx = alpha*delx; x = x + delx;
    grd = dfun(x);
    grd1 = grd(:,1); Agrd = grd(:,2:lam1)'; Am = grd(:,(lam1+1):(lam1+mu0))';
    gamj = (grd1-gj)-(Agrd-Aej)'*lamj-(Am-Aij)'*muj; qj = Hj*delx;
    dg = delx'*gamj; dq = delx'*qj;
    if dg >= 0.2*dq, theta = 1;
    else, theta = 0.8*dq/(dq-dg); end
    eta = theta*gamj + (1-theta)*qj; invdq = 1/dq; invde = 1/(delx'*eta);
    Hj = Hj + invde*(eta*eta') - invdq*(qj*qj'); Aej = Agrd; Aij = Am; gj = grd1;
    fv = fun(x);
    aj = fv(2:lam1); cj = fv((lam1+1):(lam1+mu0)); d = norm(delx); iter = iter + 1;
end
xopt = x; fopt = fv(1);
end


function alpha = linsearch(fun,xj,dx,lam,muj,crit)
% Determine step size by line search method
nmuj = length(muj); alrange = 0:0.01:1; nint = length(alrange);
hz = zeros(nint,1);
for j = 1:nint
    xdj = xj + alrange(j)*dx; fv = fun(xdj); af = fv(2:lam);
    cf = fv((lam+1):(lam+nmuj));
    hz(j) = fv(1) + 1e2*sum(af.^2)- muj'*cf;
end
[mval,indhz] = min(hz); atemp = alrange(indhz); indmu = find(muj <= crit); mc =
length(indmu);
if mc == 0, alpha = 0.95*atemp;
else
    dv = zeros(mc,1);
    for k = 1:mc
      for j = 1:nint
            aj = alrange(j); xdj = xj + aj*dx; fcj = fun(xdj);
            cj = fcj((lam+1):(lam+nmuj)); hz(j) = cj(indmu(k));
      end
    indhz = find(hz < 0); hc = length(indhz);
    if hc == 0, dv(k) = 1;
    else, dv(k) = alrange(indhz(1)-1); end
  end
      mdv = min(dv); alpha = 0.95*min(atemp,mdv);
end
end


function xqr = quadpr(Q,c,Aeq,beq,Ane,bne,x0,crit)
% minimization by quadratic programming method
nbeq = length(beq); nbne = length(bne);
rnbne = nbne + 1.5*sqrt(nbne); aone = 1-crit; x = x0(:); y = Ane*x - bne;
zeta = zeros(nbeq,1); gama = ones(nbne,1); ym = y.*gama; sumy = sum(ym); iter
= 0;
while sumy > crit
        tau = sumy/rnbne; resid = -Q*x - c + Aeq'*zeta + Ane'*gama;
        diffr = beq - Aeq*x; numt = tau - y.*gama; yj = gama./y; ysj = numt./y;
        ydm = diag(yj);
        Gr = inv(Q + Ane'*ydm*Ane); ag = Aeq*Gr*Aeq';
```

```
        ayj = resid + Ane'*ysj; diffag = diffr - Aeq*Gr*ayj;
       delz = inv(ag)*diffag;
       dx = Gr*(ayj + Aeq'*delz); dy = Ane*dx; dgama = (numt - (gama.*dy))./y;
       indny = find(dy < 0); yr = min(y(indny)./(-dy(indny)));
       indny = find(dgama < 0); gamr = min(gama(indny)./(-dgama(indny)));
       aj = aone*min([1 yr gamr]); x = x + aj*dx; gama = gama + aj*dgama;
       zeta = zeta + aj*delz;
       y = Ane*x - bne; ym = y.*gama; sumy = sum(ym);
       iter = iter + 1;
end
xqr = x;
end
```

Example 10.16 Sequential Quadratic Programming Method

Find the solution of the following constrained minimization problem using the sequential quadratic programming method:

$$\text{Minimize } f(x) = 2x_1 + x_2^2$$

$$\text{Subject to } h(x) = x_1^2 + x_2^2 = 8, \, 0 \le x_1 \le 4, \, 1 \le x_2 \le 5$$

As a starting point, use $x^0 = [2, 3]^T$. The initial values of Lagrangian multipliers can be set as 1 and 3. Set crit = 1×10^{-6}.

Solution
The inequality constraints can be rearranged as follows:

$$x_1 \ge 0, \, -x_1 + 4 \ge 0, \, x_2 - 1 \ge 0, \, -x_2 + 5 \ge 0$$

The function *fun* defines the objective function, equality constraint, and inequality constraints:

```
function fv = fun(x)
fv = [2*x(1) + x(2)^2;     % objective function
      x(1)^2 + x(2)^2 - 8; % equality constraint: h(x) = 0
      x(1);            % inequality constraint: g1(x) >= 0
      -x(1) + 4;         % inequality constraint: g2(x) >= 0
      x(2) - 1;          % inequality constraint: g3(x) >= 0
      -x(2) + 5];         % inequality constraint: g4(x) >= 0
end
```

Gradients of the objective function, equality constraint, and inequality constraints are given by

$$\nabla f(x) = [2 \, 2x_2], \, \nabla h(x) = [2x_1 \, 2x_2], \, \nabla g(x) = \begin{bmatrix} 1 & 0 \\ -1 & 0 \\ 0 & 1 \\ 0 & -1 \end{bmatrix}$$

The function *dfun* defines these gradients as follows:

```
function dfv = dfun(x)
   df = [2 2*x(2)];  % gradient of objective function
   dh = [2*x(1) 2*x(2)]; % gradient of equality constraint (h(x) = 0)
   dg = [1 0;-1 0;0 1;0 -1]; % gradient of inequality constraint (gi(x) >= 0)
   dfv = [df' dh' dg'];
end
```

Note that each gradient vector is converted to a column vector and that the gradient matrix for ∇g is transposed. The following commands implements the SQP algorithm to get solutions:

```
>> x0 = [2 3]'; lam0 = 1; mu0 = 3; crit = 1e-6;
>> [xopt,fopt,iter] = sqpopt(@fun,@dfun,x0,lam0,mu0,crit)
xopt =
      2.6458
      1.0000
fopt =
      6.2915
iter =
      8
```

10.4 DIRECT SEARCH METHODS

10.4.1 CYCLIC COORDINATE METHOD

In the cyclic coordinate search method, the search is performed along each of the coordinate directions for finding the minimum. Let u_k be the unit vector along the coordinate direction k. Then the value of α_k that minimizes $f(\alpha) = f(x + \alpha u_k)$ is determined, and a move is made to the new point $x + \alpha_k u_k$ at the end of the search along direction k. The search is conducted along all the possible directions as one stage. The search for the minimum terminates at the point where the gradient appears to be zero. For an n-dimensional problem, the cyclic coordinate algorithm can be summarized as follows[18]:

1) Choose a starting point x^1 and calculate $f_1 = f(x^1)$. Set $k = 1$, $x = x^1$, and $f = f_1$.
2) Determine α_k that minimizes $f(\alpha) = f(x + \alpha u_k)$.
3) Modify the current point by setting $x = x + \alpha_k u_k$ and calculate $f = f(x)$.
4) If $k < n$, set $k = k + 1$ and go to step 2).
5) Specify the direction as $d = x - x^1$ and determine α_d that minimizes $f(x + \alpha d)$.
6) Move to the new point by setting $x = x + \alpha_d d$ and calculate $f = f(x)$.
7) If $\|d\| > \varepsilon$ or $|f - f_1| > \varepsilon$, set $x^1 = x$, $f_1 = f$, and go to step 2).

The MATLAB function *cycopt* implements the cyclic coordinate search method. The basic syntax is

```
[xopt,fopt,iter] = cycopt(fcyc,x0,crit)
```

where fcyc is the objective function, x0 is the starting point vector, crit is the stopping criterion, xopt is the resultant optimum point, fopt is the function value at the optimum point, and iter is the number of iterations.

```
function [xopt,fopt,iter] = cycopt(fcyc,x0,crit)
% cycopt.m: minimization by the cyclic coordinate search
% Inputs:
%       fcyc: objective functions
%       x0: starting point
%       crit: stopping criterion
% Outputs:
%       xopt: optimal point
%       fopt: objective function value at x=xopt
%       iter: number of iterations
% Example:
% x0 = [-3 -1 0 1]; crit = 1e-4;
%   fc  =  @(x)   (x(1)+10*x(2))^2+5*(x(3)-x(4))^2+(x(2)-2*x(3))^4+10*(x(1)-x
(4))^4;
```

```
% [xopt,fopt,iter] = cycopt(fc,x0,crit)

stsize = 0.1; x = x0; xk = x; nv = length(x); fk = fcyc(xk); fold = fk; iter = 0;
while (1)
  iter = iter + 1;
  for j = 1:nv, h(j) = 0; end
  for k = 1:nv
    for j = 1:nv, d(j) = 0; end
    d(k) = 1; aut = 0;
    [stfit,fvfit] = quadfit(fcyc,aut,fk,stsize,xk,d,crit);
    xk(k) = xk(k) + stfit; h(k) = stfit; fk = fvfit;
  end
  aut = 0; [stfit,fvfit] = quadfit(fcyc,aut,fk,stsize,xk,h,crit);
  for j = 1:nv, xk(j) = xk(j) + stfit * h(j); end
  if (abs(fk-fold) <= crit), break; end
  fold = fk;
end
xopt = xk; fopt = fcyc(xk);
end


function [x2, f2] = quadfit(fcyc,x1,f1,stsize,x0,d,crit)
fr = 0.05;  % 0 < sfrac < 0.5
[x1,x2,x3,f1,f2,f3] = approx3pt(fcyc,x1,f1,stsize,x0,d); % 3-point pattern
tau = (sqrt(5) - 1)/2; redi = 2;
if (x3 < x1), xtemp = x1; ftemp = f1; x1 = x3; f1 = f3; x3 = xtemp; f3 = ftemp; end
iflag = 0; indc = 0; jndc = 0;
while (1)
  xdold = abs(x3 - x1); favg = (f1 + f2 + f3)/3.;
  if iflag == 0
    A = (x1-x2)*(x1-x3); B = (x2-x1)*(x2-x3); C = (x3-x1)*(x3-x2);
    x4 = (f1*(x2+x3)/A + f2*(x1+x3)/B + f3*(x1+x2)/C)/(f1/A+f2/B+f3/C)/2;
  else
    if x2 <= (x1+x3)/2, x4 = x2 + (1-tau)*(x3-x2);
    else, x4 = x3 - (1-tau)*(x2-x1);
    end
    iflag = 0;
  end
  delt = fr*min(abs(x2-x1),abs(x3-x2));
  if abs(x4-x1) < delt, x4 = x1+delt;
  elseif abs(x4 - x3) < delt, x4 = x3 - delt;
  elseif abs(x4-x2) < delt
    if x2 > (x1 + x3)/2, x4 = x2-delt; else, x4 = x2+delt; end
  end
  f4 = fcyc(x0 + x4*d);
  if (x4 > x2)
    if (f4 >= f2), x3 = x4; f3 = f4; else, x1 = x2; f1 = f2; x2 = x4; f2 = f4; end
  else
    if (f4 >= f2), x1 = x4; f1 = f4; else, x3 = x2; f3 = f2; x2 = x4; f2 = f4; end
  end
  xdnew = abs(x3-x1); fvnew = (f1 + f2 + f3)/3.;
  if abs(x3 - x1) <= crit, break; end
  if abs(fvnew - favg) <= crit
    jndc = jndc + 1; if jndc == 2, break; end
  else, jndc = 0;
  end
```

```
if xdnew/xdold > tau
   indc = indc + 1; if indc == redi, indc = 0; iflag = 1; end
else, indc = 0; iflag = 0;
end
end
end

function [x1,x2,x3,f1,f2,f3] = approx3pt(fcyc,x1,f1,stsize,x0,d)
% Three-point pattern
tau = (sqrt(5) - 1)/2; stlen = stsize; x2 = x1 + stlen; x = x0 + x2*d; f2 = fcyc(x);
if (f2 > f1)
   temp = x1; x1 = x2; x2 = temp; temp = f1; f1 = f2; f2 = temp; stlen = -stlen;
end
while (1)
   stlen = stlen/tau; x3 = x2 + stlen; x = x0 + x3*d; f3 = fcyc(x);
   if (f3 > f2), break; else, f1 = f2; x1 = x2; f2 = f3; x2 = x3; end
end
end
```

Example 10.17 Cyclic Coordinate Search Method

Find the minimum of the Powell's function using the cyclic coordinate search method:

$$f(x) = (x_1 + 10x_2)^2 + 5(x_3 - x_4)^2 + (x_2 - 2x_3)^4 + 10(x_1 - x_4)^4$$

As a starting point, use $x^0 = [-3, -1, 0, 1]$ and crit $= 1 \times 10^{-4}$.

Solution

```
>> x0 = [-3 -1 0 1]; crit = 1e-4;
>> fc = @(x) (x(1)+10*x(2))^2+5*(x(3)-x(4))^2+(x(2)-2*x(3))^4+10*(x(1)-x(4))^4;
>> [xopt,fopt,iter] = cycopt(fc,x0,crit)
xopt =
     0.1042   -0.0097   0.0543   0.0566
fopt =
     3.1864e-04
iter =
     14
```

10.4.2 HOOKE-JEEVES PATTERN SEARCH METHOD

In the Hooke-Jeeves pattern search method, an initial step size α is chosen and the search is initiated from a given starting point. Let u_k be the unit vector along the coordinate direction x_k. Then the function value at $x + \alpha u_k$ is evaluated, where α is the step size. If the function reduces, the current point x is updated according to $x + \alpha u_k$. If the function value does not decrease, calculate $f(x - \alpha u_k)$. If this function value reduces, x is updated to be $x - \alpha u_k$.

Suppose that the starting point is denoted by Q, whose coordinate is x_Q. If the search is not successful about the starting point Q, the step size is reduced as $\alpha = c\alpha$, where $c(<1)$ is a step reduction parameter. If the exploration is successful, the pattern direction $x_R - x_Q$ is established and the point R becomes the new starting point. The Hooke-Jeeves pattern search algorithm can be summarized as follows[19]:

1) Select a starting point x^1, an initial step size α, a step reduction parameter c, and an acceleration parameter β.
2) Set $x_Q = x^1$ and search about the point x_Q. If the function value reduces, define the new point generated as a new starting point x_Q and rename the old starting point as $x_{Q'}$. If the search is not successful, reduce the step size as $\alpha = c\alpha$.
3) Find a point $x_P = x_Q + \beta(x_Q - x_{Q'})$ and perform exploration about the point x_P.
4) If the function value reduces, define the new point generated as a new starting point x_Q, rename the old starting point as $x_{Q'}$, and go to step 3).
5) If the search is not successful, perform exploration about the point x_Q. Reduce the step size as $\alpha = c\alpha$ until convergence is achieved. Take the new point as a new starting point and go to step 3).

The MATLAB function *hjopt* implements the Hooke-Jeeves pattern search method. The basic syntax is

```
[xopt,fopt,iter] = hjopt(fhj,x0,crit)
```

where fhj is the objective function, x0 is the starting point vector, crit is the stopping criterion, xopt is the resultant optimum point, fopt is the function value at the optimum point, and iter is the number of iterations.

```
function [xopt,fopt,iter] = hjopt(fhj,x0,crit)
% hjopt.m: minimization by the Hooke and Jeeves pattern search method
% Inputs:
%      fhj: objective functions
%      x0: starting point
%      crit: stopping criterion
% Outputs:
%      xopt: optimal point
%      fopt: objective function value at x=xopt
%      iter: number of iterations
% Example:
% x0 = [-3 -1 0 1]; crit = 1e-4;
% f = @(x) (x(1)+10*x(2))^2+5*(x(3)-x(4))^2+(x(2)-2*x(3))^4+10*(x(1)-x(4))^4;
% [xopt,fopt,iter] = hjopt(f,x0,crit)

inist = 0.5; fr = 0.125; x = x0; xb = x; n = length(x0); stsize = inist;
nc = 0; % number of contraction steps
nb = 0; % number of base changes
np = 0; % number of pattern moves
fb = fhj(xb); iter = 1; fold = fb;
while (1)
    fk = fb; [fk, x] = search(fhj, n, stsize, fk, x);
    if (fb-fk > crit)
        while (2)
            icv = 0;
            for j = 1:n, cp = x(j); x(j) = 2*cp - xb(j); xb(j) = cp; end
            fb = fk; fk = fhj(x); [fk, x] = search(fhj, n, stsize, fk, x);
            if (fb-fk <= crit)
                x = xb;
                if (nb > 1), if abs(fk - fold) < crit, icv = 1; break; end; end
                nb = nb + 1; break;
            end
            np = np + 1;
```

```
      end
      if (icv == 1), break; end
   else
      fold = fk; if (stsize < crit), break; end; stsize = fr*stsize; nc = nc + 1;
   end
   iter = iter + 1;
end
xopt = x; fopt = fk;
end

function [fk, xk] = search(fhj, n, stsize, fk, x)
for k = 1:n
      cpt = x(k); x(k) = cpt + stsize; f = fhj(x);
      if (f < fk), fk = f;
      else
         x(k) = cpt - stsize; f = fhj(x);
         if (f < fk ), fk = f; else, x(k) = cpt; end
      end
end
xk = x;
end
```

Example 10.18 Hooke-Jeeves Pattern Search Method

Find the minimum of the Rosenbrock's function $f(x)$ using the Hooke-Jeeves pattern search method:

$$f(x) = 100(x_1^2 - x_2)^2 + (1 - x_1)^2$$

Use $x^0 = [-1, 1]$ and crit $= 1 \times 10^{-6}$.

Solution
```
>> f = @(x) 100*(x(1)^2 - x(2))^2 + (1 - x(1))^ 2; x0 = [-1 1]; crit = 1e-6;
>> [xopt,fopt,iter] = hjopt(f,x0,crit)
xopt =
  1.0078   1.0156
fopt =
 6.0860e-05
iter =
 10
```

10.4.3 ROSENBROCK'S METHOD

In Rosenbrock's method, the search is carried out in n orthogonal directions at each stage. At the next stage, new orthogonal directions are determined. Exploration of orthogonal directions presents some useful schemes for step-size decisions. Rosenbrock's method can be summarized as follows[20]:

1) Set the direction vectors as $\{v_1, v_2, \cdots, v_n\} = \{u_1, u_2, \cdots, u_n\}$ where u_i is the unit vector. Choose a starting point x^0, initial step lengths $\alpha_i (i = 1, \cdots, n)$, a step expansion parameter $\beta (>1)$, and a step reduction parameter $c (<1)$. Set $x = y = x^0$ and $k = 1$ (k: stage counter) and $j = 0$ (j: cycle counter).
2) If $j > n$, set $j = 1$. Otherwise set $j = j + 1$. Let $f_x = f(x)$ and $y = x + \alpha_j v_j$. Evaluate the

function at y: $f_y = f(y)$. If $f_y < f_x$, set $\alpha_j = \beta\alpha_j$, replace x by y ($x = y, f_x = f_y$), and go to step 3). If $f_y \geq f_x$, set $\alpha_i = -c\alpha_i$ and repeat this step.

3) Construct n linearly independent directions $w_1, w_2, \cdots, w_n$ as $w_i = \sum_{k=i}^{n} \alpha_k v_k$ ($i = 1, \cdots, n$). Calculate the magnitude of w_1: $\alpha = \sqrt{w_1^T w_1}$. If α is less than the stopping criterion, stop the procedure. If convergence is not achieved, new orthogonal directions are formed using the Gram-Schmidt procedure: $w_i' = w_i - \sum_{k=1}^{i-1}(w_i^T v_k)v_k$, $v_i = w_i'/\text{norm}(w_i')$.

4) Set $k = k + 1$ and carry out the next stage calculations. The new stage begins with n co-ordinate directions as starting vectors.

The MATLAB function *rbopt* implements Rosenbrock's method. The basic syntax is
[xopt,fopt,istag] = rbopt(fros,x0,crit)

where fros is the objective function, x0 is the starting point vector, crit is the stopping criterion, xopt is the resultant optimum point, fopt is the function value at the optimum point, and istag is the number of calculation stages.

```
function [xopt,fopt,istag] = rbopt(fros,x0,crit)
% rbopt.m: minimization by the Rosenbrock's method
% Inputs:
%       fros: objective functions
%       x0: starting point
%       crit: stopping criterion
% Outputs:
%       xopt: optimal point
%       fopt: objective function value at x=xopt
%       istag: number of stages
% Example:
% x0 = [-3 -1 0 1]; crit = 1e-4;
%   f  =  @(x)  (x(1)+10*x(2))^2+5*(x(3)-x(4))^2+(x(2)-2*x(3))^4+10*(x(1)-x
(4))^4;
% [xopt,fopt,istag] = rbopt(f,x0,crit)

n = length(x0); % number of variables
mi = 7; stsize = 1; alpha = 3; beta = 0.5; xt = x0;
% Initial vectors along coordinates
for j = 1:n
  for k = 1:n, v(j,k) = 0; end; v(j,j) = 1;
end
ft = fros(xt); istag = 0;
while (1)
      istag = istag + 1;
      % Initialization
      for j = 1:n, stp(j) = stsize; stj(j) = 0; sfv(j,1) = 0; sfv(j,2) = 0; end
      iter = 0; icont = 0; idn = 0;
  while (2)
        iter = iter + 1; if iter > n, iter = 1; end
        for j = 1:n, x(j) = xt(j) + stp(iter)*v(j,iter); end; f = fros(x);
        if f < ft
          stj(iter) = stj(iter) + stp(iter); % successful
          for j = 1:n, xt(j) = x(j); end
          ft = f; idn = 0; stp(iter) = alpha*stp(iter);
          if sfv(iter,1) == 0, sfv(iter,1) = 1; icont = icont + 1; end
        else % failure
          stp(iter) = -beta*stp(iter);
```

```
        if sfv(iter,2) == 0, sfv(iter,2) = 1; icont = icont + 1; end
        idn = idn + 1;
    end
    if icont == 2*n, break; end
    if idn > 50, mi = mi + 1; return; end
end
  if istag == 1, fold = ft;
else
  if abs(ft - fold) < crit, mi = mi + 1; display('Convergence achieved.');
break; end
end
fold = ft;
% Construct new vectors
for k = 1:n
        for j = 1:n, u(k,j) = 0; end
end
for i = 1:n
        for j = 1:n, for k = 1:n, u(k,i) = u(k,i) + stj(j)*v(k,j); end; end
end
% Orthogonalization by Gram-Schmidt procedure
for i = 1:n
        if i > 1
            for j = 1:n, v(j,i) = 0; end
            for k = 1:i-1
    c = 0; for j = 1:n, c = c + u(j,i)*v(j,k); end; for j = 1:n, v(j,i) = v(j,i) +
c*v(j,k); end
        end
        for j = 1:n, u(j,i) = u(j,i) - v(j,i); end
    end
    c = 0; for j = 1:n, c = c + u(j,i)*u(j,i); end
    c = sqrt(c);
    % Take new step length as the length of the first vector
    if i == 1
            stsize = c;
            if stsize < crit, mi = mi + 1; break; end
            if c < crit % Orthogonality is lost: reset vectors.
                for j = 1:n
                    for k = 1:n, v(j,k) = 0; end
                    v(j,j) = 1; break;
            end
        end
        for j = 1:n, v(j,i) = u(j,i)/c; end
    end
  end
end
xopt = xt; fopt = ft;
end
```

Example 10.19 Rosenbrock's Method

Find the minimum of $f(x) = x_1^2 + 2x_2^2 + 2x_1x_2$ using Rosenbrock's method. Use $x^0=[0.5,1]$ and crit=1×10^{-6}.

Solution

```
>> f = @(x) x(1)^2 + 2*x(2)^2 + 2*x(1)*x(2); x0 = [0.5 1]; crit = 1e-6;
>> [xopt,fopt,istag] = rbopt(f,x0,crit)
Convergence achieved.
xopt =
        -0.3000    0.2000
fopt =
        0.0500
istag =
        6
```

10.4.4 NELDER-MEAD'S SIMPLEX METHOD

In an n-dimensional space, $n + 1$ points form a simplex. For example, a tetrahedron forms a simplex in three-dimensional space. An initial simplex in n-dimensional space can be formed by choosing the origin as one corner and n points, each marked at a distance of s from the origin along the coordinate axes. A regular simplex is constructed in space by adding the coordinates of a starting point to each of the $n + 1$ points. For minimization, function values at the $n + 1$ corner points of the simplex are evaluated first. The point with the highest function value is replaced by a newly created point that can be obtained by reflection, expansion, or contraction.

Let f_h, f_s, and f_l be the highest, second highest, and lowest function values among the $n + 1$ corners of the simplex, respectively, and let x_h, x_s, and x_l be the corresponding coordinates. Then x_m, the average of n points excluding the highest point, can be calculated by

$$x_m = \frac{1}{n} \sum_{j=1}^{n+1} x_j (j \neq h)$$

If x_m is the mean of k points and the $(k + 1)$th point is denoted by x_{k+1}, the updated mean of the $k + 1$ points is given by

$$x_m = x_m + \frac{x_{k+1} - x_m}{k + 1}$$

The reflected point x_r is given by $x_r = x_m + \gamma (x_m - x_h)$ (γ: reflection parameter, $\gamma > 0$), and the expanded point x_e is given by $x_e = x_r + \beta (x_r - x_m)$ (β: expansion parameter, $\beta > 1$). If $f_r = f(x_r) > f_h$, the contracted point x_c is given by $x_c = x_m + c(x_h - x_m)$ (c: contraction parameter, $0 < c < 1$). If $f_j < f_r \leq f_h (j \neq h)$, the contracted point x_c is given by $x_c = x_m + c(x_r - x_m)$. The point x_j is updated by $x_j = x_l + \alpha (x_j - x_l)$.[21]

Nelder-Mead's simplex method can be summarized as follows[22]:

1) Choose a starting point x_1 and evaluate $f_1 = f(x_1)$. Set $k = 0$.
2) Generate a simplex by creating n new corner points and calculate function values at these points. Set $k = k + 1$.
3) Determine the highest, second highest, and lowest function values f_h, f_s, and f_l, and the corresponding coordinates x_h, x_s, and x_l, respectively. If $|f_h - f_l| < \varepsilon$, go to step 13).
4) Determine x_m, the average of n points excluding the highest point, and the reflected point x_r. Calculate $f_r = f(x_r)$. If $f_r \geq f_l$, go to step 7).
5) Evaluate x_e and $f_e = f(x_e)$. If $f_e \geq f_l$, go to step 11). Replace x_h and f_h by x_e and f_e, respectively, and go to step 2).
6) Replace x_h and f_h by x_r and f_r, respectively, and go to step 2).
7) If $f_r > f_h$, go to step 9).

8) Replace x_h and f_h by x_r and f_r, respectively.

9) If $f_r \leq f_s$, go to step 2).

10) Evaluate x_c and $f_c = f(x_c)$. If $f_c > f_h$, go to step 11). Replace x_h and f_h by x_c and f_c, respectively, and go to step 2).

11) Perform an update to find x_j and $f_j = f(x_j)$.

12) Calculate the function values at the new corners of the simplex and go to step 2).

13) If $k > 1$ and $|f_{min} - f_l| < \varepsilon$, the convergence is achieved; stop the calculation procedure.

14) Set $f_{min} = f_l$ and replace x_1 and f_1 by x_l and f_l, respectively. Reduce the step reduction parameter and go to step 2).

The MATLAB function *nmopt* implements Nelder-Mead's simplex method. The basic syntax is

```
[xopt,fopt,iter] = nmopt(fun,x0,crit)
```

where fun is the objective function, x0 is the starting point vector, crit is the stopping criterion, xopt is the resultant optimum point, fopt is the function value at the optimum point, and iter is the number of iterations.

```
function [xopt,fopt,iter] = nmopt(fun,x0,crit)
% nmopt.m: minimization by the Nelder and Mead's simplex method
% Inputs:
%        fun: objective functions
%        x0: starting point
%        crit: stopping criterion
% Outputs:
%        xopt: optimal point
%        fopt: objective function value at x=xopt
%        iter: number of iterations
% Example:
% x0 = [-3 -1 0 1]; crit = 1e-4;
% f = @(x) (x(1)+10*x(2))^2+5*(x(3)-x(4))^2+(x(2)-2*x(3))^4+10*(x(1)-x(4))^4;
% [xopt,fopt,iter] = nmopt(f,x0,crit)

rf = 1; % reflection parameter
ef = 2; % expansion parameter
cf = 0.5; % contraction parameter
sf = 0.5; % scale parameter
n = length(x0); beta = 1; n1 = n+1; iter = 0;
x = x0; fv(1) = fun(x);
for j = 1:n, pt(1,j) = x(j); end
% Create simplex and evaluate function values
for k = 2:n1
    u = zeros(1,n); u(1,k-1) = 1; pt(k,:) = pt(1,:) + beta*u; x = pt(k,:);
    fv(k) = fun(x);
end
while (1)
    nloop = 0;
    while (2)
        nloop = nloop + 1; if (nloop == 2), break; end
        iter = iter + 1;
        % Find highest, lease, and second highest values
        [fh, indh] = max(fv); [fl, indl] = min(fv); fs = fv(indl); inds = indl;
        for k = 1:n1
            if (k ~= indh), if (fs < fv(k)), fs = fv(k); inds = k; end; end
        end
```

```
                % Find the average of n points xm
                for k = 1:n
                  xm(k) = 0;
                  for j = 1:n1, if (j ~= indh), xm(k) = xm(k) + pt(j,k); end; end
                  xm(k) = xm(k)/n;
                end
                % Reflection procedure
                xr = xm + rf*(xm - pt(indh,:)); fr = fun(xr);
                if (fr >= fl & fr <= fs) % accept reflection
                  fv(indh) = fr; pt(indh,:) = xr; break;
                end
                % Expansion procedure
                if (fr < fl )
                  xe = xr + ef*(xr - xm); fe = fun(xe);
                  if (fe < fl) % accept expansion
                    fv(indh) = fe; pt(indh,:) = xe; break;
                  else % accept reflection
                    fv(indh) = fr; pt(indh,:) = xr; break;
                  end
                end
                % Contraction procedure
                if (fr > fh)
                  xc = xm + cf*(pt(indh,:) - xm); [fc] = fun(xc);
                  if (fc <= fh) % accept contraction
                    fv(indh) = fc; pt(indh,:) = xc; break;
                  end
                elseif (fr > fs & fr <= fh)
                  xc = xm + cf*(xr - xm); fc = fun(xc);
                  if (fc <= fr) % accept contraction
                    fv(indh) = fc; pt(indh,:) = xc; break;
                  end
                end
                % Scaling
                for k = 1:n1
                  if (k ~= indl)
                    for j = 1:n, x(j) = sf *pt(k,j) + (1 - sf) *pt(indl,j);
                    pt(k,j) = x(j); end
                    fv(k) = fun(x);
          end
        end
  end
  sigma = std(fv); avg = mean(fv);
  if (sigma <= crit), inc = inc + 1; if (inc == 2), break; end
  else, inc = 0; end
end
xopt = pt(indl,:); fopt = fl;
end
```

Example 10.20 Nelder-Mead's Simplex Method

Find the minimum of the Powell function using Nelder-Merad's simplex method:
$$f(x) = (x_1 + 10x_2)^2 + 5(x_3 - x_4)^2 + (x_2 - 2x_3)^4 + 10(x_1 - x_4)^4$$

As a starting point, use $x^0 = [-3, -1, 0, 1]$ and crit $= 1 \times 10^{-6}$.

Solution
```
>> f = @(x) (x(1)+10*x(2))^2+5*(x(3)-x(4))^2+(x(2)-2*x(3))^4+10*(x(1)-x
(4))^4;
>> x0 = [-3 -1 0 1]; crit = 1e-6; [xopt,fopt,iter] = nmopt(f,x0,crit)
xopt =
    -0.0244    0.0025   -0.0097   -0.0100
fopt =
    1.3627e-06
iter =
    80
```

10.4.5 SIMULATED ANNEALING (SA) METHOD

Metals have phase transformations when the temperature changes. Annealing is a process where stresses are relieved from a previously hardened metal. More specifically, annealing is the physical process of heating up a metal above its melting point, followed by cooling it down so slowly that the excited metal atoms can settle into a minimum energy level, yielding a crystal with a regular structure. Energy in the annealing process sometimes may increase even while the trend is a net decrease. This property can be applied to optimization problems.

Consider a simple minimization problem of the form

$$\text{Minimize } f(x)$$

$$\text{Subject to } l_i \le x_i \le u_i (i = 1, \cdots, n)$$

where $x = [x_1, x_2, \cdots, x_n]^T$ is a column vector of n real-valued variables. In optimization applications, we begin with an initial state of temperature T set at a high level. The simulated annealing process can be implemented using Boltzmann's probability distribution function given by

$$p(\Delta E) = e^{-\frac{\Delta E}{kT}}$$

where k is Boltzmann's constant, which can be chosen as equal to 1 in optimization applications. The change Δf in the objective function value is accepted whenever it represents a decrease. If it is an increase, we accept it with a probability of $p(\Delta f) = e^{-\Delta f/T}$ ($k = 1$). The solution of minimization problem by the simulated annealing method can be summarized as follows[23]:

1) Choose a starting point x and evaluate $f = f(x)$. Set $x_{min} = x$ and $f_{min} = f$.
2) Choose a vector s with step size s_i along the coordinate direction u_i. The initial value of each s_i may be set equal to a step size $s_T (=1)$ and a step reduction parameter r_s.
3) Define a vector a of acceptance ratios with each element equal to 1.
4) Choose a starting temperature T and a temperature reduction factor r_T. Set $k = 1$.
5) Set the temperature as $r_T T$ and the step size as $r_s s_T$. At each temperature, N_T iterations are carried out, and each iteration consists of N_C cycles.
6) Generate a random number $r(-1 \le r \le 1)$ and evaluate a new point $x_s = x + r s_i u_i$. If this point lies outside the bounds, the ith component of x_s is adjusted to be a random point in the interval l_i to u_i.
7) Evaluate $f_s = f(x_s)$. If $f_s \le f$, the point is accepted by setting $x = x_s$. If $f_s \le f_{min}$, update f_{min} and x_{min}. If $f_s > f$, the point is accepted with a probability of $p = e^{-(f-f_s)/T}$.

8) Generate a random number r. If $r < p$, accept f_s. If a rejection takes place, update the acceptance ratio a_i as $a_i = a_i - 1/N_C$.

9) Set $k = k + 1$ and go to step 5).

10) At the end of N_C cycles, use a_i to update the step size for the direction. A low value of a_i implies that there are more rejections, suggesting that the step size should be reduced. A high rate of a_i implies more acceptances, which may be due to a small step size. In this case, the step size is to be increased.

The MATLAB function *smopt* implements the simulated annealing method. The basic syntax is [xopt,fopt,iter] = saopt(fun,T,xl,xu,rp,rs,crit)

where fun is the objective function, T is the initial temperature, xl and xu are the lower and upper bounds, rp is the temperature reduction factor, rs is the step reduction parameter, crit is the stopping criterion, xopt is the resultant optimum point, fopt is the function value at the optimum point, and iter is the number of cycle iterations.

```
function [xopt,fopt,iter] = saopt(fun,T,xl,xu,rp,rs,crit)
% saopt.m: minimization by simulated annealing (SA) method
% Inputs:
%      fun: objective function
%      T: initial temperature
%      xl,xu: lower and upper bounds on x
%      rp: temperature reduction factor
%      rs: step reduction parameter
%      crit: stopping criterion
% Outputs:
%      xopt: optimal point
%      fopt: objective function value at x=xopt
%      iter: number of cycle iterations
% Example:
%      f = @(x) -cos(5*sqrt(sum(x-5).^2)) + 0.1*sum(x-5).^2;
%      T = 100; xl = -10*[1 1]; xu = 10*[1 1]; rp = 0.8; rs = 0.9; crit = 1e-8;
%      [xopt,fopt,iter] = saopt(f,T,xl,xu,rp,rs,crit)

% Initialization
sf = 2; stp = 1; np = 10; nc = 20; nt = 2e4; ir = 16; rcrit = 1e-10; n = length(xu);
% Feasible starting point
for j = 1:n, x(j) = xl(j) + rand*(xu(j) - xl(j)); xs(j) = x(j); xmin(j) = x(j); end
f = fun(x); fmin = f; fold = f; citer = 0;
% Set step sizes, step factors and acceptance ratios
for j = 1:n, stsize(j) = stp; ar(j) = 1; end
while (1)
   for piter = 1:np % temperature step loop
      for ic = 1:nc % % search cycles: search along coordinate direction
         for k = 1:n
            xs(k) = x(k) + (2*rand - 1)*stsize(k);
            if (xs(k) < xl(k)) | (xs(k) > xu(k)), xs(k) = xl(k) + rand*(xu(k) - xl(k));
            end
            fs = fun(xs);
               if fs <= f % point is accepted: update xmin and fmin
               x(k) = xs(k); f = fs;
               if fs < fmin, xmin = xs; fmin = fs; end
               else
                  p = exp((f - fs)/T);
```

```
            if rand < p, x(k) = xs(k); f = fs;
            else % point rejected
                xs(k) = x(k); ar(k) = ar(k) - 1/nc;
            end
         end
      end
   end
   % Adjust step so that about half the points are accepted.
   for j = 1:n
      if ar(j) > 0.6, stsize(j) = stsize(j)*(1 + sf*(ar(j) - 0.6)/0.4);
      elseif ar(j) < 0.4, stsize(j) = stsize(j)/(1 + sf*(0.4 - ar(j))/0.4); end
      if stsize(j) > xu(j) - xl(j), stsize(j) = xu(j) - xl(j); end
      ar(j) = 1;
   end
end
stsize = stp*(xu - xl); fcrit = crit + rcrit*abs(fmin);
if (fmin <= fold) & (fold - fmin < fcrit)
   citer = citer + 1; if citer >= ir, break; end
else, citer = 0;
end
% Reduce temperature
T = rp*T; stp = rs*stp; x = xmin; f = fmin; fold = f;
end
xopt = xmin; fopt = fmin; iter = citer;
end
```

Example 10.21 Simulated Annealing Method

Find the minimum of the corrugated spring function $f(x)$ using the simulated annealing method[24]:

$$f(x) = -\cos(kR) + 0.1R^2, R = \sqrt{(x_1 - c)^2 + (x_2 - c)^2}, -10 \le x_1, x_2 \le 10$$

Set the initial temperature as T = 100, the temperature reduction factor rp = 0.8, the step reduction parameter rs = 0.9, and crit = 1×10^{-8}.

Solution

The corrugated spring function is defined by the function *fcy* as follows:
```
function fv = fcy(x)
% Corrugated spring function
C = 0; for j = 1:2, C = C + (x(j) - 5)^2; end
fv = -cos(5*sqrt(C)) + 0.1*C;
end
```

The following commands carry out the SA calculation procedure:
```
>> T = 100; xl = -10*[1 1]; xu = 10*[1 1]; rp = 0.8; rs = 0.9; crit = 1e-8;
>> [xopt,fopt,iter] = saopt(@fcy,T,xl,xu,rp,rs,crit)
xopt =
    4.9999    5.0001
fopt =
   -1.0000
iter =
    16
```

10.4.6 Genetic Algorithm (GA)

A natural process takes place through the mutation and recombination of chromosomes in a population to yield a better gene structure. The genetic algorithm is a typical direct search method modeled on natural evolution and selection processes toward the survival of the fittest. One of the significant characteristics of the genetic algorithm is the use of stochastic information in its implementation, and therefore this method may be considered a global optimization method. A basic genetic algorithm can be summarized as follows[25]:

1) Set a set of solutions or chromosomes (population) $P(0)$. Evaluate the initial solution for fitness. Set $t = 0$.
2) Generate the set of children (crossover, mutation) using the genetic operators. Add a new set of randomly generated population and evaluate again the population fitness.
3) Determine which members will be part of the next generation (competitive selection). Select population $P(t + 1)$.
4) If convengence is not achieved, set $t = t + 1$ and go to step 2).

Figure 10.5 shows the flowchart for a genetic algorithm.[26]

In general, a genetic algorithm problem is posed in the form

$$\text{Maximize } f(x)$$

$$\text{Subject to } l_i \leq x_i \leq u_i (i = 1, \cdots, n)$$

where $x = [x_1, x_2, \cdots, x_n]^T$ is a column vector of n real-valued variables. Each variable is represented as a binary number, say, of m bits. This is performed by dividing the feasible interval of variable x_i into $2^m - 1$ intervals. The interval step s_i for variable x_i is given by

$$s_i = \frac{u_i - l_i}{2^m - 1}$$

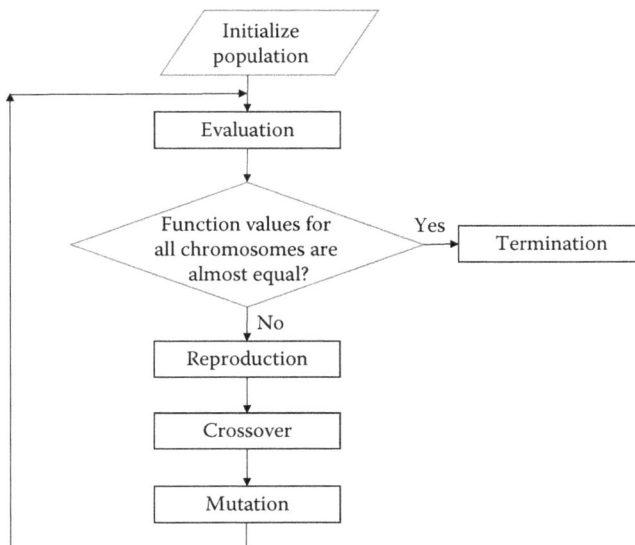

FIGURE 10.5 Flowchart for a genetic algorithm. (From Yang, W.Y. et al., *Applied Numerical Methods Using MATLAB*, Wiley Interscience, Hoboken, NJ, 2005, p. 338.)

In the creation of the initial population, each member of the population is a string of size $n * m$ bits. In the evaluation phase, the values of variables for each member are extracted. The function values $f_1, f_2, \cdots$ are evaluated. These values are referred to as fitness values. The reproduction takes the steps of creation of a mating pool, crossover, and mutation operations. The sum of the scaled fitness value S is calculated as

$$S = \sum_{j=1}^{z} f_j', \quad f_j' = \frac{f_j + Q}{R}, \quad Q = 0.1 f_h - 1.1 f_l, \quad R = \max(1, f_h + Q)$$

Roulette wheel selection is used to make copies of the fittest members for reproduction. The crossover operation is carried out on the mating parent pool to produce offspring. There is a chance that the parent selection and crossover operations may result in nearly identical individuals that may not be the fittest. Mutation is performed, in which a random operation is carried out to change the expected fitness. The population is evaluated again, and the highest fitness value and the corresponding variable values are stored as f_{max} and x_{max}. This completes a generation. If the predetermined number of generations is complete, the calculation process is terminated.

The MATLAB function *gaopt* implements the genetic algorithm. The basic syntax is

```
[xopt,fopt,iter] = gaopt(fun,xl,xu,nb,ps,ng,mp)
```

where fun is the objective function, xl and xu are the lower and upper bounds, nb is the number of binary digits, ps is the population size, ng is the number of generations, mp is the mutation probability, xopt is the resultant optimum point, fopt is the function value at the optimum point, and iter is the number of function evaluations.

```
function [xopt,fopt,iter] = gaopt(fun,xl,xu,nb,ps,ng,mp)
% gaopt.m: maximization by the genetic algorithm
% Inputs:
%       fun: objective functions
%       xl,xu: lower and upper bounds on x
%       nb: number of binary digits
%       ps: population size
%       ng: number of generations
%       mp: mutation probability
% Outputs:
%       xopt: optimal point
%       fopt: objective function value at x=xopt
%       iter: number of function evaluations
% Example:
%    xl=-6*[1 1 1 1]; xu=6*[1 1 1 1]; nb=8; ps=40; ng=50; mp=0.02;
%    f = @(x) (x(1)+10*x(2))^2+5*(x(3)-x(4))^2+(x(2)-2*x(3))^4+10*(x(1)-x(4))^4;
%    [xopt,fopt,iter] = gaopt(f,xl,xu,nb,ps,ng,mp)

n = length(xu);
Pn = round(rand(ps,n*nb)); % initial population
for kg = 1:ng
        ft = fitness(fun,ps,nb,n,xl,xu,Pn); [fmax,kmax] = max(ft);
        if (kg == 1)
                fbest = fmax;
                for k = 1:n
                        ab = num2str(Pn(kmax,nb*(k-1)+1:nb*k));
                        adec = bin2dec(ab); % convert binary string to decimal integer
                        xbest(k) = xl(k) + (xu(k)-xl(k))/(2^nb - 1)*adec;
                end
```

```
            else
                if (fmax > fbest)
                    fbest = fmax;
                    for k = 1:n
                        ab = num2str(Pn(kmax,nb*(k-1)+1:nb*k)); adec = bin2dec(ab);
                         xbest(k) = xl(k) + (xu(k)-xl(k))/(2^nb - 1)*adec;
            end
          end
        end

    [frk,ft] = frank(ps, ft); % scaling
    for k = 1: ps % shuffling: roulette wheel
        ik = frk(k); ptemp = ps; ktemp = k; am(ik) = 2*(ptemp + 1 - ktemp)/
        (ptemp*(ptemp + 1));
    end
    for k = 2: ps, ptemp = am(k - 1); am(k) = ptemp + am(k); end
    %
    for k = 1:ps % selection
      kstr1 = roul(ps,am); kstr2 = roul(ps,am); kchd = cros(Pn,nb,n,kstr1, kstr2);
        kchd = mutat(kchd,mp,nb,n);
        for j = 1: nb*n, Pn(k, j) = kchd(j); end
    end
end
xopt = xbest; fopt = fbest; iter = ng*ps;
end

function ft = fitness(fun,ps,nb,n,xl,xu,Pn)  % function evaluation
for k = 1:ps
    for j = 1:n
        a = num2str(Pn(k,nb*(j-1) + 1:nb*j)); adec = bin2dec(a);
        x(j) = xl(j) + (xu(j)-xl(j))/(2^nb - 1)*adec;
    end
    f = fun(x); ft(k) = f;
end
end

function [frk,ft] = frank(ps, ft)
for k = 1:ps, frk(k) = k; end
for k = 1:ps-1
    temp = ft(k); ktemp = k;
    for j = k+1:ps
        if (ft(j) > temp), temp = ft(j); ktemp = j; end
    end
    jtemp = frk(ktemp); frk(ktemp) = frk(k); frk(k) = jtemp;
    ftemp = ft(ktemp); ft(ktemp) = ft(k); ft(k) = ftemp;
end
end

function kstr = roul(ps,am)  % shuffling operation
temp = rand;
for k = 1:ps
    while (1)
        if (temp > am(k)), break; end
        kstr = k; return;
    end
```

```
end
end

function kchd = cros(Pn,nb,n,kstr1, kstr2)  % crossover operation
for j = 1:n
        jp = nb*(j - 1); nc = floor(rand*nb + 0.5);
        if (nc == 0)
            for k = 1:nb, kchd(k+jp) = Pn(kstr2, k+jp); end
    elseif (nc == nb)
            for k = 1:nb, kchd(k+jp) = Pn(kstr1, k+jp); end
    else
            for k = 1: nc, kchd(k+jp) = Pn(kstr1, k+jp); end
            for k = nc+1:nb, kchd(k+jp) = Pn(kstr2, k+jp); end
    end
end
end

function [kchd] = mutat(kchd,mp,nb,n)  % mutation operation
for k = 1:nb*n
        if (rand <= mp)
        if (kchd(k) == 1), kchd(k) = 0;
        else, kchd(k) = 1;
    end
    end
end
end
```

Example 10.22 Genetic Algorithm

Find the minimum of the corrugated spring function $f(x)$ using the genetic algorithm[27]:

$$f(x) = -\cos(kR) + 0.1R^2, \; R = \sqrt{(x_1 - c)^2 + (x_2 - c)^2}, \; -10 \le x_1, x_2 \le 10$$

Set nb = 8, ps = 50, ng = 60 and mp = 0.05.

Solution
Note that the function *gaopt* carries out the maximization problem. Thus, the corrugated spring function defined in the previous example (function *fcy*) should be multiplied by −1 as follows:
```
function fv = fcy2(x)
% Corrugated spring function
C = 0; for j = 1:2, C = C + (x(j) - 5)^2; end
fv = - (-cos(5*sqrt(C)) + 0.1*C);
end
```

The following commands carry out the GA calculation procedure:
```
>> xl = -10*[1 1]; xu = 10*[1 1]; nb = 8; ps = 50; ng = 60; mp = 0.05;
>> [xopt,fopt,iter] = gaopt(@fcy2,xl,xu,nb,ps,ng,mp)
xopt =
      4.9020    4.9804
fopt =
      0.8766
iter =
        3000
```

10.5 MIXED-INTEGER PROGRAMMING

10.5.1 ZERO-ONE PROGRAMMING METHOD

Many engineering problems can be formulated as zero-one problems. For example, selection of optimal locations of actuators may be formulated as a zero-one problem. Moreover, a certain class of integer programming problems may be converted into equivalent zero-one linear programming problems. Consider a simple minimization problem of the form

$$\text{Minimize } f(x)$$

$$\text{Subject to } g_i(x) \le b_i, \ x_i = 0 \text{ or } 1 \ (i = 1, \cdots, n)$$

where $x = [x_1, x_2, \cdots, x_n]^T$ is a vector of n real-valued variables. If there are inequality constraints of "greater than" ($\ge$) type, these can be converted into constraints of "less than" ($\le$) type. The search for the solution starts with an initial point $x = 0$. If this initial solution is also feasible, then we are done. If it is not feasible, the variable values are adjusted so that feasibility is attained.

The MATLAB function *ozopt* implements the zero-one programming method. The basic syntax is

```
[xopt,fopt,iter] = ozopt(A,c,nl,ne)
```

where A is the coefficient matrix for constraints, c is a row vector of coefficients of the objective function, nl is the number of "less than" ($\le$) type constraints, ne is the number of equality ($=$) constraints, xopt is the resultant optimum point, fopt is the function value at the optimum point, and iter is the number of iterations. The last column of matrix A contains the right-hand sides of "less than" inequality constraints.

```
function [xopt,fopt,iter] = ozopt(A,c,nl,ne)
% ozopt.m: zero-one programming method for inter minimization problem
%        inequality constraints should be rearranged to <= constraints.
% Inputs:
%     A: coefficient matrix for constraints including right hand sides
%     c: coefficient vector for objective function (row vector)
%     nl: number of <= constraints
%     ne: number of = constraints
% Outputs:
%     xopt: optimal point (0 or 1)
%     fopt: function value at the optimal point (=f(xopt))
%     iter: number of iterations
% Example
%     nl = 3; ne = 0;
%     A = [-1 3 6 6; -2 -3 -3 -2; -1 0 -1 -1]; c = [4 5 3];
%     [xopt,fopt,iter] = ozopt(A,c,nl,ng,ne)

% nl = 3; ne = 3;
%A = [1 0 0 1 0 1 0 0 0 1;0 1 0 0 0 0 1 0 0 1;0 0 0 1 0 0 0 0 1 1;1 1 1 0 0 0 0 0 0 1;0 0 0 1 1 0 0 0 0
1;0 0 0 0 0 0 0 1 1 1];
%c = -[5 3 1 3 5 2 5 5 2];

nv = length(c); m = nl + ne; x = zeros(1,nv); Am = zeros(m+1,nv+1); xvec =
zeros(1,nv);
indv = zeros(1,nv); temA = zeros(1,m+1); xmin = zeros(1,nv);
[rAm,cAm] = size(A); Am(1:rAm,1:cAm) = A;
for k = 1:m, Am(k,nv+1) = -Am(k,nv + 1); end
% Convert <= constraints to >=
```

```
for k = 1:nl, for j = 1:nv+1, Am(k,j) = -Am(k,j); end; end
% Convert = constraints to >= by adding one >= constraint
if ne > 0
    m = m + 1;
    for k = nl+1:m-1
        for j = 1:nv+1, Am(m,j) = Am(m,j) - Am(k,j); end
    end
end
% Convert variables if c(j) < 0
check0 = 0;
for j = 1:nv
    if c(j) < 0
        indv(j) = 1; check0 = check0 + c(j); c(j) = -c(j);
        for k = 1:m, Am(k,nv+1) = Am(k,nv+1) + Am(k,j); Am(k,j) = -Am(k,j); end
    end
end
for k = 1:nv, xvec(k) = k; x(k) = 0; end
for k = 1:m, temA(k) = Am(k,nv+1); end
indf = 1; inds = 0; f = 0; iter = 0;
while (1)
    sflag = 0; iter = iter + 1;
    if xvec(indf) > -1
       sflag = 1;
       for k = 1:m, if temA(k) < 0, sflag = 0; break; end; end
    end
    if sflag == 1 % feasible solution
       inds = inds + 1;
       if inds == 1 || (inds > 1 && f < fmin)
            fmin = f;
            for k = 1:nv, xmin(k) = x(k); end
       end
    end
  cflag = 0; nflag = 0;
  if sflag == 0 % infeasible solution
        for k = 1:m
        indk = temA(k);
        if indk < 0
            for j = indf+1:nv
                if Am(k,xvec(j)) > 0, indk = indk + Am(k,xvec(j)); end
            end
          end
          if indk < 0, cflag = 1; break; end
      end
      if inds > 0
          nflag = 1;
          for k = indf+1:nv, if f + c(xvec(k)) < fmin, nflag = 0; break; end; end
      end
  end
  if sflag == 1 || cflag == 1 || nflag == 1
        while xvec(indf) < 0
            xvec(indf) = -xvec(indf); indf = indf - 1;
            if indf == 0, break; end
        end
        if indf == 0, break; end
        x(xvec(indf)) = 0;
```

```
                % Update constraints and function value (f)
                 for k = 1:m, temA(k) = temA(k) - Am(k,xvec(indf)); end
                 f = f - c(xvec(indf)); xvec(indf) = -xvec(indf);
            else
                 indf = indf + 1;
                 for k = indf:nv
                     difa = 0;
                     for j = 1:m
                         temg = temA(j) + Am(j,xvec(k)); if temg < 0,
                         difa = difa - temg; end
                     end
                     if k == indf, delf = difa; ink = k;
                     else, if delf > difa, delf = difa; ink = k; end; end
                 end
                 inj = xvec(indf); xvec(indf) = xvec(ink); xvec(ink) = inj;
                 % Update x and f
                 x(xvec(indf)) = 1;
                 for k = 1:m, temA(k) = temA(k) + Am(k,xvec(indf)); end
                 f = f + c(xvec(indf));
             end
        end
if inds == 0, display('No feasible solution.'); return; end
% Assign outputs
for k = 1:nv, if (indv(k) == 1), xmin(k) = 1 - xmin(k); end; end
xopt = xmin; fopt = fmin;
end
```

Example 10.23 Zero-One Programming Method

Find the solution of the following minimization problem using the zero-one programming method:

$$\text{Minimize } f(x) = 4x_1 + 3x_2$$

$$\text{Subject to } 2x_1 - 5x_2 \le 10, \ 3x_1 + 2x_2 \le 9, \ 0 \le x_1, \ x_2 \ (x_i = 0 \text{ or } 1)$$

Solution

```
>> nl = 2; ne = 0; A = [2 -5 10;3 2 9]; c = [4 3]; [xopt,fopt,iter] = ozopt
(A,c,nl,ne)
xopt =
      0    1
fopt =
     -1
iter =
      4
```

10.5.2 Branch-and-Bound Method

The branch-and-bound method is useful to solve mixed-integer linear programming problems. In this method, a continuous relaxed linear programming problem is solved at every solution stage. If the solution of the continuous problem happens to be an integer solution, it represents the optimum solution of the integer problem. Otherwise, at least one of the integer variables must assume a noninteger value. If a variable x_i is not an integer, it can be represented by

$$I_x < x_i < I_x + 1$$

where I_x is the largest integer bounded by x_i. For example, if $x_i = 4.7$, $I_x = 4$. In this case, two subproblems are formulated, one with the additional upper-bound constraint and another with the lower-bound constraint, as follows:

$$x_i \leq I_x, \ x_i \geq I_x + 1$$

In this branching process, some portion of the continuous space that is not feasible for the integer problem is eliminated, while all the integer feasible solutions are retained. The solution of a continuous problem forms a node, and the linear programming problem is solved at each node k. The process of branching and solving a sequence of continuous problems is continued until an integer feasible solution is found for one of the two continuous problems.[28] The branch-and-bound algorithm can be summarized as follows[29]:

1) Set the root node with all discrete variables free. Set the current incumbent solution to $f^* = \infty$, $x^* = [\infty]$.
2) If any active feasible partial solutions remain, choose one as x^s. If none exist and there exists an incumbent solution, it is done. If none exist and there is no incumbent solution, the problem is infeasible.
3) Obtain the completion of the partial solution x^s using a continuous relaxation of the original problem.
4) If there are no feasible completions of the partial solution x^s, set $s + 1 \rightarrow s$ and go to step 2).
5) If the best feasible completion of the partial solution x^s cannot improve the incumbent solution, set $s + 1 \rightarrow s$ and go to step 2).
6) If the best feasible completion is better than the incumbent, update the incumbent solution as $f^s \rightarrow f^*$ and $x^s \rightarrow x^*$. Set $s + 1 \rightarrow s$ and go to step 2).

The MATLAB function *bnbopt* implements the branch-and-bound algorithm. The basic syntax is
`[xopt,fopt,iter] = bnbopt(A,b,c,nl,ng,ne,ibd)`

where A is the coefficient matrix for constraints, b is a row vector representing the right-hand side of the constraints, c is a row vector of coefficients of the objective function, nl is the number of "less than" ($\leq$) type constraints, ng is the number of "greater than" ($\geq$) type constraints, ne is the number of equality (=) constraints, ibd is the variable index vector, xopt is the resultant optimum point, fopt is the function value at the optimum point, and iter is the number of iterations.

```
function [xopt,fopt,iter] = bnbopt(A,b,c,nl,ng,ne,ibd)
% bnbopt.m: branch and bound method for mixed inter minimization problem
% Inputs:
%      A: coefficient matrix for constraints
%      b: right hand side of constraints (row vector)
%      c: coefficient vector for objective function (row vector)
%      nl: number of <= constraints
%      ng: number of >= constraints
%      ne: number of = constraints
%      ibd: variable index vector
% Outputs:
%      xopt: optimal point (0 or 1)
%      fopt: function value at the optimal point (=f(xopt))
```

```
%     iter: number of iterations

critd = 1e-4; critf = 5e-2; kmax = 1e4; ksol = 0; kf = 0; fmin = 1e30;
nv = length(c); nbd = length(ibd);
for j = 1:length(b)
  if (b(j) < 0), disp('b(j) must be non-negative.'); return; end
end
for k = 1:nv, vtype(k) = 1; end
for k = 1:nbd, vtype(ibd(k)) = 2; end
for j = 1:nv, kvar(j) = j; end
iter = 0; kconv = 0;
while (1)
    iter = iter + 1;
    if (iter > kmax), disp('Maximum possible iterations exceeded'); break; end
    ktrac = 0; [f,x,kflag] = linpm(nv,nl,ng,ne,kf,kvar,A,b,c);
    if (nbd == 0)
       if (kflag > 0), disp('No feasible solution.'); return;
       else, fmin = f; xmin = x; break; end
  end
  jc = 0;
  while (2)
      jc = jc+1; if (jc == 2), break; end
      if (kflag > 0)
        if (kf == 0), disp('No feasible solution.'); return; end
        ktrac = 1; break; % backtrack operation
    else
      if (iter == 1), flb = f; end  % best low bound
        if (f >= fmin), ktrac = 1; break; end
        vmax = 0;
        for j = 1:nv
           if(vtype(j) == 2)
              diffx = abs(x(j)-round(x(j)));
              if (diffx > vmax), vmax = diffx; jv = j; end
         end
      end
      if (vmax < critd)
          ksol = 1;
          if (f < fmin), fmin = f; for j = 1:nv, xmin(j) = x(j); end; end
          gapf = abs(flb - fmin)/abs(flb);
          if (gapf <= critf), kconv = 1; break; end
          ktrac = 1; % backtrack operation
          break
        else
          kf = kf + 1; jt = kvar(kf); kvar(kf) = kvar(jv); kvar(jv) = jt;
      end
    end
  end
  while (ktrac == 1)
      if (kvar(kf) > 0), kvar(kf) = -kvar(kf); break;
      else
        kvar(kf) = -kvar(kf); kf = kf - 1;
        if (kf == 0)
           if(ksol == 0), disp('No feasible solution.'); return; end
           kconv = 1; break;
        end
```

```
            end
        end
        if (kconv == 1), break; end
    end
xopt = xmin; fopt = fmin;
end

function [f,x,kflag] = linpm(nv,nl,ng,ne,kf,kvar,A,B,C)
if (kf > 0)
  idn = [1:nv];
  for k = 1:kf, jk = abs(kvar(k)); idn(jk) = 0; end
  difnv = nv - kf; nej = ne; Aj = A; Bj = B; f0 = 0; cj = 0; jm = 0;
  for k = 1:nv
      if idn(k) > 0, cj = cj+1; Ck(cj) = C(k);
      else
          jm = jm+1;
          if (kvar(jm) > 0), cm = 1; else, cm = 0;
          end
          f0 = f0 + C(k)*cm;
      end
  end
  for k = 1:nl, conk(k) = -1; end
  for k = 1:ng, conk(nl+k) = 1; end
  for k = 1:nl+ng+ne
      cj = 0; jm = 0;
      for j = 1:nv
          if (idn(j)) > 0, cj = cj + 1; Aj(k,cj) = A(k,j);
          else
              jm = jm+1;
              if (kvar(jm) > 0), cm = 1; else, cm = 0; end
              Bj(k) = Bj(k) - A(k,j)*cm;
          end
      end
    if (k <= nl+ng & Bj(k) < 0), conk(k) = -conk(k); Aj(k,:) = -Aj(k,:); Bj(k) = -Bj
(k); end
  end
  nlk = 0; ngk = 0;
  for k = 1:nl+ng
      if (conk(k) == -1), nlk = nlk + 1;
      elseif (conk(k)==1), ngk = ngk + 1;
      end
  end
  [ids,xk] = sort(conk); A = Aj; B = Bj;
  for k = 1:nl+ng, A(k,:) = Aj(xk(k),:); B(k) = Bj(xk(k)); end
  nk = nlk + ngk + nej; Aj = A(1:nk,1:difnv); Bj = B(1:nk);
else
  difnv = nv; nlk = nl; ngk = ng; nej = ne; Aj = A; Bj = B; Ck = C;
end
[fk,xs,kflag] = linsimx(difnv,nlk,ngk,nej,Aj,Bj,Ck);
if (kflag > 0), x = xs; f = fk; return; end
if (kf > 0)
  cj = 0; jm = 0;
  for j = 1:nv
      if (idn(j)) > 0. cj = cj+1; x(j) = xs(cj);
      else
```

```
        jm = jm+1;
        if (kvar(jm)>0), cm = 1; else, cm = 0; end
        x(j) = cm;
      end
    end
    f = fk + f0;
  else
    f = fk; x = xs; return;
  end
end

function [f,x,kflag] = linsimx(nv,nl,ng,ne,A,B,C)
% linear programming by simplex method
mnp = 20; bigm = 0; kflag = 0;
nc = nv + nl + ne + 2*ng; nr = nl + ne + ng + 1;
% Initialization
Aj(1:nr,1:nc) = 0; Bj(1:nr) = 0; Aj(1:nr-1,1:nv) = A(1:nr-1,1:nv);
Aj(nr,1:nv) = C(1:nv); Bj(1:nr-1) = B(1:nr-1); A = Aj; B = Bj;
for j = 1:nv, A(nr,j) = C(j); end
for j = 1:nv, bigm = bigm + mnp*abs(A(nr,j)); end
if (nl > 0) % slack variables
  for k = 1:nl, A(k,nv+k) = 1; Bs(k) = nv + k; end
end
if (ng > 0)
  for k = 1:ng
    A(nl+k,nv+nl+k) = -1; A(nl+k, nv+nl+ng+k) = 1; Bs(nl+k) = nv + nl + ng + k;
  end
end
if (ne > 0)
  for k = 1:ne, A(nl+ng+k,nv+nl+2*ng+k) = 1; Bs(nl+ng+k) = nv + nl + 2*ng + k; end
end
mge = ng + ne;
if (mge > 0), for k = 1:mge, A(nr,nv+nl+ng+k) = bigm; end
end
if (mge > 0) % Remove artificial variables
  for k = 1:mge
    C = A(nr,nv+nl+ng+k); for j = 1:nc, A(nr,j) = A(nr,j) - C*A(nl+k,j); end
    B(nr) = B(nr) - C*B(nl+k);
  end
end
% Simplex method
while (1)
  jflag = 0;
  for j = 1:nc, if (A(nr,j) < 0), jflag = 1; break; end; end
  if (jflag == 0), break; end
  C = bigm;
  for j = 1:nc, if (A(nr,j) < C), C = A(nr,j); idv = j; end; end
  jn = 0; jk = 0;
  for k = 1:nr-1
    if (A(k,idv) > 0)
      jk = jk + 1; Dm = B(k)/(A(k,idv) + 1e-10);
      if (jk == 1), C = Dm; jp = k;
      else, if (Dm < C), C = Dm; jp = k; end
      end
      jn = 1;
```

```
      end
   end
   Bs(jp) = idv; if (jn == 0), kflag = 1; f = 0; x = 0; return; end
   Dm = 1/A(jp,idv); B(jp) = Dm*B(jp);
   for j = 1:nc, A(jp,j) = Dm*A(jp,j); end
   for k = 1:nr
      if (k ~= jp)
         Em = A(k,idv); for j = 1:nc, A(k,j) = A(k,j) - Em*A(jp,j); end
         B(k) = B(k) - Em*B(jp);
      end
   end
end
x(1:nv) = 0; nt = nv + nl + ng;
for k = 1:nr-1, if (Bs(k) > nt), kflag = 1; f = 0; x = 0; return; end; end
for k = 1:nv
   for j = 1:nr-1, if (Bs(j) == k), x(k) = B(j); break; end; end
end
f = B(nr); f = -f; % minimization
end
```

Example 10.24 Branch-and-Bound Algorithm (1)

Find the solution of the following minimization problem using the branch-and-bound algorithm:

$$\text{Minimize } f(x) = 12x_1 + 10x_2 + 6x_3 + 4x_4$$

Subject to $x_1 + 3x_2 + 2x_3 + 4x_4 \le 8$, $x_1 + x_2 + x_3 = 2$, $x_3 + x_4 = 1 (x_i = 0 \text{ or } 1)$

Solution
```
>> A = [1 3 2 4;1 1 1 0;0 0 1 1]; b = [8 2 1]; c = [12 10 6 4]; nl = 1; ne = 2; ng = 0; ibd =
[1 2 3 4];
>> [xopt,fopt,iter] = bnbopt(A,b,c,nl,ng,ne,ibd)
xopt =
      0    1    1    0
fopt =
     16
iter =
      1
```

Example 10.25 Branch-and-Bound Algorithm (2)

Find the solution of the following maximization problem using the branch-and-bound algorithm[30]:

$$\text{Maximize } f(x) = 5x_1 + 3x_2 + x_3 + 3x_4 + 5x_5 + 2x_6 + 5x_7 + 5x_8 + 2x_9$$

Subject to $x_1 + x_4 + x_6 \le 1$, $x_2 + x_7 \le 1$, $x_4 + x_9 \le 1$,

$$x_1 + x_2 + x_3 = 1, x_4 + x_5 = 1, x_8 + x_9 = 1 (x_i = 0 \text{ or } 1)$$

Solution
```
>> A = [1 0 0 1 0 1 0 0 0;0 1 0 0 0 0 1 0 0;0 0 0 1 0 0 0 0 1;1 1 1 0 0 0 0 0 0;0 0 0 1 1 0 0 0 0;0
0 0 0 0 0 0 1 1];
>> b = ones(1,6); c = -[5 3 1 3 5 2 5 5 2]; nl = 3; ne = 3; ng = 0; ibd = [1:9];
```

```
>> [xopt,fopt,iter] = bnbopt(A,b,c,nl,ng,ne,ibd)
xopt =
      1    0    0    0    1    0    1    1    0
fopt =
    -20
iter =
   1
```

We can see that the maximum function value is $f_{max} = 20$.

10.6 USE OF MATLAB BUILT-IN FUNCTIONS

MATLAB provides many useful functions that allow solution of various optimization problems. Table 10.4 lists the names of MATLAB built-in optimization routines.

10.6.1 UNCONSTRAINED OPTIMIZATION

10.6.1.1 fminsearch

The function *fminsearch* can be used to solve a multidimensional unconstrained nonlinear minimization problem. This function uses Nelder-Mead's simplex (direct search) method. The basic syntax is
`[x, f] = fminsearch(fun, x0)`

where f is the function value at the optimum point, x0 is the initial search point, fun is a function handle, and x is a local minimizer of fun.

10.6.1.2 fminunc

The function *fminunc* finds a local minimum of a function of several variables. This function can be used as
`[x, f] = fminunc(fun, x0)`

where f is the function value at the optimum point, x0 is the initial search point, fun is a function handle, and x is a local minimizer of fun.

TABLE 10.4
MATLAB Built-In Optimization Routines

Problem Type	Function Name	Description
Unconstrained optimization	*fminbnd*	Bounded nonlinear minimization
	fminsearch	Unconstrained nonlinear minimization
	fminunc	Local minimum of a function of several variables
	lsqnonlin	Nonlinear least-squares problem
Constrained optimization	*lsqlin*	Constrained linear least-squares problem
	fmincon	Constrained minimum of a function
	fminimax	Minimax solution of a function
	quadprog	Quadratic programming problem
Linear programming	*linprog*	Linear programming problem
Mixed-integer programming	*intlinprog*	Mixed-integer linear programming problem

10.6.1.3 fminbnd

The function *fminbnd* solves a single-variable bounded nonlinear function minimization problem. If we want to find a local minimum of the function fun in the interval $x_1 < x < x_2$, we can execute the following commands:

```
[x, f] = fminbnd(fun, x1 ,x2)
```

where x1 and x2 denote the lower and upper bounds, fun is a function handle, and x is a local minimizer of fun.

10.6.1.4 lsqnonlin

The function *lsqnonlin* attempts to solve nonlinear least-squares problems of the form

$$\text{Minimize } \sum f(x)^2$$

The basic syntax is
```
x = lsqnonlin(fun, x0, lb, ub)
```

where x0 is the starting point, fun is a function handle, lb and ub are the lower and upper bounds ($lb < x < ub$), and x is a local minimizer of fun. If no bounds exist, we may use empty matrices for lb and ub or just omit them. If x_i is unbounded below, set lb(i) = -Inf, and set ub(i) = Inf if x_i is unbounded above.

Example 10.26 Unconstrained Optimization by Built-In Functions

(1) Find the minimum of $f(x) = 2x^2\sin(x) + e^{-x}$ using the built-in functions *fminsearch* and *fminunc*.

(2) Minimize $f(x) = 2x^2\sin(x) + e^{-x}$ using the built-in function *fminbnd* in the interval $-4 \leq x \leq 0$.

(3) Minimize $f(x) = 2x^2\sin(x) + e^{-x}$ using the built-in function *lsqnonlin* in the interval $-4 \leq x \leq 0$. Use $x_0 = -1$ as a starting point. Check the result using the function *fminbnd*.

Solution

(1) Let the starting point be $x_0 = 1$:
```
>> f = @(x) 2*x^2*sin(x) + exp(-x); x0 = 1; [x, fv] = fminsearch(f, x0)
x =
      0.3488
fv =
      0.7887
>> [x, fv] = fminunc(f, x0)
x =
      0.3488
fv =
      0.7887
```

(2) For the interval $-4 \leq x \leq 0$, x1 = lb = −4 and x2 = ub = 0.

```
>> f = @(x) 2*x^2*sin(x) + exp(-x); lb = -4; ub = 0; [x,fv] = fminbnd(f, lb ,ub)
x =
     -1.7540
fv =
     -0.2724
```

(3) For the interval $-4 \le x \le 0$, x1 = lb = -4 and x2 = ub = 0, and x0 = -1.

```
>> f = @(x) 2*x^2*sin(x) + exp(-x); lb = -4; ub = 0; x0 = -1; x = lsqnonlin(f, x0,
lb, ub)
x =
    -1.4962
>> f = @(x) (2*x^2*sin(x) + exp(-x))^2; lb = -4; ub = 0; x0 = -1; [x,fv] = fminbnd
(f, lb ,ub)
x =
    -1.4962
fv =
    6.5941e-11
```

Example 10.27 Parameters of the Michaelis-Menten Model

The enzyme reaction $E + S \rightarrow E + P$ can be described by the Michaelis-Menten kinetics

$$r_m = \frac{r_{max}S}{k_m + S}$$

where

r_m is the reaction rate
r_{max} is the maximum reaction rate
k_m is a constant
S is the substrate concentration

Table 10.5 shows experimental data for S and r. The parameters $r_{\max}$ and k_{m} in the reaction model can be found from the optimization of the squared residual given by

$$J = \sum_{k=1}^{n} (r - r_m)^2 = \sum_{k=1}^{n} \left(r - \frac{r_{max}S}{k_m + S} \right)^2$$

Use the built-in function *fminsearch* or *fminunc* to find the optimal value of r_{max} and k_m. Plot reaction rates calculated by the model equation using the optimal parameters.

Solution

The script *optmm* uses the built-in function *fminsearch* to find the optimal parameters. The function *fminunc* gives similar results. Figure 10.6 shows the observed rates as well as those calculated from the model equation using the optimal parameters.

TABLE 10.5
Observed Reaction Rate versus Substrate Concentration

Subject	$S\,(ng/ml)$	$r\,(nmol/min)$	Subject	$S\,(ng/ml)$	$r\,(nmol/min)$
1	0.74	0.10	9	8.32	0.82
2	1.19	0.21	10	10.10	1.26
3	1.46	0.21	11	11.10	0.55
4	1.62	0.14	12	20.10	3.34
5	1.63	0.34	13	21.73	2.49
6	5.27	0.49	14	25.10	2.01
7	5.87	0.36	15	27.78	1.80
8	6.26	0.48	16	35.70	1.68

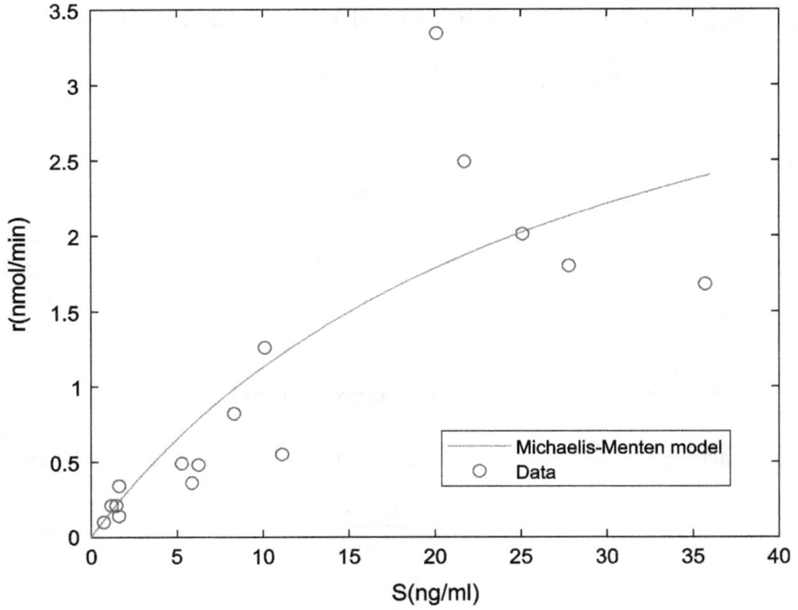

FIGURE 10.6   Observed rates (r) and rates obtained from the model (r_m).

```
% optmm.m: determination of optimal parameters for Michaelis-Menten model
clear all;
S = [0.74, 1.19, 1.46, 1.62, 1.63, 5.27, 5.87, 6.26, 8.32, 10.10, 11.10,...
    20.10, 21.73, 25.10, 27.78, 35.70]; % substrate concentration
r = [0.10, 0.21, 0.21, 0.14, 0.34, 0.49, 0.36, 0.48, 0.82, 1.26, 0.55,...
    3.34, 2.49, 2.01, 1.80, 1.68]; % reaction rate
rmax0 = 1.7; km0 = 11; % initial guesses
J = @(x) sum((r - x(1)*S./(x(2) + S)).^2); % x(1) = rmax, x(2) = km
x = fminsearch(J,[rmax0, km0]); rmax = x(1); km = x(2); % optimization
fprintf('Maximum reaction rate (rm) = %g\n',rmax);
fprintf('Michaelis constant (km) = %g\n',km);
% Compare data and model
Si = linspace(floor(min(S)),ceil(max(S)),1000);
ri = rmax*Si./(km + Si); % model
plot(Si,ri,S,r,'o'), xlabel('S(ng/ml)'), ylabel('r(nmol/min)')
legend('Michaelis-Menten model','Data','location','best')

>> optmm
Maximum reaction rate (rm) = 4.23704
Michaelis constant (km) = 27.4776
```

Example 10.28 Optimal Reflux Ratio in a Binary Column[31]

We consider determination of the optimal reflux ratio (L/D) for the operation of a binary distillation column. The rule of thumb about the minimum reflux ratio is to operate 1.2 times the minimum value given by Fenske equation. The minimum reflux ratio (L/D)$_{min}$ by the Fenske equation is as follows:

$$\left(\frac{L}{D}\right)_{min} = \frac{1}{\alpha - 1}\left\{\frac{x_D}{x_F} - \alpha\left(\frac{1 - x_D}{1 - x_F}\right)\right\}$$

The objective function consists of the cost of the column (C_{col}) and the energy cost (C_{eng}) in the reboiler:

$$J(L/D) = \frac{1}{3}C_{col} + C_{eng}$$

The column cost is calculated as the steel cost:

$$C_{col} = \left\{A + \pi N\left(\frac{d}{2}\right)^2\right\}w\rho_s C_{st}$$

where

$A(m^2)$ is the area
N is the number of trays
$d(m)$ is the column diameter
$w(m)$ is the width of the steel
$\rho_s(kg/m^3)$ is the steel density
$C_{st}(\$)$ is the steel cost
A and w are given by

$$A = 4\pi\left(\frac{d}{2}\right)^2 + \pi dh, \quad w = \frac{rP}{\xi S - 0.6P} + 0.0032$$

where

$h(m)$ is the height
$P(psi)$ is the operating pressure
$S(psi)$ is the tensor stress for the material
$r(m)$ is the column radius
ξ is the welding efficiency
h is given by

$$h = 0.6\left(\frac{N - 1}{\eta} + 1\right) + 2$$

where η is the efficiency of the theoretical trays. The column diameter d is determined as follows:

$$d = \sqrt{\frac{4V \times 22.4 \times 760 \times (T + 273.15)}{273\pi PK\sqrt{(\rho_L - \rho_G)/\rho_L}}}$$

The McCabe-Thiele method can be used to determine the number of theoretical trays N. We assume that the equilibrium composition is given by

$$y^* = \frac{\alpha x^*}{1 + (\alpha - 1)x^*}$$

The operating lines for the rectifying and stripping sections can be represented as

$$y_m = \left(\frac{L}{V}\right)x_{m-1} + \left(\frac{D}{V}\right)x_D, \quad y_{n+1} = \left(\frac{L'}{V'}\right)x_n + \left(\frac{B}{V'}\right)x_B$$

V can be obtained from the following equation:

$$V = F\left(1 + \frac{L}{D}\right)\left(\frac{x_F - x_B}{x_D - x_B}\right)$$

The energy cost C_{eng} in the reboiler is given by

$$Q = \lambda F\left(1 + \frac{L}{D}\right)\left(\frac{x_F - x_B}{x_D - x_B}\right), \; C_{eng} = \frac{Q}{\lambda_s}C_{ss}$$

where C_{ss} is the steam cost. Find the optimal reflux ratio using the built-in function *fminsearch*. Use the given data.

$\xi = 0.8$, $\eta = 0.75$, $S = 12000\,psi$, $\rho_G = 2\,kg/m^3$, $\rho_L = 850\,kg/m^3$, $\rho_s = 8000\,kg/m^3$,

$K = 0.05\,m/sec$, $\lambda = 800\,kJ/kmol$, $\lambda_s = 1800\,kJ/kg$, $C_{ss} = \$0.05/kg$, $C_{st} = \$10/kg$,

$F = 100\,kmol/sec$, $T = 70°C$, $P = 760\,mmHg$, $x_B = 0.05$, $x_D = 0.85$, $x_F = 0.4$, $\alpha = 2.3$.

Solution

The optimization problem can be solved by the following procedure:

1) Determine the number of theoretical trays N.
2) Find the vapor rate V.
3) Compute d, h, w, and A.
4) Find the energy cost in the reboiler C_{eng}.
5) Evaluate the objective function $J(L/D)$.

The objective function is defined by the function *rfobj* as follows:

```
function C = rfobj(rf,opdat)
% Retrieve data
we = opdat.we; eta = opdat.eta; S = opdat.S; K = opdat.K;
rhoG = opdat.rhoG; rhoL = opdat.rhoL; rhos = opdat.rhos;
lamb = opdat.lamb; lambs = opdat.lambs; Css = opdat.Css; Cst = opdat.Cst;
F = opdat.F; T = opdat.T; P = opdat.P; alpa = opdat.alpa;
xB = opdat.xB; xD = opdat.xD; xF = opdat.xF;
% Set parameters
rLV = rf/(rf + 1); rDV = 1/(1 + rf);
rpLV = (rf + ((xD-xB)/(xF-xB)))/(1 + rf);
rBV = (((xD-xB)/(xF-xB)) - 1)/(1 + rf);
% Determine the number of trays
N = 1; ye(1) = xD; xe(1) = ye(1)/(alpa*(1-ye(1)) + ye(1));
while xe(N) > xB
  if xe(N) > xF
    ye(N+1) = rLV*xe(N) + rDV*xD; % rectifying section line
  else
    ye(N+1) = rpLV*xe(N) - rBV*xB; % stripping section line
  end
  xe(N+1) = ye(N+1)/(alpa*(1-ye(N+1)) + ye(N+1));
  N = N + 1;
end
V = F*(1 + rf)*(xF-xB)/(xD-xB); % vapor rate
```

```
d = sqrt((4*V*22.4*760*(T+273.15))/(273*pi*P*K*sqrt((rhoL-rhoG)/rhoL)));
h = 0.6*((N-1)/eta + 1) + 2; % height
w = 14.7*(P/760)*d/2/(we*S - 0.6*14.7*(P/760)) + 0.0032; % steel width
A = 4*pi*(d/2)^2 + pi*d*h; % area
% Cost and objective function
Ceng = lamb*F*(1 + rf)*Css*(xF-xB)/(xD-xB)/lambs; % Energy cost
Ccol = (A + pi*N*(d/2)^2)*w*rhos*Cst; % column cost
C = Ccol/3 + Ceng;
end
```

The script *optrf* defines the data and uses the built-in function *fminsearch* to find the optimal reflux ratio.

```
% optrf.m: optimal reflux ratio in a binary distillation column
clear all;
% Data
opdat.we = 0.8; opdat.eta = 0.75; opdat.S = 12000; opdat.K = 0.05;
opdat.rhoG = 2; opdat.rhoL = 850; opdat.rhos = 8000;
opdat.lamb = 800; opdat.lambs = 1800; opdat.Css = 0.05; opdat.Cst = 10;
opdat.F = 100; opdat.T = 70; opdat.P = 760; opdat.alpa = 2.3;
opdat.xB = 0.05; opdat.xD = 0.85; opdat.xF = 0.4;
% Assign data
alpa = opdat.alpa; xD = opdat.xD; xF = opdat.xF;
% Find optimal reflux ratio rfopt
rf0 = 1.5; % initial guess
rfopt = fminsearch(@rfobj,rf0,[],opdat);
minrf = (xD/xF - alpa*(1-xD)/(1-xF))/(alpa-1); % minimum rf by Fenske eqn.
rfpr = 1.2*minrf; % reflux ratio by the rule of thumb (=1.2*minrf)
fprintf('Optimal reflux ratio = %g\n', rfopt);
fprintf('Reflux ratio by the rule of thumb = %g\n', rfpr);
```

Execution of the script *optrf* produces the following results:
```
>> optrf
Optimal reflux ratio = 1.55125
Reflux ratio by the rule of thumb = 1.43077
```

10.6.2 CONSTRAINED OPTIMIZATION

10.6.2.1 lsqlin

The function *lsqlin* solves the constrained linear least-squares problem of the form

$$\text{Minimize } \frac{1}{2}\|Cx - d\|^2$$

$$\text{Subject to } Ax \leq b, \, A_{eq}x = b_{eq}$$

The basic syntax is
```
x = lsqlin(C, d, A, b, Aeq, beq, lb, ub, x0)
```

where $C \in R^{m \times n}$ is a coefficient matrix, x0 is the starting point, fun is a function handle, lb and ub are the lower and upper bounds (lb < x < ub), and x is a local minimizer of fun. If no bounds exist, we may use empty matrices for lb and ub or just omit them. If x_i is unbounded below,

set lb(i) = -Inf, and set ub(i) = Inf if x_i is unbounded above. Depending on the problem type or conditions, the corresponding input arguments may be omitted. For example, if only inequality constraints exist, we can use
```
x = lsqlin(C, d, A, b)
```

10.6.2.2 fmincon

The function *fmincon* finds a constrained minimum of a function of several variables of the form

$$\text{Minimize } f(x)$$

$$\text{Subject to } Ax \le b, \ A_e qx = b_e q \text{ (linear constraints), } \text{lb} \le x \le \text{ub}.$$

The basic syntax is
```
[x, f] = fmincon(fun, x0, A, b, Aeq, Beq, lb, ub)
```

where f is the function value at the optimum point, x0 is the starting point, and fun is a function handle. If no bounds exist, we may use empty matrices for lb and ub or just omit them. If x_i is unbounded below, set lb(i) = -Inf, and set ub(i) = Inf if x_i is unbounded above. Depending on the problem type or conditions, the corresponding input arguments may be omitted. For example, if only inequality constraints exist, we can use
```
x = fmincon(fun, x0, A, b)
```

10.6.2.3 fminimax

The function *fminimax* finds a minimax solution of a function of several variables of the form

$$\text{Minimize (Max. } f(x))$$

$$\text{Subject to } Ax \le b, A_{eq}x = b_{eq} \text{ (linear constraints), } \text{lb} \le x \le \text{ub}.$$

The basic syntax is
```
x = fminimax(fun, x0, A, b, Aeq, Beq, lb, ub)
```

where x0 is the starting point and fun is a function handle. If no bounds exist, we may use empty matrices for lb and ub or just omit them. For example, if there are no constraints, we can use
```
x = fminimax(fun, x0)
```

10.6.2.4 quadprog

The function *quadprog* solves quadratic programming problems of the form

$$\text{Minimize } \frac{1}{2}x^T Hx + c^T x$$

$$\text{Subject to } Ax \le b, A_{eq}x = b_{eq} \text{ (linear constraints), } lb \le x \le ub.$$

The basic syntax is
```
[x, f] = quadprog(H, c, A, b, Aeq, beq, lb, ub, x0)
```

where f is the function value at the optimum point. If no bounds exist, we may use empty matrices for lb and ub or just omit them. For example, if only inequality constraints exist, we can use
```
x = quadprog(H, c, A, b)
```

Example 10.29 Constrained Optimization by Built-In Functions

(1) Solve the constrained linear least-squares problem given by

$$\text{Minimize } \frac{1}{2}\|Cx - d\|^2$$

$$\text{Subject to } Ax \le b$$

where $A = \begin{bmatrix} -2 & 1 \\ 3 & 5 \end{bmatrix}$, $b = \begin{bmatrix} 6 \\ 8 \end{bmatrix}$, $C = \begin{bmatrix} 2 & 0 \\ 0 & 3 \end{bmatrix}$, $d = \begin{bmatrix} 4 \\ 4 \end{bmatrix}$

(2) Minimize

$$f(x) = \left(x_1 - \frac{1}{2}\right)^2 (x_1 + 1)^2 + 2(x_2 + 1)^2 (x_2 - 1)^2$$

$$\text{subject to } 2x_1 + 4x_2 \le 7, \; -3x_1 + x_2 \le 3$$

Set $x_0 = [00]$.

(3) Find a minimax solution of

$$f(x) = \frac{1}{(x - 0.3)^2 + 0.01} + \frac{1}{(x - 0.9)^2 + 0.04} - 5$$

Use $x_0 = 1$.

(4) Minimize

$$f(x) = -4x_1 + x_1^2 - 2x_1 x_2 + 2x_2^2$$

Subject to $2x_1 + x_2 \le 6$, $x_1 - 4x_2 \le 0$, $x_1 \ge 0$, $x_2 \ge 0$

Solution

(1) We use the function *lsqlin*:
```
>> C = [2 0;0 3]; d = [4; 4]; A = [-2 1;3 5]; b = [6; 8]; x = lsqlin(C, d, A, b)
Optimization terminated.
x =
  1.3039
  0.8177
```

(2) We can see that $A = \begin{bmatrix} 2 & 4 \\ -3 & 1 \end{bmatrix}$, $b = \begin{bmatrix} 7 \\ 3 \end{bmatrix}$. The function *fmincon* may be used:
```
>> f = @(x) (x(1)-1/2)^2*(x(1)+1)^2 + 2*(x(2)+1)^2*(x(2)-1)^2;
>> A = [2 4;-3 1]; b = [7; 3]; x0 = [0 0]; [x fv] = fmincon(f, x0, A, b)
x =
     0.5000  -1.0000
fv =
   2.8902e-16
```

(3) We use the function *fminimax*:
```
>> f = @(x) 1/((x-0.3)^2+0.01) + 1/((x-0.9)^2+0.04) - 5; x0 = 1; [x,fv] = fmi-
nimax(f, x0)
x =
   2.2928e+07
```

```
fv =
    -5.0000
```

(4) The function and the constraints can be expressed as

$$f(x) = \frac{1}{2}[x_1 \ x_2]\begin{bmatrix} 2 & -2 \\ -2 & 4 \end{bmatrix}\begin{bmatrix} x_1 \\ x_2 \end{bmatrix} + [-40]\begin{bmatrix} x_1 \\ x_2 \end{bmatrix}, \begin{bmatrix} 2 & 1 \\ 1 & -4 \\ -1 & 0 \\ 0 & -1 \end{bmatrix} x \leq \begin{bmatrix} 6 \\ 0 \\ 0 \\ 0 \end{bmatrix}$$

We use the function *quadprog:*
```
>> H = [2 -2;-2 4]; c = [-4; 0]; A = [2 1;1 -4;-1 0;0 -1]; b = [6 0 0 0]'; [x,f] =
quadprog(H, c, A, b)
x =
    2.4615
    1.0769
f =
    -6.7692
```

10.6.3 LINEAR AND MIXED-INTEGER PROGRAMMING PROBLEMS

10.6.3.1 linprog
The function *linprog* solves the linear programming problem of the form

$$\text{Minimize } f^T x$$

$$\text{Subject to } Ax \leq b, \ A_e qx = b_e q \text{ (linear constraints)}, \ \text{lb} \leq x \leq \text{ub}$$

The basic syntax is
```
[x, fv] = linprog(f, A, b, Aeq, Beq, lb, ub, x0)
```

where fv is the function value at the optimum point x. If no bounds exist, we may use empty matrices for lb and ub or just omit them. If x_i is unbounded below, set lb(i) = -Inf, and set ub(i) = Inf if x_i is unbounded above. Depending on the problem type or conditions, the corresponding input arguments may be omitted. For example, if only inequality constraints exist, we can use
```
[x, fv] = linprog (f, A, b)
```

10.6.3.2 intlinprog
The function intlinprog solves mixed-integer linear programming problems of the form

$$\text{Minimize } f^T x$$

$$\text{Subject to } Ax \leq b, A_e qx = b_e q \text{ (linear constraints)}, \ lb \leq x \leq ub, \ x_i: \text{ integers.}$$

The basic syntax is
```
[x, fv] = intlinprog (f, intcon, A, b, Aeq, Beq, lb, ub)
```

where intcon is the index vector containing integer variables. If no bounds exist, we may use empty matrices for lb and ub or just omit them. If x_i is unbounded below, set lb(i) = -Inf, and set ub(i) = Inf if x_i is unbounded above. Depending on the problem type or conditions, the corresponding input arguments may be omitted. For example, if only inequality constraints exist, we can use

```
[x, fv] = linprog (f, A, b)
```

Example 10.30 Linear Programming Problem by a Built-In Function

$$\text{Minimize } f(x) = -50x_1 - 98x_2 - 25x_3 - 43x_4$$

$$\text{Subject to } 6x_1 + 12x_2 + 3x_3 + 8x_4 \leq 1150, \ 4x_1 + 30x_2 + 2x_3 + x_4 \leq 750, \ x_i \geq 0 (i = 1, \cdots, 4)$$

Solution

The coefficient vector is given by $f = [-50 - 98 - 25 - 43]^T$ and the constraints can be expressed as

$$\begin{bmatrix} 6 & 12 & 3 & 8 \\ 4 & 30 & 2 & 1 \\ -1 & 0 & 0 & 0 \\ 0 & -1 & 0 & 0 \\ 0 & 0 & -1 & 0 \\ 0 & 0 & 0 & -1 \end{bmatrix} x \leq \begin{bmatrix} 1150 \\ 750 \\ 0 \\ 0 \\ 0 \\ 0 \end{bmatrix}$$

```
>> f = [-50 -98 -25 -43]'; A = [6 12 3 8;4 30 2 1;-1 0 0 0;0 -1 0 0;0 0 -1 0;0 0 0 -1]; b =
[1150 750 0 0 0 0]';
>> [x, fv] = linprog(f, A, b)
Optimal solution found.
x =
 186.5385
      0
      0
   3.8462
fv =
 -9.4923e+03
```

Example 10.31 Mixed-Integer Programming Problem by a Built-In Function

$$\text{Minimize } f(x) = -2x_1 - 3x_2 \ (x_1, x_2: \text{ integer variables})$$

$$\text{Subject to } 4x_1 + 10x_2 \leq 45, \ 4x_1 + 4x_2 \leq 23, \ x_i \geq 0 (i = 1,2)$$

Solution

The coefficient vector is given by $f = [-2 - 3]^T$ and the constraints can be expressed as

$$\begin{bmatrix} 4 & 10 \\ 4 & 4 \\ -1 & 0 \\ 0 & -1 \end{bmatrix} x \leq \begin{bmatrix} 45 \\ 23 \\ 0 \\ 0 \end{bmatrix}$$

Since x_1 and x_2 are integer variables, intcon = [1 2].

```
>> c = [-2 -3]; A = [4 10;4 4;-1 0;0 -1]; b = [45 23 0 0]; intcon = [1 2];
>> [x, fv] = intlinprog(c, intcon, A, b)
LP: Optimal objective value is -15.166667.
Optimal solution found.
```

```
x =
   1.0000
   4.0000
fv =
  -14
```

PROBLEMS

10.1 A metallic ball released from a height h at an angle θ and velocity v with respect to the horizontal in a gravitational field g travels a distance d when it hits the ground. The distance d is given by

$$d = \left(\frac{v\sin\theta}{g} + \sqrt{\frac{2h}{g} + \left(\frac{v\sin\theta}{g}\right)^2} \right) v\cos\theta$$

Use the Fibonacci search method to find θ for which the distance d is a maximum and to determine the maximum distance d.

Data: $h = 60\,m$, $v = 80\,m$, $g = 9.8\,m/sec^2$, $n = 30$, $a = 0$, $b = 1.6$.

10.2 A company plans to borrow $S \times 1000$ dollars to build a new plant on equal yearly installments over the next n years. The interest charged is $r_e = e_1 + e_2 S$. The money earned can be reinvested at a rate of m_r per year. The expected future value of the return is given by

$$f_R = P(1 - e^{-wS})$$

and the future value of the payment is given by

$$f_P = \left[\frac{(1 + m_r)^n - 1}{m_r} \right]\left[\frac{r_e(1 + r_e)^n}{(1 + r_e)^n - 1} \right] S$$

Use the golden section search method to determine the value of S to be borrowed for the maximum future value of profit F, which is given by $F = f_R - f_P$.

Data: $e_1 = 0.046$, $e_2 = 0.00028$, $m_r = 0.061$, $P = 500 \times 10^3\$$, $w = 0.012/\$1000$, $n = 6$.

10.3 Use Brent's quadratic fit method to find the minimum of $f(x) = 87.5(1 - x^3)^2 + (1 - x^2) + 3(1 - x)^2$. The initial point can be specified as $x_1 = 0.01$. Let the step size be $h = 0.2$ and the stopping criterion be 1×10^{-6}.

10.4 Use the Shubert-Piyavskii method to find the maximum of

$$f(x) = \sum_{j=1}^{5} j\sin((j + 1)x + j)$$

Use the following conditions: initial interval $= [-10, 10]$, Lipschitz constant $C = 80$, stopping criterion $= 1 \times 10^{-6}$, maximum number of function evaluations $= 2000$.[32]

10.5 A desired product C is produced by decomposition of A in a CSTR. The decomposition reaction is accompanied by two other side reactions from which undesired by-products B and D are produced.

The reaction rates are given by[33]

$$r_B = 1,\ r_C = 0.6C_A{}^2,\ r_D = 0.22C_A$$

where

r_B, r_C and r_D are the production rates of species B, C, and D, respectively

C_A is the concentration of reactant A in the effluent stream

The concentration of C in the effluent stream, C_C, is given by

$$C_C = \frac{r_C}{r_B + r_C + r_D}(C_{A0} - C_A)$$

where the initial concentration of the reactant A is $C_{A0} = 4\,mol/liter$. What is the optimal concentration of A in the effluent stream that maximizes the desired product C? What is the maximum value of C_C?

10.6 Use the steepest descent method to find the minimum of the function given by

$$f(x) = x_1^2 - x_1 x_2 - 4x_1 + x_2^2 - x_2$$

The initial point can be specified as $[0, 0]$. Use an initial step size of $\alpha_0 = 2$. The stopping criterion and the maximum number of function evaluations can be set as 1×10^{-6} and 10,000, respectively.

10.7 Use the steepest descent method to find the minimum of Wood's function, given by

$$f(x) = 100(x_2 - x_1^2)^2 + (1 - x_1)^2 + 90(x_4 - x_3^2)^2 + (1 - x_3)^2$$

$$+ 10.1[(x_2 - 1)^2 + (x_4 - 1)^2] + 19.8(x_2 - 1)(x_4 - 1)$$

The initial point can be specified as $[-3, -1, -3, -1]$. Use an initial step size of $\alpha_0 = 2$. The stopping criterion and the maximum number of function evaluations can be set as 1×10^{-6} and 10,000, respectively.

10.8 Use Newton's method to find the minimum of the function given by

$$f(x) = \begin{bmatrix} 2x_1 - x_2 - 4 \\ - x_1 + 2x_2 - 1 \end{bmatrix}$$

The initial point can be specified as $[0, 0]$. The stopping criterion and the maximum number of function evaluations can be set as 1×10^{-6} and 1000, respectively.

10.9 Wood's function is given by

$$f(x) = 100(x_2 - x_1^2)^2 + (1 - x_1)^2 + 90(x_4 - x_3^2)^2 + (1 - x_3)^2$$

$$+ 10.1[(x_2 - 1)^2 + (x_4 - 1)^2] + 19.8(x_2 - 1)(x_4 - 1)$$

Use the conjugate gradient method to find the minimum of Wood's function. The initial point can be specified as $[-3, -1, -3, -1]$. Use an initial step size of $\alpha_0 = 2$. The stopping criterion and the maximum number of function evaluations can be set as 1×10^{-6} and 10,000, respectively.

10.10 Use the Davidon-Fletcher-Powell (DFP) method to find the minimum of Rosenbrock's function, given by

$$f(x) = 100(x_2 - x_1^2)^2 + (1 - x_1)^2$$

The initial point can be specified as $[-1.2, 1]$. Use an initial step size of $\alpha_0 = 1$. The stopping criterion and the maximum number of function evaluations can be set as 1×10^{-6} and 1000, respectively.

10.11 Himmelblau's function is given by

$$f(x) = (x_1^2 + x_2 - 11)^2 + (x_1 + x_2^2 - 7)^2.$$

1) Construct mesh and contour plots for this function in the range of $-6 \leq x_1, x_2 \leq 6$.
2) Determine the minimum using the Davidon-Fletcher-Powell (DFP) method. Use an initial step size of $\alpha_0 = 1$ and set the initial point as $[0, 0]$. The stopping criterion and the maximum number of function evaluations can be set as 1×10^{-6} and 1000, respectively.

10.12 Find the solution of the following LP problem using the two-phase method:

$$\text{Minimize } f(x) = 2x_1 + 6x_2$$

$$\text{Subject to } -x_1 + x_2 \leq 1, \ 2x_1 + x_2 \leq 2, \ x_1 \geq 0, \ x_2 \geq 0$$

10.13 A farmer has the choice of producing tomatoes, green peppers, or cucumbers on a 200 acre farm. A total of 480 workdays of labor is available, as shown in Table P10.13. Assuming fertilizer cost is the same for all products, determine the optimum crop combination using the two-phase simplex method.

10.14 Find the solution of the following LP problem using the interior point method:

$$\text{Minimize } f(x) = -x_1 - 2x_2 - x_3$$

$$\text{Subject to } 2x_1 + x_2 - x_3 \leq 2, \ 2x_1 - x_2 + 5x_3 \leq 6, \ 4x_1 + x_2 + x_3 \leq 9, \ x_1 \geq 0, \ x_2 \geq 0, \ x_3 \geq 0$$

10.15 A company manufactures three chemical products (product 1, product 2 and product 3). Table P10.15 shows the needed resources for each product. There are 100 hours of engineering, 600 hours of labor, and 300 lb of material available per day. The unit profits on these products are $10, $6, and $4, respectively. Find the amount of each product to be produced to achieve the maximum profit per day.[34] What is the maximum profit?

TABLE P10.13
Yield and Labor

	Yield ($/acre)	Labor (Workdays/Acre)
Tomatoes	410	6
Green peppers	350	8
Cucumbers	390	7

TABLE P10.15
Resources for Each Product

Product	Engineering Service (hr)	Direct Labor (hr)	Raw Material (lb)
1	1	10	3
2	2	4	2
3	1	5	1

10.16 Find the solution of the following constrained minimization problem using Rosen's gradient projection method:

$$\text{Minimize } f(x) = -x_1 x_2 x_3$$

$$\text{Subject to } 0 \leq x_1 + 2x_2 + x_3 \leq 70, \ 0 \leq x_1, \ x_2, \ x_3 \leq 40$$

As a starting point, use $x^0 = [5, 5, 5]$. Note that this point satisfies all constraints.

10.17 Find the solution of the following constrained minimization problem using the generalized reduced gradient method:

$$\text{Minimize } f(x) = -(1.2x_1 + 3x_2)$$

$$\text{Subject to } h(x) = x_1^2 + 6x_2^2 - 1 \leq 0, \ 0 \leq x_1, \ x_2 \leq 10$$

The inequality constraint can be converted to an equality by introducing a slack variable x_3 as follows:

$$g(x) = x_1^2 + 6x_2^2 + x_3 - 1 = 0$$

As a starting point, use $x^0 = [0, 0, 1]$. Note that this point satisfies all constraints. Set kmax = 1000 and crit = 1×10^{-4}.

10.18 Find the solution of the following constrained minimization problem using the sequential quadratic programming method[35]:

$$\text{Minimize } f(x) = x_1^2 + x_2$$

$$\text{Subject to } h(x) = x_1^2 + x_2^2 = 9, \ 1 \leq x_1 \leq 5, \ 2 \leq x_2 \leq 4$$

As a starting point, use $x^0 = [3, 3]$. The initial values of Lagrangian multipliers can be set to 1 and 4. Set crit = 1×10^{-6}.

10.19 Find the solution of the foll owing constrained minimization problem using the sequential quadratic programming method:

$$\text{Minimize } f(x) = x_1^2 + x_2^2 + 2.3x_3^2 - 1.2x_4^2 - 4x_1 - 6x_2 - 20x_3 + 6x_4 + 100$$

Subject to

$$h_1(x) = x_1^2 + x_2^2 + x_3^2 + x_4^2 + x_1 - x_2 + x_3 - x_4 \leq 7$$

$$h_2(x) = x_1^2 + 2x_2^2 + x_3^2 + 2x_4^2 - x_1 - x_4 \leq 11$$

$$h_3(x) = 2x_1^2 + x_2^2 + x_3^2 + 2x_1 - x_2 - x_4 \leq 6, \ -100 \leq x_1, \ x_2, \ x_3, \ x_4 \leq 100$$

As a starting point, we can use $x^0 = [0000]^T$. The initial values of Lagrangian multipliers can be set to 1 and 4. Set crit = 1×10^{-6}.

10.20 Find the minimum of each of the given functions using the cyclic coordinate search method:

1. $f(x) = 100(x_1^2 - x_2)^2 + (1 - x_1)^2$ (Rosenbrock's function), $x^0 = [-1, 1]$, crit = 1×10^{-4}.
2. $f(x) = x_1^2 + 2x_2^2 + 2x_1 x_2$, $x^0 = [0.5, 1]$, crit = 1×10^{-4}.

10.21 Find the minimum of $f(x) = x_1^2 + 2x_2^2 + 2x_1 x_2$ using Nelder-Mead's simplex method. Use $x^0 = [0. 5, 1]$ and crit = 1×10^{-6}.

10.22 Find the minimum of Rosenbrock's function using the simulated annealing method:

$$f(x) = 100(x_1^2 - x_2)^2 + (1 - x_1)^2 (-5 \leq x_1, x_2 \leq 5)$$

Set the initial temperature as $T = 100$, the temperature reduction factor rp = 0.8, the step reduction parameter rs = 0.9, and crit = 1×10^{-8}.

10.23 Find the minimum of the given function $f(x)$ using the simulated annealing method[36]:

$$f(x) = x_1^4 - 16x_1^2 - 5x_1 + x_2^4 - 16x_2^2 - 5x_2 (-5 \leq x_1, x_2 \leq 5)$$

Set the initial temperature as $T = 100$, the temperature reduction factor rp = 0.8, the step reduction parameter rs = 0.9, and crit = 1×10^{-8}.

10.24 Find the minimum of Powell's function using the genetic algorithm:

$$f(x) = (x_1 + 10x_2)^2 + 5(x_3 - x_4)^2 + (x_2 - 2x_3)^4 + 10(x_1 - x_4)^4 - 5 \leq x_1, x_2, x_3, x_4 \leq 5$$

Set nb = 8, ps = 50, ng = 50, and mp = 0.04.

10.25 Find the solution of the following minimization problem using the branch-and-bound algorithm[37]:

$$\text{Minimize } f(x) = 65x_1 + 57x_2 + 42x_3 + 20x_4$$

$$\text{Subject to } 4x_1 + 5x_2 + 6x_3 + 10x_4 \leq 12, \ x_1 + x_2 = 1, \ x_3 + x_4 = 1 \ (x_i = 0 \text{ or } 1)$$

10.26 Find the solution of the following minimization problem using the zero-one programming method:

$$\text{Minimize } f(x) = 4x_1 + 5x_2 + 3x_3$$

$$\text{Subject to } -x_1 + 3x_2 + 6x_2 \leq 6, \ -2x_1 - 3x_2 - 3x_2 \leq -2, \ -x_1 - x_3 \leq -1, \ 0 \leq x_1, x_2$$

10.27 Suppose that the annual cost $C(d)$ ($) of a pipeline can be expressed as a function of the pipe diameter d (in.) as follows:

$$C(d) = 500 + 240d^{1.5} + \frac{2}{d^{2.5}} + \frac{27}{d^4}$$

1. Plot $C(d)$ versus d.
2. Find the optimal value of d using a built-in function.
3. If d is within the interval [1, 3], what is the optimum annual cost? Use a built-in function that solves the constrained optimization problem.

REFERENCES

1. Brent, R. P., *Algorithms of Minimization without Derivatives*, Prentice-Hall, Englewood Cliffs, NJ, pp. 47–60, 1973.
2. Belegundu, A. D. and T. R. Chandrupatla, *Optimization Concepts and Applications in Engineering*, 2nd ed., Cambridge University Press, New York, NY, pp. 75–76, 2011.
3. Fletcher, R. and M. J. D. Powell, A rapidly convergent descent method for minimization, *The Computer Journal*, 6, pp. 163–168, 1963.
4. Belegundu, A. D. and T. R. Chandrupatla, *Optimization Concepts and Applications in Engineering*, 2nd ed., Cambridge University Press, New York, NY, p. 108, 2011.
5. Edgar, T. F., D. M. Himmelblau, and L. S. Lasdon, *Optimization of Chemical Processes*, 2nd ed., McGraw-Hill, New York, NY, p. 197, 2001.

6. Belegundu, A. D. and T. R. Chandrupatla, *Optimization Concepts and Applications in Engineering*, 2nd ed., Cambridge University Press, New York, NY, p. 114, 2011.

7. Rao, S. S., *Engineering Optimization: Theory and Practice*, 4th ed., John Wiley & Sons, Inc., Hoboken, NJ, pp. 150–152, 2009.

8. Belegundu, A. D. and T. R. Chandrupatla, *Optimization Concepts and Applications in Engineering*, 2nd ed., Cambridge University Press, New York, NY, pp. 132–135, 2011.

9. Karmarkar, N., A new polynomial time algorithm for linear programming, *Combinatorica*, 4, pp. 373–395, 1984.

10. Lindfield, G. G. and J. E. T. Penny, *Numerical Methods*, 3rd ed., Academic Press, Waltham, MA, pp. 374–375, 2012.

11. Adidharma, H. and V. Temyanko, *Mathcad for Chemical Engineers*, Trafford publishing, Victoria, BC, Canada, pp. 98–99, 2007.

12. Belegundu, A. D. and T. R. Chandrupatla, *Optimization Concepts and Applications in Engineering*, 2nd ed., Cambridge University Press, New York, NY, p. 220, 2011.

13. Belegundu, A. D. and T. R. Chandrupatla, *Optimization Concepts and Applications in Engineering*, 2nd ed., Cambridge University Press, New York, NY, p. 226, 2011.

14. Rao, S. S., *Engineering Optimization*, 4th ed., John Wiley & Sons, Inc., Hoboken, NJ, pp. 413–414, 2009.

15. Lasdon, L. et al., Design and testing of a generalized relaxed gradient code for nonlinear programming, *ACM Transactions on Mathematical Software*, 4(1), pp. 34–50, 1978.

16. Antoniou, A. and W.-S. Lu, *Practical Optimization*, Springer, New York, NY, pp. 514–515, 2007.

17. Antoniou, A. and W.-S. Lu, *Practical Optimization*, Springer, New York, NY, p. 516, 2007.

18. Belegundu, A. D. and T. R. Chandrupatla, *Optimization Concepts and Applications in Engineering*, 2nd ed., Cambridge University Press, New York, NY, pp. 295–296, 2011.

19. Hooke, R. and T. A. Jeeves, Direct search solution of numerical and statistical problems, *Journal of the ACM*, 8, pp. 212–229, 1961.

20. Rosenbrock, H. H., An automatic method for finding the greatest or least value of a function, *The Computer Journal*, 3, pp. 175–184, 1960.

21. Belegundu, A. D. and T. R. Chandrupatla, *Optimization Concepts and Applications in Engineering*, 2nd ed., Cambridge University Press, New York, NY, pp. 309–311, 2011.

22. Nelder, J. A. and R. Mead, A simplex method for function minimization, *The Computer Journal*, 7, pp. 308–313, 1965.

23. Corana, A., M. Marchesi, C. Martini, and S. Ridella, Minimizing multimodal functions of continuous variables with simulated annealing algorithm, *ACM Transactions on Mathematical Software*, 13, pp. 262–280, 1987.

24. Belegundu, A. D. and T. R. Chandrupatla, *Optimization Concepts and Applications in Engineering*, 2nd ed., Cambridge University Press, New York, NY, pp. 317–318, 2011.

25. Venkataraman, P., *Applied Optimization with MATLAB Programming*, 2nd ed., John Wiley & Sons, Inc., Hoboken, NJ, pp. 458–459, 2009.

26. Yang, W. Y., W. Cao, T.-S. Chung, and J. Morris, *Applied Numerical Methods Using MATLAB*, Wiley Interscience, Hoboken, NJ, p. 338, 2005.

27. Belegundu, A. D. and T. R. Chandrupatla, *Optimization Concepts and Applications in Engineering*, 2nd ed., Cambridge University Press, New York, NY, p. 322, 2011.

28. Rao, S. S., *Engineering Optimization: Theory and Practice*, 4th ed., John Wiley & Sons, Inc., Hoboken, NJ, p. 610, 2009.

29. Venkataraman, P., *Applied Optimization with MATLAB Programming*, 2nd ed., John Wiley & Sons, Inc., Hoboken, NJ, pp. 416–417, 2009.

30. Belegundu, A. D. and T. R. Chandrupatla, *Optimization Concepts and Applications in Engineering*, 2nd ed., Cambridge University Press, New York, NY, p. 368, 2011.

31. Mariano M. Martin (Editor), *Introduction to Software for Chemical Engineers*, CRC Press, Taylor & Francis Group, Boca Raton, FL, pp. 146–148, 2015.

32. Belegundu, A. D. and T. R. Chandrupatla, *Optimization Concepts and Applications in Engineering*, 2nd ed., Cambridge University Press, New York, NY, p. 77, 2011.

33. Adidharma, H. and V. Temyanko, *Mathcad for Chemical Engineers*, Trafford publishing, Victoria, BC, Canada, p. 96, 2007.

34. Adidharma, H. and V. Temyanko, *Mathcad for Chemical Engineers*, Trafford publishing, Victoria, BC, Canada, p. 99, 2007.

35. Antoniou, A. and W.-S. Lu, *Practical Optimization*, Springer, New York, NY, p. 517, 2007.

36. Yang, W. Y., W. Cao, T.-S. Chung, and J. Morris, *Applied Numerical Methods Using MATLAB*, Wiley Interscience, Hoboken, NJ, pp. 337–338, 2005.
37. Belegundu, A. D. and T. R. Chandrupatla, *Optimization Concepts and Applications in Engineering*, 2nd ed., Cambridge University Press, New York, NY, pp. 366–367, 2011.

11 Computational Intelligence

11.1 FUZZY SYSTEMS

Fuzzy logic is a method of reasoning that resembles human reasoning. The fuzzy logic approach imitates the method of decision making in humans, which involves all intermediate possibilities between the digital values of "yes" and "no". Fuzzy logic is an extension of multivalued logic, and is related to classes of objects with unsharp boundaries in which membership is a matter of degree. Thus, a critical step in applying fuzzy logic strategies is the membership function assessment of a variable, which is proper for demonstrating the decision maker's modeling preferences. To use fuzzy logic, we first need a fuzzy set. In a fuzzy set, the transition from membership to non-membership is not well defined. We quantify the degree of membership with values between 0 (not a member) and 1 (definitely a member). Fuzzy logic and fuzzy sets allow approximate reasoning. An element of a fuzzy set belongs to the set with a certain degree of certainty. Fuzzy logic allows reasoning with uncertain facts to infer new facts, with a degree of certainty associated with each fact. The uncertainty in fuzzy systems is referred to as nonstatistical uncertainty, and should not be confused with statistical uncertainty.[1]

In recent years, the number and variety of applications of fuzzy logic have increased significantly. The applications range from consumer products such as cameras, camcorders, washing machines, and microwave ovens to industrial process control, medical instrumentation, decision-support systems, and portfolio selection.

The Fuzzy Logic Toolbox™ provides MATLAB® functions, apps, and a Simulink® block for analyzing, designing, and simulating systems based on fuzzy logic. It guides you through the steps of designing fuzzy inference systems. Functions are provided for many common methods, including fuzzy clustering and adaptive neuro-fuzzy learning. The toolbox lets you model complex system behaviors using simple logic rules, and then implement these rules in a fuzzy inference system.

11.1.1 MEMBERSHIP FUNCTIONS

A membership function defines the fuzzy set. The function is used to associate a degree of membership for each of the elements of the domain to the corresponding fuzzy set. Membership functions for fuzzy sets can be of any shape or type, as determined by experts in the domain over which the sets are defined.[2] For each input and output variable in a fuzzy inference system, one or more membership functions define the possible linguistic sets for that variable. MATLAB provides many default membership functions, as shown in Table 11.1.

trimf is the triangular membership function. y = trimf(x, [a b c]) returns a matrix which is the triangular membership function evaluated at x. Parameters [a b c] determine the break points of this membership function ($a \leq b \leq c$).

trapmf is the trapezoidal membership function. y = trapmf(x, [a b c d]) returns a matrix which is the trapezoidal membership function evaluated at x. Parameters [a b c d] determine the break points of this membership function. It is required that $a \leq b$ and $c \leq d$.

sigmf is the sigmoid curve membership function. y = sigmf(x, [a b]) returns a matrix which is the sigmoid membership function evaluated at x. Parameters [a b] determine the shape and position of this membership function. The output y is given by

TABLE 11.1
Default Membership Functions Provided by Fuzzy Logic Toolbox

Function Type	Function Name	Synopsis
Triangular	*trimf*	y = trimf(x, [a b c])
Trapezoidal	*trapmf*	y = trapmf(x, [a b c d])
Sigmoid	*sigmf*	y = sigmf(x, [a b])
Gaussian	*gaussmf*	y = gaussmf(x, [σ, c])
Z-membership	*zmf*	y = zmf(x, [a b])
S-membership	*smf*	y = smf(x, [a b])

$$y = \frac{1}{1 + e^{-a(x-b)}}$$

gaussmf is the Gaussian curve membership function. y = gaussmf(x, [σ, c]) returns a matrix which is the Gaussian membership function evaluated at x. Parameters [σ, c] determine the shape and position of this membership function. The output y is given by

$$y = e^{-\frac{1}{2\sigma^2}(x-c)^2}$$

zmf is the Z-shaped curve membership function. y = zmf(x, [a b]) returns a matrix which is the Z-shaped membership function evaluated at x. Parameters [a b] determine the break points of this membership function. When $a > b$, *zmf* is a smooth transition from 1 (at b) to 0 (at a). When $a \leq b$, *zmf* becomes a reverse step function which jumps from 1 to 0 at $(a + b)/2$.

smf is the S-shaped curve membership function. y = smf(x, [a b]) returns a matrix which is the S-shaped membership function evaluated at x. Parameters [a b] determine the break points of this membership function. When $a < b$, *smf* is a smooth transition from 0 (at a) to 1 (at b). When $a \geq b$, *smf* becomes a step function which jumps from 0 to 1 at $(a + b)/2$.

Example 11.1   Plots of Membership Functions

Let y1, y2, y3, and y4 be fuzzy sets defined as
y1 = trimf(x, [3 6 8]); y2 = sigmf(x, [2 4]); y3 = gaussmf(x, [2 5]);
y4 = smf(x, [1 8]);
Plot these fuzzy sets for $0 \leq x \leq 10$.

Solution

The script *mfplots* produces plots of fuzzy sets defined by some of the membership functions (triangular, sigmoid, Gaussian, and S-membership), shown in Figure 11.1.

```
% mfplots.m
x = 0:0.1:10;
y1 = trimf(x,[3 6 8]); % Triangular
subplot(2,2,1), plot(x,y1), xlabel('trimf, P = [3 6 8]'), ylim([-0.05 1.05]), title
('Triangular')
y2 = sigmf(x,[2 4]); % Sigmoid
subplot(2,2,2), plot(x,y2), xlabel('sigmf, P = [2 4]'), ylim([-0.05 1.05]), title
('Sigmoidal')
y3 = gaussmf(x,[2 5]); % Gaussian
subplot(2,2,3), plot(x,y3), xlabel('gaussmf, P = [2 5]'), title('Gaussian')
```

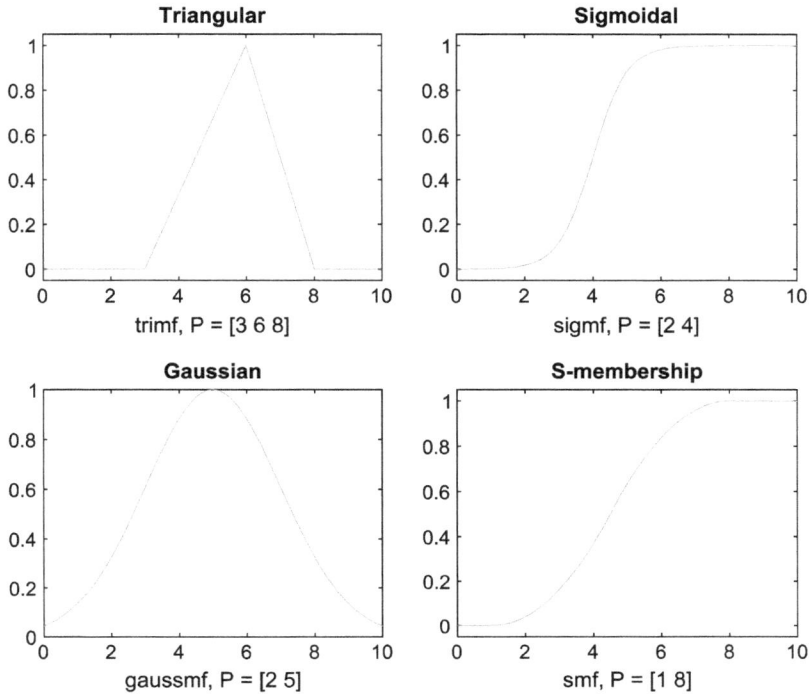

FIGURE 11.1   Plots of some membership functions.

```
y4 = smf(x,[1 8]); % S-membership
subplot(2,2,4), plot(x,y4), xlabel('smf, P = [1 8]'), ylim([-0.05 1.05]), title('S-
membership')
```

11.1.2 FUZZY ARITHMETIC OPERATIONS

The default function *fuzarith* performs fuzzy arithmetic operations. The basic syntax is

```
C = fuzarith(X,A,B,operator)
```

where A and B are convex fuzzy sets of universe X. A, B, and X should be vectors of the same dimension. Membership grades of A and B outside of X are zero. Operator is one of the following, specified as a character row vector or string scalar: 'sum', 'sub', 'prod', or 'div'. C is the returned fuzzy set, which is a column vector with the same length as X.

Example 11.2   Fuzzy Arithmetic Operations

Suppose that fuzzy sets A and B are given by A = trapmf(x,[-10 -2 1 3]) and B = gaussmf(x,[2 5]) in the range of $-20 \leq x \leq 20$. Evaluate the sum, difference, and product of A and B and plot the results.

Solution

The script *fzopn* performs the desired arithmetic operations and creates the plots shown in Figure 11.2.

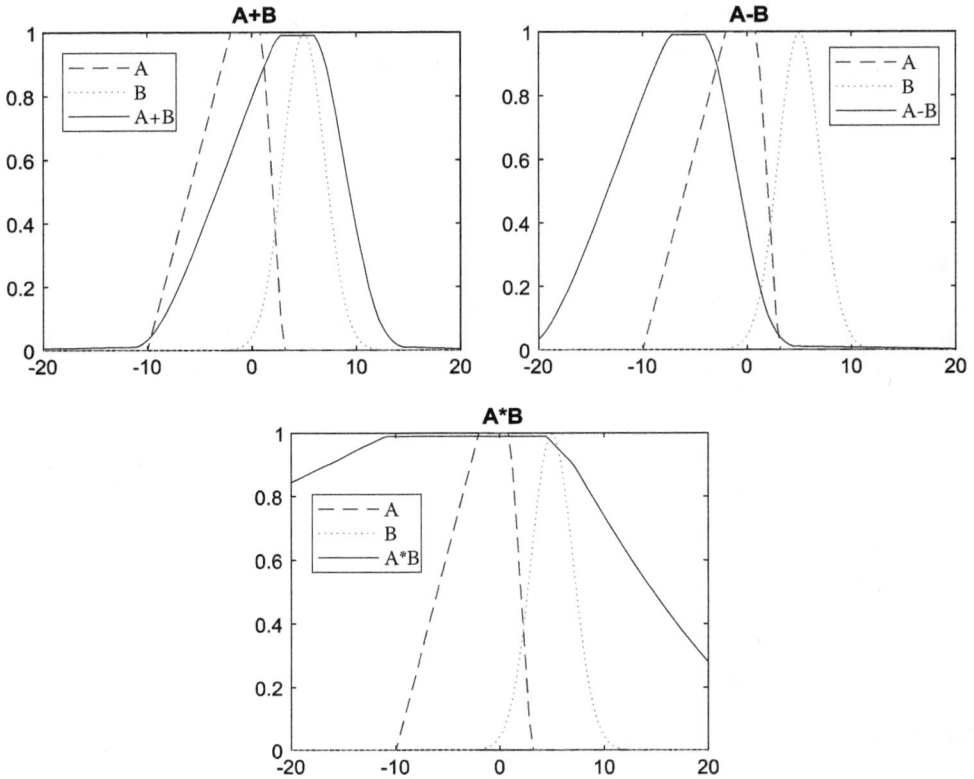

FIGURE 11.2   Fuzzy arithmetic operations.

```
% fzopn.m: fuzzy arithmetic operations using fuzarith function
N = 101; minx = -20; maxx = 20;
x = linspace(minx,maxx,N); % generate N points between minx and maxx
A = trapmf(x,[-10 -2 1 3]); B = gaussmf(x,[2 5]); % define A and B using membership functions
Csum = fuzarith(x,A,B,'sum'); % A+B
Csub = fuzarith(x,A,B,'sub'); % A-B
Cprod = fuzarith(x,A,B,'prod'); % A*B
subplot(2,2,1),   plot(x,A,'b--',x,B,'m:',x,Csum,'k-'),   title('   A+B'),   le-
gend('A','B','A+B')
subplot(2,2,2),   plot(x,A,'b--',x,B,'m:',x,Csub,'k-'),   title('   A-B'),   legend
('A','B','A-B')
subplot(2,2,3),   plot(x,A,'b--',x,B,'m:',x,Cprod,'k-'),   title('   A*B'),   le-
gend('A','B','A*B')
```

11.1.3 DEFUZZIFICATION

Defuzzification of the membership function can be performed by the built-in function *defuzz*. The basic syntax is y = defuzz(x,mf,type) where y is a defuzzified value out of a membership function mf positioned at associated variable value x, using one of several defuzzification strategies provided by the argument type. Table 11.2 shows the types of defuzzification strategies available in MATLAB.

TABLE 11.2
Defuzzification Types

Type	Description
centroid	Centroid of the area under the output fuzzy set (default for Mamdani systems)
bisector	Bisector of the area under the output fuzzy set
mom	Mean of the values for which the output fuzzy set is maximum
lom	Largest value for which the output fuzzy set is maximum
som	Smallest value for which the output fuzzy set is maximum

For example, suppose that a fuzzy set A is defined by the trapezoidal membership function as A = `trapmf(x,[-10 -8 -4 7])` in the range of $-10 \le x \le 10$. Defuzzification using the centroid method can be performed as follows:

```
>> x = linspace(-10,10,100); A = trapmf(x,[-10 -8 -4 7]); dfzout = defuzz(x, A,
'centroid')
dfzout =
 -3.2850
```

Example 11.3 Comparison of Defuzzification Methods[3]
The fuzzy sets A and B are defined by triangular and trapezoidal membership functions as A = `trimf(x,[-5 -4 -2])` and B = `trapmf(x,[-5 -3 2 5])` in the range of $-5 \le x \le 5$. The fuzzy set C is given by

$$C = \max(0.7A, 0.5B)$$

Perform defuzzification of C using centroid, bisector, and mom methods.

Solution
The script *compdfz* performs defuzzification using centroid, bisector, and mom methods. It produces line plots to compare the results, shown in Figure 11.3.

```
% compdfz.m
x = -5:0.1:5;
A = trimf(x,[-5 -4 -2]); B = trapmf(x,[-5 -3 2 5]); C = max(0.7*A, 0.5*B); %
produce fuzzy sets
plot(x,C,'LineWidth',4), ylim([-1 1])
x1 = defuzz(x,C,'centroid')
h1   =   line([x1   x1],[-0.2   1.2],'Color','k');   t1   =   text(x1,-0.2,'
centroid','FontWeight','bold');
x2 = defuzz(x,C,'bisector')
h2   =   line([x2   x2],[-0.4   1.2],'Color','k');   t2   =   text(x2,-0.4,'
bisector','FontWeight','bold');
x3 = defuzz(x,C,'mom')
h3   =   line([x3   x3],[-0.7   1.2],'Color','k');   t3   =   text(x3,-0.7,'mom',
'FontWeight','bold');

>> compdfz
x1 =
 -0.5896
```

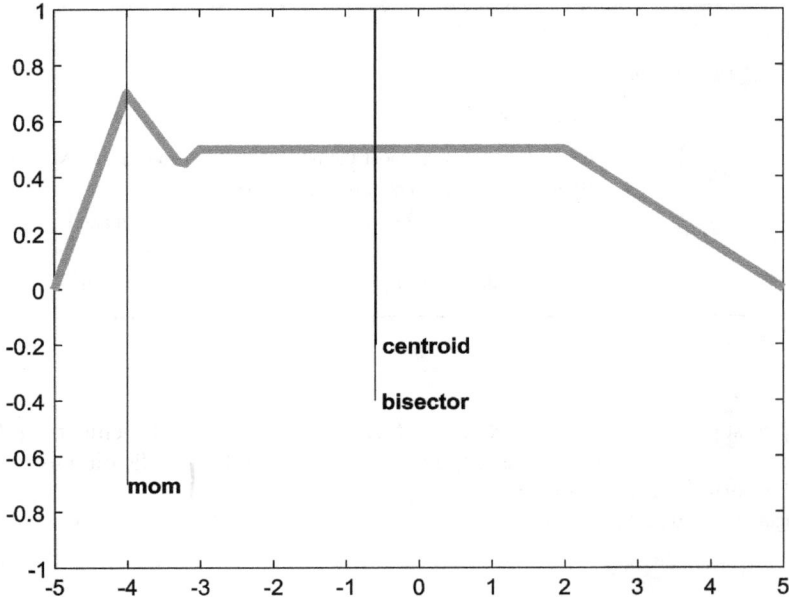

FIGURE 11.3 Defuzzification by centroid, bisector, and mom methods.

```
x2 =
 -0.6000
x3 =
  -4
```

Usually the centroid defuzzification method generates the most accurate and reliable results.

11.1.4 Fuzzy Inference Systems

11.1.4.1 Mamdani Fuzz Inference System (FIS) Objects

Mamdani fuzzy inference is the most commonly used fuzzy methodology, due to its simple min-max structure. This method was proposed by Ebrahim Mamdani[4] as an attempt to control a combined steam engine and boiler by synthesizing a set of linguistic control rules obtained from experienced human operators. This method is based on the fuzzy algorithms proposed by Lotfi Zadeh to handle complex systems and decision processes.[5] In Mamdani-type inference, the output membership functions are expected to be fuzzy sets, and the fuzzy set for each output variable should be defuzzified after the aggregation process.

MATLAB provides a sample FIS object named *demofis* that includes two inputs, one output, and three rules. This FIS object can be uploaded and referenced as follows:

```
>> load demofis, demofis
```

The first input object of demofis and its membership functions can be referenced by

```
>> demofis.Inputs(1)
>> demofis.Inputs(1).MembershipFunctions
```

The built-in function *plotfis* produces a block diagram representing the system and creates an input-output display of the FIS. Inputs and their membership functions appear to the left of the FIS structural characteristics, while outputs and their membership functions appear on the right. The built-in function *plotmf* displays all of the membership functions for a given variable. This function plots all of the membership functions in the FIS called *fismat* associated with a variable with a given type and index. The built-in function *gensurf* generates a plot of the output surface of a given FIS object with the first two inputs and the first output. If the FIS object has more than two inputs, *gensurf* uses midrange values of the nondisplayed inputs for plot generation.

The script *showdemofis* loads the sample FIS object *demofis* and plots membership functions and the output surface, as shown in Figures 11.4 and 11.5.

```
% showdemofis.m
load demofis;
demofis.Rules % display all rules
figure(1), plotfis(demofis) % construct a block diagram representing the system
figure(2)
subplot(2,2,1), plotmf(demofis,'input',1) % the membership function for the 1st
input
subplot(2,2,2), plotmf(demofis,'input',2) % the membership function for the 2nd
input
subplot(2,2,3), plotmf(demofis,'output',1) % the membership function for the
1st output
subplot(2,2,4), gensurf(demofis) % FIS outputs for input variables

>> showdemofis
ans =
 1x3 fisrule array with properties:
  Description
  Antecedent
  Consequent
  Weight
  Connection
 Details:
```

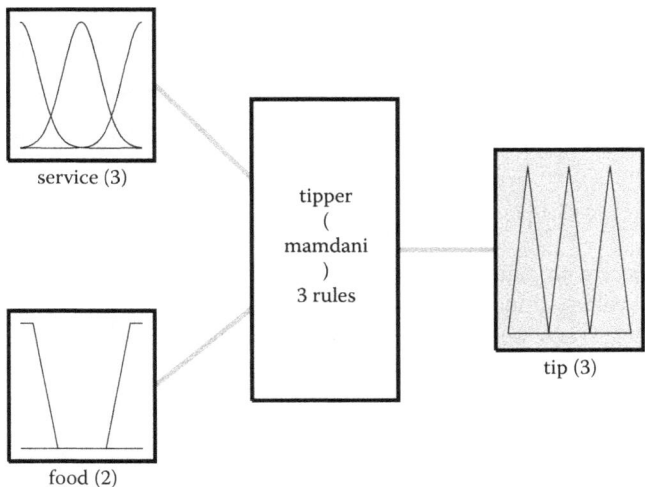

System tipper: 2 inputs, 1 outputs, 3 rules

FIGURE 11.4 Block diagram of the sample FIS object (*demofis*).

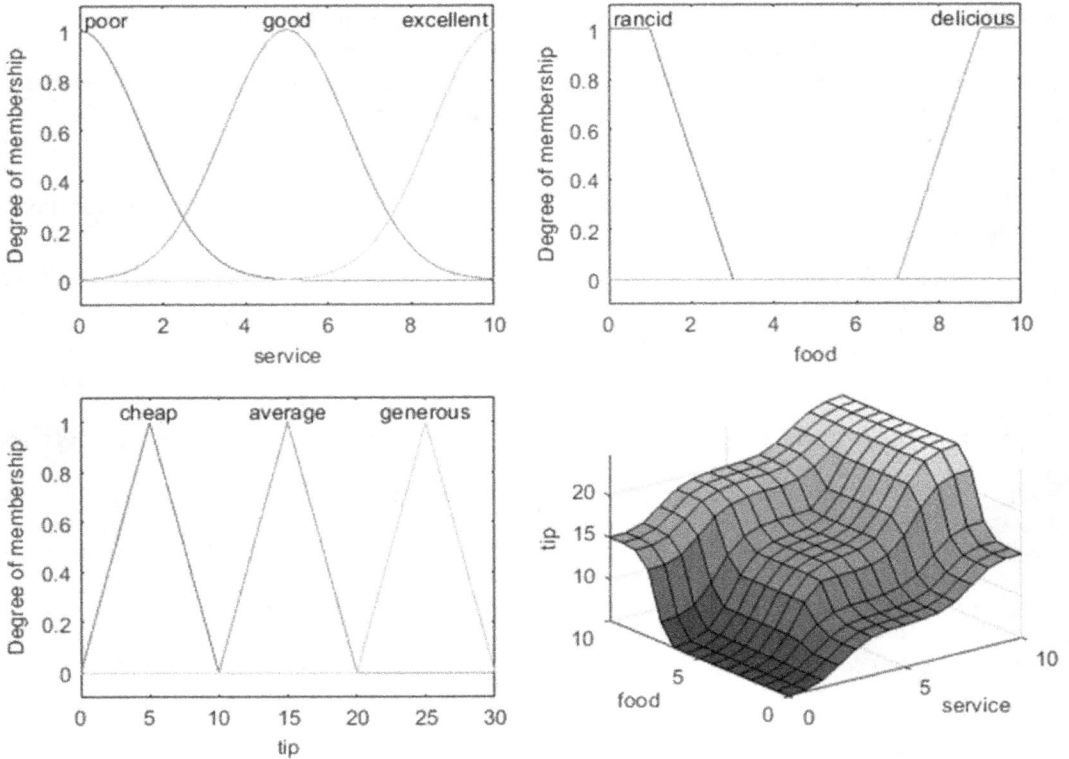

FIGURE 11.5 Plots of membership functions and surface plot of outputs.

```
                    Description
 ─────────────────────────────────────────────────────────────
 1    "service==poor | food==rancid => tip=cheap (1) "
 2    "service==good => tip=average (1) "
 3    "service==excellent | food==delicious => tip=generous (1) "
```

11.1.4.2 Creation of an FIS Object Using the *mamfis* Function

The built-in function *mamfis* can be used to create a Mamdani object. The basic syntax is

```
fis1 = mamfis
fis2 = mamfis(name1, value1, ...)
```

fis1 is an FIS object with default property values, and fis2 is an FIS with the specified pairs of parameter names and values. If input-output data are available, the built-in function *genfis* can be used to create an FIS object. When we call the existing.*fis* files, the built-in function *readfis* may be used.

Example 11.4   Creation of an FIS Object

Use the built-in function *mamfis* to create a Mamdani FIS object with three inputs and one output.

Solution

```
>> fis = mamfis ("NumInputs",3,"NumOutputs",1)
fis =
```

```
mamfis with properties:
              Name: "fis"
         AndMethod: "min"
          OrMethod: "max"
     ImplicationMethod: "min"
     AggregationMethod: "max"
   DefuzzificationMethod: "centroid"
            Inputs: [1×3 fisvar]
           Outputs: [1×1 fisvar]
             Rules: [1×27 fisrule]
  DisableStructuralChecks: 0
    See 'getTunableSettings' method for parameter optimization.
fis = mamfis("NumInputs",3,"NumOutputs",1);
```

As an example, we now create an FIS object that determines the flow rate of cooling water of an exothermic reactor. First, create an FIS object *rxnfis*, specifying its name as "cwrate". There are two input variables, representing the reaction state in the reactor (color and pressure). Let the range of each input variable be [0 10]. Add the first input variable, for the color, using the *addInput* function. Add membership functions for each of the color levels using addMF. In this case, we use Gaussian membership functions. Add the second input variable, for the pressure, and add two trapezoidal membership functions. Add the output variable for the coolant rate (coolant), and add two trapezoidal membership functions and one triangular membership function. We specify the following three rules for the FIS object *rxnfis* as a numeric array:

1. If (color is red) or (pressure is low), then (coolant is low).
2. If (color is orange), then (coolant is average).
3. If (color is yellow) or (pressure is high), then (coolant is high).

In this case, the array is a 3×5 array. Each row of the array contains one rule, in the following format:

Column 1: Index of membership function for first input
Column 2: Index of membership function for second input
Column 3: Index of membership function for output
Column 4: Rule weight (from 0 to 1)
Column 5: Fuzzy operator (1 for AND, 2 for OR)

For the membership function induces, indicate a NOT condition using a negative value. For example, the rule (the first row of the array) [1 1 1 1 2] represents "If input 1 is MF 1 (the first membership function associated with input 1) or if input 2 is MF 1, then output 1 should be MF 1 (the first membership function associated with output 1)." We can add rules using the function *addRule*. To evaluate the output of the FIS object for a given input combination, we can use the built-in *evolves* command. If we want to evaluate three different types of input variables, we supply input variables as a numeric array (each row denoting a combination of an input variable).

The script *muffs* creates the desired FIS object, evaluates it according to the procedure described, and generates the surface plot shown in Figure 11.6.

```
% muffs.m
% Create an FIS object
rxnfis = mamfis('Name',"cwrate");
% Add the first input variable (color) and membership functions
rxnfis = addInput(rxnfis,[0 10],'Name',"color");
```

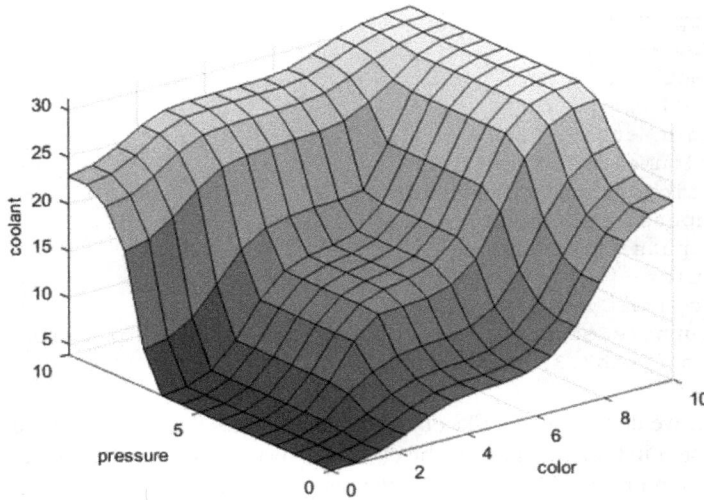

FIGURE 11.6   Surface plot of outputs from the FIS *rxnfis*.

```
rxnfis = addMF(rxnfis,"color","gaussmf",[1.5 0],'Name',"red");
rxnfis = addMF(rxnfis,"color","gaussmf",[1.5 5],'Name',"orange");
rxnfis = addMF(rxnfis,"color","gaussmf",[1.5 10],'Name',"yellow");
% Add the second input variable (pressure) and trapezoidal membership functions
rxnfis = addInput(rxnfis,[0 10],'Name',"pressure");
rxnfis = addMF(rxnfis,"pressure","trapmf",[-2 0 1 3],'Name',"low");
rxnfis = addMF(rxnfis,"pressure","trapmf",[7 9 10 12],'Name',"high");
% Add the output variable (coolant) and triangular membership functions
rxnfis = addOutput(rxnfis,[0 40],'Name',"coolant");
rxnfis = addMF(rxnfis,"coolant","trapmf",[-2 0 5 10],'Name',"low");
rxnfis = addMF(rxnfis,"coolant","trimf",[10 15 20],'Name',"average");
rxnfis = addMF(rxnfis,"coolant","trapmf",[20 25 40 50],'Name',"high");
% Create and add a numeric array representing rules
numrule = [1 1 1 1 2; 2 0 2 1 1; 3 2 3 1 2];
rxnfis = addRule(rxnfis,numrule);
% Evaluation of the created FIS object
in1 = evalfis(rxnfis, [1 2])
vals = [3 5; 2 7; 3 1]; % three different types of input variables
in3 = evalfis(rxnfis,vals)
gensurf(rxnfis)

>> myfis
in1 =
  4.4216
in3 =
 12.0223
  7.0500
  7.1236
```

11.1.4.3 Creation of an FIS Object Using the *genfis* Function

When input and output data are available, we can use the built-in function *genfis* to create an FIS object. The basic syntax is

```
fis = genfis(indat,outdat)
```

where indat and outdate are the files containing the input and output data, respectively. When we create the FIS object using grid partitioning, we use the default *genfisOptions* to create the option as follows:

```
opt = genfisOptions('GridPartition');
```

The script *datfis* generates input and output data sets and uses *genfisOptions* to apply the grid partitioning. The first input variable is associated with three Gaussian membership functions, and the second input variable is associated with five triangular membership functions. The script produces a plot for each input membership function, as shown in Figure 11.7.

```
% datfis.m: creation of a FIS object with options
indat = [rand(10,1) 10*rand(10,1)-5]; % input data (10x2)
outdat = rand(10,1); % output data (10x1)
opt = genfisOptions('GridPartition'); % create a FIS object using grid partitioning
opt.NumMembershipFunctions = [3 5]; % 3 gaussian and 5 triangular mf's for the
1st and 2nd input variable
opt.InputMembershipFunctionType = ["gaussmf" "trimf"];
exfis = genfis(indat,outdat,opt); % generate a FIS object
[x1,mf1] = plotmf(exfis,'input',1); subplot(2,1,1), plot(x1,mf1), xlabel('input 1
(gaussmf)')
[x2,mf2] = plotmf(exfis,'input',2); subplot(2,1,2), plot(x2,mf2), xlabel('input 2
(trimf)')
```

The FIS object *exfis* created by the script *datfis* is a structure with 10 property fields, and the Rules field contains 15 rules. The built-in function *showrule* is used to display the rules as follows:

```
>> showrule(exfis)
ans =
```

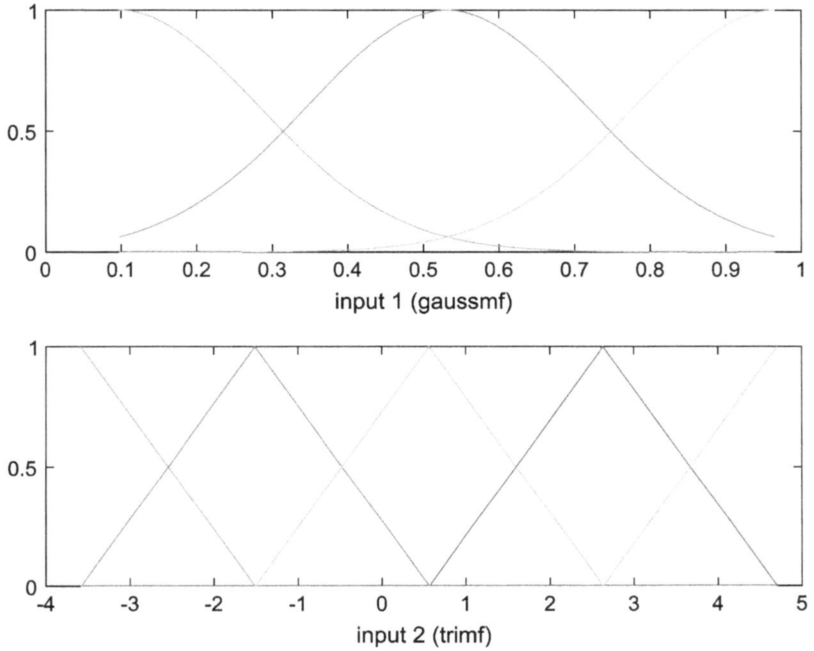

FIGURE 11.7 Plots of membership functions.

```
15×78 char array
'1. If (input1 is in1mf1) and (input2 is in2mf1) then (output is out1mf1) (1) '
'2. If (input1 is in1mf1) and (input2 is in2mf2) then (output is out1mf2) (1) '
'3. If (input1 is in1mf1) and (input2 is in2mf3) then (output is out1mf3) (1) '
'4. If (input1 is in1mf1) and (input2 is in2mf4) then (output is out1mf4) (1) '
'5. If (input1 is in1mf1) and (input2 is in2mf5) then (output is out1mf5) (1) '
'6. If (input1 is in1mf2) and (input2 is in2mf1) then (output is out1mf6) (1) '
'7. If (input1 is in1mf2) and (input2 is in2mf2) then (output is out1mf7) (1) '
'8. If (input1 is in1mf2) and (input2 is in2mf3) then (output is out1mf8) (1) '
'9. If (input1 is in1mf2) and (input2 is in2mf4) then (output is out1mf9) (1) '
'10. If (input1 is in1mf2) and (input2 is in2mf5) then (output is out1mf10) (1)'
'11. If (input1 is in1mf3) and (input2 is in2mf1) then (output is out1mf11) (1)'
'12. If (input1 is in1mf3) and (input2 is in2mf2) then (output is out1mf12) (1)'
'13. If (input1 is in1mf3) and (input2 is in2mf3) then (output is out1mf13) (1)'
'14. If (input1 is in1mf3) and (input2 is in2mf4) then (output is out1mf14) (1)'
'15. If (input1 is in1mf3) and (input2 is in2mf5) then (output is out1mf15) (1)'
```

11.1.4.4 Creation of an FIS Object Using Fuzzy Logic Designer

Fuzzy Logic Designer lets you design and test fuzzy inference systems for modeling complex system behaviors. As an example, we now create the FIS object *rxnfi* for the cooling water example using Fuzzy Logic Designer. As before, this example creates a Mamdani fuzzy inference system for a two-input, one-output cooling water rate problem. Given a number between 0 and 10 that represents the color of the reactant in the reactor (where 10 is bright yellow and 0 is dark red), and another number between 0 and 10 that represents the reactor pressure (where 10 is very high and 0 is very low), what should the flow rate of the cooling water be? The starting point is to write down the three heuristic rules of operation:

1. If (color is red) or (pressure is low), then (coolant rate is low).
2. If (color is orange), then (coolant rate is average).
3. If (color is yellow) or (pressure is high), then (coolant rate is high).

Now that we know the rules and have an idea of what the output should look like, use the UI tools to construct a fuzzy inference system for this decision process. To open Fuzzy Logic Designer, type the following command at the MATLAB prompt:

```
>> fuzzyLogicDesigner
```

Fuzzy Logic Designer opens and displays a diagram of the fuzzy inference system, with the names of the input variables on the left and those of the output variables on the right, as shown in Figure 11.8(a). The sample membership functions shown in the boxes are just icons and do not depict the actual shapes of the membership functions.

In this example, we construct a two-input, one output system. The two inputs are reactant "color" and reactor "pressure," and the one output is "coolant" flow rate. To add a second input variable and change the variable names to reflect these designations, select Edit > Add variable > Input. A second yellow box labeled "input2" appears in the editor window. The editing procedure for input variables is as follows:

1. Click the yellow box "input1". This box is highlighted with a red outline. Edit the Name field from "input1" to "color," and press Enter.
2. Click the yellow box "input2," edit the Name field from "input2" to "pressure," and press Enter.
3. Click the blue box "output1," edit the Name field from "output1" to "coolant," and press Enter.
4. Select File > Export > To Workspace, enter the Workspace variable name "cwrate," and click OK.

(a)   (b)

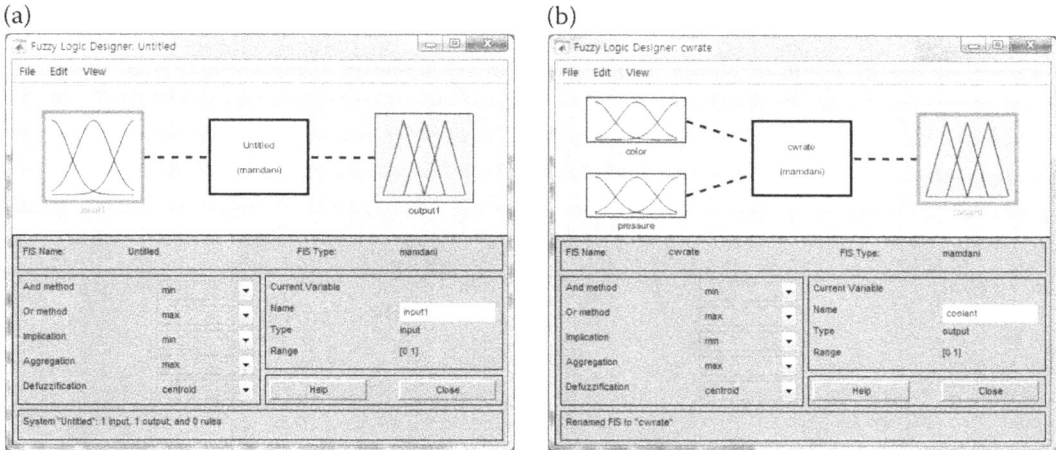

FIGURE 11.8   Editor window (a) of Fuzzy Logic Designer and (b) for the cwrate FIS object.

The diagram is updated to reflect the new names of the input and output variables. There is now a new variable in the Workspace called cwrate that contains all the information about this system. The editor window looks like Figure 11.8(b).

Next, define the membership functions associated with each of the variables. To do this, we have to open the Membership Function Editor. To open the Membership Function Editor, select Edit > Membership Functions within the Fuzzy Logic Designer window. Or you can type and enter *mfedit* at the command line. The process of specifying the membership functions for the two-input cooling water example, cwrate, is as follows:

1. Double-click the input variable "color" to open the Membership Function Editor.
2. In the Membership Function Editor, enter "[0 10]" in the Range and Display Range fields.
3. Select Edit > Remove All MFs to remove the default membership functions for the input variable "color."
4. Select Edit > Add MFs to open the Membership Functions dialog box.
5. In the Membership Functions dialog box, select "gaussmf" as the MF Type. Verify that 3 is selected as the Number of MFs. Click OK to add three Gaussian curves to the input variable "color."
6. Click on the curve named "mf1" to select it, and in the Current Membership Function (click on MF to select) area, enter "red" in the Name field and "[1.5 0]" in the Params field.
7. Click on the curve named "mf2" to select it, and in the Current Membership Function (click on MF to select) area, enter "orange" in the Name field and "[1.5 5]" in the Params field.
8. Click on the curve named "mf3," and in the Current Membership Function (click on MF to select) area, enter "yellow" in the Name field and "[1.5 10]" in the Params field. The Membership Function Editor window looks similar to Figure 11.9(a).
9. In the FIS Variables area, click the input variable "pressure" to select it. Enter "[0 10]" in the Range and Display Range fields.
10. Create the membership functions for the input variable "pressure."
11. Select Edit > Remove All MFs to remove the default Membership Functions for the input variable "pressure." Select Edit > Add MFs to open the Membership Functions dialog box.
12. In the Membership Functions dialog box, select "trapmf" as the MF Type, and select 2 in the Number of MFs drop-down list. Click OK to add two trapezoidal curves to the input variable "pressure."

(a)   (b)

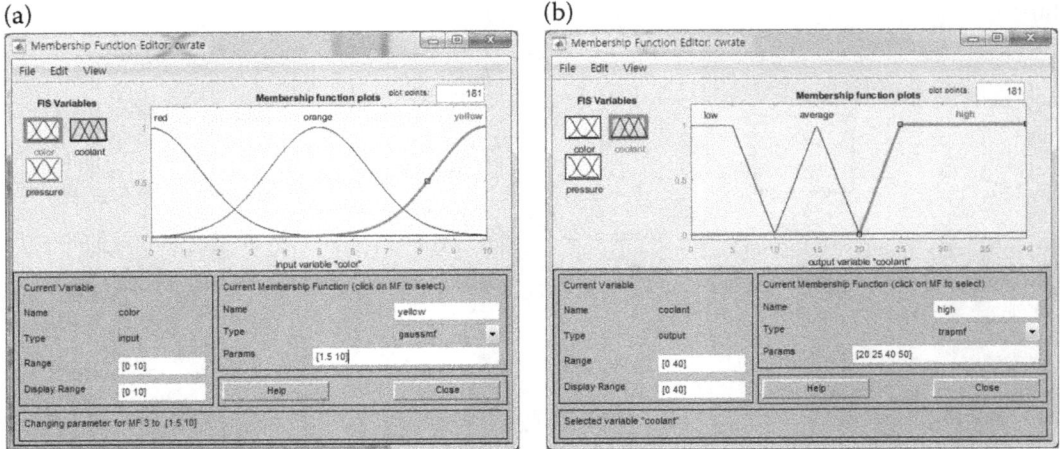

FIGURE 11.9   Membership Function Editor for (a) input variables and (b) output variable.

13. Rename the membership functions for the input variable "pressure," and click the input variable "pressure" in the FIS Variables area to select it.
14. Click on the curve named "mf1," and in the Current Membership Function (click on MF to select) area, enter "low" in the Name field and "[-2 0 1 3]" in the Params field.
15. Click on the curve named "mf2" to select it, and enter "high" in the Name field. Reset the associated parameters if desired.
16. Click on the output variable "coolant" to select it. Enter "[0 40]" in the Range and Display Range fields to cover the output range.
17. Click the curve named "mf1" to select it, and in the Current Membership Function (click on MF to select) area, enter "low" in the Name field. In the Type field, select "trapmf" as the MF Type, and in the Params field, enter "[-2 0 5 10]."
18. Click the curve named "mf2" to select it, and in the Current Membership Function (click on MF to select) area, enter "average" in the Name field. In the Type field, select "trimf" as the MF Type. In the Params field, enter "[10 15 20]."
19. Click the curve named "mf3" to select it, and enter "high" in the Name field. In the Type field, select "trapmf" as the MF Type, and enter "[20 25 40 50]" in the Params field.

The Membership Function Editor looks similar to Figure 11.9(b).

To insert rules, we open the Rule Editor by selecting Edit > Rules. To insert the first rule in the Rule Editor, select "red" under the variable "color," "low" under the variable "pressure," the "or" radio button in the Connection block, "low" under the output variable "coolant," and then click Add rule. Follow a similar procedure to insert the second and third rules in the Rule Editor, to get the rules shown in Figure 11.10(a). The numbers in parentheses represent weights.

Now the fuzzy inference system has been completely defined, in that the variables, membership functions, and rules necessary to calculate the coolant flow rate are in place. We can use the Rule Viewer by selecting Rules from the View menu. The Rule Viewer displays a road map of the whole fuzzy inference process, as shown in Figure 11.10(b). We can see a single window with 10 plots nested in it. The three plots across the top of the figure represent the antecedent and consequent of the first rule. Each rule is a row of plots, and each column is a variable. The rule number is displayed on the left of each row. We can click on a rule number to view the rule in the status line. The first two columns of plots (the six yellow plots) show the membership functions referenced by the antecedent, or the "if" part of each rule. The third column of plots (the three blue plots) shows the membership functions referenced

(a)

(b)

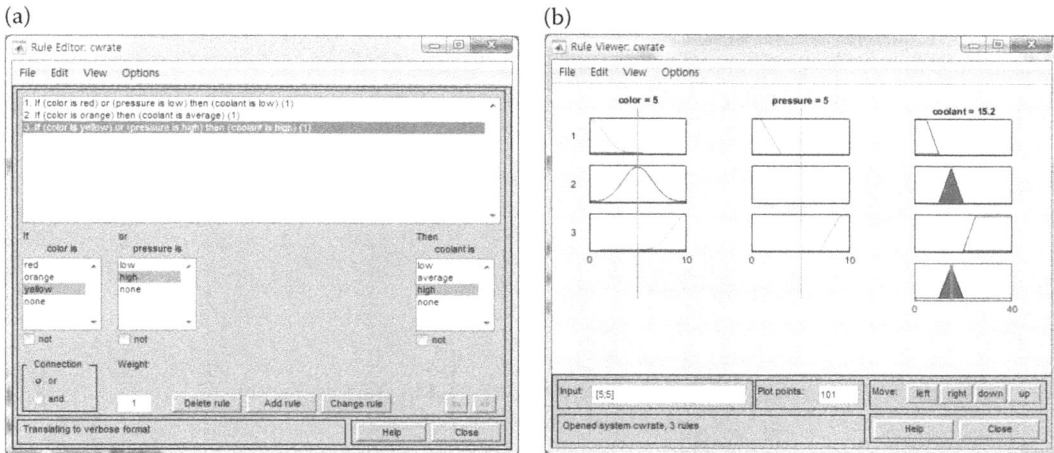

FIGURE 11.10   (a) Rule Editor window; (b) Rule Viewer window.

by the consequent, or the "then" part of each rule. The fourth plot in the third column represents the aggregate weighted decision for the given inference system. This decision will depend on the input values for the system. The defuzzified output is displayed as a bold vertical line on this plot.

The variables and their current values are displayed above the columns. In the lower left, there is a text field Input in which we can enter specific input values. For the two-input system, we can enter an input vector—[4 7], for example—and then press Enter. We can also adjust these input values by clicking on any of the three plots for each input. This will move the red index line horizontally, to the point where we have clicked. The defuzzified output value is shown by the thick line passing through the aggregate fuzzy set. We can shift the plots using "left," "right," "down," and "up." The Rule Viewer allows us to interpret the entire fuzzy inference process at once, and shows how the shape of certain membership functions influences the overall result.

The Rule Viewer shows one calculation at a time and in great detail. If we want to see the entire output surface of the system (that is, the entire span of the output set based on the entire span of the input set), we need to open the Surface Viewer. To open the Surface Viewer, select Surface from the View menu. Upon opening the Surface Viewer, we can see a three-dimensional curve that represents the mapping from color and pressure to the coolant flow rate, as shown in Figure 11.11.

11.2 ARTIFICIAL NEURAL NETWORKS

11.2.1 NEURAL NETWORK MODELS

Artificial neural networks are a set of algorithms, modeled loosely after the human brain, that are designed to recognize patterns. They interpret sensory data through a kind of machine perception, labeling or clustering raw input. The patterns they recognize are numerical, contained in vectors into which all real-world data, be it images, sound, text, or time series, must be translated. Neural networks help us cluster and classify. They help group unlabeled data according to similarities among the example inputs, and they classify data when they have a labeled data set to train on. Neural networks belong to a group of information processing techniques which can be used to find knowledge, patterns, or models from a large amount of data.

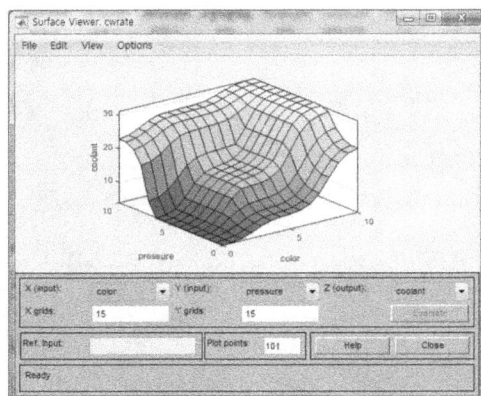

FIGURE 11.11   Surface Viewer.

Neural networks are normally based on mathematical regression to correlate input and output streams to and from process units. Such models principally rely on a large number of experimental data. Neural networks differ from traditional regression because they have more potential for finding the unseen structure.

Artificial neural networks are composed of simple elements operating in parallel. These elements are inspired by biological systems. The connections between these elements largely determine the function of the neural network. Artificial neural networks can be considered as a kind of nonlinear mapping process of two different subspaces. A neural network can be trained to perform a particular function by adjusting the values of the weights between elements. It has been found theoretically that a three-layer feed-forward network can estimate any continuous functions after training. The Deep Learning Toolbox provided by MATLAB contains algorithms, functions, and applications to create, train, visualize, and simulate neural networks.

In the neural network model, the set of inputs to the hidden layer, $\{S_j\}$, is computed after the sample $A_k = (a_1^k, a_2^k, \cdots, a_n^k)$ is supplied to the input layer, and the output vector from the hidden layer, $\{B_j\}$, is obtained through the activation function (usually the sigmoid function is used), σ, as follows:

$$B_j = \sigma\left(S_j\right) = \sigma\left(\sum_{i=1}^{N} W_{ij} \cdot a_j - \theta_j\right)$$

where

W_{ij} is the connection weight of the jth input layer to the ith hidden layer
a_j is the input vector to the internal recurrent neural network
θ_j is the output layer threshold
N represents the number of neurons used in the hidden layer

Next, the input vector to the output layer, $\{L_t\}$, is computed, followed by computation of the output vector of the output layer, $\{C_i^k\}$, through the sigmoid function σ as follows:

$$C_t^k = \sigma(L_t) = \sigma\left(\sum_{j=1}^{p} V_{jt} \cdot b_j - \gamma_t\right)$$

where

V_{jt} is the connection weight of the tth hidden layer to the jth output layer
b_j is the output vector of the hidden layer
γ_t is the threshold of the output layer
p is the number of neurons used in the output layer

The gradient drop law is used to backwardly compute the error of the network node, and the backward propagation of the accumulated error is usually employed to update the connection weights of the network. This procedure is repeated constantly. The output vector from the neural network, including the bias node, can be expressed as follows:

$$Y(k) = \sum_{j=1}^{p} WO_j \cdot \sigma(S_j(k)) + WO_{bias}, \; S_j(k) = \sum_{i=1}^{p} WR_{ij} \cdot \sigma(S_j(k-1)) + \sum_{i=1}^{n} WI_{ij} \cdot I_{ij}(k) + WI_{j,bias}$$

where
WI is the weight coefficient matrix connecting the input layer to the intermediate hidden layer
WR is the weight coefficient matrix connecting the feedback layer backward to the hidden layer
WO is the weight coefficient matrix connecting the hidden layer to the output layer
I is the input vector for the bias node
WI_{bias} is the weight coefficients connecting bias node 1 to the intermediate hidden layer
WO_{bias} is the weight coefficients connecting bias node 2 to the output layer
The standard steps for designing neural networks include data collection, network creation, network configuration, initialization of the weights and biases, network training, network validation, and network application. Typical areas of application for are pattern recognition, clustering, time-series analysis, and function fitting.

11.2.2 FUNCTION FITTING NEURAL NETWORKS

11.2.2.1 Construction and Training of a Function Fitting Network

The built-in function *fitnet* can be used to design a function fitting network. The basic syntax is summarized in Table 11.3. The name of the training function, *trfn* in the syntax, can be specified as one of the functions shown in Table 11.4. For example, the Fletcher-Powell conjugate gradient algorithm can be specified as the training algorithm by using "traincgf."

As an example, suppose that the 1×20 input data set x and 1×20 target data set t are given as follows:

```
>> x = [4.99 5.15 5.30 5.45 5.59 5.73 5.89 6.07 6.24 6.39 6.53 6.65 6.74 6.82 6.90
6.97 7.03 7.09 7.15 7.20];
>> t = [7.31 7.43 7.57 7.71 7.85 7.99 8.11 8.18 8.14 8.01 7.76 7.46 7.16 6.84 6.52
6.20 5.89 5.57 5.26 4.95];
```

We now construct and view a function fitting neural network with one hidden layer of size 10. If we want to create a network with three hidden layers and the numbers of elements in the layers are 10, 8, and 5 (that is, the size of the first hidden layer is 10, that of the second hidden layer is 8, and

TABLE 11.3
Syntax of fitnet

Syntax	Description
	Returns a function fitting network with a hidden layer size of hid
net = fitnet(hid)	
	Returns a function fitting network with a hidden layer size of hid and a training
net = fitnet(hid,trfn)	function specified by trfn

TABLE 11.4
Training Functions

Training Function	Algorithm
trainlm	Levenberg-Marquardt (default)
trainbr	Bayesian regularization
trainbfg	BFGS quasi-Newton
trainrp	Resilient back propagation
trainscg	Scaled conjugate gradient
traincgb	Conjugate gradient with Powell-Beale restarts
traincgf	Fletcher-Powell conjugate gradient
traincgp	Polak-Ribiére conjugate gradient
trainoss	One-step secant
traingdx	Variable learning rate gradient descent
traingdm	Gradient descent with momentum
traingd	Gradient descent

that of the third hidden layer is 5), we specify hid =[10,8,5]. The function fitting neural network created is displayed in Figure 11.12.

```
>> mynet = fitnet(10);
>> view(mynet)
```

Example 11.5 Creation of a Multilayer Network

Create a network with four hidden layers. The numbers of nodes of the layers are 4, 7, 8, and 6 for the first, second, third, and fourth hidden layer, respectively.

Solution

A network with four hidden layers can be created by fitnet([a b c d]) where a, b, c, and d represent the number of nodes of each hidden layer. Figure 11.13 shows the network created.

```
>> exnet = fitnet([4 7 8 6]); view(exnet)
```

As indicated in Figure 11.13, the sizes of the input and output are zero. These sizes are adjusted during the training according to the training data.

We use the built-in function *train* to train the network using the training data. The basic syntax is

```
net = train(net,x,t)
[net,tr] = train(net,x,t)
```

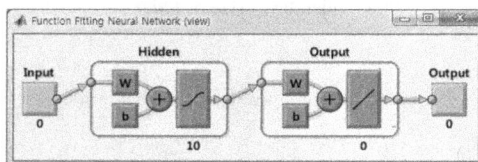

FIGURE 11.12 A function fitting neural network with one hidden layer of size 10.

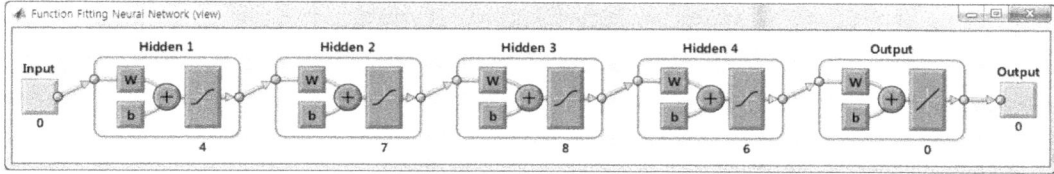

FIGURE 11.13 A network with four hidden layers.

where x is the input data set ($n \times q$ matrix), t is the target data set ($m \times q$ matrix), n is the number of input elements, q is the number of sample data, m is the number of network output elements, and tr is the structure containing the information on the training results. This structure consists of the field for the index on the data classified as training and validation, the field for the number of epochs, and the field for the list on the training status. In our example, we can see that the sizes of the input and output are 1, as shown in Figure 11.14.

```
>> mynet = train(mynet,x,t);
>> view(mynet)
```

We now estimate the targets using the trained network and assess the performance of the network. The default performance function is mean squared error.

```
>> y = mynet(x);
>> pf = perform(mynet,y,t)
pf =
  0.0375
```

The default training algorithm for a function fitting network is the Levenberg-Marquardt (trainlm) algorithm shown in Table 11.4. We now use the Bayesian regularization algorithm (trainbr) and compare the performance results.

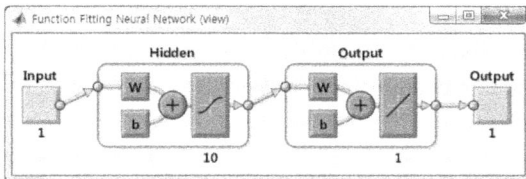

FIGURE 11.14 The trained function fitting neural network.

```
>> mynet = fitnet(10,'trainbr'); mynet = train(mynet,x,t); y = mynet(x);
>> pf = perform(mynet,y,t)
pf =
  6.2013e-05
```

We can see that the Bayesian regularization algorithm improves the performance of the network in terms of estimating the target values.

11.2.2.2 Creation and Training of a Feed-forward Network

Feed-forward neural networks consist of a series of layers. The first layer has a connection from the network input, and each subsequent layer has a connection from the previous layer. The final layer produces the network's output. Feed-forward networks can be used for any kind of input-output mapping. A feed-forward network with one hidden layer containing enough neurons can fit any finite input-output mapping problem.[6] The built-in function *feedforwardnet* is used to construct a feed-forward network. The basic syntax is

```
net = feedforwardnet(hid,trfn)
```

where hid is a row vector of sizes of one or more hidden layers (default value = 10) and trfn is the training, which that can be selected from the list shown in Table 11.4 (default = 'trainlm').

As an example, we use the input and target data sets defined before to construct a feed-forward network. The script *myffnet* creates a feed-forward network, trains it, and assesses its performance.

```
% myffnet.m
x = [4.99 5.15 5.30 5.45 5.59 5.73 5.89 6.07 6.24 6.39 6.53 6.65 6.74 6.82 6.90
6.97 7.03 7.09 7.15 7.20]; % input
t = [7.31 7.43 7.57 7.71 7.85 7.99 8.11 8.18 8.14 8.01 7.76 7.46 7.16 6.84 6.52
6.20 5.89 5.57 5.26 4.95]; % target
ffnet = feedforwardnet(10); % one hidden layer with 10 neurons
ffnet = train(ffnet,x,t); % train the feedforward network
view(ffnet)
y = ffnet(x);
perf = perform(ffnet,y,t)  % assess performance

>> myffnet
perf =
  0.0634
```

Example 11.6   Training and Validation

Create a feed-forward network with one hidden layer containing three nodes using the Levenberg-Marquardt algorithm (trainlm). Train and test the network using the data given in Table 11.5.

Solution

The script *testffnet* produces the desired feed-forward network and plots the results of validation, as shown in Figure 11.15.

```
% testffnet.m
clear all;
x = [0 1.99 0.10 2.28 0.21 2.56 0.32 2.85 0.44 3.13 0.58 3.43 0.74 3.77...
    0.94 4.12 1.21 4.47 1.55 4.83]; % train data: input
t = [5.05 9.47 5.66 8.97 6.32 8.46 6.94 8.01 7.57 7.65 8.21 7.38 8.84...
    7.19 9.43 7.10 9.90 7.10 9.98 7.22]; % train data: target
```

TABLE 11.5
Data Set for Training and Validation

Training Data Input (x)	Target (t)	Testing Data Input (x)	Target (t)
0	5.05	1.85	9.69
1.99	9.47	0.05	5.36
0.10	5.66	2.13	9.23
2.28	8.97	0.16	5.99
0.21	6.32	2.42	8.71
2.56	8.46	0.27	6.63
0.32	6.94	2.70	8.22
2.85	8.01	0.38	7.26
0.44	7.57	2.99	7.82
3.13	7.65	0.51	7.90
0.58	8.21	3.28	7.51
3.43	7.38	0.66	8.52
0.74	8.84	3.59	7.28
3.77	7.19	0.83	9.14
0.94	9.43	3.95	7.13
4.12	7.10	1.07	9.70
1.21	9.90	4.30	7.09
4.47	7.10	1.37	9.99
1.55	9.98	4.65	7.14
4.83	7.22	1.70	9.86

```
xv = [1.85 0.05 2.13 0.16 2.42 0.27 2.70 0.38 2.99 0.51 3.28 0.66 3.59...
   0.83 3.95 1.07 4.30 1.37 4.65 1.70]; % test data: input
tv = [9.69 5.36 9.23 5.99 8.71 6.63 8.22 7.26 7.82 7.90 7.51 8.52 7.28...
   9.14 7.13 9.70 7.09 9.99 7.14 9.86]; % test data: target
samnet = feedforwardnet(3,'trainlm'); % create a network having one hidden
layer with 3 nodes
samnet = train(samnet,x,t); % train the network
ty = samnet(xv); % output from the trained network
plot(xv,tv,'o',xv,ty,'*'),   xlabel('Input'),   ylabel('Target'),   legend
('Target data','Network output')
perf = perform(samnet,ty,tv) % assess performance

>> testffnet
perf =
 1.8428e-04
```

11.2.2.3 Creation and Training of a Cascade Network

Cascade-forward networks are similar to feed-forward networks but have extra connections, including a connection from the input and every previous layer to the following layers. As with feed-forward networks, cascade-forward networks of two (or more) layers can learn any finite input-output function arbitrarily, well given enough hidden neurons. The basic syntax is

```
net = cascadeforwardnet(hid,trfn)
```

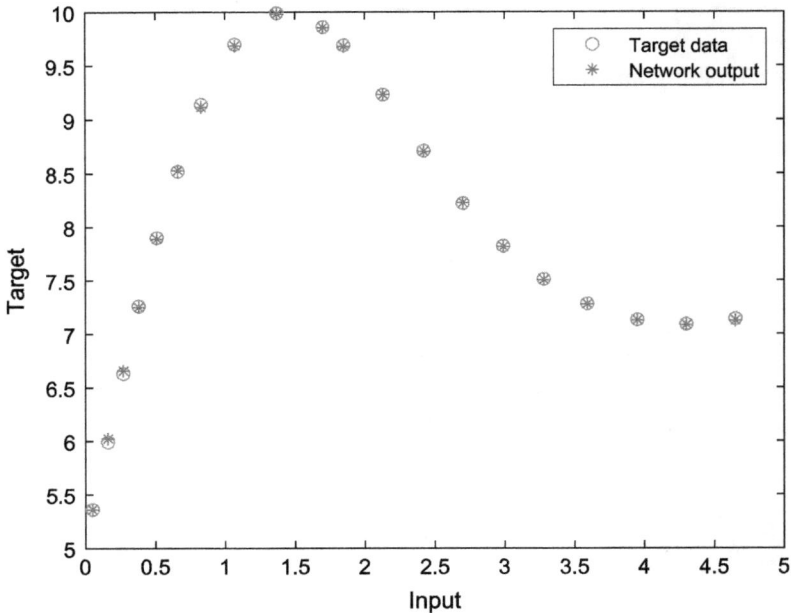

FIGURE 11.15   Results of validation of a feed-forward network with one hidden layer.

where hid is a row vector representing sizes of n hidden layers, trfn is a back propagation training function, and net is a cascade-forward neural network with $n + 1$ layers. The sizes of the input and output layers are set to 0. These sizes will automatically be configured to match particular data by training.

As an example, we use the input and target data sets defined before to construct a cascade-forward network. The script *ccfnet* creates a cascade-forward network, trains it, and assesses its performance. Figure 11.16 shows the cascade-forward network created by the script.

```
% ccfnet.m
x = [4.99 5.15 5.30 5.45 5.59 5.73 5.89 6.07 6.24 6.39 6.53 6.65 6.74 6.82 6.90
6.97 7.03 7.09 7.15 7.20]; % input
t = [7.31 7.43 7.57 7.71 7.85 7.99 8.11 8.18 8.14 8.01 7.76 7.46 7.16 6.84 6.52
6.20 5.89 5.57 5.26 4.95]; % target
cfnet = cascadeforwardnet(10); % one hidden layer with 10 neurons
cfnet = train(cfnet,x,t); % train the cascade-forward network
view(cfnet)
y = cfnet(x);
perf = perform(cfnet,y,t)  % assess performance

>> ccfnet
perf =
  0.0634
```

11.2.3 REGRESSION FITTING

11.2.3.1 Regression Models

The built-in function *regression* can be used to perform regression between targets and outputs. We first create a neural network, train it, and plot the regression between its targets and outputs. The basic syntax of *regression* is $[r,m,b]$ = regression(t,y) where t is the target matrix or

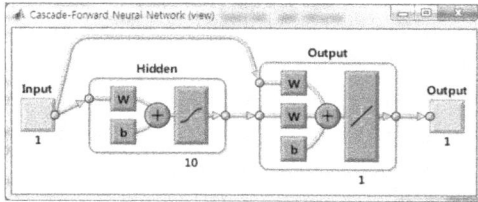

FIGURE 11.16 A cascade-forward network with 10 nodes.

cell array data with n matrix rows, y is the output matrix or cell array data of the same size, r is the regression values for each of the n matrix rows, m is the slope of the regression fit for each of the n matrix rows, and b is the offset of the regression fit for each of the n matrix rows. The built-in function *plotregression* plots the linear regression of targets relative to outputs.

The built-in function *plotfit* plots the output function of a network across the range of the inputs, and also plots targets and output data points associated with input values. The basic syntax is

```
plotfit(net,x,t)
```

The plot appears only for neural networks with one input. If the network has more than one output, only the first output or target appears. The error bars in the plot represent the difference between network outputs and targets.

Example 11.7   Regression Fit

Create a feed-forward network with one hidden layer containing 10 nodes using the Levenberg-Marquardt algorithm (trainlm). Train the network using the training data given in Table 11.6, and calculate and plot the regression between its targets and outputs. Use the built-in function *plotfit* to plot the output function of the network.

Solution

The script *regfit* creates a feed-forward network with one hidden layer containing 10 nodes, trains it using the training data given in Table 11.6, and calculates and plots the regression between its targets and outputs, as shown in Figure 11.17.

```
% regfit.m
clear all;
x = [1.85 0.05 2.13 0.16 2.42 2.56 0.32 0.38 2.99 0.51 3.28 0.66 0.74 3.77...
   0.94 4.12 4.30 1.37 4.65 1.70 0 1.99 0.10 2.28 0.21 0.27 2.70 2.85 0.44...
   3.13 0.58 3.43 3.59 0.83 3.95 1.07 1.21 4.47 1.55 4.83]; % train data: input
t = [9.69 5.36 9.23 5.99 8.71 8.46 6.94 7.26 7.82 7.90 7.51 8.52 8.84 7.19...
   9.43 7.10 7.09 9.99 7.14 9.86 5.05 9.47 5.66 8.97 6.32 6.63 8.22 8.01...
   7.57 7.65 8.21 7.38 7.28 9.14 7.13 9.70 9.90 7.10 9.98 7.22]; % train data:
target
rgnet = feedforwardnet(10,'trainlm'); % create a network having one hidden
layer with 10 nodes
rgnet = train(rgnet,x,t); % train the network
y = rgnet(x); % output from the trained network
[r,m,b] = regression(t,y) % calculate regression between targets and outputs
figure(1), plotregression(t,y) % plot the regression
figure(2), plotfit(rgnet,x,t) % plot the output function

>> regfit
r =
   0.9996
```

TABLE 11.6
Data Set for Training

Input (x)	Target (t)	Input (x)	Target (t)
1.85	9.69	0	5.05
0.05	5.36	1.99	9.47
2.13	9.23	0.10	5.66
0.16	5.99	2.28	8.97
2.42	8.71	0.21	6.32
2.56	8.46	0.27	6.63
0.32	6.94	2.70	8.22
0.38	7.26	2.85	8.01
2.99	7.82	0.44	7.57
0.51	7.90	3.13	7.65
3.28	7.51	0.58	8.21
0.66	8.52	3.43	7.38
0.74	8.84	3.59	7.28
3.77	7.19	0.83	9.14
0.94	9.43	3.95	7.13
4.12	7.10	1.07	9.70
4.30	7.09	1.21	9.90
1.37	9.99	4.47	7.10
4.65	7.14	1.55	9.98
1.70	9.86	4.83	7.22

```
m =
   0.9959
b =
   0.0318
```

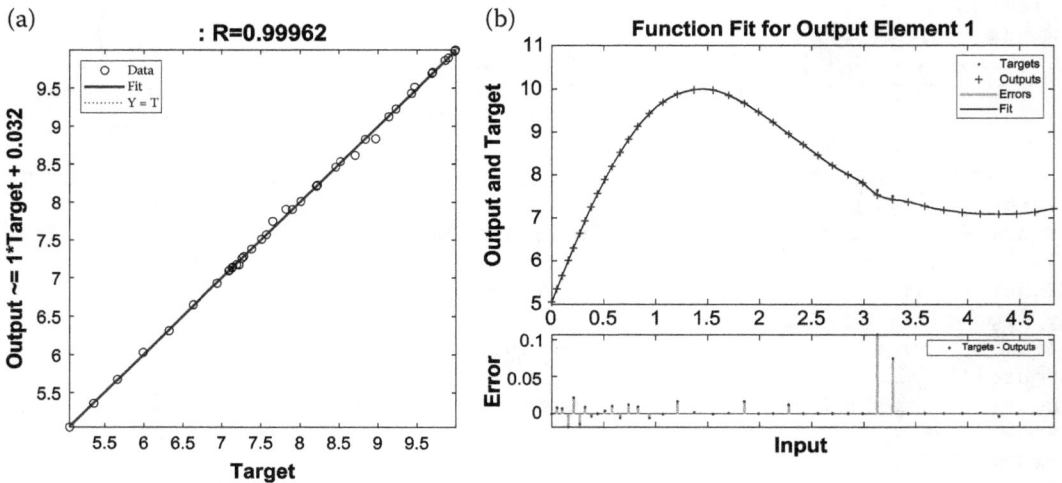

FIGURE 11.17　(a) Regression between targets and outputs; (b) plot of the output function.

11.2.3.2 Data Fitting

Neural networks are very good at function fitting problems. A neural network with enough hidden layers and neurons can fit any data with arbitrary accuracy. Neural networks are particularly well suited for addressing nonlinear problems. Data for function fitting problems are set up for a neural network by organizing the data into two matrices, the input matrix x and the target matrix t.

As an example, we now consider the data obtained from the operation of a paper manufacturing plant. Table 11.7 shows operation data acquired during the operation of the plant. The input variables are Stockflow (stock flow rate, *liter/min*), Talcflow (talc flow rate, *liter/min*), and Pressure (kg/cm^2), and the output variable is Ash (ash content, %). We use a neural network to represent the output variable as a function of input variables.

We first define input and target matrices, followed by creating the function fitting network using the built-in function *fitnet*. The default network for our example is a feed-forward network with the default tan-sigmoid transfer function in the hidden layer and linear transfer function in the output layer. The network has one hidden layer with 10 neurons and one output neuron, because there is only one target value (Ash) associated with each input vector.

```
ppnet = fitnet(10); % one hidden layer with 10 neurons
```

We can randomly divide data for training, validation, and testing. Suppose that the input and target vectors are divided with 70% used for training, 15% for validation, and 15% for testing. The division of data can be performed as follows:

```
ppnet.divideParam.trainRatio = 0.7;
ppnet.divideParam.valRatio = 0.15;
ppnet.divideParam.testRatio = 0.15;
```

Then we can train the network and assess the performance of the network as before.

```
[ppnet,tr] = train(ppnet,x,t);
y = ppnet(x); % output calculation by network
neterr = gsubtract(y,t); % error
perm = perform(ppnet,t,y) % performance
```

The script *nnfitash* performs the whole procedure and produces the data fitting results shown in Figures 11.18 and 11.19.

```
% nnfitash.m
clear all;
xc1 = [0.0015 0.0076 0.0214 0.0273 0.0304 0.0263 0.0131 0.0087 0.0097 0.0069...
    0.0015 0.0143 0.0133 0.0083 0.0047 0.0039 0.0043 0.0032 0.0006 0.0056...
    0.0169 0.0240 0.0208 0.0125 0.0077 0.0086 0.0097 0.0117 0.0127 0.0083...
    0.0123 0.0243 0.0555 0.0619 0.0666 0.0763 0.0796 0.0820 0.0901 0.1051...
    0.1232 0.1313 0.1319 0.1387 0.1531 0.1636 0.1766 0.1940 0.2035 0.2092];
xc2 = [0.0365 0.0363 0.0361 0.0359 0.0357 0.0355 0.0353 0.0351 0.0350 0.0319...
    0.0169 0.0121 0.0119 0.0160 0.0227 0.0226 0.0224 0.0222 0.0220 0.0218...
    0.0216 0.0214 0.0192 0.0137 0.0149 0.0225 0.0224 0.0195 0.0120 0.0045...
    0.0009 0.0013 0.0024 0.0027 0.0030 0.0033 0.0036 0.0038 0.0041 0.0044...
    0.0046 0.0049 0.0052 0.0055 0.0057 0.0060 0.0063 0.0065 0.0068 0.0071];
xc3 = [0.0003 0.0007 0.0010 0.0013 0.0017 0.0020 0.0024 0.0027 0.0031 0.0034...
    0.0041 0.0044 0.0048 0.0051 0.0054 0.0058 0.0061 0.0065 0.0068 0.0072...
    0.0075 0.0078 0.0082 0.0085 0.0089 0.0092 0.0095 0.0099 0.0102 0.0106...
    0.0109 0.0112 0.0119 0.0123 0.0126 0.0129 0.0133 0.0136 0.0140 0.0143...
```

TABLE 11.7
Paper Plant Operation Data

Stockflow (*liter/min*)	Talcflow (*liter/min*)	Pressure (kg/cm^2)	Ash (%)
0.0015	0.0365	0.0003	0.1113
0.0076	0.0363	0.0007	0.1131
0.0214	0.0361	0.0010	0.1173
0.0273	0.0359	0.0013	0.1182
0.0304	0.0357	0.0017	0.1169
0.0263	0.0355	0.0020	0.1190
0.0131	0.0353	0.0024	0.1227
0.0087	0.0351	0.0027	0.1257
0.0097	0.0350	0.0031	0.1327
0.0069	0.0319	0.0034	0.1350
0.0015	0.0169	0.0041	0.1346
0.0143	0.0121	0.0044	0.1327
0.0133	0.0119	0.0048	0.1308
0.0083	0.0160	0.0051	0.1302
0.0047	0.0227	0.0054	0.1310
0.0039	0.0226	0.0058	0.1324
0.0043	0.0224	0.0061	0.1330
0.0032	0.0222	0.0065	0.1318
0.0006	0.0220	0.0068	0.1316
0.0056	0.0218	0.0072	0.1320
0.0169	0.0216	0.0075	0.1305
0.0240	0.0214	0.0078	0.1306
0.0208	0.0192	0.0082	0.1277
0.0125	0.0137	0.0085	0.1254
0.0077	0.0149	0.0089	0.1256
0.0086	0.0225	0.0092	0.1230
0.0097	0.0224	0.0095	0.1225
0.0117	0.0195	0.0099	0.1207
0.0127	0.0120	0.0102	0.1223
0.0083	0.0045	0.0106	0.1279
0.0123	0.0009	0.0109	0.1316
0.0243	0.0013	0.0112	0.1366
0.0555	0.0024	0.0119	0.1373
0.0619	0.0027	0.0123	0.1379
0.0666	0.0030	0.0126	0.1369
0.0763	0.0033	0.0129	0.1374
0.0796	0.0036	0.0133	0.1331
0.0820	0.0038	0.0136	0.1303
0.0901	0.0041	0.0140	0.1239
0.1051	0.0044	0.0143	0.1174
0.1232	0.0046	0.0147	0.1135
0.1313	0.0049	0.0150	0.1120
0.1319	0.0052	0.0153	0.1095
0.1387	0.0055	0.0157	0.1078
0.1531	0.0057	0.0161	0.1050

(*continued*)

TABLE 11.7 (continued)
Paper Plant Operation Data

Stockflow (*liter/min*)	Talcflow (*liter/min*)	Pressure (*kg/cm²*)	Ash (%)
0.1636	0.0060	0.0160	0.1016
0.1766	0.0063	0.0167	0.0964
0.1940	0.0065	0.0170	0.0940
0.2035	0.0068	0.0174	0.0885
0.2092	0.0071	0.0177	0.0871

```
   0.0147 0.0150 0.0153 0.0157 0.0161 0.0160 0.0167 0.0170 0.0174 0.0177];
t = [0.1113 0.1131 0.1173 0.1182 0.1169 0.1190 0.1227 0.1257 0.1327 0.1350...
   0.1346 0.1327 0.1308 0.1302 0.1310 0.1324 0.1330 0.1318 0.1316 0.1320...
   0.1305 0.1306 0.1277 0.1254 0.1256 0.1230 0.1225 0.1207 0.1223 0.1279...
```

(a)

(b)

(c)

(d)

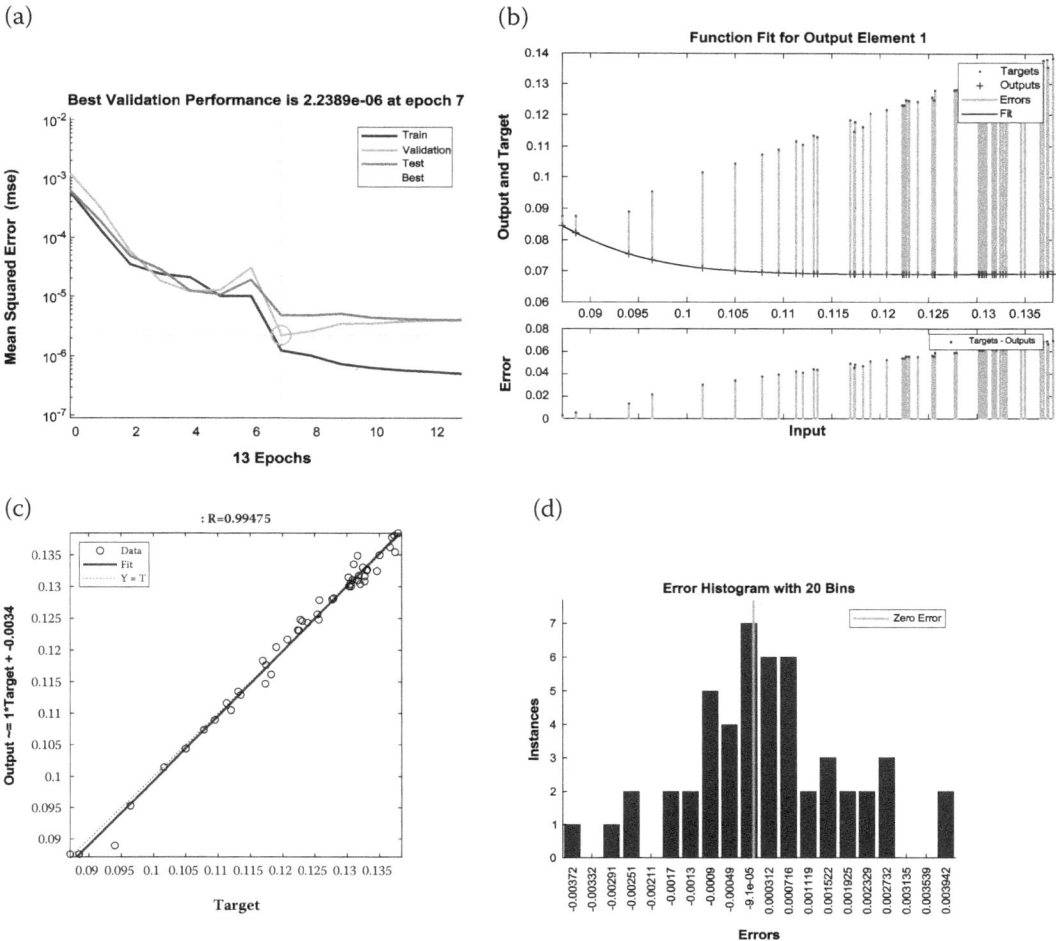

FIGURE 11.18 Data fitting by network: (a) mean squared error, (b) function fit, (c) target output, (d) error histogram.

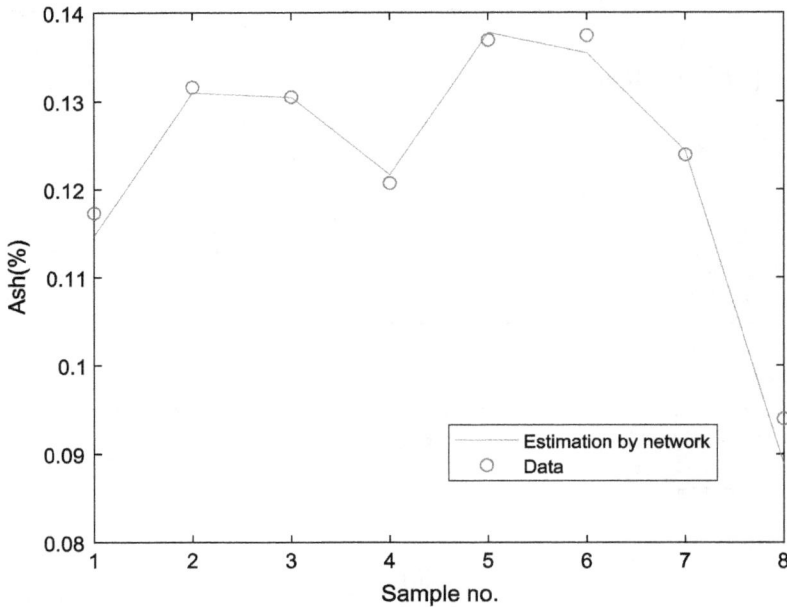

FIGURE 11.19 Data fitting by the function fitting network.

```
   0.1316 0.1366 0.1373 0.1379 0.1369 0.1374 0.1331 0.1303 0.1239 0.1174...
   0.1135 0.1120 0.1095 0.1078 0.1050 0.1016 0.0964 0.0940 0.0885 0.0871];
x = [xc1; xc2; xc3]; % input data matrix
ppnet = fitnet(10); % one hidden layer with 10 neurons
ppnet.divideParam.trainRatio = 0.7; % use 70% for training
ppnet.divideParam.valRatio = 0.15; % use 15% for validation
ppnet.divideParam.testRatio = 0.15; % use 15% for testing
[ppnet,tr] = train(ppnet,x,t);
y = ppnet(x); % output calculation by network
neterr = gsubtract(y,t); % error
perm = perform(ppnet,t,y) % performance
view(ppnet)
figure(1), plotperform(tr)
figure(2), plotfit(ppnet,t,y)
figure(3), plotregression(t,y)
figure(4), ploterrhist(neterr)
% Compare data and network output
testx = x(:,tr.testInd); % input data set for test
testt = t(:,tr.testInd); % target data set for test
testy = ppnet(testx); % network output for test input data
k = 1:length(tr.testInd);
figure(5), plot(k,testy,k,testt,'o')
xlabel('Sample no.'),ylabel('Ash(%)'),legend('Estimation by
network','Data','Location','best')
```

Example 11.8 Estimation of Moisture Content

Table 11.8 shows operation data acquired during the operation of a paper manufacturing plant. The input variables are Stockflow (stock flow rate, *liter/min*), Talcflow (talc flow rate, *liter/min*), and Pressure (kg/cm^2), and the output variable is Moisture (moisture content, %). Create a function fitting neural network, train it, and plot test data and moisture contents estimated by the network.

TABLE 11.8
Paper Plant Operation Data

Stockflow (*liter/min*)	Talcflow (*liter/min*)	Pressure (kg/cm^2)	Moisture (%)
0.0015	0.0365	0.0003	0.0924
0.0076	0.0363	0.0007	0.1665
0.0214	0.0361	0.0010	0.2198
0.0273	0.0359	0.0013	0.2121
0.0304	0.0357	0.0017	0.2093
0.0263	0.0355	0.0020	0.2271
0.0131	0.0353	0.0024	0.2195
0.0087	0.0351	0.0027	0.2074
0.0097	0.0350	0.0031	0.1973
0.0069	0.0319	0.0034	0.2503
0.0015	0.0169	0.0041	0.2534
0.0143	0.0121	0.0044	0.2769
0.0133	0.0119	0.0048	0.2854
0.0083	0.0160	0.0051	0.2886
0.0047	0.0227	0.0054	0.2897
0.0039	0.0226	0.0058	0.2924
0.0043	0.0224	0.0061	0.2818
0.0032	0.0222	0.0065	0.3950
0.0006	0.0220	0.0068	0.3943
0.0056	0.0218	0.0072	0.3899
0.0169	0.0216	0.0075	0.3875
0.0240	0.0214	0.0078	0.3863
0.0208	0.0192	0.0082	0.3849
0.0125	0.0137	0.0085	0.3872
0.0077	0.0149	0.0089	0.3879
0.0086	0.0225	0.0092	0.3913
0.0097	0.0224	0.0095	0.3963
0.0117	0.0195	0.0099	0.4031
0.0127	0.0120	0.0102	0.4054
0.0083	0.0045	0.0106	0.3968
0.0123	0.0009	0.0109	0.3780
0.0243	0.0013	0.0112	0.3733
0.0555	0.0024	0.0119	0.3732
0.0619	0.0027	0.0123	0.3828
0.0666	0.0030	0.0126	0.3916
0.0763	0.0033	0.0129	0.4053
0.0796	0.0036	0.0133	0.4149
0.0820	0.0038	0.0136	0.4196
0.0901	0.0041	0.0140	0.4207
0.1051	0.0044	0.0143	0.4237
0.1232	0.0046	0.0147	0.4193
0.1313	0.0049	0.0150	0.4007
0.1319	0.0052	0.0153	0.3919
0.1387	0.0055	0.0157	0.3726
0.1531	0.0057	0.0161	0.3670

(continued)

TABLE 11.8 (continued)
Paper Plant Operation Data

Stockflow (*liter/min*)	Talcflow (*liter/min*)	Pressure (*kg/cm²*)	Moisture (%)
0.1636	0.0060	0.0160	0.3785
0.1766	0.0063	0.0167	0.3822
0.1940	0.0065	0.0170	0.3932
0.2035	0.0068	0.0174	0.4069
0.2092	0.0071	0.0177	0.4109

The network should have one hidden layer with eight neurons. Divide the data so that 60% is used for training, 20% is used for validation, and 20% is used for testing.

Solution

The script *nnfitmoist* defines data set, creates and trains the function fitting network, and generates the plot of the data and the moisture values estimated by the network, as shown in Figure 11.20.

```
% nnfitmoist.m
clear all;
xc1 = [0.0015 0.0076 0.0214 0.0273 0.0304 0.0263 0.0131 0.0087 0.0097 0.0069...
  0.0015 0.0143 0.0133 0.0083 0.0047 0.0039 0.0043 0.0032 0.0006 0.0056...
  0.0169 0.0240 0.0208 0.0125 0.0077 0.0086 0.0097 0.0117 0.0127 0.0083...
  0.0123 0.0243 0.0555 0.0619 0.0666 0.0763 0.0796 0.0820 0.0901 0.1051...
  0.1232 0.1313 0.1319 0.1387 0.1531 0.1636 0.1766 0.1940 0.2035 0.2092];
xc2 = [0.0365 0.0363 0.0361 0.0359 0.0357 0.0355 0.0353 0.0351 0.0350 0.0319...
  0.0169 0.0121 0.0119 0.0160 0.0227 0.0226 0.0224 0.0222 0.0220 0.0218...
  0.0216 0.0214 0.0192 0.0137 0.0149 0.0225 0.0224 0.0195 0.0120 0.0045...
```

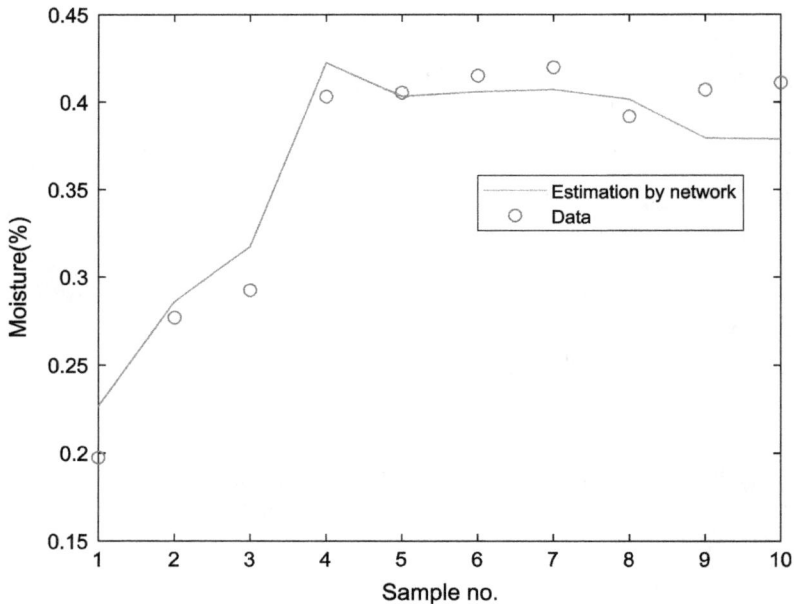

FIGURE 11.20 Moisture estimation by the network.

```
      0.0009 0.0013 0.0024 0.0027 0.0030 0.0033 0.0036 0.0038 0.0041 0.0044...
      0.0046 0.0049 0.0052 0.0055 0.0057 0.0060 0.0063 0.0065 0.0068 0.0071];
xc3 = [0.0003 0.0007 0.0010 0.0013 0.0017 0.0020 0.0024 0.0027 0.0031 0.0034...
      0.0041 0.0044 0.0048 0.0051 0.0054 0.0058 0.0061 0.0065 0.0068 0.0072...
      0.0075 0.0078 0.0082 0.0085 0.0089 0.0092 0.0095 0.0099 0.0102 0.0106...
      0.0109 0.0112 0.0119 0.0123 0.0126 0.0129 0.0133 0.0136 0.0140 0.0143...
      0.0147 0.0150 0.0153 0.0157 0.0161 0.0160 0.0167 0.0170 0.0174 0.0177];
t = [0.0924 0.1665 0.2198 0.2121 0.2093 0.2271 0.2195 0.2074 0.1973 0.2503...
      0.2534 0.2769 0.2854 0.2886 0.2897 0.2924 0.2818 0.3950 0.3943 0.3899...
      0.3875 0.3863 0.3849 0.3872 0.3879 0.3913 0.3963 0.4031 0.4054 0.3968...
      0.3780 0.3733 0.3732 0.3828 0.3916 0.4053 0.4149 0.4196 0.4207 0.4237...
      0.4193 0.4007 0.3919 0.3726 0.3670 0.3785 0.3822 0.3932 0.4069 0.4109];
x = [xc1; xc2; xc3]; % input data matrix
ppnet = fitnet(8); % one hidden layer with 8 neurons
ppnet.divideParam.trainRatio = 0.6;
ppnet.divideParam.valRatio = 0.2;
ppnet.divideParam.testRatio = 0.2;
[ppnet,tr] = train(ppnet,x,t);
y = ppnet(x); % output calculation by network
neterr = gsubtract(y,t); % error
perm = perform(ppnet,t,y) % performance
% Compare data and network output
testx = x(:,tr.testInd); % input data set for test
testt = t(:,tr.testInd); % target data set for test
testy = ppnet(testx); % network output for test input data
k = 1:length(tr.testInd);
plot(k,testy,k,testt,'o'), xlabel('Sample no.'),ylabel('Moisture(%)')
legend('Estimation by network','Data','Location','best')
```

11.3 SUPPORT VECTOR MACHINES

11.3.1 FUNDAMENTALS OF SUPPORT VECTOR MACHINES

Support vector machines (SVMs), introduced by Vapnik et al., are a modeling technique that produces mathematical expressions based on input and output data.[7] SVMs were initially related to the classification of data. Recently, they have been extended to solve nonlinear regression problems by introducing an alternative loss function.[8] The SVM for regression is formulated to solve a convex optimization problem usually known as quadratic programming. The basic idea of support vector regression is to find a function that can agree with the training data with the smallest predicting error.[9] However, SVMs have the drawback of higher computational load due to the computation of the constrained optimization problem. This drawback could be overcome by the introduction of a least-squares SVM (LSSVM). In an LSSVM, the inequality constraints are transformed into equality constraints and the objective function consists of a sum of squared error, as in the training of a typical neural network. With this reformulation, the calculation procedure becomes very simple and the solution is obtained as a linear system. Iterative methods including the conjugate gradient algorithm can be effectively used to solve the linear system.

LSSVMs have enjoyed successful application in many areas of nonlinear modeling due to their good generalization and lower computational burden. Given a certain individual learning subset, $L_k (k = 1, ..., T)$ which consists of data samples $\{x_i, y_i\}_{i=1}^n$, an LSSVM regression model is transformed into an optimization programming model as follows:

$$\min J(\omega, \xi) = \frac{1}{2}\omega^T\omega + \frac{1}{2}\gamma \sum_{i=1}^n \xi_i^2$$

$$\text{Subject to } y_i = \omega^T \varphi(x_i) + b + \xi_i (i = 1, \ldots, n)$$

where

φ is a nonlinear function that maps the input data into a higher-dimensional subspace

b is the bias

ω is the weight vector

$\xi = [\xi_1, \ldots, \xi_n]^T$ is the error variable vector

γ is the penalty factor

If the data are noisy, a smaller γ is used to prevent overfitting. The well-known Lagrange function and Karush-Kuhn-Tucker conditions are employed in the solution of the optimization problem. The resultant LSSVM model for function estimation can be represented as follows:

$$y(x) = \sum_{i=1}^{n} \alpha_i K(x, x_i) + b$$

where

$\alpha = [\alpha_1, \ldots, \alpha_n]^T$ is the Lagrange multiplier vector

K is the kernel function used to substitute the mapping procedure and avoid calculation of φ

Usually, the kernel K is a symmetric function satisfying Mercer's theorem. The following Gaussian radial basis function is commonly used as the kernel function:

$$K(x, x_i) = \exp(-\|x - x_i\|^2 / \sigma^2)$$

where σ is a tuning kernel parameter.

Statistics and Machine Learning Toolbox™ provides supervised and unsupervised machine learning algorithms including SVMs. The built-in function *fitrsvm* trains or validates an SVM regression model on a low- to moderate-dimensional data set, while *fitrlinear* trains a linear SVM regression model on a high-dimensional data set.

11.3.2 CREATION OF REGRESSION MODELS USING *FITRSVM*

The built-in function *fitrsvm* trains an SVM regression model on a low- to moderate-dimensional data set and uses the quadratic programming method to minimize the cross-validation objective function. The basic syntax is summarized in Table 11.9.

As an example, consider the paper plant operation data given in Table 11.8. The function *datgen* generates the data set to be used in the development of an SVM regression model.

```
function [X Y] = datgen % datgen.m: generate data set for SVM regression practices
xc1 = [0.0015 0.0076 0.0214 0.0273 0.0304 0.0263 0.0131 0.0087 0.0097 0.0069...
    0.0015 0.0143 0.0133 0.0083 0.0047 0.0039 0.0043 0.0032 0.0006 0.0056...
    0.0169 0.0240 0.0208 0.0125 0.0077 0.0086 0.0097 0.0117 0.0127 0.0083...
    0.0123 0.0243 0.0555 0.0619 0.0666 0.0763 0.0796 0.0820 0.0901 0.1051...
    0.1232 0.1313 0.1319 0.1387 0.1531 0.1636 0.1766 0.1940 0.2035 0.2092];
xc2 = [0.0365 0.0363 0.0361 0.0359 0.0357 0.0355 0.0353 0.0351 0.0350 0.0319...
    0.0169 0.0121 0.0119 0.0160 0.0227 0.0226 0.0224 0.0222 0.0220 0.0218...
    0.0216 0.0214 0.0192 0.0137 0.0149 0.0225 0.0224 0.0195 0.0120 0.0045...
    0.0009 0.0013 0.0024 0.0027 0.0030 0.0033 0.0036 0.0038 0.0041 0.0044...
    0.0046 0.0049 0.0052 0.0055 0.0057 0.0060 0.0063 0.0065 0.0068 0.0071];
xc3 = [0.0003 0.0007 0.0010 0.0013 0.0017 0.0020 0.0024 0.0027 0.0031 0.0034...
    0.0041 0.0044 0.0048 0.0051 0.0054 0.0058 0.0061 0.0065 0.0068 0.0072...
    0.0075 0.0078 0.0082 0.0085 0.0089 0.0092 0.0095 0.0099 0.0102 0.0106...
    0.0109 0.0112 0.0119 0.0123 0.0126 0.0129 0.0133 0.0136 0.0140 0.0143...
    0.0147 0.0150 0.0153 0.0157 0.0161 0.0160 0.0167 0.0170 0.0174 0.0177];
```

TABLE 11.9
Basic Syntax of *fitrsvm*

Syntax	Description
modl = fitrsvm(Tbl,ResponseVarName)	Creates an SVM regression model using the predictor values in the table Tbl and the response values in Tbl.ResponseVarName.
modl = fitrsvm(Tbl,formula)	Creates an SVM regression model using the predictor values in the table Tbl. formula is an explanatory model of the response and a subset of predictor variables in Tbl used to fit modl.
modl = fitrsvm(Tbl,Y)	Creates an SVM regression model trained using the predictor values in the table Tbl and the response values in the vector Y.
modl = fitrsvm(X,Y)	Creates an SVM regression model trained using the predictor values in the matrix X and the response values in the vector Y.

```
t = [0.0924 0.1665 0.2198 0.2121 0.2093 0.2271 0.2195 0.2074 0.1973 0.2503...
   0.2534 0.2769 0.2854 0.2886 0.2897 0.2924 0.2818 0.3950 0.3943 0.3899...
   0.3875 0.3863 0.3849 0.3872 0.3879 0.3913 0.3963 0.4031 0.4054 0.3968...
   0.3780 0.3733 0.3732 0.3828 0.3916 0.4053 0.4149 0.4196 0.4207 0.4237...
   0.4193 0.4007 0.3919 0.3726 0.3670 0.3785 0.3822 0.3932 0.4069 0.4109];
X = [xc1' xc2' xc3']; % input data matrix
Y = t'; % response (output) vector
end
```

The SVM regression model in the following is trained using the predictor values in the matrix X and the response values in the vector Y generated by the function *datgen*.

```
>> Tbl = datgen; X = Tbl(:,1:3); Y = Tbl(:,4); modl = fitrsvm(X,Y)
modl =
RegressionSVM
          ResponseName: 'Y'
  CategoricalPredictors: []
    ResponseTransform: 'none'
              Alpha: [42×1 double]
               Bias: 0.3733
     KernelParameters: [1×1 struct]
     NumObservations: 50
      BoxConstraints: [50×1 double]
      ConvergenceInfo: [1×1 struct]
      IsSupportVector: [50×1 logical]
             Solver: 'SMO'
Properties, Methods
>> modl.ConvergenceInfo.Converged
ans =
 logical
 1
```

We can see that the value of the *Converged* field of the structure *modl.ConvergenceInfo* is 1, which means that convergence is achieved. If this value is 0, we can standardize and retrain the data set by using the Standardize option, as follows:

```
>> modl = fitrsvm(X,Y,'Standardize',true)
```

For the purpose of cross validation, we generate two SVM regression models using five-fold cross validation. We use the default linear kernel for one model (linearmodl) and the Gaussian kernel function for the other model (gaussmodl). The options for kernel functions are shown in Table 11.10.

```
>> Tbl = datgen; X = Tbl(:,1:3); Y = Tbl(:,4);
>> linearmodl = fitrsvm(X,Y, 'Standardize',true,'KFold',5);
>> gaussmodl = fitrsvm(X,Y, 'Standardize',true,'KFold',5, 'KernelFunction',
'gaussian');
```

The built-in function *kfoldLoss* gives the generalization error (mean squared error) for each model as follows:

```
>> mselinearmodl = kfoldLoss(linearmodl), msegaussmodl = kfoldLoss(gaussmodl)
mselinearmodl =
  0.0010
msegaussmodl =
  7.3967e-04
```

We can see that the SVM regression model using the Gaussian kernel performs better than the one using the linear kernel function.

We use the built-in function *predict* to predict responses using an SVM regression model. The basic syntax is

```
Y = predict(modl, X)
```

The script *estmoist* estimates output values using the SVM regression model. The script generates the data set and rearranges it using the *table* command. Suitable names can be assigned to each data column by using the VariableNames option. The whole data set can be divided into two parts: one for training and the other for testing. When 70% of the whole data set is to be used in training, we set the value of Holdout to 0.3 in the built-in *cvpartition* function (which indicates that 30% of the whole data will be used in testing). Results of estimation are shown in Figure 11.21.

```
% estmoist.m: predict output by SVM regression model
clear all; D = datgen; % generate data set
[n,m] = size(D); % data set size (n: number of rows, m: number of columns)
X = [D(:,1) D(:,2) D(:,3)]; Y = D(:,4);
rng(10); % specify positive integer seed value
testdat = cvpartition(n, 'Holdout',0.3); % hold out 30% of the data for testing
indtrain = training(testdat); % training set indices
indtest = test(testdat); % test set indices
modl = fitrsvm(X(indtrain,:),Y(indtrain),'Standardize',true); % generation of
regression model
```

TABLE 11.10

Options for Kernel Functions

Option	Kernel Function	Function Definition
gaussian or rbf	Gaussian or radial basis function (RBF)	$F(x_j, x_k) = e^{-\lVert x_j - x_k \rVert^2}$
linear	Linear	$F(x_j, x_k) = x_j^T x_k$
polynomial	Polynomial	$F(x_j, x_k) = (1 + x_j^T x_k)^q$

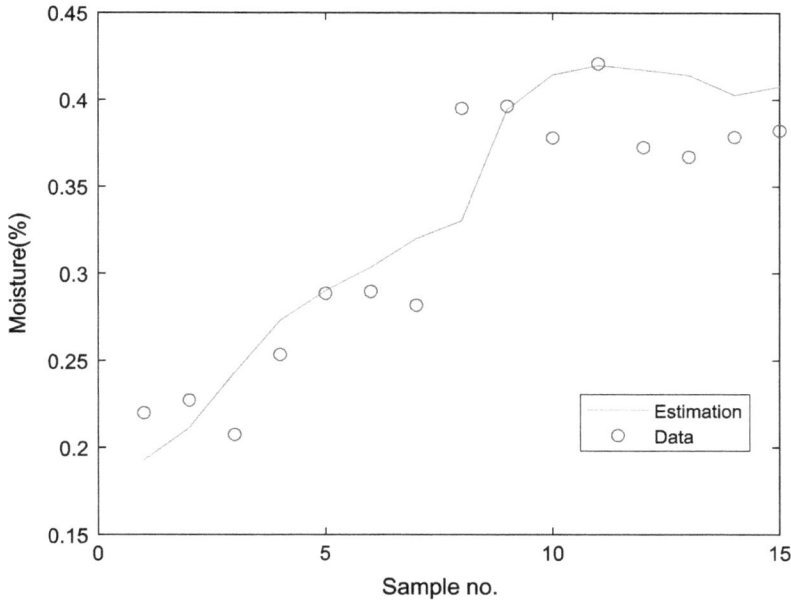

FIGURE 11.21 Moisture estimation by the SVM regression model.

```
yp = predict(modl,X(indtest,:)); % predict outputs
yd = D(indtest,4); k = 1:length(yd);
plot(k,yp,k,yd,'o')  % plot data and estimation results
xlabel('Sample no.'),ylabel('Moisture(%)'),legend
('Estimation','Data','Location','best')
```

Example 11.9 Estimation of Ash Content

Consider the paper plant operation data shown in Table 11.7. Construct the SVM regression model and predict the Ash content by the model. 75% of the whole data set should be used in training and the remaining 25% in testing. Produce the plot representing the estimation results with data points.

Solution

The script *estash* defines the data set, creates and trains the SVM regression model, predicts the ash content, and produces the plot showing the estimation results and data shown in Figure 11.22.

```
% estash.m: predict ash content by SVM regression model
clear all;
xc1 = [0.0015 0.0076 0.0214 0.0273 0.0304 0.0263 0.0131 0.0087 0.0097 0.0069...
  0.0015 0.0143 0.0133 0.0083 0.0047 0.0039 0.0043 0.0032 0.0006 0.0056...
  0.0169 0.0240 0.0208 0.0125 0.0077 0.0086 0.0097 0.0117 0.0127 0.0083...
  0.0123 0.0243 0.0555 0.0619 0.0666 0.0763 0.0796 0.0820 0.0901 0.1051...
  0.1232 0.1313 0.1319 0.1387 0.1531 0.1636 0.1766 0.1940 0.2035 0.2092];
xc2 = [0.0365 0.0363 0.0361 0.0359 0.0357 0.0355 0.0353 0.0351 0.0350 0.0319...
  0.0169 0.0121 0.0119 0.0160 0.0227 0.0226 0.0224 0.0222 0.0220 0.0218...
  0.0216 0.0214 0.0192 0.0137 0.0149 0.0225 0.0224 0.0195 0.0120 0.0045...
  0.0009 0.0013 0.0024 0.0027 0.0030 0.0033 0.0036 0.0038 0.0041 0.0044...
  0.0046 0.0049 0.0052 0.0055 0.0057 0.0060 0.0063 0.0065 0.0068 0.0071];
xc3 = [0.0003 0.0007 0.0010 0.0013 0.0017 0.0020 0.0024 0.0027 0.0031 0.0034...
  0.0041 0.0044 0.0048 0.0051 0.0054 0.0058 0.0061 0.0065 0.0068 0.0072...
```

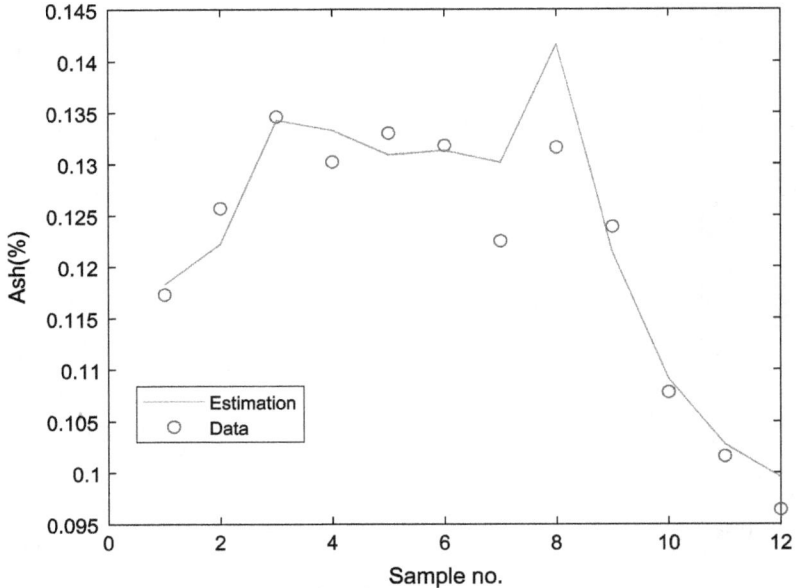

FIGURE 11.22   Estimation of ash content by the SVM regression model.

```
   0.0075 0.0078 0.0082 0.0085 0.0089 0.0092 0.0095 0.0099 0.0102 0.0106...
   0.0109 0.0112 0.0119 0.0123 0.0126 0.0129 0.0133 0.0136 0.0140 0.0143...
   0.0147 0.0150 0.0153 0.0157 0.0161 0.0160 0.0167 0.0170 0.0174 0.0177];
t = [0.1113 0.1131 0.1173 0.1182 0.1169 0.1190 0.1227 0.1257 0.1327 0.1350...
   0.1346 0.1327 0.1308 0.1302 0.1310 0.1324 0.1330 0.1318 0.1316 0.1320...
   0.1305 0.1306 0.1277 0.1254 0.1256 0.1230 0.1225 0.1207 0.1223 0.1279...
   0.1316 0.1366 0.1373 0.1379 0.1369 0.1374 0.1331 0.1303 0.1239 0.1174...
   0.1135 0.1120 0.1095 0.1078 0.1050 0.1016 0.0964 0.0940 0.0885 0.0871];
X = [xc1' xc2' xc3']; Y = t'; % define input and output(ash) data
[n,m] = size(X); % data set size (n: number of data samples)
rng(10); % specify positive integer seed value
testdat = cvpartition(n,'Holdout',0.25); % hold out 25% of the data for testing
indtrain = training(testdat); % training set indices
indtest = test(testdat); % test set indices
modl = fitrsvm(X(indtrain,:),Y(indtrain),'Standardize',true); % generate of
regression model
yp = predict(modl,X(indtest,:)); % predict outputs
yd = Y(indtest); k = 1:length(yd);
plot(k,yp,k,yd,'o')  % plot data and estimation results
xlabel('Sample
no.'),ylabel('Ash(%)'),legend('Estimation','Data','Location','best')
```

11.3.3 Creation of Regression Models Using *fitrlinear*

The built-in function *fitrlinear* trains linear regression models with high-dimensional, full or sparse predictor data. Available linear regression models include regularized SVM and least-squares regression methods. This function minimizes the objective function using schemes that reduce computing time. The basic syntax is shown in Table 11.11.

As an input data set, we consider a 10000×1000 sparse matrix $X = [x_1, x_2, \cdots, x_{1000}]$. Assume that 10% of all the elements of X are nonzero. Suppose that the output Y is defined by

TABLE 11.11

Basic Syntax of *fitrlinear*

Syntax	Description
modl = fitrlinear(X,Y)	Creates a trained regression model object that contains the results of fitting an SVM regression model to the predictor X and response Y.
modl = fitrlinear(X,Y,Name,Value)	Creates a trained linear regression model with additional options specified by one or more pairs of Name and Value.
[modl,FitInfo] = fitrlinear(_)	Returns optimization details using any one of the previous syntaxes.

$$Y = x_{100} + 2x_{200} + \epsilon$$

where ϵ is a vector of random normalized error with mean 0 and standard deviation 0.3. We can create a normal distributed sparse random matrix by using the built-in function *sprandn*. The SVM regression model can be constructed by the function *fitrlinear*. When 30% of the sample data set is used in validation, we set the 'Holdout option' to 0.3 as before.

```
>> n = 1e4; m = 1e3;  % dimension of input data set X
>> nz = 0.1; rng(1); X = sprandn(n, m, nz); Y = X(:,100) + 2*X(:,200) +
0.3*randn(n,1);
>> Tmd = fitrlinear(X,Y, 'Holdout',0.3); % create SVM regression model (30% of
data are used in testing)
```

The object *Tmd* contains the property *Trained*, which is a 1×1 cell array holding a regression linear model trained using the training data set. Now we predict the output values using the built-in function *predict*.

```
>> modl = Tmd.Trained{1};
>> indtrain = training(Tmd.Partition); indtest = test(Tmd.Partition);
>> ytrain = predict(modl,X(indtrain,:)); ytest = predict(modl,X(indtest,:));
```

We plot the estimation results and data values as shown in Figure 11.23.

```
>> y = Y(indtest); % data
>> k = 1: length(y); plot(k,ytest,k,y,'o'), xlabel('Sample no.'),ylabel
('Y'),legend('Estimation','Data')
```

Example 11.10   Regression of Ash Content

Consider the paper plant operation data shown in Table 11.7. Construct the SVM regression model by using the function *fitrlinear* and predict the ash content by the model. 75% of the whole data set should be used in training and the remaining 25% in testing. Produce the plot representing the estimation results with data points.

Solution

The script *regash* defines the data set, creates the SVM regression model using *fitrlinear*, trains the model, predicts the ash content, and produces the plot showing the estimation results and data shown in Figure 11.24.

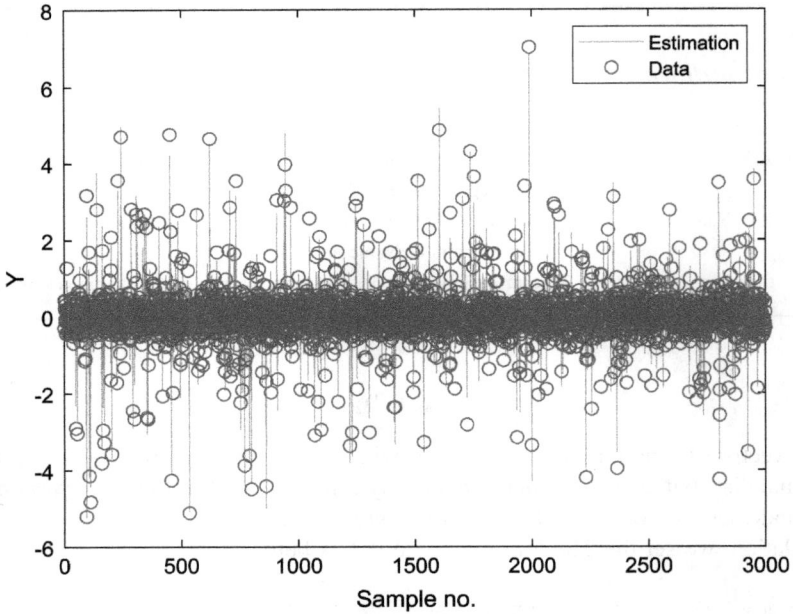

FIGURE 11.23 Estimation of output values by the SVM regression model.

```
% regash.m: predict ash content by SVM regression model
clear all;
xc1 = [0.0015 0.0076 0.0214 0.0273 0.0304 0.0263 0.0131 0.0087 0.0097 0.0069...
    0.0015 0.0143 0.0133 0.0083 0.0047 0.0039 0.0043 0.0032 0.0006 0.0056...
```

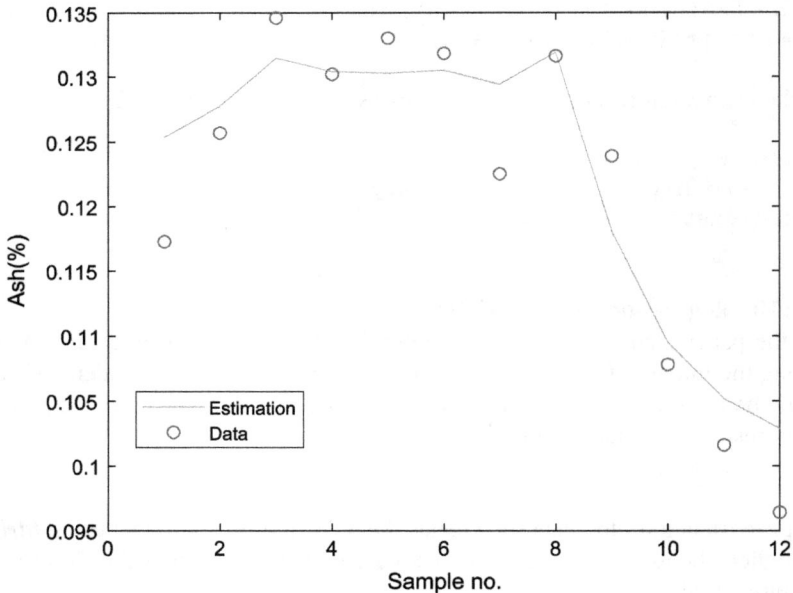

FIGURE 11.24 Estimation of ash content by the regression model.

```
      0.0169 0.0240 0.0208 0.0125 0.0077 0.0086 0.0097 0.0117 0.0127 0.0083...
      0.0123 0.0243 0.0555 0.0619 0.0666 0.0763 0.0796 0.0820 0.0901 0.1051...
      0.1232 0.1313 0.1319 0.1387 0.1531 0.1636 0.1766 0.1940 0.2035 0.2092];
xc2 = [0.0365 0.0363 0.0361 0.0359 0.0357 0.0355 0.0353 0.0351 0.0350 0.0319...
      0.0169 0.0121 0.0119 0.0160 0.0227 0.0226 0.0224 0.0222 0.0220 0.0218...
      0.0216 0.0214 0.0192 0.0137 0.0149 0.0225 0.0224 0.0195 0.0120 0.0045...
      0.0009 0.0013 0.0024 0.0027 0.0030 0.0033 0.0036 0.0038 0.0041 0.0044...
      0.0046 0.0049 0.0052 0.0055 0.0057 0.0060 0.0063 0.0065 0.0068 0.0071];
xc3 = [0.0003 0.0007 0.0010 0.0013 0.0017 0.0020 0.0024 0.0027 0.0031 0.0034...
      0.0041 0.0044 0.0048 0.0051 0.0054 0.0058 0.0061 0.0065 0.0068 0.0072...
      0.0075 0.0078 0.0082 0.0085 0.0089 0.0092 0.0095 0.0099 0.0102 0.0106...
      0.0109 0.0112 0.0119 0.0123 0.0126 0.0129 0.0133 0.0136 0.0140 0.0143...
      0.0147 0.0150 0.0153 0.0157 0.0161 0.0160 0.0167 0.0170 0.0174 0.0177];
t = [0.1113 0.1131 0.1173 0.1182 0.1169 0.1190 0.1227 0.1257 0.1327 0.1350...
      0.1346 0.1327 0.1308 0.1302 0.1310 0.1324 0.1330 0.1318 0.1316 0.1320...
      0.1305 0.1306 0.1277 0.1254 0.1256 0.1230 0.1225 0.1207 0.1223 0.1279...
      0.1316 0.1366 0.1373 0.1379 0.1369 0.1374 0.1331 0.1303 0.1239 0.1174...
      0.1135 0.1120 0.1095 0.1078 0.1050 0.1016 0.0964 0.0940 0.0885 0.0871];
X = [xc1' xc2' xc3']; Y = t'; % define input and output(ash) data
[n,m] = size(X); % data set size (n: number of data samples)
rng(10); % specify positive integer seed value
Tmd = fitrlinear(X,Y,'Holdout',0.25); % create regression model (hold out 25% of
the data for testing)
modl = Tmd.Trained{1};
indtrain = training(Tmd.Partition); % training set indices
indtest = test(Tmd.Partition); % test set indices
ytrain = predict(modl,X(indtrain,:));
yp = predict(modl,X(indtest,:)); % predict outputs
yd = Y(indtest); k = 1:length(yd);
plot(k,yp,k,yd,'o')  % plot data and estimation results
xlabel('Sample no.'),ylabel('Ash(%)'),legend
('Estimation','Data','Location','best')
```

Example 11.11   Regression of Output Values

The input data set is defined as a 100×50 sparse matrix $X = [x_1, x_2, \cdots, x_{50}]$. Assume that 10% of all the elements of X are nonzero. Suppose that the output Y is defined by

$$Y = 1.2*x_{20} + 0.5*\sin(x_{40}) + \epsilon$$

where ϵ is a vector of random normalized error with mean 0 and standard deviation 0.3. We can create a normal distributed sparse random matrix by using the built-in function *sprandn*. Create the SVM regression model by using the function *fitrlinear*. 30% of the data set is to be used in validation.

Solution

The script *linsvmr* defines the data set, creates the SVM regression model using *ftirlinear*, trains the model, predicts the output values, and produces the plot showing the estimation results and data shown in Figure 11.25.

```
% linsvmr.m
n = 100; m = 50;  % dimension of input data set X
nz = 0.1; rng(1); X = sprandn(n,m,nz); Y = 1.2*X(:,20) + 0.5*sin(X(:,40)) +
0.3*randn(n,1);
```

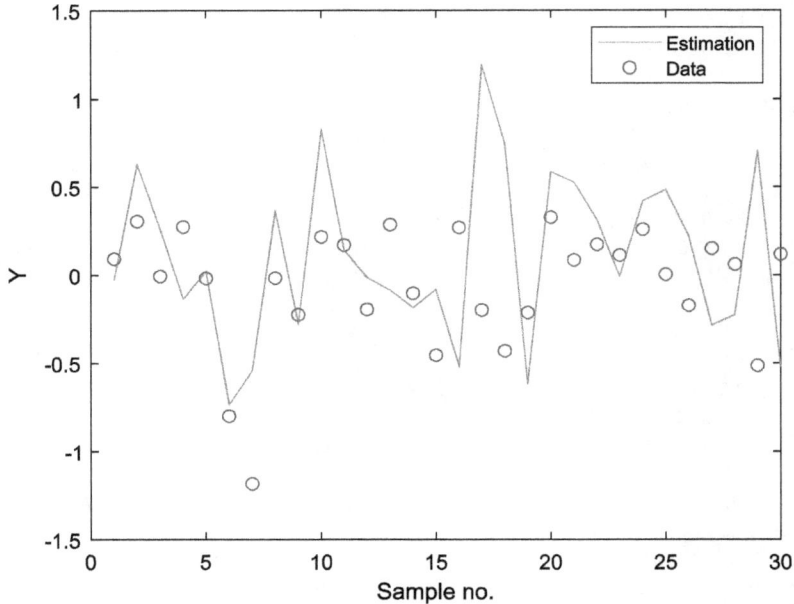

FIGURE 11.25   Estimation of output values by the SVM regression model.

```
Tmd = fitrlinear(X,Y, 'Holdout',0.3); % create SVM regression model (30% of data
are used in testing)
modl = Tmd.Trained{1};
indtrain = training(Tmd.Partition); indtest = test(Tmd.Partition);
ytrain = predict(modl,X(indtrain,:)); ytest = predict(modl,X(indtest,:));
y = Y(indtest); % data
k  =  1: length(y); plot(k,ytest,k,y,'o'),  xlabel('Sample  no.'),ylabel
('Y'),legend('Estimation','Data')
```

PROBLEMS

11.1 Plot the fuzzy sets y1 and y2 defined by trapezoidal and Z-membership functions as y1 = trapmf(x, [1 5 7 8]) and y2 = zmf(x, [3 7]) for $0 \leq x \leq 10$.

11.2 Suppose that fuzzy sets A and B are given by A = trimf(x, [-6 6 8]) and B = sigmf (x, [2 5]) in the range of $-10 \leq x \leq 10$. Evaluate the sum, difference, and product of A and B and plot the results.

11.3 Fuzzy sets A and B are defined by sigmoid and Z-membership functions as A = sigmf(x, [-42 4]) and B = zmf(x, [-2 5]) in the range of $-5 \leq x \leq 5$. The fuzzy set C is given by

$$C = \max(0.4A, 0.6B)$$

Perform defuzzification of C using centroid, bisector, and mom methods.

11.4 Create a function fitting network with three hidden layers. The numbers of nodes of the layers are 5, 7, and 6 for the first, second, and third hidden layer, respectively.

11.5 Create a function fitting network with one hidden layer containing three nodes using the Levenberg-Marquardt algorithm (trainlm). Train and test the network using the data given in Table P11.5.

11.6 Table P11.6 shows operation data acquired during the operation of a paper manufacturing plant. The input variables are Stockflow (stock flow rate, *liter/min*) and Pressure (kg/cm^2), and the output variable is Moisture (moisture content, %). Create and train a function fitting neural network and plot the test data and the moisture contents estimated by the network. The network should have two hidden layer with six and five neurons. Divide the data so that 60% is used for training, 20% is used for validation, and 20% is used for testing.

11.7 Consider the data shown in Table P11.5. Combine the training data and testing data as a single data set. Use this data set to construct an SVM regression model and predict the target data. 70% of the whole data set should be used in training, and the remaining 30% should be used in testing. Produce the plot representing the estimation results with data points. Note that the data should be sorted in ascending order prior to constructing an SVM regression model.

11.8 Consider the paper plant operation data shown in Table P11.6. Construct an SVM regression model by using the function *fitrlinear* and predict the moisture content. 70% of the whole data set should be used in the training, and the remaining 30% should be used in testing. Produce the plot representing the estimation results with data points.

TABLE P11.5
Data Set for Training and Testing

Training Data Input (x)	Target (t)	Testing Data Input (x)	Target (t)
1.85	9.69	0	5.05
0.05	5.36	1.99	9.47
2.13	9.23	0.10	5.66
0.16	5.99	2.28	8.97
2.42	8.71	0.21	6.32
2.56	8.46	0.27	6.63
0.32	6.94	2.70	8.22
0.38	7.26	2.85	8.01
2.99	7.82	0.44	7.57
0.51	7.90	3.13	7.65
3.28	7.51	0.58	8.21
0.66	8.52	3.43	7.38
0.74	8.84	3.59	7.28
3.77	7.19	0.83	9.14
0.94	9.43	3.95	7.13
4.12	7.10	1.07	9.70
4.30	7.09	1.21	9.90
1.37	9.99	4.47	7.10
4.65	7.14	1.55	9.98
1.70	9.86	4.83	7.22

TABLE P11.6
Paper Plant Operation Data

Stockflow (*liter/min*)	Pressure (*kg/cm²*)	Moisture (%)
0.0015	0.0003	0.0924
0.0076	0.0007	0.1665
0.0214	0.0010	0.2198
0.0273	0.0013	0.2121
0.0304	0.0017	0.2093
0.0263	0.0020	0.2271
0.0131	0.0024	0.2195
0.0087	0.0027	0.2074
0.0097	0.0031	0.1973
0.0069	0.0034	0.2503
0.0015	0.0041	0.2534
0.0143	0.0044	0.2769
0.0133	0.0048	0.2854
0.0083	0.0051	0.2886
0.0047	0.0054	0.2897
0.0039	0.0058	0.2924
0.0043	0.0061	0.2818
0.0032	0.0065	0.3950
0.0006	0.0068	0.3943
0.0056	0.0072	0.3899
0.0169	0.0075	0.3875
0.0240	0.0078	0.3863
0.0208	0.0082	0.3849
0.0125	0.0085	0.3872
0.0077	0.0089	0.3879
0.0086	0.0092	0.3913
0.0097	0.0095	0.3963
0.0117	0.0099	0.4031
0.0127	0.0102	0.4054
0.0083	0.0106	0.3968
0.0123	0.0109	0.3780
0.0243	0.0112	0.3733
0.0555	0.0119	0.3732
0.0619	0.0123	0.3828
0.0666	0.0126	0.3916
0.0763	0.0129	0.4053
0.0796	0.0133	0.4149
0.0820	0.0136	0.4196
0.0901	0.0140	0.4207
0.1051	0.0143	0.4237
0.1232	0.0147	0.4193
0.1313	0.0150	0.4007
0.1319	0.0153	0.3919
0.1387	0.0157	0.3726
0.1531	0.0161	0.3670

(*continued*)

TABLE P11.6 (continued)
Paper Plant Operation Data

Stockflow (*liter/min*)	Pressure (*kg/cm²*)	Moisture (%)
0.1636	0.0160	0.3785
0.1766	0.0167	0.3822
0.1940	0.0170	0.3932
0.2035	0.0174	0.4069
0.2092	0.0177	0.4109

REFERENCES

1. Engelbrecht, A. P., *Computational Intelligence*, 2nd ed., John Wiley &Sons, Ltd., West Sussex, England, p. 10, 2007.
2. Engelbrecht, A. P., *Computational Intelligence*, 2nd ed., John Wiley &Sons, Ltd., West Sussex, England, p. 454, 2007.
3. Sumathi, S. and P. Surekha, *Computational Intelligence Paradigms*, CRC Press, Taylor & Francis Group, Boca Raton, FL, pp. 274–276, 2010.
4. Mamdani, E. H. and S. Assilian, An experiment in linguistic synthesis with a fuzzy logic controller, *Int. J. Man-Machine Stud.*, 7, pp. 1–13, 1975.
5. Zadeh, L. A., Outline of a new approach to the analysis of complex systems and decision processes, *IEEE Transaction on System Man and Cybernetics*, 3(1), pp. 28–44, 1973.
6. Engelbrecht, A. P., *Computational Intelligence*, 2nd ed., John Wiley &Sons, Ltd., West Sussex, England, p. 28, 2007.
7. Vapnik, V. N., *The Nature of Statistical Learning Theory*, Springer-Verlag, New York, NY, pp. 35–40, 1995.
8. Zeng, L., H. Zhou, K. Cen, and C. Wang, A comparative study of optimization algorithms for low NOx combustion modification at a coal-fired utility boiler, *Expert System Application*, 36, pp. 2780–2793, 2009.
9. Wei, Z., X. Li, L. Xu, and C. Yanting, Comparative study of computational intelligence approached for NOx reduction of coal-fired boiler, *Energy*, 55, pp. 683–692, 2013.

Index

For Product Safety Concerns and Information please contact our EU
representative GPSR@taylorandfrancis.com
Taylor & Francis Verlag GmbH, Kaufingerstraße 24, 80331 München, Germany

www.ingramcontent.com/pod-product-compliance
Lightning Source LLC
Chambersburg PA
CBHW080334220326
41598CB00030B/4501